CONTENTS

REVIEW AND PREVIEW 1

1 Functions and Their Graphs 1
2 Types of Functions; Shifting and Scaling 7
3 Graphing Calculators and Computers 10
4 Principles of Problem Solving 15

1 INTRODUCTION TO VECTORS AND VECTOR FUNCTIONS 17

1.1 Vectors 17
1.2 The Dot Product 19
1.3 Vector Functions 23
Review 26

2 LIMITS AND RATES OF CHANGE 31

2.1 The Tangent and Velocity Problems 31
2.2 The Limit of a Function 33
2.3 Calculating Limits using the Limit Laws 36
2.4 The Precise Definition of a Limit 41
2.5 Continuity 45
2.6 Limits at Infinity; Horizontal Asymptotes 49
2.7 Tangents, Velocities, and Other Rates of Change 52
Review 55

3 DERIVATIVES 59

3.1 Derivatives 59
3.2 Differentiation Formulas 65
3.3 Rates of Change in the Natural and Social Sciences 69
3.4 Derivatives of Trigonometric Functions 71
3.5 The Chain Rule 74
3.6 Implicit Differentiation 77

3.7 Derivatives of Vector Functions 80
3.8 Higher Derivatives 82
3.9 Slopes and Tangents of Parametric Curves 86
3.10 Related Rates 90
3.11 Differentials; Linear and Quadratic Approximations 93
3.12 Newton's Method 96
Review 99

PROBLEMS PLUS 105

4 INVERSE FUNCTIONS: Exponential, Logarithmic, and Inverse Trigonometric Functions 111

4.1 Exponential Functions and Their Derivatives 111
4.2 Inverse Functions 114
4.3 Logarithmic Functions 117
4.4 Derivatives of Logarithmic Functions 120
4.5 Exponential Growth and Decay 122
4.6 Inverse Trigonometric Functions 124
4.7 Hyperbolic Functions 128
4.8 Indeterminate Forms and l'Hospital's Rule 131
Review 135

APPLICATIONS PLUS 139

5 APPLICATIONS OF DIFFERENTIATION 141

5.1 What does f' say about f? 141
5.2 Maximum and Minimum Values 143
5.3 Derivatives and the Shapes of Curves 148
5.4 Graphing with Calculus *and* Calculators 155
5.5 Applied Maximum and Minimum Problems 163
5.6 Applications to Economics 169
5.7 Antiderivatives 171
Review 176

PROBLEMS PLUS 185

CONTENTS

6 INTEGRALS 191

6.1 Sigma Notation 191
6.2 Area 193
6.3 The Definite Integral 198
6.4 The Fundamental Theorem of Calculus 202
6.5 The Substitution Rule 207
6.6 The Logarithm Defined as an Integral 211
Review 212

APPLICATIONS PLUS 217

7 APPLICATIONS OF INTEGRATION 219

7.1 Areas Between Curves 219
7.2 Volume 225
7.3 Volumes by Cylindrical Shells 230
7.4 Work 232
7.5 Average Value of a Function 233
Review 234

PROBLEMS PLUS 237

8 TECHNIQUES OF INTEGRATION 241

8.1 Integration by Parts 241
8.2 Trigonometric Integrals 244
8.3 Trigonometric Substitution 247
8.4 Integration of Rational Functions by Partial Fractions 251
8.5 Rationalizing Substitutions 256
8.6 Strategy for Integration 258
8.7 Using Tables of Integrals and Computer Algebra Systems 262
8.8 Approximate Integration 265
8.9 Improper Integrals 270
Review 276

APPLICATIONS PLUS 281

9 **FURTHER APPLICATIONS OF INTEGRATION** 285

9.1 Differential Equations 285
9.2 First-Order Linear Equations 289
9.3 Arc Length 292
9.4 Area of a Surface of Revolution 295
9.5 Moments and Centers of Mass 298
9.6 Hydrostatic Pressure and Force 302
9.7 Applications to Economics and Biology 304
Review 305

PROBLEMS PLUS 309

10 **INFINITE SEQUENCES AND SERIES** 313

10.1 Sequences 313
10.2 Series 316
10.3 The Integral and Comparison Tests; Estimating Sums 322
10.4 Other Convergence Tests 325
10.5 Power Series 328
10.6 Representations of Functions as Power Series 331
10.7 Taylor and Maclaurin Series 335
10.8 The Binomial Series 340
10.9 Applications of Taylor Polynomials 343
Review 349

PROBLEMS PLUS 353

11 **THREE-DIMENSIONAL ANALYTIC GEOMETRY AND VECTORS** 357

11.1 Three-Dimensional Coordinate Systems 357
11.2 Vectors and the Dot Product in Three Dimensions 360
11.3 The Cross Product 363
11.4 Equations of Lines and Planes 366
11.5 Quadric Surfaces 371
11.6 Vector Functions and Space Curves 375
11.7 Arc Length and Curvature 381
11.8 Motion in Space: Velocity and Acceleration 385
Review 387

APPLICATIONS PLUS 393

12 PARTIAL DERIVATIVES 397

12.1 Functions of Several Variables 397

12.2 Limits and Continuity 403

12.3 Partial Derivatives 406

12.4 Tangent Planes and Differentials 412

12.5 The Chain Rule 414

12.6 Directional Derivatives and the Gradient Vector 417

12.7 Maximum and Minimum Values 421

12.8 Lagrange Multipliers 429

Review 433

PROBLEMS PLUS 439

13 MULTIPLE INTEGRALS 443

13.1 Double Integrals over Rectangles 443

13.2 Iterated Integrals 444

13.3 Double Integrals over General Regions 446

13.4 Polar Coordinates 450

13.5 Double Integrals in Polar Coordinates 456

13.6 Applications of Double Integrals 458

13.7 Surface Area 460

13.8 Triple Integrals 462

13.9 Cylindrical and Spherical Coordinates 467

13.10 Triple Integrals in Cylindrical and Spherical Coordinates 469

13.11 Change of Variables in Multiple Integrals 473

Review 475

APPLICATIONS PLUS 481

14 VECTOR CALCULUS 485

14.1 Vector Fields 485

14.2 Line Integrals 487

14.3 The Fundamental Theorem for Line Integrals 490

14.4 Green's Theorem 492

14.5 Curl and Divergence 495

14.6 Parametric Surfaces and Their Areas 498

14.7 Surface Integrals 502

14.8 Stokes' Theorem 505

14.9 The Divergence Theorem 507

Review 508

PROBLEMS PLUS 511

15 SECOND-ORDER DIFFERENTIAL EQUATIONS 515

15.1 Second-Order Linear Equations 515

15.2 Nonhomogeneous Linear Equations 517

15.3 Applications of Second-Order Differential Equations 520

15.4 Using Series to Solve Differential Equations 521

Review 523

APPENDIXES 525

A Numbers, Inequalities and Absolute Values 525

B Coordinate Geometry and Lines 528

C Graphs of Second-Degree Equations 531

D Trigonometry 533

E Mathematical Induction 537

G Lies My Calculator and Computer Told Me 538

H Complex Numbers 541

I Conic Sections 543

J Conic Sections in Polar Coordinates 546

REVIEW AND PREVIEW

EXERCISES 1

1. $f(x) = 2x^2 + 3x - 4$, so $f(0) = 2(0)^2 + 3(0) - 4 = -4$, $f(2) = 2(2)^2 + 3(2) - 4 = 10$,

$f(\sqrt{2}) = 2(\sqrt{2})^2 + 3(\sqrt{2}) - 4 = 3\sqrt{2}$,

$f(1+\sqrt{2}) = 2(1+\sqrt{2})^2 + 3(1+\sqrt{2}) - 4 = 2(1+2+2\sqrt{2}) + 3 + 3\sqrt{2} - 4 = 5 + 7\sqrt{2}$,

$f(-x) = 2(-x)^2 + 3(-x) - 4 = 2x^2 - 3x - 4$,

$f(x+1) = 2(x+1)^2 + 3(x+1) - 4 = 2(x^2 + 2x + 1) + 3x + 3 - 4 = 2x^2 + 7x + 1$,

$2f(x) = 2(2x^2 + 3x - 4) = 4x^2 + 6x - 8$, and

$f(2x) = 2(2x)^2 + 3(2x) - 4 = 2(4x^2) + 6x - 4 = 8x^2 + 6x - 4$.

3. $f(x) = x - x^2$, so $f(2+h) = 2 + h - (2+h)^2 = 2 + h - 4 - 4h - h^2 = -(h^2 + 3h + 2)$,

$f(x+h) = x + h - (x+h)^2 = x + h - x^2 - 2xh - h^2$, and

$$\frac{f(x+h) - f(x)}{h} = \frac{x + h - x^2 - 2xh - h^2 - x + x^2}{h} = \frac{h - 2xh - h^2}{h} = 1 - 2x - h.$$

5. $f(x) = \sqrt{x}$, $0 \le x \le 4$

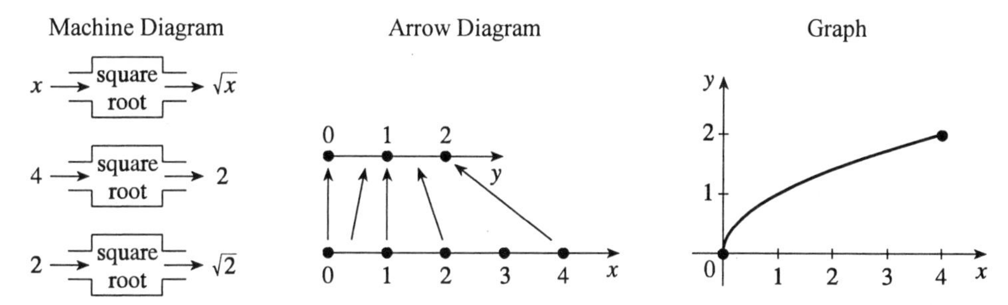

7. The range of f is the set of values of f, $\{0, 1, 2, 4\}$.

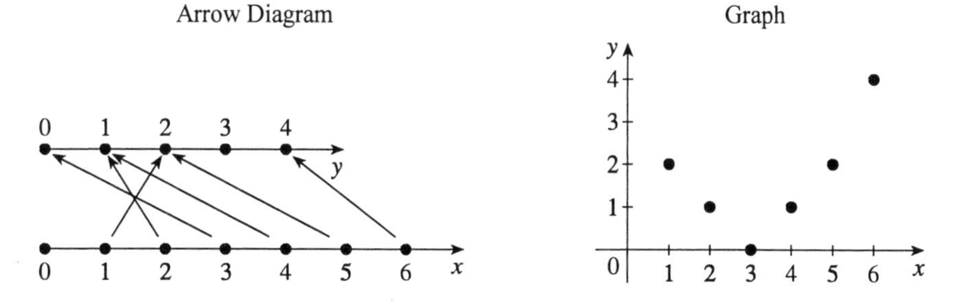

9. $f(x) = 6 - 4x$, $-2 \le x \le 3$. The domain is $[-2, 3]$. If $-2 \le x \le 3$, then

$14 = 6 - 4(-2) \ge 6 - 4x \ge 6 - 4(3) = -6$, so the range is $[-6, 14]$.

11. $h(x) = \sqrt{2x-5}$ is defined when $2x - 5 \geq 0$ or $x \geq \frac{5}{2}$, so the domain is $\left[\frac{5}{2}, \infty\right)$ and the range is $[0, \infty)$.

13. $F(x) = \sqrt{1-x^2}$ is defined when $1 - x^2 \geq 0 \quad \Leftrightarrow \quad x^2 \leq 1 \quad \Leftrightarrow \quad |x| \leq 1 \quad \Leftrightarrow \quad -1 \leq x \leq 1$, so the domain is $[-1, 1]$ and the range is $[0, 1]$.

15. $f(x) = \dfrac{x+2}{x^2-1}$ is defined for all x except when $x^2 - 1 = 0 \quad \Leftrightarrow \quad x = 1$ or $x = -1$, so the domain is $\{x \mid x \neq \pm 1\}$.

17. $g(x) = \sqrt[4]{x^2 - 6x}$ is defined when $0 \leq x^2 - 6x = x(x-6) \quad \Leftrightarrow \quad x \geq 6$ or $x \leq 0$, so the domain is $(-\infty, 0] \cup [6, \infty)$.

19. $\phi(x) = \sqrt{\dfrac{x}{\pi - x}}$ is defined when $\dfrac{x}{\pi - x} \geq 0$. This fraction is positive when the numerator and denominator have the *same* sign. So either $x \leq 0$ and $\pi - x < 0$ ($\Leftrightarrow$ $x > \pi$), which is impossible, or $x \geq 0$ and $\pi - x > 0$ ($\Leftrightarrow$ $x < \pi$), and so the domain is $[0, \pi)$.

21. $f(t) = \sqrt[3]{t-1}$ is defined for every t, since every real number has a cube root. The domain is the set of all real numbers.

23. $f(x) = 3 - 2x$. Domain is $\mathbb{R}$.

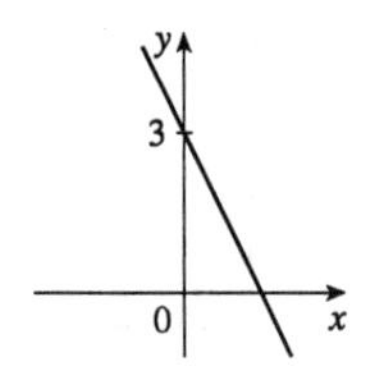

25. $f(x) = x^2 + 2x - 1 = (x^2 + 2x + 1) - 2 = (x+1)^2 - 2$, so the graph is a parabola with vertex at $(-1, -2)$. The domain is $\mathbb{R}$.

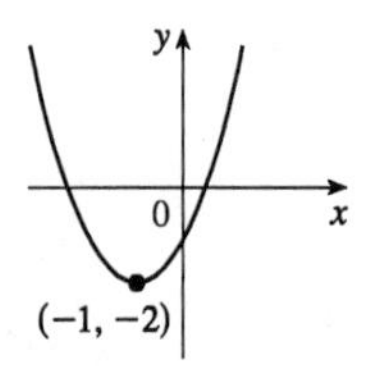

27. $g(x) = \sqrt{-x}$. The domain is $\{x \mid -x \geq 0\} = (-\infty, 0]$.

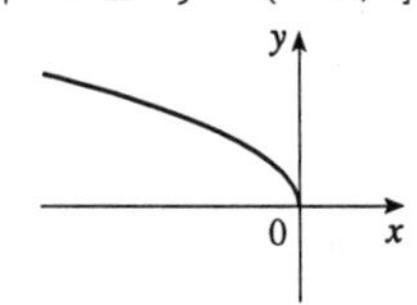

29. $h(x) = \sqrt{4 - x^2}$. Now $y = \sqrt{4 - x^2} \quad \Rightarrow \quad y^2 = 4 - x^2 \quad \Leftrightarrow \quad x^2 + y^2 = 4$, so the graph is the top half of a circle of radius 2. The domain is $\{x \mid 4 - x^2 \geq 0\} = [-2, 2]$.

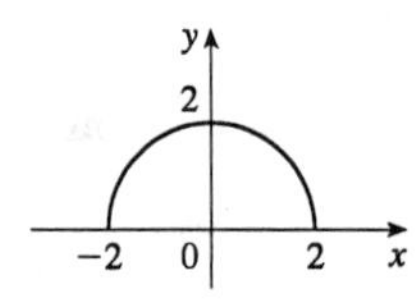

31. $F(x) = \dfrac{1}{x}$. The domain is $\{x \mid x \neq 0\}$.

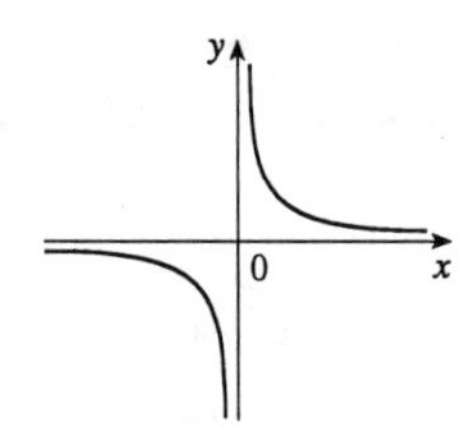

33. $G(x) = |x| + x = \begin{cases} 2x & \text{if } x \geq 0 \\ 0 & \text{if } x < 0 \end{cases}$

Domain is $\mathbb{R}$.

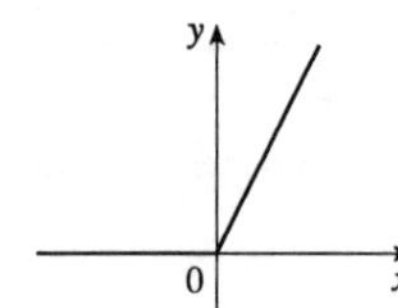

35. $H(x) = |2x| = \begin{cases} 2x & \text{if } x \geq 0 \\ -2x & \text{if } x < 0 \end{cases}$

Domain is $\mathbb{R}$.

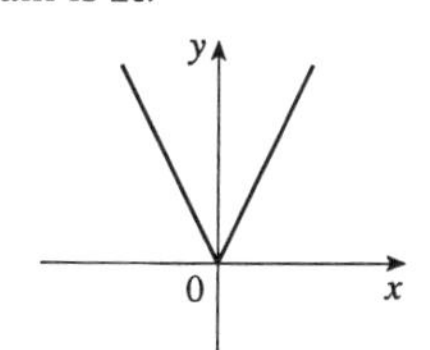

37. $f(x) = \dfrac{x}{|x|} = \begin{cases} 1 & \text{if } x > 0 \\ -1 & \text{if } x < 0 \end{cases}$

Domain is $\{x \mid x \neq 0\}$.

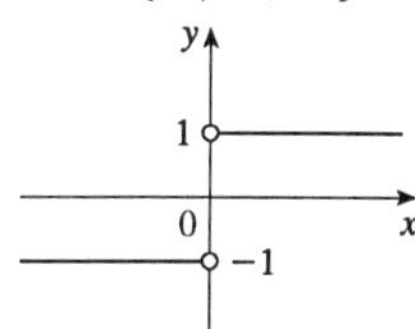

39. $f(x) = \dfrac{x^2 - 1}{x - 1} = \dfrac{(x+1)(x-1)}{x-1}$, so for $x \neq 1$, $f(x) = x + 1$. Domain is $\{x \mid x \neq 1\}$.

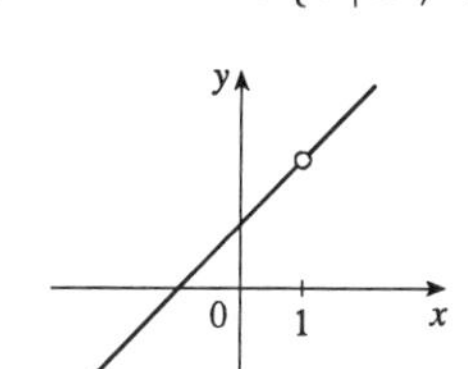

41. $f(x) = \begin{cases} 0 & \text{if } x < 2 \\ 1 & \text{if } x \geq 2 \end{cases}$ Domain is $\mathbb{R}$.

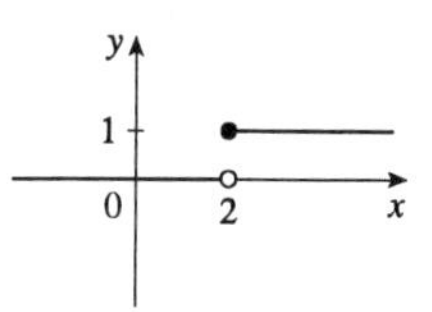

43. $f(x) = \begin{cases} x & \text{if } x \leq 0 \\ x + 1 & \text{if } x > 0 \end{cases}$ Domain is $\mathbb{R}$.

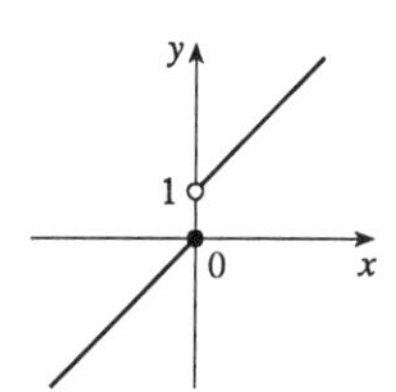

45. $f(x) = \begin{cases} -1 & \text{if } x < -1 \\ x & \text{if } -1 \leq x \leq 1 \\ 1 & \text{if } x > 1 \end{cases}$ Domain is $\mathbb{R}$.

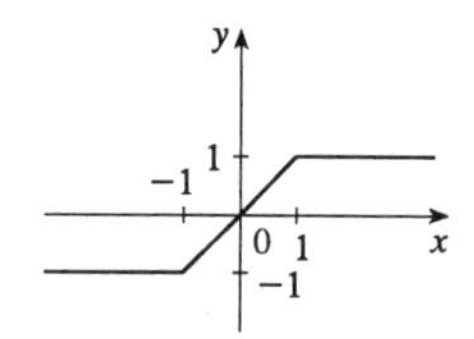

47. $f(x) = \begin{cases} x + 2 & \text{if } x \leq -1 \\ x^2 & \text{if } x > -1 \end{cases}$ Domain is $\mathbb{R}$.

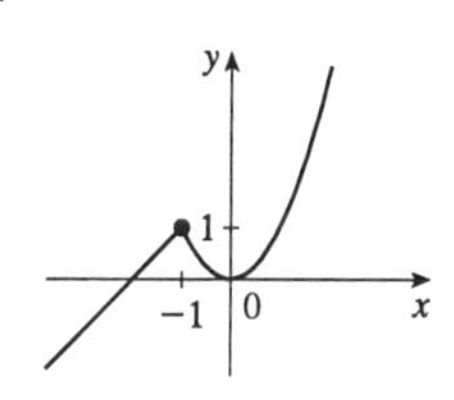

49. $f(x) = \begin{cases} -1 & \text{if } x \leq -1 \\ 3x + 2 & \text{if } -1 < x < 1 \\ 7 - 2x & \text{if } x \geq 1 \end{cases}$

Domain is $\mathbb{R}$.

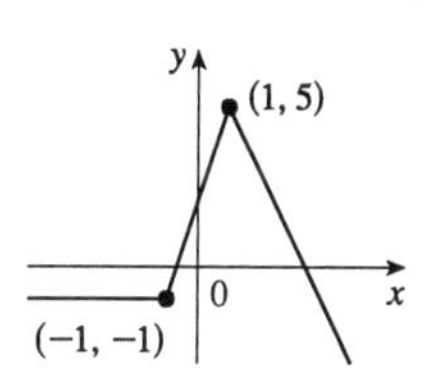

51. Yes, the curve is the graph of a function. The domain is $[-3, 2]$ and the range is $[-2, 2]$.

53. No, this is not the graph of a function since for $x = -1$ there are infinitely many points on the curve.

55. The slope of this line segment is $\dfrac{-6 - 1}{4 - (-2)} = -\dfrac{7}{6}$, so an equation is $y - 1 = -\frac{7}{6}(x + 2)$. The function is $f(x) = -\frac{7}{6}x - \frac{4}{3}$, $-2 \leq x \leq 4$.

57. $x + (y - 1)^2 = 0 \quad \Leftrightarrow \quad y - 1 = \pm\sqrt{-x}$. The bottom half is given by the function $f(x) = 1 - \sqrt{-x}$, $x \leq 0$.

59. For $-1 \le x \le 2$, the graph is the line with slope 1 and y-intercept 1, that is, the line $y = x + 1$. For $2 < x \le 4$, the graph is the line with slope $-\frac{3}{2}$ and x-intercept 4, so $y = -\frac{3}{2}(x - 4) = -\frac{3}{2}x + 6$. So the function is
$$f(x) = \begin{cases} x + 1 & \text{if } -1 \le x \le 2 \\ -\frac{3}{2}x + 6 & \text{if } 2 < x \le 4 \end{cases}$$

61. Let the length and width of the rectangle be L and W respectively. Then the perimeter is $2L + 2W = 20$, and the area is $A = LW$. Solving the first equation for W in terms of L gives $W = \dfrac{20 - 2L}{2} = 10 - L$. Thus $A(L) = L(10 - L) = 10L - L^2$. Since lengths are positive, the domain of A is $0 < L < 10$.

63. Let the length of a side of the equilateral triangle be x. Then by the Pythagorean Theorem, the height y of the triangle satisfies $y^2 + \left(\frac{1}{2}x\right)^2 = x^2$, so that $y = \frac{\sqrt{3}}{2}x$. Thus the area of the triangle is $A = \frac{1}{2}xy$ and so $A(x) = \frac{1}{2}\left(\frac{\sqrt{3}}{2}x\right)x = \frac{\sqrt{3}}{4}x^2$, with domain $x > 0$.

65. Let each side of the base of the box have length x, and let the height of the box be h. Since the volume is 2, we know that $2 = hx^2$, so that $h = 2/x^2$, and the surface area is $S = x^2 + 4xh$. Thus $S(x) = x^2 + 4x(2/x^2) = x^2 + 8/x$, with domain $x > 0$.

67. The height of the box is x and the length and width are $L = 20 - 2x$, $W = 12 - 2x$. Then $V = LWx$ and so $V(x) = (20 - 2x)(12 - 2x)(x) = 4(10 - x)(6 - x)(x) = 4x\left(60 - 16x + x^2\right) = 4x^3 - 64x^2 + 240x$, with domain $0 < x < 6$.

69. **(a)** T is a linear function of h, so $T = mh + b$ with m and b constants. We know two points on the graph of T as a function of h: $(h, T) = (0,20)$ and $(h,T) = (1,10)$. The slope of the line passing through these two points is $(10 - 20)/(1 - 0) = -10$ and the line's T-intercept is 20, so the slope-intercept form of the equation of the line is $T = -10h + 20$.

(b) The slope is $m = -10°\,\text{C/km}$, and it represents the rate of change of temperature with respect to height.

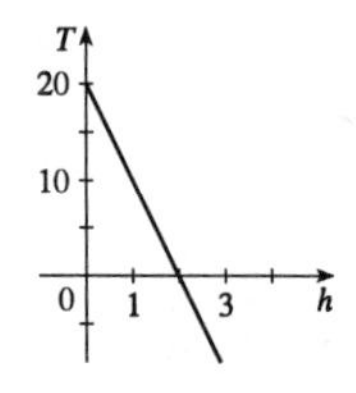

(c) At a height of $h = 2.5$ km, the temperature is
$$\begin{aligned} T &= -10(2.5) + 20 \\ &= -25 + 20 \\ &= -5°\,\text{C}. \end{aligned}$$

71. The water will cool down almost to freezing as the ice melts. Then, when the ice has melted, the water will slowly warm up to room temperature.

73. Of course, this graph depends strongly on the geographical location!

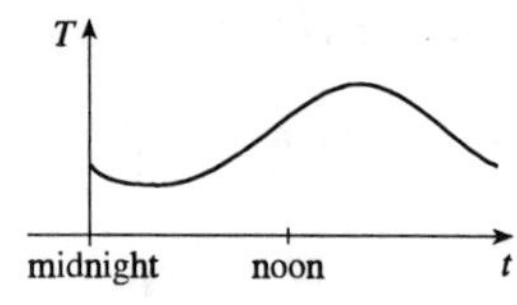

75. **(a)**

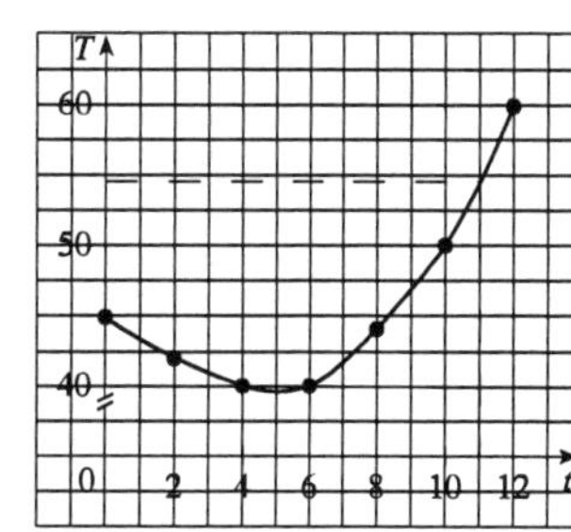

(b) $T(11) \approx 54^\circ\text{F}$

77. $f(-x) = \dfrac{1}{(-x)^2} = \dfrac{1}{x^2} = f(x)$, so f is an even function and symmetric with respect to the y-axis.

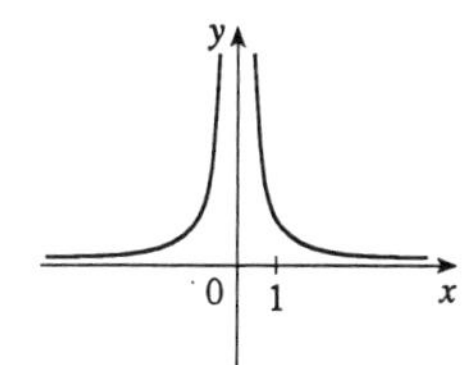

79. $f(-x) = (-x)^2 + (-x) = x^2 - x$. Since this is neither $f(x)$ nor $-f(x)$, the function f is neither even nor odd.

81. $f(-x) = (-x)^3 - (-x) = -x^3 + x = -(x^3 - x)$
$= -f(x)$, so f is odd and symmetric with respect to the origin.

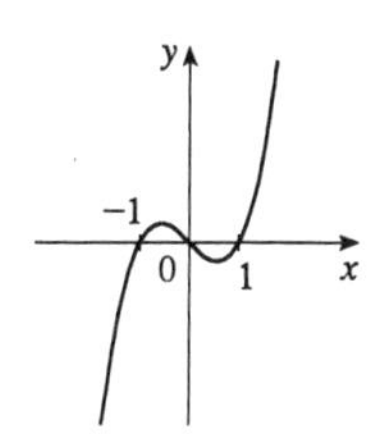

NOTE: For the rest of this section, "$D =$ " stands for "The domain of the function is".

83. $f(x) = x^3 + 2x^2$; $g(x) = 3x^2 - 1$. $D = \mathbb{R}$ for both f and g.

$(f+g)(x) = x^3 + 2x^2 + 3x^2 - 1 = x^3 + 5x^2 - 1$, $D = \mathbb{R}$.

$(f-g)(x) = x^3 + 2x^2 - (3x^2 - 1) = x^3 - x^2 + 1$, $D = \mathbb{R}$.

$(fg)(x) = (x^3 + 2x^2)(3x^2 - 1) = 3x^5 + 6x^4 - x^3 - 2x^2$, $D = \mathbb{R}$.

$(f/g)(x) = (x^3 + 2x^2)/(3x^2 - 1)$, $D = \left\{x \mid x \neq \pm\frac{1}{\sqrt{3}}\right\}$.

85. $f(x) = x$, $g(x) = 1/x$

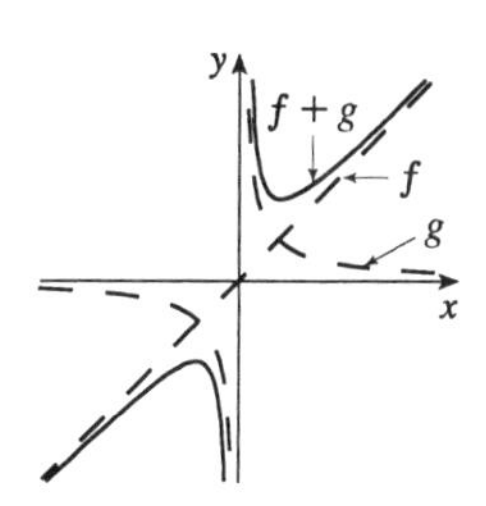

87. $f(x) = 2x^2 - x$, $g(x) = 3x + 2$. $D = \mathbb{R}$ for both f and g, and hence for their composites.

$(f \circ g)(x) = f(g(x)) = f(3x+2) = 2(3x+2)^2 - (3x+2) = 18x^2 + 21x + 6$.

$(g \circ f)(x) = g(f(x)) = g(2x^2 - x) = 3(2x^2 - x) + 2 = 6x^2 - 3x + 2$.

$(f \circ f)(x) = f(f(x)) = f(2x^2 - x) = 2(2x^2 - x)^2 - (2x^2 - x) = 8x^4 - 8x^3 + x$.

$(g \circ g)(x) = g(g(x)) = g(3x+2) = 3(3x+2) + 2 = 9x + 8$.

89. $f(x) = 1/x$, $D = \{x \mid x \neq 0\}$; $g(x) = x^3 + 2x$, $D = \mathbb{R}$.

$(f \circ g)(x) = f(g(x)) = f(x^3 + 2x) = 1/(x^3 + 2x)$, $D = \{x \mid x^3 + 2x \neq 0\} = \{x \mid x \neq 0\}$.

$(g \circ f)(x) = g(f(x)) = g(1/x) = 1/x^3 + 2/x$, $D = \{x \mid x \neq 0\}$.

$(f \circ f)(x) = f(f(x)) = f(1/x) = \dfrac{1}{1/x} = x$, $D = \{x \mid x \neq 0\}$.

$(g \circ g)(x) = g(g(x)) = g(x^3 + 2x) = (x^3 + 2x)^3 + 2(x^3 + 2x) = x^9 + 6x^7 + 12x^5 + 10x^3 + 4x$, $D = \mathbb{R}$.

91. $f(x) = \sqrt[3]{x}$, $D = \mathbb{R}$; $g(x) = 1 - \sqrt{x}$, $D = [0, \infty)$.

$(f \circ g)(x) = f(g(x)) = f\left(1 - \sqrt{x}\right) = \sqrt[3]{1 - \sqrt{x}}$, $D = [0, \infty)$.

$(g \circ f)(x) = g(f(x)) = g\left(\sqrt[3]{x}\right) = 1 - x^{1/6}$, $D = [0, \infty)$.

$(f \circ f)(x) = f(f(x)) = f\left(\sqrt[3]{x}\right) = x^{1/9}$, $D = \mathbb{R}$.

$(g \circ g)(x) = g(g(x)) = g\left(1 - \sqrt{x}\right) = 1 - \sqrt{1 - \sqrt{x}}$, $D = \left\{x \geq 0 \mid 1 - \sqrt{x} \geq 0\right\} = [0, 1]$.

93. $f(x) = \dfrac{x + 2}{2x + 1}$, $D = \left\{x \mid x \neq -\frac{1}{2}\right\}$; $g(x) = \dfrac{x}{x - 2}$, $D = \{x \mid x \neq 2\}$.

$(f \circ g)(x) = f(g(x)) = f\left(\dfrac{x}{x - 2}\right) = \dfrac{x/(x - 2) + 2}{2x/(x - 2) + 1} = \dfrac{3x - 4}{3x - 2}$, $D = \left\{x \mid x \neq 2, \frac{2}{3}\right\}$.

$(g \circ f)(x) = g(f(x)) = g\left(\dfrac{x + 2}{2x + 1}\right) = \dfrac{(x + 2)/(2x + 1)}{(x + 2)/(2x + 1) - 2} = \dfrac{-x - 2}{3x}$, $D = \left\{x \mid x \neq 0, -\frac{1}{2}\right\}$.

$(f \circ f)(x) = f(f(x)) = f\left(\dfrac{x + 2}{2x + 1}\right) = \dfrac{(x + 2)/(2x + 1) + 2}{2(x + 2)/(2x + 1) + 1} = \dfrac{5x + 4}{4x + 5}$, $D = \left\{x \mid x \neq -\frac{1}{2}, -\frac{5}{4}\right\}$.

$(g \circ g)(x) = g(g(x)) = g\left(\dfrac{x}{x - 2}\right) = \dfrac{x/(x - 2)}{x/(x - 2) - 2} = \dfrac{x}{4 - x}$, $D = \{x \mid x \neq 2, 4\}$.

95. $(f \circ g \circ h)(x) = f(g(h(x))) = f(g(x - 1)) = f\left(\sqrt{x - 1}\right) = \sqrt{x - 1} - 1$

97. $(f \circ g \circ h)(x) = f(g(h(x))) = f\left(g\left(\sqrt{x}\right)\right) = f\left(\sqrt{x} - 5\right) = \left(\sqrt{x} - 5\right)^4 + 1$

99. Let $g(x) = x - 9$ and $f(x) = x^5$. Then $(f \circ g)(x) = (x - 9)^5 = F(x)$.

101. Let $g(x) = x^2$ and $f(x) = \dfrac{x}{x + 4}$. Then $(f \circ g)(x) = \dfrac{x^2}{x^2 + 4} = G(x)$.

103. Let $h(x) = x^2$, $g(x) = x + 1$ and $f(x) = \dfrac{1}{x}$. Then $(f \circ g \circ h)(x) = \dfrac{1}{x^2 + 1} = H(x)$.

105. Let r be the radius of the ripple in cm. The area of the ripple is $A = \pi r^2$ but, as a function of time, $r = 60t$. Thus, $A = \pi(60t)^2 = 3600\pi t^2$.

107. We need a function g so that $f(g(x)) = 3(g(x)) + 5 = h(x) = 3x^2 + 3x + 2 = 3(x^2 + x) + 2$ $= 3(x^2 + x - 1) + 5$. So we see that $g(x) = x^2 + x - 1$.

109. The function $g(x) = x$ has domain $(-\infty, \infty)$. However, the function $f \circ f$, where $f(x) = 1/x$, has for its domain $(-\infty, 0) \cup (0, \infty)$ even though the rule is the same: $(f \circ f)(x) = f(1/x) = x$.

EXERCISES 2

1. **(a)** $f(x) = \sqrt[5]{x}$ is a root function.

(b) $g(x) = \sqrt{1 - x^2}$ is an algebraic function because it is a root of a polynomial.

(c) $h(x) = x^9 + x^4$ is a polynomial of degree 9.

(d) $r(x) = \dfrac{x^2 + 1}{x^3 + x}$ is a rational function because it is a ratio of polynomials.

(e) $s(x) = \tan 2x$ is a trigonometric function.

(f) $t(x) = \log_{10} x$ is a logarithmic function.

3. **(a)** To graph $y = f(2x)$ we compress the graph of f horizontally by a factor of 2.

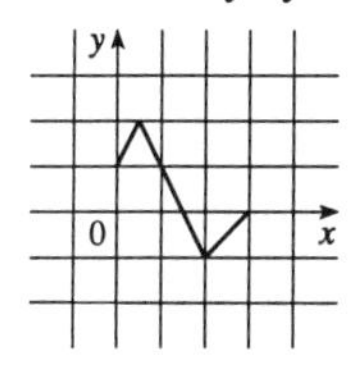

(b) To graph $y = f\left(\frac{1}{2}x\right)$ we expand the graph of f horizontally by a factor of 2.

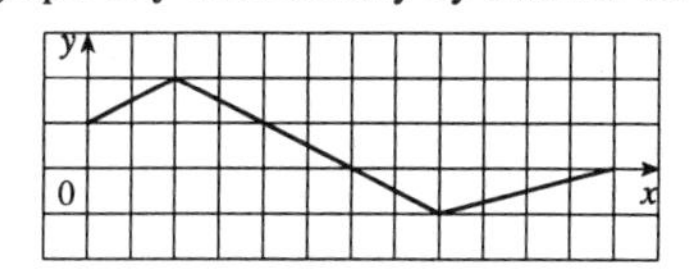

(c) To graph $y = f(-x)$ we reflect the graph of f about the y-axis.

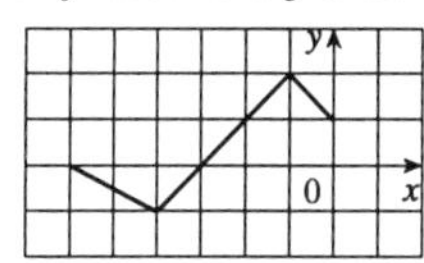

(d) To graph $y = -f(-x)$ we reflect the graph of f about the y-axis, then about the x-axis.

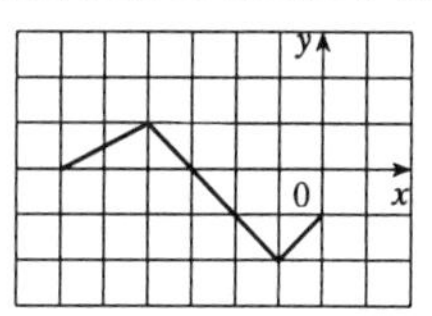

5. $y = -1/x$

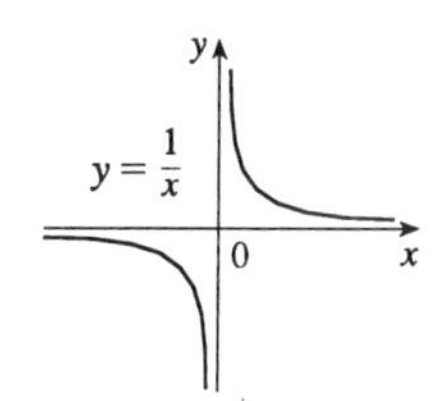

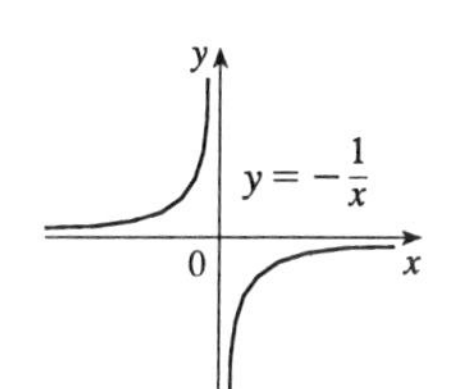

7. $y = 2\sin x$

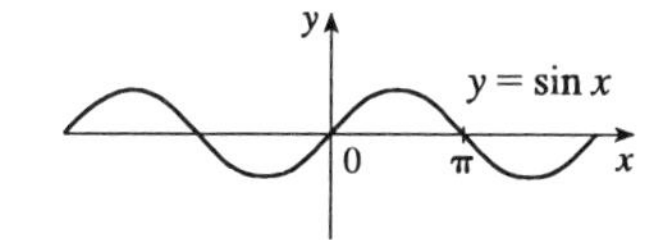

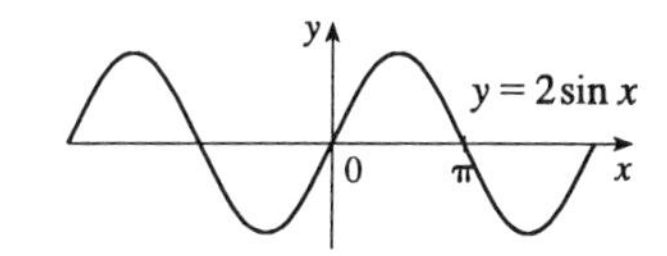

9. $y = (x-1)^3 + 2$

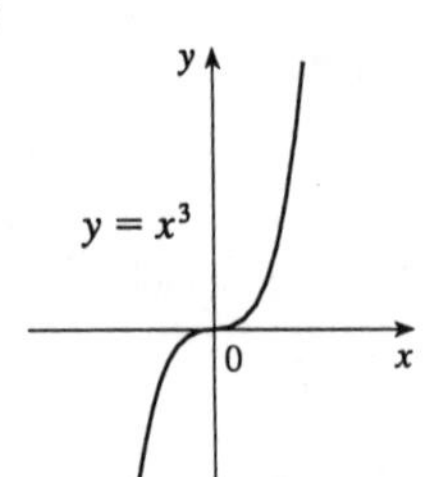

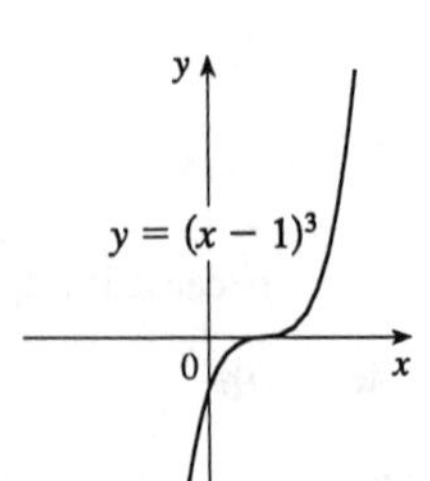

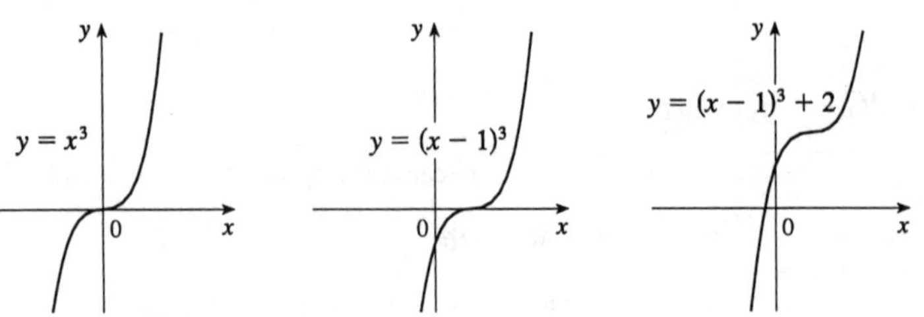

11. $y = \tan 2x$

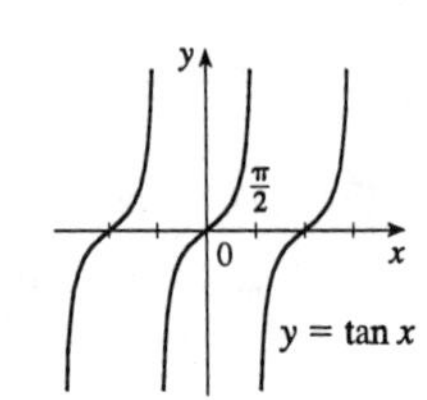

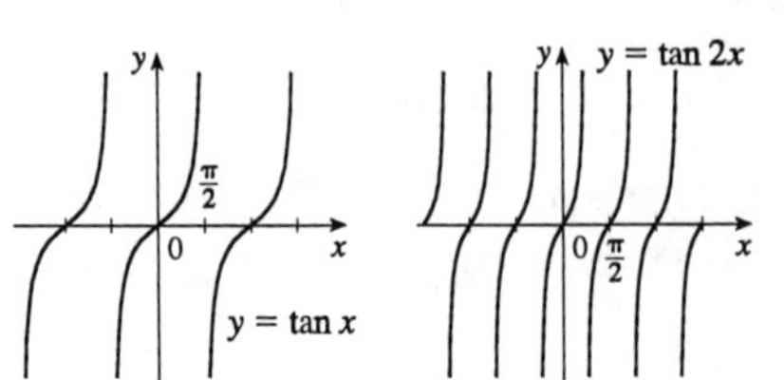

13. $y = \cos(x/2)$

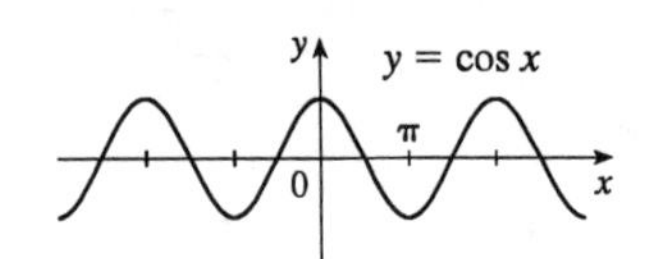

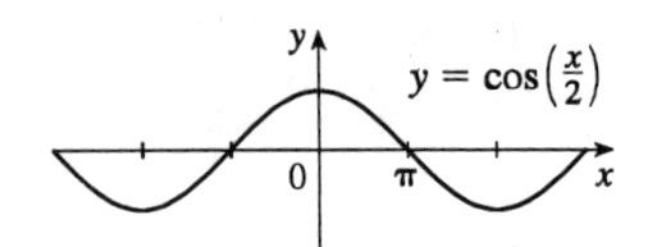

15. $y = \dfrac{1}{x-3}$

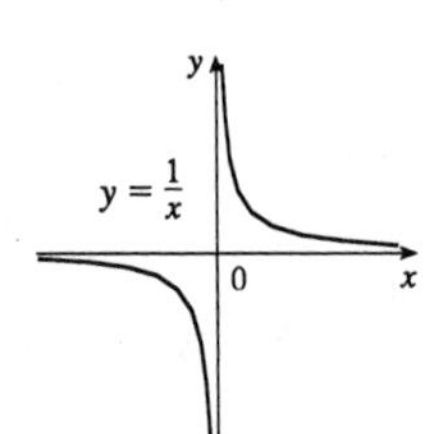

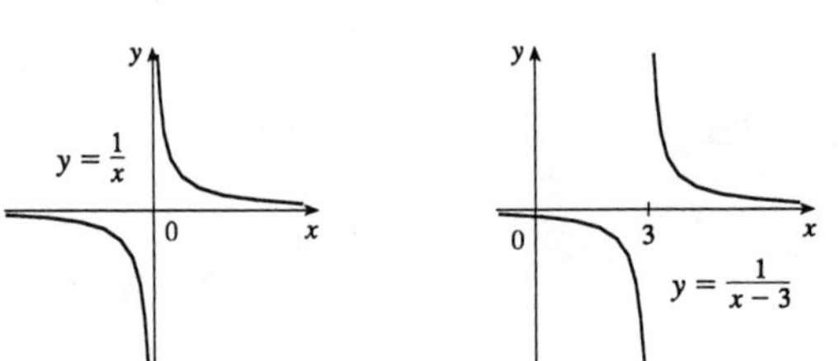

17. $y = \frac{1}{3}\sin\left(x - \frac{\pi}{6}\right)$

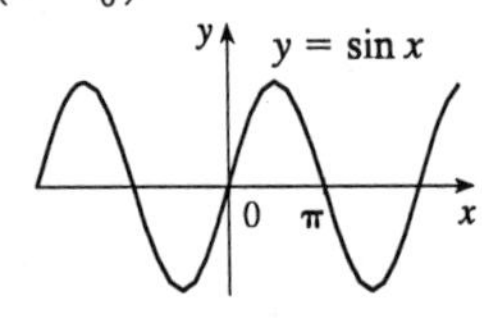

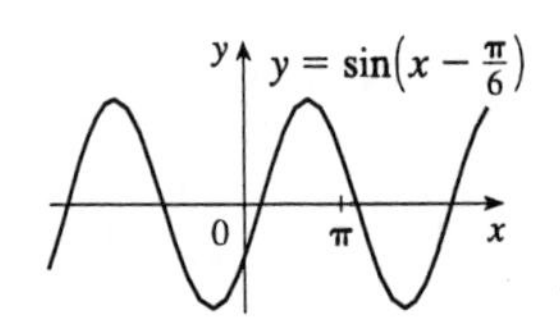

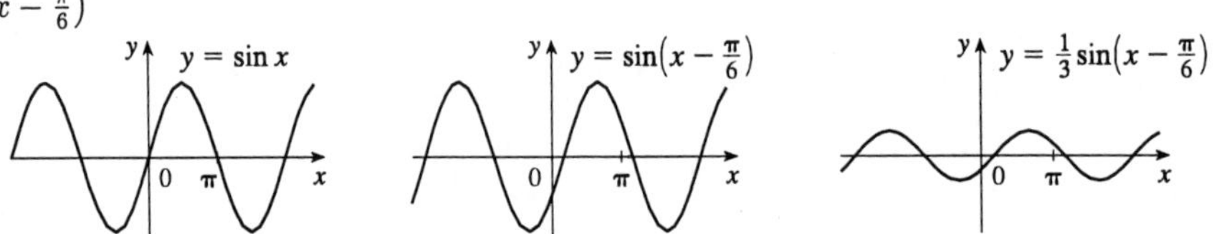

19. $y = 1 + 2x - x^2 = -(x-1)^2 + 2$

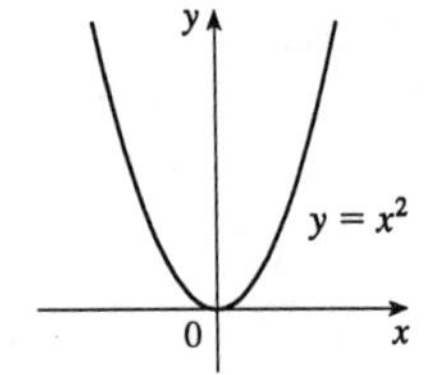

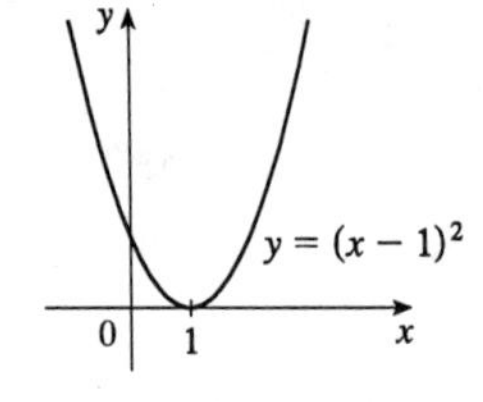

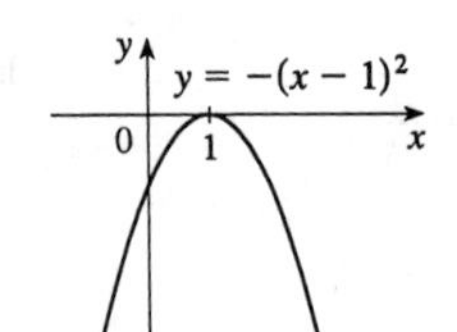

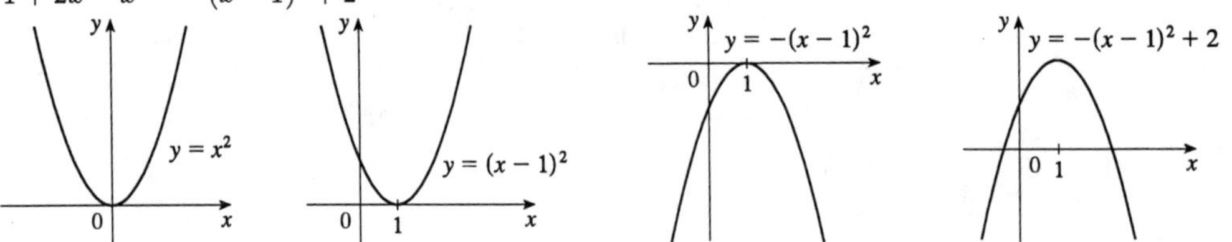

21. $y = 2 - \sqrt{x+1}$

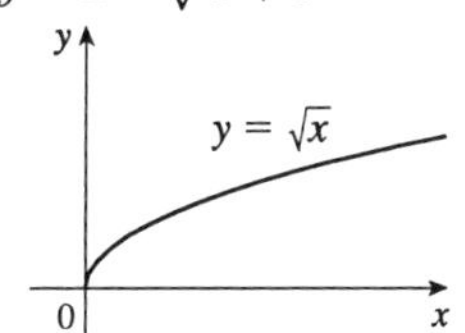

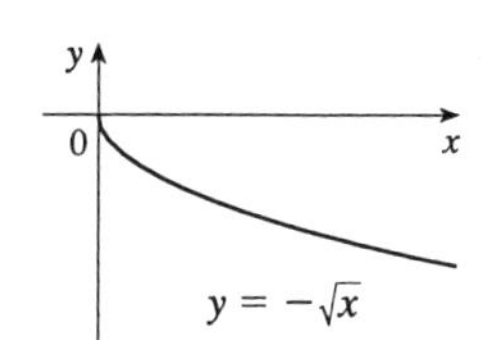

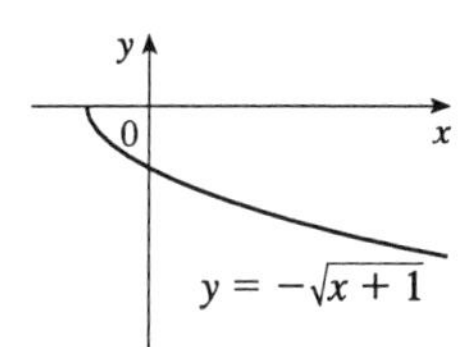

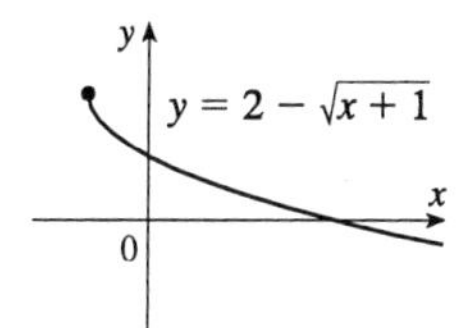

23. $y = x^2 - 2x = (x-1)^2 - 1$ $\qquad$ $y = |x^2 - 2x|$

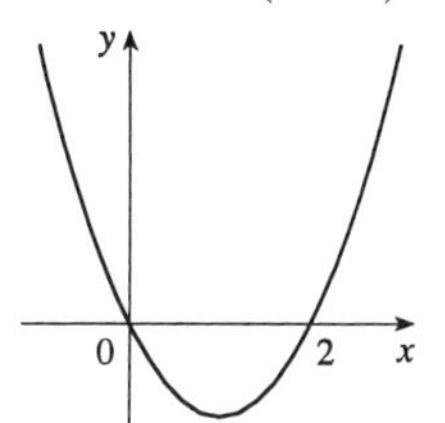

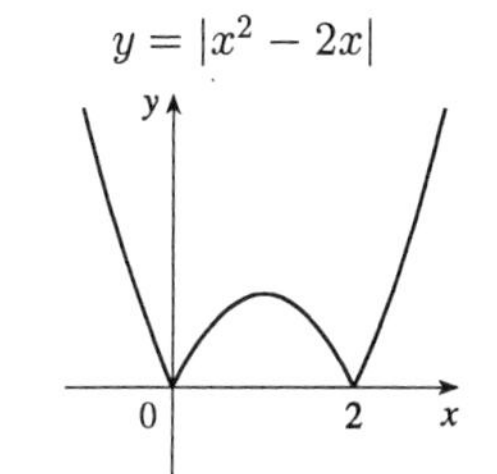

25. $y = ||x| - 1|$

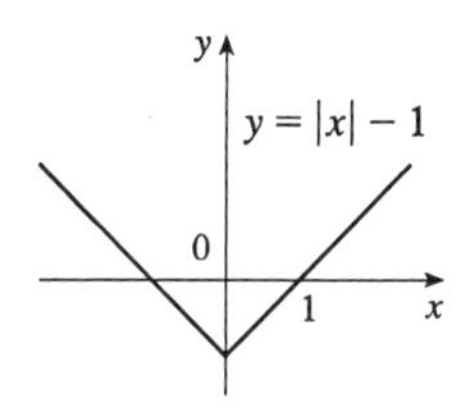

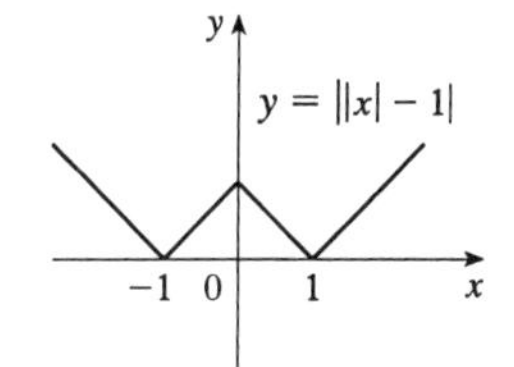

27. **(a)** To obtain $y = f(|x|)$, the portion of $y = f(x)$ right of the y-axis is reflected in the y-axis.

(b) $y = \sin|x|$

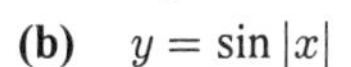

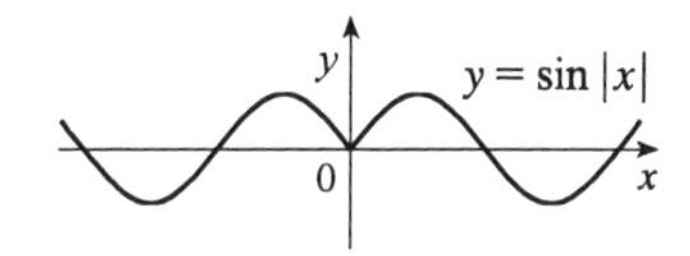

29. Note that there are vertical asymptotes wherever $f(x) = 0$, since division by 0 is impossible.

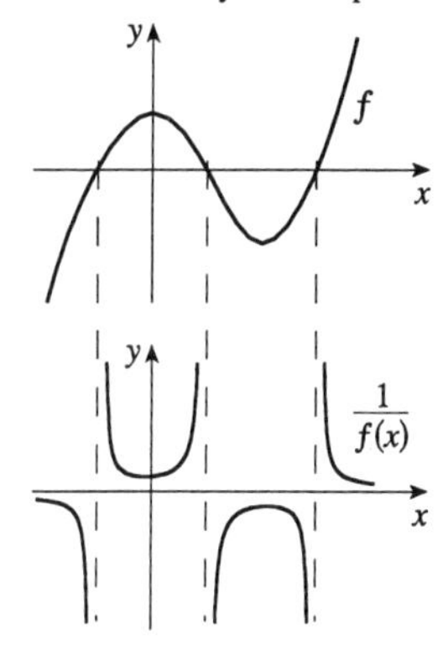

EXERCISES 3

1. $f(x) = x^4 + 2$

(a) $[-2, 2]$ by $[-2, 2]$

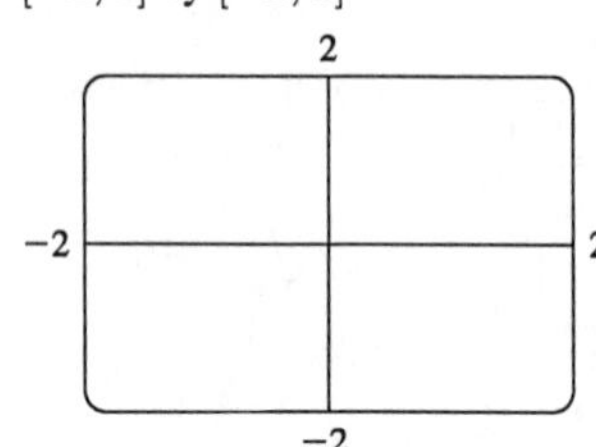

(b) $[0, 4]$ by $[0, 4]$

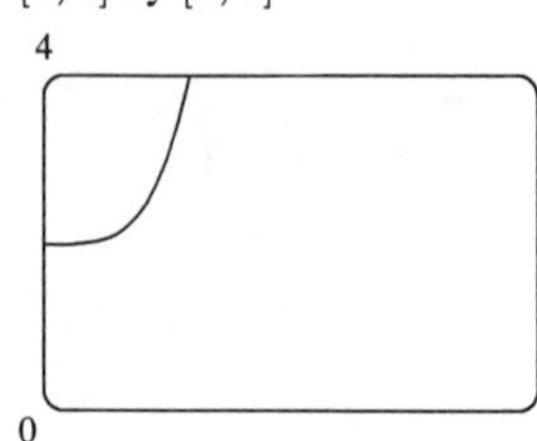

(c) $[-4, 4]$ by $[-4, 4]$

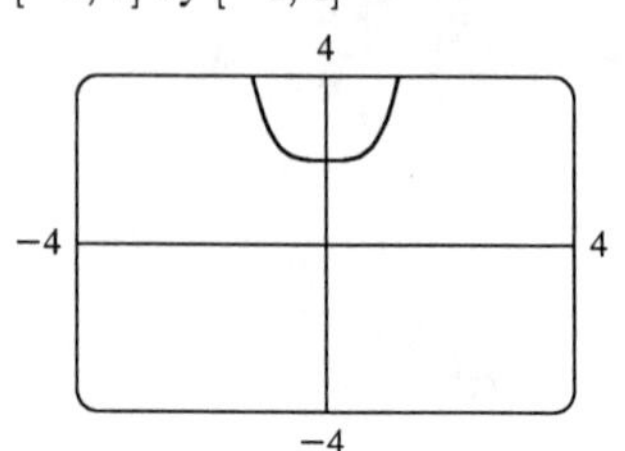

(d) $[-8, 8]$ by $[-4, 40]$

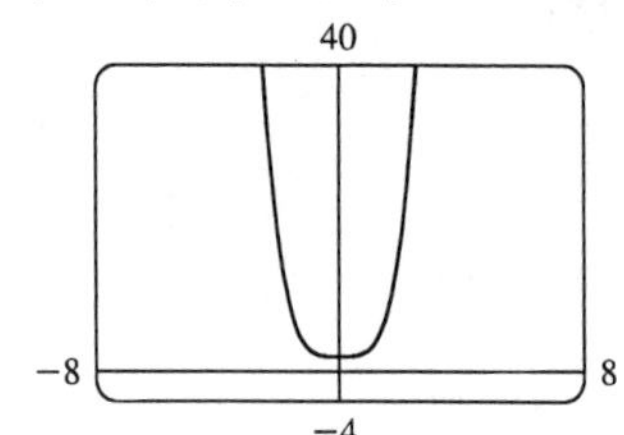

(e) $[-40, 40]$ by $[-80, 800]$

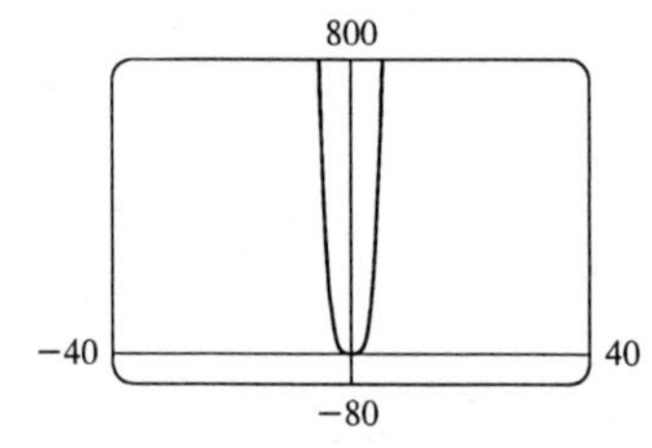

The most appropriate graph is produced in viewing rectangle (d).

3. $f(x) = 10 + 25x - x^3$

(a) $[-4, 4]$ by $[-4, 4]$

4

−4 4

−4

(b) $[-10, 10]$ by $[-10, 10]$

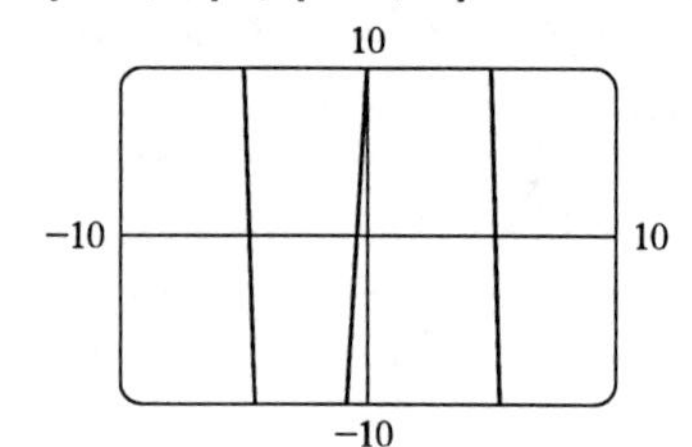

(c) $[-20, 20]$ by $[-100, 100]$

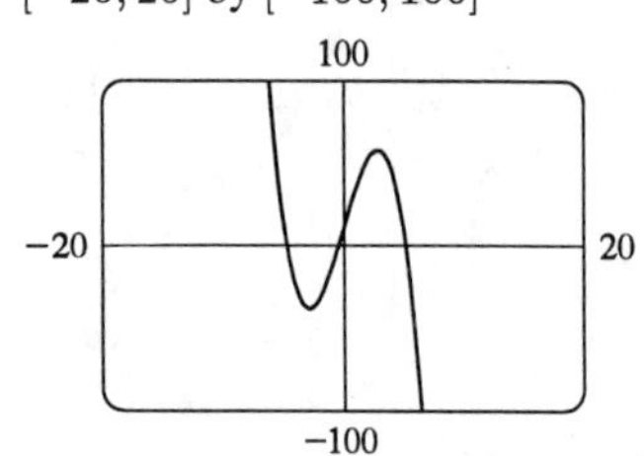

(d) $[-100, 100]$ by $[-200, 200]$

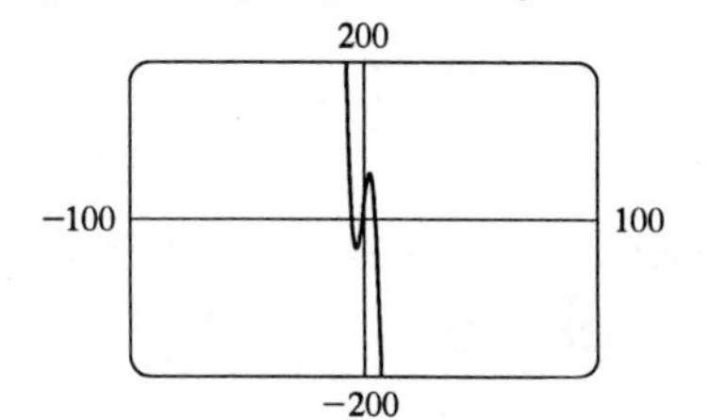

The most appropriate graph is produced in viewing rectangle (c).

5. $f(x) = 4 + 6x - x^2$
Note that many similar rectangles give equally good views of the function.

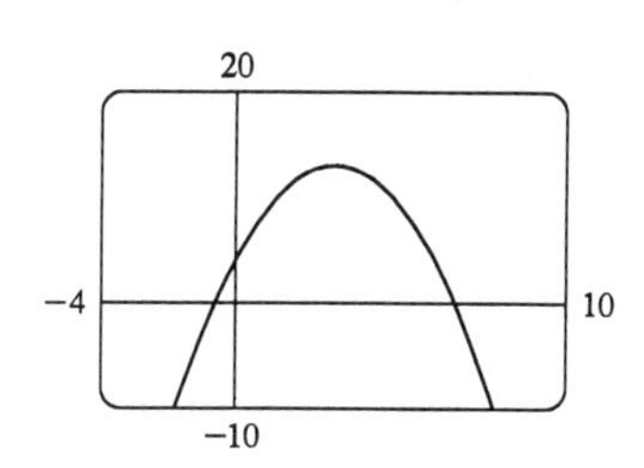

7. $f(x) = \sqrt[4]{256 - x^2}$ To find an appropriate viewing rectangle, we calculate f's domain and range: $256 - x^2 \geq 0 \quad \Leftrightarrow \quad x^2 \leq 256 \quad \Leftrightarrow \quad |x| \leq 16 \quad \Leftrightarrow \quad -16 \leq x \leq 16$, so the domain is $[-16, 16]$. Also, $0 \leq \sqrt[4]{256 - x^2} \leq \sqrt[4]{256} = 4$, so the range is $[0, 4]$. Thus we choose the viewing rectangle to be $[-20, 20]$ by $[-2, 6]$.

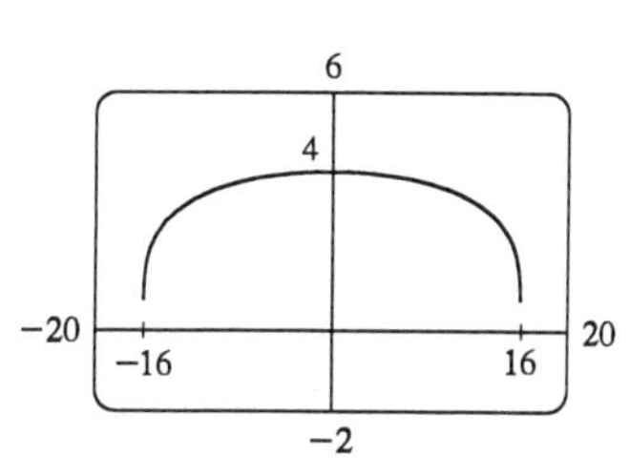

9. $f(x) = 0.01x^3 - x^2 + 5$

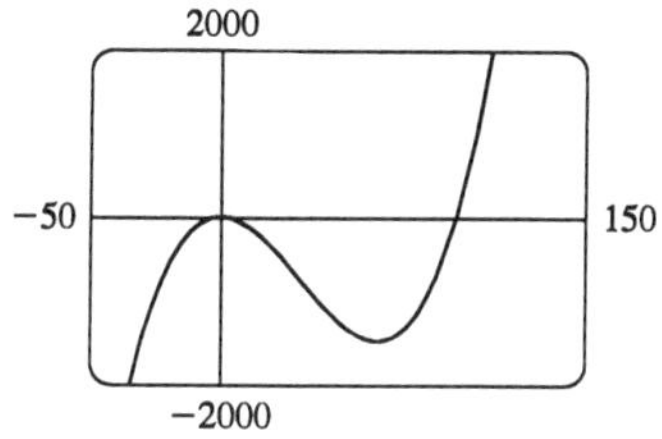

2000
−50
150
−2000

11. $y = \dfrac{1}{x^2 + 25}$

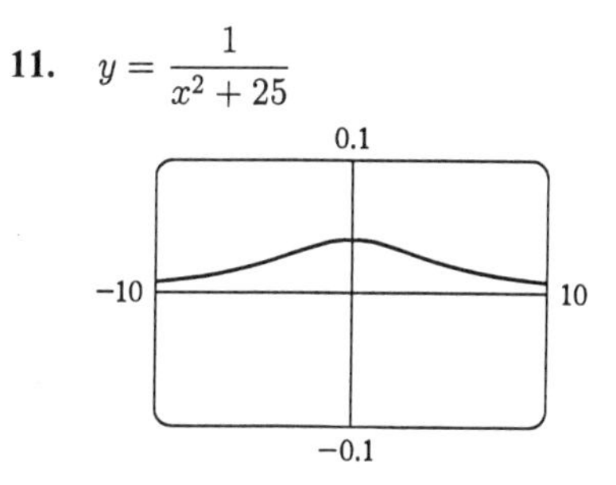

13. $y = x^4 - 4x^3$

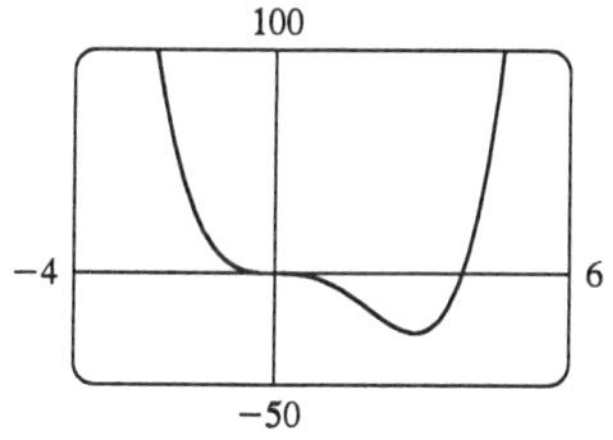

100
−4
6
−50

15. $y = \dfrac{2x - 1}{x + 3}$

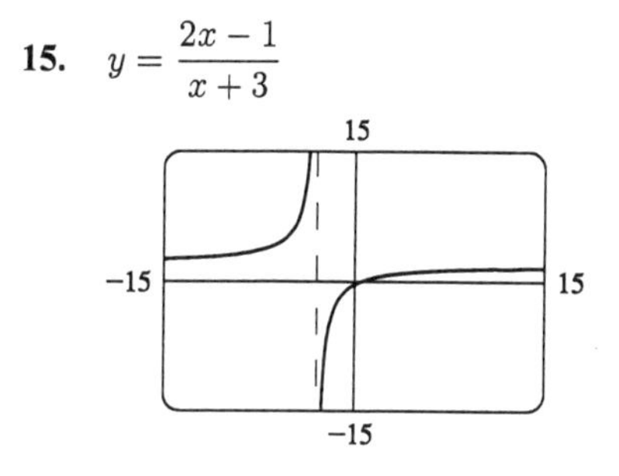

17. $f(x) = \cos(100x)$

1.5
−0.1
0.1
−1.5

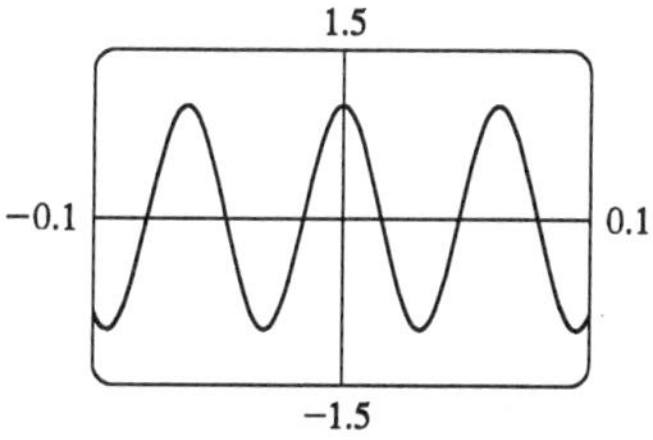

19. $f(x) = \sin(x/40)$

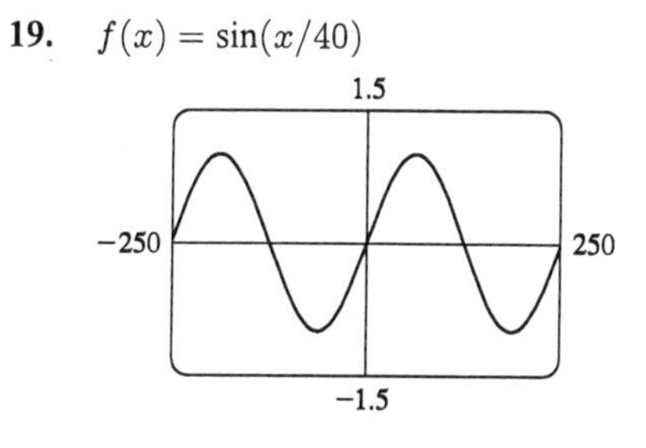

21. $y = 3^{\cos(x^2)}$

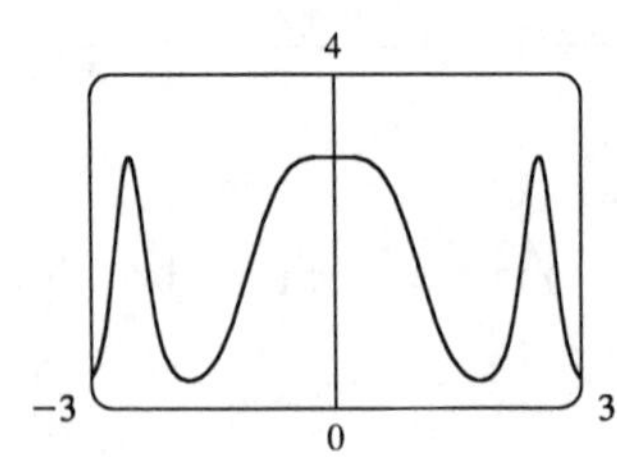

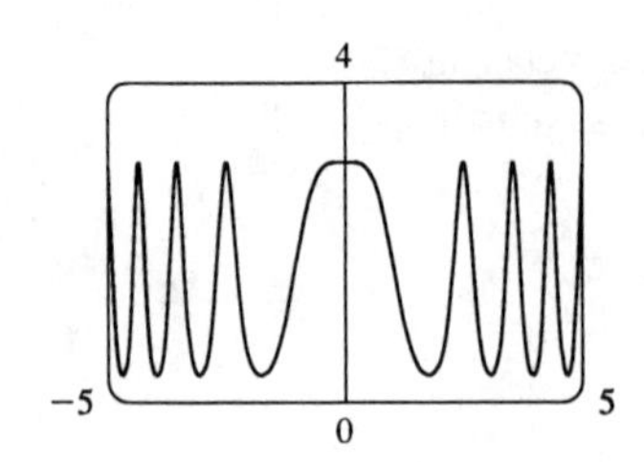

23. $y = \pm\sqrt{\dfrac{1-4x^2}{2}}$

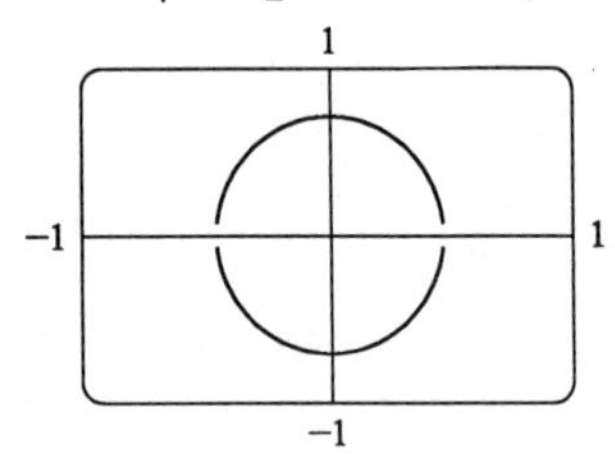

24. $y = \pm\sqrt{1+9x^2}$

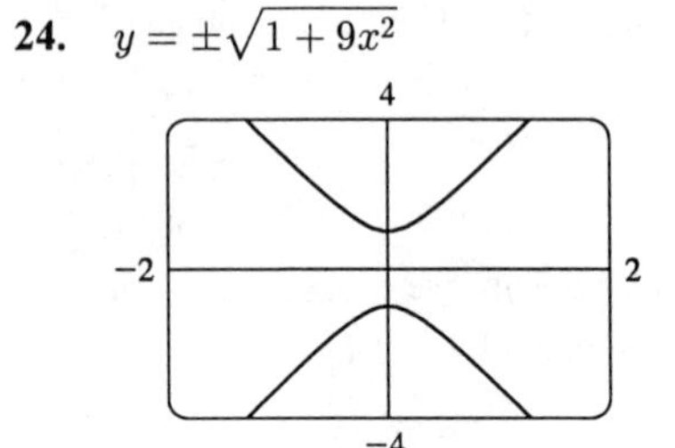

25. In Maple, we can use the procedure

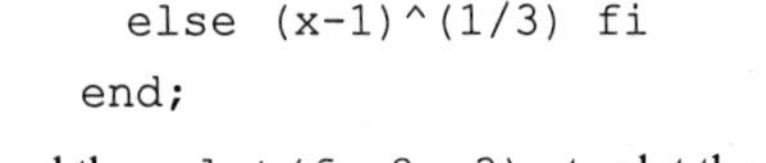

```
f:=proc(x)
    if x<=1 then x^3-2*x+1
    else (x-1)^(1/3) fi
  end;
```

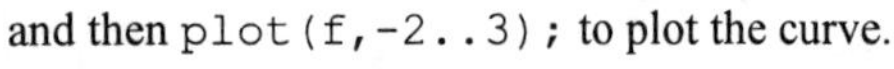

and then `plot(f,-2..3);` to plot the curve.
To define f in Mathematica, we can use
`f[x_]:=If[x<=1,x^3-2*x+1,(x-1)^(1/3)].`

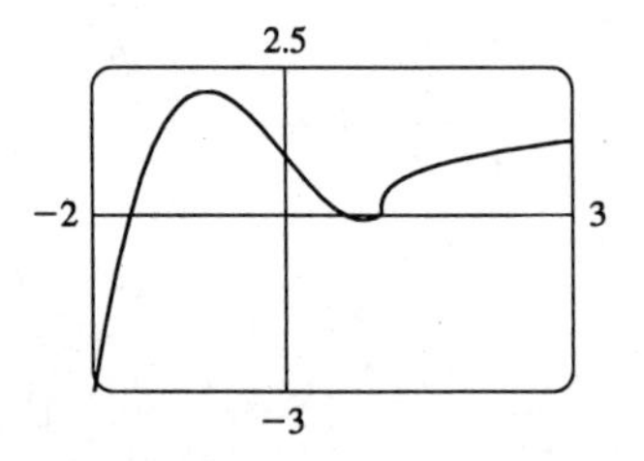

27. We first graph $f(x) = 3x^3 + x^2 + x - 2$ in the viewing rectangle $[-2, 2] \times [-30, 30]$ to find the approximate value of the root. The only root appears to be between $x = 0.5$ and $x = 1$, so we graph f again in the rectangle $[0.5, 1] \times [-1, 1]$ (or use the cursor on a graphing calculator). From the second graph, it appears that the only solution to the equation $f(x) = 0$ is 0.67, to 2 decimal places.

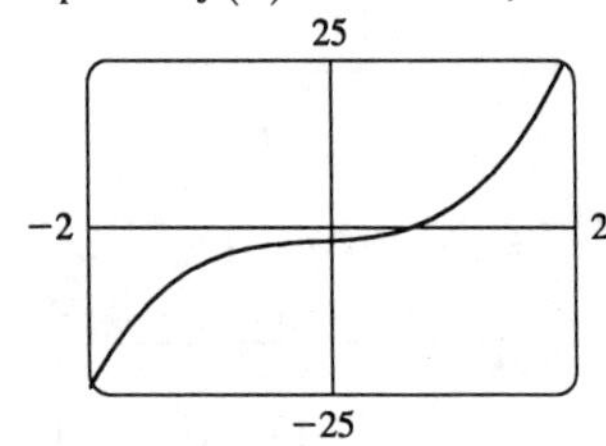

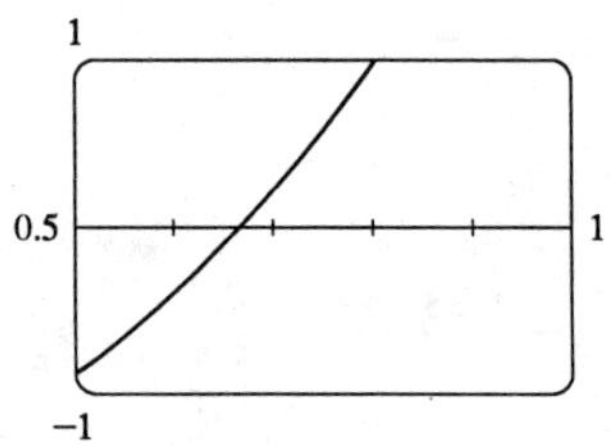

29. From the first graph, it appears that the roots lie near ± 2. We zoom in (or use the cursor) and find that the solutions are -1.90, 0 and 1.90, to two decimal places.

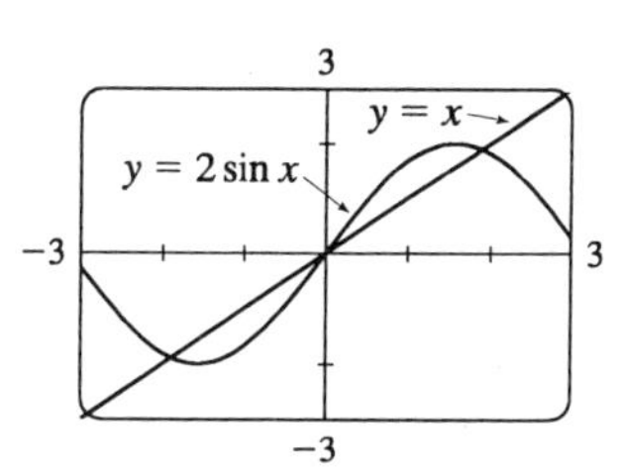

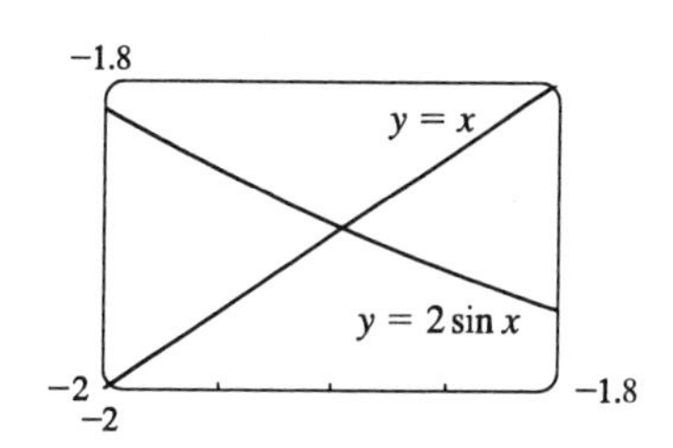

31. $g(x) = x^3/10$ is eventually larger than $f(x) = 10x^2$.

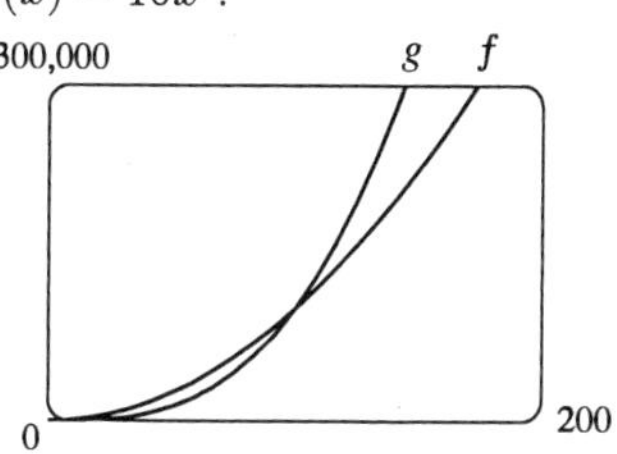

33. **(a)** **(i)** $[0, 5]$ by $[0, 20]$ **(ii)** $[0, 25]$ by $[0, 10^7]$ **(iii)** $[0, 50]$ by $[0, 10^8]$

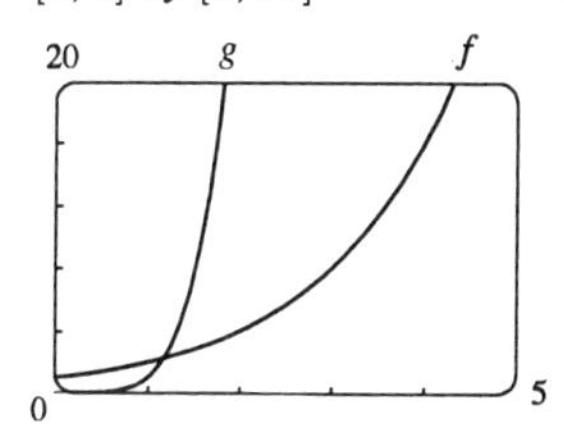

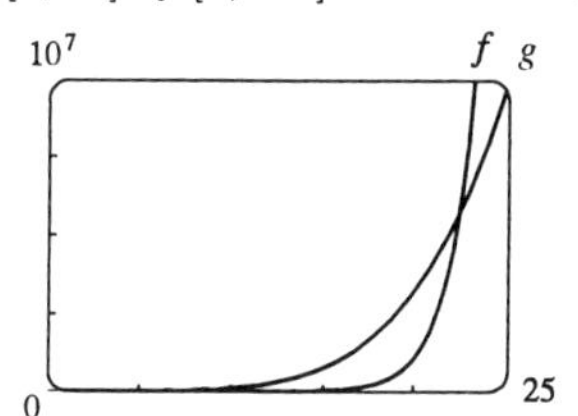

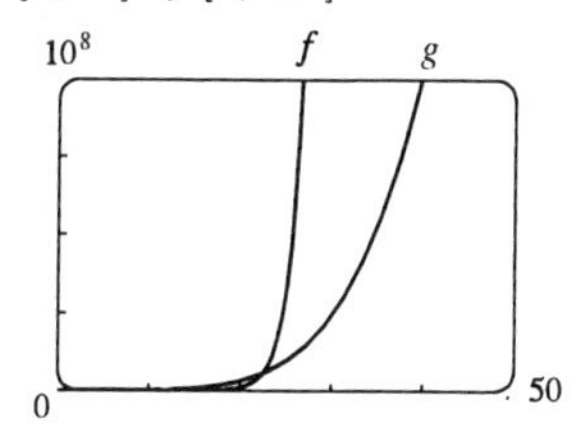

As x gets large, f grows much more quickly than g.

(b) From the graphs in part (a), it appears that the two solutions are $x \approx 1.2$ and 22.4.

35. We see from the graph of $y = |\sin x - x|$ that there are two solutions to the equation $|\sin x - x| = 0.1$: $x \approx -0.85$ and $x \approx 0.85$. The condition $|\sin x - x| < 0.1$ holds for any x lying between these two values.

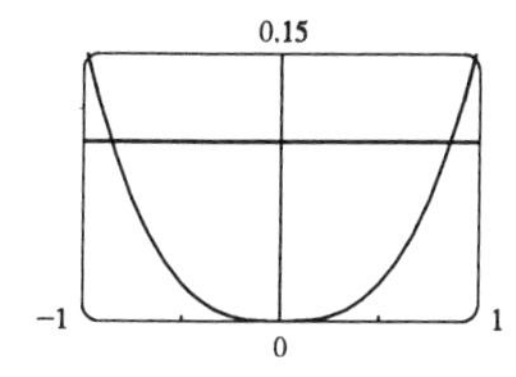

37. **(a)** The root functions $y = \sqrt{x}$, $y = \sqrt[4]{x}$ and $y = \sqrt[6]{x}$

3

√x ∜x

⁶√x

−1 4

−1

(b) The root functions $y = x$, $y = \sqrt[3]{x}$ and $y = \sqrt[5]{x}$

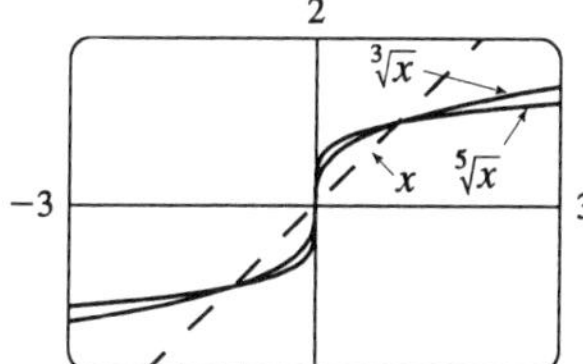

(c) The root functions
$y = \sqrt{x}$, $y = \sqrt[3]{x}$, $y = \sqrt[4]{x}$ and $y = \sqrt[5]{x}$

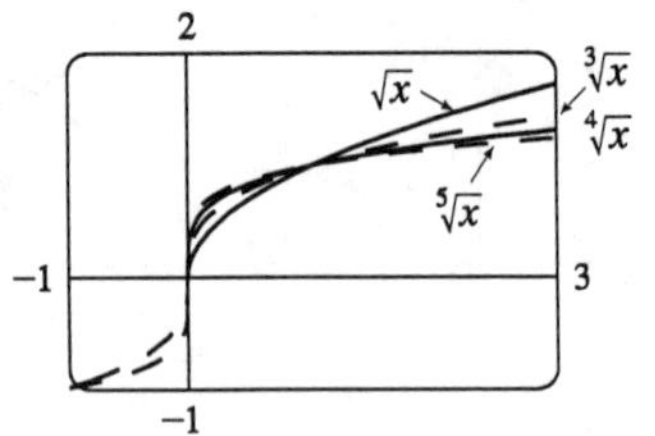

(d)
- For any n, the nth root of 0 is 0 and the nth root of 1 is 1, that is, all nth root functions pass through the points $(0, 0)$ and $(1, 1)$.
- For odd n, the domain of the nth root function is $\mathbb{R}$, while for even n, it is $\{x \in \mathbb{R} \mid x \geq 0\}$.
- Graphs of even root functions look similar to that of $\sqrt{x}$, while those of odd root functions resemble that of $\sqrt[3]{x}$.
- As n increases, the graph of $\sqrt[n]{x}$ becomes steeper near 0 and flatter for $x > 1$.

39. $f(x) = x^4 + cx^2 + x$
If $c < 0$, there are three humps: two minimum points and a maximum point. These humps get flatter as c increases, until at $c = 0$ two of the humps disappear and there is only one minimum point. This single hump then moves to the right and approaches the origin as c increases.

41. $y = x^n 2^{-x}$
As n increases, the maximum of the function moves further from the origin, and gets larger. Note, however, that regardless of n, the function approaches 0 as $x \to \infty$.

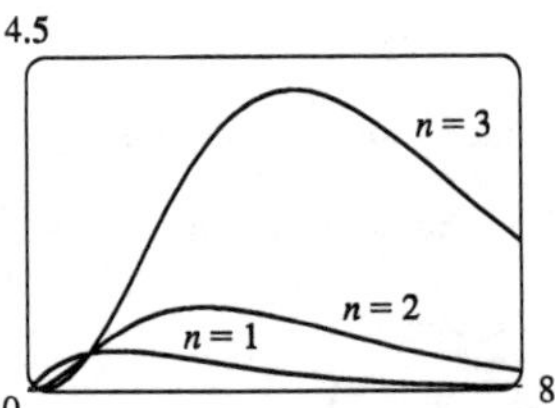
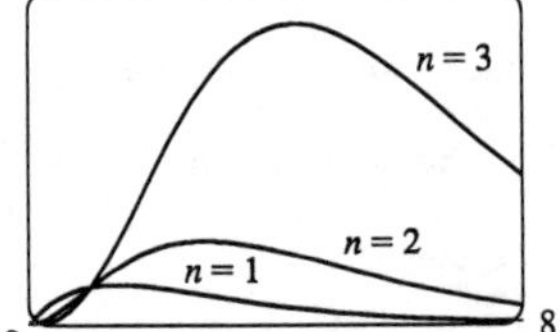

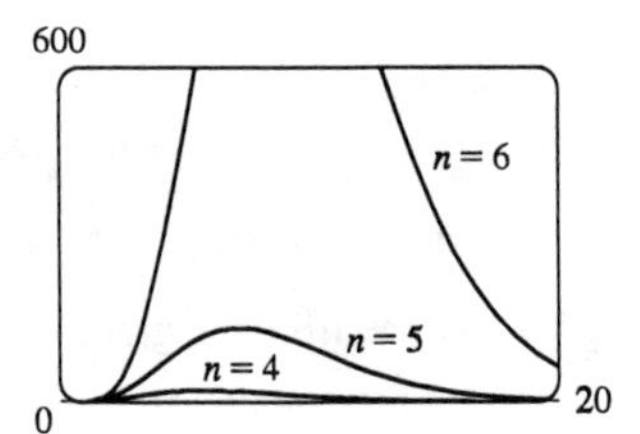

43. $y^2 = cx^3 + x^2$
If $c < 0$, the loop is to the right of the origin, and if c is positive, it is to the left. In both cases, the closer c is to 0, the larger the loop is. (In the limiting case, $c = 0$, the loop is "infinite," that is, it doesn't close.) Also, the larger $|c|$ is, the steeper the slope is on the loopless side of the origin.

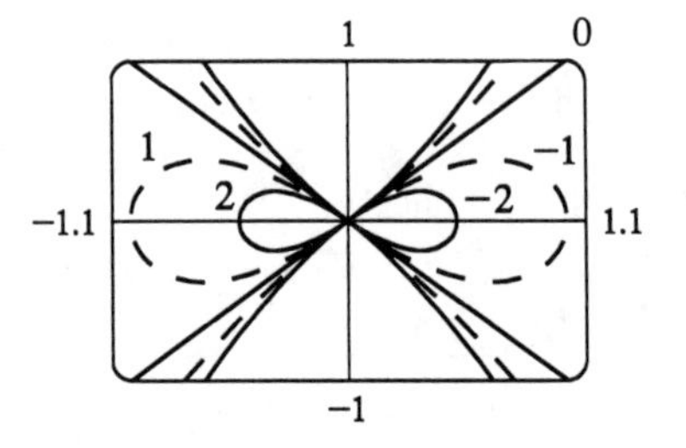

EXERCISES 4

1. As in Example 2, $|x+1| = \begin{cases} -x-1 & \text{if } x < -1 \\ x+1 & \text{if } x \geq 1 \end{cases}$ and $|x+4| = \begin{cases} -x-4 & \text{if } x < -4 \\ x+4 & \text{if } x \geq -4 \end{cases}$

Therefore we consider three cases: $x < -4$, $-4 \leq x < -1$, and $x \geq -1$.

If $x < -4$, we must have $-x-1-x-4 \leq 5 \quad \Leftrightarrow \quad x \geq -5$.

If $-4 \leq x < -1$, we must have $-x-1+x+4 \leq 5 \quad \Leftrightarrow \quad 3 \leq 5$.

If $x \geq -1$, we must have $x+1+x+4 \leq 5 \quad \Leftrightarrow \quad x \leq 0$.

These conditions together imply that $-5 \leq x \leq 0$.

3. $|2x-1| = \begin{cases} 1-2x & \text{if } x < \frac{1}{2} \\ 2x-1 & \text{if } x \geq \frac{1}{2} \end{cases}$ and $|x+5| = \begin{cases} -x-5 & \text{if } x < -5 \\ x+5 & \text{if } x \geq -5 \end{cases}$

Therefore we consider the cases $x < -5$, $-5 \leq x < \frac{1}{2}$, and $x \geq \frac{1}{2}$.

If $x < -5$, we must have $1-2x-(-x-5) = 3 \;\Leftrightarrow\; x = 3$, which is false, since we are considering $x < -5$.

If $-5 \leq x < \frac{1}{2}$, we must have $1-2x-(x+5) = 3 \;\Leftrightarrow\; x = -\frac{7}{3}$.

If $x \geq \frac{1}{2}$, we must have $2x-1-(x+5) = 3 \;\Leftrightarrow\; x = 9$.

So the two solutions of the equation are $x = -\frac{7}{3}$ and $x = 9$.

5. The final digit in 947^{362} is determined by 7^{362}, since $947^{362} = (900+40+7)^{362}$, and every term in the expansion of this expression is the product of powers of either 40 or 900 or both, the last digits of which are all zero, except the term 7^{362}. Looking at the first few powers of 7 we see: $7^1 = 7$, $7^2 = 49$, $7^3 = 343$, $7^4 = 2401$, $7^5 = 16{,}807$, $7^6 = 117{,}649$ and it appears that the final digit follows a cyclical pattern, namely $7 \to 9 \to 3 \to 1 \to 7 \to 9$, of length 4. Since $362 \div 4 = 90$ with remainder 2, the final digit is the second in the cycle, that is, 9.

7. $f_0(x) = x^2$ and $f_{n+1}(x) = f_0(f_n(x))$ for $n = 0, 1, 2, \ldots$.

$f_1(x) = f_0(f_0(x)) = f_0(x^2) = (x^2)^2 = x^4$, $f_2(x) = f_0(f_1(x)) = f_0(x^4) = (x^4)^2 = x^8$,

$f_3(x) = f_0(f_2(x)) = f_0(x^8) = (x^8)^2 = x^{16}, \ldots$. Thus a general formula is $f_n(x) = x^{2^{n+1}}$.

9. $f(x) = |x^2 - 4|x| + 3|$. If $x \geq 0$, then $f(x) = |x^2-4x+3| = |(x-1)(x-3)|$.

Case (i): If $0 < x \leq 1$, then $f(x) = x^2 - 4x + 3$.

Case (ii): If $1 < x \leq 3$, then $f(x) = -(x^2-4x+3) = -x^2+4x-3$.

Case (iii): If $x > 3$, then $f(x) = x^2-4x+3$.

This enables us to sketch the graph for $x \geq 0$. Then we use the fact that f is an even function to reflect this part of the graph about the y-axis to obtain the entire graph. Or, we could consider also the cases $x < -3$, $-3 \leq x < -1$, and $-1 \leq x < 0$.

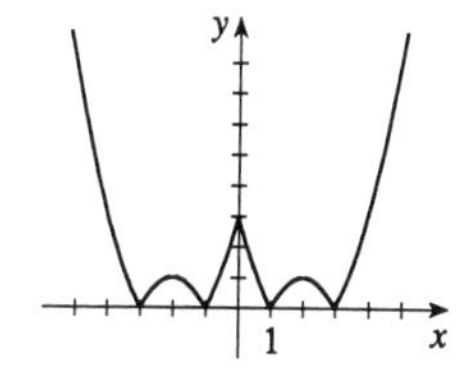

11. $\left[\sqrt{3+2\sqrt{2}}-\sqrt{3-2\sqrt{2}}\right]^2 = \left(3+2\sqrt{2}\right)+\left(3-2\sqrt{2}\right)-2\sqrt{\left(3+2\sqrt{2}\right)\left(3-2\sqrt{2}\right)}$

$= 6-2\sqrt{9+6\sqrt{2}-6\sqrt{2}-4\sqrt{2}\sqrt{2}} = 6-2=4$. So the given expression is $\sqrt{4}=2$.

13. 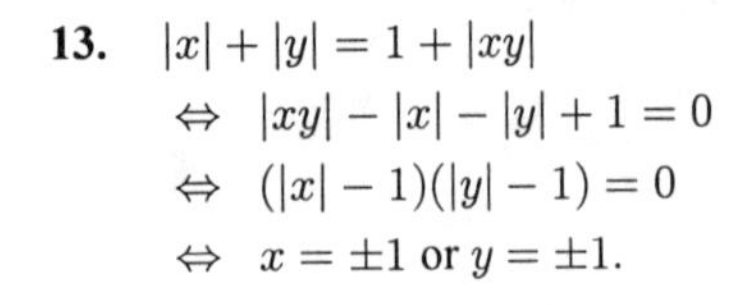

$|x|+|y| = 1+|xy|$

$\Leftrightarrow \; |xy|-|x|-|y|+1=0$

$\Leftrightarrow \; (|x|-1)(|y|-1)=0$

$\Leftrightarrow \; x=\pm 1$ or $y=\pm 1$.

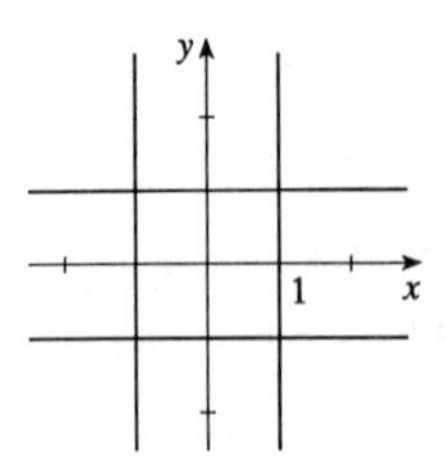

15. $|x|+|y|\leq 1$. The boundary of the region has equation $|x|+|y|=1$. In quadrants I, II, III, IV, this becomes the lines $x+y=1$, $-x+y=1$, $-x-y=1$, and $x-y=1$ respectively.

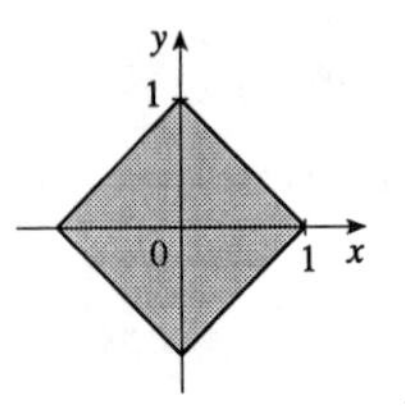

17. **(a)** The amount of ribbon needed is equal to the circumference of the earth, which is about $2\pi r = 2\pi(3960)\approx 24{,}880$ mi.

(b) The additional ribbon needed is $2\pi(r+1)-2\pi(r)=2\pi(1)\approx 6.3$ ft. (!)

19. Let x represent the length of the base in cm and h the length of the altitude in cm. By the Pythagorean Theorem, $3^2+x^2=5^2 \quad\Leftrightarrow\quad x^2=16 \quad\Leftrightarrow\quad x=4$ and so the area of the triangle is $\frac{1}{2}\cdot 4\cdot 3=6$. But the area of the triangle is also $\frac{1}{2}\cdot 5\cdot h=6 \quad\Leftrightarrow\quad h=2.4$ and hence the length of the altitude is 2.4 cm.

21. We use a proof by contradiction. Assume that $\sqrt{3}$ is rational. So $\sqrt{3}=p/q$ for some integers p and q. This fraction is assumed to be in lowest form (that is, p and q have no common factors). Then $3q^2=p^2$. We see that 3 must be one of the prime factors of p^2, and thus of p itself. So $p=3k$ for some integer k. Substituting this into the previous equation, we get $3q^2=(3k)^2=9k^2 \quad\Leftrightarrow\quad q^2=3k^2$. So 3 is one of the prime factors of q^2, and thus of q itself. So 3 divides both p and q. But this contradicts our assumption that the fraction p/q was in lowest form So $\sqrt{3}$ is not expressible as a ratio of integers, that is, $\sqrt{3}$ is irrational.

23. The statement is false. Here is one particular counterexample:

	First Half	Second Half	Whole Season
Player A	1/99	1/1	$2/100=.020$
Player B	0/1	98/99	$98/100=.980$

25. The odometer reading is proportional to the number of tire revolutions. Let r_1 represent the number of revolutions made by the tire on the 400 mi trip and r_2 represent the number of revolutions made by the tire on the 390 mi trip. Then $r_1=400k$ and $r_2=390k$ for some constant k. Let d be the actual distance traveled, R_1 be the radius of normal tires, and R_2 be the radius of snow tires. Then $d=2\pi R_1 r_1=2\pi R_2 r_2 \;\Leftrightarrow\; R_1 r_1=R_2 r_2$

$\Leftrightarrow \quad R_2=\dfrac{R_1 r_1}{r_2}=\dfrac{15\cdot 400k}{390k}=\dfrac{15\cdot 40}{39}\approx 15.38$. So the radius of the snow tires is about 15.4 in.

CHAPTER ONE

EXERCISES 1.1

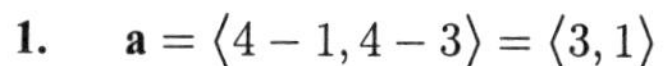

1. $\mathbf{a} = \langle 4-1, 4-3 \rangle = \langle 3, 1 \rangle$

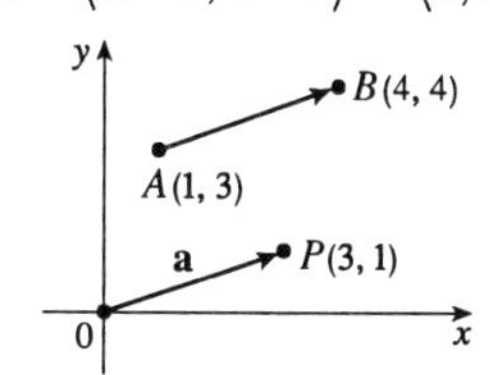

3. $\mathbf{a} = \langle 3-3, -3-(-1) \rangle = \langle 0, -2 \rangle$

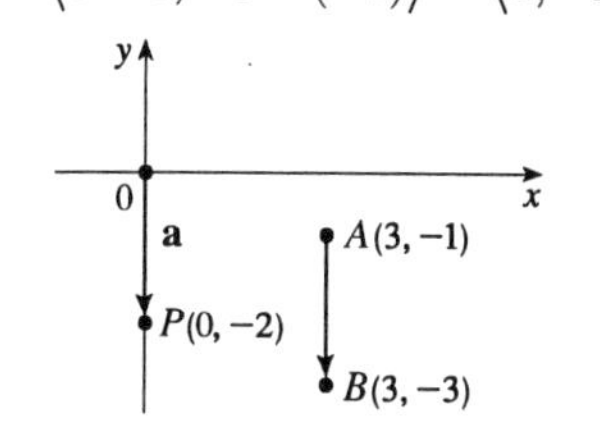

5. $\langle 2, 3 \rangle + \langle 3, -4 \rangle = \langle 2+3, 3+(-4) \rangle = \langle 5, -1 \rangle$

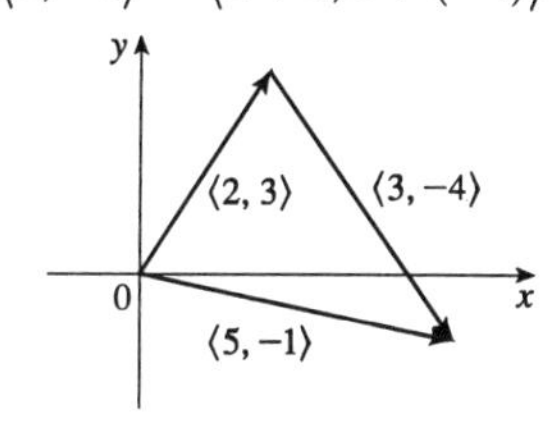

(using the triangle law)

7. $\langle 1, 0 \rangle + \langle 0, 1 \rangle = \langle 1, 1 \rangle$

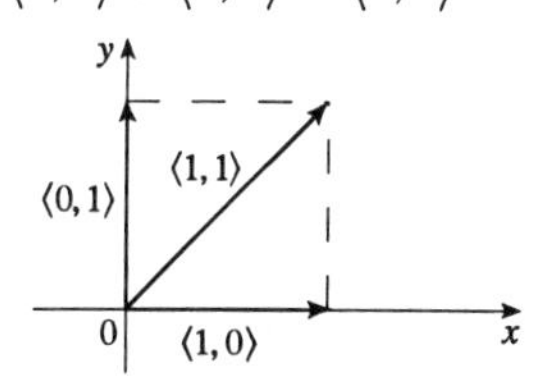

(using the parallelogram law)

9. $\mathbf{a} = \langle 5, -12 \rangle, \mathbf{b} = \langle -2, 8 \rangle$

$|\mathbf{a}| = \sqrt{5^2 + (-12)^2} = \sqrt{169} = 13$ $\qquad$ $\mathbf{a} + \mathbf{b} = \langle 5-2, -12+8 \rangle = \langle 3, -4 \rangle$

$\mathbf{a} - \mathbf{b} = \langle 5-(-2), -12-8 \rangle = \langle 7, -20 \rangle$ $\qquad$ $2\mathbf{a} = \langle 2(5), 2(-12) \rangle = \langle 10, -24 \rangle$

$3\mathbf{a} + 4\mathbf{b} = 3\langle 5, -12 \rangle + 4\langle -2, 8 \rangle = \langle 15, -36 \rangle + \langle -8, 32 \rangle = \langle 7, -4 \rangle$

11. $\mathbf{a} = \langle 2, -3 \rangle, \mathbf{b} = \langle 1, 4 \rangle$

$|\mathbf{a}| = \sqrt{2^2 + (-3)^2} = \sqrt{13}$ $\qquad$ $\mathbf{a} + \mathbf{b} = \langle 2+1, -3+4 \rangle = \langle 3, 1 \rangle$

$\mathbf{a} - \mathbf{b} = \langle 2-1, -3-4 \rangle = \langle 1, -7 \rangle$ $\qquad$ $2\mathbf{a} = \langle 2(2), 2(-3) \rangle = \langle 4, -6 \rangle$

$3\mathbf{a} + 4\mathbf{b} = \langle 6, -9 \rangle + \langle 4, 16 \rangle = \langle 10, 7 \rangle$

13. $\mathbf{a} = \mathbf{i} - \mathbf{j}, \mathbf{b} = \mathbf{i} + \mathbf{j}$

$|\mathbf{a}| = \sqrt{1^2 + (-1)^2} = \sqrt{2}$ $\qquad$ $\mathbf{a} + \mathbf{b} = (\mathbf{i} - \mathbf{j}) + (\mathbf{i} + \mathbf{j}) = 2\mathbf{i}$

$\mathbf{a} - \mathbf{b} = (\mathbf{i} - \mathbf{j}) - (\mathbf{i} + \mathbf{j}) = -2\mathbf{j}$ $\qquad$ $2\mathbf{a} = 2(\mathbf{i} - \mathbf{j}) = 2\mathbf{i} - 2\mathbf{j}$

$3\mathbf{a} + 4\mathbf{b} = 3(\mathbf{i} - \mathbf{j}) + 4(\mathbf{i} + \mathbf{j}) = 3\mathbf{i} - 3\mathbf{j} + 4\mathbf{i} + 4\mathbf{j} = 7\mathbf{i} + \mathbf{j}$

15. $\mathbf{a} = \mathbf{i} + \mathbf{j}, \mathbf{b} = 2\mathbf{i} - \mathbf{j}$

$|\mathbf{a}| = \sqrt{1^2 + 1^2} = \sqrt{2}$ $\qquad$ $\mathbf{a} + \mathbf{b} = (\mathbf{i} + \mathbf{j}) + (2\mathbf{i} - \mathbf{j}) = 3\mathbf{i}$

$\mathbf{a} - \mathbf{b} = (\mathbf{i} + \mathbf{j}) - (2\mathbf{i} - \mathbf{j}) = -\mathbf{i} + 2\mathbf{j}$ $\qquad$ $2\mathbf{a} = 2(\mathbf{i} + \mathbf{j}) = 2\mathbf{i} + 2\mathbf{j}$

$3\mathbf{a} + 4\mathbf{b} = 3(\mathbf{i} + \mathbf{j}) + 4(2\mathbf{i} - \mathbf{j}) = 3\mathbf{i} + 3\mathbf{j} + 8\mathbf{i} - 4\mathbf{j} = 11\mathbf{i} - \mathbf{j}$

17. $|\langle 1, 2 \rangle| = \sqrt{1^2 + 2^2} = \sqrt{5}$. Thus, $\mathbf{u} = \frac{1}{\sqrt{5}}\langle 1, 2 \rangle = \left\langle \frac{1}{\sqrt{5}}, \frac{2}{\sqrt{5}} \right\rangle$.

19. $|\mathbf{i}+\mathbf{j}| = \sqrt{1^2+1^2} = \sqrt{2}$. Thus, $\mathbf{u} = \frac{1}{\sqrt{2}}(\mathbf{i}+\mathbf{j}) = \frac{1}{\sqrt{2}}\mathbf{i} + \frac{1}{\sqrt{2}}\mathbf{j}$.

21. We can find an expression for **i** by eliminating **j**. To do this, we will add 3 times **b** to **a**.
$\mathbf{a} = 2\mathbf{i}+3\mathbf{j}, \mathbf{b} = \mathbf{i}-\mathbf{j} \quad \Rightarrow \quad \mathbf{a}+3\mathbf{b} = 2\mathbf{i}+3\mathbf{j}+3\mathbf{i}-3\mathbf{j} = 5\mathbf{i} \quad \Rightarrow \quad \mathbf{i} = \frac{1}{5}\mathbf{a}+\frac{3}{5}\mathbf{b}$. Substituting this expression for **i** into $\mathbf{a} = 2\mathbf{i}+3\mathbf{j}$ gives $\mathbf{a} = 2\left(\frac{1}{5}\mathbf{a}+\frac{3}{5}\mathbf{b}\right)+3\mathbf{j} \quad \Rightarrow \quad \frac{3}{5}\mathbf{a}-\frac{6}{5}\mathbf{b} = 3\mathbf{j} \quad \Rightarrow \quad \mathbf{j} = \frac{1}{5}\mathbf{a}-\frac{2}{5}\mathbf{b}$.

23. By the triangle law, $\overrightarrow{AB}+\overrightarrow{BC} = \overrightarrow{AC}$, and $\overrightarrow{AC} = -\overrightarrow{CA} \quad \Rightarrow \quad \overrightarrow{AB}+\overrightarrow{BC}+\overrightarrow{CA} = -\overrightarrow{CA}+\overrightarrow{AC} = \mathbf{0}$.

25. **(a), (b)**

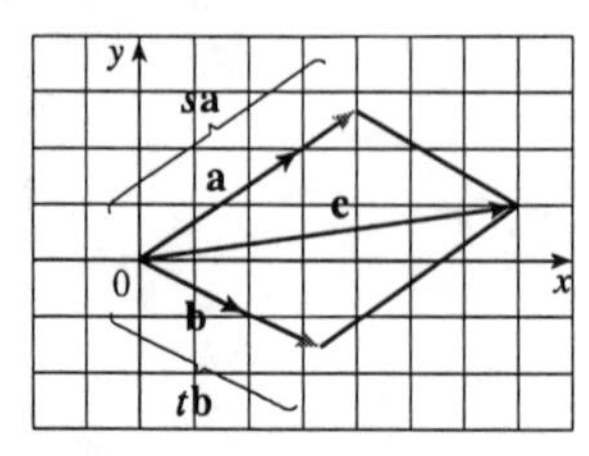

(c) From the sketch, we estimate that $s \approx 1.3$ and $t \approx 1.6$.

(d) $\mathbf{c} = s\mathbf{a}+t\mathbf{b} \quad \Leftrightarrow \quad \langle 7,1\rangle = \langle 3s,2s\rangle + \langle 2t,-t\rangle \quad \Leftrightarrow \quad 7 = 3s+2t$ and $1 = 2s-t$. Solving these equations gives $s = \frac{9}{7}$ and $t = \frac{11}{7}$.

27. $|\mathbf{F}_1| = 10$ lb and $|\mathbf{F}_2| = 12$ lb.
$\mathbf{F}_1 = -|\mathbf{F}_1|\cos 45°\,\mathbf{i} + |\mathbf{F}_1|\sin 45°\,\mathbf{j} = -10\cos 45°\,\mathbf{i} + 10\sin 45°\,\mathbf{j} = -5\sqrt{2}\mathbf{i}+5\sqrt{2}\mathbf{j}$
$\mathbf{F}_2 = |\mathbf{F}_2|\cos 30°\,\mathbf{i} + |\mathbf{F}_2|\sin 30°\,\mathbf{j} = 12\cos 30°\,\mathbf{i} + 12\sin 30°\,\mathbf{j} = 6\sqrt{3}\mathbf{i}+6\mathbf{j}$
$\mathbf{F} = \mathbf{F}_1+\mathbf{F}_2 = \left(6\sqrt{3}-5\sqrt{2}\right)\mathbf{i} + \left(6+5\sqrt{2}\right)\mathbf{j} \approx 3.32\mathbf{i}+13.07\mathbf{j}$
$|\mathbf{F}| \approx \sqrt{(3.32)^2+(13.07)^2} \approx 13.5$ lb. $\tan\theta = \dfrac{6+5\sqrt{2}}{6\sqrt{3}-5\sqrt{2}} \quad \Rightarrow \quad \theta = \tan^{-1}\dfrac{6+5\sqrt{2}}{6\sqrt{3}-5\sqrt{2}} \approx 76°$.

29. With respect to the water's surface, the woman's velocity is the vector sum of the velocity of the ship with respect to the water, and her velocity with respect to the ship. If we let north be the positive y-direction, then $\mathbf{v} = \langle 0,22\rangle + \langle -3,0\rangle = \langle -3,22\rangle$. The woman's speed is $|\mathbf{v}| = \sqrt{9+484} \approx 22.2$ mi/h. The vector **v** makes an angle θ with the east, where $\theta = \tan^{-1}\frac{22}{-3} \approx 98°$. Therefore, the woman's direction is about N$(98-90)°$ W $=$ N $8°$ W.

31. $|\mathbf{r}-\mathbf{r}_0|$ is the distance between the points (x,y) and (x_0,y_0), so the set of points is a circle with radius 1 and center (x_0,y_0).
Alternate Method: $|\mathbf{r}-\mathbf{r}_0| = 1 \quad \Leftrightarrow \quad \sqrt{(x-x_0)^2+(y-y_0)^2} = 1 \quad \Leftrightarrow \quad (x-x_0)^2+(y-y_0)^2 = 1$, which is the equation of a circle with radius 1 and center (x_0,y_0).

33.
$$\begin{aligned}\mathbf{a}+(\mathbf{b}+\mathbf{c}) &= \langle a_1,a_2\rangle + (\langle b_1,b_2\rangle + \langle c_1,c_2\rangle) = \langle a_1,a_2\rangle + \langle b_1+c_1,b_2+c_2\rangle \\ &= \langle a_1+b_1+c_1, a_2+b_2+c_2\rangle = \langle (a_1+b_1)+c_1, (a_2+b_2)+c_2\rangle \\ &= \langle a_1+b_1, a_2+b_2\rangle + \langle c_1,c_2\rangle = (\langle a_1,a_2\rangle + \langle b_1,b_2\rangle) + \langle c_1,c_2\rangle = (\mathbf{a}+\mathbf{b})+\mathbf{c}\end{aligned}$$

35.
$$\begin{aligned}(c+d)\mathbf{a} &= (c+d)\langle a_1,a_2\rangle = \langle (c+d)a_1, (c+d)a_2\rangle \\ &= \langle ca_1+da_1, ca_2+da_2\rangle = \langle ca_1,ca_2\rangle + \langle da_1,da_2\rangle = c\mathbf{a}+d\mathbf{a}.\end{aligned}$$

37. Consider quadrilateral $ABCD$ with sides AB and CD parallel and of equal length; that is, $\overrightarrow{AB} = \overrightarrow{DC}$. Thus, $\overrightarrow{AD} = \overrightarrow{AB}+\overrightarrow{BD} = \overrightarrow{DC}+\overrightarrow{BD}$ $\left(\text{since } \overrightarrow{AB} = \overrightarrow{DC}\right) = \overrightarrow{BD}+\overrightarrow{DC} = \overrightarrow{BC}$. This shows that sides AD and BC are parallel and have equal lengths.

EXERCISES 1.2

1. **(a)** $\mathbf{a}\cdot\mathbf{b}$ is a scalar, and the dot product is defined only for vectors, so $(\mathbf{a}\cdot\mathbf{b})\cdot\mathbf{c}$ has no meaning.

(b) $(\mathbf{a}\cdot\mathbf{b})\mathbf{c}$ is a scalar multiple of a vector, so it does have meaning.

(c) Both $|\mathbf{a}|$ and $\mathbf{b}\cdot\mathbf{c}$ are scalars, so $|\mathbf{a}|(\mathbf{b}\cdot\mathbf{c})$ is an ordinary product of real numbers, and has meaning.

(d) Both $\mathbf{a}$ and $\mathbf{b}+\mathbf{c}$ are scalars, so the dot product $\mathbf{a}\cdot(\mathbf{b}+\mathbf{c})$ has meaning.

(e) $\mathbf{a}\cdot\mathbf{b}$ is a scalar, but $\mathbf{c}$ is a vector, and so the two quantities cannot be added and this expression has no meaning.

(f) $\mathbf{a}$ is a scalar, and the dot product is defined only for vectors, so $|\mathbf{a}|\cdot(\mathbf{b}+\mathbf{c})$ has no meaning.

3. $|\mathbf{a}|=2$, $|\mathbf{b}|=3$, $\theta=\frac{\pi}{3}\quad\Rightarrow\quad\mathbf{a}\cdot\mathbf{b}=(2)(3)\cos\frac{\pi}{3}=6\cdot\frac{1}{2}=3$

5. $\mathbf{a}=\langle 4,7\rangle$, $\mathbf{b}=\langle -2,1\rangle\quad\Rightarrow\quad\mathbf{a}\cdot\mathbf{b}=4(-2)+7(1)=-1$

7. $\mathbf{a}=2\mathbf{i}+3\mathbf{j}$, $\mathbf{b}=\mathbf{i}-3\mathbf{j}\quad\Rightarrow\quad\mathbf{a}\cdot\mathbf{b}=2(1)+3(-3)=-7$

9. $\mathbf{u}$, $\mathbf{v}$, and $\mathbf{w}$ are all unit vectors, so the triangle is an equilateral triangle. Thus, the angle between $\mathbf{u}$ and $\mathbf{v}$ is 60° and $\mathbf{u}\cdot\mathbf{v}=|\mathbf{u}||\mathbf{v}|\cos 60^\circ=(1)(1)\left(\frac{1}{2}\right)=\frac{1}{2}$. If $\mathbf{w}$ is moved so it has the same initial point as $\mathbf{u}$, we can see that the angle between them is 120°, and we have $\mathbf{u}\cdot\mathbf{w}=|\mathbf{u}||\mathbf{w}|\cos 120^\circ=(1)(1)\left(-\frac{1}{2}\right)=-\frac{1}{2}$.

11. **(a)** $\mathbf{i}\cdot\mathbf{j}=\langle 1,0\rangle\cdot\langle 0,1\rangle=(1)(0)+(0)(1)=0$.

Another Method: Because $\mathbf{i}$ and $\mathbf{j}$ are perpendicular, the cosine factor in the dot product is $\cos\frac{\pi}{2}=0$.

(b) $\mathbf{i}\cdot\mathbf{i}=|\mathbf{i}|^2=1^2=1$ since $\mathbf{i}$ is a unit vector. Similarly, $\mathbf{j}\cdot\mathbf{j}=|\mathbf{j}|^2=1$.

13. $\mathbf{a}=\langle 1,2\rangle$, $\mathbf{b}=\langle 3,4\rangle\quad\Rightarrow\quad\cos\theta=\dfrac{\mathbf{a}\cdot\mathbf{b}}{|\mathbf{a}||\mathbf{b}|}=\dfrac{1(3)+2(4)}{\sqrt{1^2+2^2}\sqrt{3^2+4^2}}=\dfrac{11}{\sqrt{5}\sqrt{25}}\quad\Rightarrow$

$\theta=\cos^{-1}\left(\dfrac{11}{5\sqrt{5}}\right)\approx 10^\circ$.

15. $\mathbf{a}=\langle 1,2\rangle$, $\mathbf{b}=\langle 12,-5\rangle\quad\Rightarrow\quad\cos\theta=\dfrac{\mathbf{a}\cdot\mathbf{b}}{|\mathbf{a}||\mathbf{b}|}=\dfrac{1(12)+2(-5)}{\sqrt{1^2+2^2}\sqrt{12^2+(-5)^2}}\quad\Rightarrow$

$\theta=\cos^{-1}\left(\dfrac{2}{\sqrt{5}(13)}\right)\approx 86^\circ$.

17. $\mathbf{a}=6\mathbf{i}-2\mathbf{j}$, $\mathbf{b}=\mathbf{i}+\mathbf{j}\quad\Rightarrow\quad\cos\theta=\dfrac{6(1)+(-2)(1)}{\sqrt{6^2+(-2)^2}\sqrt{1^2+1^2}}\quad\Rightarrow\quad\theta=\cos^{-1}\left(\dfrac{4}{\sqrt{40}\sqrt{2}}\right)\approx 63^\circ$.

19. $\angle A = \cos^{-1}\left(\dfrac{\overrightarrow{AB}\cdot\overrightarrow{AC}}{\left|\overrightarrow{AB}\right|\left|\overrightarrow{AC}\right|}\right) = \cos^{-1}\left(\dfrac{\langle 5,-1\rangle\cdot\langle -2,-4\rangle}{|\langle 5,-1\rangle||\langle -2,-4\rangle|}\right) = \cos^{-1}\left(\dfrac{-6}{\sqrt{26}\sqrt{20}}\right) \approx 105.26° \approx 105°$

$\angle B = \cos^{-1}\left(\dfrac{\overrightarrow{BA}\cdot\overrightarrow{BC}}{\left|\overrightarrow{BA}\right|\left|\overrightarrow{BC}\right|}\right) = \cos^{-1}\left(\dfrac{\langle -5,1\rangle\cdot\langle -7,-3\rangle}{|\langle -5,1\rangle||\langle -7,-3\rangle|}\right) = \cos^{-1}\left(\dfrac{32}{\sqrt{26}\sqrt{58}}\right) \approx 34.51° \approx 35°$

$\angle C = \cos^{-1}\left(\dfrac{\overrightarrow{CA}\cdot\overrightarrow{CB}}{\left|\overrightarrow{CA}\right|\left|\overrightarrow{CB}\right|}\right) = \cos^{-1}\left(\dfrac{\langle 2,4\rangle\cdot\langle 7,3\rangle}{|\langle 2,4\rangle||\langle 7,3\rangle|}\right) = \cos^{-1}\left(\dfrac{26}{\sqrt{20}\sqrt{58}}\right) \approx 40.24° \approx 40°$

(Check: $\angle A + \angle B + \angle C \approx 180°$)

21. $\mathbf{a} = \langle 2,-4\rangle$ and $\mathbf{b} = \langle -1,2\rangle \quad\Rightarrow\quad \mathbf{a} = -2\mathbf{b}$. Since $\mathbf{a}$ is a scalar multiple of $\mathbf{b}$, the two vectors are *parallel*.

23. $\mathbf{a} = \langle 2,8\rangle$ and $\mathbf{b} = \langle -1,2\rangle \quad\Rightarrow\quad \mathbf{a}\cdot\mathbf{b} = 2(-1)+8(2) = -2+16 = 14 \neq 0$, so $\mathbf{a}$ and $\mathbf{b}$ are not orthogonal. Since $\mathbf{a}$ is not a scalar multiple of $\mathbf{b}$, the given vectors are *neither* orthogonal nor parallel.

25. $\mathbf{a} = 3\mathbf{i}+\mathbf{j}$ and $\mathbf{b} = -3\mathbf{i}+9\mathbf{j} \quad\Rightarrow\quad \mathbf{a}\cdot\mathbf{b} = 3(-3)+1(9) = -9+9 = 0 \quad\Rightarrow\quad$ $\mathbf{a}$ and $\mathbf{b}$ are *orthogonal.*

27. $(x\mathbf{i}-2\mathbf{j})$ and $(x\mathbf{i}+8\mathbf{j})$ are orthogonal if and only if $(x\mathbf{i}-2\mathbf{j})\cdot(x\mathbf{i}+8\mathbf{j}) = 0 \quad\Leftrightarrow\quad x^2-16=0 \quad\Leftrightarrow$ $x^2 = 16 \quad\Leftrightarrow\quad x = \pm 4$.

29. $\langle x,1\rangle$ and $\langle 4,x\rangle$ are orthogonal if and only if $\langle x,1\rangle\cdot\langle 4,x\rangle = 0 \quad\Leftrightarrow\quad 4x+x=0 \quad\Leftrightarrow\quad 5x=0 \quad\Leftrightarrow$ $x = 0$.

31. An arbitrary vector $a\mathbf{i}+b\mathbf{j}$ is orthogonal to $\mathbf{i}+3\mathbf{j}$ if and only if $(a\mathbf{i}+b\mathbf{j})\cdot(\mathbf{i}+3\mathbf{j}) = 0 \quad\Leftrightarrow\quad a+3b=0 \quad\Leftrightarrow$ $a = -3b$. So $a\mathbf{i}+b\mathbf{j} = -3b\mathbf{i}+b\mathbf{j}$ and $|-3b\mathbf{i}+b\mathbf{j}| = \sqrt{(-3b)^2+b^2} = \sqrt{10b^2} = \sqrt{10}|b|$. To be a unit vector, $|b|\sqrt{10}$ must equal 1. Thus, $|b| = 1/\sqrt{10}$ so $b = \pm 1/\sqrt{10}$ and the vectors orthogonal to $\mathbf{i}+3\mathbf{j}$ are $-\frac{3}{\sqrt{10}}\mathbf{i}+\frac{1}{\sqrt{10}}\mathbf{j}$ and $\frac{3}{\sqrt{10}}\mathbf{i}-\frac{1}{\sqrt{10}}\mathbf{j}$.

33. $\mathbf{a} = \langle 2,3\rangle$ and $\mathbf{b} = \langle 4,1\rangle \quad\Rightarrow\quad$ the scalar projection of $\mathbf{b}$ onto $\mathbf{a}$ is $\text{comp}_{\mathbf{a}}\,\mathbf{b} = \dfrac{\mathbf{a}\cdot\mathbf{b}}{|\mathbf{a}|} = \dfrac{2\cdot 4+3\cdot 1}{\sqrt{2^2+3^2}} = \dfrac{11}{\sqrt{13}}$.

The vector projection of $\mathbf{b}$ onto $\mathbf{a}$ is $\text{proj}_{\mathbf{a}}\,\mathbf{b} = \dfrac{\mathbf{a}\cdot\mathbf{b}}{|\mathbf{a}|^2}\mathbf{a} = \frac{11}{\sqrt{13}}\cdot\frac{1}{\sqrt{13}}\langle 2,3\rangle = \frac{11}{13}\langle 2,3\rangle = \left\langle \frac{22}{13},\frac{33}{13}\right\rangle$.

35. $\mathbf{a} = \langle 4,2\rangle$ and $\mathbf{b} = \langle 1,1\rangle \quad\Rightarrow\quad$ the scalar projection of $\mathbf{b}$ onto $\mathbf{a}$ is $\text{comp}_{\mathbf{a}}\,\mathbf{b} = \dfrac{\mathbf{a}\cdot\mathbf{b}}{|\mathbf{a}|} = \dfrac{4\cdot 1+2\cdot 1}{\sqrt{4^2+2^2}} = \dfrac{6}{\sqrt{20}} = \dfrac{3}{\sqrt{5}}$. The vector projection of $\mathbf{b}$ onto $\mathbf{a}$ is $\text{proj}_{\mathbf{a}}\,\mathbf{b} = \dfrac{\mathbf{a}\cdot\mathbf{b}}{|\mathbf{a}|^2}\mathbf{a} = \frac{3}{\sqrt{5}}\cdot\frac{1}{2\sqrt{5}}\langle 4,2\rangle = \frac{1}{5}\langle 6,3\rangle = \left\langle \frac{6}{5},\frac{3}{5}\right\rangle$.

37. $\mathbf{a} = \mathbf{i}$ and $\mathbf{b} = \mathbf{i}-\mathbf{j} \quad\Rightarrow\quad$ the scalar projection of $\mathbf{b}$ onto $\mathbf{a}$ is $\text{comp}_{\mathbf{a}}\,\mathbf{b} = \dfrac{\mathbf{a}\cdot\mathbf{b}}{|\mathbf{a}|} = \dfrac{1(1)+0(-1)}{\sqrt{1^2+0^2}} = \dfrac{1}{1} = 1$.

The vector projection of $\mathbf{b}$ onto $\mathbf{a}$ is $\text{proj}_{\mathbf{a}}\,\mathbf{b} = \dfrac{\mathbf{a}\cdot\mathbf{b}}{|\mathbf{a}|^2}\mathbf{a} = 1\cdot\dfrac{1}{1}(\mathbf{i}) = \mathbf{i}$.

39. $\mathbf{a} = \langle 2,1 \rangle \quad \Rightarrow \quad \mathbf{a}^{\perp} = \langle -1,2 \rangle$ by Definition 1.2.5. The vector projection of $\mathbf{b}$ onto $\mathbf{a}^{\perp}$ is

$$\operatorname{proj}_{\mathbf{a}^{\perp}} \mathbf{b} = \frac{\mathbf{a}^{\perp} \cdot \mathbf{b}}{|\mathbf{a}^{\perp}|^2}\mathbf{a}^{\perp} = \frac{\langle -1,2 \rangle \cdot \langle 1,1 \rangle}{|\langle -1,2 \rangle|^2}\langle -1,2 \rangle = \frac{-1(1)+2(1)}{\left[\sqrt{(-1)^2+2^2}\right]^2}\langle -1,2 \rangle = \frac{1}{\left(\sqrt{5}\right)^2}\langle -1,2 \rangle = \langle -\tfrac{1}{5}, \tfrac{2}{5} \rangle.$$

41. $\mathbf{a} = \langle -2,3 \rangle \quad \Rightarrow \quad \mathbf{a}^{\perp} = \langle -3,-2 \rangle$.

$$\operatorname{proj}_{\mathbf{a}^{\perp}} \mathbf{b} = \frac{\mathbf{a}^{\perp} \cdot \mathbf{b}}{|\mathbf{a}^{\perp}|^2}\mathbf{a}^{\perp} = \frac{\langle -3,-2 \rangle \cdot \langle 0,2 \rangle}{|\langle -3,-2 \rangle|^2}\langle -3,-2 \rangle = \frac{-4}{\left(\sqrt{13}\right)^2}\langle -3,-2 \rangle = \langle \tfrac{12}{13}, \tfrac{8}{13} \rangle.$$

43. As in Example 9, the points $(0,0)$ and $(1,2)$ lie on the line $y = 2x$, so the displacement vector is $\mathbf{a} = \langle 1,2 \rangle - \langle 0,0 \rangle = \langle 1,2 \rangle$. Now $\mathbf{a}$ is parallel to the line $y = 2x$ and $\mathbf{a}^{\perp} = \langle -2,1 \rangle$ is perpendicular to the line. Next we form the displacement vector $\mathbf{b}$ from $(0,0)$ to the point $(3,7)$. $\mathbf{b} = \langle 3,7 \rangle - \langle 0,0 \rangle = \langle 3,7 \rangle$. $|\operatorname{comp}_{\mathbf{a}^{\perp}} \mathbf{b}|$ gives the distance from the point $(3,7)$ to the line $y = 2x$.

$$|\operatorname{comp}_{\mathbf{a}^{\perp}} \mathbf{b}| = \left|\frac{\mathbf{a}^{\perp} \cdot \mathbf{b}}{|\mathbf{a}^{\perp}|}\right| = \left|\frac{\langle -2,1 \rangle \cdot \langle 3,7 \rangle}{|\langle -2,1 \rangle|}\right| = \left|\frac{-2(3)+1(7)}{\sqrt{(-2)^2+1^2}}\right| = \left|\frac{1}{\sqrt{5}}\right| = \frac{1}{\sqrt{5}}.$$

45. As in Exercise 43, we have the line $3x - 4y + 5 = 0$ or $y = \frac{1}{4}(3x+5)$. Letting $x = 1$ and $x = 5$ (to make $3x+5$ a multiple of 4) gives us the points $(1,2)$ and $(5,5)$. So $\mathbf{a} = \langle 5,5 \rangle - \langle 1,2 \rangle = \langle 4,3 \rangle$, $\mathbf{a}^{\perp} = \langle -3,4 \rangle$, and $\mathbf{b} = \langle -2,3 \rangle - \langle 1,2 \rangle = \langle -3,1 \rangle$. $|\operatorname{comp}_{\mathbf{a}^{\perp}} \mathbf{b}| = \left|\frac{\mathbf{a}^{\perp} \cdot \mathbf{b}}{|\mathbf{a}^{\perp}|}\right| = \left|\frac{\langle -3,4 \rangle \cdot \langle -3,1 \rangle}{|\langle -3,4 \rangle|}\right| = \left|\frac{9+4}{\sqrt{25}}\right| = \frac{13}{5}$.

47. By using the point $(0,9)$ on the line $y - 2x = 9$, we can treat this problem just like Exercise 43 by finding the distance from the point $(0,9)$ to the line $y = 2x+3$. The points $(0,3)$ and $(1,5)$ are on the line $y = 2x+3$. So $\mathbf{a} = \langle 1,5 \rangle - \langle 0,3 \rangle = \langle 1,2 \rangle$, $\mathbf{a}^{\perp} = \langle -2,1 \rangle$, and $\mathbf{b} = \langle 0,9 \rangle - \langle 0,3 \rangle = \langle 0,6 \rangle$.

$$|\operatorname{comp}_{\mathbf{a}^{\perp}} \mathbf{b}| = \left|\frac{\mathbf{a}^{\perp} \cdot \mathbf{b}}{|\mathbf{a}^{\perp}|}\right| = \left|\frac{\langle -2,1 \rangle \cdot \langle 0,6 \rangle}{|\langle -2,1 \rangle|}\right| = \left|\frac{6}{\sqrt{5}}\right| = \frac{6}{\sqrt{5}}.$$

49. To show that $\operatorname{orth}_{\mathbf{a}} \mathbf{b}$ is orthogonal to $\mathbf{a}$, we examine the dot product

$$(\operatorname{orth}_{\mathbf{a}} \mathbf{b}) \cdot \mathbf{a} = (\mathbf{b} - \operatorname{proj}_{\mathbf{a}}\mathbf{b}) \cdot \mathbf{a} = \mathbf{b} \cdot \mathbf{a} - (\operatorname{proj}_{\mathbf{a}} \mathbf{b}) \cdot \mathbf{a} = \mathbf{b} \cdot \mathbf{a} - \frac{\mathbf{a} \cdot \mathbf{b}}{|\mathbf{a}|^2}\mathbf{a} \cdot \mathbf{a} = \mathbf{b} \cdot \mathbf{a} - \frac{\mathbf{a} \cdot \mathbf{b}}{|\mathbf{a}|^2}|\mathbf{a}|^2$$

$= \mathbf{b} \cdot \mathbf{a} - \mathbf{a} \cdot \mathbf{b} = 0$. Thus, $\operatorname{orth}_{\mathbf{a}} \mathbf{b}$ is orthogonal to $\mathbf{a}$.

To find a relationship between $\operatorname{orth}_{\mathbf{a}} \mathbf{b}$ and $\operatorname{proj}_{\mathbf{a}^{\perp}} \mathbf{b}$, we could use $\mathbf{a} = \langle a_1, a_2 \rangle$ and $\mathbf{b} = \langle b_1, b_2 \rangle$ and show that

$\operatorname{orth}_{\mathbf{a}} \mathbf{b} = \operatorname{proj}_{\mathbf{a}^{\perp}} \mathbf{b} = \dfrac{-a_2 b_1 + a_1 b_2}{a_1^2 + a_2^2}\langle -a_2, a_1 \rangle$, proving that $\operatorname{orth}_{\mathbf{a}} \mathbf{b}$ and $\operatorname{proj}_{\mathbf{a}^{\perp}} \mathbf{b}$ are equal. The figure shows that fact.

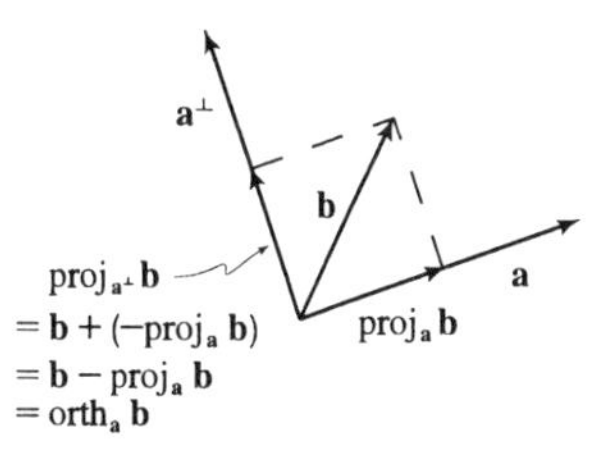

51. $\operatorname{comp}_{\mathbf{a}} \mathbf{b} = \dfrac{\mathbf{a} \cdot \mathbf{b}}{|\mathbf{a}|} \quad \Leftrightarrow \quad 2 = \dfrac{\langle 3,-1 \rangle \cdot \langle b_1, b_2 \rangle}{\sqrt{10}} \quad \Leftrightarrow \quad 2\sqrt{10} = 3b_1 - b_2$. One solution is obtained by taking $b_1 = 0$ so that $b_2 = -2\sqrt{10}$. Thus, $b = \langle 0, -2\sqrt{10} \rangle$ is a specific solution. If we let $b_1 = s$ (s some real number), then $b_2 = 3s - 2\sqrt{10}$. So the general solution is $\mathbf{b} = \langle s, 3s - 2\sqrt{10} \rangle$, $s \in \mathbb{R}$.

53. The displacement vector is $\mathbf{D} = \langle 4,9 \rangle - \langle 2,3 \rangle = \langle 2,6 \rangle$. Since $\mathbf{F} = 10\mathbf{i} + 18\mathbf{j}$, the work done is $\mathbf{W} = \mathbf{F} \cdot \mathbf{D} = \langle 10,18 \rangle \cdot \langle 2,6 \rangle = 20 + 108 = 128$ J.

55. $W = |\mathbf{F}||\mathbf{D}|\cos\theta = (25)(10)\cos 20° \approx 235$ ft-lb

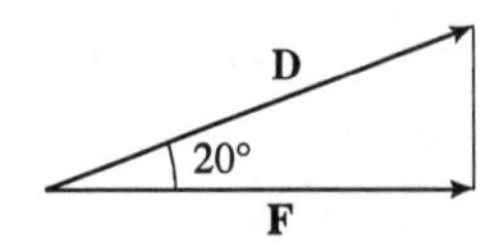

57. **(a)** First note that $\mathbf{n} = \langle a, b\rangle$ is perpendicular to the line, because if $Q_1 = (a_1, b_1)$ and $Q_2 = (a_2, b_2)$ lie on the line, then $\mathbf{n} \cdot \overrightarrow{Q_1Q_2} = aa_2 - aa_1 + bb_2 - bb_1 = 0$, since $aa_2 + bb_2 = -c = aa_1 + bb_1$ from the equation of the line. Let $P_2 = (x_2, y_2)$ lie on the line. Then the distance from P_1 to the line is the absolute value of the scalar projection of $\overrightarrow{P_1P_2}$ onto $\mathbf{n}$.

$\text{comp}_{\mathbf{n}}\left(\overrightarrow{P_1P_2}\right) = \dfrac{|\mathbf{n}\cdot\langle x_2 - x_1, y_2 - y_1\rangle|}{|\mathbf{n}|} = \dfrac{|ax_2 - ax_1 + by_2 - by_1|}{\sqrt{a^2+b^2}} = \dfrac{|ax_1 + by_1 + c|}{\sqrt{a^2+b^2}}$ since $ax_2 + by_2 = -c$.

(b) The distance from $P_1(x_1, y_1) = (-2, 3)$ to the line $ax + by + c = 0$ $(3x - 4y + 5 = 0)$ is $\dfrac{|3\cdot -2 + -4\cdot 3 + 5|}{\sqrt{3^2 + (-4)^2}} = \dfrac{13}{5}$.

59. Let $\mathbf{a} = \langle a_1, a_2\rangle$ and $\mathbf{b} = \langle b_1, b_2\rangle$.

Property 2: $\mathbf{a}\cdot\mathbf{b} = \langle a_1, a_2\rangle \cdot \langle b_1, b_2\rangle = a_1b_1 + a_2b_2$

$= b_1a_1 + b_2a_2 = \langle b_1, b_2\rangle \cdot \langle a_1, a_2\rangle = \mathbf{b}\cdot\mathbf{a}$

Property 5: $\mathbf{0}\cdot\mathbf{a} = \langle 0, 0\rangle \cdot \langle a_1, a_2\rangle = (0)(a_1) + (0)(a_2) = 0$

61. $|\mathbf{a}\cdot\mathbf{b}| = ||\mathbf{a}||\mathbf{b}|\cos\theta| = |\mathbf{a}||\mathbf{b}||\cos\theta|$. Since $|\cos\theta| \le 1$, $|\mathbf{a}\cdot\mathbf{b}| = |\mathbf{a}||\mathbf{b}||\cos\theta| \le |\mathbf{a}||\mathbf{b}|$.

Note: We have equality in the case of $\cos\theta = \pm 1$, so $\theta = 0$ or $\theta = \pi$, thus, equality when **a** and **b** are parallel.

63. **(a)**

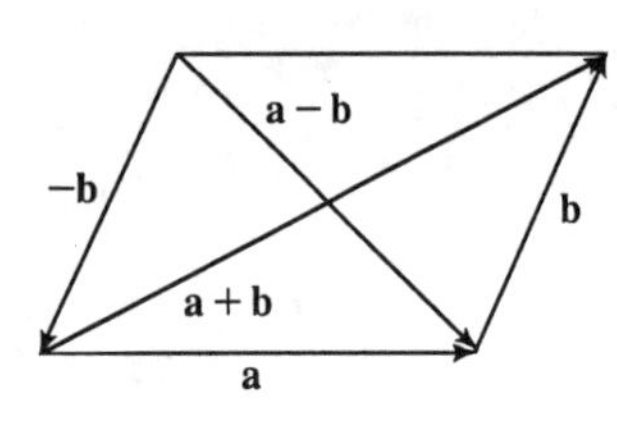

The Parallelogram Law states that the sum of the squares of the lengths of the diagonals of a parallelogram equals the sum of the squares of its (four) sides.

(b) $|\mathbf{a}+\mathbf{b}|^2 = (\mathbf{a}+\mathbf{b})\cdot(\mathbf{a}+\mathbf{b}) = |\mathbf{a}|^2 + 2(\mathbf{a}\cdot\mathbf{b}) + |\mathbf{b}|^2$ and

$|\mathbf{a}-\mathbf{b}|^2 = (\mathbf{a}-\mathbf{b})\cdot(\mathbf{a}-\mathbf{b}) = |\mathbf{a}|^2 - 2(\mathbf{a}\cdot\mathbf{b}) + |\mathbf{b}|^2$. Adding these two equations gives

$|\mathbf{a}+\mathbf{b}|^2 + |\mathbf{a}-\mathbf{b}|^2 = 2|\mathbf{a}|^2 + 2|\mathbf{b}|^2$.

EXERCISES 1.3

1. (a)

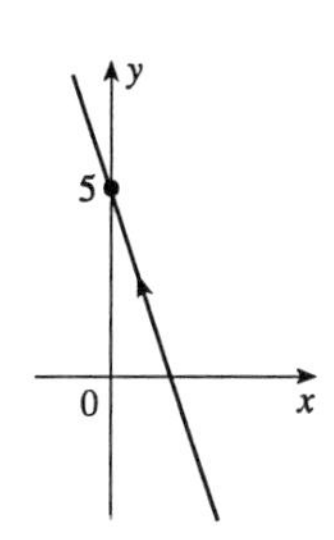

(b) $x = 1 - t, y = 2 + 3t$

$y = 2 + 3(1 - x) = 5 - 3x$, so

$3x + y = 5$

3. (a)

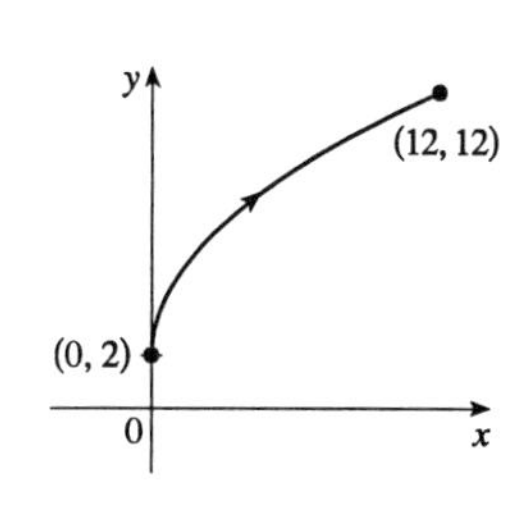

(b) $x = 3t^2, y = 2 + 5t, 0 \le t \le 2$

$x = 3\left(\dfrac{y-2}{5}\right)^2 = \frac{3}{25}(y-2)^2$,

$2 \le y \le 12$

5. (a)

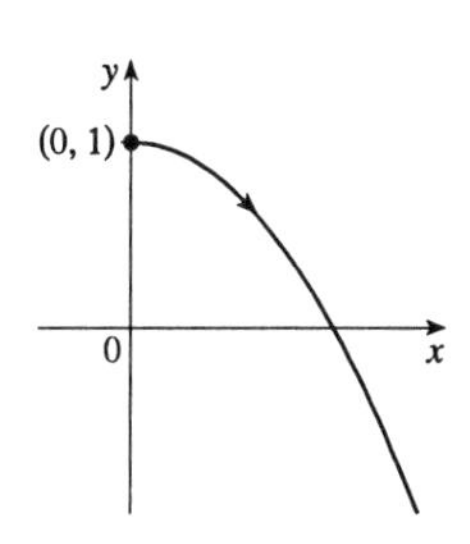

(b) $x = \sqrt{t}, y = 1 - t$

$y = 1 - t = 1 - x^2, x \ge 0$

7. (a)

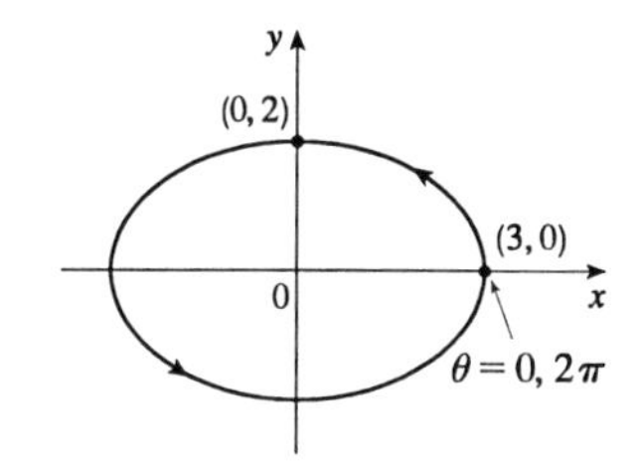

(b) $x = 3\cos\theta, y = 2\sin\theta, 0 \le \theta \le 2\pi$

$\left(\dfrac{x}{3}\right)^2 + \left(\dfrac{y}{2}\right)^2 = \cos^2\theta + \sin^2\theta = 1$, or

$\frac{1}{9}x^2 + \frac{1}{4}y^2 = 1$

9. (a)

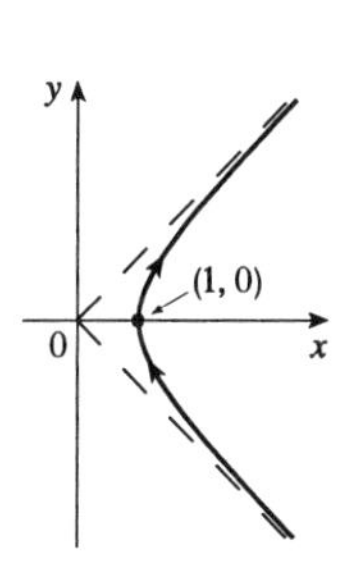

(b) $x = \sec\theta, y = \tan\theta, -\frac{\pi}{2} < \theta < \frac{\pi}{2}$

$x^2 - y^2 = \sec^2\theta - \tan^2\theta = 1, x \ge 1$,

or $x = \sqrt{y^2 + 1}$

11. (a)

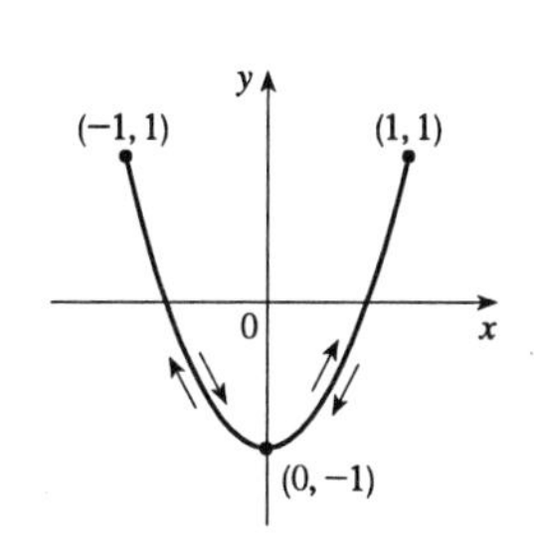

(b) $x = \cos t, y = \cos 2t$

$y = \cos 2t = 2\cos^2 t - 1 = 2x^2 - 1$,

so $y + 1 = 2x^2, -1 \le x \le 1$

13. **(a)**

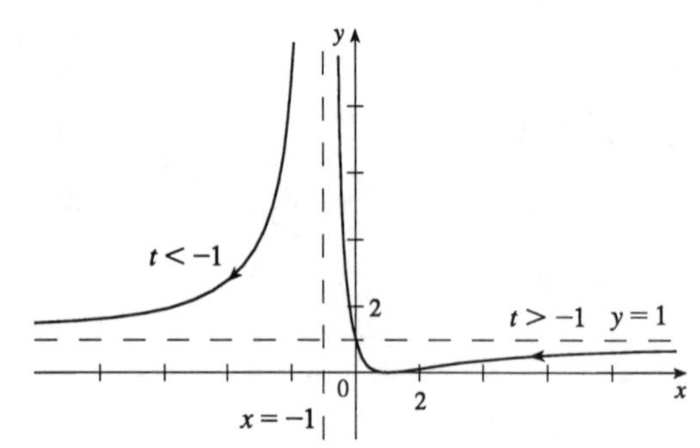

For values of t near -1, $|x|$ is large and y is near 1 ($y = 1$ is a horizontal asymptote). x is never equal to -1 and $y \geq 0$ always.

(b) $x = \dfrac{1-t}{1+t} \quad \Rightarrow \quad (1+t)x = 1-t \quad \Rightarrow \quad x + xt = 1 - t \quad \Rightarrow \quad xt + t = 1 - x \quad \Rightarrow$

$t(x+1) = 1 - x \quad \Rightarrow \quad t = \dfrac{1-x}{1+x}$, so $y = t^2 = \left(\dfrac{1-x}{1+x}\right)^2$, or $y = \left(\dfrac{x-1}{x+1}\right)^2$.

15. $x = 2 + \cos t$, $y = 3 + \sin t \quad \Rightarrow \quad (x-2)^2 + (y-3)^2 = \cos^2 t + \sin^2 t = 1$, so the motion takes place on a unit circle centered at $(2, 3)$. As t goes from 0 to 2π, the particle makes one complete counterclockwise rotation around the circle, starting and ending at $(3, 3)$.

17. $x = \cos^2 t = y^2$, so the particle moves along the parabola $x = y^2$. As t goes from 0 to 4π, the particle moves from $(1, 1)$ down to $(1, -1)$ (at $t = \pi$), back up to $(1, 1)$ again (at $t = 2\pi$), and then repeats this entire cycle between $t = 2\pi$ and $t = 4\pi$.

19. $y = \csc t = 1/\sin t = 1/x$. The particle slides down the first quadrant branch of the hyperbola $xy = 1$ from $\left(\frac{1}{2}, 2\right)$ to $(\sin 1, \csc 1) \approx (0.84147, 1.1884)$ as t goes from $\frac{\pi}{6}$ to 1.

21. $x = t \cos t$, $y = t \sin t$

Graph with $-6\pi \leq t \leq 6\pi$

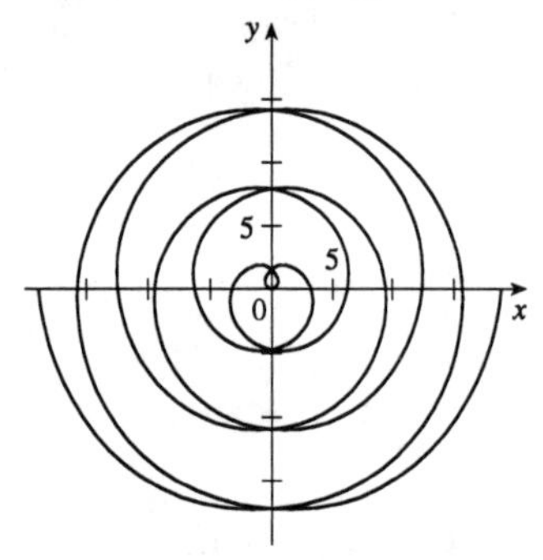

23. $x = \dfrac{3t}{1+t^3}$, $y = \dfrac{3t^2}{1+t^3}$

Graph with $-10 \leq t \leq 10$

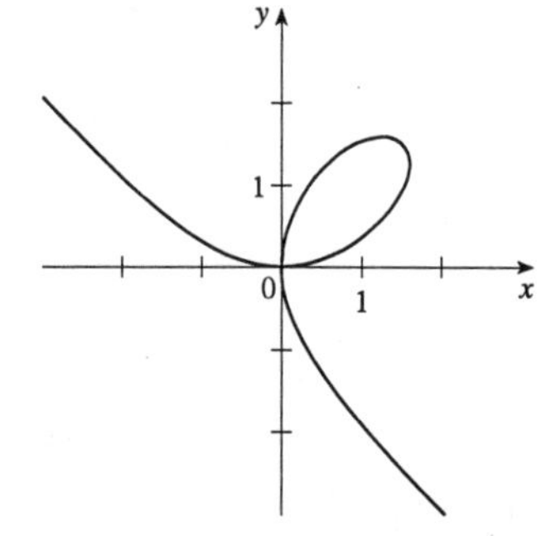

25. **(a)** We proceed as in Example 6 with $\mathbf{v} = \langle -2, 7 \rangle - \langle 1, 3 \rangle = \langle -3, 4 \rangle$. Now $\mathbf{r}(t) = \mathbf{r}_0 + t\mathbf{v} = \langle 1, 3 \rangle + t\langle -3, 4 \rangle = \langle 1 - 3t, 3 + 4t \rangle$.

(b) $x(t) = 1 - 3t$, $y(t) = 3 + 4t$

27. **(a)** We proceed as in Example 6 with $\mathbf{v} = \langle -2, 5 \rangle - \langle 4, -1 \rangle = \langle -6, 6 \rangle$. Now $\mathbf{r}(t) = \mathbf{r}_0 + t\mathbf{v} = \langle 4, -1 \rangle + t\langle -6, 6 \rangle = \langle 4 - 6t, -1 + 6t \rangle$.

(b) $x(t) = 4 - 6t$, $y(t) = -1 + 6t$

29. **(a)** $\mathbf{r}(t) = \overrightarrow{OP} + t\mathbf{a} = \langle -4, 5 \rangle + t\langle -2, 6 \rangle$

(b) $\mathbf{r}(t) = \langle -4 - 2t, 5 + 6t \rangle \quad \Rightarrow \quad x = -4 - 2t,\ y = 5 + 6t$

(c) $x = -4 - 2t \quad \Rightarrow \quad t = -\frac{1}{2}(x + 4) \quad \Rightarrow \quad y = 5 + 6\left[-\frac{1}{2}(x + 4)\right] \quad \Rightarrow \quad y = -3x - 7$

31. If we write L_1: $\mathbf{r}_1(t) = (2 + 3t)\mathbf{i} + (-1 + 4t)\mathbf{j}$ as $\mathbf{r}_1(t) = \langle 2, -1 \rangle + t\langle 3, 4 \rangle$ and L_2: $\mathbf{r}_2(t) = (1 + 4t)\mathbf{i} + (2 - 2t)\mathbf{j}$ as $\mathbf{r}_2(t) = \langle 1, 2 \rangle + t\langle 4, -2 \rangle$, then we see that L_1 is parallel to $\langle 3, 4 \rangle$ and L_2 is parallel to $\langle 4, -2 \rangle$. These vectors are not parallel since $\langle 4, -2 \rangle$ isn't a scalar multiple of $\langle 3, 4 \rangle$, so L_1 and L_2 must intersect. Since $\langle 3, 4 \rangle \cdot \langle 4, -2 \rangle = 6 \neq 0$, L_1 isn't perpendicular to L_2. To find the point of intersection, we seek numbers s and t such that $\mathbf{r}_1(s) = \mathbf{r}_2(t)$; that is, $(2 + 3s)\mathbf{i} + (-1 + 4s)\mathbf{j} = (1 + 4t)\mathbf{i} + (2 - 2t)\mathbf{j}$. Equating **i**- and **j**-components gives us $\left\{ \begin{array}{c} 2 + 3s = 1 + 4t \\ -1 + 4s = 2 - 2t \end{array} \right\} \quad \Leftrightarrow \quad \left\{ \begin{array}{c} 3s - 4t = -1 \\ 4s + 2t = 3 \end{array} \right\} \quad \Leftrightarrow \quad \left\{ \begin{array}{c} s = \frac{5}{11} \\ t = \frac{13}{22} \end{array} \right\}$ so $\mathbf{r}_1(s) = \mathbf{r}_2(t) = \frac{37}{11}\mathbf{i} + \frac{9}{11}\mathbf{j}$. The point of intersection is $\left(\frac{37}{11}, \frac{9}{11}\right)$. Another way to find the point of intersection would be to find Cartesian equations for L_1 and L_2 and solve them simultaneously for x and y. We will use this method in the solution to Exercise 33.

33. Let $\mathbf{r}_1(t) = (2 - t)\mathbf{i} + (-3 + 5t)\mathbf{j} = \langle 2, -3 \rangle + t\langle -1, 5 \rangle$ and $\mathbf{r}_2(t) = (8 + 10t)\mathbf{i} + (2 + 2t)\mathbf{j} = \langle 8, 2 \rangle + t\langle 10, 2 \rangle$. $\mathbf{r}_1$ and $\mathbf{r}_2$ are not parallel since $\langle 10, 2 \rangle$ is not a scalar multiple of $\langle -1, 5 \rangle$. $\mathbf{r}_1$ and $\mathbf{r}_2$ are perpendicular since $\langle -1, 5 \rangle \cdot \langle 10, 2 \rangle = 0$. To find the point of intersection, we will find Cartesian equations for L_1 and L_2 and then solve them simultaneously for x and y. L_1: $x = 2 - t \quad \Rightarrow \quad t = 2 - x$ so $y = -3 + 5t = -3 + 5(2 - x)$ gives us $y = -5x + 7$. L_2: $x = 8 + 10t \quad \Rightarrow \quad t = \frac{1}{10}(x - 8)$ so $y = 2 + 2t = 2 + 2\left[\frac{1}{10}(x - 8)\right]$ gives us $y = \frac{1}{5}x + \frac{2}{5}$. (Note that the slopes, -5 and $\frac{1}{5}$, are negative reciprocals of one another, confirming that L_1 and L_2 are perpendicular.) Solving $-5x + 7 = \frac{1}{5}x + \frac{2}{5}$ gives us $-25x + 35 = x + 2 \quad \Leftrightarrow \quad 33 = 26x \quad \Leftrightarrow \quad x = \frac{33}{26}$ and so $y = -5\left(\frac{33}{26}\right) + 7 = \frac{17}{26}$. The point of intersection is $\left(\frac{33}{26}, \frac{17}{26}\right)$.

35. **(a)** $\mathbf{r}(t) = (t^2 + 2t)\mathbf{i} + (t + 4)\mathbf{j} \quad \Rightarrow \quad \mathbf{r}(2) = 8\mathbf{i} + 6\mathbf{j}$, so the position of the object at time $t = 2$ is $(8, 6)$.

(b) $t + 4$ must equal 7, so $t = 3$ seconds. Checking, $\mathbf{r}(3) = 15\mathbf{i} + 7\mathbf{j}$.

(c) $t + 4$ must equal 9, so $t = 5$. But $\mathbf{r}(5) = 35\mathbf{i} + 9\mathbf{j}$, so the object does *not* pass through the point $(20, 9)$.

(d) $y = t + 4 \quad \Rightarrow \quad t = y - 4$, so $x = t^2 + 2t = (y - 4)^2 + 2(y - 4)$, or, equivalently, $x = y^2 - 6y + 8$. Assuming $t \geq 0$, $x \geq 0$ and $y \geq 4$.

37. $x = t^2, y = t^3 - ct$. We use a graphing device to produce the graphs for various values of c. Note that all the members of the family are symmetric about the x-axis. For $c < 0$, the graph does not cross itself, but for $c = 0$ it has a cusp at $(0, 0)$ and for $c > 0$ the graph crosses itself at $x = c$, so the loop grows larger as c increases.

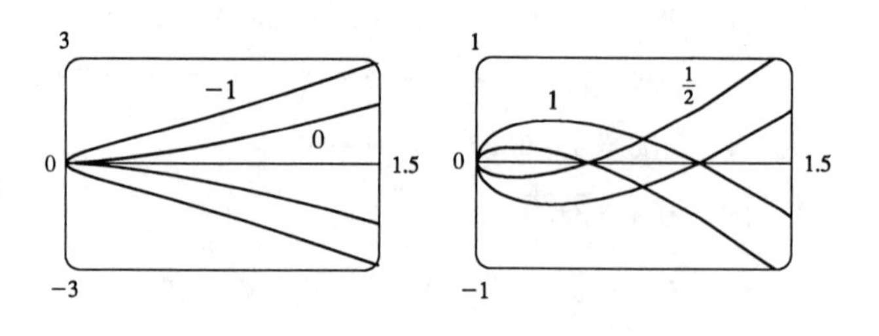

39. Note that all the Lissajous figures are symmetric about the x-axis. The parameters a and b simply stretch the graph in the x- and y-directions respectively. For $a = b = n = 1$ the graph is simply a circle with radius 1. For $n = 2$ the graph crosses itself at the origin and there are loops above and below the x-axis. In general, the figures have $n - 1$ points of intersection, all of which are on the y-axis, and a total of n closed loops.

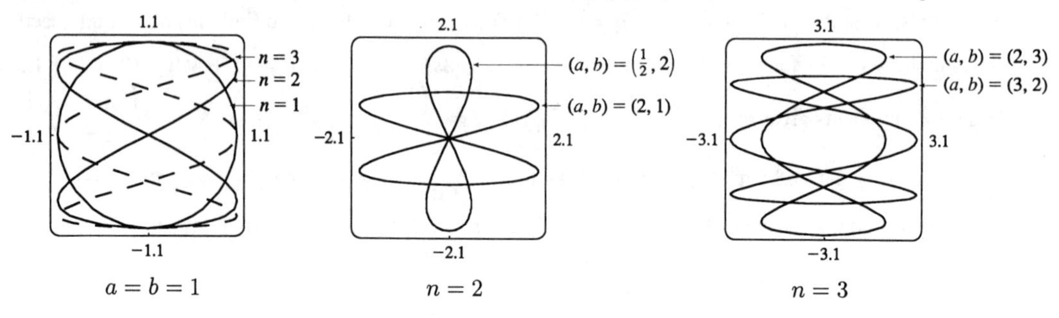

REVIEW EXERCISES FOR CHAPTER 1

1. $\mathbf{a} = \overrightarrow{OB} - \overrightarrow{OA} = \langle 3, -1 \rangle - \langle 2, 1 \rangle = \langle 1, -2 \rangle$

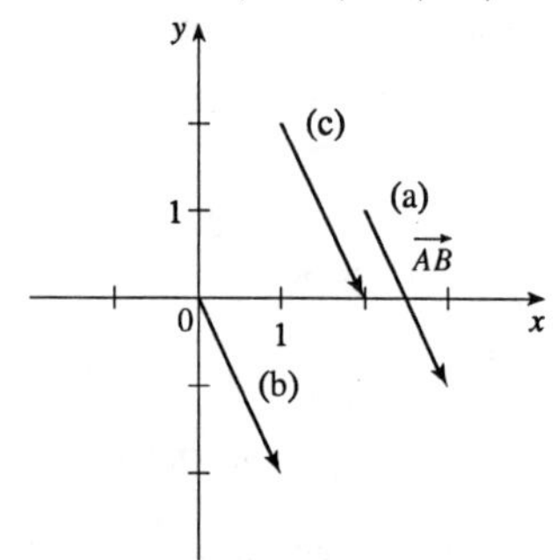

3. sum $= (2\mathbf{i} - 3\mathbf{j}) + (-4\mathbf{i} + \mathbf{j}) = -2\mathbf{i} - 2\mathbf{j}$

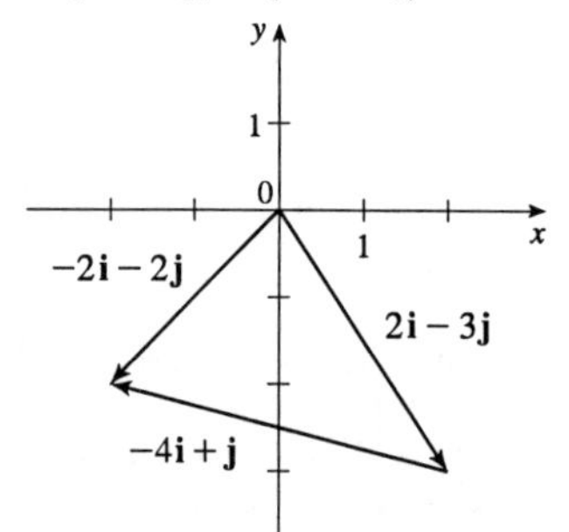

5. sum $= \langle 4,3\rangle + \langle -1,2\rangle = \langle 3,5\rangle$

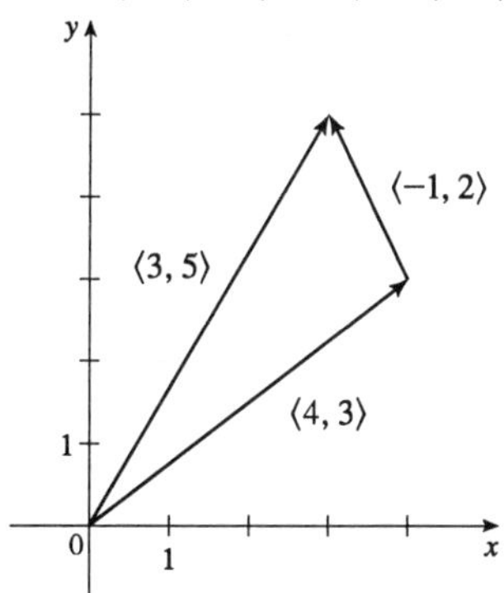

7. $\overrightarrow{OB} = \overrightarrow{OA} + \mathbf{a} = (1\mathbf{i} + 0\mathbf{j}) + (-2\mathbf{i} + 3\mathbf{j}) = -\mathbf{i} + 3\mathbf{j}$, so $B = (-1, 3)$.

9. $\overrightarrow{OB} = \overrightarrow{OA} + \mathbf{a} \quad \Rightarrow \quad \overrightarrow{OA} = \overrightarrow{OB} - \mathbf{a} = \langle 1,5\rangle - \langle -3,4\rangle = \langle 4,1\rangle$, so $A = (4, 1)$.

11. $\mathbf{a} = \langle 3,-4\rangle$, $\mathbf{b} = \langle 1,3\rangle$, $\mathbf{c} = \langle 2,1\rangle$

(a) The length of $\mathbf{a}$ is $|\mathbf{a}| = \sqrt{3^2 + (-4)^2} = \sqrt{25} = 5$.

(b) A unit vector having the same direction as $\mathbf{a}$ is $\mathbf{a}/|\mathbf{a}| = \frac{1}{5}\langle 3,-4\rangle = \langle \frac{3}{5}, -\frac{4}{5}\rangle$.

(c) The vectors that are orthogonal to $\mathbf{a}$ are $\pm\mathbf{a}^{\perp}$, or $\pm\langle 4,3\rangle$. Thus, there are two unit vectors that are orthogonal to $\mathbf{a}$, $\pm\frac{1}{5}\langle 4,3\rangle = \langle \frac{4}{5}, \frac{3}{5}\rangle$ or $\langle -\frac{4}{5}, -\frac{3}{5}\rangle$.

(d) $-4\mathbf{a} + 3\mathbf{b} = -4\langle 3,-4\rangle + 3\langle 1,3\rangle = \langle -12,16\rangle + \langle 3,9\rangle = \langle -9,25\rangle$

(e) $\mathbf{c} = s\mathbf{a} + t\mathbf{b} \quad \Leftrightarrow \quad \langle 2,1\rangle = \langle 3s,-4s\rangle + \langle t,3t\rangle \quad \Leftrightarrow \quad \langle 2,1\rangle = \langle 3s+t, -4s+3t\rangle \quad \Leftrightarrow$

$$\left\{ \begin{array}{r} 3s + t = 2 \\ -4s + 3t = 1 \end{array} \right\} \quad \Leftrightarrow \quad \left\{ \begin{array}{l} s = \frac{5}{13} \\ t = \frac{11}{13} \end{array} \right\}$$

13. $\mathbf{a} = \mathbf{i} - 7\mathbf{j}$, $\mathbf{b} = 2\mathbf{i} + 3\mathbf{j}$, $\mathbf{c} = 4\mathbf{j}$

(a) $|\mathbf{a}| = \sqrt{1^2 + (-7)^2} = \sqrt{50}$ or $5\sqrt{2}$

(b) $\mathbf{a}/|\mathbf{a}| = \frac{1}{\sqrt{50}}(\mathbf{i} - 7\mathbf{j}) = \frac{1}{\sqrt{50}}\mathbf{i} - \frac{7}{\sqrt{50}}\mathbf{j}$

(c) $\pm\mathbf{a}^{\perp}/|\mathbf{a}^{\perp}| = \pm\frac{1}{\sqrt{50}}(7\mathbf{i} + \mathbf{j}) = \pm\left(\frac{7}{\sqrt{50}}\mathbf{i} + \frac{1}{\sqrt{50}}\mathbf{j}\right)$

(d) $-4\mathbf{a} + 3\mathbf{b} = -4(\mathbf{i} - 7\mathbf{j}) + 3(2\mathbf{i} + 3\mathbf{j}) = 2\mathbf{i} + 37\mathbf{j}$

(e) $\mathbf{c} = s\mathbf{a} + t\mathbf{b} \quad \Leftrightarrow \quad 4\mathbf{j} = s(\mathbf{i} - 7\mathbf{j}) + t(2\mathbf{i} + 3\mathbf{j}) \quad \Leftrightarrow \quad 4\mathbf{j} = (s + 2t)\mathbf{i} + (-7s + 3t)\mathbf{j} \quad \Leftrightarrow$

$$\left\{ \begin{array}{r} s + 2t = 0 \\ -7s + 3t = 4 \end{array} \right\} \quad \Leftrightarrow \quad \left\{ \begin{array}{l} s = -\frac{8}{17} \\ t = \frac{4}{17} \end{array} \right\}$$

15. $\mathbf{a} = \langle 1, 2 \rangle$, $\mathbf{b} = \langle -2, 3 \rangle$

(a) $\mathbf{a} \cdot \mathbf{b} = \langle 1, 2 \rangle \cdot \langle -2, 3 \rangle = 1(-2) + 2(3) = 4$

(b) $\cos\theta = \dfrac{\mathbf{a} \cdot \mathbf{b}}{|\mathbf{a}||\mathbf{b}|} \quad \Rightarrow \quad \theta = \cos^{-1}\left(\dfrac{4}{\sqrt{5}\sqrt{13}}\right) \approx 60^\circ$

(c) $\text{comp}_{\mathbf{a}}\, \mathbf{b} = \dfrac{\mathbf{a} \cdot \mathbf{b}}{|\mathbf{a}|} = \dfrac{4}{\sqrt{5}}$

(d) $\text{proj}_{\mathbf{a}}\, \mathbf{b} = \dfrac{\mathbf{a} \cdot \mathbf{b}}{|\mathbf{a}|^2}\mathbf{a} = \dfrac{4}{\sqrt{5}} \cdot \dfrac{1}{\sqrt{5}}\langle 1, 2 \rangle = \frac{4}{5}\langle 1, 2 \rangle = \langle \frac{4}{5}, \frac{8}{5} \rangle$

17. $\mathbf{a} = -3\mathbf{i} + 4\mathbf{j}$, $\mathbf{b} = -2\mathbf{j}$

(a) $\mathbf{a} \cdot \mathbf{b} = (-3\mathbf{i} + 4\mathbf{j}) \cdot (-2\mathbf{j}) = -3(0) + 4(-2) = -8$

(b) $\cos\theta = \dfrac{\mathbf{a} \cdot \mathbf{b}}{|\mathbf{a}||\mathbf{b}|} \quad \Rightarrow \quad \theta = \cos^{-1}\left(\dfrac{-8}{5 \cdot 2}\right) \approx 143^\circ$

(c) $\text{comp}_{\mathbf{a}}\, \mathbf{b} = \dfrac{\mathbf{a} \cdot \mathbf{b}}{|\mathbf{a}|} = -\dfrac{8}{5}$

(d) $\text{proj}_{\mathbf{a}}\, \mathbf{b} = \dfrac{\mathbf{a} \cdot \mathbf{b}}{|\mathbf{a}|^2}\mathbf{a} = -\dfrac{8}{5} \cdot \dfrac{1}{5}(-3\mathbf{i} + 4\mathbf{j}) = \frac{24}{25}\mathbf{i} - \frac{32}{25}\mathbf{j}$

19. $\mathbf{a} = \langle 7, -6 \rangle$, $\mathbf{b} = \langle 6, 7 \rangle$

(a) $\mathbf{a} \cdot \mathbf{b} = \langle 7, -6 \rangle \cdot \langle 6, 7 \rangle = 42 - 42 = 0$

(b) $\cos\theta = \dfrac{\mathbf{a} \cdot \mathbf{b}}{|\mathbf{a}||\mathbf{b}|} \quad \Rightarrow \quad \theta = \cos^{-1}(0) = 90^\circ$

(c) $\text{comp}_{\mathbf{a}}\, \mathbf{b} = \dfrac{\mathbf{a} \cdot \mathbf{b}}{|\mathbf{a}|} = 0$

(d) $\text{proj}_{\mathbf{a}}\, \mathbf{b} = \dfrac{\mathbf{a} \cdot \mathbf{b}}{|\mathbf{a}|^2}\mathbf{a} = 0\mathbf{a} = \langle 0, 0 \rangle$

21. Using Exercise 57 from Section 1.2, the distance from the point $(2, 3)$ to the line $2x - y - 7 = 0$ is $\dfrac{|2(2) + (-1)(3) + (-7)|}{\sqrt{2^2 + (-1)^2}} = \dfrac{|-6|}{\sqrt{5}} = \dfrac{6}{\sqrt{5}}$.

23. Find a point on the line $2x + 4y = 7$, say $\left(\frac{7}{2}, 0\right)$ (let $y = 0$). Now find the distance d from $\left(\frac{7}{2}, 0\right)$ to the line $2x + 4y - 9 = 0$ as we did in Exercise 21: $d = \dfrac{\left|2\left(\frac{7}{2}\right) + 4(0) + (-9)\right|}{\sqrt{2^2 + 4^2}} = \dfrac{|-2|}{\sqrt{20}} = \dfrac{1}{\sqrt{5}}$.

25. Let $\mathbf{D} = \langle 2, 3 \rangle - \langle -1, 2 \rangle = \langle 3, 1 \rangle$. Then the work done by $\mathbf{F}$ is $\mathbf{F} \cdot \mathbf{D} = (5\mathbf{i} + 6\mathbf{j}) \cdot (3\mathbf{i} + \mathbf{j}) = 21$.

27. $x = 1 - t^2, y = 1 - t \quad (-1 \le t \le 1)$

$x = 1 - (1 - y)^2 = 2y - y^2 \ (0 \le y \le 2)$

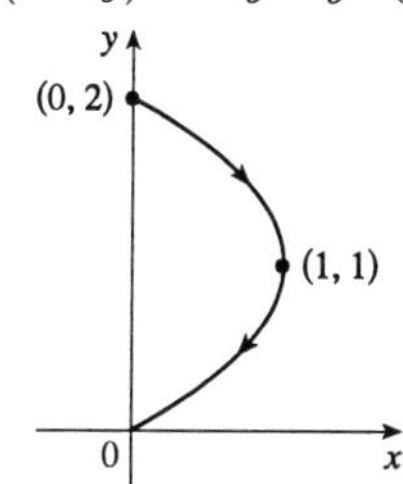

29. $x = 1 + \sin t, y = 2 + \cos t \quad \Rightarrow$

$(x - 1)^2 + (y - 2)^2 = \sin^2 t + \cos^2 t = 1$

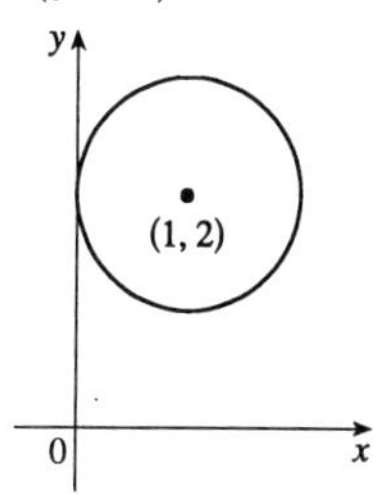

31. **(a)** In Equation 1 of Section 1.3, we use $\mathbf{r}_0 = \langle 1, 2 \rangle$ and $\mathbf{v} = \langle -3, 4 \rangle - \langle 1, 2 \rangle = \langle -4, 2 \rangle$ to obtain $\mathbf{r}(t) = \mathbf{r}_0 + t\mathbf{v} = \langle 1, 2 \rangle + t\langle -4, 2 \rangle$.

(b) $\mathbf{r}(t) = \langle 1 - 4t, 2 + 2t \rangle \quad \Rightarrow \quad x = 1 - 4t, y = 2 + 2t$

(c) $x = 1 - 4t \quad \Rightarrow \quad t = \frac{1}{4}(1 - x)$, so $y = 2 + 2t = 2 + 2\left[\frac{1}{4}(1 - x)\right] = 2 + \frac{1}{2} - \frac{1}{2}x = -\frac{1}{2}x + \frac{5}{2}$.

33. **(a)** Take $\mathbf{r}_0 = \langle 2, 3 \rangle$ and $\mathbf{v} = \langle 1, -5 \rangle$ in Equation 1.3.1 to get $\mathbf{r}(t) = \mathbf{r}_0 + t\mathbf{v} = \langle 2, 3 \rangle + t\langle 1, -5 \rangle$.

(b) $\mathbf{r}(t) = \langle 2 + t, 3 - 5t \rangle \quad \Rightarrow \quad x = 2 + t, y = 3 - 5t$

(c) $x = 2 + t \quad \Rightarrow \quad t = x - 2$, so $y = 3 - 5(x - 2) = -5x + 13$.

35. **(a)** $\mathbf{r}(t) = (t^2 + t)\mathbf{i} + (t - 4)\mathbf{j} \quad \Rightarrow \quad \mathbf{r}(2) = 6\mathbf{i} - 2\mathbf{j}$, so the position of the object at time $t = 2$ is $(6, -2)$.

(b) $t - 4$ must equal -1, so $t = 3$. Checking, $\mathbf{r}(3) = 12\mathbf{i} - \mathbf{j}$.

(c) $t - 4$ must equal 8, so $t = 12$. But $\mathbf{r}(12) = 156\mathbf{i} + 8\mathbf{j}$, so the object does *not* pass through the point $(4, 8)$.

37. The points $(5, 1)$ and $(2, -1)$ are on the line $2x - 3y = 7$, so the vector $\mathbf{a} = \langle 5, 1 \rangle - \langle 2, -1 \rangle = \langle 3, 2 \rangle$ is parallel to it and $\mathbf{a}^\perp = \langle -2, 3 \rangle$ is perpendicular to it. Now $2\mathbf{i} - 3\mathbf{j} = -\mathbf{a}^\perp$, so $2\mathbf{i} - 3\mathbf{j}$ is also perpendicular to the line. In general, $a\mathbf{i} + b\mathbf{j}$ is perpendicular to the line $ax + by = c$. To see this, let (x_1, y_1) and (x_2, y_2) be points on the line. Then $ax_1 + by_1 = c$ and $ax_2 + by_2 = c$. Subtracting the first relation from the second, we get $a(x_2 - x_1) + b(y_2 - y_1) = 0$ or, equivalently, $(a\mathbf{i} + b\mathbf{j}) \cdot [(x_2 - x_1)\mathbf{i} + (y_2 - y_1)\mathbf{j}] = 0$. The vector in brackets is parallel to the line since it is the difference of the position vectors $x_2\mathbf{i} + y_2\mathbf{j}$ and $x_1\mathbf{i} + y_1\mathbf{j}$. Since $a\mathbf{i} + b\mathbf{j}$ is perpendicular to the vector in brackets, it is perpendicular to the line $ax + by = c$.

CHAPTER TWO

EXERCISES 2.1

1. **(a)** Slopes of the secant lines:

x	m_{PQ}
0	$\dfrac{2.6-1.3}{0-3} \approx -0.43$
1	$\dfrac{2.0-1.3}{1-3} = -0.35$
2	$\dfrac{1.1-1.3}{2-3} = 0.2$
4	$\dfrac{2.1-1.3}{4-3} = 0.8$
5	$\dfrac{3.5-1.3}{5-3} = 1.1$

(b) The slope of the tangent line at P is about $\dfrac{2.5-0}{5-0.6} \approx 0.57$.

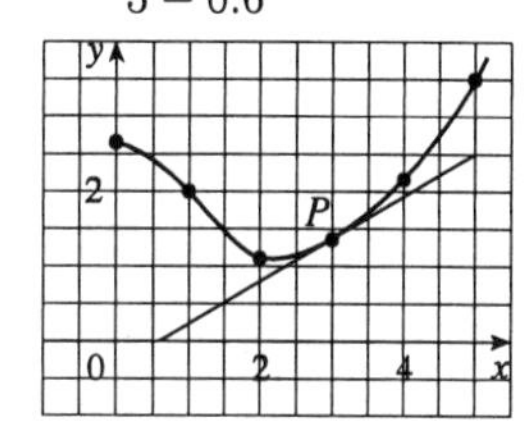

3. **(a)** $\overrightarrow{PQ} = \langle t, t^2-1\rangle - \langle 2, 3\rangle = \langle t-2, t^2-4\rangle$. The secant vector $\mathbf{s}(t) = \dfrac{1}{t-2}\overrightarrow{PQ} = \dfrac{1}{t-2}\langle t-2, t^2-4\rangle = \langle 1, t+2\rangle$ for $t \neq 2$.

(i)

t	3	2.5	2.1	2.01	2.001
$\mathbf{s}(t)$	$\langle 1,5\rangle$	$\langle 1,4.5\rangle$	$\langle 1,4.1\rangle$	$\langle 1,4.01\rangle$	$\langle 1,4.001\rangle$

(ii)

t	1	1.5	1.9	1.99	1.999
$\mathbf{s}(t)$	$\langle 1,3\rangle$	$\langle 1,3.5\rangle$	$\langle 1,3.9\rangle$	$\langle 1,3.99\rangle$	$\langle 1,3.999\rangle$

(b) The tangent vector at $P(2,3)$ appears to be $\langle 1, 4\rangle$.

(c) $\mathbf{L}(t) = \mathbf{r}(2) + t\langle 1,4\rangle = \langle 2,3\rangle + t\langle 1,4\rangle$

5. **(a)** The average velocity $\mathbf{v}_{\text{av}}$ at $t=2$ is $\dfrac{1}{h}[\mathbf{r}(2+h) - \mathbf{r}(2)]$, where $\mathbf{r}(t) = (3t)\mathbf{i} + (45t - 16t^2)\mathbf{j}$.

(i) $h = 0.5 \Rightarrow \mathbf{v}_{\text{av}} = \frac{1}{0.5}[\mathbf{r}(2.5) - \mathbf{r}(2)] = 2[\langle 7.5, 12.5\rangle - \langle 6, 26\rangle] = 2\langle 1.5, -13.5\rangle = \langle 3, -27\rangle$,

where the components are measured in ft/s.

(ii) $h = 0.1 \Rightarrow \mathbf{v}_{\text{av}} = \frac{1}{0.1}[\mathbf{r}(2.1) - \mathbf{r}(2)] = 10[\langle 6.3, 23.94\rangle - \langle 6, 26\rangle] = 10\langle 0.3, -2.06\rangle = \langle 3, -20.6\rangle$

(iii) $h = 0.05 \Rightarrow$

$\mathbf{v}_{\text{av}} = \frac{1}{0.05}[\mathbf{r}(2.05) - \mathbf{r}(2)] = 20[\langle 6.15, 25.01\rangle - \langle 6, 26\rangle] = 20\langle 0.15, -0.99\rangle = \langle 3, -19.8\rangle$

(iv) $h = 0.01 \Rightarrow$

$\mathbf{v}_{\text{av}} = \frac{1}{0.01}[\mathbf{r}(2.01) - \mathbf{r}(2)] = 100[\langle 6.03, 25.8084\rangle - \langle 6, 26\rangle] = 100\langle 0.03, -0.1916\rangle = \langle 3, -19.16\rangle$

(b) As h gets small, it appears that $\mathbf{v}_{\text{av}}$ approaches $\langle 3, -19\rangle$.

7. For the curve $y = \sqrt{x}$ and the point $P(4, 2)$:

(a)

	x	Q	m_{PQ}
(i)	5	$(5, 2.236068)$	0.236068
(ii)	4.5	$(4.5, 2.121320)$	0.242641
(iii)	4.1	$(4.1, 2.024846)$	0.248457
(iv)	4.01	$(4.01, 2.002498)$	0.249844
(v)	4.001	$(4.001, 2.000250)$	0.249984

	x	Q	m_{PQ}
(vi)	3	$(3, 1.732051)$	0.267949
(vii)	3.5	$(3.5, 1.870829)$	0.258343
(viii)	3.9	$(3.9, 1.974842)$	0.251582
(ix)	3.99	$(3.99, 1.997498)$	0.250156
(x)	3.999	$(3.999, 1.999750)$	0.250016

(b) The slope appears to be $\frac{1}{4}$.

(c) $y - 2 = \frac{1}{4}(x - 4)$ or $x - 4y + 4 = 0$

9. **(a)** At $t = 2$, $y = 40(2) - 16(2)^2 = 16$. The average velocity between times 2 and $2 + h$ is

$$\frac{40(2+h) - 16(2+h)^2 - 16}{h} = \frac{-24h - 16h^2}{h} = -24 - 16h, \text{ if } h \neq 0.$$

(i) $h = 0.5$, -32 ft/s

(ii) $h = 0.1$, -25.6 ft/s

(iii) $h = 0.05$, -24.8 ft/s

(iv) $h = 0.01$, -24.16 ft/s

(b) The instantaneous velocity when $t = 2$ is -24 ft/s.

11. Average velocity between times 1 and $1 + h$ is

$$\frac{s(1+h) - s(1)}{h} = \frac{(1+h)^3/6 - 1/6}{h} = \frac{h^3 + 3h^2 + 3h}{6h} = \frac{h^2 + 3h + 3}{6} \text{ if } h \neq 0.$$

(a) **(i)** $v_{\text{av}} = \dfrac{2^2 + 3(2) + 3}{6} = \dfrac{13}{6}$ ft/s

(ii) $v_{\text{av}} = \dfrac{1^2 + 3(1) + 3}{6} = \dfrac{7}{6}$ ft/s

(iii) $v_{\text{av}} = \dfrac{(0.5)^2 + 3(0.5) + 3}{6} = \dfrac{19}{24}$ ft/s

(iv) $v_{\text{av}} = \dfrac{(0.1)^2 + 3(0.1) + 3}{6} = \dfrac{331}{600}$ ft/s

(b) As h approaches 0, the velocity approaches $\frac{1}{2}$ ft/s.

(c) & (d)

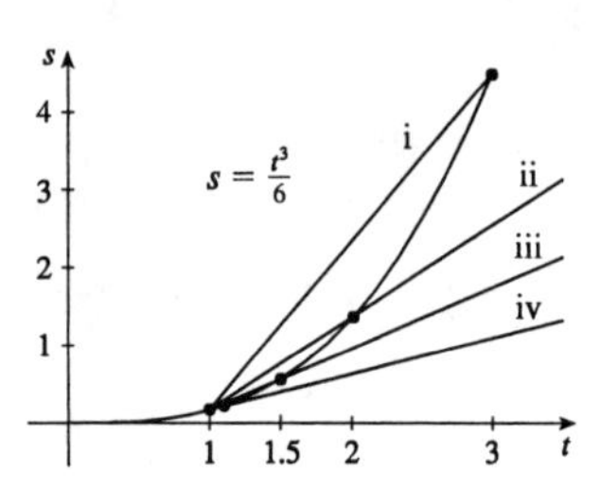

EXERCISES 2.2

1. **(a)** $\lim_{x\to 1} f(x) = 3$ **(b)** $\lim_{x\to 3^-} f(x) = 2$ **(c)** $\lim_{x\to 3^+} f(x) = -2$

(d) $\lim_{x\to 3} f(x)$ doesn't exist **(e)** $f(3) = 1$ **(f)** $\lim_{x\to -2^-} f(x) = -1$

(g) $\lim_{x\to -2^+} f(x) = -1$ **(h)** $\lim_{x\to -2} f(x) = -1$ **(i)** $f(-2) = -3$

3. **(a)** $\lim_{x\to 3} f(x) = 2$ **(b)** $\lim_{x\to 1} f(x) = -1$ **(c)** $\lim_{x\to -3} f(x) = 1$

(d) $\lim_{x\to 2^-} f(x) = 1$ **(e)** $\lim_{x\to 2^+} f(x) = 2$ **(f)** $\lim_{x\to 2} f(x)$ doesn't exist

5. **(a)** $\lim_{x\to 3} f(x) = \infty$ **(b)** $\lim_{x\to 7} f(x) = -\infty$ **(c)** $\lim_{x\to -4} f(x) = -\infty$

(d) $\lim_{x\to -9^-} f(x) = \infty$ **(e)** $\lim_{x\to -9^+} f(x) = -\infty$

(f) The equations of the vertical asymptotes: $x = -9$, $x = -4$, $x = 3$, $x = 7$

7. **(a)**

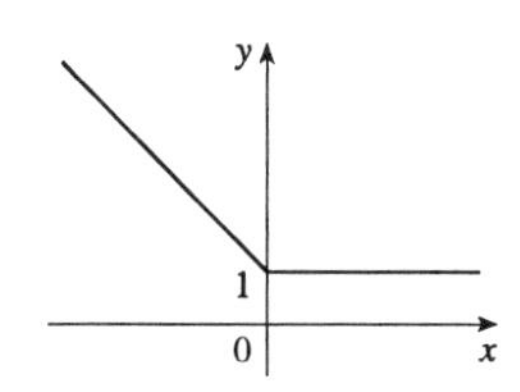

(b) **(i)** $\lim_{x\to 0^-} f(x) = 1$

(ii) $\lim_{x\to 0^+} f(x) = 1$

(iii) $\lim_{x\to 0} f(x) = 1$

9. For $g(x) = \dfrac{x-1}{x^3-1}$:

x	$g(x)$
0.2	0.806452
0.4	0.641026
0.6	0.510204
0.8	0.409836
0.9	0.369004
0.99	0.336689

x	$g(x)$
1.8	0.165563
1.6	0.193798
1.4	0.229358
1.2	0.274725
1.1	0.302115
1.01	0.330022

It appears that $\lim_{x\to 1} \dfrac{x-1}{x^3-1} = 0.\overline{3} = \dfrac{1}{3}$.

11. For $F(x) = \dfrac{(1/\sqrt{x}) - \frac{1}{5}}{x - 25}$:

x	$F(x)$
26	-0.003884
25.5	-0.003941
25.1	-0.003988
25.05	-0.003994
25.01	-0.003999

x	$F(x)$
24	-0.004124
24.5	-0.004061
24.9	-0.004012
24.95	-0.004006
24.99	-0.004001

It appears that $\lim_{x\to 25} F(x) = -0.004$.

13. For $f(x) = \dfrac{1 - \cos x}{x^2}$:

x	$f(x)$
1	0.459698
0.5	0.489670
0.4	0.493369
0.3	0.496261
0.2	0.498336
0.1	0.499583
0.05	0.499896
0.01	0.499996

It appears that $\lim\limits_{x\to 0} \dfrac{1 - \cos x}{x^2} = 0.5$.

15. $\mathbf{r}(t) = (2t - 1)\mathbf{i} + \left(\dfrac{t^2 - 1}{t - 1}\right)\mathbf{j} = (2t - 1)\mathbf{i} + (t + 1)\mathbf{j}$ for $t \neq 1$.

t	0.2	0.4	0.6	0.8	0.9	0.99
$\mathbf{r}(t)$	$-0.6\mathbf{i} + 1.2\mathbf{j}$	$-0.2\mathbf{i} + 1.4\mathbf{j}$	$0.2\mathbf{i} + 1.6\mathbf{j}$	$0.6\mathbf{i} + 1.8\mathbf{j}$	$0.8\mathbf{i} + 1.9\mathbf{j}$	$0.98\mathbf{i} + 1.99\mathbf{j}$

t	1.8	1.6	1.4	1.2	1.1	1.01
$\mathbf{r}(t)$	$2.6\mathbf{i} + 2.8\mathbf{j}$	$2.2\mathbf{i} + 2.6\mathbf{j}$	$1.8\mathbf{i} + 2.4\mathbf{j}$	$1.4\mathbf{i} + 2.2\mathbf{j}$	$1.2\mathbf{i} + 2.1\mathbf{j}$	$1.02\mathbf{i} + 2.01\mathbf{j}$

It appears that $\lim\limits_{t\to 1} \mathbf{r}(t) = \mathbf{i} + 2\mathbf{j}$.

17. $\lim\limits_{x\to 5^+} \dfrac{6}{x - 5} = \infty$ since $(x - 5) \to 0$ as $x \to 5^+$ and $\dfrac{6}{x - 5} > 0$ for $x > 5$.

19. $\lim\limits_{x\to 3} \dfrac{1}{(x - 3)^8} = \infty$ since $(x - 3) \to 0$ as $x \to 3$ and $\dfrac{1}{(x - 3)^8} > 0$.

21. $\lim\limits_{x\to -2^+} \dfrac{x - 1}{x^2(x + 2)} = -\infty$ since $(x + 2) \to 0$ as $x \to 2^+$ and $\dfrac{x - 1}{x^2(x + 2)} < 0$ for $-2 < x < 0$.

23. **(a)**

x	$f(x)$
0.5	-1.14
0.9	-3.69
0.99	-33.7
0.999	-333.7
0.9999	-3333.7
0.99999	$-33{,}333.7$

x	$f(x)$
1.5	0.42
1.1	3.02
1.01	33.0
1.001	333.0
1.0001	3333.0
1.00001	33,333.3

From these calculations, it seems that $\lim\limits_{x\to 1^-} f(x) = -\infty$ and $\lim\limits_{x\to 1^+} f(x) = \infty$.

(b) If x is slightly smaller than 1, then $x^3 - 1$ will be a negative number close to 0, and the reciprocal of $x^3 - 1$, that is, $f(x)$, will be a negative number with large absolute value. So $\lim_{x \to 1^-} f(x) = -\infty$.

If x is slightly larger than 1, then $x^3 - 1$ will be a small positive number, and its reciprocal, $f(x)$, will be a large positive number.

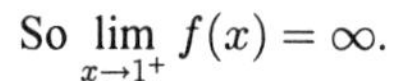

So $\lim_{x \to 1^+} f(x) = \infty$.

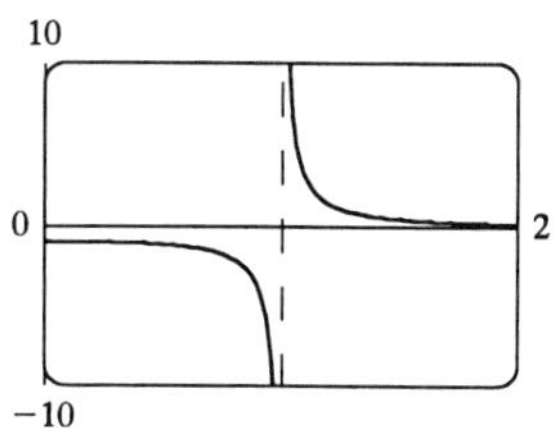

(c) It appears from the graph of f that $\lim_{x \to 1^-} f(x) = -\infty$ and $\lim_{x \to 1^+} f(x) = \infty$.

25. Let $h(x) = (1 + x)^{1/x}$.

x	$h(x)$
1.0	2.0
0.1	2.593742
0.01	2.704814
0.001	2.716924
0.0001	2.718146
0.00001	2.718268
0.000001	2.718280
0.0000001	2.718282
0.00000001	2.718282
0.000000001	2.718282

It appears that $\lim_{x \to 0}(1 + x)^{1/x} \approx 2.71828$.

27. For $f(x) = x^2 - (2^x/1000)$:

(a)

x	$f(x)$
1	0.998000
0.8	0.638259
0.6	0.358484
0.4	0.158680
0.2	0.038851
0.1	0.008928
0.05	0.001465

It appears that $\lim_{x \to 0} f(x) = 0$.

(b)

x	$f(x)$
0.04	0.000572
0.02	−0.000614
0.01	−0.000907
0.005	−0.000978
0.003	−0.000993
0.001	−0.001000

It appears that $\lim_{x \to 0} f(x) = -0.001$.

29. From the following graphs, it seems that $\lim\limits_{x\to 0} \dfrac{\tan(4x)}{x} = 4$.

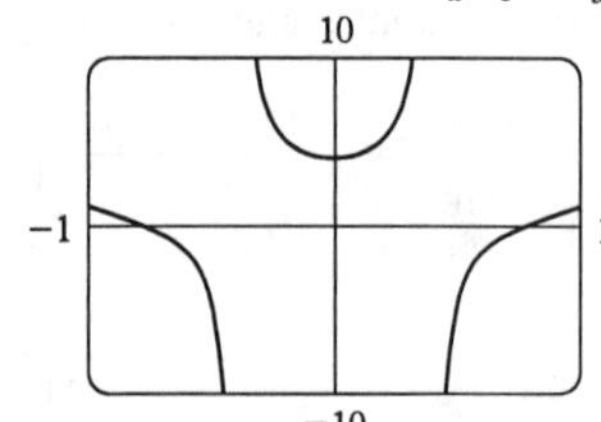

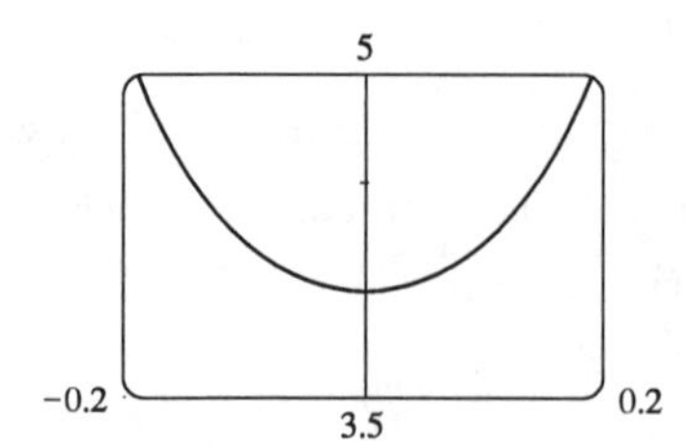

31. No matter how many times we zoom in towards the origin, the graphs appear to consist of almost-vertical lines. This indicates more and more frequent oscillations as $x \to 0$.

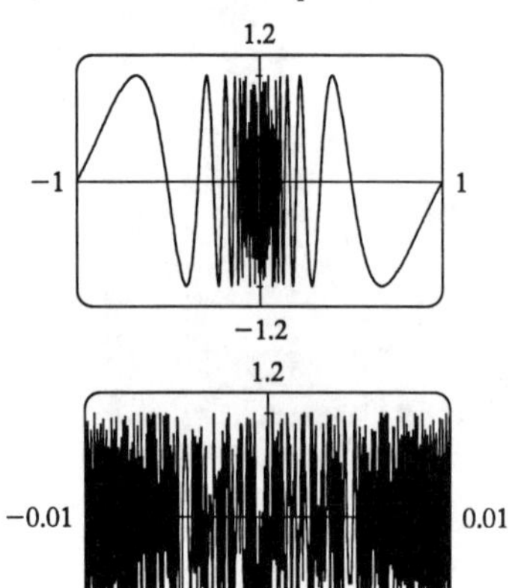

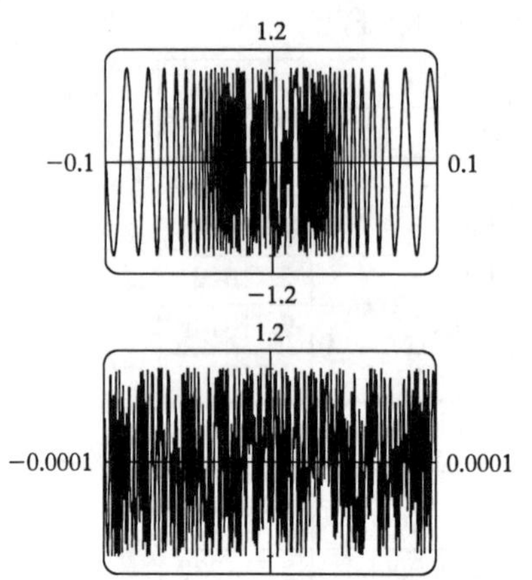

EXERCISES 2.3

1.
$$\begin{aligned}\lim_{x\to 4}\left(5x^2 - 2x + 3\right) &= \lim_{x\to 4} 5x^2 - \lim_{x\to 4} 2x + \lim_{x\to 4} 3 && \text{(Limit Laws 2 \& 1)}\\ &= 5\lim_{x\to 4} x^2 - 2\lim_{x\to 4} x + 3 && \text{(3 \& 7)}\\ &= 5(4)^2 - 2(4) + 3 = 75 && \text{(9 \& 8)}\end{aligned}$$

3.
$$\begin{aligned}\lim_{x\to 2}\left(x^2+1\right)\left(x^2+4x\right) &= \lim_{x\to 2}\left(x^2+1\right)\lim_{x\to 2}\left(x^2+4x\right) && \text{(4)}\\ &= \left(\lim_{x\to 2} x^2 + \lim_{x\to 2} 1\right)\left(\lim_{x\to 2} x^2 + 4\lim_{x\to 2} x\right) && \text{(1 \& 3)}\\ &= \left[(2)^2+1\right]\left[(2)^2+4(2)\right] = 60 && \text{(9, 7 \& 8)}\end{aligned}$$

5. $$\lim_{x\to -1}\frac{x-2}{x^2+4x-3}=\frac{\lim\limits_{x\to -1}(x-2)}{\lim\limits_{x\to -1}(x^2+4x-3)} \qquad (5)$$
$$=\frac{\lim\limits_{x\to -1}x-\lim\limits_{x\to -1}2}{\lim\limits_{x\to -1}x^2+4\lim\limits_{x\to -1}x-\lim\limits_{x\to -1}3} \qquad (2,\ 1\ \&\ 3)$$
$$=\frac{(-1)-2}{(-1)^2+4(-1)-3}=\frac{1}{2} \qquad (8,\ 7\ \&\ 9)$$

7. $$\lim_{x\to -1}\sqrt{x^3+2x+7}=\sqrt{\lim_{x\to -1}(x^3+2x+7)} \qquad (11)$$
$$=\sqrt{\lim_{x\to -1}x^3+2\lim_{x\to -1}x+\lim_{x\to -1}7} \qquad (1\ \&\ 3)$$
$$=\sqrt{(-1)^3+2(-1)+7}=2 \qquad (9,\ 8\ \&\ 6)$$

9. $$\lim_{t\to -2}(t+1)^9(t^2-1)=\lim_{t\to -2}(t+1)^9\lim_{t\to -2}(t^2-1) \qquad (4)$$
$$=\left[\lim_{t\to -2}(t+1)\right]^9\lim_{t\to -2}(t^2-1) \qquad (6)$$
$$=\left[\lim_{t\to -2}t+\lim_{t\to -2}1\right]^9\left[\lim_{t\to -2}t^2-\lim_{t\to -2}1\right] \qquad (1\ \&\ 2)$$
$$=[(-2)+1]^9\left[(-2)^2-1\right]=-3 \qquad (8,\ 7\ \&\ 9)$$

11. $$\lim_{w\to -2}\sqrt[3]{\frac{4w+3w^3}{3w+10}}=\sqrt[3]{\lim_{w\to -2}\frac{4w+3w^3}{3w+10}} \qquad (11)$$
$$=\sqrt[3]{\frac{\lim\limits_{w\to -2}(4w+3w^3)}{\lim\limits_{w\to -2}(3w+10)}} \qquad (5)$$
$$=\sqrt[3]{\frac{4\lim\limits_{w\to -2}w+3\lim\limits_{w\to -2}w^3}{3\lim\limits_{w\to -2}w+\lim\limits_{w\to -2}10}} \qquad (1\ \&\ 3)$$
$$=\sqrt[3]{\frac{4(-2)+3(-2)^3}{3(-2)+10}}=-2 \qquad (8,\ 9\ \&\ 7)$$

13. **(a)** $\lim\limits_{x\to a}[f(x)+h(x)]=\lim\limits_{x\to a}f(x)+\lim\limits_{x\to a}h(x)=-3+8=5$

(b) $\lim\limits_{x\to a}[f(x)]^2=\left[\lim\limits_{x\to a}f(x)\right]^2=(-3)^2=9$

(c) $\lim\limits_{x\to a}\sqrt[3]{h(x)}=\sqrt[3]{\lim\limits_{x\to a}h(x)}=\sqrt[3]{8}=2$

(d) $\lim\limits_{x\to a}\dfrac{1}{f(x)}=\dfrac{1}{\lim\limits_{x\to a}f(x)}=\dfrac{1}{-3}=-\dfrac{1}{3}$

(e) $\lim\limits_{x\to a}\dfrac{f(x)}{h(x)}=\dfrac{\lim\limits_{x\to a}f(x)}{\lim\limits_{x\to a}h(x)}=\dfrac{-3}{8}=-\dfrac{3}{8}$

(f) $\lim\limits_{x\to a}\dfrac{g(x)}{f(x)}=\dfrac{\lim\limits_{x\to a}g(x)}{\lim\limits_{x\to a}f(x)}=\dfrac{0}{-3}=0$

(g) The limit does not exist, since $\lim\limits_{x\to a}g(x)=0$ but $\lim\limits_{x\to a}f(x)\neq 0$.

(h) $\lim\limits_{x\to a}\dfrac{2f(x)}{h(x)-f(x)}=\dfrac{2\lim\limits_{x\to a}f(x)}{\lim\limits_{x\to a}h(x)-\lim\limits_{x\to a}f(x)}=\dfrac{2(-3)}{8-(-3)}=-\dfrac{6}{11}$

15. $\lim_{x\to-3}\dfrac{x^2-x+12}{x+3}$ does not exist since $x+3\to0$ but $x^2-x+12\to24$ as $x\to-3$.

17. $\lim_{x\to-1}\dfrac{x^2-x-2}{x+1}=\lim_{x\to-1}\dfrac{(x+1)(x-2)}{x+1}=\lim_{x\to-1}(x-2)=-3$

19. $\lim_{t\to1}\dfrac{t^3-t}{t^2-1}=\lim_{t\to1}\dfrac{t(t^2-1)}{t^2-1}=\lim_{t\to1}t=1$

21. $\lim_{h\to0}\dfrac{(h-5)^2-25}{h}=\lim_{h\to0}\dfrac{(h^2-10h+25)-25}{h}=\lim_{h\to0}\dfrac{h^2-10h}{h}=\lim_{h\to0}(h-10)=-10$

23. $\lim_{h\to0}\dfrac{(1+h)^4-1}{h}=\lim_{h\to0}\dfrac{(1+4h+6h^2+4h^3+h^4)-1}{h}=\lim_{h\to0}\dfrac{4h+6h^2+4h^3+h^4}{h}$

$$=\lim_{h\to0}\left(4+6h+4h^2+h^3\right)=4$$

25. $\lim_{x\to-2}\dfrac{x+2}{x^2-x-6}=\lim_{x\to-2}\dfrac{x+2}{(x-3)(x+2)}=\lim_{x\to-2}\dfrac{1}{x-3}=-\dfrac{1}{5}$

27. $\lim_{t\to9}\dfrac{9-t}{3-\sqrt{t}}=\lim_{t\to9}\dfrac{(3+\sqrt{t})(3-\sqrt{t})}{3-\sqrt{t}}=\lim_{t\to9}\left(3+\sqrt{t}\right)=3+\sqrt{9}=6$

29. $\lim_{t\to0}\dfrac{\sqrt{2-t}-\sqrt{2}}{t}=\lim_{t\to0}\dfrac{\sqrt{2-t}-\sqrt{2}}{t}\cdot\dfrac{\sqrt{2-t}+\sqrt{2}}{\sqrt{2-t}+\sqrt{2}}=\lim_{t\to0}\dfrac{-t}{t\left(\sqrt{2-t}+\sqrt{2}\right)}=\lim_{t\to0}\dfrac{-1}{\sqrt{2-t}+\sqrt{2}}$

$$=-\frac{1}{2\sqrt{2}}=-\frac{\sqrt{2}}{4}$$

31. $\lim_{x\to9}\dfrac{x^2-81}{\sqrt{x}-3}=\lim_{x\to9}\dfrac{(x-9)(x+9)}{\sqrt{x}-3}=\lim_{x\to9}\dfrac{\left(\sqrt{x}-3\right)\left(\sqrt{x}+3\right)(x+9)}{\sqrt{x}-3}$

$$=\lim_{x\to9}\left(\sqrt{x}+3\right)(x+9)=\lim_{x\to9}\left(\sqrt{x}+3\right)\lim_{x\to9}(x+9)=\left(\sqrt{9}+3\right)(9+9)=108$$

33. $\lim_{t\to0}\left[\dfrac{1}{t\sqrt{1+t}}-\dfrac{1}{t}\right]=\lim_{t\to0}\dfrac{1-\sqrt{1+t}}{t\sqrt{1+t}}=\lim_{t\to0}\dfrac{(1-\sqrt{1+t})(1+\sqrt{1+t})}{t\sqrt{t+1}(1+\sqrt{1+t})}=\lim_{t\to0}\dfrac{-t}{t\sqrt{1+t}(1+\sqrt{1+t})}$

$$=\lim_{t\to0}\frac{-1}{\sqrt{1+t}(1+\sqrt{1+t})}=\frac{-1}{\sqrt{1+0}(1+\sqrt{1+0})}=-\frac{1}{2}$$

35. $\lim_{x\to0}\dfrac{x}{\sqrt{1+3x}-1}=\lim_{x\to0}\dfrac{x\left(\sqrt{1+3x}+1\right)}{\left(\sqrt{1+3x}-1\right)\left(\sqrt{1+3x}+1\right)}=\lim_{x\to0}\dfrac{x\left(\sqrt{1+3x}+1\right)}{3x}$

$$=\lim_{x\to0}\frac{\sqrt{1+3x}+1}{3}=\frac{\sqrt{1}+1}{3}=\frac{2}{3}$$

37. $\lim_{x\to2}\dfrac{x-\sqrt{3x-2}}{x^2-4}=\lim_{x\to2}\dfrac{\left(x-\sqrt{3x-2}\right)\left(x-\sqrt{3x-2}\right)}{\left(x^2-4\right)\left(x-\sqrt{3x-2}\right)}=\lim_{x\to2}\dfrac{x^2-3x+2}{\left(x^2-4\right)\left(x+\sqrt{3x-2}\right)}$

$$=\lim_{x\to2}\frac{(x-2)(x-1)}{(x-2)(x+2)\left(x+\sqrt{3x-2}\right)}=\lim_{x\to2}\frac{(x-1)}{(x+2)\left(x+\sqrt{3x-2}\right)}=\frac{1}{4\left(2+\sqrt{4}\right)}=\frac{1}{16}$$

39. $\lim_{t\to1}\mathbf{r}(t)=\lim_{t\to1}\left[(2t-3)\mathbf{i}+\left(\dfrac{t^2-t}{t-1}\right)\mathbf{j}\right]=\left[\lim_{t\to1}(2t-3)\right]\mathbf{i}+\left[\lim_{t\to1}\dfrac{t(t-1)}{t-1}\right]\mathbf{j}$

$$=\left[\lim_{t\to1}(2t-3)\right]\mathbf{i}+\left[\lim_{t\to1}(t)\right]\mathbf{j}=(-1)\mathbf{i}+(1)\mathbf{j}=-\mathbf{i}+\mathbf{j}$$

41. $\lim_{t\to2}\mathbf{r}(t)=\lim_{t\to2}\left\langle\dfrac{4-t}{2-\sqrt{t}},\dfrac{t^2-4}{t-2}\right\rangle=\left\langle\lim_{t\to2}\dfrac{(2+\sqrt{t})(2-\sqrt{t})}{2-\sqrt{t}},\lim_{t\to2}\dfrac{(t+2)(t-2)}{t-2}\right\rangle$

$$=\left\langle\lim_{t\to2}\left(2+\sqrt{t}\right),\lim_{t\to2}(t+2)\right\rangle=\langle2+\sqrt{2},4\rangle$$

43. The anticipated position of the particle at time $t = 2$ is

$$\lim_{t\to 2}\mathbf{r}(t) = \lim_{t\to 2}\left[\left(\frac{t^2-4}{t^2-t-2}\right)\mathbf{i} + (2t)\mathbf{j}\right] = \left[\lim_{t\to 2}\frac{(t+2)(t-2)}{(t-2)(t+1)}\right]\mathbf{i} + \left[\lim_{t\to 2}(2t)\right]\mathbf{j}$$
$$= \left[\lim_{t\to 2}\left(\frac{t+2}{t+1}\right)\right]\mathbf{i} + \left[\lim_{t\to 2}(2t)\right]\mathbf{j} = \tfrac{4}{3}\mathbf{i} + 4\mathbf{j}$$

45. Let $f(x) = -x^2$, $g(x) = x^2\cos 20\pi x$ and $h(x) = x^2$. Then $-1 \le \cos 20\pi x \le 1 \quad\Rightarrow\quad f(x) \le g(x) \le h(x)$. So since $\lim_{x\to 0} f(x) = \lim_{x\to 0} h(x) = 0$, by the Squeeze Theorem we have $\lim_{x\to 0}(x^2\cos 20\pi x) = \lim_{x\to 0} g(x) = 0$.

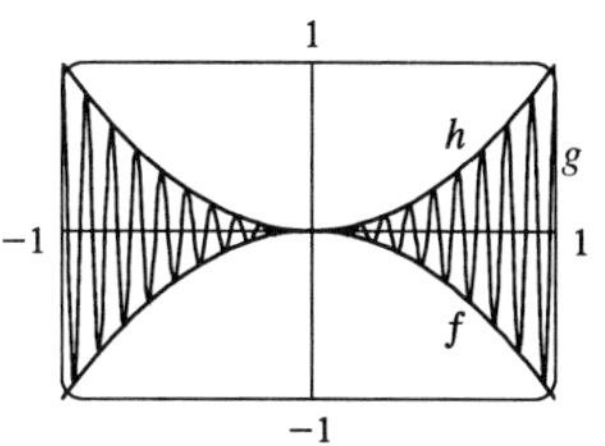

47. $1 \le f(x) \le x^2 + 2x + 2$ for all x. But $\lim_{x\to -1} 1 = 1$ and $\lim_{x\to -1}(x^2+2x+2) = \lim_{x\to -1} x^2 + 2\lim_{x\to -1} x + \lim_{x\to -1} 2$ $= (-1)^2 + 2(-1) + 2 = 1$. Therefore, by the Squeeze Theorem, $\lim_{x\to -1} f(x) = 1$.

49. $-1 \le \sin(1/x) \le 1 \quad\Rightarrow\quad -x^2 \le x^2\sin(1/x) \le x^2$. Since $\lim_{x\to 0}(-x^2) = 0$ and $\lim_{x\to 0} x^2 = 0$, we have $\lim_{x\to 0} x^2\sin(1/x) = 0$ by the Squeeze Theorem.

51. $\lim_{x\to 4^-}\sqrt{16-x^2} = \sqrt{\lim_{x\to 4^-} 16 - \lim_{x\to 4^-} x^2} = \sqrt{16-4^2} = 0$

53. If $x > -4$, then $|x+4| = x+4$, so $\lim_{x\to -4^+}|x+4| = \lim_{x\to -4^+}(x+4) = -4+4 = 0$.

If $x < -4$, then $|x+4| = -(x+4)$, so $\lim_{x\to -4^-}|x+4| = \lim_{x\to -4^-} -(x+4) = 4-4 = 0$.

Therefore $\lim_{x\to -4}|x+4| = 0$.

55. If $x > 2$, then $|x-2| = x-2$, so $\lim_{x\to 2^+}\frac{|x-2|}{x-2} = \lim_{x\to 2^+}\frac{x-2}{x-2} = \lim_{x\to 2^+} 1 = 1$.

If $x < 2$, then $|x-2| = -(x-2)$ so $\lim_{x\to 2^-}\frac{|x-2|}{x-2} = \lim_{x\to 2^-}\frac{-(x-2)}{x-2} = \lim_{x\to 2^-} -1 = -1$.

The right and left limits are different, so $\lim_{x\to 2}\frac{|x-2|}{x-2}$ does not exist.

57. $[\![x]\!] = -2$ for $-2 \le x < -1$, so $\lim_{x\to -2^+}[\![x]\!] = \lim_{x\to -2^+}(-2) = -2$

59. $[\![x]\!] = -3$ for $-3 \le x < -2$, so $\lim_{x\to -2.4}[\![x]\!] = \lim_{x\to -2.4}(-3) = -3$.

61. $\lim_{x\to 1^+}\sqrt{x^2+x-2} = \sqrt{\lim_{x\to 1^+} x^2 + \lim_{x\to 1^+} x - \lim_{x\to 1^+} 2} = \sqrt{1^2+1-2} = 0$

Notice that the domain of $\sqrt{x^2+x-2}$ is $(-\infty, -2] \cup [1, \infty)$.

63. Since $|x| = -x$ for $x < 0$, we have $\lim_{x\to 0^-}\left(\frac{1}{x} - \frac{1}{|x|}\right) = \lim_{x\to 0^-}\left(\frac{1}{x} - \frac{1}{-x}\right) = \lim_{x\to 0^-}\frac{2}{x}$, which does not exist since the denominator $\to 0$ and the numerator does not.

65. **(a)**

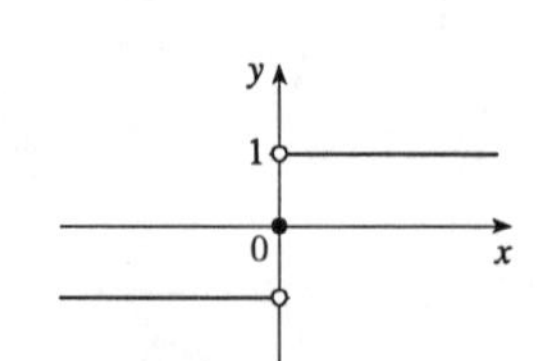

(b) **(i)** Since $\operatorname{sgn} x = 1$ for $x > 0$, $\lim_{x\to 0^+} \operatorname{sgn} x = \lim_{x\to 0^+} 1 = 1$.

(ii) Since $\operatorname{sgn} x = -1$ for $x < 0$, $\lim_{x\to 0^-} \operatorname{sgn} x = \lim_{x\to 0^-} -1 = -1$.

(iii) Since $\lim_{x\to 0^-} \operatorname{sgn} x \neq \lim_{x\to 0^+} \operatorname{sgn} x$, $\lim_{x\to 0} \operatorname{sgn} x$ does not exist.

(iv) Since $|\operatorname{sgn} x| = 1$ for $x \neq 0$, $\lim_{x\to 0} |\operatorname{sgn} x| = \lim_{x\to 0} 1 = 1$.

67. **(a)** $\lim_{x\to -1^-} g(x) = \lim_{x\to -1^-} (-x^3) = -(-1)^3 = 1$,

$\lim_{x\to -1^+} g(x) = \lim_{x\to -1^+} (x+2)^2 = (-1+2)^2 = 1$

(b) By part (a), $\lim_{x\to -1} g(x) = 1$.

(c)

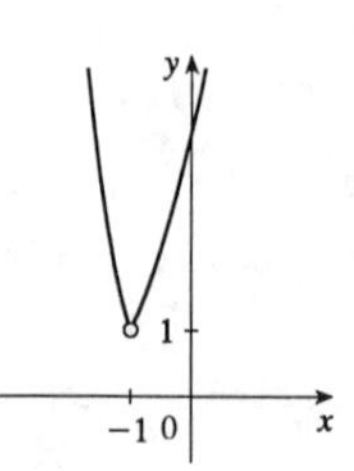

69. **(a)** **(i)** $[\![x]\!] = n-1$ for $n-1 \leq x < n$, so $\lim_{x\to n^-} [\![x]\!] = \lim_{x\to n^-} (n-1) = n-1$.

(ii) $[\![x]\!] = n$ for $n \leq x < n+1$, so $\lim_{x\to n^+} [\![x]\!] = \lim_{x\to n^+} n = n$.

(b) $\lim_{x\to a} [\![x]\!]$ exists $\Leftrightarrow$ a is not an integer.

71. **(a)** **(i)** $\displaystyle\lim_{x\to 1^+} \frac{x^2-1}{|x-1|} = \lim_{x\to 1^+} \frac{x^2-1}{x-1} = \lim_{x\to 1^+} (x+1) = 2$

(ii) $\displaystyle\lim_{x\to 1^-} \frac{x^2-1}{|x-1|} = \lim_{x\to 1^-} \frac{x^2-1}{-(x-1)} = \lim_{x\to 1^-} -(x+1) = -2$

(b) No, $\lim_{x\to 1} F(x)$ does not exist since $\lim_{x\to 1^+} F(x) \neq \lim_{x\to 1^-} F(x)$.

(c)

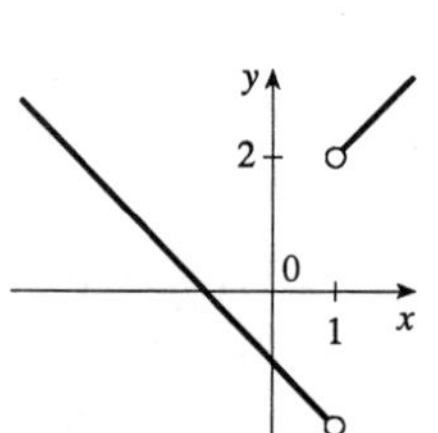

73. Since $p(x)$ is a polynomial, $p(x) = a_0 + a_1x + a_2x^2 + \cdots + a_nx^n$. Thus, by the Limit Laws,

$$\lim_{x\to a} p(x) = \lim_{x\to a}\left(a_0 + a_1x + a_2x^2 + \cdots + a_nx^n\right) = a_0 + a_1 \lim_{x\to a} x + a_2 \lim_{x\to a} x^2 + \cdots + a_n \lim_{x\to a} x^n$$

$= a_0 + a_1a + a_2a^2 + \cdots + a_na^n = p(a)$. Thus, for any polynomial p, $\lim_{x\to a} p(x) = p(a)$.

75. Observe that $0 \leq f(x) \leq x^2$ for all x, and $\lim_{x\to 0} 0 = 0 = \lim_{x\to 0} x^2$. So, by the Squeeze Theorem, $\lim_{x\to 0} f(x) = 0$.

77. Let $f(x) = H(x)$ and $g(x) = 1 - H(x)$, where H is the Heaviside function defined in Example 2.2.6. Then $\lim_{x\to 0} f(x)$ and $\lim_{x\to 0} g(x)$ do not exist but $\lim_{x\to 0}[f(x)g(x)] = \lim_{x\to 0} 0 = 0$.

79. Let $t = \sqrt[3]{1+cx}$. Then $t \to 1$ as $x \to 0$ and $t^3 = 1 + cx$ $\Rightarrow$ $x = (t^3-1)/c$. (If $c = 0$, then the limit is obviously 0.) Therefore

$$\lim_{x\to 0} \frac{\sqrt[3]{1+cx}-1}{x} = \lim_{t\to 1} \frac{t-1}{(t^3-1)/c} = \lim_{t\to 1} \frac{c(t-1)}{(t-1)(t^2+t+1)} = \lim_{t\to 1} \frac{c}{t^2+t+1} = \frac{c}{1^2+1+1} = \frac{c}{3}.$$

Another Method: Multiply numerator and denominator by $(1+cx)^{2/3} + (1+cx)^{1/3} + 1$.

81. Since the denominator approaches 0 as $x \to -2$, the limit will exist only if the numerator also approaches 0 as $x \to -2$. In order for this to happen, we need $\lim\limits_{x\to -2}(3x^2 + ax + a + 3) = 0 \quad \Leftrightarrow$

$3(-2)^2 + a(-2) + a + 3 = 0 \quad \Leftrightarrow \quad 12 - 2a + a + 3 = 0 \quad \Leftrightarrow \quad a = 15$. With $a = 15$, the limit becomes

$\lim\limits_{x\to -2} \dfrac{3x^2 + 15x + 18}{x^2 + x - 2} = \lim\limits_{x\to -2} \dfrac{3(x+2)(x+3)}{(x-1)(x+2)} = \dfrac{3(-2+3)}{-2-1} = -1.$

83. $y - 1 < [\![y]\!] \le y$, so $x^2\left(\dfrac{1}{4x^2} - 1\right) < x^2\left[\!\!\left[\dfrac{1}{4x^2}\right]\!\!\right] \le x^2\left(\dfrac{1}{4x^2}\right) = \dfrac{1}{4} \quad (x \ne 0)$. But

$\lim\limits_{x\to 0} x^2\left(\dfrac{1}{4x^2} - 1\right) = \lim\limits_{x\to 0}\left(\dfrac{1}{4} - x^2\right) = \dfrac{1}{4}$ and $\lim\limits_{x\to 0} \dfrac{1}{4} = \dfrac{1}{4}$. So by the Squeeze Theorem, $\lim\limits_{x\to 0} x^2\left[\!\!\left[\dfrac{1}{4x^2}\right]\!\!\right] = \dfrac{1}{4}$.

EXERCISES 2.4

1. **(a)** $|(6x+1) - 19| < 0.1 \quad \Leftrightarrow \quad |6x - 18| < 0.1 \quad \Leftrightarrow \quad 6|x-3| < 0.1 \quad \Leftrightarrow \quad |x - 3| < (0.1)/6 = \frac{1}{60}$

(b) $|(6x+1) - 19| < 0.01 \quad \Leftrightarrow \quad |x - 3| < (0.01)/6 = \frac{1}{600}$

3. On the left side, we need $|x - 2| < \left|\frac{10}{7} - 2\right| = \frac{4}{7}$. On the right side, we need $|x - 2| < \left|\frac{10}{3} - 2\right| = \frac{4}{3}$. For both of these conditions to be satisfied at once, we need the more restrictive of the two to hold, that is, $|x - 2| < \frac{4}{7}$. So we can choose $\delta = \frac{4}{7}$, or any smaller positive number.

5. $\left|\sqrt{4x+1} - 3\right| < 0.5 \quad \Leftrightarrow \quad 2.5 < \sqrt{4x+1} < 3.5.$

We plot the three parts of this inequality on the same screen and identify the x-coordinates of the points of intersection using the cursor. It appears that the inequality holds for $1.32 \le x \le 2.81$. Since $|2 - 1.32| = 0.68$ and $|2 - 2.81| = 0.81$, we choose $0 < \delta \le \min\{0.68, 0.81\} = 0.68$.

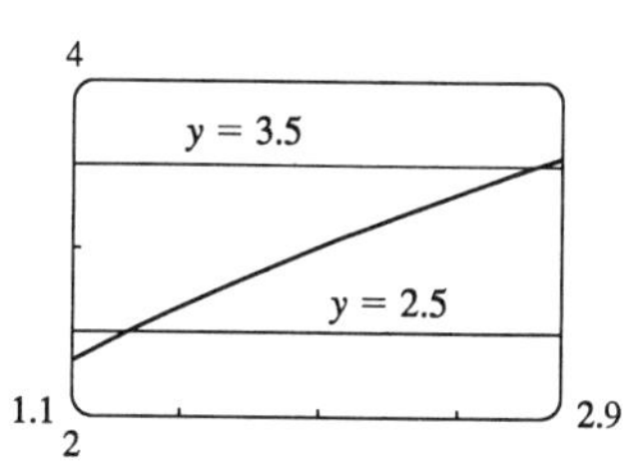

7. For $\epsilon = 1$, the definition of a limit requires that we find δ such that $|(4 + x - 3x^3) - 2| < 1 \quad \Leftrightarrow$ $1 < 4 + x - 3x^3 < 3$ whenever $|x - 1| < \delta$. If we plot the graphs of $y = 1$, $y = 4 + x - 3x^3$ and $y = 3$ on the same screen, we see that we need $0.86 \le x \le 1.11$. So since $|1 - 0.86| = 0.14$ and $|1 - 1.11| = 0.11$, we choose $\delta = 0.11$ (or any smaller positive number). For $\epsilon = 0.1$, we must find δ such that $|(4 + x - 3x^3) - 2| < 0.1 \quad \Leftrightarrow \quad 1.9 < 4 + x - 3x^3 < 2.1$ whenever $|x - 1| < \delta$. From the graph, we see that we need $0.988 \le x \le 1.012$. So since $|1 - 0.988| = 0.012$ and $|1 - 1.012| = 0.012$, we must choose $\delta = 0.012$ (or any smaller positive number) for the inequality to hold.

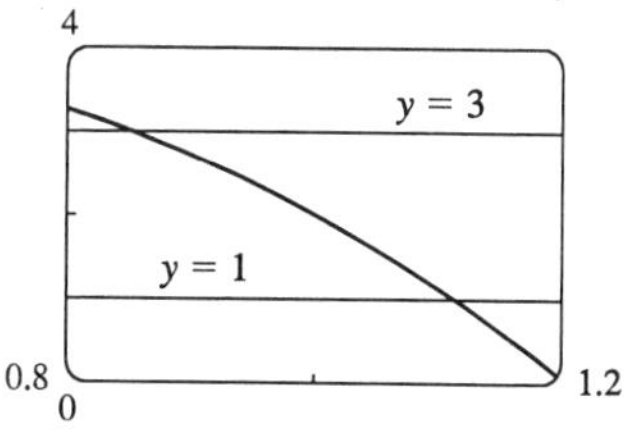

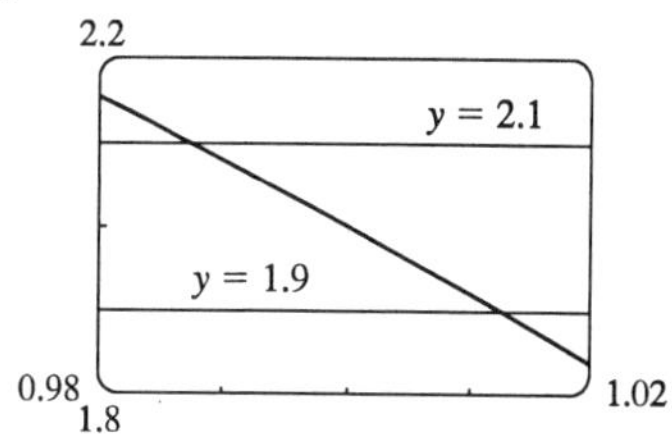

9. From the graph, we see that $\dfrac{x}{(x^2+1)(x-1)^2} > 100$ whenever $0.93 \le x \le 1.07$. So since $|1-0.93| = 0.7$ and $|1-1.07| = 0.7$, we can take $\delta = 0.07$ (or any smaller positive number).

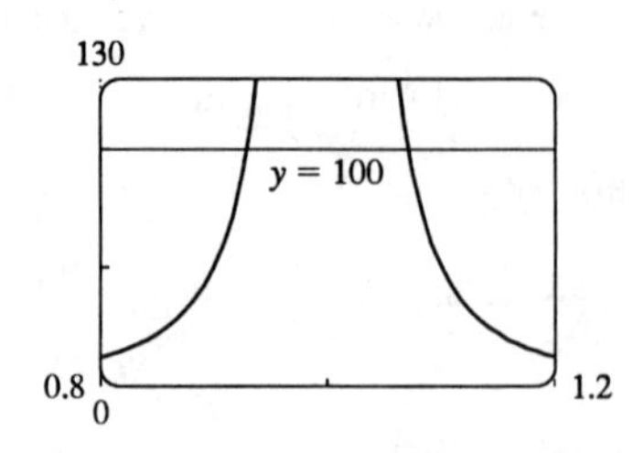

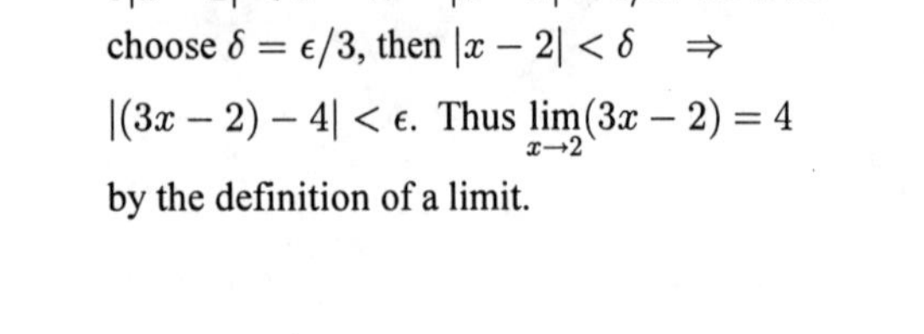

11. Given $\epsilon > 0$, we need $\delta > 0$ such that if $|x-2| < \delta$, then $|(3x-2)-4| < \epsilon \quad\Leftrightarrow\quad |3x-6| < \epsilon \quad\Leftrightarrow$ $3|x-2| < \epsilon \quad\Leftrightarrow\quad |x-2| < \epsilon/3$. So if we choose $\delta = \epsilon/3$, then $|x-2| < \delta \quad\Rightarrow$ $|(3x-2)-4| < \epsilon$. Thus $\lim\limits_{x\to 2}(3x-2) = 4$ by the definition of a limit.

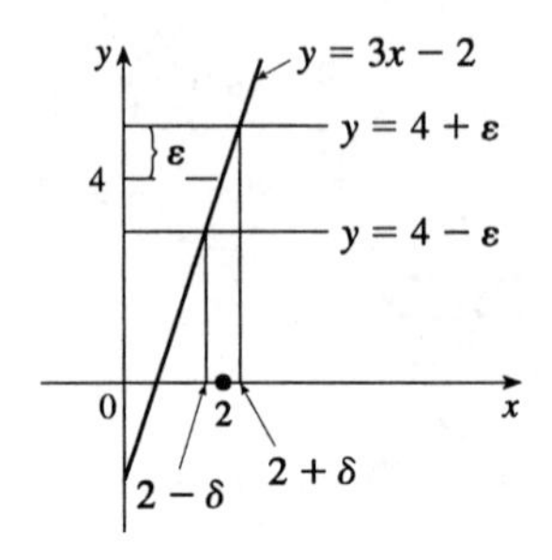

13. Given $\epsilon > 0$, we need $\delta > 0$ such that if $|x-(-1)| < \delta$, then $|(5x+8)-3| < \epsilon \;\Leftrightarrow\; |5x+5| < \epsilon \;\Leftrightarrow$ $5|x+1| < \epsilon \quad\Leftrightarrow\quad |x-(-1)| < \epsilon/5$. So if we choose $\delta = \epsilon/5$, then $|x-(-1)| < \delta \;\Rightarrow\; |(5x+8)-3| < \epsilon$. Thus $\lim\limits_{x\to -1}(5x+8) = 3$ by the definition of a limit.

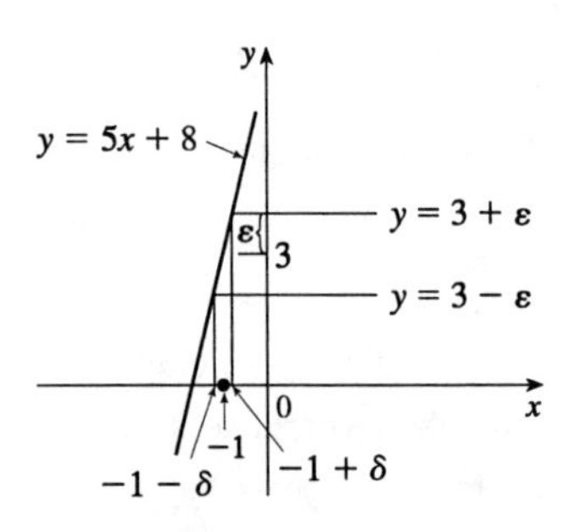

15. Given $\epsilon > 0$, we need $\delta > 0$ such that if $|x-2| < \delta$ then $\left|\dfrac{x}{7} - \dfrac{2}{7}\right| < \epsilon \;\Leftrightarrow\; \frac{1}{7}|x-2| < \epsilon \;\Leftrightarrow\; |x-2| < 7\epsilon$.

So take $\delta = 7\epsilon$. Then $|x-2| < \delta \quad\Rightarrow\quad \left|\dfrac{x}{7} - \dfrac{2}{7}\right| < \epsilon$. Thus $\lim\limits_{x\to 2}\dfrac{x}{7} = \dfrac{2}{7}$ by the definition of a limit.

17. Given $\epsilon > 0$, we need $\delta > 0$ such that if $|x-(-5)| < \delta$ then $\left|\left(4 - \frac{3}{5}x\right) - 7\right| < \epsilon \quad\Leftrightarrow\quad \frac{3}{5}|x+5| < \epsilon \quad\Leftrightarrow$ $|x-(-5)| < \frac{5}{3}\epsilon$. So take $\delta = \frac{5}{3}\epsilon$. Then $|x-(-5)| < \delta \quad\Rightarrow\quad \left|\left(4-\frac{3}{5}x\right) - 7\right| < \epsilon$. Thus $\lim\limits_{x\to -5}\left(4 - \frac{3}{5}x\right) = 7$ by the definition of a limit.

19. Given $\epsilon > 0$, we need $\delta > 0$ such that if $|x-a| < \delta$ then $|x-a| < \epsilon$. So $\delta = \epsilon$ will work.

21. Given $\epsilon > 0$, we need $\delta > 0$ such that if $|x| < \delta$ then $|x^2 - 0| < \epsilon \quad\Leftrightarrow\quad x^2 < \epsilon \quad\Leftrightarrow\quad |x| < \sqrt{\epsilon}$. Take $\delta = \sqrt{\epsilon}$. Then $|x-0| < \delta \quad\Rightarrow\quad |x^2 - 0| < \epsilon$. Thus $\lim\limits_{x\to 0} x^2 = 0$ by the definition of a limit.

23. Given $\epsilon > 0$, we need $\delta > 0$ such that if $|x-0| < \delta$ then $||x| - 0| < \epsilon$. But $||x|| = |x|$. So this is true if we pick $\delta = \epsilon$.

25. Given $\epsilon > 0$, we need $\delta > 0$ such that if $|x - 2| < \delta$, then $|(x^2 - 4x + 5) - 1| < \epsilon \quad \Leftrightarrow \quad |x^2 - 4x + 4| < \epsilon$ $\quad \Leftrightarrow \quad |(x-2)^2| < \epsilon$. So take $\delta = \sqrt{\epsilon}$. Then $|x - 2| < \delta \quad \Leftrightarrow \quad |x - 2| < \sqrt{\epsilon} \quad \Leftrightarrow \quad |(x-2)^2| < \epsilon$. So $\lim\limits_{x \to 2} (x^2 - 4x + 5) = 1$ by the definition of a limit.

27. Given $\epsilon > 0$, we need $\delta > 0$ such that if $|x - (-2)| < \delta$ then $|(x^2 - 1) - 3| < \epsilon$ or upon simplifying we need $|x^2 - 4| < \epsilon$ whenever $|x + 2| < \delta$. Notice that if $|x + 2| < 1$, then $-1 < x + 2 < 1 \quad \Rightarrow \quad -5 < x - 2 < -3$ $\quad \Rightarrow \quad |x - 2| < 5$. So take $\delta = \min\{\epsilon/5, 1\}$. Then $|x - 2| < 5$ and $|x + 2| < \epsilon/5$, so $|(x^2 - 1) - 3| = |(x + 2)(x - 2)| = |x + 2||x - 2| < (\epsilon/5)(5) = \epsilon$.
Therefore, by the definition of a limit, $\lim\limits_{x \to -2} (x^2 - 1) = 3$.

29. Given $\epsilon > 0$, we let $\delta = \min\left\{2, \dfrac{\epsilon}{8}\right\}$. If $0 < |x - 3| < \delta$ then $|x - 3| < 2 \quad \Rightarrow \quad 1 < x < 5 \quad \Rightarrow$ $|x + 3| < 8$. Also $|x - 3| < \dfrac{\epsilon}{8}$, so $|x^2 - 9| = |x + 3||x - 3| < 8 \cdot \dfrac{\epsilon}{8} = \epsilon$. Thus $\lim\limits_{x \to 3} x^2 = 9$.

31. *1. Guessing a value for* δ Given $\epsilon > 0$, we must find $\delta > 0$ such that $\left|\sqrt{x} - \sqrt{a}\right| < \epsilon$ whenever $0 < |x - a| < \delta$. But $\left|\sqrt{x} - \sqrt{a}\right| = \dfrac{|x - a|}{\sqrt{x} + \sqrt{a}} < \epsilon$ (from the hint). Now if we can find a positive constant C such that $\sqrt{x} + \sqrt{a} > C$ then $\dfrac{|x - a|}{\sqrt{x} + \sqrt{a}} < \dfrac{|x - a|}{C} < \epsilon$, and we take $|x - a| < C\epsilon$. We can find this number by restricting x to lie in some interval centered at a. If $|x - a| < \frac{1}{2}a$, then $\frac{1}{2}a < x < \frac{3}{2}a \quad \Rightarrow$ $\sqrt{x} + \sqrt{a} > \sqrt{\frac{1}{2}a} + \sqrt{a}$, and so $C = \sqrt{\frac{1}{2}a} + \sqrt{a}$ is a suitable choice for the constant. So $|x - a| < \left(\sqrt{\frac{1}{2}a} + \sqrt{a}\right)\epsilon$. This suggests that we let $\delta = \min\left\{\frac{1}{2}a, \left(\sqrt{\frac{1}{2}a} + \sqrt{a}\right)\epsilon\right\}$.

2. Showing that δ *works* Given $\epsilon > 0$, we let $\delta = \min\left\{\frac{1}{2}a, \left(\sqrt{\frac{1}{2}a} + \sqrt{a}\right)\epsilon\right\}$. If $0 < |x - a| < \delta$, then $|x - a| < \frac{1}{2}a \quad \Rightarrow \quad \sqrt{x} + \sqrt{a} > \sqrt{\frac{1}{2}a} + \sqrt{a}$ (as in part 1). Also $|x - a| < \left(\sqrt{\frac{1}{2}a} + \sqrt{a}\right)\epsilon$, so $\left|\sqrt{x} + \sqrt{a}\right| = \dfrac{|x - a|}{\sqrt{x} + \sqrt{a}} < \dfrac{\left(\sqrt{a/2} + \sqrt{a}\right)\epsilon}{\left(\sqrt{a/2} + \sqrt{a}\right)} = \epsilon$. Therefore $\lim\limits_{x \to a} \sqrt{x} = \sqrt{a}$ by the definition of a limit.

33. Suppose that $\lim\limits_{x \to 0} f(x) = L$. Given $\epsilon = \frac{1}{2}$, there exists $\delta > 0$ such that $0 < |x| < \delta \quad \Rightarrow \quad |f(x) - L| < \frac{1}{2}$. Take any rational number r with $0 < |r| < \delta$. Then $f(r) = 0$, so $|0 - L| < \frac{1}{2}$, so $L \leq |L| < \frac{1}{2}$. Now take any irrational number s with $0 < |s| < \delta$. Then $f(s) = 1$, so $|1 - L| < \frac{1}{2}$. Hence $1 - L < \frac{1}{2}$, so $L > \frac{1}{2}$. This contradicts $L < \frac{1}{2}$, so $\lim\limits_{x \to 0} f(x)$ does not exist.

35. $\dfrac{1}{(x+3)^4} > 10{,}000 \quad \Leftrightarrow \quad (x+3)^4 < \dfrac{1}{10{,}000} \quad \Leftrightarrow \quad |x - (-3)| = |x + 3| < \dfrac{1}{10}$

37. Let $N < 0$ be given. Then, for $x < -1$, we have $\dfrac{5}{(x+1)^3} < N \quad \Leftrightarrow \quad \dfrac{5}{N} < (x+1)^3 \quad \Leftrightarrow \quad \sqrt[3]{\dfrac{5}{N}} < x+1.$

Let $\delta = -\sqrt[3]{\dfrac{5}{N}}$. Then $-1-\delta < x < -1 \quad \Rightarrow \quad \sqrt[3]{\dfrac{5}{N}} < x+1 < 0 \quad \Rightarrow \quad \dfrac{5}{(x+1)^3} < N$, so

$\displaystyle\lim_{x \to -1^-} \frac{5}{(x+1)^3} = -\infty.$

39. $\displaystyle\lim_{t \to a^+} \mathbf{r}(t) = \mathbf{v}$ if for every number $\epsilon > 0$, there is a corresponding number $\delta > 0$ such that $|\mathbf{r}(t) - \mathbf{v}| < \epsilon$ whenever $a < t < a + \delta$.

41. $|\mathbf{r}(t) - (7\mathbf{i} + 5\mathbf{j})| = |(2t+1)\mathbf{i} + (t+2)\mathbf{j} - (7\mathbf{i} + 5\mathbf{j})| = |(2t-6)\mathbf{i} + (t-3)\mathbf{j}| = |t-3||2\mathbf{i} + \mathbf{j}|$
$= |t-3|\sqrt{2^2+1^2} = \sqrt{5}|t-3|$

Now $\sqrt{5}|t-3| < \epsilon \quad \Rightarrow \quad |t-3| < \epsilon/\sqrt{5}$.

(a) If $\epsilon = 0.01$, then t must be within $0.01/\sqrt{5}$ of 3. ($0.01/\sqrt{5} \approx 0.00447$)

(b) Proceeding as in part (a), $0.001/\sqrt{5} \approx 0.000447$.

(c) Proceeding as in part (a), $0.0001/\sqrt{5} \approx 0.0000447$.

43. Let $\mathbf{r}(t) = \langle f(t), g(t) \rangle$ and $\mathbf{L} = \langle \ell_1, \ell_2 \rangle$. We will show that $\displaystyle\lim_{t \to a} \mathbf{r}(t) = \mathbf{L} \quad \Leftrightarrow \quad \lim_{t \to a} f(t) = \ell_1$ and $\displaystyle\lim_{t \to a} g(t) = \ell_2$.

To prove the forward implication, suppose that $\displaystyle\lim_{t \to a} \mathbf{r}(t) = \mathbf{L}$ and let $\epsilon > 0$. Then there is a corresponding $\delta > 0$ such that $|\mathbf{r}(t) - \mathbf{L}| < \epsilon$ whenever $0 < |t-a| < \delta$. Notice that $|\mathbf{r}(t) - \mathbf{L}| = |\langle f(t), g(t) \rangle - \langle \ell_1, \ell_2 \rangle| = |\langle f(t) - \ell_1, g(t) - \ell_2 \rangle| = \sqrt{[f(t) - \ell_1]^2 + [g(t) - \ell_2]^2}$. It follows that $|\mathbf{r}(t) - \mathbf{L}| \geq \sqrt{[f(t) - \ell_1]^2} = |f(t) - \ell_1|$, and similarly $|\mathbf{r}(t) - \mathbf{L}| \geq |g(t) - \ell_2|$. If $0 < |t-a| < \delta$, then $|\mathbf{r}(t) - \mathbf{L}| < \epsilon$, so $|f(t) - \ell_1| < \epsilon$ and $|g(t) - \ell_2| < \epsilon$. This proves that $\displaystyle\lim_{t \to a} f(t) = \ell_1$ and $\displaystyle\lim_{t \to a} g(t) = \ell_2$.

For the reverse implication, suppose that $\displaystyle\lim_{t \to a} f(t) = \ell_1$ and $\displaystyle\lim_{t \to a} g(t) = \ell_2$. Suppose $\epsilon > 0$. Then there exist numbers $\delta_1 > 0$ and $\delta_2 > 0$ such that $|f(t) - \ell_1| < \epsilon/\sqrt{2}$ whenever $0 < |t-a| < \delta_1$ and $|g(t) - \ell_2| < \epsilon/\sqrt{2}$ whenever $0 < |t-a| < \delta_2$. Let $\delta = \min(\delta_1, \delta_2)$. If $0 < |t-a| < \delta$, then $0 < |t-a| < \delta_1$ and $0 < |t-a| < \delta_2$, so $|f(t) - \ell_1| < \epsilon/\sqrt{2}$ and $|g(t) - \ell_2| < \epsilon/\sqrt{2}$. It follows that $|\mathbf{r}(t) - \mathbf{L}| = \sqrt{|f(t) - \ell_1|^2 + |g(t) - \ell_2|^2} < \sqrt{\left(\epsilon/\sqrt{2}\right)^2 + \left(\epsilon/\sqrt{2}\right)^2} = \sqrt{\epsilon^2/2 + \epsilon^2/2} = \sqrt{\epsilon^2} = \epsilon$. This proves that $\displaystyle\lim_{t \to a} \mathbf{r}(t) = \mathbf{L}$.

EXERCISES 2.5

1. **(a)** f is discontinuous at -5, -3, -1, 3, 5, 8 and 10.

(b) f is continuous from the left at -5 and -3, and continuous from the right at 8.

It is continuous on neither side at -1, 3, 5, and 10.

3. $\lim\limits_{x\to 3}(x^4 - 5x^3 + 6) = \lim\limits_{x\to 3} x^4 - 5\lim\limits_{x\to 3} x^3 + \lim\limits_{x\to 3} 6 = 3^4 - 5(3^3) + 6 = -48 = f(3)$. Thus f is continuous at 3.

5. $\lim\limits_{x\to 5} f(x) = \lim\limits_{x\to 5}\left(1 + \sqrt{x^2 - 9}\right) = \lim\limits_{x\to 5} 1 + \sqrt{\lim\limits_{x\to 5} x^2 - \lim\limits_{x\to 5} 9} = 1 + \sqrt{5^2 - 9} = 5 = f(5)$. Thus f is continuous at 5.

7. $$\lim_{t\to -8} g(t) = \lim_{t\to -8}\frac{\sqrt[3]{t}}{(t+1)^4} = \frac{\sqrt[3]{\lim\limits_{t\to -8} t}}{\left(\lim\limits_{t\to -8} t + 1\right)^4} = \frac{\sqrt[3]{-8}}{(-8+1)^4} = -\frac{2}{2401} = g(-8).$$ Thus g is continuous at -8.

9. For $-4 < a < 4$ we have $\lim\limits_{x\to a} f(x) = \lim\limits_{x\to a} x\sqrt{16 - x^2} = \lim\limits_{x\to a} x\sqrt{\lim\limits_{x\to a} 16 - \lim\limits_{x\to a} x^2} = a\sqrt{16 - a^2} = f(a)$, so f is continuous on $(-4, 4)$. Similarly, we get $\lim\limits_{x\to 4^-} f(x) = 0 = f(4)$ and $\lim\limits_{x\to -4^+} f(x) = 0 = f(-4)$, so f is continuous from the left at 4 and from the right at -4. Thus f is continuous on $[-4, 4]$.

11. For any $a \in \mathbb{R}$ we have $\lim\limits_{x\to a} f(x) = \lim\limits_{x\to a}(x^2 - 1)^8 = \left(\lim\limits_{x\to a} x^2 - \lim\limits_{x\to a} 1\right)^8 = (a^2 - 1)^8 = f(a)$. Thus f is continuous on $(-\infty, \infty)$.

13. $f(x) = -\dfrac{1}{(x-1)^2}$ is discontinuous at 1 since $f(1)$ is not defined.

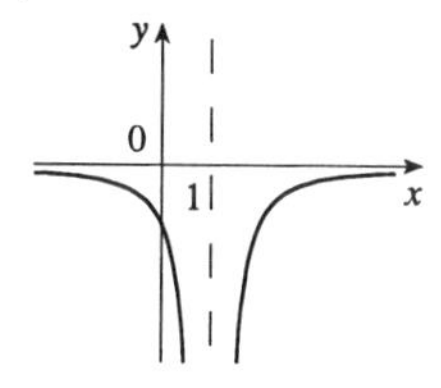

15. $\lim\limits_{x\to 1} f(x) = \lim\limits_{x\to 1}\left[-\dfrac{1}{(x-1)^2}\right]$ does not exist.

Therefore f is discontinuous at 1.

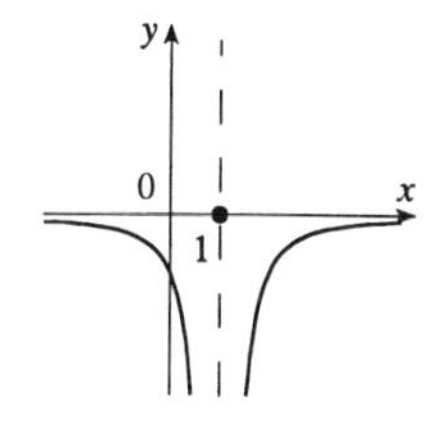

17. Since $f(x) = x^2 - 2$ for $x \neq -3$,

$\lim\limits_{x\to -3} f(x) = \lim\limits_{x\to -3}(x^2 - 2) = (-3)^2 - 2 = 7$.

But $f(-3) = 5$, so $\lim\limits_{x\to -3} f(x) \neq f(-3)$.

Therefore f is discontinuous at -3.

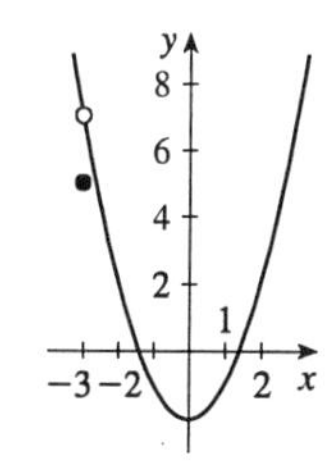

19. $f(x) = (x+1)(x^3+8x+9)$ is a polynomial, so by Theorem 5 it is continuous on $\mathbb{R}$.

21. $g(x) = x+1$, a polynomial, is continuous (by Theorem 5) and $f(x) = \sqrt{x}$ is continuous on $[0,\infty)$ by Theorem 6, so $f(g(x)) = \sqrt{x+1}$ is continuous on $[-1,\infty)$ by Theorem 8. By Theorem 4 #5, $H(x) = 1/\sqrt{x+1}$ is continuous on $(-1,\infty)$.

23. $g(x) = x-1$ and $G(x) = x^2-2$ are both polynomials, so by Theorem 5 they are continuous. Also $f(x) = \sqrt[5]{x}$ is continuous by Theorem 6, so $f(g(x)) = \sqrt[5]{x-1}$ is continuous on $\mathbb{R}$ by Theorem 8. Thus the product $h(x) = \sqrt[5]{x-1}(x^2-2)$ is continuous on $\mathbb{R}$ by Theorem 4 #4.

25. Since the discriminant of t^2+t+1 is negative, t^2+t+1 is always positive. So the domain of $F(t)$ is $\mathbb{R}$. By Theorem 5 the polynomial $(t^2+t+1)^3$ is continuous. By Theorems 6 and 8 the composition $F(t) = \sqrt{(t^2+t+1)^3}$ is continuous on $\mathbb{R}$.

27. $g(x) = x^3 - x$ is continuous on $\mathbb{R}$ since it is a polynomial [Theorem 5(a)], and $f(x) = |x|$ is continuous on $\mathbb{R}$ by Example 9(a). So $L(x) = |x^3 - x|$ is continuous on $\mathbb{R}$ by Theorem 8.

29. f is continuous on $(-\infty,3)$ and $(3,\infty)$ since on each of these intervals it is a polynomial. Also $\lim\limits_{x\to 3^+} f(x) = \lim\limits_{x\to 3^+}(5-x) = 2$ and $\lim\limits_{x\to 3^-} f(x) = \lim\limits_{x\to 3^-}(x-1) = 2$, so $\lim\limits_{x\to 3} f(x) = 2$. Since $f(3) = 5-3 = 2$, f is also continuous at 3. Thus f is continuous on $(-\infty,\infty)$.

31. f is continuous on $(-\infty,0)$ and $(0,\infty)$ since on each of these intervals it is a polynomial. Now $\lim\limits_{x\to 0^-} f(x) = \lim\limits_{x\to 0^-}(x-1)^3 = -1$ and $\lim\limits_{x\to 0^+} f(x) = \lim\limits_{x\to 0^+}(x+1)^3 = 1$. Thus $\lim\limits_{x\to 0} f(x)$ does not exist, so f is discontinuous at 0. Since $f(0) = 1$, f is continuous from the right at 0.

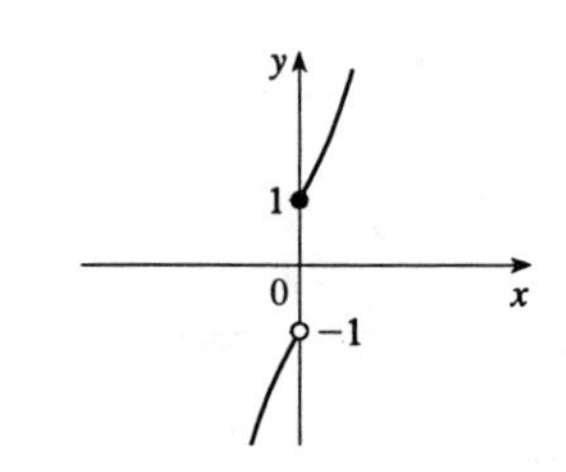

33. f is continuous on $(-\infty,-1)$, $(-1,1)$ and $(1,\infty)$. Now $\lim\limits_{x\to -1^-} f(x) = \lim\limits_{x\to -1^-}\dfrac{1}{x} = -1$ and $\lim\limits_{x\to -1^+} f(x) = \lim\limits_{x\to -1^+} x = -1$, so $\lim\limits_{x\to -1} f(x) = -1 = f(-1)$ and f is continuous at -1.

Also $\lim\limits_{x\to 1^-} f(x) = \lim\limits_{x\to 1^-} x = 1$ and $\lim\limits_{x\to 1^+} f(x) = \lim\limits_{x\to 1^+}\dfrac{1}{x^2} = 1$, so $\lim\limits_{x\to 1} f(x) = 1 = f(1)$ and f is continuous at 1. Thus f has no discontinuities.

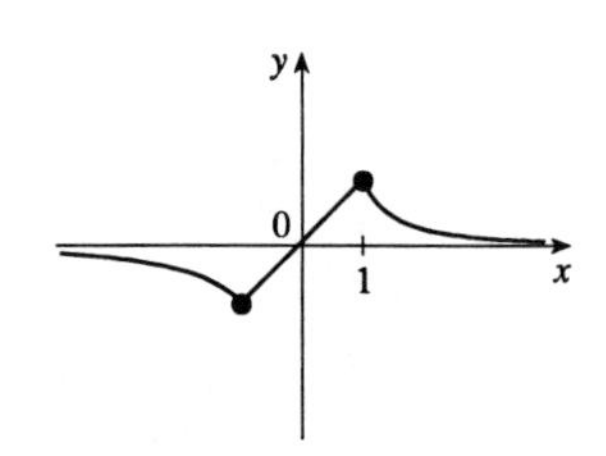

35. $f(x) = [\![2x]\!]$ is continuous except when $2x = n \quad \Leftrightarrow$ $x = n/2$, n an integer. In fact, $\lim\limits_{x \to n/2^-} [\![2x]\!] = n - 1$ and $\lim\limits_{x \to n/2^+} [\![2x]\!] = n = f(n)$, so f is continuous only from the right at $n/2$.

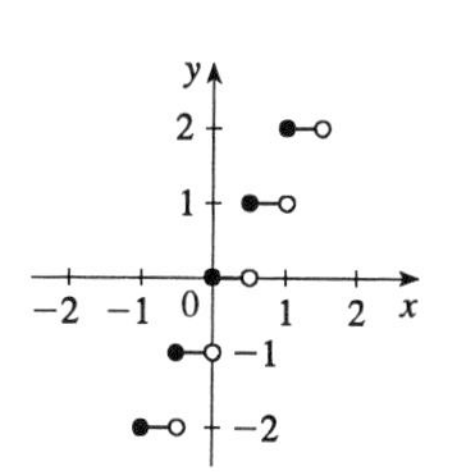

37. f is continuous on $(-\infty, 3)$ and $(3, \infty)$. Now $\lim\limits_{x \to 3^-} f(x) = \lim\limits_{x \to 3^-} (cx + 1) = 3c + 1$ and $\lim\limits_{x \to 3^+} f(x) = \lim\limits_{x \to 3^+} (cx^2 - 1) = 9c - 1$. So f is continuous $\quad \Leftrightarrow \quad 3c + 1 = 9c - 1 \quad \Leftrightarrow \quad 6c = 2 \quad \Leftrightarrow$ $c = \frac{1}{3}$. Thus for f to be continuous on $(-\infty, \infty)$, $c = \frac{1}{3}$.

39. The functions $2x$, $cx^2 + d$ and $4x$ are continuous on their own domains, so the only possible problems occur at $x = 1$ and $x = 2$. The left- and right-hand limits at these points must be the same in order for $\lim\limits_{x \to 1} h(x)$ and $\lim\limits_{x \to 2} h(x)$ to exist. So we must have $2 \cdot 1 = c(1)^2 + d$ and $c(2)^2 + d = 4 \cdot 2$. From the first of these equations we get $d = 2 - c$. Substituting this into the second, we get $4c + (2 - c) = 8 \quad \Leftrightarrow \quad c = 2$. Back-substituting into the first to get d, we find that $d = 0$.

41. **(a)** $\lim\limits_{x \to 1^-} f(x) = \lim\limits_{x \to 1^-} (1 - x^2) = 0$ and $\lim\limits_{x \to 1^+} f(x) = \lim\limits_{x \to 1^+} (1 + x/2) = \frac{3}{2}$. Thus $\lim\limits_{x \to 1} f(x)$ does not exist, so f is not continuous at 1.

(b) $f(0) = 1$ and $f(2) = 2$. For $0 \leq x \leq 1$, f takes the values in $[0, 1]$. For $1 < x \leq 2$, f takes the values in $(1.5, 2]$. Thus f does not take on the value 1.5 $\big($or any other value in $(1, 1.5]\big)$.

43. $f(x) = x^3 - x^2 + x$ is continuous on $[2, 3]$ and $f(2) = 6$, $f(3) = 21$. Since $6 < 10 < 21$, there is a number c in $(2, 3)$ such that $f(c) = 10$ by the Intermediate Value Theorem.

45. $f(x) = x^3 - 3x + 1$ is continuous on $[0, 1]$ and $f(0) = 1$, $f(1) = -1$. Since $-1 < 0 < 1$, there is a number c in $(0, 1)$ such that $f(c) = 0$ by the Intermediate Value Theorem. Thus there is a root of the equation $x^3 - 3x + 1 = 0$ in the interval $(0, 1)$.

47. $f(x) = x^3 + 2x - (x^2 + 1) = x^3 + 2x - x^2 - 1$ is continuous on $[0, 1]$ and $f(0) = -1$, $f(1) = 1$. Since $-1 < 0 < 1$, there is a number c in $(0, 1)$ such that $f(c) = 0$ by the Intermediate Value Theorem. Thus there is a root of the equation $x^3 + 2x - x^2 - 1 = 0$, or equivalently, $x^3 + 2x = x^2 + 1$, in the interval $(0, 1)$.

49. **(a)** $f(x) = x^3 - x + 1$ is continuous on $[-2, -1]$ and $f(-2) = -5$, $f(-1) = 1$. Since $-5 < 0 < 1$, there is a number c in $(-2, 1)$ such that $f(c) = 0$ by the Intermediate Value Theorem. Thus there is a root of the equation $x^3 - x + 1 = 0$ in the interval $(-2, -1)$.

(b) $f(-1.33) \approx -0.0226$ and $f(-1.32) \approx 0.0200$, so there is a root between -1.33 and -1.32.

51. **(a)** Let $f(x) = x^5 - x^2 - 4$. Then $f(1) = 1^5 - 1^2 - 4 = -4 < 0$ and $f(2) = 2^5 - 2^2 - 4 = 24 > 0$. So by the Intermediate Value Theorem, there is a number c in $(1, 2)$ such that $c^5 - c^2 - 4 = 0$.

(b) We can see from the graphs that, correct to three decimal places, the root is $x \approx 1.434$.

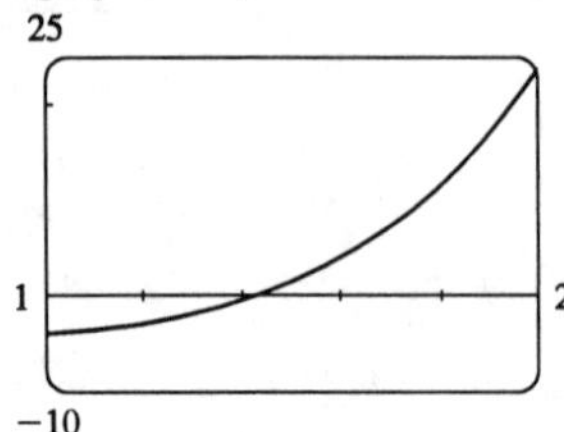

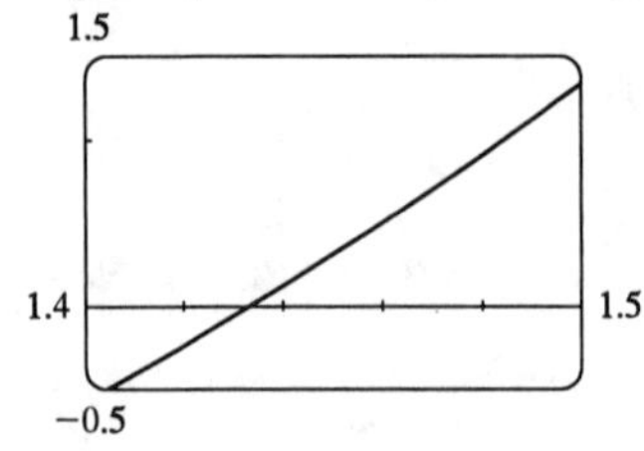

53. ($\Rightarrow$) If f is continuous at a, then by Theorem 7 with $g(h) = a + h$, we have

$$\lim_{h\to0} f(a+h) = f\left(\lim_{h\to0}(a+h)\right) = f(a).$$

($\Leftarrow$) Let $\epsilon > 0$. Since $\lim_{h\to0} f(a+h) = f(a)$, there exists $\delta > 0$ such that $|h| < \delta \Rightarrow$ $|f(a+h) - f(a)| < \epsilon$. So if $|x - a| < \delta$, then $|f(x) - f(a)| = |f(a + (x-a)) - f(a)| < \epsilon$. Thus $\lim_{x\to a} f(x) = f(a)$ and so f is continuous at a.

55. $f(x) = \begin{cases} 0 & \text{if } x \text{ is rational} \\ 1 & \text{if } x \text{ is irrational} \end{cases}$ is continuous nowhere. For, given any number a and any $\delta > 0$, the interval $(a - \delta, a + \delta)$ contains both infinitely many rational and infinitely many irrational numbers. Since $f(a) = 0$ or 1, there are infinitely many numbers x with $|x - a| < \delta$ and $|f(x) - f(a)| = 1$. Thus $\lim_{x\to a} f(x) \neq f(a)$. [In fact, $\lim_{x\to a} f(x)$ does not even exist.]

57. If there is such a number, it satisfies the equation $x^3 + 1 = x \Leftrightarrow x^3 - x + 1 = 0$.
Let the LHS of this equation be called $f(x)$. Now $f(-2) = (-2)^3 - (-2) + 1 = -5 < 0$, and $f(-1) = (-1)^3 - (-1) + 1 = 1 > 0$. Note also that $f(x)$ is a polynomial, and thus continuous. So by the Intermediate Value Theorem, there is a number c between -2 and -1 such that $f(c) = 0$, so that $c = c^3 + 1$.

59. Define $u(t)$ to be the monk's distance from the monastery, as a function of time, on the first day, and define $d(t)$ to be his distance from the monastery, as a function of time, on the second day. Let D be the distance from the monastery to the top of the mountain. From the given information we know that $u(0) = 0$, $u(12) = D$, $d(0) = D$ and $d(12) = 0$. Now consider the function $u - d$, which is clearly continuous (assuming that the monk does not use his mental powers to instantaneously transport himself). We calculate that $(u - d)(0) = -D$ and $(u - d)(12) = D$. So by the Intermediate Value Theorem there must be some time t_0 between 0 and 12 such that $(u - d)(t_0) = 0 \Leftrightarrow u(t_0) = d(t_0)$. So at time t_0 after 7:00 A.M., the monk will be at the same place on both days.

61. $\mathbf{r}(t)$ is continuous at $t = a \Leftrightarrow \lim_{t\to a} \mathbf{r}(t) = \mathbf{r}(a) = \langle f(a), g(a) \rangle \Leftrightarrow$ (by Exercise 2.4.43) $\lim_{t\to a} f(t) = f(a)$ and $\lim_{t\to a} g(t) = g(a) \Leftrightarrow f$ and g are continuous at $t = a$.

EXERCISES 2.6

1. $\displaystyle \lim_{x\to\infty}\frac{1}{x\sqrt{x}}=\lim_{x\to\infty}\frac{1}{x^{3/2}}=0$ by Theorem 4.

3. $$\lim_{x\to\infty}\frac{x+4}{x^2-2x+5}=\lim_{x\to\infty}\frac{\dfrac{1}{x}+\dfrac{4}{x^2}}{1-\dfrac{2}{x}+\dfrac{5}{x^2}}\overset{(5)}{=}\frac{\displaystyle\lim_{x\to\infty}\left(\frac{1}{x}+\frac{4}{x^2}\right)}{\displaystyle\lim_{x\to\infty}\left(1-\frac{2}{x}+\frac{5}{x^2}\right)}\overset{(1,2,3)}{=}\frac{\displaystyle\lim_{x\to\infty}\frac{1}{x}+4\lim_{x\to\infty}\frac{1}{x^2}}{\displaystyle\lim_{x\to\infty}1-2\lim_{x\to\infty}\frac{1}{x}+5\lim_{x\to\infty}\frac{1}{x^2}}$$
$$=\frac{0+4(0)}{1-2(0)+5(0)}=0 \text{ by (7) and Theorem 4.}$$

5. $$\lim_{x\to-\infty}\frac{(1-x)(2+x)}{(1+2x)(2-3x)}=\lim_{x\to-\infty}\frac{\left[\dfrac{1}{x}-1\right]\left[\dfrac{2}{x}+1\right]}{\left[\dfrac{1}{x}+2\right]\left[\dfrac{2}{x}-3\right]}=\frac{\left[\displaystyle\lim_{x\to-\infty}\frac{1}{x}-1\right]\left[\displaystyle\lim_{x\to-\infty}\frac{2}{x}+1\right]}{\left[\displaystyle\lim_{x\to-\infty}\frac{1}{x}+2\right]\left[\displaystyle\lim_{x\to-\infty}\frac{2}{x}-3\right]}$$
$$\overset{(5,4,1,2,7)}{=}\frac{(0-1)(0+1)}{(0+2)(0-3)}=\frac{1}{6}$$

7. $$\lim_{x\to\infty}\frac{1}{3+\sqrt{x}}=\lim_{x\to\infty}\frac{1/\sqrt{x}}{(3/\sqrt{x})+1}\overset{(5,1,3)}{=}\frac{\displaystyle\lim_{x\to\infty}(1/\sqrt{x})}{3\displaystyle\lim_{x\to\infty}(1/\sqrt{x})+\lim_{x\to\infty}1}=\frac{0}{3(0)+1}=0 \text{ (by Theorem 4 with } r=\tfrac{1}{2}.)$$

Or: Note that $0<\dfrac{1}{3+\sqrt{x}}<\dfrac{1}{\sqrt{x}}$ and use the Squeeze Theorem.

9. $$\lim_{r\to\infty}\frac{r^4-r^2+1}{r^5+r^3-r}=\lim_{r\to\infty}\frac{\dfrac{1}{r}-\dfrac{1}{r^3}+\dfrac{1}{r^5}}{1+\dfrac{1}{r^2}-\dfrac{1}{r^4}}=\frac{\displaystyle\lim_{r\to\infty}\frac{1}{r}-\lim_{r\to\infty}\frac{1}{r^3}+\lim_{r\to\infty}\frac{1}{r^5}}{\displaystyle\lim_{r\to\infty}1+\lim_{r\to\infty}\frac{1}{r^2}-\lim_{r\to\infty}\frac{1}{r^4}}=\frac{0-0+0}{1+0-0}=0$$

11. $$\lim_{x\to\infty}\frac{\sqrt{1+4x^2}}{4+x}=\lim_{x\to\infty}\frac{\sqrt{(1/x^2)+4}}{(4/x)+1}=\frac{\sqrt{0+4}}{0+1}=2$$

13. $$\lim_{x\to\infty}\frac{1-\sqrt{x}}{1+\sqrt{x}}=\lim_{x\to\infty}\frac{(1/\sqrt{x})-1}{(1/\sqrt{x})+1}=\frac{0-1}{0+1}=-1$$

15. $$\lim_{x\to\infty}\left(\sqrt{x^2+1}-\sqrt{x^2-1}\right)=\lim_{x\to\infty}\left(\sqrt{x^2+1}-\sqrt{x^2-1}\right)\frac{\sqrt{x^2+1}+\sqrt{x^2-1}}{\sqrt{x^2+1}+\sqrt{x^2-1}}$$
$$=\lim_{x\to\infty}\frac{(x^2+1)-(x^2-1)}{\sqrt{x^2+1}+\sqrt{x^2-1}}=\lim_{x\to\infty}\frac{2}{\sqrt{x^2+1}+\sqrt{x^2-1}}$$
$$=\lim_{x\to\infty}\frac{2/x}{\sqrt{1+(1/x^2)}+\sqrt{1-(1/x^2)}}=\frac{0}{\sqrt{1+0}+\sqrt{1-0}}=0$$

17. $$\lim_{x\to\infty}\left(\sqrt{1+x}-\sqrt{x}\right)=\lim_{x\to\infty}\left(\sqrt{1+x}-\sqrt{x}\right)\left(\frac{\sqrt{1+x}+\sqrt{x}}{\sqrt{1+x}+\sqrt{x}}\right)=\lim_{x\to\infty}\frac{(1+x)-x}{\sqrt{1+x}+\sqrt{x}}$$
$$=\lim_{x\to\infty}\frac{1}{\sqrt{1+x}+\sqrt{x}}=\lim_{x\to\infty}\frac{1/\sqrt{x}}{\sqrt{(1/x)+1}+1}=\frac{0}{\sqrt{0+1}+1}=0$$

19. $\lim\limits_{x\to-\infty}\left(\sqrt{x^2+x+1}+x\right)=\lim\limits_{x\to-\infty}\left(\sqrt{x^2+x+1}+x\right)\left[\dfrac{\sqrt{x^2+x+1}-x}{\sqrt{x^2+x+1}-x}\right]$

$$=\lim_{x\to-\infty}\frac{x+1}{\left(\sqrt{x^2+x+1}-x\right)}=\lim_{x\to-\infty}\frac{1+(1/x)}{-\sqrt{1+(1/x)+(1/x^2)}-1}=\frac{1+0}{-\sqrt{1+0+0}-1}=-\frac{1}{2}$$

21. $\sqrt{x}$ is large when x is large, so $\lim\limits_{x\to\infty}\sqrt{x}=\infty$.

23. $\lim\limits_{x\to\infty}(x-\sqrt{x})=\lim\limits_{x\to\infty}\sqrt{x}(\sqrt{x}-1)=\infty$ since $\sqrt{x}\to\infty$ and $\sqrt{x}-1\to\infty$ as $x\to\infty$.

25. $\lim\limits_{x\to-\infty}(x^3-5x^2)=-\infty$ since $x^3\to-\infty$ and $-5x^2\to-\infty$ as $x\to-\infty$.

Or: $\lim\limits_{x\to-\infty}(x^3-5x^2)=\lim\limits_{x\to-\infty}x^2(x-5)=-\infty$ since $x^2\to\infty$ and $x-5\to-\infty$.

27. $\lim\limits_{x\to\infty}\dfrac{x^7-1}{x^6-1}=\lim\limits_{x\to\infty}\dfrac{1-1/x^7}{(1/x)-(1/x^7)}=\infty$ since $1-\dfrac{1}{x^7}\to1$ while $\dfrac{1}{x}-\dfrac{1}{x^7}\to0^+$ as $x\to\infty$.

Or: Divide numerator and denominator by x^6 instead of x^7.

29. $\lim\limits_{x\to\infty}\dfrac{\sqrt{x}+3}{x+3}=\lim\limits_{x\to\infty}\dfrac{(1/\sqrt{x})+(3/x)}{1+3/x}=\dfrac{0+0}{1+0}=0$

31. If $f(x)=x^2/2^x$, then a calculator gives $f(0)=0$, $f(1)=0.5$, $f(2)=1$, $f(3)=1.125$, $f(4)=1$, $f(5)=0.78125$, $f(6)=0.5625$, $f(7)=0.3828125$, $f(8)=0.25$, $f(9)=0.158203125$, $f(10)=0.09765625$, $f(20)\approx0.00038147$, $f(50)\approx2.2204\times10^{-12}$, $f(100)\approx7.8886\times10^{-27}$. It appears that $\lim\limits_{x\to\infty}(x^2/2^x)=0$.

33. $\lim\limits_{x\to\pm\infty}\dfrac{x}{x+4}=\lim\limits_{x\to\pm\infty}\dfrac{1}{1+4/x}=\dfrac{1}{1+0}=1$, so $y=1$ is a horizontal asymptote.

$\lim\limits_{x\to-4^-}\dfrac{x}{x+4}=\infty$ and $\lim\limits_{x\to-4^+}\dfrac{x}{x+4}=-\infty$, so $x=-4$ is a vertical asymptote. The graph confirms these calculations.

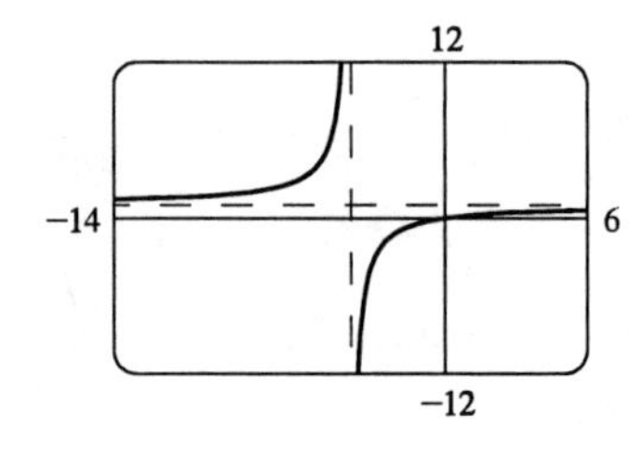

35. $\lim\limits_{x\to\pm\infty}\dfrac{x^3}{x^2+3x-10}=\lim\limits_{x\to\pm\infty}\dfrac{x}{1+(3/x)-(10/x^2)}=\pm\infty$, so there is no horizontal asymptote. $\lim\limits_{x\to2^+}\dfrac{x^3}{x^2+3x-10}=\lim\limits_{x\to2^+}\dfrac{x^3}{(x+5)(x-2)}=\infty$, since $\dfrac{x^3}{(x+5)(x-2)}>0$ for $x>2$. Similarly, $\lim\limits_{x\to2^-}\dfrac{x^3}{x^2+3x-10}=-\infty$ and $\lim\limits_{x\to-5^-}\dfrac{x^3}{x^2+3x-10}=-\infty$, $\lim\limits_{x\to-5^+}\dfrac{x^3}{x^2+3x-10}=\infty$, so $x=2$ and $x=-5$ are vertical asymptotes. The graph confirms these calculations.

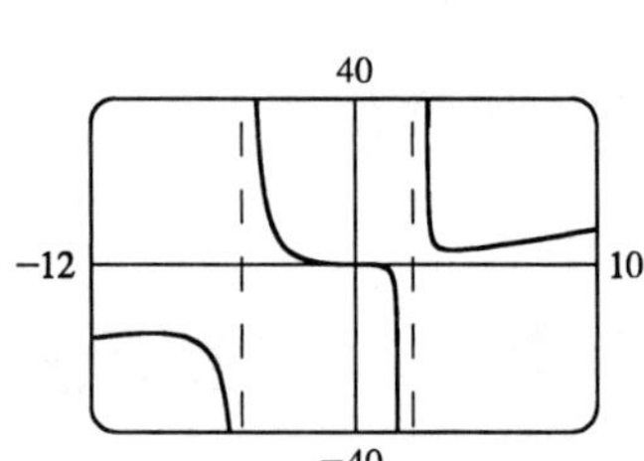

37. $\lim\limits_{x\to\infty}\dfrac{x}{\sqrt[4]{x^4+1}}=\lim\limits_{x\to\infty}\dfrac{1}{\sqrt[4]{1+(1/x^4)}}=\dfrac{1}{\sqrt[4]{1+0}}=1$ and

$\lim\limits_{x\to-\infty}\dfrac{x}{\sqrt[4]{x^4+1}}=\lim\limits_{x\to-\infty}\dfrac{1}{-\sqrt[4]{1+(1/x^4)}}=\dfrac{1}{-\sqrt[4]{1+0}}=-1,$

so $y=\pm1$ are horizontal asymptotes. There are no vertical asymptotes.

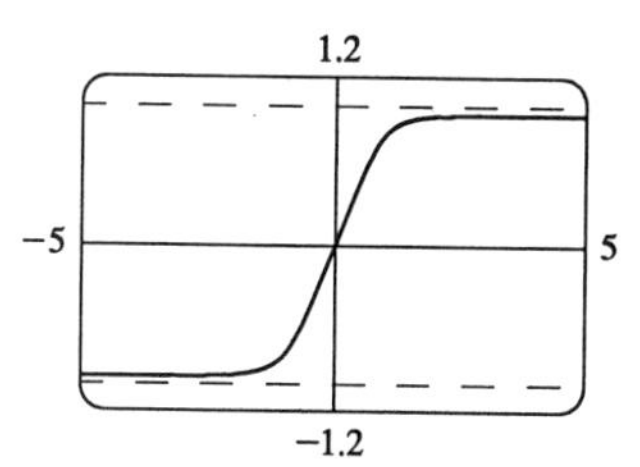

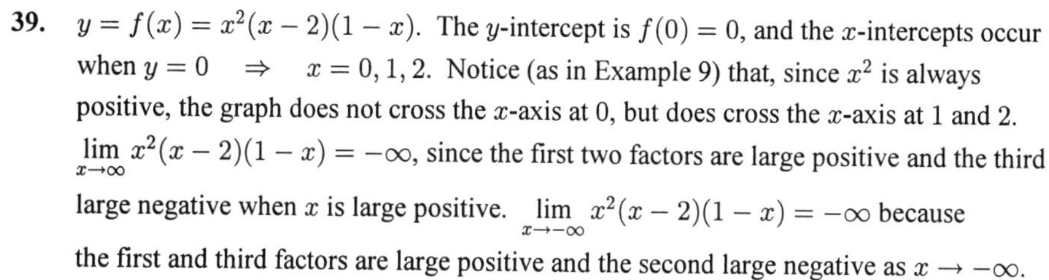

39. $y=f(x)=x^2(x-2)(1-x)$. The y-intercept is $f(0)=0$, and the x-intercepts occur when $y=0 \quad\Rightarrow\quad x=0,1,2$. Notice (as in Example 9) that, since x^2 is always positive, the graph does not cross the x-axis at 0, but does cross the x-axis at 1 and 2. $\lim\limits_{x\to\infty}x^2(x-2)(1-x)=-\infty$, since the first two factors are large positive and the third large negative when x is large positive. $\lim\limits_{x\to-\infty}x^2(x-2)(1-x)=-\infty$ because the first and third factors are large positive and the second large negative as $x\to-\infty$.

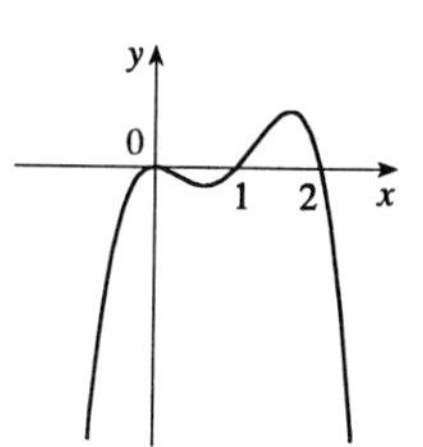

41. $y=f(x)=(x+4)^5(x-3)^4$. The y-intercept is $f(0)=4^5(-3)^4=82{,}944$. The x-intercepts occur when $y=0 \quad\Rightarrow\quad x=-4,3$. Notice (as in Example 9) that the graph does not cross the x-axis at 3 because $(x-3)^4$ is always positive, but does cross the x-axis at -4.

$\lim\limits_{x\to\infty}(x+4)^5(x-3)^4=\infty$ since both factors are large positive when x is large positive. $\lim\limits_{x\to-\infty}(x+4)^5(x-3)^4=-\infty$ since the first factor is large negative and the second factor is large positive when x is large negative.

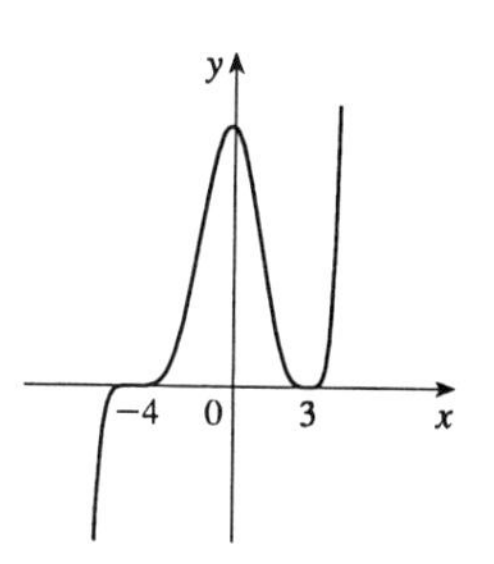

43. Divide numerator and denominator by the highest power of x in $Q(x)$.

(a) If $\deg(P)<\deg(Q)$, then numerator $\to0$ but denominator doesn't. So $\lim\limits_{x\to\infty}\dfrac{P(x)}{Q(x)}=0$.

(b) If $\deg(P)>\deg(Q)$, then numerator $\to\pm\infty$ but denominator doesn't, so $\lim\limits_{x\to\infty}\dfrac{P(x)}{Q(x)}=\pm\infty$ (depending on the ratio of the leading coefficients of P and Q.)

45. $\lim\limits_{x\to\infty}\dfrac{4x-1}{x}=\lim\limits_{x\to\infty}\left(4-\dfrac{1}{x}\right)=4$, and $\lim\limits_{x\to\infty}\dfrac{4x^2+3x}{x^2}=\lim\limits_{x\to\infty}\left(4+\dfrac{3}{x}\right)=4$. Therefore by the Squeeze Theorem, $\lim\limits_{x\to\infty}f(x)=4$.

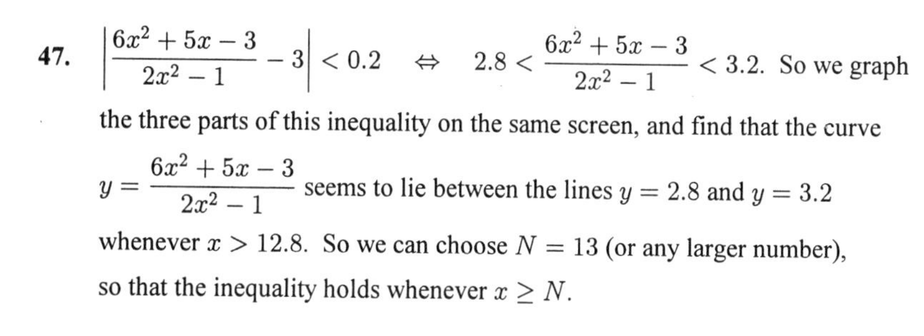

47. $\left|\dfrac{6x^2+5x-3}{2x^2-1}-3\right|<0.2 \quad\Leftrightarrow\quad 2.8<\dfrac{6x^2+5x-3}{2x^2-1}<3.2$. So we graph the three parts of this inequality on the same screen, and find that the curve $y=\dfrac{6x^2+5x-3}{2x^2-1}$ seems to lie between the lines $y=2.8$ and $y=3.2$ whenever $x>12.8$. So we can choose $N=13$ (or any larger number), so that the inequality holds whenever $x\geq N$.

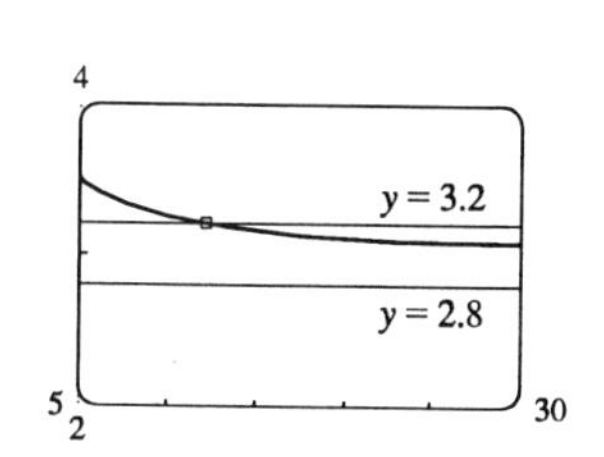

49. For $\epsilon = 0.5$, we need to find N such that

$$\left|\frac{\sqrt{4x^2+1}}{x+1} - (-2)\right| < 0.5 \quad \Leftrightarrow \quad -2.5 < \frac{\sqrt{4x^2+1}}{x+1} < -1.5$$

whenever $x \le N$. We graph the three parts of this inequality on the same screen, and see that the inequality holds for $x \le -6$. So we choose $N = -6$ (or any smaller number).

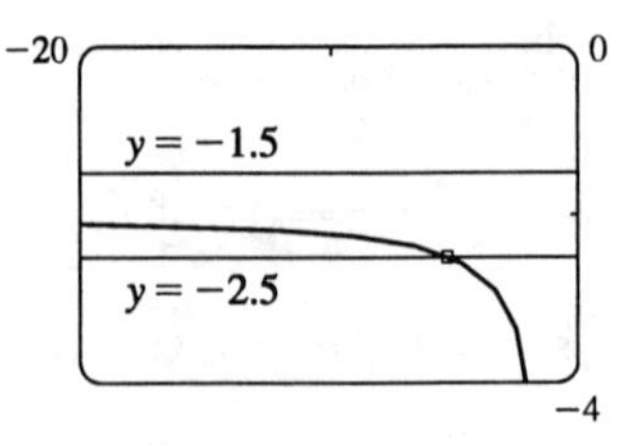

For $\epsilon = 0.1$, we need $-2.1 < \dfrac{\sqrt{4x^2+1}}{x+1} < -1.9$ whenever $x \le N$. From the graph, it seems that this inequality holds for $x \le -22$. So we choose any $N = -22$ (or any smaller number).

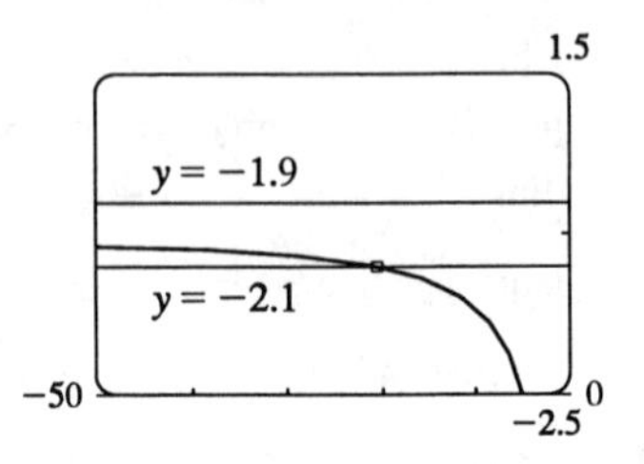

51. **(a)** $1/x^2 < 0.0001 \quad \Leftrightarrow \quad x^2 > 1/0.0001 = 10{,}000 \quad \Leftrightarrow \quad x > 100 \quad (x > 0)$

(b) If $\epsilon > 0$ is given, then $1/x^2 < \epsilon \quad \Leftrightarrow \quad x^2 > 1/\epsilon \quad \Leftrightarrow \quad x > 1/\sqrt{\epsilon}$. Let $N = 1/\sqrt{\epsilon}$.

Then $x > N \quad \Rightarrow \quad x > \dfrac{1}{\sqrt{\epsilon}} \quad \Rightarrow \quad \left|\dfrac{1}{x^2} - 0\right| = \dfrac{1}{x^2} < \epsilon$ so $\displaystyle\lim_{x\to\infty} \frac{1}{x^2} = 0$.

53. For $x < 0$, $|1/x - 0| = -1/x$. If $\epsilon > 0$ is given, then $-1/x < \epsilon \quad \Leftrightarrow \quad x < -1/\epsilon$.

Take $N = -\dfrac{1}{\epsilon}$. Then $x < N \quad \Rightarrow \quad x < -\dfrac{1}{\epsilon} \quad \Rightarrow \quad \left|\dfrac{1}{x} - 0\right| = -\dfrac{1}{x} < \epsilon$, so $\displaystyle\lim_{x\to-\infty} \frac{1}{x} = 0$.

55. Suppose that $\displaystyle\lim_{x\to\infty} f(x) = L$ and let $\epsilon > 0$ be given. Then there exists $N > 0$ such that $x > N \quad \Rightarrow$ $|f(x) - L| < \epsilon$. Let $\delta = 1/N$. Then $0 < t < \delta \quad \Rightarrow \quad t < 1/N \quad \Rightarrow \quad 1/t > N \quad \Rightarrow \quad |f(1/t) - L| < \epsilon$. So $\displaystyle\lim_{t\to 0^+} f(1/t) = L = \lim_{x\to\infty} f(x)$. Now suppose that $\displaystyle\lim_{x\to-\infty} f(x) = L$ and let $\epsilon > 0$ be given. Then there exists $N < 0$ such that $x < N \quad \Rightarrow \quad |f(x) - L| < \epsilon$. Let $\delta = -1/N$. Then $-\delta < t < 0 \quad \Rightarrow \quad t > 1/N \quad \Rightarrow$ $1/t < N \quad \Rightarrow \quad |f(1/t) - L| < \epsilon$. So $\displaystyle\lim_{x\to 0^-} f(1/t) = L = \lim_{x\to-\infty} f(x)$.

EXERCISES 2.7

1. **(a)** **(i)** $m = \displaystyle\lim_{x\to-3} \frac{x^2+2x-3}{x-(-3)} = \lim_{x\to-3} \frac{(x+3)(x-1)}{x+3} = \lim_{x\to-3}(x-1) = -4$

(ii)

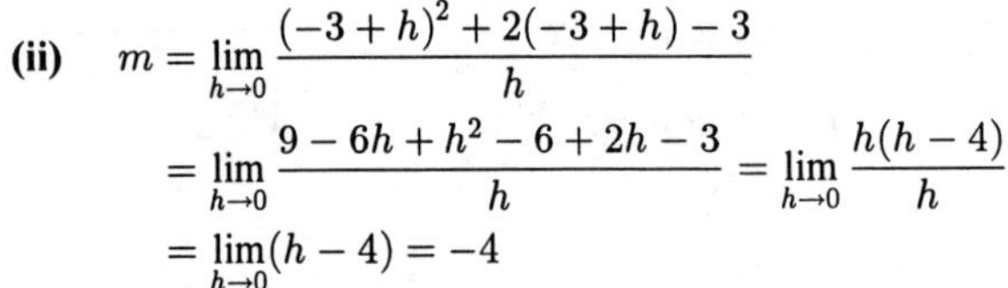

$$\begin{aligned} m &= \lim_{h\to 0} \frac{(-3+h)^2 + 2(-3+h) - 3}{h} \\ &= \lim_{h\to 0} \frac{9 - 6h + h^2 - 6 + 2h - 3}{h} = \lim_{h\to 0} \frac{h(h-4)}{h} \\ &= \lim_{h\to 0}(h-4) = -4 \end{aligned}$$

(b) The equation of the tangent line is $y - 3 = -4(x+3)$ or $y = -4x - 9$.

(c)

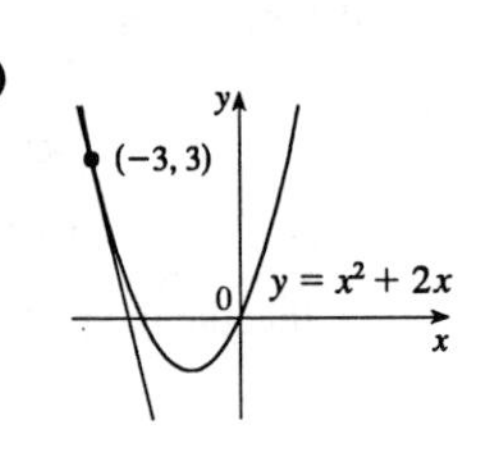

3. Using (1), $m = \lim\limits_{x \to -2} \dfrac{1 - 2x - 3x^2 + 7}{x + 2} = \lim\limits_{x \to -2} \dfrac{-3x^2 - 2x + 8}{x + 2} = \lim\limits_{x \to -2} \dfrac{(-3x + 4)(x + 2)}{x + 2}$

$= \lim\limits_{x \to -2} (-3x + 4) = 10$. Thus the equation of the tangent is $y + 7 = 10(x + 2)$ or $y = 10x + 13$.

5. Using (1), $m = \lim\limits_{x \to -2} \dfrac{1/x^2 - \frac{1}{4}}{x + 2} = \lim\limits_{x \to -2} \dfrac{4 - x^2}{4x^2(x + 2)} = \lim\limits_{x \to -2} \dfrac{(2 - x)(2 + x)}{4x^2(x + 2)} = \lim\limits_{x \to -2} \dfrac{2 - x}{4x^2} = \dfrac{1}{4}$. Thus the equation of the tangent is $y - \frac{1}{4} = \frac{1}{4}(x + 2)$ or $x - 4y + 3 = 0$.

7. **(a)** $m = \lim\limits_{x \to a} \dfrac{2/(x + 3) - 2/(a + 3)}{x - a} = \lim\limits_{x \to a} \dfrac{2(a - x)}{(x - a)(x + 3)(a + 3)} = \lim\limits_{x \to a} \dfrac{-2}{(x + 3)(a + 3)} = \dfrac{-2}{(a + 3)^2}$

(b) **(i)** $a = -1 \;\Rightarrow\; m = \frac{-2}{(-1+3)^2} = -\frac{1}{2}$ **(ii)** $a = 0 \;\Rightarrow\; m = \frac{-2}{(0+3)^2} = -\frac{2}{9}$

(iii) $a = 1 \;\Rightarrow\; m = \frac{-2}{(1+3)^2} = -\frac{1}{8}$

9. **(a)** Using (1), $m = \lim\limits_{x \to a} \dfrac{(x^3 - 4x + 1) - (a^3 - 4a + 1)}{x - a} = \lim\limits_{x \to a} \dfrac{(x^3 - a^3) - 4(x - a)}{x - a}$

$= \lim\limits_{x \to a} \dfrac{(x - a)(x^2 + ax + a^2) - 4(x - a)}{x - a} = \lim\limits_{x \to a} \left(x^2 + ax + a^2 - 4\right) = 3a^2 - 4$.

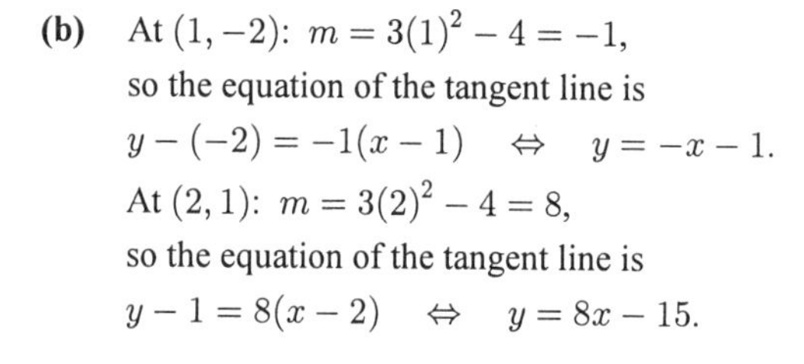

(b) At $(1, -2)$: $m = 3(1)^2 - 4 = -1$,
so the equation of the tangent line is
$y - (-2) = -1(x - 1) \;\Leftrightarrow\; y = -x - 1$.
At $(2, 1)$: $m = 3(2)^2 - 4 = 8$,
so the equation of the tangent line is
$y - 1 = 8(x - 2) \;\Leftrightarrow\; y = 8x - 15$.

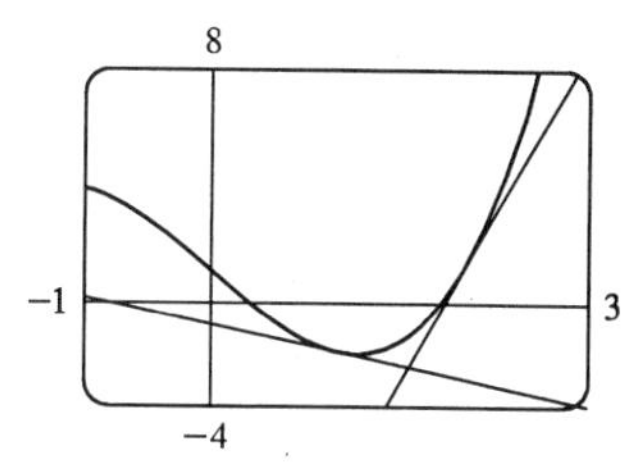

11. Let $s(t) = 40t - 16t^2$. $v(2) = \lim\limits_{t \to 2} \dfrac{s(t) - s(2)}{t - 2} = \lim\limits_{t \to 2} \dfrac{40t - 16t^2 - 16}{t - 2} = \lim\limits_{t \to 2} \dfrac{8(t - 2)(-2t + 1)}{t - 2}$

$= \lim\limits_{t \to 2} 8(-2t + 1) = -24$. Thus, the instantaneous velocity when $t = 2$ is -24 ft/s.

13. $v(a) = \lim\limits_{h \to 0} \dfrac{s(a + h) - s(a)}{h} = \lim\limits_{h \to 0} \dfrac{4(a + h)^3 + 6(a + h) + 2 - (4a^3 + 6a + 2)}{h}$

$= \lim\limits_{h \to 0} \dfrac{4a^3 + 12a^2h + 12ah^2 + 4h^3 + 6a + 6h + 2 - 4a^3 - 6a - 2}{h}$

$= \lim\limits_{h \to 0} \dfrac{12a^2h + 12ah^2 + 4h^3 + 6h}{h} = \lim\limits_{h \to 0} \left(12a^2 + 12ah + 4h^2 + 6\right) = 12a^2 + 6$

So $v(1) = 12(1)^2 + 6 = 18$ m/s, $v(2) = 12(2)^2 + 6 = 54$ m/s, and $v(3) = 12(3)^2 + 6 = 114$ m/s.

15. **(a)** Since the slope of the tangent at $s = 0$ is 0, the car's initial velocity was 0.

(b) The slope of the tangent is greater at C than at B, so the car was going faster at C.

(c) Near A, the tangent lines are becoming steeper as x increases, so the velocity was increasing, so the car was speeding up. Near B, the tangent lines are becoming less steep, so the car was slowing down. The steepest tangent near C is the one right at C, so at C the car had just finished speeding up, and was about to start slowing down.

(d) Between D and E, the slope of the tangent is 0, so the car did not move during that time.

17. **(a)** **(i)** $[8,11]$: $\dfrac{7.9-11.5}{3}=-1.2°/\text{h}$ **(ii)** $[8,10]$: $\dfrac{9.0-11.5}{2}=-1.25°/\text{h}$

(iii) $[8,9]$: $\dfrac{10.2-11.5}{1}=-1.3°/\text{h}$

(b) The instantaneous rate of change is approximately $-1.6°/\text{h}$ at 8 P.M.

19. **(a)** **(i)** $\dfrac{\Delta C}{\Delta x}=\dfrac{C(105)-C(100)}{5}=\dfrac{6601.25-6500}{5}=\$20.25/\text{unit}.$

(ii) $\dfrac{\Delta C}{\Delta x}=\dfrac{C(101)-C(100)}{1}=\dfrac{6520.05-6500}{1}=\$20.05/\text{unit}.$

(b) $\dfrac{C(100+h)-C(100)}{h}=\dfrac{5000+10(100+h)+0.05(100+h)^2-6500}{h}=20+0.05h,\ h\neq 0$. So as h approaches 0, the rate of change of C approaches $\$20/\text{unit}$.

21. **(a)** $\mathbf{v}=\lim\limits_{t\to 2}\dfrac{1}{t-2}[\mathbf{r}(t)-\mathbf{r}(2)]=\lim\limits_{t\to 2}\dfrac{1}{t-2}\left[(t^2+t)\mathbf{i}+\dfrac{2}{t}\mathbf{j}-(6\mathbf{i}+\mathbf{j})\right]$

$=\lim\limits_{t\to 2}\dfrac{1}{t-2}\left[(t^2+t-6)\mathbf{i}+\left(\dfrac{2}{t}-1\right)\mathbf{j}\right]=\lim\limits_{t\to 2}\dfrac{1}{t-2}\left[(t+3)(t-2)\mathbf{i}-\dfrac{1}{t}(t-2)\mathbf{j}\right]$

$=\lim\limits_{t\to 2}\left[(t+3)\mathbf{i}-\dfrac{1}{t}\mathbf{j}\right]=5\mathbf{i}-\frac{1}{2}\mathbf{j}$

(b) The tangent line to the curve at $t=2$ is given by

$\mathbf{L}(t)=\mathbf{r}(2)+t\mathbf{v}=(6\mathbf{i}+\mathbf{j})+t\left(5\mathbf{i}-\frac{1}{2}\mathbf{j}\right)=(6+5t)\mathbf{i}+\left(1-\frac{1}{2}t\right)\mathbf{j}$. Parametric equations for the tangent line are $x=6+5t$, $y=1-\frac{1}{2}t$.

23. **(a)** The velocity of the ball when $t=1$ is

$\mathbf{v}(1)=\lim\limits_{t\to 1}\dfrac{1}{t-1}[\mathbf{r}(t)-\mathbf{r}(1)]=\lim\limits_{t\to 1}\dfrac{1}{t-1}\left[10t\mathbf{i}+(30t-16t^2)\mathbf{j}-(10\mathbf{i}+14\mathbf{j})\right]$

$=\lim\limits_{t\to 1}\dfrac{1}{t-1}\left[(10t-10)\mathbf{i}+(-16t^2+30t-14)\mathbf{j}\right]$

$=\lim\limits_{t\to 1}\dfrac{1}{t-1}[10(t-1)\mathbf{i}+(-2)(8t-7)(t-1)\mathbf{j}]$

$=\lim\limits_{t\to 1}[10\mathbf{i}-2(8t-7)\mathbf{j}]=(10\mathbf{i}-2\mathbf{j})$ ft/s

(b) The speed of the ball is $|10\mathbf{i}-2\mathbf{j}|=\sqrt{10^2+(-2)^2}=\sqrt{104}\approx 10.2$ ft/s.

REVIEW EXERCISES FOR CHAPTER 2

1. False, since $\lim\limits_{x\to 4}\dfrac{2x}{x-4}$ and $\lim\limits_{x\to 4}\dfrac{8}{x-4}$ do not exist.

3. True by Limit Law #5, since $\lim\limits_{x\to 1}(x^2+2x-4)=-1\neq 0$.

5. False. For example, let $f(x)=\begin{cases} x^2+1 & \text{if } x\neq 0 \\ 2 & \text{if } x=0\end{cases}$

Then $f(x)>1$ for all x, but $\lim\limits_{x\to 0}f(x)=\lim\limits_{x\to 0}(x^2+1)=1$.

7. True, by the definition of a limit with $\epsilon=1$.

9. True. See Exercise 1.3.67.

11. True by Theorem 1.5.7 with $a=2$, $b=5$, and $g(x)=4x^2-11$.

13. $\lim\limits_{x\to 4}\sqrt{x+\sqrt{x}}=\sqrt{4+\sqrt{4}}=\sqrt{6}$ since the function is continuous.

15. $\lim\limits_{t\to -1}\dfrac{t+1}{t^3-t}=\lim\limits_{t\to -1}\dfrac{t+1}{t(t+1)(t-1)}=\lim\limits_{t\to -1}\dfrac{1}{t(t-1)}=\dfrac{1}{(-1)(-2)}=\dfrac{1}{2}$

17. $\lim\limits_{h\to 0}\dfrac{(1+h)^2-1}{h}=\lim\limits_{h\to 0}\dfrac{1+2h+h^2-1}{h}=\lim\limits_{h\to 0}\dfrac{2h+h^2}{h}=\lim\limits_{h\to 0}(2+h)=2$

19. $\lim\limits_{x\to -1}\dfrac{x^2-x-2}{x^2+3x-2}=\dfrac{\lim\limits_{x\to -1}(x^2-x-2)}{\lim\limits_{x\to -1}(x^2+3x-2)}=\dfrac{(-1)^2-(-1)-2}{(-1)^2+3(-1)-2}=\dfrac{0}{-4}=0$

21. $\lim\limits_{t\to 6}\dfrac{17}{(t-6)^2}=\infty$ since $(t-6)^2\to 0$ and $\dfrac{17}{(t-6)^2}>0$.

23. $\lim\limits_{s\to 16}\dfrac{4-\sqrt{s}}{s-16}=\lim\limits_{s\to 16}\dfrac{4-\sqrt{s}}{(\sqrt{s}+4)(\sqrt{s}-4)}=\lim\limits_{s\to 16}\dfrac{-1}{\sqrt{s}+4}=\dfrac{-1}{\sqrt{16}+4}=-\dfrac{1}{8}$

25. $\lim\limits_{x\to 8^-}\dfrac{|x-8|}{x-8}=\lim\limits_{x\to 8^-}\dfrac{-(x-8)}{x-8}=\lim\limits_{x\to 8^-}(-1)=-1$

27. $\lim\limits_{x\to 0}\dfrac{1-\sqrt{1-x^2}}{x}\cdot\dfrac{1+\sqrt{1-x^2}}{1+\sqrt{1-x^2}}=\lim\limits_{x\to 0}\dfrac{1-(1-x^2)}{x\left(1+\sqrt{1-x^2}\right)}=\lim\limits_{x\to 0}\dfrac{x^2}{x\left(1+\sqrt{1-x^2}\right)}=\lim\limits_{x\to 0}\dfrac{x}{1+\sqrt{1-x^2}}=0$

29. $\lim\limits_{x\to\infty}\dfrac{1+2x-x^2}{1-x+2x^2}=\lim\limits_{x\to\infty}\dfrac{(1/x^2)+(2/x)-1}{(1/x^2)-(1/x)+2}=\dfrac{0+0-1}{0-0+2}=-\dfrac{1}{2}$

31. $\lim\limits_{x\to\infty}\dfrac{\sqrt{x^2-9}}{2x-6}=\lim\limits_{x\to\infty}\dfrac{\sqrt{1-9/x^2}}{2-6/x}=\dfrac{\sqrt{1-0}}{2-0}=\dfrac{1}{2}$

33. $\lim\limits_{x\to\infty}\left(\sqrt[3]{x}-\frac{1}{3}x\right)=\lim\limits_{x\to\infty}\sqrt[3]{x}\left(1-\frac{1}{3}x^{2/3}\right)=-\infty$, since $\sqrt[3]{x}\to\infty$ and $1-\frac{1}{3}x^{2/3}\to-\infty$.

35. $\lim\limits_{t\to 3}\left(\dfrac{2t^2-18}{t-3}\mathbf{i}+t^2\mathbf{j}\right)=\left[\lim\limits_{t\to 3}\dfrac{2(t-3)(t+3)}{t-3}\right]\mathbf{i}+\left[\lim\limits_{t\to 3}t^2\right]\mathbf{j}=\left[\lim\limits_{t\to 3}2(t+3)\right]\mathbf{i}+\left[\lim\limits_{t\to 3}t^2\right]\mathbf{j}=12\mathbf{i}+9\mathbf{j}$

37. $\lim\limits_{t\to 1}\left\langle\dfrac{1-\sqrt{t}}{1-t},t^3-4t\right\rangle=\left\langle\lim\limits_{t\to 1}\dfrac{1-\sqrt{t}}{(1-\sqrt{t})(1+\sqrt{t})},\lim\limits_{t\to 1}(t^3-4t)\right\rangle=\left\langle\lim\limits_{t\to 1}\dfrac{1}{1+\sqrt{t}},1^3-4\cdot 1\right\rangle=\left\langle\frac{1}{2},-3\right\rangle$

39. Given $\epsilon > 0$, we need $\delta > 0$ so that if $|x-5| < \delta$ then $|(7x-27)-8| < \epsilon \Leftrightarrow |7x-35| < \epsilon \Leftrightarrow |x-5| < \epsilon/7$. So take $\delta = \epsilon/7$. Then $|x-5| < \delta \Rightarrow |(7x-27)-8| < \epsilon$. Thus $\lim\limits_{x\to5}(7x-27) = 8$ by the definition of a limit.

41. Given $\epsilon > 0$, we need $\delta > 0$ so that if $|x-2| < \delta$ then $|x^2-3x-(-2)| < \epsilon$. First, note that if $|x-2| < 1$, then $-1 < x-2 < 1$, so $0 < x-1 < 2 \Rightarrow |x-1| < 2$. Now let $\delta = \min\{\epsilon/2, 1\}$. Then $|x-2| < \delta$ $\Rightarrow |x^2-3x-(-2)| = |(x-2)(x-1)| = |x-2||x-1| < (\epsilon/2)(2) = \epsilon$.

Thus $\lim\limits_{x\to2}(x^2-3x) = -2$ by the definition of a limit.

43. Since $2x-1 \le f(x) \le x^2$ for $0 < x < 3$ and $\lim\limits_{x\to1}(2x-1) = 1 = \lim\limits_{x\to1}x^2$, we have $\lim\limits_{x\to1}f(x) = 1$ by the Squeeze Theorem.

45. **(a)** $f(x) = \sqrt{-x}$ if $x < 0$, $f(x) = 3-x$ if $0 \le x < 3$, $f(x) = (x-3)^2$ if $x > 3$. So

(i) $\lim\limits_{x\to0^+}f(x) = \lim\limits_{x\to0^+}(3-x) = 3$

(ii) $\lim\limits_{x\to0^-}f(x) = \lim\limits_{x\to0^-}\sqrt{-x} = 0$

(iii) Because of (i) and (ii), $\lim\limits_{x\to0}f(x)$ does not exist.

(iv) $\lim\limits_{x\to3^-}f(x) = \lim\limits_{x\to3^-}(3-x) = 0$

(v) $\lim\limits_{x\to3^+}f(x) = \lim\limits_{x\to3^+}(x-3)^2 = 0$

(vi) Because of (iv) and (v), $\lim\limits_{x\to3}f(x) = 0$.

(b) f is discontinuous at 0 since $\lim\limits_{x\to0}f(x)$ does not exist.

f is discontinuous at 3 since $f(3)$ does not exist.

(c)

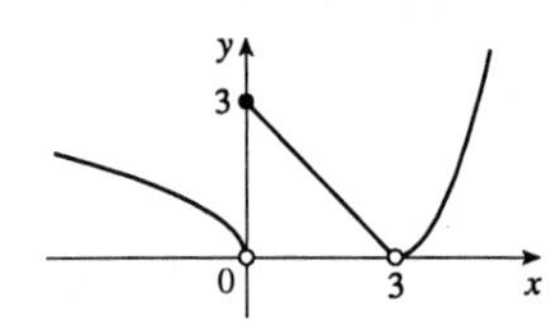

47. $f(x) = \dfrac{x+1}{x^2+x+1}$ is rational so it is continuous on its domain which is $\mathbb{R}$. ($x^2+x+1=0$ has no real roots.)

49. $f(x) = 2x^3+x^2+2$ is a polynomial, so it is continuous on $[-2,-1]$ and $f(-2) = -10 < 0 < 1 = f(-1)$. So by the Intermediate Value Theorem there is a number c in $(-2,-1)$ such that $f(c) = 0$, that is, the equation $2x^3+x^2+2 = 0$ has a root in $(-2,-1)$.

51. **(a)** The slope of the tangent line at $(2,1)$ is $\lim\limits_{x\to2}\dfrac{f(x)-f(2)}{x-2} = \lim\limits_{x\to2}\dfrac{9-2x^2-1}{x-2}$

$$= \lim_{x\to2}\frac{8-2x^2}{x-2} = \lim_{x\to2}\frac{-2(x^2-4)}{x-2} = \lim_{x\to2}\frac{-2(x-2)(x+2)}{x-2} = \lim_{x\to2}-2(x+2) = -8.$$

(b) The equation of this tangent line is $y-1 = -8(x-2)$ or $8x+y = 17$.

53. **(a)** $\mathbf{v} = \lim\limits_{t\to2}\dfrac{1}{t-2}[\mathbf{r}(t)-\mathbf{r}(2)] = \lim\limits_{t\to2}\dfrac{1}{t-2}\left[(3t^2-4t)\mathbf{i}+(5t+2)\mathbf{j}-(4\mathbf{i}+12\mathbf{j})\right]$

$$= \lim_{t\to2}\frac{1}{t-2}\left[(3t^2-4t-4)\mathbf{i}+(5t-10)\mathbf{j}\right] = \lim_{t\to2}\frac{1}{t-2}[(t-2)(3t+2)\mathbf{i}+5(t-2)\mathbf{j}]$$

$$= \lim_{t\to2}[(3t+2)\mathbf{i}+5\mathbf{j}] = 8\mathbf{i}+5\mathbf{j}$$

(b) $\mathbf{L}(t) = \mathbf{r}(2)+t\mathbf{v} = \langle 4,12\rangle + t\langle 8,5\rangle = \langle 4+8t, 12+5t\rangle$

(c) Parametric equations for the tangent line at $\mathbf{r}(2)$ are $x = 4+8t$, $y = 12+5t$.

55. **(a)** For $\mathbf{r}(t) = \langle t^2 + 3t, 4t^2 \rangle$, the average velocity $\mathbf{v}_{\text{av}}$ over the interval $[2, 2+h]$ is

$$\begin{aligned}\frac{\mathbf{r}(2+h) - \mathbf{r}(2)}{h} &= \frac{1}{h}\left[\left\langle (2+h)^2 + 3(2+h), 4(2+h)^2 \right\rangle - \langle 10, 16 \rangle\right] \\ &= \frac{1}{h}\left\langle h^2 + 7h, 4h^2 + 16h \right\rangle = \langle h+7, 4h+16 \rangle \text{ ft/s.}\end{aligned}$$

(i) When $h = 2$, $\mathbf{v}_{\text{av}} = \langle 9, 24 \rangle$ ft/s.

(ii) When $h = 1$, $\mathbf{v}_{\text{av}} = \langle 8, 20 \rangle$ ft/s.

(iii) When $h = 0.5$, $\mathbf{v}_{\text{av}} = \langle 7.5, 18 \rangle$ ft/s.

(iv) When $h = 0.1$, $\mathbf{v}_{\text{av}} = \langle 7.1, 16.4 \rangle$ ft/s.

(b) The instantaneous velocity when $t = 2$ s is $\displaystyle\lim_{h\to 0} \frac{\mathbf{r}(2+h) - \mathbf{r}(2)}{h} = \lim_{h\to 0} \langle h+7, 4h+16 \rangle = \langle 7, 16 \rangle$ ft/s.

57. **(a)** $s = 1 + 2t + t^2/4$. The average velocity over the time interval $[1, 1+h]$ is
$\displaystyle\frac{s(1+h) - s(1)}{h} = \frac{1 + 2(1+h) + (1+h)^2/4 - 13/4}{h} = \frac{10h + h^2}{4h} = \frac{10+h}{4}$. So for the following intervals the average velocities are:

(i) $[1, 3]$: $(10+2)/4 = 3\text{ m/s}$

(ii) $[1, 2]$: $(10+1)/4 = 2.75\text{ m/s}$

(iii) $[1, 1.5]$: $(10+0.5)/4 = 2.625\text{ m/s}$

(iv) $[1, 1.1]$: $(10+0.1)/4 = 2.525\text{ m/s}$

(b) When $t = 1$ the velocity is $\displaystyle\lim_{h\to 0} \frac{s(1+h) - s(1)}{h} = \lim_{h\to 0} \frac{10+h}{4} = 2.5\text{ m/s}$.

59. The inequality $\left|\dfrac{x+1}{x-1} - 3\right| < 0.2$ is equivalent to the double inequality $2.8 < \dfrac{x+1}{x-1} < 3.2$. Graphing the functions $y = 2.8$, $y = |(x+1)/(x-1)|$ and $y = 3.2$ on the interval $[1.9, 2.15]$, we see that the inequality holds whenever $1.91 < x < 2.11$ (approximately). So since $|2 - 1.91| = 0.09$ and $|2 - 2.15| = 0.15$, any positive $\delta \le 0.09$ will do.

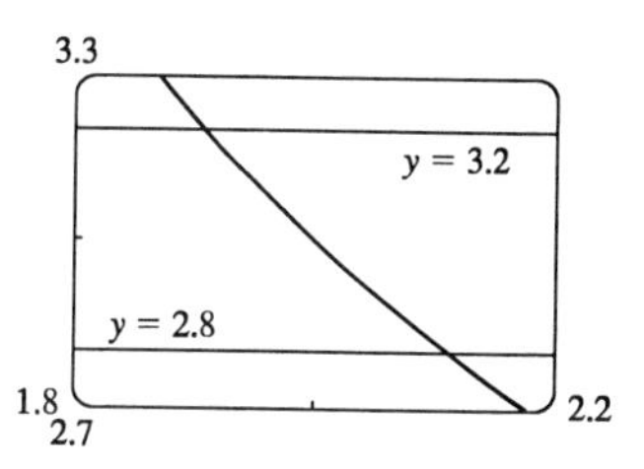

61. $|f(x)| \le g(x) \quad \Leftrightarrow \quad -g(x) \le f(x) \le g(x)$ and $\displaystyle\lim_{x\to a} g(x) = 0 = \lim_{x\to a} -g(x)$.

Thus, by the Squeeze Theorem, $\displaystyle\lim_{x\to a} f(x) = 0$.

63. $\displaystyle\lim_{x\to a} f(x) = \lim_{x\to a}\left(\tfrac{1}{2}[f(x)+g(x)] + \tfrac{1}{2}[f(x)-g(x)]\right) = \tfrac{1}{2}\lim_{x\to a}[f(x)+g(x)] + \tfrac{1}{2}\lim_{x\to a}[f(x)-g(x)] = \tfrac{1}{2}\cdot 2 + \tfrac{1}{2}\cdot 1$
$= \tfrac{3}{2}$, and $\displaystyle\lim_{x\to a} g(x) = \lim_{x\to a}([f(x)+g(x)] - f(x)) = \lim_{x\to a}[f(x)+g(x)] - \lim_{x\to a} f(x) = 2 - \tfrac{3}{2} = \tfrac{1}{2}$.

So $\displaystyle\lim_{x\to a}[f(x)g(x)] = \left[\lim_{x\to a} f(x)\right]\left[\lim_{x\to a} g(x)\right] = \tfrac{3}{2}\cdot\tfrac{1}{2} = \tfrac{3}{4}$.

Alternate Solution: Since $\displaystyle\lim_{x\to a}[f(x)+g(x)]$ and $\displaystyle\lim_{x\to a}[f(x)-g(x)]$ exist, we must have

$\displaystyle\lim_{x\to a}[f(x)+g(x)]^2 = \left(\lim_{x\to a}[f(x)+g(x)]\right)^2$ and $\displaystyle\lim_{x\to a}[f(x)-g(x)]^2 = \left(\lim_{x\to a}[f(x)-g(x)]\right)^2$, so

$$\begin{aligned}\lim_{x\to a}[f(x)\,g(x)] &= \lim_{x\to a} \tfrac{1}{4}\left([f(x)+g(x)]^2 - [f(x)-g(x)]^2\right) \text{ (because all of the } f^2 \text{ and } g^2 \text{ cancel)} \\ &= \tfrac{1}{4}\left(\lim_{x\to a}[f(x)+g(x)]^2 - \lim_{x\to a}[f(x)-g(x)]^2\right) = \tfrac{1}{4}(2^2 - 1^2) = \tfrac{3}{4}.\end{aligned}$$

CHAPTER THREE

EXERCISES 3.1

1. $f'(2) = \lim_{h\to 0} \dfrac{f(2+h)-f(2)}{h} = \lim_{h\to 0} \dfrac{3(2+h)^2 - 5(2+h) - \left[3(2)^2 - 5(2)\right]}{h}$

$= \lim_{h\to 0} \dfrac{12+12h+3h^2-10-5h-12+10}{h} = \lim_{h\to 0} \dfrac{3h^2+7h}{h} = \lim_{h\to 0}(3h+7) = 7$

So the equation of the tangent line at $(2,2)$ is $y-2=7(x-2)$ or $7x-y=12$.

3. **(a)** $F'(1) = \lim_{x\to 1} \dfrac{F(x)-F(1)}{x-1} = \lim_{x\to 1} \dfrac{x^3-5x+1-(-3)}{x-1} = \lim_{x\to 1} \dfrac{x^3-5x+4}{x-1} = \lim_{x\to 1} \dfrac{(x-1)(x^2+x-4)}{x-1}$

$= \lim_{x\to 1}\left(x^2+x-4\right) = -2.$

So the equation of the tangent line at $(1,-3)$ is $y-(-3) = -2(x-1)$ $\Leftrightarrow$ $y=-2x-1$.

Note: Instead of using Equation 3 to compute $F'(1)$, we could have used Equation 1.

(b)

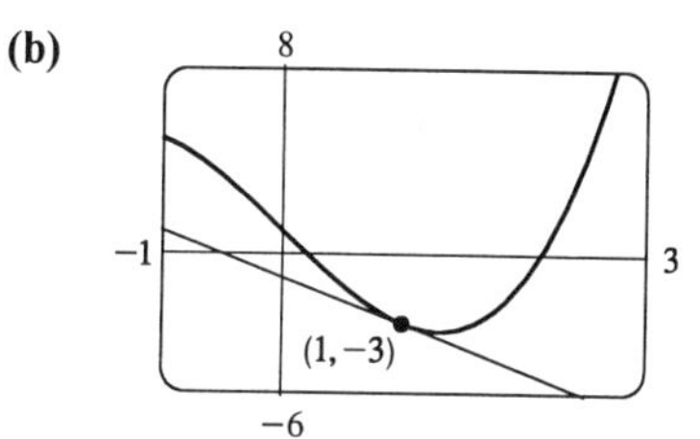

5. $v(2) = f'(2) = \lim_{h\to 0} \dfrac{f(2+h)-f(2)}{h} = \lim_{h\to 0} \dfrac{(2+h)^2-6(2+h)-5-(2^2-6(2)-5)}{h}$

$= \lim_{h\to 0} \dfrac{4+4h+h^2-12-6h-5-4+12+5}{h} = \lim_{h\to 0} \dfrac{h^2-2h}{h} = \lim_{h\to 0}(h-2) = -2\text{ m/s}$

7. $f'(a) = \lim_{h\to 0} \dfrac{f(a+h)-f(a)}{h} = \lim_{h\to 0} \dfrac{1+(a+h)-2(a+h)^2-(1+a-2a^2)}{h}$

$= \lim_{h\to 0} \dfrac{h-4ah-2h^2}{h} = \lim_{h\to 0}(1-4a-2h) = 1-4a$

9. $f'(a) = \lim_{h\to 0} \dfrac{f(a+h)-f(a)}{h} = \lim_{h\to 0} \dfrac{\dfrac{a+h}{2(a+h)-1} - \dfrac{a}{2a-1}}{h}$

$= \lim_{h\to 0} \dfrac{(a+h)(2a-1)-a(2a+2h-1)}{h(2a+2h-1)(2a-1)} = \lim_{h\to 0} \dfrac{-h}{h(2a+2h-1)(2a-1)}$

$= \lim_{h\to 0} \dfrac{-1}{(2a+2h-1)(2a-1)} = -\dfrac{1}{(2a-1)^2}$

11. $f'(a) = \lim\limits_{h \to 0} \dfrac{f(a+h) - f(a)}{h} = \lim\limits_{h \to 0} \dfrac{\dfrac{2}{\sqrt{3-(a+h)}} - \dfrac{2}{\sqrt{3-a}}}{h} = \lim\limits_{h \to 0} \dfrac{2\left(\sqrt{3-a} - \sqrt{3-a-h}\right)}{h\sqrt{3-a-h}\,\sqrt{3-a}}$

$$= \lim_{h \to 0} \frac{2\left(\sqrt{3-a} - \sqrt{3-a-h}\right)}{h\sqrt{3-a-h}\,\sqrt{3-a}} \cdot \frac{\sqrt{3-a} + \sqrt{3-a-h}}{\sqrt{3-a} + \sqrt{3-a-h}}$$

$$= \lim_{h \to 0} \frac{2[3-a-(3-a-h)]}{h\sqrt{3-a-h}\,\sqrt{3-a}\left(\sqrt{3-a} + \sqrt{3-a-h}\right)}$$

$$= \lim_{h \to 0} \frac{2}{\sqrt{3-a-h}\,\sqrt{3-a}\left(\sqrt{3-a} + \sqrt{3-a-h}\right)}$$

$$= \frac{2}{\sqrt{3-a}\,\sqrt{3-a}\left(2\sqrt{3-a}\right)} = \frac{1}{(3-a)^{3/2}}$$

13. By Equation 1, $\lim\limits_{h \to 0} \dfrac{\sqrt{1+h} - 1}{h} = f'(1)$ where $f(x) = \sqrt{x}$. [Or $f'(0)$ where $f(x) = \sqrt{1+x}$; the answers to Exercises 13-18 are not unique.]

15. $\lim\limits_{x \to 1} \dfrac{x^9 - 1}{x - 1} = f'(1)$ where $f(x) = x^9$. (See Equation 3.)

17. $\lim\limits_{t \to 0} \dfrac{\sin\left(\frac{\pi}{2} + t\right) - 1}{t} = f'\left(\frac{\pi}{2}\right)$ where $f(x) = \sin x$.

19. $f'(x) = \lim\limits_{h \to 0} \dfrac{f(x+h) - f(x)}{h} = \lim\limits_{h \to 0} \dfrac{5(x+h) + 3 - (5x+3)}{h} = \lim\limits_{h \to 0} \dfrac{5h}{h} = \lim\limits_{h \to 0} 5 = 5.$

Domain of f = domain of $f' = \mathbb{R}$.

21. $f'(x) = \lim\limits_{h \to 0} \dfrac{f(x+h) - f(x)}{h} = \lim\limits_{h \to 0} \dfrac{(x+h)^3 - (x+h)^2 + 2(x+h) - (x^3 - x^2 + 2x)}{h}$

$$= \lim_{h \to 0} \frac{3x^2h + 3xh^2 + h^3 - 2xh - h^2 + 2h}{h} = \lim_{h \to 0} \left(3x^2 + 3xh + h^2 - 2x - h + 2\right) = 3x^2 - 2x + 2$$

Domain of f = domain of $f' = \mathbb{R}$.

23. $g'(x) = \lim\limits_{h \to 0} \dfrac{g(x+h) - g(x)}{h} = \lim\limits_{h \to 0} \dfrac{\sqrt{1+2(x+h)} - \sqrt{1+2x}}{h} \left[\dfrac{\sqrt{1+2(x+h)} + \sqrt{1+2x}}{\sqrt{1+2(x+h)} + \sqrt{1+2x}}\right]$

$$= \lim_{h \to 0} \frac{1 + 2x + 2h - (1+2x)}{h\left[\sqrt{1+2(x+h)} + \sqrt{1+2x}\right]} = \lim_{h \to 0} \frac{2}{\sqrt{1+2(x+h)} + \sqrt{1+2x}} = \frac{1}{\sqrt{1+2x}}$$

Domain of $g = \left[-\frac{1}{2}, \infty\right)$, domain of $g' = \left(-\frac{1}{2}, \infty\right)$.

25. $G'(x) = \lim\limits_{h \to 0} \dfrac{G(x+h) - G(x)}{h} = \lim\limits_{h \to 0} \dfrac{\dfrac{4 - 3(x+h)}{2 + (x+h)} - \dfrac{4 - 3x}{2 + x}}{h}$

$$= \lim_{h \to 0} \frac{(4 - 3x - 3h)(2 + x) - (4 - 3x)(2 + x + h)}{h(2 + x + h)(2 + x)} = \lim_{h \to 0} \frac{-10h}{h(2 + x + h)(2 + x)}$$

$$= \lim_{h \to 0} \frac{-10}{(2 + x + h)(2 + x)} = \frac{-10}{(2 + x)^2}$$

Domain of G = domain of $G' = \{x \mid x \neq -2\}$.

27. $f'(x) = \lim_{h\to 0} \dfrac{f(x+h) - f(x)}{h} = \lim_{h\to 0} \dfrac{(x+h)^4 - x^4}{h} = \lim_{h\to 0} \dfrac{4x^3h + 6x^2h^2 + 4xh^3 + h^4}{h}$
$= \lim_{h\to 0} \left(4x^3 + 6x^2h + 4xh^2 + h^3\right) = 4x^3$

Domain of f = domain of $f' = \mathbb{R}$.

29. $f(x) = x \quad \Rightarrow \quad f'(x) = \lim_{h\to 0} \dfrac{x+h-x}{h} = \lim_{h\to 0} 1 = 1$

$f(x) = x^2 \quad \Rightarrow \quad f'(x) = \lim_{h\to 0} \dfrac{(x+h)^2 - x^2}{h} = \lim_{h\to 0} \dfrac{2xh + h^2}{h} = \lim_{h\to 0}(2x + h) = 2x$

$f(x) = x^3 \quad \Rightarrow \quad f'(x) = \lim_{h\to 0} \dfrac{(x+h)^3 - x^3}{h} = \lim_{h\to 0} \dfrac{3x^2h + 3xh^2 + h^3}{h} = \lim_{h\to 0}\left(3x^2 + 3xh + h^2\right) = 3x^2$

$f(x) = x^4 \quad \Rightarrow \quad f'(x) = 4x^3$ from Exercise 27.

Guess: The derivative of $f(x) = x^n$ is $f'(x) = nx^{n-1}$. Test for $n = 5$: $f(x) = x^5 \quad \Rightarrow$

$f'(x) = \lim_{h\to 0} \dfrac{(x+h)^5 - x^5}{h} = \lim_{h\to 0} \dfrac{5x^4h + 10x^3h^2 + 10x^2h^3 + 5xh^4 + h^5}{h}$
$= \lim_{h\to 0}\left(5x^4 + 10x^3h + 10x^2h^2 + 5xh^3 + h^4\right) = 5x^4$

31. **(a)** $f'(x) = \lim_{h\to 0} \dfrac{f(x+h) - f(x)}{h} = \lim_{h\to 0} \dfrac{\left[x + h - \left(\dfrac{2}{x+h}\right)\right] - \left[x - \left(\dfrac{2}{x}\right)\right]}{h}$

$= \lim_{h\to 0} \left[1 + \dfrac{\dfrac{2}{x} - \dfrac{2}{(x+h)}}{h}\right]$

$= \lim_{h\to 0} \left[1 + \dfrac{2(x+h) - 2x}{h(x)(x+h)}\right]$

$= 1 + 2x^{-2}$

Notice that when f has steep tangent lines, $f'(x)$ is very large. When f is flatter, $f'(x)$ is smaller.

(b)

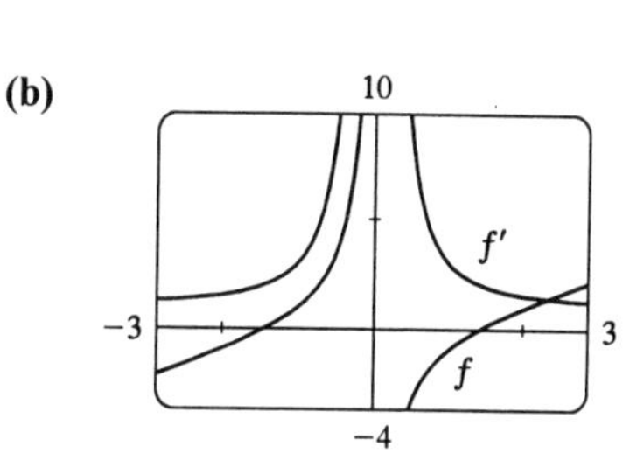

33. From the graph of f, it appears that

(a) $f'(1) \approx -2$ **(b)** $f'(2) \approx 0.8$
(c) $f'(3) \approx -1$ **(d)** $f'(4) \approx -0.5$

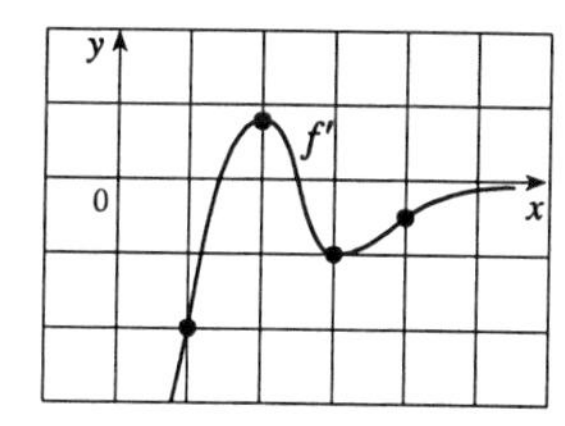

35.

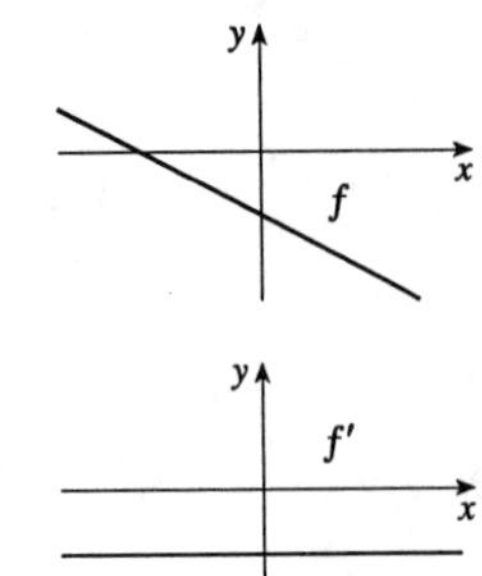

37.

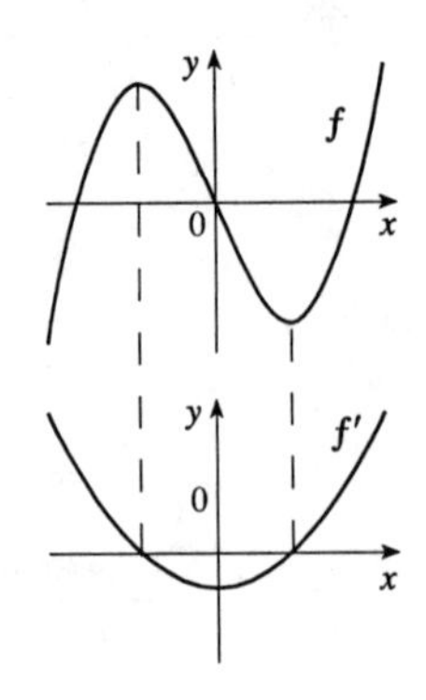

39.

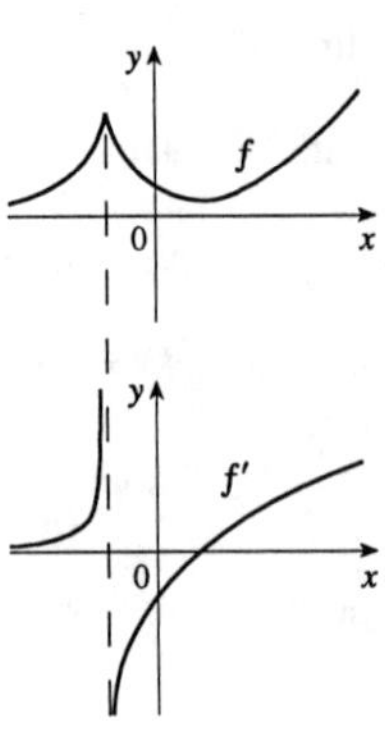

41.

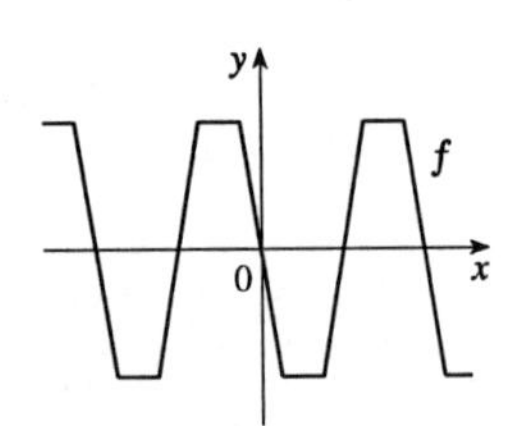

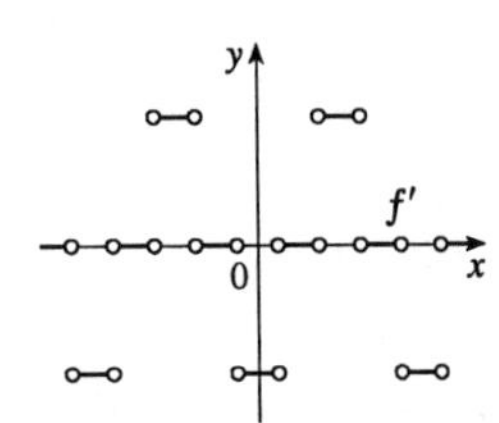

43.

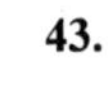

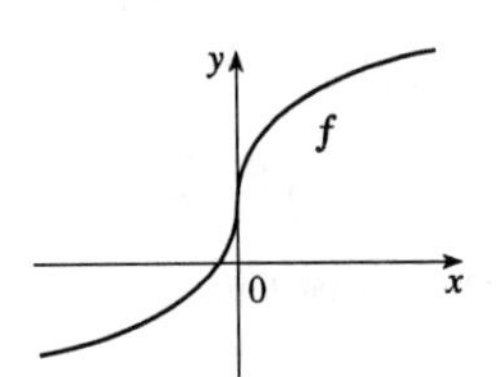

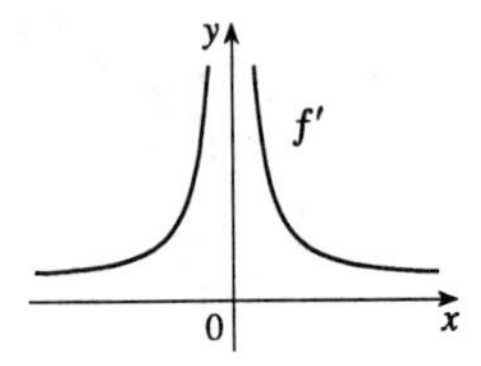

45. $f'(1) = \lim_{h\to 0} \frac{f(1+h) - f(1)}{h} = \lim_{h\to 0} \frac{3^{1+h} - 3^1}{h}$ So let $F(h) = \frac{3^{1+h} - 3}{h}$. We calculate:

h	$F(h)$
0.1	3.48
0.01	3.314
0.001	3.297
0.0001	3.296
−0.1	3.12
−0.01	3.278
−0.001	3.294
−0.0001	3.296

We estimate that $f'(1) \approx 3.296$.

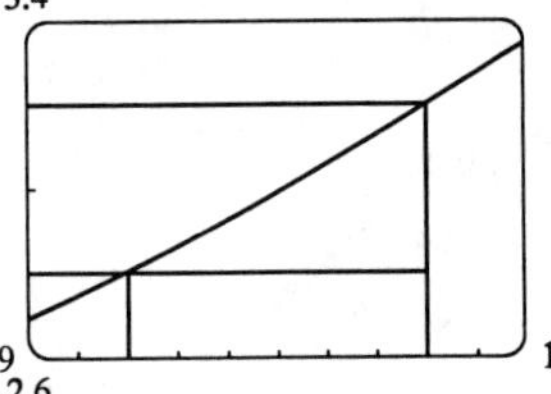

From the graph we estimate that the slope of the tangent is about $\frac{3.2 - 2.8}{1.06 - 0.94} = \frac{0.4}{0.12} \approx 3.3$.

47. We plot the points given by the data in the table, then sketch the rough shape of the curve. To estimate the derivative $f'(x)$, we draw the tangent line to the curve at x. It appears that $f'(0.1) \approx -5$, $f'(0.2) \approx 4$, $f'(0.3) \approx 8$, $f'(0.4) \approx 9$, $f'(0.5) \approx 5$, $f'(0.6) \approx -0.5$, and $f'(0.7) \approx -8$.

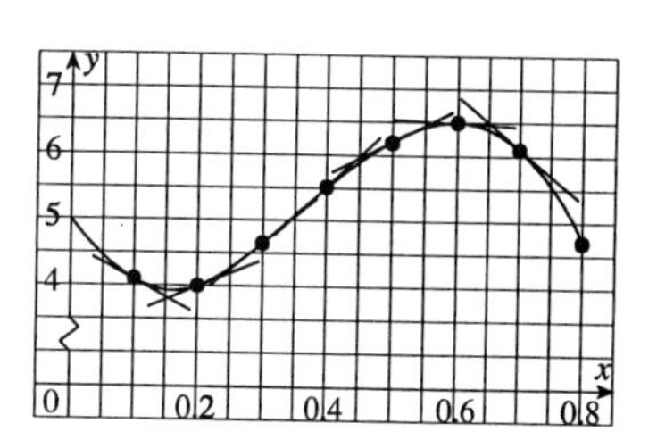

49. **(a)** $f'(a) = \lim\limits_{x\to a} \dfrac{f(x)-f(a)}{x-a} = \lim\limits_{x\to a} \dfrac{x^{1/3}-a^{1/3}}{x-a} = \lim\limits_{x\to a} \dfrac{x^{1/3}-a^{1/3}}{\left(x^{1/3}-a^{1/3}\right)\left(x^{2/3}+a^{2/3}+x^{1/3}a^{1/3}\right)}$

$= \lim\limits_{x\to a} \dfrac{1}{x^{2/3}+a^{2/3}+x^{1/3}a^{1/3}} = \lim\limits_{x\to a} \dfrac{1}{3x^{2/3}} = \dfrac{1}{3a^{2/3}}$

(b) $f'(0) = \lim\limits_{h\to 0} \dfrac{f(0+h)-f(0)}{h} = \lim\limits_{h\to 0} \dfrac{\sqrt[3]{h}-0}{h} = \lim\limits_{h\to 0} \dfrac{1}{h^{2/3}}$. This limit does not exist, and therefore $f'(0)$ does not exist.

(c) $\lim\limits_{x\to 0}|f'(x)| = \lim\limits_{x\to 0} \dfrac{1}{3x^{2/3}} = \infty$. Also f is continuous at $x=0$ (root function), so f has a vertical tangent at $x=0$.

51. f is not differentiable at $x=-1$ or at $x=11$ because the graph has vertical tangents at those points; at $x=4$, because there is a discontinuity there; and at $x=8$, because the graph has a corner there.

53. $f(x) = |x-6| = \begin{cases} 6-x & \text{if } x<6 \\ x-6 & \text{if } x\geq 6. \end{cases}$ $\lim\limits_{x\to 6^+} \dfrac{f(x)-f(6)}{x-6} = \lim\limits_{x\to 6^+} \dfrac{|x-6|-0}{x-6} = \lim\limits_{x\to 6^+} \dfrac{x-6}{x-6} = \lim\limits_{x\to 6^+} 1 = 1.$

But $\lim\limits_{x\to 6^-} \dfrac{f(x)-f(6)}{x-6} = \lim\limits_{x\to 6^-} \dfrac{|x-6|-0}{x-6}$

$= \lim\limits_{x\to 6^-} \dfrac{6-x}{x-6} = \lim\limits_{x\to 6^-}(-1) = -1.$

So $f'(6) = \lim\limits_{x\to 6} \dfrac{f(x)-f(6)}{x-6}$ does not exist.

However $f'(x) = \begin{cases} -1 & \text{if } x<6 \\ 1 & \text{if } x>6. \end{cases}$

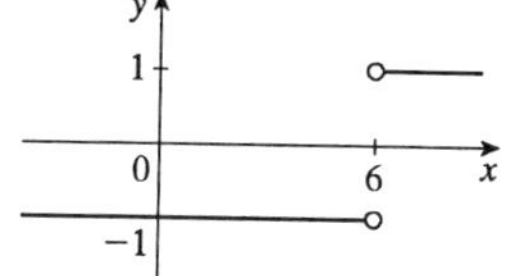

55. **(a)** $f(x) = x|x| = \begin{cases} x^2 & \text{if } x\geq 0 \\ -x^2 & \text{if } x<0 \end{cases}$

(b) Since $f(x)=x^2$ for $x\geq 0$, we have $f'(x)=2x$ for $x>0$. Since $f(x)=-x^2$ for $x<0$, we have $f'(x)=-2x$ for $x<0$. At $x=0$, we have

$f'(0) = \lim\limits_{x\to 0} \dfrac{f(x)-f(0)}{x-0} = \lim\limits_{x\to 0} \dfrac{x|x|}{x} = \lim\limits_{x\to 0}|x| = 0$

(by Example 2.3.9). So f is differentiable at 0. Thus f is differentiable for all x.

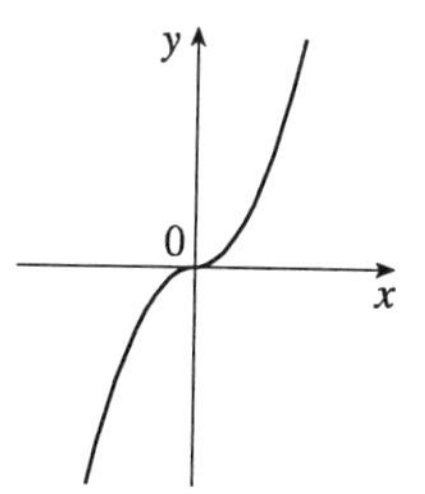

(c) From part (b) we have $f'(x) = \left\{\begin{matrix} 2x & x\geq 0 \\ -2x & x<0 \end{matrix}\right\} = 2|x|.$

57. **(a)** $f'_-(0.6) = \lim_{h \to 0^-} \frac{f(0.6+h) - f(0.6)}{h} = \lim_{h \to 0^-} \frac{|5(0.6+h) - 3| - |3-3|}{h}$

$= \lim_{h \to 0^-} \frac{|3 + 5h - 3|}{h} = \lim_{h \to 0^-} \frac{-5h}{h} = -5$

$f'_+(0.6) = \lim_{h \to 0^+} \frac{f(0.6+h) - f(0.6)}{h} = \lim_{h \to 0^+} \frac{|5(0.6+h) - 3| - |3-3|}{h}$

$= \lim_{h \to 0^+} \frac{5h}{h} = 5$

(b) Since $f'_-(0.6) \neq f'_+(0.6)$, $f'(0.6)$ does not exist.

59. Since $f(x) = x \sin(1/x)$ when $x \neq 0$ and $f(0) = 0$, we have

$f'(0) = \lim_{h \to 0} \frac{f(0+h) - f(0)}{h} = \lim_{h \to 0} \frac{h \sin(1/h) - 0}{h} = \lim_{h \to 0} \sin(1/h)$. This limit does not exist since $\sin(1/h)$ takes the values -1 and 1 on any interval containing 0. (Compare with Example 2.2.4.)

61. **(a)** If f is even, then $f'(-x) = \lim_{h \to 0} \frac{f(-x+h) - f(-x)}{h} = \lim_{h \to 0} \frac{f(x-h) - f(x)}{h}$

$= -\lim_{h \to 0} \frac{f(x-h) - f(x)}{-h}$ [let $\Delta x = -h$] $= -\lim_{\Delta x \to 0} \frac{f(x + \Delta x) - f(x)}{\Delta x} = -f'(x)$. Therefore f' is odd.

(b) If f is odd, then $f'(-x) = \lim_{h \to 0} \frac{f(-x+h) - f(-x)}{h} = \lim_{h \to 0} \frac{-f(x-h) + f(x)}{h}$

$= \lim_{h \to 0} \frac{f(x-h) - f(x)}{-h}$ [let $\Delta x = -h$] $= \lim_{\Delta x \to 0} \frac{f(x + \Delta x) - f(x)}{\Delta x} = f'(x)$. Therefore f' is even.

63.

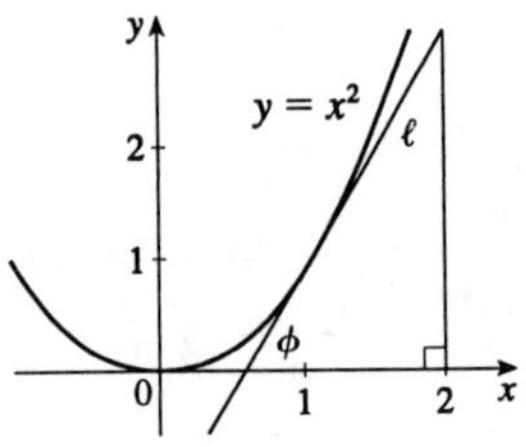

From the diagram, we see that the slope of the tangent is equal to $\tan \phi$, and also that $0 < \phi < \frac{\pi}{2}$. We know (see Exercise 29) that the derivative of $f(x) = x^2$ is $f'(x) = 2x$. So the slope of the tangent to the curve at the point $(1, 1)$ is 2. So ϕ is the angle between 0 and $\frac{\pi}{2}$ whose tangent is 2, that is, $\phi = \tan^{-1} 2 \approx 63°$.

EXERCISES 3.2

1. $f(x) = x^2 - 10x + 100 \quad \Rightarrow \quad f'(x) = 2x - 10$

3. $V(r) = \frac{4}{3}\pi r^3 \quad \Rightarrow \quad V'(r) = \frac{4}{3}\pi(3r^2) = 4\pi r^2$

5. $F(x) = (16x)^3 = 4{,}096x^3 \quad \Rightarrow \quad F'(x) = 4{,}096(3x^2) = 12{,}288x^2$

7. $Y(t) = 6t^{-9} \quad \Rightarrow \quad Y'(t) = 6(-9)t^{-10} = -54t^{-10}$

9. $g(x) = x^2 + \dfrac{1}{x^2} = x^2 + x^{-2} \quad \Rightarrow \quad g'(x) = 2x + (-2)x^{-3} = 2x - \dfrac{2}{x^3}$

11. $h(x) = \dfrac{x+2}{x-1} \quad \Rightarrow \quad h'(x) = \dfrac{(x-1)D(x+2) - (x+2)D(x-1)}{(x-1)^2} = \dfrac{x-1-(x+2)}{(x-1)^2} = \dfrac{-3}{(x-1)^2}$

13. $G(s) = (s^2 + s + 1)(s^2 + 2) \quad \Rightarrow \quad G'(s) = (2s+1)(s^2+2) + (s^2+s+1)(2s) = 4s^3 + 3s^2 + 6s + 2$

15. $y = \dfrac{x^2 + 4x + 3}{\sqrt{x}} = x^{3/2} + 4x^{1/2} + 3x^{-1/2} \quad \Rightarrow$

$y' = \frac{3}{2}x^{1/2} + 4\left(\frac{1}{2}\right)x^{-1/2} + 3\left(-\frac{1}{2}\right)x^{-3/2} = \frac{3}{2}\sqrt{x} + \dfrac{2}{\sqrt{x}} - \dfrac{3}{2x\sqrt{x}}$. *Another Method:* Use the Quotient Rule.

17. $y = \sqrt{5x} = \sqrt{5}x^{1/2} \quad \Rightarrow \quad y' = \sqrt{5}\left(\frac{1}{2}\right)x^{-1/2} = \dfrac{\sqrt{5}}{2\sqrt{x}}$

19. $y = \dfrac{1}{x^4 + x^2 + 1} \quad \Rightarrow \quad y' = \dfrac{(x^4+x^2+1)(0) - 1(4x^3+2x)}{(x^4+x^2+1)^2} = -\dfrac{4x^3+2x}{(x^4+x^2+1)^2}$

21. $y = ax^2 + bx + c \quad \Rightarrow \quad y' = 2ax + b$

23. $y = \dfrac{3t-7}{t^2+5t-4} \quad \Rightarrow \quad y' = \dfrac{(t^2+5t-4)(3) - (3t-7)(2t+5)}{(t^2+5t-4)^2} = \dfrac{-3t^2+14t+23}{(t^2+5t-4)^2}$

25. $y = x + \sqrt[5]{x^2} = x + x^{2/5} \quad \Rightarrow \quad y' = 1 + \frac{2}{5}x^{-3/5} = 1 + \dfrac{2}{5\sqrt[5]{x^3}}$

27. $u = x^{\sqrt{2}} \quad \Rightarrow \quad u' = \sqrt{2}\,x^{\sqrt{2}-1}$

29. $v = x\sqrt{x} + \dfrac{1}{x^2\sqrt{x}} = x^{3/2} + x^{-5/2} \quad \Rightarrow \quad v' = \frac{3}{2}x^{1/2} - \frac{5}{2}x^{-7/2} = \frac{3}{2}\sqrt{x} - \dfrac{5}{2x^3\sqrt{x}}$

31. $f(x) = \dfrac{x}{x + c/x} \quad \Rightarrow \quad f'(x) = \dfrac{(x + c/x)(1) - x(1 - c/x^2)}{(x + c/x)^2} = \dfrac{2cx}{(x^2+c)^2}$

33. $f(x) = \dfrac{x^5}{x^3 - 2} \quad \Rightarrow \quad f'(x) = \dfrac{(x^3-2)(5x^4) - x^5(3x^2)}{(x^3-2)^2} = \dfrac{2x^4(x^3-5)}{(x^3-2)^2}$

35. $P(x) = a_n x^n + a_{n-1}x^{n-1} + \cdots + a_2x^2 + a_1x + a_0 \quad \Rightarrow$

$P'(x) = na_nx^{n-1} + (n-1)a_{n-1}x^{n-2} + \cdots + 2a_2x + a_1$

37. $y = f(x) = x + \dfrac{4}{x} \quad \Rightarrow \quad f'(x) = 1 - \dfrac{4}{x^2}$. So the slope of the tangent line at $(2, 4)$ is $f'(2) = 0$ and its equation is $y - 4 = 0$ or $y = 4$.

39. $y = f(x) = x + \sqrt{x} \quad \Rightarrow \quad f'(x) = 1 + \frac{1}{2}x^{-1/2}$. So the slope of the tangent line at $(1, 2)$ is $f'(1) = 1 + \frac{1}{2}(1) = \frac{3}{2}$ and its equation is $y - 2 = \frac{3}{2}(x - 1)$ or $y = \frac{3}{2}x + \frac{1}{2}$ or $3x - 2y + 1 = 0$.

41. **(a)** $y = f(x) = \dfrac{1}{1+x^2} \quad \Rightarrow \quad f'(x) = \dfrac{-2x}{(1+x^2)^2}$.

So the slope of the tangent line at the point $\left(-1, \frac{1}{2}\right)$ is

$$f'(-1) = \frac{-2(-1)}{\left[1+(-1)^2\right]^2} = \frac{1}{2} \text{ and its equation is}$$

$y - \frac{1}{2} = \frac{1}{2}(x+1)$ or $y = \frac{1}{2}x + 1$ or $x - 2y + 2 = 0$.

(b)

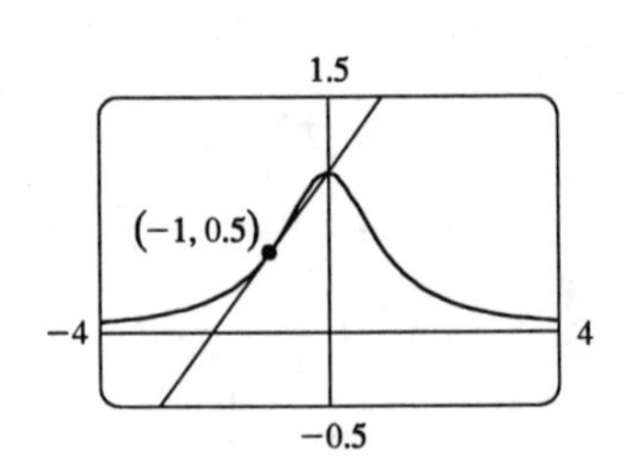

43. **(a)** $f(x) = 3x^{15} - 5x^3 + 3 \quad \Rightarrow$

$$f'(x) = 3 \cdot 15x^{14} - 5 \cdot 3x^2 = 45x^{14} - 15x^2$$

(b)

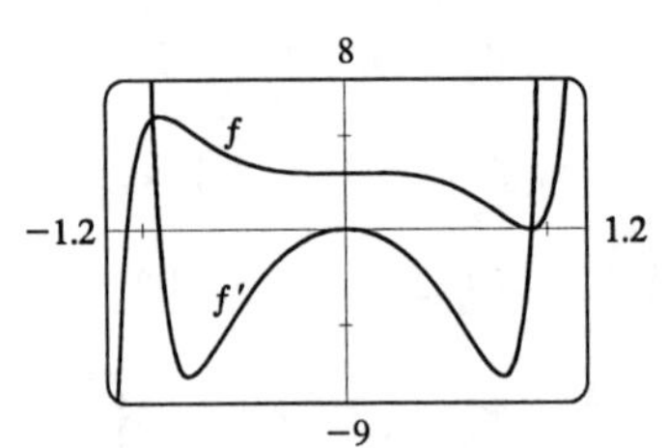

Notice that $f'(x) = 0$ when f has a horizontal tangent.

45. $y = x\sqrt{x} = x^{3/2} \quad \Rightarrow \quad y' = \frac{3}{2}\sqrt{x}$ so the tangent line is parallel to $3x - y + 6 = 0$ when $\frac{3}{2}\sqrt{x} = 3 \quad \Leftrightarrow$ $\sqrt{x} = 2 \quad \Leftrightarrow \quad x = 4$. So the point is $(4, 8)$.

47. $y = x^3 - x^2 - x + 1$ has a horizontal tangent when $y' = 3x^2 - 2x - 1 = 0 \quad (3x+1)(x-1) = 0 \Leftrightarrow x = 1$ or $-\frac{1}{3}$. Therefore the points are $(1, 0)$ and $\left(-\frac{1}{3}, \frac{32}{27}\right)$.

49. If $y = f(x) = \dfrac{x}{x+1}$ then $f'(x) = \dfrac{(x+1)(1) - x(1)}{(x+1)^2} = \dfrac{1}{(x+1)^2}$. When $x = a$, the equation of the tangent line is $y - \dfrac{a}{a+1} = \dfrac{1}{(a+1)^2}(x - a)$. This line passes through $(1, 2)$ when $2 - \dfrac{a}{a+1} = \dfrac{1}{(a+1)^2}(1-a) \quad \Leftrightarrow$ $2(a+1)^2 = a(a+1) + (1-a) = a^2 + 1 \quad \Leftrightarrow \quad a^2 + 4a + 1 = 0$. The quadratic formula gives the roots of this equation as $-2 \pm \sqrt{3}$, so there are two such tangent lines, which touch the curve at $\left(-2+\sqrt{3}, \frac{1-\sqrt{3}}{2}\right)$ and $\left(-2-\sqrt{3}, \frac{1+\sqrt{3}}{2}\right)$.

51. $y = 6x^3 + 5x - 3 \quad \Rightarrow \quad m = y' = 18x^2 + 5$, but $x^2 \geq 0$ for all x so $m \geq 5$ for all x.

53. $y = f(x) = 1 - x^2 \quad \Rightarrow \quad f'(x) = -2x$, so the tangent line at $(2, -3)$ has slope $f'(2) = -4$. The normal line has slope $-1/(-4) = \frac{1}{4}$ and equation $y + 3 = \frac{1}{4}(x - 2)$ or $x - 4y = 14$.

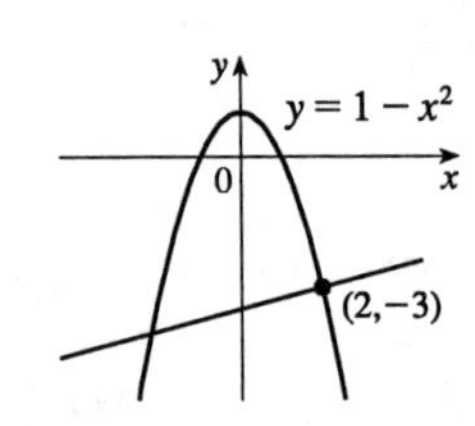

55. $y = f(x) = \sqrt[3]{x} = x^{1/3} \quad \Rightarrow \quad f'(x) = \frac{1}{3}x^{-2/3}$,
so the tangent line at $(-8, -2)$ has slope $f'(-8) = \frac{1}{12}$.
The normal line has slope $-1/\left(\frac{1}{12}\right) = -12$ and
equation $y + 2 = -12(x + 8) \quad \Leftrightarrow \quad 12x + y + 98 = 0$.

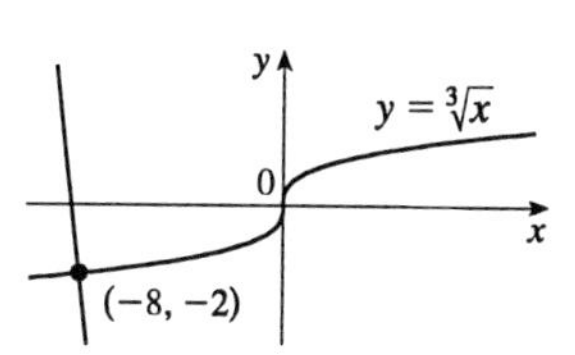

57. If the normal line has slope 16, then the tangent has slope $-\frac{1}{16}$, so $y' = 4x^3 = -\frac{1}{16} \quad \Rightarrow \quad x^3 = -\frac{1}{64} \quad \Rightarrow$
$x = -\frac{1}{4}$. The point is $\left(-\frac{1}{4}, \frac{1}{256}\right)$.

59. **(a)** $(fg)'(5) = f'(5)g(5) + f(5)g'(5) = 6(-3) + 1(2) = -16$

(b) $$\left(\frac{f}{g}\right)'(5) = \frac{f'(5)g(5) - f(5)g'(5)}{(g(5))^2} = \frac{6(-3) - 1(2)}{(-3)^2} = -\frac{20}{9}$$

(c) $$\left(\frac{g}{f}\right)'(5) = \frac{g'(5)f(5) - g(5)f'(5)}{(f(5))^2} = \frac{2(1) - (-3)(6)}{1^2} = 20$$

61. **(a)** $u(x) = f(x)g(x)$, so $u'(1) = f(1)g'(1) + g(1)f'(1) = 2 \cdot (-1) + 1 \cdot 2 = 0$

(b) $v(x) = f(x)/g(x)$, so $v'(5) = \dfrac{g(5)f'(5) - f(5)g'(5)}{[g(5)]^2} = \dfrac{2\left(-\frac{1}{3}\right) - 3 \cdot \frac{2}{3}}{2^2} = -\dfrac{2}{3}$

63. **(a)** $(fgh)' = [(fg)h]' = (fg)'h + (fg)h' = (f'g + fg')h + (fg)h' = f'gh + fg'h + fgh'$

(b) Putting $f = g = h$ in part (a), we have
$$\frac{d}{dx}[f(x)]^3 = (fff)' = f'ff + ff'f + fff' = 3fff' = 3[f(x)]^2 f'(x).$$

65. $y = \sqrt{x}(x^4 + x + 1)(2x - 3)$. Using Exercise 63(a), we have
$$\begin{aligned} y' &= \frac{1}{2\sqrt{x}}\left(x^4 + x + 1\right)(2x - 3) + \sqrt{x}\left(4x^3 + 1\right)(2x - 3) + \sqrt{x}\left(x^4 + x + 1\right)(2) \\ &= \left(x^4 + x + 1\right)\frac{2x - 3}{2\sqrt{x}} + \sqrt{x}\left[\left(4x^3 + 1\right)(2x - 3) + 2\left(x^4 + x + 1\right)\right]. \end{aligned}$$

67. $f(x) = 2 - x$ if $x \le 1$ and $f(x) = x^2 - 2x + 2$ if $x > 1$. Now we compute the right- and left-hand derivatives defined in Exercise 3.1.57:
$$f'_-(1) = \lim_{h \to 0^-} \frac{f(1 + h) - f(1)}{h} = \lim_{h \to 0^-} \frac{2 - (1 + h) - 1}{h} = \lim_{h \to 0^-} \frac{-h}{h} = \lim_{h \to 0^-} -1 = -1 \text{ and}$$
$$f'_+(1) = \lim_{h \to 0^+} \frac{f(1 + h) - f(1)}{h} = \lim_{h \to 0^+} \frac{(1 + h)^2 - 2(1 + h) + 2 - 1}{h} = \lim_{h \to 0^+} \frac{h^2}{h} = \lim_{h \to 0^+} h = 0.$$
Thus $f'(1)$ does not exist since $f'_-(1) \neq f'_+(1)$, so f is not differentiable at 1. But $f'(x) = -1$ for $x < 1$ and $f'(x) = 2x - 2$ if $x > 1$.

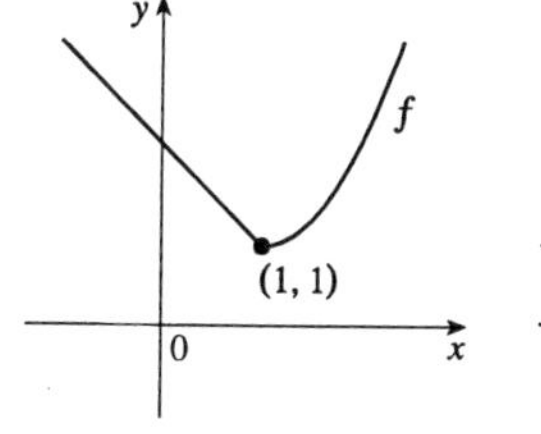

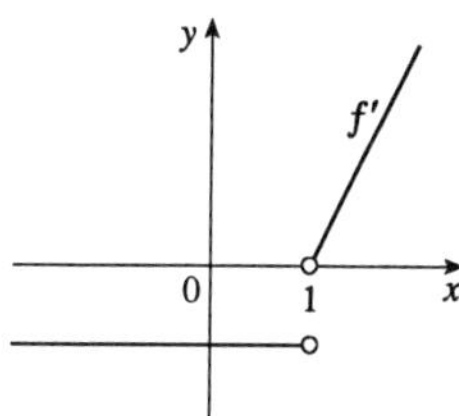

69. **(a)** Note that $x^2 - 9 < 0$ for $x^2 < 9 \quad \Leftrightarrow \quad |x| < 3 \quad \Leftrightarrow \quad -3 < x < 3$. So

$$f(x) = \begin{cases} x^2 - 9 & \text{if } x \le -3 \\ -x^2 + 9 & \text{if } -3 < x < 3 \\ x^2 - 9 & \text{if } x \ge 3 \end{cases} \quad \Rightarrow \quad f'(x) = \begin{cases} 2x & \text{if } x < -3 \\ -2x & \text{if } -3 < x < 3 \\ 2x & \text{if } x > 3. \end{cases}$$

To show that $f'(3)$ does not exist we investigate $\lim\limits_{h \to 0} \dfrac{f(3+h) - f(3)}{h}$ by computing the left- and right-hand derivatives defined in Exercise 3.1.57.

$$f'_-(3) = \lim_{h \to 0^-} \frac{f(3+h) - f(3)}{h} = \lim_{h \to 0^-} \frac{\left(-(3+h)^2 + 9\right) - 0}{h} = \lim_{h \to 0^-} (-6 + h) = -6 \text{ and}$$

$$f'_+(3) = \lim_{h \to 0^+} \frac{f(3+h) - f(3)}{h} = \lim_{h \to 0^+} \frac{\left[(3+h)^2 + 9\right] - 0}{h} = \lim_{h \to 0^+} \frac{6h + h^2}{h} = \lim_{h \to 0^+} (6 + h) = 6.$$

Since the left and right limits are different, $\lim\limits_{h \to 0} \dfrac{f(3+h) - f(3)}{h}$ does not exist, that is, $f'(3)$ does not exist. Similarly, $f'(-3)$ does not exist. Therefore f is not differentiable at 3 or at -3.

(b)

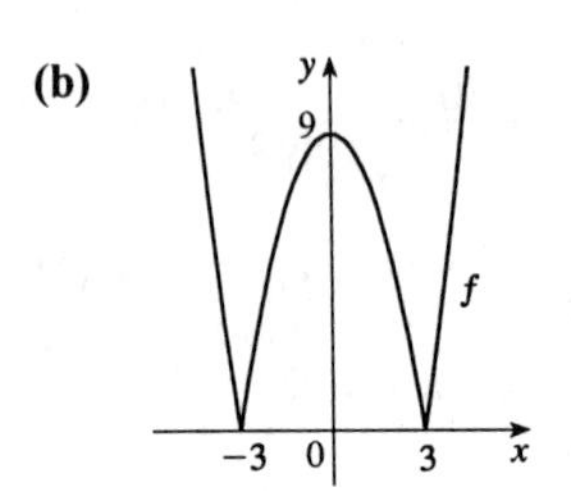

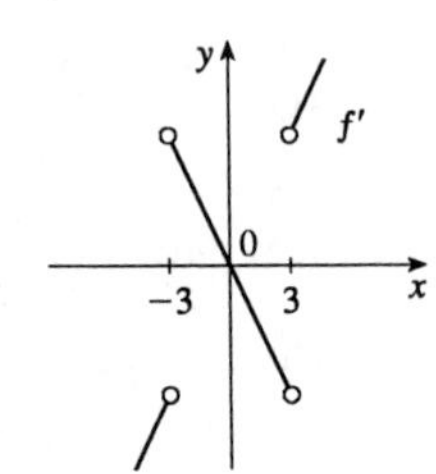

71. $y = f(x) = ax^2 \quad \Rightarrow \quad f'(x) = 2ax$. So the slope of the tangent to the parabola at $x = 2$ is $m = 2a(2) = 4a$. The slope of the given line is seen to be -2, so we must have $4a = -2 \quad \Leftrightarrow \quad a = -\frac{1}{2}$. So the point in question has y-coordinate $-\frac{1}{2} \cdot 2^2 = -2$. Now we simply require that the given line, whose equation is $2x + y = b$, pass through the point $(2, -2)$: $2(2) + (-2) = b \quad \Leftrightarrow \quad b = 2$. So we must have $a = -\frac{1}{2}$ and $b = 2$.

73. $F = f/g \quad \Rightarrow \quad f = Fg \quad \Rightarrow \quad f' = F'g + Fg' \quad \Rightarrow \quad F' = \dfrac{f' - Fg'}{g} = \dfrac{f' - (f/g)g'}{g} = \dfrac{f'g - fg'}{g^2}$

75. *Solution 1:* Let $f(x) = x^{1000}$. Then, by the definition of the derivative,

$f'(1) = \lim\limits_{x \to 1} \dfrac{f(x) - f(1)}{x - 1} = \lim\limits_{x \to 1} \dfrac{x^{1000} - 1}{x - 1}$. But this is just the limit we want to find, and we know (from the Power Rule) that $f'(x) = 1000x^{999}$, so $f'(1) = 1000(1)^{999} = 1000$. So $\lim\limits_{x \to 1} \dfrac{x^{1000} - 1}{x - 1} = 1000$.

Solution 2: Note that $(x^{1000} - 1) = (x - 1)(x^{999} + x^{998} + x^{997} + \cdots + x^2 + x + 1)$. So

$$\lim_{x \to 1} \frac{x^{1000} - 1}{x - 1} = \lim_{x \to 1} \frac{(x-1)(x^{999} + x^{998} + x^{997} + \cdots + x^2 + x + 1)}{x - 1}$$
$$= \lim_{x \to 1} \left(x^{999} + x^{998} + x^{997} + \cdots + x^2 + x + 1\right) = 1 + 1 + 1 + \cdots + 1 + 1 + 1 = 1000 \text{ as above.}$$

EXERCISES 3.3

1. **(a)** $v(t) = f'(t) = 2t - 6$ **(b)** $v(2) = 2(2) - 6 = -2\text{ ft/s}$

(c) It is at rest when $v(t) = 2t - 6 = 0 \quad \Leftrightarrow \quad t = 3.$

(d) It moves in the positive direction when $2t - 6 > 0 \quad \Leftrightarrow \quad t > 3.$

(e) Distance in positive direction $= |f(4) - f(3)| = |1 - 0| = 1\text{ ft}$

Distance in negative direction $= |f(3) - f(0)| = |0 - 9| = 9\text{ ft}$

Total distance traveled $= 1 + 9 = 10\text{ ft}$

(f)

$t = 3$ $t = 0$

0 9 s

3. **(a)** $v(t) = f'(t) = 6t^2 - 18t + 12$ **(b)** $v(2) = 6(2)^2 - 18(2) + 12 = 0\text{ ft/s}$

(c) It is at rest when $v(t) = 6t^2 - 18t + 12 = 6(t-1)(t-2) = 0 \quad \Leftrightarrow \quad t = 1$ or 2.

(d) It moves in the positive direction when $6(t-1)(t-2) > 0 \quad \Leftrightarrow \quad 0 \le t < 1$ or $t > 2$.

(e) Distance in positive direction $= |f(4) - f(2)| + |f(1) - f(0)| = |33 - 5| + |6 - 1| = 33\text{ ft}$

Distance in negative direction $= |f(2) - f(1)| = |5 - 6| = 1\text{ ft}$

Total distance traveled $= 33 + 1 = 34\text{ ft}$

(f)

$t = 2$ $t = 0$ $t = 1$

0 1 5 6 s

5. **(a)** $v(t) = s'(t) = \dfrac{(t^2+1)(1) - t(2t)}{(t^2+1)^2} = \dfrac{1 - t^2}{(t^2+1)^2}$ **(b)** $v(2) = \dfrac{1 - (2)^2}{(2^2+1)^2} = -\dfrac{3}{25}\text{ ft/s}$

(c) It is at rest when $v = 0 \quad \Leftrightarrow \quad 1 - t^2 = 0 \quad \Leftrightarrow \quad t = 1.$

(d) It moves in the positive direction when $v > 0 \quad \Leftrightarrow \quad 1 - t^2 > 0 \quad \Leftrightarrow \quad t^2 < 1 \quad \Leftrightarrow \quad 0 \le t < 1.$

(e) Distance in positive direction $= |s(1) - s(0)| = \left|\frac{1}{2} - 0\right| = \frac{1}{2}\text{ ft}$

Distance in negative direction $= |s(4) - s(1)| = \left|\frac{4}{17} - \frac{1}{2}\right| = \frac{9}{34}\text{ ft}$

Total distance traveled $= \frac{1}{2} + \frac{9}{34} = \frac{13}{17}\text{ ft}$

(f)

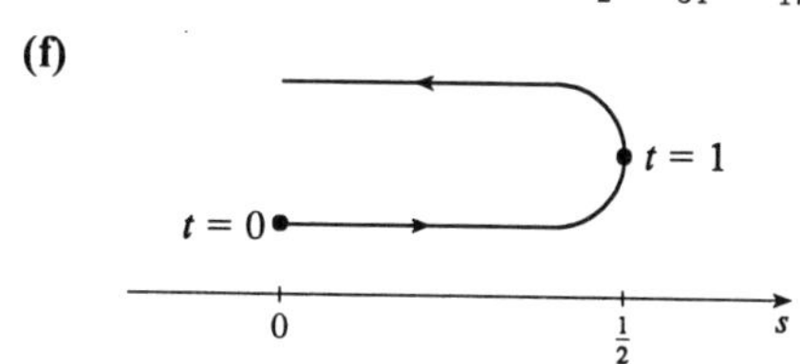

7. $s(t) = t^3 - 4.5t^2 - 7t \quad \Rightarrow \quad v(t) = s'(t) = 3t^2 - 9t - 7 = 5 \quad \Leftrightarrow \quad 3t^2 - 9t - 12 = 0 \quad \Leftrightarrow$
$3(t-4)(t+1) = 0 \quad \Leftrightarrow \quad t = 4$ or -1. Since $t \geq 0$, the particle reaches a velocity of 5 m/s at $t = 4\text{ s}$.

9. **(a)** $V(x) = x^3$, so the average rate of change is:

(i) $\dfrac{V(6) - V(5)}{6 - 5} = 6^3 - 5^3 = 216 - 125 = 91$ **(ii)** $\dfrac{V(5.1) - V(5)}{5.1 - 5} = \dfrac{(5.1)^3 - 5^3}{0.1} = 76.51$

(iii) $\dfrac{V(5.01) - V(5)}{5.01 - 5} = \dfrac{(5.01)^3 - 5^3}{0.01} = 75.1501$

(b) $V'(x) = 3x^2$, $V'(5) = 75$

(c) The surface area is $S(x) = 6x^2$, so $V'(x) = 3x^2 = \frac{1}{2}(6x^2) = \frac{1}{2}S(x)$.

11. After t seconds the radius is $r = 60t$, so the area is $A(t) = \pi(60t)^2 = 3600\pi t^2 \quad \Rightarrow \quad A'(t) = 7200\pi t \quad \Rightarrow$

(a) $A'(1) = 7200\pi\text{ cm}^2/\text{s}$ **(b)** $A'(3) = 21{,}600\pi\text{ cm}^2/\text{s}$ **(c)** $A'(5) = 36{,}000\pi\text{ cm}^2/\text{s}$

13. $S(r) = 4\pi r^2 \quad \Rightarrow \quad S'(r) = 8\pi r$

(a) $S'(1) = 8\pi\text{ ft}^2/\text{ft}$ **(b)** $S'(2) = 16\pi\text{ ft}^2/\text{ft}$ **(c)** $S'(3) = 24\pi\text{ ft}^2/\text{ft}$

15. $f(x) = 3x^2$, so the linear density at x is $\rho(x) = f'(x) = 6x$.

(a) $\rho(1) = 6\text{ kg/m}$

(b) $\rho(2) = 12\text{ kg/m}$

(c) $\rho(3) = 18\text{ kg/m}$

17. $Q(t) = t^3 - 2t^2 + 6t + 2$, so the current is $Q'(t) = 3t^2 - 4t + 6$.

(a) $Q'(0.5) = 3(0.5)^2 - 4(0.5) + 6 = 4.75\text{ A}$

(b) $Q'(1) = 3(1)^2 - 4(1) + 6 = 5\text{ A}$

19. **(a)** $PV = C \quad \Rightarrow \quad V = \dfrac{C}{P} \quad \Rightarrow \quad \dfrac{dV}{dP} = -\dfrac{C}{P^2}$

(b) $\beta = -\dfrac{1}{V}\dfrac{dV}{dP} = -\dfrac{1}{V}\left(-\dfrac{C}{P^2}\right) = \dfrac{C}{(PV)P} = \dfrac{C}{CP} = \dfrac{1}{P}$

21. **(a)** rate of reaction $= \dfrac{d[C]}{dt} = \dfrac{a^2k(akt+1) - (a^2kt)(ak)}{(akt+1)^2} = \dfrac{a^2k(akt + 1 - akt)}{(akt+1)^2} = \dfrac{a^2k}{(akt+1)^2}$

(b) $a - x = a - \dfrac{a^2kt}{akt+1} = \dfrac{a^2kt + a - a^2kt}{akt+1} = \dfrac{a}{akt+1}$.

So $k(a-x)^2 = k\left(\dfrac{a}{akt+1}\right)^2 = \dfrac{a^2k}{(akt+1)^2} = \dfrac{dx}{dt}$.

23. $m(t) = 5 - 0.02t^2 \quad \Rightarrow \quad m'(t) = -0.04t \quad \Rightarrow \quad m'(1) = -0.04$

25. $v(r) = \dfrac{P}{4\eta\ell}\left(R^2 - r^2\right) \quad \Rightarrow \quad v'(r) = \dfrac{P}{4\eta\ell}(-2r) = -\dfrac{Pr}{2\eta\ell}$.

When $\ell = 3$, $P = 3000$ and $\eta = 0.027$, we have $v'(0.005) = -\dfrac{3000(0.005)}{2(0.027)(3)} \approx -92.6\,\dfrac{\text{cm/s}}{\text{cm}}$.

27. $C(x) = 420 + 1.5x + 0.002x^2 \quad \Rightarrow \quad C'(x) = 1.5 + 0.004x \quad \Rightarrow$

$C'(100) = 1.5 + (0.004)(100) = \$1.90/\text{item}$

$C(101) - C(100) = (420 + 151.5 + 20.402) - (420 + 150 + 20) = \$1.902/\text{item}$

29. $C(x) = 2000 + 3x + 0.01x^2 + 0.0002x^3 \quad \Rightarrow \quad C'(x) = 3 + 0.02x + 0.0006x^2 \quad \Rightarrow$

$C'(100) = 3 + 0.02(100) + 0.0006(10{,}000) = 3 + 2 + 6 = \$11/\text{item}$

$$C(101) - C(100) = (2000 + 303 + 102.01 + 206.0602) - (2000 + 300 + 100 + 200)$$
$$= 11.0702 \approx \$11.07/\text{item}$$

EXERCISES 3.4

1. $\lim\limits_{x\to 0}(x^2 + \cos x) = \lim\limits_{x\to 0} x^2 + \lim\limits_{x\to 0} \cos x = 0^2 + \cos 0 = 0 + 1 = 1$

3. $\lim\limits_{x\to\pi/3}(\sin x - \cos x) = \sin\frac{\pi}{3} - \cos\frac{\pi}{3} = \frac{\sqrt{3}}{2} - \frac{1}{2}$

5. $\lim\limits_{x\to\pi/4} \dfrac{\sin x}{3x} = \dfrac{\sin(\pi/4)}{3\pi/4} = \dfrac{1/\sqrt{2}}{3\pi/4} = \dfrac{2\sqrt{2}}{3\pi}$

7. $\lim\limits_{t\to 0} \dfrac{\sin 5t}{t} = \lim\limits_{t\to 0} \dfrac{5\sin 5t}{5t} = 5\lim\limits_{t\to 0} \dfrac{\sin 5t}{5t} = 5\cdot 1 = 5$

9. $\lim\limits_{\theta\to 0} \dfrac{\sin(\cos\theta)}{\sec\theta} = \dfrac{\sin\left(\lim\limits_{\theta\to 0}\cos\theta\right)}{\lim\limits_{\theta\to 0}\sec\theta} = \dfrac{\sin 1}{1} = \sin 1$

11. $\lim\limits_{x\to\pi/4} \dfrac{\tan x}{4x} = \dfrac{\tan(\pi/4)}{4(\pi/4)} = \dfrac{1}{\pi}$

13. $\lim\limits_{\theta\to 0} \dfrac{\sin^2\theta}{\theta} = \lim\limits_{\theta\to 0}\left(\dfrac{\sin\theta}{\theta}\right)\sin\theta = \lim\limits_{\theta\to 0}\dfrac{\sin\theta}{\theta}\lim\limits_{\theta\to 0}\sin\theta = 1\cdot 0 = 0$

15. $\lim\limits_{x\to 0} \dfrac{\tan 3x}{3\tan 2x} = \lim\limits_{x\to 0} \dfrac{\dfrac{\tan 3x}{3x}}{2\dfrac{\tan 2x}{2x}} = \dfrac{1}{2}\,\dfrac{\lim\limits_{x\to 0}\dfrac{\sin 3x}{3x}\cdot\dfrac{1}{\cos 3x}}{\lim\limits_{x\to 0}\dfrac{\sin 2x}{2x}\cdot\lim\limits_{x\to 0}\dfrac{1}{\cos 2x}} = \dfrac{1}{2}\,\dfrac{1\cdot 1}{1\cdot 1} = \dfrac{1}{2}$

17. $\dfrac{d}{dx}(\csc x) = \dfrac{d}{dx}\left(\dfrac{1}{\sin x}\right) = \dfrac{(\sin x)(0) - 1(\cos x)}{\sin^2 x} = \dfrac{-\cos x}{\sin^2 x} = -\dfrac{1}{\sin x}\cdot\dfrac{\cos x}{\sin x} = -\csc x\cot x$

19. $\dfrac{d}{dx}(\cot x) = \dfrac{d}{dx}\left(\dfrac{\cos x}{\sin x}\right) = \dfrac{(\sin x)(-\sin x) - (\cos x)(\cos x)}{\sin^2 x} = -\dfrac{\sin^2 x + \cos^2 x}{\sin^2 x} = -\dfrac{1}{\sin^2 x} = -\csc^2 x$

21. $y = \sin x + \cos x \quad \Rightarrow \quad dy/dx = \cos x - \sin x$

23. $y = \csc x\cot x \quad \Rightarrow \quad dy/dx = (-\csc x\cot x)\cot x + \csc x\,(-\csc^2 x) = -\csc x\,(\cot^2 x + \csc^2 x)$

25. $y = \dfrac{\tan x}{x} \quad \Rightarrow \quad \dfrac{dy}{dx} = \dfrac{x\sec^2 x - \tan x}{x^2}$

27. $y = \dfrac{x}{\sin x + \cos x} \quad\Rightarrow$

$$\frac{dy}{dx} = \frac{(\sin x + \cos x) - x(\cos x - \sin x)}{(\sin x + \cos x)^2} = \frac{(1+x)\sin x + (1-x)\cos x}{\sin^2 x + \cos^2 x + 2\sin x \cos x} = \frac{(1+x)\sin x + (1-x)\cos x}{1 + \sin 2x}$$

29. $y = x^{-3} \sin x \tan x \quad\Rightarrow$

$$\frac{dy}{dx} = -3x^{-4} \sin x \tan x + x^{-3} \cos x \tan x + x^{-3} \sin x \sec^2 x = x^{-4} \sin x \left(-3 \tan x + x + x \sec^2 x\right)$$

31. $y = \dfrac{x^2 \tan x}{\sec x} \quad\Rightarrow$

$$\frac{dy}{dx} = \frac{\sec x\,(2x \tan x + x^2 \sec^2 x) - x^2 \tan x \sec x \tan x}{\sec^2 x} = \frac{2x \tan x + x^2(\sec^2 x - \tan^2 x)}{\sec x} = \frac{2x \tan x + x^2}{\sec x}.$$

Another Method: Write $y = x^2 \sin x$. Then $y' = 2x \sin x + x^2 \cos x$.

33. $y = \tan x \quad\Rightarrow\quad y' = \sec^2 x \quad\Rightarrow$ The slope of the tangent line at $\left(\frac{\pi}{4}, 1\right)$ is $\sec^2 \frac{\pi}{4} = 2$ and the equation is $y - 1 = 2\left(x - \frac{\pi}{4}\right)$ or $4x - 2y = \pi - 2$.

35. **(a)** $y = x \cos x \quad\Rightarrow$
$y' = x(-\sin x) + \cos x\,(1) = \cos x - x \sin x$
So the slope of the tangent at the point $(\pi, -\pi)$ is $\cos \pi - \pi \sin \pi = -1 - \pi(0) = -1$, and its equation is $y + \pi = -(x - \pi)$
$\Leftrightarrow\quad y = -x$.

(b)

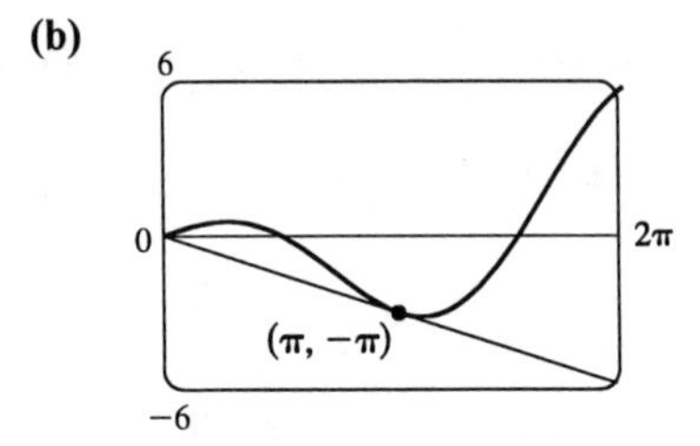

37. $y = x + 2\sin x$ has a horizontal tangent when $y' = 1 + 2\cos x = 0 \quad\Leftrightarrow\quad \cos x = -\frac{1}{2} \quad\Leftrightarrow$
$x = (2n+1)\pi \pm \frac{\pi}{3}$, n an integer.

39.

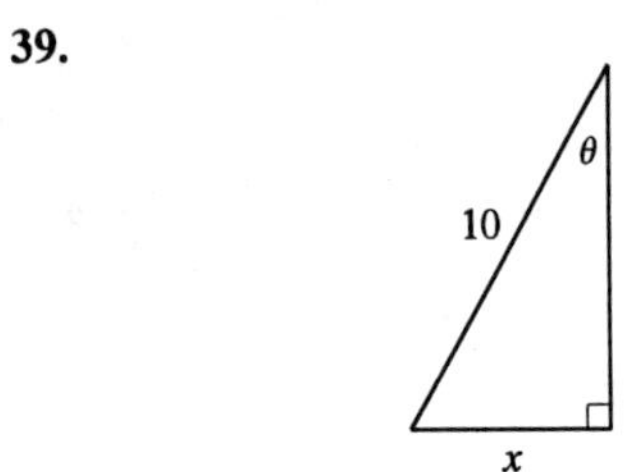

From the diagram we can see that $\sin\theta = 10/x$ $\Leftrightarrow\quad x = 10\sin\theta$. But we want to find the rate of change of x with respect to θ, that is, $dx/d\theta$. Taking the derivative of the above expression, $dx/d\theta = 10(\cos\theta)$. So when $\theta = \frac{\pi}{3}$, $dx/d\theta = 10\cos\frac{\pi}{3} = 10\left(\frac{1}{2}\right) = 5$ ft/rad.

41. $$\lim_{\theta\to 0} \frac{\cos\theta - 1}{\theta} = \lim_{\theta\to 0} \frac{1 - 2\sin^2(\theta/2) - 1}{\theta} = \lim_{\theta\to 0} \frac{-\sin^2(\theta/2)}{\theta/2} = -\lim_{\theta\to 0} \frac{\sin(\theta/2)}{\theta/2} \lim_{\theta\to 0} \sin(\theta/2) = -1 \cdot 0 = 0$$

43. $$\lim_{x\to 0} \frac{\cot 2x}{\csc x} = \lim_{x\to 0} \frac{\cos 2x \sin x}{\sin 2x} = \lim_{x\to 0} \cos 2x \left[\frac{(\sin x)/x}{(\sin 2x)/x}\right] = \lim_{x\to 0} \cos 2x \left[\frac{\lim\limits_{x\to 0}[(\sin x)/x]}{2\lim\limits_{x\to 0}[(\sin 2x)/2x]}\right] = 1 \cdot \frac{1}{2\cdot 1} = \frac{1}{2}$$

45. $$\lim_{x\to\pi} \frac{\tan x}{\sin 2x} = \lim_{x\to\pi} \frac{\sin x}{\cos x\,(2\sin x \cos x)} = \lim_{x\to\pi} \frac{1}{2\cos^2 x} = \frac{1}{2(-1)^2} = \frac{1}{2}$$

47. Divide numerator and denominator by θ. ($\sin\theta$ also works.)

$$\lim_{\theta\to 0}\frac{\sin\theta}{\theta+\tan\theta}=\lim_{\theta\to 0}\frac{\frac{\sin\theta}{\theta}}{1+\frac{\sin\theta}{\theta}\cdot\frac{1}{\cos\theta}}=\frac{\lim\limits_{\theta\to 0}\frac{\sin\theta}{\theta}}{1+\lim\limits_{\theta\to 0}\frac{\sin\theta}{\theta}\lim\limits_{\theta\to 0}\frac{1}{\cos\theta}}=\frac{1}{1+1\cdot 1}=\frac{1}{2}$$

49. $$\lim_{x\to 0}\frac{\cos x\sin x-\tan x}{x^2\sin x}=\lim_{x\to 0}\frac{\cos x\sin x-\sin x/\cos x}{x^2\sin x}=\lim_{x\to 0}\frac{\cos^2 x\sin x-\sin x}{x^2\sin x\cos x}=\lim_{x\to 0}\frac{\cos^2 x-1}{x^2\cos x}$$
$$=\lim_{x\to 0}\left(\frac{-\sin^2 x}{x^2}\right)\frac{1}{\cos x}=-\left[\lim_{x\to 0}\frac{\sin x}{x}\right]^2\left[\lim_{x\to 0}\frac{1}{\cos x}\right]=-1$$

51. $\lim\limits_{x\to 0}\dfrac{\sin(\sin x)}{\sin x}=\lim\limits_{\sin x\to 0}\dfrac{\sin(\sin x)}{\sin x}$ since as $x\to 0$, $\sin x\to 0$. So we make the substitution $y=\sin x$, and see that $\lim\limits_{x\to 0}\dfrac{\sin(\sin x)}{\sin x}=\lim\limits_{y\to 0}\dfrac{\sin y}{y}=1$.

53. If $t=1/x$ then $\lim\limits_{x\to\infty}\cos(1/x)=\lim\limits_{t\to 0^+}\cos t=\cos 0=1$.

55. **(a)** $\dfrac{d}{dx}\tan x=\dfrac{d}{dx}\dfrac{\sin x}{\cos x}\quad\Rightarrow\quad\sec^2 x=\dfrac{\cos x\cos x-\sin x\,(-\sin x)}{\cos^2 x}=\dfrac{\cos^2 x+\sin^2 x}{\cos^2 x}$. So $\sec^2 x=\dfrac{1}{\cos^2 x}$.

(b) $\dfrac{d}{dx}\sec x=\dfrac{d}{dx}\dfrac{1}{\cos x}\quad\Rightarrow\quad\sec x\tan x=\dfrac{(\cos x)(0)-1(-\sin x)}{\cos^2 x}$. So $\sec x\tan x=\dfrac{\sin x}{\cos^2 x}$.

(c) $\dfrac{d}{dx}(\sin x+\cos x)=\dfrac{d}{dx}\dfrac{1+\cot x}{\csc x}\quad\Rightarrow$

$$\cos x-\sin x=\frac{\csc x\,(-\csc^2 x)-(1+\cot x)(-\csc x\cot x)}{\csc^2 x}=\frac{-\csc^2 x+\cot^2 x+\cot x}{\csc x}$$

So $\cos x-\sin x=\dfrac{\cot x-1}{\csc x}$.

57. By the definition of radian measure, $s=r\theta$, where r is the radius of the circle. By drawing the bisector of the angle θ, we can see that $\sin\dfrac{\theta}{2}=\dfrac{d/2}{r}\quad\Rightarrow\quad d=2r\sin\dfrac{\theta}{2}$. So

$$\lim_{\theta\to 0^+}\frac{s}{d}=\lim_{\theta\to 0^+}\frac{r\theta}{2r\sin(\theta/2)}=\lim_{\theta\to 0^+}\frac{2\cdot(\theta/2)}{2\sin(\theta/2)}=\lim_{\theta\to 0}\frac{\theta/2}{\sin(\theta/2)}=1.$$

[This is just the reciprocal of the limit $\lim\limits_{x\to 0}\dfrac{\sin x}{x}=1$ combined with the fact that as $\theta\to 0$, $\dfrac{\theta}{2}\to 0$ also.]

EXERCISES 3.5

1. $y = u^2$, $u = x^2 + 2x + 3$

(a) $\dfrac{dy}{dx} = \dfrac{dy}{du}\dfrac{du}{dx} = 2u(2x+2) = 4u(x+1)$. When $x = 1$, $u = 1^2 + 2(1) + 3 = 6$, so $\left.\dfrac{dy}{dx}\right|_{x=1} = 4(6)(1+1) = 48$.

(b) $y = u^2 = (x^2+2x+3)^2 = x^4 + 4x^2 + 9 + 4x^3 + 6x^2 + 12x = x^4 + 4x^3 + 10x^2 + 12x + 9$, so $\dfrac{dy}{dx} = 4x^3 + 12x^2 + 20x + 12$ and $\left.\dfrac{dy}{dx}\right|_{x=1} = 4(1)^3 + 12(1)^2 + 20(1) + 12 = 48$.

3. $y = u^3$, $u = x + 1/x$

(a) $\dfrac{dy}{dx} = \dfrac{dy}{du}\dfrac{du}{dx} = 3u^2\left(1 - \dfrac{1}{x^2}\right)$. When $x = 1$, $u = 1 + \dfrac{1}{1} = 2$, so $\left.\dfrac{dy}{dx}\right|_{x=1} = 3(2)^2\left(1 - \dfrac{1}{1^2}\right) = 0$.

(b) $y = u^3 = \left(x + \dfrac{1}{x}\right)^3 = x^3 + 3x^2\left(\dfrac{1}{x}\right) + 3x\left(\dfrac{1}{x}\right)^2 + \left(\dfrac{1}{x}\right)^3 = x^3 + 3x + 3x^{-1} + x^{-3}$, so $\dfrac{dy}{dx} = 3x^2 + 3 - 3x^{-2} - 3x^{-4}$ and $\left.\dfrac{dy}{dx}\right|_{x=1} = 3(1)^2 + 3 - 3(1)^{-2} - 3(1)^{-4} = 0$.

5. $F(x) = (x^2+4x+6)^5 \quad\Rightarrow$

$F'(x) = 5(x^2+4x+6)^4 \dfrac{d}{dx}(x^2+4x+6) = 5(x^2+4x+6)^4(2x+4) = 10(x^2+4x+6)^4(x+2)$

7. $G(x) = (3x-2)^{10}(5x^2-x+1)^{12} \quad\Rightarrow$

$$\begin{aligned} G'(x) &= 10(3x-2)^9(3)(5x^2-x+1)^{12} + (3x-2)^{10}(12)(5x^2-x+1)^{11}(10x-1) \\ &= 30(3x-2)^9(5x^2-x+1)^{12} + 12(3x-2)^{10}(5x^2-x+1)^{11}(10x-1) \end{aligned}$$

[This can be simplified to $6(3x-2)^9(5x^2-x+1)^{11}(85x^2-51x+9)$.]

9. $f(t) = (2t^2-6t+1)^{-8} \quad\Rightarrow\quad f'(t) = -8(2t^2-6t+1)^{-9}(4t-6) = -16(2t^2-6t+1)^{-9}(2t-3)$

11. $g(x) = \sqrt{x^2-7x} = (x^2-7x)^{1/2} \quad\Rightarrow\quad g'(x) = \frac{1}{2}(x^2-7x)^{-1/2}(2x-7) = \dfrac{2x-7}{2\sqrt{x^2-7x}}$

13. $h(t) = (t - 1/t)^{3/2} \quad\Rightarrow\quad h'(t) = \frac{3}{2}(t-1/t)^{1/2}(1+1/t^2)$

15. $F(y) = \left(\dfrac{y-6}{y+7}\right)^3 \quad\Rightarrow\quad F'(y) = 3\left(\dfrac{y-6}{y+7}\right)^2 \dfrac{(y+7)(1)-(y-6)(1)}{(y+7)^2} = 3\left(\dfrac{y-6}{y+7}\right)^2 \dfrac{13}{(y+7)^2} = \dfrac{39(y-6)^2}{(y+7)^4}$

17. $f(z) = (2z-1)^{-1/5} \quad\Rightarrow\quad f'(z) = -\frac{1}{5}(2z-1)^{-6/5}(2) = -\frac{2}{5}(2z-1)^{-6/5}$

19. $y = (2x-5)^4(8x^2-5)^{-3} \quad\Rightarrow$

$$\begin{aligned} y' &= 4(2x-5)^3(2)(8x^2-5)^{-3} + (2x-5)^4(-3)(8x^2-5)^{-4}(16x) \\ &= 8(2x-5)^3(8x^2-5)^{-3} - 48x(2x-5)^4(8x^2-5)^{-4} \end{aligned}$$

[This simplifies to $8(2x-5)^3(8x^2-5)^{-4}(-4x^2+30x-5)$.]

21. $y = \tan 3x \quad \Rightarrow \quad y' = \sec^2 3x \dfrac{d}{dx}(3x) = 3\sec^2 3x$

23. $y = \cos(x^3) \quad \Rightarrow \quad y' = -\sin(x^3)(3x^2) = -3x^2 \sin(x^3)$

25. $y = (1+\cos^2 x)^6 \quad \Rightarrow \quad y' = 6(1+\cos^2 x)^5 2\cos x\,(-\sin x) = -12\cos x \sin x\,(1+\cos^2 x)^5$

27. $y = \cos(\tan x) \quad \Rightarrow \quad y' = -\sin(\tan x)\sec^2 x$

29. $y = \sec^2 2x - \tan^2 2x \quad \Rightarrow \quad y' = 2\sec 2x\,(\sec 2x \tan 2x)(2) - 2\tan 2x \sec^2(2x)(2) = 0$

Easier method: $y = \sec^2 2x - \tan^2 2x = 1 \quad \Rightarrow \quad y' = 0$

31. $y = \csc \dfrac{x}{3} \quad \Rightarrow \quad y' = -\dfrac{1}{3}\csc\dfrac{x}{3}\cot\dfrac{x}{3}$

33. $y = \sin^3 x + \cos^3 x \quad \Rightarrow \quad y' = 3\sin^2 x \cos x + 3\cos^2 x(-\sin x) = 3\sin x \cos x\,(\sin x - \cos x)$

35. $y = \sin\dfrac{1}{x} \quad \Rightarrow \quad y' = \cos\dfrac{1}{x}\left(-\dfrac{1}{x^2}\right) = -\dfrac{1}{x^2}\cos\dfrac{1}{x}$

37. $y = \dfrac{1+\sin 2x}{1-\sin 2x} \quad \Rightarrow \quad y' = \dfrac{(1-\sin 2x)(2\cos 2x) - (1+\sin 2x)(-2\cos 2x)}{(1-\sin 2x)^2} = \dfrac{4\cos 2x}{(1-\sin 2x)^2}$

39. $y = \tan^2(x^3) \quad \Rightarrow \quad y' = 2\tan(x^3)\sec^2(x^3)(3x^2) = 6x^2\tan(x^3)\sec^2(x^3)$

41. $y = \cos^2(\cos x) + \sin^2(\cos x) = 1 \quad \Rightarrow \quad y' = 0.$

43. $y = \sqrt{x+\sqrt{x}} \quad \Rightarrow \quad y' = \frac{1}{2}\left(x+\sqrt{x}\right)^{-1/2}\left(1+\frac{1}{2}x^{-1/2}\right) = \dfrac{1}{2\sqrt{x+\sqrt{x}}}\left(1+\dfrac{1}{2\sqrt{x}}\right)$

45. $f(x) = \left[x^3 + (2x-1)^3\right]^3 \quad \Rightarrow$

$f'(x) = 3\left[x^3+(2x-1)^3\right]^2\left[3x^2+3(2x-1)^2(2)\right] = 9\left[x^3+(2x-1)^3\right]^2\left[9x^2-8x+2\right]$

47. $y = \sin\left(\tan\sqrt{\sin x}\right) \quad \Rightarrow \quad y' = \cos\left(\tan\sqrt{\sin x}\right)\left(\sec^2\sqrt{\sin x}\right)\left(\dfrac{1}{2\sqrt{\sin x}}\right)(\cos x)$

49. $y = f(x) = (x^3 - x^2 + x - 1)^{10} \quad \Rightarrow \quad f'(x) = 10(x^3 - x^2 + x - 1)^9(3x^2 - 2x + 1)$. The slope of the tangent at $(1,0)$ is $f'(1) = 0$ and its equation is $y - 0 = 0(x-1)$ or $y = 0$.

51. $y = f(x) = \dfrac{8}{\sqrt{4+3x}} \quad \Rightarrow \quad f'(x) = 8\left(-\frac{1}{2}\right)(4+3x)^{-3/2}(3) = -12(4+3x)^{-3/2}$. The slope of the tangent at $(4,2)$ is $f'(4) = -\frac{3}{16}$ and its equation is $y - 2 = -\frac{3}{16}(x-4)$ or $3x + 16y = 44$.

53. **(a)**

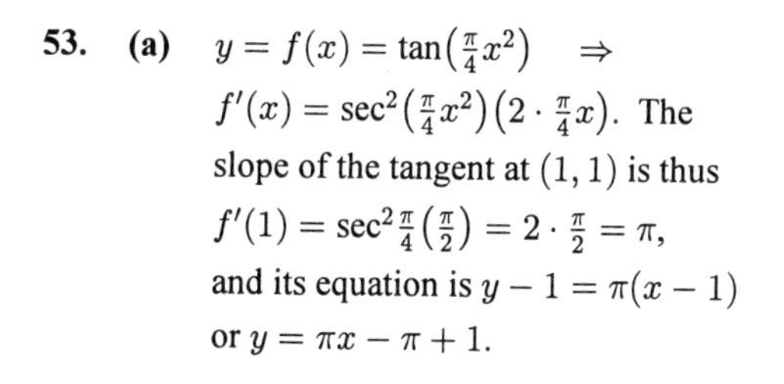

$y = f(x) = \tan\left(\frac{\pi}{4}x^2\right) \quad \Rightarrow$
$f'(x) = \sec^2\left(\frac{\pi}{4}x^2\right)\left(2\cdot\frac{\pi}{4}x\right)$. The slope of the tangent at $(1,1)$ is thus $f'(1) = \sec^2\frac{\pi}{4}\left(\frac{\pi}{2}\right) = 2\cdot\frac{\pi}{2} = \pi$, and its equation is $y - 1 = \pi(x-1)$ or $y = \pi x - \pi + 1$.

(b)

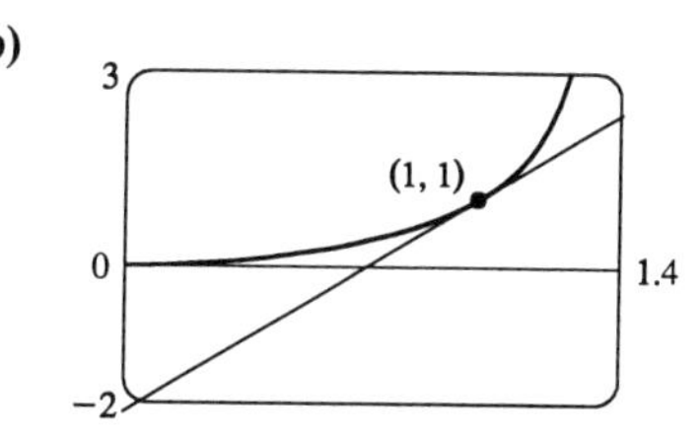

55. **(a)** $f(x) = \dfrac{\sqrt{1-x^2}}{x} \quad \Rightarrow$

$$f'(x) = \frac{x \cdot \frac{1}{2}(1-x^2)^{-1/2}(-2x) - \sqrt{1-x^2}}{x^2} = \frac{-1}{\sqrt{1-x^2}} - \frac{\sqrt{1-x^2}}{x^2} = \frac{-x^2 - \sqrt{1-x^2}\sqrt{1-x^2}}{x^2\sqrt{1-x^2}} = \frac{-1}{x^2\sqrt{1-x^2}}$$

(b)

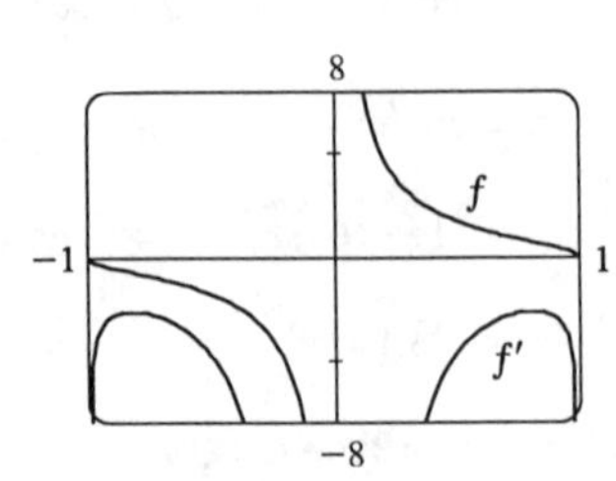

Notice that all tangents to the graph of f have negative slopes and $f'(x) < 0$ always.

57. For the tangent line to be horizontal $f'(x) = 0$. $f(x) = 2\sin x + \sin^2 x \quad \Rightarrow$

$f'(x) = 2\cos x + 2\sin x\cos x = 0 \quad \Leftrightarrow \quad 2\cos x\,(1+\sin x) = 0 \quad \Leftrightarrow \quad \cos x = 0$

or $\sin x = -1$, so $x = \left(n + \frac{1}{2}\right)\pi$ or $\left(2n + \frac{3}{2}\right)\pi$ where n is any integer. So the points on the curve with a horizontal tangent are $\left(\left(2n+\frac{1}{2}\right)\pi, 3\right)$ and $\left(\left(2n+\frac{3}{2}\right)\pi, -1\right)$ where n is any integer.

59. $F(x) = f(g(x)) \quad \Rightarrow \quad F'(x) = f'(g(x))\,g'(x)$, so $F'(3) = f'(g(3))g'(3) = f'(6)g'(3) = 7 \cdot 4 = 28$.

61. $s(t) = 10 + \frac{1}{4}\sin(10\pi t) \quad \Rightarrow \quad$ the velocity after t seconds is

$v(t) = s'(t) = \frac{1}{4}\cos(10\pi t)(10\pi) = \frac{5\pi}{2}\cos(10\pi t)$ cm/s.

63. **(a)** $\dfrac{dB}{dt} = \left(0.35\cos\dfrac{2\pi t}{5.4}\right)\left(\dfrac{2\pi}{5.4}\right) = \dfrac{7\pi}{54}\cos\dfrac{2\pi t}{5.4}$ **(b)** At $t = 1$, $\dfrac{dB}{dt} = \dfrac{7\pi}{54}\cos\dfrac{2\pi}{5.4} \approx 0.16$.

65. **(a)** Since h is differentiable on $[0, \infty)$ and $\sqrt{x}$ is differentiable on $(0, \infty)$, it follows that $G(x) = h\left(\sqrt{x}\right)$ is differentiable on $(0, \infty)$.

(b) By the Chain Rule, $G'(x) = h'\left(\sqrt{x}\right)\dfrac{d}{dx}\sqrt{x} = \dfrac{h'\left(\sqrt{x}\right)}{2\sqrt{x}}$.

67. **(a)** $F(x) = f(\cos x) \quad \Rightarrow \quad F'(x) = f'(\cos x)\dfrac{d}{dx}(\cos x) = -\sin x\, f'(\cos x)$

(b) $G(x) = \cos(f(x)) \quad \Rightarrow \quad G'(x) = -\sin(f(x))f'(x)$

69. $g(x) = f(b+mx) + f(b-mx) \quad \Rightarrow$

$g'(x) = f'(b+mx)D(b+mx) + f'(b-mx)D(b-mx) = mf'(b+mx) - mf'(b-mx)$

So $g'(0) = mf'(b) - mf'(b) = 0$.

71. **(a)** If f is even, then $f(x) = f(-x)$. Using the Chain Rule to differentiate this equation, we get $f'(x) = f'(-x)\dfrac{d}{dx}(-x) = -f'(-x)$. Thus $f'(-x) = -f'(x)$, so f' is odd.

(b) If f is odd, then $f(x) = -f(-x)$. Differentiating this equation, we get $f'(x) = -f'(-x)(-1) = f'(-x)$, so f' is even.

73. $\frac{d}{dx}(\sin^n x \cos nx) = n \sin^{n-1} x \cos x \cos nx + \sin^n x\,(-n \sin nx)$

$= n \sin^{n-1} x\,(\cos nx \cos x - \sin nx \sin x) = n \sin^{n-1} x \cos[(n+1)x]$

75. $f(x) = |x| = \sqrt{x^2} \quad \Rightarrow \quad f'(x) = \frac{1}{2}(x^2)^{-1/2}(2x) = x/\sqrt{x^2} = x/|x|.$

77. Using Exercise 75, we have $h(x) = x|2x-1| \quad \Rightarrow$

$h'(x) = |2x-1| + x\frac{2x-1}{|2x-1|}(2) = |2x-1| + \frac{2x(2x-1)}{|2x-1|}.$

79. Since $\theta^\circ = \left(\frac{\pi}{180}\right)\theta$ rad, we have $\frac{d}{d\theta}(\sin\theta^\circ) = \frac{d}{d\theta}\left(\sin\frac{\pi}{180}\theta\right) = \frac{\pi}{180}\cos\frac{\pi}{180}\theta = \frac{\pi}{180}\cos\theta^\circ.$

EXERCISES 3.6

1. **(a)** $x^2 + 3x + xy = 5 \quad \Rightarrow \quad 2x + 3 + y + xy' = 0 \quad \Rightarrow \quad y' = -\frac{2x+y+3}{x}$

(b) $x^2 + 3x + xy = 5 \quad \Rightarrow \quad y = \frac{5 - x^2 - 3x}{x} = \frac{5}{x} - x - 3 \quad \Rightarrow \quad y' = -\frac{5}{x^2} - 1$

(c) $y' = -\frac{2x+y+3}{x} = \frac{-2x - 3 - (-3 - x + 5/x)}{x} = -1 - \frac{5}{x^2}$

3. **(a)** $2y^2 + xy = x^2 + 3 \quad \Rightarrow \quad 4yy' + y + xy' = 2x \quad \Rightarrow \quad y' = \frac{2x-y}{x+4y}$

(b) Use the quadratic formula: $2y^2 + xy - (x^2+3) = 0 \quad \Rightarrow$

$y = \frac{-x \pm \sqrt{x^2 + 8(x^2+3)}}{4} = \frac{-x \pm \sqrt{9x^2+24}}{4} \quad \Rightarrow \quad y' = \frac{1}{4}\left(-1 \pm \frac{9x}{\sqrt{9x^2+24}}\right)$

(c) $y' = \frac{2x-y}{x+4y} = \frac{2x - \frac{1}{4}\left(-x \pm \sqrt{9x^2+24}\right)}{x + \left(-x \pm \sqrt{9x^2+24}\right)} = \frac{1}{4}\left(-1 \pm \frac{9x}{\sqrt{9x^2+24}}\right)$

5. $x^2 - xy + y^3 = 8 \quad \Rightarrow \quad 2x - y - xy' + 3y^2y' = 0 \quad \Rightarrow \quad y' = \frac{y-2x}{3y^2-x}$

7. $2y^2 + \sqrt[3]{xy} = 3x^2 + 17 \quad \Rightarrow \quad 4yy' + \frac{1}{3}x^{-2/3}y^{1/3} + \frac{1}{3}x^{1/3}y^{-2/3}y' = 6x \quad \Rightarrow$

$y' = \frac{6x - \frac{1}{3}x^{-2/3}y^{1/3}}{4y + \frac{1}{3}x^{1/3}y^{-2/3}} = \frac{18x - x^{-2/3}y^{1/3}}{12y + x^{1/3}y^{-2/3}}$

9. $x^4 + y^4 = 16 \quad \Rightarrow \quad 4x^3 + 4y^3y' = 0 \quad \Rightarrow \quad y' = -\frac{x^3}{y^3}$

11. $\frac{y}{x-y} = x^2 + 1 \quad \Rightarrow \quad 2x = \frac{(x-y)y' - y(1-y')}{(x-y)^2} = \frac{xy'-y}{(x-y)^2} \quad \Rightarrow \quad y' = \frac{y}{x} + 2(x-y)^2$

Another Method: Write the equation as $y = (x-y)(x^2+1) = x^3 + x - yx^2 - y$. Then $y' = \frac{3x^2 + 1 - 2xy}{x^2+2}$.

13. $\cos(x-y) = y\sin x \quad\Rightarrow\quad -\sin(x-y)(1-y') = y'\sin x + y\cos x \quad\Rightarrow\quad y' = \dfrac{\sin(x-y)+y\cos x}{\sin(x-y)-\sin x}$

15. $xy = \cot(xy) \quad\Rightarrow\quad y + xy' = -\csc^2(xy)(y+xy') \quad\Rightarrow\quad (y+xy')[1+\csc^2(xy)] = 0 \quad\Rightarrow\quad y + xy' = 0$
$\Rightarrow\quad y' = -y/x$

17. $y^4 + x^2y^2 + yx^4 = y + 1 \Rightarrow 4y^3 + 2x\dfrac{dx}{dy}y^2 + 2x^2y + x^4 + 4yx^3\dfrac{dx}{dy} = 1 \Rightarrow \dfrac{dx}{dy} = \dfrac{1-4y^3-2x^2y-x^4}{2xy^2+4yx^3}$

19. $x[f(x)]^3 + xf(x) = 6 \quad\Rightarrow\quad [f(x)]^3 + 3x[f(x)]^2f'(x) + f(x) + xf'(x) = 0 \quad\Rightarrow$
$f'(x) = -\dfrac{[f(x)]^3+f(x)}{3x[f(x)]^2+x} \quad\Rightarrow\quad f'(3) = -\dfrac{(1)^3+1}{3(3)(1)^2+3} = -\dfrac{1}{6}$

21. $\dfrac{x^2}{16} - \dfrac{y^2}{9} = 1 \quad\Rightarrow\quad \dfrac{x}{8} - \dfrac{2yy'}{9} = 0 \quad\Rightarrow\quad y' = \dfrac{9x}{16y}$. When $x = -5$ and $y = \frac{9}{4}$ we have $y' = \dfrac{9(-5)}{16(9/4)} = -\dfrac{5}{4}$
so the equation of the tangent is $y - \frac{9}{4} = -\frac{5}{4}(x+5)$ or $5x + 4y + 16 = 0$.

23. $y^2 = x^3(2-x) = 2x^3 - x^4 \quad\Rightarrow\quad 2yy' = 6x^2 - 4x^3 \quad\Rightarrow\quad y' = \dfrac{3x^2-2x^3}{y}$. When $x = y = 1$,
$y' = \dfrac{3(1)^2 - 2(1)^3}{1} = 1$, so the equation of the tangent line is $y - 1 = 1(x-1)$ or $y = x$.

25. $2(x^2+y^2)^2 = 25(x^2-y^2) \quad\Rightarrow\quad 4(x^2+y^2)(2x+2yy') = 25(2x-2yy') \quad\Rightarrow\quad y' = \dfrac{25x-4x(x^2+y^2)}{25y+4y(x^2+y^2)}$.
When $x = 3$ and $y = 1$, $y' = \dfrac{75 - 120}{25 + 40} = -\dfrac{9}{13}$ so the equation of the tangent is $y - 1 = -\frac{9}{13}(x-3)$ or
$9x + 13y = 40$.

27. **(a)** $y^2 = 5x^4 - x^2 \quad\Rightarrow\quad 2yy' = 5(4x^3) - 2x \quad\Rightarrow$
$y' = \dfrac{10x^3 - x}{y}$. So at the point $(1, 2)$ we have
$y' = \dfrac{10(1)^3 - 1}{2} = \dfrac{9}{2}$, and the equation of
the tangent line is $y - 2 = \frac{9}{2}(x-1) \quad\Leftrightarrow\quad y = \frac{9}{2}x - \frac{5}{2}$.

(b)

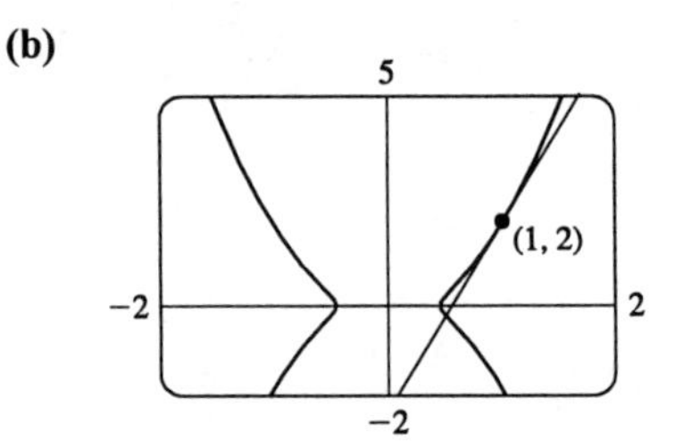

29. From Exercise 25, a tangent to the lemniscate will be horizontal $\Rightarrow\quad y' = 0 \quad\Rightarrow\quad 25x - 4x(x^2+y^2) = 0$
$\Rightarrow\quad x^2 + y^2 = \frac{25}{4}$. (Note that $x = 0 \quad\Rightarrow\quad y = 0$ and there is no horizontal tangent at the origin.) Putting this in the equation of the lemniscate, we get $x^2 - y^2 = \frac{25}{8}$. Solving these two equations we have $x^2 = \frac{75}{16}$ and $y^2 = \frac{25}{16}$, so the four points are $\left(\pm\frac{5\sqrt{3}}{4}, \pm\frac{5}{4}\right)$.

31. $\dfrac{x^2}{a^2}-\dfrac{y^2}{b^2}=1 \quad\Rightarrow\quad \dfrac{2x}{a^2}-\dfrac{2yy'}{b^2}=0 \quad\Rightarrow\quad y'=\dfrac{b^2x}{a^2y} \quad\Rightarrow$ the equation of the tangent at (x_0,y_0) is $y-y_0=\dfrac{b^2x_0}{a^2y_0}(x-x_0)$. Multiplying both sides by $\dfrac{y_0}{b^2}$ gives $\dfrac{y_0y}{b^2}-\dfrac{y_0^2}{b^2}=\dfrac{x_0x}{a^2}-\dfrac{x_0^2}{a^2}$. Since (x_0,y_0) lies on the hyperbola, we have $\dfrac{x_0x}{a^2}-\dfrac{y_0y}{b^2}=\dfrac{x_0^2}{a^2}-\dfrac{y_0^2}{b^2}=1$.

33. If the circle has radius r, its equation is $x^2+y^2=r^2 \quad\Rightarrow\quad 2x+2yy'=0 \quad\Rightarrow\quad y'=-\dfrac{x}{y}$, so the slope of the tangent line at $P(x_0,y_0)$ is $-\dfrac{x_0}{y_0}$. The slope of OP is $\dfrac{y_0}{x_0}=\dfrac{-1}{-x_0/y_0}$, so the tangent is perpendicular to OP.

35. $2x^2+y^2=3$ and $x=y^2$ intersect when $2x^2+x-3=(2x+3)(x-1)=0 \quad\Leftrightarrow\quad x=-\frac{3}{2}$ or 1, but $-\frac{3}{2}$ is extraneous. $2x^2+y^2=3 \quad\Rightarrow\quad 4x+2yy'=0 \quad\Rightarrow\quad y'=-2x/y$ and $x=y^2 \quad\Rightarrow\quad 1=2yy' \quad\Rightarrow$ $y'=1/(2y)$. At $(1,1)$ the slopes are $m_1=-2$ and $m_2=\frac{1}{2}$, so the curves are orthogonal there. By symmetry they are also orthogonal at $(1,-1)$.

37. $x^2+y^2=r^2$ is a circle with center O and $ax+by=0$ is a line through O. By Exercise 33, the curves are orthogonal.

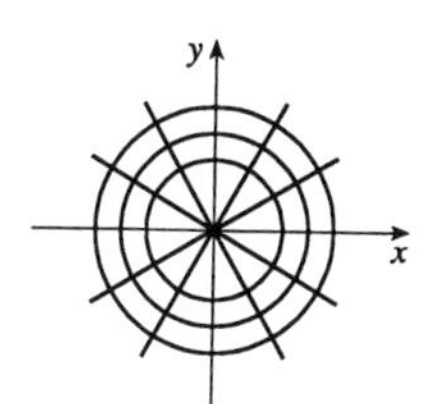

39. $y=cx^2 \quad\Rightarrow\quad y'=2cx$ and $x^2+2y^2=k$ $\Rightarrow\quad 2x+4yy'=0 \quad\Rightarrow\quad y'=-\dfrac{x}{2y}=-\dfrac{x}{2cx^2}$ $=-1/(2cx)$, so the curves are orthogonal.

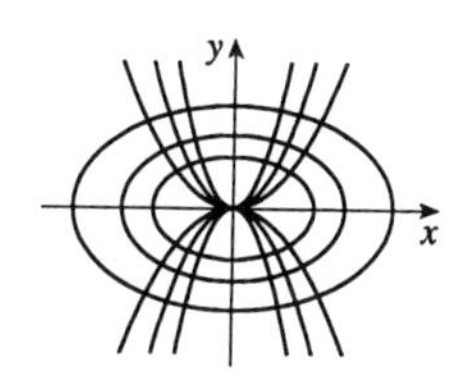

41. $y=0 \quad\Rightarrow\quad x^2+x(0)+0^2=3 \quad\Leftrightarrow\quad x=\pm\sqrt{3}$. So the graph of the ellipse crosses the x-axis at the points $(\pm\sqrt{3},0)$. Using implicit differentiation to find y', we get $2x-xy'-y+2yy'=0 \;\Rightarrow\; y'(2y-x)=y-2x$ $\Rightarrow\quad y'=\dfrac{y-2x}{2y-x}$. So $y'(\sqrt{3},0)=\dfrac{0-2\sqrt{3}}{2(0)-\sqrt{3}}=2$, and $y'(-\sqrt{3},0)=\dfrac{0+2\sqrt{3}}{2(0)+\sqrt{3}}=2=y'(\sqrt{3},0)$. So the tangent lines at these points are parallel.

43. $x^2y^2+xy=2 \quad\Rightarrow\quad 2xy^2+2x^2yy'+y+xy'=0 \quad\Leftrightarrow\quad y'(2x^2y+x)=-2xy^2-y \quad\Leftrightarrow$ $y'=-\dfrac{2xy^2+y}{2x^2y+x}$. So $-\dfrac{2xy^2+y}{2x^2y+x}=-1 \quad\Leftrightarrow\quad 2xy^2+y=2x^2y+x \quad\Leftrightarrow\quad y(2xy+1)=x(2xy+1) \quad\Leftrightarrow$ $(2xy+1)(y-x)=0 \quad\Leftrightarrow\quad y=x$ or $xy=-\frac{1}{2}$. But $xy=-\frac{1}{2} \quad\Rightarrow\quad x^2y^2+xy=\frac{1}{4}-\frac{1}{2}\neq 2$ so we must have $x=y$. Then $x^2y^2+xy=2 \quad\Rightarrow\quad x^4+x^2=2 \quad\Leftrightarrow\quad x^4+x^2-2=0 \quad\Leftrightarrow\quad (x^2+2)(x^2-1)=0$. So $x^2=-2$, which is impossible, or $x^2=1 \quad\Leftrightarrow\quad x=\pm 1$. So the points on the curve where the tangent line has a slope of -1 are $(-1,-1)$ and $(1,1)$.

45. We use implicit differentiation to find y': $2x + 4(2yy') = 0 \quad \Rightarrow \quad y' = -\frac{x}{4y}$. Now let h be the height of the lamp, and let (a, b) be the point of tangency of the line passing through the points $(3, h)$ and $(-5, 0)$. This line has slope $(h - 0)/[3 - (-5)] = \frac{1}{8}h$. But the slope of the tangent line through the point (a, b) can be expressed as $y' = -\frac{a}{4b}$, or as $\frac{b - 0}{a - (-5)} = \frac{b}{a + 5}$ [since the line passes through $(-5, 0)$ and (a, b)], so $-\frac{a}{4b} = \frac{b}{a + 5} \quad \Leftrightarrow$ $4b^2 = -a^2 - 5a \quad \Leftrightarrow \quad a^2 + 4b^2 = -5a$. But $a^2 + 4b^2 = 5$, since (a, b) is on the ellipse, so $5 = -5a \quad \Leftrightarrow$ $a = -1$. Then $4b^2 = -1 - 5(-1) = 4 \quad \Rightarrow \quad b = 1$, since the point is on the top half of the ellipse. So $\frac{h}{8} = \frac{b}{a + 5} = \frac{1}{-1 + 5} = \frac{1}{4} \quad \Rightarrow \quad h = 2$. So the lamp is located 2 units above the x-axis.

EXERCISES 3.7

1. $\mathbf{r}(t) = \langle t, 2t \rangle$ corresponds to the parametric equations $x = t$ and $y = 2t$, or the Cartesian equation $y = 2x$.

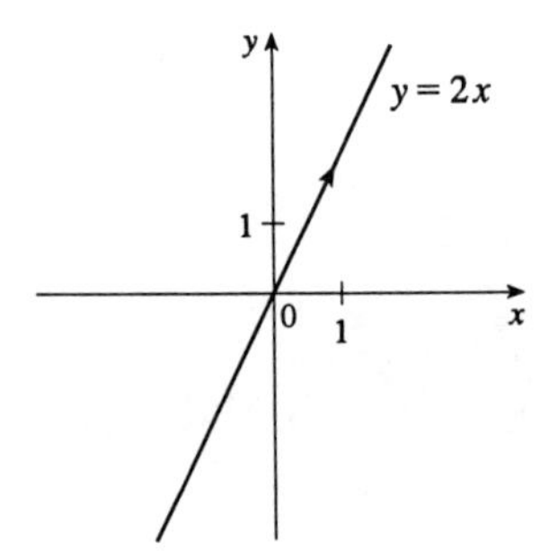

3. $\mathbf{r}(t) = \sin t\,\mathbf{i} + \cos t\,\mathbf{j}$ corresponds to the parametric equations $x = \sin t$ and $y = \cos t$, or the Cartesian equation $x^2 + y^2 = 1$.

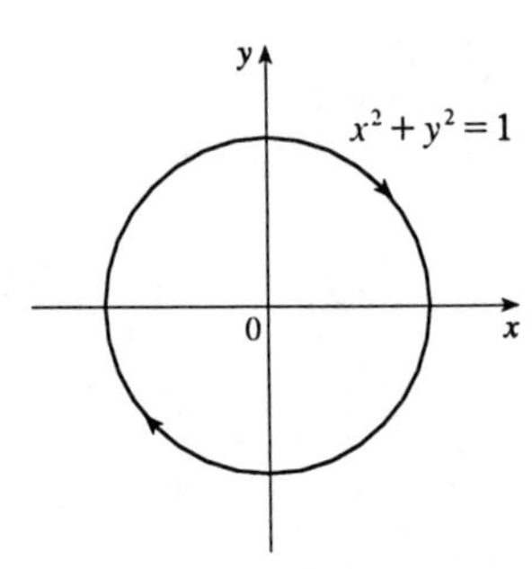

5. $\mathbf{r}(t) = \langle t^3 + 1, 2t \rangle$ corresponds to the parametric equations $x = t^3 + 1$ and $y = 2t$, or the Cartesian equation $x = \left(\frac{y}{2}\right)^3 + 1 = \frac{1}{8}y^3 + 1$.

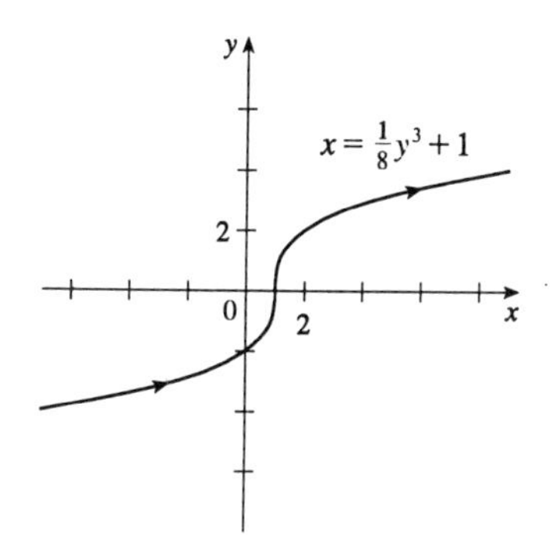

7. In general, the domain of $\mathbf{r}$ consists of all values of t for which the expression $\mathbf{r}(t)$ is defined. The domain of $\mathbf{r}(t) = \langle t^2, t \rangle$ is all real numbers since t^2 and t are just polynomials. The derivative of $\mathbf{r}(t)$ is $\mathbf{r}'(t) = \langle 2t, 1 \rangle$ and the domain of $\mathbf{r}'$ is $\mathbb{R}$.

9. The domain of $\mathbf{r}(t) = \langle t^2 - 4, \sqrt{t-4} \rangle$ is $t \geq 4$. The derivative of $\mathbf{r}(t)$ is $\mathbf{r}'(t) = \left\langle 2t, 1/\left(2\sqrt{t-4}\right) \right\rangle$ and the domain of $\mathbf{r}'$ is $t > 4$ (does not include 4).

11. $\mathbf{r}(t) = \langle 2t, 3t^2 \rangle$ and $t = 1$. $\mathbf{r}'(t) = \langle 2, 6t \rangle$ so $\mathbf{r}'(1) = \langle 2, 6 \rangle$ is a tangent vector at the point with $t = 1$. (The point is $(2, 3)$, but is not needed.) To get a tangent vector of unit length, we divide the vector by its length.
$\dfrac{\langle 2,6 \rangle}{|\langle 2,6 \rangle|} = \dfrac{\langle 2,6 \rangle}{\sqrt{2^2 + 6^2}} = \frac{1}{\sqrt{40}}\langle 2,6 \rangle = \frac{1}{2\sqrt{10}}\langle 2,6 \rangle = \langle \frac{1}{\sqrt{10}}, \frac{3}{\sqrt{10}} \rangle.$

13. $\mathbf{r}(t) = (1 + t^3)\mathbf{i} + t^2\mathbf{j}$ and $t = 1$. $\mathbf{r}'(t) = 3t^2\mathbf{i} + 2t\mathbf{j} \quad \Rightarrow \quad \mathbf{r}'(1) = 3\mathbf{i} + 2\mathbf{j}$. $|\mathbf{r}'(1)| = \sqrt{3^2 + 2^2} = \sqrt{13}$, so the desired tangent vector is $\frac{1}{\sqrt{13}}(3\mathbf{i} + 2\mathbf{j}) = \frac{3}{\sqrt{13}}\mathbf{i} + \frac{2}{\sqrt{13}}\mathbf{j}$.

15. $\mathbf{r}(t) = \langle t, 25t - 5t^2 \rangle$ and $t = 1 \quad \Rightarrow \quad \mathbf{r}'(t) = \langle 1, 25 - 10t \rangle$, so the velocity is $\mathbf{v}(1) = \mathbf{r}'(1) = \langle 1, 15 \rangle$ and the speed is $|\mathbf{r}'(1)| = \sqrt{1^2 + 15^2} = \sqrt{226}$.

17. $\mathbf{r}(t) = (t^4 + 7t)\mathbf{i} + (t^2 + 2t)\mathbf{j}$ and $t = 1 \quad \Rightarrow \quad \mathbf{r}'(t) = (4t^3 + 7)\mathbf{i} + (2t + 2)\mathbf{j}$, so the velocity is $\mathbf{v}(1) = \mathbf{r}'(1) = 11\mathbf{i} + 4\mathbf{j}$ and the speed is $|\mathbf{r}'(1)| = \sqrt{11^2 + 4^2} = \sqrt{137}$.

19. The angle of intersection of the two curves, $\mathbf{r}_1(t) = \langle t, t^2 \rangle$ and $\mathbf{r}_2(t) = \langle \sin t, \sin 2t \rangle$, is the angle between the two tangent vectors to the curves at the point of intersection. Since $\mathbf{r}_1'(t) = \langle 1, 2t \rangle$ and $t = 0$ at $(0, 0)$, $\mathbf{r}_1'(0) = \langle 1, 0 \rangle$ is a tangent vector to $\mathbf{r}_1$ at $(0, 0)$. Also $\mathbf{r}_2' = \langle \cos t, 2\cos 2t \rangle$ so $\mathbf{r}_2'(0) = \langle 1, 2 \rangle$ is a tangent vector to $\mathbf{r}_2$ at $(0, 0)$. If θ is the angle between these two tangent vectors, then
$\cos\theta = \dfrac{\mathbf{r}_1'(0) \cdot \mathbf{r}_2'(0)}{|\mathbf{r}_1'(0)||\mathbf{r}_2'(0)|} = \dfrac{1}{\sqrt{1}\sqrt{5}}\langle 1, 0 \rangle \cdot \langle 1, 2 \rangle = \dfrac{1}{\sqrt{5}}$ and $\theta = \cos^{-1}\frac{1}{\sqrt{5}} \approx 63°$.

21. $\frac{d}{dt}[\mathbf{u}(t)+\mathbf{v}(t)] = \frac{d}{dt}\left[\langle u_1, u_2\rangle \cdot \langle v_1, v_2\rangle\right] = \frac{d}{dt}[u_1 v_1 + u_2 v_2] = u_1 v_1' + v_1 u_1' + u_2 v_2' + v_2 u_2'$
$= (u_1 v_1' + u_2 v_2') + (v_1 u_1' + v_2 u_2') = \langle u_1, u_2\rangle \cdot \langle v_1', v_2'\rangle + \langle v_1, v_2\rangle \cdot \langle u_1', u_2'\rangle$
$= \mathbf{u}(t) \cdot \mathbf{v}'(t) + \mathbf{v}(t) \cdot \mathbf{u}'(t)$

23. Since $\mathbf{r}(t)$ is always perpendicular to $\mathbf{r}'(t)$, we know that $\mathbf{r}(t) \cdot \mathbf{r}'(t) = 0$. Following the hint, we differentiate the equation $|\mathbf{r}(t)|^2 = \mathbf{r}(t) \cdot \mathbf{r}(t)$. $\frac{d}{dt}|\mathbf{r}(t)|^2 = \frac{d}{dt}[\mathbf{r}(t) \cdot \mathbf{r}(t)] \Rightarrow \frac{d}{dt}|\mathbf{r}(t)|^2 = \mathbf{r}(t) \cdot \mathbf{r}'(t) + \mathbf{r}(t) \cdot \mathbf{r}'(t)$ $[= 0 + 0]$, so $|\mathbf{r}(t)|^2$, and hence $|\mathbf{r}(t)|$, is a constant. Thus, the curve lies on a circle with center the origin.

EXERCISES 3.8

1. $a = f, b = f', c = f''$. We can see this because where a has a horizontal tangent, $b = 0$, and where b has a horizontal tangent, $c = 0$. We can immediately see that c can be neither f nor f', since at the points where c has a horizontal tangent, neither a nor b is equal to 0.

3. $f(x) = x^4 - 3x^3 + 16x \quad \Leftrightarrow \quad f'(x) = 4x^3 - 9x^2 + 16 \quad \Rightarrow \quad f''(x) = 12x^2 - 18x$

5. $h(x) = \sqrt{x^2+1} \quad \Rightarrow \quad h'(x) = \frac{1}{2}(x^2+1)^{-1/2}(2x) = \frac{x}{\sqrt{x^2+1}} \quad \Rightarrow$

$$h''(x) = \frac{\sqrt{x^2+1} - x\left(x/\sqrt{x^2+1}\right)}{x^2+1} = \frac{x^2+1-x^2}{(x^2+1)^{3/2}} = \frac{1}{(x^2+1)^{3/2}}$$

7. $F(s) = (3s+5)^8 \Rightarrow F'(s) = 8(3s+5)^7(3) = 24(3s+5)^7 \Rightarrow F''(s) = 168(3s+5)^6(3) = 504(3s+5)^6$

9. $y = \frac{x}{1-x} \quad \Rightarrow \quad y' = \frac{1(1-x) - x(-1)}{(1-x)^2} = \frac{1}{(1-x)^2} \quad \Rightarrow \quad y'' = -2(1-x)^{-3}(-1) = \frac{2}{(1-x)^3}$

11. $y = (1-x^2)^{3/4} \quad \Rightarrow \quad y' = \frac{3}{4}(1-x^2)^{-1/4}(-2x) = -\frac{3}{2}x(1-x^2)^{-1/4} \quad \Rightarrow$
$y'' = -\frac{3}{2}(1-x^2)^{-1/4} - \frac{3}{2}x\left(-\frac{1}{4}\right)(1-x^2)^{-5/4}(-2x) = -\frac{3}{2}(1-x^2)^{-1/4} - \frac{3}{4}x^2(1-x^2)^{-5/4}$
$= \frac{3}{4}(1-x^2)^{-5/4}(x^2-2)$

13. $H(t) = \tan^3(2t-1) \quad \Rightarrow \quad H'(t) = 3\tan^2(2t-1)\sec^2(2t-1)(2) = 6\tan^2(2t-1)\sec^2(2t-1) \quad \Rightarrow$
$H''(t) = 12\tan(2t-1)\sec^2(2t-1)(2)\sec^2(2t-1) + 6\tan^2(2t-1)2\sec(2t-1)\sec(2t-1)\tan(2t-1)(2)$
$= 24\tan(2t-1)\sec^4(2t-1) + 24\tan^3(2t-1)\sec^2(2t-1)$

15. **(a)** $f(x) = 2\cos x + \sin^2 x \quad \Rightarrow \quad f'(x) = 2(-\sin x) + 2\sin x\,(\cos x) = \sin 2x - 2\sin x \quad \Rightarrow$
$f''(x) = 2\cos 2x - 2\cos x = 2(\cos 2x - \cos x)$

(b)

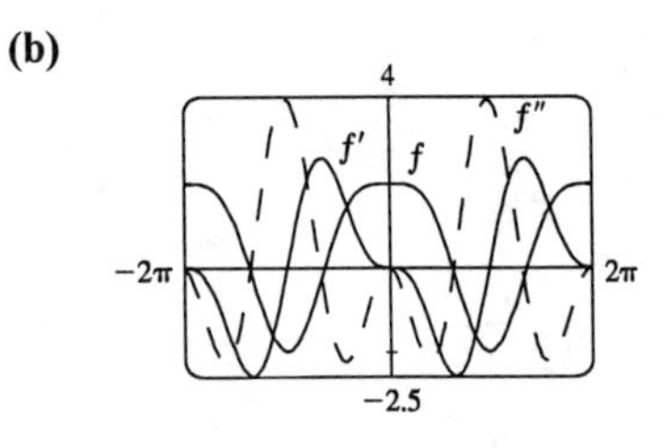

We can see that our answers are plausible, since f has horizontal tangents where $f'(x) = 0$, and f' has horizontal tangents where $f''(x) = 0$.

17. $y = \sqrt{5t-1} \quad\Rightarrow\quad y' = \frac{1}{2}(5t-1)^{-1/2}(5) = \frac{5}{2}(5t-1)^{-1/2} \quad\Rightarrow$
$y'' = -\frac{5}{4}(5t-1)^{-3/2}(5) = -\frac{25}{4}(5t-1)^{-3/2} \quad\Rightarrow\quad y''' = \frac{75}{8}(5t-1)^{-5/2}(5) = \frac{375}{8}(5t-1)^{-5/2}$

19. $f(x) = (2-3x)^{-1/2} \quad\Rightarrow\quad f(0) = 2^{-1/2} = \frac{1}{\sqrt{2}}$

$f'(x) = -\frac{1}{2}(2-3x)^{-3/2}(-3) = \frac{3}{2}(2-3x)^{-3/2} \quad\Rightarrow\quad f'(0) = \frac{3}{2}(2)^{-3/2} = \frac{3}{4\sqrt{2}}$

$f''(x) = -\frac{9}{4}(2-3x)^{-5/2}(-3) = \frac{27}{4}(2-3x)^{-5/2} \quad\Rightarrow\quad f''(0) = \frac{27}{4}(2)^{-5/2} = \frac{27}{16\sqrt{2}}$

$f'''(x) = \frac{405}{8}(2-3x)^{-7/2} \quad\Rightarrow\quad f'''(0) = \frac{405}{8}(2)^{-7/2} = \frac{405}{64\sqrt{2}}$

21. $f(\theta) = \cot\theta \quad\Rightarrow\quad f'(\theta) = -\csc^2\theta \quad\Rightarrow\quad f''(\theta) = -2\csc\theta(-\csc\theta\cot\theta) = 2\csc^2\theta\cot\theta \quad\Rightarrow$
$f'''(\theta) = 2(-2\csc^2\theta\cot\theta)\cot\theta + 2\csc^2\theta(-\csc^2\theta) = -2\csc^2\theta(2\cot^2\theta + \csc^2\theta) \quad\Rightarrow$
$f'''\left(\frac{\pi}{6}\right) = -2(2)^2\left[2\left(\sqrt{3}\right)^2 + (2)^2\right] = -80$

23. $x^3 + y^3 = 1 \quad\Rightarrow\quad 3x^2 + 3y^2y' = 0 \quad\Rightarrow\quad y' = -\dfrac{x^2}{y^2} \quad\Rightarrow$

$y'' = -\dfrac{2xy^2 - 2x^2yy'}{y^4} = -\dfrac{2xy^2 - 2x^2y(-x^2/y^2)}{y^4} = -\dfrac{2xy^3 + 2x^4}{y^5} = -\dfrac{2x(y^3+x^3)}{y^5} = -\dfrac{2x}{y^5}$, since x and y must satisfy the original equation, $x^3 + y^3 = 1$.

25. $x^2 + 6xy + y^2 = 8 \quad\Rightarrow\quad 2x + 6y + 6xy' + 2yy' = 0 \quad\Rightarrow\quad y' = -\dfrac{x+3y}{3x+y} \quad\Rightarrow$

$$y'' = -\frac{(1+3y')(3x+y) - (x+3y)(3+y')}{(3x+y)^2} = \frac{8(y-xy')}{(3x+y)^2} = \frac{8\left[y - x(-x-3y)/(3x+y)\right]}{(3x+y)^2}$$
$$= \frac{8[y(3x+y) + x(x+3y)]}{(3x+y)^3} = \frac{8(x^2+6xy+y^2)}{(3x+y)^3} = \frac{64}{(3x+y)^3},$$ since x and y must satisfy the original equation, $x^2 + 6xy + y^2 = 8$.

27. $f(x) = x - x^2 + x^3 - x^4 + x^5 - x^6 \quad\Rightarrow\quad f'(x) = 1 - 2x + 3x^2 - 4x^3 + 5x^4 - 6x^5 \quad\Rightarrow$

$f''(x) = -2 + 6x - 12x^2 + 20x^3 - 30x^4 \quad\Rightarrow\quad f'''(x) = 6 - 24x + 60x^2 - 120x^3 \quad\Rightarrow$

$f^{(4)}(x) = -24 + 120x - 360x^2 \quad\Rightarrow\quad f^{(5)}(x) = 120 - 720x \quad\Rightarrow\quad f^{(6)}(x) = -720 \quad\Rightarrow$

$f^{(n)}(x) = 0$ for $7 \le n \le 73$.

29. $f(x) = x^n \quad\Rightarrow\quad f'(x) = nx^{n-1} \quad\Rightarrow\quad f''(x) = n(n-1)\,x^{n-2} \quad\Rightarrow\quad \cdots \quad\Rightarrow$
$f^{(n)}(x) = n(n-1)(n-2)\cdots 2\cdot 1\,x^{n-n} = n!$

31. $f(x) = 1/(3x^3) = \frac{1}{3}x^{-3} \quad\Rightarrow\quad f'(x) = \frac{1}{3}(-3)x^{-4} \quad\Rightarrow\quad f''(x) = \frac{1}{3}(-3)(-4)x^{-5} \quad\Rightarrow$

$f'''(x) = \frac{1}{3}(-3)(-4)(-5)x^{-6} \quad\Rightarrow\quad \cdots \quad\Rightarrow$

$f^{(n)}(x) = \frac{1}{3}(-3)(-4)\cdots[-(n+2)]x^{-(n+3)} = \dfrac{(-1)^n\cdot 3\cdot 4\cdot 5\cdot\cdots\cdot(n+2)}{3x^{n+3}} = \dfrac{(-1)^n(n+2)!}{6x^{n+3}}$

33. In general, $Df(2x) = 2f'(2x)$, $D^2f(2x) = 4f''(2x)$, $\cdots$, $D^nf(2x) = 2^nf^{(n)}(2x)$. Since $f(x) = \cos x$ and $50 = 4(12) + 2$, we have $f^{(50)}(x) = f^{(2)}(x) = -\cos x$, so $D^{50}\cos 2x = -2^{50}\cos 2x$.

35. **(a)** $s = t^3 - 3t \quad \Rightarrow \quad v(t) = s'(t) = 3t^2 - 3 \quad \Rightarrow \quad a(t) = v'(t) = 6t$

(b) $a(1) = 6(1) = 6 \text{ m/s}^2$

(c) $v(t) = 3t^2 - 3 = 0$ when $t^2 = 1$, that is, $t = 1$ and $a(1) = 6\,\text{m/s}^2$.

37. **(a)** $s = At^2 + Bt + C \quad \Rightarrow \quad v(t) = s'(t) = 2At + B \quad \Rightarrow \quad a(t) = v'(t) = 2A$

(b) $a(1) = 2A\,\text{m/s}^2$

(c) The acceleration at these instants is $2A\,\text{m/s}^2$, since $a(t)$ is constant.

39. **(a)** $s(t) = t^4 - 4t^3 + 2 \quad \Rightarrow \quad v(t) = s'(t) = 4t^3 - 12t^2 \quad \Rightarrow$
$a(t) = v'(t) = 12t^2 - 24t = 12t(t-2) = 0$ when $t = 0$ or 2.

(b) $s(0) = 2\,\text{m}$, $v(0) = 0\,\text{m/s}$, $s(2) = -14\,\text{m}$, $v(2) = -16\,\text{m/s}$

41. **(a)** $\mathbf{r}(t) = \langle \cos t, \sin t \rangle \quad \Rightarrow \quad x = \cos t,\ y = \sin t \quad \Rightarrow \quad x^2 + y^2 = 1$

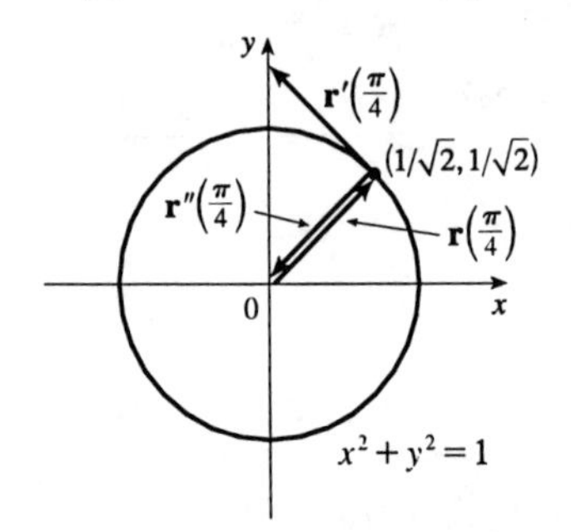

(b) $\mathbf{r}'(t) = \langle -\sin t, \cos t \rangle$ and $\mathbf{r}''(t) = \langle -\cos t, -\sin t \rangle = -\mathbf{r}(t)$.

(c) $t = \frac{\pi}{4}$, so $\mathbf{r}\left(\frac{\pi}{4}\right) = \left\langle \frac{\sqrt{2}}{2}, \frac{\sqrt{2}}{2} \right\rangle$, $\mathbf{r}'\left(\frac{\pi}{4}\right) = \left\langle -\frac{\sqrt{2}}{2}, \frac{\sqrt{2}}{2} \right\rangle$, and $\mathbf{r}''\left(\frac{\pi}{4}\right) = \left\langle -\frac{\sqrt{2}}{2}, -\frac{\sqrt{2}}{2} \right\rangle$.

43. **(a)** $\mathbf{r}(t) = (1+t)\mathbf{i} + t^2\mathbf{j} \quad \Rightarrow \quad x = 1 + t,\ y = t^2 \quad \Rightarrow \quad y = (x-1)^2$

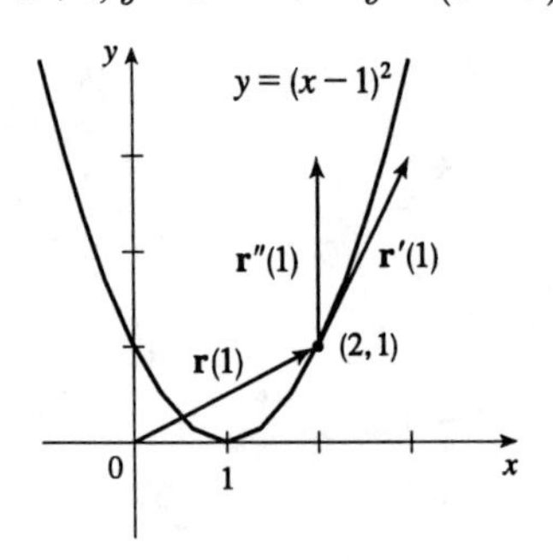

(b) $\mathbf{r}'(t) = \mathbf{i} + 2t\mathbf{j}$ and $\mathbf{r}''(t) = 2\mathbf{j}$.

(c) $t = 1$, so $\mathbf{r}(1) = 2\mathbf{i} + \mathbf{j}$, $\mathbf{r}'(1) = \mathbf{i} + 2\mathbf{j}$, and $\mathbf{r}''(1) = 2\mathbf{j}$.

45. $\mathbf{r}(t) = \langle t, 25t - 5t^2 \rangle \Rightarrow \mathbf{r}'(t) = \langle 1, 25 - 10t \rangle \Rightarrow \mathbf{r}''(t) = \langle 0, -10 \rangle$. The acceleration at any time t (including $t = 1$) is $\langle 0, -10 \rangle$.

47. $\mathbf{r}(t) = (t^4 + 7t)\mathbf{i} + (t^2 + 2t)\mathbf{j} \Rightarrow \mathbf{r}'(t) = (4t^3 + 7)\mathbf{i} + (2t + 2)\mathbf{j} \Rightarrow \mathbf{r}''(t) = (12t^2)\mathbf{i} + 2\mathbf{j}$. The acceleration at $t = 1$ is $\mathbf{r}''(1) = 12\mathbf{i} + 2\mathbf{j}$.

49. As in Example 6, the satellite's position is $\mathbf{r}(t) = \langle R\cos\omega t, R\sin\omega t \rangle$, its velocity is $\mathbf{r}'(t) = \langle -R\omega\sin\omega t, R\omega\cos\omega t \rangle$, its speed is $|\mathbf{r}'(t)| = R\omega$, and its acceleration is $\mathbf{r}''(t) = \langle -R\omega^2\cos\omega t, -R\omega^2\sin\omega t \rangle$. Here we have $R = 6600 + 1000 = 7600$ km. Also, 1 hr 46 min $= \left(1 + \frac{46}{60}\right)$ h $= 1.7\overline{6}$ h. So $2\pi/\omega = 1.7\overline{6} \Rightarrow \omega = 2\pi/1.7\overline{6} \approx 3.56$ rad/h. Substituting for R and ω gives us $\mathbf{v}(t) \approx \langle -27{,}030\sin 3.56t, 27{,}030\cos 3.56t \rangle$, $|\mathbf{v}| \approx 27{,}030$ km/h, and $\mathbf{a}(t) \approx \langle -96{,}131\cos 3.56t, -96{,}131\sin 3.56t \rangle$.

51. From Example 2 in Section 3.7, we have $\mathbf{r}(t) = \langle x, y \rangle = \langle 500t, 500\sqrt{3}t - 16t^2 \rangle$. The cannonball hits the ground when $t > 0$ and $y = 0$. $y = 0 \Leftrightarrow 500\sqrt{3}t - 16t^2 = 0 \Leftrightarrow 4t\left(125\sqrt{3} - 4t\right) = 0 \Leftrightarrow t = 0, \frac{125\sqrt{3}}{4}$. Let $t_1 = \frac{125\sqrt{3}}{4} \approx 54.13$ s. $\mathbf{r}'(t) = \langle 500, 500\sqrt{3} - 32t \rangle$ and $\mathbf{r}''(t) = \langle 0, -32 \rangle \Rightarrow$ the velocity of the cannonball is $\mathbf{r}'(t_1) = \langle 500, 500\sqrt{3} - 32 \cdot \frac{125\sqrt{3}}{4} \rangle = \langle 500, -500\sqrt{3} \rangle \approx \langle 500, -866 \rangle$ (component units are ft/s), the speed of the cannonball is $|\mathbf{v}(t)| = \sqrt{500^2 + \left(-500\sqrt{3}\right)^2} = \sqrt{1{,}000{,}000} = 1000$ ft/s, and the acceleration of the cannonball is $\mathbf{a}(t_1) = \mathbf{r}''(t_1) = \langle 0, -32 \rangle$ [component units are (ft/s)/s].

Note: $\mathbf{r}''(t) = \langle 0, -32 \rangle$ (ft/s)/s for *all* t.

53. **(a)** $y(t) = A\sin\omega t \Rightarrow v(t) = y'(t) = A\omega\cos\omega t \Rightarrow a(t) = v'(t) = -A\omega^2\sin\omega t$

(b) $a(t) = -A\omega^2\sin\omega t = -\omega^2 y(t)$

(c) $|v(t)| = A\omega|\cos\omega t|$ is a maximum when $\cos\omega t = \pm 1 \Leftrightarrow \sin\omega t = 0 \Leftrightarrow a(t) = -A\omega^2\sin^2\omega t = 0$.

55. Let $P(x) = ax^2 + bx + c$. Then $P'(x) = 2ax + b$ and $P''(x) = 2a$.
$P''(2) = 2 \Rightarrow 2a = 2 \Rightarrow a = 1$. $P'(2) = 3 \Rightarrow 4a + b = 4 + b = 3 \Rightarrow b = -1$.
$P(2) = 5 \Rightarrow 2^2 - 2 + c = 5 \Rightarrow c = 3$. So $P(x) = x^2 - x + 3$.

57. $P(x) = c_n x^n + c_{n-1}x^{n-1} + \cdots + c_1 x + c_0 \Rightarrow P'(x) = nc_n x^{n-1} + (n-1)c_{n-1}x^{n-2} + \cdots \Rightarrow$
$P''(x) = n(n-1)c_n x^{n-2} + \cdots \Rightarrow P^{(n)}(x) = n(n-1)(n-2)\cdots(1)c_n x^{n-n} = n!\,c_n$ which is a constant.
Therefore $P^{(m)}(x) = 0$ for $m > n$.

59. $f(x) = xg(x^2) \Rightarrow f'(x) = g(x^2) + xg'(x^2)2x = g(x^2) + 2x^2 g'(x^2) \Rightarrow$
$f''(x) = 2xg'(x^2) + 4xg'(x^2) + 4x^3 g''(x^2) = 6xg'(x^2) + 4x^3 g''(x^2)$

61. $f(x) = g\left(\sqrt{x}\right) \Rightarrow f'(x) = \dfrac{g'\left(\sqrt{x}\right)}{2\sqrt{x}} \Rightarrow f''(x) = \dfrac{\dfrac{g''\left(\sqrt{x}\right)}{2\sqrt{x}} \cdot 2\sqrt{x} - \dfrac{g'\left(\sqrt{x}\right)}{\sqrt{x}}}{4x} = \dfrac{\sqrt{x}\,g''\left(\sqrt{x}\right) - g'\left(\sqrt{x}\right)}{4x\sqrt{x}}$

63. **(a)** $f(x) = \dfrac{1}{x^2+x} \quad \Rightarrow \quad f'(x) = \dfrac{-(2x+1)}{(x^2+x)^2} \quad \Rightarrow$

$$f''(x) = \frac{(x^2+x)^2(-2) + (2x+1)(2)(x^2+x)(2x+1)}{(x^2+x)^4} = \frac{2(3x^2+3x+1)}{(x^2+x)^3} \quad \Rightarrow$$

$$f'''(x) = \frac{(x^2+x)^3(2)(6x+3) - 2(3x^2+3x+1)(3)(x^2+x)^2(2x+1)}{(x^2+x)^6}$$

$$= \frac{-6(4x^3+6x^2+4x+1)}{(x^2+x)^4} \quad \Rightarrow$$

$$f^{(4)}(x) = \frac{(x^2+x)^4(-6)(12x^2+12x+4) + 6(4x^3+6x^2+4x+1)(4)(x^2+x)^3(2x+1)}{(x^2+x)^8}$$

$$= \frac{24(5x^4+10x^3+10x^2+5x+1)}{(x^2+x)^5}$$

$f^{(5)}(x) = ?$

(b) $f(x) = \dfrac{1}{x(x+1)} = \dfrac{1}{x} - \dfrac{1}{x+1} \quad \Rightarrow \quad f'(x) = -x^{-2} + (x+1)^{-2} \quad \Rightarrow \quad f''(x) = 2x^{-3} - 2(x+1)^{-3}$

$\Rightarrow \; f'''(x) = (-3)(2)x^{-4} + (3)(2)(x+1)^{-4} \;\Rightarrow\; \cdots \;\Rightarrow\; f^{(n)}(x) = (-1)^n n!\left[x^{-(n+1)} - (x+1)^{-(n+1)}\right]$

65. The Chain Rule says that $\dfrac{dy}{dx} = \dfrac{dy}{du}\dfrac{du}{dx}$, so

$$\frac{d^2y}{dx^2} = \frac{d}{dx}\left(\frac{dy}{dx}\right) = \frac{d}{dx}\left(\frac{dy}{du}\frac{du}{dx}\right) = \left[\frac{d}{dx}\left(\frac{dy}{du}\right)\right]\frac{du}{dx} + \frac{dy}{du}\frac{d}{dx}\left(\frac{du}{dx}\right) \quad \text{(Product Rule)}$$

$$= \left[\frac{d}{du}\left(\frac{dy}{du}\right)\frac{du}{dx}\right]\frac{du}{dx} + \frac{dy}{du}\frac{d^2u}{dx^2} = \frac{d^2y}{du^2}\left(\frac{du}{dx}\right)^2 + \frac{dy}{du}\frac{d^2u}{dx^2}.$$

EXERCISES 3.9

1. $x = t^2 + t,\ y = t^2 - t;\ t = 0.$ $\dfrac{dy}{dt} = 2t - 1,\ \dfrac{dx}{dt} = 2t + 1$, so $\dfrac{dy}{dx} = \dfrac{dy/dt}{dx/dt} = \dfrac{2t-1}{2t+1}$. When $t = 0$, $x = y = 0$ and $\dfrac{dy}{dx} = -1$. The tangent is $y - 0 = (-1)(x - 0)$, or $y = -x$.

3. $x = t^2 + t,\ y = \sqrt{t};\ t = 4.$ $\dfrac{dy}{dt} = \dfrac{1}{2\sqrt{t}},\ \dfrac{dx}{dt} = 2t + 1$, so $\dfrac{dy}{dx} = \dfrac{dy/dt}{dx/dt} = \dfrac{1}{2\sqrt{t}(2t+1)}$. When $t = 4$, $(x, y) = (20, 2)$ and $dy/dx = 1/(2 \cdot 2 \cdot 9) = \frac{1}{36}$, so an equation of the tangent is $y - 2 = \frac{1}{36}(x - 20)$, or $y = \frac{1}{36}x + \frac{13}{9}$.

5. $\mathbf{r}(t) = t\sin t\,\mathbf{i} + t\cos t\,\mathbf{j} \quad \Rightarrow \quad x = t\sin t,\ y = t\cos t;\ t = \pi.$ $\dfrac{dy}{dx} = \dfrac{dy/dt}{dx/dt} = \dfrac{-t\sin t + \cos t}{t\cos t + \sin t}$. When $t = \pi$, $(x, y) = (0, -\pi)$ and $dy/dx = -1/(-\pi) = 1/\pi$, so an equation of the tangent is $y + \pi = \frac{1}{\pi}(x - 0)$, or $y = \frac{1}{\pi}x - \pi$.

7. **(a)** $x = 1 - t$, $y = 1 - t^2$; $(1, 1)$. $\dfrac{dy}{dt} = -2t$, $\dfrac{dx}{dt} = -1$, and $\dfrac{dy}{dx} = \dfrac{dy/dt}{dx/dt} = 2t$. When $x = 1$, $t = 0$, and $dy/dx = 0$, so an equation of the tangent is $y - 1 = 0(x - 1)$, or $y = 1$.

(b) $y = 1 - t^2 = 1 - (1 - x)^2$, so $\dfrac{dy}{dx} = -2(1 - x)(-1) = 2 - 2x$. When $x = 1$, $\dfrac{dy}{dx} = 0$, so an equation of the tangent is $y = 1$, as in part (a).

9. **(a)** $x = 5\cos t$, $y = 5\sin t$; $(3, 4)$. $\dfrac{dy}{dt} = 5\cos t$, $\dfrac{dx}{dt} = -5\sin t$, $\dfrac{dy}{dx} = \dfrac{dy/dt}{dx/dt} = -\cot t$.

At $(3, 4)$, $t = \tan^{-1}\dfrac{y}{x} = \tan^{-1}\dfrac{4}{3}$, so $\dfrac{dy}{dx} = -\dfrac{3}{4}$, and the tangent is $y - 4 = -\frac{3}{4}(x - 3)$, or $y = -\frac{3}{4}x + \frac{25}{4}$.

(b) $x^2 + y^2 = 25$, so $2x + 2y\dfrac{dy}{dx} = 0$, or $\dfrac{dy}{dx} = -\dfrac{x}{y}$. At $(3, 4)$, $\dfrac{dy}{dx} = -\dfrac{3}{4}$, and as in (a) the tangent is $y = -\frac{3}{4}x + \frac{25}{4}$.

11. $x = 1 - 2\cos t$, $y = 2 + 3\sin t$. $\dfrac{dx}{dt} = 2\sin t$, $\dfrac{dy}{dt} = 3\cos t$. Since $\dfrac{dy}{dx} = \dfrac{dy/dt}{dx/dt}$, a vertical tangent occurs when $\dfrac{dx}{dt} = 0$ and $\dfrac{dy}{dt} \neq 0$; that is, when $t = \pi n$ for some integer n. If n is even, then $(x, y) = (-1, 2)$. If n is odd, then $(x, y) = (3, 2)$. A horizontal tangent occurs when $\dfrac{dy}{dt} = 0$ and $\dfrac{dx}{dt} \neq 0$; that is, when $t = (2k + 1)\frac{\pi}{2} = \frac{\pi}{2} + k\pi$ for some integer k. If k is even, then $(x, y) = (1, 5)$. If k is odd, then $(x, y) = (1, -1)$.

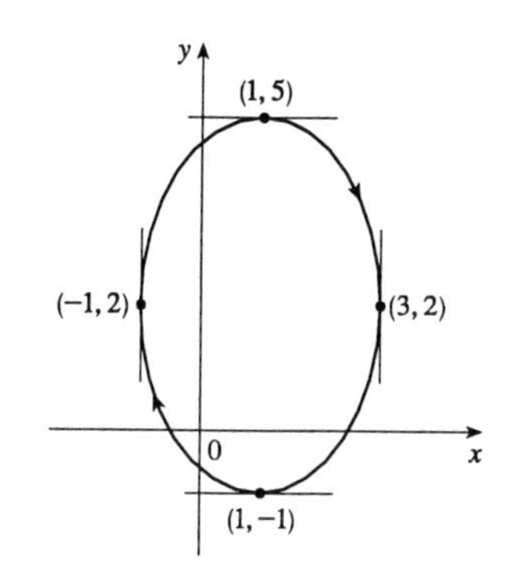

13. $x = t(t^2 - 3) = t^3 - 3t$, $y = 3(t^2 - 3)$. $\dfrac{dx}{dt} = 3t^2 - 3 = 3(t - 1)(t + 1)$, $\dfrac{dy}{dt} = 6t$. $\dfrac{dy}{dt} = 0 \Leftrightarrow t = 0 \Leftrightarrow (x, y) = (0, -9)$. $\dfrac{dx}{dt} = 0 \Leftrightarrow t = \pm 1 \Leftrightarrow (x, y) = (-2, -6)$ or $(2, -6)$. So there is a horizontal tangent at $(0, -9)$ and there are vertical tangents at $(-2, -6)$ and $(2, -6)$.

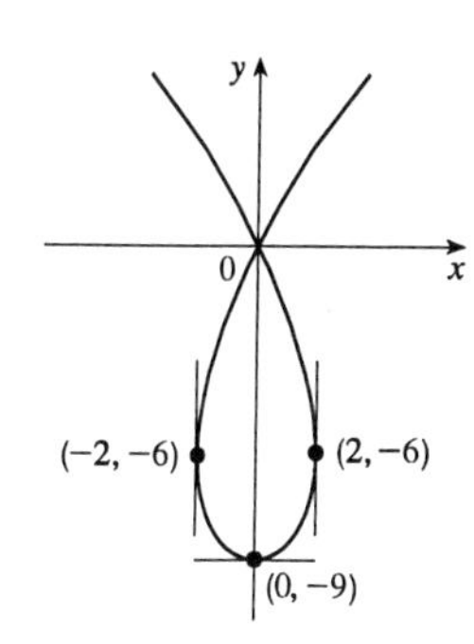

15. $x = \dfrac{3t}{1+t^3}$, $y = \dfrac{3t^2}{1+t^3}$. $\dfrac{dx}{dt} = \dfrac{(1+t^3)3 - 3t(3t^2)}{(1+t^3)^2} = \dfrac{3-6t^3}{(1+t^3)^2}$,

$\dfrac{dy}{dt} = \dfrac{(1+t^3)(6t) - 3t^2(3t^2)}{(1+t^3)^2} = \dfrac{6t - 3t^4}{(1+t^3)^2} = \dfrac{3t(2-t^3)}{(1+t^3)^2}$. $\dfrac{dy}{dt} = 0 \iff t = 0$ or $\sqrt[3]{2} \iff$

$(x,y) = (0,0)$ or $\left(\sqrt[3]{2}, \sqrt[3]{4}\right)$. $\dfrac{dx}{dt} = 0 \iff t^3 = \frac{1}{2} \iff t = 2^{-1/3} \iff$

$(x,y) = \left(\sqrt[3]{4}, \sqrt[3]{2}\right)$. There are horizontal tangents at $(0,0)$ and $\left(\sqrt[3]{2}, \sqrt[3]{4}\right)$, and there are vertical tangents at $\left(\sqrt[3]{4}, \sqrt[3]{2}\right)$ and $(0,0)$. [The vertical tangent at $(0,0)$ is undetectable by the methods of this section because that tangent corresponds to the limiting position of the point (x,y) as $t \to \pm\infty$.]

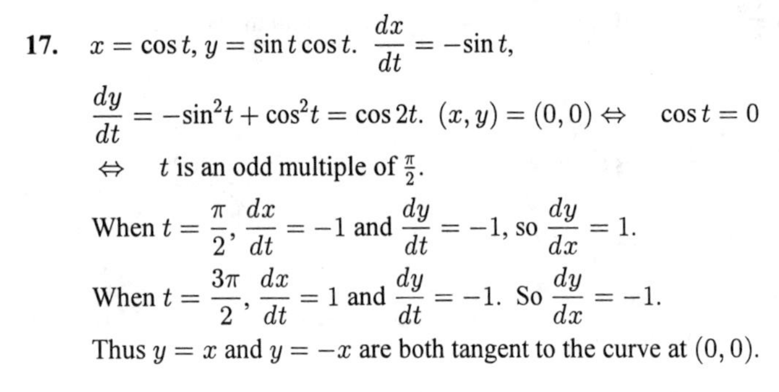

17. $x = \cos t$, $y = \sin t \cos t$. $\dfrac{dx}{dt} = -\sin t$,

$\dfrac{dy}{dt} = -\sin^2 t + \cos^2 t = \cos 2t$. $(x,y) = (0,0) \iff \cos t = 0$

$\iff t$ is an odd multiple of $\frac{\pi}{2}$.

When $t = \dfrac{\pi}{2}$, $\dfrac{dx}{dt} = -1$ and $\dfrac{dy}{dt} = -1$, so $\dfrac{dy}{dx} = 1$.

When $t = \dfrac{3\pi}{2}$, $\dfrac{dx}{dt} = 1$ and $\dfrac{dy}{dt} = -1$. So $\dfrac{dy}{dx} = -1$.

Thus $y = x$ and $y = -x$ are both tangent to the curve at $(0,0)$.

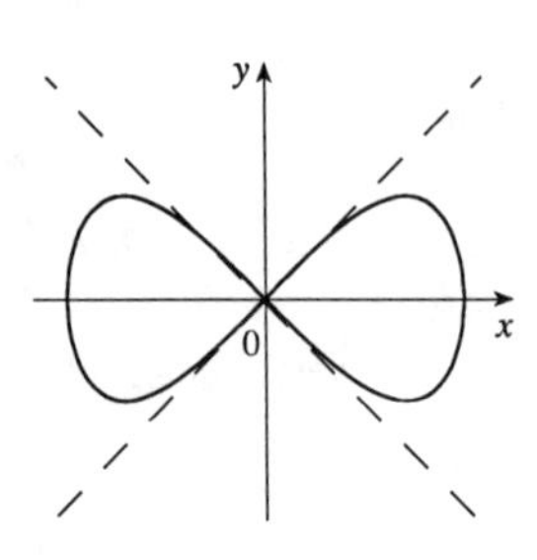

19. $x = t(t^2-3)$, $y = 3(t^2-3)$. We seek values t_1 and t_2 such that $t_1 \neq t_2$, $x(t_1) = x(t_2)$, and $y(t_1) = y(t_2)$. $y(t_1) = y(t_2) \Rightarrow 3(t_1^2 - 3) = 3(t_2^2 - 3) \Rightarrow t_1^2 = t_2^2 \Rightarrow t_1 = \pm t_2$. Since $t_1 \neq t_2$, we conclude that $t_1 = -t_2$. Substituting $-t_2$ for t_1 in $x(t_1) = x(t_2)$ gives $x(-t_2) = x(t_2) \Rightarrow -t_2(t_2^2 - 3) = t_2(t_2^2 - 3) \Rightarrow$ $0 = 2t_2(t_2^2 - 3) \Rightarrow t_2 = 0, \pm\sqrt{3}$. Since $t_1 \neq t_2$, $t_2 = \pm\sqrt{3}$ and $t_1 = \mp\sqrt{3}$. These values give the point $(0,0)$.

$\dfrac{dy}{dx} = \dfrac{dy/dt}{dx/dt} = \dfrac{6t}{3t^2-3}$, so $\left.\dfrac{dy}{dx}\right|_{t=\pm\sqrt{3}} = \dfrac{6\left(\pm\sqrt{3}\right)}{6} = \pm\sqrt{3}$ and equations of the tangents are $y = \sqrt{3}x$ and $y = -\sqrt{3}x$.

21. The line with parametric equations $x = -7t$, $y = 12t - 5$ is $y = 12\left(-\frac{1}{7}x\right) - 5$, which has slope $-\frac{12}{7}$. The curve $x = t^3 + 4t$, $y = 6t^2$ has slope $\dfrac{dy}{dx} = \dfrac{dy/dt}{dx/dt} = \dfrac{12t}{3t^2+4}$. This equals $-\dfrac{12}{7} \iff 3t^2 + 4 = -7t \iff$ $(3t+4)(t+1) = 0 \iff t = -1$ or $t = -\frac{4}{3} \iff (x,y) = (-5, 6)$ or $\left(-\frac{208}{27}, \frac{32}{3}\right)$.

23. The coordinates of T are $(r\cos\theta, r\sin\theta)$. Since TP was unwound from arc TA, TP has length $r\theta$. Also $\angle PTQ = \angle PTR - \angle QTR = \frac{1}{2}\pi - \theta$, so P has coordinates $x = r\cos\theta + r\theta\cos\left(\frac{1}{2}\pi - \theta\right) = r(\cos\theta + \theta\sin\theta)$, $y = r\sin\theta - r\theta\sin\left(\frac{1}{2}\pi - \theta\right) = r(\sin\theta - \theta\cos\theta)$.

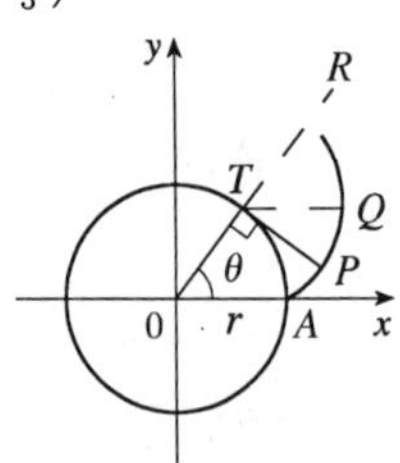

25. **(a)** The center Q of the smaller circle has coordinates $((a-b)\cos\theta, (a-b)\sin\theta)$. Arc PS on circle C has length $a\theta$ since it is equal in length to arc AS (the smaller circle rolls without slipping against the larger). Thus $\angle PQS = \frac{a}{b}\theta$ and $\angle PQT = \frac{a}{b}\theta - \theta$, so P has coordinates

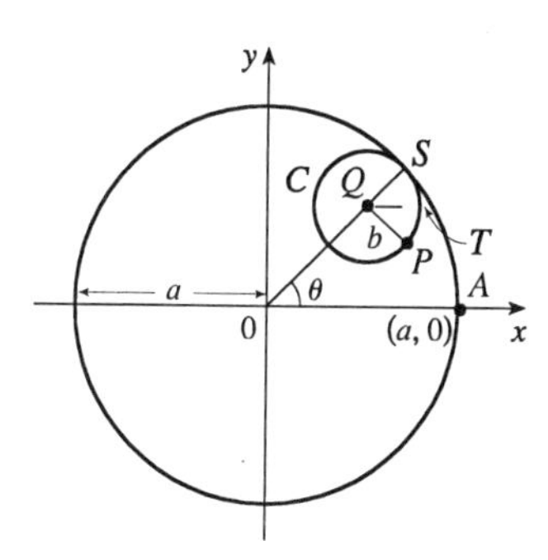

$$x = (a-b)\cos\theta + b\cos(\angle PQT) = (a-b)\cos\theta + b\cos\left(\frac{a-b}{b}\theta\right),$$

and $y = (a-b)\sin\theta - b\sin(\angle PQT) = (a-b)\sin\theta - b\sin\left(\frac{a-b}{b}\theta\right)$.

(b) If $b = \frac{a}{4}$, then $a - b = \frac{3a}{4}$ and $\frac{a-b}{b} = 3$, so

$x = \frac{3a}{4}\cos\theta + \frac{a}{4}\cos 3\theta = \frac{3a}{4}\cos\theta + \frac{a}{4}\left(4\cos^3\theta - 3\cos\theta\right) = a\cos^3\theta$ and

$y = \frac{3a}{4}\sin\theta - \frac{a}{4}\sin 3\theta = \frac{3a}{4}\sin\theta - \frac{a}{4}\left(3\sin\theta - 4\sin^3\theta\right) = a\sin^3\theta$.

The curve is symmetric about the origin.

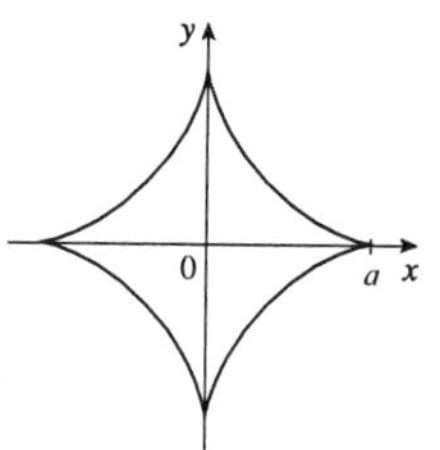

(c)

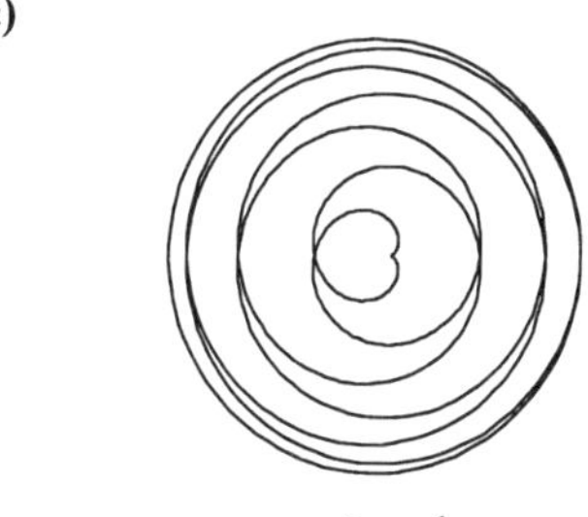

$a/b = \frac{1}{8}$

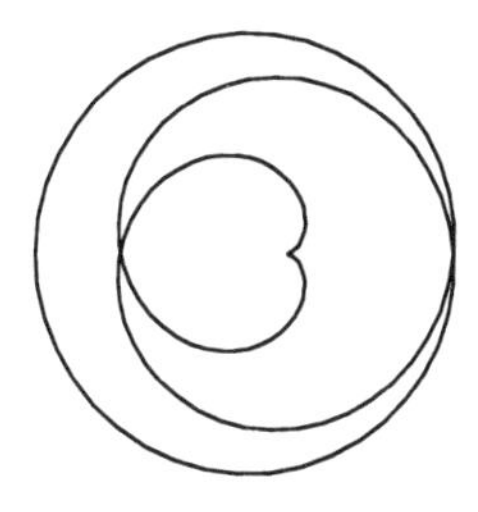

$a/b = \frac{1}{4}$

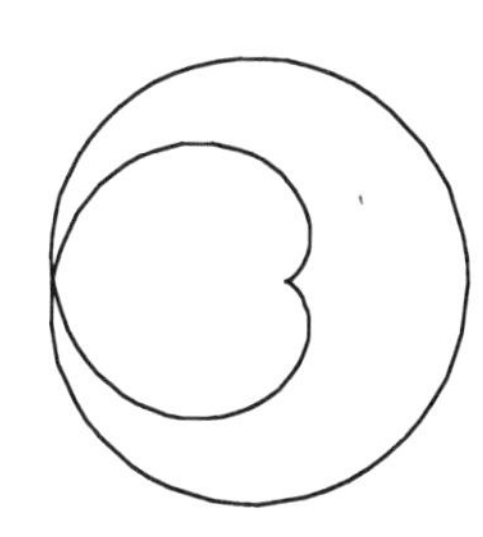

$a/b = \frac{1}{3}$

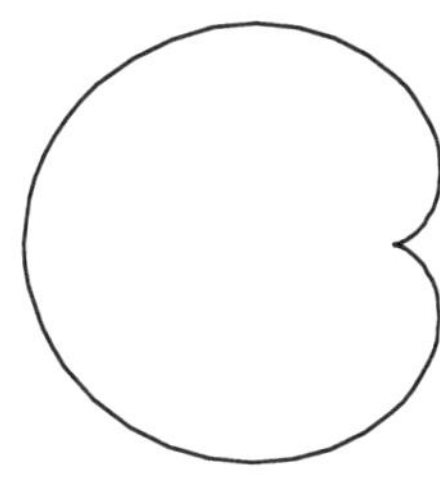

$a/b = \frac{1}{2}$

$a/b = e - 2, 0 \le t \le 446$

$a/b = \frac{7}{5}$

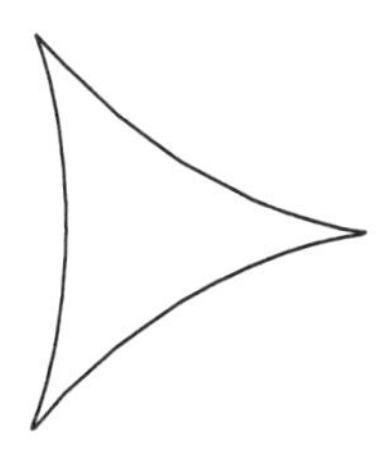

$a/b = 3$

$a/b = \frac{11}{3}$

$a/b = 23$

27. (a) $x = r\theta - d\sin\theta$, $y = r - d\cos\theta$; $\dfrac{dx}{d\theta} = r - d\cos\theta$, $\dfrac{dy}{d\theta} = d\sin\theta$. So $\dfrac{dy}{dx} = \dfrac{d\sin\theta}{r - d\cos\theta}$.

(b) If $0 < d < r$, then $|d\cos\theta| \le d < r$, so $r - d\cos\theta \ge r - d > 0$. This shows that $dx/d\theta$ never vanishes, so the trochoid can have no vertical tangents if $d < r$.

EXERCISES 3.10

1. $V = x^3 \quad \Rightarrow \quad \dfrac{dV}{dt} = 3x^2\dfrac{dx}{dt}$

3. $xy = 1 \quad \Rightarrow \quad x\dfrac{dy}{dt} + y\dfrac{dx}{dt} = 0$. If $\dfrac{dx}{dt} = 4$ and $x = 2$, then $y = \dfrac{1}{2}$, so $\dfrac{dy}{dt} = -\dfrac{y}{x}\dfrac{dx}{dt} = -\dfrac{1/2}{2}(4) = -1$.

5. If the radius is r and the diameter x, then $V = \frac{4}{3}\pi r^3 = \frac{\pi}{6}x^3 \quad \Rightarrow \quad -1 = \dfrac{dV}{dt} = \dfrac{\pi}{2}x^2\dfrac{dx}{dt} \quad \Rightarrow \quad \dfrac{dx}{dt} = -\dfrac{2}{\pi x^2}$.

When $x = 10$, $\dfrac{dx}{dt} = -\dfrac{2}{\pi(100)} = -\dfrac{1}{50\pi}$. So the rate of decrease is $\dfrac{1}{50\pi}\dfrac{\text{cm}}{\text{min}}$.

7. We are given that $dx/dt = 5$ ft/s. By similar triangles,

$\dfrac{15}{6} = \dfrac{x+y}{y} \quad \Rightarrow \quad y = \frac{2}{3}x$.

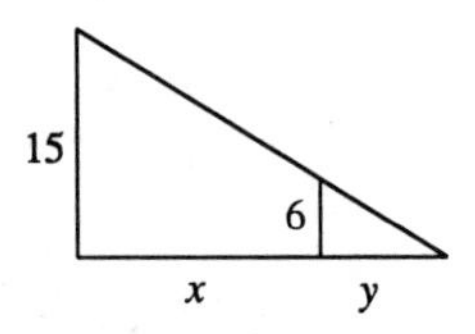

(a) The shadow moves at a rate of

$\dfrac{d}{dt}(x+y) = \dfrac{d}{dt}\left(x + \frac{2}{3}x\right) = \dfrac{5}{3}\dfrac{dx}{dt} = \frac{5}{3}(5) = \frac{25}{3}$ ft/s.

(b) The shadow lengthens at a rate of $\dfrac{dy}{dt} = \dfrac{d}{dt}\left(\frac{2}{3}x\right) = \dfrac{2}{3}\dfrac{dx}{dt} = \frac{2}{3}(5) = \dfrac{10}{3}$ ft/s.

9. We are given that $dx/dt = 500$ mi/h. By the Pythagorean Theorem,

$y^2 = x^2 + 1$, so $2y\dfrac{dy}{dt} = 2x\dfrac{dx}{dt} \quad \Rightarrow \quad \dfrac{dy}{dt} = \dfrac{x}{y}\dfrac{dx}{dt} = 500\dfrac{x}{y}$.

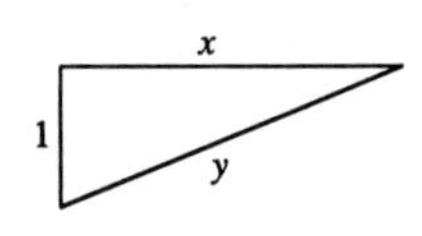

When $y = 2$, $x = \sqrt{3}$, so $\dfrac{dy}{dt} = 500\left(\frac{\sqrt{3}}{2}\right) = 250\sqrt{3}$ mi/h.

11. We are given that $\dfrac{dx}{dt} = 60$ mi/h and $\dfrac{dy}{dt} = 25$ mi/h.

$z^2 = x^2 + y^2 \quad \Rightarrow \quad 2z\dfrac{dz}{dt} = 2x\dfrac{dx}{dt} + 2y\dfrac{dy}{dt}$. After 2 hours, $x = 120$ and $y = 50 \quad \Rightarrow \quad z = 130$, so

$\dfrac{dz}{dt} = \dfrac{1}{z}\left(x\dfrac{dx}{dt} + y\dfrac{dy}{dt}\right) = \dfrac{120(60) + 50(25)}{130} = 65$ mi/h.

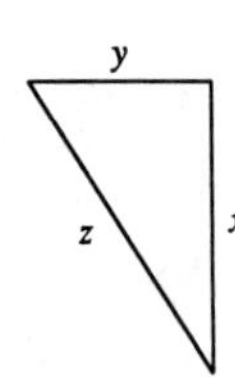

13. We are given that $\frac{dx}{dt} = 35$ km/h and $\frac{dy}{dt} = 25$ km/h.

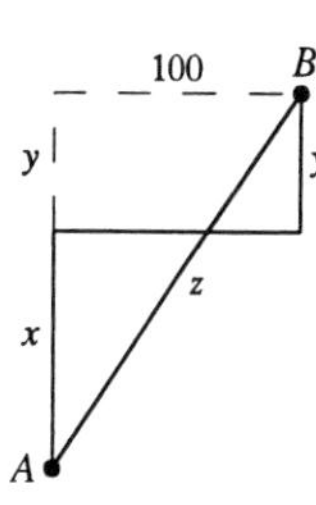

$z^2 = (x+y)^2 + 100^2 \quad\Rightarrow\quad 2z\frac{dz}{dt} = 2(x+y)\left(\frac{dx}{dt} + \frac{dy}{dt}\right)$.

At 4:00 P.M., $x = 140$ and $y = 100 \quad\Rightarrow\quad z = 260$, so

$\frac{dz}{dt} = \frac{x+y}{z}\left(\frac{dx}{dt} + \frac{dy}{dt}\right) = \frac{140+100}{260}(35+25) = \frac{720}{13} \approx 55.4$ km/h.

15. $A = \frac{bh}{2}$, where b is the base and h is the altitude. We are given that $\frac{dh}{dt} = 1$ and $\frac{dA}{dt} = 2$. So

$2 = \frac{dA}{dt} = \frac{b}{2}\frac{dh}{dt} + \frac{h}{2}\frac{db}{dt} = \frac{b}{2} + \frac{h}{2}\frac{db}{dt} \quad\Rightarrow\quad \frac{db}{dt} = \frac{4-b}{h}$. When $h = 10$ and $A = 100$, we have $b = 20$, so

$\frac{db}{dt} = \frac{4-20}{10} = -1.6$ cm/min.

17. If $C =$ the rate at which water is pumped in, then $\frac{dV}{dt} = C - 10{,}000$,

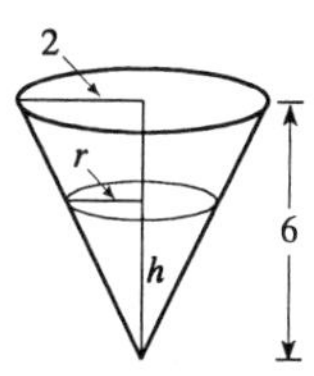

where $V = \frac{1}{3}\pi r^2 h$ is the volume at time t. By similar triangles, $\frac{r}{2} = \frac{h}{6}$

$\Rightarrow \quad r = \frac{1}{3}h \quad\Rightarrow\quad V = \frac{1}{3}\pi\left(\frac{1}{3}h\right)^2 h = \frac{\pi}{27}h^3 \quad\Rightarrow\quad \frac{dV}{dt} = \frac{\pi}{9}h^2\frac{dh}{dt}$.

When $h = 200$, $\frac{dh}{dt} = 20$, so $C - 10{,}000 = \frac{\pi}{9}(200)^2(20) \quad\Rightarrow$

$C = 10{,}000 + \frac{800{,}000}{9}\pi \approx 2.89 \times 10^5$ cm^3/min.

19. $V = \frac{1}{2}[0.3 + (0.3 + 2a)]h(10)$, where $\frac{a}{h} = \frac{0.25}{0.5} = \frac{1}{2}$ so

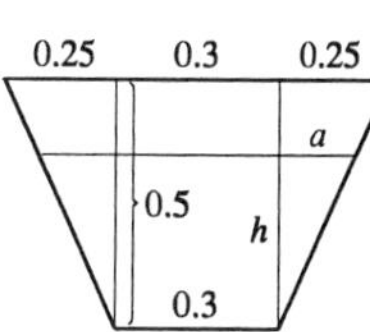
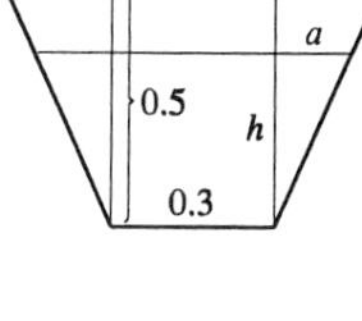

$2a = h \quad\Rightarrow\quad V = 5(0.6 + h)h = 3h + 5h^2 \quad\Rightarrow$

$0.2 = \frac{dV}{dt} = (3 + 10h)\frac{dh}{dt} \quad\Rightarrow\quad \frac{dh}{dt} = \frac{0.2}{3+10h}$. When

$h = 0.3$, $\frac{dh}{dt} = \frac{0.2}{3 + 10(0.3)} = \frac{0.2}{6}$ m/min $= \frac{10}{3}$ cm/min.

21. We are given that $\frac{dV}{dt} = 30$ ft^3/min. $V = \frac{1}{3}\pi\left(\frac{h}{2}\right)^2 h = \frac{h^3\pi}{12}$

$\Rightarrow \quad 30 = \frac{dV}{dt} = \frac{h^2\pi}{4}\frac{dh}{dt} \quad\Rightarrow\quad \frac{dh}{dt} = \frac{120}{\pi h^2}$.

When $h = 10$ ft, $\frac{dh}{dt} = \frac{120}{10^2\pi} = \frac{6}{5\pi} \approx 0.38$ ft/min.

23. $A = \frac{1}{2}bh$, but $b = 5$ m and $h = 4\sin\theta$ so $A = 10\sin\theta$.

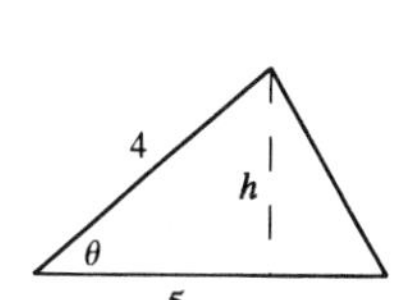

We are given $\frac{d\theta}{dt} = 0.06$ rad/s. $\frac{dA}{dt} = 10\cos\theta\,\frac{d\theta}{dt} = 0.6\cos\theta$.

When $\theta = \frac{\pi}{3}$, $\frac{dA}{dt} = 10(0.06)\left(\cos\frac{\pi}{3}\right) = (0.6)\left(\frac{1}{2}\right) = 0.3$ m^2/s.

25. $PV = C \quad\Rightarrow\quad P\frac{dV}{dt} + V\frac{dP}{dt} = 0 \quad\Rightarrow\quad \frac{dV}{dt} = -\frac{V}{P}\frac{dP}{dt}$. When $V = 600$, $P = 150$ and $\frac{dP}{dt} = 20$, we have $\frac{dV}{dt} = -\frac{600}{150}(20) = -80$, so the volume is decreasing at a rate of 80 cm^3/min.

27. **(a)** By the Pythagorean Theorem, $4000^2 + y^2 = \ell^2$. Differentiating with respect to t, we obtain $2y\dfrac{dy}{dt} = 2\ell\dfrac{d\ell}{dt}$. We know that $\dfrac{dy}{dt} = 600$, so when $y = 3000$ and $\ell = 5000$,

$$\frac{d\ell}{dt} = \frac{y(dy/dt)}{\ell} = \frac{3000(600)}{5000} = \frac{1800}{5} = 360\text{ ft/s}.$$

(b) Here $\tan\theta = y/4000$, so $\sec^2\theta\,\dfrac{d\theta}{dt} = \dfrac{1}{4000}\dfrac{dy}{dt} \Rightarrow \dfrac{d\theta}{dt} = \dfrac{\cos^2\theta}{4000}\dfrac{dy}{dt}$. When $y = 3000$, $\dfrac{dy}{dt} = 600$, $\ell = 5000$ and $\cos\theta = \dfrac{4000}{\ell} = \dfrac{4000}{5000} = \dfrac{4}{5}$, so $\dfrac{d\theta}{dt} = \dfrac{(4/5)^2}{4000}(600) = 0.096\text{ rad/s}$.

29. We are given that $\dfrac{dx}{dt} = 2\text{ ft/s}$. $x = 10\sin\theta \Rightarrow \dfrac{dx}{dt} = 10\cos\theta\,\dfrac{d\theta}{dt}$.

When $\theta = \dfrac{\pi}{4}$, $\dfrac{d\theta}{dt} = \dfrac{2}{10\left(1/\sqrt{2}\right)} = \dfrac{\sqrt{2}}{5}\text{ rad/s}$.

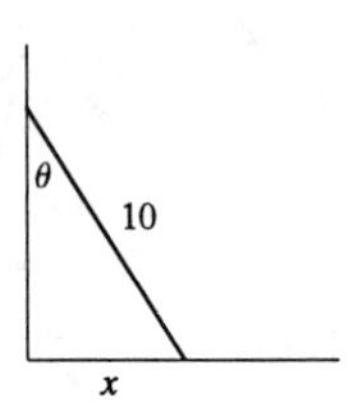

31. We are given that $\dfrac{dx}{dt} = 300\text{ km/h}$. By the Law of Cosines,

$y^2 = x^2 + 1 - 2x\cos 120^\circ = x^2 + 1 - 2x\left(-\frac{1}{2}\right) = x^2 + x + 1$,

so $2y\dfrac{dy}{dt} = 2x\dfrac{dx}{dt} + \dfrac{dx}{dt} \Rightarrow \dfrac{dy}{dt} = \dfrac{2x+1}{2y}\dfrac{dx}{dt}$. After 1 minute,

$x = \frac{300}{60} = 5 \Rightarrow y = \sqrt{31} \Rightarrow$

$$\frac{dy}{dt} = \frac{2(5)+1}{2\sqrt{31}}(300) = \frac{1650}{\sqrt{31}} \approx 296\text{ km/h}.$$

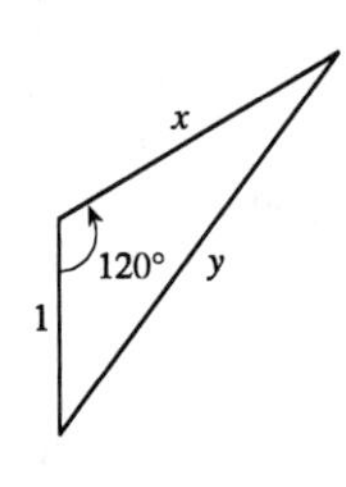

33. Let the distance between the runner and the friend be ℓ. Then by the Law of Cosines,

$\ell^2 = 200^2 + 100^2 - 2\cdot 200\cdot 100\cdot\cos\theta = 50{,}000 - 40{,}000\cos\theta$ (★).

Differentiating implicitly with respect to t, we obtain

$2\ell\dfrac{d\ell}{dt} = -40{,}000(-\sin\theta)\dfrac{d\theta}{dt}$. Now if D is the distance run when the angle is θ radians, then $D = 100\theta$, so $\theta = \frac{1}{100}D \Rightarrow$

$\dfrac{d\theta}{dt} = \dfrac{1}{100}\dfrac{dD}{dt} = \dfrac{7}{100}$. To substitute into the expression for $\dfrac{d\ell}{dt}$, we must know $\sin\theta$ at the time when $\ell = 200$, which we find from (★): $200^2 = 50{,}000 - 40{,}000\cos\theta \Leftrightarrow \cos\theta = \frac{1}{4}$

$\Rightarrow \sin\theta = \sqrt{1 - \left(\frac{1}{4}\right)^2} = \frac{\sqrt{15}}{4}$. Substituting, we get $2\ell\dfrac{d\ell}{dt} = 40{,}000\dfrac{\sqrt{15}}{4}\left(\dfrac{7}{100}\right) \Rightarrow$

$\dfrac{d\ell}{dt} = \dfrac{700\sqrt{15}}{2\cdot 200} = \dfrac{7\sqrt{15}}{4} \approx 6.78\text{ m/s}$. Whether the distance between them is increasing or decreasing depends on the direction in which the runner is running.

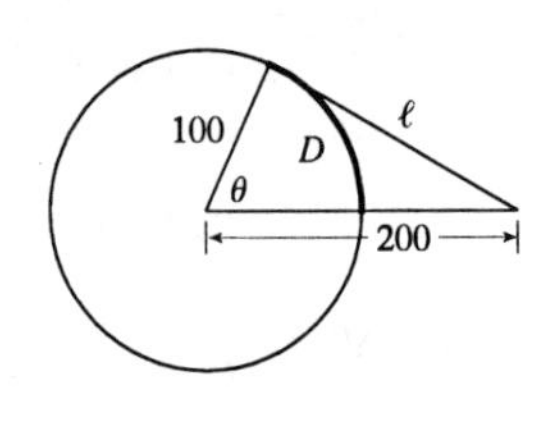

EXERCISES 3.11

1. $y = x^5 \quad \Rightarrow \quad dy = 5x^4 dx$

3. $y = \sqrt{x^4 + x^2 + 1} \quad \Rightarrow \quad dy = \frac{1}{2}(x^4 + x^2 + 1)^{-1/2}(4x^3 + 2x)dx = \dfrac{2x^3 + x}{\sqrt{x^4 + x^2 + 1}}dx$

5. $y = \sin 2x \quad \Rightarrow \quad dy = 2\cos 2x\, dx$

7. **(a)** $y = 1 - x^2 \quad \Rightarrow \quad dy = -2x\, dx$

(b) When $x = 5$ and $dx = \frac{1}{2}$, $dy = -2(5)\left(\frac{1}{2}\right) = -5$.

9. **(a)** $y = (x^2 + 5)^3 \quad \Rightarrow \quad dy = 3(x^2 + 5)^2\, 2x\, dx = 6x(x^2 + 5)^2\, dx$

(b) When $x = 1$ and $dx = 0.05$, $dy = 6(1)(1^2 + 5)^2(0.05) = 10.8$.

11. **(a)** $y = \cos x \quad \Rightarrow \quad dy = -\sin x\, dx$

(b) When $x = \frac{\pi}{6}$ and $dx = 0.05$, $dy = -\frac{1}{2}(0.05) = -0.025$.

13. $y = x^2$, $x = 1$, $\Delta x = 0.5 \quad \Rightarrow$
$\Delta y = (1.5)^2 - 1^2 = 1.25$.
$dy = 2x\, dx = 2(1)(0.5) = 1$

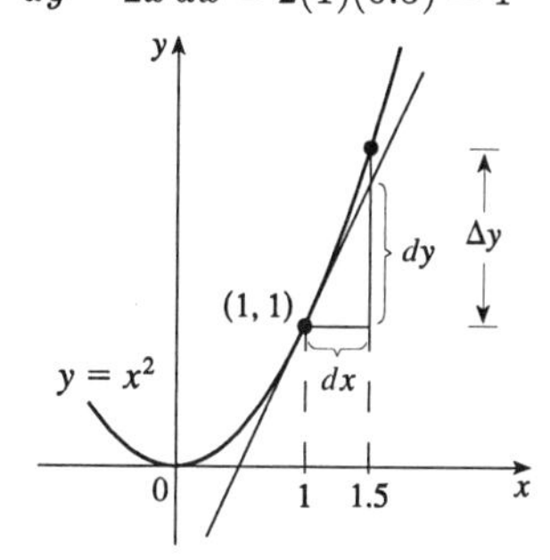

15. $y = 6 - x^2$, $x = -2$, $\Delta x = 0.4 \quad \Rightarrow$
$\Delta y = \left(6 - (-1.6)^2\right) - \left(6 - (-2)^2\right) = 1.44$
$dy = -2x\, dx = -2(-2)(0.4) = 1.6$

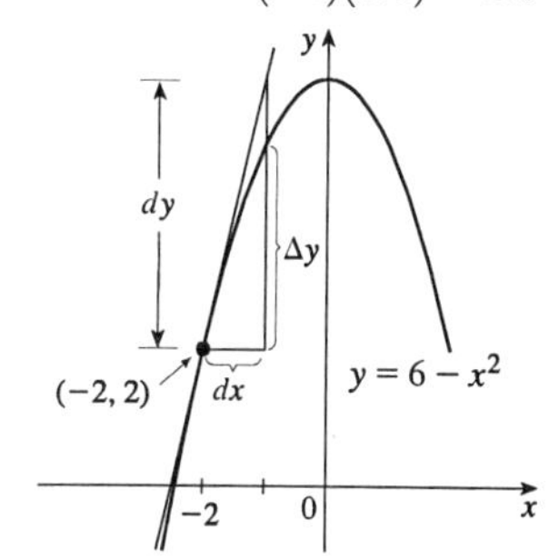

17. $y = f(x) = 2x^3 + 3x - 4$, $x = 3 \quad \Rightarrow \quad dy = (6x^2 + 3)dx = 57\, dx$

$\Delta x = 1 \quad \Rightarrow \quad \Delta y = f(4) - f(3) = 136 - 59 = 77$, $dy = 57(1) = 57$, $\Delta y - dy = 77 - 57 = 20$

$\Delta x = 0.5 \quad \Rightarrow \quad \Delta y = f(3.5) - f(3) = 92.25 - 59 = 33.25$, $dy = 57(0.5) = 28.5$,

$\Delta y - dy = 33.25 - 28.5 = 4.75$

$\Delta x = 0.1 \quad \Rightarrow \quad \Delta y = f(3.1) - f(3) = 64.882 - 59 = 5.882$, $dy = 57(0.1) = 5.7$,

$\Delta y - dy = 5.882 - 5.7 = 0.182$

$\Delta x = 0.01 \quad \Rightarrow \quad \Delta y = f(3.01) - f(3) = 59.571802 - 59 = 0.571802$, $dy = 57(0.01) = 0.57$,

$\Delta y - dy = 0.571802 - 0.57 = 0.001802$

19. $y = f(x) = \sqrt{x} \quad \Rightarrow \quad dy = \dfrac{1}{2\sqrt{x}}\,dx$. When $x = 36$ and $dx = 0.1$, $dy = \frac{1}{2\sqrt{36}}(0.1) = \frac{1}{120}$, so

$\sqrt{36.1} = f(36.1) \approx f(36) + dy = \sqrt{36} + \frac{1}{120} \approx 6.0083$.

21. $y = f(x) = 1/x \quad \Rightarrow \quad dy = (-1/x^2)dx$. When $x = 10$ and $dx = 0.1$, $dy = \left(-\frac{1}{100}\right)(0.1) = -0.001$, so

$\frac{1}{10.1} = f(10.1) \approx f(10) + dy = 0.1 - 0.001 = 0.099$.

23. $y = f(x) = \sin x \quad \Rightarrow \quad dy = \cos x\,dx$. When $x = \frac{\pi}{3}$ and $dx = -\frac{\pi}{180}$, $dy = \cos\frac{\pi}{3}\left(-\frac{\pi}{180}\right) = -\frac{\pi}{360}$, so

$\sin 59° = f\left(\frac{59}{180}\pi\right) \approx f\left(\frac{\pi}{3}\right) + dy = \frac{\sqrt{3}}{2} - \frac{\pi}{360} \approx 0.857$.

25. **(a)** If x is the edge length, then $V = x^3 \quad \Rightarrow \quad dV = 3x^2\,dx$. When $x = 30$ and $dx = 0.1$,

$dV = 3(30)^2(0.1) = 270$, so the maximum error is about 270 cm^3.

(b) $S = 6x^2 \quad \Rightarrow \quad dS = 12x\,dx$. When $x = 30$ and $dx = 0.1$, $dS = 12(30)(0.1) = 36$, so the maximum error is about 36 cm^2.

27. **(a)** For a sphere of radius r, the circumference is $C = 2\pi r$ and the surface area is $S = 4\pi r^2$, so $r = C/(2\pi)$

$\Rightarrow \quad S = 4\pi(C/2\pi)^2 = C^2/\pi \quad \Rightarrow \quad dS = (2/\pi)C\,dC$. When $C = 84$ and $dC = 0.5$,

$dS = \frac{2}{\pi}(84)(0.5) = \frac{84}{\pi}$, so the maximum error is about $\frac{84}{\pi} \approx 27\text{ cm}^2$.

(b) Relative error $\approx \dfrac{dS}{S} = \dfrac{84/\pi}{84^2/\pi} = \dfrac{1}{84} \approx 0.012$

29. **(a)** $V = \pi r^2 h \quad \Rightarrow \quad \Delta V \approx dV = 2\pi rh\,dr = 2\pi rh\,\Delta r$

(b) $\Delta V = \pi(r + \Delta r)^2 h - \pi r^2 h$, so the error is $\Delta V - dv = \pi(r + \Delta r)^2 h - \pi r^2 h - 2\pi rh\,\Delta r = \pi(\Delta r)^2 h$

31. $L(x) = f(1) + f'(1)(x - 1)$. $f(x) = x^3 \quad \Rightarrow \quad f'(x) = 3x^2$ so $f(1) = 1$ and $f'(1) = 3$. So

$L(x) = 1 + 3(x - 1) = 3x - 2$.

33. $f(x) = 1/x \quad \Rightarrow \quad f'(x) = -1/x^2$. So $f(4) = \frac{1}{4}$ and $f'(4) = -\frac{1}{16}$.

So $L(x) = f(4) + f'(4)(x - 4) = \frac{1}{4} + \left(-\frac{1}{16}\right)(x - 4) = \frac{1}{2} - \frac{1}{16}x$.

35. $f(x) = \sqrt{1 + x} \quad \Rightarrow \quad f'(x) = \dfrac{1}{2\sqrt{1 + x}}$ so $f(0) = 1$ and $f'(0) = \frac{1}{2}$.

So $f(x) \approx f(0) + f'(0)(x - 0) = 1 + \frac{1}{2}(x - 0) = 1 + \frac{1}{2}x$.

37. $f(x) = \dfrac{1}{(1 + 2x)^4} \quad \Rightarrow \quad f'(x) = \dfrac{-8}{(1 + 2x)^5}$ so $f(0) = 1$ and $f'(0) = -8$.

So $f(x) \approx f(0) + f'(0)(x - 0) = 1 + (-8)(x - 0) = 1 - 8x$.

39. $f(x) = \sqrt{1 - x} \quad \Rightarrow \quad f'(x) = \dfrac{-1}{2\sqrt{1 - x}}$ so $f(0) = 1$ and $f'(0) = -\frac{1}{2}$. Therefore

$\sqrt{1 - x} = f(x) \approx f(0) + f'(0)(x - 0) = 1 + \left(-\frac{1}{2}\right)(x - 0) = 1 - \frac{1}{2}x$.

So $\sqrt{0.9} = \sqrt{1 - 0.1} \approx 1 - \frac{1}{2}(0.1) = 0.95$ and

$\sqrt{0.99} = \sqrt{1 - 0.01} \approx 1 - \frac{1}{2}(0.01) = 0.995$.

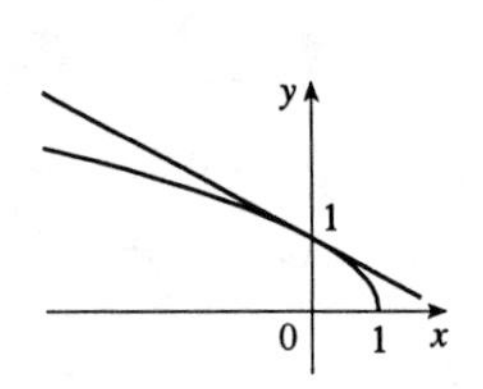

41. We need $\sqrt{1+x} - 0.1 < 1 + \frac{1}{2}x < \sqrt{1+x} + 0.1$. By zooming in or using a cursor, we see that this is true when $-0.69 < x < 1.09$.

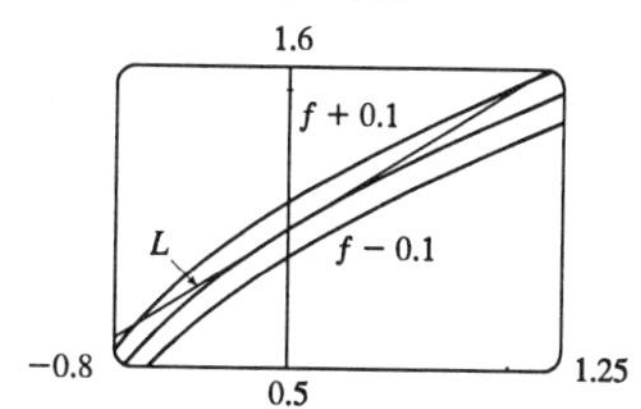

43. We need $1/(1+2x)^4 - 0.1 < 1 - 8x$ and $1 - 8x < 1/(1+2x)^4 + 0.1$, which both hold when $-0.045 < x < 0.055$.

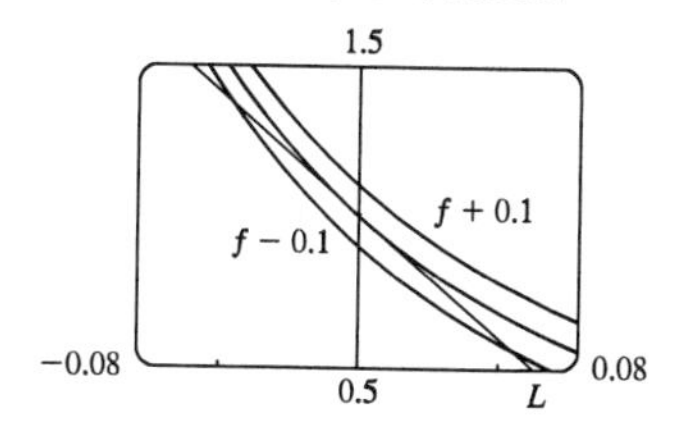

45. Using (10) with $f(x) = 1/x$, $f'(x) = -1/x^2$, and $f''(x) = 2/x^3$,

$$1/x \approx f(4) + f'(4)(x-4) + \tfrac{1}{2} f''(4)(x-4)^2 = \tfrac{1}{4} + (-1)4^{-2}(x-4) + \tfrac{1}{2}(2)4^{-3}(x-4)^2$$
$$= \tfrac{1}{4} - \tfrac{1}{16}(x-4) + \tfrac{1}{64}(x-4)^2$$

47. $f(x) = \sec x$, $f'(x) = \sec x \tan x$, and $f''(x) = \sec x \tan^2 x + \sec^3 x$, so

$$\sec x \approx f(0) + f'(0)(x) + \tfrac{1}{2} f''(x)(x)^2 = \sec 0 + \sec 0 \tan 0(x) + \tfrac{1}{2}[\sec 0(\sec^2 0) + \tan 0(\sec 0 \tan 0)]x^2$$
$$= \tfrac{1}{2}x^2 + 1.$$

49. $f(x) = \sqrt{x}$, $f'(x) = \dfrac{1}{2\sqrt{x}}$, and $f''(x) = \dfrac{1}{-4x\sqrt{x}}$,

so the linear approximation is

$$\sqrt{x} \approx f(1) + f'(1)(x-1) = \sqrt{1} + \tfrac{1}{2\sqrt{1}}(x-1) = 1 + \tfrac{1}{2}(x-1),$$

and the quadratic approximation is

$$\sqrt{x} \approx f(1) + f'(1)(x-1) + \tfrac{1}{2} f''(1)(x-1)^2 = 1 + \tfrac{1}{2}(x-1) - \tfrac{1}{8}(x-1)^2.$$

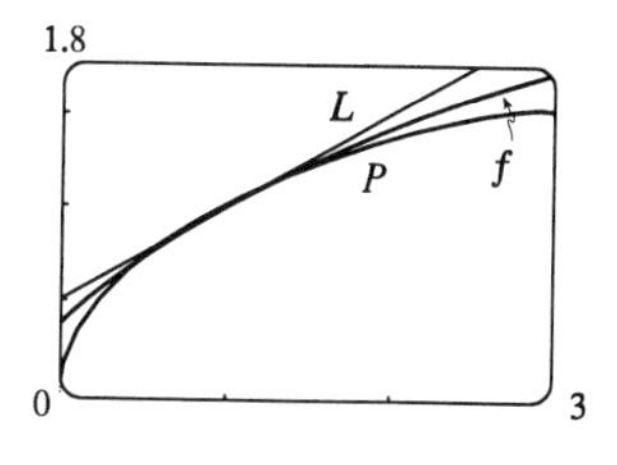

51. **(a)** $f(x) = \cos x \;\Rightarrow\; f'(x) = -\sin x \;\Rightarrow\; f''(x) = -\cos x$.

Thus the linear approximation is

$$\cos x \approx f\left(\tfrac{\pi}{6}\right) + f'\left(\tfrac{\pi}{6}\right)\left(x - \tfrac{\pi}{6}\right) = \cos\tfrac{\pi}{6} - \sin\tfrac{\pi}{6}\left(x - \tfrac{\pi}{6}\right)$$
$$= \tfrac{\sqrt{3}}{2} - \tfrac{1}{2}\left(x - \tfrac{\pi}{6}\right),$$ and the quadratic approximation is

$$\cos x \approx f\left(\tfrac{\pi}{6}\right) + f'\left(\tfrac{\pi}{6}\right)\left(x - \tfrac{\pi}{6}\right) + \tfrac{1}{2}f''\left(\tfrac{\pi}{6}\right)\left(x - \tfrac{\pi}{6}\right)^2$$
$$= \cos\tfrac{\pi}{6} - \sin\tfrac{\pi}{6}\left(x - \tfrac{\pi}{6}\right) + \tfrac{1}{2}\left(-\cos\tfrac{\pi}{6}\right)\left(x - \tfrac{\pi}{6}\right)^2$$
$$= \tfrac{\sqrt{3}}{2} - \tfrac{1}{2}\left(x - \tfrac{\pi}{6}\right) - \tfrac{\sqrt{3}}{4}\left(x - \tfrac{\pi}{6}\right)^2.$$

(b)

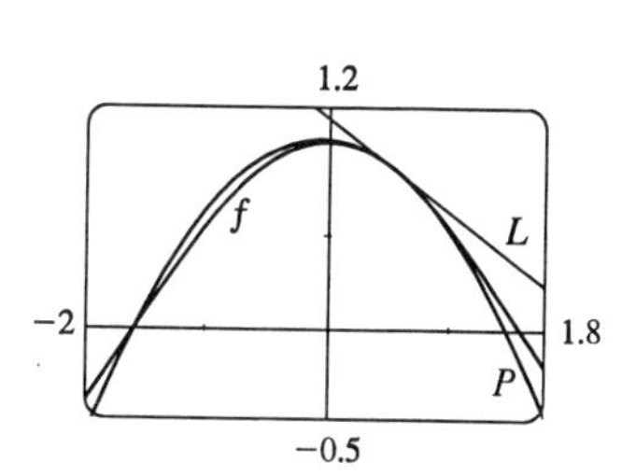

(c) We need $\cos x - 0.1 < \frac{\sqrt{3}}{2} - \frac{1}{2}\left(x - \frac{\pi}{6}\right) < \cos x + 0.1$. From the graph, it appears that the linear approximation has the required accuracy when $0.06 < x < 1.03$.

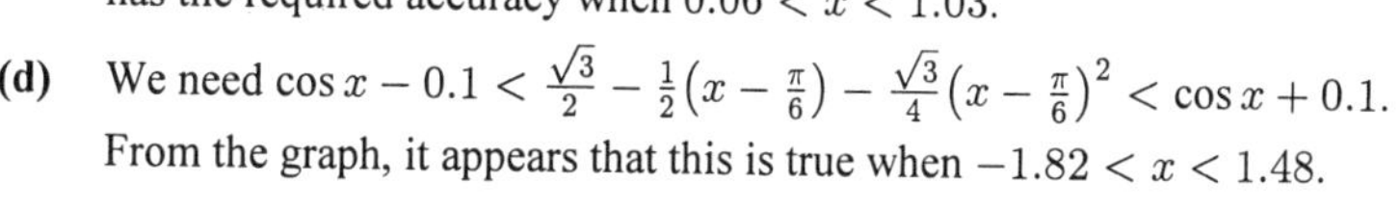

(d) We need $\cos x - 0.1 < \frac{\sqrt{3}}{2} - \frac{1}{2}\left(x - \frac{\pi}{6}\right) - \frac{\sqrt{3}}{4}\left(x - \frac{\pi}{6}\right)^2 < \cos x + 0.1$. From the graph, it appears that this is true when $-1.82 < x < 1.48$.

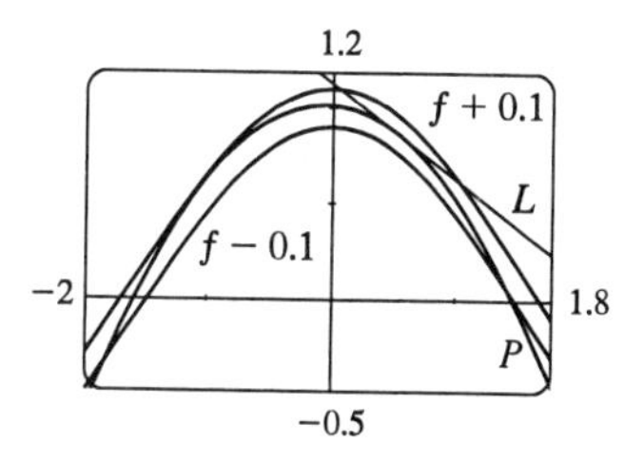

53. **(a)** $dc = \dfrac{dc}{dx}\,dx = 0\,dx = 0$

(b) $d(cu) = \dfrac{d}{dx}(cu)dx = c\dfrac{du}{dx}\,dx = c\,du$

(c) $d(u+v) = \dfrac{d}{dx}(u+v)dx = \left(\dfrac{du}{dx} + \dfrac{dv}{dx}\right)dx = \dfrac{du}{dx}\,dx + \dfrac{dv}{dx}\,dx = du + dv$

(d) $d(uv) = \dfrac{d}{dx}(uv)dx = \left(u\dfrac{dv}{dx} + v\dfrac{du}{dx}\right)dx = u\dfrac{dv}{dx}\,dx + v\dfrac{du}{dx}\,dx = u\,dv + v\,du$

(e) $d\left(\dfrac{u}{v}\right) = \dfrac{d}{dx}\left(\dfrac{u}{v}\right)dx = \dfrac{v\dfrac{du}{dx} - u\dfrac{dv}{dx}}{v^2}\,dx = \dfrac{v\dfrac{du}{dx}\,dx - u\dfrac{dv}{dx}\,dx}{v^2} = \dfrac{v\,du - u\,dv}{v^2}$

(f) $d(x^n) = \dfrac{d}{dx}(x^n)dx = nx^{n-1}\,dx$

55. $P(x) = a_0 + a_1x + a_2x^2 + a_3x^3 + \cdots + a_nx^n \quad \Rightarrow$

$P'(x) = a_1 + 2a_2x + 3a_3x^2 + \cdots \quad \Rightarrow$

$P''(x) = 2a_2 + 2\cdot 3a_3x + 3\cdot 4a_4x + \cdots \quad \Rightarrow$

$P'''(x) = 2\cdot 3a_3 + 2\cdot 3\cdot 4a_4x + \cdots \quad \Rightarrow$

$P^{(k)}(x) = 2\cdot 3\cdot 4\cdot\cdots\cdot ka_k + 2\cdot 3\cdot 4\cdot\cdots\cdot k\cdot(k+1)a_{k+1}x + \cdots \quad \Rightarrow$

$P^{(n)}(x) = n!\,a_n$. Therefore $P^{(k)}(0) = f^{(k)}(0) = k!\,a_k$, and so $a_k = \dfrac{f^{(k)}(0)}{k!}$ for $k = 1, 2, \ldots, n$.

Now let $f(x) = \sin x$. Then $f'(x) = \cos x$, $f''(x) = -\sin x$ and $f'''(x) = -\cos x$. So the Taylor polynomial of degree 3 for $\sin x$ is $P(x) = \sin 0 + \dfrac{\cos 0}{1!}x + \dfrac{-\sin 0}{2!}x^2 + \dfrac{-\cos 0}{3!}x^3 = x - \dfrac{x^3}{6}$.

EXERCISES 3.12

1.

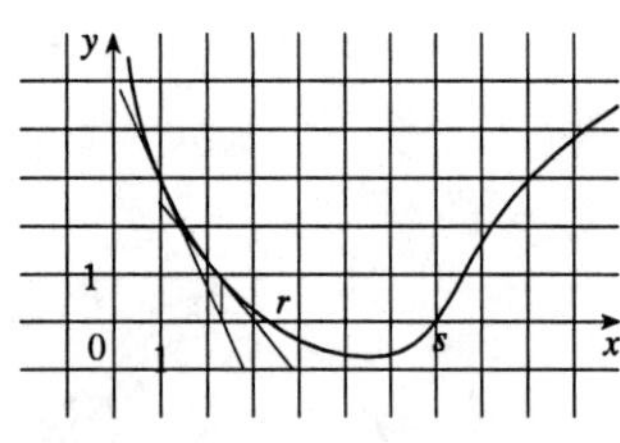

$x_2 \approx 2.3$, $x_3 \approx 3$

3. $f(x) = x^3 + x + 1 \quad \Rightarrow \quad f'(x) = 3x^2 + 1$, so $x_{n+1} = x_n - \dfrac{x_n^3 + x_n + 1}{3x_n^2 + 1}$. $x_1 = -1 \quad \Rightarrow$

$x_2 = -1 - \dfrac{-1 - 1 + 1}{3\cdot 1 + 1} = -0.75 \quad \Rightarrow \quad x_3 = -0.75 - \dfrac{(-0.75)^3 - 0.75 + 1}{3(-0.75)^2 + 1} \approx -0.6860$

5. $f(x) = x^5 - 10 \quad \Rightarrow \quad f'(x) = 5x^4$, so $x_{n+1} = x_n - \dfrac{x_n^5 - 10}{5x_n^4}$. $x_1 = 1.5 \quad \Rightarrow$

$x_2 = 1.5 - \dfrac{(1.5)^5 - 10}{5(1.5)^4} \approx 1.5951 \quad \Rightarrow \quad x_3 = 1.5951 - \dfrac{f(1.5951)}{f'(1.5951)} \approx 1.5850$

7. Finding $\sqrt[4]{22}$ is equivalent to finding the positive root of $x^4 - 22 = 0$ so we take $f(x) = x^4 - 22 \quad \Rightarrow$ $f'(x) = 4x^3$ and $x_{n+1} = x_n - \dfrac{x_n^4 - 22}{4x_n^3}$. Taking $x_1 = 2$, we get $x_2 = 2.1875$, $x_3 \approx 2.16605940$, $x_4 \approx 2.16573684$ and $x_5 \approx x_6 \approx 2.16573677$. Thus $\sqrt[4]{22} \approx 2.16573677$ to eight decimal places.

9. $f(x) = x^3 - 2x - 1 \quad \Rightarrow \quad f'(x) = 3x^2 - 2$, so $x_{n+1} = x_n - \dfrac{x_n^3 - 2x_n - 1}{3x_n^2 - 2}$. Taking $x_1 = 1.5$, we get $x_2 \approx 1.631579$, $x_3 \approx 1.618184$, $x_4 \approx 1.618034$ and $x_5 \approx 1.618034$. So the root is 1.618034 to six decimal places.

11. From the graph it appears that there is a root near 2, so we take $x_1 = 2$. Write the equation as $f(x) = 2\sin x - x = 0$. Then $f'(x) = 2\cos x - 1$, so $x_{n+1} = x_n - \dfrac{2\sin x_n - x_n}{2\cos x_n - 1} \quad \Rightarrow \quad x_1 = 2$, $x_2 \approx 1.900996$, $x_3 \approx 1.895512$, $x_4 \approx 1.895494$, and $x_5 \approx 1.895494$. So the root is 1.895494 to 6 decimal places.

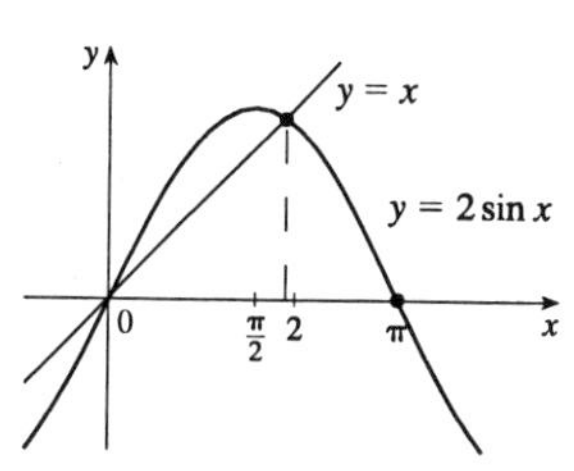

13. $f(x) = x^3 - 4x + 1 \quad \Rightarrow \quad f'(x) = 3x^2 - 4$, so $x_{n+1} = x_n - \dfrac{x_n^3 - 4x_n + 1}{3x_n^2 - 4}$. Observe that $f(-3) = -14$, $f(-2) = 1$, $f(0) = 1$, $f(1) = -2$ and $f(2) = 1$ so there are roots in $[-3, -2]$, $[0, 1]$ and $[1, 2]$.

$[-3, -2]$
$x_1 = -2$
$x_2 = -2.125$
$x_3 \approx -2.114975$
$x_4 \approx -2.114908$
$x_5 \approx -2.114908$

$[0, 1]$
$x_1 = 0$
$x_2 = 0.25$
$x_3 \approx 0.254098$
$x_4 \approx 0.254102$
$x_5 \approx 0.254102$

$[1, 2]$
$x_1 = 2$
$x_2 = 1.875$
$x_3 \approx 1.860979$
$x_4 \approx 1.860806$
$x_5 \approx 1.860806$

To six decimal places, the roots are -2.114908, 0.254102 and 1.860806.

15. $f(x) = x^4 + x^2 - x - 1 \quad \Rightarrow \quad f'(x) = 4x^3 + 2x - 1$, so $x_{n+1} = x_n - \dfrac{x_n^4 + x_n^2 - x_n - 1}{4x_n^3 + 2x_n - 1}$. Note that $f(1) = 0$, so $x = 1$ is a root. Also $f(-1) = 2$ and $f(0) = -1$, so there is a root in $[-1, 0]$. A sketch shows that these are the only roots. Taking $x_1 = -0.5$, we have $x_2 = -0.575$, $x_3 \approx -0.569867$, $x_4 \approx -0.569840$ and $x_5 \approx -0.569840$. The roots are 1 and -0.569840, to six decimal places.

17.

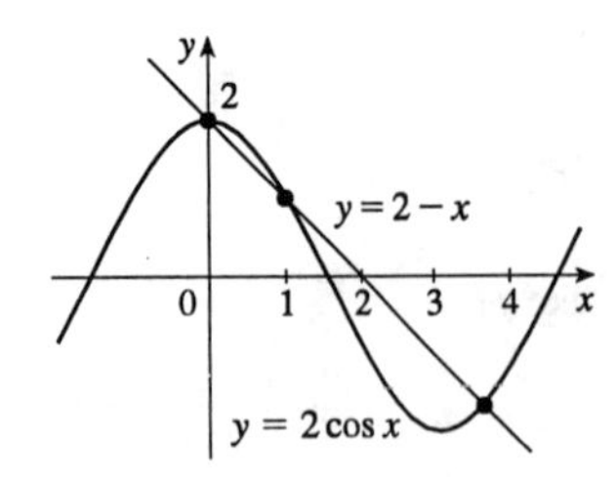

Clearly $x = 0$ is a root. From the sketch, there appear to be roots near 1 and 3.5. Write the equation as $f(x) = 2\cos x + x - 2 = 0$.

Then $f'(x) = -2\sin x + 1$, so $x_{n+1} = x_n - \dfrac{2\cos x_n + x_n - 2}{1 - 2\sin x_n}$.

Taking $x_1 = 1$, we get $x_2 \approx 1.118026$, $x_3 \approx 1.109188$, $x_4 \approx 1.109144$ and $x_5 \approx 1.109144$. Taking $x_1 = 3.5$, we get $x_2 \approx 3.719159$, $x_3 \approx 3.698331$, $x_4 \approx 3.698154$ and $x_5 \approx 3.698154$.

To six decimal places the roots are 0, 1.109144 and 3.698154.

19.

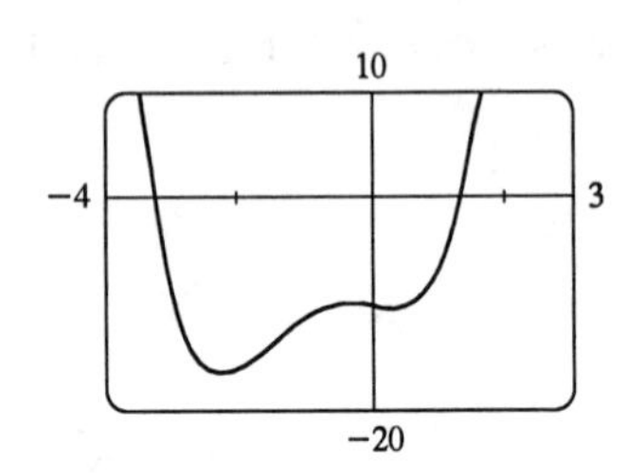

From the graph, there appear to be roots near -3.2 and 1.4.

Let $f(x) = x^4 + 3x^3 - x - 10 \quad \Rightarrow \quad f'(x) = 4x^3 + 9x^2 - 1$, so

$x_{n+1} = x_n - \dfrac{x_n^4 + 3x_n^3 - x_n - 10}{4x_n^3 + 9x_n^2 - 1}$. Taking $x_1 = -3.2$, we get

$x_2 \approx -3.20617358$, $x_3 \approx -3.20614267 \approx x_4$. Taking $x_1 = 1.4$, we get $x_2 \approx 1.37560834$, $x_3 \approx 1.37506496$, $x_4 \approx 1.37506470 \approx x_5$.

To eight decimal places, the roots are -3.20614267 and 1.37506470.

21.

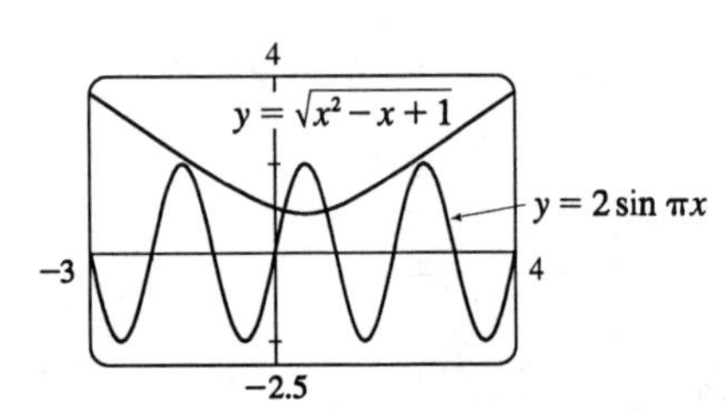

From the graph, we see that there are roots of this equation near 0.2 and 0.8. Let $f(x) = \sqrt{x^2 - x + 1} - 2\sin \pi x \quad \Rightarrow$

$$f'(x) = \frac{2x - 1}{2\sqrt{x^2 - x + 1}} - 2\pi \cos \pi x, \text{ so}$$

$$x_{n+1} = x_n - \frac{\sqrt{x_n^2 - x_n + 1} - 2\sin \pi x_n}{\dfrac{2x_n - 1}{2\sqrt{x_n^2 - x_n + 1}} - 2\pi\cos \pi x_n}.$$

Taking $x_1 = 0.2$, we get $x_2 \approx 0.15212015$, $x_3 \approx 0.15438067$, $x_4 \approx 0.15438500 \approx x_5$. Taking $x_1 = 0.8$, we get $x_2 \approx 0.84787985$, $x_3 \approx 0.84561933$, $x_4 \approx 0.845614500 \approx x_5$. So, to eight decimal places, the roots of the equation are 0.15438500 and 0.84561500.

23. **(a)** $f(x) = x^2 - a \quad \Rightarrow \quad f'(x) = 2x$, so Newton's Method gives

$$x_{n+1} = x_n - \frac{x_n^2 - a}{2x_n} = x_n - \tfrac{1}{2}x_n + \frac{a}{2x_n} = \frac{1}{2}\left(x_n + \frac{a}{x_n}\right).$$

(b) Using (a) with $x_1 = 30$, we get $x_2 \approx 31.666667$, $x_3 \approx 31.622807$, $x_4 \approx 31.622777$ and $x_5 \approx 31.622777$. So $\sqrt{1000} \approx 31.622777$.

25. If we attempt to compute x_2 we get $x_2 = x_1 - \dfrac{f(x_1)}{f'(x_1)}$, but $f(x) = x^3 - 3x + 6 \quad \Rightarrow$

$f'(x_1) = 3x_1^2 - 3 = 3(1)^2 - 3 = 0$. For Newton's Method to work $f'(x_n) \neq 0$ (no horizontal tangents).

27. For $f(x) = x^{1/3}$, $f'(x) = \frac{1}{3}x^{-2/3}$ and $x_{n+1} = x_n - \dfrac{f(x_n)}{f'(x_n)} = x_n - \dfrac{x_n^{1/3}}{\frac{1}{3}x_n^{-2/3}} = x_n - 3x_n = -2x_n$. Therefore each successive approximation becomes twice as large as the previous one in absolute value, so the sequence of approximations fails to converge to the root, which is 0.

29. The volume of the silo, in terms of its radius, is $V(r) = \pi r^2(30) + \frac{1}{2}\left(\frac{4}{3}\pi r^3\right) = 30\pi r^2 + \frac{2}{3}\pi r^3$. From a graph of V, we see that $V(r) = 15{,}000$ at $r \approx 11$ ft. Now we use Newton's Method to solve the equation $V(r) - 15{,}000 = 0$. First we must calculate $\dfrac{dV}{dr} = 60\pi r + 2\pi r^2$, so $r_{n+1} = r_n - \dfrac{30\pi r_n^2 + \frac{2}{3}\pi r_n^3 - 15{,}000}{60\pi r_n + 2\pi r_n^2}$.
Taking $r_1 = 11$, we get $r_2 = 11.2853$, $r_3 = 11.2807 \approx r_4$. So in order for the silo to hold 15,000 ft^3 of grain, its radius must be about 11.2807 ft.

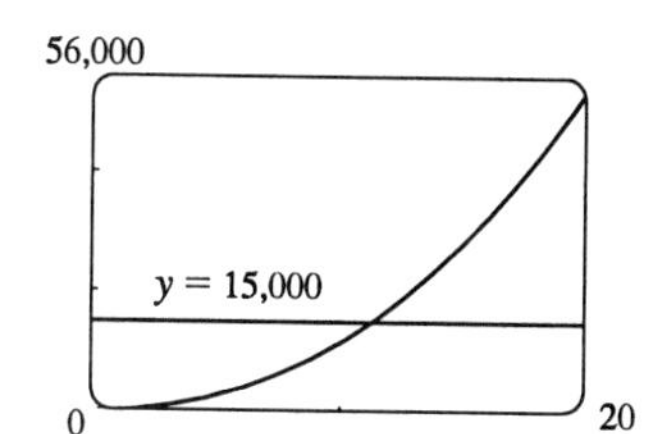

31. In this case, $A = 18{,}000$, $R = 375$, and $n = 60$. So the formula becomes $18{,}000 = \dfrac{375}{x}\left[1 - (1+x)^{-60}\right] \Leftrightarrow$ $48x = 1 - (1+x)^{-60} \Leftrightarrow 48x(1+x)^{60} - (1+x)^{60} + 1 = 0$. Let the LHS be called $f(x)$, so that $f'(x) = 48x(60)(1+x)^{59} + 48(1+x)^{60} - 60(1+x)^{59} = 12(1+x)^{59}(244x - 1)$. So we use Newton's Method with $x_{n+1} = x_n - \dfrac{48x_n(1+x_n)^{60} - (1+x_n)^{60} + 1}{12(1+x_n)^{59}(244x_n - 1)}$ and $x_1 = 1\% = 0.01$.

We get $x_2 \approx 0.0082202$, $x_3 \approx 0.0076802$, $x_4 \approx 0.0076291$, $x_5 \approx 0.0076286 \approx x_6$. So the dealer is charging a monthly interest rate of 0.76286%.

REVIEW EXERCISES FOR CHAPTER 3

1. False; see the warning after Theorem 3.1.8.

3. False. See the discussion before the Product Rule.

5. True, by the Chain Rule.

7. False. $f(x) = |x^2 + x| = x^2 + x$ for $x \ge 0$ or $x \le -1$ and $|x^2 + x| = -(x^2 + x)$ for $-1 < x < 0$. So $f'(x) = 2x + 1$ for $x > 0$ or $x < -1$ and $f'(x) = -(2x + 1)$ for $-1 < x < 0$. But $|2x + 1| = 2x + 1$ for $x \ge -\frac{1}{2}$ and $|2x + 1| = -2x - 1$ for $x < -\frac{1}{2}$.

9. True. $g(x) = x^5 \Rightarrow g'(x) = 5x^4 \Rightarrow g'(2) = 5(2)^4 = 80$, and by the definition of the derivative, $\lim\limits_{x \to 2} \dfrac{g(x) - g(2)}{x - 2} = g'(2) = 80$.

11. False. A tangent to the parabola has slope $\dfrac{dy}{dx} = 2x$, so at $(-2, 4)$ the slope of the tangent is $2(-2) = -4$ and the equation is $y - 4 = -4(x + 2)$. [The equation $y - 4 = 2x(x + 2)$ is not even linear!]

13. $f(x) = x^3 + 5x + 4 \quad \Rightarrow$

$$f'(x) = \lim_{h \to 0} \frac{f(x+h) - f(x)}{h} = \lim_{h \to 0} \frac{(x+h)^3 + 5(x+h) + 4 - (x^3 + 5x + 4)}{h}$$

$$= \lim_{h \to 0} \frac{3x^2h + 3xh^2 + h^3 + 5h}{h} = \lim_{h \to 0} \left(3x^2 + 3xh + h^2 + 5\right) = 3x^2 + 5$$

15. $f(x) = \sqrt{3 - 5x} \quad \Rightarrow$

$$f'(x) = \lim_{h \to 0} \frac{f(x+h) - f(x)}{h} = \lim_{h \to 0} \frac{\sqrt{3 - 5(x+h)} - \sqrt{3 - 5x}}{h}$$

$$= \lim_{h \to 0} \frac{\sqrt{3 - 5x - 5h} - \sqrt{3 - 5x}}{h} \left(\frac{\sqrt{3 - 5x - 5h} + \sqrt{3 - 5x}}{\sqrt{3 - 5x - 5h} + \sqrt{3 - 5x}} \right)$$

$$= \lim_{h \to 0} \frac{-5h}{h\left(\sqrt{3 - 5x - 5h} + \sqrt{3 - 5x}\right)} = \lim_{h \to 0} \frac{-5}{\sqrt{3 - 5x - 5h} + \sqrt{3 - 5x}} = \frac{-5}{2\sqrt{3 - 5x}}$$

17. $y = (x+2)^8(x+3)^6 \quad \Rightarrow \quad y' = 6(x+3)^5(x+2)^8 + 8(x+2)^7(x+3)^6 = 2(7x+18)(x+2)^7(x+3)^5$

19. $y = \dfrac{x}{\sqrt{9 - 4x}} \quad \Rightarrow \quad y' = \dfrac{\sqrt{9 - 4x} - x\left[-4/\left(2\sqrt{9 - 4x}\right)\right]}{9 - 4x} = \dfrac{9 - 4x + 2x}{(9 - 4x)^{3/2}} = \dfrac{9 - 2x}{(9 - 4x)^{3/2}}$

21. $x^2y^3 + 3y^2 = x - 4y \quad \Rightarrow \quad 2xy^3 + 3x^2y^2y' + 6yy' = 1 - 4y' \quad \Rightarrow \quad y' = \dfrac{1 - 2xy^3}{3x^2y^2 + 6y + 4}$

23. $y = \sqrt{x\sqrt{x\sqrt{x}}} = \left[x\left(x^{3/2}\right)^{1/2}\right]^{1/2} = \left[x\left(x^{3/4}\right)\right]^{1/2} = x^{7/8} \quad \Rightarrow \quad y' = \frac{7}{8}x^{-1/8}$

25. $y = \dfrac{x}{8 - 3x} \quad \Rightarrow \quad y' = \dfrac{(8 - 3x) - x(-3)}{(8 - 3x)^2} = \dfrac{8}{(8 - 3x)^2}$

27. $y = (x \tan x)^{1/5} \quad \Rightarrow \quad y' = \frac{1}{5}(x \tan x)^{-4/5}(\tan x + x \sec^2 x)$

29. $x^2 = y(y+1) = y^2 + y \quad \Rightarrow \quad 2x = 2yy' + y' \quad \Rightarrow \quad y' = 2x/(2y+1)$

31. $y = \dfrac{(x-1)(x-4)}{(x-2)(x-3)} = \dfrac{x^2 - 5x + 4}{x^2 - 5x + 6} \quad \Rightarrow$

$$y' = \frac{(x^2 - 5x + 6)(2x - 5) - (x^2 - 5x + 4)(2x - 5)}{(x^2 - 5x + 6)^2} = \frac{2(2x - 5)}{(x-2)^2(x-3)^2}$$

33. $y = \tan\sqrt{1 - x} \quad \Rightarrow \quad y' = \left(\sec^2\sqrt{1 - x}\right)\left(\dfrac{1}{2\sqrt{1 - x}}\right)(-1) = -\dfrac{\sec^2\sqrt{1 - x}}{2\sqrt{1 - x}}$

35. $y = \sin\left(\tan\sqrt{1 + x^3}\right) \quad \Rightarrow \quad y' = \cos\left(\tan\sqrt{1 + x^3}\right)\left(\sec^2\sqrt{1 + x^3}\right)\left[3x^2/\left(2\sqrt{1 + x^3}\right)\right]$

37. $y = \cot(3x^2 + 5) \quad \Rightarrow \quad y' = -\csc^2(3x^2 + 5)(6x) = -6x\csc^2(3x^2 + 5)$

39. $y = \cos^2(\tan x) \quad \Rightarrow \quad y' = 2\cos(\tan x)[-\sin(\tan x)]\sec^2 x = -\sin(2\tan x)\sec^2 x$

41. $f(x) = (2x - 1)^{-5} \quad \Rightarrow \quad f'(x) = -5(2x - 1)^{-6}(2) = -10(2x - 1)^{-6} \quad \Rightarrow$

$f''(x) = 60(2x - 1)^{-7}(2) = 120(2x - 1)^{-7} \quad \Rightarrow \quad f''(0) = 120(-1)^{-7} = -120$

43. $x^6 + y^6 = 1 \quad \Rightarrow \quad 6x^5 + 6y^5y' = 0 \quad \Rightarrow \quad y' = -\dfrac{x^5}{y^5} \quad \Rightarrow$

$$y'' = -\frac{5x^4y^5 - x^5(5y^4y')}{y^{10}} = -\frac{5x^4y^5 - 5x^5y^4\left(-x^5/y^5\right)}{y^{10}} = -\frac{5x^4y^6 + 5x^{10}}{y^{11}} = -\frac{5x^4(y^6 + x^6)}{y^{11}} = -\frac{5x^4}{y^{11}}$$

45. $\lim_{x\to 0} \dfrac{\sec x}{1-\sin x} = \dfrac{\sec 0}{1-\sin 0} = \dfrac{1}{1-0} = 1$

47. $y = \dfrac{x}{x^2-2} \quad\Rightarrow\quad y' = \dfrac{(x^2-2)-x(2x)}{(x^2-2)^2} = \dfrac{-x^2-2}{(x^2-2)^2}$. When $x=2$, $y' = \dfrac{-2^2-2}{(2^2-2)^2} = -\dfrac{3}{2}$, so the equation of the tangent at $(2,1)$ is $y-1 = -\frac{3}{2}(x-2)$ or $3x+2y-8=0$.

49. $y = \tan x \quad\Rightarrow\quad y' = \sec^2 x$. When $x = \frac{\pi}{3}$, $y' = 2^2 = 4$, so the equation of the tangent line at $\left(\frac{\pi}{3}, \sqrt{3}\right)$ is $y - \sqrt{3} = 4\left(x - \frac{\pi}{3}\right)$ or $y = 4x + \sqrt{3} - \frac{4}{3}\pi$.

51. $y = \sin x + \cos x \quad\Rightarrow\quad y' = \cos x - \sin x = 0 \quad\Leftrightarrow\quad \cos x = \sin x$ and $0 \le x \le 2\pi \quad\Leftrightarrow\quad x = \frac{\pi}{4}$ or $\frac{5\pi}{4}$, so the points are $\left(\frac{\pi}{4}, \sqrt{2}\right)$ and $\left(\frac{5\pi}{4}, -\sqrt{2}\right)$.

53. $f(x) = (x-a)(x-b)(x-c) \quad\Rightarrow\quad f'(x) = (x-b)(x-c) + (x-a)(x-c) + (x-a)(x-b)$. So $\dfrac{f'(x)}{f(x)} = \dfrac{(x-b)(x-c)+(x-a)(x-c)+(x-a)(x-b)}{(x-a)(x-b)(x-c)} = \dfrac{1}{x-a} + \dfrac{1}{x-b} + \dfrac{1}{x-c}$.

55. **(a)** $h'(x) = f'(x)g(x) + f(x)g'(x) \quad\Rightarrow\quad h'(2) = f'(2)g(2) + f(2)g'(2) = (-2)(5) + (3)(4) = 2$

(b) $F'(x) = f'(g(x))g'(x) \quad\Rightarrow\quad F'(2) = f'(g(2))g'(2) = f'(5)(4) = 11 \cdot 4 = 44$

57. The graph of a has tangent lines with positive slope for $x < 0$ and negative slope for $x > 0$, and the values of c fit this pattern, so c must be the graph of the derivative of the function for a. The graph of c has horizontal tangent lines to the left and right of the x-axis and b has zeros at these points. Hence b is the graph of the derivative of the function for c. Therefore a is the graph of f, c is the graph of f', and b is the graph of f''.

59. **(a)** $f(x) = x\sqrt{5-x} \quad\Rightarrow\quad f'(x) = \dfrac{-x}{2\sqrt{5-x}} + \sqrt{5-x} = \dfrac{10-3x}{2\sqrt{5-x}}$

(b) At $(1,2)$: $f'(1) = 1\left(\dfrac{-1}{2\sqrt{5-1}}\right) + \sqrt{4} = \dfrac{7}{4}$. So the equation of the tangent is

$y - 2 = \frac{7}{4}(x-1) \quad\Leftrightarrow\quad y = \frac{7}{4}x + \frac{1}{4}$.

At $(4,4)$: $f'(4) = 4\left(\frac{-1}{2\sqrt{1}}\right) + \sqrt{1} = -1$.

So the equation of the tangent is

$y - 4 = -(x-4) \quad\Leftrightarrow\quad y = -x + 8$.

(c)

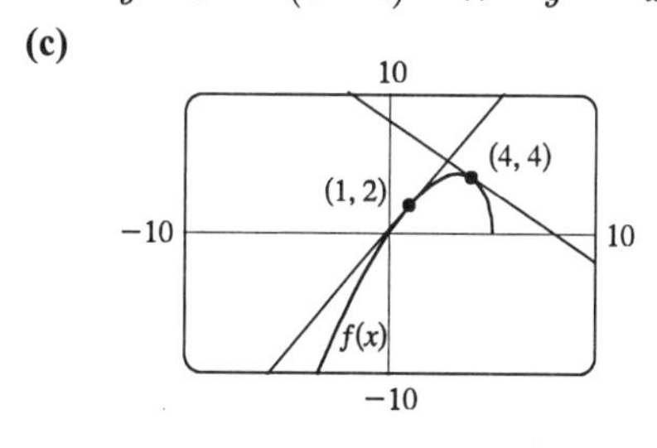

(d)

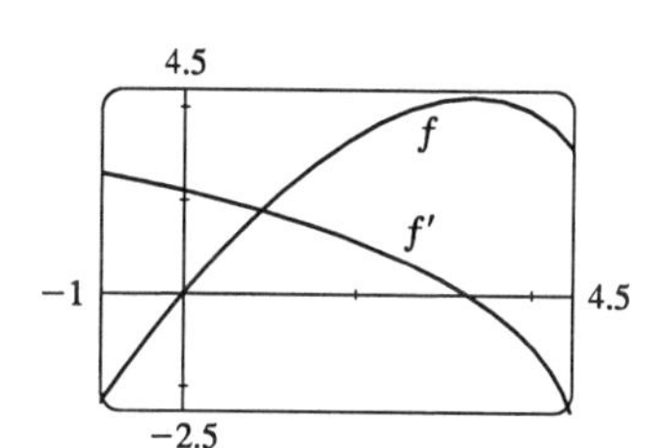

The graphs look reasonable, since f' is positive where f has tangents with positive slope, and f' is negative where f has tangents with negative slope.

61. $f(x) = x^2 g(x) \quad\Rightarrow\quad f'(x) = 2xg(x) + x^2 g'(x)$

63. $f(x) = (g(x))^2 \quad\Rightarrow\quad f'(x) = 2g(x)\,g'(x)$

65. $f(x) = g(g(x)) \quad\Rightarrow\quad f'(x) = g'(g(x))g'(x)$

67. $h(x) = \dfrac{f(x)g(x)}{f(x)+g(x)} \quad\Rightarrow$

$$h'(x) = \frac{[f'(x)g(x) + f(x)g'(x)][f(x)+g(x)] - f(x)g(x)[f'(x)+g'(x)]}{[f(x)+g(x)]^2} = \frac{f'(x)[g(x)]^2 + g'(x)[f(x)]^2}{[f(x)+g(x)]^2}$$

69. Using the Chain Rule repeatedly, $h(x) = f(g(\sin 4x)) \quad\Rightarrow$

$$h'(x) = f'(g(\sin 4x)) \cdot \frac{d}{dx}(g(\sin 4x)) = f'(g(\sin 4x)) \cdot g'(\sin 4x) \cdot \frac{d}{dx}(\sin 4x)$$
$$= f'(g(\sin 4x))g'(\sin 4x)(\cos 4x)(4)$$

71. **(a)** $y = t^3 - 12t + 3 \quad\Rightarrow\quad v(t) = y' = 3t^2 - 12 \quad\Rightarrow\quad a(t) = v'(t) = 6t$

(b) $v(t) = 3(t^2 - 4) > 0$ when $t > 2$, so it moves upward when $t > 2$ and downward when $0 \le t < 2$.

(c) Distance upward $= y(3) - y(2) = -6 - (-13) = 7$,
Distance downward $= y(0) - y(2) = 3 - (-13) = 16$. Total distance $= 7 + 16 = 23$

73. **(a)** $\mathbf{r}(t) = \langle \sin 2t, \cos 2t\rangle \quad\Rightarrow\quad x = \sin 2t,\ y = \cos 2t \quad\Rightarrow\quad x^2 + y^2 = 1$. For increasing t, the motion around the circle is clockwise.

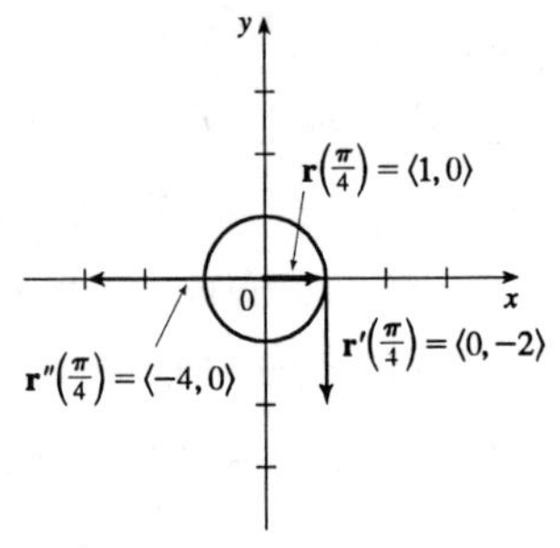

(b) $\mathbf{r}'(t) = \langle 2\cos 2t, -2\sin 2t\rangle$ and $\mathbf{r}''(t) = \langle -4\sin 2t, -4\cos 2t\rangle = -4\mathbf{r}(t)$.

(c) $t = \frac{\pi}{4}$, so $\mathbf{r}\left(\frac{\pi}{4}\right) = \langle 1, 0\rangle$, $\mathbf{r}'\left(\frac{\pi}{4}\right) = \langle 0, -2\rangle$, and $\mathbf{r}''\left(\frac{\pi}{4}\right) = \langle -4, 0\rangle$.

75. **(a)** $\mathbf{r}(t) = (4t^2 - 6t)\mathbf{i} + t^3\mathbf{j} \quad\Rightarrow\quad x = 4t^2 - 6t,\ y = t^3 \quad\Rightarrow\quad x = 4\left(y^{1/3}\right)^2 - 6\left(y^{1/3}\right) = 4y^{2/3} - 6y^{1/3}$

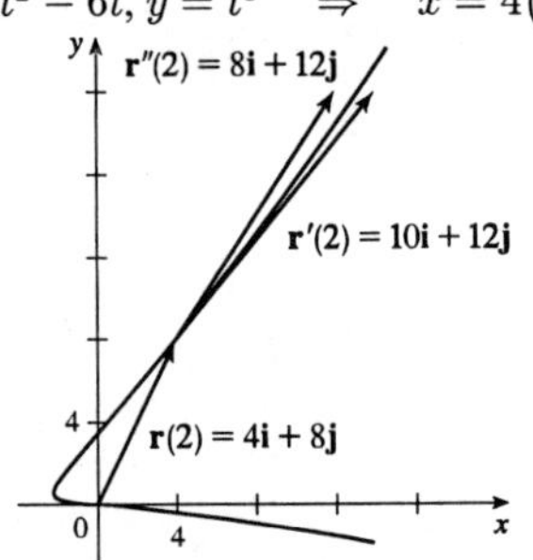

(b) $\mathbf{r}'(t) = (8t - 6)\mathbf{i} + 3t^2\mathbf{j}$ and $\mathbf{r}''(t) = 8\mathbf{i} + 6t\mathbf{j}$.

(c) $t = 2$, so $\mathbf{r}(2) = 4\mathbf{i} + 8\mathbf{j}$, $\mathbf{r}'(2) = 10\mathbf{i} + 12\mathbf{j}$, and $\mathbf{r}''(2) = 8\mathbf{i} + 12\mathbf{j}$.

77. position: $\mathbf{r}(t) = (t^2 - 4t)\mathbf{i} + (t^3 - 5t)\mathbf{j}$; $\mathbf{r}(1) = -3\mathbf{i} - 4\mathbf{j}$
velocity: $\mathbf{r}'(t) = (2t - 4)\mathbf{i} + (3t^2 - 5)\mathbf{j}$; $\mathbf{r}'(1) = -2\mathbf{i} - 2\mathbf{j}$
acceleration: $\mathbf{r}''(t) = 2\mathbf{i} + 6t\mathbf{j}$; $\mathbf{r}''(1) = 2\mathbf{i} + 6\mathbf{j}$
speed: $|\mathbf{r}'(1)| = |-2\mathbf{i} - 2\mathbf{j}| = \sqrt{(-2)^2 + (-2)^2} = \sqrt{8} = 2\sqrt{2}$

79. position: $\mathbf{r}(t) = \langle 3\cos 2t, -4\sin 2t\rangle$; $\mathbf{r}\left(\frac{\pi}{8}\right) = \langle 3\sqrt{2}/2, -2\sqrt{2}\rangle$
velocity: $\mathbf{r}'(t) = \langle -6\sin 2t, -8\cos 2t\rangle$; $\mathbf{r}'\left(\frac{\pi}{8}\right) = \langle -3\sqrt{2}, -4\sqrt{2}\rangle$
acceleration: $\mathbf{r}''(t) = \langle -12\cos 2t, 16\sin 2t\rangle$; $\mathbf{r}''\left(\frac{\pi}{8}\right) = \langle -6\sqrt{2}, 8\sqrt{2}\rangle$
speed: $\left|\mathbf{r}'\left(\frac{\pi}{8}\right)\right| = \left|\langle -3\sqrt{2}, -4\sqrt{2}\rangle\right| = \sqrt{\left(-3\sqrt{2}\right)^2 + \left(-4\sqrt{2}\right)^2} = \sqrt{18+32} = 5\sqrt{2}$

81. $x = t^2 + 2t$, $y = t^3 - t$; $t = 1$. $\dfrac{dy}{dt} = 3t^2 - 1$, $\dfrac{dx}{dt} = 2t + 2$, so $\dfrac{dy}{dx} = \dfrac{dy/dt}{dx/dt} = \dfrac{3t^2-1}{2t+2}$. When $t = 1$, $(x, y) = (3, 0)$ and $\dfrac{dy}{dx} = \dfrac{2}{4} = \dfrac{1}{2}$, so an equation of the tangent is $y - 0 = \frac{1}{2}(x - 3)$, or $y = \frac{1}{2}x - \frac{3}{2}$.

83. $\mathbf{r}(t) = t\cos t\,\mathbf{i} + t\sin t\,\mathbf{j} \Rightarrow x = t\cos t$, $y = t\sin t$; $t = \frac{\pi}{4}$. $\dfrac{dy}{dx} = \dfrac{dy/dt}{dx/dt} = \dfrac{t\cos t + \sin t}{-t\sin t + \cos t}$. When $t = \frac{\pi}{4}$, $(x, y) = \left(\frac{\pi}{4}\cdot\frac{\sqrt{2}}{2}, \frac{\pi}{4}\cdot\frac{\sqrt{2}}{2}\right) = \left(\frac{\pi\sqrt{2}}{8}, \frac{\pi\sqrt{2}}{8}\right)$ and $\dfrac{dy}{dx} = \dfrac{\frac{\pi}{4}\cdot\frac{\sqrt{2}}{2} + \frac{\sqrt{2}}{2}}{-\frac{\pi}{4}\cdot\frac{\sqrt{2}}{2} + \frac{\sqrt{2}}{2}} = \dfrac{\frac{\sqrt{2}}{2}\left(\frac{\pi}{4}+1\right)}{\frac{\sqrt{2}}{2}\left(-\frac{\pi}{4}+1\right)} = \dfrac{\pi+4}{-\pi+4}$, so an equation of the tangent is $y - \frac{\pi\sqrt{2}}{8} = \frac{4+\pi}{4-\pi}\left(x - \frac{\pi\sqrt{2}}{8}\right)$, or $y = \left(\frac{4+\pi}{4-\pi}\right)x - \frac{\pi\sqrt{2}}{8}\left(\frac{8+\pi}{4-\pi}\right)$.

85. $m = x\left(1 + \sqrt{x}\right) = x + x^{3/2} \Rightarrow \rho = dm/dx = 1 + \frac{3}{2}\sqrt{x}$, so the density when $x = 4$ is $1 + \frac{3}{2}\sqrt{4} = 4\text{ kg/m}$.

87. If x = edge length, then $V = x^3 \Rightarrow dV/dt = 3x^2\,dx/dt = 10 \Rightarrow dx/dt = 10/(3x^2)$ and $S = 6x^2$ $\Rightarrow dS/dt = (12x)dx/dt = 12x\left[10/(3x^2)\right] = 40/x$. When $x = 30$, $dS/dt = \frac{40}{30} = \frac{4}{3}\text{ cm}^2/\text{min}$.

89. Given $dh/dt = 5$ and $dx/dt = 15$, find dz/dt. $z^2 = x^2 + h^2 \Rightarrow$
$2z\dfrac{dz}{dt} = 2x\dfrac{dx}{dt} + 2h\dfrac{dh}{dt} \Rightarrow \dfrac{dz}{dt} = \dfrac{1}{z}(15x + 5h)$.
When $t = 3$, $h = 45 + 3(5) = 60$ and $x = 15(3) = 45 \Rightarrow$
$z = 75$, so $\dfrac{dz}{dt} = \frac{1}{75}[15(45) + 5(60)] = 13\text{ ft/s}$.

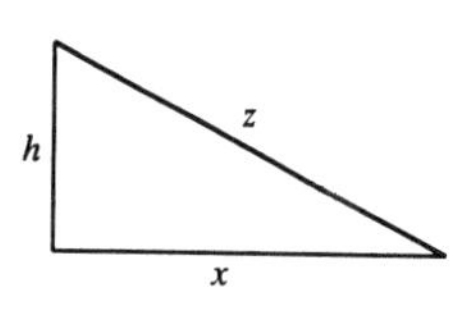

91. We are given $d\theta/dt = -0.25$ rad/h.
$x = 400\cot\theta \Rightarrow \dfrac{dx}{dt} = -400\csc^2\theta\,\dfrac{d\theta}{dt}$.
When $\theta = \dfrac{\pi}{6}$, $\dfrac{dx}{dt} = -400(2)^2(-0.25) = 400\text{ ft/h}$.

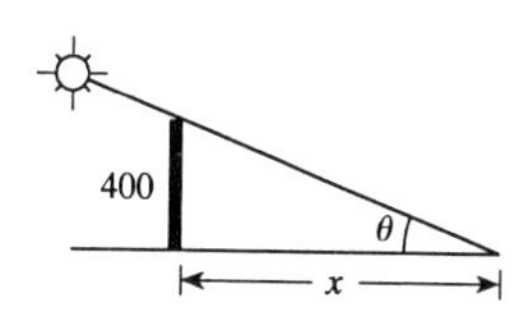

93. $y = x^3 - 2x^2 + 1 \Rightarrow dy = (3x^2 - 4x)dx$. When $x = 2$ and $dx = 0.2$, $dy = \left[3(2)^2 - 4(2)\right](0.2) = 0.8$.

95. $f(x) = \sqrt[3]{1+3x} = (1+3x)^{1/3} \Rightarrow f'(x) = (1+3x)^{-2/3}$ so
$L(x) = f(0) + f'(0)(x - 0) = 1^{1/3} + 1^{-2/3}x = 1 + x$. Thus $\sqrt[3]{1+3x} \approx 1 + x \Rightarrow$
$\sqrt[3]{1.03} = \sqrt[3]{1+3(0.01)} \approx 1 + (0.01) = 1.01$.

97. The linear approximation is $\sqrt[3]{1+3x} \approx 1 + x$, so for the required accuracy we want
$\sqrt[3]{1+3x} - 0.1 < 1 + x < \sqrt[3]{1+3x} + 0.1$.
From the graph, it appears that this is true when $-0.23 < x < 0.40$.

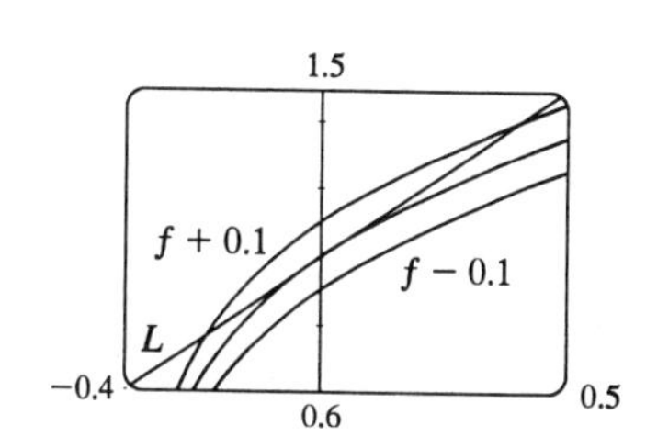

99. $f(x) = x^4 + x - 1 \quad \Rightarrow \quad f'(x) = 4x^3 + 1 \quad \Rightarrow \quad x_{n+1} = x_n - \dfrac{x_n^4 + x_n - 1}{4x_n^3 + 1}$. If $x_1 = 0.5$ then $x_2 \approx 0.791667$, $x_3 \approx 0.729862$, $x_4 \approx 0.724528$, $x_5 \approx 0.724492$ and $x_6 \approx 0.724492$, so, to six decimal places, the root is 0.724492.

101. $y = x^6 + 2x^2 - 8x + 3$ has a horizontal tangent when $y' = 6x^5 + 4x - 8 = 0$. Let $f(x) = 6x^5 + 4x - 8$. Then $f'(x) = 30x^4 + 4$, so $x_{n+1} = x_n - \dfrac{6x_n^5 + 4x_n - 8}{30\,x_n^4 + 4}$. A sketch shows that the root is near 1, so we take $x_1 = 1$. Then $x_2 \approx 0.9412$, $x_3 \approx 0.9341$, $x_4 \approx 0.9340$ and $x_5 \approx 0.9340$. Thus, to four decimal places, the point is $(0.9340, -2.0634)$.

103. $\displaystyle\lim_{h\to 0} \frac{(2+h)^6 - 64}{h} = \frac{d}{dx}x^6\bigg|_{x=2} = 6(2)^5 = 192$

105. Differentiating the expression for $g(x)$ and using the Chain Rule repeatedly, we obtain

$$\begin{aligned} g(x) &= f(x^3 + f(x^2 + f(x))) \quad \Rightarrow \\ g'(x) &= f'(x^3 + f(x^2 + f(x)))\,(x^3 + f(x^2 + f(x)))' \\ &= f'(x^3 + f(x^2 + f(x)))\left(3x^2 + f'(x^2 + f(x))\,[x^2 + f(x)]'\right) \\ &= f'(x^3 + f(x^2 + f(x)))\,(3x^2 + f'(x^2 + f(x))[2x + f'(x)]). \text{ So} \end{aligned}$$

$$\begin{aligned} g'(1) &= f'(1^3 + f(1^2 + f(1)))\,(3\cdot 1^2 + f'(1^2 + f(1))[2\cdot 1 + f'(1)]) \\ &= f'(1 + f(1+1))(3 + f'(1+1)[2 + f'(1)]) = f'(1+2)[3 + f'(2)(2+1)] = 3(3 + 2\cdot 3) = 27. \end{aligned}$$

107.

$$\begin{aligned} \lim_{x\to 0} \frac{\sqrt{1+\tan x} - \sqrt{1+\sin x}}{x^3} &= \lim_{x\to 0} \frac{\left(\sqrt{1+\tan x} - \sqrt{1+\sin x}\right)\left(\sqrt{1+\tan x} + \sqrt{1+\sin x}\right)}{x^3\left(\sqrt{1+\tan x} + \sqrt{1+\sin x}\right)} \\ &= \lim_{x\to 0} \frac{(1+\tan x) - (1+\sin x)}{x^3\left(\sqrt{1+\tan x} + \sqrt{1+\sin x}\right)} = \lim_{x\to 0} \frac{\sin x(1/\cos x - 1)\cos x}{x^3\left(\sqrt{1+\tan x} + \sqrt{1+\sin x}\right)\cos x} \\ &= \lim_{x\to 0} \frac{\sin x\,(1-\cos x)(1+\cos x)}{x^3\left(\sqrt{1+\tan x} + \sqrt{1+\sin x}\right)\cos x\,(1+\cos x)} \\ &= \lim_{x\to 0} \frac{\sin x \cdot \sin^2 x}{x^3\left(\sqrt{1+\tan x} + \sqrt{1+\sin x}\right)\cos x\,(1+\cos x)} \\ &= \left(\lim_{x\to 0} \frac{\sin x}{x}\right)^3 \lim_{x\to 0} \frac{1}{\left(\sqrt{1+\tan x} + \sqrt{1+\sin x}\right)\cos x\,(1+\cos x)} \\ &= 1^3 \cdot \frac{1}{\left(\sqrt{1} + \sqrt{1}\right)\cdot 1 \cdot (1+1)} = \frac{1}{4} \end{aligned}$$

109. We are given that $|f(x)| \le x^2$ for all x. In particular, $|f(0)| \le 0$, but $|a| \ge 0$ for all a. The only conclusion is that $f(0) = 0$. Now $\left|\dfrac{f(x) - f(0)}{x - 0}\right| = \left|\dfrac{f(x)}{x}\right| = \dfrac{|f(x)|}{|x|} \le \dfrac{x^2}{|x|} = \dfrac{|x^2|}{|x|} \quad \Rightarrow \quad -|x| \le \dfrac{f(x) - f(0)}{x - 0} \le |x|$. But $\displaystyle\lim_{x\to 0} -|x| = 0 = \lim_{x\to 0}|x|$, so by the Squeeze Theorem, $\displaystyle\lim_{x\to 0} \frac{f(x) - f(0)}{x - 0} = 0$. So by the definition of the derivative, f is differentiable at 0 and, furthermore, $f'(0) = 0$.

PROBLEMS PLUS (after Chapter 3)

1. Let a be the x-coordinate of Q. Then $y = 1 - x^2 \quad \Rightarrow \quad$ the slope at $Q = y'(a) = -2a$. But since the triangle is equilateral, $\angle ACB = 60°$, so that the slope at Q is $\tan 120° = -\sqrt{3}$. Therefore we must have that $-2a = -\sqrt{3} \quad \Rightarrow \quad a = \frac{\sqrt{3}}{2}$. Therefore the point Q has coordinates $\left(\frac{\sqrt{3}}{2}, 1 - \left(\frac{\sqrt{3}}{2}\right)^2\right) = \left(\frac{\sqrt{3}}{2}, \frac{1}{4}\right)$ and by symmetry P has coordinates $\left(-\frac{\sqrt{3}}{2}, \frac{1}{4}\right)$.

3. $1 + x + x^2 + \cdots + x^{100} = \dfrac{1 - x^{101}}{1 - x}$ $(x \neq 1)$. If $x = 1$, then the sum is clearly equal to $101 > 0$. If $x \geq 0$, then we have a sum of positive terms which is clearly positive. And if $x < 0$ then $x^{101} < 0 \quad \Rightarrow \quad 1 - x > 0$ and $1 - x^{101} > 0 \quad \Rightarrow \quad \dfrac{1 - x^{101}}{1 - x} > 0$. Therefore $1 + x + x^2 + \cdots + x^{100} = \dfrac{1 - x^{101}}{1 - x} \geq 0$ for all x.

5. For $-\frac{1}{2} < x < \frac{1}{2}$ we have $2x - 1 < 0$, so $|2x - 1| = -(2x - 1)$ and $2x + 1 > 0 \quad \Rightarrow \quad |2x + 1| = 2x + 1$. Therefore, $\displaystyle\lim_{x \to 0} \frac{|2x - 1| - |2x + 1|}{x} = \lim_{x \to 0} \frac{-(2x - 1) - (2x + 1)}{x} = \lim_{x \to 0} \frac{-4x}{x} = \lim_{x \to 0}(-4) = -4$.

7. We use mathematical induction. Let S_n be the statement that $\dfrac{d^n}{dx^n}(\sin^4 x + \cos^4 x) = 4^{n-1}\cos(4x + n\pi/2)$.

S_1 is true because

$$\begin{aligned}
\frac{d}{dx}(\sin^4 x + \cos^4 x) &= 4\sin^3 x\cos x - 4\cos^3 x\sin x = 4\sin x\cos x(\sin^2 x - \cos^2 x) \\
&= -4\sin x\cos x\cos 2x = -2\sin 2x\cos 2x = -\sin 4x \\
&= \cos\left[\tfrac{\pi}{2} - (-4x)\right] = \cos\left(\tfrac{\pi}{2} + 4x\right)4^{n-1}\cos\left(4x + n\tfrac{\pi}{2}\right) \text{ when } n = 1.
\end{aligned}$$

Now assume S_k is true, that is, $\dfrac{d^k}{dx^k}(\sin^4 x + \cos^4 x) = 4^{k-1}\cos\left(4x + k\frac{\pi}{2}\right)$. Then

$$\begin{aligned}
\frac{d^{k+1}}{dx^{k+1}}(\sin^4 x + \cos^4 x) &= \frac{d}{dx}\left[\frac{d^k}{dx^k}(\sin^4 x + \cos^4 x)\right] = \frac{d}{dx}\left[4^{k-1}\cos\left(4x + k\tfrac{\pi}{2}\right)\right] \\
&= -4^{k-1}\sin\left(4x + k\tfrac{\pi}{2}\right) \cdot \frac{d}{dx}\left(4x + k\tfrac{\pi}{2}\right) = -4^k\sin\left(4x + k\tfrac{\pi}{2}\right) \\
&= 4^k\sin\left(-4x - k\tfrac{\pi}{2}\right) = 4^k\cos\left[\tfrac{\pi}{2} - \left(-4x - k\tfrac{\pi}{2}\right)\right] \\
&= 4^k\cos\left[4x + (k+1)\tfrac{\pi}{2}\right] \text{ which shows that } S_{k+1} \text{ is true.}
\end{aligned}$$

Therefore $\dfrac{d_n}{dx_n}(\sin^4 x + \cos^4 x) = 4^{n-1}\cos\left(4x + n\frac{\pi}{2}\right)$ for every positive integer n by mathematical induction.

Another Proof: First write

$\sin^4 x + \cos^4 x = (\sin^2 x + \cos^2 x)^2 - 2\sin^2 x\cos^2 x = 1 - \frac{1}{2}\sin^2 2x = 1 - \frac{1}{4}(1 - \cos 4x) = \frac{3}{4} + \frac{1}{4}\cos 4x$.

Then we have $\dfrac{d^n}{dx^n}(\sin^4 x + \cos^4 x) = \dfrac{d^n}{dx^n}\left(\frac{3}{4} + \frac{1}{4}\cos 4x\right) = \frac{1}{4} \cdot 4^n\cos\left(4x + n\frac{\pi}{2}\right) = 4^{n-1}\cos\left(4x + n\frac{\pi}{2}\right)$.

9. It seems from the figure that as P approaches the point $(0, 2)$ from the right, $x_T \to \infty$ and $y_T \to 2^+$. As P approaches the point $(3, 0)$ from the left, it appears that $x_T \to 3^+$ and $y_T \to \infty$. So we guess that $x_T \in (3, \infty)$ and $y_T \in (2, \infty)$. It is more difficult to estimate the range of values for x_N and y_N. We might perhaps guess that $x_N \in (0, 3)$, and $y_N \in (-\infty, 0)$ or $(-2, 0)$.

In order to actually solve the problem, we implicitly differentiate the equation of the ellipse to find the equation of the tangent line: $\frac{x^2}{9} + \frac{y^2}{4} = 1 \quad \Rightarrow \quad \frac{2x}{9} + \frac{2y}{4}y' = 0$, so $y' = -\frac{4}{9}\frac{x}{y}$. So at the point (x_0, y_0) on the ellipse, the equation of the tangent line is $y - y_0 = -\dfrac{4}{9}\dfrac{x_0}{y_0}(x - x_0)$ or $4x_0x + 9y_0y = 4x_0^2 + 9y_0^2$. This can be written as $\dfrac{x_0x}{9} + \dfrac{y_0y}{4} = \dfrac{x_0^2}{9} + \dfrac{y_0^2}{4} = 1$, because (x_0, y_0) lies on the ellipse. So an equation of the tangent line is $\dfrac{x_0x}{9} + \dfrac{y_0y}{4} = 1$.

Therefore the x-intercept x_T for the tangent line is given by $\dfrac{x_0x_T}{9} = 1 \quad \Leftrightarrow \quad x_T = \dfrac{9}{x_0}$, and the y-intercept y_T is given by $\dfrac{y_0y_T}{4} = 1 \quad \Leftrightarrow \quad y_T = \dfrac{4}{y_0}$.

So as x_0 takes on all values in $(0, 3)$, x_T takes on all values in $(3, \infty)$, and as y_0 takes on all values in $(0, 2)$, y_T takes on all values in $(2, \infty)$.

At the point (x_0, y_0) on the ellipse, the slope of the normal line is $-\dfrac{1}{y'(x_0, y_0)} = \dfrac{9}{4}\dfrac{y_0}{x_0}$, and its equation is $y - y_0 = \dfrac{9}{4}\dfrac{y_0}{x_0}(x - x_0)$. So the x-intercept x_N for the normal line is given by $0 - y_0 = \dfrac{9}{4}\dfrac{y_0}{x_0}(x_N - x_0) \quad \Rightarrow$ $x_N = -\dfrac{4x_0}{9} + x_0 = \dfrac{5x_0}{9}$, and the y-intercept y_N is given by $y_N - y_0 = \dfrac{9}{4}\dfrac{y_0}{x_0}(0 - x_0) \quad \Rightarrow$ $y_N = -\dfrac{9y_0}{4} + y_0 = -\dfrac{5y_0}{4}$.

So as x_0 takes on all values in $(0, 3)$, x_N takes on all values in $\left(0, \frac{5}{3}\right)$, and as y_0 takes on all values in $(0, 2)$, y_N takes on all values in $\left(-\frac{5}{2}, 0\right)$.

11. **(a)**
$$\begin{aligned} D &= \left\{x \mid 3 - x \geq 0, 2 - \sqrt{3 - x} \geq 0, 1 - \sqrt{2 - \sqrt{3 - x}} \geq 0\right\} \\ &= \left\{x \mid 3 \geq x, 2 \geq \sqrt{3 - x}, 1 \geq \sqrt{2 - \sqrt{3 - x}}\right\} = \left\{x \mid 3 \geq x, 4 \geq 3 - x, 1 \geq 2 - \sqrt{3 - x}\right\} \\ &= \left\{x \mid x \leq 3, x \geq -1, 1 \leq \sqrt{3 - x}\right\} = \{x \mid x \leq 3, x \geq -1, 1 \leq 3 - x\} \\ &= \{x \mid x \leq 3, x \geq -1, x \leq 2\} = \{x \mid -1 \leq x \leq 2\} = [-1, 2] \end{aligned}$$

(b) $f(x) = \sqrt{1 - \sqrt{2 - \sqrt{3 - x}}} \quad \Rightarrow$
$$\begin{aligned} f'(x) &= \frac{1}{2\sqrt{1 - \sqrt{2 - \sqrt{3 - x}}}}\frac{d}{dx}\left(1 - \sqrt{2 - \sqrt{3 - x}}\right) \\ &= \frac{1}{2\sqrt{1 - \sqrt{2 - \sqrt{3 - x}}}} \cdot \frac{-1}{2\sqrt{2 - \sqrt{3 - x}}}\frac{d}{dx}\left(2 - \sqrt{3 - x}\right) \\ &= -\frac{1}{8\sqrt{1 - \sqrt{2 - \sqrt{3 - x}}}\sqrt{2 - \sqrt{3 - x}}\sqrt{3 - x}} \end{aligned}$$

13. (a) For $x \geq 0$,

$$h(x) = |x^2 - 6|x| + 8| = |x^2 - 6x + 8| = |(x-2)(x-4)| = \begin{cases} x^2 - 6x + 8 & \text{if } 0 \leq x \leq 2 \\ -x^2 + 6x - 8 & \text{if } 2 < x < 4 \\ x^2 - 6x + 8 & \text{if } x \geq 4 \end{cases}$$

and for $x < 0$,

$$h(x) = |x^2 - 6|x| + 8| = |x^2 + 6x + 8|$$

$$= |(x+2)(x+4)| = \begin{cases} x^2 + 6x + 8 & \text{if } x \leq -4 \\ -x^2 - 6x - 8 & \text{if } -4 < x < -2 \\ x^2 + 6x + 8 & \text{if } -2 \leq x < 0 \end{cases}$$

Or: Use the fact that h is an even function and reflect the part of the graph for $x \geq 0$ about the y-axis.

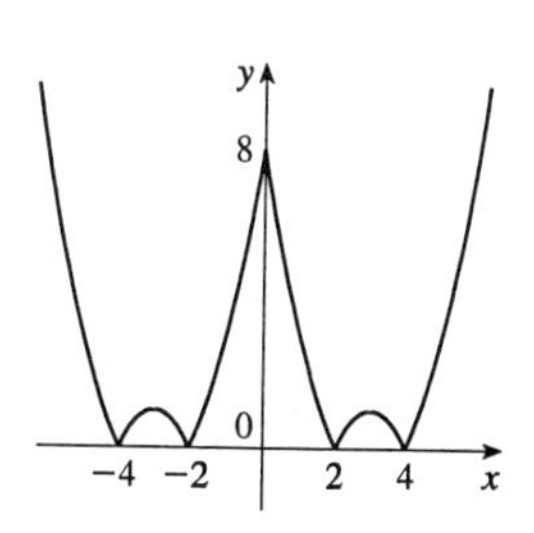

(b) To find where h is differentiable we check the points 0, 2 and 4 by computing the left- and right-hand derivatives:

$$h'_-(0) = \lim_{t \to 0^-} \frac{h(0+t) - h(0)}{t} = \lim_{t \to 0^-} \frac{(0+t)^2 + 6(0+t) + 8 - 8}{t} = \lim_{t \to 0^-} (t + 6) = 6$$

$$h'_+(0) = \lim_{t \to 0^+} \frac{h(0+t) - h(0)}{t} = \lim_{t \to 0^+} \frac{(0+t)^2 - 6(0+t) + 8 - 8}{t} = \lim_{t \to 0^+} (t - 6) = -6$$

$\neq h'_-(0)$, so h is not differentiable at 0. Similarly h is not differentiable at ± 2 or ± 4. This can also be seen from the graph.

15.

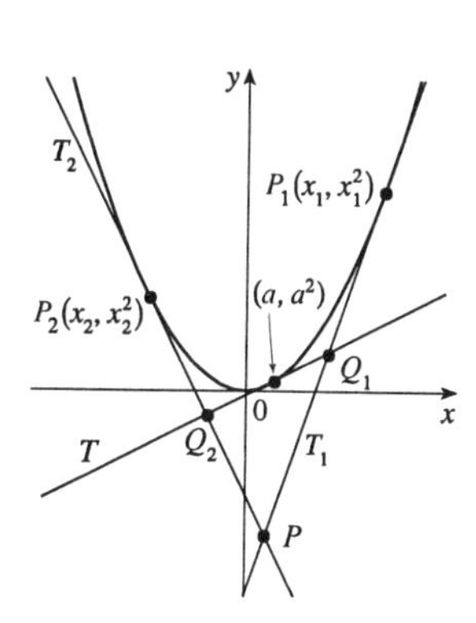

The equation of T_1 is $y - x_1^2 = 2x_1(x - x_1) = 2x_1 x - 2x_1^2$ or $y = 2x_1 x - x_1^2$. The equation of T_2 is $y = 2x_2 x - x_2^2$. Solving for the point of intersection, we get $2x(x_1 - x_2) = x_1^2 - x_2^2 \quad \Rightarrow x = \frac{1}{2}(x_1 + x_2)$. Therefore the coordinates of P are $\left(\frac{1}{2}(x_1 + x_2), x_1 x_2\right)$. So if the point of contact of T is (a, a^2), then Q_1 is $\left(\frac{1}{2}(a + x_1), a x_1\right)$ and Q_2 is $\left(\frac{1}{2}(a + x_2), a x_2\right)$. Therefore $|PQ_1|^2 = \frac{1}{4}(a - x_2)^2 + x_1^2(a - x_2)^2 = (a - x_2)^2\left(\frac{1}{4} + x_1^2\right)$ and $|PP_1|^2 = \frac{1}{4}(x_1 - x_2)^2 + x_1^2(x_1 - x_2)^2 = (x_1 - x_2)^2\left(\frac{1}{4} + x_1^2\right)$.

So $\dfrac{|PQ_1|^2}{|PP_1|^2} = \dfrac{(a - x_2)^2}{(x_1 - x_2)^2}$, and similarly $\dfrac{|PQ_2|^2}{|PP_2|^2} = \dfrac{(x_1 - a)^2}{(x_1 - x_2)^2}$.

Finally, $\dfrac{|PQ_1|}{|PP_1|} + \dfrac{|PQ_2|}{|PP_2|} = \dfrac{a - x_2}{x_1 - x_2} + \dfrac{x_1 - a}{x_1 - x_2} = 1$.

17. (a) Since f is differentiable at 0, f is continuous at 0 so $f(0) = \lim\limits_{x \to 0} f(x) = \lim\limits_{x \to 0} \dfrac{f(x)}{x} \cdot x = \lim\limits_{x \to 0} \dfrac{f(x)}{x} \cdot \lim\limits_{x \to 0} x = 4 \cdot 0 = 0$.

(b) $f'(0) = \lim\limits_{x \to 0} \dfrac{f(x) - f(0)}{x - 0} = \lim\limits_{x \to 0} \dfrac{f(x)}{x} = 4 \quad$ [since $f(0) = 0$ from (a)]

(c) $\lim\limits_{x \to 0} \dfrac{g(x)}{f(x)} = \lim\limits_{x \to 0} \dfrac{g(x)/x}{f(x)/x} = \dfrac{\lim\limits_{x \to 0}[g(x)/x]}{\lim\limits_{x \to 0}[f(x)/x]} = \dfrac{2}{4} = \dfrac{1}{2}$

19. $\displaystyle\lim_{x\to0}\frac{\sin(a+2x)-2\sin(a+x)+\sin a}{x^2}=\lim_{x\to0}\frac{\sin a\cos 2x+\cos a\sin 2x-2\sin a\cos x-2\cos a\sin x+\sin a}{x^2}$

$$=\lim_{x\to0}\frac{\sin a(\cos 2x-2\cos x+1)+\cos a(\sin 2x-2\sin x)}{x^2}$$

$$=\lim_{x\to0}\frac{\sin a(2\cos^2x-1-2\cos x+1)+\cos a(2\sin x\cos x-2\sin x)}{x^2}$$

$$=\lim_{x\to0}\frac{\sin a(2\cos x)(\cos x-1)+\cos a(2\sin x)(\cos x-1)}{x^2}$$

$$=\lim_{x\to0}\frac{2(\cos x-1)[\sin a\cos x+\cos a\sin x](\cos x+1)}{x^2(\cos x+1)}$$

$$=\lim_{x\to0}\frac{-2\sin^2x\,[\sin(a+x)]}{x^2(\cos x+1)}=-2\lim_{x\to0}\left(\frac{\sin x}{x}\right)^2\cdot\frac{\sin(a+x)}{\cos x+1}$$

$$=-2(1)^2\frac{\sin(a+0)}{\cos 0+1}=-\sin a$$

21. **(a)** If the two lines L_1 and L_2 have slopes m_1 and m_2 and angles of inclination ϕ_1 and ϕ_2, then $m_1=\tan\phi_1$ and $m_2=\tan\phi_2$. The figure shows that $\phi_2=\phi_1+\alpha$ and so $\alpha=\phi_2-\phi_1$. Therefore, using the identity for $\tan(x-y)$, we have

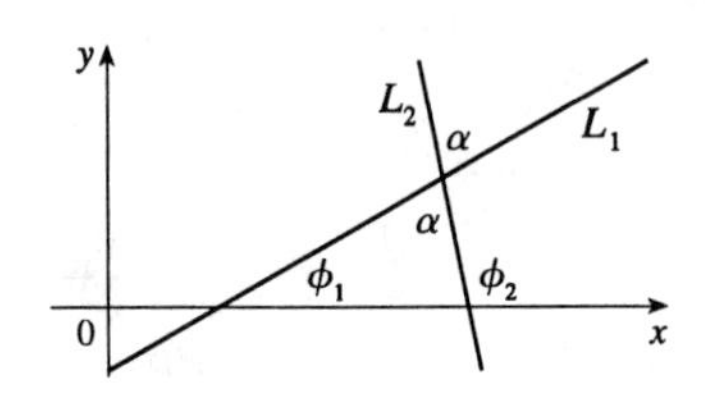

$$\tan\alpha=\tan(\phi_2-\phi_1)=\frac{\tan\phi_2-\tan\phi_1}{1+\tan\phi_2\tan\phi_1}\text{ and so }\tan\alpha=\frac{m_2-m_1}{1+m_1m_2}.$$

(b) **(i)** The parabolas intersect when $x^2=(x-2)^2\Rightarrow x=1$. If $y=x^2$, then $y'=2x$, so the slope of the tangent to $y=x^2$ at $(1,1)$ is $m_1=2(1)=2$. If $y=(x-2)^2$, then $y'=2(x-2)$, so the slope of the tangent to $y=(x-2)^2$ at $(1,1)$ is $m_2=2(1-2)=-2$. Therefore

$$\tan\alpha=\frac{m_2-m_1}{1+m_1m_2}=\frac{-2-2}{1+2(-2)}=\frac{4}{3}\text{ and so }\alpha=\tan^{-1}\tfrac{4}{3}\approx 53^\circ.$$

(ii) $x^2-y^2=3$ and $x^2-4x+y^2+3=0$ intersect when $x^2-4x+x^2=0\Leftrightarrow 2x(x-2)=0$ $\Rightarrow x=0$ or 2, but 0 is extraneous. If $x^2-y^2=3$ then $2x-2yy'=0\Rightarrow y'=x/y$ and $x^2-4x+y^2+3=0\Rightarrow 2x-4+2yy'=0\Rightarrow y'=\dfrac{2-x}{y}$. At $(2,1)$ the slopes are $m_1=2$ and $m_2=0$, so $\tan\alpha=\dfrac{0-2}{1+2\cdot 0}=-2\Rightarrow\alpha\approx117^\circ$. At $(2,-1)$ the slopes are $m_1=-2$ and $m_2=0$, so $\tan\alpha=\dfrac{0-(-2)}{1+(-2)(0)}=2\Rightarrow\alpha\approx63^\circ$.

23. Since $\angle ROQ=\angle OQP=\theta$, the triangle QOR is isosceles, so $|QR|=|RO|=x$. By the Law of Cosines, $x^2=x^2+r^2-2rx\cos\theta$. Hence $2rx\cos\theta=r^2$, so $x=\dfrac{r^2}{2r\cos\theta}=\dfrac{r}{2\cos\theta}$. Note that as $y\to0^+$, $\theta\to0^+$ (since $\sin\theta=y/r$), and hence $x\to\dfrac{r}{2\cos 0}=\dfrac{r}{2}$. Thus as P is taken closer and closer to the x-axis, the point R approaches the midpoint of the radius AO.

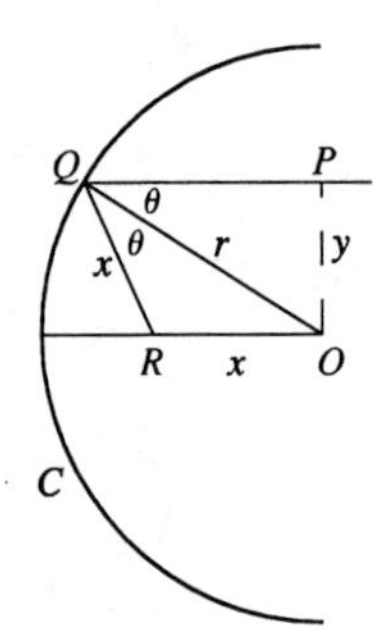

25. $y = x^4 - 2x^2 - x \Rightarrow y' = 4x^3 - 4x - 1$. The equation of the tangent line at $x = a$ is $y - (a^4 - 2a^2 - a) = (4a^3 - 4a - 1)(x - a)$ or $y = (4a^3 - 4a - 1)x + (-3a^4 + 2a^2)$ and similarly for $x = b$. So if at $x = a$ and $x = b$ we have the same tangent line, then $4a^3 - 4a - 1 = 4b^3 - 4b - 1$ and $-3a^4 + 2a^2 = -3b^4 + 2b^2$. The first equation gives $a^3 - b^3 = a - b \Rightarrow (a - b)(a^2 + ab + b^2) = (a - b)$. Assuming $a \neq b$, we have $1 = a^2 + ab + b^2$. The second equation gives $3(a^4 - b^4) = 2(a^2 - b^2) \Rightarrow 3(a^2 - b^2)(a^2 + b^2) = 2(a^2 - b^2)$ which is true if $a = -b$. Substituting into $1 = a^2 + ab + b^2$ gives $1 = a^2 - a^2 + a^2 \Rightarrow a = \pm 1$ so that $a = 1$ and $b = -1$ or vice versa. It is easily verified that the points $(1, -2)$ and $(-1, 0)$ have a common tangent line.

As long as there are only two such points, we are done. So we show that these are in fact the only two such points. Suppose that $a^2 - b^2 \neq 0$. Then $3(a^2 - b^2)(a^2 + b^2) = 2(a^2 - b^2)$ gives $3(a^2 + b^2) = 2$ or $a^2 + b^2 = \frac{2}{3}$. Thus $ab = (a^2 + ab + b^2) - (a^2 + b^2) = 1 - \frac{2}{3} = \frac{1}{3}$, so $b = \dfrac{1}{3a}$. Hence $a^2 + \dfrac{1}{9a^2} = \dfrac{2}{3}$, so $9a^4 + 1 = 6a^2 \Rightarrow 0 = 9a^4 - 6a^2 + 1 = (3a^2 - 1)^2$. So $3a^2 - 1 = 0$, so $a^2 = \dfrac{1}{3} \Rightarrow b^2 = \dfrac{1}{9a^2} = \dfrac{1}{3} = a^2$, contradicting our assumption that $a^2 \neq b^2$.

27. Because of the periodic nature of the lattice points, it suffices to consider the points in the 5×2 grid shown. We can see that the minimum value of r occurs when there is a line with slope $\frac{2}{5}$ which touches the circle centered at $(3, 1)$ and the circles centered at $(0, 0)$ and $(5, 2)$. To find P, the point at which the line is tangent to the circle at $(0, 0)$, we simultaneously solve $x^2 + y^2 = r^2$ and $y = -\frac{5}{2}x \Rightarrow x^2 + \frac{25}{4}x^2 = r^2 \Rightarrow x^2 = \frac{4}{29}r^2 \Rightarrow x = \frac{2}{\sqrt{29}}r$, $y = -\frac{5}{\sqrt{29}}r$. To find Q, we either use symmetry or solve $(x - 3)^2 + (y - 1)^2 = r^2$ and $y - 1 = -\frac{5}{2}(x - 3)$. As above, we get $x = 3 - \frac{2}{\sqrt{29}}r$, $y = 1 + \frac{5}{\sqrt{29}}r$. Now the slope of the line PQ is $\frac{2}{5}$, so

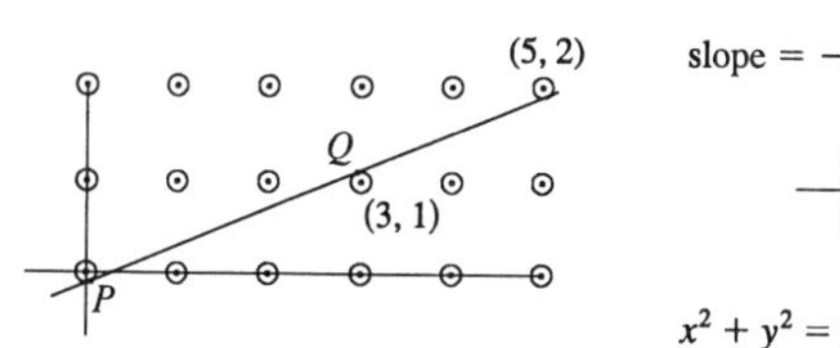

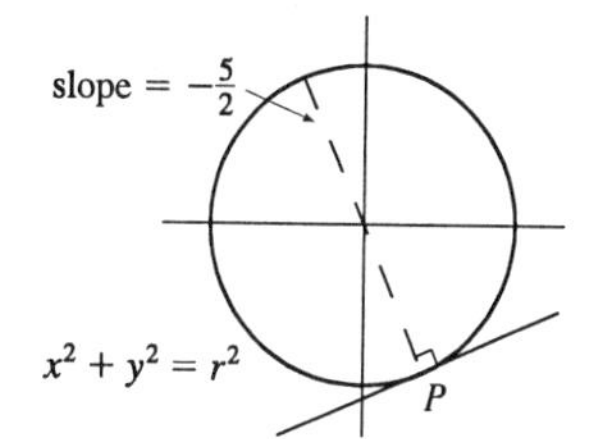

$$m_{PQ} = \frac{1 + \frac{5}{\sqrt{29}}r - \left(-\frac{5}{\sqrt{29}}r\right)}{3 - \frac{2}{\sqrt{29}}r - \frac{2}{\sqrt{29}}r} = \frac{1 + \frac{10}{\sqrt{29}}r}{3 - \frac{4}{\sqrt{29}}r} = \frac{\sqrt{29} + 10r}{3\sqrt{29} - 4r} = \frac{2}{5} \Rightarrow 5\sqrt{29} + 50r = 6\sqrt{29} - 8r \Leftrightarrow$$

$58r = \sqrt{29} \Leftrightarrow r = \frac{\sqrt{29}}{58}$. So the minimum value of r for which any line with slope $\frac{2}{5}$ intersects circles with radius r centered at the lattice points on the plane is $r = \frac{\sqrt{29}}{58}$.

CHAPTER FOUR

EXERCISES 4.1

1.

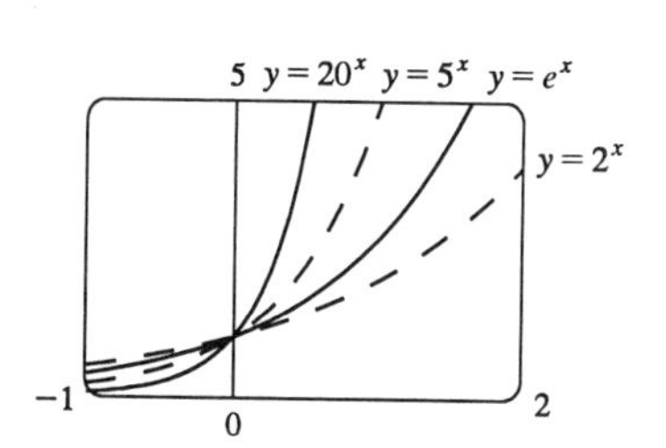

All of these graphs approach 0 as $x \to -\infty$, all of them pass through the point $(0, 1)$, and all of them are increasing and approach ∞ as $x \to \infty$. The larger the base, the faster the function increases for $x > 0$, and the faster it approaches 0 as $x \to -\infty$.

3.

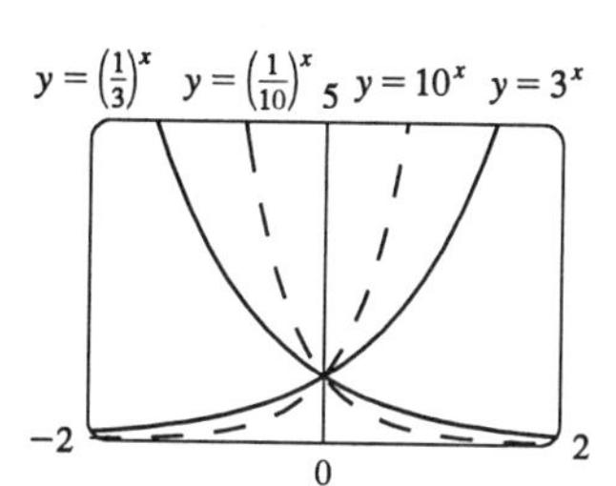

The functions with bases greater than 1 (3^x and 10^x) are increasing, while those with bases less than 1 $\left[\left(\frac{1}{3}\right)^x \text{ and } \left(\frac{1}{10}\right)^x\right]$ are decreasing. The graph of $\left(\frac{1}{3}\right)^x$ is the reflection of that of 3^x about the y-axis, and the graph of $\left(\frac{1}{10}\right)^x$ is the reflection of that of 10^x about the y-axis. The graph of 10^x increases more quickly than that of 3^x for $x > 0$, and approaches 0 faster as $x \to -\infty$.

5.

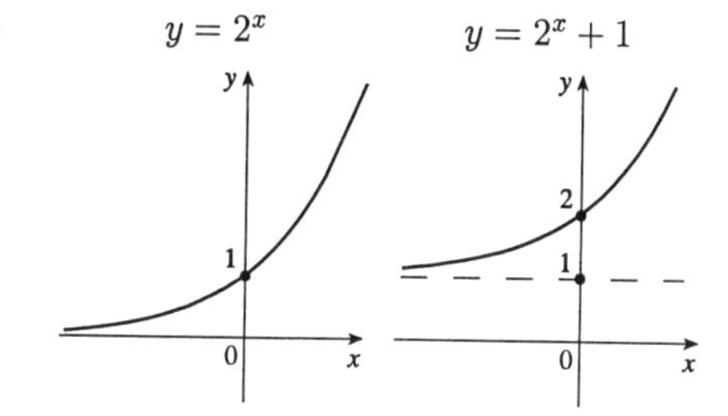

7.

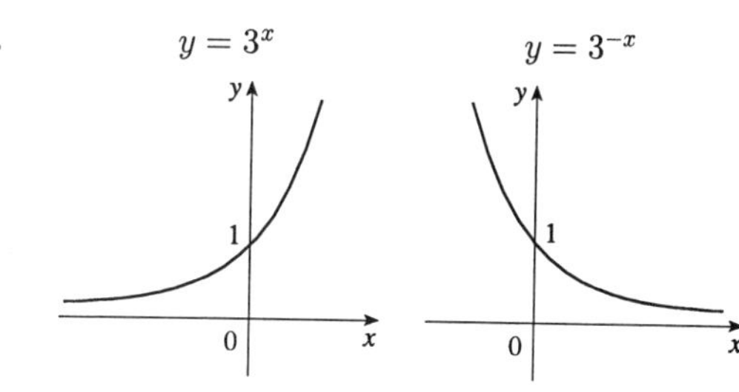

9.

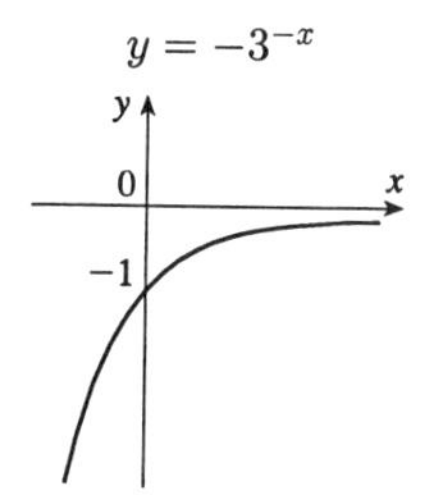

11.

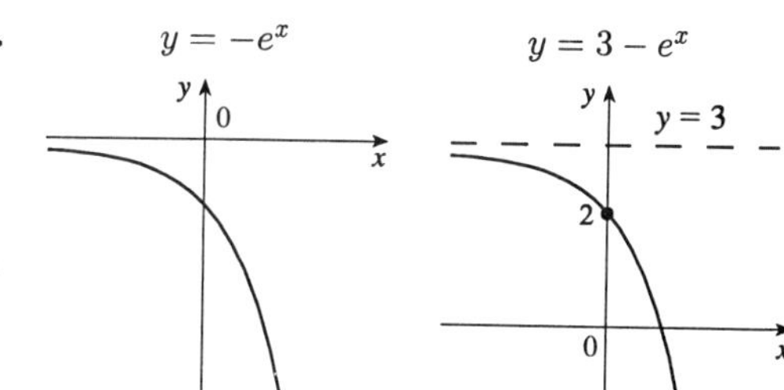

13. $\lim\limits_{x\to\infty} (1.1)^x = \infty$ by Equation 3 since $1.1 > 1$.

15. $\lim\limits_{x\to-\infty} \left(\frac{\pi}{4}\right)^x = \infty$ since $0 < \frac{\pi}{4} < 1$.

17. Divide numerator and denominator by e^{3x}: $\lim\limits_{x\to\infty} \dfrac{e^{3x} - e^{-3x}}{e^{3x} + e^{-3x}} = \lim\limits_{x\to\infty} \dfrac{1 - e^{-6x}}{1 + e^{-6x}} = \dfrac{1 - 0}{1 + 0} = 1$

19. $\lim_{x\to 1^-} e^{2/(x-1)} = 0$ since $\dfrac{2}{x-1} \to -\infty$ as $x \to 1^-$.

21. $\lim_{x\to\pi/2^-} \dfrac{2}{1+e^{\tan x}} = 0$ since $\tan x \to \infty \quad \Rightarrow \quad e^{\tan x} \to \infty$.

23. $2\text{ ft} = 24\text{ in}$, $f(24) = 24^2\text{ in} = 576\text{ in} = 48\text{ ft}$. $g(24) = 2^{24}\text{ in} = 2^{24}/(12\cdot 5280)\text{ mi} \approx 265\text{ mi}$

25. **(a)** Let $f(h) = \dfrac{4^h - 1}{h}$. Then $f(0.1) \approx 1.487$, $f(0.01) \approx 1.396$, $f(0.001) \approx 1.387$, and $f(0.0001) \approx 1.386$.

These quantities represent the slopes of secant lines to the curve $y = 4^x$, through the points $(0,1)$ and $(h, 4^h)$ $\left(\text{since they are of the form } \dfrac{4^h - 4^0}{h - 0}\right)$.

(b) The value of the limit $\lim_{h\to 0} \dfrac{4^h - 1}{h}$ is about 1.39, judging from the calculations in part (a).

(c) The limit in part (b) represents the slope of the tangent line to the curve $y = 4^x$ at $(0,1)$.

27. $f(x) = e^{\sqrt{x}} \quad \Rightarrow \quad f'(x) = e^{\sqrt{x}}\big/\left(2\sqrt{x}\right)$

29. $y = xe^{2x} \quad \Rightarrow \quad y' = e^{2x} + xe^{2x}(2) = e^{2x}(1+2x)$

31. $h(t) = \sqrt{1-e^t} \quad \Rightarrow \quad h'(t) = -e^t\Big/\left(2\sqrt{1-e^t}\right)$

33. $y = e^{x\cos x} \quad \Rightarrow \quad y' = e^{x\cos x}(\cos x - x\sin x)$

35. $y = e^{-1/x} \quad \Rightarrow \quad y' = e^{-1/x}/x^2$

37. $y = \tan(e^{3x-2}) \quad \Rightarrow \quad y' = 3e^{3x-2}\sec^2(e^{3x-2})$

39. $y = \dfrac{e^{3x}}{1+e^x} \quad \Rightarrow \quad y' = \dfrac{3e^{3x}(1+e^x) - e^{3x}(e^x)}{(1+e^x)^2} = \dfrac{3e^{3x} + 3e^{4x} - e^{4x}}{(1+e^x)^2} = \dfrac{3e^{3x} + 2e^{4x}}{(1+e^x)^2}$

41. $y = x^e \quad \Rightarrow \quad y' = ex^{e-1}$

43. $y = f(x) = e^{-x}\sin x \quad \Rightarrow \quad f'(x) = -e^{-x}\sin x + e^{-x}\cos x \quad \Rightarrow \quad f'(\pi) = e^{-\pi}(\cos\pi - \sin\pi) = -e^{-\pi}$, so the equation of the tangent at $(\pi, 0)$ is $y - 0 = -e^{-\pi}(x - \pi)$ or $x + e^{\pi}y = \pi$.

45. $\cos(x-y) = xe^x \quad \Rightarrow \quad -\sin(x-y)(1-y') = e^x + xe^x \quad \Rightarrow \quad y' = 1 + \dfrac{e^x(1+x)}{\sin(x-y)}$

47. $y = e^{2x} + e^{-3x} \quad \Rightarrow \quad y' = 2e^{2x} - 3e^{-3x} \quad \Rightarrow \quad y'' = 4e^{2x} + 9e^{-3x}$, so

$y'' + y' - 6y = (4e^{2x} + 9e^{-3x}) + (2e^{2x} - 3e^{-3x}) - 6(e^{2x} + e^{-3x}) = 0.$

49. $y = e^{rx} \quad \Rightarrow \quad y' = re^{rx} \quad \Rightarrow \quad y'' = r^2e^{rx}$, so $y'' + 5y' - 6y = r^2e^{rx} + 5re^{rx} - 6e^{rx}$

$= e^{rx}(r^2 + 5r - 6) = e^{rx}(r+6)(r-1) = 0 \quad \Rightarrow \quad (r+6)(r-1) = 0 \quad \Rightarrow \quad r = 1 \text{ or } -6.$

51. $f(x) = e^{-2x} \quad \Rightarrow \quad f'(x) = -2e^{-2x} \quad \Rightarrow \quad f''(x) = (-2)^2e^{-2x} \quad \Rightarrow \quad f'''(x) = (-2)^3e^{-2x} \quad \Rightarrow \quad \cdots$

$\Rightarrow \quad f^{(8)}(x) = (-2)^8e^{-2x} = 256e^{-2x}$

53. **(a)** $f(x) = e^x + x$ is continuous on $\mathbb{R}$ and $f(-1) = e^{-1} - 1 < 0 < 1 = f(0)$, so by the Intermediate Value Theorem, $e^x + x = 0$ has a root in $(-1, 0)$.

(b) $f(x) = e^x + x \quad \Rightarrow \quad f'(x) = e^x + 1$, so $x_{n+1} = x_n - \dfrac{e^{x_n} + x_n}{e^{x_n} + 1}$. From Exercise 31 we know that there is a root between -1 and 0, so we take $x_1 = -0.5$. Then $x_2 \approx -0.566311$, $x_3 \approx -0.567143$, and $x_4 \approx -0.567143$, so the root is -0.567143 to six decimal places.

55. **(a)** $\displaystyle\lim_{t\to\infty} p(t) = \lim_{t\to\infty} \frac{1}{1 + ae^{-kt}} = \frac{1}{1 + a \cdot 0} = 1,$ since $k > 0 \quad \Rightarrow \quad -kt \to -\infty$.

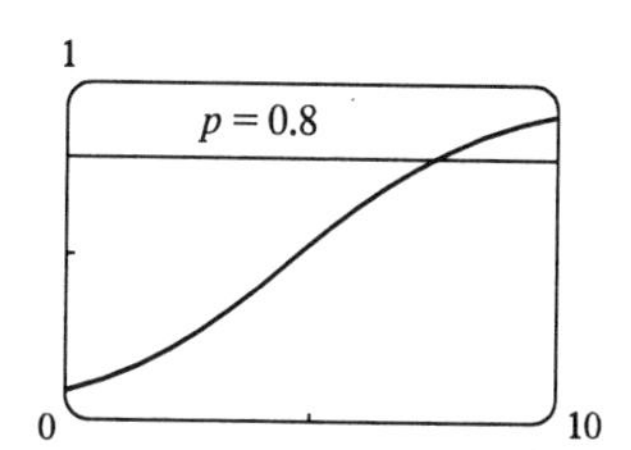

(b) $\dfrac{dp}{dt} = -\left(1 + ae^{-kt}\right)^{-2}\left(-kae^{-kt}\right) = \dfrac{kae^{-kt}}{(1 + ae^{-kt})^2}$

(c) From the graph, it seems that $p(t) = 0.8$ (indicating that 80% of the population has heard the rumor) when $t \approx 7.4$ hours.

57. We recognize this limit as the definition of the derivative of the function $f(x) = e^{\sin x}$ at $x = \pi$, since it is of the form $\displaystyle\lim_{x\to\pi} \frac{f(x) - f(\pi)}{x - \pi}$. Therefore, the limit is equal to $f'(\pi) = (\cos \pi)e^{\sin \pi} = -1 \cdot e^0 = -1$.

59. $x = e^{t/3} \cos t$, $y = e^{t/3} \sin t$ for $-\pi \le t \le \pi$. Note that $x^2 + y^2 = e^{2t/3} \cos^2 t + e^{2t/3} \sin^2 t = e^{2t/3} \left(\cos^2 t + \sin^2 t\right) = e^{2t/3}$, so $\sqrt{x^2 + y^2} = \sqrt{e^{2t/3}} = e^{t/3}$. Since $\sqrt{x^2 + y^2}$ is the distance to the origin, we see that this distance from the origin to the curve increases as t increases. When $t = -\pi$, $(x, y) = \left(-e^{-\pi/3}, 0\right) \approx (-0.35, 0)$, and when $t = \pi$, $(x, y) = \left(-e^{\pi/3}, 0\right) \approx (-2.85, 0)$. The shape of the graph is a portion of an exponential spiral.

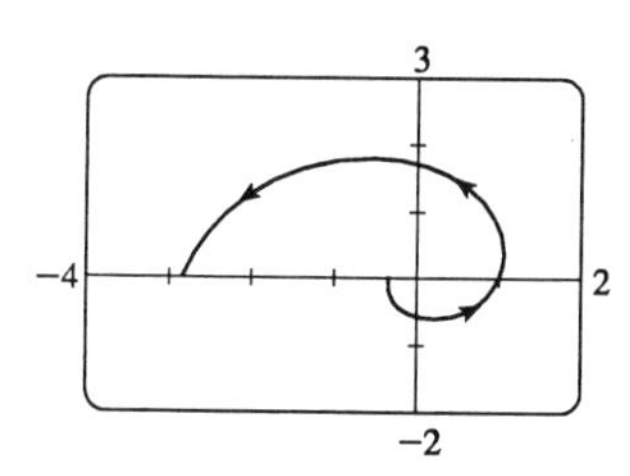

61. $x = e^{-t}$ and $y = te^{2t} \quad \Rightarrow \quad \dfrac{dy}{dx} = \dfrac{dy/dt}{dx/dt} = \dfrac{t \cdot 2e^{2t} + e^{2t}}{-e^{-t}} = \dfrac{e^{2t}(2t + 1)}{-e^{-t}} = -e^{3t}(2t + 1)$.

EXERCISES 4.2

1. The diagram shows that there is a horizontal line which intersects the graph more than once, so the function is not one-to-one.

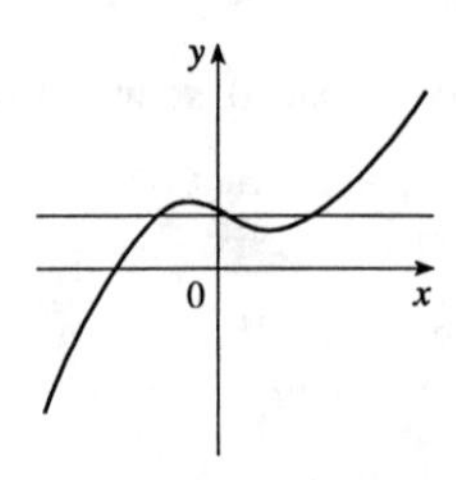

3. The function is one-to-one because no horizontal line intersects the graph more than once.

5. The diagram shows that there is a horizontal line which intersects the graph more than once, so the function is not one-to-one.

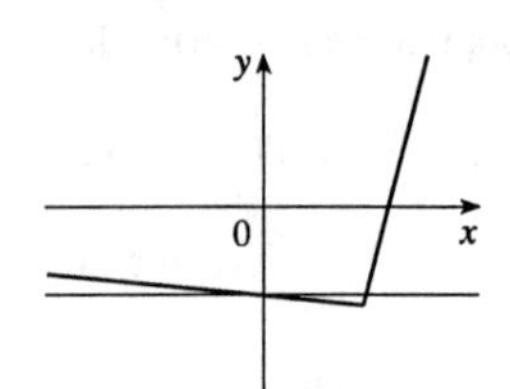

7. $x_1 \neq x_2 \Rightarrow 7x_1 \neq 7x_2 \Rightarrow 7x_1 - 3 \neq 7x_2 - 3 \Rightarrow f(x_1) \neq f(x_2)$, so f is 1-1.

9. $x_1 \neq x_2 \Rightarrow \sqrt{x_1} \neq \sqrt{x_2} \Rightarrow g(x_1) \neq g(x_2)$, so g is 1-1.

11. $h(x) = x^4 + 5 \Rightarrow h(1) = 6 = h(-1)$, so h is not 1-1.

13. $x_1 \neq x_2 \Rightarrow 4x_1 \neq 4x_2 \Rightarrow 4x_1 + 7 \neq 4x_2 + 7 \Rightarrow f(x_1) \neq f(x_2)$, so f is 1-1. $y = 4x + 7 \Rightarrow 4x = y - 7 \Rightarrow x = (y-7)/4$. Interchange x and y: $y = (x-7)/4$. So $f^{-1}(x) = (x-7)/4$.

15. $f(x) = \dfrac{1+3x}{5-2x}$. If $f(x_1) = f(x_2)$, then $\dfrac{1+3x_1}{5-2x_1} = \dfrac{1+3x_2}{5-2x_2} \Rightarrow$
$5 + 15x_1 - 2x_2 - 6x_1x_2 = 5 - 2x_1 + 15x_2 - 6x_1x_2 \Rightarrow 17x_1 = 17x_2 \Rightarrow x_1 = x_2$, so f is one-to-one.
$y = \dfrac{1+3x}{5-2x} \Rightarrow 5y - 2xy = 1 + 3x \Rightarrow x(3+2y) = 5y - 1 \Rightarrow x = \dfrac{5y-1}{2y+3}$.
Interchange x and y: $y = \dfrac{5x-1}{2x+3}$. So $f^{-1}(x) = \dfrac{5x-1}{2x+3}$.

17. $x_1 \neq x_2 \Rightarrow 5x_1 \neq 5x_2 \Rightarrow 2 + 5x_1 \neq 2 + 5x_2 \Rightarrow \sqrt{2+5x_1} \neq \sqrt{2+5x_2} \Rightarrow$
$f(x_1) \neq f(x_2)$, so f is 1-1. $y = \sqrt{2+5x} \Rightarrow y^2 = 2 + 5x$ and $y \geq 0 \Rightarrow 5x = y^2 - 2 \Rightarrow$
$x = \dfrac{y^2-2}{5}$, $y \geq 0$. Interchange x and y: $y = \dfrac{x^2-2}{5}$, $x \geq 0$. So $f^{-1}(x) = \dfrac{x^2-2}{5}$, $x \geq 0$.

19. **(a)** $x_1 \neq x_2 \Rightarrow 2x_1 \neq 2x_2 \Rightarrow 2x_1 + 1 \neq 2x_2 + 1 \Rightarrow f(x_1) \neq f(x_2)$, so f is 1-1.

(b) $f(1) = 3 \quad \Rightarrow \quad g(3) = 1$. Also $f'(x) = 2$, so $g'(3) = 1/f'(3) = \frac{1}{2}$.

(c) $y = 2x + 1 \quad \Rightarrow \quad x = \frac{1}{2}(y - 1)$.
Interchanging x and y gives
$y = \frac{1}{2}(x - 1)$, so $f^{-1}(x) = \frac{1}{2}(x - 1)$.
Domain(g) = range$(f) = \mathbb{R}$.
Range(g) = domain$(f) = \mathbb{R}$.

(d) $g(x) = \frac{1}{2}(x - 1) \quad \Rightarrow \quad g'(x) = \frac{1}{2}$
$\Rightarrow \quad g'(3) = \frac{1}{2}$ as in (b).

(e)

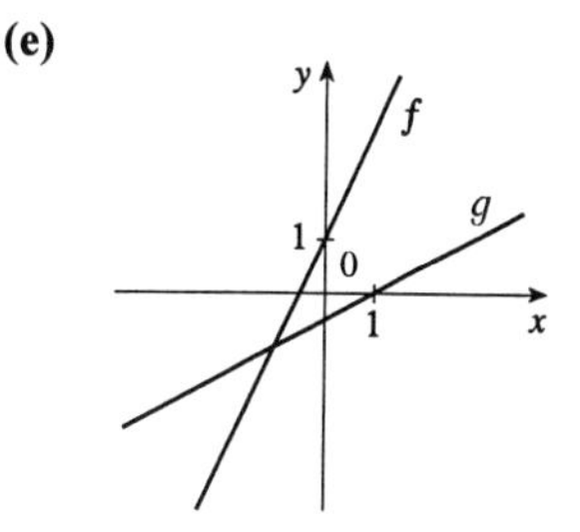

21. **(a)** $x_1 \neq x_2 \quad \Rightarrow \quad x_1^3 \neq x_2^3 \quad \Rightarrow \quad f(x_1) \neq f(x_2)$, so f is one-to-one.

(b) $f'(x) = 3x^2$ and $f(2) = 8 \quad \Rightarrow \quad g(8) = 2$, so $g'(8) = 1/f'(g(8)) = 1/f'(2) = \frac{1}{12}$.

(c) $y = x^3 \quad \Rightarrow \quad x = y^{1/3}$. Interchanging
x and y gives $y = x^{1/3}$, so $f^{-1}(x) = x^{1/3}$.
Domain(g) = range$(f) = \mathbb{R}$.
Range(g) = domain$(f) = \mathbb{R}$.

(d) $g(x) = x^{1/3} \quad \Rightarrow$
$g'(x) = \frac{1}{3}x^{-2/3} \quad \Rightarrow$
$g'(8) = \frac{1}{3}\left(\frac{1}{4}\right) = \frac{1}{12}$ as in part (b).

(e)

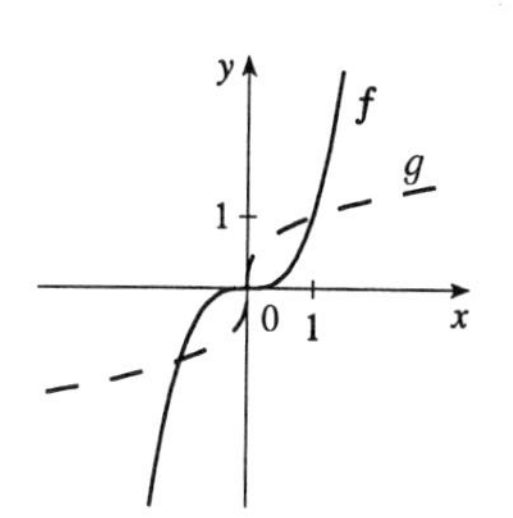

23. **(a)** Since $x \geq 0$, $x_1 \neq x_2 \Rightarrow x_1^2 \neq x_2^2 \Rightarrow 9 - x_1^2 \neq 9 - x_2^2 \Rightarrow f(x_1) \neq f(x_2)$, so f is 1-1.

(b) $f'(x) = -2x$ and $f(1) = 8 \Rightarrow g(8) = 1$, so $g'(8) = \dfrac{1}{f'(g(8))} = \dfrac{1}{f'(1)} = \dfrac{1}{(-2)} = -\dfrac{1}{2}$.

(c) $y = 9 - x^2 \quad \Rightarrow \quad x^2 = 9 - y \quad \Rightarrow$
$x = \sqrt{9 - y}$. Interchange x and y:
$y = \sqrt{9 - x}$, so $f^{-1}(x) = \sqrt{9 - x}$.
Domain(g) = range$(f) = [0, 9]$.
Range(g) = domain$(f) = [0, 3]$.

(d) $g'(x) = -1\big/\left(2\sqrt{9 - x}\right) \quad \Rightarrow$
$g'(8) = -\frac{1}{2}$ as in (b).

(e)

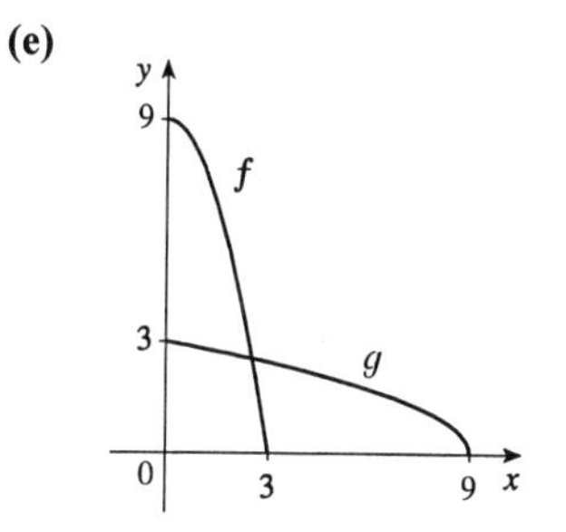

25. $f(0) = 1 \Rightarrow g(1) = 0$, and $f'(x) = 3x^2 + 1 \Rightarrow f'(0) = 1$. Therefore $g'(1) = \dfrac{1}{f'(g(1))} = \dfrac{1}{f'(0)} = \dfrac{1}{1} = 1$.

27. $f(0) = 3 \quad \Rightarrow \quad g(3) = 0$, and $f'(x) = 2x + \frac{\pi}{2}\sec^2(\pi x/2) \quad \Rightarrow \quad f'(0) = 1 \cdot \frac{\pi}{2} = \frac{\pi}{2}$. Thus $g'(3) = 1/f'(g(3)) = 1/f'(0) = 2/\pi$.

29. $f(0) = 1 \quad \Rightarrow \quad g(1) = 0$, and $f'(x) = e^x \quad \Rightarrow \quad f'(0) = 1$. Therefore $g'(1) = 1/f'(g(1)) = 1$.

31. $f(4) = 5 \quad \Rightarrow \quad g(5) = 4$. Therefore, $g'(5) = \dfrac{1}{f'(g(5))} = \dfrac{1}{f'(4)} = \dfrac{1}{2/3} = \dfrac{3}{2}$.

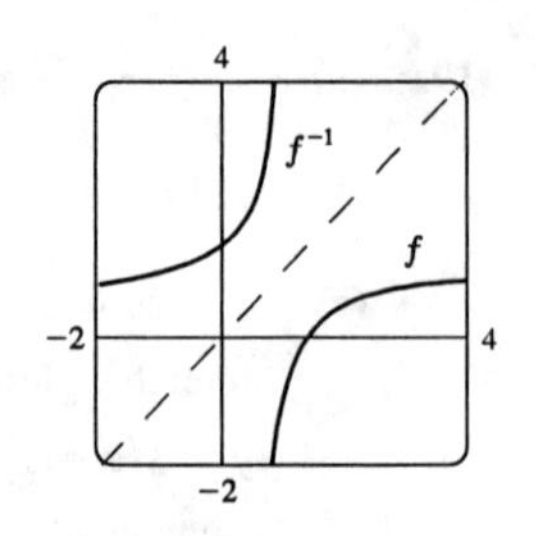

33. $y = 1 - 2/x^2 \quad \Rightarrow \quad 1 - y = 2/x^2$

$\Rightarrow \quad x^2 = 2/(1-y) \quad \Rightarrow \quad x = \sqrt{\dfrac{2}{1-y}}$,

since $x > 0$. Interchange x and y:

$y = \sqrt{\dfrac{2}{1-x}}$. So $f^{-1}(x) = \sqrt{\dfrac{2}{1-x}}$.

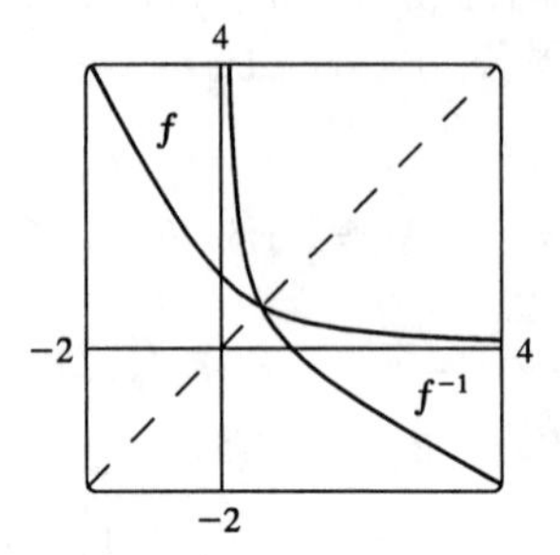

35. Since $f'(x) = \dfrac{2x}{2\sqrt{x^2+1}} - 1 = \dfrac{x - \sqrt{x^2+1}}{\sqrt{x^2+1}}$ is negative for all x, we know that f is a decreasing function on $\mathbb{R}$, and hence is 1-1. We could also use the Horizontal Line Test to show that f is 1-1.
The parametric equations for the graph of f are $x = t$, $y = \sqrt{t^2+1} - t$; for the graph of f^{-1} they are $x = \sqrt{t^2+1} - t$, $y = t$.

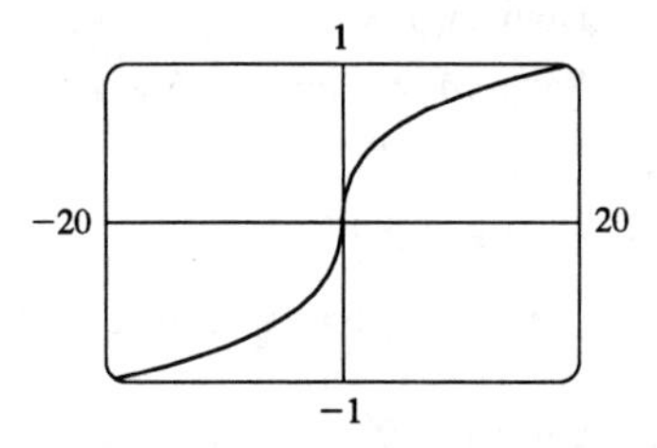

37. **(a)** $\sqrt[5]{x} - \sqrt[5]{y} = y \quad \Rightarrow \quad \sqrt[5]{x} = y + \sqrt[5]{y} \quad \Rightarrow$

$x = \left(y + \sqrt[5]{y}\right)^5$. Interchange x and y:

$y = \left(x + \sqrt[5]{x}\right)^5$. So $f^{-1}(x) = \left(x + \sqrt[5]{x}\right)^5$.

(b) The parametric equations for the graph of f^{-1} are $x = t$, $y = \left(t + \sqrt[5]{t}\right)^5$.
So the parametric equations for the graph of $f = (f^{-1})^{-1}$ are $x = \left(t + \sqrt[5]{t}\right)^5$, $y = t$.

39. $y = \sqrt[n]{x} \quad \Rightarrow \quad y^n = x \quad \Rightarrow \quad ny^{n-1}y' = 1 \quad \Rightarrow \quad y' = \dfrac{1}{ny^{n-1}} = \dfrac{1}{n\left(\sqrt[n]{x}\right)^{n-1}} = \dfrac{1}{n}x^{(1/n)-1}$

41. We know that $g'(x) = \dfrac{1}{f'(g(x))}$. Thus $g''(x) = -\dfrac{g'(x)f''(g(x))}{[f'(g(x))]^2} = -\dfrac{f''(g(x))}{f'(g(x))[f'(g(x))]^2} = -\dfrac{f''(g(x))}{[f'(g(x))]^3}$.

EXERCISES 4.3

1. $\log_2 64 = 6$ since $2^6 = 64$.

3. $\log_8 2 = \frac{1}{3}$ since $8^{1/3} = 2$.

5. $\log_3 \frac{1}{27} = -3$ since $3^{-3} = \frac{1}{27}$.

7. $\ln e^{\sqrt{2}} = \sqrt{2}$

9. $\log_{10} 1.25 + \log_{10} 80 = \log_{10}(1.25 \cdot 80) = \log_{10} 100 = 2$

11. $\log_8 6 - \log_8 3 + \log_8 4 = \log_8 \frac{6 \cdot 4}{3} = \log_8 8 = 1$

13. $2^{(\log_2 3 + \log_2 5)} = 2^{\log_2 15} = 15$

15. $\log_5 a + \log_5 b - \log_5 c = \log_5(ab/c)$

17. $2\ln 4 - \ln 2 = \ln 4^2 - \ln 2 = \ln 16 - \ln 2 = \ln \frac{16}{2} = \ln 8$

Or: $2\ln 4 - \ln 2 = 2\ln 2^2 - \ln 2 = 4\ln 2 - \ln 2 = 3\ln 2$

19. $\frac{1}{3}\ln x - 4\ln(2x+3) = \ln\left(x^{1/3}\right) - \ln(2x+3)^4 = \ln\left(x^{1/3}/(2x+3)^4\right)$

21. **(a)** $\log_2 5 = \dfrac{\ln 5}{\ln 2} \approx 2.321928$

(b) $\log_5 26.05 = \dfrac{\ln 26.05}{\ln 5} \approx 2.025563$

(c) $\log_3 e = \dfrac{1}{\ln 3} \approx 0.910239$

(d) $\log_{0.7} 14 = \dfrac{\ln 14}{\ln 0.7} \approx -7.399054$

23.

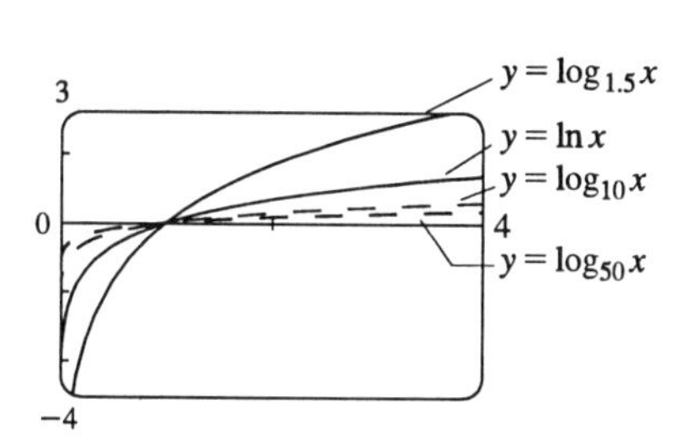

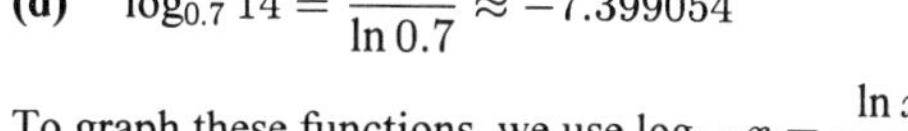

To graph these functions, we use $\log_{1.5} x = \dfrac{\ln x}{\ln 1.5}$ and $\log_{50} x = \dfrac{\ln x}{\ln 50}$. These graphs all approach $-\infty$ as $x \to 0^+$, and they all pass through the point $(1, 0)$. Also, they are all increasing, and all approach ∞ as $x \to \infty$. The functions with larger bases increase extremely slowly, and the ones with smaller bases do so somewhat more quickly. The functions with large bases approach the y-axis more closely as $x \to 0^+$.

25. $y = \log_{10} x$ $\quad$ $y = \log_{10}(x+5)$

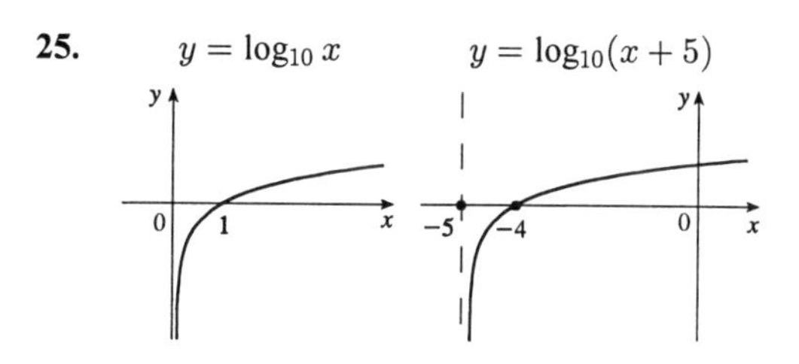

27. $y = \ln x$ $\quad$ $y = -\ln x$

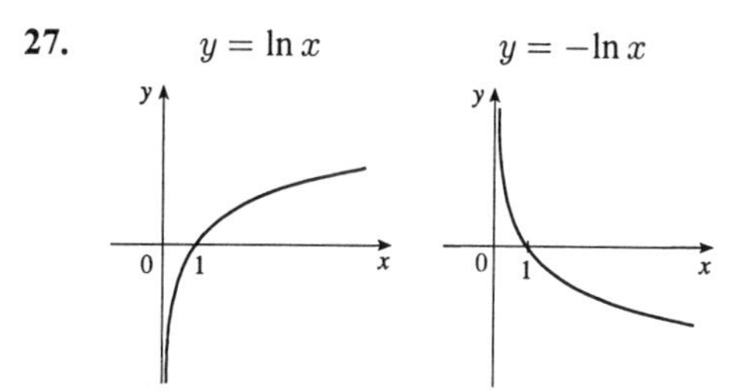

29. $y = \ln(-x)$ $\quad$ $y = -\ln(-x)$

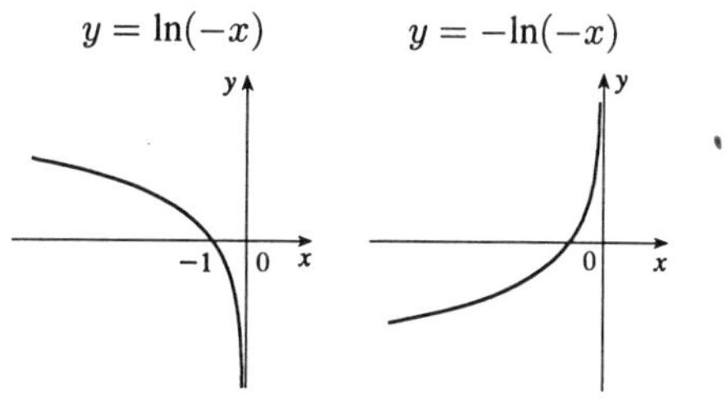

31. $y = \ln(x^2) = 2\ln|x|$

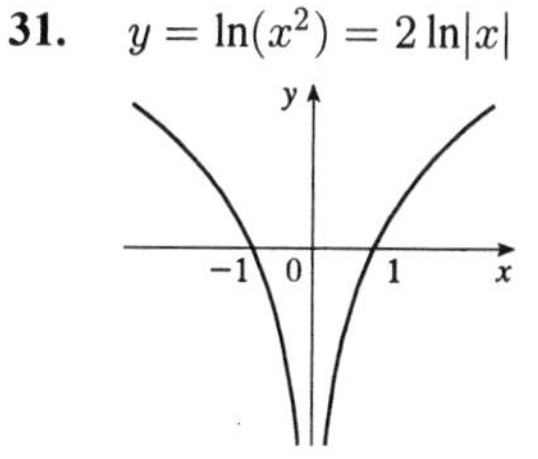

33. $y = \ln x$ $\quad$ $y = \ln(x+3)$

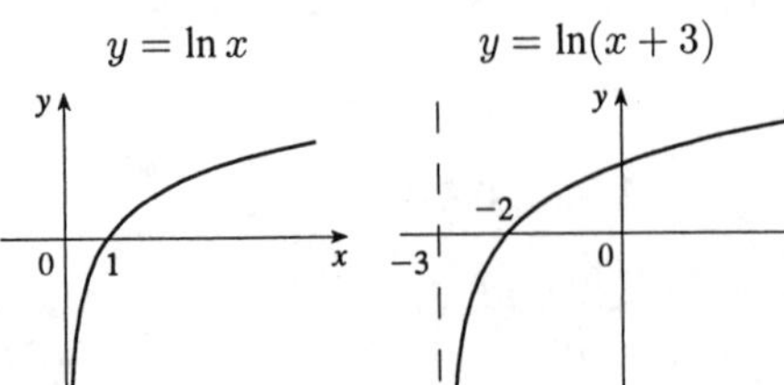

35. $\log_2 x = 3 \quad \Leftrightarrow \quad x = 2^3 = 8$

37. $e^x = 16 \quad \Leftrightarrow \quad \ln e^x = \ln 16 \quad \Leftrightarrow \quad x = \ln 16 \quad \Leftrightarrow \quad x = 4\ln 2$

39. $\ln(2x-1) = 3 \quad \Leftrightarrow \quad e^{\ln(2x-1)} = e^3 \quad \Leftrightarrow \quad 2x - 1 = e^3 \quad \Leftrightarrow \quad x = \frac{1}{2}(e^3+1)$

41. $3^{x+2} = m \quad \Leftrightarrow \quad \log_3 m = x+2 \quad \Leftrightarrow \quad x = \log_3 m - 2$

43. $\ln x = \ln 5 + \ln 8 = \ln 40 \quad \Leftrightarrow \quad x = 40$

45. $5 = \ln(e^{2x-1}) = 2x - 1 \quad \Leftrightarrow \quad x = 3$

47. $\ln(\ln x) = 1 \quad \Leftrightarrow \quad e^{\ln(\ln x)} = e^1 \quad \Leftrightarrow \quad \ln x = e^1 = e \quad \Leftrightarrow \quad e^{\ln x} = e^e \quad \Leftrightarrow \quad x = e^e$

49. $2^{3^x} = 5 \quad \Leftrightarrow \quad 3^x = \log_2 5 \quad \Leftrightarrow \quad \log_3(\log_2 5) = x$. *Or:* $2^{3^x} = 5 \quad \Leftrightarrow \quad \ln 2^{3^x} = \ln 5 \quad \Leftrightarrow \quad 3^x \ln 2 = \ln 5$ $\quad \Leftrightarrow \quad 3^x = \dfrac{\ln 5}{\ln 2}$. Hence $\ln 3^x = x \ln 3 = \ln\left(\dfrac{\ln 5}{\ln 2}\right) \quad \Leftrightarrow \quad x = \dfrac{\ln(\ln 5/\ln 2)}{\ln 3}$.

51. $\ln(x+6) + \ln(x-3) = \ln 5 + \ln 2 \quad \Leftrightarrow \quad \ln[(x+6)(x-3)] = \ln 10 \quad \Leftrightarrow \quad (x+6)(x-3) = 10 \quad \Leftrightarrow$ $x^2 + 3x - 18 = 10 \quad \Leftrightarrow \quad x^2 + 3x - 28 = 0 \quad \Leftrightarrow \quad (x+7)(x-4) = 0 \quad \Leftrightarrow \quad x = -7$ or 4. However, $x = -7$ is not a solution since $\ln(-7+6)$ is not defined. So $x = 4$ is the only solution.

53. $e^{ax} = Ce^{bx} \quad \Leftrightarrow \quad \ln e^{ax} = \ln(Ce^{bx}) \quad \Leftrightarrow \quad ax = \ln C + bx \quad \Leftrightarrow \quad (a-b)x = \ln C \quad \Leftrightarrow \quad x = \dfrac{\ln C}{a-b}$

55. $\ln(x-5) = 3 \quad \Rightarrow \quad x - 5 = e^3 \quad \Rightarrow \quad x = e^3 + 5 \approx 25.0855$

57. $e^{2-3x} = 20 \quad \Rightarrow \quad 2 - 3x = \ln 20 \quad \Rightarrow \quad x = \frac{1}{3}(2 - \ln 20) \approx -0.3319$

59. 3 ft = 36 in, so we need x such that $\log_2 x = 36 \quad \Leftrightarrow \quad x = 2^{36} = 68{,}719{,}476{,}736$. In miles, this is $68{,}719{,}476{,}736 \text{ in} \cdot \dfrac{1 \text{ ft}}{12 \text{ in}} \cdot \dfrac{1 \text{ mi}}{5280 \text{ ft}} \approx 1{,}084{,}587.7 \text{ mi}$.

61. If I is the intensity of the 1989 San Francisco earthquake, then $\log_{10}(I/S) = 7.1 \quad \Rightarrow$ $\log_{10}(16I/S) = \log_{10}16 + \log_{10}(I/S) = \log_{10}16 + 7.1 \approx 8.3$.

63. $\lim\limits_{x\to 5^+} \ln(x-5) = -\infty$ since $x - 5 \to 0^+$ as $x \to 5^+$.

65. $\lim\limits_{x\to\infty} \log_2(x^2 - x) = \infty$ since $x^2 - x \to \infty$ as $x \to \infty$.

67. $\lim\limits_{x\to\pi/2^-} \log_{10}(\cos x) = -\infty$ since $\cos x \to 0^+$ as $x \to \frac{\pi}{2}^-$.

69. $\lim_{x\to\infty} \ln(1+e^{-x^2}) = \ln\left(1+\lim_{x\to\infty} e^{-x^2}\right) = \ln(1+0) = 0$

71. $f(x) = \log_{10}(1-x)$ Domain$(f) = \{x \mid 1-x>0\} = \{x \mid x<1\} = (-\infty, 1)$. Range$(f) = \mathbb{R}$.

73. $F(t) = \sqrt{t}\ln(t^2-1)$ Domain$(F) = \{t \mid t \geq 0 \text{ and } t^2-1>0\} = \{t \mid t>1\} = (1,\infty)$. Range$(F) = \mathbb{R}$.

75. $y = \ln(x+3) \quad \Rightarrow \quad e^y = e^{\ln(x+3)} = x+3 \quad \Rightarrow \quad x = e^y - 3$.
Interchange x and y: the inverse function is $y = e^x - 3$.

77. $y = e^{\sqrt{x}} \quad \Rightarrow \quad \ln y = \ln e^{\sqrt{x}} = \sqrt{x} \quad \Rightarrow \quad x = (\ln y)^2$. Also note that $\sqrt{x} \geq 0 \quad \Rightarrow \quad y = e^{\sqrt{x}} \geq 1$.
Interchange x and y: the inverse function is $y = (\ln x)^2$, $x \geq 1$.

79. $y = \dfrac{10^x}{10^x+1} \quad \Rightarrow \quad 10^x y + y = 10^x \quad \Rightarrow \quad 10^x(1-y) = y \quad \Rightarrow \quad 10^x = \dfrac{y}{1-y} \quad \Rightarrow$
$x = \log_{10}\left(\dfrac{y}{1-y}\right)$. Interchange x and y: $y = \log_{10}\left(\dfrac{x}{1-x}\right)$ is the inverse function.

81. **(a)** We have to show that $-f(x) = f(-x)$.

$$-f(x) = -\ln\left(x+\sqrt{x^2+1}\right) = \ln\left[\left(x+\sqrt{x^2+1}\right)^{-1}\right] = \ln\frac{1}{x+\sqrt{x^2+1}}$$
$$= \ln\left(\frac{1}{x+\sqrt{x^2+1}}\cdot\frac{x-\sqrt{x^2+1}}{x-\sqrt{x^2+1}}\right) = \ln\frac{x-\sqrt{x^2+1}}{x^2-x^2-1}$$
$$= \ln\left(\sqrt{x^2+1}-x\right) = f(-x).$$ Thus, f is an odd function.

(b) Let $y = \ln\left(x+\sqrt{x^2+1}\right)$, then $e^y = x+\sqrt{x^2+1} \quad \Leftrightarrow \quad (e^y-x)^2 = x^2+1 \quad \Leftrightarrow$
$e^{2y} - 2xe^y + x^2 = x^2+1 \quad \Leftrightarrow \quad 2xe^y = e^{2y}-1 \quad \Leftrightarrow \quad x = \dfrac{e^{2y}-1}{2e^y} = \frac{1}{2}(e^y - e^{-y})$. Thus, the inverse function is $f^{-1}(x) = \frac{1}{2}(e^x - e^{-x})$.

83. Let $x = \log_{10} 99$, $y = \log_9 82$. Then $10^x = 99 < 10^2 \quad \Rightarrow \quad x < 2$, and $9^y = 82 > 9^2 \quad \Rightarrow \quad y > 2$.
Therefore $y = \log_9 82$ is larger.

85. **(a)** Let $\epsilon > 0$ be given. We need N such that $|a^x - 0| < \epsilon$ when $x < N$. But $a^x < \epsilon \quad \Leftrightarrow \quad x < \log_a \epsilon$. Let $N = \log_a \epsilon$. Then $x < N \quad \Rightarrow \quad x < \log_a \epsilon \quad \Rightarrow \quad |a^x - 0| = a^x < \epsilon$, so $\lim_{x\to-\infty} a^x = 0$.

(b) Let $M > 0$ be given. We need N such that $a^x > M$ when $x > N$. But $a^x > M \quad \Leftrightarrow \quad x > \log_a M$. Let $N = \log_a M$. Then $x > N \quad \Rightarrow \quad x > \log_a M \quad \Rightarrow \quad a^x > M$, so $\lim_{x\to\infty} a^x = \infty$.

87. $\ln(x^2-2x-2) \leq 0 \quad \Rightarrow \quad 0 < x^2-2x-2 \leq 1$. Now $x^2-2x-2 \leq 1$ gives $x^2-2x-3 \leq 0$ and hence $(x-3)(x+1) \leq 0$. So $-1 \leq x \leq 3$. Now $0 < x^2-2x-2 \quad \Rightarrow \quad x < 1-\sqrt{3}$ or $x > 1+\sqrt{3}$. Therefore $\ln(x^2-2x-2) \leq 0 \quad \Leftrightarrow \quad -1 \leq x < 1-\sqrt{3}$ or $1+\sqrt{3} < x \leq 3$.

EXERCISES 4.4

1. $f(x) = \ln(x+1) \quad \Rightarrow \quad f'(x) = 1/(x+1)$, $\text{Dom}(f) = \text{Dom}(f') = \{x \mid x+1>0\}$
$= \{x \mid x > -1\} = (-1, \infty)$ [Note that, in general, $\text{Dom}(f') \subset \text{Dom}(f)$.]

3. $f(x) = x^2 \ln(1-x^2) \quad \Rightarrow \quad f'(x) = 2x\ln(1-x^2) + \dfrac{x^2(-2x)}{1-x^2} = 2x\ln(1-x^2) - \dfrac{2x^3}{1-x^2}$,
$\text{Dom}(f) = \text{Dom}(f') = \{x \mid 1-x^2>0\} = \{x \mid |x|<1\} = (-1,1)$

5. $f(x) = \log_3(x^2-4) \quad \Rightarrow \quad f'(x) = \dfrac{2x}{(x^2-4)\ln 3}$,
$\text{Dom}(f) = \text{Dom}(f') = \{x \mid x^2-4>0\} = \{x \mid |x|>2\} = (-\infty,-2)\cup(2,\infty)$

7. $y = x\ln x \quad \Rightarrow \quad y' = \ln x + x(1/x) = \ln x + 1 \quad \Rightarrow \quad y'' = 1/x$

9. $y = \log_{10} x \quad \Rightarrow \quad y' = \dfrac{1}{x\ln 10} \quad \Rightarrow \quad y'' = -\dfrac{1}{x^2\ln 10}$

11. $f(x) = \sqrt{x}\ln x \quad \Rightarrow \quad f'(x) = \dfrac{1}{2\sqrt{x}}\ln x + \sqrt{x}\left(\dfrac{1}{x}\right) = \dfrac{\ln x + 2}{2\sqrt{x}}$

13. $g(x) = \ln\dfrac{a-x}{a+x} = \ln(a-x) - \ln(a+x) \quad \Rightarrow \quad g'(x) = \dfrac{-1}{a-x} - \dfrac{1}{a+x} = \dfrac{-2a}{a^2-x^2}$

15. $F(x) = \ln\sqrt{x} = \frac{1}{2}\ln x \quad \Rightarrow \quad F'(x) = \dfrac{1}{2}\left(\dfrac{1}{x}\right) = \dfrac{1}{2x}$

17. $f(t) = \log_2(t^4 - t^2 + 1) \quad \Rightarrow \quad f'(t) = \dfrac{4t^3 - 2t}{(t^4 - t^2 + 1)\ln 2}$

19. $g(u) = \dfrac{1-\ln u}{1+\ln u} \quad \Rightarrow \quad g'(u) = \dfrac{(1+\ln u)(-1/u) - (1-\ln u)(1/u)}{(1+\ln u)^2} = -\dfrac{2}{u(1+\ln u)^2}$

21. $y = (\ln\sin x)^3 \quad \Rightarrow \quad y' = 3(\ln\sin x)^2\dfrac{\cos x}{\sin x} = 3(\ln\sin x)^2\cot x$

23. $y = \dfrac{\ln x}{1+x^2} \quad \Rightarrow \quad y' = \dfrac{(1+x^2)(1/x) - 2x\ln x}{(1+x^2)^2} = \dfrac{1+x^2-2x^2\ln x}{x(1+x^2)^2}$

25. $y = \ln|x^3 - x^2| \quad \Rightarrow \quad y' = \dfrac{1}{x^3-x^2}(3x^2-2x) = \dfrac{x(3x-2)}{x^2(x-1)} = \dfrac{3x-2}{x(x-1)}$

27. $F(x) = e^x\ln x \quad \Rightarrow \quad F'(x) = e^x\ln x + e^x\left(\dfrac{1}{x}\right) = e^x\left(\ln x + \dfrac{1}{x}\right)$

29. $f(t) = \pi^{-t} \quad \Rightarrow \quad f'(t) = \pi^{-t}(\ln\pi)(-1) = -\pi^{-t}\ln\pi$

31. $h(t) = t^3 - 3^t \quad \Rightarrow \quad h'(t) = 3t^2 - 3^t\ln 3$

33. $y = \ln[e^{-x}(1+x)] = \ln(e^{-x}) + \ln(1+x) = -x + \ln(1+x) \quad \Rightarrow \quad y' = -1 + \dfrac{1}{1+x} = -\dfrac{x}{1+x}$

35. $y = x^{\sin x} \quad \Rightarrow \quad \ln y = \sin x\ln x \quad \Rightarrow \quad \dfrac{y'}{y} = \cos x\ln x + \dfrac{\sin x}{x} \quad \Rightarrow \quad y' = x^{\sin x}\left(\cos x\ln x + \dfrac{\sin x}{x}\right)$

37. $y = x^{e^x} \quad \Rightarrow \quad \ln y = e^x\ln x \quad \Rightarrow \quad \dfrac{y'}{y} = e^x\ln x + \dfrac{e^x}{x} \quad \Rightarrow \quad y' = x^{e^x}e^x\left(\ln x + \dfrac{1}{x}\right)$

39. $y = (\ln x)^x \quad \Rightarrow \quad \ln y = x\ln\ln x \quad \Rightarrow \quad \dfrac{y'}{y} = \ln\ln x + x\cdot\dfrac{1}{\ln x}\cdot\dfrac{1}{x} \quad \Rightarrow \quad y' = (\ln x)^x\left(\ln\ln x + \dfrac{1}{\ln x}\right)$

41. $y = x^{1/\ln x} \quad \Rightarrow \quad \ln y = \left(\dfrac{1}{\ln x}\right)\ln x = 1 \quad \Rightarrow \quad y = e \quad \Rightarrow \quad y' = 0$

43. $y = \cos\left(x^{\sqrt{x}}\right) \quad \Rightarrow \quad y' = -\sin\left(x^{\sqrt{x}}\right)x^{\sqrt{x}}\left(\dfrac{\ln x + 2}{2\sqrt{x}}\right)$ by Example 16

45. $f(x) = \dfrac{x}{\ln x} \quad \Rightarrow \quad f'(x) = \dfrac{\ln x - x(1/x)}{(\ln x)^2} = \dfrac{\ln x - 1}{(\ln x)^2} \quad \Rightarrow \quad f'(e) = \dfrac{1-1}{1^2} = 0$

47. $f(x) = \sin x + \ln x \quad \Rightarrow$

$f'(x) = \cos x + \dfrac{1}{x}$

This is reasonable, because the graph shows that f increases when $f'(x)$ is positive.

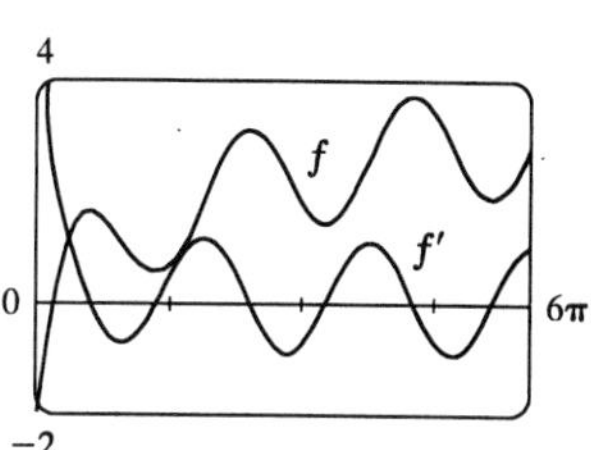

49. $y = f(x) = \ln \ln x \quad \Rightarrow \quad f'(x) = \dfrac{1}{\ln x}\left(\dfrac{1}{x}\right) \quad \Rightarrow \quad f'(e) = \dfrac{1}{e}$, so the equation of the tangent at $(e, 0)$ is $y - 0 = \dfrac{1}{e}(x - e)$ or $x - ey = e$.

51. $x = \ln t$ and $y = te^t$. Since $(0, e)$ is on the curve, $x = 0 \quad \Rightarrow \quad \ln t = 0 \quad \Rightarrow \quad t = 1$.
$\dfrac{dy}{dx} = \dfrac{dy/dt}{dx/dt} = \dfrac{te^t + e^t}{1/t} = \dfrac{e^t(t+1)}{1/t} = te^t(t+1)$. When $t = 1$, $\dfrac{dy}{dx} = 2e$, so an equation of the tangent line is $y - e = 2e(x - 0)$, or $y = 2ex + e$.

53. $y = \ln(x^2 + y^2) \quad \Rightarrow \quad y' = \dfrac{1}{x^2 + y^2}\dfrac{d}{dx}(x^2 + y^2) \quad \Rightarrow \quad (x^2 + y^2)y' = 2x + 2yy' \quad \Rightarrow$
$(x^2 + y^2)y' - 2yy' = 2x \quad \Rightarrow \quad (x^2 + y^2 - 2y)y' = 2x \quad \Rightarrow \quad y' = \dfrac{2x}{x^2 + y^2 - 2y}$

55. $f(x) = \ln(x - 1) \quad \Rightarrow \quad f'(x) = 1/(x-1) = (x-1)^{-1} \quad \Rightarrow \quad f''(x) = -(x-1)^{-2} \quad \Rightarrow$
$f'''(x) = 2(x-1)^{-3} \quad \Rightarrow \quad f^{(4)}(x) = -2 \cdot 3(x-1)^{-4} \quad \Rightarrow \quad \cdots \quad \Rightarrow$
$f^{(n)}(x) = (-1)^{n-1} \cdot 2 \cdot 3 \cdot 4 \cdot \cdots \cdot (n-1)(x-1)^{-n} = (-1)^{n-1}\dfrac{(n-1)!}{(x-1)^n}$

57.

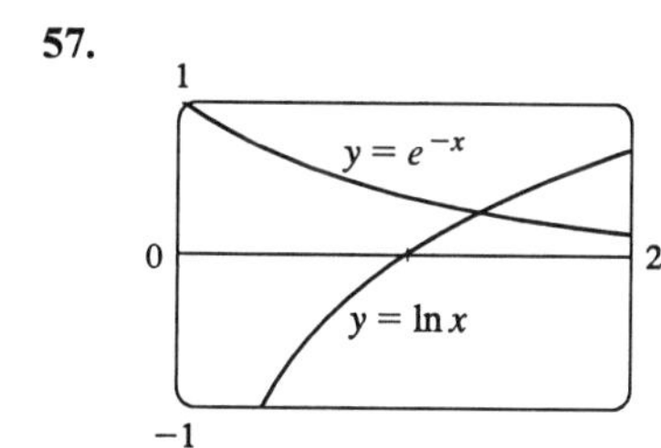

From the graph, it appears that the only root of the equation occurs at about $x = 1.3$. So we use Newton's Method with this as our initial approximation, and with $f(x) = \ln x - e^{-x} \quad \Rightarrow \quad f'(x) = 1/x + e^{-x}$. The formula is $x_{n+1} = x_n - f(x_n)/f'(x_n)$, and we calculate $x_1 = 1.3$, $x_2 \approx 1.309760$, $x_3 \approx x_4 \approx 1.309800$. So, correct to six decimal places, the root of the equation $\ln x = e^{-x}$ is $x = 1.309800$.

59. $y = (3x - 7)^4(8x^2 - 1)^3 \quad \Rightarrow \quad \ln|y| = 4\ln|3x - 7| + 3\ln|8x^2 - 1| \quad \Rightarrow \quad \dfrac{y'}{y} = \dfrac{12}{3x - 7} + \dfrac{48x}{8x^2 - 1} \quad \Rightarrow$
$y' = (3x - 7)^4(8x^2 - 1)^3\left(\dfrac{12}{3x - 7} + \dfrac{48x}{8x^2 - 1}\right)$

61. $y = \dfrac{(x+1)^4(x-5)^3}{(x-3)^8} \quad\Rightarrow\quad \ln|y| = 4\ln|x+1| + 3\ln|x-5| - 8\ln|x-3| \quad\Rightarrow$

$\dfrac{y'}{y} = \dfrac{4}{x+1} + \dfrac{3}{x-5} - \dfrac{8}{x-3} \quad\Rightarrow\quad y' = \dfrac{(x+1)^4(x-5)^3}{(x-3)^8}\left(\dfrac{4}{x+1} + \dfrac{3}{x-5} - \dfrac{8}{x-3}\right)$

63. $y = \dfrac{e^x\sqrt{x^5+2}}{(x+1)^4(x^2+3)^2} \quad\Rightarrow\quad \ln y = x + \frac{1}{2}\ln(x^5+2) - 4\ln|x+1| - 2\ln(x^2+3) \quad\Rightarrow$

$\dfrac{y'}{y} = 1 + \dfrac{5x^4}{2(x^5+2)} - \dfrac{4}{x+1} - \dfrac{4x}{x^2+3}$. So $y' = \dfrac{e^x\sqrt{x^5+2}}{(x+1)^4(x^2+3)^2}\left[1 + \dfrac{5x^4}{2(x^5+2)} - \dfrac{4}{x+1} - \dfrac{4x}{x^2+3}\right]$.

65. $f(x) = 2x + \ln x \quad\Rightarrow\quad f'(x) = 2 + 1/x$. If $g = f^{-1}$, then $f(1) = 2 \quad\Rightarrow\quad g(2) = 1$, so $g'(2) = 1/f'(g(2)) = 1/f'(1) = \frac{1}{3}$.

67. If $f(x) = \ln(1+x)$, then $f'(x) = 1/(1+x)$, so $f'(0) = 1$. Thus

$\lim\limits_{x\to 0}\dfrac{\ln(1+x)}{x} = \lim\limits_{x\to 0}\dfrac{f(x)}{x} = \lim\limits_{x\to 0}\dfrac{f(x)-f(0)}{x-0} = f'(0) = 1.$

69. Let $m = n/x$. Then $n = xm$, and as $n \to \infty$, $m \to \infty$. Therefore,

$\lim\limits_{n\to\infty}\left(1 + \dfrac{x}{n}\right)^n = \lim\limits_{m\to\infty}\left(1 + \dfrac{1}{m}\right)^{mx} = \left[\lim\limits_{m\to\infty}\left(1 + \dfrac{1}{m}\right)^m\right]^x = e^x$ by Equation 7.

EXERCISES 4.5

1. **(a)** By Theorem 2, $y(t) = y(0)e^{kt} = 100e^{kt} \quad\Rightarrow\quad y\left(\frac{1}{3}\right) = 100e^{k/3} = 200 \quad\Rightarrow$ $k/3 = \ln(200/100) = \ln 2 \quad\Rightarrow\quad k = 3\ln 2$. So $y(t) = 100e^{(3\ln 2)t} = 100\cdot 2^{3t}$.

(b) $y(10) = 100\cdot 2^{30} \approx 1.07\times 10^{11}$ cells

(c) $y(t) = 100\cdot 2^{3t} = 10{,}000 \quad\Rightarrow\quad 2^{3t} = 100 \quad\Rightarrow\quad 3t\ln 2 = \ln 100 \quad\Rightarrow\quad t = (\ln 100)/(3\ln 2) \approx 2.2\,\text{h}$

3. **(a)** $y(t) = y(0)e^{kt} = 500\,e^{kt} \quad\Rightarrow\quad y(3) = 500e^{3k} = 8000 \quad\Rightarrow\quad e^{3k} = 16 \quad\Rightarrow\quad 3k = \ln 16 \quad\Rightarrow$ $y(t) = 500e^{(\ln 16)t/3} = 500\cdot 16^{t/3}$

(b) $y(4) = 500\cdot 16^{4/3} \approx 20{,}159$

(c) $y(t) = 500\cdot 16^{t/3} = 30{,}000 \quad\Rightarrow\quad 16^{t/3} = 60 \quad\Rightarrow\quad \frac{1}{3}t\ln 16 = \ln 60 \quad\Rightarrow$ $t = 3(\ln 60)/(\ln 16) \approx 4.4\,\text{h}$

5. **(a)** Let the population (in millions) in the year t be $P(t)$. Since the initial time is the year 1750, we substitute $t - 1750$ for t in Theorem 2, so the exponential model gives $P(t) = P(1750)e^{k(t-1750)}$. Then $P(1800) = 906 = 728e^{k(1800-1750)} \quad \Rightarrow \quad \ln \frac{906}{728} = k(50) \quad \Rightarrow \quad k = \frac{1}{50} \ln \frac{906}{728} \approx 0.0043748$. So with this model, we estimate $P(1900) \approx P(1750)e^{k(1900-1750)} \approx 728e^{150(0.0043748)} \approx 1403$ million, and $P(1950) \approx 728e^{200(0.0043748)} \approx 1746$ million. Both of these estimates are much too low.

(b) In this case, the exponential model gives $P(t) = P(1850)e^{k(t-1850)} \quad \Rightarrow$ $P(1900) = 1608 = 1171e^{k(1900-1850)} \quad \Rightarrow \quad \ln \frac{1608}{1171} = k(50) \quad \Rightarrow \quad k = \frac{1}{50} \ln \frac{1608}{1171} \approx 0.006343$. So with this model, we estimate $P(1950) \approx 1171e^{100(0.006343)} \approx 2208$ million. This is still too low, but closer than the estimate of $P(1950)$ in part (a).

(c) The exponential model gives $P(t) = P(1900)e^{k(t-1900)} \quad \Rightarrow \quad P(1950) = 2517 = 1608e^{k(1950-1900)}$ $\Rightarrow \quad \ln \frac{2517}{1608} = k(50) \quad \Rightarrow \quad k = \frac{1}{50} \ln \frac{2517}{1608} \approx 0.008962$. With this model, we estimate $P(1992) \approx 1608e^{0.008962(1992-1900)} \approx 3667$ million. This is much too low.

The discrepancy is explained by the fact that the world birth rate (average yearly number of births per person) is about the same as always, whereas the mortality rate (especially the infant mortality rate) is much lower, owing mostly to advances in medical science and to the wars in the first part of the twentieth century. The exponential model assumes, among other things, that the birth and mortality rates will remain constant.

7. **(a)** If $y = [N_2O_5]$ then $\dfrac{dy}{dt} = -0.0005y \quad \Rightarrow \quad y(t) = y(0)e^{-0.0005t} = Ce^{-0.0005t}$.

(b) $y(t) = Ce^{-0.0005t} = 0.9C \quad \Rightarrow \quad e^{-0.0005t} = 0.9 \quad \Rightarrow \quad -0.0005t = \ln 0.9 \quad \Rightarrow$ $t = -2000 \ln 0.9 \approx 211$ s

9. **(a)** If $y(t)$ is the mass remaining after t days, then $y(t) = y(0)e^{kt} = 50e^{kt} \quad \Rightarrow$ $y(0.00014) = 50e^{0.00014k} = 25 \quad \Rightarrow \quad e^{0.00014k} = \frac{1}{2} \quad \Rightarrow \quad k = -(\ln 2)/0.00014 \quad \Rightarrow \quad y(t) = 50$ $e^{-(\ln 2)t/0.00014} = 50 \cdot 2^{-t/0.00014}$

(b) $y(0.01) = 50 \cdot 2^{-0.01/0.00014} \approx 1.57 \times 10^{-20}$ mg

(c) $50e^{-(\ln 2)t/0.00014} = 40 \quad \Rightarrow \quad -(\ln 2)t/0.00014 = \ln 0.8 \quad \Rightarrow \quad t = -0.00014 \dfrac{\ln 0.8}{\ln 2} \approx 4.5 \times 10^{-5}$ s

11. Let $y(t)$ be the level of radioactivity. Thus, $y(t) = y(0)e^{-kt}$ and k is determined by using the half-life: $\frac{1}{2} = e^{-5730k} \quad \Rightarrow \quad k = -\dfrac{\ln \frac{1}{2}}{5730} = \dfrac{\ln 2}{5730}$. If 0.74 of the ^{14}C remains, then we know that $0.74 = e^{-t(\ln 2)/5730} \quad \Rightarrow$ $\ln 0.74 = -\dfrac{t \ln 2}{5730} \quad \Rightarrow \quad t = -\dfrac{5730(\ln 0.74)}{\ln 2} \approx 2489 \approx 2500$ years.

13. Let $y(t) =$ temperature after t minutes. Then $\dfrac{dy}{dt} = -\frac{1}{10}[y(t) - 21]$. If $u(t) = y(t) - 21$, then $\dfrac{du}{dt} = -\dfrac{u}{10}$ $\Rightarrow \quad u(t) = u(0)\, e^{-t/10} = 12\, e^{-t/10} \quad \Rightarrow \quad y(t) = 21 + u(t) = 21 + 12\, e^{-t/10}$.

15. **(a)** Let $y(t)$ = temperature after t minutes. Newton's Law of Cooling implies that $\frac{dy}{dt} = k(y - 75)$. Let $u(t) = y(t) - 75$. Then $\frac{du}{dt} = ku$, so $u(t) = u(0)e^{kt} = 110e^{kt} \Rightarrow y(t) = 75 + 110e^{kt} \Rightarrow$ $y(30) = 75 + 110e^{30k} = 150 \Rightarrow e^{30k} = \frac{75}{110} = \frac{15}{22} \Rightarrow k = \frac{1}{30}\ln\frac{15}{22}$, so $y(t) = 75 + 110e^{\frac{1}{30}t\ln\left(\frac{15}{22}\right)}$ and $y(45) = 75 + 110e^{\frac{45}{30}\ln\left(\frac{15}{22}\right)} \approx 137\,°\text{F}$.

(b) $y(t) = 75 + 110e^{\frac{1}{30}t\ln\left(\frac{15}{22}\right)} = 100 \Rightarrow e^{\frac{1}{30}t\ln\left(\frac{15}{22}\right)} = \frac{25}{110} \Rightarrow \frac{1}{30}t\ln\frac{15}{22} = \ln\frac{25}{110} \Rightarrow t = \dfrac{30\ln\frac{25}{110}}{\ln\frac{15}{22}} \approx 116$ min

17. **(a)** Let $P(h)$ be the pressure at altitude h. Then $dP/dh = kP \Rightarrow P(h) = P(0)e^{kh} = 101.3e^{kh} \Rightarrow$ $P(1000) = 101.3e^{1000k} = 87.14 \Rightarrow 1000k = \ln\left(\frac{87.14}{101.3}\right) \Rightarrow P(h) = 101.3\,e^{\frac{1}{1000}h\ln\left(\frac{87.14}{101.3}\right)}$, so $P(3000) = 101.3e^{3\ln\left(\frac{87.14}{101.3}\right)} \approx 64.5$ kPa.

(b) $P(6187) = 101.3\,e^{\frac{6187}{1000}\ln\left(\frac{87.14}{101.3}\right)} \approx 39.9$ kPa

19. With the notation of Example 4, $A_0 = 3000$, $i = 0.05$, and $t = 5$.

(a) $n = 1$: $A = 3000(1.05)^5 = \$3828.84$

(b) $n = 2$: $A = 3000\left(1 + \frac{0.05}{2}\right)^{10} = \$3840.25$

(c) $n = 12$: $A = 3000\left(1 + \frac{0.05}{12}\right)^{60} = \$3850.08$

(d) $n = 52$: $A = 3000\left(1 + \frac{0.05}{52}\right)^{5\cdot 52} = \$3851.61$

(e) $n = 365$: $A = 3000\left(1 + \frac{0.05}{365}\right)^{5\cdot 365} = \$3852.01$

(f) continuously: $A = 3000e^{(0.05)5} = \$3852.08$

21. **(a)** If $y(t)$ is the amount of salt at time t, then $y(0) = 1500(0.3) = 450$ kg. The rate of change of y is $\frac{dy}{dt} = -\left(\frac{y(t)}{1500}\frac{\text{kg}}{\text{L}}\right)\left(20\,\frac{\text{L}}{\text{min}}\right) = -\frac{1}{75}y(t)\,\frac{\text{kg}}{\text{min}}$, so $y(t) = y(0)e^{-t/75} = 450e^{-t/75} \Rightarrow$ $y(30) = 450e^{-0.4} \approx 301.6$ kg.

(b) When the concentration is 0.2 kg/L, the amount of salt is $1500(0.2) = 300$ kg. So $y(t) = 450e^{-t/75} = 300$ $\Rightarrow e^{-t/75} = \frac{2}{3} \Rightarrow -t/75 = \ln\frac{2}{3} \Rightarrow t = -75\ln\frac{2}{3} \approx 30.41$ min.

EXERCISES 4.6

1. $\cos^{-1}(-1) = \pi$ since $\cos\pi = -1$.

3. $\tan^{-1}\sqrt{3} = \frac{\pi}{3}$ since $\tan\frac{\pi}{3} = \sqrt{3}$.

5. $\csc^{-1}\sqrt{2} = \frac{\pi}{4}$ since $\csc\frac{\pi}{4} = \sqrt{2}$.

7. $\cot^{-1}\left(-\sqrt{3}\right) = \frac{5\pi}{6}$ since $\cot\frac{5\pi}{6} = -\sqrt{3}$.

9. $\sin(\sin^{-1}0.7) = 0.7$

11. $\tan^{-1}\left(\tan\frac{4\pi}{3}\right) = \tan^{-1}\sqrt{3} = \frac{\pi}{3}$

13. Let $\theta = \cos^{-1}\frac{4}{5}$, so $\cos\theta = \frac{4}{5}$. Then $\sin\left(\cos^{-1}\frac{4}{5}\right) = \sin\theta = \sqrt{1 - \left(\frac{4}{5}\right)^2} = \sqrt{\frac{9}{25}} = \frac{3}{5}$.

15. $\arcsin\left(\sin\frac{5\pi}{4}\right) = \arcsin\left(-\frac{1}{\sqrt{2}}\right) = -\frac{\pi}{4}$

17. Let $\theta = \sin^{-1}\frac{5}{13}$. Then $\sin\theta = \frac{5}{13}$, so $\cos\left(2\sin^{-1}\frac{5}{13}\right) = \cos 2\theta = 1 - 2\sin^2\theta = 1 - 2\left(\frac{5}{13}\right)^2 = \frac{119}{169}$.

19. Let $y = \sin^{-1}x$. Then $-\frac{\pi}{2} \le y \le \frac{\pi}{2} \quad \Rightarrow \quad \cos y \ge 0$, so $\cos(\sin^{-1}x) = \cos y = \sqrt{1 - \sin^2 y} = \sqrt{1 - x^2}$

21. Let $y = \tan^{-1}x$. Then $\tan y = x$, so from the triangle we see that

$$\sin(\tan^{-1}x) = \sin y = \frac{x}{\sqrt{1+x^2}}.$$

$\sqrt{1+x^2}$ x y 1

23.

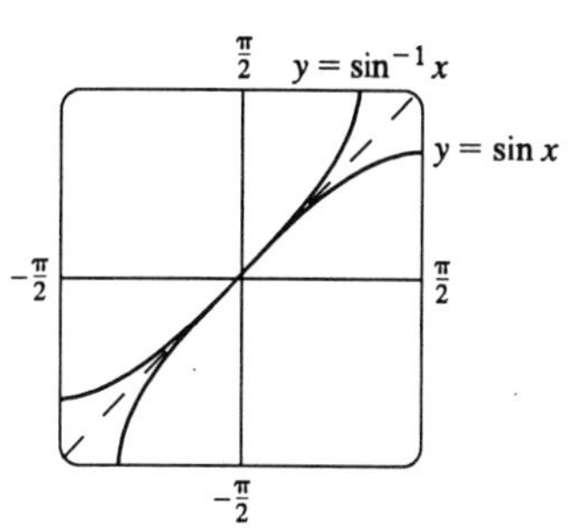

The graph of $\sin^{-1}x$ is the reflection of the graph of $\sin x$ about the line $y = x$.

25. Let $y = \cos^{-1}x$. Then $\cos y = x$ and $0 \le y \le \pi \quad \Rightarrow \quad -\sin y \dfrac{dy}{dx} = 1 \quad \Rightarrow$

$$\frac{dy}{dx} = -\frac{1}{\sin y} = -\frac{1}{\sqrt{1-\cos^2 y}} = -\frac{1}{\sqrt{1-x^2}}$$ (Note that $\sin y \ge 0$ for $0 \le y \le \pi$.)

27. Let $y = \cot^{-1}x$. Then $\cot y = x \quad \Rightarrow \quad -\csc^2 y \dfrac{dy}{dx} = 1 \quad \Rightarrow \quad \dfrac{dy}{dx} = -\dfrac{1}{\csc^2 y} = -\dfrac{1}{1+\cot^2 y} = -\dfrac{1}{1+x^2}$.

29. Let $y = \csc^{-1}x$. Then $\csc y = x \quad \Rightarrow \quad -\csc y \cot y \dfrac{dy}{dx} = 1 \quad \Rightarrow$

$$\frac{dy}{dx} = -\frac{1}{\csc y \cot y} = -\frac{1}{\csc y\sqrt{\csc^2 y - 1}} = -\frac{1}{x\sqrt{x^2-1}}.$$ Note that $\cot y \ge 0$ on the domain of $\csc^{-1}x$.

31. $g(x) = \tan^{-1}(x^3) \quad \Rightarrow \quad g'(x) = \dfrac{1}{1+(x^3)^2}(3x^2) = \dfrac{3x^2}{1+x^6}$

33. $y = \sin^{-1}(x^2) \quad \Rightarrow \quad y' = \dfrac{1}{\sqrt{1-(x^2)^2}}(2x) = \dfrac{2x}{\sqrt{1-x^4}}$

35. $H(x) = (1+x^2)\arctan x \quad \Rightarrow \quad H'(x) = (2x)\arctan x + (1+x^2)\dfrac{1}{1+x^2} = 1 + 2x\arctan x$

37. $g(t) = \sin^{-1}\left(\dfrac{4}{t}\right) \quad \Rightarrow \quad g'(t) = \dfrac{1}{\sqrt{1-(4/t)^2}}\left(-\dfrac{4}{t^2}\right) = -\dfrac{4}{\sqrt{t^4 - 16t^2}}$

39. $G(t) = \cos^{-1}\sqrt{2t-1} \quad \Rightarrow \quad G'(t) = -\dfrac{1}{\sqrt{1-(2t-1)}}\dfrac{2}{2\sqrt{2t-1}} = -\dfrac{1}{\sqrt{2(-2t^2+3t-1)}}$

41. $y = \sec^{-1}\sqrt{1+x^2} \quad \Rightarrow \quad y' = \left[\dfrac{1}{\sqrt{1+x^2}\sqrt{(1+x^2)-1}}\right]\left[\dfrac{2x}{2\sqrt{1+x^2}}\right] = \dfrac{x}{(1+x^2)\sqrt{x^2}} = \dfrac{x}{(1+x^2)|x|}$

43. $y = \tan^{-1}(\sin x) \quad \Rightarrow \quad y' = \dfrac{\cos x}{1+\sin^2 x}$

45. $y = (\tan^{-1}x)^{-1} \quad \Rightarrow \quad y' = -(\tan^{-1}x)^{-2}\left(\dfrac{1}{1+x^2}\right) = -\dfrac{1}{(1+x^2)(\tan^{-1}x)^2}$

47. $y = x^2\cot^{-1}(3x) \quad \Rightarrow \quad y' = 2x\cot^{-1}(3x) + x^2\left[-\dfrac{1}{1+(3x)^2}\right](3) = 2x\cot^{-1}(3x) - \dfrac{3x^2}{1+9x^2}$

49. $y = \arccos\left(\dfrac{b + a\cos x}{a + b\cos x}\right) \quad \Rightarrow$

$$y' = -\frac{1}{\sqrt{1 - \left(\dfrac{b + a\cos x}{a + b\cos x}\right)^2}} \frac{(a + b\cos x)(-a\sin x) - (b + a\cos x)(-b\sin x)}{(a + b\cos x)^2}$$

$$= \frac{1}{\sqrt{a^2 + b^2\cos^2 x - b^2 - a^2\cos^2 x}} \frac{(a^2 - b^2)\sin x}{|a + b\cos x|}$$

$$= \frac{1}{\sqrt{a^2 - b^2}\sqrt{1 - \cos^2 x}} \frac{(a^2 - b^2)\sin x}{|a + b\cos x|} = \frac{\sqrt{a^2 - b^2}}{|a + b\cos x|} \frac{\sin x}{|\sin x|}$$

But $0 \le x \le \pi$, so $|\sin x| = \sin x$. Also $a > b > 0 \quad \Rightarrow \quad b\cos x \ge -b > -a$, so $a + b\cos x > 0$.

Thus $y' = \dfrac{\sqrt{a^2 - b^2}}{a + b\cos x}$.

51. $g(x) = \sin^{-1}(3x + 1) \quad \Rightarrow \quad g'(x) = \dfrac{3}{\sqrt{1 - (3x + 1)^2}} = \dfrac{3}{\sqrt{-9x^2 - 6x}}$,

$\text{Dom}(g) = \{x \mid -1 \le 3x + 1 \le 1\} = \{x \mid -\frac{2}{3} \le x \le 0\} = [-\frac{2}{3}, 0]$,

$\text{Dom}(g') = \{x \mid -1 < 3x + 1 < 1\} = (-\frac{2}{3}, 0)$

53. $S(x) = \sin^{-1}(\tan^{-1}x) \quad \Rightarrow \quad S'(x) = \left[\sqrt{1 - (\tan^{-1}x)^2}(1 + x^2)\right]^{-1}$,

$\text{Dom}(S) = \{x \mid -1 \le \tan^{-1}x \le 1\} = \{x \mid \tan(-1) \le x \le \tan 1\} = [-\tan 1, \tan 1]$,

$\text{Dom}(S') = \{x \mid -1 < \tan^{-1}x < 1\} = (-\tan 1, \tan 1)$

55. $U(t) = 2^{\arctan t} \quad \Rightarrow \quad U'(t) = 2^{\arctan t}(\ln 2)/(1 + t^2)$, $\text{Dom}(U) = \text{Dom}(U') = \mathbb{R}$

57. $g(x) = x\sin^{-1}\left(\dfrac{x}{4}\right) + \sqrt{16 - x^2} \quad \Rightarrow \quad g'(x) = \sin^{-1}\left(\dfrac{x}{4}\right) + \dfrac{x}{4\sqrt{1 - (x/4)^2}} - \dfrac{x}{\sqrt{16 - x^2}} = \sin^{-1}\left(\dfrac{x}{4}\right)$

$\Rightarrow \quad g'(2) = \sin^{-1}\frac{1}{2} = \frac{\pi}{6}$

59. $f(x) = e^x - x^2\arctan x \quad \Rightarrow$

$$f'(x) = e^x - \left[x^2\left(\frac{1}{1 + x^2}\right) + 2x\arctan x\right]$$

$$= e^x - \frac{x^2}{1 + x^2} - 2x\arctan x$$

This is reasonable because the graphs show that f is increasing when $f'(x)$ is positive.

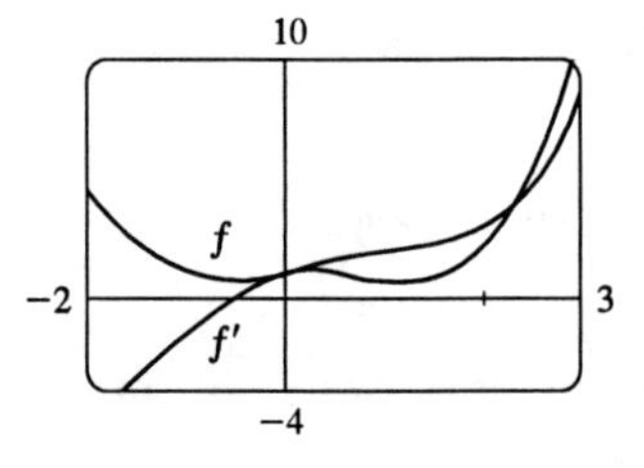

61. $\displaystyle\lim_{x \to -1^+} \sin^{-1}x = \sin^{-1}(-1) = -\frac{\pi}{2}$

63. $\displaystyle\lim_{x \to \infty} \tan^{-1}(x^2) = \frac{\pi}{2}$ since $x^2 \to \infty$ as $x \to \infty$.

65.

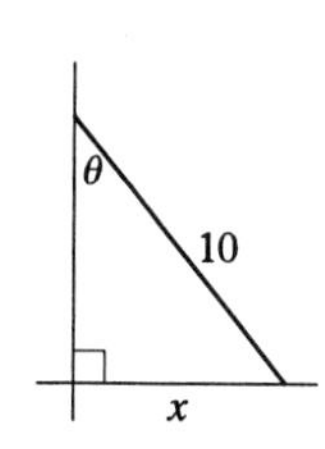

$$\frac{dx}{dt} = 2\text{ ft/s}, \sin\theta = \frac{x}{10} \quad \Rightarrow \quad \theta = \sin^{-1}\left(\frac{x}{10}\right),$$

$$\frac{d\theta}{dx} = \frac{1/10}{\sqrt{1-(x/10)^2}}, \ \frac{d\theta}{dt} = \frac{d\theta}{dx}\frac{dx}{dt} = \frac{1/10}{\sqrt{1-(x/10)^2}}(2)\text{ rad/s},$$

$$\left.\frac{d\theta}{dt}\right|_{x=6} = \frac{2/10}{\sqrt{1-(6/10)^2}}\text{ rad/s} = \frac{1}{4}\text{ rad/s}$$

67. By reflecting the graph of $y = \sec x$ (see Figure 11) about the line $y = x$, we get the graph of $y = \sec^{-1}x$.

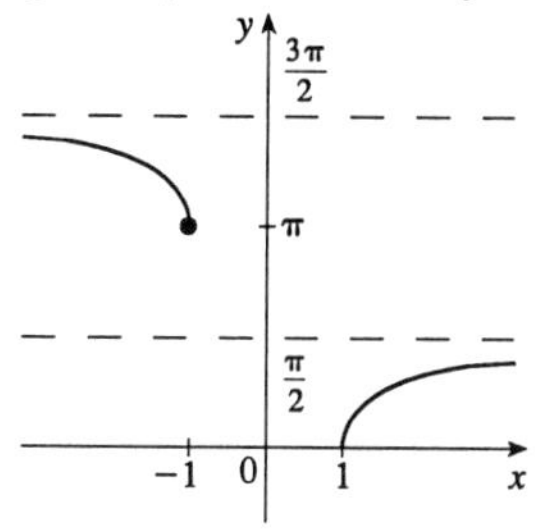

69. **(a)** $\arctan\frac{1}{2} + \arctan\frac{1}{3} = \arctan\left(\dfrac{\frac{1}{2}+\frac{1}{3}}{1-\frac{1}{2}\cdot\frac{1}{3}}\right) = \arctan 1 = \dfrac{\pi}{4}$

(b) $2\arctan\frac{1}{3} + \arctan\frac{1}{7} = \left(\arctan\frac{1}{3} + \arctan\frac{1}{3}\right) + \arctan\frac{1}{7} = \arctan\left(\dfrac{\frac{1}{3}+\frac{1}{3}}{1-\frac{1}{3}\cdot\frac{1}{3}}\right) + \arctan\frac{1}{7}$

$$= \arctan\tfrac{3}{4} + \arctan\tfrac{1}{7} = \arctan\left(\frac{\frac{3}{4}+\frac{1}{7}}{1-\frac{3}{4}\cdot\frac{1}{7}}\right) = \arctan 1 = \frac{\pi}{4}$$

71. $y = \sec^{-1}x \quad \Rightarrow \quad \sec y = x \quad \Rightarrow \quad \sec y\tan y\,\dfrac{dy}{dx} = 1 \quad \Rightarrow \quad \dfrac{dy}{dx} = \dfrac{1}{\sec y\tan y}$. Now $\tan^2 y = \sec^2 y - 1 = x^2 - 1$, so $\tan y = \pm\sqrt{x^2-1}$. For $y \in \left[0, \frac{\pi}{2}\right)$, $x \geq 1$, so $\sec y = x = |x|$ and $\tan y \geq 0$ $\Rightarrow$ $\dfrac{dy}{dx} = \dfrac{1}{x\sqrt{x^2-1}} = \dfrac{1}{|x|\sqrt{x^2-1}}$. For $y \in \left(\frac{\pi}{2}, \pi\right]$, $x \leq -1$, so $|x| = -x$ and $\tan y = -\sqrt{x^2-1}$ $\Rightarrow$

$$\frac{dy}{dx} = \frac{1}{\sec y\tan y} = \frac{1}{x\left(-\sqrt{x^2-1}\right)} = \frac{1}{(-x)\sqrt{x^2-1}} = \frac{1}{|x|\sqrt{x^2-1}}.$$

EXERCISES 4.7

1. **(a)** $\sinh 0 = \frac{1}{2}(e^0 - e^0) = 0$ **(b)** $\cosh 0 = \frac{1}{2}(e^0 + e^0) = \frac{1}{2}(1+1) = 1$

3. **(a)** $\sinh(\ln 2) = \dfrac{e^{\ln 2} - e^{-\ln 2}}{2} = \dfrac{2 - \frac{1}{2}}{2} = \dfrac{3}{4}$ **(b)** $\sinh 2 = \frac{1}{2}(e^2 - e^{-2}) \approx 3.62686$

5. **(a)** $\operatorname{sech} 0 = \dfrac{1}{\cosh 0} = \dfrac{1}{1} = 1$ **(b)** $\cosh^{-1} 1 = 0$ because $\cosh 0 = 1$.

7. $\sinh(-x) = \frac{1}{2}\left[e^{-x} - e^{-(-x)}\right] = \frac{1}{2}(e^{-x} - e^x) = -\frac{1}{2}(e^x - e^{-x}) = -\sinh x$

9. $\cosh x + \sinh x = \frac{1}{2}(e^x + e^{-x}) + \frac{1}{2}(e^x - e^{-x}) = \frac{1}{2}(2e^x) = e^x$

11. $\sinh x \cosh y + \cosh x \sinh y = \left[\frac{1}{2}(e^x - e^{-x})\right]\left[\frac{1}{2}(e^y + e^{-y})\right] + \left[\frac{1}{2}(e^x + e^{-x})\right]\left[\frac{1}{2}(e^y - e^{-y})\right]$

$= \frac{1}{4}[(e^{x+y} + e^{x-y} - e^{-x+y} - e^{-x-y}) + (e^{x+y} - e^{x-y} + e^{-x+y} - e^{-x-y})]$

$= \frac{1}{4}(2e^{x+y} - 2e^{-x-y}) = \frac{1}{2}\left[e^{x+y} - e^{-(x+y)}\right] = \sinh(x+y)$

13. Divide both sides of the identity $\cosh^2 x - \sinh^2 x = 1$ by $\sinh^2 x$:

$$\frac{\cosh^2 x}{\sinh^2 x} - 1 = \frac{1}{\sinh^2 x} \quad \Leftrightarrow \quad \coth^2 x - 1 = \operatorname{csch}^2 x.$$

15. By Exercise 11, $\sinh 2x = \sinh(x+x) = \sinh x \cosh x + \cosh x \sinh x = 2 \sinh x \cosh x$.

17. $\tanh(\ln x) = \dfrac{\sinh(\ln x)}{\cosh(\ln x)} = \dfrac{(e^{\ln x} - e^{-\ln x})/2}{(e^{\ln x} + e^{-\ln x})/2} = \dfrac{x - 1/x}{x + 1/x} = \dfrac{x^2 - 1}{x^2 + 1}$

19. By Exercise 9, $(\cosh x + \sinh x)^n = (e^x)^n = e^{nx} = \cosh nx + \sinh nx$.

21. $\tanh x = \frac{4}{5} > 0$, so $x > 0$. $\coth x = 1/\tanh x = \frac{5}{4}$, $\operatorname{sech}^2 x = 1 - \tanh^2 x = 1 - \left(\frac{4}{5}\right)^2 = \frac{9}{25} \Rightarrow \operatorname{sech} x = \frac{3}{5}$ (since $\operatorname{sech} x > 0$), $\cosh x = 1/\operatorname{sech} x = \frac{5}{3}$, $\sinh x = \tanh x \cosh x = \frac{4}{5} \cdot \frac{5}{3} = \frac{4}{3}$, and $\operatorname{csch} x = 1/\sinh x = \frac{3}{4}$.

23. **(a)** $\displaystyle\lim_{x\to\infty} \tanh x = \lim_{x\to\infty} \frac{e^x - e^{-x}}{e^x + e^{-x}} = \lim_{x\to\infty} \frac{1 - e^{-2x}}{1 + e^{-2x}} = \frac{1-0}{1+0} = 1$

(b) $\displaystyle\lim_{x\to-\infty} \tanh x = \lim_{x\to-\infty} \frac{e^x - e^{-x}}{e^x + e^{-x}} = \lim_{x\to-\infty} \frac{e^{2x} - 1}{e^{2x} + 1} = \frac{0-1}{0+1} = -1$

(c) $\displaystyle\lim_{x\to\infty} \sinh x = \lim_{x\to\infty} \frac{e^x - e^{-x}}{2} = \infty$

(d) $\displaystyle\lim_{x\to-\infty} \sinh x = \lim_{x\to-\infty} \frac{e^x - e^{-x}}{2} = -\infty$

(e) $\displaystyle\lim_{x\to\infty} \operatorname{sech} x = \lim_{x\to\infty} \frac{2}{e^x + e^{-x}} = 0$

(f) $\displaystyle\lim_{x\to\infty} \coth x = \lim_{x\to\infty} \frac{e^x + e^{-x}}{e^x - e^{-x}} = \lim_{x\to\infty} \frac{1 + e^{-2x}}{1 - e^{-2x}} = \frac{1+0}{1-0} = 1$ [*Or:* Use part (a)]

(g) $\displaystyle\lim_{x\to 0^+} \coth x = \lim_{x\to 0^+} \frac{\cosh x}{\sinh x} = \infty$, since $\sinh x \to 0$ and $\coth x > 0$.

(h) $\displaystyle\lim_{x\to 0^-} \coth x = \lim_{x\to 0^-} \frac{\cosh x}{\sinh x} = -\infty$, since $\sinh x \to 0$ and $\coth x < 0$.

(i) $\displaystyle\lim_{x\to-\infty} \operatorname{csch} x = \lim_{x\to-\infty} \frac{2}{e^x - e^{-x}} = 0$

25. Let $y = \sinh^{-1}x$. Then $\sinh y = x$ and, by Example 1(a), $\cosh y = \sqrt{1 + \sinh^2 y} = \sqrt{1 + x^2}$. So by Exercise 9, $e^y = \sinh y + \cosh y = x + \sqrt{1 + x^2} \quad \Rightarrow \quad y = \ln\left(x + \sqrt{1 + x^2}\right)$.

27. **(a)** Let $y = \tanh^{-1}x$. Then $x = \tanh y = \dfrac{e^y - e^{-y}}{e^y + e^{-y}} = \dfrac{e^{2y} - 1}{e^{2y} + 1} \quad \Rightarrow \quad xe^{2y} + x = e^{2y} - 1 \quad \Rightarrow$

$$e^{2y} = \frac{1 + x}{1 - x} \quad \Rightarrow \quad 2y = \ln\left(\frac{1 + x}{1 - x}\right) \quad \Rightarrow \quad y = \frac{1}{2}\ln\left(\frac{1 + x}{1 - x}\right).$$

(b) Let $y = \tanh^{-1}x$. Then $x = \tanh y$, so from Exercise 18 we have $e^{2y} = \dfrac{1 + \tanh y}{1 - \tanh y} = \dfrac{1 + x}{1 - x} \quad \Rightarrow$

$$2y = \ln\left(\frac{1 + x}{1 - x}\right) \quad \Rightarrow \quad y = \frac{1}{2}\ln\left(\frac{1 + x}{1 - x}\right).$$

29. **(a)** Let $y = \cosh^{-1}x$. Then $\cosh y = x$ and $y \geq 0 \quad \Rightarrow \quad \sinh y \dfrac{dy}{dx} = 1 \quad \Rightarrow$

$$\frac{dy}{dx} = \frac{1}{\sinh y} = \frac{1}{\sqrt{\cosh^2 y - 1}} = \frac{1}{\sqrt{x^2 - 1}}$$ (since $\sinh y \geq 0$ for $y \geq 0$). *Or:* Use Formula 4.

(b) Let $y = \tanh^{-1}x$. Then $\tanh y = x \;\Rightarrow\; \operatorname{sech}^2 y \dfrac{dy}{dx} = 1 \;\Rightarrow\; \dfrac{dy}{dx} = \dfrac{1}{\operatorname{sech}^2 y} = \dfrac{1}{1 - \tanh^2 y} = \dfrac{1}{1 - x^2}$.

Or: Use Formula 5.

(c) Let $y = \operatorname{csch}^{-1}x$. Then $\operatorname{csch} y = x \quad \Rightarrow \quad -\operatorname{csch} y \coth y \dfrac{dy}{dx} = 1 \quad \Rightarrow \quad \dfrac{dy}{dx} = -\dfrac{1}{\operatorname{csch} y \coth y}$.

By Exercise 13, $\coth y = \pm\sqrt{\operatorname{csch}^2 y + 1} = \pm\sqrt{x^2 + 1}$. If $x > 0$, then $\coth y > 0$, so $\coth y = \sqrt{x^2 + 1}$. If $x < 0$, then $\coth y < 0$, so $\coth y = -\sqrt{x^2 + 1}$. In either case we have

$$\frac{dy}{dx} = -\frac{1}{\operatorname{csch} y \coth y} = -\frac{1}{|x|\sqrt{x^2 + 1}}.$$

(d) Let $y = \operatorname{sech}^{-1}x$. Then $\operatorname{sech} y = x \quad \Rightarrow \quad -\operatorname{sech} y \tanh y \dfrac{dy}{dx} = 1 \quad \Rightarrow$

$$\frac{dy}{dx} = -\frac{1}{\operatorname{sech} y \tanh y} = -\frac{1}{\operatorname{sech} y\sqrt{1 - \operatorname{sech}^2 y}} = -\frac{1}{x\sqrt{1 - x^2}}.$$ (Note that $y > 0$ and so $\tanh y > 0$.)

(e) Let $y = \coth^{-1}x$. Then $\coth y = x \;\Rightarrow\; -\operatorname{csch}^2 y \dfrac{dy}{dx} = 1 \;\Rightarrow\; \dfrac{dy}{dx} = -\dfrac{1}{\operatorname{csch}^2 y} = \dfrac{1}{1 - \coth^2 y} = \dfrac{1}{1 - x^2}$

by Exercise 13.

31. $f(x) = \tanh 3x \quad \Rightarrow \quad f'(x) = 3\operatorname{sech}^2 3x$

33. $h(x) = \cosh(x^4) \quad \Rightarrow \quad h'(x) = \sinh(x^4)4x^3 = 4x^3\sinh(x^4)$

35. $G(x) = x^2 \operatorname{sech} x \quad \Rightarrow \quad G'(x) = 2x \operatorname{sech} x - x^2 \operatorname{sech} x \tanh x$

37. $H(t) = \tanh(e^t) \quad \Rightarrow \quad H'(t) = \operatorname{sech}^2(e^t)[e^t] = e^t \operatorname{sech}^2(e^t)$

39. $y = x^{\cosh x} \;\Rightarrow\; \ln y = \cosh x \ln x \;\Rightarrow\; \dfrac{y'}{y} = \sinh x \ln x + \dfrac{\cosh x}{x} \;\Rightarrow\; y' = x^{\cosh x}\left(\sinh x \ln x + \dfrac{\cosh x}{x}\right)$

41. $y = \cosh^{-1}(x^2) \quad \Rightarrow \quad y' = \left(1\Big/\sqrt{(x^2)^2 - 1}\right)(2x) = 2x/\sqrt{x^4 - 1}$

43. $y = x\ln(\operatorname{sech} 4x) \quad \Rightarrow \quad y' = \ln(\operatorname{sech} 4x) + x\,\dfrac{-\operatorname{sech} 4x \tanh 4x}{\operatorname{sech} 4x}(4) = \ln(\operatorname{sech} 4x) - 4x\tanh 4x$

45. $y = x\sinh^{-1}(x/3) - \sqrt{9+x^2} \quad \Rightarrow$

$$y' = \sinh^{-1}\left(\frac{x}{3}\right) + x\frac{1/3}{\sqrt{1+(x/3)^2}} - \frac{2x}{2\sqrt{9+x^2}} = \sinh^{-1}\left(\frac{x}{3}\right) + \frac{x}{\sqrt{9+x^2}} - \frac{x}{\sqrt{9+x^2}} = \sinh^{-1}\left(\frac{x}{3}\right)$$

47. $y = \coth^{-1}\sqrt{x^2+1} \quad \Rightarrow \quad y' = \dfrac{1}{1-(x^2+1)}\dfrac{2x}{2\sqrt{x^2+1}} = -\dfrac{1}{x\sqrt{x^2+1}}$

49. $x = 4\cosh t$ and $y = 3\sinh t$, so $\cosh^2 t - \sinh^2 t = 1 \quad \Rightarrow \quad \left(\frac{x}{4}\right)^2 - \left(\frac{y}{3}\right)^2 = 1$, which is a hyperbola with right and left branches. Since $x = \cosh t \geq 4$, we have $\left(\frac{x}{4}\right)^2 - \left(\frac{y}{3}\right)^2 = 1 \quad \Rightarrow \quad \left(\frac{x}{4}\right)^2 = 1 + \left(\frac{y}{3}\right)^2 \quad \Rightarrow$ $\dfrac{x}{4} = \sqrt{1+\left(\frac{y}{3}\right)^2} \quad \Rightarrow \quad x = 4\sqrt{1+y^2/9}$, which is the right branch of the hyperbola.

51. $x = 4\cosh t$, $y = 3\sinh t$; $(4,0)$. $y = 0 \quad \Rightarrow \quad 3\sinh t = 0 \quad \Rightarrow \quad \sinh t = 0 \quad \Rightarrow \quad t = 0$. Let $\mathbf{r}(t) = \langle 4\cosh t, 3\sinh t\rangle$ so that $\mathbf{r}'(t) = \langle 4\sinh t, 3\cosh t\rangle$ and $\mathbf{r}'(0) = \langle 0, 3\rangle$. This vector is tangent to the curve at $(4,0)$ and has length 3. Thus, we divide $\langle 0,3\rangle$ by 3 to obtain the tangent vector of unit length, namely, $\langle 0,1\rangle$.

53. **(a)** $y = A\sinh mx + B\cosh mx \quad \Rightarrow \quad y' = mA\cosh mx + mB\sinh mx \quad \Rightarrow$
$y'' = m^2A\sinh mx + m^2B\cosh mx = m^2y$

(b) From part (a), a solution of $y'' = 9y$ is $y(x) = A\sinh 3x + B\cosh 3x$. So
$-4 = y(0) = A\sinh 0 + B\cosh 0 = B$, so $B = -4$. Now $y'(x) = 3A\cosh 3x - 12\sinh 3x \quad \Rightarrow$
$6 = y'(0) = 3A \quad \Rightarrow \quad A = 2$, so $y = 2\sinh 3x - 4\cosh 3x$.

55. The tangent to $y = \cosh x$ has slope 1 when $y' = \sinh x = 1 \quad \Rightarrow \quad x = \sinh^{-1}1 = \ln\left(1+\sqrt{2}\right)$, by Equation 3. Since $\sinh x = 1$ and $y = \cosh x = \sqrt{1+\sinh^2 x}$, we have $\cosh x = \sqrt{2}$. The point is $\left(\ln\left(1+\sqrt{2}\right), \sqrt{2}\right)$.

EXERCISES 4.8

NOTE: The use of l'Hospital's Rule is indicated by an H above the equal sign: $\overset{\text{H}}{=}$

1. $\lim_{x\to 2}\frac{x-2}{x^2-4}=\lim_{x\to 2}\frac{x-2}{(x-2)(x+2)}=\lim_{x\to 2}\frac{1}{x+2}=\frac{1}{4}$

3. $\lim_{x\to -1}\frac{x^6-1}{x^4-1}\overset{\text{H}}{=}\lim_{x\to -1}\frac{6x^5}{4x^3}=\frac{-6}{-4}=\frac{3}{2}$

5. $\lim_{x\to 0}\frac{e^x-1}{\sin x}\overset{\text{H}}{=}\lim_{x\to 0}\frac{e^x}{\cos x}=\frac{1}{1}=1$

7. $\lim_{x\to 0}\frac{\sin x}{x^3}\overset{\text{H}}{=}\lim_{x\to 0}\frac{\cos x}{3x^2}=\infty$

9. $\lim_{x\to 0}\frac{\tan x}{x+\sin x}\overset{\text{H}}{=}\lim_{x\to 0}\frac{\sec^2 x}{1+\cos x}=\frac{1}{1+1}=\frac{1}{2}$

11. $\lim_{x\to\infty}\frac{\ln x}{x}\overset{\text{H}}{=}\lim_{x\to\infty}\frac{1/x}{1}=0$

13. $\lim_{x\to\infty}\frac{e^x}{x^3}\overset{\text{H}}{=}\lim_{x\to\infty}\frac{e^x}{3x^2}\overset{\text{H}}{=}\lim_{x\to\infty}\frac{e^x}{6x}\overset{\text{H}}{=}\lim_{x\to\infty}\frac{e^x}{6}=\infty$

15. $\lim_{x\to a}\frac{x^{1/3}-a^{1/3}}{x-a}\overset{\text{H}}{=}\lim_{x\to a}\frac{(1/3)x^{-2/3}}{1}=\frac{1}{3a^{2/3}}$

17. $\lim_{x\to 0}\frac{e^x-1-x}{x^2}\overset{\text{H}}{=}\lim_{x\to 0}\frac{e^x-1}{2x}\overset{\text{H}}{=}\lim_{x\to 0}\frac{e^x}{2}=\frac{1}{2}$

19. $\lim_{x\to 0}\frac{\sin x}{e^x}=\frac{0}{1}=0$

21. $\lim_{x\to 0}\frac{1-\cos x}{x^2}\overset{\text{H}}{=}\lim_{x\to 0}\frac{\sin x}{2x}\overset{\text{H}}{=}\lim_{x\to 0}\frac{\cos x}{2}=\frac{1}{2}$

23. $\lim_{x\to 2^-}\frac{\ln x}{\sqrt{2-x}}=\infty$ since $\sqrt{2-x}\to 0$ but $\ln x\to\ln 2$

25. $\lim_{x\to\infty}\frac{\ln\ln x}{\sqrt{x}}\overset{\text{H}}{=}\lim_{x\to\infty}\frac{1/(x\ln x)}{1/(2\sqrt{x})}=\lim_{x\to\infty}\frac{2}{\sqrt{x}\ln x}=0$

27. $\lim_{x\to 0}\frac{\tan^{-1}(2x)}{3x}\overset{\text{H}}{=}\lim_{x\to 0}\frac{2/(1+4x^2)}{3}=\frac{2}{3}$

29. $\lim_{x\to 0}\frac{\tan\alpha x}{x}\overset{\text{H}}{=}\lim_{x\to 0}\frac{\alpha\sec^2\alpha x}{1}=\alpha$

31. $\lim_{x\to 0}\frac{\tan 2x}{\tanh 3x}\overset{\text{H}}{=}\lim_{x\to 0}\frac{2\sec^2 2x}{3\,\text{sech}^2 3x}=\frac{2}{3}$

33. $\lim_{x\to 0}\frac{x+\sin 3x}{x-\sin 3x}\overset{\text{H}}{=}\lim_{x\to 0}\frac{1+3\cos 3x}{1-3\cos 3x}=\frac{1+3}{1-3}=-2$

35. $\lim_{x\to 0}\frac{e^{4x}-1}{\cos x}=\frac{0}{1}=0$

37. $\lim_{x\to 0} \frac{\tan x - \sin x}{x^3} \overset{\text{H}}{=} \lim_{x\to 0} \frac{\sec^2 x - \cos x}{3x^2} \overset{\text{H}}{=} \lim_{x\to 0} \frac{2\sec^2 x \tan x + \sin x}{6x}$

$\overset{\text{H}}{=} \lim_{x\to 0} \frac{4\sec^2 x \tan^2 x + 2\sec^4 x + \cos x}{6} = \frac{0+2+1}{6} = \frac{1}{2}$

39. $\lim_{x\to 0^+} \sqrt{x}\ln x = \lim_{x\to 0^+} \frac{\ln x}{x^{-1/2}} \overset{\text{H}}{=} \lim_{x\to 0^+} \frac{1/x}{-\frac{1}{2}x^{-3/2}} = \lim_{x\to 0^+}\left(-2\sqrt{x}\right) = 0$

41. $\lim_{x\to\infty} e^{-x}\ln x = \lim_{x\to\infty} \frac{\ln x}{e^x} \overset{\text{H}}{=} \lim_{x\to\infty} \frac{1/x}{e^x} = \lim_{x\to\infty} \frac{1}{xe^x} = 0$

43. $\lim_{x\to\infty} x^3 e^{-x^2} = \lim_{x\to\infty} \frac{x^3}{e^{x^2}} \overset{\text{H}}{=} \lim_{x\to\infty} \frac{3x^2}{2xe^{x^2}} = \lim_{x\to\infty} \frac{3x}{2e^{x^2}} \overset{\text{H}}{=} \lim_{x\to\infty} \frac{3}{4xe^{x^2}} = 0$

45. $\lim_{x\to\pi}(x-\pi)\cot x = \lim_{x\to\pi} \frac{x-\pi}{\tan x} \overset{\text{H}}{=} \lim_{x\to\pi} \frac{1}{\sec^2 x} = \frac{1}{(-1)^2} = 1$

47. $\lim_{x\to 0}\left(\frac{1}{x^4} - \frac{1}{x^2}\right) = \lim_{x\to 0} \frac{1-x^2}{x^4} = \infty$

49. $\lim_{x\to 0}\left(\frac{1}{x} - \csc x\right) = \lim_{x\to 0}\left(\frac{1}{x} - \frac{1}{\sin x}\right) = \lim_{x\to 0} \frac{\sin x - x}{x\sin x}$

$\overset{\text{H}}{=} \lim_{x\to 0} \frac{\cos x - 1}{\sin x + x\cos x} \overset{\text{H}}{=} \lim_{x\to 0} \frac{-\sin x}{2\cos x - x\sin x} = \frac{0}{2} = 0$

51. $\lim_{x\to\infty}\left(x - \sqrt{x^2-1}\right) = \lim_{x\to\infty}\left(x - \sqrt{x^2-1}\right)\frac{x+\sqrt{x^2-1}}{x+\sqrt{x^2-1}} = \lim_{x\to\infty} \frac{x^2-(x^2-1)}{x+\sqrt{x^2-1}} = \lim_{x\to\infty} \frac{1}{x+\sqrt{x^2-1}} = 0$

53. $\lim_{x\to\infty}\left(\frac{x^3}{x^2-1} - \frac{x^3}{x^2+1}\right) = \lim_{x\to\infty} \frac{x^3(x^2+1) - x^3(x^2-1)}{(x^2-1)(x^2+1)} = \lim_{x\to\infty} \frac{2x^3}{x^4-1} = \lim_{x\to\infty} \frac{2/x}{1-1/x^4} = 0$

55. $y = x^{\sin x} \quad\Rightarrow\quad \ln y = \sin x \ln x$, so $\lim_{x\to 0^+} \ln y = \lim_{x\to 0^+} \sin x \ln x = \lim_{x\to 0^+} \frac{\ln x}{\csc x} \overset{\text{H}}{=} \lim_{x\to 0^+} \frac{1/x}{-\csc x \cot x}$

$= -\left(\lim_{x\to 0^+} \frac{\sin x}{x}\right)\left(\lim_{x\to 0^+} \tan x\right) = -1\cdot 0 = 0 \quad\Rightarrow\quad \lim_{x\to 0^+} x^{\sin x} = \lim_{x\to 0^+} e^{\ln y} = e^0 = 1.$

57. $y = (1-2x)^{1/x} \quad\Rightarrow\quad \ln y = \frac{1}{x}\ln(1-2x) \quad\Rightarrow\quad \lim_{x\to 0} \ln y = \lim_{x\to 0} \frac{\ln(1-2x)}{x} \overset{\text{H}}{=} \lim_{x\to 0} \frac{-2/(1-2x)}{1} = -2$

$\Rightarrow\quad \lim_{x\to 0}(1-2x)^{1/x} = \lim_{x\to 0} e^{\ln y} = e^{-2}$

59. $y = \left(1 + \frac{3}{x} + \frac{5}{x^2}\right)^x \quad\Rightarrow\quad \ln y = x\ln\left(1 + \frac{3}{x} + \frac{5}{x^2}\right) \quad\Rightarrow$

$\lim_{x\to\infty} \ln y = \lim_{x\to\infty} \frac{\ln\left(1 + \frac{3}{x} + \frac{5}{x^2}\right)}{1/x} \overset{\text{H}}{=} \lim_{x\to\infty} \frac{\left(-\frac{3}{x^2} - \frac{10}{x^3}\right)\Big/\left(1 + \frac{3}{x} + \frac{5}{x^2}\right)}{-1/x^2} = \lim_{x\to\infty} \frac{3+10/x}{1+3/x+5/x^2} = 3$, so

$\lim_{x\to\infty}\left(1 + \frac{3}{x} + \frac{5}{x^2}\right)^x = \lim_{x\to\infty} e^{\ln y} = e^3.$

61. $y = x^{1/x} \quad\Rightarrow\quad \ln y = (1/x)\ln x \quad\Rightarrow\quad \lim_{x\to\infty} \ln y = \lim_{x\to\infty} \frac{\ln x}{x} \overset{\text{H}}{=} \lim_{x\to\infty} \frac{1/x}{1} = 0 \quad\Rightarrow$

$\lim_{x\to\infty} x^{1/x} = \lim_{x\to\infty} e^{\ln y} = e^0 = 1$

63. $y = (\cot x)^{\sin x} \quad \Rightarrow \quad \ln y = \sin x \ln(\cot x) \quad \Rightarrow$

$$\lim_{x\to 0^+} \ln y = \lim_{x\to 0^+} \frac{\ln(\cot x)}{\csc x} \overset{\text{H}}{=} \lim_{x\to 0^+} \frac{(-\csc^2 x)/\cot x}{-\csc x \cot x} = \lim_{x\to 0^+} \frac{\csc x}{\cot^2 x}$$

$$= \lim_{x\to 0^+} \frac{\sin x}{\cos^2 x} = 0, \text{ so } \lim_{x\to 0^+} (\cot x)^{\sin x} = \lim_{x\to 0^+} e^{\ln y} = e^0 = 1.$$

65. $y = \left(\dfrac{x}{x+1}\right)^x \quad \Rightarrow \quad \ln y = x \ln\left(\dfrac{x}{x+1}\right) \quad \Rightarrow$

$$\lim_{x\to\infty} \ln y = \lim_{x\to\infty} x \ln\left(\frac{x}{x+1}\right) = \lim_{x\to\infty} \frac{\ln x - \ln(x+1)}{1/x} \overset{\text{H}}{=} \lim_{x\to\infty} \frac{1/x - 1/(x+1)}{-1/x^2}$$

$$= \lim_{x\to\infty} \left(-x + \frac{x^2}{x+1}\right) = \lim_{x\to\infty} \frac{-x}{x+1} = -1, \text{ so } \lim_{x\to\infty} \left(\frac{x}{x+1}\right)^x = \lim_{x\to\infty} e^{\ln y} = e^{-1}$$

Or: $\displaystyle \lim_{x\to\infty} \left(\frac{x}{x+1}\right)^x = \lim_{x\to\infty} \left[\left(\frac{x+1}{x}\right)^{-1}\right]^x = \left[\lim_{x\to\infty} \left(1 + \frac{1}{x}\right)^x\right]^{-1} = e^{-1}$

67. Let $y = (-\ln x)^x$. Then $\ln y = x \ln(-\ln x) \quad \Rightarrow \quad \displaystyle \lim_{x\to 0^+} \ln y = \lim_{x\to 0^+} x \ln(-\ln x) = \lim_{x\to 0^+} \frac{\ln(-\ln x)}{1/x}$

$$\overset{\text{H}}{=} \lim_{x\to 0^+} \frac{(1/-\ln x)(-1/x)}{-1/x^2} = \lim_{x\to 0^+} \frac{-x}{\ln x} = 0 \quad \Rightarrow \quad \lim_{x\to 0^+} (-\ln x)^x = e^0 = 1.$$

69.

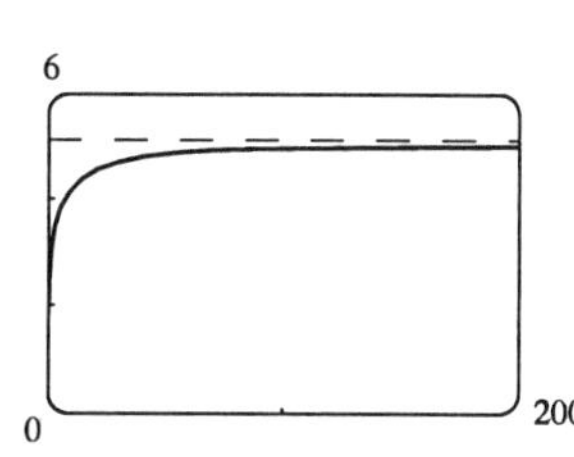

From the graph, it appears that $\displaystyle \lim_{x\to\infty} x[\ln(x+5) - \ln x] = 5$.

Now $\displaystyle \lim_{x\to\infty} x[\ln(x+5) - \ln x] = \lim_{x\to\infty} \frac{\ln(x+5) - \ln x}{1/x}$

$$\overset{\text{H}}{=} \lim_{x\to\infty} \frac{1/(x+5) - 1/x}{-1/x^2} = \lim_{x\to\infty} \frac{5x^2}{x(x+5)} = 5.$$

71.

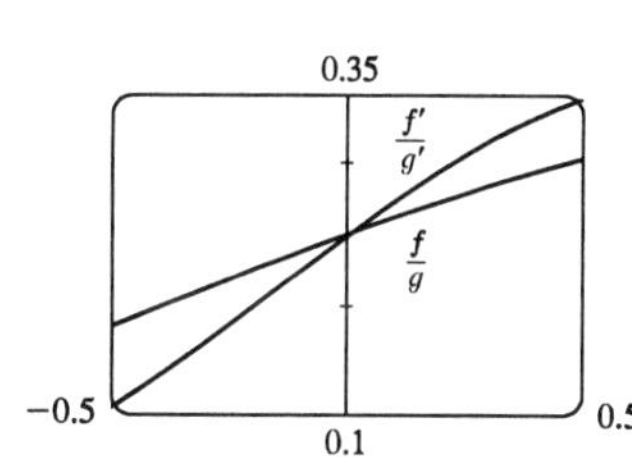

From the graph, it appears that

$$\lim_{x\to 0} \frac{f(x)}{g(x)} = \lim_{x\to 0} \frac{f'(x)}{g'(x)} \approx 0.25.$$

We calculate $\displaystyle \lim_{x\to 0} \frac{f(x)}{g(x)} = \lim_{x\to 0} \frac{e^x - 1}{x^3 + 4x}$

$$\overset{\text{H}}{=} \lim_{x\to 0} \frac{e^x}{3x^2 + 4} = \frac{1}{4}.$$

73. We see that both numerator and denominator approach 0, so we can use l'Hospital's Rule:

$$\lim_{x\to a} \frac{\sqrt{2a^3x - x^4} - a\sqrt[3]{aax}}{a - \sqrt[4]{ax^3}} \overset{\text{H}}{=} \lim_{x\to a} \frac{\frac{1}{2}(2a^3x - x^4)^{-1/2}(2a^3 - 4x^3) - a\left(\frac{1}{3}\right)(aax)^{-2/3}a^2}{-\frac{1}{4}(ax^3)^{-3/4}(3ax^2)}$$

$$= \frac{\frac{1}{2}(2a^3a - a^4)^{-1/2}(2a^3 - 4a^3) - \frac{1}{3}a^3(a^2a)^{-2/3}}{-\frac{1}{4}(aa^3)^{-3/4}(3aa^2)}$$

$$= \frac{(a^4)^{-1/2}(-a^3) - \frac{1}{3}a^3(a^3)^{-2/3}}{-\frac{3}{4}a^3(a^4)^{-3/4}} = \frac{-a - \frac{1}{3}a}{-3/4} = \frac{4}{3}\left(\frac{4}{3}a\right) = \frac{16}{9}a$$

75. Since $\lim_{h\to 0}[f(x+h)-f(x-h)] = f(x)-f(x) = 0$ (f is differentiable and hence continuous) and $\lim_{h\to 0} 2h = 0$, we use l'Hospital's Rule:

$$\lim_{h\to 0}\frac{f(x+h)-f(x-h)}{2h} \overset{\text{H}}{=} \lim_{h\to 0}\frac{f'(x+h)-f'(x-h)(-1)}{2} = \frac{f'(x)+f'(x)}{2} = \frac{2f'(x)}{2} = f'(x)$$

77. $\displaystyle\lim_{x\to\infty}\frac{e^x}{x^n} \overset{\text{H}}{=} \lim_{x\to\infty}\frac{e^x}{nx^{n-1}} \overset{\text{H}}{=} \lim_{x\to\infty}\frac{e^x}{n(n-1)x^{n-2}} \overset{\text{H}}{=} \cdots \overset{\text{H}}{=} \lim_{x\to\infty}\frac{e^x}{n!} = \infty$

79. **(a)** We show that $\displaystyle\lim_{x\to 0}\frac{f(x)}{x^n} = 0$ for every integer $n \geq 0$. Let $y = \dfrac{1}{x^2}$. Then

$$\lim_{x\to 0}\frac{f(x)}{x^{2n}} = \lim_{x\to 0}\frac{e^{-1/x^2}}{(x^2)^n} = \lim_{y\to\infty}\frac{y^n}{e^y} \overset{\text{H}}{=} \lim_{y\to\infty}\frac{ny^{n-1}}{e^y} \overset{\text{H}}{=} \cdots \overset{\text{H}}{=} \lim_{y\to\infty}\frac{n!}{e^y} = 0 \quad\Rightarrow$$

$$\lim_{x\to 0}\frac{f(x)}{x^n} = \lim_{x\to 0} x^n\frac{f(x)}{x^{2n}} = \lim_{x\to 0} x^n \lim_{x\to 0}\frac{f(x)}{x^{2n}} = 0. \text{ Thus } f'(0) = \lim_{x\to 0}\frac{f(x)-f(0)}{x-0} = \lim_{x\to 0}\frac{f(x)}{x} = 0.$$

(b) Using the Chain Rule and the Quotient Rule we see that $f^{(n)}(x)$ exists for $x \neq 0$. In fact, we prove by induction that for each $n \geq 0$, there is a polynomial p_n and a non-negative integer k_n with $f^{(n)}(x) = p_n(x)f(x)/x^{k_n}$ for $x \neq 0$. This is true for $n = 0$; suppose it is true for the nth derivative. Then

$$\begin{aligned} f^{(n+1)}(x) &= \left[x^{k_n}[p_n'(x)f(x) + p_n(x)f'(x)] - k_n x^{k_n-1}\,p_n(x)f(x)\right]x^{-2k_n} \\ &= \left[x^{k_n}\,p_n'(x) + p_n(x)\left(2/x^3\right) - k_n x^{k_n-1}\,p_n(x)\right]f(x)x^{-2k_n} \\ &= \left[x^{k_n+3}\,p_n'(x) + 2\,p_n(x) - k_n x^{k_n+2}\,p_n(x)\right]f(x)x^{-(2k_n+3)}, \end{aligned}$$

which has the desired form.

Now we show by induction that $f^{(n)}(0) = 0$ for all n. By (a), $f'(0) = 0$. Suppose that $f^{(n)}(0) = 0$. Then

$$\begin{aligned} f^{(n+1)}(0) &= \lim_{x\to 0}\frac{f^{(n)}(x)-f^{(n)}(0)}{x-0} = \lim_{x\to 0}\frac{f^{(n)}(x)}{x} = \lim_{x\to 0}\frac{p_n(x)f(x)/x^{k_n}}{x} = \lim_{x\to 0}\frac{p_n(x)f(x)}{x^{k_n+1}} \\ &= \lim_{x\to 0} p_n(x)\lim_{x\to 0}\frac{f(x)}{x^{k_n+1}} = p_n(0)\cdot 0 = 0. \end{aligned}$$

REVIEW EXERCISES FOR CHAPTER 4

1. False. For example, $\cos\frac{\pi}{2} = \cos\left(-\frac{\pi}{2}\right) = 0$, so $\cos x$ is not 1-1.

3. True, since $\ln x$ is an increasing function on $(0, \infty)$.

5. True, since $e^x \neq 0$ for all x.

7. False. For example, $(\ln e)^6 = 1^6 = 1$, but $6\ln e = 6$. In fact $\ln(x^6) = 6\ln x$.

9. False. $\ln 10$ is a constant, so its derivative is 0.

11. False. The "-1" is not an exponent; it is an indication of an inverse function. See Equation 4.6.4.

13. True. See Figure 2 in Section 4.7.

15. True. By Equation 4.4.4, $\dfrac{d}{dx}\log_8 x = \dfrac{1}{x\ln 8} = \dfrac{1}{\ln(2^3)\,x} = \dfrac{1}{(3\ln 2)x}$.

17. $y = e^x$

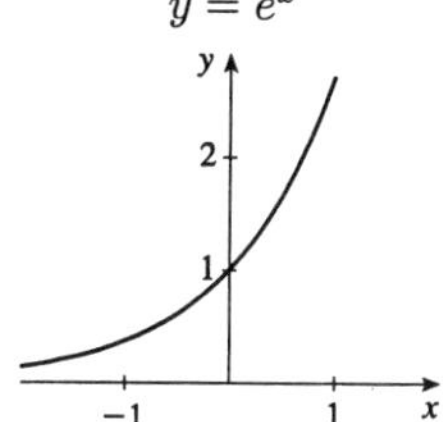

19. $y = e^{-x}$

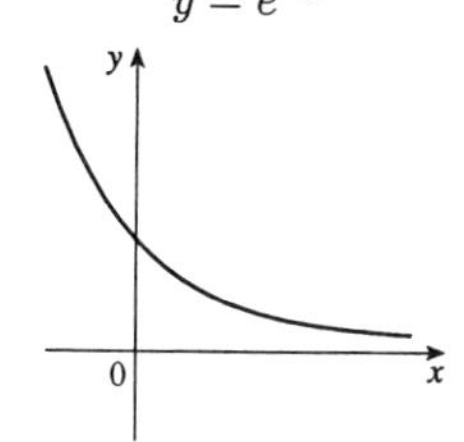

$y = -e^{-x}$

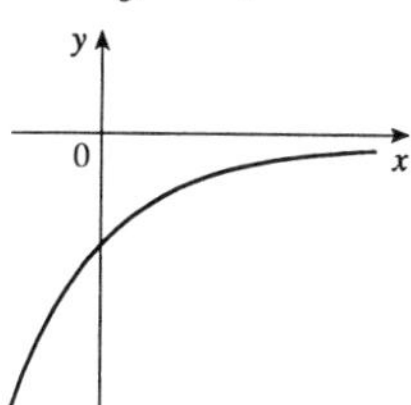

21. $y = \ln x$

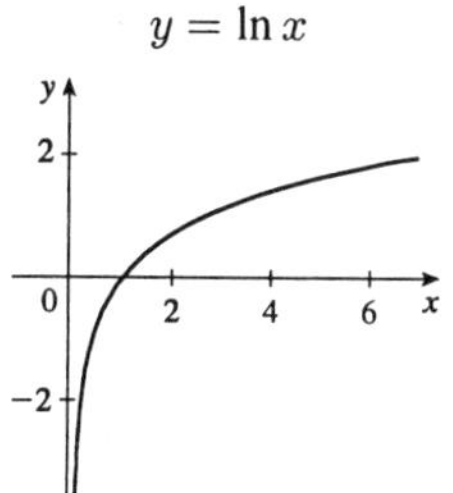

23. $y = 2 - \ln x$

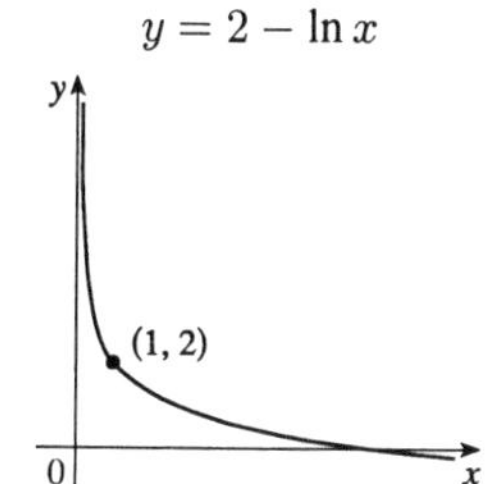

25. $y = \tan^{-1}x$

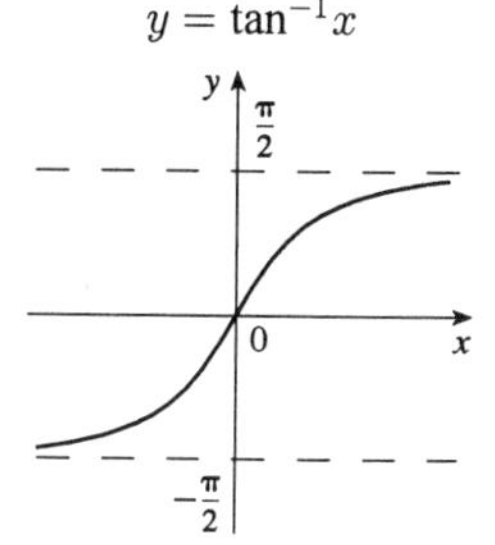

27. $e^x = 5 \quad\Rightarrow\quad x = \ln(e^x) = \ln 5$

29. $\log_{10}(e^x) = 1 \quad\Rightarrow\quad e^x = 10 \quad\Rightarrow\quad x = \ln(e^x) = \ln 10$

Or: $1 = \log_{10}(e^x) = x\log_{10} e \quad\Rightarrow\quad x = 1/\log_{10} e = \ln 10$

31. $2 = \ln(x^\pi) = \pi\ln x \quad\Rightarrow\quad \ln x = 2/\pi \quad\Rightarrow\quad x = e^{2/\pi}$

33. $\tan x = 4 \quad\Rightarrow\quad x = \tan^{-1}4 + n\pi = \arctan 4 + n\pi$, n an integer

35. $y = \log_{10}(x^2 - x) \quad\Rightarrow\quad y' = \dfrac{1}{x^2 - x}(\log_{10} e)(2x - 1) = \dfrac{2x - 1}{(\ln 10)(x^2 - x)}$

37. $y = \dfrac{\sqrt{x+1}\,(2-x)^5}{(x+3)^7} \quad\Rightarrow\quad \ln|y| = \frac{1}{2}\ln(x+1) + 5\ln|2-x| - 7\ln(x+3) \quad\Rightarrow$

$$\frac{y'}{y} = \frac{1}{2(x+1)} + \frac{-5}{2-x} - \frac{7}{x+3} \quad\Rightarrow\quad y' = \frac{\sqrt{x+1}\,(2-x)^5}{(x+3)^7}\left[\frac{1}{2(x+1)} - \frac{5}{2-x} - \frac{7}{x+3}\right]$$

39. $y = e^{cx}(c\sin x - \cos x) \quad \Rightarrow \quad y' = ce^{cx}(c\sin x - \cos x) + e^{cx}(c\cos x + \sin x) = (c^2+1)e^{cx}\sin x$

41. $y = \ln(\sec^2 x) = 2\ln|\sec x| \quad \Rightarrow \quad y' = (2/\sec x)(\sec x \tan x) = 2\tan x$

43. $y = xe^{-1/x} \quad \Rightarrow \quad y' = e^{-1/x} + xe^{-1/x}(1/x^2) = e^{-1/x}(1 + 1/x)$

45. $y = (\cos^{-1}x)^{\sin^{-1}x} \quad \Rightarrow \quad \ln y = \sin^{-1}x \ln(\cos^{-1}x) \quad \Rightarrow$

$$\frac{y'}{y} = \frac{1}{\sqrt{1-x^2}}\ln(\cos^{-1}x) + (\sin^{-1}x)\left(\frac{1}{\cos^{-1}x}\right)\left(-\frac{1}{\sqrt{1-x^2}}\right) \quad \Rightarrow$$

$$y' = (\cos^{-1}x)^{\sin^{-1}x - 1}\left[\frac{\cos^{-1}x\ln(\cos^{-1}x) - \sin^{-1}x}{\sqrt{1-x^2}}\right]$$

47. $y = e^{e^x} \quad \Rightarrow \quad y' = e^{e^x}e^x = e^{x+e^x}$

49. $y = \ln\dfrac{1}{x} + \dfrac{1}{\ln x} = -\ln x + (\ln x)^{-1} \quad \Rightarrow \quad y' = -\dfrac{1}{x} - \dfrac{1}{x(\ln x)^2}$

51. $y = 7^{\sqrt{2x}} \quad \Rightarrow \quad y' = 7^{\sqrt{2x}}(\ln 7)\left[1/\left(2\sqrt{2x}\right)\right](2) = 7^{\sqrt{2x}}(\ln 7)/\sqrt{2x}$

53. $y = \ln(\cosh 3x) \quad \Rightarrow \quad y' = (1/\cosh 3x)(\sinh 3x)(3) = 3\tanh 3x$

55. $y = \cosh^{-1}(\sinh x) \quad \Rightarrow \quad y' = (\cosh x)/\sqrt{\sinh^2 x - 1}$

57. $y = \ln\sin x - \frac{1}{2}\sin^2 x \quad \Rightarrow \quad y' = \dfrac{\cos x}{\sin x} - \sin x\cos x = \cot x - \sin x \cos x$

59. $y = \sin^{-1}\left(\dfrac{x-1}{x+1}\right) \quad \Rightarrow$

$$y' = \frac{1}{\sqrt{1 - [(x-1)/(x+1)]^2}}\,\frac{(x+1)-(x-1)}{(x+1)^2} = \frac{1}{\sqrt{(x+1)^2 - (x-1)^2}}\left(\frac{2}{x+1}\right)$$

$$= \frac{2}{\sqrt{4x}(x+1)} = \frac{1}{\sqrt{x}(x+1)}.$$ [Note that the domain of y is $x \geq 0$.]

61. $y = \frac{1}{4}\left[\ln(x^2+x+1) - \ln(x^2-x+1)\right] + \dfrac{1}{2\sqrt{3}}\left[\tan^{-1}\left(\dfrac{2x+1}{\sqrt{3}}\right) + \tan^{-1}\left(\dfrac{2x-1}{\sqrt{3}}\right)\right] \quad \Rightarrow$

$$y' = \frac{1}{4}\left[\frac{2x+1}{x^2+x+1} - \frac{2x-1}{x^2-x+1}\right] + \frac{1}{2\sqrt{3}}\left[\frac{2/\sqrt{3}}{1+\left[(2x+1)/\sqrt{3}\right]^2} + \frac{2/\sqrt{3}}{1+\left[(2x-1)/\sqrt{3}\right]^2}\right]$$

$$= \frac{1}{4}\left[\frac{2x+1}{x^2+x+1} - \frac{2x-1}{x^2-x+1}\right] + \frac{1}{4(x^2+x+1)} + \frac{1}{4(x^2-x+1)}$$

$$= \frac{1}{2}\left[\frac{x+1}{x^2+x+1} - \frac{x-1}{x^2-x+1}\right] = \frac{1}{x^4+x^2+1}$$

63. $f(x) = 2^x \quad \Rightarrow \quad f'(x) = 2^x\ln 2 \quad \Rightarrow \quad f''(x) = 2^x(\ln 2)^2 \quad \Rightarrow \quad \cdots \quad \Rightarrow \quad f^{(n)}(x) = 2^x(\ln 2)^n$

65. We first show it is true for $n = 1$: $f'(x) = e^x + xe^x = (x+1)e^x$. We now assume it is true for $n = k$: $f^{(k)}(x) = (x+k)e^x$. With this assumption, we must show it is true for $n = k+1$:

$$f^{(k+1)}(x) = \frac{d}{dx}\left[f^{(k)}(x)\right] = \frac{d}{dx}[(x+k)e^x] = e^x + (x+k)e^x = [x+(k+1)]e^x.$$

Therefore $f^{(n)}(x) = (x+n)e^x$ by mathematical induction.

67. $y = f(x) = \ln(e^x + e^{2x}) \quad \Rightarrow \quad f'(x) = \dfrac{e^x + 2e^{2x}}{e^x + e^{2x}} \quad \Rightarrow \quad f'(0) = \frac{3}{2}$, so the tangent line at $(0, \ln 2)$ is $y - \ln 2 = \frac{3}{2}x$ or $3x - 2y + \ln 4 = 0$.

69.

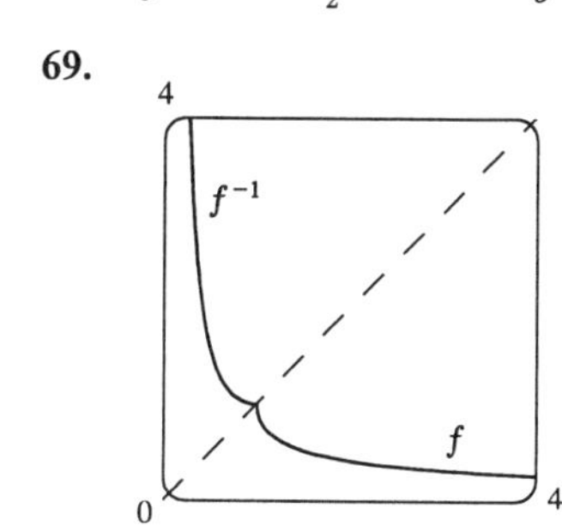

$f(x) = \sqrt{x} - \sqrt{x-1}$ has domain $[1, \infty)$. To see that f is 1-1, we can either graph the function and use the Horizontal Line Test, or we can calculate $f'(x) = \dfrac{1}{2\sqrt{x}} - \dfrac{1}{2\sqrt{x-1}} = \dfrac{\sqrt{x-1} - \sqrt{x}}{2\sqrt{x(x-1)}} < 0$, so f is decreasing and hence 1-1. The parametric equations of the graph of f are $x = t$, $y = \sqrt{t} - \sqrt{t-1}$, and so the parametric equations of the graph of f^{-1} are $x = \sqrt{t} - \sqrt{t-1}$, $y = t$.

71. $y = [\ln(x+4)]^2 \quad \Rightarrow \quad y' = 2\dfrac{\ln(x+4)}{x+4} = 0 \quad \Leftrightarrow \quad \ln(x+4) = 0 \quad \Leftrightarrow \quad x + 4 = 1 \quad \Leftrightarrow \quad x = -3$, so the tangent is horizontal at $(-3, 0)$.

73. The slope of the tangent at the point (a, e^a) is $\left[\dfrac{d}{dx}e^x\right]_{x=a} = e^a$. The equation of the tangent line is thus $y - e^a = e^a(x - a)$. We substitute $x = 0$, $y = 0$ into this equation, since we want the line to pass through the origin: $0 - e^a = e^a(0 - a) \quad \Leftrightarrow \quad -e^a = e^a(-a) \quad \Leftrightarrow \quad a = 1$. So the equation of the tangent is $y - e = e(x - 1)$, or $y = ex$.

75. $\lim\limits_{x \to -\infty} 10^{-x} = \infty$ since $-x \to \infty$ as $x \to -\infty$.

77. $\lim\limits_{x \to 0^+} \ln(\tan x) = -\infty$ since $\tan x \to 0^+$ as $x \to 0^+$.

79. $\lim\limits_{x \to -4^+} e^{1/(x+4)} = \infty$ since $\dfrac{1}{x+4} \to \infty$ as $x \to -4^+$.

81. $\lim\limits_{x \to \infty} \dfrac{e^x}{e^{2x} + e^{-x}} = \lim\limits_{x \to \infty} \dfrac{e^{-x}}{1 + e^{-3x}} = \dfrac{0}{1+0} = 0$

83. $\lim\limits_{x \to 1} \cos^{-1}\left(\dfrac{x}{x+1}\right) = \cos^{-1}\frac{1}{2} = \frac{\pi}{3}$

85. $\lim\limits_{x \to \pi} \dfrac{\sin x}{x^2 - \pi^2} \overset{\text{H}}{=} \lim\limits_{x \to \pi} \dfrac{\cos x}{2x} = -\dfrac{1}{2\pi}$

87. $\lim\limits_{x \to \infty} \dfrac{\ln(\ln x)}{\ln x} \overset{\text{H}}{=} \lim\limits_{x \to \infty} \dfrac{1/(x \ln x)}{1/x} = \lim\limits_{x \to \infty} \dfrac{1}{\ln x} = 0$

89. $\lim\limits_{x \to 0} \dfrac{\ln(1-x) + x + \frac{1}{2}x^2}{x^3} \overset{\text{H}}{=} \lim\limits_{x \to 0} \dfrac{-\dfrac{1}{1-x} + 1 + x}{3x^2} \overset{\text{H}}{=} \lim\limits_{x \to 0} \dfrac{-\dfrac{1}{(1-x)^2} + 1}{6x} \overset{\text{H}}{=} \lim\limits_{x \to 0} \dfrac{-\dfrac{2}{(1-x)^3}}{6} = -\dfrac{2}{6} = -\dfrac{1}{3}$

91. $\lim\limits_{x \to 0^+} \sin x (\ln x)^2 = \lim\limits_{x \to 0^+} \dfrac{(\ln x)^2}{\csc x} \overset{\text{H}}{=} \lim\limits_{x \to 0^+} \dfrac{2 \ln x / x}{-\csc x \cot x} = -2 \lim\limits_{x \to 0} \dfrac{\sin x}{x} \lim\limits_{x \to 0} \dfrac{\ln x}{\cot x} = -2 \lim\limits_{x \to 0} \dfrac{\ln x}{\cot x}$

$$\overset{\text{H}}{=} -2 \lim_{x \to 0} \frac{1/x}{-\csc^2 x} = 2 \lim_{x \to 0} \frac{\sin^2 x}{x} = 2 \lim_{x \to 0} \frac{\sin x}{x} \lim_{x \to 0} \sin x = 2 \cdot 1 \cdot 0 = 0$$

93. $\lim\limits_{x \to 1} (\ln x)^{\sin x} = (\ln 1)^{\sin 1} = 0^{\sin 1} = 0$

95. $\lim\limits_{x \to 0^+} \dfrac{x^{1/3} - 1}{x^{1/4} - 1} = \dfrac{0 - 1}{0 - 1} = 1$

97. **(a)** $y(t) = y(0)e^{kt} = 1000e^{kt} \Rightarrow y(2) = 1000e^{2k} = 9000 \Rightarrow e^{2k} = 9 \Rightarrow 2k = \ln 9 \Rightarrow$ $k = \frac{1}{2}\ln 9 = \ln 3 \Rightarrow y(t) = 1000e^{(\ln 3)t} = 1000 \cdot 3^t$

(b) $y(3) = 1000 \cdot 3^3 = 27{,}000$

(c) $1000 \cdot 3^t = 2000 \Rightarrow 3^t = 2 \Rightarrow t\ln 3 = \ln 2 \Rightarrow t = (\ln 2)/(\ln 3) \approx 0.63\text{ h}$

99. Using the formula in Example 4.5.4, $A(t) = A_0\left(1 + \dfrac{i}{n}\right)^{nt}$, where $A_0 = 10{,}000.00$ and $i = 0.06$ then for:

(a) $n = 1$: $A(4) = 10{,}000(1 + 0.06)^{1 \cdot 4} = \$12{,}624.77$

(b) $n = 2$: $A(4) = 10{,}000\left(1 + \frac{0.06}{2}\right)^{2 \cdot 4} = \$12{,}667.70$

(c) $n = 4$: $A(4) = 10{,}000\left(1 + \frac{0.06}{4}\right)^{4 \cdot 4} = \$12{,}689.86$

(d) $n = 12$: $A(4) = 10{,}000\left(1 + \frac{0.06}{12}\right)^{12 \cdot 4} = \$12{,}704.89$

(e) $n = 365$: $A(4) = 10{,}000\left(1 + \frac{0.06}{365}\right)^{365 \cdot 4} = \$12{,}712.24$

(f) Using the formula for continuous interest, $A(t) = A_0e^{it}$, we have $A(4) = 10{,}000 \cdot e^{0.06 \cdot 4} = \$12{,}712.49$.

101. **(a)** $C'(t) = -kC(t) \Rightarrow C(t) = C(0)e^{-kt}$ by Theorem 4.5.2. But $C(0) = C_0$. Thus $C(t) = C_0e^{-kt}$.

(b) $C(30) = \frac{1}{2}C_0$ since the concentration is reduced by half. Thus, $\frac{1}{2}C_0 = C_0e^{-30k} \Rightarrow \ln\frac{1}{2} = -30k \Rightarrow$ $k = -\frac{1}{30}\ln\frac{1}{2} = \frac{1}{30}\ln 2$. Since 10% of the original concentration remains if 90% is eliminated, we want the value of t such that $C(t) = \frac{1}{10}C_0$. Therefore, $\frac{1}{10}C_0 = C_0e^{-t(\ln 2)/30} \Rightarrow t = -\dfrac{30}{\ln 2}\ln 0.1 \approx 100\text{ h}$.

103. $s(t) = Ae^{-ct}\cos(\omega t + \delta) \Rightarrow$

$$v(t) = s'(t) = -cAe^{-ct}\cos(\omega t + \delta) + Ae^{-ct}[-\omega\sin(\omega t + \delta)] = -Ae^{-ct}[c\cos(\omega t + \delta) + \omega\sin(\omega t + \delta)] \Rightarrow$$

$$a(t) = v'(t) = cAe^{-ct}[c\cos(\omega t + \delta) + \omega\sin(\omega t + \delta)] + \left(-Ae^{-ct}\right)\left[-\omega c\sin(\omega t + \delta) + \omega^2\cos(\omega t + \delta)\right]$$
$$= Ae^{-ct}\left[\left(c^2 - \omega^2\right)\cos(\omega t + \delta) + 2c\omega\sin(\omega t + \delta)\right]$$

105. $f(x) = e^{g(x)} \Rightarrow f'(x) = e^{g(x)}g'(x)$

107. $f(x) = \ln|g(x)| \Rightarrow f'(x) = g'(x)/g(x)$

109. $f(x) = \ln g(e^x) \Rightarrow f'(x) = \dfrac{1}{g(e^x)}\,g'(e^x)e^x$

111. $f(x) = \ln x + \tan^{-1}x \Rightarrow f(1) = \ln 1 + \tan^{-1}1 = \frac{\pi}{4} \Rightarrow g\left(\frac{\pi}{4}\right) = 1$.

$f'(x) = \dfrac{1}{x} + \dfrac{1}{1 + x^2}$, so $g'\left(\frac{\pi}{4}\right) = \dfrac{1}{f'(1)} = \dfrac{1}{3/2} = \dfrac{2}{3}$.

113. Let $y = \tan^{-1}x$. Then $\tan y = x$, so from the triangle we see that $\sin\left(\tan^{-1}x\right) = \sin y = \dfrac{x}{\sqrt{1 + x^2}}$.

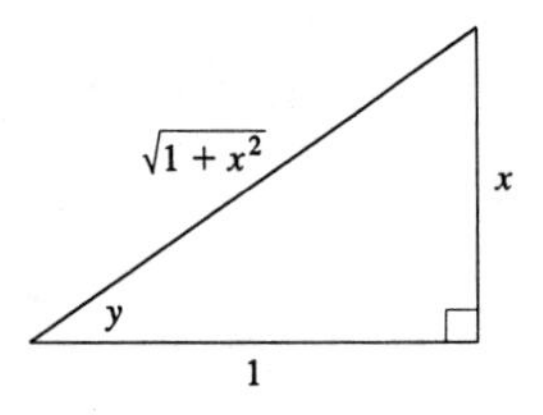

Using this fact we have that

$$\sin\left(\tan^{-1}(\sinh x)\right) = \frac{\sinh x}{\sqrt{1 + \sinh^2 x}} = \frac{\sinh x}{\cosh x} = \tanh x.$$

Hence $\sin^{-1}(\tanh x) = \sin^{-1}(\sin(\tan^{-1}(\sinh x))) = \tan^{-1}(\sinh x)$.

APPLICATIONS PLUS (after Chapter 4)

1. We can assume without loss of generality that $\theta = 0$ at time $t = 0$, so that $\theta = 12\pi t$ rad. (The angular velocity of the wheel is 360 rpm $= 360 \cdot 2\pi\text{ rad}/60\text{ s} = 12\pi\text{ rad/s}$.) Then the position of A as a function of time is $A = (40\cos\theta, 40\sin\theta) = (40\cos 12\pi t, 40\sin 12\pi t)$, so $\sin\alpha = \dfrac{40\sin\theta}{120} = \dfrac{\sin\theta}{3} = \frac{1}{3}\sin 12\pi t$.

(a) Differentiating the expression for $\sin\alpha$, we get $\cos\alpha \cdot \dfrac{d\alpha}{dt} = \frac{1}{3} \cdot 12\pi \cdot \cos 12\pi t = 4\pi\cos\theta$. When $\theta = \frac{\pi}{3}$, we have $\sin\alpha = \frac{1}{3}\sin\theta = \frac{\sqrt{3}}{6}$, so $\cos\alpha = \sqrt{1 - \left(\frac{\sqrt{3}}{6}\right)^2} = \sqrt{\frac{11}{12}}$ and

$$\frac{d\alpha}{dt} = \frac{4\pi\cos\frac{\pi}{3}}{\cos\alpha} = \frac{2\pi}{\sqrt{11/12}} = \frac{4\pi\sqrt{3}}{\sqrt{11}}\text{ rad/s}.$$

(b) By the Law of Cosines, $|AP|^2 = |OA|^2 + |OP|^2 - 2|OA||OP|\cos\theta \Rightarrow$

$$120^2 = 40^2 + |OP|^2 - 2 \cdot 40|OP|\cos\theta \Rightarrow |OP|^2 - (80\cos\theta)|OP| - 12{,}800 = 0 \Rightarrow$$

$$|OP| = \tfrac{1}{2}\left(80\cos\theta \pm \sqrt{6400\cos^2\theta + 51{,}200}\right) = 40\cos\theta \pm 40\sqrt{\cos^2\theta + 8}$$

$$= 40\left(\cos\theta + \sqrt{8 + \cos^2\theta}\right)\text{cm} \quad (\text{since } |OP| > 0).$$

As a check, note that $|OP| = 160$ cm when $\theta = 0$ and $|OP| = 80\sqrt{2}$ cm when $\theta = \frac{\pi}{2}$.

(c) By part (b), the x-coordinate of P is given by $x = 40\left(\cos\theta + \sqrt{8 + \cos^2\theta}\right)$, so

$$\frac{dx}{dt} = \frac{dx}{d\theta}\frac{d\theta}{dt} = 40\left(-\sin\theta - \frac{2\cos\theta\sin\theta}{2\sqrt{8 + \cos^2\theta}}\right) \cdot 12\pi = -480\pi\sin\theta\left(1 + \frac{\sin\theta\cos\theta}{\sqrt{8 + \cos^2\theta}}\right)\text{cm/s}.$$

In particular, $dx/dt = 0$ cm/s when $\theta = 0$ and $dx/dt = -480\pi$ cm/s when $\theta = \frac{\pi}{2}$.

3. (a) $T = 2\pi\sqrt{\dfrac{L}{g}} = \dfrac{2\pi}{\sqrt{g}}L^{1/2} \Rightarrow dT = \dfrac{2\pi}{\sqrt{g}}\dfrac{1}{2}L^{-1/2}dL \Rightarrow \dfrac{dT}{2\pi/\sqrt{g}} = \dfrac{dL}{2\sqrt{L}} \Rightarrow$

$$\frac{dT}{\left(2\pi/\sqrt{g}\right)\sqrt{L}} = \frac{dL}{2L} \Rightarrow \frac{dT}{T} = \frac{dL}{2L}$$

(b) $dL = \dfrac{2L}{T}dT$. Set $dT = -15$ s, $T = 3600$ s. Then $dL = \dfrac{2L}{3600} \cdot (-15) = -\frac{30}{3600}L = -\frac{1}{120}L$. Thus shorten the pendulum by $\frac{1}{120}L$.

(c) $T = \dfrac{2\pi\sqrt{L}}{\sqrt{g}} = 2\pi\sqrt{L}g^{-1/2} \Rightarrow dT = 2\pi\sqrt{L}\left(-\frac{1}{2}\right)t^{-3/2}dg$. Therefore, $dg = -\dfrac{g\sqrt{g}\,dT}{\pi\sqrt{L}}$.

5. $V = \frac{4}{3}\pi r^3 \quad\Rightarrow\quad \frac{dV}{dt} = 4\pi r^2 \frac{dr}{dt}$. But $\frac{dV}{dt}$ is proportional to the surface area, so $\frac{dV}{dt} = k \cdot 4\pi r^2$ for some constant k. Therefore $4\pi r^2 \frac{dr}{dt} = k \cdot 4\pi r^2 \quad\Rightarrow\quad \frac{dr}{dt} = k = \text{constant} \quad\Rightarrow\quad r = kt + r_0$. To find k we use the fact that when $t = 3$, $r = 3k + r_0$ and $V = \frac{1}{2}V_0 \quad\Rightarrow\quad \frac{4}{3}\pi(3k + r_0)^3 = \frac{1}{2} \cdot \frac{4}{3}\pi r_0^3 \quad\Rightarrow\quad (3k + r_0)^3 = \frac{1}{2}r_0^3$ $\Rightarrow\quad (3k + r_0) = \frac{1}{\sqrt[3]{2}}r_0 \quad\Rightarrow\quad k = \frac{1}{3}r_0\left(\frac{1}{\sqrt[3]{2}} - 1\right)$. Therefore $r = \frac{1}{3}r_0\left(\frac{1}{\sqrt[3]{2}} - 1\right)t + r_0$. When the snowball has melted completely we have $r = 0 \quad\Rightarrow\quad \frac{1}{3}r_0\left(\frac{1}{\sqrt[3]{2}} - 1\right)t + r_0 = 0$ which gives $t = \frac{3\sqrt[3]{2}}{\sqrt[3]{2} - 1}$. Therefore it takes $\frac{3\sqrt[3]{2}}{\sqrt[3]{2} - 1} - 3 = \frac{3}{\sqrt[3]{2} - 1} \approx 11$ hours and 33 minutes longer.

7. **(a)** Let $S(t)$ be sales (in \$ millions) at time t. Then $\frac{dS}{dt} = k(25 - S)$ for some constant k.

(b) $\frac{dS}{dt} = k(25 - S)$. Let $\sigma = 25 - S$, so $\frac{d\sigma}{dt} = -\frac{dS}{dt}$ and the differential equation becomes $\frac{d\sigma}{dt} = -k\sigma$.

By Theorem 4.5.2, the solution to this equation is $\sigma(t) = \sigma(0)e^{-kt} \quad\Rightarrow$

$25 - S(t) = [25 - S(0)]e^{-kt} = 25e^{-kt} \quad\Rightarrow\quad S(t) = 25(1 - e^{-kt})$. To find k, we substitute

$S(3) = 3.5 = 25(1 - e^{-3k}) \quad\Rightarrow\quad -3k = \ln 0.86 \quad\Rightarrow\quad k = -\frac{1}{3}\ln 0.86$. So

$S(t) = 25\left[1 - (0.86)^{t/3}\right]$.

(c) $S(8) = 25(1 - 0.86^{8/3}) \approx \$8.28$ (million)

(d) $12.5 = S = 25\left[1 - (0.86)^{t/3}\right] \quad\Rightarrow\quad \frac{1}{2} = 1 - (0.86)^{t/3} \quad\Rightarrow\quad (0.86)^{8/3} = \frac{1}{2} \quad\Rightarrow\quad (0.86)^t = \frac{1}{8}$

$\Rightarrow\quad t\ln(0.86) = -\ln 8 \quad\Rightarrow\quad t = \frac{\ln 8}{-\ln(0.86)} \approx 13.8$ years

CHAPTER FIVE

EXERCISES 5.1

1. **(a)** Since $f'(x) > 0$ on $(-\infty, 0)$ and $(3, \infty)$, f is increasing on the same intervals. $f'(x) < 0$ and f is decreasing on $(0, 3)$.

(b) Since $f'(x) = 0$ at $x = 0$ and f' changes from positive to negative there, f changes from increasing to decreasing and has a local maximum at $x = 0$. Since $f'(x) = 0$ at $x = 3$ and changes from negative to positive there, f changes from decreasing to increasing and has a local minimum at $x = 3$.

(c)

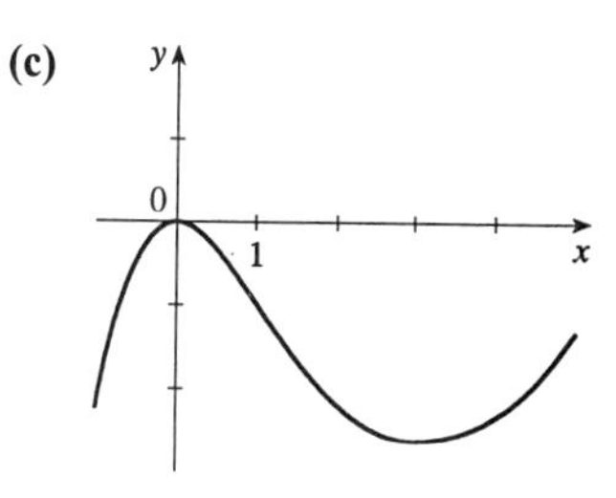

3. The derivative f' is increasing when the slopes of the tangent lines of f are becoming larger as x increases. This seems to be the case on the interval $(2, 5)$. The derivative is decreasing when the slopes of the tangent lines of f are becoming smaller as x increases, and this seems to be the case on $(-\infty, 2)$ and $(5, \infty)$. So f' is increasing on $(2, 5)$ and decreasing on $(-\infty, 2)$ and $(5, \infty)$.

5. If $D(t)$ is the size of the deficit as a function of time, then at the time of the speech $D'(t) > 0$, but $D''(t) < 0$ because $D''(t) = (D')'(t)$ is the rate of change of $D'(t)$.

7. **(a)** The rate of increase of the population is initially very small, then increases rapidly until about 1932 when it starts decreasing. The rate becomes negative by 1936, peaks in magnitude in 1937, and approaches 0 in 1940.

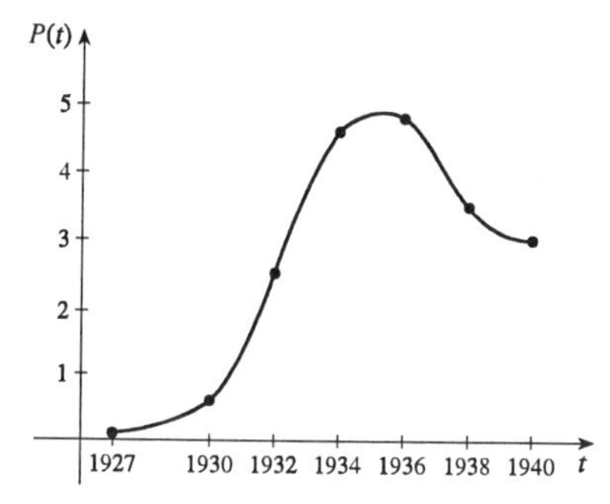

(b) Inflection points (IP) appear to be at $(1932, 2.5)$ and $(1937, 4.3)$. The rates of population increase and decrease have their maximum values at those points.

9. Most students learn more in the third hour of studying than in the eighth hour, so $K(3) - K(2)$ is larger than $K(8) - K(7)$. In other words, as you begin studying for a test, the rate of knowledge gain is large and then starts to taper off, so $K'(t)$ decreases and the graph of K is concave downward.

11. **(a)** f is increasing where f' is positive, that is, on $(0,2)$, $(4,6)$, and $(8,\infty)$; and decreasing where f' is negative, that is, on $(2,4)$ and $(6,8)$.

(b) f has local maxima where f' changes from positive to negative, at $x=2$ and at $x=6$, and local minima where f' changes from negative to positive, at $x=4$ and at $x=8$.

(c) f is concave upward where f' is increasing, that is, on $(3,6)$ and $(6,\infty)$, and concave downward where f' is decreasing, that is, on $(0,3)$.

(d) There is a point of inflection where f changes from being CD to being CU, that is, at $x=3$.

(e)

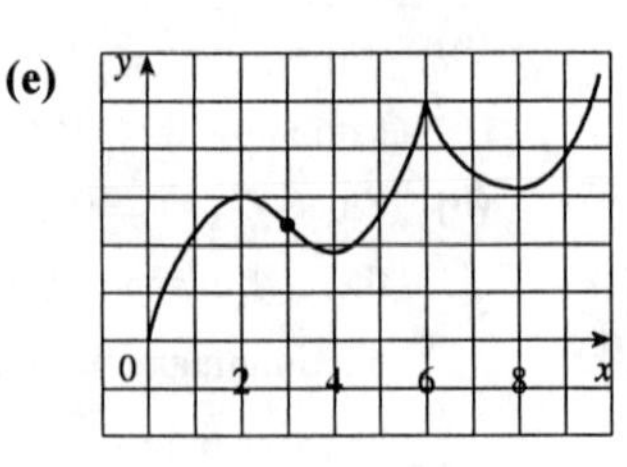

13. The function must be always decreasing and concave downward.

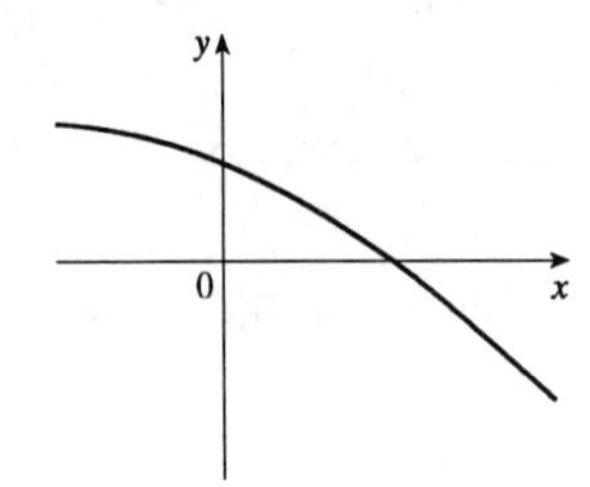

15. $f(-1)=4$ and $f(1)=0$ gives us two points to start with.

$f'(-1)=f'(1)=0 \Longrightarrow$ horizontal tangents at $x=\pm 1$. $f'(x)<0$ if $|x|<1 \Longrightarrow f$ is decreasing on $(-1,1)$. $f'(x)>0$ if $|x|>1 \Longrightarrow f$ is increasing on $(-\infty,-1)$ and $(1,\infty)$. $f''(x)<0$ if $x<0 \Longrightarrow f$ is concave downward on $(-\infty,0)$. $f''(x)>0$ if $x>0 \Longrightarrow f$ is concave upward on $(0,\infty)$ and there is an inflection point at $x=0$.

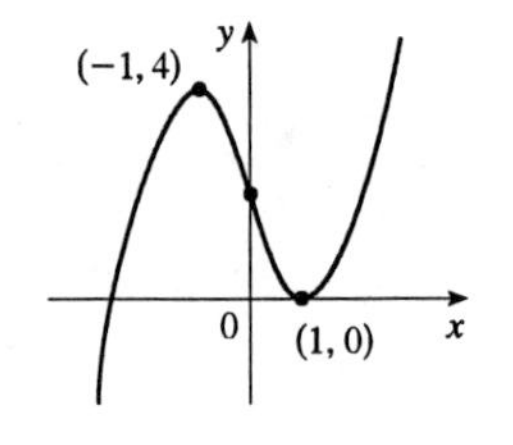

17. First we plot the points which are known to be on the graph: $(2,-1)$ and $(0,0)$. We can also draw a short line segment of slope 0 at $x=2$, since we are given that $f'(2)=0$. Now we know that $f'(x)<0$ (that is, the function is decreasing) on $(0,2)$, and that $f''(x)<0$ on $(0,1)$ and $f''(x)>0$ on $(1,2)$. So we must join the points $(0,0)$ and $(2,-1)$ in such a way that the curve is concave down on $(0,1)$ and concave up on $(1,2)$. The curve must be concave up and increasing on $(2,4)$ and concave down and increasing on $(4,\infty)$. Now we just need to reflect the curve in the y-axis, since we are given that f is an even function.

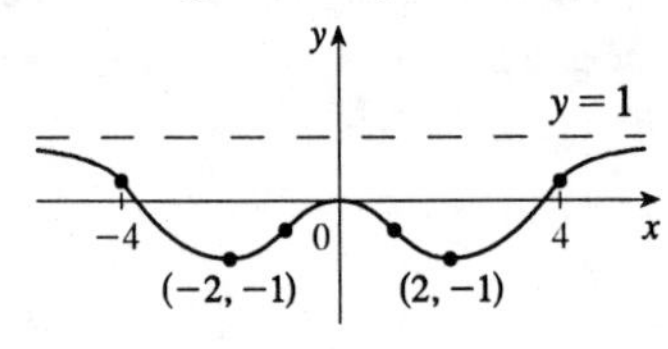

19. **(a)** Since e^{-x^2} is positive for all x, $f'(x) = xe^{-x^2}$ is positive where $x > 0$ and negative where $x < 0$. Thus, f is increasing on $(0, \infty)$ and decreasing on $(-\infty, 0)$.

(b) Since f changes from decreasing to increasing at $x = 0$, f has a minimum value there.

21. For small x, f is negative, so the graph of its antiderivative must be decreasing. So only b can be f's antiderivative.

EXERCISES 5.2

1. A function f has an *absolute minimum* at $x = c$ if $f(c)$ is the smallest function value on the entire domain of f, whereas f has a *local minimum* at c if $f(c)$ is the smallest function value when x is near c.

3. Absolute maximum at b; absolute minimum at d; local maxima at b, e; local minima at d, s; neither a maximum nor a minimum at a, c, r, and t.

5. Absolute maximum value is $f(4) = 4$; absolute minimum value is $f(7) = 0$; local maximum values are $f(4) = 4$ and $f(6) = 3$; local minimum values are $f(2) = 1$ and $f(5) = 2$.

7.

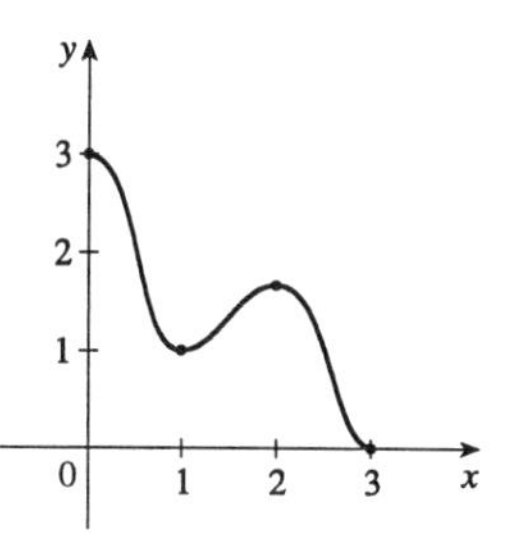

9.

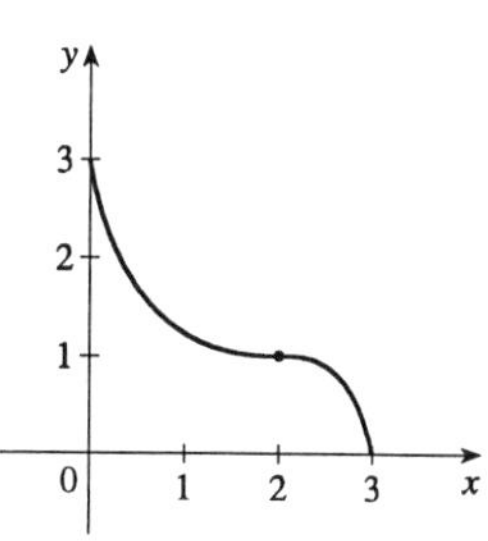

11. **(a)**

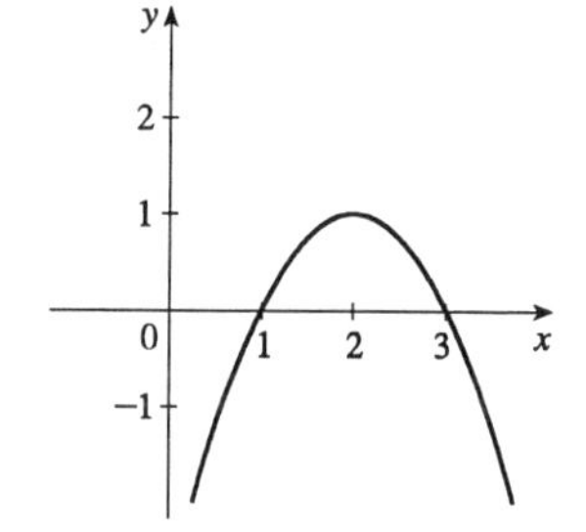

(b)

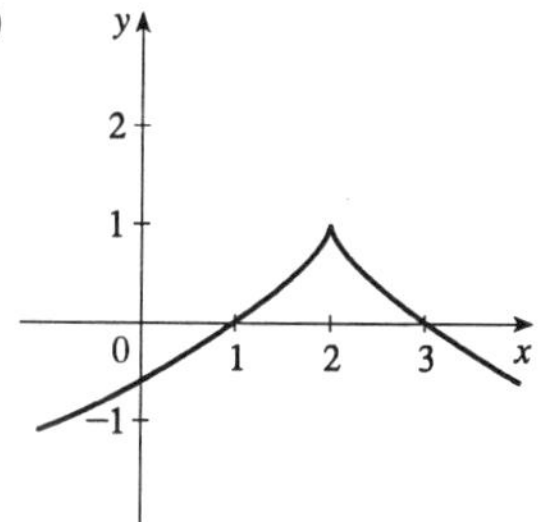

(c)

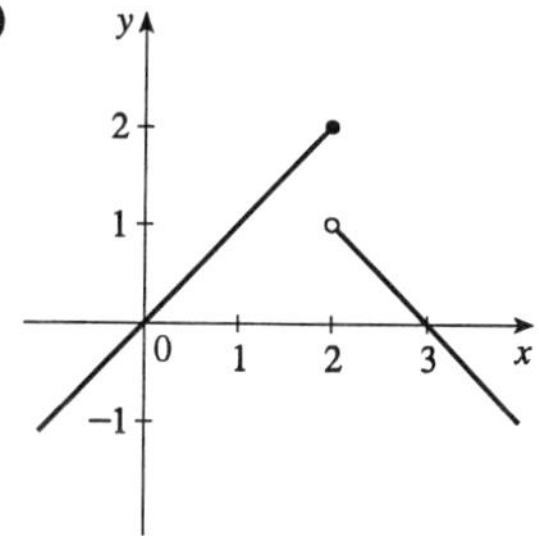

13. **(a)** *Note:* By the Extreme Value Theorem, f must *not* be continuous; because if it were, it would attain an absolute minimum.

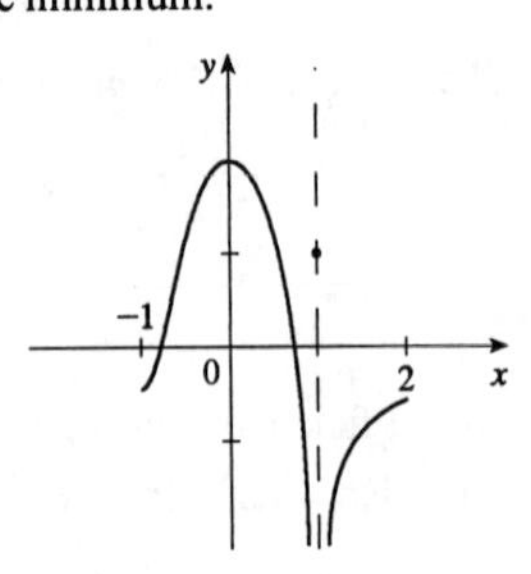

(b)

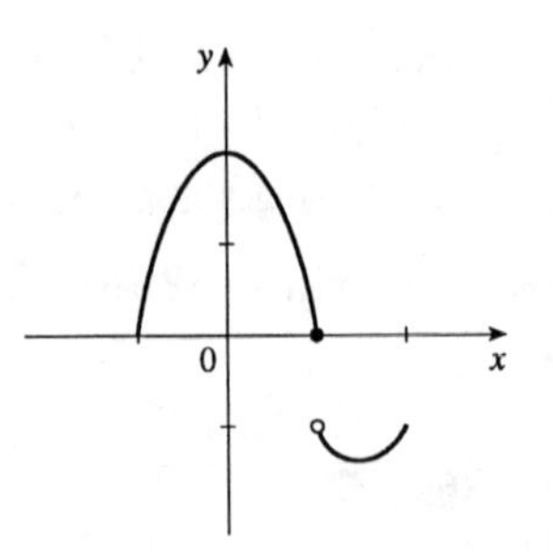

15. $f(x) = 1 + 2x$, $x \geq -1$. Absolute minimum $f(-1) = -1$; no local minimum. No local or absolute maximum.

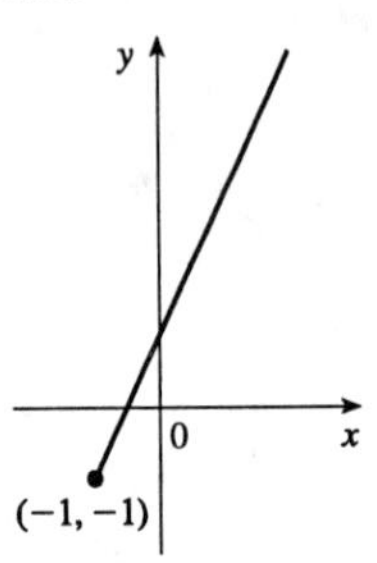

17. $f(x) = 1 - x^2$, $-2 \leq x \leq 1$. Absolute and local maximum $f(0) = 1$. Absolute minimum $f(-2) = -3$; no local minimum.

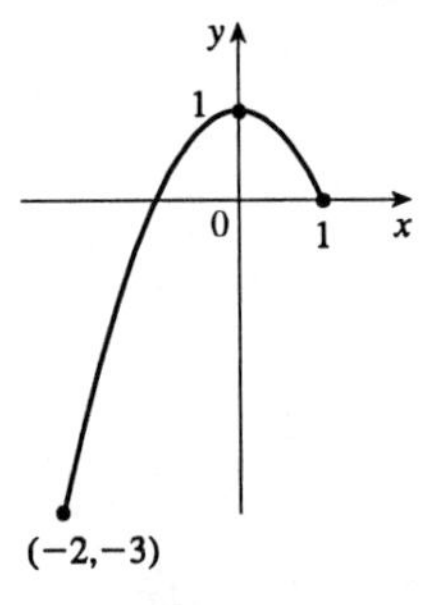

19. $f(\theta) = \sin\theta$, $-2\pi \leq \theta \leq 2\pi$. Absolute and local maxima $f\left(-\frac{3\pi}{2}\right) = f\left(\frac{\pi}{2}\right) = 1$. Absolute and local minima $f\left(-\frac{\pi}{2}\right) = f\left(\frac{3\pi}{2}\right) = -1$.

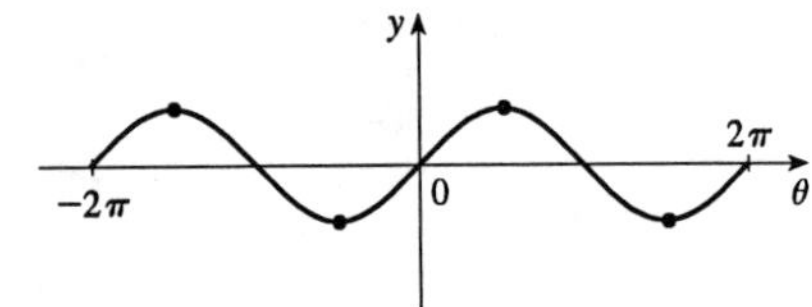

21. $f(x) = x^5$. No maximum or minimum.

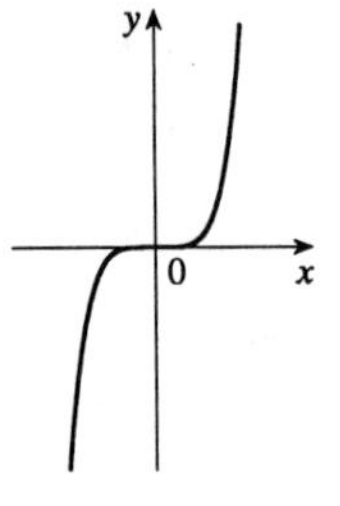

23. $f(x) = 1 - e^{-x}$, $x \geq 0$. Absolute minimum $f(0) = 0$; no local minimum. No absolute or local maximum.

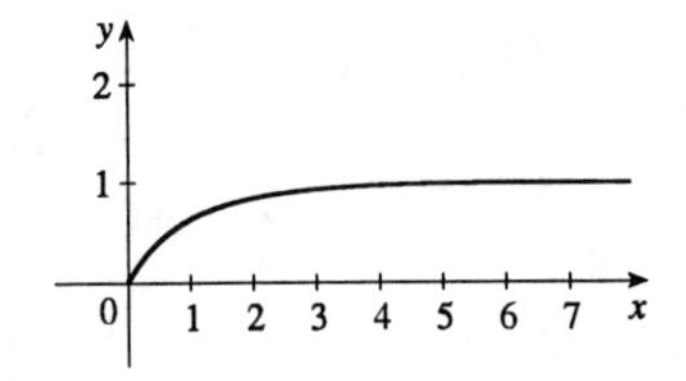

25. $f(x) = 4x^3 - 9x^2 - 12x + 3 \Longrightarrow f'(x) = 12x^2 - 18x - 12 = 6\left(2x^2 - 3x - 2\right) = 6(2x+1)(x-2)$.

$f'(x) = 0 \Longrightarrow x = -\frac{1}{2}, 2$; so the critical numbers are $x = -\frac{1}{2}, 2$.

27. $s(t) = t^4 + 4t^3 + 2t^2 \Longrightarrow s'(t) = 4t^3 + 12t^2 + 4t = 4t\left(t^2 + 3t + 1\right) = 0$ when $t = 0$ or $t^2 + 3t + 1 = 0$. By the quadratic formula, the critical numbers are $t = 0, \frac{-3 \pm \sqrt{5}}{2}$.

29. $f(r) = \dfrac{r}{r^2+1} \Longrightarrow f'(r) = \dfrac{\left(r^2+1\right)1 - r(2r)}{\left(r^2+1\right)^2} = \dfrac{-r^2+1}{\left(r^2+1\right)^2} = 0 \Longleftrightarrow r^2 = 1 \Longleftrightarrow r = \pm 1$, so these are the critical numbers. Note that $f'(x)$ always exists since $r^2 + 1 \neq 0$.

31. $F(x) = x^{4/5}(x-4)^2 \Longrightarrow$

$$F'(x) = \tfrac{4}{5}x^{-1/5}(x-4)^2 + 2x^{4/5}(x-4) = \tfrac{1}{5}x^{-1/5}(x-4)\left[4(x-4) + 10x\right]$$

$$= \frac{(x-4)(14x-16)}{5x^{1/5}} = \frac{2(x-4)(7x-8)}{5x^{1/5}} = 0 \text{ when } x = 4, \tfrac{8}{7}\text{; and } F'(0) \text{ does not exist.}$$

Critical numbers are $0, \frac{8}{7}, 4$.

33. $f(\theta) = \sin^2(2\theta) \Longrightarrow$

$f'(\theta) = 2\sin(2\theta)\cos(2\theta)(2) = 2(2\sin 2\theta\cos 2\theta) = 2\left[\sin(2 \cdot 2\theta)\right] = 2\sin 4\theta = 0 \Longleftrightarrow \sin 4\theta = 0 \Longleftrightarrow$ $4\theta = n\pi$, n an integer. So $\theta = n\pi/4$ are the critical numbers.

35. $f(x) = x\ln x \Longrightarrow f'(x) = x(1/x) + \ln x = \ln x + 1 = 0 \Longleftrightarrow \ln x = -1 \Longleftrightarrow x = e^{-1} = 1/e$. Therefore, the only critical number is $x = 1/e$.

37. $f(x) = x^2 - 2x + 2$, $[0, 3]$. $f'(x) = 2x - 2 = 0 \Longleftrightarrow x = 1$. $f(0) = 2$, $f(1) = 1$, $f(3) = 5$. So $f(3) = 5$ is the absolute maximum and $f(1) = 1$ is the absolute minimum.

39. $f(x) = 3x^5 - 5x^3 - 1$, $[-2, 2]$. $f'(x) = 15x^4 - 15x^2 = 15x^2(x+1)(x-1) = 0 \Longleftrightarrow x = -1, 0, 1$.

$f(-2) = -57$, $f(-1) = 1$, $f(0) = -1$, $f(1) = -3$, $f(2) = 55$. So $f(-2) = -57$ is the absolute minimum and $f(2) = 55$ is the absolute maximum.

41. $f(x) = x^2 + \dfrac{2}{x}$, $\left[\frac{1}{2}, 2\right]$. $f'(x) = 2x - \dfrac{2}{x^2} = 2\dfrac{x^3 - 1}{x^2} = 0 \Longleftrightarrow x^3 - 1 = 0 \Longleftrightarrow (x-1)\left(x^2 + x + 1\right) = 0$,

but $x^2 + x + 1 \neq 0$, so $x = 1$. The denominator is 0 at $x = 0$, but not in the desired interval. $f\left(\frac{1}{2}\right) = \frac{17}{4}$, $f(1) = 3$, $f(2) = 5$. So $f(1) = 3$ is the absolute minimum and $f(2) = 5$ is the absolute maximum.

43. $f(x) = \sin x + \cos x$, $\left[0, \frac{\pi}{3}\right]$. $f'(x) = \cos x - \sin x = 0 \iff \sin x = \cos x \Longrightarrow \dfrac{\sin x}{\cos x} = 1 \Longrightarrow$

$\tan x = 1 \Longrightarrow x = \frac{\pi}{4}$. $f(0) = 1$, $f\left(\frac{\pi}{4}\right) = \sqrt{2} \approx 1.41$, $f\left(\frac{\pi}{3}\right) = \frac{\sqrt{3}+1}{2} \approx 1.37$. So $f(0) = 1$ is the absolute

minimum and $f\left(\frac{\pi}{4}\right) = \sqrt{2}$ is the absolute maximum.

45. $f(x) = xe^{-x}$, $[0, 2]$. $f'(x) = x\left(-e^{-x}\right) + e^{-x} = e^{-x}(1 - x) = 0 \iff x = 1$. Now $f(0) = 0$, $f(1) = e^{-1} = 1/e$, and $f(2) = 2/e^2 \approx 0.27$, so $f(0) = 0$ is the absolute minimum and $f(1) = 1/e$ is the absolute maximum.

47. (a) From the graph, it appears that the absolute maximum value is about $f(-1.63) = 9.71$, and the absolute minimum value is about $f(1.63) = -7.71$. These values make sense because the graph is symmetric about the point $(0, 1)$. [$y = x^3 - 8x$ is symmetric about the origin.]

(b) $f(x) = x^3 - 8x + 1 \Longrightarrow f'(x) = 3x^2 - 8$. So $f'(x) = 0 \Longrightarrow x = \pm\sqrt{\frac{8}{3}}$.

$$f\left(\pm\sqrt{\tfrac{8}{3}}\right) = \left(\pm\sqrt{\tfrac{8}{3}}\right)^3 - 8\left(\pm\sqrt{\tfrac{8}{3}}\right) + 1 = \pm\tfrac{8}{3}\sqrt{\tfrac{8}{3}} \mp 8\sqrt{\tfrac{8}{3}} + 1$$

$$= -\tfrac{16}{3}\sqrt{\tfrac{8}{3}} + 1 = 1 - \tfrac{32\sqrt{6}}{9} \text{ (minimum) or } \tfrac{16}{3}\sqrt{\tfrac{8}{3}} + 1 = 1 + \tfrac{32\sqrt{6}}{9} \text{ (maximum).}$$

(From the graph, we see that the extreme values do not occur at the endpoints.)

49. (a)

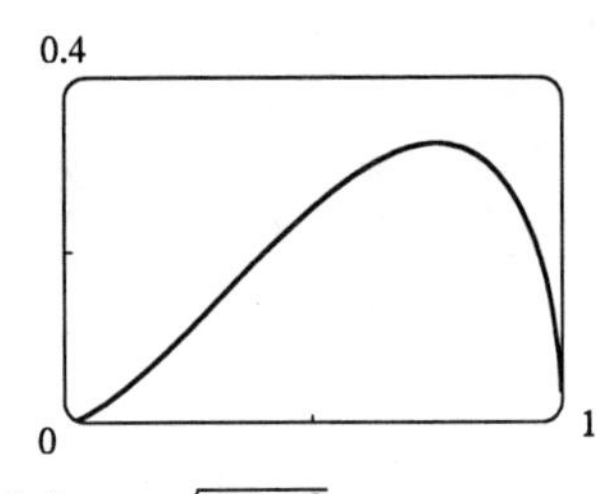

From the graph, it appears that the absolute maximum value is about $f(0.75) = 0.32$, and the absolute minimum value is $f(0) = f(1) = 0$, that is, at both endpoints.

(b) $f(x) = x\sqrt{x - x^2} \Longrightarrow$

$$f'(x) = x \cdot \frac{1 - 2x}{2\sqrt{x - x^2}} + \sqrt{x - x^2} = \frac{\left(x - 2x^2\right) + \left(2x - 2x^2\right)}{2\sqrt{x - x^2}} = \frac{3x - 4x^2}{2\sqrt{x - x^2}}. \text{ So } f'(x) = 0 \Longrightarrow$$

$3x - 4x^2 = 0 \Longrightarrow x = 0$ or $\frac{3}{4}$. $f(0) = f(1) = 0$ (minima), and $f\left(\frac{3}{4}\right) = \frac{3}{4}\sqrt{\frac{3}{4} - \left(\frac{3}{4}\right)^2} = \frac{3\sqrt{3}}{16}$

(maximum).

51. The density is defined as $\rho = \dfrac{\text{mass}}{\text{volume}} = \dfrac{1000}{V(T)}$ (in g/cm^3). But a critical point of ρ will also be a critical point of V $\left[\text{since } \dfrac{d\rho}{dT} = -1000V^{-2}\dfrac{dV}{dT} \text{ and } V \text{ is never } 0\right]$, and V is easier to differentiate than ρ. $V(T) = 999.87 - 0.06426T + 0.0085043T^2 - 0.0000679T^3 \implies$ $V'(T) = -0.06426 + 0.0170086T - 0.0002037T^2$. Setting this equal to 0 and using the quadratic formula to find T, we get $T = \dfrac{-0.0170086 \pm \sqrt{0.0170086^2 - 4 \cdot 0.0002037 \cdot 0.06426}}{2(-0.0002037)} \approx 3.9665^\circ$ or 79.5318°. Since we are only interested in the region $0^\circ \le T \le 30^\circ$, we check the density ρ at the endpoints and at 3.9665°: $\rho(0) \approx \dfrac{1000}{999.87} \approx 1.00013$; $\rho(30) \approx \dfrac{1000}{1003.7641} \approx 0.99625$; $\rho(3.9665) \approx \dfrac{1000}{999.7447} \approx 1.000255$. So water has its maximum density at about 3.9665°C.

53. $v(t) = 0.001302t^3 - 0.09029t^2 + 23.61t - 3.083 \implies$ $a(t) = v'(t) = 0.003906t^2 - 0.18058t + 23.61 \implies a'(t) = 0.007812t - 0.18058$. $a'(t) = 0 \implies$ $t_1 = \frac{0.18058}{0.007812} \approx 23.12$. Evaluating $a(t)$ at the critical number and the endpoints gives us: $a(0) = 23.61$, $a(t_1) \approx 21.52$, and $a(126) \approx 62.87$. The absolute maximum is about $62.87\ \text{ft/s}^2$ and the absolute minimum is about $21.52\ \text{ft/s}^2$.

55. **(a)** $v(r) = k(r_0 - r)r^2 = kr_0r^2 - kr^3 \implies$ $v'(r) = 2kr_0r - 3kr^2$. $v'(r) = 0 \implies$ $kr(2r_0 - 3r) = 0 \implies r = 0$ or $\frac{2}{3}r_0$ (but 0 is not in the interval). Evaluating v at $\frac{1}{2}r_0$, $\frac{2}{3}r_0$, and r_0, we get $v\left(\frac{1}{2}r_0\right) = \frac{1}{8}kr_0^3$, $v\left(\frac{2}{3}r_0\right) = \frac{4}{27}kr_0^3$, and $v(r_0) = 0$. Since $\frac{4}{27} > \frac{1}{8}$, v attains its maximum value at $r = \frac{2}{3}r_0$. This supports the statement in the text.

(b) From part (a), the maximum value of v is $\frac{4}{27}kr_0^3$.

(c)

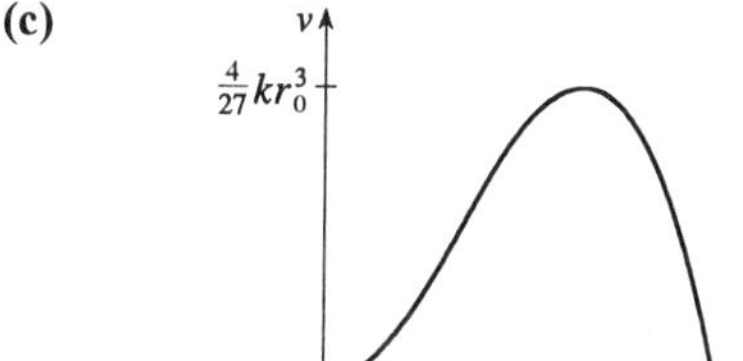

EXERCISES 5.3

Note: We will use the following abbreviations:

D	the domain of f	VA	vertical asymptote(s)
HA	horizontal asymptote(s)	IP	inflection point(s)
CU	concave upward	CD	concave downward

1. Geometrically, we are looking for the x-coordinates at which the slope of the tangent line equals the slope of the line segment connecting the endpoints. $\frac{f(8)-f(0)}{8-0}=\frac{6-4}{8}=\frac{1}{4}$. The values of c which satisfy $f'(c)=\frac{1}{4}$ seem to be about $c=0.8$, 3.2, 4.4, and 6.1.

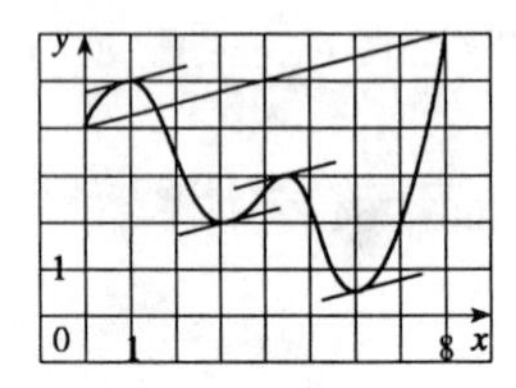

3. **(a)** Use the Increasing/Decreasing (I/D) Test on page 316.

(b) Use the Concavity Test on page 318.

(c) At any value of x where the concavity changes, we have an inflection point at $(x, f(x))$.

5. There is an inflection point at $x=1$ because $f''(x)$ changes from negative to positive there, and one at $x=7$ because $f''(x)$ changes from positive to negative there.

7. **(a)** $f(x)=x^6+192x+17 \Longrightarrow f'(x)=6x^5+192=6\left(x^5+32\right)$. So $f'(x)>0 \Longleftrightarrow x^5>-32 \Longleftrightarrow x>-2$ and $f'(x)<0 \Longleftrightarrow x<-2$. So f is increasing on $(-2,\infty)$ and decreasing on $(-\infty,-2)$.

(b) f changes from decreasing to increasing at its only critical number, $x=-2$. Thus, $f(-2)=-303$ is a local minimum.

(c) $f''(x)=30x^4\geq 0$ for all x, so the concavity of f doesn't change and there are no inflection points. f is concave upward on $(-\infty,\infty)$.

9. **(a)** $y=f(x)=xe^x \Longrightarrow f'(x)=xe^x+e^x=e^x(x+1)$. So $f'(x)>0 \Longleftrightarrow x+1>0 \Longleftrightarrow x>-1$. Thus, f is increasing on $(-1,\infty)$ and decreasing on $(-\infty,-1)$.

(b) f changes from decreasing to increasing at its only critical number, $x=-1$. Thus, $f(-1)=-e^{-1}$ is a local minimum.

(c) $f'(x)=e^x(x+1) \Longrightarrow f''(x)=e^x(1)+(x+1)e^x=e^x(x+2)$. So $f''(x)>0 \Longleftrightarrow x+2>0 \Longleftrightarrow x>-2$. Thus, f is concave upward on $(-2,\infty)$ and concave downward on $(-\infty,-2)$. Since the concavity changes direction at $x=-2$, the point $\left(-2,-2e^{-2}\right)$ is an inflection point.

11. **(a)** $y=f(x)=\frac{\ln x}{\sqrt{x}}$. (Note that f is only defined for $x>0$.)

$$f'(x)=\frac{\sqrt{x}\,(1/x)-\ln x\left(\frac{1}{2}x^{-1/2}\right)}{x}=\frac{\frac{1}{\sqrt{x}}-\frac{\ln x}{2\sqrt{x}}}{x}\cdot\frac{2\sqrt{x}}{2\sqrt{x}}=\frac{2-\ln x}{2x^{3/2}}>0 \Longleftrightarrow \ln x<2 \Longleftrightarrow$$

$x<e^2$. Therefore f is increasing on $\left(0,e^2\right)$ and decreasing on $\left(e^2,\infty\right)$.

(b) f changes from increasing to decreasing at $x = e^2$, so $f\left(e^2\right) = \dfrac{\ln e^2}{\sqrt{e^2}} = \dfrac{2}{e}$ is a local maximum.

(c) $$f''(x) = \frac{2x^{3/2}(-1/x) - (2 - \ln x)\left(3x^{1/2}\right)}{(2x^{3/2})^2} = \frac{-2x^{1/2} + 3x^{1/2}(\ln x - 2)}{4x^3}$$
$$= \frac{x^{1/2}(-2 + 3\ln x - 6)}{4x^3} = \frac{3\ln x - 8}{4x^{5/2}}$$

$f''(x) = 0 \iff \ln x = \frac{8}{3} \iff x = e^{8/3}$. $f''(x) > 0 \iff x > e^{8/3}$, so f is concave upward on $\left(e^{8/3}, \infty\right)$ and concave downward on $\left(0, e^{8/3}\right)$. There is an inflection point at $\left(e^{8/3}, 8/\left(3e^{4/3}\right)\right) \approx (14.39, 0.70)$.

13. **(a)** $f(x) = 1 - 3x + 5x^2 - x^3 \implies f'(x) = -3 + 10x - 3x^2 = -\left(3x^2 - 10x + 3\right) = -(3x - 1)(x - 3)$.

$f'(x) = 0 \iff x = \frac{1}{3}, 3$. $f'(x) > 0 \iff \frac{1}{3} < x < 3$ [the graph of f' is a parabola opening down]. Thus, f is increasing on $\left(\frac{1}{3}, 3\right)$ and decreasing on $\left(-\infty, \frac{1}{3}\right)$ and $(3, \infty)$.

(b) f changes from decreasing to increasing at $x = \frac{1}{3}$, so $f\left(\frac{1}{3}\right) = \frac{14}{27}$ is a local minimum. f changes from increasing to decreasing at $x = 3$, so $f(3) = 10$ is a local maximum.

(c) $f'(x) = -3x^2 + 10x - 3 \implies f''(x) = -6x + 10 = 0 \iff x = \frac{5}{3}$.

$f''(x) > 0 \iff x < \frac{5}{3}$, so f is CU on $\left(-\infty, \frac{5}{3}\right)$ and CD on $\left(\frac{5}{3}, \infty\right)$.

There is an IP at $\left(\frac{5}{3}, \frac{142}{27}\right) \approx (1.67, 5.26)$.

(d)

15. **(a)** $f(x) = \left(x^2 - 1\right)^3 \implies f'(x) = 6x\left(x^2 - 1\right)^2 \geq 0 \iff x > 0\ (x \neq 1)$, so f is increasing on $(0, \infty)$ and decreasing on $(-\infty, 0)$.

(b) $f(0) = -1$ is a local minimum since f changes from decreasing to increasing at $x = 0$.

(c) $f''(x) = 6\left(x^2 - 1\right)^2 + 24x^2\left(x^2 - 1\right) = 6\left(x^2 - 1\right)\left(5x^2 - 1\right)$. The roots ± 1 and $\pm\frac{1}{\sqrt{5}}$ divide $\mathbb{R}$ into five intervals.

Interval	$x^2 - 1$	$5x^2 - 1$	$f''(x)$	Concavity
$x < -1$	$+$	$+$	$+$	upward
$-1 < x < -\frac{1}{\sqrt{5}}$	$-$	$+$	$-$	downward
$-\frac{1}{\sqrt{5}} < x < \frac{1}{\sqrt{5}}$	$-$	$-$	$+$	upward
$\frac{1}{\sqrt{5}} < x < 1$	$-$	$+$	$-$	downward
$x > 1$	$+$	$+$	$+$	upward

From the table, we see that f is CU on $(-\infty, -1)$, $\left(-\frac{1}{\sqrt{5}}, \frac{1}{\sqrt{5}}\right)$ and $(1, \infty)$, and CD on $\left(-1, -\frac{1}{\sqrt{5}}\right)$ and $\left(\frac{1}{\sqrt{5}}, 1\right)$. There are inflection points at $x = \pm 1$, $\pm\frac{1}{\sqrt{5}}$.

(d)

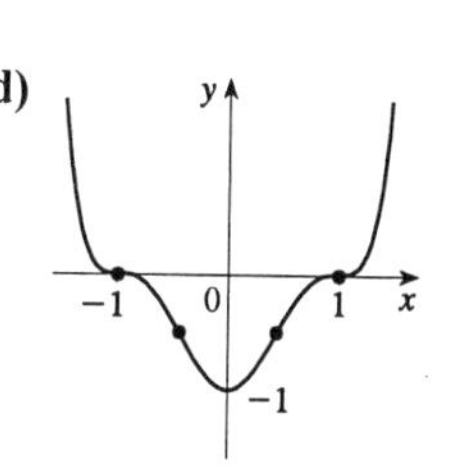

17. **(a)** $f(x) = x^{1/3}(x+3)^{2/3} \implies$

$$f'(x) = \tfrac{1}{3}x^{-2/3}(x+3)^{2/3} + x^{1/3}\left(\tfrac{2}{3}\right)(x+3)^{-1/3}$$
$$= \frac{(x+3)^{2/3}}{3x^{2/3}} + \frac{2x^{1/3}}{3(x+3)^{1/3}} = \frac{(x+3)+2x}{3x^{2/3}(x+3)^{1/3}} = \frac{x+1}{x^{2/3}(x+3)^{1/3}}.$$

The critical numbers are -3, -1, and 0. Note that $x^{2/3} \geq 0$ for all x. So $f'(x) > 0$ when $x < -3$ or $x > -1$ and $f'(x) < 0$ when $-3 < x < -1 \implies f$ is increasing on $(-\infty, -3)$ and $(-1, \infty)$ and decreasing on $(-3, -1)$.

(b) At $x = -3$, f changes from increasing to decreasing and at $x = -1$, vice versa, so $f(-3) = 0$ is a local maximum and $f(-1) = -4^{1/3} \approx -1.6$ is a local minimum.

(c)
$$f''(x) = \frac{x^{2/3}(x+3)^{1/3} \cdot 1 - (x+1)\left[x^{2/3}\tfrac{1}{3}(x+3)^{-2/3} + (x+3)^{1/3}\tfrac{2}{3}x^{-1/3}\right]}{\left[x^{2/3}(x+3)^{1/3}\right]^2}$$
$$= \frac{x^{2/3}(x+3)^{1/3} - (x+1)\left[\dfrac{x^{2/3}}{3(x+3)^{2/3}} + \dfrac{2(x+3)^{1/3}}{3x^{1/3}}\right]}{x^{4/3}(x+3)^{2/3}}$$
$$= \frac{x^{2/3}(x+3)^{1/3} - (x+1)\left[\dfrac{x+2(x+3)}{3x^{1/3}(x+3)^{2/3}}\right]}{x^{4/3}(x+3)^{2/3}} = \frac{3x(x+3) - 3(x+1)(x+2)}{3x^{5/3}(x+3)^{4/3}}$$
$$= -\frac{2}{x^{5/3}(x+3)^{4/3}}$$

Note that $(x+3)^{4/3} > 0$ for $x \neq -3$, so the sign of $f''(x)$ is the same as the sign of $-x^{5/3}$. Thus, $f''(x) > 0$ when $x < 0$, so f is CU on $(-\infty, -3)$ and $(-3, 0)$ and CD on $(0, \infty)$. There is an IP at $x = 0$.

(d)

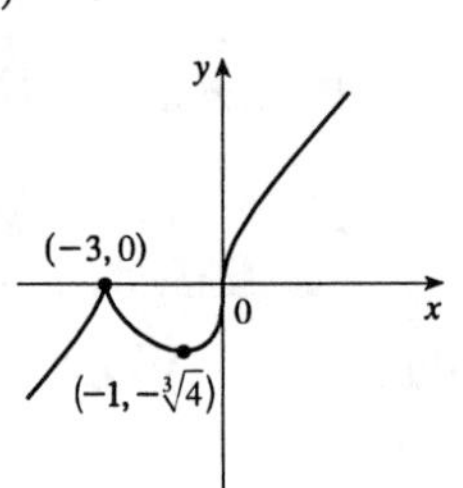

19. **(a)** $f(x) = 2\cos x + \sin^2 x \implies f'(x) = -2\sin x + 2\sin x \cos x = 2\sin x(\cos x - 1)$. Since $\cos x \leq 1$, $\cos x - 1 \leq 0$, so the sign of $f'(x)$ is the opposite of the sign of $\sin x$. Thus, $f'(x) > 0 \iff \sin x < 0 \iff (2n-1)\pi < x < 2n\pi$, so f is increasing on $((2n-1)\pi, 2n\pi)$ and decreasing on $(2n\pi, (2n+1)\pi)$.

(b) f changes from increasing to decreasing at $x = 2n\pi$, so $f(2n\pi) = 2$ are local maxima. f changes from decreasing to increasing when $x = 2n\pi + \pi$, so $f((2n+1)\pi) = -2$ are local minima.

(c) $f'(x) = -2\sin x + 2\sin x\cos x = -2\sin x + \sin 2x \implies$
$f''(x) = -2\cos x + 2\cos 2x = 2\left(2\cos^2 x - \cos x - 1\right)$
$= 2(2\cos x + 1)(\cos x - 1) > 0 \iff$
$\cos x < -\frac{1}{2} \iff x \in \left(2n\pi + \frac{2\pi}{3}, 2n\pi + \frac{4\pi}{3}\right)$, so f is CU on these intervals and CD on $\left(2n\pi - \frac{2\pi}{3}, 2n\pi + \frac{2\pi}{3}\right)$. There are IP at $\left(2n\pi \pm \frac{2\pi}{3}, -\frac{1}{4}\right)$.

(d)

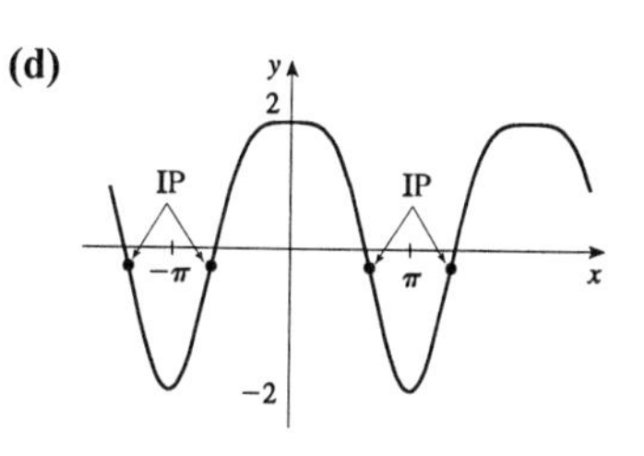

21. **(a)** $\lim\limits_{x\to\pm\infty} \dfrac{1+x^2}{1-x^2} = \lim\limits_{x\to\pm\infty} \dfrac{(1/x^2)+1}{(1/x^2)-1} = -1$, so $y = -1$ is a HA. $\lim\limits_{x\to 1^-} \dfrac{1+x^2}{1-x^2} = \infty$,

$\lim\limits_{x\to 1^+} \dfrac{1+x^2}{1-x^2} = -\infty$, $\lim\limits_{x\to -1^-} \dfrac{1+x^2}{1-x^2} = -\infty$, $\lim\limits_{x\to -1^+} \dfrac{1+x^2}{1-x^2} = \infty$. So $x = 1$ and $x = -1$ are VA.

(b) $f(x) = \dfrac{1+x^2}{1-x^2} = -1 + \dfrac{2}{1-x^2} \implies f'(x) = \dfrac{4x}{(1-x^2)^2} > 0 \iff x > 0\ (x \neq 1)$, so f increases on $(0,1)$, $(1,\infty)$ and decreases on $(-\infty,-1)$, $(-1,0)$.

(c) $f(0) = 1$ is a local minimum.

(d) $f''(x) = \dfrac{4\left(1-x^2\right)^2 - 4x\cdot 2\left(1-x^2\right)(-2x)}{(1-x^2)^4} = \dfrac{4\left(1+3x^2\right)}{(1-x^2)^3}$. Since the numerator is always positive, the sign of $f''(x)$ is the same as the sign of $1-x^2$. Thus, $f''(x) > 0 \iff 1 - x^2 > 0 \iff x^2 < 1 \iff$
$-1 < x < 1$, so f is CU on $(-1,1)$ and CD on $(-\infty,-1)$ and $(1,\infty)$. There is no IP since $x = \pm 1$ are not in the domain of f.

(e)

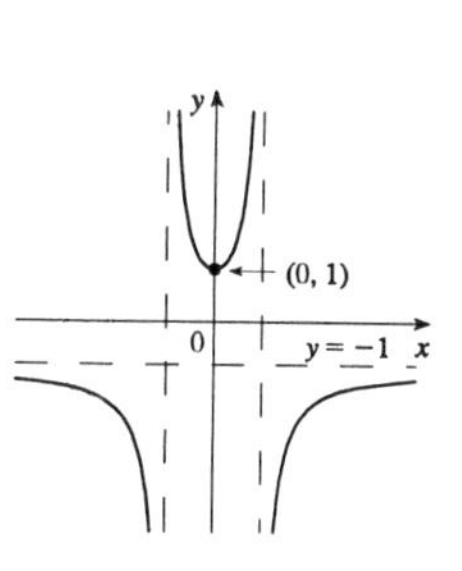

23. **(a)** $\lim\limits_{x\to -\infty} \left(\sqrt{x^2+1} - x\right) = \infty$ and

$\lim\limits_{x\to\infty} \left(\sqrt{x^2+1} - x\right) = \lim\limits_{x\to\infty} \left(\sqrt{x^2+1} - x\right) \dfrac{\sqrt{x^2+1}+x}{\sqrt{x^2+1}+x} = \lim\limits_{x\to\infty} \dfrac{1}{\sqrt{x^2+1}+x} = 0$, so $y = 0$ is a HA.

(b) $f(x) = \sqrt{x^2+1} - x \implies f'(x) = \dfrac{x}{\sqrt{x^2+1}} - 1$. Since $\dfrac{x}{\sqrt{x^2+1}} < 1$ for all x, $f'(x) < 0$, so f is decreasing on $\mathbb{R}$.

(c) No minimum or maximum

(d)

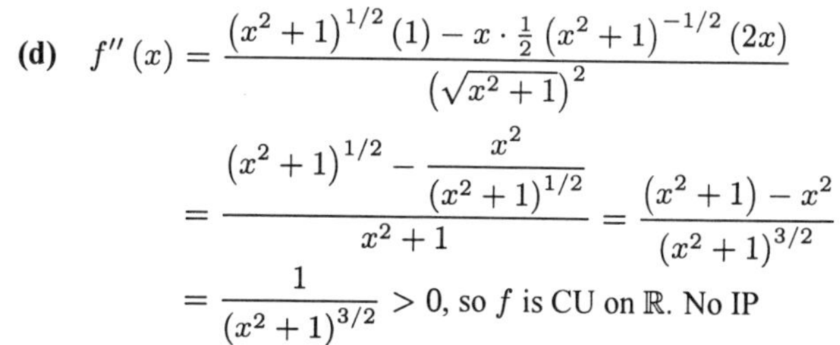

$$f''(x) = \frac{\left(x^2+1\right)^{1/2}(1) - x\cdot\frac{1}{2}\left(x^2+1\right)^{-1/2}(2x)}{\left(\sqrt{x^2+1}\right)^2}$$
$$= \frac{\left(x^2+1\right)^{1/2} - \dfrac{x^2}{\left(x^2+1\right)^{1/2}}}{x^2+1} = \frac{\left(x^2+1\right) - x^2}{\left(x^2+1\right)^{3/2}}$$
$$= \frac{1}{\left(x^2+1\right)^{3/2}} > 0$$, so f is CU on $\mathbb{R}$. No IP

(e)

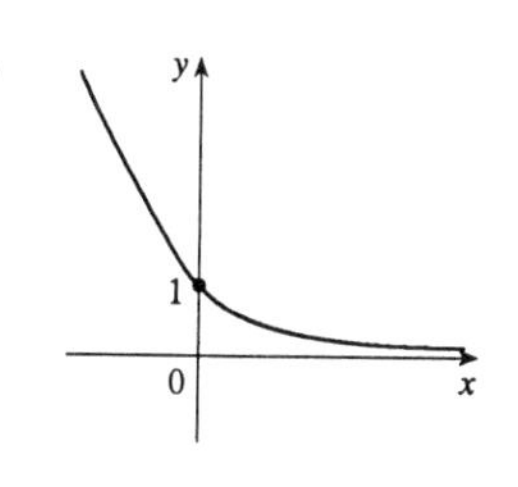

25. **(a)** $\lim_{x\to\pm\infty} e^{-1/(x+1)} = 1$ since $-1/(x+1) \to 0$,

so $y = 1$ is a HA. $\lim_{x\to-1^+} e^{-1/(x+1)} = 0$ since $-1/(x+1) \to -\infty$, $\lim_{x\to-1^-} e^{-1/(x+1)} = \infty$ since $-1/(x+1) \to \infty$, so $x = -1$ is a VA.

(b) $f(x) = e^{-1/(x+1)} \implies f'(x) = e^{-1/(x+1)}/(x+1)^2 \implies f'(x) > 0$ for all x except -1, so f is increasing on $(-\infty, -1)$ and $(-1, \infty)$.

(c) No local maximum or minimum

(d) $$f''(x) = \frac{(x+1)^2 e^{-1/(x+1)}\left[1/(x+1)^2\right] - e^{-1/(x+1)}\left[2(x+1)\right]}{\left[(x+1)^2\right]^2} = \frac{e^{-1/(x+1)}\left[1-(2x+2)\right]}{(x+1)^4}$$
$$= -\frac{e^{-1/(x+1)}(2x+1)}{(x+1)^4} \implies$$

$f''(x) > 0 \iff 2x+1 < 0 \iff x < -\frac{1}{2}$, so f is CU on $(-\infty, -1)$ and $\left(-1, -\frac{1}{2}\right)$, and CD on $\left(-\frac{1}{2}, \infty\right)$. f has an IP at $\left(-\frac{1}{2}, e^{-2}\right)$.

(e)

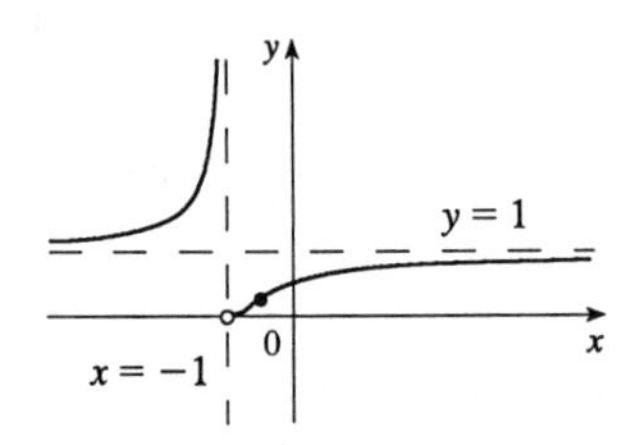

27. **(a)** From the graphs of $f(x) = 3x^5 - 40x^3 + 30x^2$, it seems that f is concave upward on $(-2, 0.25)$ and $(2, \infty)$, and concave downward on $(-\infty, -2)$ and $(0.25, 2)$, with inflection points at about $(-2, 350)$, $(0.25, 1)$, and $(2, -100)$.

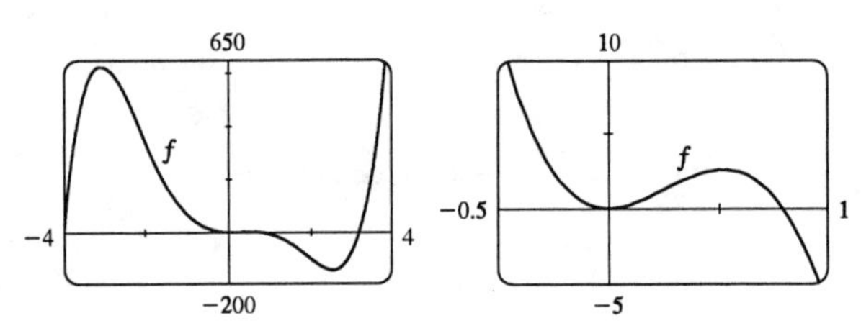

(b) From the graph of $f''(x) = 60x^3 - 240x + 60$, it seems that f is CU on $(-2.1, 0.25)$ and $(1.9, \infty)$, and CD on $(-\infty, -2.1)$ and $(0.25, 2)$, with inflection points at about $(-2.1, 386)$, $(0.25, 1.3)$ and $(1.9, -87)$. (We have to check back on the graph of f to find the y-coordinates of the inflection points.)

29. (a)

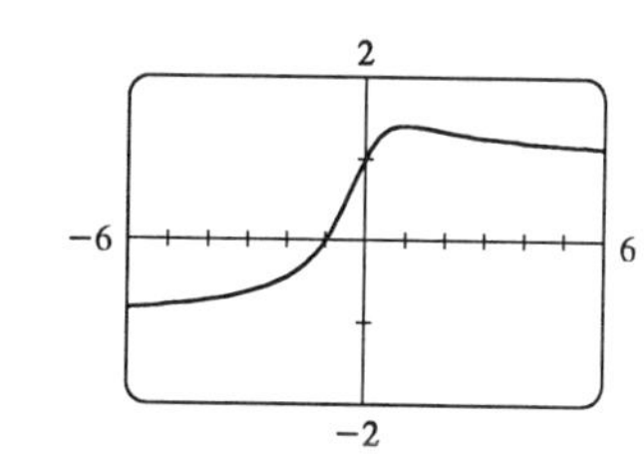

Tracing the graph gives us an estimate of $f(1) = 1.41$ as a local maximum, and no local minimum. $f(x) = \dfrac{x+1}{\sqrt{x^2+1}} \Longrightarrow$

$f'(x) = \dfrac{1-x}{(x^2+1)^{3/2}}$. $f'(x) = 0 \Longleftrightarrow x = 1$.

$f(1) = \frac{2}{\sqrt{2}} = \sqrt{2}$ is the exact value.

(b) From the graph in part (a), f increases most rapidly somewhere between $x = -\frac{1}{2}$ and $x = -\frac{1}{4}$. To find the exact value, we need to find the maximum value of f', which we can do by finding the critical numbers of f'. $f''(x) = \dfrac{2x^2 - 3x - 1}{(x^2+1)^{5/2}} = 0 \Longleftrightarrow x = \dfrac{3 \pm \sqrt{17}}{4}$. $x = \dfrac{3+\sqrt{17}}{4}$ corresponds to the *minimum* value of f'. The maximum value of f' is at $\left(\frac{3-\sqrt{17}}{4}, \sqrt{\frac{7}{6} - \frac{\sqrt{17}}{6}}\right) \approx (-0.28, 0.69)$.

31.

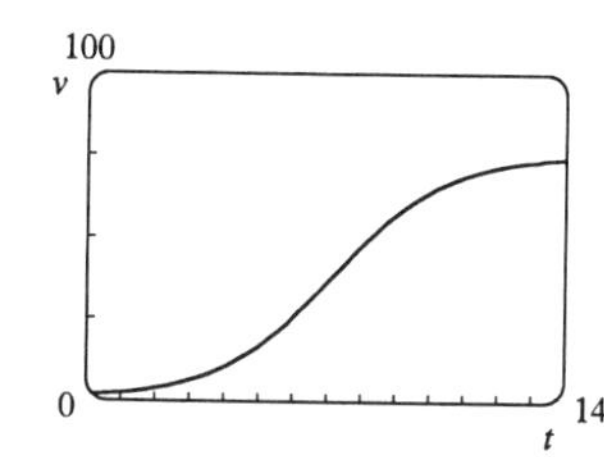

From the graph, we estimate that the most rapid increase in the number of VCRs occurs at about $t = 7$. To maximize the first derivative, we need to determine the values for which the second derivative is 0.

$$V(t) = \frac{75}{1+74e^{-0.6t}} \Longrightarrow V'(t) = -\frac{75\left[74e^{-0.6t}(-0.6)\right]}{(1+74e^{-0.6t})^2} = \frac{3330e^{-0.6t}}{(1+74e^{-0.6t})^2} \Longrightarrow$$

$$V''(t) = \frac{(1+74e^{-0.6t})^2\left[3330e^{-0.6t}(-0.6)\right] - (3330e^{-0.6t})\,2\,(1+74e^{-0.6t})\left[74e^{-0.6t}(-0.6)\right]}{\left[(1+74e^{-0.6t})^2\right]^2}$$

$$= \frac{(1+74e^{-0.6t})\left[3330e^{-0.6t}(-0.6)\right]\left[(1+74e^{-0.6t}) - 2(74e^{-0.6t})\right]}{(1+74e^{-0.6t})^4}$$

$$= \frac{-1998e^{-0.6t}(1-74e^{-0.6t})}{(1+74e^{-0.6t})^3}$$

$V''(t) = 0 \Longleftrightarrow 1 = 74e^{-0.6t} \Longleftrightarrow e^{0.6t} = 74 \Longleftrightarrow 0.6t = \ln 74 \Longleftrightarrow t = \frac{5}{3}\ln 74 \approx 7.173$ years, which corresponds to early September 1987.

33. In Maple, we define f and then use the command `plot(diff(diff(f,x),x),x=-3..3);`. In Mathematica, we define f and then use `Plot[Dt[Dt[f,x],x],{x,-3,3}]`.We see that $f'' > 0$ for $x > 0.1$ and $f'' < 0$ for $x < 0.1$. So f is concave up on $(0.1, \infty)$ and concave down on $(-\infty, 0.1)$.

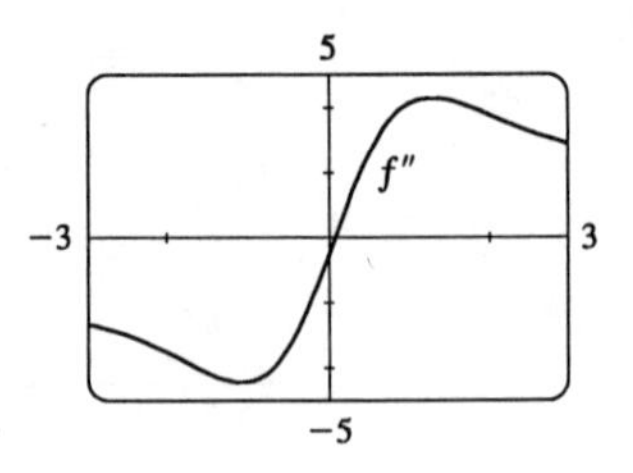

35. $f(x) = ax^3 + bx^2 + cx + d \Longrightarrow f(1) = a + b + c + d = 0$, $f(-2) = -8a + 4b - 2c + d = 3$, and $f'(x) = 3ax^2 + 2bx + c$. Also $f'(1) = 3a + 2b + c = 0$ and $f'(-2) = 12a - 4b + c = 0$ by Fermat's Theorem. Solving these four equations, we get $a = \frac{2}{9}$, $b = \frac{1}{3}$, $c = -\frac{4}{3}$, $d = \frac{7}{9}$, so the function is $f(x) = \frac{1}{9}\left(2x^3 + 3x^2 - 12x + 7\right)$.

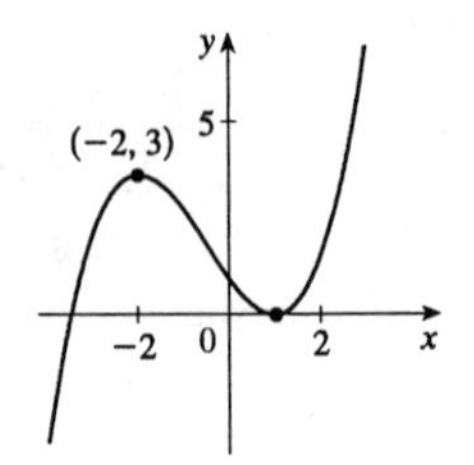

37. If f and g are CU on I, then $f'' > 0$ and $g'' > 0$ on I, so $(f + g)'' = f'' + g'' > 0$ on $I \Longrightarrow f + g$ is CU on I.

39. Since f and g are positive, increasing, and CU on I, we have $f > 0$, $f' > 0$, $f'' > 0$, $g > 0$, $g' > 0$, $g'' > 0$ on I. Then $(fg)' = fg' + f'g \Longrightarrow (fg)'' = (fg'' + g'f') + (gf'' + f'g') = fg'' + 2f'g' + f''g > 0 \Longrightarrow fg$ is CU on I.

41. $f(x) = \tan x - x \Longrightarrow f'(x) = \sec^2 x - 1 > 0$ for $0 < x < \frac{\pi}{2}$ since $\sec^2 x > 1$ for $0 < x < \frac{\pi}{2}$. So f is increasing on $\left(0, \frac{\pi}{2}\right)$. Thus, $f(x) > f(0) = 0$ for $0 < x < \frac{\pi}{2} \Longrightarrow \tan x - x > 0 \Longrightarrow \tan x > x$ for $0 < x < \frac{\pi}{2}$.

43. We are given that f is differentiable (and therefore continuous) everywhere. In particular, we can apply the Mean Value Theorem on the interval $[0, 4]$. There exists a number c in $(0, 4)$ such that $f(4) - f(0) = f'(c)(4 - 0)$, so $f(4) = f(0) + 4f'(c) = -3 + 4f'(c)$. We are given that $f'(x) \le 5$ for all x, so in particular we know that $f'(c) \le 5$. Multiplying both sides of this inequality by 4, we have $4f'(c) \le 20$, so $f(4) = -3 + 4f'(c) \le -3 + 20 = 17$. The largest possible value for $f(4)$ is 17.

45. Let $g(t)$ and $h(t)$ be the position functions of the two runners and let $f(t) = g(t) - h(t)$. By hypothesis, $f(0) = g(0) - h(0) = 0$ and $f(b) = g(b) - h(b) = 0$, where b is the finishing time. Then by the Mean Value Theorem, there is a time c, with $0 < c < b$, such that $f'(c) = \dfrac{f(b) - f(0)}{b - 0}$. But $f(b) = f(0) = 0$, so $f'(c) = 0$. Since $f'(c) = g'(c) - h'(c) = 0$, we have $g'(c) = h'(c)$. So at time c, both runners have the same velocity $g'(c) = h'(c)$.

47. Let the cubic function be $f(x) = ax^3 + bx^2 + cx + d \implies f'(x) = 3ax^2 + 2bx + c \implies$

$f''(x) = 6ax + 2b$. So f is CU when $6ax + 2b > 0 \iff x > -\frac{b}{3a}$, and CD when

$x < -\frac{b}{3a}$, and so the only point of inflection occurs when $x = -\frac{b}{3a}$. If the graph

has three x-intercepts x_1, x_2 and x_3, then the equation of $f(x)$ must factor as

$f(x) = a(x - x_1)(x - x_2)(x - x_3) = a\left[x^3 - (x_1 + x_2 + x_3)x^2 + (x_1x_2 + x_1x_3 + x_2x_3)x - x_1x_2x_3\right]$.

So $b = -a(x_1 + x_2 + x_3)$. Hence, the x-coordinate of the point of inflection is

$$-\frac{b}{3a} = -\frac{-a(x_1 + x_2 + x_3)}{3a} = \frac{x_1 + x_2 + x_3}{3}.$$

EXERCISES 5.4

Abbreviations:

HA	horizontal asymptote(s)	VA	vertical asymptote(s)
CU	concave upward	CD	concave downward
IP	inflection point(s)	FDT	First Derivative Test

1. $f(x) = 4x^4 - 7x^2 + 4x + 6 \implies f'(x) = 16x^3 - 14x + 4 \implies f''(x) = 48x^2 - 14$

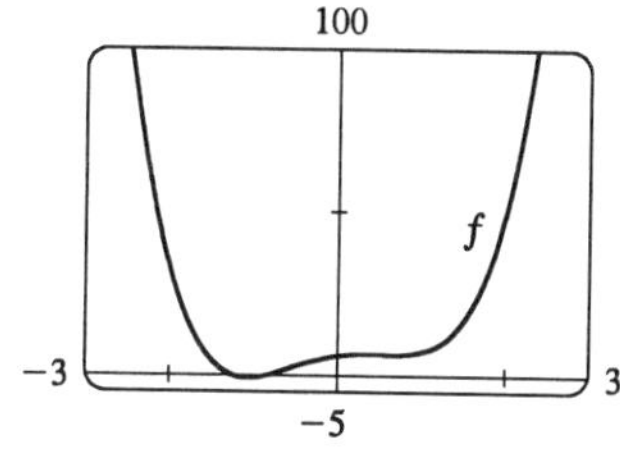

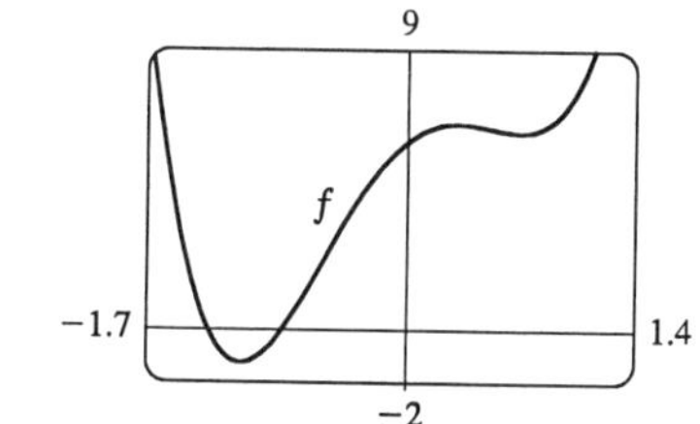

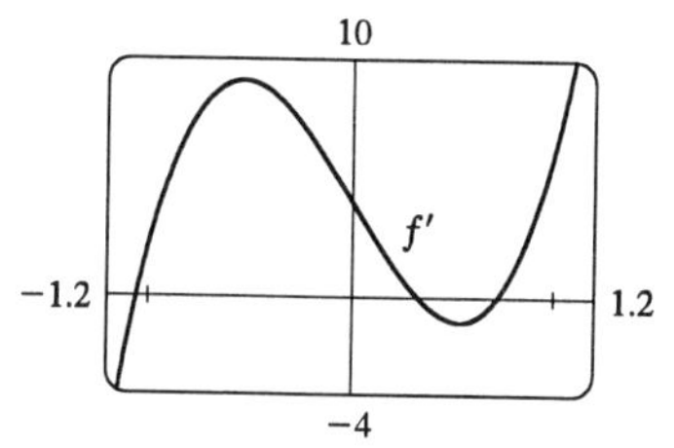

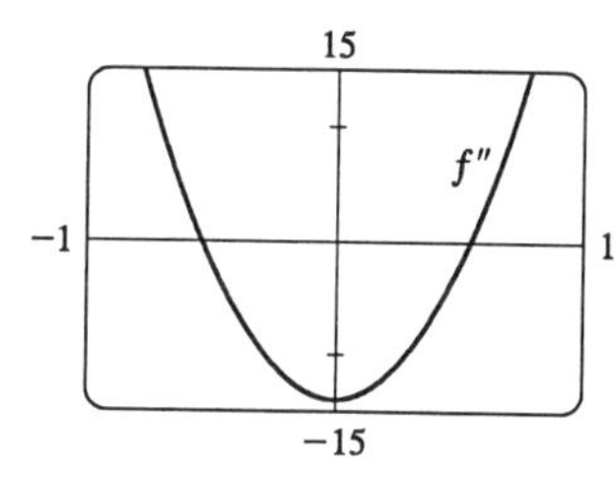

After finding suitable viewing rectangles (by ensuring that we have located all of the x-values where either $f' = 0$ or $f'' = 0$) we estimate from the graph of f' that f is increasing on $(-1.1, 0.3)$ and $(0.7, \infty)$ and decreasing on $(-\infty, -1.1)$ and $(0.3, 0.7)$, with a local maximum of $f(0.3) \approx 6.6$ and minima of $f(-1.1) \approx -1.0$ and $f(0.7) \approx 6.3$. We estimate from the graph of f'' that f is CU on $(-\infty, -0.5)$ and $(0.5, \infty)$ and CD on $(-0.5, 0.5)$, and that f has inflection points at about $(-0.5, 2.5)$ and $(0.5, 6.5)$.

3. $f(x) = \sqrt[3]{x^2 - 3x - 5} \Longrightarrow f'(x) = \dfrac{1}{3}\dfrac{2x - 3}{(x^2 - 3x - 5)^{2/3}} \Longrightarrow f''(x) = -\dfrac{2}{9}\dfrac{x^2 - 3x + 24}{(x^2 - 3x - 5)^{5/3}}$

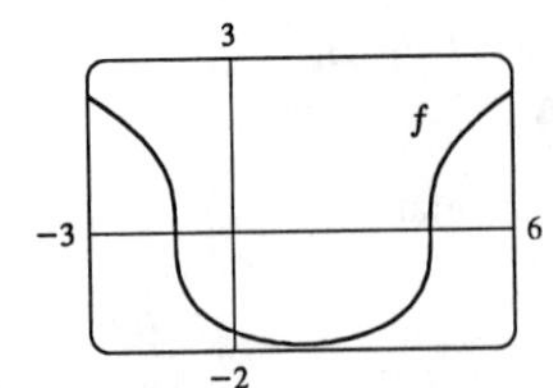

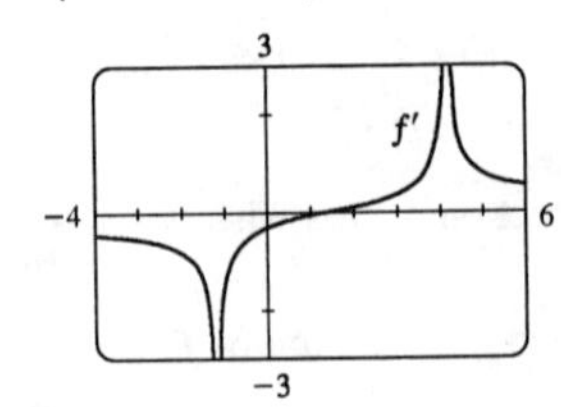

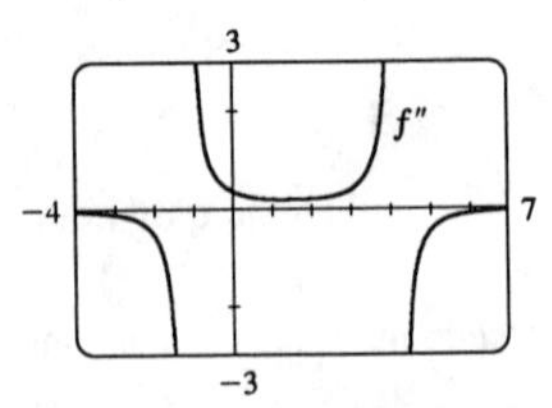

Note: With some CAS's, including Maple, it is necessary to define $f(x) = \dfrac{x^2 - 3x - 5}{|x^2 - 3x - 5|}\left|x^2 - 3x - 5\right|^{1/3}$, since the CAS does not compute real cube roots of negative numbers. We estimate from the graph of f' that f is increasing on $(1.5, 4.2)$ and $(4.2, \infty)$, and decreasing on $(-\infty, -1.2)$ and $(-1.2, 1.5)$. f has no maximum. Minimum: $f(1.5) \approx -1.9$. From the graph of f'', we estimate that f is CU on $(-1.2, 4.2)$ and CD on $(-\infty, -1.2)$ and $(4.2, \infty)$. IP $(-1.2, 0)$ and $(4.2, 0)$.

5. $f(x) = x^2 \sin x \Longrightarrow f'(x) = 2x \sin x + x^2 \cos x \Longrightarrow f''(x) = 2 \sin x + 4x \cos x - x^2 \sin x$

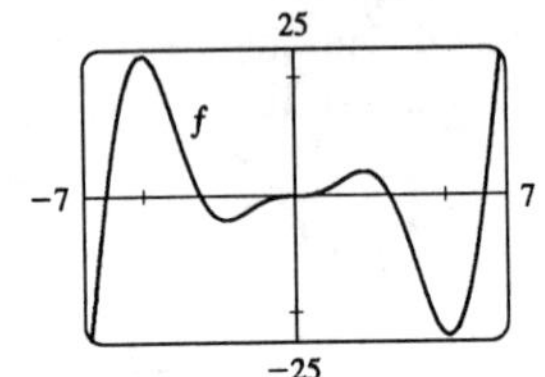

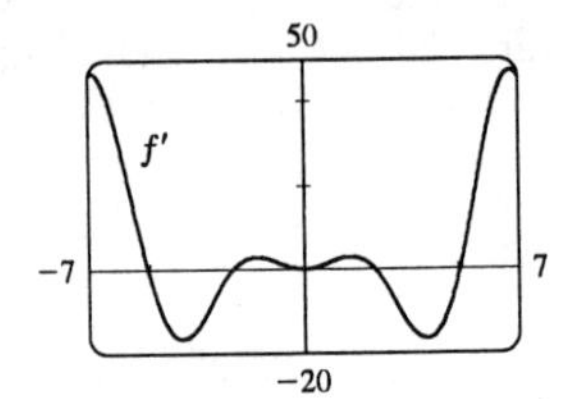

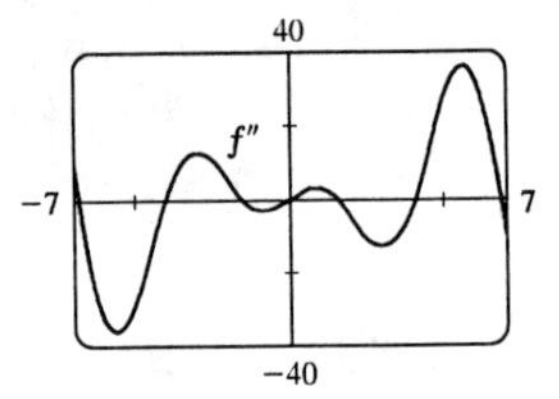

We estimate from the graph of f' that f is increasing on $(-7, -5.1)$, $(-2.3, 2.3)$, and $(5.1, 7)$ and decreasing on $(-5.1, -2.3)$, and $(2.3, 5.1)$. Local maxima: $f(-5.1) \approx 24.1$, $f(2.3) \approx 3.9$. Local minima: $f(-2.3) \approx -3.9$, $f(5.1) \approx -24.1$. From the graph of f'', we estimate that f is CU on $(-7, -6.8)$, $(-4.0, -1.5)$, $(0, 1.5)$, and $(4.0, 6.8)$, and CD on $(-6.8, -4.0)$, $(-1.5, 0)$, $(1.5, 4.0)$, and $(6.8, 7)$. f has IP at $(-6.8, -24.4)$, $(-4.0, 12.0)$, $(-1.5, -2.3)$, $(0, 0)$, $(1.5, 2.3)$, $(4.0, -12.0)$ and $(6.8, 24.4)$.

7.

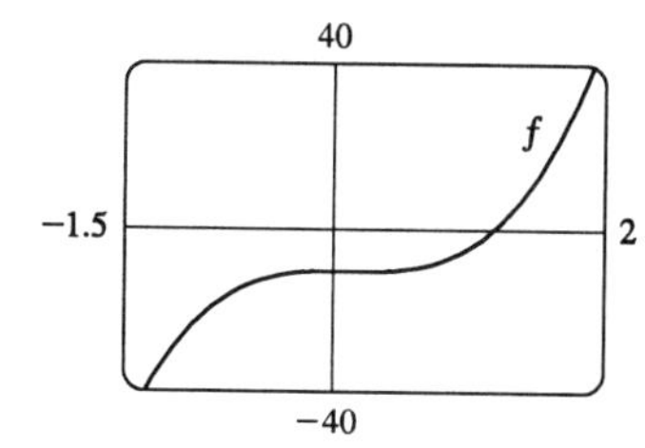

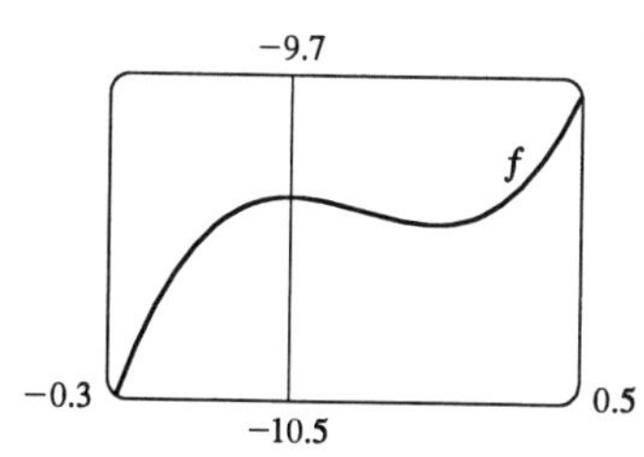

From the graphs, it appears that $f(x) = 8x^3 - 3x^2 - 10$ increases on $(-\infty, 0)$ and $(0.25, \infty)$ and decreases on $(0, 0.25)$; that f has a local maximum of $f(0) = -10.0$ and a local minimum of $f(0.25) \approx -10.1$; that f is CU on $(0.1, \infty)$ and CD on $(-\infty, 0.1)$; and that f has an IP at $(0.1, -10)$. $f(x) = 8x^3 - 3x^2 - 10 \Longrightarrow$ $f'(x) = 24x^2 - 6x = 6x(4x-1)$, which is positive ($f$ is increasing) for $(-\infty, 0)$ and $\left(\frac{1}{4}, \infty\right)$, and negative ($f$ is decreasing) on $\left(0, \frac{1}{4}\right)$. By the FDT, f has a local maximum at $x = 0$: $f(0) = -10$; and f has a local minimum at $\frac{1}{4}$: $f\left(\frac{1}{4}\right) = \frac{1}{8} - \frac{3}{16} - 10 = -\frac{161}{16}$. $f'(x) = 24x^2 - 6x \Longrightarrow f''(x) = 48x - 6 = 6(8x-1)$, which is positive ($f$ is CU) on $\left(\frac{1}{8}, \infty\right)$, and negative ($f$ is CD) on $\left(-\infty, \frac{1}{8}\right)$. f has an IP at $\left(\frac{1}{8}, f\left(\frac{1}{8}\right)\right) = \left(\frac{1}{8}, -\frac{321}{32}\right)$.

9.

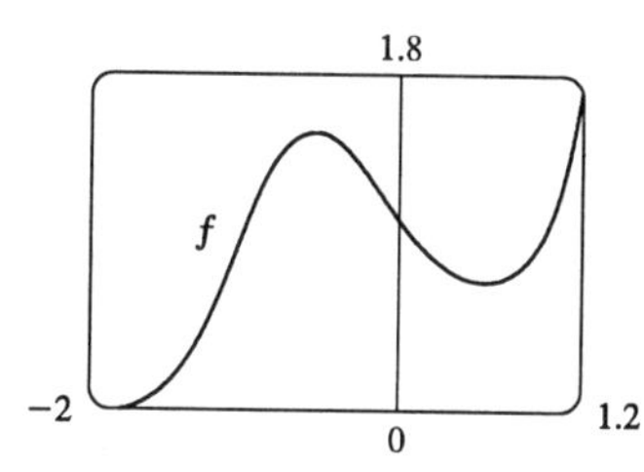

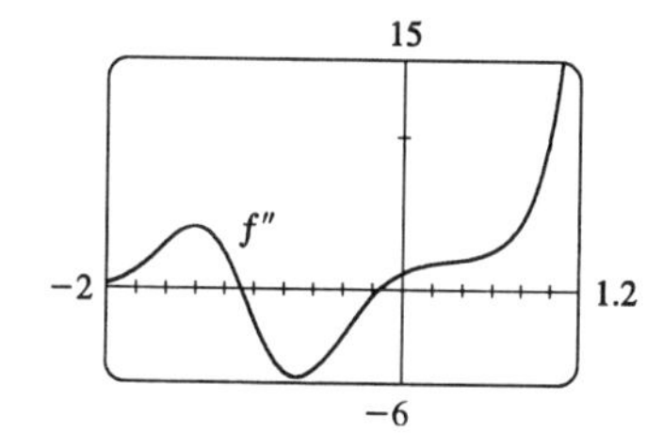

$f(x) = e^{x^3 - x} \to 0$ as $x \to -\infty$, and $f(x) \to \infty$ as $x \to \infty$. From the graph, it appears that f has a local minimum of about $f(0.58) = 0.68$, and a local maximum of about $f(-0.58) = 1.47$.

To find the exact values, we calculate $f'(x) = \left(3x^2 - 1\right)e^{x^3 - x}$, which is 0 when $3x^2 - 1 = 0 \Longleftrightarrow x = \pm\frac{1}{\sqrt{3}}$. The negative root corresponds to the local maximum $f\left(-\frac{1}{\sqrt{3}}\right) = e^{(-1/\sqrt{3})^3 - (-1/\sqrt{3})} = e^{2\sqrt{3}/9}$, and the positive root corresponds to the local minimum $f\left(\frac{1}{\sqrt{3}}\right) = e^{(1/\sqrt{3})^3 - (1/\sqrt{3})} = e^{-2\sqrt{3}/9}$. To estimate the inflection points, we calculate and graph

$$f''(x) = \frac{d}{dx}\left[\left(3x^2 - 1\right)e^{x^3 - x}\right] = \left(3x^2 - 1\right)e^{x^3 - x}\left(3x^2 - 1\right) + e^{x^3 - x}(6x) = e^{x^3 - x}\left(9x^4 - 6x^2 + 6x + 1\right).$$

From the graph, it appears that $f''(x)$ changes sign (and thus f has inflection points) at $x \approx -0.15$ and $x \approx -1.09$. From the graph of f, we see that these x-values correspond to inflection points at about $(-0.15, 1.15)$ and $(-1.09, 0.82)$.

11. 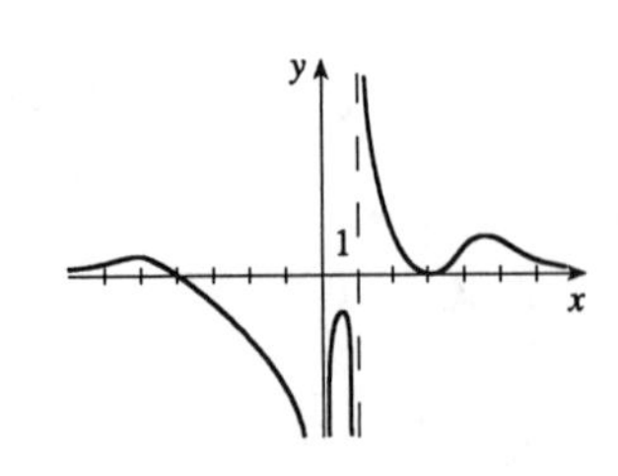

$f(x) = \dfrac{(x+4)(x-3)^2}{x^4(x-1)}$ has VA at $x = 0$ and at $x = 1$ since

$\lim\limits_{x\to 0} f(x) = -\infty$, $\lim\limits_{x\to 1^-} f(x) = -\infty$ and $\lim\limits_{x\to 1^+} f(x) = \infty$.

$f(x) = \dfrac{(1+4/x)(1-3/x)^2}{x(x-1)} \to 0^+$ as $x \to \pm\infty$, so f is asymptotic to the x-axis. Since f is undefined at $x = 0$, it has no y-intercept.

$f(x) = 0 \Longrightarrow (x+4)(x-3)^2 = 0 \Longrightarrow x = -4$ or $x = 3$, so f has x-intercepts -4 and 3. Note, however, that the graph of f is only tangent to the x-axis and does not cross it at $x = 3$, since f is positive as $x \to 3^-$ and as $x \to 3^+$.

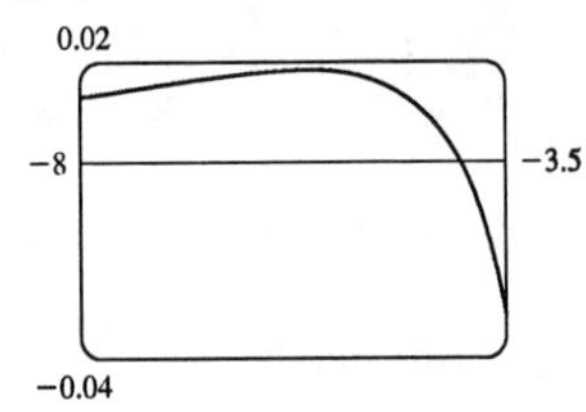

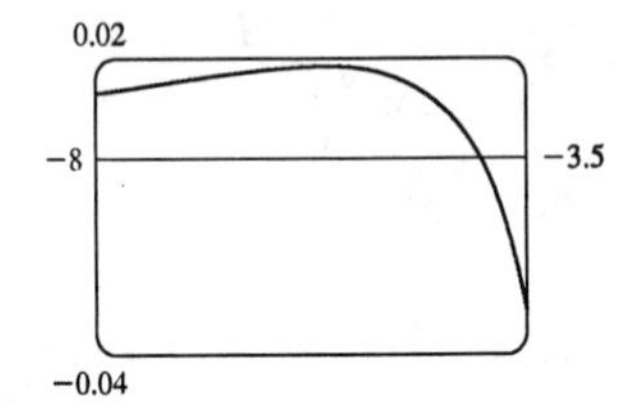

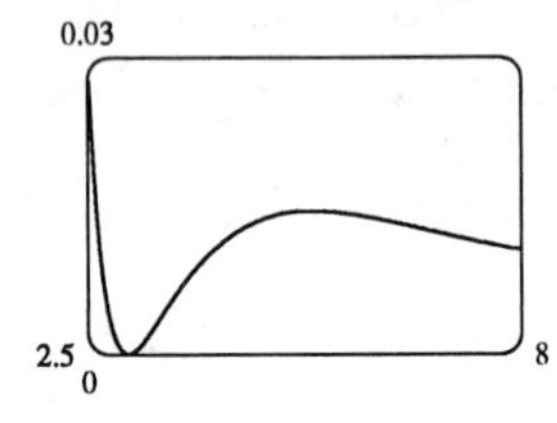

From these graphs, it appears that f has three maxima and one minimum. The maxima are approximately $f(-5.6) = 0.0182$, $f(0.82) = -281.5$ and $f(5.2) = 0.0145$ and we know (since the graph is tangent to the x-axis at $x = 3$) that the minimum is $f(3) = 0$.

13. $f(x) = \dfrac{x^2(x+1)^3}{(x-2)^2(x-4)^4} \Longrightarrow f'(x) = -\dfrac{x(x+1)^2(x^3+18x^2-44x-16)}{(x-2)^3(x-4)^5}$ (from CAS).

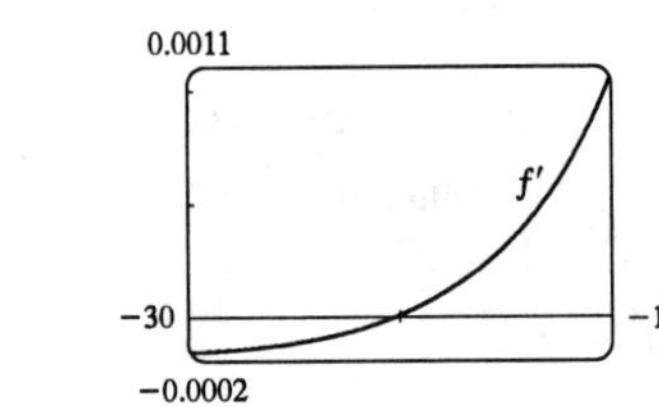

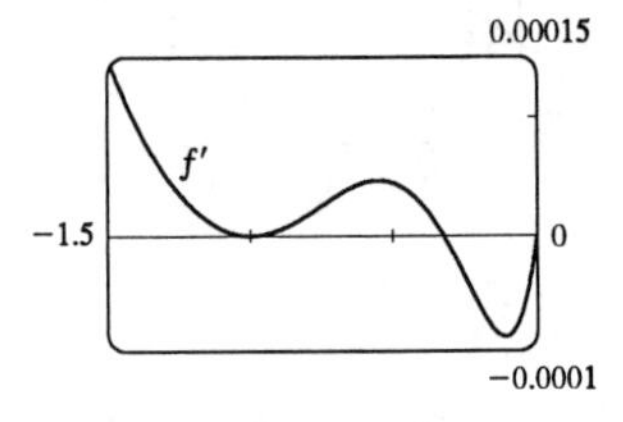

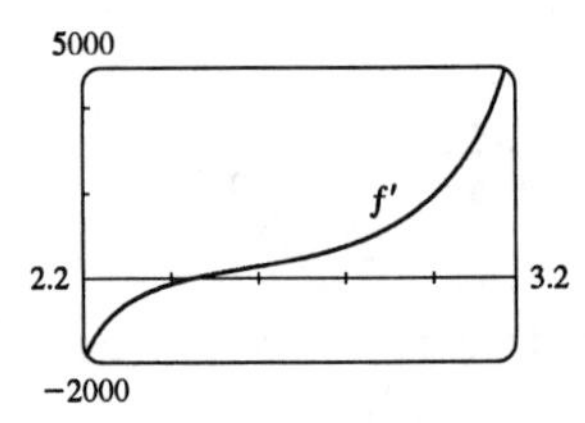

From the graphs of f', it seems that the critical points which indicate extrema occur at $x \approx -20$, -0.3, and 2.5, as estimated in Example 3. (There is another critical point at $x = -1$, but the sign of f' does not change there.)

We differentiate again, obtaining $f''(x) = 2\dfrac{(x+1)(x^6+36x^5+6x^4-628x^3+684x^2+672x+64)}{(x-2)^4(x-4)^6}$.

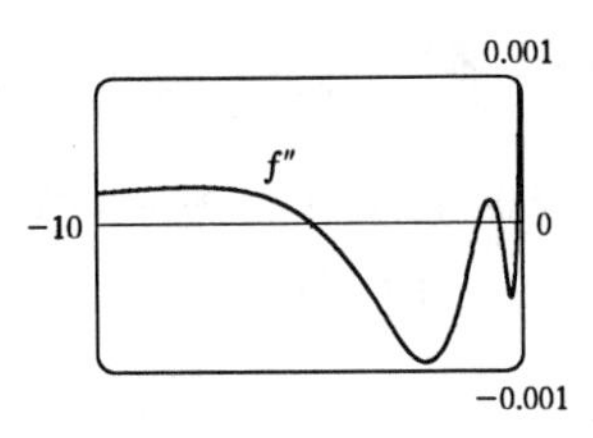

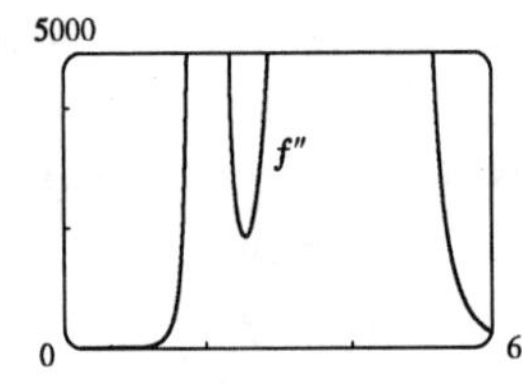

From the graphs of f'', it appears that f is CU on $(-\infty, -5.0)$, $(-1.0, -0.5)$, $(-0.1, 2.0)$, $(2.0, 4.0)$ and $(4.0, \infty)$ and CD on $(-5.0, -1.0)$ and $(-0.5, -0.1)$. We check back on the graphs of f to find the y-coordinates of the inflection points, and find that these points are approximately $(-5, -0.005)$, $(-1, 0)$, $(-0.5, 0.00001)$, and $(-0.1, 0.0000066)$.

15.

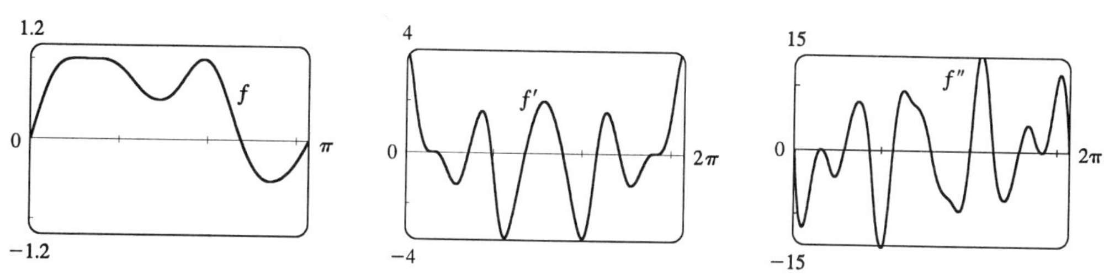

From the graph of $f(x) = \sin(x + \sin 3x)$ in the viewing rectangle $[0, \pi]$ by $[-1.2, 1.2]$, it looks like f has two maxima and two minima. If we calculate and graph $f'(x) = [\cos(x + \sin 3x)](1 + 3\cos 3x)$ on $[0, 2\pi]$, we see that the graph of f' appears to be almost tangent to the x-axis at about $x = 0.7$. The graph of $f'' = -[\sin(x + \sin 3x)](1 + 3\cos 3x)^2 + \cos(x + \sin 3x)(-9\sin 3x)$ is even more interesting near this x-value: it seems to just touch the x-axis.

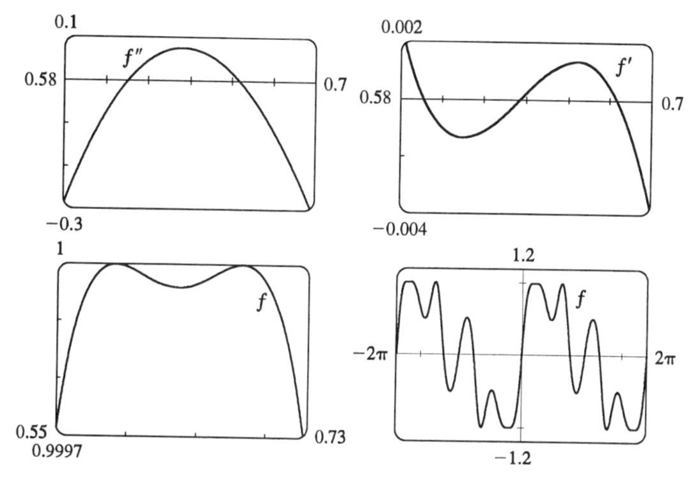

If we zoom in on this place on the graph of f'', we see that f'' actually does cross the axis twice near $x = 0.65$, indicating a change in concavity for a very short interval. If we look at the graph of f' on the same interval, we see that it changes sign three times near $x = 0.65$, indicating that what we had thought was a broad extremum at about $x = 0.7$ actually consists of three extrema (two maxima and a minimum). These maxima are roughly $f(0.59) = 1$ and $f(0.68) = 1$, and the minimum is roughly $f(0.64) = 0.99996$. There are also a maximum of about $f(1.96) = 1$ and minima of about $f(1.46) = 0.49$ and $f(2.73) = -0.51$. The points of inflection on $(0, \pi)$ are about $(0.61, 0.99998)$, $(0.66, 0.99998)$, $(1.17, 0.72)$, $(1.75, 0.77)$, and $(2.28, 0.34)$. On $(\pi, 2\pi)$, they are about $(4.01, -0.34)$, $(4.54, -0.77)$, $(5.11, -0.72)$, $(5.62, -0.99998)$, and $(5.67, -0.99998)$. There are also IP at $(0, 0)$ and $(\pi, 0)$. Note that the function is odd and periodic with period 2π, and it is also rotationally symmetric about all points of the form $((2n + 1)\pi, 0)$, n an integer.

17.

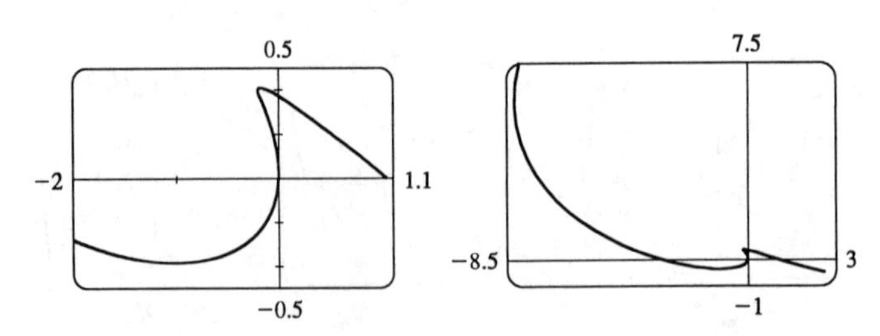

We graph the curve $x = t^4 - 2t^3 - 2t^2$, $y = t^3 - t$ in the viewing rectangle $[-2, 1.1]$ by $[-0.5, 0.5]$. This rectangle corresponds approximately to $t \in [-1, 0.8]$. We estimate that the curve has horizontal tangents at about $(-1, -0.4)$ and $(-0.17, 0.39)$ and vertical tangents at about $(0, 0)$ and $(-0.19, 0.37)$. We calculate $\frac{dy}{dx} = \frac{dy/dt}{dx/dt} = \frac{3t^2 - 1}{4t^3 - 6t^2 - 4t}$. The horizontal tangents occur when $dy/dt = 3t^2 - 1 = 0 \iff t = \pm\frac{1}{\sqrt{3}}$, so both horizontal tangents are shown in our graph. $t = \frac{1}{\sqrt{3}}$ corresponds to the point $\left(\frac{-2\sqrt{3}-5}{9}, \frac{-2\sqrt{3}}{9}\right) \approx (-0.94, -0.38)$ and $t = -\frac{1}{\sqrt{3}}$ corresponds to $\left(\frac{2\sqrt{3}-5}{9}, \frac{2\sqrt{3}}{9}\right) \approx (-0.17, 0.38)$. The vertical tangents occur when $dx/dt = 2t\left(2t^2 - 3t - 2\right) = 0 \iff 2t\left(2t + 1\right)\left(t - 2\right) = 0 \iff t = 0, -\frac{1}{2}$ or 2. It seems that we have missed one vertical tangent, and indeed if we plot the curve on the t-interval $[-1.2, 2.2]$ we see that there is another vertical tangent at $(-8, 6)$. The t-values and points at which there are vertical tangents are $t = 0$, $(0, 0)$; $t = -\frac{1}{2}$, $\left(-\frac{3}{16}, \frac{3}{8}\right)$; and $t = 2$, $(-8, 6)$.

19. $x = t^3 - ct$, $y = t^2$. For $c = 0$, there is a cusp at $(0, 0)$. For $c < 0$, there is a local minimum at $(0, 0)$. For $c > 0$, there is a loop whose size increases as c increases ($c = \frac{1}{2}$ and $c = 1$ are shown in the figure). The curve intersects itself at the point $(0, c)$ when $t = \pm\sqrt{c}$ (solve $x = 0$.)

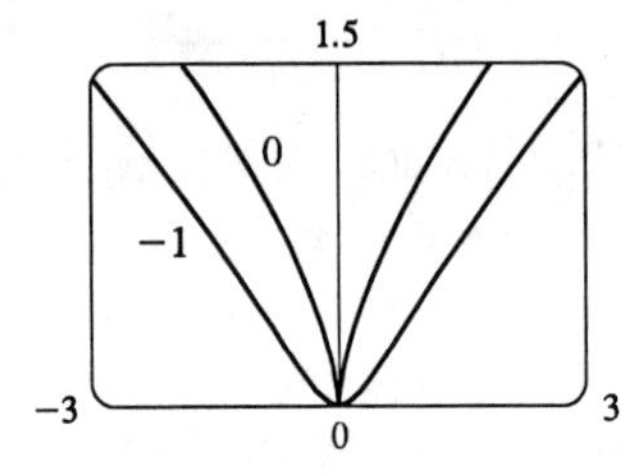

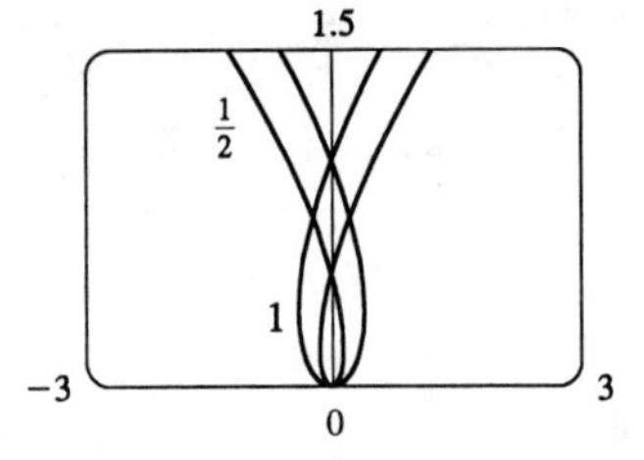

From the second figure, we see that the left- and rightmost points of the loop occur when there are vertical tangent lines. $dx/dt = 0 \implies 3t^2 - c = 0 \implies t = \pm\sqrt{c/3}$. The rightmost point occurs when $t = -\sqrt{c/3}$ and has coordinates $\left(\frac{2c\sqrt{3c}}{9}, \frac{c}{3}\right)$. The leftmost point occurs when $t = \sqrt{c/3}$ and has coordinates $\left(-\frac{2c\sqrt{3c}}{9}, \frac{c}{3}\right)$.

21. Note that $c = 0$ is a transitional value at which the graph consists of the x-axis. Also, we can see that if we substitute $-c$ for c, the function $f(x) = \dfrac{cx}{1 + c^2x^2}$ will be reflected in the x-axis, so we investigate only positive values of c (except $c = -1$, as a demonstration of this reflective property). Also, f is an odd function. $\lim\limits_{x \to \pm\infty} f(x) = 0$, so $y = 0$ is a horizontal asymptote for all c. We calculate

$$f'(x) = \frac{c\left(1 + c^2x^2\right) - cx\left(2c^2x\right)}{\left(1 + c^2x^2\right)^2} = -\frac{c\left(c^2x^2 - 1\right)}{\left(1 + c^2x^2\right)^2}.\ f'(x) = 0 \iff c^2x^2 - 1 = 0 \iff x = \pm 1/c.\ \text{So}$$

there is an absolute maximum of $f(1/c) = \frac{1}{2}$ and an absolute minimum of $f(-1/c) = -\frac{1}{2}$. These extrema have the same value regardless of c, but the maximum points move closer to the y-axis as c increases.

$$\begin{aligned} f''(x) &= \frac{\left(-2c^3x\right)\left(1 + c^2x^2\right)^2 - \left(-c^3x^2 + c\right)\left[2\left(1 + c^2x^2\right)\left(2c^2x\right)\right]}{\left(1 + c^2x^2\right)^4} \\ &= \frac{\left(-2c^3x\right)\left(1 + c^2x^2\right) + \left(c^3x^2 - c\right)\left(4c^2x\right)}{\left(1 + c^2x^2\right)^3} = \frac{2c^3x\left(c^2x^2 - 3\right)}{\left(1 + c^2x^2\right)^3}. \end{aligned}$$

$f''(x) = 0 \iff x = 0$ or $\pm\sqrt{3}/c$, so there are inflection points at $(0, 0)$ and at $\left(\pm\sqrt{3}/c, \pm\sqrt{3}/4\right)$. Again, the y-coordinate of the inflection points does not depend on c, but as c increases, both inflection points approach the y-axis.

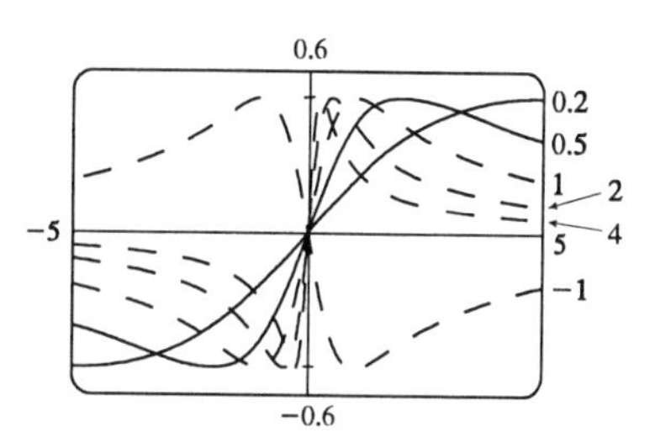

23.

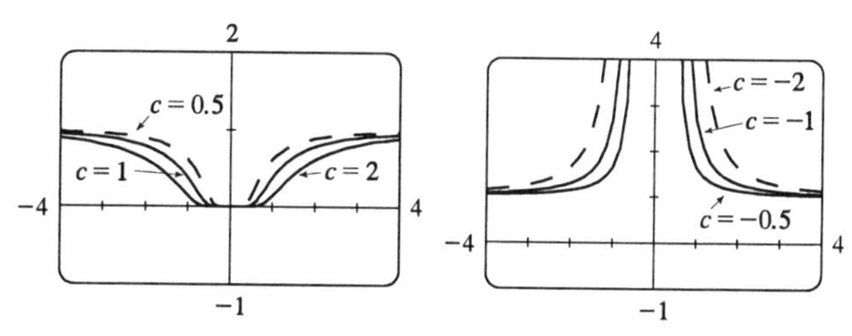

$c = 0$ is a transitional value — we get the graph of $y = 1$. For $c > 0$, we see that there is a HA at $y = 1$, and that the graph spreads out as c increases. At first glance there appears to be a minimum at $(0, 0)$, but $f(0)$ is undefined, so there is no minimum or maximum.For $c < 0$, we still have the HA at $y = 1$, but the range is $(1, \infty)$ rather than $(0, 1)$. We also have a VA at $x = 0$. $f(x) = e^{-c/x^2} \implies$

$f'(x) = e^{-c/x^2}\left(-2c/x^3\right) \implies f''(x) = \dfrac{2c\left(2c - 3x^2\right)}{x^6e^{c/x^2}}$. $f'(x) \neq 0$ and $f'(x)$ exists for all $x \neq 0$ (and 0 is not in the domain of f), so there are no maxima or minima. $f''(x) = 0 \implies x = \pm\sqrt{2c/3}$, so if $c > 0$, the inflection points spread out as c increases, and if $c < 0$, there are no IP. For $c > 0$, there are IP at $\left(\pm\sqrt{2c/3}, e^{-3/2}\right)$. Note that the y-coordinate of the IP is constant.

25. $f(x) = x^4 + cx^2 = x^2(x^2 + c)$. Note that f is an even function. For $c \geq 0$, the only x-intercept is the point $(0, 0)$. We calculate $f'(x) = 4x^3 + 2cx = 4x\left(x^2 + \frac{1}{2}c\right) \Longrightarrow f''(x) = 12x^2 + 2c$. If $c \geq 0$, $x = 0$ is the only critical point and there are no inflection points. As we can see from the examples, there is no change in the basic shape of the graph for $c \geq 0$; it merely becomes steeper as c increases. For $c = 0$, the graph is the simple curve $y = x^4$.

For $c < 0$, there are x-intercepts at 0 and at $\pm\sqrt{-c}$. Also, there is a maximum at $(0, 0)$, and there are minima at $\left(\pm\sqrt{-\frac{1}{2}c}, -\frac{1}{4}c^2\right)$. As $c \to -\infty$, the x-coordinates of these minima get larger in absolute value, and the minimum points move downward. There are inflection points at $\left(\pm\sqrt{-\frac{1}{6}c}, -\frac{5}{36}c^2\right)$, which also move away from the origin as $c \to -\infty$.

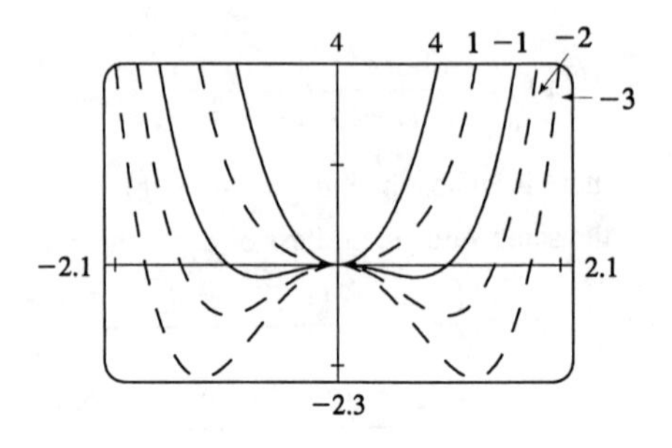

EXERCISES 5.5

1. If x is one number, the other is $100 - x$. Maximize $f(x) = x(100 - x) = 100x - x^2$. $f'(x) = 100 - 2x = 0$ $\Rightarrow$ $x = 50$. Now $f''(x) = -2 < 0$, so there is an absolute maximum at $x = 50$. The numbers are 50 and 50.

3. The two numbers are x and $\dfrac{100}{x}$ where $x > 0$. Minimize $f(x) = x + \dfrac{100}{x}$. $f'(x) = 1 - \dfrac{100}{x^2} = \dfrac{x^2 - 100}{x^2}$. The critical number is $x = 10$. Since $f'(x) < 0$ for $0 < x < 10$ and $f'(x) > 0$ for $x > 10$, there is an absolute minimum at $x = 10$. The numbers are 10 and 10.

5. Let p be the perimeter and x and y the lengths of the sides, so $p = 2x + 2y$ $\Rightarrow$ $y = \frac{1}{2}p - x$. The area is $A(x) = x\left(\frac{1}{2}p - x\right) = \frac{1}{2}px - x^2$. Now $0 = A'(x) = \frac{1}{2}p - 2x$ $\Rightarrow$ $x = \frac{1}{4}p$. Since $A''(x) = -2 < 0$, there is an absolute maximum where $x = \frac{1}{4}p$. The sides of the rectangle are $\frac{1}{4}p$ and $\frac{1}{2}p - \frac{1}{4}p = \frac{1}{4}p$, so the rectangle is a square.

7. Here $5x + 2y = 750$ so $y = (750 - 5x)/2$. Maximize $A = xy = x(750 - 5x)/2 = 375x - \frac{5}{2}x^2$. Now $A'(x) = 375 - 5x = 0$ $\Rightarrow$ $x = 75$. Since $A''(x) = -5 < 0$ there is an absolute maximum when $x = 75$. Then $y = \frac{375}{2}$. The largest area is $75\left(\frac{375}{2}\right) = 14{,}062.5\text{ ft}^2$.

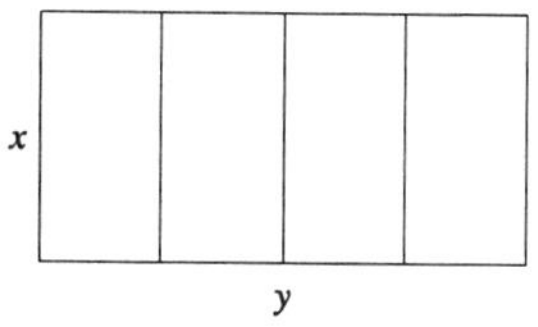

9. Let b be the base of the box and h the height. The surface area is $1200 = b^2 + 4hb$ $\Rightarrow$ $h = (1200 - b^2)/(4b)$. The volume is $V = b^2h = b^2(1200 - b^2)/4b = 300b - b^3/4$ $\Rightarrow$ $V'(b) = 300 - \frac{3}{4}b^2$. $V'(b) = 0$ $\Rightarrow$ $b = \sqrt{400} = 20$. Since $V'(b) > 0$ for $0 < b < 20$ and $V'(b) < 0$ for $b > 20$, there is an absolute maximum when $b = 20$. Then $h = 10$, so the largest possible volume is $(20)^2(10) = 4000\text{ cm}^3$.

11. $10 = (2w)(w)\,h = 2w^2h$, so $h = 5/w^2$. The cost is
$$C(w) = 10(2w^2) + 6[2(2wh) + 2hw] + 6(2w^2) = 32w^2 + 36wh = 32w^2 + 180/w.$$
$C'(w) = 64w - 180/w^2 = 4(16w^3 - 45)/w^2$ $\Rightarrow$ $w = \sqrt[3]{\frac{45}{16}}$ is the critical number. $C'(w) < 0$ for $0 < w < \sqrt[3]{\frac{45}{16}}$ and $C'(w) > 0$ for $w > \sqrt[3]{\frac{45}{16}}$. The minimum cost is $C\left(\sqrt[3]{\frac{45}{16}}\right) = 32(2.8125)^{2/3} + 180/\sqrt[3]{2.8125} \approx \$191.28$.

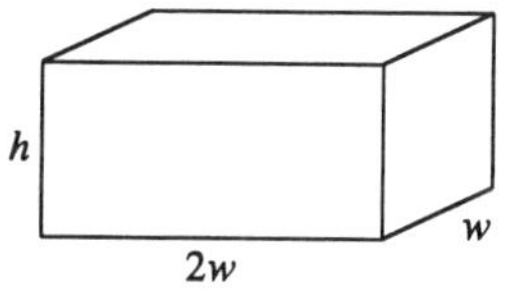

13. For (x, y) on the line $y = 2x - 3$, the distance to the origin is $\sqrt{(x-0)^2 + (2x-3)^2}$. We minimize the square of the distance, that is, $x^2 + (2x-3)^2 = 5x^2 - 12x + 9 = D(x)$. $D'(x) = 10x - 12 = 0$ $\Rightarrow$ $x = \frac{6}{5}$. Since there is a point closest to the origin, $x = \frac{6}{5}$ and hence $y = -\frac{3}{5}$. So the point is $\left(\frac{6}{5}, -\frac{3}{5}\right)$.

15. By symmetry, the points are (x, y) and $(x, -y)$, where $y > 0$. The square of the distance is $D(x) = (x-2)^2 + y^2 = (x-2)^2 + 4 + x^2 = 2x^2 - 4x + 8$. So $D'(x) = 4x - 4 = 0 \quad \Rightarrow \quad x = 1$ and $y = \pm\sqrt{4+1} = \pm\sqrt{5}$. The points are $\left(1, \pm\sqrt{5}\right)$.

17. $\mathbf{r}(t) = (t - \sin t)\mathbf{i} + (1 - \cos t)\mathbf{j} \quad \Rightarrow \quad \mathbf{r}'(t) = (1 - \cos t)\mathbf{i} + \sin t\,\mathbf{j}$.
speed $= |\mathbf{r}'(t)| = \sqrt{(1-\cos t)^2 + (\sin t)^2} = \sqrt{1 - 2\cos t + \cos^2 t + \sin^2 t} = \sqrt{2 - 2\cos t}$. The speed will be a maximum when $2 - 2\cos t$ is a maximum. Since $\cos t = -1$ when $t = \pi + 2\pi n$, $2 - 2\cos t$ has a maximum value of 4 when $t = (2n+1)\pi$. Thus, the maximum speed is $\sqrt{4} = 2$.

19.

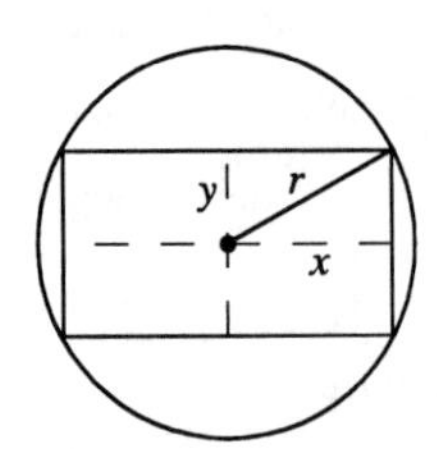

Area of rectangle is $4xy$. Also $r^2 = x^2 + y^2$ so $y = \sqrt{r^2 - x^2}$, so the area is $A(x) = 4x\sqrt{r^2 - x^2}$. Now

$$A'(x) = 4\left(\sqrt{r^2 - x^2} - \frac{x^2}{\sqrt{r^2 - x^2}}\right) = 4\frac{r^2 - 2x^2}{\sqrt{r^2 - x^2}}.$$

The critical number is $x = \frac{1}{\sqrt{2}}r$. Clearly this gives a maximum. The dimensions are $2x = \sqrt{2}r$ and $2y = \sqrt{2}r$.

21.

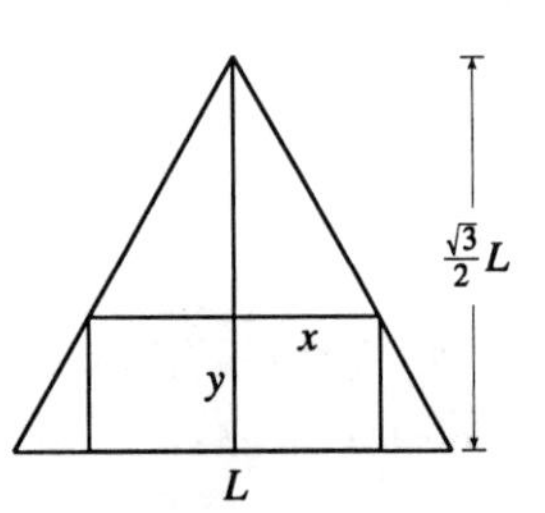

$$\frac{\frac{\sqrt{3}}{2}L - y}{x} = \frac{\frac{\sqrt{3}}{2}L}{L/2} = \sqrt{3} \text{ (similar triangles)} \Rightarrow$$

$\sqrt{3}x = \frac{\sqrt{3}}{2}L - y \quad \Rightarrow \quad y = \frac{\sqrt{3}}{2}(L - 2x)$. The area of the inscribed rectangle is $A(x) = (2x)y = \sqrt{3}x(L - 2x)$ where $0 \le x \le L/2$. Now $0 = A'(x) = \sqrt{3}L - 4\sqrt{3}x \quad \Rightarrow$ $x = \sqrt{3}L/(4\sqrt{3}) = L/4$. Since $A(0) = A(L/2) = 0$, the maximum occurs when $x = L/4$, and $y = \frac{\sqrt{3}}{2}L - \frac{\sqrt{3}}{4}L = \frac{\sqrt{3}}{4}L$, so the dimensions are $L/2$ and $\frac{\sqrt{3}}{4}L$.

23.

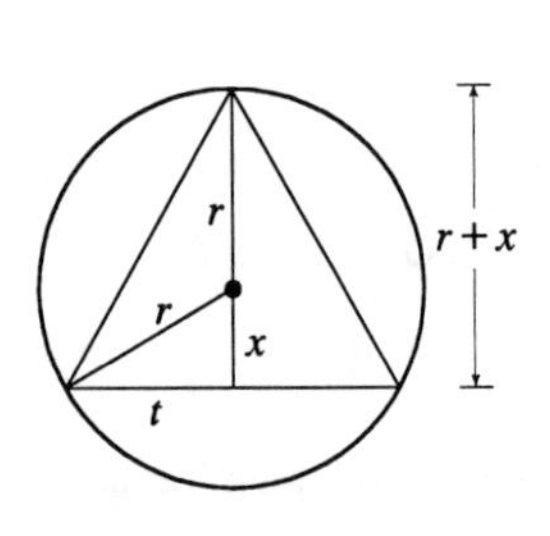

The area of the triangle is

$A(x) = \frac{1}{2}(2t)(r + x) = t(r + x) = \sqrt{r^2 - x^2}(r + x)$. Then

$$0 = A'(x) = r\frac{-2x}{2\sqrt{r^2 - x^2}} + \sqrt{r^2 - x^2} + x\frac{-2x}{2\sqrt{r^2 - x^2}}$$

$$= -\frac{x^2 + rx}{\sqrt{r^2 - x^2}} + \sqrt{r^2 - x^2} \quad \Rightarrow \quad \frac{x^2 + rx}{\sqrt{r^2 - x^2}} = \sqrt{r^2 - x^2}$$

$\Rightarrow \quad x^2 + rx = r^2 - x^2 \quad \Rightarrow \quad 0 = 2x^2 + rx - r^2 = (2x - r)(x + r)$

$\Rightarrow \quad x = \frac{1}{2}r$ or $x = -r$. Now $A(r) = 0 = A(-r) \quad \Rightarrow$ the maximum occurs where $x = \frac{1}{2}r$, so the triangle has height $r + \frac{1}{2}r = \frac{3}{2}r$ and base $2\sqrt{r^2 - \left(\frac{1}{2}r\right)^2} = 2\sqrt{\frac{3}{4}r^2} = \sqrt{3}r$.

25.

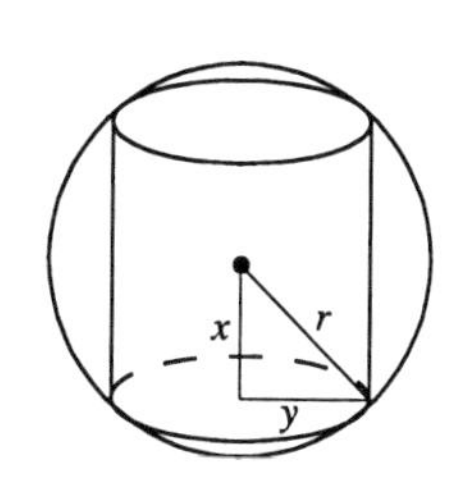

The cylinder has volume $V = \pi y^2(2x)$. Also $x^2 + y^2 = r^2 \Rightarrow y^2 = r^2 - x^2$, so $V(x) = \pi(r^2 - x^2)(2x) = 2\pi\,(r^2x - x^3)$, where $0 \le x \le r$. $V'(x) = 2\pi(r^2 - 3x^2) = 0 \Rightarrow x = r/\sqrt{3}$. Now $V(0) = V(r) = 0$, so there is a maximum when $x = r/\sqrt{3}$ and $V\left(r/\sqrt{3}\right) = 4\pi r^3/\left(3\sqrt{3}\right)$.

27.

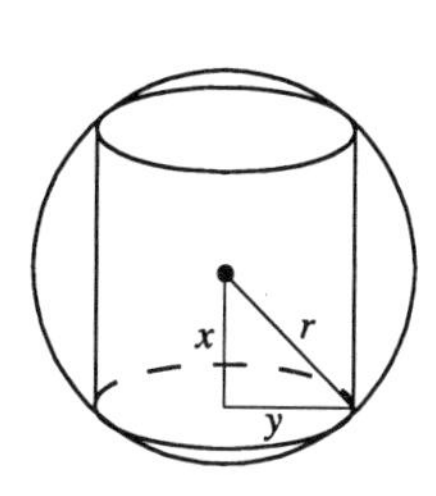

The cylinder has surface area $2\pi y^2 + 2\pi y(2x)$. Now $x^2 + y^2 = r^2 \Rightarrow y = \sqrt{r^2 - x^2}$, so the surface area is $S(x) = 2\pi(r^2 - x^2) + 4\pi x\sqrt{r^2 - x^2}$, $0 \le x \le r$. $S'(x) = -4\pi x + 4\pi\sqrt{r^2 - x^2} - 4\pi x^2/\sqrt{r^2 - x^2}$

$$= \frac{4\pi\left(r^2 - 2x^2 - x\sqrt{r^2 - x^2}\right)}{\sqrt{r^2 - x^2}} = 0 \Rightarrow x\sqrt{r^2 - x^2} = r^2 - 2x^2 \quad (\bigstar)$$

$$\Rightarrow x^2(r^2 - x^2) = r^4 - 4r^2x^2 + 4x^4 \Rightarrow 5x^4 - 5r^2x^2 + r^4 = 0.$$

By the quadratic formula, $x^2 = \frac{5 \pm \sqrt{5}}{10}r^2$, but we reject the root with the $+$ sign since it doesn't satisfy ($\bigstar$). So $x = \sqrt{\frac{5 - \sqrt{5}}{10}}\,r$. Since $S(0) = S(r) = 0$, the maximum occurs at the critical number and $x^2 = \frac{5 - \sqrt{5}}{10}r^2 \Rightarrow y^2 = \frac{5 + \sqrt{5}}{10}r^2 \Rightarrow$ the surface area is $2\pi\left(\frac{5 + \sqrt{5}}{10}\right)r^2 + 4\pi\sqrt{\frac{5 - \sqrt{5}}{10}}\sqrt{\frac{5 + \sqrt{5}}{10}}\,r^2 = \pi r^2\left(1 + \sqrt{5}\right)$.

29.

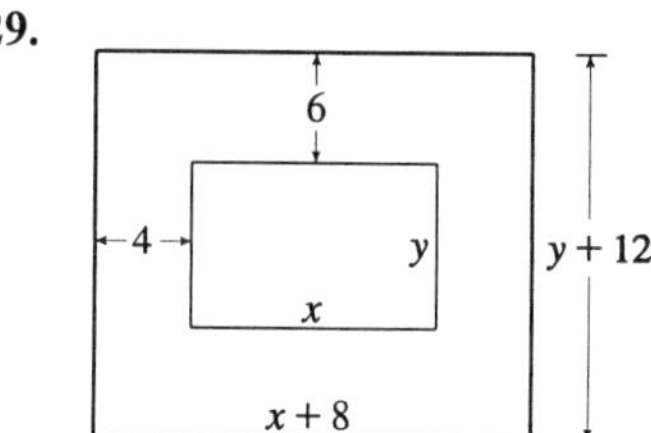

$xy = 384 \Rightarrow y = 384/x$. Total area is $A(x) = (8 + x)(12 + 384/x) = 12(40 + x + 256/x)$, so $A'(x) = 12\left(1 - 256/x^2\right) = 0 \Rightarrow x = 16$. There is an absolute minimum when $x = 16$ since $A'(x) < 0$ for $0 < x < 16$ and $A'(x) > 0$ for $x > 16$. When $x = 16$, $y = 384/16 = 24$, so the dimensions are 24 cm and 36 cm.

31.

Let x be the length of the wire used for the square. The total area is $A(x) = \frac{1}{16}x^2 + \frac{\sqrt{3}}{36}(10 - x)^2$, $0 \le x \le 10$.

$A'(x) = \frac{1}{8}x - \frac{\sqrt{3}}{18}(10 - x) = 0 \Leftrightarrow x = \dfrac{40\sqrt{3}}{9 + 4\sqrt{3}}$. Now

$A(0) = \left(\frac{\sqrt{3}}{36}\right)100 \approx 4.81$, $A(10) = \frac{100}{16} = 6.25$ and $A\left(\frac{40\sqrt{3}}{9 + 4\sqrt{3}}\right) \approx 2.72$, so

(a) The maximum occurs when $x = 10$ m, and all the wire is used for the square.

(b) The minimum occurs when $x = \frac{40\sqrt{3}}{9 + 4\sqrt{3}} \approx 4.35$ m.

33.

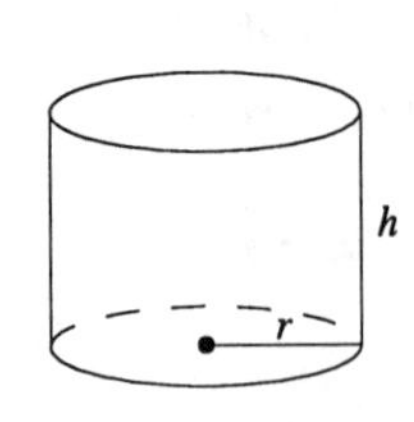

The volume is $V = \pi r^2 h$ and the surface area is

$$S(r) = \pi r^2 + 2\pi r h = \pi r^2 + 2\pi r\left(\frac{V}{\pi r^2}\right) = \pi r^2 + \frac{2V}{r}.$$

$S'(r) = 2\pi r - 2V/r^2 = 0 \quad \Rightarrow \quad 2\pi r^3 = 2V \quad \Rightarrow \quad r = \sqrt[3]{V/\pi}$ cm.

This gives an absolute minimum since $S'(r) < 0$ for $0 < r < \sqrt[3]{V/\pi}$ and $S'(r) > 0$ for $r > \sqrt[3]{\dfrac{V}{\pi}}$. When $r = \sqrt[3]{\dfrac{V}{\pi}}$,

$$h = \frac{V}{\pi r^2} = \frac{V}{\pi (V/\pi)^{2/3}} = \sqrt[3]{\frac{V}{\pi}} \text{ cm.}$$

35.

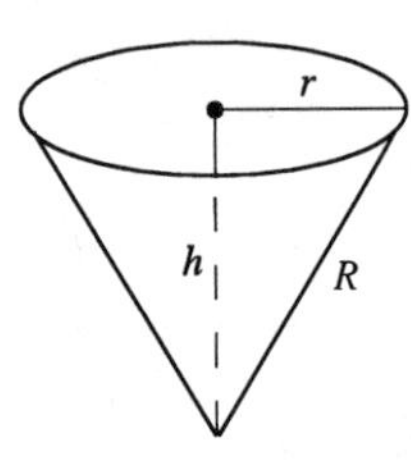

$h^2 + r^2 = R^2 \quad \Rightarrow \quad V = \frac{\pi}{3} r^2 h = \frac{\pi}{3}(R^2 - h^2)h = \frac{\pi}{3}(R^2 h - h^3)$.

$V'(h) = \frac{\pi}{3}(R^2 - 3h^2) = 0$ when $h = \frac{1}{\sqrt{3}}R$. This gives an absolute maximum since $V'(h) > 0$ for $0 < h < \frac{1}{\sqrt{3}}R$ and $V'(h) < 0$ for $h > \frac{1}{\sqrt{3}}R$. Maximum volume is $V\left(\frac{1}{\sqrt{3}}R\right) = \frac{2}{9\sqrt{3}}\pi R^3$.

37. $S = 6sh - \frac{3}{2}s^2 \cot\theta + 3s^2 \frac{\sqrt{3}}{2} \csc\theta$

(a) $\dfrac{dS}{d\theta} = \frac{3}{2}s^2 \csc^2\theta - 3s^2 \frac{\sqrt{3}}{2} \csc\theta \cot\theta$ or $\frac{3}{2}s^2 \csc\theta \left(\csc\theta - \sqrt{3}\cot\theta\right)$.

(b) $\dfrac{dS}{d\theta} = 0$ when $\csc\theta - \sqrt{3}\cot\theta = 0 \quad \Rightarrow \quad \dfrac{1}{\sin\theta} - \sqrt{3}\dfrac{\cos\theta}{\sin\theta} = 0 \quad \Rightarrow \quad \cos\theta = \frac{1}{\sqrt{3}}$. The First Derivative Test shows that the minimum surface area occurs when $\theta = \cos^{-1}\frac{1}{\sqrt{3}} \approx 55^\circ$.

(c)

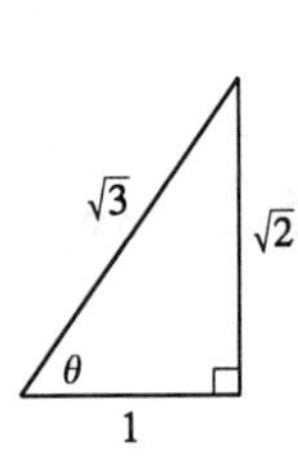

If $\cos\theta = \frac{1}{\sqrt{3}}$, then $\cot\theta = \frac{1}{\sqrt{2}}$ and $\csc\theta = \frac{\sqrt{3}}{\sqrt{2}}$, so the surface area is

$$S = 6sh - \tfrac{3}{2}s^2 \tfrac{1}{\sqrt{2}} + 3s^2 \tfrac{\sqrt{3}}{2} \tfrac{\sqrt{3}}{\sqrt{2}} = 6sh - \tfrac{3}{2\sqrt{2}}s^2 + \tfrac{9}{2\sqrt{2}}s^2 = 6s\left(h + \tfrac{1}{2\sqrt{2}}s\right)$$

39. Here $T(x) = \dfrac{\sqrt{x^2 + 25}}{6} + \dfrac{5 - x}{8}, 0 \le x \le 5, \quad \Rightarrow \quad T'(x) = \dfrac{x}{6\sqrt{x^2 + 25}} - \dfrac{1}{8} = 0 \quad \Leftrightarrow \quad 8x = 6\sqrt{x^2 + 25}$

$\Leftrightarrow \quad 16x^2 = 9(x^2 + 25) \quad \Leftrightarrow \quad x = \frac{15}{\sqrt{7}}$. But $\frac{15}{\sqrt{7}} > 5$, so T has no critical number. Since $T(0) \approx 1.46$ and $T(5) \approx 1.18$, he should row directly to B.

41.

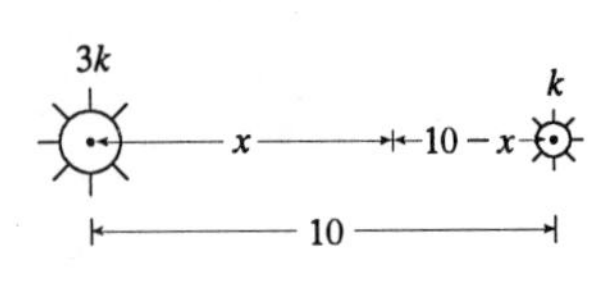

The total illumination is $I(x) = \dfrac{3k}{x^2} + \dfrac{k}{(10 - x)^2}, \; 0 < x < 10$.

Then $I'(x) = \dfrac{-6k}{x^3} + \dfrac{2k}{(10 - x)^3} = 0 \quad \Rightarrow \quad 6k(10 - x)^3 = 2kx^3$

$\Rightarrow \quad \sqrt[3]{3}(10 - x) = x \quad \Rightarrow \quad x = 10\sqrt[3]{3}/\left(1 + \sqrt[3]{3}\right) \approx 5.9$ ft.

This gives a minimum since there is clearly no maximum.

43.

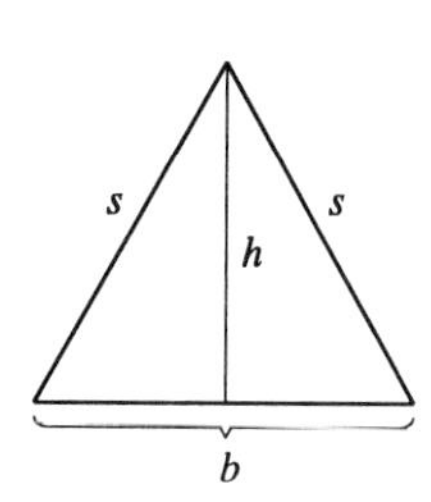

Here $s^2 = h^2 + b^2/4$ so $h^2 = s^2 - b^2/4$. The area is $A = \frac{1}{2}b\sqrt{s^2 - b^2/4}$. Let the perimeter be p, so $2s + b = p$ or $s = (p-b)/2 \Rightarrow$ $A(b) = \frac{1}{2}b\sqrt{(p-b)^2/4 - b^2/4} = b\sqrt{p^2 - 2pb}/4$. Now $A'(b) = \dfrac{\sqrt{p^2 - 2pb}}{4} - \dfrac{bp/4}{\sqrt{p^2 - 2pb}} = \dfrac{-3pb + p^2}{4\sqrt{p^2 - 2pb}}$. Therefore $A'(b) = 0$ $\Rightarrow \quad -3pb + p^2 = 0 \quad \Rightarrow \quad b = p/3$. Since $A'(b) > 0$ for $b < p/3$ and $A'(b) < 0$ for $b > p/3$, there is an absolute maximum when $b = p/3$. But then $2s + p/3 = p$ so $s = p/3 \quad \Rightarrow$ $s = b \quad \Rightarrow \quad$ the triangle is equilateral.

45.

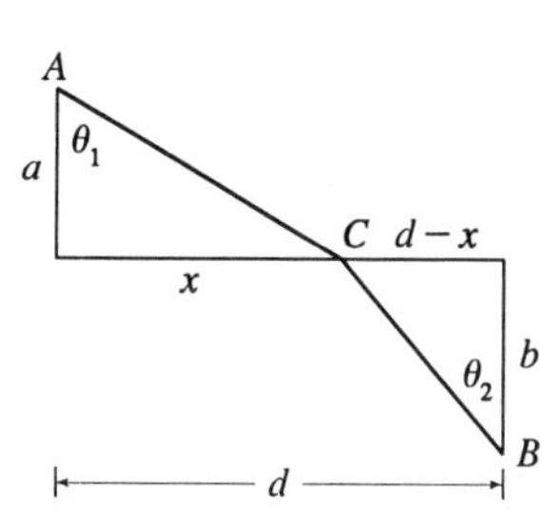

The total time is $T(x) = (\text{time from } A \text{ to } C) + (\text{time from } C \text{ to } B)$

$= \sqrt{a^2 + x^2}/v_1 + \sqrt{b^2 + (d-x)^2}/v_2,\ 0 < x < d.$

$$T'(x) = \frac{x}{v_1\sqrt{a^2 + x^2}} - \frac{d - x}{v_2\sqrt{b^2 + (d-x)^2}} = \frac{\sin\theta_1}{v_1} - \frac{\sin\theta_2}{v_2}.$$

The minimum occurs when $T'(x) = 0 \quad \Rightarrow \quad \dfrac{\sin\theta_1}{v_1} = \dfrac{\sin\theta_2}{v_2}$.

47.

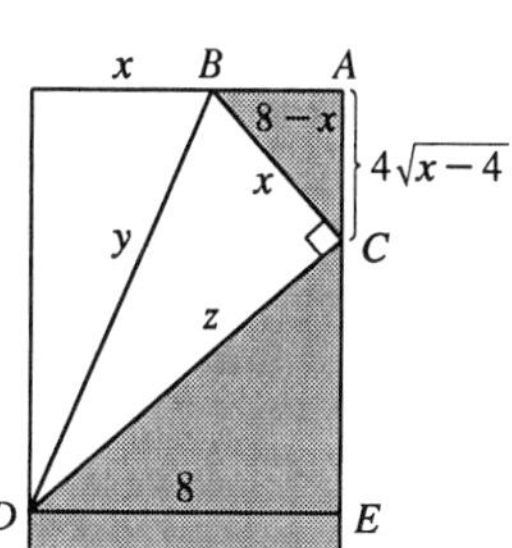

$y^2 = x^2 + z^2$, but triangles CDE and BCA are similar, so $z/8 = x\big/\left(4\sqrt{x-4}\right)$. Thus we minimize

$f(x) = y^2 = x^2 + 4x^2/(x-4) = x^3/(x-4),\ 4 < x \le 8.$

$$f'(x) = \frac{3x^2(x-4) - x^3}{(x-4)^2} = \frac{2x^2(x-6)}{(x-4)^2} = 0 \text{ when } x = 6.$$

$f'(x) < 0$ when $x < 6$, $f'(x) > 0$ when $x > 6$, so the minimum occurs when $x = 6$ in.

49.

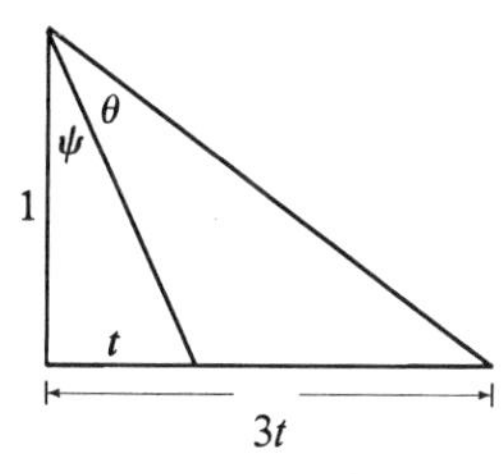

It suffices to maximize $\tan\theta$. Now $\dfrac{3t}{1} = \tan(\psi + \theta) = \dfrac{\tan\psi + \tan\theta}{1 - \tan\psi\tan\theta}$

(see endpapers) $= \dfrac{t + \tan\theta}{1 - t\tan\theta}$. So $3t(1 - t\tan\theta) = t + \tan\theta$

$\Rightarrow \quad 2t = (1 + 3t^2)\tan\theta \quad \Rightarrow \quad \tan\theta = \dfrac{2t}{1 + 3t^2}$.

Let $f(t) = \tan\theta = \dfrac{2t}{1 + 3t^2} \quad \Rightarrow$

$f'(t) = \dfrac{2(1 + 3t^2) - 2t(6t)}{(1 + 3t^2)^2} = \dfrac{2(1 - 3t^2)}{(1 + 3t^2)^2} = 0 \iff 1 - 3t^2 = 0 \iff t = \frac{1}{\sqrt{3}}$ since $t \ge 0$. Now $f'(t) > 0$ for $0 \le t < \frac{1}{\sqrt{3}}$ and $f'(t) < 0$ for $t > \frac{1}{\sqrt{3}}$, so f has an absolute maximum when $t = \frac{1}{\sqrt{3}}$ and

$$\tan\theta = \frac{2\left(1/\sqrt{3}\right)}{1 + 3\left(1/\sqrt{3}\right)^2} = \frac{1}{\sqrt{3}} \quad \Rightarrow \quad \theta = \frac{\pi}{6}.$$

51.

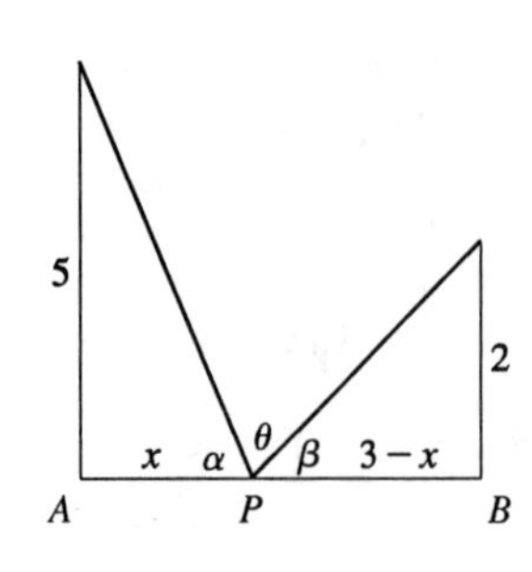

$x = 5\cot\alpha,\ 3 - x = 2\cot\beta \quad\Rightarrow$

$$\theta = \pi - \cot^{-1}\left(\frac{x}{5}\right) - \cot^{-1}\left(\frac{3-x}{2}\right) \quad\Rightarrow$$

$$\frac{d\theta}{dx} = \frac{1}{1+\left(\frac{x}{5}\right)^2}\left(\frac{1}{5}\right) + \frac{1}{1+\left(\frac{3-x}{2}\right)^2}\left(-\frac{1}{2}\right) = 0 \quad\Rightarrow$$

$$5\left(1+\frac{x^2}{25}\right) = 2\left(1+\frac{9-6x+x^2}{4}\right)$$

$\Rightarrow \quad 50 + 2x^2 = 65 - 30x + 5x^2 \quad\Rightarrow\quad x^2 - 10x + 5 = 0 \quad\Rightarrow\quad x = 5 \pm 2\sqrt{5}$. We reject the root with the $+$ sign, since it is larger than 3. $d\theta/dx > 0$ for $x < 5 - 2\sqrt{5}$ and $d\theta/dx < 0$ for $x > 5 - 2\sqrt{5}$, so θ is maximized when $|AP| = x = 5 - 2\sqrt{5}$.

53.

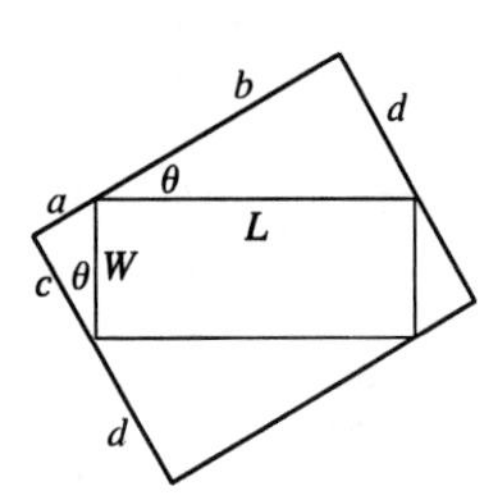

$a = W\sin\theta,\ c = W\cos\theta,\ b = L\cos\theta,\ d = L\sin\theta$, so the area of the circumscribed rectangle is

$$\begin{aligned} A(\theta) &= (a+b)(c+d) = (W\sin\theta + L\cos\theta)(W\cos\theta + L\sin\theta) \\ &= LW\sin^2\theta + LW\cos^2\theta + \left(L^2+W^2\right)\sin\theta\cos\theta \\ &= LW + \tfrac{1}{2}(L^2+W^2)\sin 2\theta,\ 0 \le \theta \le \tfrac{\pi}{2}. \end{aligned}$$

This expression shows, without calculus, that the maximum value of $A(\theta)$ occurs when $\sin 2\theta = 1 \;\Leftrightarrow\; 2\theta = \frac{\pi}{2} \;\Leftrightarrow\; x = \frac{\pi}{4}$. So the maximum area is $A\left(\frac{\pi}{4}\right) = LW + \frac{1}{2}(L^2+W^2) = \frac{1}{2}(L+W)^2$.

55. $L(x) = |AP| + |BP| + |CP| = x + \sqrt{(5-x)^2 + 2^2} + \sqrt{(5-x)^2 + 3^2}$

$$= x + \sqrt{x^2 - 10x + 29} + \sqrt{x^2 - 10x + 34} \quad\Rightarrow$$

$$L'(x) = 1 + \frac{x-5}{\sqrt{x^2-10x+29}} + \frac{x-5}{\sqrt{x^2-10x+34}}$$

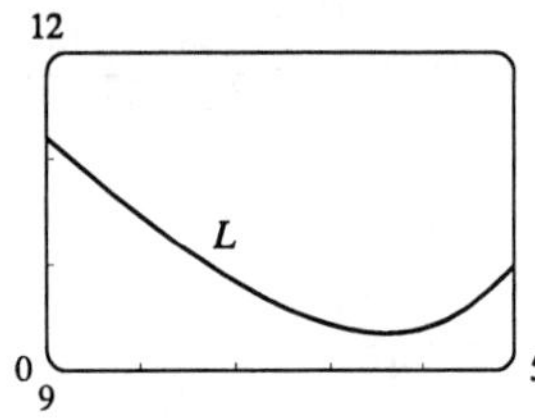

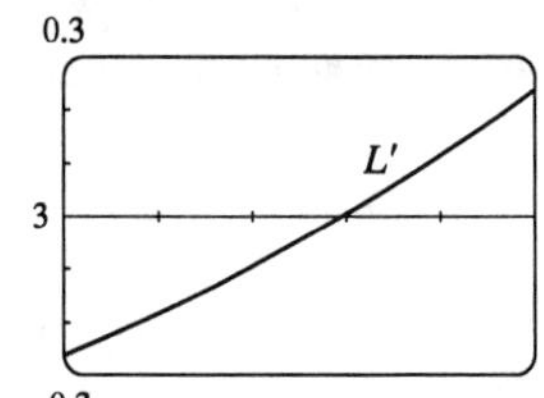

From the graphs of L and L', it seems that the minimum value of L is about $L(3.59) = 9.35$ m.

EXERCISES 5.6

1. **(a)** $C(0)$ represents the fixed costs of production, such as rent, utilities, machinery etc., which are incurred even when nothing is produced.

(b) The inflection point is the point at which $C''(x)$ changes from negative to positive, that is, the marginal cost $C'(x)$changes from decreasing to increasing. So the marginal cost is minimized.

(c) The marginal cost function is $C'(x)$.

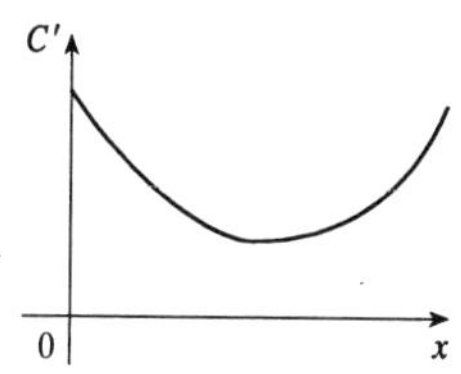

3. **(a)** $C(x) = 10{,}000 + 25x + x^2$, $C(1000) = \$1{,}035{,}000$, $c(x) = \dfrac{C(x)}{x} = \dfrac{10{,}000}{x} + 25 + x$, $c(1000) = \$1035$. $C'(x) = 25 + 2x$, $C'(1000) = \$2025/\text{unit}$.

(b) We must have $c(x) = C'(x) \Rightarrow 10{,}000/x + 25 + x = 25 + 2x \Rightarrow 10{,}000/x = x \Rightarrow x^2 = 10{,}000 \Rightarrow x = 100$. This is a minimum since $c''(x) = 20{,}000/x^3 > 0$.

(c) The minimum average cost is $c(100) = \$225$.

5. **(a)** $C(x) = 45 + \dfrac{x}{2} + \dfrac{x^2}{560}$, $C(1000) = \$2330.71$. $c(x) = \dfrac{45}{x} + \dfrac{1}{2} + \dfrac{x}{560}$, $c(1000) = \$2.33$.
$C'(x) = \dfrac{1}{2} + \dfrac{x}{280}$, $C'(1000) = \$4.07/\text{unit}$

(b) We must have $C'(x) = c(x) \Rightarrow \dfrac{1}{2} + \dfrac{x}{280} = \dfrac{45}{x} + \dfrac{1}{2} + \dfrac{x}{560} \Rightarrow \dfrac{45}{x} = \dfrac{x}{560} \Rightarrow x^2 = (45)(560) \Rightarrow x = \sqrt{25{,}200} \approx 159$. This is a minimum since $c''(x) = 90/x^2 > 0$.

(c) The minimum average cost is $c(159) = \$1.07$.

7. **(a)** $C(x) = 2\sqrt{x} + \dfrac{x^2}{8000}$, $C(1000) = \$188.25$. $c(x) = \dfrac{2}{\sqrt{x}} + \dfrac{x}{8000}$, $c(1000) = \$0.19$.
$C'(x) = \dfrac{1}{\sqrt{x}} + \dfrac{x}{4000}$, $C'(1000) = \$0.28/\text{unit}$.

(b) We must have $C'(x) = c(x) \Rightarrow \dfrac{1}{\sqrt{x}} + \dfrac{x}{4000} = \dfrac{2}{\sqrt{x}} + \dfrac{x}{8000} \Rightarrow \dfrac{x}{8000} = \dfrac{1}{\sqrt{x}} \Rightarrow$
$x^{3/2} = 8000 \Rightarrow x = (8000)^{2/3} = 400$. This is a minimum since $c''(x) = \frac{3}{2}x^{-5/2} > 0$.

(c) The minimum average cost is $c(400) = \$0.15$.

9. $C(x) = 680 + 4x + 0.01x^2$, $p(x) = 12 \Rightarrow R(x) = xp(x) = 12x$. If the profit is maximum, then $R'(x) = C'(x) \Rightarrow 12 = 4 + 0.02x \Rightarrow 0.02x = 8 \Rightarrow x = 400$. Now $R''(x) = 0 < 0.02 = C''(x)$, so $x = 400$ gives a maximum.

11. $C(x) = 1200 + 25x - 0.0001x^2$, $p(x) = 55 - x/1000$. Then $R(x) = xp(x) = 55x - x^2/1000$. If the profit is maximum, then $R'(x) = C'(x) \quad \Leftrightarrow \quad 55 - x/500 = 25 - 0.0002x \quad \Rightarrow \quad 30 = 0.0018x \quad \Rightarrow$ $x = 30/0.0018 \approx 16{,}667$. Now $R''(x) = -\frac{1}{500} < -0.0002 = C''(x)$, so $x = 16{,}667$ gives a maximum.

13. $C(x) = 1450 + 36x - x^2 + 0.001x^3$, $p(x) = 60 - 0.01x$. Then $R(x) = xp(x) = 60x - 0.01x^2$. If the profit is maximum, then $R'(x) = C'(x) \quad \Leftrightarrow \quad 60 - 0.02x = 36 - 2x + 0.003x^2 \quad \Rightarrow \quad 0.003x^2 - 1.98x - 24 = 0$.

By the quadratic formula, $x = \dfrac{1.98 \pm \sqrt{(-1.98)^2 + 4(0.003)(24)}}{2(0.003)} = \dfrac{1.98 \pm \sqrt{4.2084}}{0.006}$. Since $x > 0$,

$x \approx (1.98 + 2.05)/0.006 \approx 672$. Now $R''(x) = -0.02$ and $C''(x) = -2 + 0.006x \quad \Rightarrow \quad C''(672) = 2.032$ $\Rightarrow \quad R''(672) < C''(672) \quad \Rightarrow \quad$ there is a maximum at $x = 672$.

15. $C(x) = 0.001x^3 - 0.3x^2 + 6x + 900$. The marginal cost is $C'(x) = 0.003x^2 - 0.6x + 6$. $C'(x)$ is increasing when $C''(x) > 0 \quad \Leftrightarrow \quad 0.006x - 0.6 > 0 \quad \Leftrightarrow \quad x > 0.6/0.006 = 100$. So $C'(x)$ starts to increase when $x = 100$.

17. **(a)** We are given that the demand function p is linear and $p(27{,}000) = 10$, $p(33{,}000) = 8$, so the slope is $\dfrac{10 - 8}{27{,}000 - 33{,}000} = -\dfrac{1}{3000}$ and the equation of the graph is $y - 10 = \left(-\frac{1}{3000}\right)(x - 27{,}000) \quad \Rightarrow$ $p(x) = 19 - x/3000$.

(b) The revenue is $R(x) = xp(x) = 19x - x^2/3000 \quad \Rightarrow \quad R'(x) = 19 - x/1500 = 0$ when $x = 28{,}500$. Since $R''(x) = -1/1500 < 0$, the maximum revenue occurs when $x = 28{,}500 \quad \Rightarrow \quad$ the price is $p(28{,}500) = \$9.50$.

19. **(a)** $p(x) = 450 - \frac{1}{10}(x - 1000) = 550 - x/10$.

(b) $R(x) = xp(x) = 500x - x^2/10$. $R'(x) = 550 - x/5 = 0$ when $x = 5(550) = 2750$. $p(2750) = 275$, so the rebate should be $450 - 275 = \$175$.

(c) $P(x) = R(x) - C(x) = 550x - x^2/10 - 68{,}000 - 150x = 400x - x^2/10 - 68{,}000$, $P'(x) = 400 - x/5 = 0$ when $x = 2000$. $p(2000) = 550 - 200 = 350$. Therefore the rebate to maximize profits should be $450 - 350 = \$100$.

EXERCISES 5.7

1. $f(x) = 12x^2 + 6x - 5 \quad \Rightarrow \quad F(x) = 12\left(\frac{1}{3}x^3\right) + 6\left(\frac{1}{2}x^2\right) - 5x + C = 4x^3 + 3x^2 - 5x + C$

3. $f(x) = 6x^9 - 4x^7 + 3x^2 + 1 \quad \Rightarrow$
$F(x) = 6\left(\frac{1}{10}x^{10}\right) - 4\left(\frac{1}{8}x^8\right) + 3\left(\frac{1}{3}x^3\right) + x + C = \frac{3}{5}x^{10} - \frac{1}{2}x^8 + x^3 + x + C$

5. $f(x) = \sqrt{x} + \sqrt[3]{x} = x^{1/2} + x^{1/3} \quad \Rightarrow \quad F(x) = \frac{1}{3/2}x^{3/2} + \frac{1}{4/3}x^{4/3} + C = \frac{2}{3}x^{3/2} + \frac{3}{4}x^{4/3} + C$

7. $f(x) = 6/x^5 = 6x^{-5} \quad \Rightarrow \quad F(x) = \begin{cases} 6x^{-4}/(-4) + C_1 = -3/(2x^4) + C_1 & \text{if } x < 0 \\ -3/(2x^4) + C_2 & \text{if } x > 0 \end{cases}$

9. $g(t) = \dfrac{t^3 + 2t^2}{\sqrt{t}} = t^{5/2} + 2t^{3/2} \quad \Rightarrow \quad G(t) = \dfrac{t^{7/2}}{7/2} + \dfrac{2t^{5/2}}{5/2} + C = \frac{2}{7}t^{7/2} + \frac{4}{5}t^{5/2} + C$

11. $h(x) = \sin x - 2\cos x \quad \Rightarrow \quad H(x) = -\cos x - 2\sin x + C$

13. $f(t) = \sec^2 t + t^2 \quad \Rightarrow \quad F(t) = \tan t + \frac{1}{3}t^3 + C_n$ on the interval $\left(n\pi - \frac{\pi}{2}, n\pi + \frac{\pi}{2}\right)$.

15. $f(x) = 2x + \dfrac{5}{\sqrt{1 - x^2}} \quad \Rightarrow \quad F(x) = x^2 + 5\sin^{-1}x + C$

17. $f''(x) = x^2 + x^3 \quad \Rightarrow \quad f'(x) = \frac{1}{3}x^3 + \frac{1}{4}x^4 + C \quad \Rightarrow \quad f(x) = \frac{1}{12}x^4 + \frac{1}{20}x^5 + Cx + D$

19. $f''(x) = 1 \quad \Rightarrow \quad f'(x) = x + C \quad \Rightarrow \quad f(x) = \frac{1}{2}x^2 + Cx + D$

21. $f'''(x) = 24x \quad \Rightarrow \quad f''(x) = 12x^2 + C \quad \Rightarrow \quad f'(x) = 4x^3 + Cx + D \quad \Rightarrow$
$f(x) = x^4 + \frac{1}{2}Cx^2 + Dx + E$

23. $f'(x) = 4x + 3 \quad \Rightarrow \quad f(x) = 2x^2 + 3x + C \quad \Rightarrow \quad -9 = f(0) = C \quad \Rightarrow \quad f(x) = 2x^2 + 3x - 9$

25. $f'(x) = 3\sqrt{x} - 1/\sqrt{x} = 3x^{1/2} - x^{-1/2} \quad \Rightarrow \quad f(x) = 3\left(\frac{1}{3/2}\right)x^{3/2} - \frac{1}{1/2}x^{1/2} + C \quad \Rightarrow$
$2 = f(1) = 2 - 2 + C = C \quad \Rightarrow \quad f(x) = 2x^{3/2} - 2x^{1/2} + 2$

27. $f'(x) = 3\cos x + 5\sin x \quad \Rightarrow \quad f(x) = 3\sin x - 5\cos x + C \quad \Rightarrow \quad 4 = f(0) = -5 + C \quad \Rightarrow \quad C = 9$
$\Rightarrow \quad f(x) = 3\sin x - 5\cos x + 9.$

29. $f'(x) = 2/x \quad \Rightarrow \quad f(x) = 2\ln|x| + C = 2\ln(-x) + C$ (since $x < 0$). Now $f(-1) = 2\ln 1 + C = 7 \quad \Rightarrow$
$C = 7$. Therefore $f(x) = 2\ln(-x) + 7$, $x < 0$.

31. $f''(x) = x \quad \Rightarrow \quad f'(x) = \frac{1}{2}x^2 + C \quad \Rightarrow \quad 2 = f'(0) = C \quad \Rightarrow \quad f'(x) = \frac{1}{2}x^2 + 2 \quad \Rightarrow$
$f(x) = \frac{1}{6}x^3 + 2x + D \quad \Rightarrow \quad -3 = f(0) = D \quad \Rightarrow \quad f(x) = \frac{1}{6}x^3 + 2x - 3$

33. $f''(x) = x^2 + 3\cos x \quad \Rightarrow \quad f'(x) = \frac{1}{3}x^3 + 3\sin x + C \quad \Rightarrow \quad 3 = f'(0) = C \quad \Rightarrow$
$f'(x) = \frac{1}{3}x^3 + 3\sin x + 3 \quad \Rightarrow \quad f(x) = \frac{1}{12}x^4 - 3\cos x + 3x + D \quad \Rightarrow \quad 2 = f(0) = -3 + D \quad \Rightarrow$
$D = 5 \quad \Rightarrow \quad f(x) = \frac{1}{12}x^4 - 3\cos x + 3x + 5$

35. $f''(x) = 6x + 6 \quad \Rightarrow \quad f'(x) = 3x^2 + 6x + C \quad \Rightarrow \quad f(x) = x^3 + 3x^2 + Cx + D \quad \Rightarrow \quad 4 = f(0) = D$ and $3 = f(1) = 1 + 3 + C + D = 4 + C + 4 \quad \Rightarrow \quad C = -5 \quad \Rightarrow \quad f(x) = x^3 + 3x^2 - 5x + 4$

37. $f''(x) = x^{-3} \quad \Rightarrow \quad f'(x) = -\frac{1}{2}x^{-2} + C \quad \Rightarrow \quad f(x) = \frac{1}{2}x^{-1} + Cx + D \quad \Rightarrow \quad 0 = f(1) = \frac{1}{2} + C + D$ and $0 = f(2) = \frac{1}{4} + 2C + D$. Solving these equations, we get $C = \frac{1}{4}$, $D = -\frac{3}{4}$, so $f(x) = 1/(2x) + \frac{1}{4}x - \frac{3}{4}$.

39. $f''(x) = x^{-2}$, $x > 0 \quad \Rightarrow \quad f'(x) = -1/x + C \quad \Rightarrow \quad f(x) = -\ln x + Cx + D$. $0 = f(1) = C + D$ and $0 = f(2) = -\ln 2 + 2C + D = -\ln 2 + 2C - C = -\ln 2 + C \quad \Rightarrow \quad C = \ln 2$ and $D = -\ln 2$. So $f(x) = -\ln x + (\ln 2)x - \ln 2$.

41. We have that $f'(x) = 2x + 1 \quad \Rightarrow \quad f(x) = x^2 + x + C$. But f passes through $(1, 6)$ so that $6 = f(1) = 1^2 + 1 + C \quad \Rightarrow \quad C = 4$. Therefore $f(x) = x^2 + x + 4 \quad \Rightarrow \quad f(2) = 2^2 + 2 + 4 = 10$.

43. b is the antiderivative of f. For small x, f is negative, so the graph of its antiderivative must be decreasing. But both a and c are increasing for small x, so only b can be f's antiderivative. Also, f is positive where b is increasing, which supports our conclusion.

45. The graph of F will have a minimum at 0 and a maximum at 2, since $f = F'$ goes from negative to positive at $x = 0$, and from positive to negative at $x = 2$.

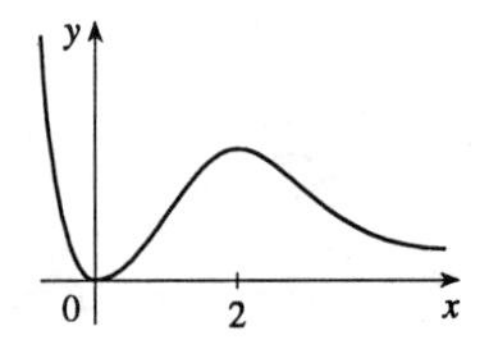

47.

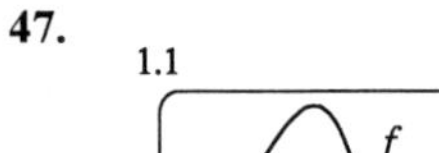
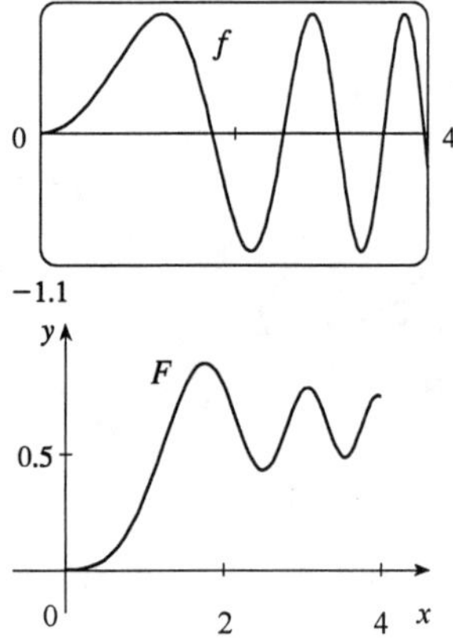

49.

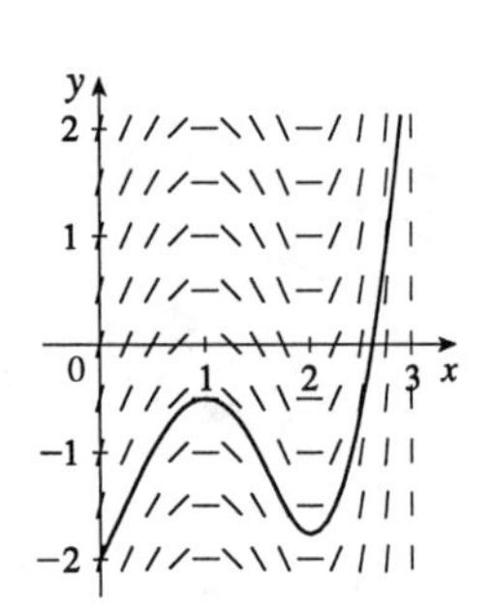

51.

x	$f(x)$
0	1
0.5	0.959
1.0	0.841
1.5	0.665
2.0	0.455
2.5	0.239
3.0	0.047
3.5	−0.100
4.0	−0.189
4.5	−0.217
5.0	−0.192
5.5	−0.128
6.0	−0.047

We compute slopes as in the table and draw a direction field as in Example 6. Then we use the direction field to graph F starting at $(0,0)$.

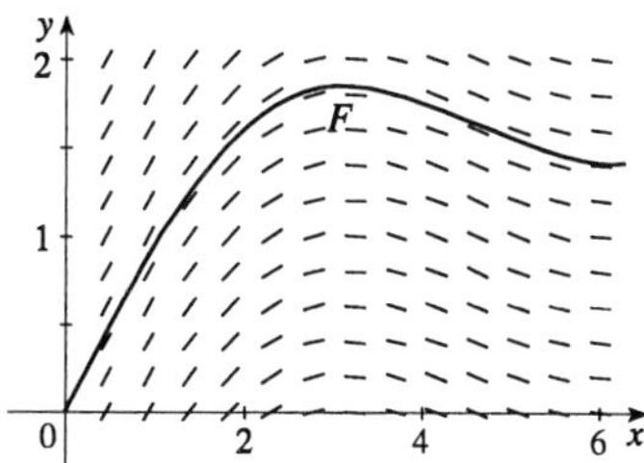

53.

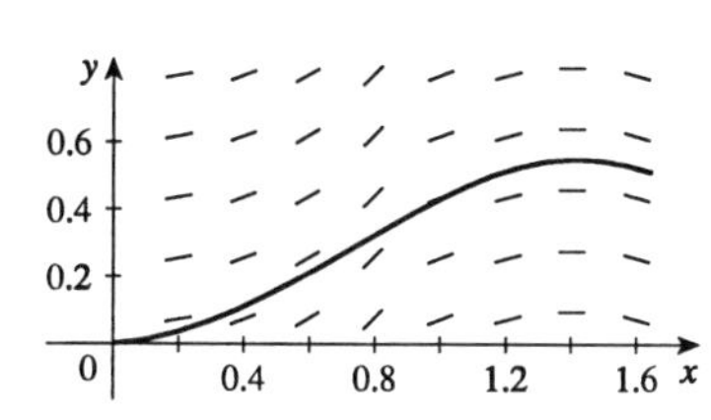

55. $v(t) = s'(t) = 3 - 2t \quad \Rightarrow \quad s(t) = 3t - t^2 + C \quad \Rightarrow \quad 4 = s(0) = C \quad \Rightarrow \quad s(t) = 3t - t^2 + 4$

57. $a(t) = v'(t) = 3t + 8 \quad \Rightarrow \quad v(t) = \frac{3}{2}t^2 + 8t + C \quad \Rightarrow \quad -2 = v(0) = C \quad \Rightarrow \quad v(t) = \frac{3}{2}t^2 + 8t - 2 \quad \Rightarrow$
$s(t) = \frac{1}{2}t^3 + 4t^2 - 2t + D \quad \Rightarrow \quad 1 = s(0) = D \quad \Rightarrow \quad s(t) = \frac{1}{2}t^3 + 4t^2 - 2t + 1$

59. $a(t) = v'(t) = t^2 - t \quad \Rightarrow \quad v(t) = \frac{1}{3}t^3 - \frac{1}{2}t^2 + C \quad \Rightarrow \quad s(t) = \frac{1}{12}t^4 - \frac{1}{6}t^3 + Ct + D \quad \Rightarrow \quad 0 = s(0) = D$
and $12 = s(6) = 108 - 36 + 6C + 0 \quad \Rightarrow \quad C = -10 \quad \Rightarrow \quad s(t) = \frac{1}{12}t^4 - \frac{1}{6}t^3 - 10t$

61. **(a)** $v'(t) = a(t) = -9.8 \quad \Rightarrow \quad v(t) = -9.8t + C$, but $C = v(0) = 0$, so $v(t) = -9.8t \quad \Rightarrow$
$s(t) = -4.9t^2 + D \quad \Rightarrow \quad D = s(0) = 450 \quad \Rightarrow \quad s(t) = 450 - 4.9t^2$

(b) It reaches the ground when $0 = s(t) = 450 - 4.9t^2 \quad \Rightarrow \quad t^2 = 450/4.9 \quad \Rightarrow \quad t = \sqrt{450/4.9} \approx 9.58$ s.

(c) $v = -9.8\sqrt{450/4.9} \approx -93.9\,\text{m/s}$

63. **(a)** $v'(t) = -9.8 \quad \Rightarrow \quad v(t) = -9.8t + C \quad \Rightarrow \quad 5 = v(0) = C$, so $v(t) = 5 - 9.8t \quad \Rightarrow$
$s(t) = 5t - 4.9t^2 + D \quad \Rightarrow \quad D = s(0) = 450 \quad \Rightarrow \quad s(t) = 450 + 5t - 4.9t^2$

(b) It reaches the ground when $450 + 5t - 4.9t^2 = 0$. By the quadratic formula, the positive root of this equation is $t = \dfrac{5 + \sqrt{8845}}{9.8} \approx 10.1$ s

(c) $v = 5 - 9.8 \cdot \dfrac{5 + \sqrt{8845}}{9.8} \approx -94.0\,\text{m/s}$

65. By Exercise 64, $s(t) = -4.9t^2 + v_0 t + s_0$ and $v(t) = s'(t) = -9.8t + v_0$. So $[v(t)]^2 = (9.8)^2 t^2 - 19.6 v_0 t + v_0^2$ and $v_0^2 - 19.6[s(t) - s_0] = v_0^2 - 19.6[-4.9t^2 + v_0 t] = v_0^2 + (9.8)^2 t^2 - 19.6 v_0 t = [v(t)]^2$

67. Marginal cost $= 1.92 - 0.002x = C'(x) \quad \Rightarrow \quad C(x) = 1.92x - 0.001x^2 + K$. But $C(1) = 1.92 - 0.001 + K = 562 \quad \Rightarrow \quad K = 560.081$. Therefore $C(x) = 1.92x - 0.001x^2 + 560.081 \quad \Rightarrow$ $C(100) = 1.92(100) - 0.001(100)^2 + 560.081 = 742.081$, so the cost of producing 100 items is \$742.08.

69. Taking the upward direction to be positive we have that for $0 \le t \le 10$ (using the subscript 1 to refer to $0 \le t \le 10$), $a_1(t) = -9 + 0.9t = v_1'(t) \quad \Rightarrow \quad v_1(t) = -9t + 0.45t^2 + v_0$, but $v_1(0) = v_0 = -10 \quad \Rightarrow$ $v_1(t) = -9t + 0.45t^2 - 10 = s_1'(t) \quad \Rightarrow \quad s_1(t) = -\frac{9}{2}t^2 + 0.15t^3 - 10t + s_0$. But $s_1(0) = 500 = s_0 \quad \Rightarrow$ $s_1(t) = -\frac{9}{2}t^2 + 0.15t^3 - 10t + 500$. Now for $t > 10$, $a(t) = 0 = v'(t) \quad \Rightarrow$ $v(t) = \text{constant} = v_1(10) = -9(10) + 0.45(10)^2 - 10 = -55 \quad \Rightarrow \quad v(t) = -55 = s'(t) \quad \Rightarrow$ $s(t) = -55t + s_{10}$. But $s(10) = s_1(10) \quad \Rightarrow \quad -55(10) + s_{10} = 100 \quad \Rightarrow \quad s_{10} = 650 \quad \Rightarrow$ $s(t) = -55t + 650$. When the raindrop hits the ground we have that $s(t) = 0 \quad \Rightarrow \quad -55t + 650 = 0 \quad \Rightarrow$ $t = \frac{650}{55} = \frac{130}{11} \approx 11.8$ s.

71. $a(t) = a$ and the initial velocity is $30 \text{ mi/h} = 30 \cdot \frac{5280}{3600} = 44$ ft/s and final velocity $50 \text{ mi/h} = 50 \cdot \frac{5280}{3600} = \frac{220}{3}$ ft/s. So $v(t) = at + 44 \quad \Rightarrow \quad \frac{220}{3} = v(5) = 5a + 44 \quad \Rightarrow \quad a = \frac{88}{15} \approx 5.87 \text{ ft/s}^2$.

73. The height at time t is $s(t) = -16t^2 + h$, where $h = s(0)$ is the height of the cliff. $v(t) = -32t = -120$ when $t = 3.75$, so $0 = s(3.75) = -16(3.75)^2 + h \quad \Rightarrow \quad h = 16(3.75)^2 = 225$ ft.

75. $\mathbf{a}(t) = \mathbf{v}'(t) = -10\mathbf{j} \quad \Rightarrow \quad \mathbf{v}(t) = -10t\mathbf{j} + \mathbf{C}$ ($\mathbf{C}$ is a constant vector). Now $\mathbf{v}(0) = -10(0)\mathbf{j} + \mathbf{C} = \mathbf{C}$ and $\mathbf{v}(0) = \mathbf{i} + \mathbf{j}$ (given) $\quad \Rightarrow \quad \mathbf{C} = \mathbf{i} + \mathbf{j}$, so $\mathbf{v}(t) = \mathbf{r}'(t) = -10t\mathbf{j} + (\mathbf{i} + \mathbf{j}) = \mathbf{i} + (1 - 10t)\mathbf{j}$ (the velocity of the particle), and hence, $\mathbf{r}(t) = t\mathbf{i} + (t - 5t^2)\mathbf{j} + \mathbf{D}$ ($\mathbf{D}$ is a constant vector). But $\mathbf{r}(0) = \mathbf{D}$ and $\mathbf{r}(0) = \mathbf{0}$ (given) $\Rightarrow \quad \mathbf{D} = \mathbf{0}$. Thus, $\mathbf{r}(t) = t\mathbf{i} + (-5t^2 + t)\mathbf{j}$ (the position of the particle).

77. **(a)** $\mathbf{a}(t) = t^2\mathbf{i} + \cos 2t\,\mathbf{j} \quad \Rightarrow \quad \mathbf{v}(t) = \frac{1}{3}t^3\mathbf{i} + \frac{1}{2}\sin 2t\,\mathbf{j} + \mathbf{C}$. Now $\mathbf{v}(0) = \mathbf{C}$ and $\mathbf{v}(0) = \mathbf{j} \quad \Rightarrow \quad \mathbf{C} = \mathbf{j}$, so $\mathbf{v}(t) = \frac{1}{3}t^3\mathbf{i} + \frac{1}{2}\sin 2t\,\mathbf{j} + (\mathbf{j}) = \frac{1}{3}t^3\mathbf{i} + \left(\frac{1}{2}\sin 2t + 1\right)\mathbf{j}$, and hence, $\mathbf{r}(t) = \frac{1}{12}t^4\mathbf{i} + \left(t - \frac{1}{4}\cos 2t\right)\mathbf{j} + \mathbf{D}$. But $\mathbf{r}(0) = -\frac{1}{4}\mathbf{j} + \mathbf{D}$ and $\mathbf{r}(0) = \mathbf{i} \quad \Rightarrow \quad \mathbf{D} = \mathbf{i} + \frac{1}{4}\mathbf{j}$. Thus,

$$\mathbf{r}(t) = \tfrac{1}{12}t^4\mathbf{i} + \left(t - \tfrac{1}{4}\cos 2t\right)\mathbf{j} + \left(\mathbf{i} + \tfrac{1}{4}\mathbf{j}\right) = \left(\tfrac{1}{12}t^4 + 1\right)\mathbf{i} + \left(t - \tfrac{1}{4}\cos 2t + \tfrac{1}{4}\right)\mathbf{j}.$$

(b) Graph $x = \frac{1}{12}t^4 + 1$, $y = t - \frac{1}{4}\cos 2t + \frac{1}{4}$; $t \ge 0$.

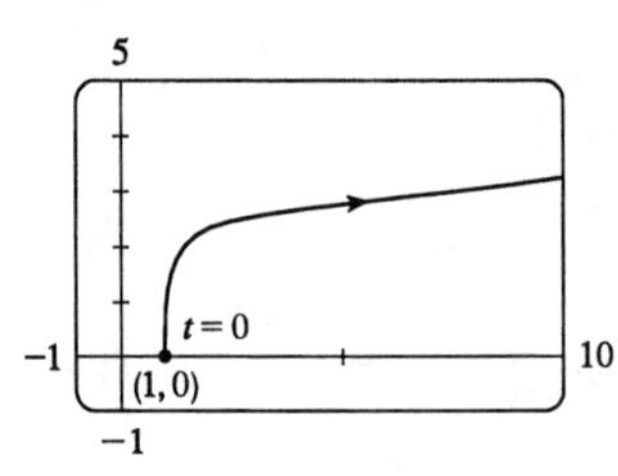

79. **(a)** From Example 11, the range d of the projectile is

$$\frac{v_0^2 \sin 2\alpha}{g} = \frac{(500)^2 \sin(2 \cdot 30°)}{g} \approx \frac{(500)^2 \left(\sqrt{3}/2\right)}{9.8} \approx 22{,}092.5 \text{ m}.$$

(b) The maximum height is reached when the vertical component of the velocity is 0. From Equation 7, $y = (v_0 \sin \alpha)t - \frac{1}{2}gt^2 \quad \Rightarrow \quad y' = v_0 \sin \alpha - gt$, which is the vertical component of the velocity. $y' = 0$ $\Rightarrow \quad v_0 \sin \alpha - gt = 0 \quad \Rightarrow \quad t = \dfrac{v_0 \sin \alpha}{g} = \dfrac{500 \sin 30°}{g} = \dfrac{250}{g}$. Substituting $\dfrac{250}{g}$ for t in the expression for y gives $(500 \sin 30°)\left(\dfrac{250}{g}\right) - \frac{1}{2}g\left(\dfrac{250}{g}\right)^2 = \dfrac{1}{2}\dfrac{(250)^2}{g} \approx 3188.8 \text{ m}$.

(c) From Example 11, the projectile lands when $t = \dfrac{2v_0 \sin \alpha}{g}$. Now

$$\begin{aligned}\mathbf{v}(t) = \mathbf{r}'(t) &= (v_0 \cos \alpha)\mathbf{i} + (v_0 \sin \alpha - gt)\mathbf{j} = (v_0 \cos \alpha)\mathbf{i} + (v_0 \sin \alpha - 2v_0 \sin \alpha)\mathbf{j} \\ &= v_0(\cos \alpha \, \mathbf{i} - \sin \alpha \, \mathbf{j}),\end{aligned}$$

so the speed at impact is $|\mathbf{v}(t)| = v_0 = 500 \text{ m/s}$. Note that the final velocity is equal to the initial velocity.

81. From Example 11, the horizontal distance d travelled is given by $d = \dfrac{v_0^2 \sin 2\alpha}{g}$. Solving for v_0 gives us $v_0^2 = \dfrac{dg}{\sin 2\alpha} \quad \Rightarrow \quad v_0 = \sqrt{\dfrac{dg}{\sin 2\alpha}}$ $(v_0 > 0)$. Substituting 90 for d, 9.8 for g, and 45° for α, we get $v_0 = \sqrt{90(9.8)} \approx 29.7 \text{ m/s}$.

83. From $x = (v_0 \cos \alpha)t$ and $y = (v_0 \sin \alpha)t - \frac{1}{2}gt^2$, we get $t = \dfrac{x}{v_0 \cos \alpha}$ and so $y = (v_0 \sin \alpha)\dfrac{x}{v_0 \cos \alpha} - \dfrac{g}{2}\left(\dfrac{x}{v_0 \cos \alpha}\right)^2 = (\tan \alpha)x - \left(\dfrac{g}{2v_0^2 \cos^2 \alpha}\right)x^2$. This equation for y is of the form $y = ax^2 + bx$, so it describes a parabola passing through the origin with vertex $\left(-b/(2a), -b^2/(4a)\right)$. The trajectory of the projectile is *part* of this parabola.

85. **(a)** The Mean Value Theorem says that there exists a number c in the interval (x_1, x_2) such that $H'(c) = \dfrac{H(x_2) - H(x_1)}{x_2 - x_1}$. Since $H = G - F$ and G and F are antiderivatives of f, $H'(c) = G'(c) - F'(c) = f(c) - f(c) = 0$. So now $\dfrac{H(x_2) - H(x_1)}{x_2 - x_1} = 0 \quad \Rightarrow \quad H(x_2) - H(x_1) = 0$ $(x_2 \neq x_1) \quad \Rightarrow \quad H(x_2) = H(x_1)$. Since this is true for any $x_1 < x_2$ in I, H must be a constant function.

(b) We have $H = G - F$ and $H(x) = C$, so $C = G - F \quad \Rightarrow \quad G(x) = F(x) + C$. Thus, any antiderivative G can be expressed as $F(x) + C$.

REVIEW EXERCISES FOR CHAPTER 5

1. False. For example, take $f(x) = x^3$, then $f'(x) = 3x^2$ and $f'(0) = 3(0)^2 = 0$, but $f(0) = 0$ is not a maximum or minimum; $(0, 0)$ is an inflection point.

3. False. For example, $f(x) = x$ is continuous on $(0, 1)$ but attains neither a maximum nor a minimum value on $(0, 1)$.

5. True, by the Test for Monotonic Functions.

7. False. For example, $f(x) = x + 2$, $g(x) = x + 1 \quad \Rightarrow \quad f'(x) = g'(x) = 1$, but $f(x) \neq g(x)$.

9. True. The graph of one such function is sketched. [An example is $f(x) = e^{-x}$.]

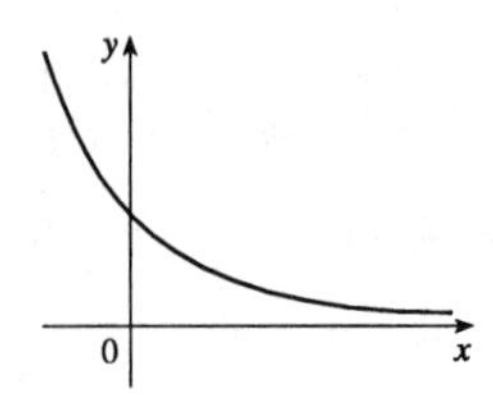

11. True. Let $x_1 < x_2$ where $x_1, x_2 \in I$. Then $f(x_1) < f(x_2)$ and $g(x_1) < g(x_2)$ (since f and g are increasing on I), so $(f + g)(x_1) = f(x_1) + g(x_1) < f(x_2) + g(x_2) = (f + g)(x_2)$.

13. False. Take $f(x) = x$ and $g(x) = x - 1$. Then both f and g are increasing on $[0, 1]$. But $f(x)g(x) = x(x - 1)$ is not increasing on $[0, 1]$.

15. True. Let $x_1, x_2 \in I$ and $x_1 < x_2$, then $f(x_1) < f(x_2)$ (f is increasing) $\Rightarrow$ $\dfrac{1}{f(x_1)} > \dfrac{1}{f(x_2)}$ (f is positive) $\Rightarrow \quad g(x_1) > g(x_2) \quad \Rightarrow \quad g(x) = \dfrac{1}{f(x)}$ is decreasing on I.

17. **(a)** $f'(x) > 0$ on $(-2, 0)$ and $(2, \infty) \quad \Rightarrow \quad f$ is increasing on those intervals. $f'(x) < 0$ on $(-\infty, -2)$ and $(0, 2) \quad \Rightarrow \quad f$ is decreasing on those intervals.

(b) $f'(x) = 0$ at $x = -2$, 0, and 2, so these are where local maxima or minima will occur. At $x = \pm 2$, f' changes from negative to positive, so f has local minima at those values. At $x = 0$, f' changes from positive to negative, so f has a local maximum there.

(c) f' is increasing on $(-\infty, -1)$ and $(1, \infty) \quad \Rightarrow \quad f'' > 0$ and f is concave upward on those intervals. f' is decreasing on $(-1, 1) \quad \Rightarrow \quad f'' < 0$ and f is concave downward on this interval.

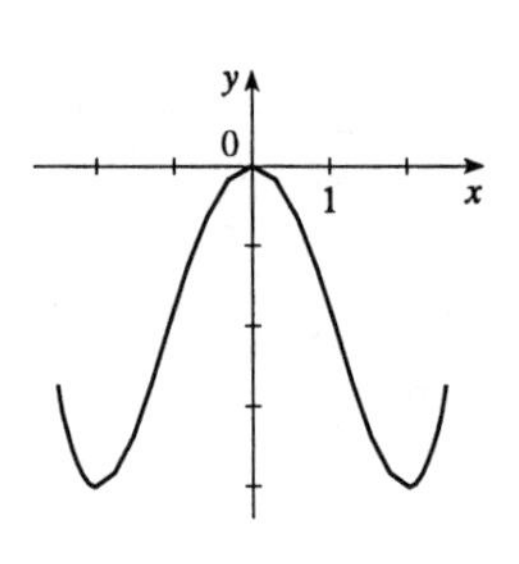

19. $f(x) = x^3 - 12x + 5$, $-5 \le x \le 3$. $f'(x) = 3x^2 - 12 = 0 \quad \Rightarrow \quad x^2 = 4 \quad \Rightarrow \quad x = \pm 2$. $f''(x) = 6x \quad \Rightarrow$ $f''(-2) = -12 < 0$, so $f(-2) = 21$ is a local maximum, and $f''(2) = 12 > 0$, so $f(2) = -11$ is a local minimum. Also $f(-5) = -60$ and $f(3) = -4$, so $f(-2) = 21$ is the absolute maximum and $f(-5) = -60$ is the absolute minimum.

21. $f(x) = \dfrac{x-2}{x+2}$, $0 \le x \le 4$. $f'(x) = \dfrac{(x+2)-(x-2)}{(x+2)^2} = \dfrac{4}{(x+2)^2} > 0 \quad \Rightarrow \quad f$ is increasing on $[0, 4]$, so f has no local extremum and $f(0) = -1$ is the absolute minimum and $f(4) = \frac{1}{3}$ is the absolute maximum.

23. $f(x) = x - \sqrt{2}\sin x$, $0 \le x \le \pi$. $f'(x) = 1 - \sqrt{2}\cos x = 0 \quad \Rightarrow \quad \cos x = \frac{1}{\sqrt{2}} \quad \Rightarrow \quad x = \frac{\pi}{4}$.
$f''\left(\frac{\pi}{4}\right) = \sqrt{2}\sin\frac{\pi}{4} = 1 > 0$, so $f\left(\frac{\pi}{4}\right) = \frac{\pi}{4} - 1$ is a local minimum. Also $f(0) = 0$ and $f(\pi) = \pi$, so the absolute minimum is $f\left(\frac{\pi}{4}\right) = \frac{\pi}{4} - 1$, the absolute maximum is $f(\pi) = \pi$.

25. **(a)** $y = f(x) = 1 + x + x^3$ is a polynomial, so there is no asymptote.

(b) $f'(x) = 1 + 3x^2 > 0$, so f is increasing on $\mathbb{R}$.

(c) Since $f' \neq 0$ and f' exists for all $\mathbb{R}$, there is no local maximum or minimum.

(d) $f''(x) = 6x \quad \Rightarrow \quad f''(x) > 0$ if $x > 0$ and $f''(x) < 0$ if $x < 0$, so f is CU on $(0, \infty)$ and CD on $(-\infty, 0)$. IP $(0, 1)$

(e)

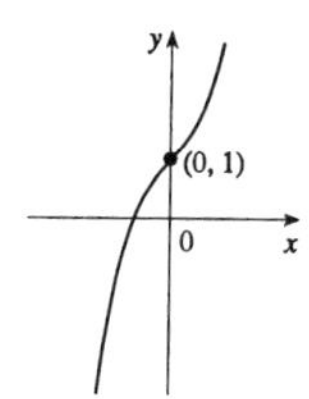

27. **(a)** $\displaystyle\lim_{x\to\pm\infty} \frac{1}{x(x-3)^2} = 0$, so $y = 0$ is a HA. $\displaystyle\lim_{x\to 0^+} \frac{1}{x(x-3)^2} = \infty$, $\displaystyle\lim_{x\to 0^-} \frac{1}{x(x-3)^2} = -\infty$, $\displaystyle\lim_{x\to 3} \frac{1}{x(x-3)^2} = \infty$, so $x = 0$ and $x = 3$ are VA.

(b) $y = f(x) = \dfrac{1}{x(x-3)^2} \quad \Rightarrow \quad f'(x) = -\dfrac{(x-3)^2 + 2x(x-3)}{x^2(x-3)^4} = \dfrac{3(1-x)}{x^2(x-3)^3} \quad \Rightarrow \quad f'(x) > 0 \quad \Leftrightarrow$ $1 < x < 3$, so f is increasing on $[1, 3)$ and decreasing on $(-\infty, 0)$, $(0, 1]$, and $(3, \infty)$.

(c) $f(1) = \frac{1}{4}$ is a local minimum.

(d) $f''(x) = \dfrac{6(2x^2 - 4x + 3)}{x^3(x-3)^4}$. Note that $2x^2 - 4x + 3 > 0$ for all x. So $f''(x) > 0 \quad \Leftrightarrow \quad x > 0 \quad \Rightarrow \quad f$ is CU on $(0, 3)$ and $(3, \infty)$ and CD on $(-\infty, 0)$. No IP

(e)

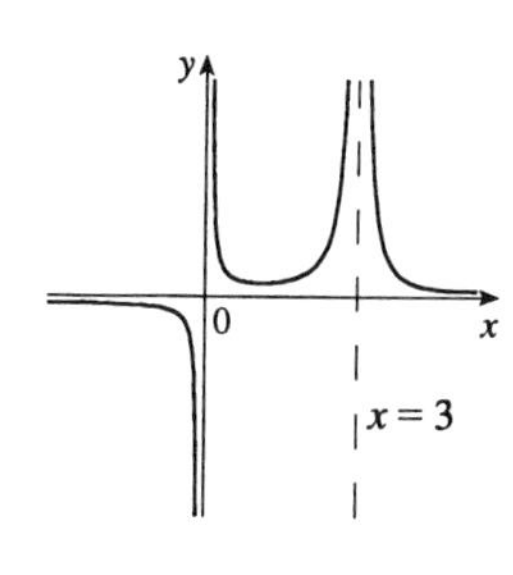

29. **(a)** $\lim\limits_{x\to-\infty} x\sqrt{5-x} = -\infty$, no asymptote

(b) $y = f(x) = x\sqrt{5-x} \quad\Rightarrow\quad f'(x) = \sqrt{5-x} - \frac{x}{2\sqrt{5-x}} = \frac{10-3x}{2\sqrt{5-x}} > 0 \quad\Leftrightarrow\quad x < \frac{10}{3}$. So f is increasing on $\left(-\infty, \frac{10}{3}\right]$ and decreasing on $\left[\frac{10}{3}, 5\right]$.

(c) $f\left(\frac{10}{3}\right) = \frac{10\sqrt{5}}{3\sqrt{3}}$ is a local and absolute maximum.

(d) $f''(x) = \dfrac{-6\sqrt{5-x} - (10-3x)\left(-1/\sqrt{5-x}\right)}{4(5-x)} = \dfrac{3x-20}{4(5-x)^{3/2}} < 0$ for all x in D, so f is CD on $(-\infty, 5)$.

(e)

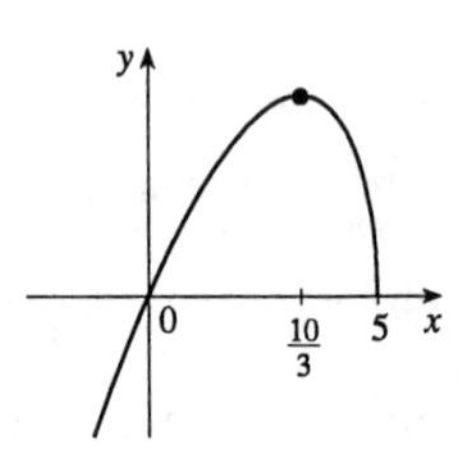

31. **(a)** $\lim\limits_{x\to 0^+} \frac{\sqrt{1-x^2}}{x} = \infty$, $\lim\limits_{x\to 0^-} \frac{\sqrt{1-x^2}}{x} = -\infty$, so $x = 0$ is a VA.

(b) $y = f(x) = \dfrac{\sqrt{1-x^2}}{x} \quad\Rightarrow\quad f'(x) = \dfrac{\left(-x^2/\sqrt{1-x^2}\right) - \sqrt{1-x^2}}{x^2} = -\dfrac{1}{x^2\sqrt{1-x^2}} < 0$, so f is decreasing on $[-1, 0)$ and $(0, 1]$.

(c) No extremum

(d) $f''(x) = \dfrac{2-3x^2}{x^3(1-x^2)^{3/2}} > 0 \quad\Leftrightarrow\quad -1 < x < -\sqrt{\frac{2}{3}}$ or $0 < x < \sqrt{\frac{2}{3}}$, so f is CU on $\left(-1, -\sqrt{\frac{2}{3}}\right)$ and $\left(0, \sqrt{\frac{2}{3}}\right)$ and CD on $\left(-\sqrt{\frac{2}{3}}, 0\right)$ and $\left(\sqrt{\frac{2}{3}}, 1\right)$. IP $\left(\pm\sqrt{\frac{2}{3}}, \pm\frac{1}{\sqrt{2}}\right)$.

(e)

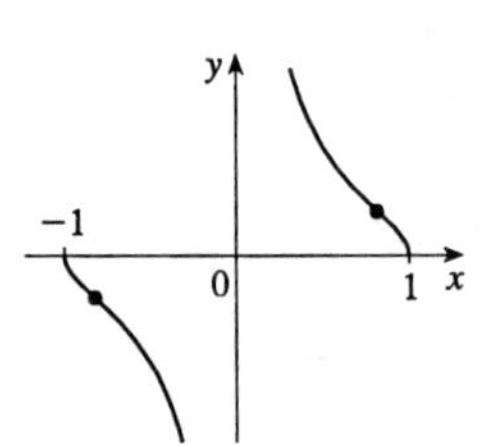

33. **(a)** $\lim\limits_{x\to\infty} \left(x^{1/2} - x^{1/3}\right) = \lim\limits_{x\to\infty} \left[x^{1/3}\left(x^{1/6} - 1\right)\right] = \infty$, no asymptote

(b) $y = f(x) = \sqrt{x} - \sqrt[3]{x} \quad\Rightarrow\quad f'(x) = \frac{1}{2}x^{-1/2} - \frac{1}{3}x^{-2/3} = \dfrac{3x^{1/6} - 2}{6x^{2/3}} > 0 \quad\Leftrightarrow\quad 3x^{1/6} > 2 \quad\Leftrightarrow$ $x > \left(\frac{2}{3}\right)^6$, so f is increasing on $\left[\left(\frac{2}{3}\right)^6, \infty\right)$ and decreasing on $\left[0, \left(\frac{2}{3}\right)^6\right]$.

(c) $f\left(\left(\frac{2}{3}\right)^6\right) = -\frac{4}{27}$ is a local minimum.

(d) $f''(x) = -\frac{1}{4}x^{-3/2} + \frac{2}{9}x^{-5/3} = \dfrac{8 - 9x^{1/6}}{36x^{5/3}} > 0 \quad \Leftrightarrow \quad x^{1/6} < \frac{8}{9} \quad \Leftrightarrow \quad x < \left(\frac{8}{9}\right)^6$, so f is CU on $\left(0, \left(\frac{8}{9}\right)^6\right)$ and CD on $\left(\left(\frac{8}{9}\right)^6, \infty\right)$. IP $\left(\frac{8}{9}, -\frac{64}{729}\right)$

(e)

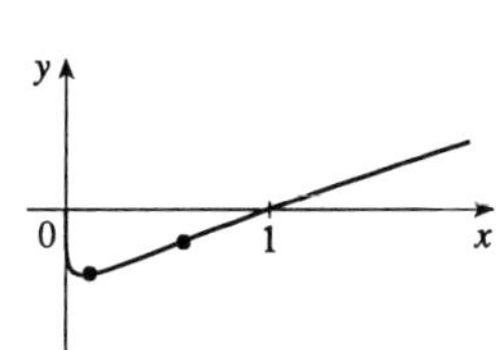

35. **(a)** $\lim\limits_{x\to\pm\infty} \sin^{-1}(1/x) = \sin^{-1}(0) = 0$, so $y = 0$ is a HA.

(b) $y = f(x) = \sin^{-1}(1/x) \quad \Rightarrow \quad f'(x) = \dfrac{1}{\sqrt{1 - (1/x)^2}}\left(-\dfrac{1}{x^2}\right) = \dfrac{-1}{\sqrt{x^4 - x^2}} < 0$, so f is decreasing on $(-\infty, -1]$ and $[1, \infty)$.

(c) No local maximum or minimum, but $f(1) = \frac{\pi}{2}$ is the absolute maximum and $f(-1) = -\frac{\pi}{2}$ is the absolute minimum.

(d) $f''(x) = \dfrac{4x^3 - 2x}{2(x^4 - x^2)^{3/2}} = \dfrac{x(2x^2 - 1)}{(x^4 - x^2)^{3/2}} > 0$ for $x > 1$ and $f''(x) < 0$ for $x < -1$, so f is CU on $(1, \infty)$ and CD on $(-\infty, -1)$.

(e)

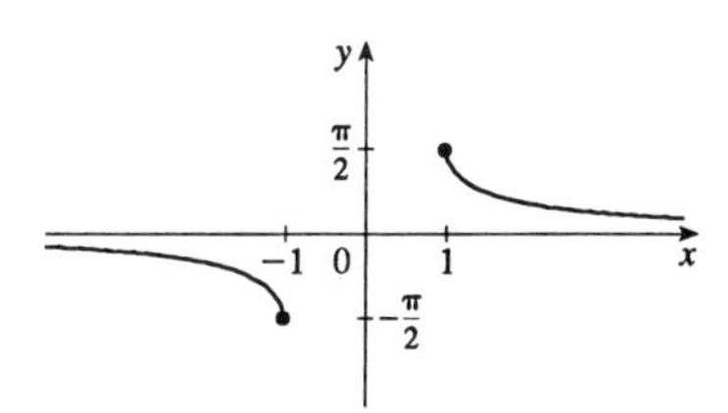

37. **(a)** $\lim\limits_{x\to\pm\infty} (e^x + e^{-3x}) = \infty$, no asymptote

(b) $y = f(x) = e^x + e^{-3x} \quad \Rightarrow \quad f'(x) = e^x - 3e^{-3x} = e^{-3x}(e^{4x} - 3) > 0 \quad \Leftrightarrow \quad e^{4x} > 3 \quad \Leftrightarrow$ $4x > \ln 3 \quad \Leftrightarrow \quad x > \frac{1}{4}\ln 3$, so f is increasing on $\left[\frac{1}{4}\ln 3, \infty\right)$ and decreasing on $\left(-\infty, \frac{1}{4}\ln 3\right]$.

(c) Absolute minimum $f\left(\frac{1}{4}\ln 3\right) = 3^{1/4} + 3^{-3/4}$

(d) $f''(x) = e^x + 9e^{-3x} > 0$, so f is CU on $(-\infty, \infty)$.

(e)

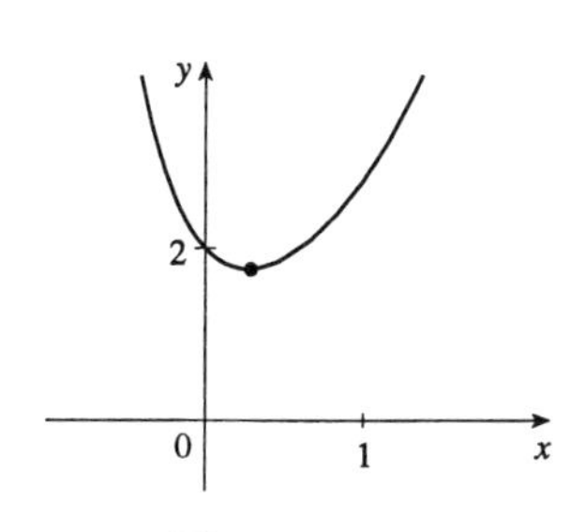

39. $f(x) = \dfrac{x^2 - 1}{x^3} \quad \Rightarrow \quad f'(x) = \dfrac{x^3(2x) - (x^2 - 1)3x^2}{x^6} = \dfrac{3 - x^2}{x^4} \quad \Rightarrow$

$f''(x) = \dfrac{x^4(-2x) - (3 - x^2)4x^3}{x^8} = \dfrac{2x^2 - 12}{x^5}$

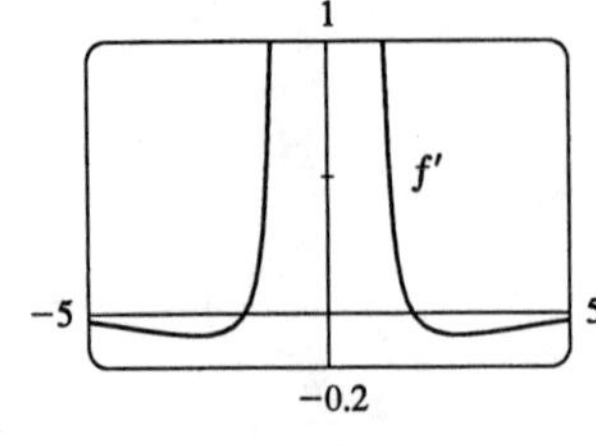

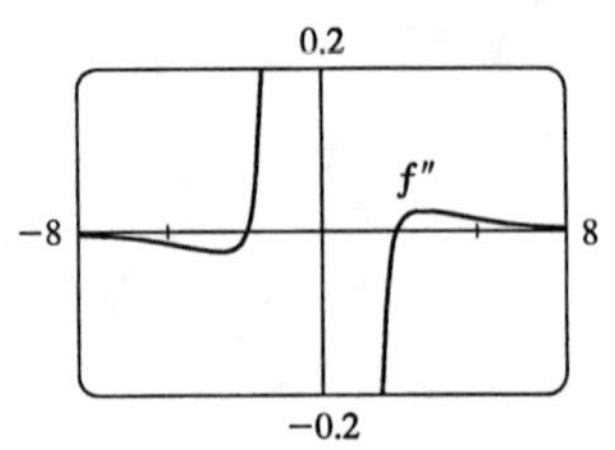

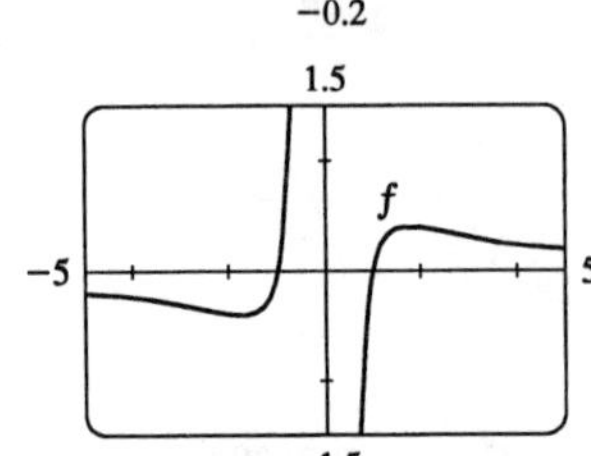

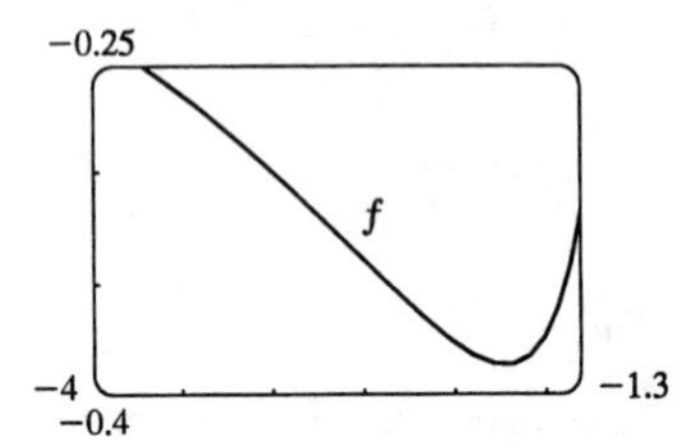

From the graphs of f' and f'', it appears that f is increasing on $[-1.73, 0)$ and $(0, 1.73]$ and decreasing on $(-\infty, -1.73]$ and $[1.73, \infty)$; f has a local maximum of about $f(1.73) = 0.38$ and a local minimum of about $f(-1.7) = -0.38$; f is CU on $(-2.45, 0)$ and $(2.45, \infty)$, and CD on $(-\infty, -2.45)$ and $(0, 2.45)$; and f has inflection points at about $(-2.45, -0.34)$ and $(2.45, 0.34)$. Now $f'(x) = \dfrac{3 - x^2}{x^4}$ is positive for $0 < x^2 < 3$, that is, f is increasing on $\left[-\sqrt{3}, 0\right)$ and $\left(0, \sqrt{3}\right]$; and $f'(x)$ is negative (and so f is decreasing) on $\left(-\infty, -\sqrt{3}\right]$ and $\left[\sqrt{3}, \infty\right)$. $f'(x) = 0$ when $x = \pm\sqrt{3}$. f' goes from positive to negative at $x = \sqrt{3}$, so f has a local maximum of $f\left(\sqrt{3}\right) = \dfrac{\left(\sqrt{3}\right)^2 - 1}{\left(\sqrt{3}\right)^3} = \dfrac{2\sqrt{3}}{9}$; and since f is odd, we know that maxima on the interval $[0, \infty)$ correspond to minima on $(-\infty, 0]$, so f has a local minimum of $f\left(-\sqrt{3}\right) = -\dfrac{2\sqrt{3}}{9}$. Also, $f''(x) = \dfrac{2x^2 - 12}{x^5}$ is positive (so f is CU) on $\left(-\sqrt{6}, 0\right)$ and $\left(\sqrt{6}, \infty\right)$, and negative (so f is CD) on $\left(-\infty, -\sqrt{6}\right)$ and $\left(0, \sqrt{6}\right)$. There are IP at $\left(\sqrt{6}, \frac{5\sqrt{6}}{36}\right)$ and $\left(-\sqrt{6}, -\frac{5\sqrt{6}}{36}\right)$.

41. $f(x) = 3x^6 - 5x^5 + x^4 - 5x^3 - 2x^2 + 2$, $f'(x) = 18x^5 - 25x^4 + 4x^3 - 15x^2 - 4x$,

$f''(x) = 90x^4 - 100x^3 + 12x^2 - 30x - 4$

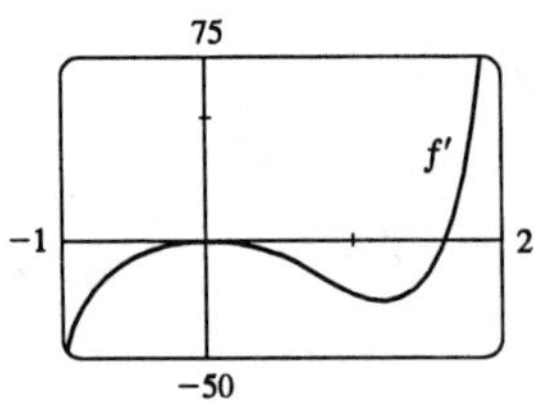

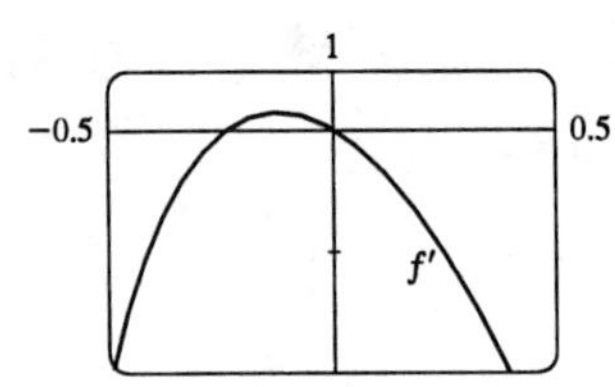

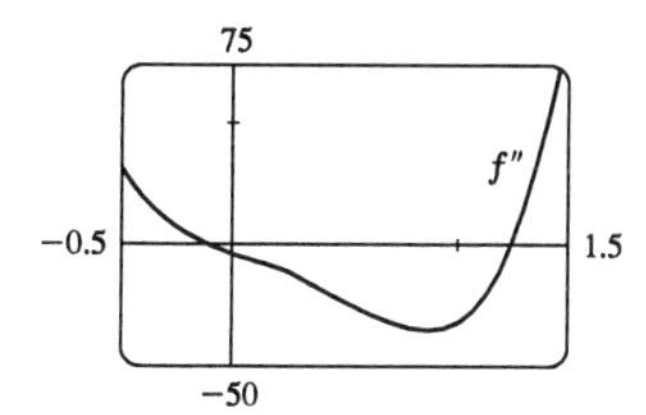

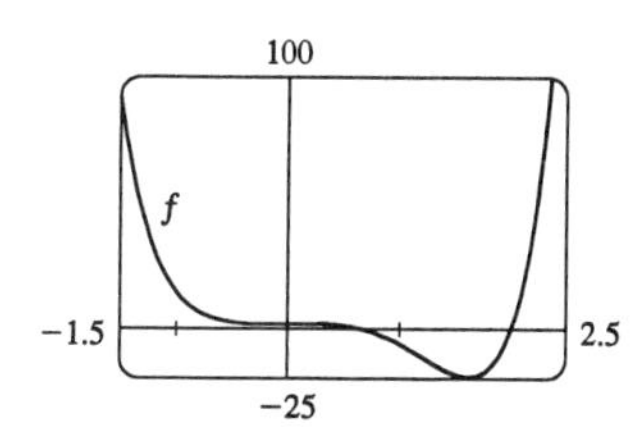

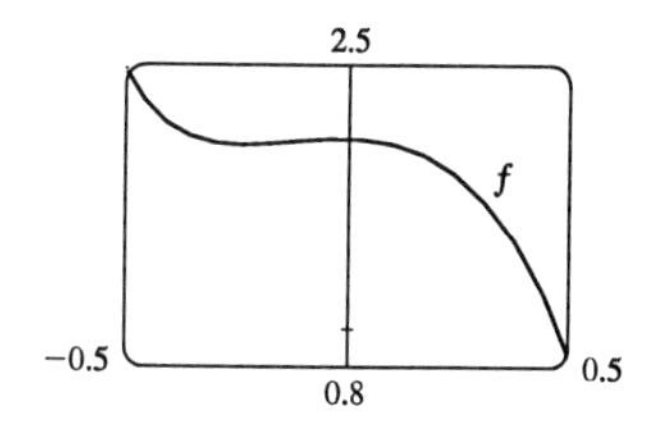

From the graphs of f' and f'', it appears that f is increasing on $[-0.23, 0]$ and $[1.62, \infty)$ and decreasing on $(-\infty, -0.23]$ and $[0, 1.62]$; f has a local maximum of about $f(0) = 2$ and local minima of about $f(-0.23) = 1.96$ and $f(1.62) = -19.2$; f is CU on $(-\infty, -0.12)$ and $(1.24, \infty)$ and CD on $(-0.12, 1.24)$; and f has inflection points at about $(-0.12, 1.98)$ and $(1.2, -12.1)$.

43. 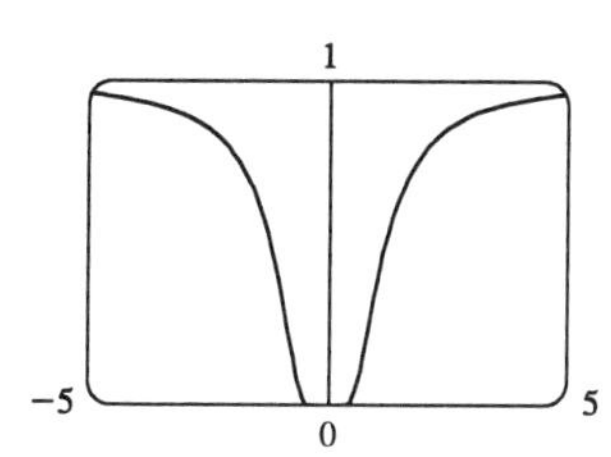

From the graph, we estimate the points of inflection to be about $(\pm 0.8, 0.2)$. $f(x) = e^{-1/x^2} \Rightarrow f'(x) = 2x^{-3}e^{-1/x^2} \Rightarrow$

$$f''(x) = 2\left[x^{-3}(2x^{-3})e^{-1/x^2} + e^{-1/x^2}(-3x^{-4})\right]$$
$$= 2x^{-6}e^{-1/x^2}(2 - 3x^2).$$ This is 0 when $2 - 3x^2 = 0$ $\Leftrightarrow$ $x = \pm\sqrt{\frac{2}{3}}$, so the inflection points are $\left(\pm\sqrt{\frac{2}{3}}, e^{-3/2}\right)$.

45. $f(x) = \arctan(\cos(3 \arcsin x))$. We use a CAS to compute f' and f'', and to graph f, f', and f'':

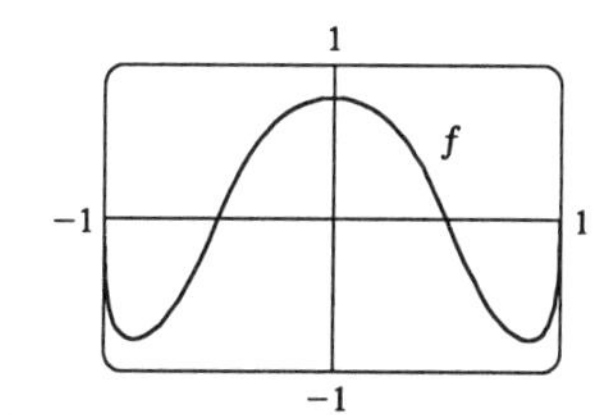

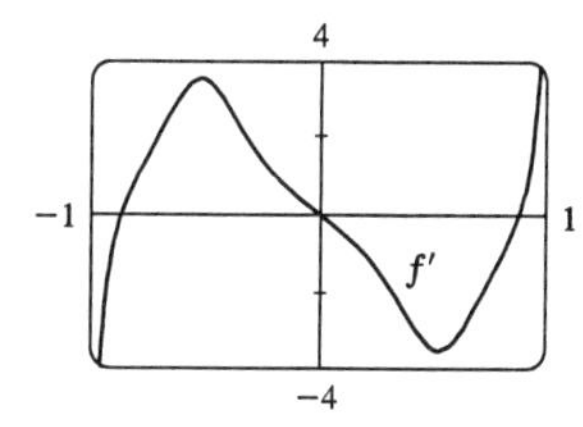

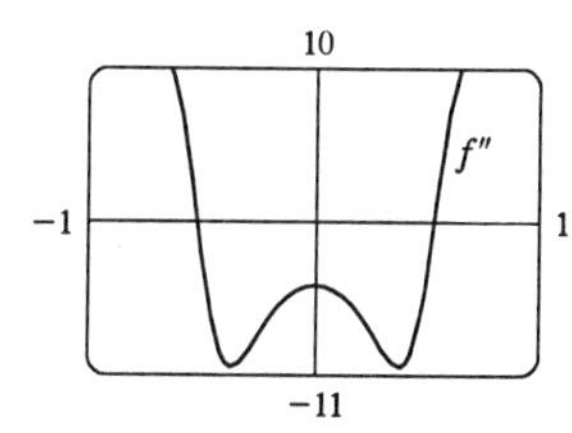

From the graph of f', it appears that the only maximum occurs at $x = 0$ and there are minima at $x = \pm 0.87$. From the graph of f'', it appears that there are inflection points at $x = \pm 0.52$.

47. $f(x) = x^{101} + x^{51} + x - 1 = 0$. Since f is continuous and $f(0) = -1$ and $f(1) = 2$, the equation has at least one root in $(0, 1)$, by the Intermediate Value Theorem. Suppose the equation has two roots, a and b, with $a < b$. Then $f(a) = 0 = f(b)$, so by the Mean Value Theorem, $f'(x) = \dfrac{f(b) - f(a)}{b - a} = \dfrac{0}{b - a} = 0$, so $f'(x)$ has a root in (a, b). But this is impossible since $f'(x) = 101x^{100} + 51x^{50} + 1 \geq 1$ for all x.

49. Since f is continuous on $[32, 33]$ and differentiable on $(32, 33)$, then by the Mean Value Theorem there exists a number c in $(32, 33)$ such that $f'(c) = \frac{1}{5}c^{-4/5} = \dfrac{\sqrt[5]{33} - \sqrt[5]{32}}{33 - 32} = \sqrt[5]{33} - 2$, but $\frac{1}{5}c^{-4/5} > 0 \Rightarrow$ $\sqrt[5]{33} - 2 > 0 \Rightarrow \sqrt[5]{33} > 2$. Also f' is decreasing, so that $f'(c) < f'(32) = \frac{1}{5}(32)^{-4/5} = 0.0125 \Rightarrow$ $0.0125 > f'(c) = \sqrt[5]{33} - 2 \Rightarrow \sqrt[5]{33} < 2.0125$. Therefore $2 < \sqrt[5]{33} < 2.0125$.

51.

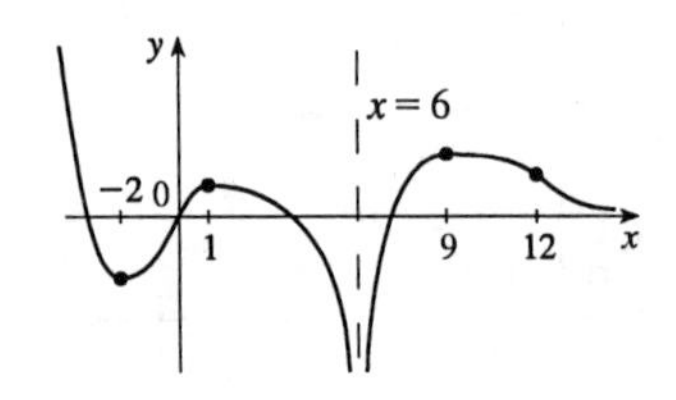

53. For $(1, 6)$ to be on the curve, we have that $6 = y(1) = 1^3 + a(1)^2 + b(1) + 1 = a + b + 2 \quad \Rightarrow \quad b = 4 - a$. Now $y' = 3x^2 + 2ax + b$ and $y'' = 6x + 2a$. Also, for $(1, 6)$ to be an inflection point it must be true that $y''(1) = 6(1) + 2a = 0 \quad \Rightarrow \quad a = -3 \quad \Rightarrow \quad b = 4 - (-3) = 7$.

55. If $B = 0$, the line is vertical and the distance from $x = -\dfrac{C}{A}$ to (x_1, y_1) is $\left|x_1 + \dfrac{C}{A}\right| = \dfrac{|Ax_1 + By_1 + C|}{\sqrt{A^2 + B^2}}$, so assume $B \neq 0$. The square of the distance from (x_1, y_1) to the line is $f(x) = (x - x_1)^2 + (y - y_1)^2$ where $Ax + By + C = 0$, so we minimize $f(x) = (x - x_1)^2 + \left(-\dfrac{A}{B}x - \dfrac{C}{B} - y_1\right)^2 \quad \Rightarrow$

$f'(x) = 2(x - x_1) + 2\left(-\dfrac{A}{B}x - \dfrac{C}{B} - y_1\right)\left(-\dfrac{A}{B}\right)$. $f'(x) = 0 \quad \Rightarrow \quad x = \dfrac{B^2 x_1 - ABy_1 - AC}{A^2 + B^2}$ and this gives a minimum since $f''(x) = 2\left(1 + \dfrac{A^2}{B^2}\right) > 0$. Substituting this value of x and simplifying gives

$f(x) = \dfrac{(Ax_1 + By_1 + C)^2}{A^2 + B^2}$, so the minimum distance is $\dfrac{|Ax_1 + By_1 + C|}{\sqrt{A^2 + B^2}}$.

57. By similar triangles, $\dfrac{y}{x} = \dfrac{r}{\sqrt{x^2 - 2rx}}$, so the area of the triangle is

$A(x) = \frac{1}{2}(2y)x = xy = \dfrac{rx^2}{\sqrt{x^2 - 2rx}} \quad \Rightarrow$

$$A'(x) = \frac{2rx\sqrt{x^2 - 2rx} - rx^2(x - r)/\sqrt{x^2 - 2rx}}{x^2 - 2rx} = \frac{rx^2(x - 3r)}{(x^2 - 2rx)^{3/2}} = 0$$

when $x = 3r$. $A'(x) < 0$ when $2r < x < 3r$, $A'(x) > 0$ when $x > 3r$. So $x = 3r$ gives a minimum and $A(3r) = r(9r^2)/\left(\sqrt{3}r\right) = 3\sqrt{3}r^2$.

59.

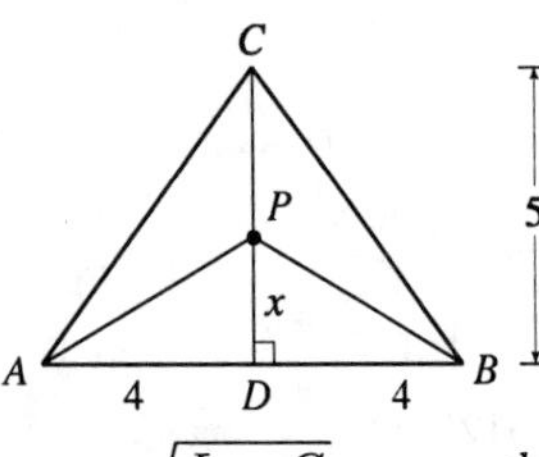

We minimize

$L(x) = |PA| + |PB| + |PC| = 2\sqrt{x^2 + 16} + (5 - x)$, $0 \le x \le 5$.

$L'(x) = 2x/\sqrt{x^2 + 16} - 1 = 0 \quad \Leftrightarrow \quad 2x = \sqrt{x^2 + 16} \quad \Leftrightarrow$

$4x^2 = x^2 + 16 \quad \Leftrightarrow \quad x = \frac{4}{\sqrt{3}}$. $L(0) = 13$, $L\left(\frac{4}{\sqrt{3}}\right) \approx 11.9$,

$L(5) \approx 12.8$, so the minimum occurs when $x = \frac{4}{\sqrt{3}}$.

61. $v = K\sqrt{\dfrac{L}{C} + \dfrac{C}{L}} \quad \Rightarrow \quad \dfrac{dv}{dL} = \dfrac{K}{2\sqrt{(L/C) + (C/L)}}\left(\dfrac{1}{C} - \dfrac{C}{L^2}\right) = 0 \quad \Leftrightarrow \quad \dfrac{1}{C} = \dfrac{C}{L^2} \quad \Leftrightarrow \quad L^2 = C^2$ $\Leftrightarrow \quad L = C$. This gives the minimum velocity since $v' < 0$ for $0 < L < C$ and $v' > 0$ for $L > C$.

63. Let x = selling price of ticket. Then $12 - x$ is the amount the ticket price has been lowered, so the number of tickets sold is $11{,}000 + 1000(12 - x) = 23{,}000 - 1000x$. The revenue is $R(x) = x(23{,}000 - 1000x) = 23{,}000x - 1000x^2$, so $R'(x) = 23{,}000 - 2000x = 0$ when $x = 11.5$. Since $R''(x) = -2000 < 0$, the maximum revenue occurs when the ticket prices are \$11.50.

65. $f'(x) = x - \sqrt[4]{x} = x - x^{1/4} \quad \Rightarrow \quad f(x) = \frac{1}{2}x^2 - \frac{4}{5}x^{5/4} + C$

67. $f'(x) = (1 + x)/\sqrt{x} = x^{-1/2} + x^{1/2} \quad \Rightarrow \quad f(x) = 2x^{1/2} + \frac{2}{3}x^{3/2} + C \quad \Rightarrow \quad 0 = f(1) = 2 + \frac{2}{3} + C \quad \Rightarrow$ $C = -\frac{8}{3} \quad \Rightarrow \quad f(x) = 2x^{1/2} + \frac{2}{3}x^{3/2} - \frac{8}{3}$

69. $f''(x) = x^3 + x \quad \Rightarrow \quad f'(x) = \frac{1}{4}x^4 + \frac{1}{2}x^2 + C \quad \Rightarrow \quad 1 = f'(0) = C \quad \Rightarrow \quad f'(x) = \frac{1}{4}x^4 + \frac{1}{2}x^2 + 1 \quad \Rightarrow$ $f(x) = \frac{1}{20}x^5 + \frac{1}{6}x^3 + x + D \quad \Rightarrow \quad -1 = f(0) = D \quad \Rightarrow \quad f(x) = \frac{1}{20}x^5 + \frac{1}{6}x^3 + x - 1$

71. $f'(x) = 2/(1 + x^2) \quad \Rightarrow \quad f(x) = 2\arctan x + C$. $f(0) = 2\arctan 0 + C = -1 \quad \Rightarrow \quad C = -1$. Therefore $f(x) = 2\arctan x - 1$.

73. $\mathbf{a}(t) = \mathbf{v}'(t) = \mathbf{i} + 2\mathbf{j} \quad \Rightarrow \quad \mathbf{v}(t) = t\mathbf{i} + 2t\mathbf{j} + \mathbf{C}$ ($\mathbf{C}$ is a constant vector). Now $\mathbf{v}(0) = \mathbf{C}$ and $\mathbf{v}(0) = \mathbf{j} \quad \Rightarrow$ $\mathbf{C} = \mathbf{j}$, so $\mathbf{v}(t) = \mathbf{r}'(t) = t\mathbf{i} + 2t\mathbf{j} + (\mathbf{j}) = t\mathbf{i} + (2t + 1)\mathbf{j}$, and hence, $\mathbf{r}(t) = \frac{1}{2}t^2\mathbf{i} + (t^2 + t)\mathbf{j} + \mathbf{D}$. But $\mathbf{r}(0) = \mathbf{D}$ and $\mathbf{r}(0) = -\mathbf{i} + \mathbf{j} \quad \Rightarrow \quad \mathbf{D} = -\mathbf{i} + \mathbf{j}$. Thus, $\mathbf{r}(t) = \frac{1}{2}t^2\mathbf{i} + (t^2 + t)\mathbf{j} + (-\mathbf{i} + \mathbf{j}) = \left(\frac{1}{2}t^2 - 1\right)\mathbf{i} + (t^2 + t + 1)\mathbf{j}$.

75. $\mathbf{a}(t) = \mathbf{v}'(t) = 2t\mathbf{i} + 3\mathbf{j} \quad \Rightarrow \quad \mathbf{v}(t) = t^2\mathbf{i} + 3t\mathbf{j} + \mathbf{C}$. Now $\mathbf{v}(0) = \mathbf{C}$ and $\mathbf{v}(0) = \mathbf{i} - \mathbf{j} \quad \Rightarrow \quad \mathbf{C} = \mathbf{i} - \mathbf{j}$, so $\mathbf{v}(t) = \mathbf{r}'(t) = t^2\mathbf{i} + 3t\mathbf{j} + (\mathbf{i} - \mathbf{j}) = (t^2 + 1)\mathbf{i} + (3t - 1)\mathbf{j}$, and hence, $\mathbf{r}(t) = \left(\frac{1}{3}t^3 + t\right)\mathbf{i} + \left(\frac{3}{2}t^2 - t\right)\mathbf{j} + \mathbf{D}$. But $\mathbf{r}(0) = \mathbf{D}$ and $\mathbf{r}(0) = \mathbf{i} + 2\mathbf{j} \quad \Rightarrow \quad \mathbf{D} = \mathbf{i} + 2\mathbf{j}$. Thus,
$\mathbf{r}(t) = \left(\frac{1}{3}t^3 + t\right)\mathbf{i} + \left(\frac{3}{2}t^2 - t\right)\mathbf{j} + (\mathbf{i} + 2\mathbf{j}) = \left(\frac{1}{3}t^3 + t + 1\right)\mathbf{i} + \left(\frac{3}{2}t^2 - t + 2\right)\mathbf{j}$.

77. Assume that the particle travels forward and does not reverse direction. $\mathbf{a}(t) = 3\mathbf{i} - \mathbf{j} \quad \Rightarrow$ $\mathbf{v}(t) = 3t\mathbf{i} - t\mathbf{j} + \mathbf{C}$, where $\mathbf{C} = \mathbf{v}(0)$. Since the unit vector on the line through $(1, 2)$ and $(4, 1)$ is $\frac{1}{\sqrt{10}}[(4 - 1)\mathbf{i} + (1 - 2)\mathbf{j}]$ and the initial speed of the particle is $|\mathbf{v}(0)| = 2$, we have
$\mathbf{v}(0) = 2 \cdot \frac{1}{\sqrt{10}}(3\mathbf{i} - \mathbf{j}) = \frac{6}{\sqrt{10}}\mathbf{i} - \frac{2}{\sqrt{10}}\mathbf{j}$, so $\mathbf{v}(t) = \left(3t + \frac{6}{\sqrt{10}}\right)\mathbf{i} + \left(-t - \frac{2}{\sqrt{10}}\right)\mathbf{j}$. Now
$\mathbf{r}(t) = \left(\frac{3}{2}t^2 + \frac{6}{\sqrt{10}}t\right)\mathbf{i} + \left(-\frac{1}{2}t^2 - \frac{2}{\sqrt{10}}t\right)\mathbf{j} + \mathbf{D}$ and $\mathbf{r}(0) = \mathbf{D} = \mathbf{i} + 2\mathbf{j} \quad \Rightarrow$
$\mathbf{r}(t) = \left(\frac{3}{2}t^2 + \frac{6}{\sqrt{10}}t + 1\right)\mathbf{i} + \left(-\frac{1}{2}t^2 - \frac{2}{\sqrt{10}}t + 2\right)\mathbf{j}$.

79.

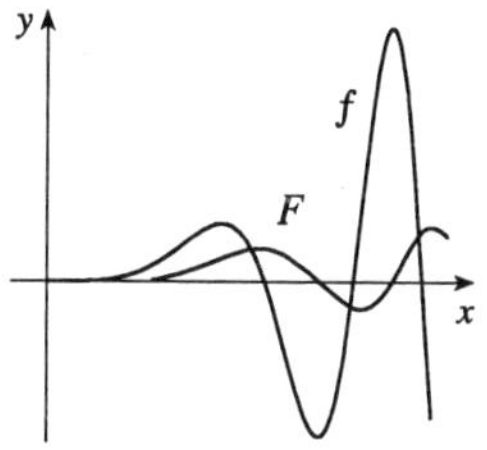

y
f
F
x

81. Choosing the positive direction to be upward, we have $a(t) = -9.8 \quad \Rightarrow \quad v(t) = -9.8t + v_0$, but $v(0) = 0 = v_0 \quad \Rightarrow \quad v(t) = -9.8t = s'(t) \quad \Rightarrow \quad s(t) = -4.9t^2 + s_0$, but $s(0) = s_0 = 500 \quad \Rightarrow$ $s(t) = -4.9t^2 + 500$. When $s = 0$, $-4.9t^2 + 500 = 0 \quad \Rightarrow \quad t = \sqrt{\frac{500}{4.9}} \quad \Rightarrow$ $v = -9.8\sqrt{\frac{500}{4.9}} \approx -98.995\,\text{m/s}$. Therefore the canister will not burst.

83. **(a)** The cross-sectional area is

$A = 2x \cdot 2y = 4xy = 4x\sqrt{100 - x^2}$, $0 \le x \le 10$, so

$$\frac{dA}{dx} = 4x\left(\tfrac{1}{2}\right)\left(100 - x^2\right)^{-1/2}(-2x) + \left(100 - x^2\right)^{1/2} \cdot 4$$

$$= \frac{-4x^2}{\left(100 - x^2\right)^{1/2}} + 4\left(100 - x^2\right)^{1/2} = 0 \text{ when}$$

$-4x^2 + 4(100 - x^2) = 0 \quad \Rightarrow \quad -8x^2 + 400 = 0$

$\Rightarrow \quad x^2 = 50 \quad \Rightarrow \quad x = \sqrt{50} \quad \Rightarrow \quad y = \sqrt{50}$. And $A(0) = A(10) = 0$. Therefore, the rectangle of maximum area is a square.

(b) $y = \sqrt{100 - x^2}$. The cross-sectional area of each rectangular plank is

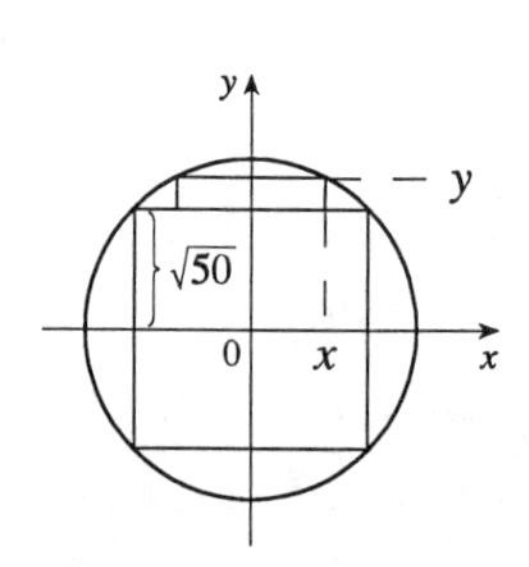

$A = 2x\left(y - \sqrt{50}\right) = 2x\left[\sqrt{100 - x^2} - \sqrt{50}\right]$, $0 \le x \le \sqrt{50}$, so

$$\frac{dA}{dx} = 2\left[\sqrt{100 - x^2} - \sqrt{50}\right] + 2x\left(\tfrac{1}{2}\right)\left(100 - x^2\right)^{-1/2}(-2x)$$

$$= 2\left(100 - x^2\right)^{1/2} - 2\sqrt{50} - \frac{2x^2}{\left(100 - x^2\right)^{1/2}}.$$

Set $\dfrac{dA}{dx} = 0$: $\left(100 - x^2\right) - \sqrt{50}\left(100 - x^2\right)^{1/2} - x^2 = 0 \quad \Rightarrow \quad 100 - 2x^2 = \sqrt{50}\left(100 - x^2\right)^{1/2}$

$\Rightarrow \quad 10{,}000 - 400x^2 + 4x^4 = 50(100 - x^2) \quad \Rightarrow \quad 2500 - 175x^2 + 2x^4 = 0 \quad \Rightarrow$

$x^2 = \dfrac{175 \pm \sqrt{10{,}625}}{4} \approx 69.5 \text{ or } 17.98 \quad \Rightarrow \quad x \approx 8.3 \text{ or } 4.2$. But $8.3 > \sqrt{50}$, so $x \approx 4.2 \quad \Rightarrow$

$y - \sqrt{50} \approx \sqrt{100 - (4.2)^2} - \sqrt{50} \approx 2.0$. Each plank should have dimensions about 8.5 inches by 2 inches.

(c) The strength is $S = k\,(2x)(2y)^2 = 8kxy^2 = 8kx(100 - x^2)$, $0 \le x \le 10$.

$\dfrac{dS}{dx} = 800k - 24kx^2 = 0$ when $24kx^2 = 800k \quad \Rightarrow \quad x^2 = \frac{100}{3} \quad \Rightarrow \quad x = \frac{10}{\sqrt{3}} \quad \Rightarrow$

$y = \sqrt{\frac{200}{3}} = \frac{10\sqrt{2}}{\sqrt{3}}$ and $S(0) = S(10) = 0$, so the maximum occurs when $x = \frac{10}{\sqrt{3}}$. The dimensions should be $\frac{20}{\sqrt{3}}$ inches by $\frac{20\sqrt{2}}{\sqrt{3}}$ inches.

PROBLEMS PLUS (after Chapter 5)

1. Let $y = f(x) = e^{-x^2}$. The area of the rectangle under the curve from $-x$ to x is $A(x) = 2xe^{-x^2}$ where $x \geq 0$. We maximize $A(x)$: $A'(x) = 2e^{-x^2} - 4x^2e^{-x^2} = 2e^{-x^2}(1 - 2x^2) = 0 \quad \Rightarrow \quad x = \frac{1}{\sqrt{2}}$. This gives a maximum since $A'(x) > 0$ for $0 \leq x < \frac{1}{\sqrt{2}}$ and $A'(x) < 0$ for $x > \frac{1}{\sqrt{2}}$. We next determine the points of inflection of $f(x)$. Now $f'(x) = -2xe^{-x^2} = -A(x)$. So $f''(x) = -A'(x)$. So $f''(x) < 0$ for $-\frac{1}{\sqrt{2}} < x < \frac{1}{\sqrt{2}}$ and $f''(x) > 0$ for $x < -\frac{1}{\sqrt{2}}$ and $x > \frac{1}{\sqrt{2}}$. So $f(x)$ changes concavity at $x = \pm\frac{1}{\sqrt{2}}$. So the two vertices of the rectangle of largest area are at the inflection points.

3. First notice that if we can prove the simpler inequality $\dfrac{x^2+1}{x} \geq 2$ for $x > 0$, then the desired inequality follows because $\dfrac{(x^2+1)(y^2+1)(z^2+1)}{xyz} = \left(\dfrac{x^2+1}{x}\right)\left(\dfrac{y^2+1}{y}\right)\left(\dfrac{z^2+1}{z}\right) \geq 2 \cdot 2 \cdot 2 = 8$. So we let $f(x) = \dfrac{x^2+1}{x} = x + \dfrac{1}{x}$, $x > 0$. Then $f'(x) = 1 - \dfrac{1}{x^2} = 0$ if $x = 1$, and $f'(x) < 0$ for $0 < x < 1$, $f'(x) > 0$ for $x > 1$. Thus the absolute minimum value of $f(x)$ for $x > 0$ is $f(1) = 2$. Therefore $\dfrac{x^2+1}{x} \geq 2$ for all positive x. $\Big($Or, without calculus, $\dfrac{x^2+1}{x} \geq 2 \quad \Leftrightarrow \quad x^2 + 1 \geq 2x \quad \Leftrightarrow \quad x^2 - 2x + 1 \geq 0 \quad \Leftrightarrow$ $(x-1)^2 \geq 0$, which is true.$\Big)$

5. Differentiating $x^2 + xy + y^2 = 12$ implicitly with respect to x gives $2x + y + x\dfrac{dy}{dx} + 2y\dfrac{dy}{dx} = 0$, so $\dfrac{dy}{dx} = -\dfrac{2x+y}{x+2y}$. At a highest or lowest point, $\dfrac{dy}{dx} = 0 \quad \Leftrightarrow \quad y = -2x$. Substituting this into the original equation gives $x^2 + x(-2x) + (-2x)^2 = 12$, so $3x^2 = 12$ and $x = \pm 2$. If $x = 2$, then $y = -2x = -4$, and if $x = -2$ then $y = 4$. Thus the highest and lowest points are $(-2, 4)$ and $(2, -4)$.

7. Consider the statement that $\dfrac{d^n}{dx^n}(e^{ax}\sin bx) = r^n e^{ax}\sin(bx + n\theta)$. For $n = 1$,

$\dfrac{d}{dx}(e^{ax}\sin bx) = ae^{ax}\sin bx + be^{ax}\cos bx$, and

$$re^{ax}\sin(bx+\theta) = re^{ax}[\sin bx\cos\theta + \cos bx\sin\theta] = re^{ax}\left(\frac{a}{r}\sin bx + \frac{b}{r}\cos bx\right)$$
$$= ae^{ax}\sin bx + be^{ax}\cos bx, \text{ since } \tan\theta = b/a \quad \Rightarrow \quad \sin\theta = b/r \text{ and } \cos\theta = a/r.$$

So the statement is true for $n = 1$.

Assume that the statement is true for $n = k$. Then

$$\frac{d^{k+1}}{dx^{k+1}}(e^{ax}\sin bx) = \frac{d}{dx}\left[r^k e^{ax}\sin(bx+k\theta)\right] = r^k a e^{ax}\sin(bx+k\theta) + r^k e^{ax} b\cos(bx+k\theta)$$

$$= r^k e^{ax}[a\sin(bx+k\theta)+b\cos(bx+k\theta)].\text{ But}$$

$$\sin[bx+(k+1)\theta] = \sin[(bx+k\theta)+\theta] = \sin(bx+k\theta)\cos\theta + \sin\theta\cos(bx+k\theta)$$

$$= \frac{a}{r}\sin(bx+k\theta) + \frac{b}{r}\cos(bx+k\theta).$$

Hence $a\sin(bx+k\theta)+b\cos(bx+k\theta) = r\sin[bx+(k+1)\theta]$. So

$$\frac{d^{k+1}}{dx^{k+1}}(e^{ax}\sin bx) = r^k e^{ax}[a\sin(bx+k\theta)+b\cos(bx+k\theta)] = r^k e^{ax}[r\sin(bx+(k+1)\theta)]$$

$$= r^{k+1}e^{ax}[\sin(bx+(k+1)\theta)].$$

Therefore the statement is true for all n by mathematical induction.

9. $f(x) = \dfrac{1}{1+|x|} + \dfrac{1}{1+|x-2|}$

$$= \begin{cases} \dfrac{1}{1-x} + \dfrac{1}{1-(x-2)} & \text{if } x<0 \\ \dfrac{1}{1+x} + \dfrac{1}{1-(x-2)} & \text{if } 0\le x<2 \\ \dfrac{1}{1+x} + \dfrac{1}{1+(x-2)} & \text{if } x\ge 2 \end{cases} \quad\Rightarrow\quad f'(x) = \begin{cases} \dfrac{1}{(1-x)^2} + \dfrac{1}{(3-x)^2} & \text{if } x<0 \\ \dfrac{-1}{(1+x)^2} + \dfrac{1}{(3-x)^2} & \text{if } 0<x<2 \\ \dfrac{-1}{(1+x)^2} - \dfrac{1}{(x-1)^2} & \text{if } x>2 \end{cases}$$

Clearly $f'(x) > 0$ for $x < 0$ and $f'(x) < 0$ for $x > 2$. For $0 < x < 2$, we have

$f'(x) = \dfrac{1}{(3-x)^2} - \dfrac{1}{(x+1)^2} = \dfrac{(x^2+2x+1)-(x^2-6x+9)}{(3-x)^2(x+1)^2} = \dfrac{8(x-1)}{(3-x)^2(x+1)^2}$, so $f'(x) < 0$ for $x < 1$, $f'(1) = 0$ and $f'(x) > 0$ for $x > 1$. We have shown that $f'(x) > 0$ for $x < 0$; $f'(x) < 0$ for $0 < x < 1$; $f'(x) > 0$ for $1 < x < 2$; and $f'(x) < 0$ for $x > 2$. Therefore by the First Derivative Test, the local maxima of f are at $x = 0$ and $x = 2$, where f takes the value $\frac{4}{3}$. Therefore $\frac{4}{3}$ is the absolute maximum value of f.

11. We must find a value x_0 such that the normal lines to the parabola $y = x^2$ at $x = \pm x_0$ intersect at a point one unit from the points $(\pm x_0, x_0^2)$. The normals to $y = x^2$ at $x = \pm x_0$ have slopes $-\dfrac{1}{\pm 2x_0}$ and pass through $(\pm x_0, x_0^2)$ respectively, so the normals have the equations $y - x_0^2 = -\dfrac{1}{2x_0}(x - x_0)$ and $y - x_0^2 = \dfrac{1}{2x_0}(x + x_0)$. The common y-intercept is $x_0^2 + \frac{1}{2}$. We want to find the value of x_0 for which the distance from $\left(0, x_0^2 + \frac{1}{2}\right)$ to (x_0, x_0^2) equals 1. The square of the distance is $(x_0 - 0)^2 + \left[x_0^2 - \left(x_0^2 + \frac{1}{2}\right)\right]^2 = x_0^2 + \frac{1}{4} = 1 \quad\Leftrightarrow\quad x_0 = \pm\frac{\sqrt{3}}{2}$. For these values of x_0, the y-intercept is $x_0^2 + \frac{1}{2} = \frac{5}{4}$, so the center of the circle is at $\left(0, \frac{5}{4}\right)$.

Alternate Solution: Let the center of the circle be $(0, a)$. Then the equation of the circle is $x^2 + (y-a)^2 = 1$. Solving with the equation of the parabola, $y = x^2$, we get $x^2 + (x^2 - a)^2 = 1 \quad\Leftrightarrow\quad x^2 + x^4 - 2ax^2 + a^2 = 1 \quad\Leftrightarrow\quad x^4 + (1-2a)x^2 + a^2 - 1 = 0$. The parabola and the circle will be tangent to each other when this quadratic equation in x^2 has equal roots, that is, when the discriminant is 0. Thus $(1-2a)^2 - 4(a^2-1) = 0 \quad\Leftrightarrow\quad 1 - 4a + 4a^2 - 4a^2 + 4 = 0 \quad\Leftrightarrow\quad 4a = 5$, so $a = \frac{5}{4}$. The center of the circle is $\left(0, \frac{5}{4}\right)$.

13. We first show that $\dfrac{x}{1+x^2} < \tan^{-1}x$ for $x > 0$. Let $f(x) = \tan^{-1}x - \dfrac{x}{1+x^2}$. Then

$f'(x) = \dfrac{1}{1+x^2} - \dfrac{1(1+x^2) - x(2x)}{(1+x^2)^2} = \dfrac{(1+x^2)-(1-x^2)}{(1+x^2)^2} = \dfrac{2x^2}{(1+x^2)^2} > 0$ for $x > 0$. So $f(x)$ is increasing on $[0, \infty)$. Hence $0 < x \quad \Rightarrow \quad 0 = f(0) < f(x) = \tan^{-1}x - \dfrac{x}{1+x^2}$. So $\dfrac{x}{1+x^2} < \tan^{-1}x$ for $0 < x$. We next show that $\tan^{-1}x < x$ for $x > 0$. Let $h(x) = x - \tan^{-1}x$. Then

$h'(x) = 1 - \dfrac{1}{1+x^2} = \dfrac{x^2}{1+x^2} > 0$. Hence $h(x)$ is increasing on $[0, \infty)$. So for $0 < x$,

$0 = h(0) < h(x) = x - \tan^{-1}x$. Hence $\tan^{-1}x < x$ for $x > 0$.

15. $f(x) = [\![x]\!] + \sqrt{x - [\![x]\!]}$. On each interval of the form $[n, n+1)$, where n is an integer, we have $f(x) = n + \sqrt{x - n}$. It is easy to see that this function is continuous and increasing on $[n, n+1)$. Also, the left-hand limit

$$\lim_{x\to(n+1)^-} f(x) = \lim_{x\to(n+1)^-}\left[[\![x]\!] + \sqrt{x - [\![x]\!]}\right] = \lim_{x\to(n+1)^-} [\![x]\!] + \sqrt{\lim_{x\to(n+1)^-} x - \lim_{x\to(n+1)^-} [\![x]\!]}$$
$$= n + \sqrt{n+1-n} = n+1 = f(n+1) = \lim_{x\to(n+1)^+} f(x),$$

so f is continuous and increasing everywhere.

17. $A = (x_1, x_1^2)$ and $B = (x_2, x_2^2)$, where x_1 and x_2 are the solutions of the quadratic equation $x^2 = mx + b$. Let $P = (x, x^2)$ and set $A_1 = (x_1, 0)$, $B_1 = (x_2, 0)$, and $P_1 = (x, 0)$. Let $f(x)$ denote the area of triangle PAB. Then $f(x)$ can be expressed in terms of the areas of three trapezoids as follows:

$$f(x) = \text{area}(A_1ABB_1) - \text{area}(A_1APP_1) - \text{area}(B_1BPP_1)$$
$$= \tfrac{1}{2}(x_2 - x_1)(x_1^2 + x_2^2) - \tfrac{1}{2}(x - x_1)(x_1^2 + x^2) - \tfrac{1}{2}(x_2 - x)(x^2 + x_2^2).$$

After expansion, canceling of terms, and factoring, we find that $f(x) = \frac{1}{2}(x_2 - x_1)(x - x_1)(x_2 - x)$.

Note: Another way to get an expression for $f(x)$ is to use the formula for an area of a triangle in terms of the coordinates of the vertices: $f(x) = \frac{1}{2}\left[(x_2x_1^2 - x_1x_2^2) + (x_1x^2 - xx_1^2) + (xx_2^2 - x_2x^2)\right]$. From our expression for $f(x)$, it follows that $f'(x) = \frac{1}{2}(x_2 - x_1)(x_1 + x_2 - 2x)$ and $f''(x) = -(x_2 - x_1) < 0$. Thus the area $f(x)$ is maximized when $x = \frac{1}{2}(x_1 + x_2)$, and $f\left(\frac{1}{2}(x_1 + x_2)\right) = \frac{1}{2}(x_2 - x_1)\frac{1}{2}(x_2 - x_1)\frac{1}{2}(x_2 - x_1) = \frac{1}{8}(x_2 - x_1)^3$. In terms of m and b, $x_1 = \frac{1}{2}\left(m - \sqrt{m^2 + 4b}\right)$ and $x_2 = \frac{1}{2}\left(m + \sqrt{m^2 + 4b}\right)$, so the maximal area is $\frac{1}{8}(m^2 + 4b)^{3/2}$ and it is attained at the point $P\left(\frac{1}{2}m, \frac{1}{4}m^2\right)$.

19. Since $[\![x]\!] \le x < [\![x]\!] + 1$, we have $1 \le \dfrac{x}{[\![x]\!]} \le 1 + \dfrac{1}{[\![x]\!]}$ for $x > 0$. As $x \to \infty$, $[\![x]\!] \to \infty$, so $\dfrac{1}{[\![x]\!]} \to 0$ and $1 + \dfrac{1}{[\![x]\!]} \to 1$. Thus $\displaystyle\lim_{x\to\infty} \frac{x}{[\![x]\!]} = 1$ by the Squeeze Theorem.

21. $f(x) = (a^2 + a - 6)\cos 2x + (a-2)x + \cos 1 \quad \Rightarrow \quad f'(x) = -(a^2 + a - 6)\sin 2x(2) + (a-2)$.

The derivative exists for all x, so the only possible critical points will occur where $f'(x) = 0 \quad \Leftrightarrow$

$2(a-2)(a+3)\sin 2x = a - 2 \quad \Leftrightarrow \quad$ either $a = 2$ or $2(a+3)\sin 2x = 1$, with the latter implying that

$\sin 2x = \dfrac{1}{2(a+3)}$. Since the range of $\sin 2x$ is $[-1, 1]$, this equation has no solution whenever either

$\dfrac{1}{2(a+3)} > 1$ or $\dfrac{1}{2(a+3)} > 1$. Solving these inequalities, we get $-\frac{7}{2} < a < -\frac{5}{2}$.

23. **(a)**

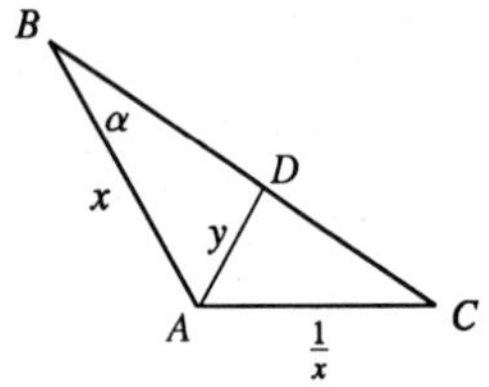

Let $y = |AD|$, $x = |AB|$, and $1/x = |AC|$, so that $|AB| \cdot |AC| = 1$.
We compute the area $\mathcal{A}$ of $\triangle ABC$ in two ways.

First, $\mathcal{A} = \frac{1}{2}|AB||AC|\sin\frac{2\pi}{3} = \frac{1}{2} \cdot 1 \cdot \frac{\sqrt{3}}{2} = \frac{\sqrt{3}}{4}$.

Second, $\mathcal{A} = (\text{area of } \triangle ABD) + (\text{area of } \triangle ACD)$

$$= \tfrac{1}{2}|AB||AD|\sin\tfrac{\pi}{3} + \tfrac{1}{2}|AD||AC|\sin\tfrac{\pi}{3}$$

$$= \tfrac{1}{2}xy\tfrac{\sqrt{3}}{2} + \tfrac{1}{2}y(1/x)\tfrac{\sqrt{3}}{2} = \tfrac{\sqrt{3}}{4}y(x + 1/x).$$

Equating the two expressions for the area, we get $\frac{\sqrt{3}}{4}y(x + 1/x) = \frac{\sqrt{3}}{4} \quad \Leftrightarrow$

$y = \dfrac{1}{x + 1/x} = \dfrac{x}{x^2+1}$, $x > 0$.

Another Method: Use the Law of Sines on the triangles ABD and ABC. In $\triangle ABD$, we have

$\angle A + \angle B + \angle D = 180° \quad \Leftrightarrow \quad 60° + \alpha + \angle D = 180° \quad \Leftrightarrow \quad \angle D = 120° - \alpha$. Thus,

$$\frac{x}{y} = \frac{\sin(120° - \alpha)}{\sin\alpha} = \frac{\sin 120°\cos\alpha - \cos 120°\sin\alpha}{\sin\alpha} = \frac{\frac{\sqrt{3}}{2}\cos\alpha + \frac{1}{2}\sin\alpha}{\sin\alpha} \quad \Rightarrow \quad \frac{x}{y} = \frac{\sqrt{3}}{2}\cot\alpha + \frac{1}{2}$$

and, by a similar argument with $\triangle ABC$, $\frac{\sqrt{3}}{2}\cot\alpha = x^2 + \frac{1}{2}$. Eliminating $\cot\alpha$ gives $\dfrac{x}{y} = \left(x^2 + \frac{1}{2}\right) + \frac{1}{2}$

$\Rightarrow \quad y = x/\left(x^2+1\right)$, $x > 0$.

(b) We differentiate our expression for y with respect to x to find the maximum:

$\dfrac{dy}{dx} = \dfrac{(x^2+1) - x(2x)}{(x^2+1)^2} = \dfrac{1-x^2}{(x^2+1)^2} = 0$ when $x = 1$. This indicates a maximum by the First Derivative Test, since $y'(x) > 0$ for $0 < x < 1$ and $y'(x) < 0$ for $x > 1$, so the maximum value of y is $y(1) = \frac{1}{2}$.

25.
$$\lim_{n\to\infty}\left(\frac{1}{\sqrt{n}\sqrt{n+1}} + \frac{1}{\sqrt{n}\sqrt{n+2}} + \cdots + \frac{1}{\sqrt{n}\sqrt{n+n}}\right) = \lim_{n\to\infty}\frac{1}{n}\left(\sqrt{\frac{n}{n+1}} + \sqrt{\frac{n}{n+2}} + \cdots + \sqrt{\frac{n}{n+n}}\right)$$

$$= \lim_{n\to\infty}\frac{1}{n}\left(\frac{1}{\sqrt{1+1/n}} + \frac{1}{\sqrt{1+2/n}} + \cdots + \frac{1}{\sqrt{1+1}}\right) = \lim_{n\to\infty}\frac{1}{n}\sum_{i=1}^{n} f\!\left(\frac{i}{n}\right) \quad \left(\text{where } f(x) = \frac{1}{\sqrt{1+x}}\right)$$

$$= \int_0^1 \frac{1}{\sqrt{1+x}}\,dx = \left[2\sqrt{1+x}\right]_0^1 = 2\left(\sqrt{2} - 1\right)$$

27. Suppose that the curve $y = a^x$ intersects the line $y = x$. Then $a^{x_0} = x_0$ for some $x_0 > 0$, and hence $a = x_0^{1/x_0}$. We find the maximum value of $g(x) = x^{1/x}$, > 0, because if a is larger than the maximum value of this function, then the curve $y = a^x$ does not intersect the line $y = x$.

$g'(x) = e^{(1/x)\ln x}\left(-\frac{1}{x^2}\ln x + \frac{1}{x}\cdot\frac{1}{x}\right) = x^{1/x}\left(\frac{1}{x^2}\right)(1 - \ln x)$. This is 0 only where $x = e$, and for $0 < x < e$, $f'(x) > 0$, while for $x > e$, $f'(x) < 0$, so g has an absolute maximum of $g(e) = e^{1/e}$. So if $y = a^x$ intersects $y = x$, we must have $0 < a \le e^{1/e}$. Conversely, suppose that $0 < a \le e^{1/e}$. Then $a^e \le e$, so the graph of $y = a^x$ lies below or touches the graph of $y = x$ at $x = e$. Also $a^0 = 1 > 0$, so the graph of $y = a^x$ lies above that of $y = x$ at $x = 0$. Therefore, by the Intermediate Value Theorem, the graphs of $y = a^x$ and $y = x$ must intersect somewhere between $x = 0$ and $x = e$.

29. Both sides of the inequality are positive, so $\cosh(\sinh x) < \sinh(\cosh x)$ $\Leftrightarrow$ $\cosh^2(\sinh x) < \sinh^2(\cosh x)$ $\Leftrightarrow$ $\sinh^2(\sinh x) + 1 < \sinh^2(\cosh x)$ $\Leftrightarrow$ $1 < [\sinh(\cosh x) - \sinh(\sinh x)][\sinh(\cosh x) + \sinh(\sinh x)]$

$\Leftrightarrow\quad 1 < \left[\sinh\left(\frac{e^x + e^{-x}}{2}\right) - \sinh\left(\frac{e^x - e^{-x}}{2}\right)\right]\left[\sinh\left(\frac{e^x + e^{-x}}{2}\right) + \sinh\left(\frac{e^x - e^{-x}}{2}\right)\right]$

$\Leftrightarrow\quad 1 < \left[2\cosh(e^x/2)\sinh(e^{-x}/2)\right]\left[2\sinh(e^x/2)\cosh(e^{-x}/2)\right]$ (use the addition formulas and cancel)

$\Leftrightarrow\quad 1 < \left[2\sinh(e^x/2)\cosh(e^x/2)\right]\left[2\sinh(e^{-x}/2)\cosh(e^{-x}/2)\right]$ $\Leftrightarrow$ $1 < \sinh e^x \sinh e^{-x}$, by the half-angle formula. Now both e^x and e^{-x} are positive, and $\sinh y > y$ for $y > 0$, since $\sinh 0 = 0$ and $(\sinh y - y)' = \cosh y - 1 > 0$ for $x > 0$, so $1 = e^x e^{-x} < \sinh e^x \sinh e^{-x}$. So, following this chain of reasoning backward, we arrive at the desired result.

Another Method: Using Formula 4.7.3, we have

$\sinh^{-1}(\cosh(\sinh x)) = \ln\left(\cosh(\sinh x) + \sqrt{1 + \cosh^2(\sinh x)}\right) = \ln(\cosh(\sinh x) + \sinh(\cosh x))$

$= \ln(e^{\sinh x}) = \sinh x$. But $\sinh x < \cosh x$, so $\sinh^{-1}(\cosh(\sinh x)) < \cosh x$. Since sinh is an increasing function, we can apply it to both sides of the inequality and get $\cosh(\sinh x) < \sinh(\cosh x)$.

31. Note that $f(0) = 0$, so for $x \ne 0$, $\left|\frac{f(x) - f(0)}{x - 0}\right| = \left|\frac{f(x)}{x}\right| = \frac{|f(x)|}{|x|} \le \frac{|\sin x|}{|x|} = \frac{\sin x}{x}$. Therefore

$|f'(0)| = \left|\lim\limits_{x\to 0}\frac{f(x) - f(0)}{x - 0}\right| = \lim\limits_{x\to 0}\left|\frac{f(x) - f(0)}{x - 0}\right| \le \lim\limits_{x\to 0}\frac{\sin x}{x} = 1$. But

$f'(x) = a_1\cos x + 2a_2\cos 2x + \cdots + na_n\cos nx$, so $|f'(0)| = |a_1 + 2a_2 + \cdots + na_n| \le 1$.

Another Solution: We are given that $\left|\sum\limits_{k=1}^{n} a_k \sin kx\right| \le |\sin x|$. So for x close to 0, and $x \ne 0$, we have

$\left|\sum\limits_{k=1}^{n} a_k \frac{\sin kx}{\sin x}\right| \le 1 \;\Rightarrow\; \lim\limits_{x\to 0}\left|\sum\limits_{k=1}^{n} a_k \frac{\sin kx}{\sin x}\right| \le 1 \;\Rightarrow\; \left|\sum\limits_{k=1}^{n} a_k \lim\limits_{x\to 0}\frac{\sin kx}{\sin x}\right| \le 1$. But by l'Hospital's Rule,

$\lim\limits_{x\to 0}\frac{\sin kx}{\sin x} = \lim\limits_{x\to 0}\frac{k\cos kx}{\cos x} = k$, so $\left|\sum\limits_{k=1}^{n} ka_k\right| \le 1$.

CHAPTER SIX

EXERCISES 6.1

1. $\sum_{i=1}^{5} \sqrt{i} = \sqrt{1} + \sqrt{2} + \sqrt{3} + \sqrt{4} + \sqrt{5}$

3. $\sum_{i=4}^{6} 3^i = 3^4 + 3^5 + 3^6$

5. $\sum_{k=0}^{4} \frac{2k-1}{2k+1} = -1 + \frac{1}{3} + \frac{3}{5} + \frac{5}{7} + \frac{7}{9}$

7. $\sum_{i=1}^{n} i^{10} = 1^{10} + 2^{10} + 3^{10} + \cdots + n^{10}$

9. $\sum_{j=0}^{n-1} (-1)^j = 1 - 1 + 1 - 1 + \cdots + (-1)^{n-1}$

11. $1 + 2 + 3 + 4 + \cdots + 10 = \sum_{i=1}^{10} i$

13. $\frac{1}{2} + \frac{2}{3} + \frac{3}{4} + \frac{4}{5} + \cdots + \frac{19}{20} = \sum_{i=1}^{19} \frac{i}{i+1}$

15. $2 + 4 + 6 + 8 + \cdots + 2n = \sum_{i=1}^{n} 2i$

17. $1 + 2 + 4 + 8 + 16 + 32 = \sum_{i=0}^{5} 2^i$

19. $x + x^2 + x^3 + \cdots + x^n = \sum_{i=1}^{n} x^i$

21. $\sum_{i=4}^{8} (3i - 2) = 10 + 13 + 16 + 19 + 22 = 80$

23. $\sum_{j=1}^{6} 3^{j+1} = 3^2 + 3^3 + 3^4 + 3^5 + 3^6 + 3^7 = 9 + 27 + 81 + 243 + 729 + 2187 = 3276$

(For a more general method, see Exercise 47.)

25. $\sum_{n=1}^{20} (-1)^n = -1 + 1 - 1 + 1 - 1 + 1 - 1 + 1 - 1 + 1 - 1 + 1 - 1 + 1 - 1 + 1 - 1 + 1 - 1 + 1 = 0$

27. $\sum_{i=0}^{4} (2^i + i^2) = (1 + 0) + (2 + 1) + (4 + 4) + (8 + 9) + (16 + 16) = 61$

29. $\sum_{i=1}^{n} 2i = 2\sum_{i=1}^{n} i = n(n+1)$

31. $\sum_{i=1}^{n} (i^2 + 3i + 4) = \sum_{i=1}^{n} i^2 + 3\sum_{i=1}^{n} i + \sum_{i=1}^{n} 4 = \frac{n(n+1)(2n+1)}{6} + \frac{3n(n+1)}{2} + 4n$

$= \frac{1}{6}[(2n^3 + 3n^2 + n) + (9n^2 + 9n) + 24n] = \frac{1}{6}(2n^3 + 12n^2 + 34n) = \frac{1}{3}n(n^2 + 6n + 17)$

33. $\sum_{i=1}^{n} (i+1)(i+2) = \sum_{i=1}^{n} (i^2 + 3i + 2) = \sum_{i=1}^{n} i^2 + 3\sum_{i=1}^{n} i + \sum_{i=1}^{n} 2$

$= \frac{n(n+1)(2n+1)}{6} + \frac{3n(n+1)}{2} + 2n = \frac{n(n+1)}{6}[(2n+1) + 9] + 2n$

$= \frac{n(n+1)}{3}(n+5) + 2n = \frac{n}{3}[(n+1)(n+5) + 6] = \frac{n}{3}(n^2 + 6n + 11)$

35. $\sum_{i=1}^{n}(i^3 - i - 2) = \sum_{i=1}^{n} i^3 - \sum_{i=1}^{n} i - \sum_{i=1}^{n} 2 = \left[\frac{n(n+1)}{2}\right]^2 - \frac{n(n+1)}{2} - 2n$

$= \frac{1}{4}n(n+1)[n(n+1) - 2] - 2n = \frac{1}{4}n(n+1)(n+2)(n-1) - 2n$

$= \frac{1}{4}n[(n+1)(n-1)(n+2) - 8] = \frac{1}{4}n[(n^2-1)(n+2) - 8] = \frac{1}{4}n(n^3 + 2n^2 - n - 10)$

37. By Theorem 2(a) and Example 3, $\sum_{i=1}^{n} c = c\sum_{i=1}^{n} 1 = cn$.

39. $\sum_{i=1}^{n}\left[(i+1)^4 - i^4\right] = (2^4 - 1^4) + (3^4 - 2^4) + (4^4 - 3^4) + \cdots + \left[(n+1)^4 - n^4\right]$

$= (n+1)^4 - 1^4 = n^4 + 4n^3 + 6n^2 + 4n$

On the other hand, $\sum_{i=1}^{n}\left[(i+1)^4 - i^4\right] = \sum_{i=1}^{n}(4i^3 + 6i^2 + 4i + 1) = 4\sum_{i=1}^{n} i^3 + 6\sum_{i=1}^{n} i^2 + 4\sum_{i=1}^{n} i + \sum_{i=1}^{n} 1$

$= 4S + n(n+1)(2n+1) + 2n(n+1) + n$ (where $S = \sum_{i=1}^{n} i^3$)

$= 4S + 2n^3 + 3n^2 + n + 2n^2 + 2n + n = 4S + 2n^3 + 5n^2 + 4n$

Thus $n^4 + 4n^3 + 6n^2 + 4n = 4S + 2n^3 + 5n^2 + 4n$, from which it follows that

$4S = n^4 + 2n^3 + n^2 = n^2(n^2 + 2n + 1) = n^2(n+1)^2$ and $S = \left[\frac{n(n+1)}{2}\right]^2$.

41. **(a)** $\sum_{i=1}^{n}\left(i^4 - (i-1)^4\right) = (1^4 - 0^4) + (2^4 - 1^4) + (3^4 - 2^4) + \cdots + \left[n^4 - (n-1)^4\right] = n^4 - 0 = n^4$

(b) $\sum_{i=1}^{100}(5^i - 5^{i-1}) = (5^1 - 5^0) + (5^2 - 5^1) + (5^3 - 5^2) + \cdots + (5^{100} - 5^{99}) = 5^{100} - 5^0 = 5^{100} - 1$

(c) $\sum_{i=3}^{99}\left(\frac{1}{i} - \frac{1}{i+1}\right) = \left(\frac{1}{3} - \frac{1}{4}\right) + \left(\frac{1}{4} - \frac{1}{5}\right) + \left(\frac{1}{5} - \frac{1}{6}\right) + \cdots + \left(\frac{1}{99} - \frac{1}{100}\right) = \frac{1}{3} - \frac{1}{100} = \frac{97}{300}$

(d) $\sum_{i=1}^{n}(a_i - a_{i-1}) = (a_1 - a_0) + (a_2 - a_1) + (a_3 - a_2) + \cdots + (a_n - a_{n-1}) = a_n - a_0$

43. $\lim_{n\to\infty}\sum_{i=1}^{n}\frac{1}{n}\left(\frac{i}{n}\right)^2 = \lim_{n\to\infty}\frac{1}{n^3}\sum_{i=1}^{n} i^2 = \lim_{n\to\infty}\frac{1}{n^3}\frac{n(n+1)(2n+1)}{6} = \lim_{n\to\infty}\frac{1}{6}\left(1 + \frac{1}{n}\right)\left(2 + \frac{1}{n}\right) = \frac{1}{6}(1)(2) = \frac{1}{3}$

45. $\lim_{n\to\infty}\sum_{i=1}^{n}\frac{2}{n}\left[\left(\frac{2i}{n}\right)^3 + 5\left(\frac{2i}{n}\right)\right] = \lim_{n\to\infty}\sum_{i=1}^{n}\left[\frac{16}{n^4}i^3 + \frac{20}{n^2}i\right] = \lim_{n\to\infty}\left[\frac{16}{n^4}\sum_{i=1}^{n} i^3 + \frac{20}{n^2}\sum_{i=1}^{n} i\right]$

$= \lim_{n\to\infty}\left[\frac{16}{n^4}\frac{n^2(n+1)^2}{4} + \frac{20}{n^2}\frac{n(n+1)}{2}\right] = \lim_{n\to\infty}\left[\frac{4(n+1)^2}{n^2} + \frac{10n(n+1)}{n^2}\right]$

$= \lim_{n\to\infty}\left[4\left(1 + \frac{1}{n}\right)^2 + 10\left(1 + \frac{1}{n}\right)\right] = 4 \cdot 1 + 10 \cdot 1 = 14$

47. Let $S = \sum_{i=1}^{n} ar^{i-1} = a + ar + ar^2 + \cdots + ar^{n-1}$. Then $rS = ar + ar^2 + \cdots + ar^{n-1} + ar^n$. Subtracting the first equation from the second, we find $(r-1)S = ar^n - a = a(r^n - 1)$, so $S = \frac{a(r^n - 1)}{r - 1}$.

49. $\displaystyle\sum_{i=1}^{n}(2i+2^i)=2\sum_{i=1}^{n}i+\sum_{i=1}^{n}2\cdot 2^{i-1}=2\,\frac{n(n+1)}{2}+\frac{2(2^n-1)}{2-1}=2^{n+1}+n^2+n-2.$

For the first sum we have used Theorem 3(c), and for the second, Exercise 47 with $a=r=2$.

51. By Theorem 3(c) we have that $\displaystyle\sum_{i=1}^{n}i=\frac{n(n+1)}{2}=78 \;\Leftrightarrow\; n(n+1)=156 \;\Leftrightarrow\; n^2+n-156=0 \;\Leftrightarrow$

$(n+13)(n-12)=0 \;\Leftrightarrow\; n=12$ or -13. But $n=-13$ produces a negative answer for the sum, so $n=12$.

53. From Formula 18c in Appendix D, $\sin x\sin y=\frac{1}{2}[\cos(x-y)-\cos(x+y)]$, so

$2\sin u\sin v=\cos(u-v)-\cos(u+v)$ ($\bigstar$). Taking $u=\frac{1}{2}x$ and $v=ix$, we get

$2\sin\left(\frac{1}{2}x\right)\sin ix=\cos\left(\left(\frac{1}{2}-i\right)x\right)-\cos\left(\left(\frac{1}{2}+i\right)x\right)=\cos\left(\left(i-\frac{1}{2}\right)x\right)-\cos\left(\left(i+\frac{1}{2}\right)x\right)$. Thus

$$\begin{aligned}2\sin\left(\tfrac{1}{2}x\right)\sum_{i=1}^{n}\sin ix&=\sum_{i=1}^{n}2\sin\left(\tfrac{1}{2}x\right)\sin ix=\sum_{i=1}^{n}\left[\cos\left(\left(i-\tfrac{1}{2}\right)x\right)-\cos\left(\left(i+\tfrac{1}{2}\right)x\right)\right]\\&=-\sum_{i=1}^{n}\left[\cos\left(\left(i+\tfrac{1}{2}\right)x\right)-\cos\left(\left(i-\tfrac{1}{2}\right)x\right)\right]=-\left[\cos\left(\left(n+\tfrac{1}{2}\right)x\right)-\cos\left(\tfrac{1}{2}x\right)\right]\quad\text{(telescoping sum)}\\&=\cos\left(\tfrac{1}{2}(n+1)x-\tfrac{1}{2}nx\right)-\cos\left(\tfrac{1}{2}(n+1)x+\tfrac{1}{2}nx\right)\\&=2\sin\left(\tfrac{1}{2}(n+1)x\right)\sin\left(\tfrac{1}{2}nx\right)\quad\left[\text{by }(\bigstar)\text{ with }u=\tfrac{1}{2}(n+1)x\text{ and }v=\tfrac{1}{2}nx\right]\end{aligned}$$

If x is not an integer multiple of 2π, then $\sin\left(\frac{1}{2}x\right)\neq 0$, so we can divide by $2\sin\left(\frac{1}{2}x\right)$ and get

$$\sum_{i=1}^{n}\sin ix=\frac{\sin\left(\frac{1}{2}nx\right)\sin\left(\frac{1}{2}(n+1)x\right)}{\sin\left(\frac{1}{2}x\right)}.$$

EXERCISES 6.2

1. **(a)** $\|P\|=\max\{1,1,1,1\}=1$

(b) $\displaystyle\sum_{i=1}^{n}f(x_i^*)\Delta x_i=\sum_{i=1}^{4}f(i-1)\cdot 1=16+15+12+7=50$

(c)

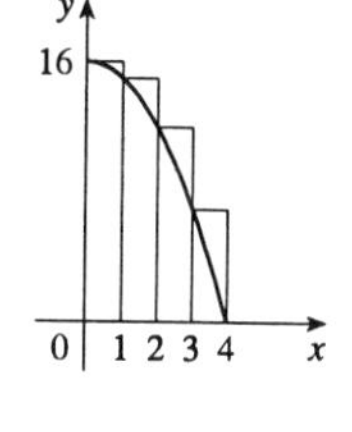

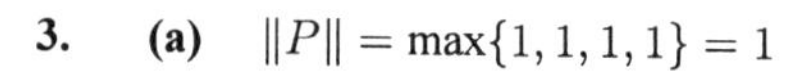

3. **(a)** $\|P\|=\max\{1,1,1,1\}=1$

(b) $\displaystyle\sum_{i=1}^{n}f(x_i^*)\Delta x_i=\sum_{i=1}^{4}f\left(i-\tfrac{1}{2}\right)\cdot 1=15.75+13.75+9.75+3.75=43$

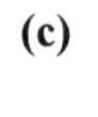
(c)

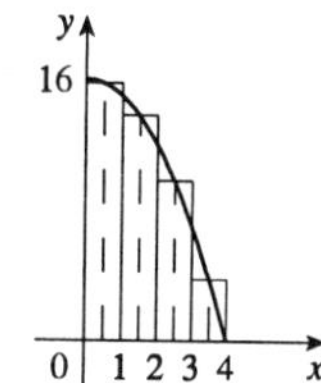

5. **(a)** $\|P\| = \max\{0.5, 0.5, 0.5, 0.5, 0.5, 0.5\} = 0.5$

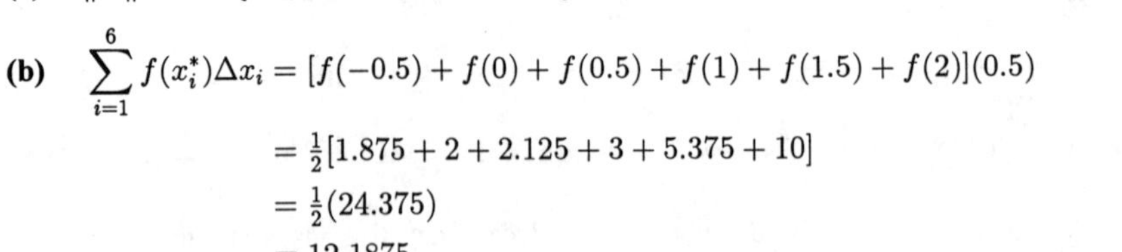

(b) $$\sum_{i=1}^{6} f(x_i^*)\Delta x_i = [f(-0.5) + f(0) + f(0.5) + f(1) + f(1.5) + f(2)](0.5)$$
$$= \tfrac{1}{2}[1.875 + 2 + 2.125 + 3 + 5.375 + 10]$$
$$= \tfrac{1}{2}(24.375)$$
$$= 12.1875$$

(c)

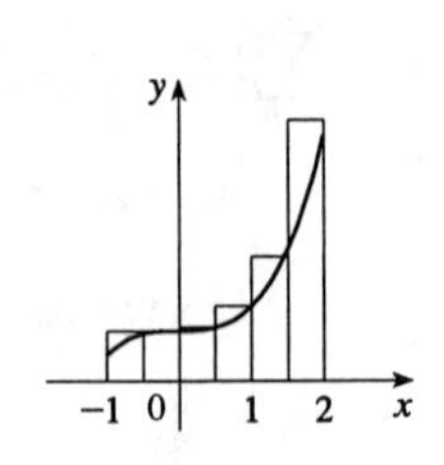

7. **(a)** $\|P\| = \max\{\frac{\pi}{4}, \frac{\pi}{4}, \frac{\pi}{4}, \frac{\pi}{4}\} = \frac{\pi}{4}$

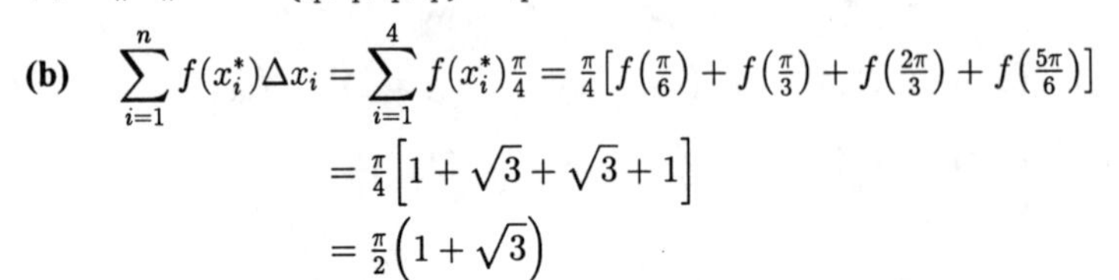

(b) $$\sum_{i=1}^{n} f(x_i^*)\Delta x_i = \sum_{i=1}^{4} f(x_i^*)\tfrac{\pi}{4} = \tfrac{\pi}{4}\left[f\left(\tfrac{\pi}{6}\right) + f\left(\tfrac{\pi}{3}\right) + f\left(\tfrac{2\pi}{3}\right) + f\left(\tfrac{5\pi}{6}\right)\right]$$
$$= \tfrac{\pi}{4}\left[1 + \sqrt{3} + \sqrt{3} + 1\right]$$
$$= \tfrac{\pi}{2}\left(1 + \sqrt{3}\right)$$

(c)

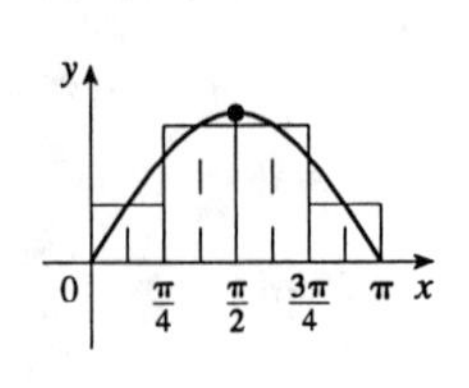

9. **(a)**

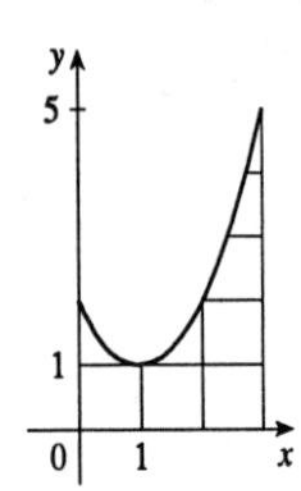

$y = x^2 - 2x + 2 = (x-1)^2 + 1$. By counting squares, we estimate that the area under the curve is between 5 and 7 — perhaps near 6.

(b) $f(x) = y = x^2 - 2x + 2$ on $[0, 3]$ with partition points $x_i = 0 + \dfrac{3i}{n} = \dfrac{3i}{n}$, $\Delta x_i = \dfrac{3}{n}$ and $x_i^* = x_i$, so

$$R_n = \sum_{i=1}^{n} f(x_i^*)\Delta x_i = \frac{3}{n}\sum_{i=1}^{n} f\left(\frac{3i}{n}\right) = \frac{3}{n}\sum_{i=1}^{n}\left[\left(\frac{3i}{n}\right)^2 - 2\left(\frac{3i}{n}\right) + 2\right] = \frac{27}{n^3}\sum_{i=1}^{n} i^2 - \frac{18}{n^2}\sum_{i=1}^{n} i + 6$$
$$= \frac{27}{6}\left[\frac{n(n+1)(2n+1)}{n^3}\right] - \frac{18}{2}\left[\frac{n(n+1)}{n^2}\right] + 6 = \frac{9}{2}\left(2 + \frac{3}{n} + \frac{1}{n^2}\right) - 9\left(1 + \frac{1}{n}\right) + 6$$
$$= 6 + \frac{9}{2n} + \frac{9}{2n^2}.$$

(c) $R_6 = 6 + \dfrac{9}{2\cdot 6} + \dfrac{9}{2\cdot 36} = \dfrac{55}{8} = 6.875$, $R_{12} = 6 + \dfrac{9}{2\cdot 12} + \dfrac{9}{2\cdot 144} = \dfrac{205}{32} = 6.40625$,

$R_{24} = 6 + \dfrac{9}{2\cdot 24} + \dfrac{9}{2\cdot 576} = \dfrac{793}{128} = 6.1953125$

(d) Since $\|P\| \to 0$ as $n \to \infty$, the area is $A = \lim\limits_{n\to\infty} R_n = \lim\limits_{n\to\infty}\left(6 + \dfrac{9}{2n} + \dfrac{9}{2n^2}\right) = 6.$

11. $f(x) = x^2 + 1$ on $[0, 2]$ with partition points $x_i = 2i/n \quad (i = 0, 1, 2, \ldots, n)$, so $\Delta x_1 = \Delta x_2 = \cdots = \Delta x_n = \frac{2}{n}$.

$\|P\| = \max\{\Delta x_i\} = \frac{2}{n}$, so $\|P\| \to 0$ is equivalent to $n \to \infty$. Taking x_i^* to be the midpoint of

$[x_{i-1}, x_i] = \left[2\frac{i-1}{n}, 2\frac{i}{n}\right]$, we get $x_i^* = \frac{2i-1}{n}$. Thus

$$A = \lim_{\|P\| \to 0} \sum_{i=1}^{n} f(x_i^*)\Delta x_i = \lim_{n \to \infty} \sum_{i=1}^{n} \left[\left(\frac{2i-1}{n}\right)^2 + 1\right]\frac{2}{n}$$

$$= \lim_{n \to \infty} \sum_{i=1}^{n} \left[\frac{8i^2}{n^3} - \frac{8i}{n^3} + \frac{2}{n^3} + \frac{2}{n}\right]$$

$$= \lim_{n \to \infty} \left[\frac{8}{n^3}\sum_{i=1}^{n} i^2 - \frac{8}{n^3}\sum_{i=1}^{n} i + \left(\frac{2}{n^3} + \frac{2}{n}\right)\sum_{i=1}^{n} 1\right]$$

$$= \lim_{n \to \infty} \left[\frac{8}{n^3}\frac{n(n+1)(2n+1)}{6} - \frac{8}{n^3}\frac{n(n+1)}{2} + \left(\frac{2}{n^3} + \frac{2}{n}\right)n\right]$$

$$= \lim_{n \to \infty} \left[\frac{4}{3} \cdot 1\left(1 + \frac{1}{n}\right)\left(2 + \frac{1}{n}\right) - \frac{4}{n} \cdot 1\left(1 + \frac{1}{n}\right) + \frac{2}{n^2} + 2\right]$$

$$= \left(\tfrac{4}{3} \cdot 1 \cdot 1 \cdot 2\right) - (0 \cdot 1 \cdot 1) + 0 + 2 = \tfrac{8}{3} + 2 = \tfrac{14}{3}.$$

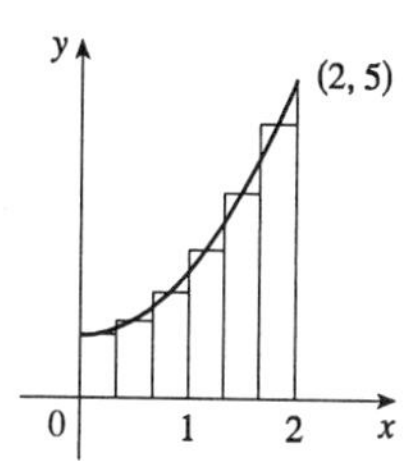

13. $f(x) = 2x + 1$ on $[0, 5]$. $x_i^* = x_i = \frac{5i}{n}$ for $i = 1, \ldots, n$ and $\Delta x_i = \frac{5}{n}$.

$$A = \lim_{n \to \infty} \sum_{i=1}^{n} \left[2\left(\frac{5i}{n}\right) + 1\right]\frac{5}{n} = \lim_{n \to \infty} \left[\frac{50}{n^2}\sum_{i=1}^{n} i + \frac{5}{n}\sum_{i=1}^{n} 1\right]$$

$$= \lim_{n \to \infty} \left[\frac{50}{n^2}\frac{n(n+1)}{2} + \frac{5}{n}n\right] = \lim_{n \to \infty} \left[25 \cdot 1\left(1 + \frac{1}{n}\right) + 5\right]$$

$$= 25 + 5 = 30$$

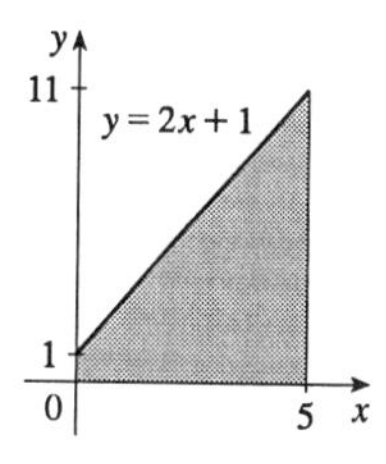

15. $f(x) = 2x^2 - 4x + 5$ on $[-3, 2]$. $x_i^* = x_i = -3 + \frac{5i}{n} \quad (i = 1, 2, \ldots, n)$.

$$A = \lim_{n \to \infty} \sum_{i=1}^{n} \left[2\left(-3 + \frac{5i}{n}\right)^2 - 4\left(-3 + \frac{5i}{n}\right) + 5\right]\frac{5}{n}$$

$$= \lim_{n \to \infty} \sum_{i=1}^{n} \left[\frac{50i^2}{n^2} - \frac{80i}{n} + 35\right]\frac{5}{n}$$

$$= \lim_{n \to \infty} \left[\frac{250}{n^3}\sum_{i=1}^{n} i^2 - \frac{400}{n^2}\sum_{i=1}^{n} i + \frac{175}{n}\sum_{i=1}^{n} 1\right]$$

$$= \lim_{n \to \infty} \left[\frac{250}{n^3}\frac{n(n+1)(2n+1)}{6} - \frac{400}{n^2}\frac{n(n+1)}{2} + \frac{175}{n}n\right]$$

$$= \lim_{n \to \infty} \left[\frac{125}{3} \cdot 1\left(1 + \frac{1}{n}\right)\left(2 + \frac{1}{n}\right) - 200 \cdot 1\left(1 + \frac{1}{n}\right) + 175\right]$$

$$= \left(\tfrac{125}{3} \cdot 1 \cdot 1 \cdot 2\right) - (200 \cdot 1 \cdot 1) + 175 = \tfrac{175}{3}$$

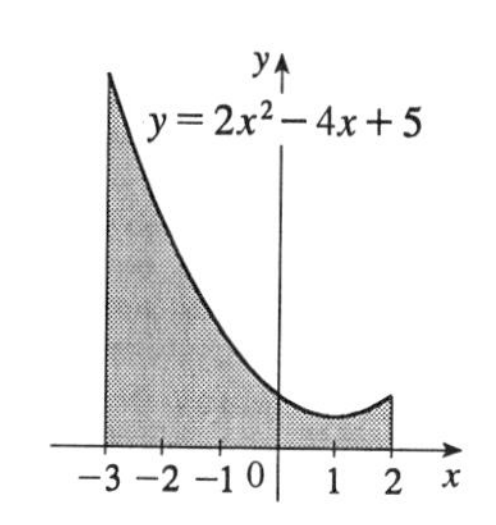

17. $f(x) = x^3 + 2x$ on $[0, 2]$. $x_i^* = x_i = \dfrac{2i}{n}$ for $i = 1, 2, \ldots, n$.

$$A = \lim_{n\to\infty} \sum_{i=1}^{n} \left[\left(\frac{2i}{n}\right)^3 + \frac{4i}{n} \right] \frac{2}{n} = \lim_{n\to\infty} \left[\frac{16}{n^4} \sum_{i=1}^{n} i^3 + \frac{8}{n^2} \sum_{i=1}^{n} i \right]$$

$$= \lim_{n\to\infty} \left[\frac{16}{n^4} \frac{n^2(n+1)^2}{4} + \frac{8}{n^2} \frac{n(n+1)}{2} \right]$$

$$= \lim_{n\to\infty} \left[4\left(1 + \frac{1}{n}\right)^2 + 4\left(1 + \frac{1}{n}\right) \right] = 4 \cdot 1 + 4 \cdot 1 = 8$$

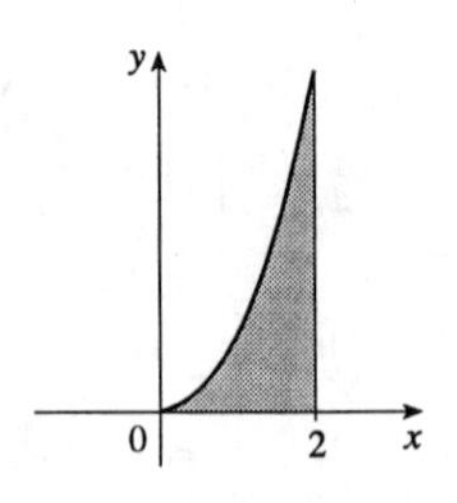

19. Here is one possible algorithm (ordered sequence of operations) for calculating the sums:

1 Let SUM = 0, let X-MIN = 0, let X-MAX = π, let STEP-SIZE = $\frac{\pi}{10}$ (or $\frac{\pi}{30}$ or $\frac{\pi}{50}$, depending on which sum we are calculating), and let RIGHT-ENDPOINT = X-MIN + STEP-SIZE.

2 Repeat steps 2a, 2b in sequence until RIGHT-ENDPOINT > X-MAX.

2a Add sin(RIGHT-ENDPOINT) to SUM.

2b Add STEP-SIZE to RIGHT-ENDPOINT.

At the end of this procedure, the variable SUM is equal to the answer we are looking for. We find that

$R_{10} = \dfrac{\pi}{10} \displaystyle\sum_{i=1}^{10} \sin\left(\frac{i\pi}{10}\right) \approx 1.9835$, $R_{30} = \dfrac{\pi}{30} \displaystyle\sum_{i=1}^{30} \sin\left(\frac{i\pi}{30}\right) \approx 1.9982$, and $R_{50} = \dfrac{\pi}{50} \displaystyle\sum_{i=1}^{50} \sin\left(\frac{i\pi}{50}\right) \approx 1.9993$.

It appears that the exact area is 2.

21. In Maple, we have to perform a number of steps before getting a numerical answer. After loading the `student` package [command: `with(student);`] we use the command `leftsum(x^(1/2),x=1..4,10` [*or* `30`, *or* `50`]`);` which gives us the expression in summation notation. To get a numerical approximation to the sum, we use `evalf(");`.

Mathematica does not have a special command for these sums, so we must type them in manually. For example, the first left sum is given by `(3/10)*Sum[Sqrt[1+3(i-1)/10],{i,1,10}]`, and we use the `N` command on the resulting output to get a numerical approximation.

In Derive, we use the `LEFT_RIEMANN` command to get the left sums, but must define the right sums ourselves.

(a) The left sums are $L_{10} = \dfrac{3}{10} \displaystyle\sum_{i=1}^{10} \sqrt{1 + \frac{3(i-1)}{10}} \approx 4.5148$, $L_{30} = \dfrac{3}{30} \displaystyle\sum_{i=1}^{30} \sqrt{1 + \frac{3(i-1)}{30}} \approx 4.6165$,

$L_{50} = \dfrac{3}{50} \displaystyle\sum_{i=1}^{50} \sqrt{1 + \frac{3(i-1)}{50}} \approx 4.6366$. The right sums are $R_{10} = \dfrac{3}{10} \displaystyle\sum_{i=1}^{10} \sqrt{1 + \frac{3i}{10}} \approx 4.8148$,

$R_{30} = \dfrac{3}{30} \displaystyle\sum_{i=1}^{30} \sqrt{1 + \frac{3i}{30}} \approx 4.7165$, $R_{50} = \dfrac{3}{50} \displaystyle\sum_{i=1}^{50} \sqrt{1 + \frac{3i}{50}} \approx 4.6966$.

(b) In Maple, we use the `leftbox` and `rightbox` commands (with the same arguments as `leftsum` and `rightsum` above) to generate the graphs.

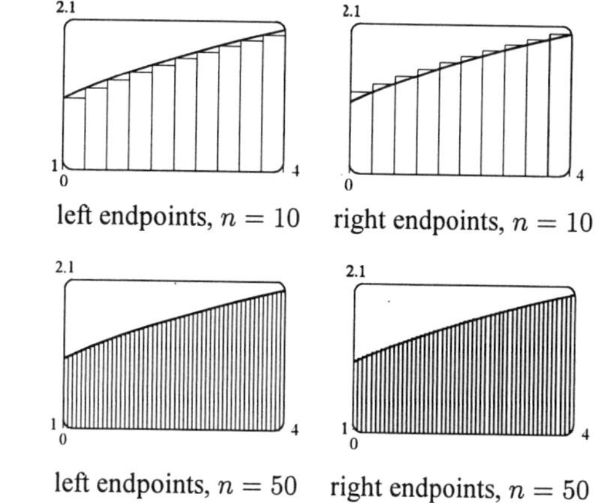

left endpoints, $n = 10$ right endpoints, $n = 10$

left endpoints, $n = 50$ right endpoints, $n = 50$

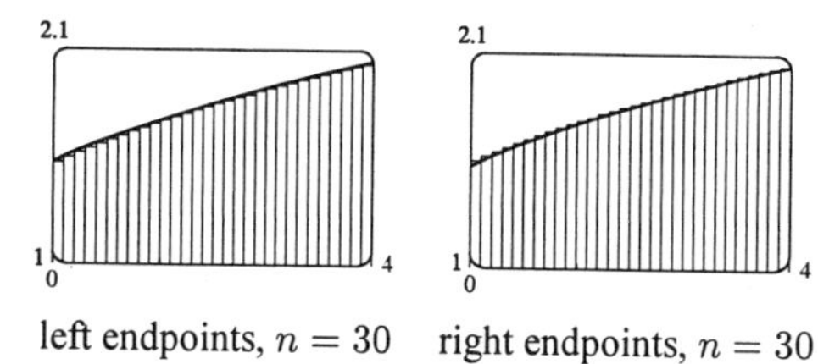

left endpoints, $n = 30$ right endpoints, $n = 30$

(c) We know that since $\sqrt{x}$ is an increasing function on $[1, 4]$, all of the left sums are smaller than the actual area, and all of the right sums are larger than the actual area (see Example 3). Since the left sum with $n = 50$ is about $4.637 > 4.6$ and the right sum with $n = 50$ is about $4.697 < 4.7$, we conclude that $4.6 < L_{50} <$ actual area $< R_{50} < 4.7$, so the actual area is between 4.6 and 4.7.

23. $\lim\limits_{n\to\infty} \sum\limits_{i=1}^{n} \frac{\pi}{4n} \tan \frac{\pi i}{4n}$ can be interpreted as the area of the region lying under the graph of $y = \tan x$ on the interval $\left[0, \frac{\pi}{4}\right]$, since for $y = \tan x$ on $\left[0, \frac{\pi}{4}\right]$ with partition points $x_i = \left(\frac{\pi}{4}\right)\frac{i}{n}$, $\Delta x = \frac{\pi}{4n}$ and $x_i^* = x_i$, the expression for the area is $A = \lim\limits_{n\to\infty} \sum\limits_{i=1}^{n} f(x_i^*)\Delta x = \lim\limits_{n\to\infty} \sum\limits_{i=1}^{n} \tan\left(\frac{\pi i}{4n}\right)\frac{\pi}{4n}$. Note that this answer is not unique, since the expression for the area is the same for the function $y = \tan(k\pi + x)$ on the interval $\left[k\pi, k\pi + \frac{\pi}{4}\right]$ where k is any integer.

25. $f(x) = \sin x$. $\Delta x_i = \frac{\pi}{n}$ and $x_i^* = x_i = \frac{i\pi}{n}$. So

$$A = \lim_{n\to\infty} \sum_{i=1}^{n} \left[\sin \frac{i\pi}{n}\left(\frac{\pi}{n}\right)\right] = \lim_{n\to\infty} \frac{\pi}{n} \sum_{i=1}^{n} \sin \frac{i\pi}{n} = \lim_{n\to\infty} \frac{\pi}{n} \frac{\sin\left(\frac{n}{2}\cdot\frac{\pi}{n}\right)\sin\left(\frac{n+1}{2}\cdot\frac{\pi}{n}\right)}{\sin\left(\frac{1}{2}\frac{\pi}{n}\right)} \quad \text{(see Exercise 6.1.53)}$$

$$= \lim_{n\to\infty} \frac{\pi}{n} \frac{\sin\left(\frac{\pi}{2} + \frac{\pi}{2n}\right)}{\sin\left(\frac{\pi}{2n}\right)} = \lim_{n\to\infty} 2 \frac{\frac{\pi}{2n}}{\sin\left(\frac{\pi}{2n}\right)} \cdot \lim_{n\to\infty} \cos\left(\frac{\pi}{2n}\right) = 2 \cdot 1 \cdot 1 = 2.$$

Here we have used the identity $\sin\left(\frac{\pi}{2} + x\right) = \cos x$.

EXERCISES 6.3

1. $f(x) = 7 - 2x$ **(a)** $\|P\| = \max\{0.6, 0.6, 0.8, 1.2, 0.8\} = 1.2$

(b) $\displaystyle\sum_{i=1}^{5} f(x_i^*)\Delta x_i = f(1.3)(0.6) + f(1.9)(0.6) + f(2.6)(0.8) + f(3.6)(1.2) + f(4.6)(0.8)$

$= (4.4)(0.6) + (3.2)(0.6) + (1.8)(0.8) + (-0.2)(1.2) + (-2.2)(0.8) = 4.$

3. $f(x) = 2 - x^2$ **(a)** $\|P\| = \max\{0.6, 0.4, 1, 0.8, 0.6, 0.6\} = 1$

(b) $\displaystyle\sum_{i=1}^{6} f(x_i^*)\Delta x_i = f(-1.4)(0.6) + f(-1)(0.4) + f(0)(1) + f(0.8)(0.8) + f(1.4)(0.6) + f(2)(0.6)$

$= (0.04)(0.6) + (1)(0.4) + (2)(1) + (1.36)(0.8) + (0.04)(0.6) + (-2)(0.6) = 2.336$

5. $f(x) = x^3$ **(a)** $\|P\| = \max\{0.5, 0.5, 0.5, 0.5\} = 0.5$

(b) $\displaystyle\sum_{i=1}^{n} f(x_i^*)\Delta x_i = \frac{1}{2}\sum_{i=1}^{4} f(x_i^*) = \tfrac{1}{2}\left[(-1)^3 + (-0.4)^3 + (0.2)^3 + 1^3\right] = -0.028$

7. **(a)** Using the right endpoints, we calculate

$$\int_0^8 f(x)\,dx \approx \sum_{i=1}^{4} f(x_i)\Delta x_i = 2[f(2) + f(4) + f(6) + f(8)] = 2(1 + 2 - 2 + 1) = 4$$

(b) Using the left endpoints, we calculate

$$\int_0^8 f(x)\,dx \approx \sum_{i=1}^{4} f(x_{i-1})\Delta x_i = 2[f(0) + f(2) + f(4) + f(6)] = 2(2 + 1 + 2 - 2) = 6$$

(c) Using the midpoint of each interval, we calculate

$$\int_0^8 f(x)\,dx \approx \sum_{i=1}^{4} f\left(\frac{x_i + x_{i-1}}{2}\right)\Delta x_i = 2[f(1) + f(3) + f(5) + f(7)] = 2(3 + 2 + 1 - 1) = 10$$

9. The width of the intervals is $\Delta x = (5 - 0)/5 = 1$ so the partition points are $0, 1, 2, 3, 4, 5$ and the midpoints are $0.5, 1.5, 2.5, 3.5, 4.5$. The Midpoint Rule gives

$\int_0^5 x^3\,dx \approx \sum_{i=1}^{5} f(\overline{x}_i)\Delta x = (0.5)^3 + (1.5)^3 + (2.5)^3 + (3.5)^3 + (4.5)^3 = 153.125.$

11. $\Delta x = (2 - 1)/10 = 0.1$ so the partition points are $1.0, 1.1, \ldots, 2.0$ and the midpoints are $1.05, 1.15, \ldots, 1.95$.

$$\int_1^2 \sqrt{1 + x^2}\,dx \approx \sum_{i=1}^{10} f(\overline{x}_i)\Delta x = 0.1\left[\sqrt{1 + (1.05)^2} + \sqrt{1 + (1.15)^2} + \cdots + \sqrt{1 + (1.95)^2}\right] \approx 1.8100$$

13. In Maple, we use the command `with(student);` to load the `sum` and `box` commands, then `m:=middlesum(sqrt(1+x^2),x=1..2,10);` which gives us the sum in summation notation, then `M:=evalf(m);` which gives $M_{10} \approx 1.81001414$, confirming the result of Exercise 11. The command `middlebox(sqrt(1+x^2),x=1..2,10);` generates the graph. Repeating for $n = 20$ and $n = 30$ gives $M_{20} \approx 1.81007263$ and $M_{30} \approx 1.81008347$.

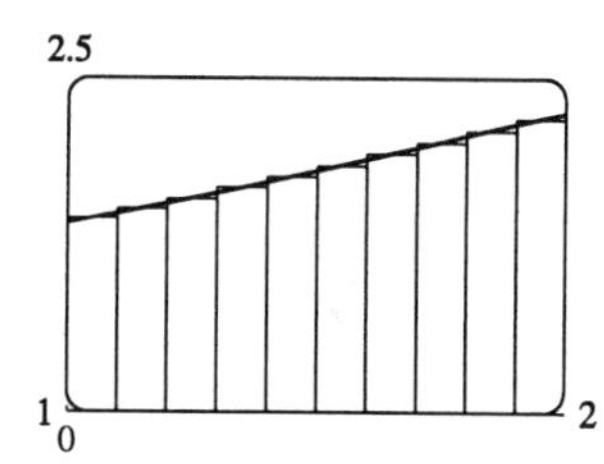

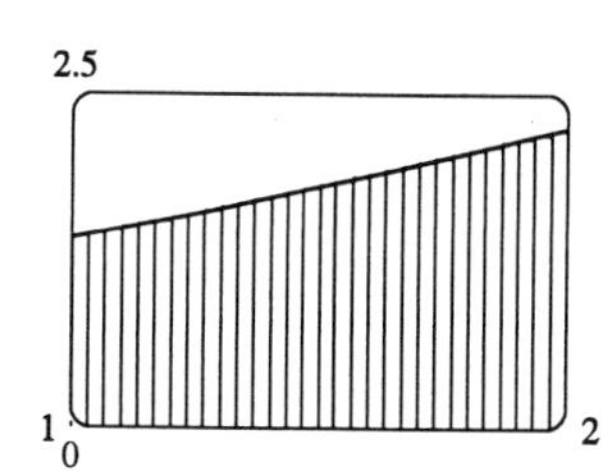

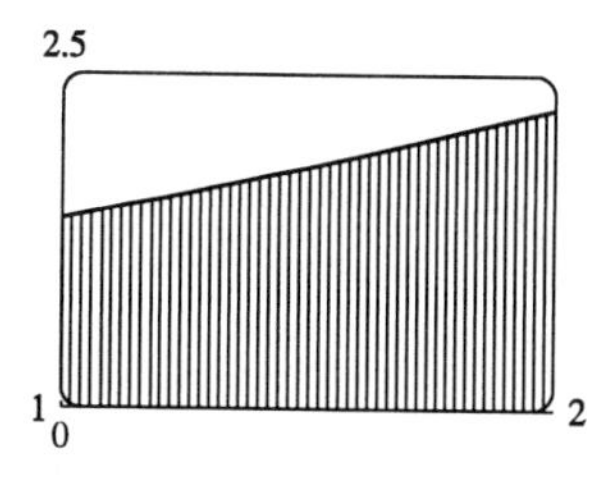

15. $\displaystyle\int_a^b c\,dx = \lim_{n\to\infty}\frac{b-a}{n}\sum_{i=1}^{n} c = \lim_{n\to\infty}\frac{b-a}{n}nc$ (by Theorem 6.1.3) $\displaystyle = \lim_{n\to\infty}(b-a)c = (b-a)c$

17. $\displaystyle\int_1^4 (x^2-2)dx = \lim_{n\to\infty}\frac{3}{n}\sum_{i=1}^{n}\left[\left(1+\frac{3i}{n}\right)^2 - 2\right] = \lim_{n\to\infty}\frac{3}{n}\sum_{i=1}^{n}\left[\frac{9i^2}{n^2}+\frac{6i}{n}-1\right]$

$$= \lim_{n\to\infty}\left[\frac{27}{n^3}\sum_{i=1}^{n} i^2 + \frac{18}{n^2}\sum_{i=1}^{n} i - \frac{3}{n}\sum_{i=1}^{n} 1\right] = \lim_{n\to\infty}\left[\frac{27}{n^3}\frac{n(n+1)(2n+1)}{6} + \frac{18}{n^2}\frac{n(n+1)}{2} - \frac{3}{n}n\right]$$

$$= \lim_{n\to\infty}\left[\frac{9}{2}\cdot 1\left(1+\frac{1}{n}\right)\left(2+\frac{1}{n}\right) + 9\cdot 1\left(1+\frac{1}{n}\right) - 3\right] = \left(\tfrac{9}{2}\cdot 2\right) + 9 - 3 = 15$$

19. $\displaystyle\int_0^b (x^3+4x)dx = \lim_{n\to\infty}\frac{b}{n}\sum_{i=1}^{n}\left[\left(\frac{bi}{n}\right)^3 + 4\left(\frac{bi}{n}\right)\right] = \lim_{n\to\infty}\left[\frac{b^4}{n^4}\sum_{i=1}^{n} i^3 + 4\frac{b^2}{n^2}\sum_{i=1}^{n} i\right]$

$$= \lim_{n\to\infty}\left[\frac{b^4}{n^4}\frac{n^2(n+1)^2}{4} + \frac{4b^2}{n^2}\frac{n(n+1)}{2}\right] = \lim_{n\to\infty}\left[\frac{b^4}{4}\cdot 1^2\left(1+\frac{1}{n}\right)^2 + 2b^2\cdot 1\left(1+\frac{1}{n}\right)\right]$$

$$= \frac{b^4}{4} + 2b^2$$

21. $\displaystyle\int_a^b x\,dx = \lim_{n\to\infty}\frac{b-a}{n}\sum_{i=1}^{n}\left[a+\frac{b-a}{n}i\right] = \lim_{n\to\infty}\left[\frac{a(b-a)}{n}\sum_{i=1}^{n} 1 + \frac{(b-a)^2}{n^2}\sum_{i=1}^{n} i\right]$

$$= \lim_{n\to\infty}\left[\frac{a(b-a)}{n}n + \frac{(b-a)^2}{n^2}\cdot\frac{n(n+1)}{2}\right] = a(b-a) + \lim_{n\to\infty}\frac{(b-a)^2}{2}\left(1+\frac{1}{n}\right)$$

$$= a(b-a) + \tfrac{1}{2}(b-a)^2 = (b-a)\left(a+\tfrac{1}{2}b-\tfrac{1}{2}a\right) = (b-a)\tfrac{1}{2}(b+a) = \tfrac{1}{2}(b^2-a^2)$$

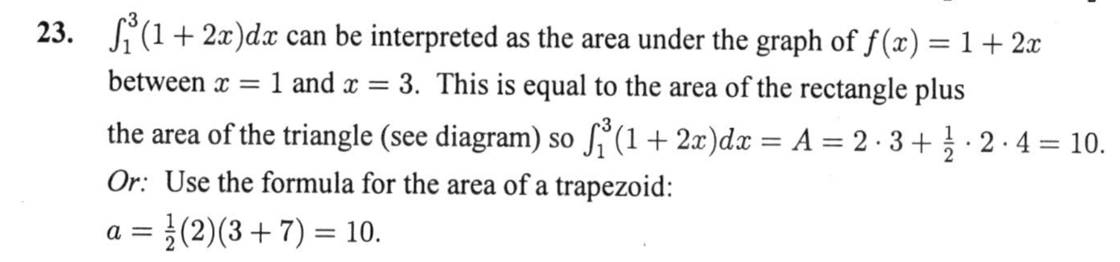

23. $\int_1^3 (1+2x)dx$ can be interpreted as the area under the graph of $f(x) = 1+2x$ between $x=1$ and $x=3$. This is equal to the area of the rectangle plus the area of the triangle (see diagram) so $\int_1^3 (1+2x)dx = A = 2\cdot 3 + \frac{1}{2}\cdot 2\cdot 4 = 10$.
Or: Use the formula for the area of a trapezoid:
$a = \frac{1}{2}(2)(3+7) = 10$.

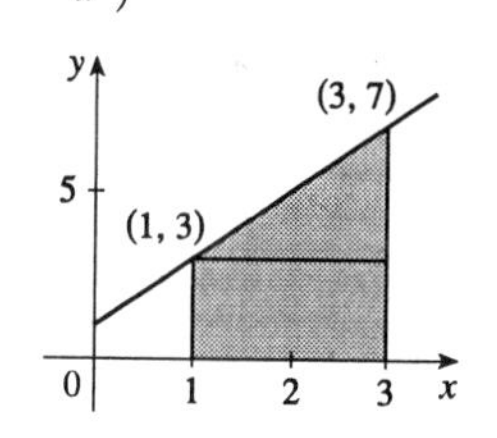

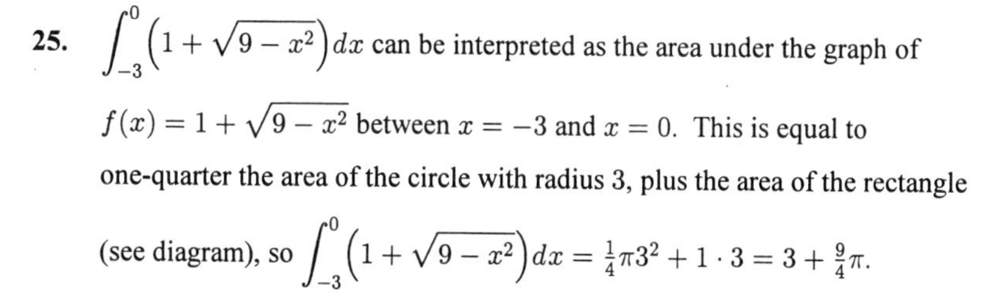

25. $\displaystyle\int_{-3}^{0}\left(1+\sqrt{9-x^2}\right)dx$ can be interpreted as the area under the graph of $f(x) = 1+\sqrt{9-x^2}$ between $x=-3$ and $x=0$. This is equal to one-quarter the area of the circle with radius 3, plus the area of the rectangle (see diagram), so $\displaystyle\int_{-3}^{0}\left(1+\sqrt{9-x^2}\right)dx = \tfrac{1}{4}\pi 3^2 + 1\cdot 3 = 3+\tfrac{9}{4}\pi$.

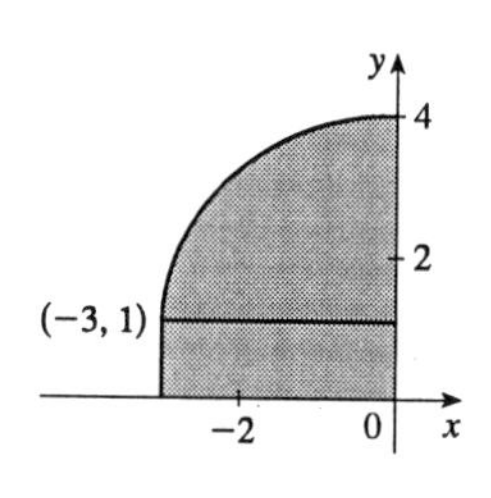

27. $\int_{-2}^{2}(1-|x|)dx$ can be interpreted as the area of the central triangle minus the areas of the outside ones (see diagram), so $\int_{-2}^{2}(1-|x|)dx = \frac{1}{2}\cdot 2\cdot 1 - 2\cdot\frac{1}{2}\cdot 1\cdot 1 = 0.$

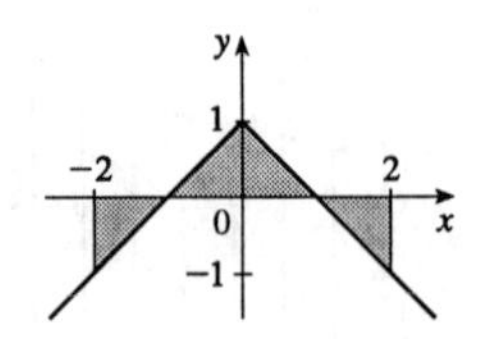

29. $\lim\limits_{\|P\|\to 0}\sum\limits_{i=1}^{n}\left[2(x_i^*)^2 - 5x_i^*\right]\Delta x_i = \int_0^1 (2x^2-5x)dx$ by Definition 2.

31. $\lim\limits_{\|P\|\to 0}\sum\limits_{i=1}^{n}\cos x_i\,\Delta x_i = \int_0^{\pi}\cos x\,dx$

33. $\lim\limits_{n\to\infty}\sum\limits_{i=1}^{n}\frac{i^4}{n^5} = \lim\limits_{n\to\infty}\frac{1}{n}\sum\limits_{i=1}^{n}\left(\frac{i}{n}\right)^4 = \int_0^1 x^4\,dx$

35. $\lim\limits_{n\to\infty}\sum\limits_{i=1}^{n}\left[3\left(1+\frac{2i}{n}\right)^5 - 6\right]\frac{2}{n} = \int_1^3 (3x^5-6)dx$

Note: To get started, notice that $\frac{2}{n} = \Delta x = \frac{b-a}{n}$ and $1+\frac{2i}{n} = a + \frac{b-a}{n}i$.

37. By the definition in Note 6, $\int_9^4 \sqrt{t}\,dt = -\int_4^9 \sqrt{t}\,dt = -\frac{38}{3}$.

39. $\int_{-4}^{-1}\sqrt{3}\,dx = \sqrt{3}(-1+4) = 3\sqrt{3}$

41. $\int_1^4(2x^2-3x+1)dx = 2\int_1^4 x^2\,dx - 3\int_1^4 x\,dx + \int_1^4 1\,dx$
$= 2\cdot\frac{1}{3}(4^3-1^3) - 3\cdot\frac{1}{2}(4^2-1^2) + 1(4-1) = \frac{45}{2} = 22.5$

43. $\int_{-1}^{1} f(x)dx = \int_{-1}^{0} f(x)dx + \int_0^1 f(x)dx = \int_{-1}^{0}(-2x)dx + \int_0^1 3x^2\,dx$
$= -2\int_{-1}^{0} x\,dx + 3\int_0^1 x^2\,dx = -2\cdot\frac{1}{2}\left[0^2-(-1)^2\right] + 3\cdot\frac{1}{3}[1^3-0^3] = 2$

45. $\int_1^3 f(x)dx + \int_3^6 f(x)dx + \int_6^{12} f(x)dx = \int_1^6 f(x)dx + \int_6^{12} f(x)dx = \int_1^{12} f(x)dx$

47. $\int_2^{10} f(x)dx - \int_2^7 f(x)dx = \int_2^7 f(x)dx + \int_7^{10} f(x)dx - \int_2^7 f(x)dx = \int_7^{10} f(x)dx$

49. $0\le \sin x < 1$ on $\left[0,\frac{\pi}{4}\right]$, so $\sin^3 x \le \sin^2 x$ on $\left[0,\frac{\pi}{4}\right]$. Hence $\int_0^{\pi/4}\sin^3 x\,dx \le \int_0^{\pi/4}\sin^2 x\,dx$ (Property 7).

51. $x\ge 4\ge 8-x$ on $[4,6]$, so $\frac{1}{x}\le\frac{1}{8-x}$ on $[4,6]$, and $\int_4^6 \frac{1}{x}\,dx \le \int_4^6 \frac{1}{8-x}\,dx.$

53. If $-1\le x\le 1$, then $0\le x^2\le 1$ and $1\le 1+x^2\le 2$, so $1\le\sqrt{1+x^2}\le\sqrt{2}$ and $1[1-(-1)] \le \int_{-1}^{1}\sqrt{1+x^2}\,dx \le \sqrt{2}[1-(-1)]$ [Property 8]; that is, $2\le\int_{-1}^{1}\sqrt{1+x^2}\,dx\le 2\sqrt{2}$.

55. If $1\le x\le 2$, then $\frac{1}{2}\le\frac{1}{x}\le 1$, so $\frac{1}{2}(2-1)\le\int_1^2\frac{1}{x}\,dx\le 1(2-1)$ or $\frac{1}{2}\le\int_1^2\frac{1}{x}\,dx\le 1$.

57. If $f(x) = x^2+2x$, $-3\le x\le 0$, then $f'(x) = 2x+2 = 0$ when $x=-1$, and $f(-1) = -1$. At the endpoints, $f(-3) = 3$, $f(0) = 0$. Thus the absolute minimum is $m=-1$ and the absolute maximum is $M=3$. Thus $-1[0-(-3)]\le\int_{-3}^{0}(x^2+2x)dx\le 3[0-(-3)]$ or $-3\le\int_{-3}^{0}(x^2+2x)dx\le 9$.

59. For $-1\le x\le 1$, $0\le x^4\le 1$ and $1\le\sqrt{1+x^4}\le\sqrt{2}$, so $1[1-(-1)]\le\int_{-1}^{1}\sqrt{1+x^4}\,dx\le\sqrt{2}[1-(-1)]$ or $2\le\int_{-1}^{1}\sqrt{1+x^4}\,dx\le 2\sqrt{2}$.

61. $\sqrt{x^4+1} \geq \sqrt{x^4} = x^2$, so $\int_1^3 \sqrt{x^4+1}\,dx \geq \int_1^3 x^2\,dx = \frac{1}{3}(3^3 - 1^3) = \frac{26}{3}$.

63. $0 \leq \sin x \leq 1$ for $0 \leq x \leq \frac{\pi}{2}$, so $x\sin x \leq x \quad \Rightarrow \quad \int_0^{\pi/2} x\sin x\,dx \leq \int_0^{\pi/2} x\,dx = \frac{1}{2}\left[\left(\frac{\pi}{2}\right)^2 - 0^2\right] = \frac{\pi^2}{8}$.

65. Using a regular partition and right endpoints as in the proof of Property 2, we calculate

$$\int_a^b cf(x)dx = \lim_{n\to\infty}\sum_{i=1}^n cf(x_i)\Delta x_i = \lim_{n\to\infty} c\sum_{i=1}^n f(x_i)\Delta x_i = c\lim_{n\to\infty}\sum_{i=1}^n f(x_i)\Delta x_i = c\int_a^b f(x)dx.$$

67. By Property 7, the inequalities $-|f(x)| \leq f(x) \leq |f(x)|$ imply that $\int_a^b(-|f(x)|)dx \leq \int_a^b f(x)dx \leq \int_a^b |f(x)|dx$. By Property 3, the left-hand integral equals $-\int_a^b |f(x)|dx$. Thus $-M \leq \int_a^b f(x)dx \leq M$, where $M = \int_a^b |f(x)|dx$. [Notice that $M \geq 0$ by Property 6.] It follows that $\left|\int_a^b f(x)dx\right| \leq M = \int_a^b |f(x)|dx$.

69. **(a)** $f(x) = x^2\sin x$ is continuous on $[0,2]$ and hence integrable by Theorem 4.

(b) $f(x) = \sec x$ is unbounded on $[0,2]$, so f is not integrable (see the remarks following Theorem 4.)

(c) $f(x)$ is piecewise continuous on $[0,2]$ with a single jump discontinuity at $x = 1$, so f is integrable.

(d) $f(x)$ has an infinite discontinuity at $x = 1$, so f is not integrable on $[0,2]$.

71. f is bounded since $|f(x)| \leq 1$ for all x in $[a,b]$. To see that f is not integrable on $[a,b]$, notice that $\sum_{i=1}^n f(x_i^*)\Delta x_i = 0$ if $x_1^*, \ldots, x_n^*$ are all chosen to be rational numbers, but $\sum_{i=1}^n f(x_i^*)\Delta x_i = \sum_{i=1}^n \Delta x_i = b - a$ if $x_1^*, \ldots, x_n^*$ are all chosen to be irrational numbers. This is true no matter how small $\|P\|$ is, since every interval $[x_{i-1}, x_i]$ with $x_{i-1} < x_i$ contains both rational and irrational numbers. $\sum_{i=1}^n f(x_i^*)\Delta x_i$ cannot approach both 0 and $b - a$ as $\|P\| \to 0$, so it has no limit as $\|P\| \to 0$.

73. Choose $x_i = 1 + \dfrac{i}{n}$ and $x_i^* = \sqrt{x_{i-1}\,x_i} = \sqrt{\left(1 + \dfrac{i-1}{n}\right)\left(1 + \dfrac{i}{n}\right)}$. Then

$$\int_1^2 x^{-2}\,dx = \lim_{n\to\infty}\frac{1}{n}\sum_{i=1}^n \frac{1}{\left(1+\dfrac{i-1}{n}\right)\left(1+\dfrac{i}{n}\right)} = \lim_{n\to\infty} n\sum_{i=1}^n \frac{1}{(n+i-1)(n+i)}$$

$$= \lim_{n\to\infty} n\sum_{i=1}^n\left[\frac{1}{n+i-1} - \frac{1}{n+1}\right] \quad \text{(by the hint)}$$

$$= \lim_{n\to\infty} n\left[\sum_{i=0}^{n-1}\frac{1}{n+i} - \sum_{i=1}^n \frac{1}{n+i}\right] = \lim_{n\to\infty} n\left[\frac{1}{n} - \frac{1}{2n}\right]$$

$$= \lim_{n\to\infty}\left[1 - \frac{1}{2}\right] = \frac{1}{2}.$$

EXERCISES 6.4

1. **(a)** $g(0) = \int_0^0 f(t)dt = 0$, $g(1) = \int_0^1 f(t)dt = 1 \cdot 2 = 2$,

$g(2) = \int_0^2 f(t)dt = \int_0^1 f(t)dt + \int_1^2 f(t)dt = g(1) + \int_1^2 f(t)dt = 2 + 1 \cdot 2 + \frac{1}{2} \cdot 1 \cdot 2 = 5$,

$g(3) = \int_0^3 f(t)dt = g(2) + \int_2^3 f(t)dt = 5 + \frac{1}{2} \cdot 1 \cdot 4 = 7$,

$g(6) = g(3) + \int_3^6 f(t)dt = 7 + \left[-\left(\frac{1}{2} \cdot 2 \cdot 2 + 1 \cdot 2\right)\right] = 7 - 4 = 3$

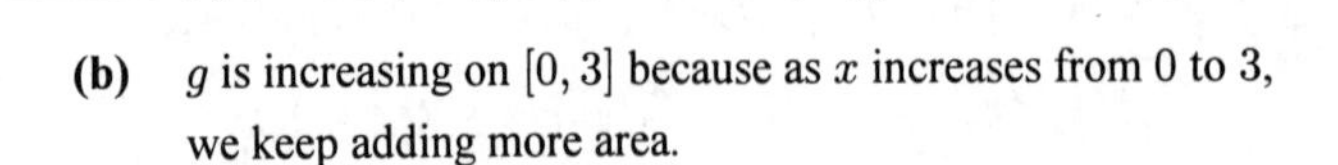

(b) g is increasing on $[0, 3]$ because as x increases from 0 to 3, we keep adding more area.

(c) g has a maximum value when we start subtracting area, that is, at $x = 3$.

(d)

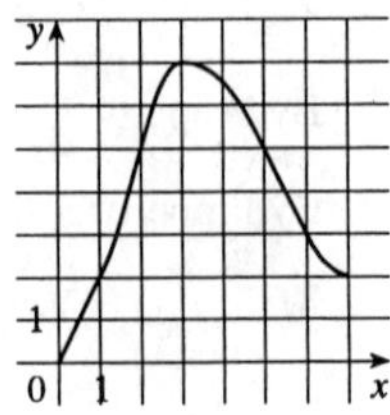

3.

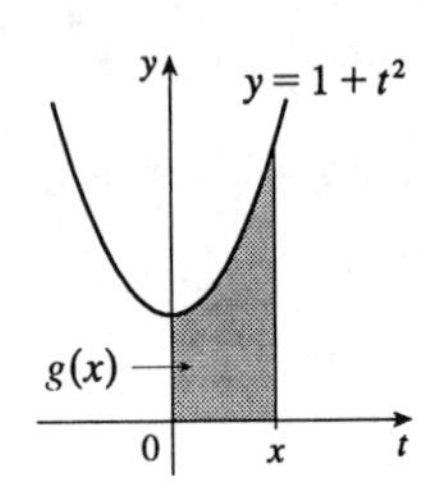

(a) By Part 1 of the Fundamental Theorem,

$g(x) = \int_0^x (1 + t^2)dt \quad \Rightarrow \quad g'(x) = f(x) = 1 + x^2$.

(b) By Part 2 of the Fundamental Theorem,

$$\begin{aligned} g(x) = \int_0^x (1 + t^2)dt &= \left[t + \tfrac{1}{3}t^3\right]_0^x \\ &= \left(x + \tfrac{1}{3}x^3\right) - \left(0 + \tfrac{1}{3}0^3\right) \\ &= x + \tfrac{1}{3}x^3 \end{aligned}$$

$\Rightarrow \quad g'(x) = 1 + x^2$.

5. $g(x) = \int_1^x (t^2 - 1)^{20} dt \quad \Rightarrow \quad g'(x) = (x^2 - 1)^{20}$

7. $g(u) = \displaystyle\int_\pi^u \frac{1}{1 + t^4}\, dt \quad \Rightarrow \quad g'(u) = \frac{1}{1 + u^4}$

9. $F(x) = \int_x^2 \cos(t^2)dt = -\int_2^x \cos(t^2)dt \quad \Rightarrow \quad F'(x) = -\cos(x^2)$

11. Let $u = \dfrac{1}{x}$. Then $\dfrac{du}{dx} = -\dfrac{1}{x^2}$, so $\dfrac{d}{dx}\displaystyle\int_2^{1/x} \sin^4 t\, dt = \frac{d}{du}\int_2^u \sin^4 t\, dt \cdot \frac{du}{dx} = \sin^4 u\, \frac{du}{dx} = \frac{-\sin^4(1/x)}{x^2}$.

13. Let $u = \tan x$. Then $\dfrac{du}{dx} = \sec^2 x$, so

$$\frac{d}{dx}\int_{\tan x}^{17} \sin(t^4)dt = -\frac{d}{dx}\int_{17}^{\tan x} \sin(t^4)dt = -\frac{d}{du}\int_{17}^{u} \sin(t^4)dt \cdot \frac{du}{dx} = -\sin(u^4)\frac{du}{dx} = -\sin(\tan^4 x)\sec^2 x.$$

15. Let $t = 5x + 1$. Then $\dfrac{dt}{dx} = 5$, so

$$\frac{d}{dx}\int_0^{5x+1} \frac{1}{u^2 - 5}\, du = \frac{d}{dt}\int_0^t \frac{1}{u^2 - 5}\, du \cdot \frac{dt}{dx} = \frac{1}{t^2 - 5}\frac{dt}{dx} = \frac{5}{25x^2 + 10x - 4}.$$

17. $\int_{-2}^4 (3x - 5)dx = \left(3 \cdot \frac{1}{2}x^2 - 5x\right)\Big|_{-2}^4 = (3 \cdot 8 - 5 \cdot 4) - [3 \cdot 2 - (-10)] = -12$

19. $\int_0^1 (1 - 2x - 3x^2)dx = \left[x - 2 \cdot \frac{1}{2}x^2 - 3 \cdot \frac{1}{3}x^3\right]_0^1 = \left[x - x^2 - x^3\right]_0^1 = (1 - 1 - 1) - 0 = -1$

21. $\int_{-3}^0 (5y^4 - 6y^2 + 14)dy = \left[5\left(\frac{1}{5}y^5\right) - 6\left(\frac{1}{3}y^3\right) + 14y\right]_{-3}^0 = \left[y^5 - 2y^3 + 14y\right]_{-3}^0 = 0 - (-243 + 54 - 42) = 231$

23. $\int_0^4 \sqrt{x}\,dx = \int_0^4 x^{1/2}\,dx = \left[\dfrac{x^{3/2}}{3/2}\right]_0^4 = \left[\dfrac{2x^{3/2}}{3}\right]_0^4 = \dfrac{2(4)^{3/2}}{3} - 0 = \dfrac{16}{3}$

25. $\int_1^3 \left[\dfrac{1}{t^2} - \dfrac{1}{t^4}\right]dt = \int_1^3 (t^{-2} - t^{-4})dt = \left[\dfrac{t^{-1}}{-1} - \dfrac{t^{-3}}{-3}\right]_1^3 = \left[\dfrac{1}{3t^3} - \dfrac{1}{t}\right]_1^3 = \left(\dfrac{1}{81} - \dfrac{1}{3}\right) - \left(\dfrac{1}{3} - 1\right) = \dfrac{28}{81}$

27. $\int_1^2 \dfrac{x^2+1}{\sqrt{x}}\,dx = \int_1^2 (x^{3/2} + x^{-1/2})dx = \left[\dfrac{x^{5/2}}{5/2} + \dfrac{x^{1/2}}{1/2}\right]_1^2 = \left[\tfrac{2}{5}x^{5/2} + 2x^{1/2}\right]_1^2$

$= \left(\tfrac{2}{5}4\sqrt{2} + 2\sqrt{2}\right) - \left(\tfrac{2}{5} + 2\right) = \dfrac{18\sqrt{2} - 12}{5} = \tfrac{6}{5}\left(3\sqrt{2} - 2\right)$

29. $\int_0^1 u(\sqrt{u} + \sqrt[3]{u})du = \int_0^1 (u^{3/2} + u^{4/3})du = \left[\dfrac{u^{5/2}}{5/2} + \dfrac{u^{7/3}}{7/3}\right]_0^1 = \left[\tfrac{2}{5}u^{5/2} + \tfrac{3}{7}u^{7/3}\right]_0^1 = \tfrac{2}{5} + \tfrac{3}{7} = \tfrac{29}{35}$

31. $\int_{-2}^3 |x^2 - 1|dx = \int_{-2}^{-1} (x^2 - 1)dx + \int_{-1}^1 (1 - x^2)dx + \int_1^3 (x^2 - 1)dx$

$= \left[\dfrac{x^3}{3} - x\right]_{-2}^{-1} + \left[x - \dfrac{x^3}{3}\right]_{-1}^1 + \left[\dfrac{x^3}{3} - x\right]_1^3$

$= \left(-\dfrac{1}{3} + 1\right) - \left(-\dfrac{8}{3} + 2\right) + \left(1 - \dfrac{1}{3}\right) - \left(-1 + \dfrac{1}{3}\right) + (9 - 3) - \left(\dfrac{1}{3} - 1\right) = \dfrac{28}{3}$

33. $\int_3^3 \sqrt{x^5 + 2}\,dx = 0$ by the definition in Note 6 in Section 6.3.

35. $\int_{-4}^2 \dfrac{2}{x^6}\,dx$ does not exist since $f(x) = \dfrac{2}{x^6}$ has an infinite discontinuity at 0.

37. $\int_1^4 \left(\sqrt{t} - \dfrac{2}{\sqrt{t}}\right)dt = \int_1^4 (t^{1/2} - 2t^{-1/2})dt = \left[\dfrac{t^{3/2}}{3/2} - 2\dfrac{t^{1/2}}{1/2}\right]_1^4 = \left[\tfrac{2}{3}t^{3/2} - 4t^{1/2}\right]_1^4$

$= \left[\tfrac{2}{3} \cdot 8 - 4 \cdot 2\right] - \left[\tfrac{2}{3} - 4\right] = \tfrac{2}{3}$

39. $\int_{-1}^0 (x+1)^3\,dx = \int_{-1}^0 (x^3 + 3x^2 + 3x + 1)dx = \left[\dfrac{x^4}{4} + 3\dfrac{x^3}{3} + 3\dfrac{x^2}{2} + x\right]_{-1}^0 = 0 - \left[\dfrac{1}{4} - 1 + \dfrac{3}{2} - 1\right]$

$= 2 - \dfrac{7}{4} = \dfrac{1}{4}$

41. $\int_{\pi/4}^{\pi/3} \sin t\,dt = [-\cos t]_{\pi/4}^{\pi/3} = -\cos\tfrac{\pi}{3} + \cos\tfrac{\pi}{4} = -\tfrac{1}{2} + \tfrac{1}{\sqrt{2}} = \tfrac{\sqrt{2} - 1}{2}$

43. $\int_{\pi/2}^{\pi} \sec x \tan x\,dx$ does not exist since $\sec x \tan x$ has an infinite discontinuity at $\tfrac{\pi}{2}$.

45. $\int_{\pi/6}^{\pi/3} \csc^2\theta\,d\theta = [-\cot\theta]_{\pi/6}^{\pi/3} = -\cot\tfrac{\pi}{3} + \cot\tfrac{\pi}{6} = -\tfrac{1}{3}\sqrt{3} + \sqrt{3} = \tfrac{2}{3}\sqrt{3}$

47. $\int_4^8 (1/x)dx = [\ln x]_4^8 = \ln 8 - \ln 4 = \ln\tfrac{8}{4} = \ln 2$

49. $\int_8^9 2^t\,dt = \left[\dfrac{1}{\ln 2}2^t\right]_8^9 = \dfrac{1}{\ln 2}(2^9 - 2^8) = \dfrac{2^8}{\ln 2}$

51. $\int_1^{\sqrt{3}} \dfrac{6}{1+x^2}\,dx = 6\left[\tan^{-1} x\right]_1^{\sqrt{3}} = 6\tan^{-1}\sqrt{3} - 6\tan^{-1} 1 = 6 \cdot \tfrac{\pi}{3} - 6 \cdot \tfrac{\pi}{4} = \tfrac{\pi}{2}$

53. $\displaystyle\int_1^e \frac{x^2+x+1}{x}\,dx = \int_1^e \left[x+1+\frac{1}{x}\right]dx = \left[\tfrac{1}{2}x^2+x+\ln x\right]_1^e$
$= \left[\tfrac{1}{2}e^2+e+\ln e\right] - \left[\tfrac{1}{2}+1+\ln 1\right] = \tfrac{1}{2}e^2+e-\tfrac{1}{2}$

55. $\displaystyle\int_0^1 \left[\sqrt[4]{x^5}+\sqrt[5]{x^4}\right]dx = \int_0^1 (x^{5/4}+x^{4/5})dx = \left[\frac{x^{9/4}}{9/4}+\frac{x^{9/5}}{9/5}\right]_0^1 = \left[\tfrac{4}{9}x^{9/4}+\tfrac{5}{9}x^{9/5}\right]_0^1 = \tfrac{4}{9}+\tfrac{5}{9}-0=1$

57. $\int_{-1}^2 (x-2|x|)dx = \int_{-1}^0 3x\,dx + \int_0^2(-x)dx = 3\left[\tfrac{1}{2}x^2\right]_{-1}^0 - \left[\tfrac{1}{2}x^2\right]_0^2 = \left(3\cdot 0 - 3\cdot\tfrac{1}{2}\right) - (2-0) = -\tfrac{7}{2} = -3.5$

59. $\int_0^2 f(x)dx = \int_0^1 x^4dx + \int_1^2 x^5dx = \tfrac{1}{5}x^5\big|_0^1 + \tfrac{1}{6}x^6\big|_1^2 = \left(\tfrac{1}{5}-0\right)+\left(\tfrac{64}{6}-\tfrac{1}{6}\right) = 10.7$

61. From the graph, it appears that the area is about 60. The actual area is
$\int_0^{27} x^{1/3}dx = \left[\tfrac{3}{4}x^{4/3}\right]_0^{27} = \tfrac{3}{4}\cdot 81 - 0$
$= \tfrac{243}{4} = 60.75$. This is $\tfrac{3}{4}$ of the area of the viewing rectangle.

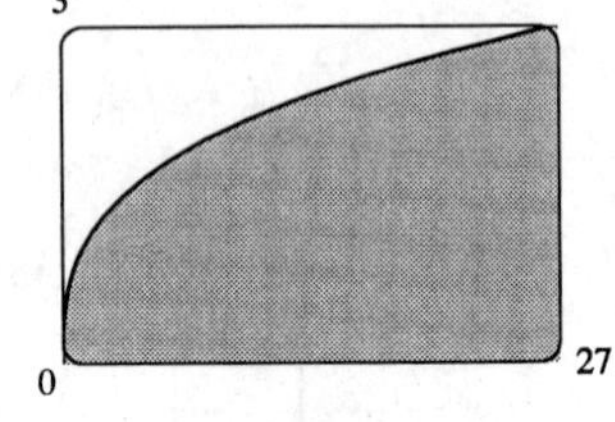

63. It appears that the area under the graph is about $\tfrac{2}{3}$ of the area of the viewing rectangle, or about $\tfrac{2}{3}\pi \approx 2.1$. The actual area is
$\int_0^\pi \sin x\,dx = [-\cos x]_0^\pi = -\cos\pi + \cos 0$
$= -(-1)+1 = 2.$

65. By zooming in on the graph of $y = x+x^2-x^4$, we see that the graph has x-intercepts at $x=0$ and at $x\approx 1.32$. So the area of the region below the curve and above the x-axis is about
$\int_0^{1.32}(x+x^2-x^4)dx = \left[\tfrac{1}{2}x^2+\tfrac{1}{3}x^3-\tfrac{1}{5}x^5\right]_0^{1.32}$
$= \left[\tfrac{1}{2}(1.32)^2+\tfrac{1}{3}(1.32)^3-\tfrac{1}{5}(1.32)^5\right]-0 \approx 0.84$

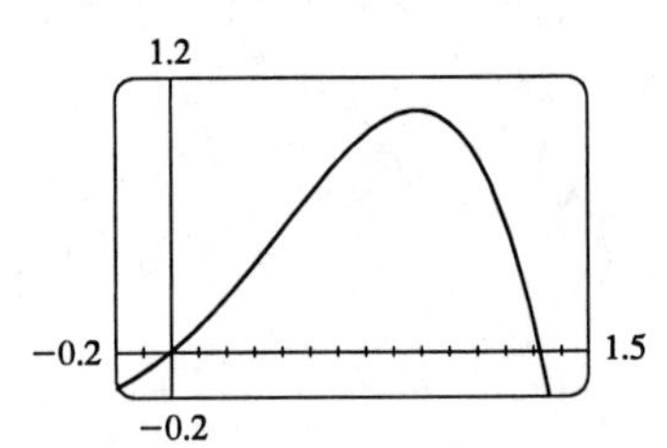

67. $\displaystyle\frac{d}{dx}\left[\frac{x}{a^2\sqrt{a^2-x^2}}+C\right] = \frac{1}{a^2}\frac{\sqrt{a^2-x^2}-x(-x)/\sqrt{a^2-x^2}}{a^2-x^2} = \frac{1}{a^2}\frac{(a^2-x^2)+x^2}{(a^2-x^2)^{3/2}} = \frac{1}{\sqrt{(a^2-x^2)^3}}$

69. $\displaystyle\frac{d}{dx}\left(\frac{x}{2}-\frac{\sin 2x}{4}+C\right) = \tfrac{1}{2}-\tfrac{1}{4}(\cos 2x)(2)+0 = \tfrac{1}{2}-\tfrac{1}{2}\cos 2x = \tfrac{1}{2}-\tfrac{1}{2}(1-2\sin^2 x) = \sin^2 x$

71. $\int x\sqrt{x}\,dx = \int x^{3/2}\,dx = \tfrac{2}{5}x^{5/2}+C$

73. $\displaystyle\int\left(2-\sqrt{x}\right)^2dx = \int\left(4-4\sqrt{x}+x\right)dx = 4x-4\frac{x^{3/2}}{3/2}+\frac{x^2}{2}+C = 4x-\tfrac{8}{3}x^{3/2}+\tfrac{1}{2}x^2+C$

75. $\int(2x+\sec x\tan x)dx = x^2+\sec x + C$

77. **(a)** displacement $= \int_0^3(3t-5)dt = \left[\tfrac{3}{2}t^2-5t\right]_0^3 = \tfrac{27}{2}-15 = -\tfrac{3}{2}$ m

(b) distance traveled $= \int_0^3|3t-5|dt = \int_0^{5/3}(5-3t)dt + \int_{5/3}^3(3t-5)dt = \left[5t-\tfrac{3}{2}t^2\right]_0^{5/3} + \left[\tfrac{3}{2}t^2-5t\right]_{5/3}^3$
$= \tfrac{25}{3}-\tfrac{3}{2}\cdot\tfrac{25}{9}+\tfrac{27}{2}-15-\left(\tfrac{3}{2}\cdot\tfrac{25}{9}-\tfrac{25}{3}\right) = \tfrac{41}{6}$ m

79. **(a)** $v'(t) = a(t) = t + 4 \quad \Rightarrow \quad v(t) = \frac{1}{2}t^2 + 4t + C \quad \Rightarrow \quad 5 = v(0) = C \quad \Rightarrow \quad v(t) = \frac{1}{2}t^2 + 4t + 5\text{ m/s}$

Or: $v(t) - v(0) = \int_0^t a(u)du = \int_0^t (u+4)du = \frac{1}{2}u^2 + 4u\Big|_0^t = \frac{1}{2}t^2 + 4t \Rightarrow v(t) = \frac{1}{2}t^2 + 4t + 5\text{ m/s}.$

(b) distance traveled $= \int_0^{10} |v(t)|\,dt = \int_0^{10} \left|\frac{1}{2}t^2 + 4t + 5\right|dt = \int_0^{10} \left(\frac{1}{2}t^2 + 4t + 5\right)dt$
$= \left[\frac{1}{6}t^3 + 2t^2 + 5t\right]_0^{10} = \frac{500}{3} + 200 + 50 = 416\frac{2}{3}\text{ m}$

81. Since $m'(x) = \rho(x)$, $m = \int_0^4 \rho(x)dx = \int_0^4 (9 + 2\sqrt{x})dx = \left[9x + \frac{4}{3}x^{3/2}\right]_0^4 = 36 + \frac{32}{3} - 0 = \frac{140}{3} = 46\frac{2}{3}\text{ kg}.$

83. Let s be the position of the car. We know from Equation 12 that $s(100) - s(0) = \int_0^{100} v(t)dt$. We use the Midpoint Rule for $0 \le t \le 100$ with $n = 5$. Note that the length of each of the five time intervals is 20 seconds $= \frac{1}{180}$ hour. So the distance traveled is

$\int_0^{100} v(t)dt \approx \frac{1}{180}[v(10) + v(30) + v(50) + v(70) + v(90)] = \frac{1}{180}(38 + 58 + 51 + 53 + 47) = \frac{247}{180} \approx 1.4\text{ miles}$

85. $g(x) = \displaystyle\int_{2x}^{3x} \frac{u-1}{u+1}du = \int_{2x}^{0} \frac{u-1}{u+1}du + \int_0^{3x} \frac{u-1}{u+1}du = -\int_0^{2x} \frac{u-1}{u+1}du + \int_0^{3x} \frac{u-1}{u+1}du \quad \Rightarrow$

$g'(x) = -\dfrac{2x-1}{2x+1} \cdot \dfrac{d}{dx}(2x) + \dfrac{3x-1}{3x+1} \cdot \dfrac{d}{dx}(3x) = -2 \cdot \dfrac{2x-1}{2x+1} + 3 \cdot \dfrac{3x-1}{3x+1}$

87. $y = \displaystyle\int_{\sqrt{x}}^{x^3} \sqrt{t}\sin t\,dt = \int_{\sqrt{x}}^{1} \sqrt{t}\sin t\,dt + \int_1^{x^3} \sqrt{t}\sin t\,dt = -\int_1^{\sqrt{x}} \sqrt{t}\sin t\,dt + \int_1^{x^3} \sqrt{t}\sin t\,dt \quad \Rightarrow$

$y' = -\sqrt[4]{x}(\sin\sqrt{x}) \cdot \dfrac{d}{dx}(\sqrt{x}) + x^{3/2}\sin(x^3) \cdot \dfrac{d}{dx}(x^3) = -\dfrac{\sqrt[4]{x}\sin\sqrt{x}}{2\sqrt{x}} + x^{3/2}\sin(x^3)(3x^2)$

$= 3x^{7/2}\sin(x^3) - (\sin\sqrt{x})/(2\sqrt[4]{x})$

89. $F(x) = \displaystyle\int_1^x f(t)dt \quad \Rightarrow \quad F'(x) = f(x) = \int_1^{x^2} \frac{\sqrt{1+u^4}}{u}du \quad \Rightarrow$

$F''(x) = f'(x) = \dfrac{\sqrt{1+(x^2)^4}}{x^2} \cdot \dfrac{d}{dx}(x^2) = \dfrac{2\sqrt{1+x^8}}{x}$. So $F''(2) = \sqrt{1+2^8} = \sqrt{257}$.

91. **(a)** The Fresnel Function $S(x) = \int_0^x \sin\left(\frac{\pi}{2}t^2\right)dt$ has local maximum values where $0 = S'(x) = \sin\left(\frac{\pi}{2}x^2\right)$ and S' changes from positive to negative. For $x > 0$, this happens when $\frac{\pi}{2}x^2 = (2n-1)\pi \quad \Leftrightarrow$ $x = \sqrt{2(2n-1)}$, n any positive integer. For $x < 0$, S' changes from positive to negative where $x = -2\sqrt{n}$, since if $x < 0$, then as x increases, x^2 decreases. S' does not change sign at $x = 0$.

(b) S is concave upward on those intervals where $S''(x) \ge 0$. Differentiating our expression for $S'(x)$, we get $S''(x) = \cos\left(\frac{\pi}{2}x^2\right)\left(2\frac{\pi}{2}x\right) = \pi x\cos\left(\frac{\pi}{2}x^2\right)$. For $x > 0$, $S''(x) > 0$ where $\cos\left(\frac{\pi}{2}x^2\right) > 0 \Leftrightarrow 0 < \frac{\pi}{2}x^2 < \frac{\pi}{2}$ or $\left(2n - \frac{1}{2}\right)\pi < \frac{\pi}{2}x^2 < \left(2n + \frac{1}{2}\right)\pi$, n any integer $\quad \Leftrightarrow \quad 0 < x < 1$ or $\sqrt{4n-1} < x < \sqrt{4n+1}$, n any positive integer. For $x < 0$, as x increases, x^2 decreases, so the intervals of upward concavity for $x < 0$ are $\left(-\sqrt{4n-1}, -\sqrt{4n-3}\right)$, n any positive integer. To summarize: S is concave upward on the intervals $(0,1)$, $\left(-\sqrt{3}, -1\right)$, $\left(\sqrt{3}, \sqrt{5}\right)$, $\left(-\sqrt{7}, -\sqrt{5}\right)$, $\left(\sqrt{7}, 3\right)$, $\ldots$.

(c)

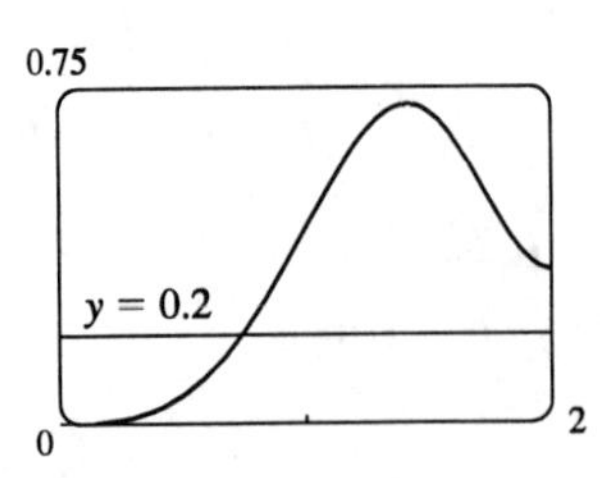

In Maple, we use `plot({int(sin(Pi*t^2/2),t=0..x),0.2},x=0..2);`. Note that Maple recognizes the Fresnel function, calling it `FresnelS(x)`. In Mathematica, we use `Plot[{Integrate[Sin[Pi*t^2/2],{t,0,x}],0.2},{x,0,2}]`. From the graphs, we see that $\int_0^x \sin\left(\frac{\pi}{2}t^2\right)dt = 0.2$ at $x \approx 0.74$.

93. **(a)** By the Fundamental Theorem of Calculus, $g'(x) = f(x)$. So $g'(x) = 0$ at $x = 1, 3, 5, 7$, and 9. g has local maxima at $x = 1$ and at $x = 5$ (since $f = g'$ changes from positive to negative there) and local minima at $x = 3$ and at $x = 7$. There is no local extremum at $x = 9$, since f is not defined for $x > 9$.

(b) We can see from the graph that $\left|\int_0^1 f\,dt\right| < \left|\int_1^3 f\,dt\right| < \cdots < \left|\int_7^9 f\,dt\right|$. So

$$g(9) = \int_0^9 f\,dt = \left|\int_0^1 f\,dt\right| - \left|\int_1^3 f\,dt\right| + \cdots - \left|\int_5^7 f\,dt\right| + \left|\int_7^9 f\,dt\right|$$
$$> g(5) = \int_0^5 f\,dt = \left|\int_0^1 f\,dt\right| - \cdots + \left|\int_3^5 f\,dt\right|,$$

which in turn is larger than $g(1)$.
So the absolute maximum of $g(x)$ occurs at $x = 9$.

(c) g is concave downward on those intervals where $g'' < 0$.
But $g'(x) = f(x)$, so $g''(x) = f'(x)$, which is negative on $\left(\frac{1}{2}, 2\right)$, $(4, 6)$ and $(8, 9)$. So g is concave down on these intervals.

(d)

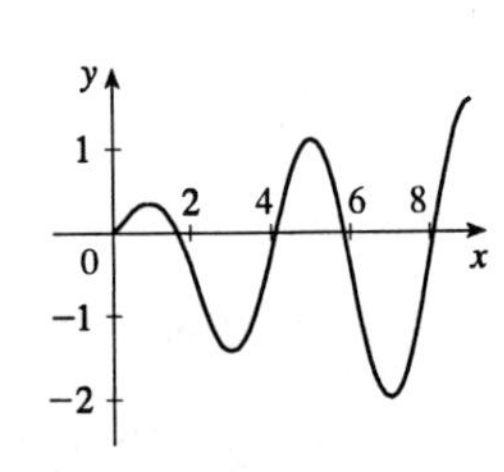

95. $$\lim_{n\to\infty}\sum_{i=1}^{n}\frac{i^3}{n^4} = \lim_{n\to\infty}\frac{1-0}{n}\sum_{i=1}^{n}\left(\frac{i}{n}\right)^3 = \int_0^1 x^3\,dx = \left.\frac{x^4}{4}\right|_0^1 = \frac{1}{4}$$

97. Suppose $h < 0$. Since f is continuous on $[x+h, x]$, the Extreme Value Theorem says that there are numbers u and v in $[x+h, x]$ such that $f(u) = m$ and $f(v) = M$, where m and M are the absolute minimum and maximum values of f on $[x+h, x]$. By Property 8 of integrals, $m(-h) \le \int_{x+h}^{x} f(t)dt \le M(-h)$; that is, $f(u)(-h) \le -\int_x^{x+h} f(t)dt \le f(v)(-h)$. Since $-h > 0$, we can divide this inequality by $-h$: $f(u) \le \frac{1}{h}\int_x^{x+h} f(t)dt \le f(v)$. By Equation 3, $\frac{g(x+h)-g(x)}{h} = \frac{1}{h}\int_x^{x+h} f(t)dt$ for $h \ne 0$, and hence $f(u) \le \frac{g(x+h)-g(x)}{h} \le f(v)$ which is Equation 4 in the case where $h < 0$.

99. **(a)** Let $f(x) = \sqrt{x} \quad \Rightarrow \quad f'(x) = 1/(2\sqrt{x}) > 0$ for $x > 0 \quad \Rightarrow \quad f$ is increasing on $[0, \infty)$. If $x \geq 0$, then $x^3 \geq 0$, so $1 + x^3 \geq 1$ and since f is increasing, this means that $f(1 + x^3) \geq f(1) \quad \Rightarrow$ $\sqrt{1 + x^3} \geq 1$ for $x \geq 0$. Next let $g(t) = t^2 - t \quad \Rightarrow \quad g'(t) = 2t - 1 \quad \Rightarrow \quad g'(t) > 0$ when $t \geq 1$. Thus g is increasing on $[1, \infty)$. And since $g(1) = 0$, $g(t) \geq 0$ when $t \geq 1$. Now let $t = \sqrt{1 + x^3}$, where $x \geq 0$. $\sqrt{1 + x^3} \geq 1$ (from above) $\quad \Rightarrow \quad t \geq 1 \quad \Rightarrow \quad g(t) \geq 0 \quad \Rightarrow \quad (1 + x^3) - \sqrt{1 + x^3} \geq 0$ for $x \geq 0$. Therefore $1 \leq \sqrt{1 + x^3} \leq 1 + x^3$ for $x \geq 0$.

(b) From part (a) and Property 7: $\int_0^1 1\,dx \leq \int_0^1 \sqrt{1 + x^3}\,dx \leq \int_0^1 (1 + x^3)dx \quad \Leftrightarrow$ $x\big|_0^1 \leq \int_0^1 \sqrt{1 + x^3}\,dx \leq \left[x + \frac{1}{4}x^4\right]_0^1 \quad \Leftrightarrow \quad 1 \leq \int_0^1 \sqrt{1 + x^3}\,dx \leq 1 + \frac{1}{4} = 1.25$.

101. If $w'(t)$ is the rate of change of weight in pounds per year, then $w(t)$ represents the weight in pounds of the child at age t. We know from the Fundamental Theorem of Calculus that $\int_5^{10} w'(t)dt = w(10) - w(5)$, so the integral represents the weight gained by the child between the ages of 5 and 10.

103. We differentiate both sides, using the Fundamental Theorem of Calculus, to get $\dfrac{f(x)}{x^2} = 2\dfrac{1}{2\sqrt{x}} \quad \Leftrightarrow$ $f(x) = x^{3/2}$. To find a, we substitute $x = a$ in the original equation to obtain $6 + \displaystyle\int_a^a \frac{f(t)}{t^2}\,dt = 2\sqrt{a} \quad \Rightarrow$ $6 + 0 = 2\sqrt{a} \quad \Rightarrow \quad a = 9$.

EXERCISES 6.5

1. Let $u = x^2 - 1$. Then $du = 2x\,dx$, so $\displaystyle\int x(x^2 - 1)^{99}\,dx = \int u^{99}\left(\tfrac{1}{2}\,du\right) = \frac{1}{2}\frac{u^{100}}{100} + C = \tfrac{1}{200}(x^2 - 1)^{100} + C$.

3. Let $u = 4x$. Then $du = 4\,dx$, so $\int \sin 4x\,dx = \int \sin u\left(\frac{1}{4}\,du\right) = \frac{1}{4}(-\cos u) + C = -\frac{1}{4}\cos 4x + C$

5. Let $u = x^2 + 6x$. Then $du = 2(x + 3)dx$, so

$$\int \frac{x + 3}{(x^2 + 6x)^2}\,dx = \frac{1}{2}\int \frac{du}{u^2} = \frac{1}{2}\int u^{-2}\,du = -\tfrac{1}{2}u^{-1} + C = -\frac{1}{2(x^2 + 6x)} + C.$$

7. Let $u = x^2 + x + 1$. Then $du = (2x + 1)dx$, so

$\int (2x + 1)(x^2 + x + 1)^3\,dx = \int u^3\,du = \frac{1}{4}u^4 + C = \frac{1}{4}(x^2 + x + 1)^4 + C$.

9. Let $u = x - 1$. Then $du = dx$, so $\int \sqrt{x - 1}\,dx = \int u^{1/2}\,du = \frac{2}{3}u^{3/2} + C = \frac{2}{3}(x - 1)^{3/2} + C$.

11. Let $u = 2 + x^4$. Then $du = 4x^3\,dx$, so $\displaystyle\int x^3\sqrt{2 + x^4}\,dx = \int u^{1/2}\left(\tfrac{1}{4}\,du\right) = \frac{1}{4}\frac{u^{3/2}}{3/2} + C = \tfrac{1}{6}(2 + x^4)^{3/2} + C$.

13. Let $u = t + 1$. Then $du = dt$, so $\displaystyle\int \frac{2}{(t + 1)^6}\,dt = 2\int u^{-6}\,du = -\frac{2}{5}u^{-5} + C = -\frac{2}{5(t + 1)^5} + C$.

15. Let $u = 1 - 2y$. Then $du = -2\,dy$, so

$$\int (1 - 2y)^{1.3}dy = \int u^{1.3}\left(-\tfrac{1}{2}\,du\right) = -\frac{1}{2}\left(\frac{u^{2.3}}{2.3}\right) + C = -\frac{(1 - 2y)^{2.3}}{4.6} + C.$$

17. Let $u = 2\theta$. Then $du = 2\,d\theta$, so $\int \cos 2\theta\, d\theta = \int \cos u\left(\frac{1}{2}\,du\right) = \frac{1}{2}\sin u + C = \frac{1}{2}\sin 2\theta + C$.

19. Let $u = x + 2$. Then $du = dx$, so $\displaystyle\int \frac{x}{\sqrt[4]{x+2}}\,dx = \int \frac{u-2}{\sqrt[4]{u}}\,du = \int \left(u^{3/4} - 2u^{-1/4}\right)du$

$= \frac{4}{7}u^{7/4} - 2\cdot\frac{4}{3}u^{3/4} + C = \frac{4}{7}(x+2)^{7/4} - \frac{8}{3}(x+2)^{3/4} + C.$

21. Let $u = t^2$. Then $du = 2t\,dt$, so $\int t\sin(t^2)dt = \int \sin u\left(\frac{1}{2}\,du\right) = -\frac{1}{2}\cos u + C = -\frac{1}{2}\cos(t^2) + C$.

23. Let $u = 1 - x^2$. Then $x^2 = 1 - u$ and $2x\,dx = -du$, so

$$\int x^3\left(1-x^2\right)^{3/2} dx = \int \left(1-x^2\right)^{3/2} x^2 \cdot x\,dx = \int u^{3/2}(1-u)\left(-\tfrac{1}{2}\right)du = \tfrac{1}{2}\int \left(u^{5/2} - u^{3/2}\right)du$$

$$= \tfrac{1}{2}\left[\tfrac{2}{7}u^{7/2} - \tfrac{2}{5}u^{5/2}\right] + C = \tfrac{1}{7}\left(1-x^2\right)^{7/2} - \tfrac{1}{5}\left(1-x^2\right)^{5/2} + C.$$

25. Let $u = 1 + \sec x$. Then $du = \sec x\tan x\,dx$, so

$\int \sec x\tan x\sqrt{1+\sec x}\,dx = \int u^{1/2}\,du = \frac{2}{3}u^{3/2} + C = \frac{2}{3}(1+\sec x)^{3/2} + C.$

27. Let $u = \cos x$. Then $du = -\sin x\,dx$, so $\int \cos^4 x\sin x\,dx = \int u^4(-du) = -\frac{1}{5}u^5 + C = -\frac{1}{5}\cos^5 x + C$.

29. Let $u = 2x + 3$. Then $du = 2\,dx$, so

$\int \sin(2x+3)dx = \int \sin u\left(\frac{1}{2}\,du\right) = -\frac{1}{2}\cos u + C = -\frac{1}{2}\cos(2x+3) + C.$

31. Let $u = 3x$. Then $du = 3\,dx$, so

$\int(\sin 3\alpha - \sin 3x)dx = \int(\sin 3\alpha - \sin u)\frac{1}{3}\,du = \frac{1}{3}[(\sin 3\alpha)\,u + \cos u] + C = (\sin 3\alpha)\,x + \frac{1}{3}\cos 3x + C.$

33. Let $u = b + cx^{a+1}$. Then $du = (a+1)cx^a\,dx$, so

$$\int x^a\sqrt{b+cx^{a+1}}\,dx = \int u^{1/2}\frac{1}{(a+1)c}\,du = \frac{1}{(a+1)c}\left(\tfrac{2}{3}u^{3/2}\right) + C = \frac{2}{3c(a+1)}\left(b+cx^{a+1}\right)^{3/2} + C.$$

35. Let $u = 2x - 1$. Then $du = 2\,dx$, so $\displaystyle\int \frac{dx}{2x-1} = \int \frac{\frac{1}{2}du}{u} = \frac{1}{2}\ln|u| + C = \frac{1}{2}\ln|2x-1| + C$.

37. Let $u = \ln x$. Then $du = \dfrac{dx}{x}$, so $\displaystyle\int \frac{(\ln x)^2}{x}\,dx = \int u^2du = \frac{1}{3}u^3 + C = \frac{1}{3}(\ln x)^3 + C$.

39. Let $u = 1 + e^x$. Then $du = e^x\,dx$, so $\int e^x(1+e^x)^{10}\,dx = \int u^{10}\,du = \frac{1}{11}u^{11} + C = \frac{1}{11}(1+e^x)^{11} + C$.

41. Let $u = \ln x$. Then $du = \dfrac{dx}{x}$, so $\displaystyle\int \frac{dx}{x\ln x} = \int \frac{du}{u} = \ln|u| + C = \ln|\ln x| + C$.

43. $\displaystyle\int \frac{e^x+1}{e^x}\,dx = \int(1+e^{-x})dx = x - e^{-x} + C$ [Substitute $u = -x$.]

45. Let $u = x^2 + 2x$. Then $du = 2(x+1)dx$, so $\displaystyle\int \frac{x+1}{x^2+2x}\,dx = \int \frac{\frac{1}{2}\,du}{u} = \frac{1}{2}\ln|u| + C = \frac{1}{2}\ln|x^2+2x| + C$.

47. $\displaystyle\int \frac{1+x}{1+x^2}\,dx = \int \frac{1}{1+x^2}\,dx + \int \frac{x}{1+x^2}\,dx = \tan^{-1}x + \frac{1}{2}\int \frac{2x\,dx}{1+x^2} = \tan^{-1}x + \frac{1}{2}\ln(1+x^2) + C$

(At the last step, we evaluate $\int du/u$ where $u = 1 + x^2$.)

49. $f(x) = \dfrac{3x-1}{(3x^2-2x+1)^4}$. Let $u = 3x^2 - 2x + 1$.

Then $du = (6x-2)dx = 2(3x-1)dx$, so

$$\int \frac{3x-1}{(3x^2-2x+1)^4}\,dx = \int \frac{1}{u^4}\left(\tfrac{1}{2}\,du\right) = \tfrac{1}{2}\int u^{-4}du$$

$$= -\tfrac{1}{6}u^{-3} + C = -\frac{1}{6(3x^2-2x+1)^3} + C.$$

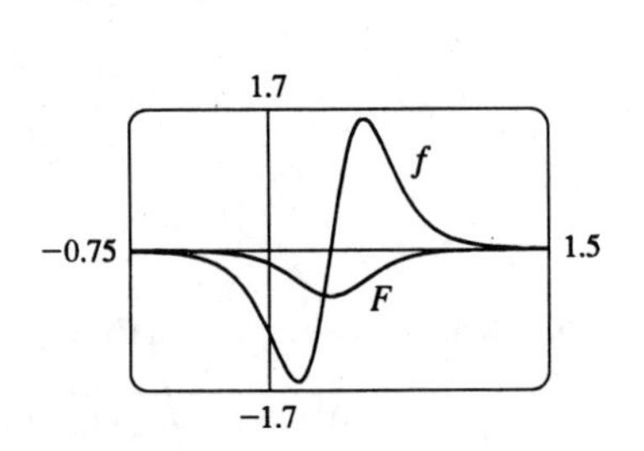

Notice that at $x = \frac{1}{3}$, the integrand goes from negative to positive, and the graph of the integral has a horizontal tangent (a local minimum).

51. $f(x) = \sin^3 x \cos x$. Let $u = \sin x$. Then $du = \cos x\,dx$, so $\int \sin^3 x \cos x\,dx = \int u^3\,du = \frac{1}{4}u^4 + C = \frac{1}{4}\sin^4 x + C$. Note that at $x = \frac{\pi}{2}$, the graph of the integrand crosses the x-axis from above, and the integral has a local maximum. Also, both f and F are periodic with period π, so at $x = 0$ and at $x = \pi$, the graph of the integrand crosses the x-axis from below, and the integral has local minima.

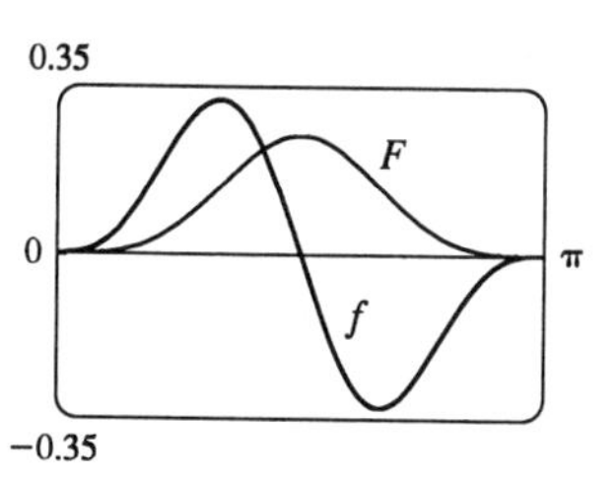

53. Let $u = 2x - 1$. Then $du = 2\,dx$, so $\int_0^1 (2x-1)^{100}dx = \int_{-1}^1 u^{100}\left(\frac{1}{2}\,du\right) = \int_0^1 u^{100}\,du$ [since the integrand is an even function] $= \left[\frac{1}{101}u^{101}\right]_0^1 = \frac{1}{101}$.

55. Let $u = x^4 + x$. Then $du = (4x^3+1)dx$, so $\displaystyle\int_0^1 (x^4+x)^5(4x^3+1)dx = \int_0^2 u^5 du = \left[\frac{u^6}{6}\right]_0^2 = \frac{2^6}{6} = \frac{32}{3}$.

57. Let $u = x - 1$. Then $du = dx$, so

$$\int_1^2 x\sqrt{x-1}\,dx = \int_0^1 (u+1)\sqrt{u}\,du = \int_0^1 \left(u^{3/2} + u^{1/2}\right)du = \left[\tfrac{2}{5}u^{5/2} + \tfrac{2}{3}u^{3/2}\right]_0^1 = \frac{2}{5} + \frac{2}{3} = \frac{16}{15}.$$

59. Let $u = \pi t$. Then $du = \pi\,dt$, so $\int_0^1 \cos \pi t\,dt = \int_0^\pi \cos u\left(\frac{1}{\pi}\,du\right) = \frac{1}{\pi}\sin u\big|_0^\pi = \frac{1}{\pi}(0-0) = 0$.

61. Let $u = 1 + \dfrac{1}{x}$. Then $du = -\dfrac{dx}{x^2}$, so

$$\int_1^4 \frac{1}{x^2}\sqrt{1+\frac{1}{x}}\,dx = \int_2^{5/4} u^{1/2}(-du) = \int_{5/4}^2 u^{1/2}\,du = \left[\tfrac{2}{3}u^{3/2}\right]_{5/4}^2 = \tfrac{2}{3}\left[2\sqrt{2} - \tfrac{5\sqrt{5}}{8}\right] = \tfrac{4\sqrt{2}}{3} - \tfrac{5\sqrt{5}}{12}.$$

63. Let $u = \cos\theta$. Then $du = -\sin\theta\,d\theta$, so

$$\int_0^{\pi/3} \frac{\sin\theta}{\cos^2\theta}\,d\theta = \int_1^{1/2} \frac{-du}{u^2} = \int_{1/2}^1 u^{-2}\,du = \left[-\frac{1}{u}\right]_{1/2}^1 = -1 + 2 = 1.$$

65. Let $u = 1 + 2x$. Then $du = 2\,dx$, so $\displaystyle\int_0^{13} \frac{dx}{\sqrt[3]{(1+2x)^2}} = \int_1^{27} u^{-2/3}\left(\tfrac{1}{2}\,du\right) = \tfrac{1}{2}\cdot 3u^{1/3}\big|_1^{27} = \tfrac{3}{2}(3-1) = 3$.

67. $\displaystyle\int_0^4 \frac{dx}{(x-2)^3}$ does not exist since $\dfrac{1}{(x-2)^3}$ has an infinite discontinuity at $x = 2$.

69. Let $u = x^2 + a^2$. Then $du = 2x\,dx$, so

$$\int_0^a x\sqrt{x^2+a^2}\,dx = \int_{a^2}^{2a^2} u^{1/2}\left(\tfrac{1}{2}\,du\right) = \tfrac{1}{2}\left[\tfrac{2}{3}\,u^{3/2}\right]_{a^2}^{2a^2} = \left[\tfrac{1}{3}u^{3/2}\right]_{a^2}^{2a^2} = \tfrac{1}{3}\left(2\sqrt{2}-1\right)a^3.$$

71. Let $u = 2x + 3$. Then $du = 2\,dx$, so

$$\int_0^3 \frac{dx}{2x+3} = \int_3^9 \frac{\frac{1}{2}\,du}{u} = \tfrac{1}{2}\ln u\big|_3^9 = \tfrac{1}{2}(\ln 9 - \ln 3) = \tfrac{1}{2}(\ln 3^2 - \ln 3) = \tfrac{1}{2}(2\ln 3 - \ln 3) = \tfrac{1}{2}\ln 3 \quad \left(\text{or } \ln\sqrt{3}\right).$$

73. Let $u = \ln x$. Then $du = \dfrac{dx}{x}$, so $\displaystyle\int_e^{e^4} \frac{dx}{x\sqrt{\ln x}} = \int_1^4 u^{-1/2}\,du = 2u^{1/2}\big|_1^4 = 2\cdot 2 - 2\cdot 1 = 2$.

75. From the graph, it appears that the area under the curve is about $1 +$ a little more than $\frac{1}{2} \cdot 1 \cdot 0.7$, or about 1.4. The exact area is given by $A = \int_0^1 \sqrt{2x+1}\,dx$. Let $u = 2x + 1$, so $du = 2\,dx$, the limits change to $2 \cdot 0 + 1 = 1$ and $2 \cdot 1 + 1 = 3$, and

$A = \int_1^3 \sqrt{u}\left(\frac{1}{2}\,du\right) = \frac{1}{3}u^{3/2}\Big|_1^3 = \frac{1}{3}\left(3\sqrt{3} - 1\right)$
$= \sqrt{3} - \frac{1}{3} \approx 1.399.$

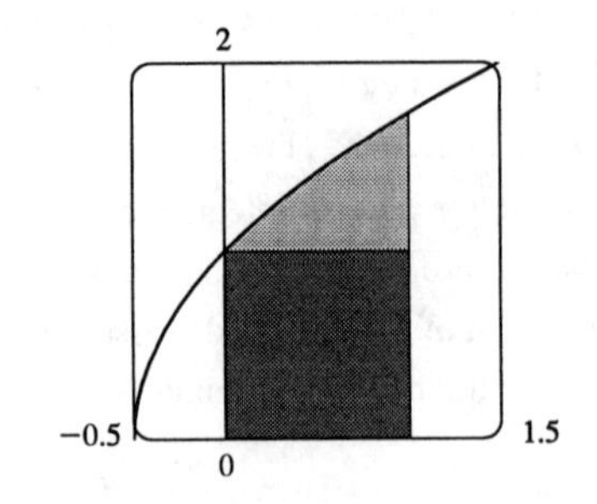

77. We split the integral: $\int_{-2}^{2}(x+3)\sqrt{4-x^2}\,dx = \int_{-2}^{2} x\sqrt{4-x^2}\,dx + \int_{-2}^{2} 3\sqrt{4-x^2}\,dx$. The first integral is 0 by Theorem 7, since $f(x) = x\sqrt{4-x^2}$ is an odd function and we are integrating from $x = -2$ to $x = 2$. The second integral we interpret as three times the area of the half-circle with radius 2, so the original integral is equal to $0 + 3 \cdot \frac{1}{2}(\pi \cdot 2^2) = 6\pi$.

79. The volume of inhaled air in the lungs at time t is

$V(t) = \int_0^t f(u)du = \int_0^t \frac{1}{2}\sin\left(\frac{2}{5}\pi u\right)du = \int_0^{2\pi t/5} \frac{1}{2}\sin v\left(\frac{5}{2\pi}\,dv\right)$ [We substitute $v = \frac{2\pi}{5}u \quad\Rightarrow\quad dv = \frac{2\pi}{5}\,du$]
$= \frac{5}{4\pi}(-\cos v)\Big|_0^{2\pi t/5} = \frac{5}{4\pi}\left[-\cos\left(\frac{2}{5}\pi t\right) + 1\right] = \frac{5}{4\pi}\left[1 - \cos\left(\frac{2}{5}\pi t\right)\right]$ liters.

81. We make the substitution $u = 2x$ in $\int_0^2 f(2x)dx$. So $du = 2dx$ and the limits become $u = 2 \cdot 0 = 2$ and $u = 2 \cdot 2 = 4$. Hence $\int_0^4 f(u)\left(\frac{1}{2}\,du\right) = \frac{1}{2}\int_0^4 f(u)\,du = \frac{1}{2}(10) = 5$.

83. Let $u = -x$. Then $du = -dx$. When $x = a$, $u = -a$; when $x = b$, $u = -b$. So

$\int_a^b f(-x)dx = \int_{-a}^{-b} f(u)(-du) = \int_{-b}^{-a} f(u)du = \int_{-b}^{-a} f(x)dx.$

From the diagram, we see that the equality follows from the fact that we are reflecting the graph of f, and the limits of integration, about the y-axis.

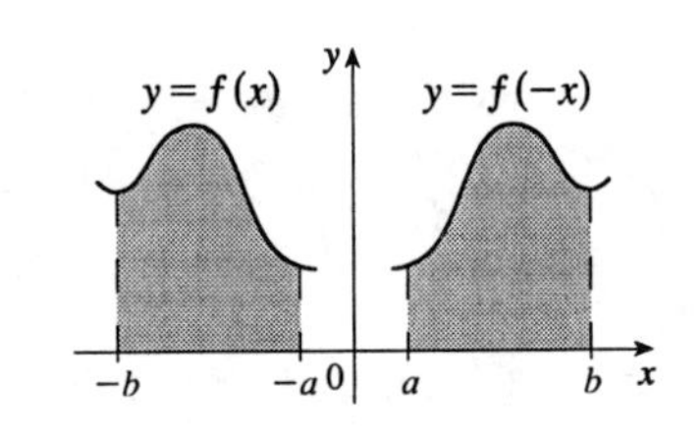

85. Let $u = 1 - x$. Then $du = -dx$. When $x = 1$, $u = 0$ and when $x = 0$, $u = 1$. So

$\int_0^1 x^a(1-x)^b\,dx = -\int_1^0 (1-u)^a u^b\,du = \int_0^1 u^b(1-u)^a\,du = \int_0^1 x^b(1-x)^a\,dx.$

87. $\dfrac{x\sin x}{1+\cos^2 x} = x \cdot \dfrac{\sin x}{2 - \sin^2 x} = xf(\sin x)$, where $f(t) = \dfrac{t}{2-t^2}$. By Exercise 86,

$$\int_0^\pi \frac{x\sin x}{1+\cos^2 x}\,dx = \int_0^\pi xf(\sin x)dx = \frac{\pi}{2}\int_0^\pi f(\sin x)dx = \frac{\pi}{2}\int_0^\pi \frac{\sin x}{1+\cos^2 x}\,dx.$$

Let $u = \cos x$. Then $du = -\sin x\,dx$. When $x = \pi$, $u = -1$ and when $x = 0$, $u = 1$. So

$$\frac{\pi}{2}\int_0^\pi \frac{\sin x}{1+\cos^2 x}\,dx = -\frac{\pi}{2}\int_1^{-1} \frac{du}{1+u^2} = \frac{\pi}{2}\int_{-1}^{1} \frac{du}{1+u^2} = \frac{\pi}{2}\left[\tan^{-1} u\right]_{-1}^{1}$$
$$= \frac{\pi}{2}\left[\tan^{-1}1 - \tan^{-1}(-1)\right] = \frac{\pi}{2}\left[\frac{\pi}{4} - \left(-\frac{\pi}{4}\right)\right] = \frac{\pi^2}{4}.$$

EXERCISES 6.6

1. **(a)**

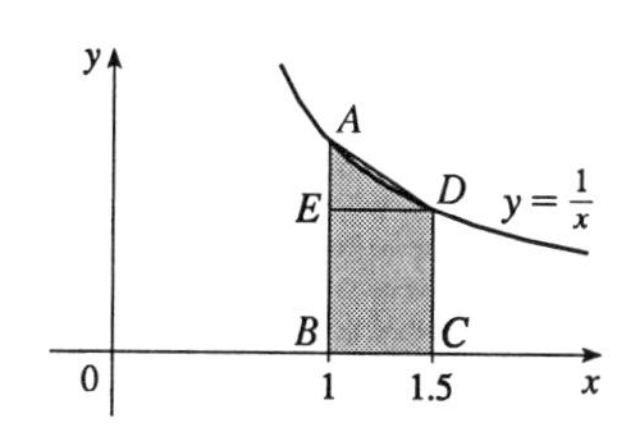

We interpret $\ln 1.5$ as the area under the curve $y = 1/x$ from $x = 1$ to $x = 1.5$. The area of the rectangle $BCDE$ is $\frac{1}{2} \cdot \frac{2}{3} = \frac{1}{3}$. The area of the trapezoid $ABCD$ is $\frac{1}{2} \cdot \frac{1}{2}\left(1 + \frac{2}{3}\right) = \frac{5}{12}$. Thus, by comparing areas, we observe that $\frac{1}{3} < \ln 1.5 < \frac{5}{12}$.

(b) With $f(t) = 1/t$, $n = 10$, and $\Delta x = 0.05$, we have

$$\ln 1.5 = \int_1^{1.5} (1/t)dt \approx (0.05)[f(1.025) + f(1.075) + \cdots + f(1.475)]$$
$$= (0.05)\left[\frac{1}{1.025} + \frac{1}{1.075} + \cdots + \frac{1}{1.475}\right] \approx 0.4054.$$

3.

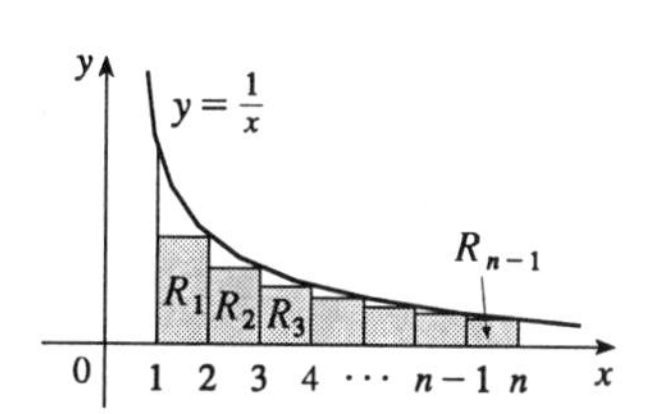

The area of R_i is $\dfrac{1}{i+1}$ and so

$$\frac{1}{2} + \frac{1}{3} + \cdots + \frac{1}{n} < \int_1^n \frac{1}{t}\,dt = \ln n.$$

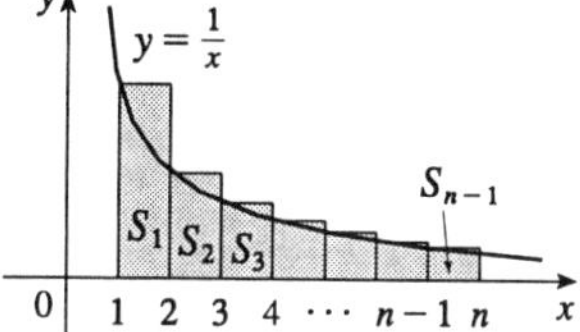

The area of S_i is $\dfrac{1}{i}$ and so

$$1 + \frac{1}{2} + \cdots + \frac{1}{n-1} > \int_1^n \frac{1}{t}\,dt = \ln n.$$

5. If $f(x) = \ln(x^r)$, then $f'(x) = (1/x^r)(rx^{r-1}) = r/x$. But if $g(x) = r\ln x$, then $g'(x) = r/x$. So f and g must differ by a constant: $\ln(x^r) = r\ln x + C$. Put $x = 1$: $\ln(1^r) = r\ln 1 + C \quad\Rightarrow\quad C = 0$, so $\ln(x^r) = r\ln x$.

7. Using the third law of logarithms and Equation 10, we have $\ln e^{rx} = rx = r\ln e^x = \ln(e^x)^r$. Since ln is a one-to-one function, it follows that $e^{rx} = (e^x)^r$.

9. Using Definition 13, the first law of logarithms, and the first law of exponents for e^x, we have $(ab)^x = e^{x\ln(ab)} = e^{x(\ln a + \ln b)} = e^{x\ln a + x\ln b} = e^{x\ln a}e^{x\ln b} = a^x b^x$.

REVIEW EXERCISES FOR CHAPTER 6

1. True by Theorem 6.1.2 (b).

3. True by repeated application of Theorem 3.2.3 (b).

5. True by Property 2 of Integrals.

7. False. For example, let $f(x) = x^2$. Then $\int_0^1 \sqrt{x^2}\,dx = \int_0^1 x\,dx = \frac{1}{2}$, but $\sqrt{\int_0^1 x^2\,dx} = \sqrt{\frac{1}{3}} = \frac{1}{\sqrt{3}}$.

9. True by Property 7 of Integrals.

11. True. The integrand is an odd function that is continuous on $[-1, 1]$, so the result follows from Equation 6.5.7 (b).

13. False. The function $f(x) = 1/x^4$ is not bounded on the interval $[-2, 1]$. It has an infinite discontinuity at $x = 0$, so it is not integrable on the interval. (If the integral were to exist, a positive value would be expected by Property 6 of Integrals.)

15. False. For example, the function $y = |x|$ is continuous on $\mathbb{R}$, but has no derivative at $x = 0$.

17. First note that either a or b must be the graph of $\int_0^x f(t)dt$, since $\int_0^0 f(t)dt = 0$, and $c(0) \neq 0$. Now notice that $b > 0$ when c is increasing, and that $c > 0$ when a is increasing. It follows that c is the graph of $f(x)$, b is the graph of $f'(x)$, and a is the graph of $\int_0^x f(t)dt$.

19. $\displaystyle\sum_{i=1}^{n} f(x_i^*)\,\Delta x_i = \sum_{i=1}^{4} f\left(\frac{i-1}{2}\right)\cdot\frac{1}{2} = \tfrac{1}{2}\left[f(0) + f\left(\tfrac{1}{2}\right) + f(1) + f\left(\tfrac{3}{2}\right)\right].$

$f(x) = 2 + (x-2)^2$, so $f(0) = 6$, $f\left(\frac{1}{2}\right) = 4.25$, $f(1) = 3$, and

$f\left(\frac{3}{2}\right) = 2.25$. Thus $\displaystyle\sum_{i=1}^{n} f(x_i^*)\,\Delta x_i = \tfrac{1}{2}(15.5) = 7.75.$

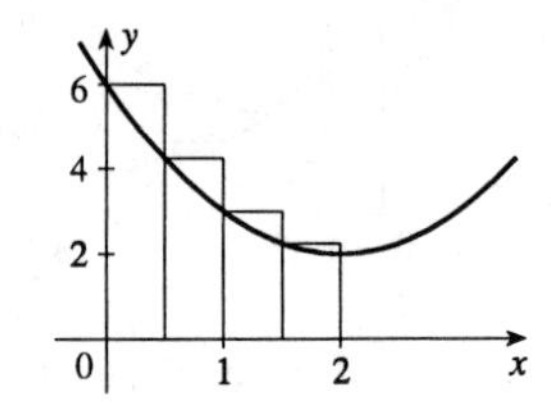

21. By Theorem 6.3.5, $\displaystyle\int_2^4 (3-4x)dx = \lim_{n\to\infty}\frac{2}{n}\sum_{i=1}^{n}\left[3 - 4\left(2 + \frac{2i}{n}\right)\right] = \lim_{n\to\infty}\frac{2}{n}\sum_{i=1}^{n}\left[-5 - \frac{8}{n}i\right]$

$$= \lim_{n\to\infty}\frac{2}{n}\left[-5n - \frac{8}{n}\frac{n(n+1)}{2}\right] = \lim_{n\to\infty}\left[-10 - 8\cdot 1\left(1 + \frac{1}{n}\right)\right] = -10 - 8 = -18.$$

23. $\int_0^5 (x^3 - 2x^2)dx = \frac{1}{4}x^4 - \frac{2}{3}x^3\Big|_0^5 = \frac{625}{4} - \frac{250}{3} = \frac{875}{12}$ (We did this integral in Exercise 22.)

25. $\int_0^1 (1 - x^9)dx = \left[x - \frac{1}{10}x^{10}\right]_0^1 = 1 - \frac{1}{10} = \frac{9}{10}$

27. $\int_1^8 \sqrt[3]{x}(x-1)dx = \int_1^8 \left(x^{4/3} - x^{1/3}\right)dx = \frac{3}{7}x^{7/3} - \frac{3}{4}x^{4/3}\Big|_1^8 = \left(\frac{3}{7}\cdot 128 - \frac{3}{4}\cdot 16\right) - \left(\frac{3}{7} - \frac{3}{4}\right) = \frac{1209}{28}$

29. Let $u = 1 + 2x^3$. Then $du = 6x^2\,dx$, so $\int_0^2 x^2(1+2x^3)^3\,dx = \int_1^{17} u^3\left(\frac{1}{6}\,du\right) = \frac{1}{24}u^4\Big|_1^{17} = \frac{1}{24}(17^4 - 1) = 3480$.

31. Let $u = 2x + 3$. Then $du = 2\,dx$, so $\displaystyle\int_3^{11} \frac{dx}{\sqrt{2x+3}} = \int_9^{25} u^{-1/2}\left(\tfrac{1}{2}\,du\right) = u^{1/2}\Big|_9^{25} = 5 - 3 = 2$.

33. $\displaystyle\int_{-2}^{-1} \frac{dx}{(2x+3)^4}$ does not exist since the integrand has an infinite discontinuity at $x = -\frac{3}{2}$.

35. Let $u = 2 + x^5$. Then $du = 5x^4\,dx$, so

$$\int \frac{x^4\,dx}{(2+x^5)^6} = \int u^{-6}\left(\tfrac{1}{5}\,du\right) = \frac{1}{5}\left(\frac{u^{-5}}{-5}\right) + C = -\frac{1}{25u^5} + C = -\frac{1}{25(2+x^5)^5} + C.$$

37. Let $u = \pi x$. Then $du = \pi\,dx$, so $\displaystyle\int \sin \pi x\,dx = \int \frac{\sin u\,du}{\pi} = \frac{-\cos u}{\pi} + C = -\frac{\cos \pi x}{\pi} + C$.

39. Let $u = \dfrac{1}{t}$. Then $du = -\dfrac{1}{t^2}\,dt$, so $\displaystyle\int \frac{\cos(1/t)}{t^2}\,dt = \int \cos u\,(-du) = -\sin u + C = -\sin\left(\frac{1}{t}\right) + C$.

41. $\displaystyle\int_0^{2\sqrt{3}} \frac{1}{x^2+4}\,dx = \left[\tfrac{1}{2}\tan^{-1}(x/2)\right]_0^{2\sqrt{3}} = \tfrac{1}{2}\left(\tan^{-1}\sqrt{3} - \tan^{-1}0\right) = \tfrac{1}{2}\cdot\tfrac{\pi}{3} = \tfrac{\pi}{6}$

43.
$$\int_2^4 \frac{1+x-x^2}{x^2}\,dx = \int_2^4\left(x^{-2} + \frac{1}{x} - 1\right)dx = \left[-\frac{1}{x} + \ln x - x\right]_2^4$$
$$= \left(-\tfrac{1}{4} + \ln 4 - 4\right) - \left(-\tfrac{1}{2} + \ln 2 - 2\right) = \ln 2 - \tfrac{7}{4}$$

45. Let $u = e^x + 1$. Then $du = e^x\,dx$, so $\displaystyle\int \frac{e^x}{e^x+1}\,dx = \int \frac{du}{u} = \ln|u| + C = \ln(e^x+1) + C$.

47. Let $u = \sqrt{x}$. Then $du = \dfrac{dx}{2\sqrt{x}}$ $\Rightarrow$ $\displaystyle\int \frac{e^{\sqrt{x}}}{\sqrt{x}}\,dx = 2\int e^u\,du = 2e^u + C = 2e^{\sqrt{x}} + C$.

49. Let $u = \ln(\cos x)$. Then $du = \dfrac{-\sin x}{\cos x}\,dx = -\tan x\,dx$ $\Rightarrow$
$\int \tan x \ln(\cos x)dx = -\int u\,du = -\frac{1}{2}u^2 + C = -\frac{1}{2}[\ln(\cos x)]^2 + C$.

51. Let $u = 1 + x^4$. Then $du = 4x^3\,dx$ $\Rightarrow$ $\displaystyle\int \frac{x^3}{1+x^4}\,dx = \frac{1}{4}\int \frac{1}{u}\,du = \tfrac{1}{4}\ln|u| + C = \tfrac{1}{4}\ln(1+x^4) + C$.

53. Let $u = 1 + \sec\theta$, so $du = \sec\theta\tan\theta\,d\theta$ $\Rightarrow$ $\displaystyle\int \frac{\sec\theta\tan\theta}{1+\sec\theta}\,d\theta = \int \frac{1}{u}\,du = \ln|u| + C = \ln|1+\sec\theta| + C$.

55. $u = 3t$ $\Rightarrow$ $\int \cosh 3t\,dt = \frac{1}{3}\int \cosh u\,du = \frac{1}{3}\sinh u + C = \frac{1}{3}\sinh 3t + C$

57. $\int_0^{2\pi}|\sin x|dx = \int_0^{\pi}\sin x\,dx - \int_{\pi}^{2\pi}\sin x\,dx = 2\int_0^{\pi}\sin x\,dx = -2\cos x\Big|_0^{\pi} = -2[(-1) - 1] = 4$

59.

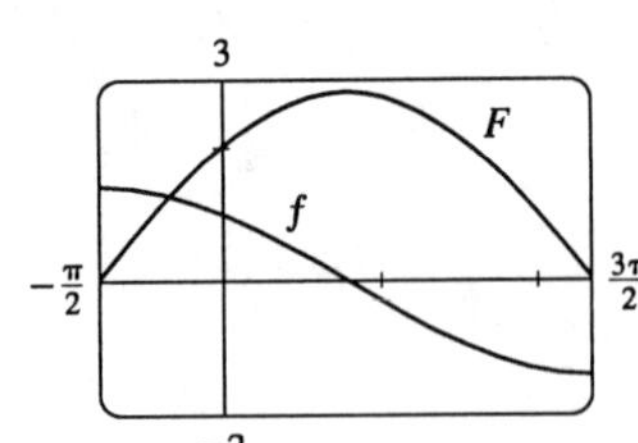

$f(x) = \dfrac{\cos x}{\sqrt{1+\sin x}}$. Let $u = 1 + \sin x$. Then $du = \cos x\,dx$, so

$$\int \frac{\cos x\,dx}{\sqrt{1+\sin x}} = \int u^{-1/2}\,du = 2u^{1/2} + C = 2\sqrt{1+\sin x} + C.$$

61.

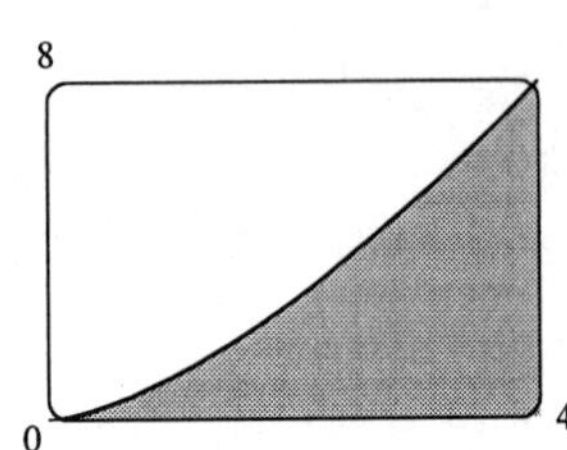

From the graph, it appears that the area under the curve $y = x\sqrt{x}$ between $x = 0$ and $x = 4$ is somewhat less than half the area of an 8×4 rectangle, so perhaps about 13 or 14. To find the exact value, we evaluate

$\int_0^4 x\sqrt{x}\,dx = \int_0^4 x^{3/2}\,dx = \frac{2}{5}x^{5/2}\Big|_0^4 = \frac{2}{5}(4)^{5/2} = \frac{64}{5} = 12.8.$

63. $F(x) = \int_1^x \sqrt{1+t^4}\,dt \quad\Rightarrow\quad F'(x) = \sqrt{1+x^4}$

65. $g(x) = \displaystyle\int_0^{x^3} \frac{t\,dt}{\sqrt{1+t^3}}$. Let $y = g(u)$ and $u = x^3$. Then

$$g'(x) = \frac{dy}{dx} = \frac{dy}{du}\frac{du}{dx} = \frac{u}{\sqrt{1+u^3}}3x^2 = \frac{x^3}{\sqrt{1+x^9}}3x^2 = \frac{3x^5}{\sqrt{1+x^9}}.$$

67. $$y = \int_{\sqrt{x}}^{x} \frac{\cos\theta}{\theta}\,d\theta = \int_1^x \frac{\cos\theta}{\theta}\,d\theta + \int_{\sqrt{x}}^{1} \frac{\cos\theta}{\theta}\,d\theta = \int_1^x \frac{\cos\theta}{\theta}\,d\theta - \int_1^{\sqrt{x}} \frac{\cos\theta}{\theta}\,d\theta \quad\Rightarrow$$

$$y' = \frac{\cos x}{x} - \frac{\cos\sqrt{x}}{\sqrt{x}}\frac{1}{2\sqrt{x}} = \frac{2\cos x - \cos\sqrt{x}}{2x}$$

69. If $1 \le x \le 3$, then $2 \le \sqrt{x^2+3} \le 2\sqrt{3}$, so $2(3-1) \le \int_1^3 \sqrt{x^2+3}\,dx \le 2\sqrt{3}(3-1)$; that is, $4 \le \int_1^3 \sqrt{x^2+3}\,dx \le 4\sqrt{3}$.

71. $0 \le x \le 1 \;\Rightarrow\; 0 \le \cos x \le 1 \;\Rightarrow\; x^2\cos x \le x^2 \;\Rightarrow\; \int_0^1 x^2\cos x\,dx \le \int_0^1 x^2\,dx = \frac{1}{3}x^3\Big|_0^1 = \frac{1}{3}$ [Property 7].

73. $\cos x \le 1 \quad\Rightarrow\quad e^x\cos x \le e^x \quad\Rightarrow\quad \int_0^1 e^x\cos x\,dx \le \int_0^1 e^x\,dx = [e^x]_0^1 = e - 1$

75. Let $f(x) = \sqrt{1+x^3}$ on $[0,1]$. The Midpoint Rule with $n = 5$ gives

$$\int_0^1 \sqrt{1+x^3}\,dx \approx \tfrac{1}{5}[f(0.1) + f(0.3) + f(0.5) + f(0.7) + f(0.9)]$$
$$= \tfrac{1}{5}\left[\sqrt{1+(0.1)^3} + \sqrt{1+(0.3)^3} + \cdots + \sqrt{1+(0.9)^3}\right] \approx 1.110.$$

77. $y = \sqrt{16-x^2}$ is a semicircle with a radius of 4. So $\int_{-4}^4 \sqrt{16-x^2}\,dx$ represents the area between the semicircle $y = \sqrt{16-x^2}$ and the x-axis and is equal to $\frac{1}{2}\pi(4)^2 = 8\pi$.

79. By the Fundamental Theorem of Calculus, we know that $F(x) = \int_a^x t^2\sin(t^2)dt$ is an antiderivative of $f(x) = x^2\sin(x^2)$. This integral cannot be expressed in any simpler form. Since $\int_a^a f\,dt = 0$ for any a, we can take $a = 1$, and then $F(1) = 0$, as required. So $F(x) = \int_1^x t^2\sin(t^2)dt$ is the desired function.

81. We want $\int_0^1 \sinh cx\,dx = 1$. To calculate the integral, we put $u = cx$, so $du = c\,dx$, the upper limit becomes c, and the equation becomes $\dfrac{1}{c}\displaystyle\int_0^c \sinh u\,du = 1 \quad\Leftrightarrow\quad \dfrac{1}{c}[\cosh c - 1] = 1$ $\Leftrightarrow\quad \cosh c - 1 = c$. We plot the function $f(c) = \cosh c - c - 1$, and see that its positive root lies at approximately $c = 1.62$. So the equation $\int_0^1 \sinh cx\,dx = 1$ holds for $c \approx 1.62$.

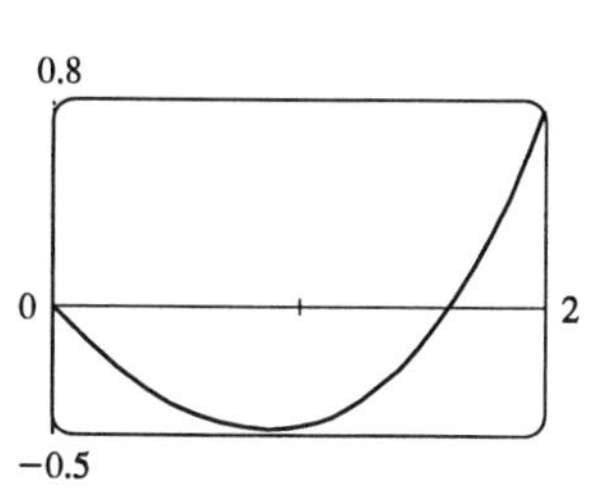

83.

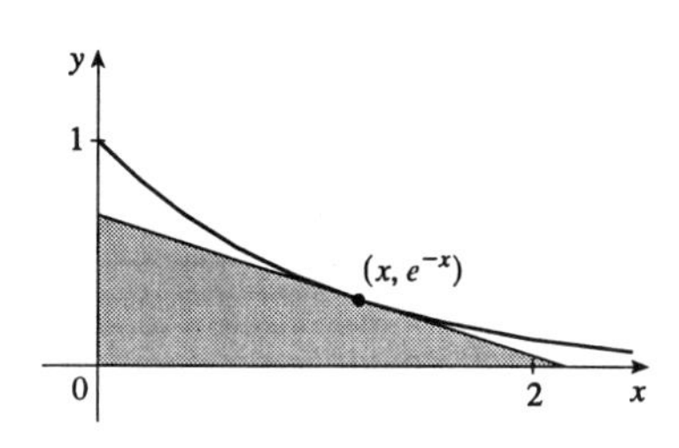

We find the equation of a tangent to the curve $y = e^{-x}$, so that we can find the x- and y-intercepts of this tangent, and then we can find the area of the triangle. The slope of the tangent at the point (a, e^{-a}) is given by $\left.\dfrac{d}{dx}e^{-x}\right|_{x=a} = -e^{-a}$, and so the equation of the tangent is $y - e^{-a} = -e^{-a}(x - a) \quad\Leftrightarrow$ $y = e^{-a}(a - x + 1)$. The y-intercept of this line is $y = e^{-a}(a - 0 + 1) = e^{-a}(a + 1)$. To find the x-intercept we set $y = 0 \quad\Rightarrow\quad e^{-a}(a - x + 1) = 0 \quad\Rightarrow$ $x = a + 1$. So the area of the triangle is $A(a) = \frac{1}{2}[e^{-a}(a+1)](a+1) = \frac{1}{2}e^{-a}(a+1)^2$. We differentiate this with respect to a: $A'(a) = \frac{1}{2}\left[e^{-a}(2)(a+1) + (a+1)^2e^{-a}(-1)\right] = \frac{1}{2}e^{-a}(1 - a^2)$. This is 0 at $a = \pm 1$, and the root $a = 1$ gives a maximum, by the First Derivative Test. So the maximum area of the triangle is $A(1) = \frac{1}{2}e^{-1}(1+1)^2 = 2e^{-1} = 2/e$.

85. Following the hint, we have $\Delta x_i = x_i - x_{i-1} = \dfrac{i^2}{n^2} - \dfrac{(i-1)^2}{n^2} = \dfrac{2i-1}{n^2}$. So

$$\int_0^1 \sqrt{x}\,dx = \lim_{n\to\infty}\sum_{i=1}^{n}\Delta x_i\,f(x_i^*) = \lim_{n\to\infty}\sum_{i=1}^{n}\left(\frac{2i-1}{n^2}\right)\sqrt{\frac{i^2}{n^2}} = \lim_{n\to\infty}\sum_{i=1}^{n}\frac{2i^2-i}{n^3}$$

$$= \lim_{n\to\infty}\frac{1}{n^3}\left[2\sum_{i=1}^{n}i^2 - \sum_{i=1}^{n}i\right] = \lim_{n\to\infty}\frac{1}{n^3}\left[\frac{n(n+1)(2n+1)}{3} - \frac{n(n+1)}{2}\right] \quad \text{[by Theorem 6.1.3(c), (d)]}$$

$$= \lim_{n\to\infty}\left[\frac{(1+1/n)(2+1/n)}{3} - \frac{1+1/n}{2n}\right] = \frac{2}{3} - 0 = \frac{2}{3}.$$

87. Let $u = f(x)$ so $du = f'(x)dx$. So $2\int_a^b f(x)f'(x)dx = 2\int_{f(a)}^{f(b)} u\,du = u^2\Big|_{f(a)}^{f(b)} = [f(b)]^2 - [f(a)]^2$.

89. Differentiating both sides of the given equation, using the Fundamental Theorem for each side, gives $f(x) = e^{2x} + 2xe^{2x} + e^{-x}f(x)$. So $f(x)(1 - e^{-x}) = e^{2x} + 2xe^{2x}$. Hence $f(x) = \dfrac{e^{2x}(1+2x)}{1 - e^{-x}}$.

91. Let $u = 1 - x$, then $du = -dx$, so $\int_0^1 f(1-x)dx = \int_1^0 f(u)(-du) = \int_0^1 f(u)du = \int_0^1 f(x)dx$.

APPLICATIONS PLUS (after Chapter 6)

1. **(a)** $I = \dfrac{k\cos\theta}{d^2} = \dfrac{k(h/d)}{d^2} = k\dfrac{h}{d^3} = k\dfrac{h}{\left(\sqrt{1600+h^2}\right)^3} = k\dfrac{h}{(1600+h^2)^{3/2}} \quad\Rightarrow$

$$\frac{dI}{dh} = k\frac{(1600+h^2)^{3/2} - h\frac{3}{2}(1600+h^2)^{1/2}\cdot 2h}{(1600+h^2)^3} = \frac{k(1600+h^2)^{1/2}(1600+h^2-3h^2)}{(1600+h^2)^3}$$

$$= \frac{k(1600-2h^2)}{(1600+h^2)^{5/2}}.\ \text{Set } \frac{dI}{dh} = 0\text{: } 1600 - 2h^2 = 0 \quad\Rightarrow\quad h^2 = 800 \quad\Rightarrow\quad h = \sqrt{800} = 20\sqrt{2}.$$

By the First Derivative Test, I has a relative maximum at $h = 20\sqrt{2} \approx 28$ ft.

(b)

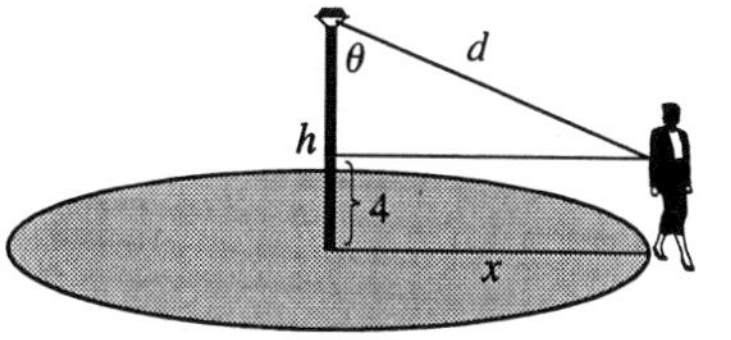

$$\frac{dx}{dt} = 4\text{ ft/s}$$

$$I = \frac{k\cos\theta}{d^2} = \frac{k[(h-4)/d]}{d^2} = \frac{k\,(h-4)}{d^3} = \frac{k(h-4)}{\left[(h-4)^2+x^2\right]^{3/2}} = k(h-4)\left([h-4]^2+x^2\right)^{-3/2}$$

$$\frac{dI}{dt} = k(h-4)\left(-\tfrac{3}{2}\right)\left[(h-4)^2+x^2\right]^{-5/2}\cdot 2x\cdot\frac{dx}{dt} = k(h-4)(-3x)\left[(h-4)^2+x^2\right]^{-5/2}\cdot 4$$

$$= \frac{-12xk(h-4)}{\left[(h-4)^2+x^2\right]^{5/2}}$$

$$\left.\frac{dI}{dt}\right|_{x=40} = -\frac{480k(h-4)}{\left[(h-4)^2+1600\right]^{5/2}}$$

3. **(a)** First note that $90\text{ mi/h} = 90\times\frac{5280}{3600}\text{ ft/s} = 132\text{ ft/s}$. Then $a(t) = 4\text{ ft/s}^2 \quad\Rightarrow\quad v(t) = 4t = 132$ when $t = \frac{132}{4} = 33$ s. It takes 33 s to reach 132 ft/s. Therefore, taking $s(0) = 0$, we have $s(t) = 2t^2$, $0 \le t \le 33$. So $s(33) = 2178$ ft.

For $33 \le t \le 933$ we have $v(t) = 132\text{ ft/s} \quad\Rightarrow\quad s(t) = 132(t-33) + C$ and $s(33) = 2178 \quad\Rightarrow$ $C = 2178$, so $s(t) = 132(t-33) + 2178$, $33 \le t \le 933$.

Therefore $s(933) = 132(900) + 2178 = 120{,}978\text{ ft} = 22.9125\text{ mi}$.

(b) As in part (a), the train accelerates for 33 s and travels 2178 ft while doing so. Similarly, it decelerates for 33 s and travels 2178 ft at the end of its trip. During the remaining $900 - 66 = 834$ s it travels at 132 ft/s, so the distance traveled is $132\cdot 834 = 110{,}088$ ft. Thus the total distance is $2178 + 110{,}088 + 2178 = 114{,}444\text{ ft} = 21.675\text{ mi}$.

5. **(a)** Let $F(t) = \int_0^t f(s)ds$. Then, by Part 1 of the Fundamental Theorem of Calculus, $F'(t) = f(t) =$ rate of depreciation, so $F(t)$ represents the loss in value over the interval $[0, t]$.

(b) $C(t) = [A + F(t)]/t$ represents the average expenditure over the interval $[0, t]$. The company wants to minimize average expenditure.

(c) $C(t) = \frac{1}{t}\left(A + \int_0^t f(s)ds\right)$. Using Part 1 of the Fundamental Theorem of Calculus, we have

$$C'(t) = -\frac{1}{t^2}\left(A + \int_0^t f(s)ds\right) + \frac{1}{t}f(t) = 0 \text{ when } tf(t) = A + \int_0^t f(s)ds \quad \Rightarrow$$

$$f(t) = \frac{1}{t}\left(A + \int_0^t f(s)ds\right) = C(t).$$

7. **(a)** $P = \dfrac{\text{area under } y = L\sin\theta}{\text{area of rectangle}} = \dfrac{\int_0^\pi L\sin\theta\, d\theta}{\pi L} = \dfrac{-L\cos\theta\big|_0^\pi}{\pi L} = \dfrac{-(-1)+1}{\pi} = \dfrac{2}{\pi}$

(b) $P = \dfrac{\text{area under } y = \frac{1}{2}L\sin\theta}{\text{area of rectangle}} = \dfrac{\int_0^\pi \frac{1}{2}L\sin\theta\, d\theta}{\pi L} = \dfrac{\int_0^\pi \sin\theta\, d\theta}{2\pi} = \dfrac{-\cos\theta\big|_0^\pi}{2\pi} = \dfrac{2}{2\pi} = \dfrac{1}{\pi}$

(c) $P = \dfrac{\text{area under } y = \frac{1}{5}L\sin\theta}{\text{area of rectangle}} = \dfrac{\int_0^\pi \frac{1}{5}L\sin\theta\, d\theta}{\pi L} = \dfrac{\int_0^\pi \sin\theta\, d\theta}{5\pi} = \dfrac{2}{5\pi}$

9. **(a)** $T_1 = \dfrac{D}{c_1}$, $T_2 = \dfrac{2|PR|}{c_1} + \dfrac{|RS|}{c_2} = \dfrac{2h\sec\theta}{c_1} + \dfrac{D - 2h\tan\theta}{c_2}$, $T_3 = \dfrac{2\sqrt{h^2 + D^2/4}}{c_1} = \dfrac{\sqrt{4h^2 + D^2}}{c_1}$.

(b) $\dfrac{dT_2}{d\theta} = \dfrac{2h}{c_1}\cdot\sec\theta\tan\theta - \dfrac{2h}{c_2}\sec^2\theta = 0$ when $2h\sec\theta\left(\dfrac{1}{c_1}\tan\theta - \dfrac{1}{c_2}\sec\theta\right) = 0 \quad \Rightarrow$

$\dfrac{1}{c_1}\dfrac{\sin\theta}{\cos\theta} - \dfrac{1}{c_2}\dfrac{1}{\cos\theta} = 0 \quad \Rightarrow \quad \sin\theta = \dfrac{c_1}{c_2}$. The First Derivative Test shows that this gives a minimum.

(c) Using part (a), we have $\dfrac{1}{4} = \dfrac{1}{c_1}$. Therefore, $c_1 = 4$. Also $\dfrac{3}{4\sqrt{5}} = \dfrac{\sqrt{4h^2+1}}{4} \quad \Rightarrow \quad \sqrt{4h^2+1} = \dfrac{3}{\sqrt{5}}$

$\Rightarrow \quad 4h^2 + 1 = \frac{9}{5} \quad \Rightarrow \quad h^2 = \frac{1}{5} \quad \Rightarrow \quad h = \frac{1}{\sqrt{5}}$. From (b), $\sin\theta = \dfrac{c_1}{c_2} = \dfrac{4}{c_2} \quad \Rightarrow$

$$\sec\theta = \frac{c_2}{\sqrt{c_2^2 - 16}} \text{ and } \tan\theta = \frac{4}{\sqrt{c_2^2 - 16}}. \text{ Thus } \frac{1}{3} = \frac{\frac{2}{\sqrt{5}}\cdot\frac{c_2}{\sqrt{c_2^2-16}}}{4} + \frac{1 - \frac{2}{\sqrt{5}}\cdot\frac{4}{\sqrt{c_2^2-16}}}{c_2} \quad \Rightarrow$$

$$4c_2 = \frac{6c_2^2}{\sqrt{5}\sqrt{c_2^2-16}} + 12\left(1 - \frac{8}{\sqrt{5}\sqrt{c_2^2-16}}\right) \quad \Rightarrow \quad 4c_2 - 12 = \frac{6(c_2^2-16)}{\sqrt{5}\sqrt{c_2^2-16}} = \frac{6}{\sqrt{5}}\sqrt{c_2^2-16}$$

$$\Rightarrow \quad 2c_2 - 6 = \frac{3}{\sqrt{5}}\sqrt{c_2^2-16} \quad \Rightarrow \quad 20c_2^2 - 120c_2 + 180 = 9c_2^2 - 144 \quad \Rightarrow$$

$11c_2^2 - 120c_2 + 324 = (c_2 - 6)(11c_2 - 54) = 0 \Rightarrow c_2 = 6$ or $\frac{54}{11}$. But the root $\frac{54}{11}$ is inadmissible because if $\tan\theta = \dfrac{4}{\sqrt{\left(\frac{54}{11}\right)^2 - 16}} \approx 1.4$, then $\theta > 45°$, which is impossible (from the diagram). So $c_2 = 6$.

CHAPTER SEVEN

EXERCISES 7.1

1. $A = \int_{-1}^{1}[(x^2+3)-x]dx = \int_{-1}^{1}(x^2-x+3)dx$
$= \left[\frac{1}{3}x^3 - \frac{1}{2}x^2 + 3x\right]_{-1}^{1} = \left(\frac{1}{3}-\frac{1}{2}+3\right) - \left(-\frac{1}{3}-\frac{1}{2}-3\right)$
$= \frac{20}{3}$

3. $A = \int_{-1}^{1}[(1-y^4)-(y^3-y)]dy = \int_{-1}^{1}(-y^4-y^3+y+1)dy$
$= \left[-\frac{1}{5}y^5 - \frac{1}{4}y^4 + \frac{1}{2}y^2 + y\right]_{-1}^{1} = \left(-\frac{1}{5}-\frac{1}{4}+\frac{1}{2}+1\right) - \left(\frac{1}{5}-\frac{1}{4}+\frac{1}{2}-1\right)$
$= \frac{8}{5}$

5. $A = \int_0^1 (x-x^2)dx = \left[\frac{1}{2}x^2 - \frac{1}{3}x^3\right]_0^1$
$= \frac{1}{2} - \frac{1}{3} = \frac{1}{6}$

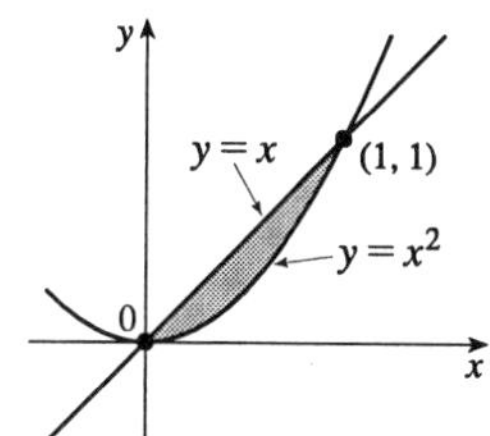

7. $A = \int_0^1 \left(\sqrt{x} - x^2\right)dx = \left[\frac{2}{3}x^{3/2} - \frac{1}{3}x^3\right]_0^1$
$= \frac{2}{3} - \frac{1}{3} = \frac{1}{3}$

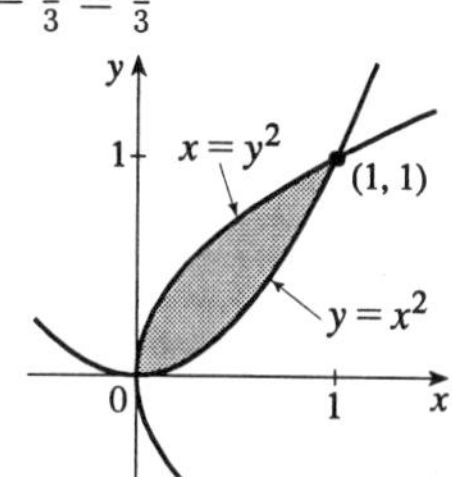

9. $A = \int_0^4 \left(\sqrt{x} - \frac{1}{2}x\right)dx = \left[\frac{2}{3}x^{3/2} - \frac{1}{4}x^2\right]_0^4$
$= \left(\frac{16}{3} - 4\right) - 0 = \frac{4}{3}$

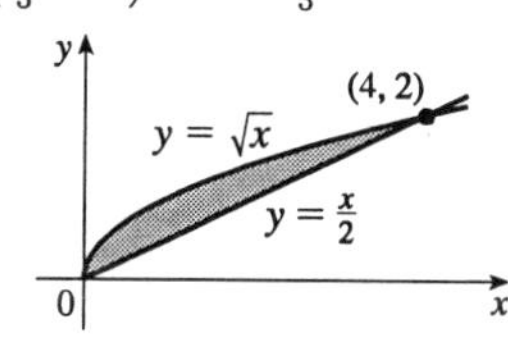

11. $A = \int_{-1}^{1}[(x^2+3)-4x^2]dx = 2\int_0^1 (3-3x^2)dx$
$= [2(3x - x^3)]_0^1 = 2(3-1) - 0 = 4$

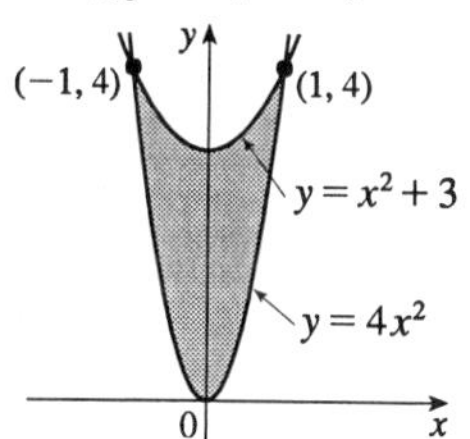

13. $A = \int_0^3 [(2x+5)-(x^2+2)]dx + \int_3^6 [(x^2+2)-(2x+5)]dx$
$= \int_0^3 (-x^2+2x+3)dx + \int_3^6 (x^2-2x-3)dx$
$= \left[-\frac{1}{3}x^3 + x^2 + 3x\right]_0^3 + \left[\frac{1}{3}x^3 - x^2 - 3x\right]_3^6$
$= (-9+9+9) - 0 + (72-36-18) - (9-9-9) = 36$

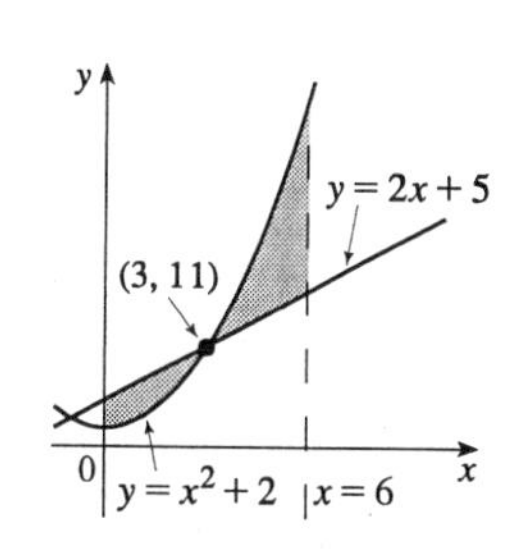

15. $A = \int_{-1}^{3}(2y + 3 - y^2)dy$
$= \left[y^2 + 3y - \frac{1}{3}y^3\right]_{-1}^{3}$
$= (9 + 9 - 9) - \left(1 - 3 + \frac{1}{3}\right) = \frac{32}{3}$

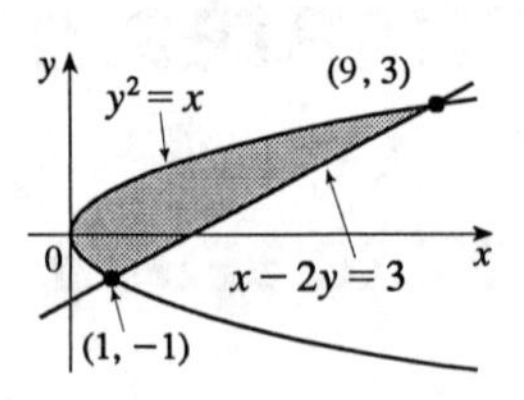

17. $A = \int_{-1}^{1}[(1 - y^2) - (y^2 - 1)]dy$
$= \int_{-1}^{1} 2(1 - y^2)dy = 4\int_0^1(1 - y^2)dy$
$= 4\left[y - \frac{1}{3}y^3\right]_0^1$
$= 4\left(1 - \frac{1}{3}\right)$
$= \frac{8}{3}$

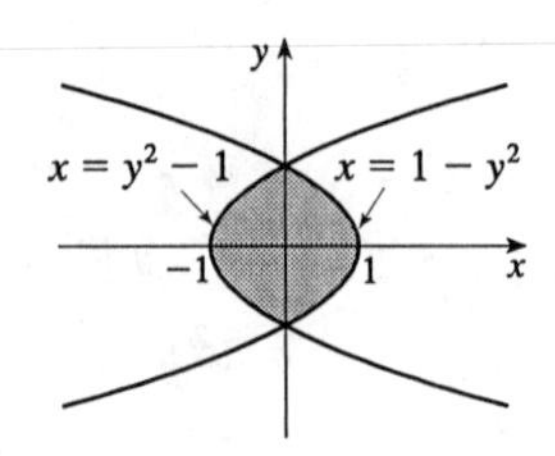

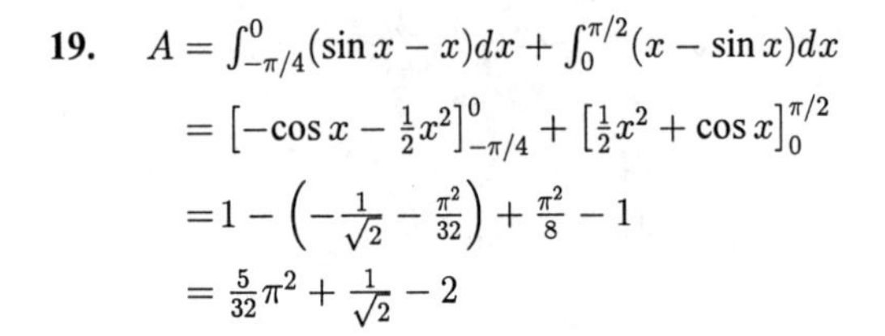

19. $A = \int_{-\pi/4}^{0}(\sin x - x)dx + \int_0^{\pi/2}(x - \sin x)dx$
$= \left[-\cos x - \frac{1}{2}x^2\right]_{-\pi/4}^{0} + \left[\frac{1}{2}x^2 + \cos x\right]_0^{\pi/2}$
$= 1 - \left(-\frac{1}{\sqrt{2}} - \frac{\pi^2}{32}\right) + \frac{\pi^2}{8} - 1$
$= \frac{5}{32}\pi^2 + \frac{1}{\sqrt{2}} - 2$

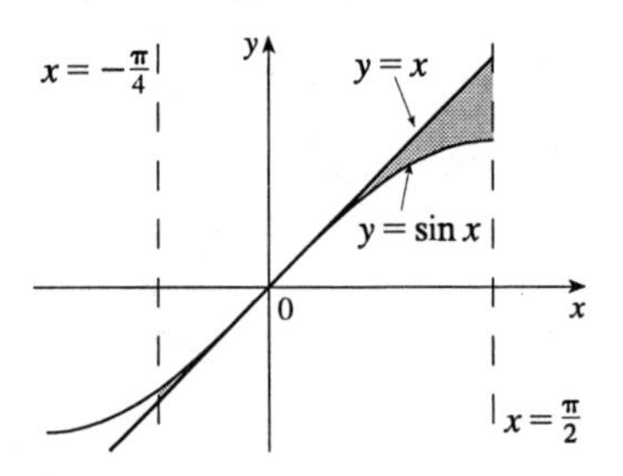

21. Notice that $\cos x = \sin 2x = 2\sin x\cos x \quad \Leftrightarrow$
$2\sin x = 1$ or $\cos x = 0 \quad \Leftrightarrow \quad x = \frac{\pi}{6}$ or $\frac{\pi}{2}$.
$A = \int_0^{\pi/6}(\cos x - \sin 2x)dx + \int_{\pi/6}^{\pi/2}(\sin 2x - \cos x)dx$
$= \left[\sin x + \frac{1}{2}\cos 2x\right]_0^{\pi/6} + \left[-\frac{1}{2}\cos 2x - \sin x\right]_{\pi/6}^{\pi/2}$
$= \frac{1}{2} + \frac{1}{2}\cdot\frac{1}{2} - \left(0 + \frac{1}{2}\cdot 1\right)$
$+ \left(\frac{1}{2} - 1\right) - \left(-\frac{1}{2}\cdot\frac{1}{2} - \frac{1}{2}\right) = \frac{1}{2}$

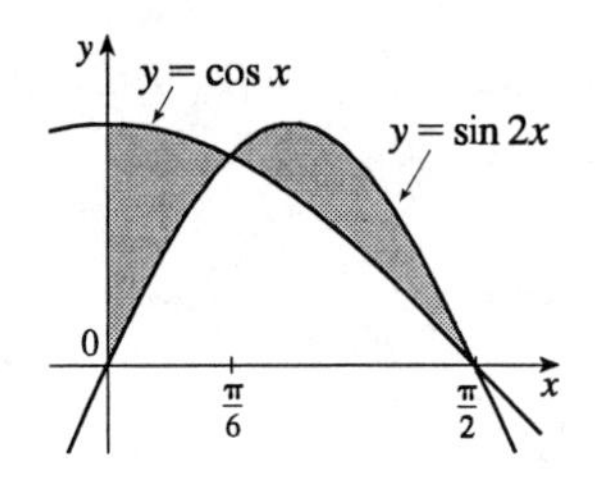

23. $\cos x = \sin 2x = 2\sin x\cos x \quad \Leftrightarrow \quad \cos x = 0$ or
$\sin x = \frac{1}{2} \quad \Leftrightarrow \quad x = \frac{\pi}{2}$ or $\frac{5\pi}{6}$.
$A = \int_{\pi/2}^{5\pi/6}(\cos x - \sin 2x)dx + \int_{5\pi/6}^{\pi}(\sin 2x - \cos x)dx$
$= \left[\sin x + \frac{1}{2}\cos 2x\right]_{\pi/2}^{5\pi/6} - \left[\sin x + \frac{1}{2}\cos 2x\right]_{5\pi/6}^{\pi}$
$= \left(\frac{1}{2} + \frac{1}{2}\cdot\frac{1}{2}\right) - \left(1 - \frac{1}{2}\right) - \left(0 + \frac{1}{2}\right) + \left(\frac{1}{2} + \frac{1}{2}\cdot\frac{1}{2}\right) = \frac{1}{2}$

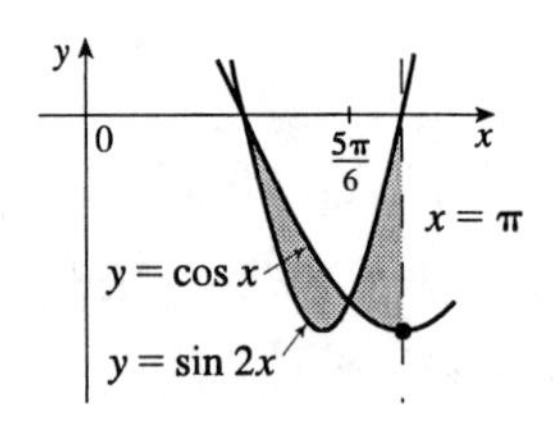

25. $A = \int_{-4}^{0}\left(-x - \left[(x+1)^2 - 7\right]\right)dx + \int_{0}^{2}\left(x - \left[(x+1)^2 - 7\right]\right)dx$
$= \int_{-4}^{0}(-x^2 - 3x + 6)dx + \int_{0}^{2}(-x^2 - x + 6)dx$
$= \left[-\frac{1}{3}x^3 - \frac{3}{2}x^2 + 6x\right]_{-4}^{0} + \left[-\frac{1}{3}x^3 - \frac{1}{2}x^2 + 6x\right]_{0}^{2}$
$= 0 - \left(\frac{64}{3} - 24 - 24\right) + \left(-\frac{8}{3} - 2 + 12\right) - 0 = 34$

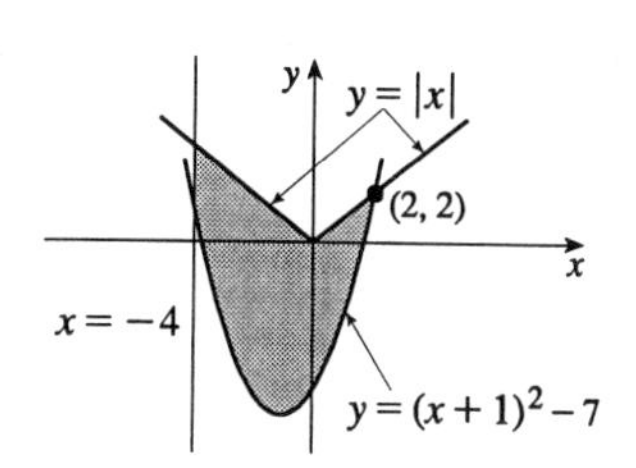

27. $A = \int_{0}^{3}\left[\frac{1}{3}x - (-x)\right]dx + \int_{3}^{6}\left[\left(8 - \frac{7}{3}x\right) - (-x)\right]dx$
$= \int_{0}^{3}\frac{4}{3}x\,dx + \int_{3}^{6}\left(-\frac{4}{3}x + 8\right)dx$
$= \left[\frac{2}{3}x^2\right]_{0}^{3} + \left[-\frac{2}{3}x^2 + 8x\right]_{3}^{6}$
$= (6 - 0) + (24 - 18) = 12$

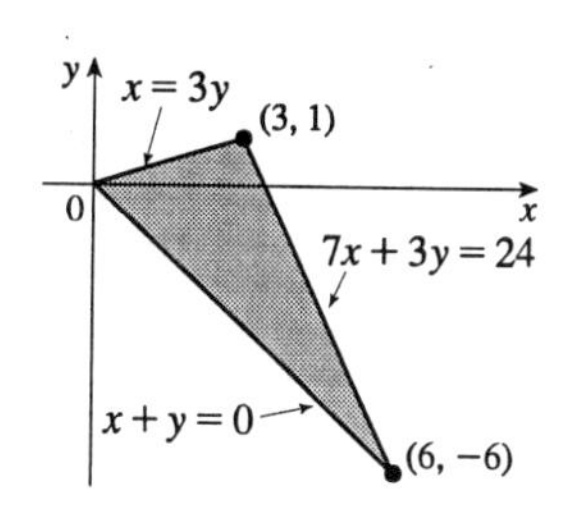

29. $A = \int_{1}^{2}\left(\frac{1}{x} - \frac{1}{x^2}\right)dx = \left[\ln x + \frac{1}{x}\right]_{1}^{2}$
$= \left(\ln 2 + \frac{1}{2}\right) - (\ln 1 + 1) = \ln 2 - \frac{1}{2}$

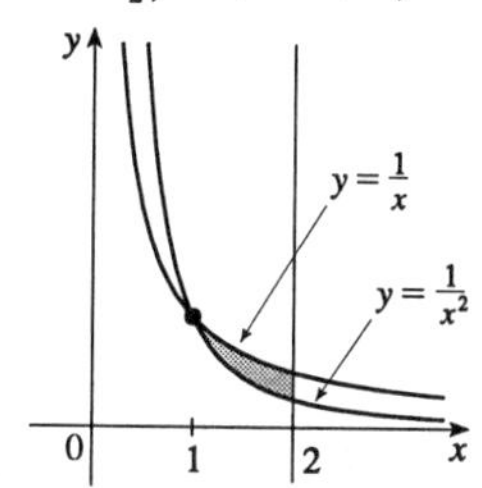

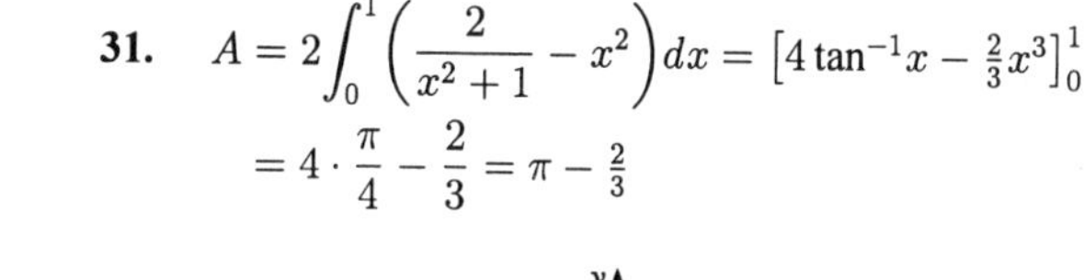

31. $A = 2\int_{0}^{1}\left(\frac{2}{x^2+1} - x^2\right)dx = \left[4\tan^{-1}x - \frac{2}{3}x^3\right]_{0}^{1}$
$= 4 \cdot \frac{\pi}{4} - \frac{2}{3} = \pi - \frac{2}{3}$

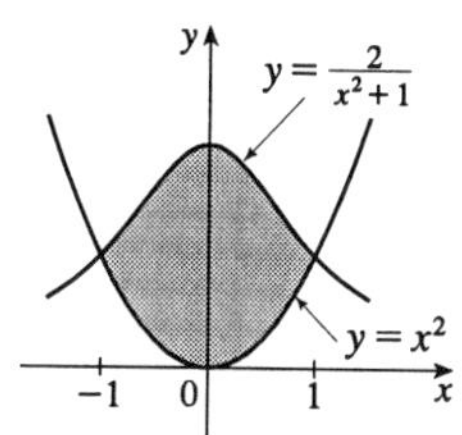

33. $A = \int_{0}^{1}(e^{3x} - e^x)dx = \left[\frac{1}{3}e^{3x} - e^x\right]_{0}^{1} = \left(\frac{1}{3}e^3 - e\right) - \left(\frac{1}{3} - 1\right)$
$= \frac{1}{3}e^3 - e + \frac{2}{3}$

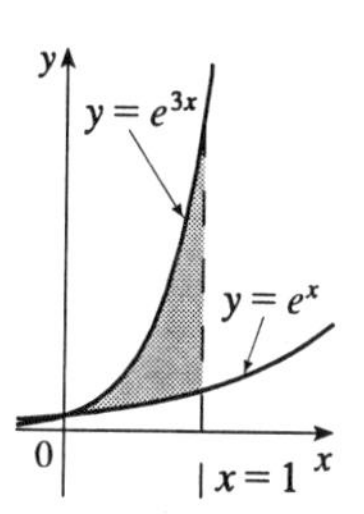

35. **(a)** $A = \int_{-4}^{-1}\left[(2x+4) + \sqrt{-4x}\right]dx + \int_{-1}^{0}2\sqrt{-4x}\,dx$
$= \left[x^2 + 4x\right]_{-4}^{-1} + 2\int_{-4}^{-1}\sqrt{-x}\,dx + 4\int_{-1}^{0}\sqrt{-x}\,dx$
$= (-3 - 0) + 2\int_{1}^{4}\sqrt{u}\,du + 4\int_{0}^{1}\sqrt{u}\,du \quad (u = -x)$
$= -3 + \left[\frac{4}{3}u^{3/2}\right]_{1}^{4} + \left[\frac{8}{3}u^{3/2}\right]_{0}^{1} = -3 + \frac{28}{3} + \frac{8}{3} = 9$

(b) $A = \int_{-4}^{2}\left[-\frac{1}{4}y^2 - \left(\frac{1}{2}y - 2\right)\right]dy = \left[-\frac{1}{12}y^3 - \frac{1}{4}y^2 + 2y\right]_{-4}^{2}$
$= \left(-\frac{2}{3} - 1 + 4\right) - \left(\frac{16}{3} - 4 - 8\right) = 9$

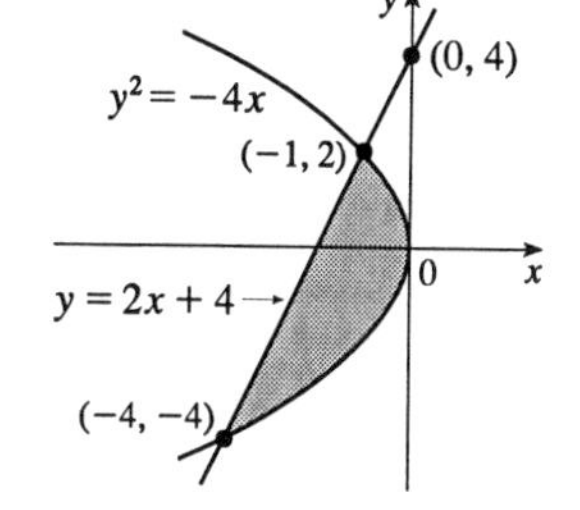

37. $A = \int_0^1 \left(8x - \frac{3}{4}x\right)dx + \int_1^4 \left[\left(-\frac{5}{3}x + \frac{29}{3}\right) - \frac{3}{4}x\right]dx$
$= \frac{29}{4}\int_0^1 x\,dx + \int_1^4 \left(-\frac{29}{12}x + \frac{29}{3}\right)dx$
$= \frac{29}{4}\left[\frac{1}{2}x^2\right]_0^1 - \frac{29}{12}\left[\frac{1}{2}x^2 - 4x\right]_1^4$
$= \frac{29}{8} - \frac{29}{12}\left(-8 - \frac{1}{2} + 4\right) = 14.5$

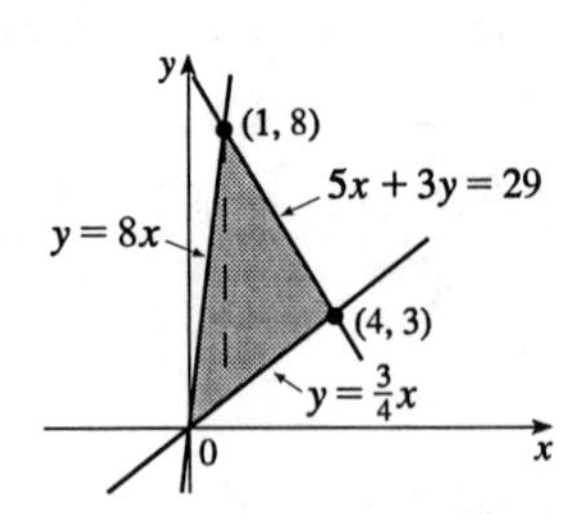

39. $\int_0^2 |x^2 - x^3|dx = \int_0^1 (x^2 - x^3)dx + \int_1^2 (x^3 - x^2)dx$
$= \left[\frac{1}{3}x^3 - \frac{1}{4}x^4\right]_0^1 + \left[\frac{1}{4}x^4 - \frac{1}{3}x^3\right]_1^2$
$= \frac{1}{3} - \frac{1}{4} + \left(4 - \frac{8}{3}\right) - \left(\frac{1}{4} - \frac{1}{3}\right) = 1.5$

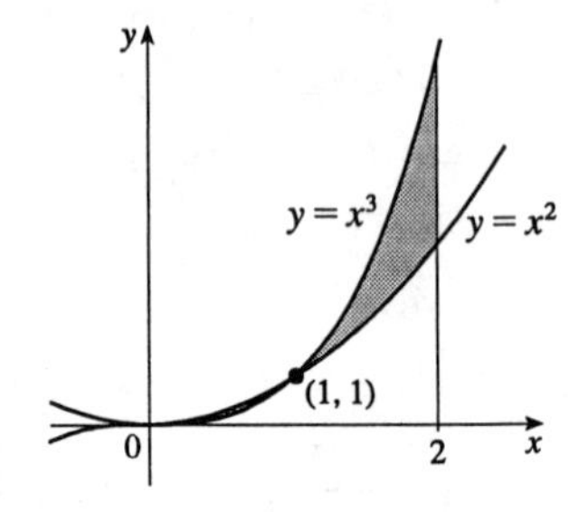

41. $\int_{-1}^2 x^3\,dx = \left[\frac{1}{4}x^4\right]_{-1}^2 = 4 - \frac{1}{4} = \frac{15}{4} = 3.75$

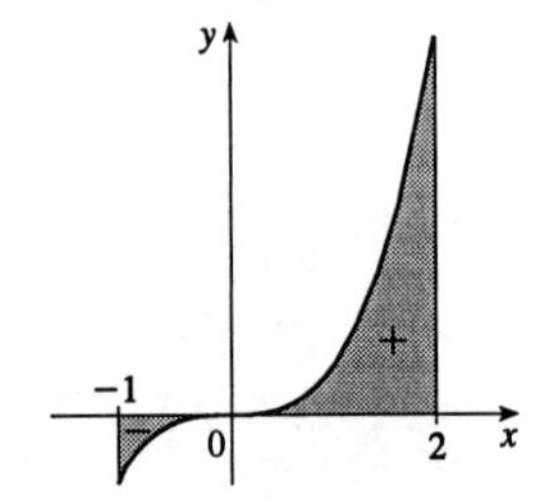

43. Let $f(x) = \sqrt{1 + x^3} - (1 - x)$, $\Delta x = \dfrac{2 - 0}{4} = \dfrac{1}{2}$.

$$A = \int_0^2 \left[\sqrt{1 + x^3} - (1 - x)\right]dx \approx \tfrac{1}{2}\left[f\left(\tfrac{1}{4}\right) + f\left(\tfrac{3}{4}\right) + f\left(\tfrac{5}{4}\right) + f\left(\tfrac{7}{4}\right)\right]$$
$$= \tfrac{1}{2}\left[\left(\tfrac{\sqrt{65}}{8} - \tfrac{3}{4}\right) + \left(\tfrac{\sqrt{91}}{8} - \tfrac{1}{4}\right) + \left(\tfrac{3\sqrt{21}}{8} + \tfrac{1}{4}\right) + \left(\tfrac{\sqrt{407}}{8} + \tfrac{3}{4}\right)\right]$$
$$= \tfrac{1}{16}\left(\sqrt{65} + \sqrt{91} + 3\sqrt{21} + \sqrt{407}\right) \approx 3.22$$

45.

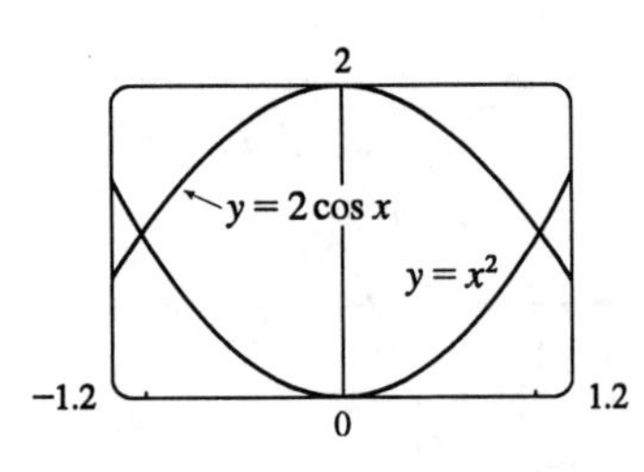

From zooming in on the graph or using a cursor, we see that the curves intersect at $x \approx \pm 1.02$, with $2\cos x > x^2$ on $[-1.02, 1.02]$. So the area between them is

$A \approx \int_{-1.02}^{1.02} (2\cos x - x^2)dx = 2\int_0^{1.02} (2\cos x - x^2)dx$
$= 2\left[2\sin x - \frac{1}{3}x^3\right]_0^{1.02} \approx 2.70.$

47.

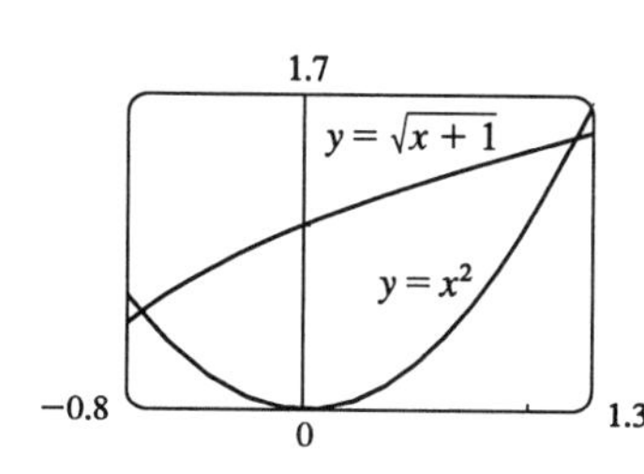

From the graph, we see that the curves intersect at $x \approx -0.72$ and at $x \approx 1.22$, with $\sqrt{x+1} > x^2$ on $[-0.72, 1.22]$. So the area between the curves is

$$A \approx \int_{-0.72}^{1.22} \left(\sqrt{x+1} - x^2\right) dx = \left[\tfrac{2}{3}(x+1)^{3/2} - \tfrac{1}{3}x^3\right]_{-0.72}^{1.22} \approx 1.38.$$

49.

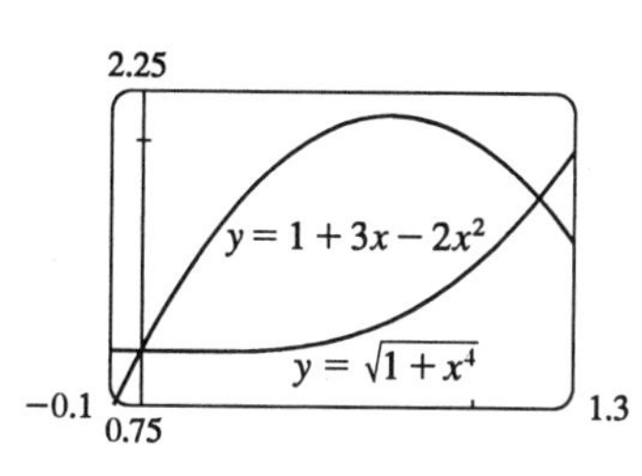

From the graph, we see that the curves intersect at $x = 0$ and at $x \approx 1.19$, with $1 + 3x - 2x^2 > \sqrt{1+x^4}$ on $[0, 1.19]$. So, using the Midpoint Rule with $f(x) = 1 + 3x - 2x^2 - \sqrt{1+x^4}$ on $[0, 1.19]$ with $n = 4$, we calculate the approximate area between the curves:

$$A \approx \int_0^{1.19} \left(1 + 3x - 2x^2 - \sqrt{1+x^4}\right) dx$$
$$\approx \tfrac{1.19}{4}\left[f\left(\tfrac{1.19}{8}\right) + f\left(\tfrac{3 \cdot 1.19}{8}\right) + f\left(\tfrac{5 \cdot 1.19}{8}\right) + f\left(\tfrac{7 \cdot 1.19}{8}\right)\right] \approx 0.83.$$

51.

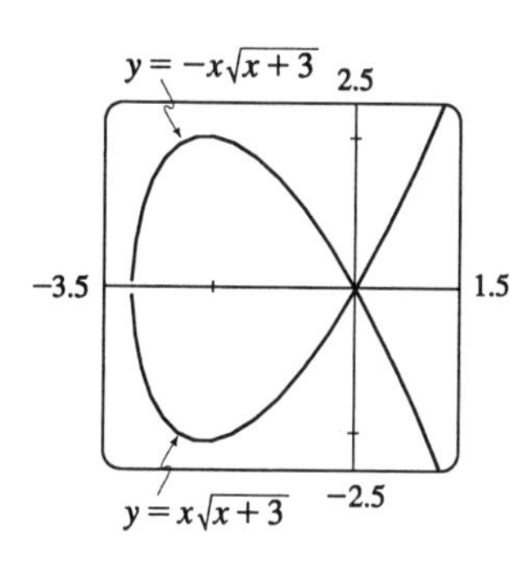

To graph this function, we must first express it as a combination of explicit functions of y; namely, $y = \pm x\sqrt{x+3}$. We can see from the graph that the loop extends from $x = -3$ to $x = 0$, and that by symmetry, the area we seek is just twice the area under the top half of the curve on this interval, the equation of the top half being $y = -x\sqrt{x+3}$. So the area is $A = 2\int_{-3}^{0}\left(-x\sqrt{x+3}\right)dx$. We substitute $u = x + 3$, so $du = dx$ and the limits change to 0 and 3, and we get

$$A = -2\int_0^3 \left[(u-3)\sqrt{u}\right] du = -2\int_0^3 \left(u^{3/2} - 3u^{1/2}\right) du$$
$$= -2\left[\tfrac{2}{5}u^{5/2} - 2u^{3/2}\right]_0^3 = -2\left[\tfrac{2}{5}\left(3^2\sqrt{3}\right) - 2\left(3\sqrt{3}\right)\right] = \tfrac{24}{5}\sqrt{3}.$$

53. We first assume that $c > 0$, since c can be replaced by $-c$ in both equations without changing the graphs, and if $c = 0$ the curves do not enclose a region. We see from the graph that the enclosed area lies between $x = -c$ and $x = c$, and by symmetry, it is equal to twice the area under the top half of the graph (whose equation is $y = c^2 - x^2$). The enclosed area is

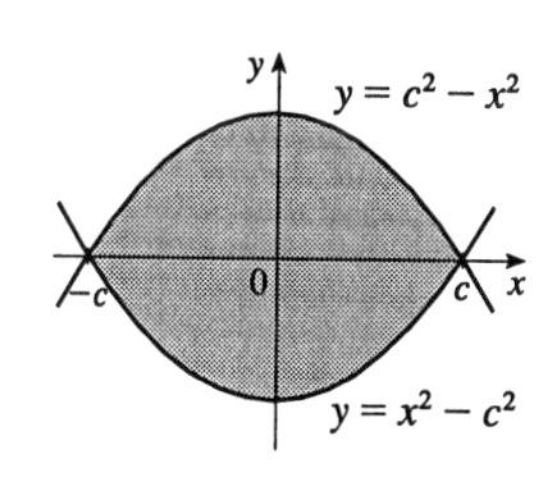

$$2\int_{-c}^{c}(c^2 - x^2)dx = 2\left[c^2x - \tfrac{1}{3}x^3\right]_{-c}^{c} = 2\left(\left[c^3 - \tfrac{1}{3}c^3\right] - \left[-c^3 - \tfrac{1}{3}(-c)^3\right]\right)$$
$$= \tfrac{8}{3}c^3, \text{ which is equal to 576 when } c = \sqrt[3]{216} = 6.$$

Note that $c = -6$ is another solution, since the graphs are the same.

55. By the symmetry of the problem, we consider only the first quadrant where $y = x^2 \Rightarrow x = \sqrt{y}$. We are looking for a number b such that $\int_0^4 x\,dy = 2\int_0^b x\,dy \Rightarrow \int_0^4 \sqrt{y}\,dy = 2\int_0^b \sqrt{y}\,dy \Rightarrow$ $\frac{2}{3}\left[y^{3/2}\right]_0^4 = \frac{4}{3}\left[y^{3/2}\right]_0^b \Rightarrow \frac{2}{3}(8-0) = \frac{4}{3}(b^{3/2}-0) \Rightarrow b^{3/2} = 4 \Rightarrow b = 4^{2/3}$.

57. We know that the area under curve A between $t=0$ and $t=x$ is $\int_0^x v_A(t)\,dt = s_A(x)$, where $v_A(t)$ is the velocity of car A and s_A is its displacement. Similarly, the area under curve B between $t=0$ and $t=x$ is $\int_0^x v_B(t)\,dt = s_B(x)$.

(a) After one minute, the area under curve A is greater than the area under curve B. So A is ahead after one minute.

(b) After two minutes, car B is traveling faster than car A and has gained some ground, but the area under curve A from $t=0$ to $t=2$ is still greater than the corresponding area for curve B, so car A is still ahead.

(c) From the graph, it appears that the area between curves A and B for $0 \le t \le 1$ (when car A is going faster), which corresponds to the distance by which car A is ahead, seems to be about 3 squares. Therefore the cars will be side by side at the time x where the area between the curves for $1 \le t \le x$ (when car B is going faster) is the same as the area for $0 \le t \le 1$. From the graph, it appears that this time is $x \approx 2.2$. So the cars are side by side when $t \approx 2.2$ minutes.

59. The curve and the line will determine a region when they intersect at two or more points. So we solve the equation $x/(x^2+1) = mx$ $\Rightarrow x = 0$ or $mx^2 + m - 1 = 0 \Rightarrow$

$$x = 0 \text{ or } x = \frac{\pm\sqrt{-4(m)(m-1)}}{2m} = \pm\sqrt{\frac{1}{m} - 1}.$$

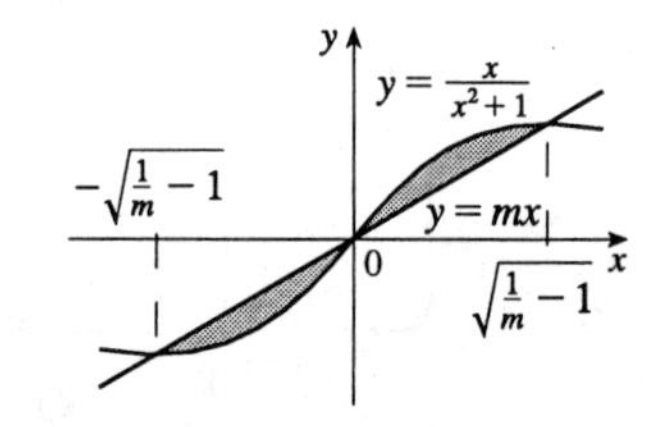

Note that if $m = 1$, this has only the solution $x = 0$, and no region is determined. But if $1/m - 1 > 0 \Leftrightarrow 1/m > 1 \Leftrightarrow 0 < m < 1$, then there are two solutions. [Another way of seeing this is to observe that the slope of the tangent to $y = x/(x^2+1)$ at the origin is $y' = 1$ and therefore we must have $0 < m < 1$.] Note that we cannot just integrate between the positive and negative roots, since the curve and the line cross at the origin. Since mx and $x/(x^2+1)$ are both odd functions, the total area is twice the area between the curves on the interval $\left[0, \sqrt{1/m - 1}\right]$. So the total area enclosed is

$$2\int_0^{\sqrt{1/m-1}} \left[\frac{x}{x^2+1} - mx\right] dx = 2\left[\tfrac{1}{2}\ln(x^2+1) - \tfrac{1}{2}mx^2\right]_0^{\sqrt{1/m-1}}$$

$$= [\ln(1/m - 1 + 1) - m(1/m - 1)] - (\ln 1 - 0) = \ln(1/m) + m - 1 = m - \ln m - 1.$$

EXERCISES 7.2

1. 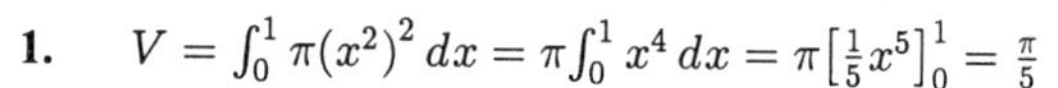

$V = \int_0^1 \pi (x^2)^2\, dx = \pi \int_0^1 x^4\, dx = \pi \left[\frac{1}{5}x^5\right]_0^1 = \frac{\pi}{5}$

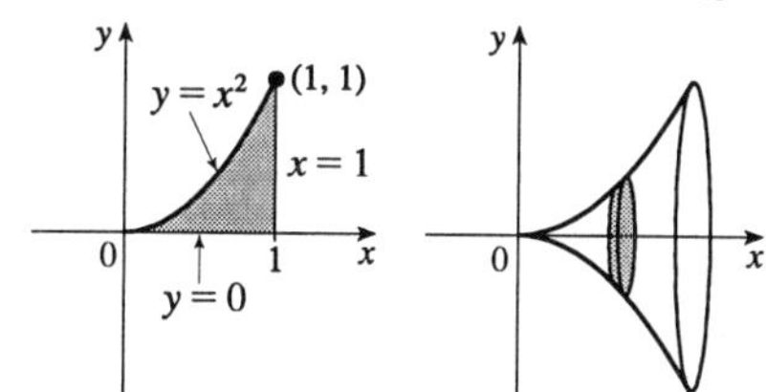

3. $V = \int_0^1 \pi(-x+1)^2\, dx = \pi \int_0^1 (x^2 - 2x + 1)dx$
$= \pi\left[\frac{1}{3}x^3 - x^2 + x\right]_0^1 = \pi\left(\frac{1}{3} - 1 + 1\right) = \frac{\pi}{3}$

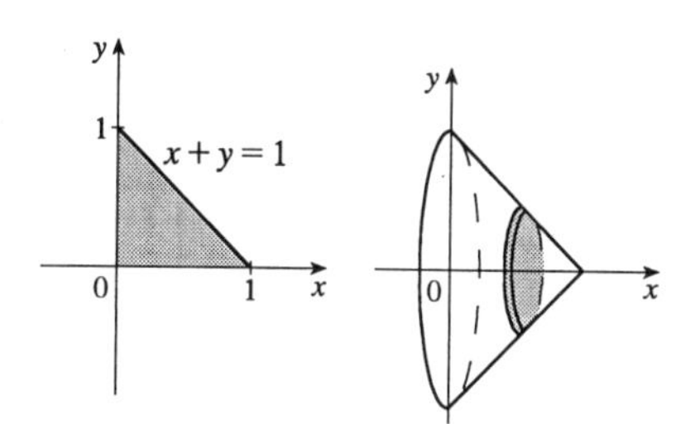

5. $V = \int_0^4 \pi\left(\sqrt{y}\right)^2 dy = \pi \int_0^4 y\, dy = \pi\left[\frac{1}{2}y^2\right]_0^4$
$= 8\pi$

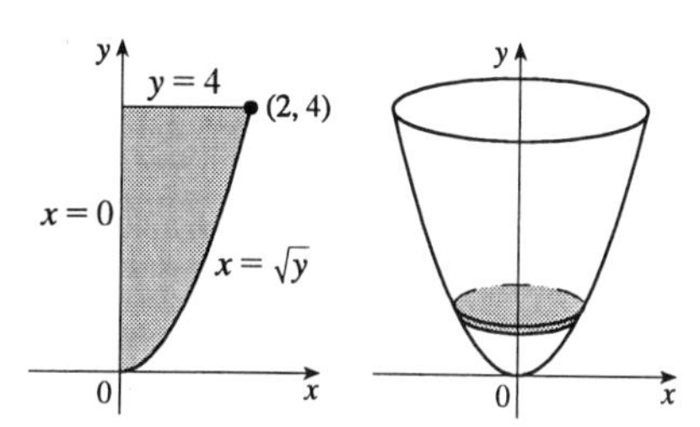

7. $V = \pi \int_0^1 \left[\left(\sqrt{x}\right)^2 - \left(x^2\right)^2\right] dx = \pi \int_0^1 (x - x^4)dx$
$= \pi\left[\frac{1}{2}x^2 - \frac{1}{5}x^5\right]_0^1 = \pi\left(\frac{1}{2} - \frac{1}{5}\right) = \frac{3\pi}{10}$

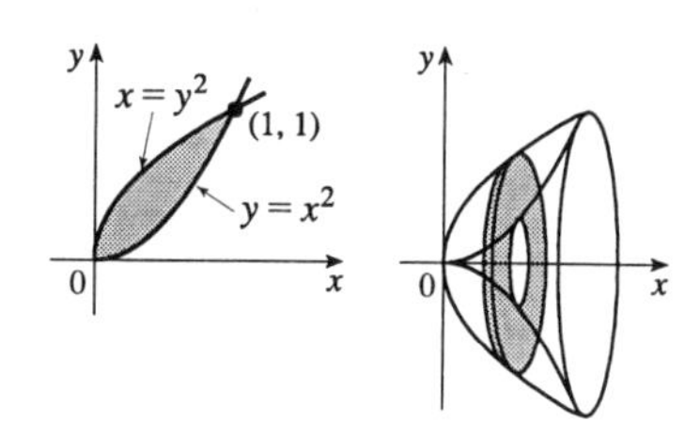

9. $V = \pi \int_0^2 \left[(2y)^2 - \left(y^2\right)^2\right] dy = \pi \int_0^2 (4y^2 - y^4)dy$
$= \pi\left[\frac{4}{3}y^3 - \frac{1}{5}y^5\right]_0^2 = \pi\left(\frac{32}{3} - \frac{32}{5}\right) = \frac{64\pi}{15}$

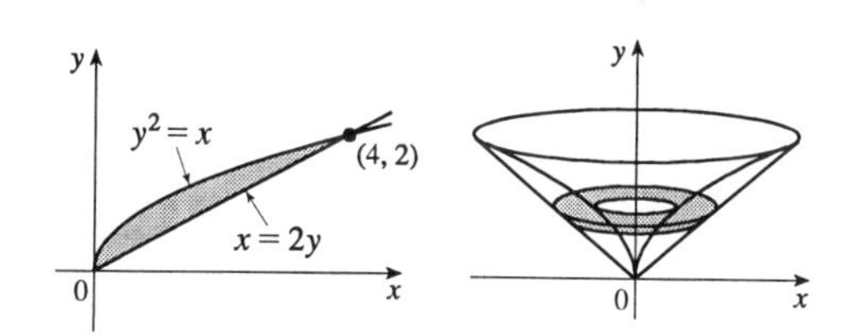

11. $V = \pi \int_{-1}^1 \left[\left(2 - x^4\right)^2 - 1^2\right] dx$
$= 2\pi \int_0^1 (3 - 4x^4 + x^8)dx$
$= 2\pi\left[3x - \frac{4}{5}x^5 + \frac{1}{9}x^9\right]_0^1$
$= 2\pi\left(3 - \frac{4}{5} + \frac{1}{9}\right) = \frac{208}{45}\pi$

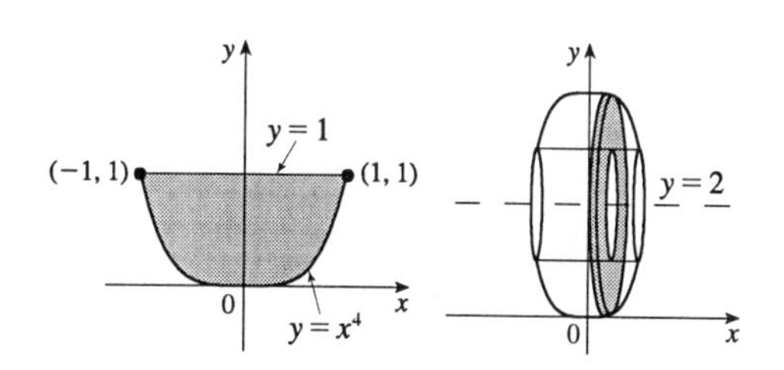

13. $V = \pi \int_0^8 \left(\frac{1}{4}x\right)^2 dx = \frac{\pi}{16}\left[\frac{1}{3}x^3\right]_0^8 = \frac{32}{3}\pi$

15. $V = \pi\int_0^2 (8-4y)^2 dy = \pi\left[64y - 32y^2 + \frac{16}{3}y^3\right]_0^2 = \pi\left(128 - 128 + \frac{128}{3}\right) = \frac{128}{3}\pi$

17. $V = \pi\int_0^8 \left[\left(\sqrt[3]{x}\right)^2 - \left(\frac{1}{4}x\right)^2\right] = \pi\int_0^8 \left(x^{2/3} - \frac{1}{16}x^2\right)dx = \pi\left[\frac{3}{5}x^{5/3} - \frac{1}{48}x^3\right]_0^8 = \pi\left(\frac{96}{5} - \frac{32}{3}\right) = \frac{128}{15}\pi$

19. $V = \pi\int_0^8 \left[\left(2 - \frac{1}{4}x\right)^2 - \left(2 - \sqrt[3]{x}\right)^2\right]dx = \pi\int_0^8 \left(-x + \frac{1}{16}x^2 + 4x^{1/3} - x^{2/3}\right)dx$

$= \pi\left[-\frac{1}{2}x^2 + \frac{1}{48}x^3 + 3x^{4/3} - \frac{3}{5}x^{5/3}\right]_0^8 = \pi\left(-32 + \frac{32}{3} + 48 - \frac{96}{5}\right) = \frac{112}{15}\pi$

21. $V = \pi\int_0^8 \left(2^2 - x^{2/3}\right)dx = \pi\left[4x - \frac{3}{5}x^{5/3}\right]_0^8 = \pi\left(32 - \frac{96}{5}\right) = \frac{64}{5}\pi$

23. $V = \pi\int_0^8 \left(2 - \sqrt[3]{x}\right)^2 dx = \pi\int_0^8 \left(4 - 4x^{1/3} + x^{2/3}\right)dx$

$= \pi\left[4x - 3x^{4/3} + \frac{3}{5}x^{5/3}\right]_0^8 = \pi\left(32 - 48 + \frac{96}{5}\right) = \frac{16}{5}\pi$

25. $V = \pi\int_0^2 (x^2-1)^2 dx = \pi\int_0^2 (x^4 - 2x^2 + 1)dx = \pi\left[\frac{1}{5}x^5 - \frac{2}{3}x^3 + x\right]_0^2 = \pi\left(\frac{32}{5} - \frac{16}{3} + 2\right) = \frac{46}{15}\pi$

27. $V = \int_0^1 \pi(e^x)^2\, dx = \int_0^1 \pi e^{2x}\, dx = \frac{1}{2}\left[\pi e^{2x}\right]_0^1 = \frac{\pi}{2}(e^2 - 1)$

29. $V = \pi\int_{-1}^1 (\sec^2 x - 1^2)dx = \pi[\tan x - x]_{-1}^1 = \pi[(\tan 1 - 1) - (-\tan 1 + 1)] = 2\pi(\tan 1 - 1)$

31. $V = \pi\int_{-3}^{-2} (-x-2)^2 dx + \pi\int_{-2}^0 (x+2)^2 dx = \pi\int_{-3}^0 (x+2)^2 dx = \left[\frac{\pi}{3}(x+2)^3\right]_{-3}^0 = \frac{\pi}{3}[8 - (-1)] = 3\pi$

33. $V = \pi\int_1^e \left[1^2 - (\ln x)^2\right]dx$

35. $x - 1 = (x-4)^2 + 1 \quad\Leftrightarrow\quad x^2 - 9x + 18 = 0 \quad\Leftrightarrow\quad x = 3$ or 6, so

$$V = \pi\int_3^6 \left[\left[6 - (x-4)^2\right]^2 - (8-x)^2\right]dx = \pi\int_3^6 \left(x^4 - 16x^3 + 83x^2 - 144x + 36\right)dx.$$

37. $V = \pi\int_0^{\pi/2} \left[(1+\cos x)^2 - 1^2\right]dx = \pi\int_0^{\pi/2} (2\cos x + \cos^2 x)dx$

39. We see from the graph in Exercise 7.1.47 that the x-coordinates of the points of intersection are $x \approx -0.72$ and $x \approx 1.22$, with $\sqrt{x+1} > x^2$ on $[-0.72, 1.22]$, so the volume of revolution is about

$$\pi\int_{-0.72}^{1.22} \left[\left(\sqrt{x+1}\right)^2 - \left(x^2\right)^2\right]dx = \pi\int_{-0.72}^{1.22} \left(x + 1 - x^4\right)dx = \pi\left[\frac{1}{2}x^2 + x - \frac{1}{5}x^5\right]_{-0.72}^{1.22} \approx 5.80.$$

41. $V = \pi\int_0^1 3^2\, dx + \pi\int_1^4 1^2\, dx + \pi\int_4^5 3^2\, dx$

$= 9\pi + 3\pi + 9\pi = 21\pi$

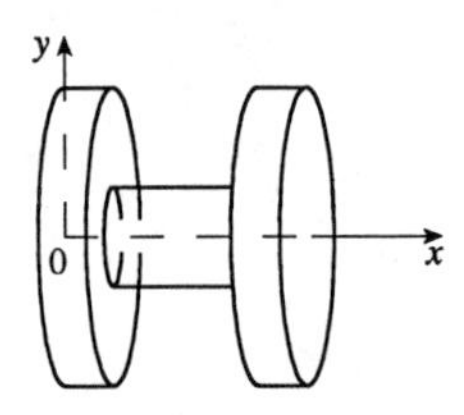

43. The solid is obtained by rotating the region under the curve $y = \tan x$, from $x = 0$ to $x = \frac{\pi}{4}$, about the x-axis.

45. The solid is obtained by rotating the region between the curves $x = y$ and $x = \sqrt{y}$ about the y-axis.

47. The solid is obtained by rotating the region between the curves $y = 5 - 2x^2$ and $y = 5 - 2x$ about the x-axis.

Or: The solid is obtained by rotating the region bounded by the curves $y = 2x$ and $y = 2x^2$ about the line $y = 5$.

49. $V = \pi \int_0^h \left(-\frac{r}{h}y + r\right)^2 dy$

$= \pi \int_0^h \left[\frac{r^2}{h^2}y^2 - \frac{2r^2}{h}y + r^2\right] dy$

$= \pi \left[\frac{r^2}{3h^2}y^3 - \frac{r^2}{h}y^2 + r^2 y\right]_0^h = \frac{1}{3}\pi r^2 h$

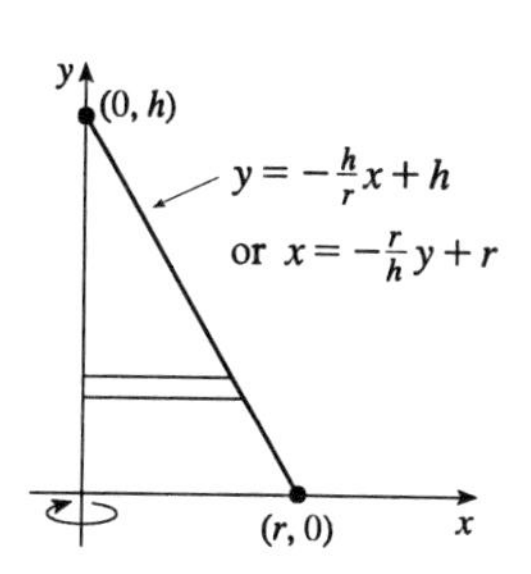

51. $V = \pi \int_{r-h}^{r} (r^2 - y^2)dy = \pi \left[r^2 y - \frac{y^3}{3}\right]_{r-h}^{r}$

$= \pi \left[\left(r^3 - \frac{r^3}{3}\right) - \left(r^2(r-h) - \frac{(r-h)^3}{3}\right)\right]$

$= \pi h^2 \left(r - \frac{h}{3}\right)$

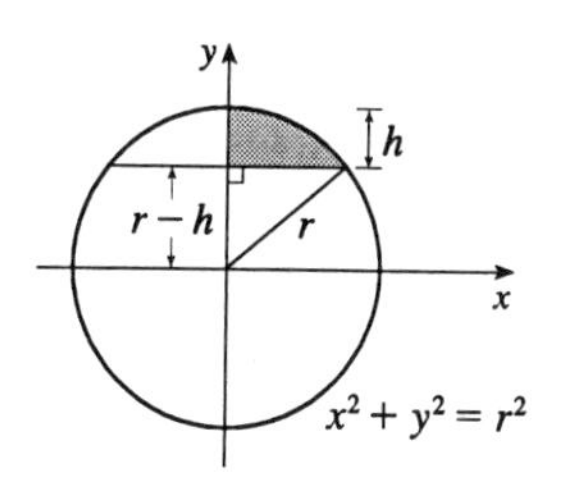

53. For a cross-section at height y, we see from similar triangles that $\frac{\alpha/2}{b/2} = \frac{h-y}{h}$, so $\alpha = b\left(1 - \frac{y}{h}\right)$. Similarly, $\beta = 2b\left(1 - \frac{y}{h}\right)$. So

$V = \int_0^h A(y)\,dy = \int_0^h 2b^2\left(1 - \frac{y}{h}\right)^2 dy = 2b^2 \int_0^h \left(1 - \frac{2y}{h} + \frac{y^2}{h^2}\right) dy$

$= 2b^2\left[y - \frac{y^2}{h} + \frac{y^3}{3h^2}\right]_0^h = 2b^2\left[h - h + \frac{1}{3}h\right] = \frac{2}{3}b^2 h$

$\left(= \frac{1}{3}Bh\right.$, where B is the area of the base, as with any pyramid.$\left.\right)$

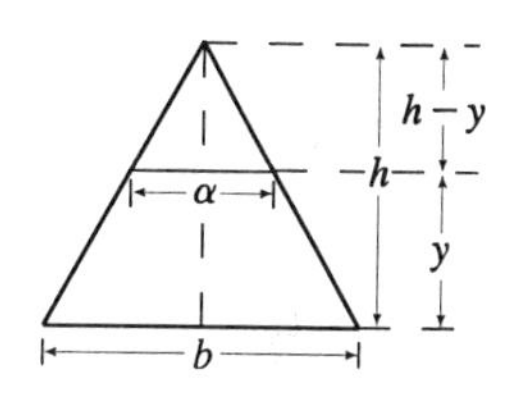

55. A cross-section at height z is a triangle similar to the base, so its area is

$A(z) = \frac{1}{2} \cdot 3\left(\frac{5-z}{5}\right) \cdot 4\left(\frac{5-z}{5}\right) = 6\left(1 - \frac{z}{5}\right)^2$, so

$V = \int_0^5 A(z)\,dz = 6\int_0^5 \left(1 - \frac{z}{5}\right)^2 dz$

$= 6\left[(-5)\frac{1}{3}\left(1 - \frac{1}{5}z\right)^3\right]_0^5 = -10(-1) = 10\text{ cm}^3.$

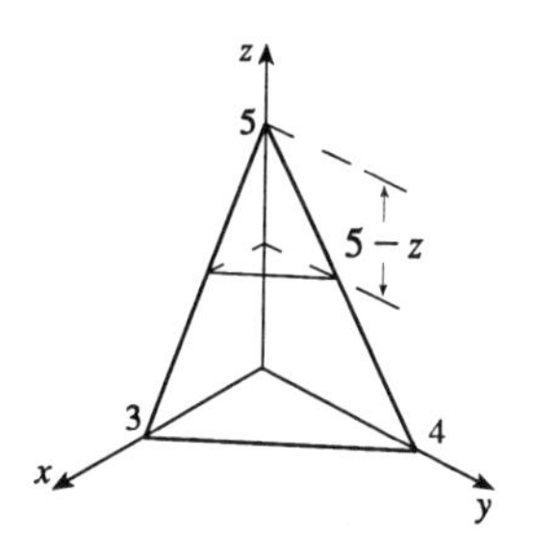

57. $V = \int_{-2}^{2} A(x)\,dx = 2\int_0^2 A(x)\,dx$

$= 2\int_0^2 \frac{1}{2}\left(\sqrt{2}y\right)^2 dx = 2\int_0^2 y^2\,dx$

$= \frac{1}{2}\int_0^2 (36 - 9x^2)dx = \frac{9}{2}\int_0^2 (4 - x^2)dx$

$= \frac{9}{2}\left[4x - \frac{x^3}{3}\right]_0^2 = \frac{9}{2}\left(8 - \frac{8}{3}\right) = 24$

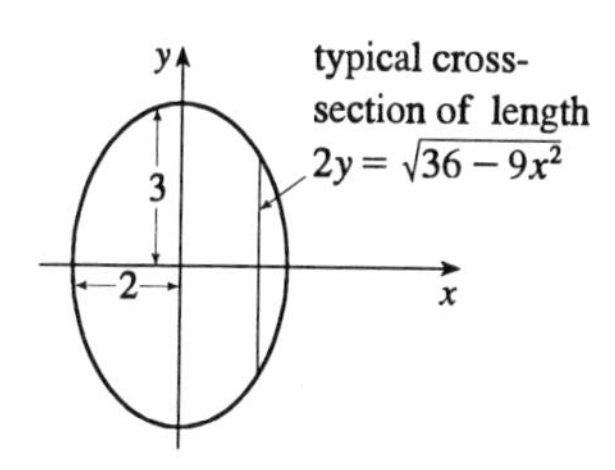

59. The cross section of the base corresponding to the coordinate y has length $2x = 2\sqrt{y}$, so

$V = \int_0^1 A(y)\,dy = \int_0^1 (2x)^2\,dy = \int_0^1 4x^2\,dy$
$= \int_0^1 4y\,dy = [2y^2]_0^1 = 2.$

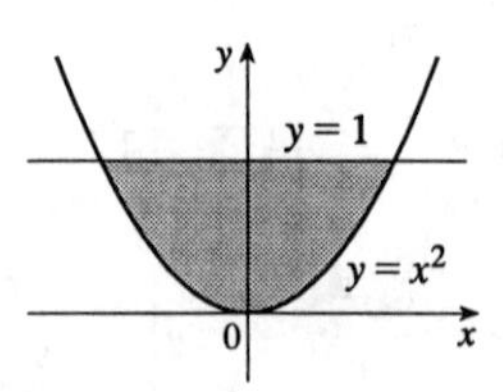

61. Assume that the base of each isosceles triangle lies in the base of S. Then its area is
$A(x) = \frac{1}{2}bh = \frac{1}{2}\left(1 - \frac{1}{2}x\right)\left(1 - \frac{1}{2}x\right) = \frac{1}{2}\left(1 - \frac{1}{2}x\right)^2$, and the volume is
$V = \int_0^2 A(x)\,dx = \int_0^2 \frac{1}{2}y^2\,dx = \frac{1}{2}\int_0^2 \left(1 - \frac{1}{2}x\right)^2 dx = \frac{1}{2}\left[\frac{2}{3}\left(\frac{1}{2}x - 1\right)^3\right]_0^2 = \frac{1}{3}.$

63. **(a)** The torus is obtained by rotating the circle $(x - R)^2 + y^2 = r^2$ about the y-axis. Solving for y, we see that the right half of the circle is given by $x = R + \sqrt{r^2 - y^2} = f(y)$ and the left half by $x = R - \sqrt{r^2 - y^2} = g(y)$.

So $V = \pi\int_{-r}^{r}\left([f(y)]^2 - [g(y)]^2\right)dy$
$= 2\pi\int_0^r 4R\sqrt{r^2 - y^2}\,dy$
$= 8\pi R\int_0^r \sqrt{r^2 - y^2}\,dy.$

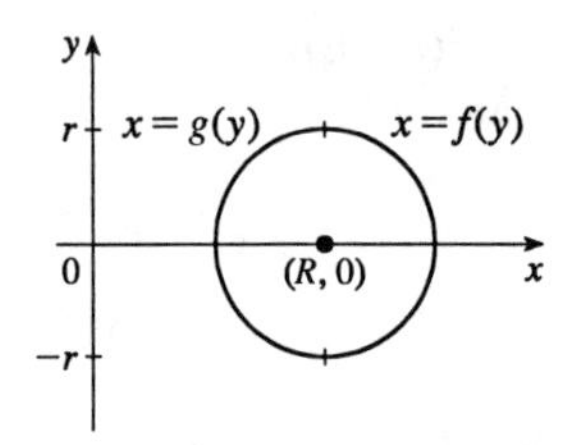

(b) Observe that the integral represents a quarter of the area of a circle with radius r, so
$8\pi R\int_0^r \sqrt{r^2 - y^2}\,dy = 8\pi R\frac{1}{4}(\pi r^2) = 2\pi^2 r^2 R.$

65. The volume is obtained by rotating the area common to two circles of radius r, as shown. The volume of the right half is

$$V_{\text{right}} = \pi\int_0^{r/2}\left[r^2 - \left(\tfrac{1}{2}r + x\right)^2\right]dx$$
$$= \pi\left[r^2 x - \tfrac{1}{3}\left(\tfrac{1}{2}r + x\right)^3\right]_0^{r/2}$$
$$= \pi\left[\left(\tfrac{1}{2}r^3 - \tfrac{1}{3}r^3\right) - \left(0 - \tfrac{1}{24}r^3\right)\right]$$
$$= \tfrac{5}{24}\pi r^3.$$

So by symmetry, the total volume is twice this, or $\frac{5}{12}\pi r^3$.

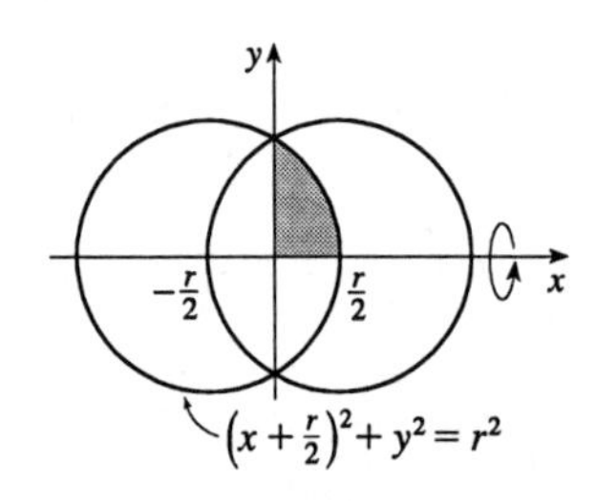

Alternate Solution: We observe that the volume is the twice the volume of a cap of a sphere, so we can use the formula from Exercise 51 with $h = \frac{1}{2}r$: $V = 2\pi\left(\frac{1}{2}r\right)^2\left(r - \dfrac{r/2}{3}\right) = \frac{5}{12}\pi r^3.$

67. The cross-sections perpendicular to the y-axis in Figure 17 are rectangles. The rectangle corresponding to the coordinate y has a base of length $2\sqrt{16-y^2}$ in the xy-plane and a height of $\frac{1}{\sqrt{3}}y$, since $\angle BAC = 30°$ and $|BC| = \frac{1}{\sqrt{3}}|AB|$. Thus $A(y) = \frac{2}{\sqrt{3}}y\sqrt{16-y^2}$ and

$$V = \int_0^4 A(y)\,dy = \tfrac{2}{\sqrt{3}}\int_0^4 \sqrt{16-y^2}\,y\,dy$$
$$= \tfrac{2}{\sqrt{3}}\int_{16}^0 u^{1/2}\left(-\tfrac{1}{2}\,du\right) \quad [\text{Put } u = 16-y^2, \text{ so } du = -2y\,dy]$$
$$= \tfrac{1}{\sqrt{3}}\int_0^{16} u^{1/2}\,du = \tfrac{1}{\sqrt{3}}\tfrac{2}{3}\left[u^{3/2}\right]_0^{16} = \tfrac{2}{3\sqrt{3}}(64) = \tfrac{128}{3\sqrt{3}}.$$

69. Take the x-axis to be the axis of the cylindrical hole of radius r. A quarter of the cross-section through y, perpendicular to the y-axis, is the rectangle shown. Using Pythagoras twice, we see that the dimensions of this rectangle are $x = \sqrt{R^2-y^2}$ and $z = \sqrt{r^2-y^2}$, so $\frac{1}{4}A(y) = xz = \sqrt{r^2-y^2}\sqrt{R^2-y^2}$,

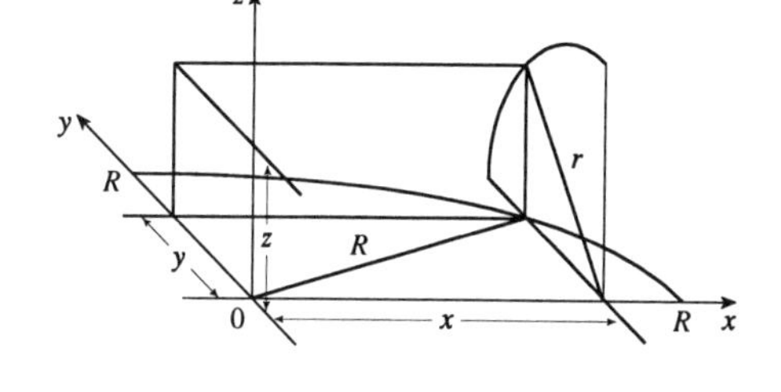

and $V = \int_{-r}^{r} A(y)\,dy = \int_{-r}^{r} 4\sqrt{r^2-y^2}\sqrt{R^2-y^2}\,dy = 8\int_0^r \sqrt{r^2-y^2}\sqrt{R^2-y^2}\,dy.$

71. **(a)** $\text{Volume}(S_1) = \int_0^h A(z)dz = \text{Volume}(S_2)$ since the cross-sectional area $A(z)$ at height z is the same for both solids.

(b) By Cavalieri's Principle, the volume of the cylinder in the figure is the same as that of a right circular cylinder with radius r and height h, that is, $\pi r^2 h$.

73. **(a)** The radius of the barrel is the same at each end by symmetry, since the function $y = R - cx^2$ is even. Since the barrel is obtained by rotating the function y about the x-axis, this radius is equal to the value of y at $x = \frac{1}{2}h$, which is $R - c\left(\frac{1}{2}h\right)^2 = R - d = r.$

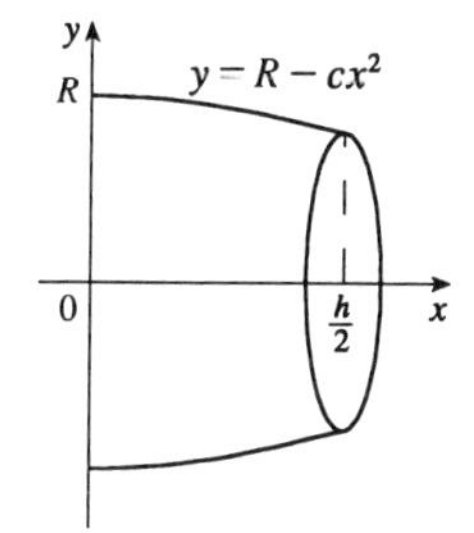

(b) The barrel is symmetric about the y-axis, so its volume is twice the volume of that part of the barrel for $x > 0$. Also, the barrel is a volume of rotation, so we use Formula 2:

$$V = 2\int_0^{h/2} \pi(y^2)dx = 2\pi\int_0^{h/2}\left(R-cx^2\right)^2 dx = 2\pi\left[R^2x - \tfrac{2}{3}Rcx^3 + \tfrac{1}{5}c^2x^5\right]_0^{h/2}$$
$$= 2\pi\left(\tfrac{1}{2}R^2h - \tfrac{1}{12}Rch^3 + \tfrac{1}{160}c^2h^5\right).$$

Trying to make this look more like the expression we want, we rewrite it as $V = \frac{1}{3}\pi h\left[2R^2 + \left(R^2 - \frac{1}{2}Rch^2 + \frac{3}{80}c^2h^4\right)\right]$. But $R^2 - \frac{1}{2}Rch^2 + \frac{3}{80}c^2h^4 = \left(R - \frac{1}{4}ch^2\right)^2 - \frac{1}{40}c^2h^4 = (R-d)^2 - \frac{2}{5}\left(\frac{1}{4}ch^2\right)^2 = r^2 - \frac{2}{5}d^2$. Substituting this back into V, we see that $V = \frac{1}{3}\pi h\left(2R^2 + r^2 - \frac{2}{5}d^2\right)$, as required.

75. We are given that the rate of change of the volume of water is $dV/dt = -kA(x)$, where k is some positive constant and $A(x)$ is the area of the surface when the water has depth x. Now we are concerned with the rate of change of the depth of the water with respect to time, that is, $\frac{dx}{dt}$. But by the Chain Rule, $\frac{dV}{dt} = \frac{dV}{dx}\frac{dx}{dt}$, so the first equation can be written $\frac{dV}{dx}\frac{dx}{dt} = -kA(x)$ (★). Also, we know that the total volume of water up to a depth x is $V(x) = \int_0^x A(s)\,ds$, where $A(s)$ is the area of a cross-section of the water at a depth s. Differentiating this equation with respect to x, we get $dV/dx = A(x)$. Substituting this into equation ★, we get $A(x)(dx/dt) = -kA(x) \quad \Rightarrow \quad dx/dt = -k$, a constant.

EXERCISES 7.3

1. $V = \int_1^2 2\pi x \cdot x^2\,dx = 2\pi \int_1^2 x^3\,dx = 2\pi\left[\frac{1}{4}x^4\right]_1^2 = 2\pi\left(\frac{15}{4}\right) = \frac{15}{2}\pi$

3. $V = \int_0^1 2\pi x e^{-x^2} dx$. Let $u = x^2$. Thus $du = 2x dx$, so $V = \pi \int_0^1 e^{-u}\,du = \pi[-e^{-u}]_0^1 = \pi(1 - 1/e)$.

5. $V = \int_0^2 2\pi x(4 - x^2)dx = 2\pi \int_0^2 (4x - x^3)dx = 2\pi\left[2x^2 - \frac{1}{4}x^4\right]_0^2 = 2\pi(8 - 4) = 8\pi$

Note: If we integrated from -2 to 2, we would be generating the volume twice.

7. $V = \int_0^1 2\pi x(x^2 - x^3)dx = 2\pi \int_0^1 (x^3 - x^4)dx = 2\pi\left[\frac{1}{4}x^4 - \frac{1}{5}x^5\right]_0^1 = 2\pi\left(\frac{1}{4} - \frac{1}{5}\right) = \frac{1}{10}\pi$

9. $V = \int_0^{16} 2\pi y \sqrt[4]{y}\,dy = 2\pi \int_0^{16} y^{5/4}\,dy = 2\pi\left[\frac{4}{9}y^{9/4}\right]_0^{16} = \frac{8}{9}\pi(512 - 0) = \frac{4096}{9}\pi$

11. $V = \int_0^9 2\pi y \cdot 2\sqrt{y}\,dy = 4\pi \int_0^9 y^{3/2}\,dy = 4\pi\left[\frac{2}{5}y^{5/2}\right]_0^9 = \frac{8}{5}\pi(243 - 0) = \frac{1944}{5}\pi$

13. $V = \int_0^1 2\pi y[(2 - y) - y^2]dy = 2\pi\left[y^2 - \frac{1}{3}y^3 - \frac{1}{4}y^4\right]_0^1 = 2\pi\left(1 - \frac{1}{3} - \frac{1}{4}\right) = \frac{5}{6}\pi$

15. $V = \int_1^4 2\pi x\sqrt{x}\,dx = 2\pi \int_1^4 x^{3/2}\,dx$
$= 2\pi\left[\frac{2}{5}x^{5/2}\right]_1^4 = \frac{4}{5}\pi(32 - 1) = \frac{124}{5}\pi$

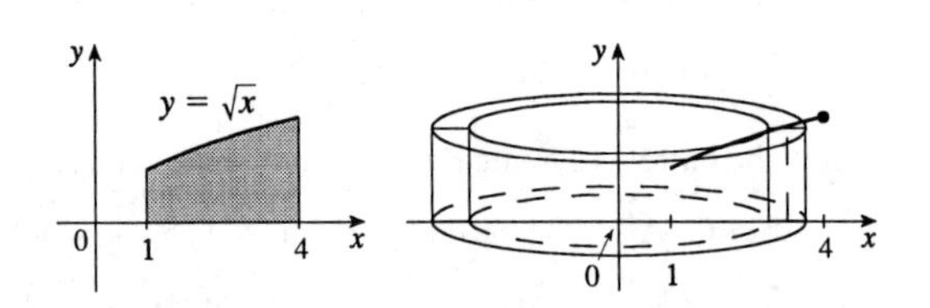

17. $V = \int_1^2 2\pi(x - 1)\,x^2\,dx = 2\pi\left[\frac{1}{4}x^4 - \frac{1}{3}x^3\right]_1^2$
$= 2\pi\left[\left(4 - \frac{8}{3}\right) - \left(\frac{1}{4} - \frac{1}{3}\right)\right] = \frac{17}{6}\pi$

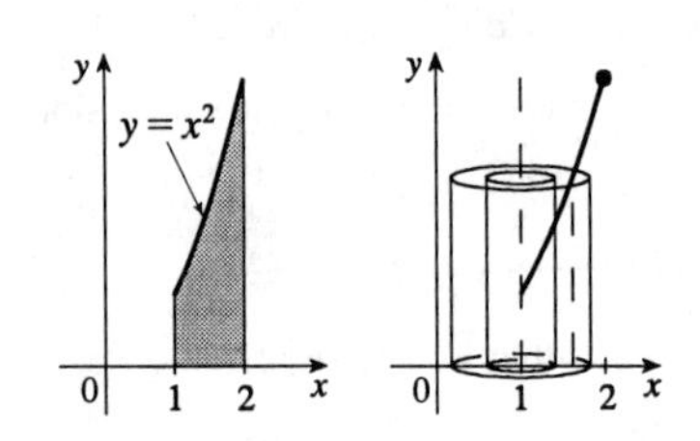

19. $V = \int_0^2 2\pi(3-y)(5-x)dy$
$= \int_0^2 2\pi(3-y)(5-y^2-1)dy$
$= \int_0^2 2\pi(12-4y-3y^2+y^3)\,dy$
$= 2\pi\left[12y - 2y^2 - y^3 + \frac{1}{4}y^4\right]_0^2$
$= 2\pi(24-8-8+4) = 24\pi$

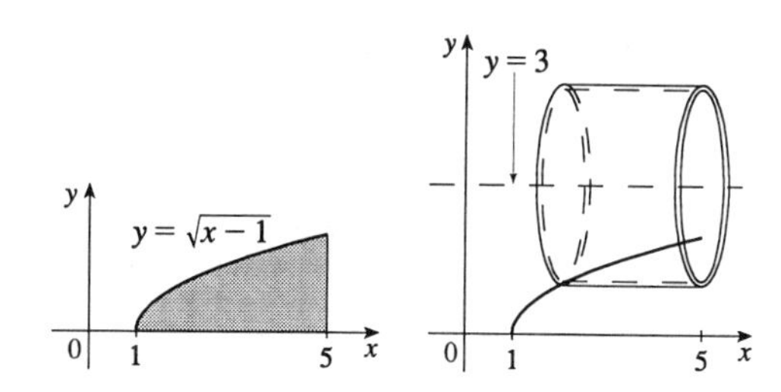

21. $V = \int_{2\pi}^{3\pi} 2\pi x \sin x\, dx$

23. $V = \int_0^{\pi/4} 2\pi y \cos y\; dy$

25. $V = \int_0^1 2\pi(x+1)\left(\sin\frac{\pi}{2}x - x^4\right)dx$

27. The solid is obtained by rotating the region bounded by the curve $y = \cos x$ and the line $y = 0$, from $x = 0$ to $x = \frac{\pi}{2}$, about the y-axis.

29. The solid is obtained by rotating the region in the first quadrant bounded by the curves $y = x^2$ and $y = x^6$ about the y-axis.

31.

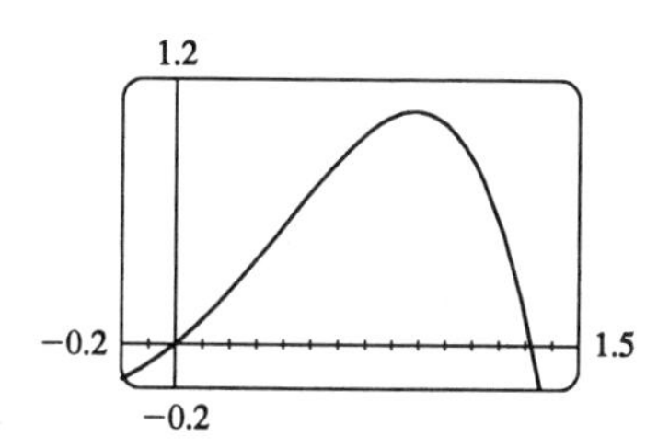

From the graph, it appears that the curves intersect at $x = 0$ and at $x \approx 1.32$, with $x + x^2 - x^4 > 0$ on $(0, 1.32)$. So the volume of the solid obtained by rotating the region about the y-axis is

$V \approx 2\pi\int_0^{1.32} x(x + x^2 - x^4)dx$
$= 2\pi\left[\frac{1}{3}x^3 + \frac{1}{4}x^4 - \frac{1}{6}x^6\right]_0^{1.32} \approx 4.05.$

33. Use disks: $V = \int_{-2}^{1} \pi(x^2+x-2)^2 dx = \pi\int_{-2}^{1}(x^4+2x^3-3x^2-4x+4)dx$
$= \pi\left[\frac{1}{5}x^5 + \frac{1}{2}x^4 - x^3 - 2x^2 + 4x\right]_{-2}^{1} = \pi\left[\left(\frac{1}{5}+\frac{1}{2}-1-2+4\right) - \left(-\frac{32}{5}+8+8-8-8\right)\right]$
$= \pi\left(\frac{33}{5}+\frac{3}{2}\right) = \frac{81}{10}\pi$

35. Use disks: $V = \pi\int_{-1}^{1}(1-y^2)^2dy = 2\pi\int_0^1(y^4-2y^2+1)dy = 2\pi\left[\frac{1}{5}y^5 - \frac{2}{3}y^3 + y\right]_0^1 = 2\pi\left(\frac{1}{5}-\frac{2}{3}+1\right) = \frac{16}{15}\pi$

37. Use disks: $V = \pi\displaystyle\int_0^2\left[\sqrt{1-(y-1)^2}\right]^2 dy = \pi\int_0^2(2y-y^2)dy = \pi\left[y^2 - \tfrac{1}{3}y^3\right]_0^2 = \pi\left(4-\tfrac{8}{3}\right) = \tfrac{4}{3}\pi$

39. $V = 2\int_0^r 2\pi x\sqrt{r^2-x^2}\,dx = -2\pi\int_0^r(r^2-x^2)^{1/2}(-2x)dx = \left[-2\pi\cdot\frac{2}{3}(r^2-x^2)^{3/2}\right]_0^r = -\frac{4}{3}\pi(0-r^3) = \frac{4}{3}\pi r^3$

41. $V = 2\pi\displaystyle\int_0^r x\left(-\frac{h}{r}x+h\right)dx = 2\pi h\int_0^r\left(-\frac{x^2}{r}+x\right)dx = 2\pi h\left[-\frac{x^3}{3r}+\frac{x^2}{2}\right]_0^r = 2\pi h\frac{r^2}{6} = \frac{\pi r^2 h}{3}$

43. If $a < b \le 0$, then a typical cylindrical shell has radius $-x$ and height $f(x)$, so
$V = \int_a^b 2\pi(-x)f(x)dx = -\int_a^b 2\pi x f(x)dx.$

45. $\Delta x = \dfrac{\pi/4 - 0}{4} = \dfrac{\pi}{16}.$
$V = \int_0^{\pi/4} 2\pi x\tan x\,dx \approx 2\pi\cdot\frac{\pi}{16}\left(\frac{\pi}{32}\tan\frac{\pi}{32} + \frac{3\pi}{32}\tan\frac{3\pi}{32} + \frac{5\pi}{32}\tan\frac{5\pi}{32} + \frac{7\pi}{32}\tan\frac{7\pi}{32}\right) \approx 1.142$

EXERCISES 7.4

1. By Equation 2, $W = Fd = (900)(8) = 7200$ J.

3. By Equation 4, $W = \int_a^b f(x)\,dx = \int_0^{10}(5x^2+1)dx = \left[\frac{5}{3}x^3 + x\right]_0^{10} = \frac{5000}{3} + 10 = \frac{5030}{3}$ ft-lb.

5. $10 = f(x) = kx = \frac{1}{3}k$ (4 inches $= \frac{1}{3}$ foot), so $k = 30$ (The units for k are pounds per foot.)

Now 6 inches $= \frac{1}{2}$ foot, so $W = \int_0^{1/2} 30x\,dx = \left[15x^2\right]_0^{1/2} = \frac{15}{4}$ ft-lb.

7. If $\int_0^{0.12} kx\,dx = 2$ J, then $2 = \left[\frac{1}{2}kx^2\right]_0^{0.12} = \frac{1}{2}k(0.0144) = 0.0072k$ and $k = \frac{2}{0.0072} = \frac{2500}{9} \approx 277.78$. Thus the work needed to stretch the spring from 35 cm to 40 cm is

$\int_{0.05}^{0.10} \frac{2500}{9} x\,dx = \left[\frac{1250}{9}x^2\right]_{1/20}^{1/10} = \frac{1250}{9}\left(\frac{1}{100} - \frac{1}{400}\right) = \frac{25}{24} \approx 1.04$ J.

9. $f(x) = kx$, so $30 = \frac{2500}{9}x$ and $x = \frac{270}{2500}$ m $= 10.8$ cm.

11. First notice that the exact height of the building does not matter. The portion of the rope from x ft to $(x + \Delta x)$ft below the top of the building weighs $\frac{1}{2}\Delta x$ lb and must be lifted x ft (approximately), so its contribution to the total work is $\frac{1}{2}x\Delta x$ ft-lb. The total work is $W = \int_0^{50} \frac{1}{2}x\,dx = \left[\frac{1}{4}x^2\right]_0^{50} = \frac{2500}{4} = 625$ ft-lb.

13. The work needed to lift the cable is $\int_0^{500} 2x\,dx = x^2\big|_0^{500} = 250{,}000$ ft-lb. The work needed to lift the coal is 800 lb $\cdot$ 500 ft $= 400{,}000$ ft-lb. Thus the total work required is $250{,}000 + 400{,}000 = 650{,}000$ ft-lb.

15. A "slice" of water Δx m thick and lying at a depth of x m (where $0 \le x \le \frac{1}{2}$) has volume $2\Delta x$ m^3, a mass of $2000\Delta x$ kg, weighs about $(9.8)(2000\Delta x) = 19{,}600\,\Delta x$ N, and thus requires about $19{,}600 x\Delta x$ J of work for its removal. So $W \approx \int_0^{1/2} 19{,}600x\,dx = 9800x^2\big|_0^{1/2} = 2450$ J.

17. A "slice" of water Δx m thick and lying x ft above the bottom has volume $8x\Delta x$ m^3 and weighs about $(9.8 \times 10^3)(8x\Delta x)$ N. It must be lifted $(5 - x)$ m by the pump, so the work needed is about $(9.8 \times 10^3)(5 - x)(8x\Delta x)$ J. The total work required is

$$
\begin{aligned}
W &\approx \int_0^3 (9.8 \times 10^3)(5 - x)8x\,dx = (9.8 \times 10^3)\int_0^3 (40x - 8x^2)dx \\
&= (9.8 \times 10^3)\left[20x^2 - \tfrac{8}{3}x^3\right]_0^3 = (9.8 \times 10^3)(180 - 72) = (9.8 \times 10^3)(108) = 1058.4 \times 10^3 \\
&\approx 1.06 \times 10^6 \text{ J.}
\end{aligned}
$$

19. Measure depth x downward from the flat top of the tank, so that $0 \le x \le 2$ ft. Then

$\Delta W = (62.5)\left(2\sqrt{4 - x^2}\right)(8\Delta x)(x + 1)$ ft-lb, so

$$
\begin{aligned}
W &\approx (62.5)(16)\int_0^2 (x+1)\sqrt{4 - x^2}\,dx = 1000\left(\int_0^2 x\sqrt{4 - x^2}\,dx + \int_0^2 \sqrt{4 - x^2}\,dx\right) \\
&= 1000\left[\int_0^4 u^{1/2}\left(\tfrac{1}{2}\,du\right) + \tfrac{1}{4}\pi(2^2)\right] \qquad \text{(Put } u = 4 - x^2\text{, so } du = -2x\,dx\text{)} \\
&= 1000\left(\tfrac{1}{2} \cdot \tfrac{2}{3}u^{3/2}\big|_0^4 + \pi\right) = 1000\left(\tfrac{8}{3} + \pi\right) \approx 5.8 \times 10^3 \text{ ft-lb.}
\end{aligned}
$$

Note: The second integral represents the area of a quarter-circle of radius 2.

21. If only 4.7×10^5 J of work is done, then only the water above a certain level (call it h) will be pumped out. So we use the same formula as in Exercise 17, except that the work is fixed, and we are trying to find the lower limit of integration: $4.7 \times 10^5 \approx \int_h^3 (9.8 \times 10^3)(5-x)8x\,dx = (9.8 \times 10^3)\left[20x^2 - \frac{8}{3}x^3\right]_h^3 \quad \Leftrightarrow$
$\frac{4.7}{9.8} \times 10^2 \approx 48 = \left(20 \cdot 3^2 - \frac{8}{3} \cdot 3^3\right) - \left(20h^2 - \frac{8}{3}h^3\right) \quad \Leftrightarrow \quad 2h^3 - 15h^2 + 45 = 0.$
To find the solution of this equation, we plot $2h^3 - 15h^2 + 45$ between $h = 0$ and $h = 3$. We see that the equation is satisfied for $h \approx 2.0$. So the depth of water remaining in the tank is about 2.0 m.

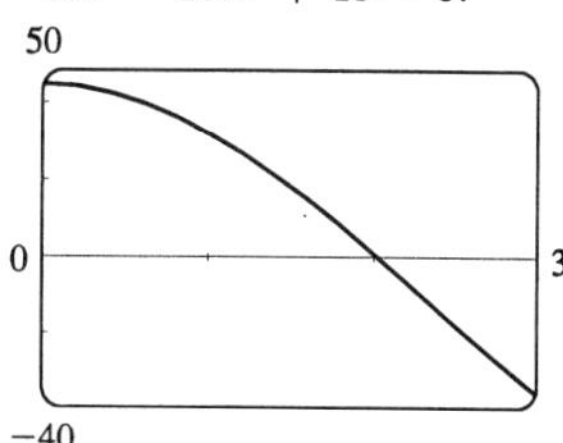

23. $V = \pi r^2 x$, so V is a function of x and P can also be regarded as a function of x. If $V_1 = \pi r^2 x_1$ and $V_2 = \pi r^2 x_2$, then $W = \int_{x_1}^{x_2} F(x)\,dx = \int_{x_1}^{x_2} \pi r^2 P(V(x))dx = \int_{x_1}^{x_2} P(V(x))dV(x)$ [Put $V(x) = \pi r^2 x$, so $dV(x) = \pi r^2\,dx$] $= \int_{V_1}^{V_2} P(V)dV$ by the Substitution Rule.

25. $$W = \int_a^b F(r)dr = \int_a^b G\frac{m_1 m_2}{r^2}\,dr = Gm_1m_2\left[\frac{-1}{r}\right]_a^b = Gm_1m_2\left(\frac{1}{a} - \frac{1}{b}\right)$$

EXERCISES 7.5

1. $f_{\text{ave}} = \frac{1}{3-0}\int_0^3 (x^2 - 2x)dx = \frac{1}{3}\left[\frac{1}{3}x^3 - x^2\right]_0^3 = \frac{1}{3}(9-9) = 0$

3. $f_{\text{ave}} = \frac{1}{1-(-1)}\int_{-1}^1 x^4\,dx = \frac{1}{2} \cdot 2\int_0^1 x^4\,dx = \left[\frac{1}{5}x^5\right]_0^1 = \frac{1}{5}$

5. $$f_{\text{ave}} = \frac{1}{\frac{\pi}{4} - \left(-\frac{\pi}{2}\right)}\int_{-\pi/2}^{\pi/4} \sin^2 x \cos x\,dx = \frac{4}{3\pi}\int_{-\pi/2}^{\pi/4} \sin^2 x \cos x\,dx$$
$= \frac{4}{3\pi}\int_{-1}^{1/\sqrt{2}} u^2\,du$ [Put $u = \sin x$, so $du = \cos x\,dx$] $= \frac{4}{3\pi}\left[\frac{1}{3}u^3\right]_{-1}^{1/\sqrt{2}}$
$= \frac{4}{9\pi}\left(\frac{1}{2\sqrt{2}} + 1\right) = \frac{4}{9\pi}\left(\frac{\sqrt{2}}{4} + 1\right) = \frac{\sqrt{2}+4}{9\pi}$

7. **(a)** $$f_{\text{ave}} = \frac{1}{2-0}\int_0^2 (4 - x^2)dx$$
$= \frac{1}{2}\left[4x - \frac{1}{3}x^3\right]_0^2 = \frac{1}{2}\left[\left(8 - \frac{8}{3}\right) - 0\right]$
$= \frac{8}{3}$

(b) $f_{\text{ave}} = f(c) \quad \Leftrightarrow \quad \frac{8}{3} = 4 - c^2$
$\Leftrightarrow \quad c^2 = \frac{4}{3} \quad \Leftrightarrow \quad c = \frac{2}{\sqrt{3}}$

(c)

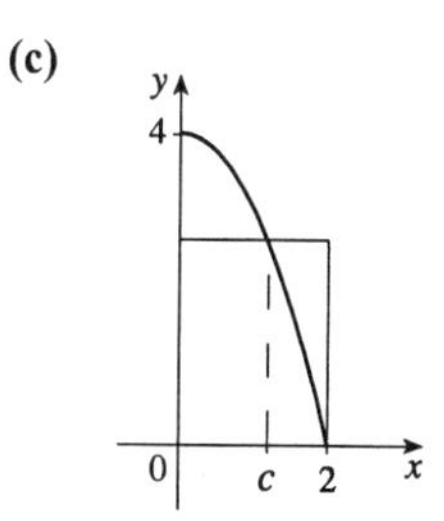

9. **(a)** $f_{\text{ave}} = \frac{1}{2-0}\int_0^2 (x^3 - x + 1)dx$
$= \frac{1}{2}\left[\frac{1}{4}x^4 - \frac{1}{2}x^2 + x\right]_0^2$
$= \frac{1}{2}(4 - 2 + 2) = 2$

(b) From the graph, it appears that $f(x) = 2$ at $x \approx 1.32$.

(c)

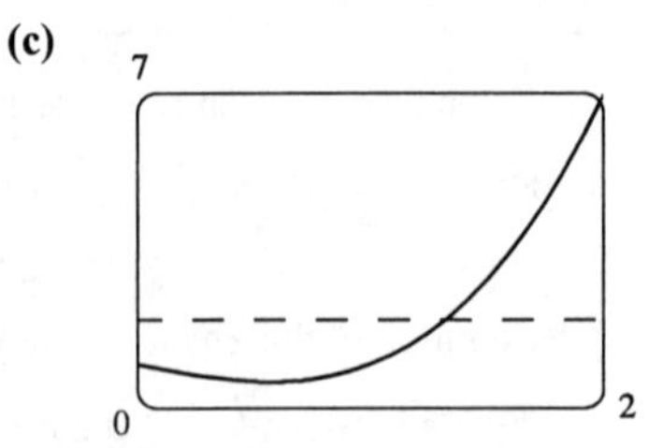

11. Since f is continuous on $[1,3]$, by the Mean Value Theorem for Integrals there exists a number c in $[1,3]$ such that $\int_1^3 f(x)\,dx = 8 = f(c)(3-1) = 2f(c)$; that is, there is a number c such that $f(c) = \frac{8}{2} = 4$.

13. $T_{\text{ave}} = \frac{1}{12}\int_0^{12}\left[50 + 14\sin\frac{1}{12}\pi t\right]dt = \frac{1}{12}\left[50\text{t} - 14 \cdot \frac{12}{\pi}\cos\frac{1}{12}\pi t\right]_0^{12}$
$= \frac{1}{12}\left[50 \cdot 12 + 14 \cdot \frac{12}{\pi} + 14 \cdot \frac{12}{\pi}\right] = \left(50 + \frac{28}{\pi}\right)^\circ\text{F} \approx 59^\circ\,\text{F}$

15. $$\rho_{\text{ave}} = \frac{1}{8}\int_0^8 \frac{12}{\sqrt{x+1}}\,dx = \frac{3}{2}\int_0^8 (x+1)^{-1/2}\,dx = \left[3\sqrt{x+1}\right]_0^8 = 9 - 3 = 6\,\text{kg/m}$$

17. $V_{\text{ave}} = \frac{1}{5}\int_0^5 V(t)dt = \frac{1}{5}\int_0^5 \frac{5}{4\pi}\left[1 - \cos\left(\frac{2}{5}\pi t\right)\right]dt = \frac{1}{4\pi}\int_0^5\left[1 - \cos\left(\frac{2}{5}\pi t\right)\right]dt$
$= \frac{1}{4\pi}\left[t - \frac{5}{2\pi}\sin\left(\frac{2}{5}\pi t\right)\right]_0^5 = \frac{1}{4\pi}[(5-0) - 0] = \frac{5}{4\pi} \approx 0.4\,\text{L}$

19. Let $F(x) = \int_a^x f(t)\,dt$ for x in $[a,b]$. Then F is continuous on $[a,b]$ and differentiable on (a,b), so by the Mean Value Theorem there is a number c in (a,b) such that $F(b) - F(a) = F'(c)(b-a)$. But $F'(x) = f(x)$ by the Fundamental Theorem of Calculus. Therefore $\int_a^b f(t)dt - 0 = f(c)(b-a)$.

REVIEW EXERCISES FOR CHAPTER 7

1. **(a)** $A = \int_a^b [f(x) - g(x)]dx$

(b) $A = \int_c^d [u(y) - v(y)]dy$

(c) Here we use disks: $V = \pi\int_a^b \left([f(x)]^2 - [g(x)]^2\right)dx$

(d) Here we use cylindrical shells: $V = 2\pi\int_a^b x[f(x) - g(x)]dx$

(e) Use shells: $V = 2\pi\int_c^d y[u(y) - v(y)]dy$

(f) Use disks: $V = \pi\int_c^d \left([u(y)]^2 - [v(y)]^2\right)dy$

3. $A = \int_0^6 [(12x - 2x^2) - (x^2 - 6x)]dx = \int_0^6 (18x - 3x^2)dx = [9x^2 - x^3]_0^6 = 9 \cdot 36 - 216 = 108$

5. By symmetry, $A = 2\int_0^1 \left(x^{1/3} - x^3\right)dx = 2\left[\frac{3}{4}x^{4/3} - \frac{1}{4}x^4\right]_0^1 = 2\left(\frac{3}{4} - \frac{1}{4}\right) = 1.$

7. $A = \int_0^{\pi} |\sin x - (-\cos x)|dx = \int_0^{3\pi/4} (\sin x + \cos x)dx - \int_{3\pi/4}^{\pi} (\sin x + \cos x)dx$

$= [\sin x - \cos x]_0^{3\pi/4} - [-\cos x + \sin x]_{3\pi/4}^{\pi}$

$= \left(\frac{1}{\sqrt{2}} + \frac{1}{\sqrt{2}}\right) - (0 - 1) - (1 + 0) + \left(\frac{1}{\sqrt{2}} + \frac{1}{\sqrt{2}}\right) = \sqrt{2} + 1 - 1 + \sqrt{2} = 2\sqrt{2}$

9. $V = \displaystyle\int_1^3 \pi\left(\sqrt{x-1}\right)^2 dx = \pi \int_1^3 (x - 1)dx = \pi\left[\tfrac{1}{2}x^2 - x\right]_1^3 = \pi\left[\left(\tfrac{9}{2} - 3\right) - \left(\tfrac{1}{2} - 1\right)\right] = 2\pi$

11. $V = \int_1^3 2\pi y(-y^2 + 4y - 3)dy = 2\pi \int_1^3 (-y^3 + 4y^2 - 3y)dy = 2\pi\left[-\tfrac{1}{4}y^4 + \tfrac{4}{3}y^3 - \tfrac{3}{2}y^2\right]_1^3$

$= 2\pi\left[\left(-\tfrac{81}{4} + 36 - \tfrac{27}{2}\right) - \left(-\tfrac{1}{4} + \tfrac{4}{3} - \tfrac{3}{2}\right)\right] = \frac{16\pi}{3}$

13. $V = \int_a^{a+h} 2\pi x \cdot 2\sqrt{x^2 - a^2}\,dx = 2\pi \int_0^{2ah+h^2} u^{1/2}\,du$ (Put $u = x^2 - a^2$, so $du = 2x\,dx$)

$= 2\pi\left[\tfrac{2}{3}u^{3/2}\right]_0^{2ah+h^2} = \tfrac{4}{3}\pi(2ah + h^2)^{3/2}$

15. $V = \displaystyle\int_0^1 \pi\left[\left(1 - x^3\right)^2 - \left(1 - x^2\right)^2\right]dx$

17. **(a)** $V = \int_0^1 \pi(x^2 - x^4)dx = \pi\left[\tfrac{1}{3}x^3 - \tfrac{1}{5}x^5\right]_0^1 = \pi\left[\tfrac{1}{3} - \tfrac{1}{5}\right] = \tfrac{2\pi}{15}$

Or: $V = \int_0^1 2\pi y\left(\sqrt{y} - y\right)dy = 2\pi\left[\tfrac{2}{5}y^{5/2} - \tfrac{1}{3}y^3\right]_0^1 = \tfrac{2\pi}{15}$

(b) $V = \displaystyle\int_0^1 \pi\left[\left(\sqrt{y}\right)^2 - y^2\right]dy = \pi\left[\tfrac{1}{2}y^2 - \tfrac{1}{3}y^3\right]_0^1 = \pi\left[\tfrac{1}{2} - \tfrac{1}{3}\right] = \tfrac{\pi}{6}$

Or: $V = \int_0^1 2\pi x(x - x^2)dx = 2\pi\left[\tfrac{1}{3}x^3 - \tfrac{1}{4}x^4\right]_0^1 = \tfrac{\pi}{6}$

(c) $V = \displaystyle\int_0^1 \pi\left[\left(2 - x^2\right)^2 - (2 - x)^2\right]dx = \int_0^1 \pi\left(x^4 - 5x^2 + 4x\right)dx$

$= \pi\left[\tfrac{1}{5}x^5 - \tfrac{5}{3}x^3 + 2x^2\right]_0^1 = \pi\left[\tfrac{1}{5} - \tfrac{5}{3} + 2\right] = \tfrac{8\pi}{15}$

Or: $V = \int_0^1 2\pi(2 - y)\left(\sqrt{y} - y\right)dy = 2\pi \int_0^1 \left(y^2 - y^{3/2} - 2y + 2y^{1/2}\right)dy$

$= 2\pi\left[\tfrac{1}{3}y^3 - \tfrac{2}{5}y^{5/2} - y^2 + \tfrac{4}{3}y^{3/2}\right]_0^1 = \tfrac{8\pi}{15}$

19. **(a)** Using the Midpoint Rule on $[0, 1]$ with $f(x) = \tan(x^2)$ and $n = 4$, we estimate

$A = \int_0^1 \tan(x^2)dx \approx \tfrac{1}{4}\left[\tan\left(\left(\tfrac{1}{8}\right)^2\right) + \tan\left(\left(\tfrac{3}{8}\right)^2\right) + \tan\left(\left(\tfrac{5}{8}\right)^2\right) + \tan\left(\left(\tfrac{7}{8}\right)^2\right)\right] \approx 0.38.$

(b) Using the Midpoint Rule on $[0, 1]$ with $f(x) = \pi\tan^2(x^2)$ (for disks) and $n = 4$, we estimate

$V = \int_0^1 f(x)dx \approx \tfrac{1}{4}\pi\left[\tan^2\left(\left(\tfrac{1}{8}\right)^2\right) + \tan^2\left(\left(\tfrac{3}{8}\right)^2\right) + \tan^2\left(\left(\tfrac{5}{8}\right)^2\right) + \tan^2\left(\left(\tfrac{7}{8}\right)^2\right)\right] \approx 0.87.$

21. The solid is obtained by rotating the region under the curve $y = \sin x$, above $y = 0$, from $x = 0$ to $x = \pi$, about the x-axis.

23. The solid is obtained by rotating the region in the first quadrant bounded by the curve $x = 4 - y^2$ and the coordinate axes about the x-axis.

25. Take the base to be the disk $x^2 + y^2 \le 9$. Then $V = \int_{-3}^{3} A(x)dx$, where $A(x_0)$ is the area of the isosceles right triangle whose hypotenuse lies along the line $x = x_0$ in the xy-plane. $A(x) = \tfrac{1}{4}\left(2\sqrt{9 - x^2}\right)^2 = 9 - x^2$, so $V = 2\int_0^3 A(x)dx = 2\int_0^3 (9 - x^2)dx = 2\left[9x - \tfrac{1}{3}x^3\right]_0^3 = 2(27 - 9) = 36.$

27. Equilateral triangles with sides measuring $\frac{1}{4}x$ meters have height $\frac{1}{4}x\sin 60^\circ = \frac{\sqrt{3}}{8}x$. Therefore,
$A(x) = \frac{1}{2}\cdot\frac{1}{4}x\cdot\frac{\sqrt{3}}{8}x = \frac{\sqrt{3}}{64}x^2$. $V = \int_0^{20} A(x)dx = \frac{\sqrt{3}}{64}\int_0^{20} x^2\,dx = \frac{\sqrt{3}}{64}\left[\frac{1}{3}x^3\right]_0^{20} = \frac{1000\sqrt{3}}{24} = \frac{125\sqrt{3}}{3}\text{ m}^3$

29. $30\text{ N} = f(x) = kx = k(0.03\text{ m})$, so $k = 30/0.03 = 1000\text{ N/m}$.
$W = \int_0^{0.08} kx\,dx = 1000\int_0^{0.08} x\,dx = 500[x^2]_0^{0.08} = 500(0.08)^2 = 3.2\text{ J}$.

31. **(a)** $W = \int_0^4 \pi\left(2\sqrt{y}\right)^2 62.5(4-y)dy = 250\pi\int_0^4 y(4-y)dy = 250\pi\left[2y^2 - \frac{1}{3}y^3\right]_0^4$
$= 250\pi\left(32 - \frac{64}{3}\right) = \frac{8000\pi}{3}$ ft-lb

(b) In part (a) we knew the final water level (0) but not the amount of work done. Here we use the same equation, except with the work fixed, and the lower limit of integration (that is, the final water level — call it h) unknown:

$W = 4000 = \int_h^4 \pi\left(2\sqrt{y}\right)^2 62.5(4-y)dy = 250\pi\int_h^4 y(4-y)dy$
$= 250\pi\left[2y^2 - \frac{1}{3}y^3\right]_h^4 = 250\pi\left[\left(2\cdot 16 - \frac{1}{3}\cdot 64\right) - \left(2h^2 - \frac{1}{3}h^3\right)\right]$
$\Leftrightarrow \quad h^3 - 6h^2 + 32 - \frac{48}{\pi} = 0$. We plot the graph of the function $f(h) = h^3 - 6h^2 + 32 - \frac{48}{\pi}$ on the interval $[0, 4]$ to see where it is 0.
From the graph, it appears that $f(h) = 0$ for $h \approx 2.1$.
So the depth of water remaining is about 2.1 ft.

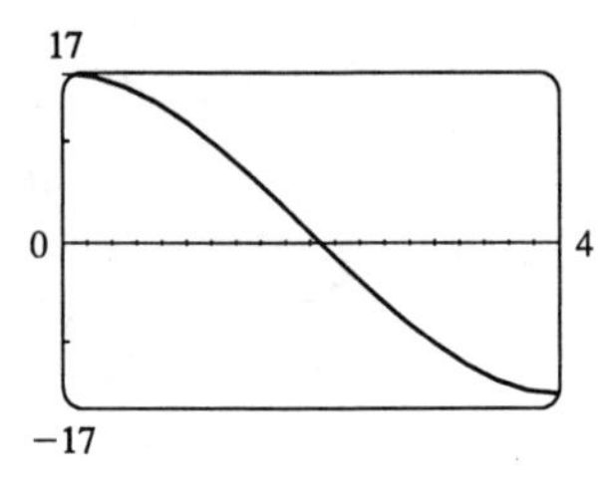

33. $\lim\limits_{h\to 0} f_{\text{ave}} = \lim\limits_{h\to 0}\frac{1}{h}\int_x^{x+h} f(t)dt = \lim\limits_{h\to 0}\frac{F(x+h)-F(x)}{h}$, where $F(x) = \int_a^x f(t)dt$.

But we recognize this limit as being $F'(x)$ by the definition of the derivative. Therefore $\lim\limits_{h\to 0} f_{\text{ave}} = F'(x) = f(x)$ by Part 1 of the Fundamental Theorem of Calculus.

PROBLEMS PLUS (after Chapter 7)

1. Let $y = f(x) = e^{-x^2}$. The area of the rectangle under the curve from $-x$ to x is $A(x) = 2xe^{-x^2}$ where $x \geq 0$. We maximize $A(x)$: $A'(x) = 2e^{-x^2} - 4x^2e^{-x^2} = 2e^{-x^2}(1 - 2x^2) = 0 \quad \Rightarrow \quad x = \frac{1}{\sqrt{2}}$. This gives a maximum since $A'(x) > 0$ for $0 \leq x < \frac{1}{\sqrt{2}}$ and $A'(x) < 0$ for $x > \frac{1}{\sqrt{2}}$. We next determine the points of inflection of $f(x)$. Now $f'(x) = -2xe^{-x^2} = -A(x)$. So $f''(x) = -A'(x)$. So $f''(x) < 0$ for $-\frac{1}{\sqrt{2}} < x < \frac{1}{\sqrt{2}}$ and $f''(x) > 0$ for $x < -\frac{1}{\sqrt{2}}$ and $x > \frac{1}{\sqrt{2}}$. So $f(x)$ changes concavity at $x = \pm\frac{1}{\sqrt{2}}$. So the two vertices of the rectangle of largest area are at the inflection points.

3. We use proof by contradiction. Suppose that $\log_2 5$ is a rational number. Then $\log_2 5 = m/n$ where m and n are positive integers $\quad \Rightarrow \quad 2^{m/n} = 5 \quad \Rightarrow \quad 2^m = 5^n$. But this is impossible since 2^m is even and 5^n is odd. So $\log_2 5$ is irrational.

5. First notice that if we can prove the simpler inequality $\dfrac{x^2+1}{x} \geq 2$ for $x > 0$, then the desired inequality follows because $\dfrac{(x^2+1)(y^2+1)(z^2+1)}{xyz} = \left(\dfrac{x^2+1}{x}\right)\left(\dfrac{y^2+1}{y}\right)\left(\dfrac{z^2+1}{z}\right) \geq 2 \cdot 2 \cdot 2 = 8$. So we let $f(x) = \dfrac{x^2+1}{x} = x + \dfrac{1}{x}$, $x > 0$. Then $f'(x) = 1 - \dfrac{1}{x^2} = 0$ if $x = 1$, and $f'(x) < 0$ for $0 < x < 1$, $f'(x) > 0$ for $x > 1$. Thus the absolute minimum value of $f(x)$ for $x > 0$ is $f(1) = 2$. Therefore $\dfrac{x^2+1}{x} \geq 2$ for all positive x. $\left(\text{Or, without calculus, } \dfrac{x^2+1}{x} \geq 2 \quad \Leftrightarrow \quad x^2 + 1 \geq 2x \quad \Leftrightarrow \quad x^2 - 2x + 1 \geq 0 \quad \Leftrightarrow \quad (x-1)^2 \geq 0\text{, which is true.}\right)$

7. Consider the statement that $\dfrac{d^n}{dx^n}(e^{ax}\sin bx) = r^n e^{ax}\sin(bx + n\theta)$. For $n = 1$,

$\dfrac{d}{dx}(e^{ax}\sin bx) = ae^{ax}\sin bx + be^{ax}\cos bx$, and

$$re^{ax}\sin(bx+\theta) = re^{ax}[\sin bx\cos\theta + \cos bx\sin\theta] = re^{ax}\left(\frac{a}{r}\sin bx + \frac{b}{r}\cos bx\right)$$
$$= ae^{ax}\sin bx + be^{ax}\cos bx, \text{ since } \tan\theta = b/a \quad \Rightarrow \quad \sin\theta = b/r \text{ and } \cos\theta = a/r.$$

So the statement is true for $n = 1$. Assume it is true for $n = k$. Then

$$\frac{d^{k+1}}{dx^{k+1}}(e^{ax}\sin bx) = \frac{d}{dx}\left[r^k e^{ax}\sin(bx+k\theta)\right] = r^k ae^{ax}\sin(bx+k\theta) + r^k e^{ax}b\cos(bx+k\theta)$$
$$= r^k e^{ax}[a\sin(bx+k\theta) + b\cos(bx+k\theta)]. \text{ But}$$
$$\sin[bx+(k+1)\theta] = \sin[(bx+k\theta)+\theta] = \sin(bx+k\theta)\cos\theta + \sin\theta\cos(bx+k\theta)$$
$$= \frac{a}{r}\sin(bx+k\theta) + \frac{b}{r}\cos(bx+k\theta).$$

Hence $a\sin(bx+k\theta) + b\cos(bx+k\theta) = r\sin[bx+(k+1)\theta]$. So

$$\frac{d^{k+1}}{dx^{k+1}}(e^{ax}\sin bx) = r^k e^{ax}[a\sin(bx+k\theta) + b\sin(bx+k\theta)] = r^k e^{ax}[r\sin(bx+(k+1)\theta)]$$
$$= r^{k+1}e^{ax}[\sin(bx+(k+1)\theta)].$$

Therefore the statement is true for all n by mathematical induction.

9. The volume generated from $x = 0$ to $x = b$ is $\int_0^b \pi[f(x)]^2\,dx$. Hence we are given that $b^2 = \int_0^b \pi[f(x)]^2\,dx$ for all $b > 0$. Differentiating both sides of this equation using the Fundamental Theorem of Calculus gives $2b = \pi[f(b)]^2 \quad\Rightarrow\quad f(b) = \sqrt{2b/\pi}$, since f is positive. Therefore $f(x) = \sqrt{2x/\pi}$.

11. Let the line through A and B have equation $y = mx + b$. Now $x^2 = mx + b$ gives $x^2 - mx - b = 0$ and hence $x = \dfrac{m \pm \sqrt{m^2 + 4b}}{2}$. So A has x-coordinate $x_1 = \dfrac{1}{2}\left(m - \sqrt{m^2+4b}\right)$ and B has x-coordinate $x_2 = \frac{1}{2}\left(m + \sqrt{m^2+4b}\right)$. So the parabolic segment has area

$$\begin{aligned}
\int_{x_1}^{x_2} [(mx+b) - x^2]dx &= \left[\tfrac{1}{2}mx^2 + bx - \tfrac{1}{3}x^3\right]_{x_1}^{x_2} = \tfrac{1}{2}m(x_2^2 - x_1^2) + b(x_2 - x_1) - \tfrac{1}{3}(x_2^3 - x_1^3)\\
&= (x_2 - x_1)\left[\tfrac{1}{2}m(x_2 + x_1) + b - \tfrac{1}{3}(x_2^2 + x_1x_2 + x_1^2)\right]\\
&= \sqrt{m^2+4b}\left(\tfrac{1}{2}m^2 + b - \tfrac{1}{3}(mx_2 + b - b + mx_1 + b)\right)\\
&= \sqrt{m^2+4b}\left[\tfrac{1}{2}m^2 + b - \tfrac{1}{3}(m^2 + b)\right]\\
&= \tfrac{1}{6}(m^2+4b)^{3/2}.
\end{aligned}$$

Now since the line through C has slope m, we see that if C has x-coordinate c, then $2c = m$ $\left[\text{since } (x^2)' = 2x\right]$ and hence $c = m/2$. So C has coordinates $\left(\frac{1}{2}m, \frac{1}{4}m^2\right)$. The line through AC has slope $\dfrac{\frac{1}{4}m^2 - x_1^2}{\frac{1}{2}m - x_1} = \dfrac{m}{2} + x_1$ and equation $y - x_1^2 = \left(\frac{1}{2}m + x_1\right)(x - x_1)$ or $y = \left(\frac{1}{2}m + x_1\right)x - \frac{1}{2}mx_1$. Similarly, the equation of the line through BC is $y = \left(\frac{1}{2}m + x_2\right)x - \frac{1}{2}mx_2$. So the area of the triangular region is

$$\begin{aligned}
&\int_{x_1}^{m/2}\left[(mx+b) - \left[\left(\tfrac{1}{2}m + x_1\right)x - \tfrac{1}{2}mx_1\right]\right]dx + \int_{m/2}^{x_2}\left[(mx+b) - \left[\left(\tfrac{1}{2}m + x_2\right)x - \tfrac{1}{2}mx_2\right]\right]dx\\
&\quad= \int_{x_1}^{m/2}\left[\left(\tfrac{1}{2}m - x_1\right)x + \left(b + \tfrac{1}{2}mx_1\right)\right]dx + \int_{m/2}^{x_2}\left[\left(\tfrac{1}{2}m - x_2\right)x + \left(b + \tfrac{1}{2}mx_2\right)\right]dx\\
&\quad= \left[\tfrac{1}{2}\left(\tfrac{1}{2}m - x_1\right)x^2 + \left(b + \tfrac{1}{2}mx_1\right)x\right]_{x_1}^{m/2} + \left[\tfrac{1}{2}\left(\tfrac{1}{2}m - x_2\right)x^2 + \left(b + \tfrac{1}{2}mx_2\right)x\right]_{m/2}^{x_2}\\
&\quad= \left[\tfrac{1}{16}m^3 - \tfrac{1}{8}m^2x_1 + \tfrac{1}{2}bm + \tfrac{1}{4}m^2x_1\right] - \left[\tfrac{1}{4}mx_1^2 - \tfrac{1}{2}x_1^3 + bx_1 + \tfrac{1}{2}mx_1^2\right]\\
&\qquad\qquad + \left[\tfrac{1}{4}mx_2^2 - \tfrac{1}{2}x_2^3 + bx_2 + \tfrac{1}{2}mx_2^2\right] - \left[\tfrac{1}{16}m^3 - \tfrac{1}{8}m^2x_2 + \tfrac{1}{2}bm + \tfrac{1}{4}m^2x_2\right]\\
&\quad= \left(b - \tfrac{1}{8}m^2\right)(x_2 - x_1) + \tfrac{3}{4}m(x_2^2 - x_1^2) - \tfrac{1}{2}(x_2^3 - x_1^3)\\
&\quad= (x_2 - x_1)\left[\left(b - \tfrac{1}{8}m^2\right) + \tfrac{3}{4}m(x_2 + x_1) - \tfrac{1}{2}(x_2^2 + x_2x_1 + x_1^2)\right]\\
&\quad= \sqrt{m^2+4b}\left[\left(b - \tfrac{1}{8}m^2\right) + \tfrac{3}{4}m^2 - \tfrac{1}{2}(m^2 + b)\right] = \tfrac{1}{8}\sqrt{m^2+4b}\,(m^2+4b) = \tfrac{1}{8}(m^2+4b)^{3/2}.
\end{aligned}$$

The result follows since $\frac{4}{3}\left[\frac{1}{8}(m^2+4b)^{3/2}\right] = \frac{1}{6}(m^2+4b)^{3/2}$, the area of the parabolic segment.

Alternate Solution: Let $A = (a, a^2)$, $B = (b, b^2)$. Then $m_{AB} = (b^2 - a^2)/(b - a) = a + b$, so the equation of AB is $y - a^2 = (a+b)(x-a)$, or $y = (a+b)x - ab$, and the area of the parabolic segment is $\int_a^b [(a+b)x - ab - x^2]dx = \left[(a+b)\frac{1}{2}x^2 - abx - \frac{1}{3}x^3\right]_a^b = \frac{1}{2}(a+b)(b^2 - a^2) - ab(b-a) - \frac{1}{3}(b^3 - a^3)$ $= \frac{1}{6}(b-a)^3$. At C, $y' = 2x = b + a$, so the x-coordinate of C is $\frac{1}{2}(a+b)$. If we calculate the area of triangle ABC as in Problems Plus #19 after Chapter 4 (by subtracting areas of trapezoids) we find that the area is $\frac{1}{2}(b-a)\left[\frac{1}{2}(a+b) - a\right]\left[b - \frac{1}{2}(a+b)\right] = \frac{1}{2}(b-a)\frac{1}{2}(b-a)\frac{1}{2}(b-a) = \frac{1}{8}(b-a)^3$. This is $\frac{3}{4}$ of the area of the parabolic segment calculated above.

13. By l'Hospital's Rule and the Fundamental Theorem, using the notation $\exp(y) = e^y$,

$$\lim_{x\to 0}\frac{\int_0^x (1-\tan 2t)^{1/t}\,dt}{x} \overset{\text{H}}{=} \lim_{x\to 0}\frac{(1-\tan 2x)^{1/x}}{1} = \exp\left[\lim_{x\to 0}\frac{\ln(1-\tan 2x)}{x}\right]$$

$$\overset{\text{H}}{=} \exp\left(\lim_{x\to 0}\frac{-2\sec^2 2x}{1-\tan 2x}\right) = \exp\left(\frac{-2\cdot 1^2}{1-0}\right) = e^{-2}.$$

15. We first show that $\dfrac{x}{1+x^2} < \tan^{-1}x$ for $x > 0$. Let $f(x) = \tan^{-1}x - \dfrac{x}{1+x^2}$. Then

$$f'(x) = \frac{1}{1+x^2} - \frac{1(1+x^2) - x(2x)}{(1+x^2)^2} = \frac{(1+x^2)-(1-x^2)}{(1+x^2)^2} = \frac{2x^2}{(1+x^2)^2} > 0 \text{ for } x > 0.$$ So $f(x)$ is increasing on $[0,\infty)$. Hence $0 < x \;\Rightarrow\; 0 = f(0) < f(x) = \tan^{-1}x - \dfrac{x}{1+x^2}$. So $\dfrac{x}{1+x^2} < \tan^{-1}x$ for $0 < x$. We next show that $\tan^{-1}x < x$ for $x > 0$. Let $h(x) = x - \tan^{-1}x$. Then

$$h'(x) = 1 - \frac{1}{1+x^2} = \frac{x^2}{1+x^2} > 0.$$ Hence $h(x)$ is increasing on $[0,\infty)$. So for $0 < x$, $0 = h(0) < h(x) = x - \tan^{-1}x$. Hence $\tan^{-1}x < x$ for $x > 0$.

17. By the Fundamental Theorem of Calculus, $f'(x) = \sqrt{1+x^3} > 0$ for $x > -1$. So f is increasing on $[-1,\infty)$ and hence is one-to-one. Note that $f(1) = 0$, so $f^{-1}(1) = 0 \;\Rightarrow\; (f^{-1})'(0) = 1/f'(1) = \frac{1}{\sqrt{2}}$.

19. Let $L = \lim\limits_{x\to\infty}\left(\dfrac{x+a}{x-a}\right)^x$, so $\ln L = \lim\limits_{x\to\infty}\ln\left(\dfrac{x+a}{x-a}\right)^x = \lim\limits_{x\to\infty} x\ln\left(\dfrac{x+a}{x-a}\right) = \lim\limits_{x\to\infty}\dfrac{\ln(x+a)-\ln(x-a)}{1/x}$

$$\overset{\text{H}}{=} \lim_{x\to\infty}\frac{\dfrac{1}{x+a}-\dfrac{1}{x-a}}{-1/x^2} = -\lim_{x\to\infty}\frac{(x-a)x^2-(x+a)x^2}{(x+a)(x-a)} = -\lim_{x\to\infty}\frac{-2ax^2}{x^2-a^2} = \lim_{x\to\infty}\frac{2a}{1-a^2/x^2} = 2a.$$ Hence $\ln L = 2a$, so $L = e^{2a}$. Hence $L = e^1 \;\Rightarrow\; 2a = 1 \;\Rightarrow\; a = \frac{1}{2}$.

21. Note that $\dfrac{d}{dx}\left(\int_0^x\left[\int_0^u f(t)dt\right]du\right) = \int_0^x f(t)dt$ by Part 1 of the Fundamental Theorem of Calculus, while

$$\frac{d}{dx}\left[\int_0^x f(u)(x-u)du\right] = \frac{d}{dx}\left[x\int_0^x f(u)du\right] - \frac{d}{dx}\left[\int_0^x f(u)u\,du\right] = \int_0^x f(u)du + xf(x) - f(x)x$$

$$= \int_0^x f(u)du.$$

Hence $\int_0^x f(u)(x-u)du = \int_0^x\left[\int_0^u f(t)dt\right]du + C$. Setting $x = 0$ gives $C = 0$.

23. $\lim\limits_{x\to\infty} x^c e^{-2x}\int_0^x e^{2t}\sqrt{t^2+1}\,dt = \lim\limits_{x\to\infty}\dfrac{\int_0^x e^{2t}\sqrt{t^2+1}\,dt}{x^{-c}e^{2x}}$. By l'Hospital's Rule, this is equal to

$$\lim_{x\to\infty}\frac{e^{2x}\sqrt{x^2+1}}{-cx^{-c-1}e^{2x}+2x^{-c}e^{2x}} = \lim_{x\to\infty}\frac{\sqrt{x^2+1}}{2x^{-c}-cx^{-c-1}} = \lim_{x\to\infty}\frac{x^{c+1}\sqrt{x^2+1}}{2x-c}.$$ This limit is finite only for $c \le -1$, and is zero for $c < -1$. If $c = -1$, the limit is $\lim\limits_{x\to\infty}\dfrac{\sqrt{x^2+1}}{2x+1} = \dfrac{1}{2}$.

25. If $f(x) = \int_0^x x^2\sin(t^2)dt = x^2\int_0^x \sin(t^2)dt$, then $f'(x) = 2x\int_0^x\sin(t^2)dt + x^2\sin(x^2)$ by the Product Rule and Part 1 of the Fundamental Theorem.

27. We must find expressions for the areas A and B, and then set them equal and see what this says about the curve C. If $P = (a, 2a^2)$, then area A is just $\int_0^a (2x^2 - x^2)dx = \int_0^a x^2\,dx = \frac{1}{3}a^3$. To find area B, we use y as the variable of integration. So we find the equation of the middle curve as a function of y: $y = 2x^2 \quad \Leftrightarrow \quad x = \sqrt{y/2}$, since we are concerned with the first quadrant only. We can express area B as $\int_0^{2a^2} \left[\sqrt{y/2} - C(y)\right] dy = \left[\frac{4}{3}(y/2)^{3/2}\right]_0^{2a^2} - \int_0^{2a^2} C(y)dy = \frac{4}{3}a^3 - \int_0^{2a^2} C(y)dy$, where $C(y)$ is the function with graph C. Setting $A = B$, we get $\frac{1}{3}a^3 = \frac{4}{3}a^3 - \int_0^{2a^2} C(y)dy \quad \Leftrightarrow \quad \int_0^{2a^2} C(y)dy = a^3$. Now we differentiate this equation with respect to a using the Chain Rule and the Fundamental Theorem: $C(2a^2)(4a) = 3a^2 \quad \Rightarrow \quad C(y) = \frac{3}{4}\sqrt{y/2}$, where $y = 2a^2$. Now we can solve for y: $x = \frac{3}{4}\sqrt{y/2} \;\Rightarrow\; x^2 = \frac{9}{16}(y/2) \;\Rightarrow\; y = \frac{32}{9}x^2$.

29. We split up the integral, and use (twice) the fact that $\int_p^q |f - g|dx \geq \int_p^q (f - g)dx$ for $q \geq p$:

$$\begin{aligned}\int_0^a |e^x - c|dx &= \int_0^{a/2} |c - e^x|dx + \int_{a/2}^a |e^x - c|dx \\ &\geq \int_0^{a/2} (c - e^x)dx + \int_{a/2}^a (e^x - c)dx = [cx - e^x]_0^{a/2} + [e^x - cx]_{a/2}^a \\ &= \left[\left(\tfrac{1}{2}ca - e^{a/2}\right) - (0 - 1)\right] + \left[(e^a - ca) - \left(e^{a/2} - \tfrac{1}{2}ca\right)\right] = e^a + 1 - 2e^{a/2} \\ &= \left(e^{a/2} - 1\right)^2.\end{aligned}$$

Another Method: Since $e^x - c \geq 0 \quad \Leftrightarrow \quad x \geq \ln c$, we can write $I(c) = \int_0^a |e^x - c|dx = \int_0^{\ln c} (c - e^x)dx + \int_{\ln c}^a (e^x - c)dx$. Evaluation gives $I(c) = 2c\ln c - 2c - ac + e^a + 1$. Then we can show that the minimum value of the function $I(c)$ is $\left(e^{a/2} - 1\right)^2$.

31. The volume is $\int_0^{\sqrt{2}} \pi r^2\,ds$, where s is measured along the line $y = x$ from the origin to P. From the figure we have $r^2 + s^2 = d^2 = x^2 + x^4$, and from the distance formula we have

$$\begin{aligned} r^2 &= \left(x - \tfrac{1}{\sqrt{2}}s\right)^2 + \left(x^2 - \tfrac{1}{\sqrt{2}}s\right)^2 \\ &= x^2 + x^4 + s^2 - \sqrt{2}s(x + x^2)\end{aligned}$$

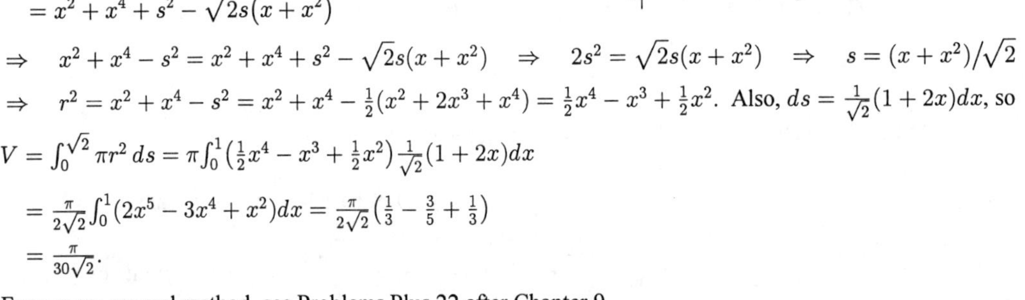

$\Rightarrow \quad x^2 + x^4 - s^2 = x^2 + x^4 + s^2 - \sqrt{2}s(x + x^2) \quad \Rightarrow \quad 2s^2 = \sqrt{2}s(x + x^2) \quad \Rightarrow \quad s = (x + x^2)/\sqrt{2}$

$\Rightarrow \quad r^2 = x^2 + x^4 - s^2 = x^2 + x^4 - \frac{1}{2}(x^2 + 2x^3 + x^4) = \frac{1}{2}x^4 - x^3 + \frac{1}{2}x^2$. Also, $ds = \frac{1}{\sqrt{2}}(1 + 2x)dx$, so

$$\begin{aligned} V &= \int_0^{\sqrt{2}} \pi r^2\,ds = \pi \int_0^1 \left(\tfrac{1}{2}x^4 - x^3 + \tfrac{1}{2}x^2\right) \tfrac{1}{\sqrt{2}}(1 + 2x)dx \\ &= \tfrac{\pi}{2\sqrt{2}} \int_0^1 (2x^5 - 3x^4 + x^2)dx = \tfrac{\pi}{2\sqrt{2}}\left(\tfrac{1}{3} - \tfrac{3}{5} + \tfrac{1}{3}\right) \\ &= \tfrac{\pi}{30\sqrt{2}}.\end{aligned}$$

For a more general method, see Problems Plus 22 after Chapter 9.

CHAPTER EIGHT

EXERCISES 8.1

1. Let $u = x$, $dv = e^{2x}\,dx \quad \Rightarrow \quad du = dx$, $v = \frac{1}{2}e^{2x}$. Then by Equation 2,
$\int xe^{2x}\,dx = \frac{1}{2}xe^{2x} - \int \frac{1}{2}e^{2x}\,dx = \frac{1}{2}xe^{2x} - \frac{1}{4}e^{2x} + C$.

3. Let $u = x$, $dv = \sin 4x\,dx \quad \Rightarrow \quad du = dx$, $v = -\frac{1}{4}\cos 4x$. Then
$\int x\sin 4x\,dx = -\frac{1}{4}x\cos 4x - \int\left(-\frac{1}{4}\cos 4x\right)dx = -\frac{1}{4}x\cos 4x + \frac{1}{16}\sin 4x + C$.

5. Let $u = x^2$, $dv = \cos 3x\,dx \quad \Rightarrow \quad du = 2x\,dx$, $v = \frac{1}{3}\sin 3x$. Then
$I = \int x^2\cos 3x\,dx = \frac{1}{3}x^2\sin 3x - \frac{2}{3}\int x\sin 3x\,dx$ by Equation 2. Next let $U = x$, $dV = \sin 3x\,dx \quad \Rightarrow$
$dU = dx$, $V = -\frac{1}{3}\cos 3x$ to get $\int x\sin 3x\,dx = -\frac{1}{3}x\cos 3x + \frac{1}{3}\int\cos 3x\,dx = -\frac{1}{3}x\cos 3x + \frac{1}{9}\sin 3x + C_1$.
Substituting for $\int x\sin 3x\,dx$, we get $I = \frac{1}{3}x^2\sin 3x - \frac{2}{3}\left(-\frac{1}{3}x\cos 3x + \frac{1}{9}\sin 3x + C_1\right)$
$= \frac{1}{3}x^2\sin 3x + \frac{2}{9}x\cos 3x - \frac{2}{27}\sin 3x + C$, where $C = -\frac{2}{3}C_1$.

7. Let $u = (\ln x)^2$, $dv = dx \quad \Rightarrow \quad du = 2\ln x \cdot \frac{1}{x}\,dx$, $v = x$. Then $I = \int(\ln x)^2 dx = x(\ln x)^2 - 2\int\ln x\,dx$.
Taking $U = \ln x$, $dV = dx \quad \Rightarrow \quad dU = 1/x\,dx$, $V = x$, we find that
$\int\ln x\,dx = x\ln x - \displaystyle\int x\cdot\frac{1}{x}\,dx = x\ln x - x + C_1$. Thus $I = x(\ln x)^2 - 2x\ln x + 2x + C$, where $C = -2C_1$.

9. $I = \int\theta\sin\theta\cos\theta\,d\theta = \frac{1}{4}\int 2\theta\sin 2\theta\,d\theta = \frac{1}{8}\int t\sin t\,dt \quad$ (Put $t = 2\theta \quad \Rightarrow \quad dt = 2\,d\theta$.)
Let $u = t$, $dv = \sin t\,dt \quad \Rightarrow \quad du = dt$, $v = -\cos t$. Then
$I = \frac{1}{8}\left(-t\cos t + \int\cos t\,dt\right) = \frac{1}{8}(-t\cos t + \sin t) + C = \frac{1}{8}(\sin 2\theta - 2\theta\cos 2\theta) + C$.

11. Let $u = \ln t$, $dv = t^2\,dt \quad \Rightarrow \quad du = dt/t$, $v = \frac{1}{3}t^3$. Then
$\int t^2\ln t\,dt = \frac{1}{3}t^3\ln t - \int\frac{1}{3}t^3(1/t)\,dt = \frac{1}{3}t^3\ln t - \frac{1}{9}t^3 + C = \frac{1}{9}t^3(3\ln t - 1) + C$.

13. First let $u = \sin 3\theta$, $dv = e^{2\theta}d\theta \quad \Rightarrow \quad du = 3\cos 3\theta\,d\theta$, $v = \frac{1}{2}e^{2\theta}$. Then
$I = \int e^{2\theta}\sin 3\theta\,d\theta = \frac{1}{2}e^{2\theta}\sin 3\theta - \frac{3}{2}\int e^{2\theta}\cos 3\theta\,d\theta$. Next let $U = \cos 3\theta$, $dU = -3\sin 3\theta\,d\theta$, $dV = e^{2\theta}d\theta$,
$V = \frac{1}{2}e^{2\theta}$ to get $\int e^{2\theta}\cos 3\theta\,d\theta = \frac{1}{2}e^{2\theta}\cos 3\theta + \frac{3}{2}\int e^{2\theta}\sin 3\theta\,d\theta$. Substituting in the previous formula gives
$I = \frac{1}{2}e^{2\theta}\sin 3\theta - \frac{3}{4}e^{2\theta}\cos 3\theta - \frac{9}{4}\int e^{2\theta}\sin 3\theta\,d\theta$ or $\frac{13}{4}\int e^{2\theta}\sin 3\theta\,d\theta = \frac{1}{2}e^{2\theta}\sin 3\theta - \frac{3}{4}e^{2\theta}\cos 3\theta + C_1$. Hence
$\int e^{2\theta}\sin 3\theta\,d\theta = \frac{1}{13}e^{2\theta}(2\sin 3\theta - 3\cos 3\theta) + C$, where $C = \frac{4}{13}C_1$.

15. Let $u = y$, $dv = \sinh y\,dy \quad \Rightarrow \quad du = dy$, $v = \cosh y$. Then
$\int y\sinh y\,dy = y\cosh y - \int\cosh y\,dy = y\cosh y - \sinh y + C$.

17. Let $u = t$, $dv = e^{-t}\,dt \quad \Rightarrow \quad du = dt$, $v = -e^{-t}$. Then Formula 6 says $\int_0^1 te^{-t}\,dt = \left[-te^{-t}\right]_0^1 + \int_0^1$
$e^{-t}\,dt = -1/e + \left[-e^{-t}\right]_0^1 = -1/e - 1/e + 1 = 1 - 2/e$.

19. Let $u = x$, $dv = \cos 2x\,dx \quad \Rightarrow \quad du = dx$, $v = \frac{1}{2}\sin 2x\,dx$. Then
$\int_0^{\pi/2} x\cos 2x\,dx = \left[\frac{1}{2}x\sin 2x\right]_0^{\pi/2} - \frac{1}{2}\int_0^{\pi/2}\sin 2x\,dx = 0 + \left[\frac{1}{4}\cos 2x\right]_0^{\pi/2} = \frac{1}{4}(-1 - 1) = -\frac{1}{2}$.

21. Let $u = \cos^{-1}x$, $dv = dx \quad \Rightarrow \quad du = -\dfrac{dx}{\sqrt{1-x^2}}$, $v = x$. Then

$I = \displaystyle\int_0^{1/2} \cos^{-1}x\,dx = \left[x\cos^{-1}x\right]_0^{1/2} + \int_0^{1/2} \frac{x\,dx}{\sqrt{1-x^2}} = \tfrac{1}{2}\cdot\tfrac{\pi}{3} + \int_1^{3/4} t^{-1/2}\left[-\tfrac{1}{2}\,dt\right]$, where $t = 1 - x^2 \quad \Rightarrow$

$dt = -2x\,dx$. Thus $I = \frac{\pi}{6} + \frac{1}{2}\int_{3/4}^{1} t^{-1/2}\,dt = \left[\sqrt{t}\right]_{3/4}^{1} = \frac{\pi}{6} + 1 - \frac{\sqrt{3}}{2} = \frac{1}{6}\left(\pi + 6 - 3\sqrt{3}\right)$.

23. Let $u = \ln(\sin x)$, $dv = \cos x\,dx \quad \Rightarrow \quad du = \dfrac{\cos x}{\sin x}\,dx$, $v = \sin x$. Then

$I = \int \cos x \ln(\sin x)dx = \sin x\ln(\sin x) - \int \cos x\,dx = \sin x\ln(\sin x) - \sin x + C$.

Another Method: Substitute $t = \sin x$, so $dt = \cos x\,dx$. Then $I = \int \ln t\,dt = t\ln t - t + C$ (see Example 2) and so $I = \sin x(\ln\sin x - 1) + C$.

25. Let $u = 2x + 3$, $dv = e^x\,dx \quad \Rightarrow \quad du = 2\,dx$, $v = e^x$. Then

$\int (2x+3)e^x\,dx = (2x+3)e^x - \int e^x \cdot 2\,dx = (2x+3)e^x - 2e^x + C = (2x+1)e^x + C$.

27. Let $w = \ln x \quad \Rightarrow \quad dw = dx/x$. Then $x = e^w$ and $dx = e^w\,dw$, so

$\int \cos(\ln x)dx = \int e^w\cos w\,dw = \frac{1}{2}e^w(\sin w + \cos w) + C$ (by the method of Example 4)

$= \frac{1}{2}x[\sin(\ln x) + \cos(\ln x)] + C$.

29. $I = \int_1^4 \ln\sqrt{x}\,dx = \frac{1}{2}\int_1^4 \ln x\,dx = \frac{1}{2}[x\ln x - x]_1^4$ as in Example 2. So

$I = \frac{1}{2}[(4\ln 4 - 4) - (0 - 1)] = 4\ln 2 - \frac{3}{2}$.

31. Let $w = \sqrt{x}$, so that $x = w^2$ and $dx = 2w\,dw$. Then use $u = 2w$, $dv = \sin w\,dw$. Thus

$\int \sin\sqrt{x}\,dx = \int 2w\sin w\,dw = -2w\cos w + \int 2\cos w\,dw = -2w\cos w + 2\sin w + C$

$= -2\sqrt{x}\cos\sqrt{x} + 2\sin\sqrt{x} + C$.

33. $\int x^5 e^{x^2}\,dx = \int (x^2)^2 e^{x^2} x\,dx = \int t^2 e^t\,\frac{1}{2}\,dt$ (where $t = x^2 \Rightarrow \frac{1}{2}\,dt = x\,dx$)

$= \frac{1}{2}(t^2 - 2t + 2)e^t + C$ (by Example 3) $= \frac{1}{2}(x^4 - 2x^2 + 2)e^{x^2} + C$.

35. Let $u = x$, $dv = \cos\pi x\,dx \quad \Rightarrow \quad du = dx$,

$v = \displaystyle\int \cos\pi x\,dx = \frac{\sin\pi x}{\pi}$. Thus

$\displaystyle\int x\cos\pi x\,dx = x\cdot\frac{\sin\pi x}{\pi} - \int \frac{\sin\pi x}{\pi}\,dx = \frac{x\sin\pi x}{\pi} + \frac{\cos\pi x}{\pi^2} + C$.

We see from the graph that this is reasonable, since the antiderivative has extrema where the original function is 0.

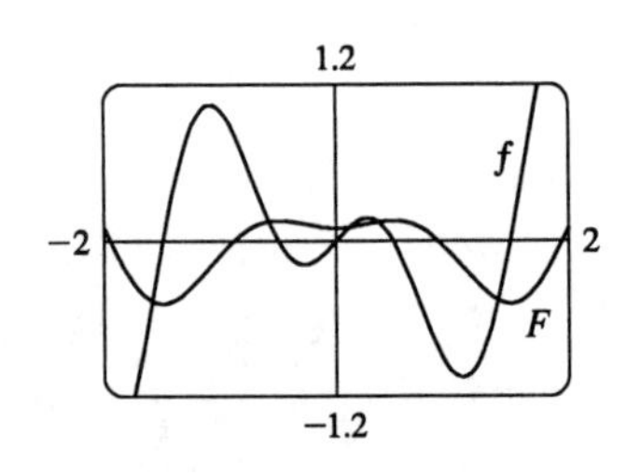

37. **(a)** Take $n = 2$ in Example 6 to get $\int \sin^2 x\,dx = -\frac{1}{2}\cos x\sin x + \frac{1}{2}\int 1\,dx = \dfrac{x}{2} - \dfrac{\sin 2x}{4} + C$.

(b) $\int \sin^4 x\,dx = -\frac{1}{4}\cos x\sin^3 x + \frac{3}{4}\int \sin^2 x\,dx = -\frac{1}{4}\cos x\sin^3 x + \frac{3}{8}x - \frac{3}{16}\sin 2x + C$.

39. **(a)** $\displaystyle\int_0^{\pi/2} \sin^n x\,dx = \left[-\frac{\cos x\sin^{n-1}x}{n}\right]_0^{\pi/2} + \frac{n-1}{n}\int_0^{\pi/2} \sin^{n-2}x\,dx = \frac{n-1}{n}\int_0^{\pi/2} \sin^{n-2}x\,dx$

(b) $\int_0^{\pi/2} \sin^3 x\,dx = \frac{2}{3}\int_0^{\pi/2} \sin x\,dx = \left[-\frac{2}{3}\cos x\right]_0^{\pi/2} = \frac{2}{3}$; $\int_0^{\pi/2} \sin^5 x\,dx = \frac{4}{5}\int_0^{\pi/2} \sin^3 x\,dx = \frac{4}{5}\cdot\frac{2}{3} = \frac{8}{15}$

(c) The formula holds for $n = 1$ (that is, $2n + 1 = 3$) by (b). Assume it holds for some $k \geq 1$. Then
$\int_0^{\pi/2} \sin^{2k+1} x\,dx = \dfrac{2 \cdot 4 \cdot 6 \cdot \cdots \cdot (2k)}{3 \cdot 5 \cdot 7 \cdot \cdots \cdot (2k+1)}$.
By part (a), $\int_0^{\pi/2} \sin^{2k+3} x\,dx = \dfrac{2k+2}{2k+3}\int_0^{\pi/2} \sin^{2k+1} x\,dx = \dfrac{2 \cdot 4 \cdot 6 \cdot \cdots \cdot [2(k+1)]}{3 \cdot 5 \cdot 7 \cdot \cdots \cdot [2(k+1)+1]}$ as desired.
By induction, the formula holds for all $n \geq 1$.

41. Let $u = (\ln x)^n$, $dv = dx \quad \Rightarrow \quad du = n(\ln x)^{n-1}(dx/x)$, $v = x$. Then
$\int (\ln x)^n\,dx = x(\ln x)^n - n\int (\ln x)^{n-1}\,dx$, by Equation 2.

43. Let $u = (x^2 + a^2)^n$, $dv = dx \quad \Rightarrow \quad du = n(x^2 + a^2)^{n-1}2x\,dx$, $v = x$. Then

$$\begin{aligned}\int (x^2+a^2)^n\,dx &= x(x^2+a^2)^n - 2n\int x^2(x^2+a^2)^{n-1}\,dx \\ &= x(x^2+a^2)^n - 2n\left[\int (x^2+a^2)^n\,dx - a^2\int (x^2+a^2)^{n-1}\,dx\right] \quad [\text{since } x^2 = (x^2+a^2) - a^2]\end{aligned}$$

$\Rightarrow \quad (2n+1)\int (x^2+a^2)^n\,dx = x(x^2+a^2)^n + 2na^2\int (x^2+a^2)^{n-1}\,dx$, and

$$\int (x^2+a^2)^n\,dx = \frac{x(x^2+a^2)^n}{2n+1} + \frac{2na^2}{2n+1}\int (x^2+a^2)^{n-1}\,dx \text{ (provided } 2n+1 \neq 0).$$

45. Take $n = 3$ in Exercise 41 to get
$\int (\ln x)^3\,dx = x(\ln x)^3 - 3\int (\ln x)^2\,dx = x(\ln x)^3 - 3x(\ln x)^2 + 6x\ln x - 6x + C$ (by Exercise 7).

47. Let $u = \sin^{-1}x$, $dv = dx \quad \Rightarrow \quad du = \dfrac{dx}{\sqrt{1-x^2}}$, $v = x$. Then

$$\begin{aligned}\text{area} &= \int_0^{1/2} \sin^{-1}x\,dx = \left[x\sin^{-1}x\right]_0^{1/2} - \int_0^{1/2} \frac{x}{\sqrt{1-x^2}}\,dx = \tfrac{1}{2}\left(\tfrac{\pi}{6}\right) + \left[\sqrt{1-x^2}\right]_0^{1/2} \\ &= \tfrac{\pi}{12} + \tfrac{\sqrt{3}}{2} - 1 = \tfrac{1}{12}\left(\pi + 6\sqrt{3} - 12\right).\end{aligned}$$

49.

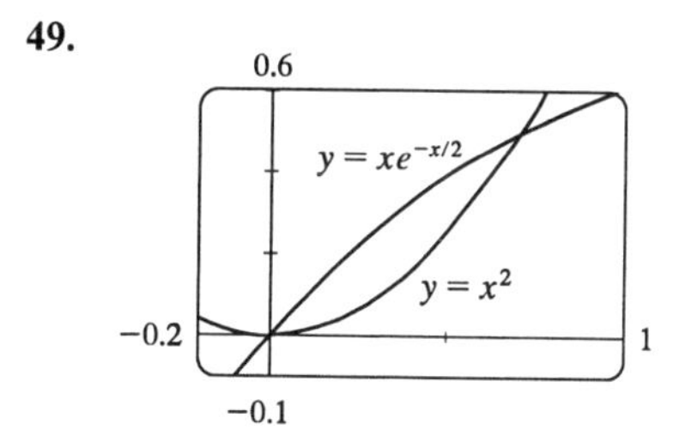

From the graph, we see that the curves intersect at approximately $x = 0$ and $x = 0.70$, with $xe^{-x/2} > x^2$ on $(0, 0.70)$. So the area bounded by the curves is approximately $A = \int_0^{0.70} \left(xe^{-x/2} - x^2\right)dx$. We separate this into two integrals, and evaluate the first one by parts with $u = x$, $dv = e^{-x/2}\,dx \quad \Rightarrow \quad du = dx$, $v = -2e^{-x/2}$:

$$\begin{aligned}A &= \left[-2xe^{-x/2}\right]_0^{0.70} - \int_0^{0.70}\left(-2e^{-x/2}\right)dx - \left[\tfrac{1}{3}x^3\right]_0^{0.70} = \left[-2(0.70)e^{-0.35} - 0\right] - \left[4e^{-x/2}\right]_0^{0.70} - \tfrac{1}{3}\left[0.70^3 - 0\right] \\ &\approx 0.080.\end{aligned}$$

51. Volume $= \int_{2\pi}^{3\pi} 2\pi x\sin x\,dx$. Let $u = x$, $dv = \sin x\,dx \quad \Rightarrow \quad du = dx$, $v = -\cos x \quad \Rightarrow$
$V = 2\pi[-x\cos x + \sin x]_{2\pi}^{3\pi} = 2\pi[(3\pi + 0) - (-2\pi + 0)] = 2\pi(5\pi) = 10\pi^2$.

53. Volume $= \int_{-1}^{0} 2\pi(1-x)e^{-x}\,dx$. Let $u = 1 - x$, $dv = e^{-x}\,dx \quad \Rightarrow \quad du = -\,dx$, $v = -e^{-x} \quad \Rightarrow$
$V = 2\pi\left[xe^{-x}\right]_{-1}^{0} = 2\pi(0 + e) = 2\pi e$

55. Since $v(t) > 0$ for all t, the desired distance $s(t) = \int_0^t v(w)dw = \int_0^t w^2 e^{-w}\,dw$. Let $u = w^2$, $dv = e^{-w}\,dw$ $\Rightarrow$ $du = 2w\,dw$, $v = -e^{-w}$. Then $s(t) = \left[-w^2 e^{-w}\right]_0^t + 2\int_0^t we^{-w}\,dw$.

Now let $U = w$, $dV = e^{-w}\,dw$ $\Rightarrow$ $dU = dw$, $V = -e^{-w}$. Then
$s(t) = -t^2e^{-t} + 2\left(\left[-we^{-w}\right]_0^t + \int_0^t e^{-w}\,dw\right) = -t^2e^{-t} - 2te^{-t} - 2e^{-t} + 2 = 2 - e^{-t}(t^2 + 2t + 2)$ meters.

57. Take $g(x) = x$ in Equation 1.

59. By Exercise 58, $\int_1^e \ln x\,dx = e\ln e - 1\ln 1 - \int_{\ln 1}^{\ln e} e^y\,dy = e - \int_0^1 e^y\,dy = e - \left[e^y\right]_0^1 = e - (e-1) = 1$.

61. Using the formula for volumes of rotation (Equation 7.2.3) and the figure, we see that
Volume $= \int_0^d \pi b^2\,dy - \int_0^c \pi a^2\,dy - \int_c^d \pi[g(y)]^2\,dy = \pi b^2 d - \pi a^2 c - \int_c^d \pi[g(y)]^2\,dy$.
Let $y = f(x)$, which gives $dy = f'(x)dx$ and $g(y) = x$, so that $V = \pi b^2 d - \pi a^2 c - \pi\int_a^b x^2 f'(x)dx$. Now integrate by parts with $u = x^2$, and $dv = f'(x)dx$ $\Rightarrow$ $du = 2x\,dx$, $v = f(x)$, and
$\int_a^b x^2 f'(x)dx = \left[x^2 f(x)\right]_a^b - \int_a^b 2x\,f(x)dx = b^2 f(b) - a^2 f(a) - \int_a^b 2x\,f(x)dx$, but $f(a) = c$ and $f(b) = d$
$\Rightarrow$ $V = \pi b^2 d - \pi a^2 c - \pi\left[b^2 d - a^2 c - \int_a^b 2x f(x)dx\right] = \int_a^b 2\pi x f(x)dx$.

EXERCISES 8.2

1. $\int_0^{\pi/2} \sin^2 3x\,dx = \int_0^{\pi/2} \frac{1}{2}(1 - \cos 6x)dx = \left[\frac{1}{2}x - \frac{1}{12}\sin 6x\right]_0^{\pi/2} = \frac{\pi}{4}$

3. $\int \cos^4 x\,dx = \int\left[\frac{1}{2}(1 + \cos 2x)\right]^2 dx = \frac{1}{4}\int(1 + 2\cos 2x + \cos^2 2x)dx$
$= \frac{1}{4}x + \frac{1}{4}\sin 2x + \frac{1}{4}\int\frac{1}{2}(1 + \cos 4x)dx = \frac{1}{4}\left[x + \sin 2x + \frac{1}{2}x + \frac{1}{8}\sin 4x\right] + C$
$= \frac{3}{8}x + \frac{1}{4}\sin 2x + \frac{1}{32}\sin 4x + C$

5. Let $u = \cos x$ $\Rightarrow$ $du = -\sin x\,dx$. Then $\int \sin^3 x\cos^4 x\,dx = \int \cos^4 x(1 - \cos^2 x)\sin x\,dx$
$= \int u^4(1 - u^2)(-du) = \int(u^6 - u^4)du = \frac{1}{7}u^7 - \frac{1}{5}u^5 + C = \frac{1}{7}\cos^7 x - \frac{1}{5}\cos^5 x + C$.

7. $\int_0^{\pi/4} \sin^4 x\cos^2 x\,dx = \int_0^{\pi/4} \sin^2 x(\sin x\cos x)^2\,dx = \int_0^{\pi/4} \frac{1}{2}(1 - \cos 2x)\left(\frac{1}{2}\sin 2x\right)^2 dx$
$= \frac{1}{8}\int_0^{\pi/4}(1 - \cos 2x)\sin^2 2x\,dx = \frac{1}{8}\int_0^{\pi/4} \sin^2 2x\,dx - \frac{1}{8}\int_0^{\pi/4} \sin^2 2x\cos 2x\,dx$
$= \frac{1}{16}\int_0^{\pi/4}(1 - \cos 4x)dx - \frac{1}{16}\left[\frac{1}{3}\sin^3 2x\right]_0^{\pi/4} = \frac{1}{16}\left[x - \frac{1}{4}\sin 4x - \frac{1}{3}\sin^3 2x\right]_0^{\pi/4}$
$= \frac{1}{16}\left(\frac{\pi}{4} - 0 - \frac{1}{3}\right) = \frac{1}{192}(3\pi - 4)$

9. $\int(1 - \sin 2x)^2\,dx = \int(1 - 2\sin 2x + \sin^2 2x)dx = \int\left[1 - 2\sin 2x + \frac{1}{2}(1 - \cos 4x)\right]dx$
$= \int\left[\frac{3}{2} - 2\sin 2x - \frac{1}{2}\cos 4x\right]dx = \frac{3}{2}x + \cos 2x - \frac{1}{8}\sin 4x + C$

11. Let $u = \sin x \quad \Rightarrow \quad du = \cos x\,dx$. Then

$$\int \cos^5 x \sin^5 x\,dx = \int u^5(1-u^2)^2\,du = \int u^5(1-2u^2+u^4)du = \int(u^5 - 2u^7 + u^9)du$$
$$= \tfrac{1}{10}u^{10} - \tfrac{1}{4}u^8 + \tfrac{1}{6}u^6 + C = \tfrac{1}{10}\sin^{10}x - \tfrac{1}{4}\sin^8 x + \tfrac{1}{6}\sin^6 x + C.$$

Or: Let $v = \cos x$, $dv = -\sin x\,dx$. Then

$$\int \cos^5 x \sin^5 x\,dx = \int v^5(1-v^2)^2(-dv) = \int(-v^5 + 2v^7 - v^9)dv$$
$$= -\tfrac{1}{10}v^{10} + \tfrac{1}{4}v^8 - \tfrac{1}{6}v^6 + C = -\tfrac{1}{10}\cos^{10}x + \tfrac{1}{4}\cos^8 x - \tfrac{1}{6}\cos^6 x + C.$$

13. Let $u = \cos x$, $du = -\sin x\,dx$. Then $\int \sin^3 x\,\sqrt{\cos x}\,dx = \int(1-\cos^2 x)\sqrt{\cos x}\,\sin x\,dx$
$= \int(1-u^2)u^{1/2}(-du) = \int(u^{5/2} - u^{1/2})du = \tfrac{2}{7}u^{7/2} - \tfrac{2}{3}u^{3/2} + C$
$= \tfrac{2}{7}(\cos x)^{7/2} - \tfrac{2}{3}(\cos x)^{3/2} + C = \left[\tfrac{2}{7}\cos^3 x - \tfrac{2}{3}\cos x\right]\sqrt{\cos x} + C.$

15. Let $u = \cos x \quad \Rightarrow \quad du = -\sin x\,dx$. Then $\displaystyle\int \cos^2 x \tan^3 x\,dx = \int \frac{\sin^3 x}{\cos x}\,dx$
$\displaystyle = \int \frac{(1-u^2)(-du)}{u} = \int\left[\frac{-1}{u} + u\right]du = -\ln|u| + \tfrac{1}{2}u^2 + C = \tfrac{1}{2}\cos^2 x - \ln|\cos x| + C.$

17. $\displaystyle\int \frac{1-\sin x}{\cos x}\,dx = \int(\sec x - \tan x)dx = \ln|\sec x + \tan x| - \ln|\sec x| + C$ (by Example 8)
$= \ln|(\sec x + \tan x)\cos x| + C = \ln|1 + \sin x| + C = \ln(1 + \sin x) + C,$

since $1 + \sin x \geq 0$.

Or: $\displaystyle\int \frac{1-\sin x}{\cos x}\,dx = \int \frac{1-\sin x}{\cos x}\cdot\frac{1+\sin x}{1+\sin x}\,dx = \int \frac{(1-\sin^2 x)dx}{\cos x(1+\sin x)} = \int \frac{\cos x\,dx}{1+\sin x} = \int \frac{dw}{w}$
(where $w = 1 + \sin x$, $dw = \cos x\,dx$) $= \ln|w| + C = \ln|1+\sin x| + C = \ln(1+\sin x) + C.$

19. $\int \tan^2 x\,dx = \int(\sec^2 x - 1)dx = \tan x - x + C.$

21. $\int \sec^4 x\,dx = \int(\tan^2 x + 1)\sec^2 x\,dx = \int \tan^2 x \sec^2 x\,dx + \int \sec^2 x\,dx = \tfrac{1}{3}\tan^3 x + \tan x + C$

23. Let $u = \tan x \quad \Rightarrow \quad du = \sec^2 x\,dx$. Then $\int_0^{\pi/4}\tan^4 x \sec^2 x\,dx = \int_0^1 u^4\,du = \left[\tfrac{1}{5}u^5\right]_0^1 = \tfrac{1}{5}.$

25. Let $u = \sec x \quad \Rightarrow \quad du = \sec x \tan x\,dx$. Then
$\int \tan x \sec^3 x\,dx = \int \sec^2 x \sec x \tan x\,dx = \int u^2\,du = \tfrac{1}{3}u^3 + C = \tfrac{1}{3}\sec^3 x + C.$

27. $\int \tan^5 x\,dx = \int(\sec^2 x - 1)^2 \tan x\,dx = \int \sec^4 x \tan x\,dx - 2\int \sec^2 x \tan x\,dx + \int \tan x\,dx$
$= \int \sec^3 x \sec x \tan x\,dx - 2\int \tan x \sec^2 x\,dx + \int \tan x\,dx$
$= \tfrac{1}{4}\sec^4 x - \tan^2 x + \ln|\sec x| + C.$ *Or:* $\tfrac{1}{4}\sec^4 x - \sec^2 x + \ln|\sec x| + C.$

29. Let $u = \sec x \quad \Rightarrow \quad du = \sec x \tan x\,dx$. Then
$\int_0^{\pi/3}\tan^5 x \sec x\,dx = \int_0^{\pi/3}(\sec^2 x - 1)^2 \sec x \tan x\,dx = \int_1^2 (u^2-1)^2\,du$
$= \int_1^2(u^4 - 2u^2 + 1)du = \left[\tfrac{1}{5}u^5 - \tfrac{2}{3}u^3 + u\right]_1^2 = \left[\tfrac{32}{5} - \tfrac{16}{3} + 2\right] - \left[\tfrac{1}{5} - \tfrac{2}{3} + 1\right] = \tfrac{38}{15}.$

31. Let $u = \tan x \quad \Rightarrow \quad du = \sec^2 x\,dx$. Then
$\displaystyle\int \frac{\sec^2 x}{\cot x}\,dx = \int \tan x \sec^2 x\,dx = \int u\,du = \tfrac{1}{2}u^2 + C = \tfrac{1}{2}\tan^2 x + C.$

33. $\int_{\pi/6}^{\pi/2} \cot^2 x\,dx = \int_{\pi/6}^{\pi/2} (\csc^2 x - 1)dx = [-\cot x - x]_{\pi/6}^{\pi/2} = \left(0 - \frac{\pi}{2}\right) - \left(-\sqrt{3} - \frac{\pi}{6}\right) = \sqrt{3} - \frac{\pi}{3}$

35. Let $u = \cot x \Rightarrow du = -\csc^2 x\,dx$. Then

$\int \cot^4 x \csc^4 x\,dx = \int u^4(u^2+1)(-du) = -\int (u^6 + u^4)du = -\frac{1}{7}u^7 - \frac{1}{5}u^5 + C = -\frac{1}{7}\cot^7 x - \frac{1}{5}\cot^5 x + C.$

37. $I = \displaystyle\int \csc x\,dx = \int \frac{\csc x(\csc x - \cot x)}{\csc x - \cot x}dx = \int \frac{-\csc x \cot x + \csc^2 x}{\csc x - \cot x}dx$. Let $u = \csc x - \cot x \Rightarrow$

$du = (-\csc x \cot x + \csc^2 x)dx$. Then $I = \int du/u = \ln|u| = \ln|\csc x - \cot x| + C$.

39. $\int \sin 5x \sin 2x\,dx = \int \frac{1}{2}[\cos(5x - 2x) - \cos(5x + 2x)]dx = \frac{1}{2}\int (\cos 3x - \cos 7x)dx$

$= \frac{1}{6}\sin 3x - \frac{1}{14}\sin 7x + C$

41. $\int \cos 3x \cos 4x\,dx = \int \frac{1}{2}[\cos(3x - 4x) + \cos(3x + 4x)]dx = \frac{1}{2}\int(\cos x + \cos 7x)dx = \frac{1}{2}\sin x + \frac{1}{14}\sin 7x + C$

43. $\displaystyle\int \frac{1 - \tan^2 x}{\sec^2 x}dx = \int \left(\cos^2 x - \sin^2 x\right)dx = \int \cos 2x\,dx = \frac{1}{2}\sin 2x + C$

45. Let $u = \cos x \Rightarrow du = -\sin x\,dx$. Then

$\int \sin^5 x\,dx = \int (1 - \cos^2 x)^2 \sin x\,dx = \int (1 - u^2)^2(-du)$

$= \int (-1 + 2u^2 - u^4)du = -\frac{1}{5}u^5 + \frac{2}{3}u^3 - u + C$

$= -\frac{1}{5}\cos^5 x + \frac{2}{3}\cos^3 x - \cos x + C.$

Notice that F is increasing when $f(x) > 0$, so the graphs serve as a check on our work.

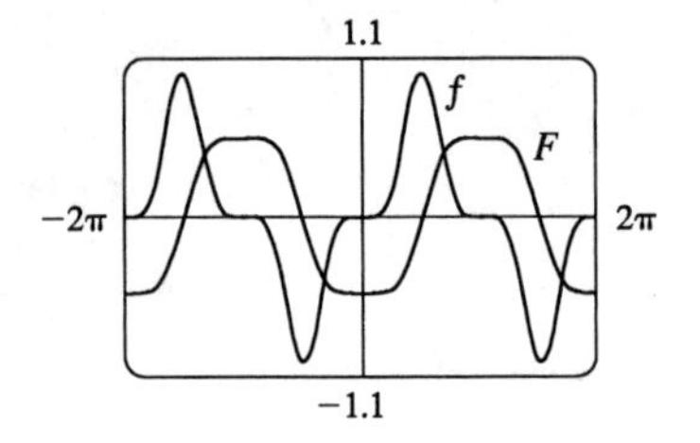

47. $f_{\text{ave}} = \frac{1}{2\pi}\int_{-\pi}^{\pi} \sin^2 x \cos^3 x\,dx = \frac{1}{2\pi}\int_{-\pi}^{\pi} \sin^2 x(1 - \sin^2 x)\cos x\,dx = \frac{1}{2\pi}\int_0^0 u^2(1 - u^2)du$ (where $u = \sin x$) $= 0$

49. For $0 < x < \frac{\pi}{2}$, we have $0 < \sin x < 1$, so $\sin^3 x < \sin x$. Hence the area is

$\int_0^{\pi/2} (\sin x - \sin^3 x)dx = \int_0^{\pi/2} \sin x(1 - \sin^2 x)dx = \int_0^{\pi/2} \cos^2 x \sin x\,dx$. Now let $u = \cos x \Rightarrow$

$du = -\sin x\,dx$. Then area $= \int_1^0 u^2(-du) = \int_0^1 u^2\,du = \left[\frac{1}{3}u^3\right]_0^1 = \frac{1}{3}$.

51.

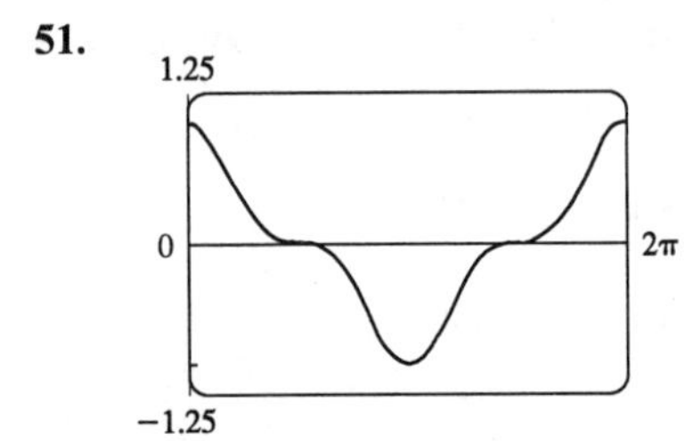

It seems from the graph that $\int_0^{2\pi} \cos^3 x\,dx = 0$, since the area below the x-axis and above the graph looks about equal to the area above the axis and below the graph. By Example 1, the integral is $\left[\sin x - \frac{1}{3}\sin^3 x\right]_0^{2\pi} = 0$. Note that due to symmetry, the integral of any odd power of $\sin x$ or $\cos x$ between limits which differ by $2n\pi$ (n any integer) is 0.

53. $V = \int_{\pi/2}^{\pi} \pi \sin^2 x\,dx = \pi\int_{\pi/2}^{\pi} \frac{1}{2}(1 - \cos 2x)dx = \pi\left[\frac{1}{2}x - \frac{1}{4}\sin 2x\right]_{\pi/2}^{\pi} = \pi\left(\frac{\pi}{2} - 0 - \frac{\pi}{4} + 0\right) = \frac{\pi^2}{4}$

55. Volume $= \pi\int_0^{\pi/2}\left[(1 + \cos x)^2 - 1^2\right]dx = \pi\int_0^{\pi/2}(2\cos x + \cos^2 x)dx$

$= \pi\left[2\sin x + \frac{1}{2}x + \frac{1}{4}\sin 2x\right]_0^{\pi/2} = \pi\left(2 + \frac{\pi}{4}\right) = 2\pi + \frac{\pi^2}{4}$

57. $s = f(t) = \int_0^t \sin\omega u \cos^2\omega u\,du$. Let $y = \cos\omega u \quad\Rightarrow\quad dy = -\omega\sin\omega u\,du$. Then

$$s = -\frac{1}{\omega}\int_1^{\cos\omega t} y^2\,dy = -\frac{1}{\omega}\left[\tfrac{1}{3}y^3\right]_1^{\cos\omega t} = \frac{1}{3\omega}\left(1 - \cos^3\omega t\right).$$

59. Just note that the integrand is odd $[f(-x) = -f(x)]$.

Or: If $m \neq n$, calculate $\int_{-\pi}^{\pi} \sin mx \cos nx\,dx = \int_{-\pi}^{\pi} \frac{1}{2}[\sin(m-n)x + \sin(m+n)x]dx$

$$= \frac{1}{2}\left[-\frac{\cos(m-n)x}{m-n} - \frac{\cos(m+n)x}{m+n}\right]_{-\pi}^{\pi} = 0.$$ If $m = n$, then the first term in each set of brackets is zero.

61. $\int_{-\pi}^{\pi} \cos mx \cos nx\,dx = \int_{-\pi}^{\pi} \frac{1}{2}[\cos(m-n)x + \cos(m+n)x]dx$. If $m \neq n$, this is equal to

$$\frac{1}{2}\left[\frac{\sin(m-n)x}{m-n} + \frac{\sin(m+n)x}{m+n}\right]_{-\pi}^{\pi} = 0.$$ If $m = n$, we get

$$\int_{-\pi}^{\pi} \tfrac{1}{2}[1 + \cos(m+n)x]dx = \left[\tfrac{1}{2}x\right]_{-\pi}^{\pi} + \left[\frac{\sin(m+n)x}{2(m+n)}\right]_{-\pi}^{\pi} = \pi + 0 = \pi.$$

EXERCISES 8.3

1. Let $x = \sin\theta$, where $-\frac{\pi}{2} \le \theta \le \frac{\pi}{2}$. Then $dx = \cos\theta\,d\theta$ and $\sqrt{1-x^2} = |\cos\theta| = \cos\theta$ (since $\cos\theta > 0$ for θ in $\left[-\frac{\pi}{2}, \frac{\pi}{2}\right]$). Thus

$$\int_{1/2}^{\sqrt{3}/2} \frac{dx}{x^2\sqrt{1-x^2}} = \int_{\pi/6}^{\pi/3} \frac{\cos\theta\,d\theta}{\sin^2\theta\cos\theta} = \int_{\pi/6}^{\pi/3} \csc^2\theta\,d\theta = [-\cot\theta]_{\pi/6}^{\pi/3}$$
$$= -\frac{1}{\sqrt{3}} - \left(-\sqrt{3}\right) = \frac{3}{\sqrt{3}} - \frac{1}{\sqrt{3}} = \frac{2}{\sqrt{3}}.$$

3. Let $u = 1 - x^2$. Then $du = -2x\,dx$, so $\displaystyle\int \frac{x}{\sqrt{1-x^2}}\,dx = -\frac{1}{2}\int \frac{du}{\sqrt{u}} = -\sqrt{u} + C = -\sqrt{1-x^2} + C.$

5. Let $2x = \sin\theta$, where $-\frac{\pi}{2} \le \theta \le \frac{\pi}{2}$. Then

$x = \frac{1}{2}\sin\theta$, $dx = \frac{1}{2}\cos\theta\,d\theta$, and $\sqrt{1-4x^2} = \sqrt{1-(2x)^2} = \cos\theta$.

$$\int\sqrt{1-4x^2}\,dx = \int \cos\theta\left(\tfrac{1}{2}\cos\theta\right)d\theta = \tfrac{1}{4}\int(1+\cos 2\theta)d\theta$$
$$= \tfrac{1}{4}\left(\theta + \tfrac{1}{2}\sin 2\theta\right) + C = \tfrac{1}{4}(\theta + \sin\theta\cos\theta) + C$$
$$= \tfrac{1}{4}\left[\sin^{-1}(2x) + 2x\sqrt{1-4x^2}\right] + C$$

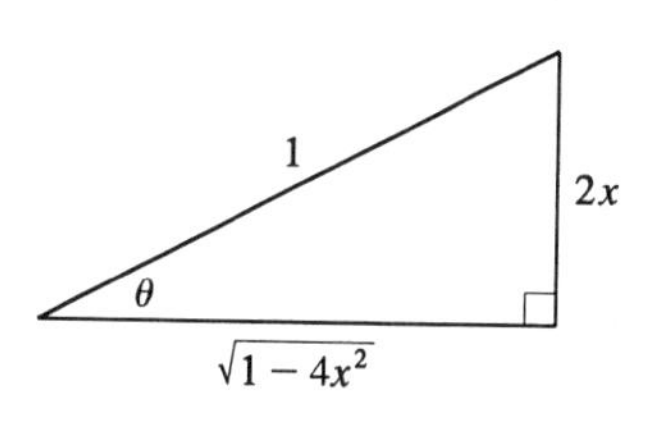

7. Let $x = 3\tan\theta$, where $-\frac{\pi}{2} < \theta < \frac{\pi}{2}$. Then $dx = 3\sec^2\theta\,d\theta$ and $\sqrt{9+x^2} = 3\sec\theta$.

$$\int_0^3 \frac{dx}{\sqrt{9+x^2}} = \int_0^{\pi/4} \frac{3\sec^2\theta\,d\theta}{3\sec\theta} = \int_0^{\pi/4} \sec\theta\,d\theta = [\ln|\sec\theta + \tan\theta|]_0^{\pi/4} = \ln\left(\sqrt{2}+1\right) - \ln 1 = \ln\left(\sqrt{2}+1\right)$$

9. Let $x = 4\sec\theta$, where $0 \le \theta < \frac{\pi}{2}$ or $\pi \le \theta < \frac{3\pi}{2}$. Then $dx = 4\sec\theta\tan\theta\,d\theta$ and $\sqrt{x^2-16} = 4|\tan\theta| = 4\tan\theta$. Thus

$$\int \frac{dx}{x^3\sqrt{x^2-16}} = \int \frac{4\sec\theta\tan\theta\,d\theta}{64\sec^3\theta\cdot 4\tan\theta} = \tfrac{1}{64}\int \cos^2\theta\,d\theta = \tfrac{1}{128}\int(1+\cos 2\theta)d\theta$$

$$= \tfrac{1}{128}\left(\theta + \tfrac{1}{2}\sin 2\theta\right) + C = \tfrac{1}{128}(\theta + \sin\theta\cos\theta) + C = \tfrac{1}{128}\left(\sec^{-1}\frac{x}{4} + \frac{4\sqrt{x^2-16}}{x^2}\right) + C$$

by the diagrams for $0 \le \theta < \frac{\pi}{2}$ and $\pi \le \theta < \frac{3\pi}{2}$, where the labels of the legs in the second diagram indicate the x-and y-coordinates of P rather than the lengths of those sides. Henceforth we omit the second diagram from our solutions.

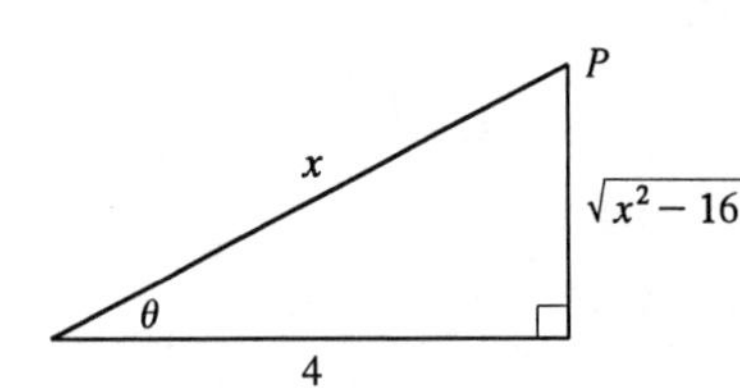

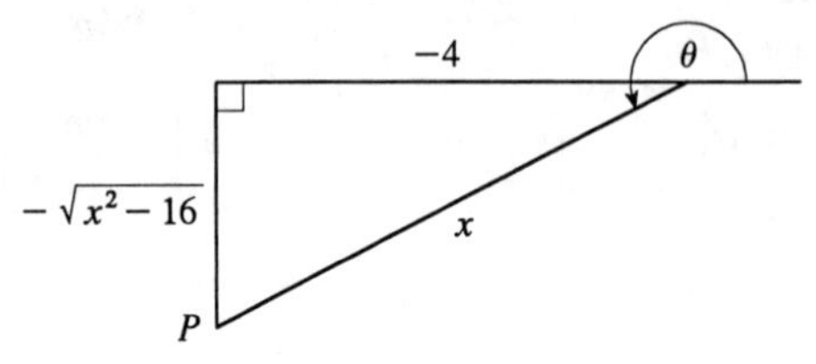

11. $9x^2 - 4 = (3x)^2 - 4$, so let $3x = 2\sec\theta$, where $0 \le \theta < \frac{\pi}{2}$ or $\pi \le \theta < \frac{3\pi}{2}$. Then $dx = \frac{2}{3}\sec\theta\tan\theta\,d\theta$ and $\sqrt{9x^2-4} = 2\tan\theta$.

$$\int \frac{\sqrt{9x^2-4}}{x}dx = \int \frac{2\tan\theta}{\frac{2}{3}\sec\theta}\cdot\tfrac{2}{3}\sec\theta\tan\theta\,d\theta$$

$$= 2\int \tan^2\theta\,d\theta = 2\int(\sec^2\theta - 1)d\theta = 2(\tan\theta - \theta) + C$$

$$= \sqrt{9x^2-4} - 2\sec^{-1}\left(\frac{3x}{2}\right) + C$$

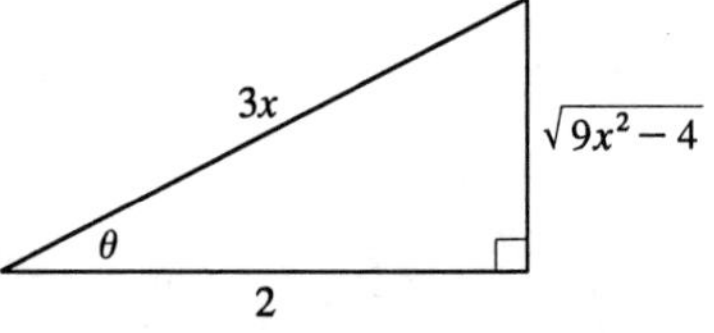

13. Let $x = a\sin\theta$, where $-\frac{\pi}{2} \le \theta \le \frac{\pi}{2}$. Then $dx = a\cos\theta\,d\theta$ and

$$\int \frac{x^2\,dx}{(a^2-x^2)^{3/2}} = \int \frac{a^2\sin^2\theta\,a\cos\theta\,d\theta}{a^3\cos^3\theta} = \int \tan^2\theta\,d\theta$$

$$= \int(\sec^2\theta - 1)d\theta = \tan\theta - \theta + C$$

$$= \frac{x}{\sqrt{a^2-x^2}} - \sin^{-1}\frac{x}{a} + C.$$

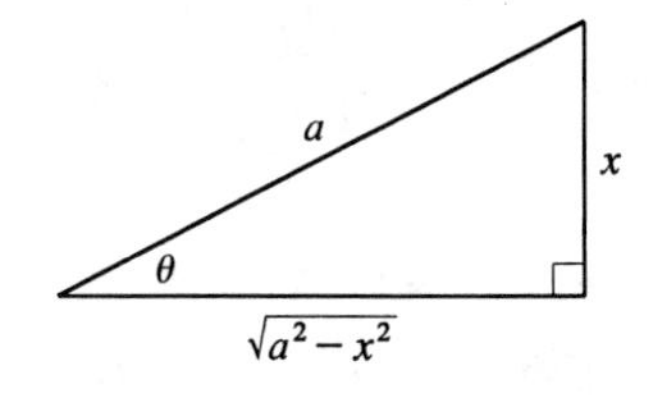

15. Let $x = \sqrt{3}\tan\theta$, where $-\frac{\pi}{2} < \theta < \frac{\pi}{2}$. Then

$$\int \frac{dx}{x\sqrt{x^2+3}} = \int \frac{\sqrt{3}\sec^2\theta\,d\theta}{\sqrt{3}\tan\theta\,\sqrt{3}\sec\theta} = \tfrac{1}{\sqrt{3}}\int \csc\theta\,d\theta$$

$$= \tfrac{1}{\sqrt{3}}\ln|\csc\theta - \cot\theta| + C = \frac{1}{\sqrt{3}}\ln\left|\frac{\sqrt{x^2+3}-\sqrt{3}}{x}\right| + C.$$

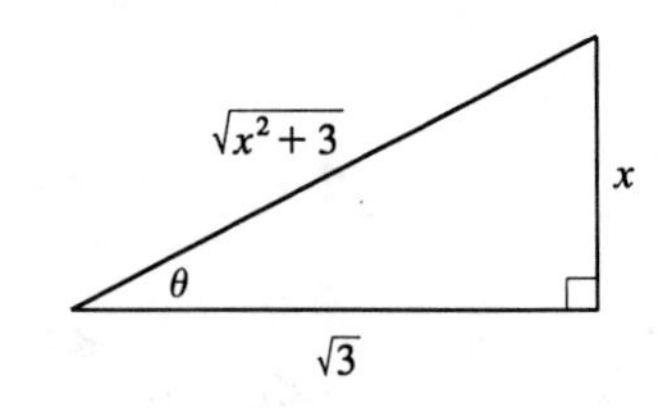

17. Let $u = 4 - 9x^2 \quad \Rightarrow \quad du = -18x\,dx$. Then $x^2 = \frac{1}{9}(4-u)$ and

$$\int_0^{2/3} x^3\sqrt{4-9x^2}\,dx = \int_4^0 \tfrac{1}{9}(4-u)u^{1/2}\left(-\tfrac{1}{18}\right)du = \tfrac{1}{162}\int_0^4\left(4u^{1/2}-u^{3/2}\right)du$$
$$= \tfrac{1}{162}\left[\tfrac{8}{3}u^{3/2} - \tfrac{2}{5}u^{5/2}\right]_0^4 = \tfrac{1}{162}\left[\tfrac{64}{3} - \tfrac{64}{5}\right] = \tfrac{64}{1215}.$$

Or: Let $3x = 2\sin\theta$, where $-\frac{\pi}{2} \le \theta \le \frac{\pi}{2}$.

19. Let $u = 1 + x^2$, $du = 2x\,dx$. Then $\int 5x\sqrt{1+x^2}\,dx = \frac{5}{2}\int u^{1/2}\,du = \frac{5}{3}u^{3/2} + C = \frac{5}{3}(1+x^2)^{3/2} + C$.

21. $2x - x^2 = -(x^2 - 2x + 1) + 1 = 1 - (x-1)^2$. Let $u = x - 1$. Then $du = dx$ and

$$\int\sqrt{2x-x^2}\,dx = \int\sqrt{1-u^2}\,du = \int\cos^2\theta\,d\theta \quad \left(\text{where } u = \sin\theta,\ -\tfrac{\pi}{2} \le \theta \le \tfrac{\pi}{2}\right)$$
$$= \tfrac{1}{2}\int(1+\cos 2\theta)d\theta = \tfrac{1}{2}\left(\theta + \tfrac{1}{2}\sin 2\theta\right) + C = \tfrac{1}{2}\left(\sin^{-1}u + u\sqrt{1-u^2}\right) + C$$
$$= \tfrac{1}{2}\left[\sin^{-1}(x-1) + (x-1)\sqrt{2x-x^2}\right] + C.$$

23. $9x^2 + 6x - 8 = (3x+1)^2 - 9$, so let $u = 3x + 1$, $du = 3\,dx$. Then $\displaystyle\int\frac{dx}{\sqrt{9x^2+6x-8}} = \int\frac{\frac{1}{3}\,du}{\sqrt{u^2-9}}$. Now let $u = 3\sec\theta$, where $0 \le \theta < \frac{\pi}{2}$ or $\pi \le \theta < \frac{3\pi}{2}$. Then $du = 3\sec\theta\tan\theta\,d\theta$ and $\sqrt{u^2-9} = 3\tan\theta$, so

$$\int\frac{\frac{1}{3}\,du}{\sqrt{u^2-9}} = \int\frac{\sec\theta\tan\theta\,d\theta}{3\tan\theta} = \tfrac{1}{3}\int\sec\theta\,d\theta = \tfrac{1}{3}\ln|\sec\theta + \tan\theta| + C_1$$
$$= \frac{1}{3}\ln\left|\frac{u+\sqrt{u^2-9}}{3}\right| + C_1 = \tfrac{1}{3}\ln\left|u + \sqrt{u^2-9}\right| + C = \tfrac{1}{3}\ln\left|3x+1+\sqrt{9x^2+6x-8}\right| + C.$$

25. $x^2 + 2x + 2 = (x+1)^2 + 1$. Let $u = x + 1$, $du = dx$. Then

$$\int\frac{dx}{(x^2+2x+2)^2} = \int\frac{du}{(u^2+1)^2} = \int\frac{\sec^2\theta\,d\theta}{\sec^4\theta} \quad \left(\begin{array}{c}\text{where } u = \tan\theta,\ du = \sec^2\theta\,d\theta,\\ \text{and } u^2+1 = \sec^2\theta\end{array}\right)$$
$$= \int\cos^2\theta\,d\theta = \tfrac{1}{2}(\theta + \sin\theta\cos\theta) + C \quad \text{(as in Exercise 21)}$$
$$= \frac{1}{2}\left[\tan^{-1}u + \frac{u}{1+u^2}\right] + C = \frac{1}{2}\left[\tan^{-1}(x+1) + \frac{x+1}{x^2+2x+2}\right] + C.$$

27. Let $u = e^t \quad \Rightarrow \quad du = e^t\,dt$. Then $\int e^t\sqrt{9-e^{2t}}\,dt = \int\sqrt{9-u^2}\,du = \int(3\cos\theta)3\cos\theta\,d\theta$ (where $u = 3\sin\theta$, $-\frac{\pi}{2} \le \theta \le \frac{\pi}{2}$) $= 9\int\cos^2\theta\,d\theta = \frac{9}{2}(\theta + \sin\theta\cos\theta) + C$ (as in Exercise 21)

$$= \frac{9}{2}\left[\sin^{-1}\left(\frac{u}{3}\right) + \frac{u}{3}\cdot\frac{\sqrt{9-u^2}}{3}\right] + C = \tfrac{9}{2}\sin^{-1}\left(\tfrac{1}{3}e^t\right) + \tfrac{1}{2}e^t\sqrt{9-e^{2t}} + C.$$

29. **(a)** Let $x = a\tan\theta$, where $-\frac{\pi}{2} < \theta < \frac{\pi}{2}$. Then $\sqrt{x^2+a^2} = a\sec\theta$ and

$$\int\frac{dx}{\sqrt{x^2+a^2}} = \int\frac{a\sec^2\theta\,d\theta}{a\sec\theta} = \int\sec\theta\,d\theta = \ln|\sec\theta + \tan\theta| + C_1 = \ln\left|\frac{\sqrt{x^2+a^2}}{a} + \frac{x}{a}\right| + C_1$$
$$= \ln\left(x + \sqrt{x^2+a^2}\right) + C, \text{ where } C = C_1 - \ln|a|$$

(b) Let $x = a\sinh t$, so that $dx = a\cosh t\,dt$ and $\sqrt{x^2+a^2} = a\cosh t$. Then

$$\int\frac{dx}{\sqrt{x^2+a^2}} = \int\frac{a\cosh t\,dt}{a\cosh t} = t + C = \sinh^{-1}(x/a) + C.$$

31. Area of $\triangle POQ = \frac{1}{2}(r\cos\theta)(r\sin\theta) = \frac{1}{2}r^2\sin\theta\cos\theta$. Area of region $PQR = \int_{r\cos\theta}^{r}\sqrt{r^2-x^2}\,dx$.

Let $x = r\cos u \quad\Rightarrow\quad dx = -r\sin u\,du$ for $\theta \le u \le \frac{\pi}{2}$. Then we obtain

$$\int\sqrt{r^2-x^2}\,dx = \int r\sin u(-r\sin u)du = -r^2\int\sin^2 u\,du$$
$$= -\tfrac{1}{2}r^2(u - \sin u\cos u) + C = -\tfrac{1}{2}r^2\cos^{-1}(x/r) + \tfrac{1}{2}x\sqrt{r^2-x^2} + C.$$ So

area of region $PQR = \frac{1}{2}\left[-r^2\cos^{-1}(x/r) + x\sqrt{r^2-x^2}\right]_{r\cos\theta}^{r} = \frac{1}{2}[0 - (-r^2\theta + r\cos\theta\, r\sin\theta)]$

$= \frac{1}{2}r^2\theta - \frac{1}{2}r^2\sin\theta\cos\theta$, so (area of sector POR) $=$ (area of $\triangle POQ$) $+$ (area of region PQR) $= \frac{1}{2}r^2\theta$.

33.

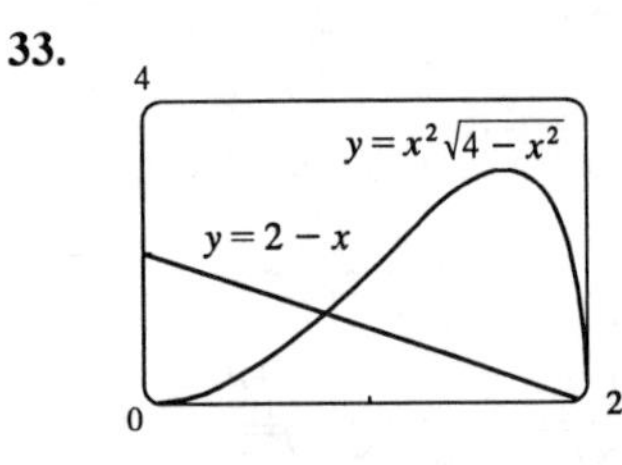

From the graph, it appears that the curve $y = x^2\sqrt{4-x^2}$ and the line $y = 2 - x$ intersect at about $x = 0.81$ and $x = 2$, with $x^2\sqrt{4-x^2} > 2 - x$ on $(0.81, 2)$. So the area bounded by the curve and the line is

$$A \approx \int_{0.81}^{2}\left[x^2\sqrt{4-x^2} - (2-x)\right]dx = \int_{0.81}^{2}x^2\sqrt{4-x^2}\,dx - \left[2x - \tfrac{1}{2}x^2\right]_{0.81}^{2}.$$

To evaluate the integral, we put $x = 2\sin\theta$, where $-\frac{\pi}{2} \le \theta \le \frac{\pi}{2}$. Then $dx = 2\cos\theta\,d\theta$, $x = 2 \quad\Rightarrow$ $\theta = \sin^{-1}1 = \frac{\pi}{2}$, and $x = 0.81 \quad\Rightarrow\quad \theta = \sin^{-1}0.405 \approx 0.417$. So

$$\int_{0.81}^{2}x^2\sqrt{4-x^2}\,dx \approx \int_{0.417}^{\pi/2}4\sin^2\theta(2\cos\theta)(2\cos\theta\,d\theta) = 4\int_{0.417}^{\pi/2}\sin^2 2\theta\,d\theta$$
$$= 4\int_{0.417}^{\pi/2}\tfrac{1}{2}(1-\cos 4\theta)d\theta = 2\left[\theta - \tfrac{1}{4}\sin 4\theta\right]_{0.417}^{\pi/2} = 2\left(\left[\tfrac{\pi}{2} - 0\right] - \left[0.417 - \tfrac{1}{4}(0.995)\right]\right) \approx 2.81.$$ So

$$A \approx 2.81 - \left[\left(2\cdot 2 - \tfrac{1}{2}\cdot 2^2\right) - \left(2\cdot 0.81 - \tfrac{1}{2}\cdot 0.81^2\right)\right] \approx 2.10.$$

35. Let the equation of the large circle be $x^2 + y^2 = R^2$. Then the equation of the small circle is $x^2 + (y-b)^2 = r^2$, where $b = \sqrt{R^2-r^2}$ is the distance between the centers of the circles. The desired area is

$$A = \int_{-r}^{r}\left[\left(b + \sqrt{r^2-x^2}\right) - \sqrt{R^2-x^2}\right]dx = 2\int_0^r\left(b + \sqrt{r^2-x^2} - \sqrt{R^2-x^2}\right)dx$$
$$= 2\int_0^r b\,dx + 2\int_0^r\sqrt{r^2-x^2}\,dx - 2\int_0^r\sqrt{R^2-x^2}\,dx.$$

The first integral is just $2br = 2r\sqrt{R^2-r^2}$. To evaluate the other two integrals, note that

$$\int\sqrt{a^2-x^2}\,dx = \int a^2\cos^2\theta\,d\theta \quad (x = a\sin\theta,\ dx = a\cos\theta\,d\theta)$$
$$= \tfrac{1}{2}\left(\tfrac{1}{2}a^2\right)\int(1+\cos 2\theta) = \tfrac{1}{2}a^2\left(\theta + \tfrac{1}{2}\sin 2\theta\right) + C = \tfrac{1}{2}a^2(\theta + \sin\theta\cos\theta) + C$$
$$= \frac{a^2}{2}\arcsin\left(\frac{x}{a}\right) + \frac{a^2}{2}\left(\frac{x}{a}\right)\frac{\sqrt{a^2-x^2}}{a} + C = \frac{a^2}{2}\arcsin\left(\frac{x}{a}\right) + \frac{x}{2}\sqrt{a^2-x^2} + C,$$ so the desired area is

$$A = 2r\sqrt{R^2-r^2} + \left[r^2\arcsin(x/r) + x\sqrt{r^2-x^2}\right]_0^r - \left[R^2\arcsin(x/R) + x\sqrt{R^2-x^2}\right]_0^r$$
$$= 2r\sqrt{R^2-r^2} + r^2\left(\tfrac{\pi}{2}\right) - \left[R^2\arcsin(r/R) + r\sqrt{R^2-r^2}\right] = r\sqrt{R^2-r^2} + \frac{\pi}{2}r^2 - R^2\arcsin(r/R).$$

37. We use cylindrical shells and assume that $R > r$. $x^2 = r^2 - (y-R)^2 \Rightarrow x = \pm\sqrt{r^2 - (y-R)^2}$, so $g(y) = 2\sqrt{r^2 - (y-R)^2}$ in Formula 7.3.3, and

$$V = \int_{R-r}^{R+r} 2\pi y \cdot 2\sqrt{r^2 - (y-R)^2}\, dy = \int_{-r}^{r} 4\pi(u+R)\sqrt{r^2 - u^2}\, du \quad (\text{where } u = y - R)$$

$$= 4\pi \int_{-r}^{r} u\sqrt{r^2 - u^2}\, du + 4\pi R \int_{-r}^{r} \sqrt{r^2 - u^2}\, du \quad \left(\begin{array}{c}\text{where } u = r\sin\theta,\ du = r\cos\theta\, d\theta \\ \text{in the second integral}\end{array}\right)$$

$$= 4\pi\left[-\tfrac{1}{3}(r^2 - u^2)^{3/2}\right]_{-r}^{r} + 4\pi R \textstyle\int_{-\pi/2}^{\pi/2} r^2\cos^2\theta\, d\theta = -\tfrac{4\pi}{3}(0-0) + 4\pi R r^2 \textstyle\int_{-\pi/2}^{\pi/2} \cos^2 d\theta$$

$$= 2\pi R r^2 \textstyle\int_{-\pi/2}^{\pi/2} (1 + \cos 2\theta)d\theta = 2\pi R r^2\left[\theta + \tfrac{1}{2}\sin 2\theta\right]_{-\pi/2}^{\pi/2} = 2\pi^2 R r^2.$$

Another Method: Use washers instead of shells, so $V = 8\pi R \int_0^r \sqrt{r^2 - y^2}\, dy$ as in Exercise 7.2.63(a), but evaluate the integral using $y = r\sin\theta$.

EXERCISES 8.4

1. $\dfrac{1}{(x-1)(x+2)} = \dfrac{A}{x-1} + \dfrac{B}{x+2}$

3. $\dfrac{x^2 + 3x - 4}{(2x-1)^2(2x+3)} = \dfrac{A}{2x-1} + \dfrac{B}{(2x-1)^2} + \dfrac{C}{2x+3}$

5. $\dfrac{1}{x^4 - x^3} = \dfrac{1}{x^3(x-1)} = \dfrac{A}{x} + \dfrac{B}{x^2} + \dfrac{C}{x^3} + \dfrac{D}{x-1}$

7. $\dfrac{x^2+1}{x^2-1} = 1 + \dfrac{2}{(x-1)(x+1)} = 1 + \dfrac{A}{x-1} + \dfrac{B}{x+1}$

9. $\dfrac{x^2 - 2}{x(x^2+2)} = \dfrac{A}{x} + \dfrac{Bx + C}{x^2+2}$

11. $\dfrac{x^4 + x^2 + 1}{(x^2+1)(x^2+4)^2} = \dfrac{Ax+B}{x^2+1} + \dfrac{Cx+D}{x^2+4} + \dfrac{Ex+F}{(x^2+4)^2}$

13. $\dfrac{x^4}{(x^2+9)^3} = \dfrac{Ax+B}{x^2+9} + \dfrac{Cx+D}{(x^2+9)^2} + \dfrac{Ex+F}{(x^2+9)^3}$

15. $\dfrac{x^3 + x^2 + 1}{x^4 + x^3 + 2x^2} = \dfrac{x^3 + x^2 + 1}{x^2(x^2 + x + 2)} = \dfrac{A}{x} + \dfrac{B}{x^2} + \dfrac{Cx+D}{x^2+x+2}$

17. $\displaystyle\int \frac{x^2}{x+1}\, dx = \int \left(x - 1 + \frac{1}{x+1}\right) dx = \tfrac{1}{2}x^2 - x + \ln|x+1| + C$

19. $\dfrac{4x-1}{(x-1)(x+2)} = \dfrac{A}{x-1} + \dfrac{B}{x+2} \Rightarrow 4x - 1 = A(x+2) + B(x-1)$ Take $x = 1$ to get $3 = 3A$, then $x = -2$ to get $-9 = -3B \Rightarrow A = 1, B = 3$. Now

$$\int_2^4 \frac{4x-1}{(x-1)(x+2)}\, dx = \int_2^4 \left[\frac{1}{x-1} + \frac{3}{x+2}\right] dx = [\ln(x-1) + 3\ln(x+2)]_2^4$$

$$= \ln 3 + 3\ln 6 - \ln 1 - 3\ln 4 = 4\ln 3 - 3\ln 2 = \ln\tfrac{81}{8}.$$

21. $\displaystyle\int \frac{6x-5}{2x+3}\,dx = \int \left[3 - \frac{14}{2x+3}\right] dx = 3x - 7\ln|2x+3| + C$

23. $\displaystyle\frac{x^2+1}{x^2-x} = 1 + \frac{x+1}{x(x-1)} = 1 - \frac{1}{x} + \frac{2}{x-1}$, so

$$\int \frac{x^2+1}{x^2-x}\,dx = x - \ln|x| + 2\ln|x-1| + C = x + \ln\frac{(x-1)^2}{|x|} + C.$$

25. $\displaystyle\frac{2x+3}{(x+1)^2} = \frac{A}{x+1} + \frac{B}{(x+1)^2} \quad\Rightarrow\quad 2x+3 = A(x+1) + B$. Take $x=-1$ to get $B=1$, and equate coefficients of x to get $A=2$. Now

$$\int_0^1 \frac{2x+3}{(x+1)^2}\,dx = \int_0^1 \left[\frac{2}{x+1} + \frac{1}{(x+1)^2}\right] dx = \left[2\ln(x+1) - \frac{1}{x+1}\right]_0^1$$
$$= 2\ln 2 - \tfrac{1}{2} - (2\ln 1 - 1) = 2\ln 2 + \tfrac{1}{2}.$$

27. $\displaystyle\frac{6x^2+5x-3}{x^3+2x^2-3x} = \frac{A}{x} + \frac{B}{x+3} + \frac{C}{x-1} \quad\Rightarrow$
$6x^2+5x-3 = A(x+3)(x-1) + B(x)(x-1) + C(x)(x+3)$.

Set $x=0$ to get $A=1$, then take $x=-3$ to get $B=3$, then set $x=1$ to get $C=2$:

$$\int_2^3 \frac{6x^2+5x-3}{x^3+2x^2-3x}\,dx = \int_2^3 \left[\frac{1}{x} + \frac{3}{x+3} + \frac{2}{x-1}\right] dx = [\ln x + 3\ln(x+3) + 2\ln(x-1)]_2^3$$
$$= (\ln 3 + 3\ln 6 + 2\ln 2) - (\ln 2 + 3\ln 5) = 4\ln 6 - 3\ln 5.$$

29. $\displaystyle\frac{1}{(x-1)^2(x+4)} = \frac{A}{x-1} + \frac{B}{(x-1)^2} + \frac{C}{x+4} \quad\Rightarrow\quad 1 = A(x-1)(x+4) + B(x+4) + C(x-1)^2$. Set $x=1$ to get $B=\frac{1}{5}$ and take $x=-4$ to get $C=\frac{1}{25}$. Now equating the coefficients of x^2, we get $0 = Ax^2 + Cx^2$ or $A = -C = -\frac{1}{25} \quad\Rightarrow$

$$\int \frac{dx}{(x-1)^2(x+4)} = \int \left[\frac{-1/25}{x-1} + \frac{1/5}{(x-1)^2} + \frac{1/25}{x+4}\right] dx = -\frac{1}{25}\ln|x-1| - \frac{1}{5}\cdot\frac{1}{x-1} + \frac{1}{25}\ln|x+4| + C$$
$$= \frac{1}{25}\left[\ln\left|\frac{x+4}{x-1}\right| - \frac{5}{x-1}\right] + C.$$

31. $\displaystyle\frac{5x^2+3x-2}{x^3+2x^2} = \frac{5x^2+3x-2}{x^2(x+2)} = \frac{A}{x} + \frac{B}{x^2} + \frac{C}{x+2}$. Multiply by $x^2(x+2)$ to get $5x^2+3x-2 = Ax(x+2) + B(x+2) + Cx^2$. Set $x=-2$ to get $C=3$, and take $x=0$ to get $B=-1$. Equating the coefficients of x^2 gives $5x^2 = Ax^2 + Cx^2$ or $A=2$. So

$$\int \frac{5x^2+3x-2}{x^3+2x^2}\,dx = \int \left[\frac{2}{x} - \frac{1}{x^2} + \frac{3}{x+2}\right] dx = 2\ln|x| + \frac{1}{x} + 3\ln|x+2| + C.$$

33. Let $u = x^3+3x^2+4$. Then $du = 3(x^2+2x)dx \Rightarrow \displaystyle\int \frac{x^2+2x}{x^3+3x^2+4}\,dx = \frac{1}{3}\int\frac{du}{u} = \tfrac{1}{3}\ln|x^3+3x^2+4| + C$.

35. $\displaystyle\frac{x^2}{(x+1)^3} = \frac{A}{x+1} + \frac{B}{(x+1)^2} + \frac{C}{(x+1)^3}$. Multiply by $(x+1)^3$ to get $x^2 = A(x+1)^2 + B(x+1) + C$. Setting $x=-1$ gives $C=1$. Equating the coefficients of x^2 gives $A=1$, and setting $x=0$ gives $B=-2$.

Now $\displaystyle\int \frac{x^2\,dx}{(x+1)^3} = \int \left[\frac{1}{x+1} - \frac{2}{(x+1)^2} + \frac{1}{(x+1)^3}\right] dx = \ln|x+1| + \frac{2}{x+1} - \frac{1}{2(x+1)^2} + C.$

37. $\dfrac{1}{x^4-x^2}=\dfrac{1}{x^2(x-1)(x+1)}=\dfrac{A}{x}+\dfrac{B}{x^2}+\dfrac{C}{x-1}+\dfrac{D}{x+1}$. Multiply by $x^2(x-1)(x+1)$ to get $1=Ax(x-1)(x+1)+B(x-1)(x+1)+Cx^2(x+1)+Dx^2(x-1)$. Setting $x=1$ gives $C=\frac{1}{2}$, taking $x=-1$ gives $D=-\frac{1}{2}$. Equating the coefficients of x^3 gives $0=A+C+D=A$. Finally, setting $x=0$ yields $B=-1$. Now $\displaystyle\int\frac{dx}{x^4-x^2}=\int\left[\frac{-1}{x^2}+\frac{1/2}{x-1}-\frac{1/2}{x+1}\right]dx=\frac{1}{x}+\frac{1}{2}\ln\left|\frac{x-1}{x+1}\right|+C.$

39. $\displaystyle\frac{x^3}{x^2+1}=\frac{(x^3+x)-x}{x^2+1}=x-\frac{x}{x^2+1}$, so $\displaystyle\int_0^1\frac{x^3}{x^2+1}\,dx=\int_0^1 x\,dx-\int_0^1\frac{x\,dx}{x^2+1}$

$$=\left[\tfrac{1}{2}x^2\right]_0^1-\frac{1}{2}\int_1^2\frac{1}{u}\,du \text{ (where } u=x^2+1,\ du=2x\,dx\text{)} =\tfrac{1}{2}-\left[\tfrac{1}{2}\ln u\right]_1^2=\tfrac{1}{2}-\tfrac{1}{2}\ln 2=\tfrac{1}{2}(1-\ln 2)$$

41. Complete the square: $x^2+x+1=\left(x+\frac{1}{2}\right)^2+\frac{3}{4}$ and let $u=x+\frac{1}{2}$. Then

$$\int_0^1\frac{x}{x^2+x+1}\,dx=\int_{1/2}^{3/2}\frac{u-1/2}{u^2+3/4}\,du=\int_{1/2}^{3/2}\frac{u}{u^2+3/4}\,du-\frac{1}{2}\int_{1/2}^{3/2}\frac{1}{u^2+3/4}\,du$$

$$=\tfrac{1}{2}\ln\left(u^2+\tfrac{3}{4}\right)-\tfrac{1}{2}\tfrac{1}{\sqrt{3}/2}\left[\tan^{-1}\left(\tfrac{2}{\sqrt{3}}u\right)\right]_{1/2}^{3/2}=\tfrac{1}{2}\ln 3-\tfrac{1}{\sqrt{3}}\left(\tfrac{\pi}{3}-\tfrac{\pi}{6}\right)=\ln\sqrt{3}-\tfrac{\pi}{6\sqrt{3}}.$$

43. $\dfrac{3x^2-4x+5}{(x-1)(x^2+1)}=\dfrac{A}{x-1}+\dfrac{Bx+C}{x^2+1}$ $\Rightarrow$ $3x^2-4x+5=A(x^2+1)+(Bx+C)(x-1)$. Take $x=1$ to get $4=2A$ or $A=2$. Now $(Bx+C)(x-1)=3x^2-4x+5-2(x^2+1)=x^2-4x+3$. Equating coefficients of x^2 and then comparing the constant terms, we get $B=1$ and $C=-3$. Hence

$$\int\frac{3x^2-4x+5}{(x-1)(x^2+1)}\,dx=\int\left[\frac{2}{x-1}+\frac{x-3}{x^2+1}\right]dx=2\ln|x-1|+\int\frac{x\,dx}{x^2+1}-3\int\frac{dx}{x^2+1}$$

$$=2\ln|x-1|+\tfrac{1}{2}\ln(x^2+1)-3\tan^{-1}x+C=\ln(x-1)^2+\ln\sqrt{x^2+1}-3\tan^{-1}x+C.$$

45. $\dfrac{1}{x^3-1}=\dfrac{1}{(x-1)(x^2+x+1)}=\dfrac{A}{x-1}+\dfrac{Bx+C}{x^2+x+1}$ $\Rightarrow$ $1=A(x^2+x+1)+(Bx+C)(x-1)$. Take $x=1$ to get $A=\frac{1}{3}$. Equate coefficients of x^2 and 1 to get $0=\frac{1}{3}+B$, $1=\frac{1}{3}-C$, so $B=-\frac{1}{3}$, $C=-\frac{2}{3}$ $\Rightarrow$

$$\int\frac{dx}{x^3-1}=\int\frac{1/3}{x-1}\,dx+\int\frac{(-1/3)x-2/3}{x^2+x+1}\,dx=\tfrac{1}{3}\ln|x-1|-\frac{1}{3}\int\frac{x+2}{x^2+x+1}\,dx$$

$$=\tfrac{1}{3}\ln|x-1|-\frac{1}{3}\int\frac{x+1/2}{x^2+x+1}\,dx-\frac{1}{3}\int\frac{(3/2)dx}{(x+1/2)^2+3/4}$$

$$=\tfrac{1}{3}\ln|x-1|-\tfrac{1}{6}\ln(x^2+x+1)-\tfrac{1}{2}\left(\tfrac{2}{\sqrt{3}}\right)\tan^{-1}\left[\left(x+\tfrac{1}{2}\right)\Big/\left(\tfrac{\sqrt{3}}{2}\right)\right]+K$$

$$=\tfrac{1}{3}\ln|x-1|-\tfrac{1}{6}\ln(x^2+x+1)-\tfrac{1}{\sqrt{3}}\tan^{-1}\left[\tfrac{1}{\sqrt{3}}(2x+1)\right]+K.$$

47. $\dfrac{x^2-2x-1}{(x-1)^2(x^2+1)}=\dfrac{A}{x-1}+\dfrac{B}{(x-1)^2}+\dfrac{Cx+D}{x^2+1}$ $\Rightarrow$

$x^2-2x-1=A(x-1)(x^2+1)+B(x^2+1)+(Cx+D)(x-1)^2$. Setting $x=1$ gives $B=-1$. Equating the coefficients of x^3 gives $A=-C$. Equating the constant terms gives $-1=-A-1+D$, so $D=A$, and setting $x=2$ gives $-1=5A-5-2A+A$ or $A=1$. We have

$$\int\frac{x^2-2x-1}{(x-1)^2(x^2+1)}\,dx=\int\left[\frac{1}{x-1}-\frac{1}{(x-1)^2}-\frac{x-1}{x^2+1}\right]dx$$

$$=\ln|x-1|+\frac{1}{x-1}-\tfrac{1}{2}\ln(x^2+1)+\tan^{-1}x+C.$$

49. $\dfrac{3x^3 - x^2 + 6x - 4}{(x^2+1)(x^2+2)} = \dfrac{Ax+B}{x^2+1} + \dfrac{Cx+D}{x^2+2} \quad \Rightarrow$
$3x^3 - x^2 + 6x - 4 = (Ax+B)(x^2+2) + (Cx+D)(x^2+1)$. Equating the coefficients gives $A + C = 3$, $B + D = -1$, $2A + C = 6$, and $2B + D = -4 \quad \Rightarrow \quad A = 3$, $C = 0$, $B = -3$, and $D = 2$. Now

$$\int \frac{3x^3 - x^2 + 6x - 4}{(x^2+1)(x^2+2)}\,dx = 3\int \frac{x-1}{x^2+1}\,dx + 2\int \frac{dx}{x^2+2} = \tfrac{3}{2}\ln(x^2+1) - 3\tan^{-1}x + \sqrt{2}\tan^{-1}\left(\frac{x}{\sqrt{2}}\right) + C.$$

51. $$\int \frac{x-3}{(x^2+2x+4)^2}\,dx = \int \frac{x-3}{\left[(x+1)^2+3\right]^2}\,dx = \int \frac{u-4}{(u^2+3)^2}\,du \quad (\text{with } u = x+1)$$

$$= \int \frac{u\,du}{(u^2+3)^2} - 4\int \frac{du}{(u^2+3)^2} = \frac{1}{2}\int \frac{dv}{v^2} - 4\int \frac{\sqrt{3}\sec^2\theta\,d\theta}{9\sec^4\theta} \quad \left[\begin{array}{l} v = u^2+3 \text{ in the first integral;} \\ u = \sqrt{3}\tan\theta \text{ in the second} \end{array}\right]$$

$$= \frac{-1}{(2v)} - \frac{4\sqrt{3}}{9}\int \cos^2\theta\,d\theta = \frac{-1}{2(u^2+3)} - \tfrac{2\sqrt{3}}{9}(\theta + \sin\theta\cos\theta) + C$$

$$= \frac{-1}{2(x^2+2x+4)} - \frac{2\sqrt{3}}{9}\left[\tan^{-1}\left(\frac{x+1}{\sqrt{3}}\right) + \frac{\sqrt{3}(x+1)}{x^2+2x+4}\right] + C$$

$$= \frac{-1}{2(x^2+2x+4)} - \frac{2\sqrt{3}}{9}\tan^{-1}\left(\frac{x+1}{\sqrt{3}}\right) - \frac{2(x+1)}{3(x^2+2x+4)} + C$$

53. Let $u = \sin^2 x - 3\sin x + 2$. Then $du = (2\sin x\cos x - 3\cos x)dx$, so

$$\int \frac{(2\sin x - 3)\cos x}{\sin^2 x - 3\sin x + 2}\,dx = \int \frac{du}{u} = \ln|u| + C = \ln\left|\sin^2 x - 3\sin x + 2\right| + C.$$

55.

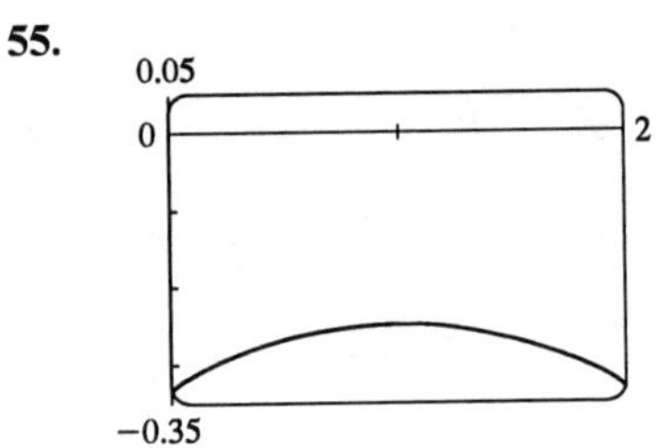

From the graph, we see that the integral will be negative, and we guess that the area is about the same as that of a rectangle with width 2 and height 0.3, so we estimate the integral to be $-(2 \cdot 0.3) = -0.6$.

Now $\dfrac{1}{x^2 - 2x - 3} = \dfrac{1}{(x-3)(x+1)} = \dfrac{A}{x-3} + \dfrac{B}{x+1} \quad \Leftrightarrow$
$1 = (A+B)x + A - 3B$, so $A = -B$ and $A - 3B = 1 \quad \Leftrightarrow$
$A = \frac{1}{4}$ and $B = -\frac{1}{4}$, so the integral becomes

$$\int_0^2 \frac{dx}{x^2-2x-3} = \frac{1}{4}\int_0^2 \frac{dx}{x-3} - \frac{1}{4}\int_0^2 \frac{dx}{x+1} = \tfrac{1}{4}\left[\ln|x-3| - \ln|x+1|\right]_0^2 = \frac{1}{4}\left[\ln\left|\frac{x-3}{x+1}\right|\right]_0^2$$
$$= \tfrac{1}{4}\left(\ln\tfrac{1}{3} - \ln 3\right) = -\tfrac{1}{2}\ln 3 \approx -0.55.$$

57. If $|x| < a$, then $\displaystyle\int \frac{dx}{a^2 - x^2} = \int \frac{a\,\text{sech}^2 u\,du}{a^2\,\text{sech}^2 u}$ (put $x = a\tanh u$) $= \dfrac{u}{a} + C = \dfrac{1}{a}\tanh^{-1}\left(\dfrac{x}{a}\right) + C.$

If $|x| > a$, then $\displaystyle\int \frac{dx}{a^2 - x^2} = \int \frac{-a\,\text{csch}^2 u\,du}{-a^2\,\text{csch}^2 u}$ (put $x = a\coth u$) $= \dfrac{u}{a} + C = \dfrac{1}{a}\coth^{-1}\left(\dfrac{x}{a}\right) + C.$

59. $$\int \frac{dx}{x^2 - 2x} = \int \frac{dx}{(x-1)^2 - 1} = \int \frac{du}{u^2 - 1} \quad (\text{put } u = x - 1)$$

$$= \frac{1}{2}\ln\left|\frac{u-1}{u+1}\right| + C \quad (\text{by Equation 6}) \quad = \frac{1}{2}\ln\left|\frac{x-2}{x}\right| + C$$

61. $\displaystyle\int \frac{x\,dx}{x^2+x-1} = \frac{1}{2}\int \frac{(2x+1)dx}{x^2+x-1} - \frac{1}{2}\int \frac{dx}{\left(x+\frac{1}{2}\right)^2 - \frac{5}{4}} = \frac{1}{2}\ln|x^2+x-1| - \frac{1}{2}\int \frac{du}{u^2 - \left(\frac{\sqrt{5}}{2}\right)^2}$

$$\left(\text{where } u = x+\tfrac{1}{2}\right) \quad = \tfrac{1}{2}\ln|x^2+x-1| - \frac{1}{2\sqrt{5}}\ln\left|\frac{u-\sqrt{5}/2}{u+\sqrt{5}/2}\right| + C$$

$$= \tfrac{1}{2}\ln|x^2+x-1| - \frac{1}{2\sqrt{5}}\ln\left|\frac{2x+1-\sqrt{5}}{2x+1+\sqrt{5}}\right| + C$$

63. $\dfrac{x+1}{x-1} = 1 + \dfrac{2}{x-1} > 0$ for $2 \le x \le 3$, so

area $= \displaystyle\int_2^3 \left[1 + \frac{2}{x-1}\right] dx = \left[x + 2\ln|x-1|\right]_2^3 = (3 + 2\ln 2) - (2 + 2\ln 1) = 1 + 2\ln 2.$

65. In this case, we use cylindrical shells, so the volume is $V = 2\pi \displaystyle\int_0^1 \frac{x\,dx}{x^2+3x+2} = 2\pi \int_0^1 \frac{x}{(x+1)(x+2)}$. We use partial fractions to simplify the integrand: $\dfrac{x}{(x+1)(x+2)} = \dfrac{A}{x+1} + \dfrac{B}{x+2} \quad\Rightarrow$

$x = (A+B)x + 2A + B$. So $A + B = 1$ and $2A + B = 0 \quad\Rightarrow\quad A = -1$ and $B = 2$. So the volume is

$$2\pi \int_0^1 \left[\frac{-1}{x+1} + \frac{2}{x+2}\right] dx = 2\pi\left[-\ln|x+1| + 2\ln|x+2|\right]_0^1$$

$$= 2\pi(-\ln 2 + 2\ln 3 + \ln 1 - 2\ln 2) = 2\pi(2\ln 3 - 3\ln 2) = 2\pi \ln \tfrac{9}{8}.$$

67. **(a)** In Maple, we define $f(x)$, and then use `convert(f,parfrac,x);` to obtain

$f(x) = \dfrac{24{,}110/4879}{5x+2} - \dfrac{668/323}{2x+1} - \dfrac{9438/80{,}155}{3x-7} + \dfrac{(22{,}098x + 48{,}935)/260{,}015}{x^2+x+5}$. In Mathematica, we use the command `Apart`, and in Derive, we use `Expand`.

(b) $\int f(x)dx = \frac{24{,}110}{4879}\cdot\frac{1}{5}\ln|5x+2| - \frac{668}{323}\cdot 2\ln|2x+1| - \frac{9438}{80{,}155}\cdot\frac{1}{3}\ln|3x-7|$

$$+ \frac{1}{260{,}015}\int \frac{22{,}098\left(x+\frac{1}{2}\right) + 37{,}886}{\left(x+\frac{1}{2}\right)^2 + \frac{19}{4}}\,dx + C$$

$$= \tfrac{24{,}110}{4879}\cdot\tfrac{1}{5}\ln|5x+2| - \tfrac{668}{323}\cdot\tfrac{1}{2}\ln|2x+1| - \tfrac{9438}{80{,}155}\cdot\tfrac{1}{3}\ln|3x-7|$$

$$+ \tfrac{1}{260{,}015}\left[22{,}098\cdot\tfrac{1}{2}\ln|x^2+x+5| + 37{,}886\cdot\sqrt{\tfrac{4}{19}}\tan^{-1}\left(\tfrac{1}{\sqrt{19/4}}\left(x+\tfrac{1}{2}\right)\right)\right] + C$$

$$= \tfrac{4822}{4879}\ln|5x+2| - \tfrac{334}{323}\ln|2x+1| - \tfrac{3146}{80{,}155}\ln|3x-7| + \tfrac{11{,}049}{260{,}015}\ln|x^2+x+5|$$

$$+ \tfrac{75{,}772}{260{,}015\sqrt{19}}\tan^{-1}\left[\tfrac{1}{\sqrt{19}}(2x+1)\right] + C.$$

If we tell Maple to `integrate(f,x);` we get

$$\frac{4822\ln(5x+2)}{4879} - \frac{334\ln(2x+1)}{323} - \frac{3146\ln(3x-7)}{80{,}155}$$

$$+ \frac{11{,}049\ln(x^2+x+5)}{260{,}015} + \frac{3988\sqrt{19}}{260{,}115}\tan^{-1}\left[\tfrac{\sqrt{19}}{19}(2x+1)\right].$$

The main difference in Maple's answer is that the absolute value signs and the constant of integration have been omitted. Also, the fractions have been reduced and the denominators rationalized.

69. There are only finitely many values of x where $Q(x)=0$ (assuming that Q is not the zero polynomial). At all other values of x, $F(x)/Q(x)=G(x)/Q(x)$, so $F(x)=G(x)$. In other words, the values of F and G agree at all except perhaps finitely many values of x. By continuity of F and G, the polynomials F and G must agree at those values of x too.

More explicitly: if a is a value of x such that $Q(a)=0$, then $Q(x)\neq 0$ for all x sufficiently close to a. Thus

$$F(a)=\lim_{x\to a}F(x)\begin{bmatrix}\text{by continuity}\\ \text{of } F\end{bmatrix}=\lim_{x\to a}G(x)\begin{bmatrix}\text{since } F(x)=G(x)\\ \text{whenever } Q(x)\neq 0\end{bmatrix}=G(a)\begin{bmatrix}\text{by continuity}\\ \text{of } G\end{bmatrix}.$$

EXERCISES 8.5

1. Let $u=\sqrt{x}$. Then $x=u^2$, $dx=2u\,du$ $\Rightarrow$

$$\int_0^1\frac{dx}{1+\sqrt{x}}=\int_0^1\frac{2u\,du}{1+u}=2\int_0^1\left[1-\frac{1}{1+u}\right]du=2[u-\ln(1+u)]_0^1=2(1-\ln 2).$$

3. Let $u=\sqrt{x}$. Then $x=u^2$, $dx=2u\,du$ $\Rightarrow$

$$\int\frac{\sqrt{x}\,dx}{x+1}=\int\frac{u\cdot 2u\,du}{u^2+1}=2\int\left[1-\frac{1}{u^2+1}\right]du=2(u-\tan^{-1}u)+C=2\left(\sqrt{x}-\tan^{-1}\sqrt{x}\right)+C.$$

5. Let $u=\sqrt[3]{x}$. Then $x=u^3$, $dx=3u^2\,du$ $\Rightarrow$

$$\int\frac{dx}{x-\sqrt[3]{x}}=\int\frac{3u^2\,du}{u^3-u}=3\int\frac{u\,du}{u^2-1}=\tfrac{3}{2}\ln\left|u^2-1\right|+C=\tfrac{3}{2}\ln\left|x^{2/3}-1\right|+C.$$

7. Let $u=\sqrt{x-1}$. Then $x=u^2+1$, $dx=2u\,du$ $\Rightarrow$ $\displaystyle\int_5^{10}\frac{x^2\,dx}{\sqrt{x-1}}=\int_2^3\frac{\left(u^2+1\right)^2 2u\,du}{u}$

$$=2\int_2^3(u^4+2u^2+1)du=2\left[\tfrac{1}{5}u^5+\tfrac{2}{3}u^3+u\right]_2^3=2\left(\tfrac{243}{5}+18+3\right)-2\left(\tfrac{32}{5}+\tfrac{16}{3}+2\right)=\tfrac{1676}{15}.$$

9. Let $u=\sqrt{x}$. Then $x=u^2$, $dx=2u\,du$ $\Rightarrow$

$$\int\frac{dx}{\sqrt{1+\sqrt{x}}}=\int\frac{2u\,du}{\sqrt{1+u}}=2\int\frac{(v^2-1)2v\,dv}{v}\quad\left(\text{put } v=\sqrt{1+u},\ u=v^2-1,\ du=2v\,dv\right)$$

$$=4\int(v^2-1)dv=\tfrac{4}{3}v^3-4v+C=\tfrac{4}{3}\left(1+\sqrt{x}\right)^{3/2}-4\sqrt{1+\sqrt{x}}+C.$$

11. Let $u=\sqrt{x}$. Then $x=u^2$, $dx=2u\,du$ $\Rightarrow$

$$\int\frac{\sqrt{x}+1}{\sqrt{x}-1}\,dx=\int\frac{u+1}{u-1}\,2u\,du=2\int\frac{u^2+u}{u-1}\,du=2\int\left[u+2+\frac{2}{u-1}\right]du$$

$$=u^2+4u+4\ln|u-1|+C=x+4\sqrt{x}+4\ln\left|\sqrt{x}-1\right|+C.$$

13. Let $u=\sqrt[3]{x^2+1}$. Then $x^2=u^3-1$, $2x\,dx=3u^2\,du$ $\Rightarrow$

$$\int\frac{x^3\,dx}{\sqrt[3]{x^2+1}}=\int\frac{(u^3-1)\tfrac{3}{2}u^2\,du}{u}=\tfrac{3}{2}\int(u^4-u)du$$

$$=\tfrac{3}{10}u^5-\tfrac{3}{4}u^2+C=\tfrac{3}{10}\left(x^2+1\right)^{5/3}-\tfrac{3}{4}\left(x^2+1\right)^{2/3}+C.$$

15. Let $u = \sqrt[4]{x}$. Then $x = u^4$, $dx = 4u^3\,du \quad \Rightarrow$

$$\int \frac{dx}{\sqrt{x}+\sqrt[4]{x}} = \int \frac{4u^3\,du}{u^2+u} = 4\int \frac{u^2\,du}{u+1} = 4\int \left[u-1+\frac{1}{u+1}\right]du$$
$$= 2u^2 - 4u + 4\ln|u+1| + C = 2\sqrt{x} - 4\sqrt[4]{x} + 4\ln\left(\sqrt[4]{x}+1\right) + C.$$

17. Let $u = \sqrt{x}$. Then $x = u^2$, $dx = 2u\,du \quad \Rightarrow$

$$\int \sqrt{\frac{1-x}{x}}\,dx = \int \frac{\sqrt{1-u^2}}{u}\,2u\,du = 2\int \sqrt{1-u^2}\,du = 2\int \cos^2\theta\,d\theta \quad (\text{put } u = \sin\theta)$$
$$= \theta + \sin\theta\cos\theta + C = \sin^{-1}\sqrt{x} + \sqrt{x(1-x)} + C.$$

Or: Let $u = \sqrt{\dfrac{1-x}{x}}$. This gives $I = \sqrt{x(1-x)} - \tan^{-1}\sqrt{\dfrac{1-x}{x}} + C$.

19. Let $u = e^x$. Then $x = \ln u$, $dx = \dfrac{du}{u} \quad \Rightarrow$

$$\int \frac{e^{2x}\,dx}{e^{2x}+3e^x+2} = \int \frac{u^2(du/u)}{u^2+3u+2} = \int \frac{u\,du}{(u+1)(u+2)} = \int \left[\frac{-1}{u+1}+\frac{2}{u+2}\right]du$$
$$= 2\ln|u+2| - \ln|u+1| + C = \ln\left[(e^x+2)^2/(e^x+1)\right] + C.$$

21. Let $u = e^x$. Then $x = \ln u$, $dx = \dfrac{du}{u} \quad \Rightarrow$

$$\int \sqrt{1-e^x}\,dx = \int \sqrt{1-u}\left(\frac{du}{u}\right) = \int \frac{\sqrt{1-u}\,du}{u} = \int \frac{v(-2v)dv}{1-v^2} \quad \left[\begin{array}{l}\text{where } v = \sqrt{1-u},\ u = 1-v^2,\\ du = -2v\,dv)\end{array}\right]$$
$$= 2\int \left[1+\frac{1}{v^2-1}\right]dv = 2\left[v + \frac{1}{2}\ln\left|\frac{v-1}{v+1}\right|\right] + C = 2\sqrt{1-e^x} + \ln\left(\frac{1-\sqrt{1-e^x}}{1+\sqrt{1-e^x}}\right) + C.$$

23. Let $t = \tan\left(\dfrac{x}{2}\right)$. Then, by Equation 1,

$$\int_0^{\pi/2} \frac{dx}{\sin x + \cos x} = \int_0^1 \frac{2\,dt}{2t+1-t^2} = -2\int_0^1 \frac{dt}{t^2-2t-1} = -2\int_0^1 \frac{dt}{(t-1)^2-2}$$
$$= -\frac{1}{\sqrt{2}}\ln\left|\frac{t-1-\sqrt{2}}{t-1+\sqrt{2}}\right|_0^1 = -\frac{1}{\sqrt{2}}\left[\ln 1 - \ln\frac{\sqrt{2}+1}{\sqrt{2}-1}\right] = \frac{1}{\sqrt{2}}\ln\left(\sqrt{2}+1\right)^2$$
$$= \sqrt{2}\ln\left(\sqrt{2}+1\right) \text{ or } -\sqrt{2}\ln\left(\sqrt{2}-1\right) \text{ since } \sqrt{2}+1 = \tfrac{1}{\sqrt{2}-1}$$
$$\text{or } \tfrac{1}{\sqrt{2}}\ln\left(3+2\sqrt{2}\right) \text{ since } \left(\sqrt{2}+1\right)^2 = 3+2\sqrt{2}.$$

25. Let $t = \tan(x/2)$. Then, by Equation 1,

$$\int \frac{dx}{3\sin x + 4\cos x} = \int \frac{2\,dt}{6t+4(1-t^2)} = \int \frac{-\,dt}{2t^2-3t-2} = -\int \left[\frac{-2/5}{2t+1}+\frac{1/5}{t-2}\right]dt$$
$$= \tfrac{1}{5}\ln\left|\frac{2t+1}{t-2}\right| + C = \tfrac{1}{5}\ln\left|\frac{2\tan(x/2)+1}{\tan(x/2)-2}\right| + C.$$

27. Let $t = \tan\left(\dfrac{x}{2}\right)$. Then $\displaystyle\int \frac{dx}{2\sin x + \sin 2x} = \frac{1}{2}\int \frac{dx}{\sin x + \sin x\cos x}$

$$= \frac{1}{2}\int \frac{2\,dt/(1+t^2)}{2t/(1+t^2)+2t(1-t^2)/(1+t^2)^2} = \frac{1}{2}\int \frac{(1+t^2)dt}{t(1+t^2)+t(1-t^2)}$$
$$= \frac{1}{4}\int \frac{(1+t^2)dt}{t} = \frac{1}{4}\int \left[\frac{1}{t}+t\right]dt = \tfrac{1}{4}\ln|t| + \tfrac{1}{8}t^2 + C = \frac{1}{4}\ln\left|\tan\frac{x}{2}\right| + \frac{1}{8}\tan^2\frac{x}{2} + C.$$

29. Let $t = \tan(x/2)$. Then

$$\int \frac{dx}{a\sin x + b\cos x} = \int \frac{2\,dt}{a(2t) + b(1-t^2)} = -\frac{2}{b}\int \frac{dt}{t^2 - 2(a/b)t - 1}$$

$$= -\frac{2}{b}\int \frac{dt}{(t-a/b)^2 - (1+a^2/b^2)} = -\frac{1}{b}\,\frac{b}{\sqrt{a^2+b^2}}\ln\left|\frac{t - a/b - \sqrt{a^2+b^2}/b}{t - a/b + \sqrt{a^2+b^2}/b}\right| + C$$

$$= \frac{1}{\sqrt{a^2+b^2}}\ln\left|\frac{b\tan(x/2) - a + \sqrt{a^2+b^2}}{b\tan(x/2) - a - \sqrt{a^2+b^2}}\right| + C.$$

31. **(a)** Let $t = \tan\left(\frac{x}{2}\right)$. Then $\int \sec x\,dx = \int \frac{dx}{\cos x} = \int \frac{2\,dt}{1-t^2} = \int \left[\frac{1}{1-t} + \frac{1}{1+t}\right]dt$

$$= \ln|1+t| - \ln|1-t| + C = \ln\left|\frac{1+t}{1-t}\right| + C = \ln\left|\frac{1+\tan(x/2)}{1-\tan(x/2)}\right| + C.$$

(b) $\tan\left(\frac{\pi}{4} + \frac{x}{2}\right) = \frac{\tan(\pi/4) + \tan(x/2)}{1 - \tan(\pi/4)\tan(x/2)} = \frac{1+\tan(x/2)}{1-\tan(x/2)}$. Substituting in the formula from part (a), we get $\int \sec x\,dx = \ln\left|\tan\left(\frac{1}{4}\pi + \frac{1}{2}x\right)\right| + C$.

33. According to Equation 8.2.1, $\int \sec x\,dx = \ln|\sec x + \tan x| + C$. Now

$$\frac{1+\tan(x/2)}{1-\tan(x/2)} = \frac{1+\sin(x/2)/\cos(x/2)}{1-\sin(x/2)/\cos(x/2)} = \frac{\cos(x/2)+\sin(x/2)}{\cos(x/2)-\sin(x/2)}$$

$$= \frac{[\cos(x/2)+\sin(x/2)]^2}{[\cos(x/2)-\sin(x/2)][\cos(x/2)+\sin(x/2)]} = \frac{1+2\cos(x/2)\sin(x/2)}{\cos^2(x/2)-\sin^2(x/2)}$$

$$= \frac{1+\sin x}{\cos x} \quad \text{(using identities from the endpapers)} \quad = \sec x + \tan x,$$

so $\ln\left|\frac{1+\tan(x/2)}{1-\tan(x/2)}\right| = \ln|\sec x + \tan x|$, and the formula in Exercise 31(a) agrees with (8.2.1).

EXERCISES 8.6

1. $$\int \frac{2x+5}{x-3}\,dx = \int \frac{(2x-6)+11}{x-3}\,dx = \int \left[2 + \frac{11}{x-3}\right]dx = 2x + 11\ln|x-3| + C$$

3. $$\int \sin^2 x\cos^3 x\,dx = \int \sin^2 x(1-\sin^2 x)\cos x\,dx = \int u^2(1-u^2)\,du \quad (\text{put } u = \sin x)$$

$$= \int (u^2 - u^4)\,du = \tfrac{1}{3}u^3 - \tfrac{1}{5}u^5 + C = \tfrac{1}{3}\sin^3 x - \tfrac{1}{5}\sin^5 x + C$$

5. Let $u = 1 - x^2$. Then $du = -2x\,dx \Rightarrow$

$$\int_0^{1/2} \frac{x\,dx}{\sqrt{1-x^2}} = -\int_1^{3/4} \frac{du}{2\sqrt{u}} = \int_{3/4}^1 \frac{du}{2\sqrt{u}} = \left[\sqrt{u}\right]_{3/4}^1 = 1 - \frac{\sqrt{3}}{2}.$$

7. Let $u = \sqrt{x-2}$. Then $x = u^2 + 2$, $dx = 2u\,du \Rightarrow \int \frac{\sqrt{x-2}}{x+2}\,dx = \int \frac{u \cdot 2u\,du}{u^2+4}$

$$= 2\int \left(1 - \frac{4}{u^2+4}\right)du = 2u - \tfrac{8}{2}\tan^{-1}\left(\tfrac{1}{2}u\right) + C = 2\sqrt{x-2} - 4\tan^{-1}\left(\tfrac{1}{2}\sqrt{x-2}\right) + C.$$

9. Use integration by parts: $u = \ln(1+x^2)$, $dv = dx \quad \Rightarrow \quad du = \dfrac{2x}{1+x^2}\,dx$, $v = x$, so

$$\int \ln(1+x^2)dx = x\ln(1+x^2) - \int x \cdot \frac{2x\,dx}{1+x^2} = x\ln(1+x^2) - 2\int\left[1 - \frac{1}{1+x^2}\right]dx$$
$$= x\ln(1+x^2) - 2x + 2\tan^{-1}x + C.$$

11. Let $u = 1 + \sqrt{x}$. Then $x = (u-1)^2$, $dx = 2(u-1)du \quad \Rightarrow$

$\int_0^1 \left(1+\sqrt{x}\right)^8 dx = \int_1^2 u^8 \cdot 2(u-1)du = 2\int_1^2 (u^9 - u^8)du = \left[\frac{1}{5}u^{10} - 2\cdot\frac{1}{9}u^9\right]_1^2 = \frac{1024}{5} - \frac{1024}{9} - \frac{1}{5} + \frac{2}{9} = \frac{4097}{45}.$

13. $$\int \frac{x\,dx}{x^2-2x+2} = \frac{1}{2}\int\frac{(2x-2)dx}{x^2-2x+2} + \int\frac{dx}{(x-1)^2+1} = \tfrac{1}{2}\ln(x^2-2x+2) + \tan^{-1}(x-1) + C$$

15. Let $u = \sqrt{9-x^2}$. Then $u^2 = 9 - x^2$, $u\,du = -x\,dx \quad \Rightarrow$

$$\int\frac{\sqrt{9-x^2}}{x}\,dx = \int\frac{\sqrt{9-x^2}}{x^2}\,x\,dx = \int\frac{u}{9-u^2}(-u)du = \int\left[1 - \frac{9}{9-u^2}\right]du$$
$$= u + 9\int\frac{du}{u^2-9} = u + \frac{9}{2\cdot 3}\ln\left|\frac{u-3}{u+3}\right| + C = \sqrt{9-x^2} + \frac{3}{2}\ln\left|\frac{\sqrt{9-x^2}-3}{\sqrt{9-x^2}+3}\right| + C$$
$$= \sqrt{9-x^2} + \frac{3}{2}\ln\frac{\left(\sqrt{9-x^2}-3\right)^2}{x^2} + C = \sqrt{9-x^2} + 3\ln\left|\frac{3-\sqrt{9-x^2}}{x}\right| + C.$$

Or: Put $x = 3\sin\theta$.

17. Integrate by parts: $u = x^2$, $dv = \cosh x\,dx \quad \Rightarrow \quad du = 2x\,dx$, $v = \sinh x$, so

$I = \int x^2\cosh x\,dx = x^2\sinh x - \int 2x\sinh x\,dx$. Now let $U = x$, $dV = \sinh x\,dx \Rightarrow dU = dx$, $V = \cosh x$.

So $I = x^2\sinh x - 2(x\cosh x - \int\cosh x\,dx)$

$\quad = x^2\sinh x - 2[x\cosh x - \sinh x] = (x^2+2)\sinh x - 2x\cosh x + C.$

19. Let $u = \sin x$. Then $\displaystyle\int\frac{\cos x\,dx}{1+\sin^2 x} = \int\frac{du}{1+u^2} = \tan^{-1}u + C = \tan^{-1}(\sin x) + C$

21. $\int_0^1 \cos\pi x\tan\pi x\,dx = \int_0^1 \sin\pi x\,dx = -\frac{1}{\pi}\int_0^1(-\pi\sin\pi x)dx = -\frac{1}{\pi}[\cos\pi x]_0^1 = -\frac{1}{\pi}(-1-1) = \frac{2}{\pi}.$

23. Integrate by parts twice, first with $u = e^{3x}$, $dv = \cos 5x\,dx$:

$\int e^{3x}\cos 5x\,dx = \frac{1}{5}e^{3x}\sin 5x - \int\frac{3}{5}e^{3x}\sin 5x\,dx = \frac{1}{5}e^{3x}\sin 5x + \frac{3}{25}e^{3x}\cos 5x - \frac{9}{25}\int e^{3x}\cos 5x\,dx$, so

$\frac{34}{25}\int e^{3x}\cos 5x\,dx = \frac{1}{25}e^{3x}(5\sin 5x + 3\cos 5x) + C_1$ and $\int e^{3x}\cos 5x\,dx = \frac{1}{34}e^{3x}(5\sin 5x + 3\cos 5x) + C.$

25. $$\int\frac{dx}{x^3+x^2+x+1} = \int\frac{dx}{(x+1)(x^2+1)} = \int\left[\frac{1/2}{x+1} - \frac{x/2-1/2}{x^2+1}\right]dx$$
$$= \frac{1}{2}\int\left(\frac{1}{x+1} - \frac{x}{x^2+1} + \frac{1}{x^2+1}\right)dx = \tfrac{1}{2}\ln|x+1| - \tfrac{1}{4}\ln(x^2+1) + \tfrac{1}{2}\tan^{-1}x + C$$

27. Let $t = x^3$. Then $dt = 3x^2\,dx \quad \Rightarrow \quad I = \int x^5 e^{-x^3}\,dx = \frac{1}{3}\int te^{-t}\,dt$. Now integrate by parts with $u = t$, $dv = e^{-t}\,dt$: $I = -\frac{1}{3}te^{-t} + \frac{1}{3}\int e^{-t}\,dt = -\frac{1}{3}te^{-t} - \frac{1}{3}e^{-t} + C = -\frac{1}{3}e^{-x^3}(x^3+1) + C.$

29. Let $u = 3x + 2$. Then

$$\int \frac{dx}{\sqrt{9x^2 + 12x - 5}} = \int \frac{dx}{\sqrt{(3x+2)^2 - 9}} = \frac{1}{3}\int \frac{du}{\sqrt{u^2 - 9}} = \tfrac{1}{3}\cosh^{-1}(u/3) + C_1$$

$$= \tfrac{1}{3}\cosh^{-1}\left[\tfrac{1}{3}(3x+2)\right] + C_1 = \tfrac{1}{3}\ln\left|3x + 2 + \sqrt{9x^2 + 12x - 5}\right| + C \quad \text{(by Equation 7.7.4).}$$

Or: Substitute $u = 3\sec\theta$.

31. $\int x^{1/3}\left(1 - x^{1/2}\right)dx = \int\left(x^{1/3} - x^{5/6}\right)dx = \frac{3}{4}x^{4/3} - \frac{6}{11}x^{11/6} + C$

33. Let $u = x^2 + 1$. Then $du = 2x\,dx$, so

$$\int \frac{x}{x^4 + 2x^2 + 10}\,dx = \int \frac{x}{(x^2+1)^2 + 9} = \frac{1}{2}\int \frac{du}{u^2 + 9} = \frac{1}{2}\cdot\frac{1}{3}\tan^{-1}\frac{u}{3} + C = \frac{1}{6}\tan^{-1}\frac{x^2+1}{3} + C.$$

35. $\int \sin^2 x\cos^4 x\,dx = \int(\sin x\cos x)^2\cos^2 x\,dx = \int \frac{1}{4}\sin^2 2x\,\frac{1}{2}(1 + \cos 2x)dx$

$$= \tfrac{1}{8}\int \sin^2 2x\,dx + \tfrac{1}{8}\int \sin^2 2x\cos 2x\,dx = \tfrac{1}{16}\int(1 - \cos 4x)dx + \tfrac{1}{16}\int \sin^2 2x(2\cos 2x)dx$$

$$= \tfrac{1}{16}x - \tfrac{1}{64}\sin 4x + \tfrac{1}{48}\sin^3 2x + C.$$

Or: Write $\int \sin^2 x\cos^4 x\,dx = \frac{1}{8}\int(1 - \cos 2x)(1 + \cos 2x)^2\,dx$.

37. Let $u = 1 - x^2$. Then $du = -2x\,dx \quad \Rightarrow$

$$\int \frac{x\,dx}{1 - x^2 + \sqrt{1 - x^2}} = -\frac{1}{2}\int \frac{du}{u + \sqrt{u}} = -\int \frac{v\,dv}{v^2 + v} \quad \left(\text{put } v = \sqrt{u},\ u = v^2,\ du = 2v\,dv\right)$$

$$= -\int \frac{dv}{v+1} = -\ln|v + 1| + C = -\ln\left(\sqrt{1 - x^2} + 1\right) + C.$$

39. Let $u = e^x$. Then $x = \ln u$, $dx = du/u \quad \Rightarrow$

$$\int \frac{e^x\,dx}{e^{2x} - 1} = \int \frac{u(du/u)}{u^2 - 1} = \int \frac{du}{u^2 - 1} = \frac{1}{2}\ln\left|\frac{u-1}{u+1}\right| + C = \tfrac{1}{2}\ln\left|\frac{e^x - 1}{e^x + 1}\right| + C.$$

41. $\displaystyle\int_{-1}^{1} x^5\cosh x\,dx = 0$ by Theorem 7.5.7, since $x^5\cosh x$ is odd.

43. $\int_{-3}^{3}|x^3 + x^2 - 2x|dx = \int_{-3}^{3}|(x+2)x(x-1)|dx$

$$= -\int_{-3}^{-2}(x^3 + x^2 - 2x)dx + \int_{-2}^{0}(x^3 + x^2 - 2x)dx - \int_{0}^{1}(x^3 + x^2 - 2x)dx + \int_{1}^{3}(x^3 + x^2 - 2x)dx.$$

Let $f(x) = \frac{1}{4}x^4 + \frac{1}{3}x^3 - x^2$. Then $f'(x) = x^3 + x^2 - 2x$, so

$\int_{-3}^{3}|x^3 + x^2 - 2x|dx = -f(-2) + f(-3) + f(0) - f(-2) - f(1) + f(0) + f(3) - f(1)$

$$= f(-3) - 2f(-2) + 2f(0) - 2f(1) + f(3) = \tfrac{9}{4} - 2\left(-\tfrac{8}{3}\right) + 2\cdot 0 - 2\left(-\tfrac{5}{12}\right) + \tfrac{81}{4} = \tfrac{86}{3}.$$

45. Let $u = \ln(\sin x)$. Then $du = \cot x\,dx \quad \Rightarrow \quad \int \cot x\ln(\sin x)dx = \int u\,du = \frac{1}{2}u^2 + C = \frac{1}{2}[\ln(\sin x)]^2 + C.$

47. $\dfrac{x}{(x^2+1)(x^2+4)} = \dfrac{Ax + B}{x^2 + 1} + \dfrac{Cx + D}{x^2 + 4} \quad \Rightarrow \quad x = (Ax + B)(x^2 + 4) + (Cx + D)(x^2 + 1) \quad \Rightarrow$

$0 = A + C$, $0 = B + D$, $1 = 4A + C$, and $0 = 4B + D \quad \Rightarrow \quad A = -C = \frac{1}{3}$, $B = D = 0$.

$$\int \frac{x\,dx}{(x^2+1)(x^2+4)} = \frac{1}{3}\int\left[\frac{x}{x^2+1} - \frac{x}{x^2+4}\right]dx = \frac{\ln(x^2+1)}{6} - \frac{\ln(x^2+4)}{6} + C = \frac{1}{6}\ln\frac{x^2+1}{x^2+4} + C.$$

49. Let $u = \sqrt[3]{x+c}$. Then $x = u^3 - c \quad \Rightarrow \quad \int x\sqrt[3]{x+c}\,dx = \int (u^3 - c)u \cdot 3u^2\,du$
$= 3\int (u^6 - cu^3)du = \frac{3}{7}u^7 - \frac{3}{4}cu^4 + C = \frac{3}{7}(x+c)^{7/3} - \frac{3}{4}c(x+c)^{4/3} + C.$

51. Let $u = \sqrt{x+1}$. Then $x = u^2 - 1 \quad \Rightarrow \quad \displaystyle\int \frac{dx}{x+4+4\sqrt{x+1}} = \int \frac{2u\,du}{u^2+3+4u} = \int \left[\frac{-1}{u+1} + \frac{3}{u+3}\right]du$
$= 3\ln|u+3| - \ln|u+1| + C = 3\ln\left(\sqrt{x+1}+3\right) - \ln\left(\sqrt{x+1}+1\right) + C.$

53. Use parts twice. First let $u = x^2 + 4x - 3$, $dv = \sin 2x\,dx \quad \Rightarrow \quad du = (2x+4)dx$, $v = -\frac{1}{2}\cos 2x$. Then
$I = \int (x^2+4x-3)\sin 2x\,dx = (x^2+4x-3)\left(-\frac{1}{2}\cos 2x\right) + \int (2x+4)\left(\frac{1}{2}\cos 2x\right)dx.$
Now let $U = 2x+4$, $dV = \frac{1}{2}\cos 2x\,dx \quad \Rightarrow \quad dU = 2\,dx$, $V = \frac{1}{4}\sin 2x$. Then
$I = (x^2+4x-3)\left(-\frac{1}{2}\cos 2x\right) + (2x+4)\left(\frac{1}{4}\sin 2x\right) - \frac{1}{2}\int \sin 2x\,dx$
$= -\frac{1}{2}(x^2+4x-3)\cos 2x + \frac{1}{2}(x+2)\sin 2x + \frac{1}{4}\cos 2x + C$
$= \frac{1}{2}(x+2)\sin 2x - \frac{1}{4}(2x^2+8x-7)\cos 2x + C.$

55. Let $u = x^2$. Then $du = 2x\,dx \quad \Rightarrow$
$$\int \frac{x\,dx}{\sqrt{16-x^4}} = \frac{1}{2}\int \frac{du}{\sqrt{16-u^2}} = \tfrac{1}{2}\sin^{-1}\left(\tfrac{1}{4}u\right) + C = \tfrac{1}{2}\sin^{-1}\left(\tfrac{1}{4}x^2\right) + C.$$

57. Let $u = \csc 2x$. Then $du = -2\cot 2x\csc 2x\,dx \quad \Rightarrow$
$\int \cot^3 2x\csc^3 2x\,dx = \int \csc^2 2x(\csc^2 2x - 1)\cot 2x\csc 2x\,dx = \int u^2(u^2-1)\left(-\frac{1}{2}\,du\right)$
$= -\frac{1}{2}\int (u^4 - u^2)du = -\frac{1}{2}\left[\frac{1}{5}u^5 - \frac{1}{3}u^3\right] + C = \frac{1}{6}\csc^3 2x - \frac{1}{10}\csc^5 2x + C.$

59. Let $u = \arctan x$. Then $du = \dfrac{dx}{1+x^2} \quad \Rightarrow \quad \displaystyle\int \frac{e^{\arctan x}}{1+x^2}\,dx = \int e^u\,du = e^u + C = e^{\arctan x} + C.$

61. Integrate by parts three times, first with $u = t^3$, $dv = e^{-2t}\,dt$:
$\int t^3 e^{-2t}\,dt = -\frac{1}{2}t^3e^{-2t} + \frac{1}{2}\int 3t^2e^{-2t}\,dt = -\frac{1}{2}t^3e^{-2t} - \frac{3}{4}t^2e^{-2t} + \frac{1}{2}\int 3te^{-2t}\,dt$
$= -e^{-2t}\left[\frac{1}{2}t^3 + \frac{3}{4}t^2\right] - \frac{3}{4}te^{-2t} + \frac{3}{4}\int e^{-2t}\,dt = -e^{-2t}\left[\frac{1}{2}t^3 + \frac{3}{4}t^2 + \frac{3}{4}t + \frac{3}{8}\right] + C$
$= -\frac{1}{8}e^{-2t}(4t^3 + 6t^2 + 6t + 3) + C.$

63. $\int \sin x\sin 2x\sin 3x\,dx = \int \sin x \cdot \frac{1}{2}[\cos(2x-3x) - \cos(2x+3x)]dx$
$= \frac{1}{2}\int (\sin x\cos x - \sin x\cos 5x)dx = \frac{1}{4}\int \sin 2x\,dx - \frac{1}{2}\int \frac{1}{2}[\sin(x+5x) + \sin(x-5x)]dx$
$= -\frac{1}{8}\cos 2x - \frac{1}{4}\int(\sin 6x - \sin 4x)dx = -\frac{1}{8}\cos 2x + \frac{1}{24}\cos 6x - \frac{1}{16}\cos 4x + C$

65. As in Example 5, $\displaystyle\int \sqrt{\frac{1+x}{1-x}}\,dx = \int \frac{1+x}{\sqrt{1-x^2}}\,dx = \int \frac{dx}{\sqrt{1-x^2}} + \int \frac{x\,dx}{\sqrt{1-x^2}} = \sin^{-1}x - \sqrt{1-x^2} + C.$
Another Method: Substitute $u = \sqrt{(1+x)/(1-x)}$.

67. $\displaystyle\int \frac{x+a}{x^2+a^2}\,dx = \frac{1}{2}\int \frac{2x\,dx}{x^2+a^2} + a\int \frac{dx}{x^2+a^2} = \frac{1}{2}\ln(x^2+a^2) + a \cdot \frac{1}{a}\tan^{-1}\left(\frac{x}{a}\right) + C$
$= \ln\sqrt{x^2+a^2} + \tan^{-1}(x/a) + C.$

69. Let $u = x^5$. Then $du = 5x^4\,dx \Rightarrow \displaystyle\int \frac{x^4\,dx}{x^{10}+16} = \int \frac{\frac{1}{5}\,du}{u^2+16} = \frac{1}{5}\cdot\frac{1}{4}\tan^{-1}\left(\tfrac{1}{4}u\right) + C = \frac{1}{20}\tan^{-1}\left(\tfrac{1}{4}x^5\right) + C.$

71. Integrate by parts with $u = x$, $dv = \sec x \tan x\,dx \quad \Rightarrow \quad du = dx$, $v = \sec x$:

$\int x \sec x \tan x\,dx = x\sec x - \int \sec x\,dx = x \sec x - \ln|\sec x + \tan x| + C.$

73. $\displaystyle\int \frac{dx}{\sqrt{x+1}+\sqrt{x}} = \int\left(\sqrt{x+1}-\sqrt{x}\right)dx = \frac{2}{3}\left[(x+1)^{3/2} - x^{3/2}\right] + C.$

75. Let $u = \sqrt{x}$. Then $du = dx/(2\sqrt{x}) \quad \Rightarrow$

$$\int \frac{\arctan\sqrt{x}}{\sqrt{x}}\,dx = \int \tan^{-1}u\,2\,du = 2u\tan^{-1}u - \int\frac{2u\,du}{1+u^2} \quad \text{(by parts)}$$

$$= 2u\tan^{-1}u - \ln(1+u^2) + C = 2\sqrt{x}\tan^{-1}\sqrt{x} - \ln(1+x) + C.$$

77. Let $u = e^x$. Then $x = \ln u$, $dx = du/u \quad \Rightarrow$

$$\int\frac{dx}{e^{3x}-e^x} = \int\frac{du/u}{u^3-u} = \int\frac{du}{(u-1)u^2(u+1)} = \int\left[\frac{1/2}{u-1} - \frac{1}{u^2} - \frac{1/2}{u+1}\right]du = \frac{1}{u} + \frac{1}{2}\ln\left|\frac{u-1}{u+1}\right| + C$$

$$= e^{-x} + \tfrac{1}{2}\ln|(e^x-1)/(e^x+1)| + C.$$

79. Let $u = \sqrt{2x-25} \quad \Rightarrow \quad u^2 = 2x - 25 \quad \Rightarrow \quad 2u\,du = 2\,dx \quad \Rightarrow$

$$\int\frac{dx}{x\sqrt{2x-25}} = \int\frac{u\,du}{\frac{1}{2}(u^2+25)\cdot u} = 2\int\frac{du}{u^2+25} = \tfrac{2}{5}\tan^{-1}\left(\tfrac{1}{5}u\right) + C = \tfrac{2}{5}\tan^{-1}\left(\tfrac{1}{5}\sqrt{2x-25}\right) + C.$$

EXERCISES 8.7

1. By Formula 99,

$$\int e^{-3x}\cos 4x\,dx = \frac{e^{-3x}}{(-3)^2+4^2}(-3\cos 4x + 4\sin 4x) + C = \frac{e^{-3x}}{25}(-3\cos 4x + 4\sin 4x) + C.$$

3. Let $u = 3x$. Then $du = 3\,dx$, so $\displaystyle\int\frac{\sqrt{9x^2-1}}{x^2}\,dx = \int\frac{\sqrt{u^2-1}}{u^2/9}\,\frac{du}{3} = 3\int\frac{\sqrt{u^2-1}}{u^2}\,du$

$$= -\frac{3\sqrt{u^2-1}}{u} + 3\ln\left|u + \sqrt{u^2-1}\right| + C \quad \text{(by Formula 42)} = -\frac{\sqrt{9x^2-1}}{x} + 3\ln\left|3x + \sqrt{9x^2-1}\right| + C.$$

5. $\int x^2 e^{3x}\,dx = \frac{1}{3}x^2e^{3x} - \frac{2}{3}\int xe^{3x}\,dx$ (Formula 97) $= \frac{1}{3}x^2e^{3x} - \frac{2}{3}\left[\frac{1}{9}(3x-1)e^{3x}\right] + C$ (Formula 96)

$= \frac{1}{27}(9x^2 - 6x + 2)e^{3x} + C$

7. Let $u = x^2$. Then $du = 2x\,dx$, so $\int x\sin^{-1}(x^2)dx = \frac{1}{2}\int\sin^{-1}u\,du = \frac{1}{2}\left(u\sin^{-1}u + \sqrt{1-u^2}\right) + C$

(Formula 87) $= \frac{1}{2}\left(x^2\sin^{-1}(x^2) + \sqrt{1-x^4}\right) + C.$

9. Let $u = e^x$. Then $du = e^x\,dx$, so $\int e^x\,\text{sech}(e^x)dx = \int\text{sech}\,u\,du = \tan^{-1}|\sinh u| + C$ (Formula 107)

$= \tan^{-1}[\sinh(e^x)] + C.$

11. Let $u = x + 2$. Then $\displaystyle\int\sqrt{5-4x-x^2}\,dx = \int\sqrt{9-(x+2)^2}\,dx = \int\sqrt{9-u^2}\,du$

$$= \frac{u}{2}\sqrt{9-u^2} + \frac{9}{2}\sin^{-1}\frac{u}{3} + C \text{ (Formula 30)} = \frac{x+2}{2}\sqrt{5-4x-x^2} + \frac{9}{2}\sin^{-1}\frac{x+2}{3} + C.$$

13. $\int \sec^5 x\,dx = \frac{1}{4}\tan x\sec^3 x + \frac{3}{4}\int \sec^3 x\,dx$ (Formula 77)

$= \frac{1}{4}\tan x\sec^3 x + \frac{3}{4}\left(\frac{1}{2}\tan x\sec x + \frac{1}{2}\int \sec x\,dx\right)$ (Formula 77 again)

$= \frac{1}{4}\tan x\sec^3 x + \frac{3}{8}\tan x\sec x + \frac{3}{8}\ln|\sec x + \tan x| + C$ (Formula 14)

15. Let $u = \sin x$. Then $du = \cos x\,dx$, so $\int \sin^2 x\cos x\ln(\sin x)dx = \int u^2\ln u\,du$

$= \frac{1}{9}u^3(3\ln u - 1) + C$ (Formula 101) $= \frac{1}{9}\sin^3 x[3\ln(\sin x) - 1] + C$.

17. $\displaystyle\int \sqrt{2+3\cos x}\tan x\,dx = -\int \frac{\sqrt{2+3\cos x}}{\cos x}(-\sin x\,dx) = -\int \frac{\sqrt{2+3u}}{u}\,du$ (where $u = \cos x$)

$\displaystyle= -2\sqrt{2+3u} - 2\int \frac{du}{u\sqrt{2+3u}}$ (Formula 58) $\displaystyle= -2\sqrt{2+3u} - 2\cdot\frac{1}{\sqrt{2}}\ln\left|\frac{\sqrt{2+3u}-\sqrt{2}}{\sqrt{2+3u}+\sqrt{2}}\right| + C$

(Formula 57) $\displaystyle= -2\sqrt{2+3\cos x} - \sqrt{2}\ln\left|\frac{\sqrt{2+3\cos x}-\sqrt{2}}{\sqrt{2+3\cos x}+\sqrt{2}}\right| + C$

19. $\int_0^{\pi/2}\cos^5 x\,dx = \frac{1}{5}[\cos^4 x\sin x]_0^{\pi/2} + \frac{4}{5}\int_0^{\pi/2}\cos^3 x\,dx$ (Formula 74)

$= 0 + \frac{4}{5}\left[\frac{1}{3}(2+\cos^2 x)\sin x\right]_0^{\pi/2}$ (Formula 68) $= \frac{4}{15}(2-0) = \frac{8}{15}$

21. Let $u = x^5$, $du = 5x^4\,dx$.

$\displaystyle\int \frac{x^4\,dx}{\sqrt{x^{10}-2}} = \frac{1}{5}\int \frac{du}{\sqrt{u^2-2}} = \frac{1}{5}\ln\left|u + \sqrt{u^2-2}\right| + C$ (Formula 43) $= \frac{1}{5}\ln\left|x^5 + \sqrt{x^{10}-2}\right| + C$.

23. Let $u = 1 + e^x$, so $du = e^x\,dx$. Then $\int e^x\ln(1+e^x)dx = \int \ln u\,du = u\ln u - u + C$ (Formula 100)

$= (1+e^x)\ln(1+e^x) - e^x - 1 + C = (1+e^x)\ln(1+e^x) - e^x + C_1$.

25. Let $u = e^x \Rightarrow \ln u = x \Rightarrow dx = \dfrac{du}{u}$. Then $\displaystyle\int \sqrt{e^{2x}-1}\,dx = \int \frac{\sqrt{u^2-1}}{u}\,du$

$= \sqrt{u^2-1} - \cos^{-1}(1/u) + C$ (Formula 41) $= \sqrt{e^{2x}-1} - \cos^{-1}(e^{-x}) + C$.

Or: Let $u = \sqrt{e^{2x}-1}$.

27. Volume $\displaystyle= \int_0^1 \frac{2\pi x}{(1+5x)^2}\,dx = 2\pi\left[\frac{1}{25(1+5x)} + \frac{1}{25}\ln|1+5x|\right]_0^1$ (Formula 51)

$= \frac{2\pi}{25}\left(\frac{1}{6} + \ln 6 - 1 - \ln 1\right) = \frac{2\pi}{25}\left(\ln 6 - \frac{5}{6}\right)$

29. **(a)** $\displaystyle\frac{d}{du}\left[\frac{1}{b^3}\left(a + bu - \frac{a^2}{a+bu} - 2a\ln|a+bu|\right) + C\right] = \frac{1}{b^3}\left[b + \frac{ba^2}{(a+bu)^2} - \frac{2ab}{(a+bu)}\right]$

$\displaystyle= \frac{1}{b^3}\left[\frac{b(a+bu)^2 + ba^2 - (a+bu)2ab}{(a+bu)^2}\right] = \frac{1}{b^3}\left[\frac{b^3u^2}{(a+bu)^2}\right] = \frac{u^2}{(a+bu)^2}$

(b) Let $t = a + bu \Rightarrow dt = b\,du$.

$\displaystyle\int \frac{u^2\,du}{(a+bu)^2} = \frac{1}{b^3}\int \frac{(t-a)^2}{t^2}\,dt = \frac{1}{b^3}\int\left(1 - \frac{2a}{t} + \frac{a^2}{t^2}\right)dt = \frac{1}{b^3}\left(t - 2a\ln|t| - \frac{a^2}{t}\right) + C$

$\displaystyle= \frac{1}{b^3}\left(a + bu - \frac{a^2}{a+bu} - 2a\ln|a+bu|\right) + C$

31. Maple, Mathematica and Derive all give $\int x^2\sqrt{5-x^2}\,dx = -\frac{1}{4}x(5-x^2)^{3/2} + \frac{5}{8}x\sqrt{5-x^2} + \frac{25}{8}\sin^{-1}\left(\frac{1}{\sqrt{5}}x\right)$. Using Formula 31, we get $\int x^2\sqrt{5-x^2}\,dx = \frac{1}{8}x(2x^2-5)\sqrt{5-x^2} + \frac{1}{8}(5^2)\sin^{-1}\left(\frac{1}{\sqrt{5}}x\right) + C$. But $-\frac{1}{4}x(5-x^2)^{3/2} + \frac{5}{8}x\sqrt{5-x^2} = \frac{1}{8}x\sqrt{5-x^2}[5-2(5-x^2)] = \frac{1}{8}x(2x^2-5)\sqrt{5-x^2}$, and the $\sin^{-1}$ terms are the same in each expression, so the answers are equivalent.

33. Maple and Derive both give $\int \sin^3 x\cos^2 x\,dx = -\frac{1}{5}\sin^2 x\cos^3 x - \frac{2}{15}\cos^3 x$ (although Derive factors the expression), and Mathematica gives $\int \sin^3 x\cos^2 x\,dx = -\frac{1}{8}\cos x - \frac{1}{48}\cos 3x + \frac{1}{80}\cos 5x$. We can use a CAS to show that both of these expressions are equal to $-\frac{1}{3}\cos^3 x + \frac{1}{5}\cos^5 x$. Using Formula 86, we write

$$\begin{aligned}\int \sin^3 x\cos^2 x\,dx &= -\tfrac{1}{5}\sin^2 x\cos^3 x + \tfrac{2}{5}\int \sin x\cos^2 x\,dx = -\tfrac{1}{5}\sin^2 x\cos^3 x + \tfrac{2}{5}\left(-\tfrac{1}{3}\cos^3 x\right) + C \\ &= -\tfrac{1}{5}\sin^2 x\cos^3 x - \tfrac{2}{15}\cos^3 x + C.\end{aligned}$$

35. Maple gives $\int x\sqrt{1+2x}\,dx = \frac{1}{10}(1+2x)^{5/2} - \frac{1}{6}(1+2x)^{3/2}$, Mathematica gives $\sqrt{1+2x}\left(\frac{2}{5}x^2 + \frac{1}{15}x - \frac{1}{15}\right)$, and Derive gives $\frac{1}{15}(1+2x)^{3/2}(3x-1)$. The first two expressions can be simplified to Derive's result. If we use Formula 54, we get $\int x\sqrt{1+2x}\,dx = \frac{2}{15(2)^2}(3\cdot 2x - 2\cdot 1)(1+2x)^{3/2} + C$

$= \frac{1}{30}(6x-2)(1+2x)^{3/2} + C = \frac{1}{15}(3x-1)(1+2x)^{3/2}$.

37. Maple gives $\int \tan^3 x\,dx = \frac{1}{2}\tan^2 x - \frac{1}{2}\ln(1+\tan^2 x)$, while Mathematica and Derive both give $\ln\cos x + \frac{1}{2}\tan^2 x$. These expressions are equivalent, since $-\frac{1}{2}\ln(1+\tan^2 x) = \ln\left[(\sec^2 x)^{-1/2}\right] = \ln\cos x$. Using Formula 69, we get $\int \tan^3 x\,dx = \frac{1}{2}\tan^2 x + \ln|\cos x| + C$.

39. Maple gives the antiderivative

$$F(x) = \int \frac{x^2-1}{x^4+x^2+1}\,dx = -\tfrac{1}{2}\ln(x^2+x+1) + \tfrac{1}{2}\ln(x^2-x+1).$$

We can see that at 0, this antiderivative is 0. From the graphs, it appears that F has a maximum at $x=-1$ and a minimum at $x=1$ [since $F'(x) = f(x)$ changes sign at these x-values], and that F has inflection points at $x\approx -1.7$, $x=0$ and $x\approx 1.7$ [since $f(x)$ has extrema at these x-values].

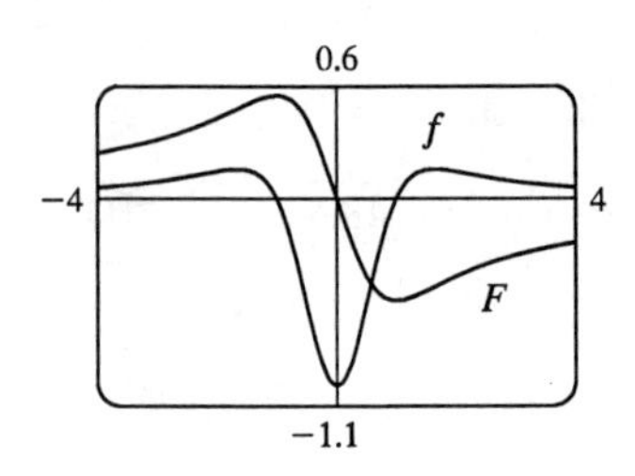

41. Since f is everywhere positive, we know that its antiderivative F is increasing. The antiderivative given by Maple is

$$\begin{aligned}\int \sin^4 x\cos^6 x\,dx = -\tfrac{1}{10}\sin^3 x\cos^7 x - \tfrac{3}{80}\sin x\cos^7 x \\ + \tfrac{1}{160}\cos^5 x\sin x + \tfrac{1}{128}\cos^3 x\sin x \\ + \tfrac{3}{256}\cos x\sin x + \tfrac{3}{256}x,\end{aligned}$$

and this is 0 at $x=0$.

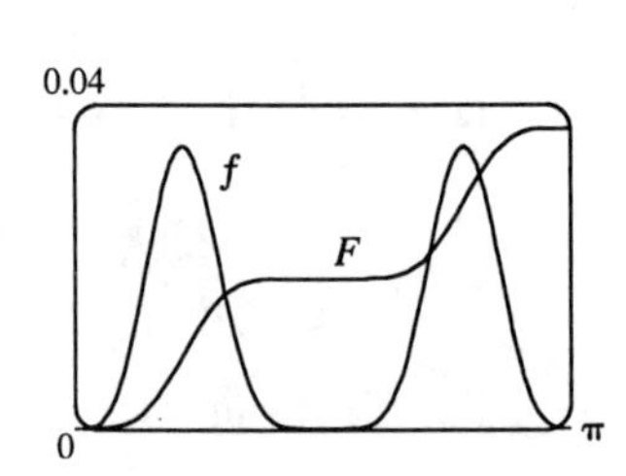

EXERCISES 8.8

1. **(a)** $L_2 = \sum_{i=1}^{2} f(x_{i-1})\Delta x = 2\,f(0) + 2\,f(2) = 2(0.5) + 2(2.5) = 6$

$R_2 = \sum_{i=1}^{2} f(x_i)\Delta x = 2\,f(2) + 2\,f(4) = 2(2.5) + 2(3.5) = 12$

$M_2 = \sum_{i=1}^{2} f(\overline{x}_i)\Delta x = 2\,f(1) + 2\,f(3) \approx 2(1.7) + 2(3.2) = 9.8$

(b)

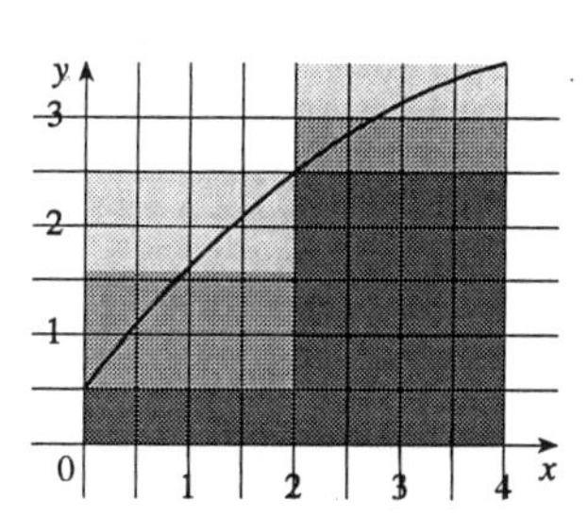

L_2 is an underestimate, since the area under the small rectangles is less than the area under the curve, and R_2 is an overestimate, since the area under the large rectangles is greater than the area under the curve. It appears that M_2 is an overestimate, though it is fairly close to I. See the solution to Exercise 37 for a proof of the fact that if f is concave down on $[a, b]$, then the Midpoint Rule is an overestimate of $\int_a^b f(x)dx$.

(c) $T_2 = \left(\frac{1}{2}\Delta x\right)[f(x_0) + 2f(x_1) + f(x_2)] = \frac{2}{2}[f(0) + 2\,f(2) + f(4)] = 0.5 + 2(2.5) + 3.5 = 9.$

This approximation is an underestimate, since the graph is concave down. See the solution to Exercise 37 for a general proof of this conclusion.

(d) For any n, we will have $L_n < T_n < I < M_n < R_n$.

3. $f(x) = \sqrt{1 + x^3}$, $\Delta x = \dfrac{1 - (-1)}{8} = \dfrac{1}{4}$

(a) $T_8 = \frac{0.25}{2}\left[f(-1) + 2f\left(-\frac{3}{4}\right) + 2f\left(-\frac{1}{2}\right) + \cdots + 2f\left(\frac{1}{2}\right) + 2f\left(\frac{3}{4}\right) + f(1)\right] \approx 1.913972$

(b) $S_8 = \frac{0.25}{3}\left[f(-1) + 4f\left(-\frac{3}{4}\right) + 2f\left(-\frac{1}{2}\right) + 4f\left(-\frac{1}{4}\right) + 2f(0) + 4f\left(\frac{1}{4}\right) + 2f\left(\frac{1}{2}\right) + 4f\left(\frac{3}{4}\right) + f(1)\right]$

≈ 1.934766

5. $f(x) = \dfrac{\sin x}{x}$, $\Delta x = \dfrac{\pi - \pi/2}{6} = \dfrac{\pi}{12}$

(a) $T_6 = \frac{\pi}{24}\left[f\left(\frac{\pi}{2}\right) + 2f\left(\frac{7\pi}{12}\right) + 2f\left(\frac{2\pi}{3}\right) + 2f\left(\frac{3\pi}{4}\right) + 2f\left(\frac{5\pi}{6}\right) + 2f\left(\frac{11\pi}{12}\right) + f(\pi)\right] \approx 0.481672$

(b) $S_6 = \frac{\pi}{36}\left[f\left(\frac{\pi}{2}\right) + 4f\left(\frac{7\pi}{12}\right) + 2f\left(\frac{2\pi}{3}\right) + 4f\left(\frac{3\pi}{4}\right) + 2f\left(\frac{5\pi}{6}\right) + 4f\left(\frac{11\pi}{12}\right) + f(\pi)\right] \approx 0.481172$

7. $f(x) = e^{-x^2}$, $\Delta x = \dfrac{1 - 0}{10} = 0.1$

(a) $T_{10} = \frac{0.1}{2}[f(0) + 2f(0.1) + 2f(0.2) + \cdots + 2f(0.8) + 2f(0.9) + f(1)] \approx 0.746211$

(b) $M_{10} = 0.1\left[f(0.05) + f(0.15) + f(0.25) + \cdots + f(0.75) + f(0.85) + f(0.95)\right] \approx 0.747131$

(c) $S_{10} = \frac{0.1}{3}\left[f(0) + 4f(0.1) + 2f(0.2) + 4f(0.3) + 2f(0.4) + 4f(0.5)\right.$

$\left. + 2f(0.6) + 4f(0.7) + 2f(0.8) + 4f(0.9) + f(1)\right] \approx 0.746825$

9. $f(x) = \cos(e^x)$, $\Delta x = \dfrac{1/2 - 0}{8} = \dfrac{1}{16}$

(a) $T_8 = \frac{1}{32}\left[f(0) + 2f\left(\frac{1}{16}\right) + 2f\left(\frac{1}{8}\right) + \cdots + 2f\left(\frac{7}{16}\right) + f\left(\frac{1}{2}\right)\right] \approx 0.132465$

(b) $M_8 = \frac{1}{16}\left[f\left(\frac{1}{32}\right) + f\left(\frac{3}{32}\right) + f\left(\frac{5}{32}\right) + \cdots + f\left(\frac{15}{32}\right)\right] \approx 0.132857$

(c) $S_8 = \frac{1}{48}\left[f(0) + 4f\left(\frac{1}{16}\right) + 2f\left(\frac{1}{8}\right) + 4f\left(\frac{3}{16}\right) + 2f\left(\frac{1}{4}\right) + 4f\left(\frac{5}{16}\right) + 2f\left(\frac{3}{8}\right) + 4f\left(\frac{7}{16}\right) + f\left(\frac{1}{2}\right)\right]$
≈ 0.132727

11. $f(x) = x^5 e^x$, $\Delta x = \dfrac{1-0}{10} = \dfrac{1}{10}$

(a) $T_{10} = \frac{0.1}{2}[f(0) + 2f(0.1) + 2f(0.2) + \cdots + 2f(0.9) + f(1)] \approx 0.409140$

(b) $M_{10} = 0.1[f(0.05) + f(0.15) + f(0.25) + \cdots + f(0.95)] \approx 0.388849$

(c) $S_{10} = \frac{0.1}{3}[f(0) + 4f(0.1) + 2f(0.2) + 4f(0.3) + 2f(0.4) + 4f(0.5)$
$+ 2f(0.6) + 4f(0.7) + 2f(0.8) + 4f(0.9) + f(1)] \approx 0.395802$

13. $f(x) = e^{1/x}$, $\Delta x = \dfrac{2-1}{4} = \dfrac{1}{4}$

(a) $T_4 = \frac{1}{4\cdot 2}[f(1) + 2f(1.25) + 2f(1.5) + 2f(1.75) + f(2)] \approx 2.031893$

(b) $M_4 = \frac{1}{4}[f(1.125) + f(1.375) + f(1.625) + f(1.875)] \approx 2.014207$

(c) $S_4 = \frac{1}{4\cdot 3}[f(1) + 4f(1.25) + 2f(1.5) + 4f(1.75) + f(2)] \approx 2.020651$

15. $f(x) = \dfrac{1}{1+x^4}$, $\Delta x = \dfrac{3-0}{6} = \dfrac{1}{2}$

(a) $T_6 = \frac{1}{2\cdot 2}[f(0) + 2f(0.5) + 2f(1) + 2f(1.5) + 2f(2) + 2f(2.5) + f(3)] \approx 1.098004$

(b) $M_6 = \frac{1}{2}[f(0.25) + f(0.75) + f(1.25) + f(1.75) + f(2.25) + f(2.75)] \approx 1.098709$

(c) $S_6 = \frac{1}{2\cdot 3}[f(0) + 4f(0.5) + 2f(1) + 4f(1.5) + 2f(2) + 4f(2.5) + f(3)] = 1.109031$

17. $f(x) = e^{-x^2}$, $\Delta x = \dfrac{2-0}{10} = \dfrac{1}{5}$

(a) $T_{10} = \frac{1}{5\cdot 2}[f(0) + 2(f(0.2) + f(0.4) + \cdots + f(1.8)) + f(2)] \approx 0.881839$
$M_{10} = \frac{1}{5}[f(0.1) + f(0.3) + f(0.5) + \cdots + f(1.7) + f(1.9)] \approx 0.882202$

(b) $f(x) = e^{-x^2}$, $f'(x) = -2xe^{-x^2}$, $f''(x) = (4x^2 - 2)e^{-x^2}$, $f'''(x) = 4x(3 - 2x^2)e^{-x^2}$.
$f'''(x) = 0 \quad \Leftrightarrow \quad x = 0$ or $x = \pm\sqrt{\frac{3}{2}}$. So to find the maximum value of $|f''(x)|$ on $[0, 2]$, we need only consider its values at $x = 0$, $x = 2$, and $x = \sqrt{\frac{3}{2}}$. $|f''(0)| = 2$, $|f''(2)| \approx 0.2564$ and $\left|f''\left(\sqrt{\frac{3}{2}}\right)\right| = 4e^{-3/2} \approx 0.8925$. Thus, taking $K = 2$, $a = 0$, $b = 2$, and $n = 10$ in Theorem 5, we get $|E_T| \le 2 \cdot 2^3/\left[12(10)^2\right] = \frac{1}{75} = 0.01\overline{3}$, and $|E_M| \le 2 \cdot 2^3/\left[24(10)^2\right] = 0.00\overline{6}$.

19. **(a)** $T_{10} = \frac{1}{10\cdot 2}[f(0) + 2(f(0.1) + f(0.2) + \cdots + f(0.9)) + f(1)] \approx 1.719713$
$S_{10} = \frac{1}{10\cdot 3}[f(0) + 4f(0.1) + 2f(0.2) + 4f(0.3) + \cdots + 4f(0.9) + f(1)] \approx 1.7182828$
Since $\int_0^1 e^x\,dx = [e^x]_0^1 = e - 1 \approx 1.71828183$, $E_T \approx -0.00143166$ and $E_S \approx -0.00000095$.

(b) $f(x) = e^x \quad \Rightarrow \quad f''(x) = e^x \le e$ for $0 \le x \le 1$. Taking $K = e$, $a = 0$, $b = 1$, and $n = 10$ in Theorem 5, we get $|E_T| \le \dfrac{e(1)^3}{12(10)^2} \approx 0.002265 > 0.00143166$ [actual $|E_T|$ from (a)]. $f^{(4)}(x) = e^x < e$ for $0 \le x \le 1$. Using Theorem 7, we have $|E_S| \le e(1)^5/\left[180(10)^4\right] \approx 0.0000015 > 0.00000095$ [actual $|E_S|$ from (a)]. We see that the actual errors are about two-thirds the size of the error estimates.

21. Take $K = 2$ (as in Exercise 17) in Theorem 5. $|E_T| \le \dfrac{K(b-a)^3}{12n^2} \le 10^{-5} \quad \Leftrightarrow \quad \dfrac{1}{6n^2} \le 10^{-5} \quad \Leftrightarrow$ $6n^2 \ge 10^5 \quad \Leftrightarrow \quad n \ge 129.099\ldots \quad \Leftrightarrow \quad n \ge 130$. Take $n = 130$ in the trapezoidal method. For E_M, again take $K = 2$ in Theorem 5 to get $|E_M| \le 2(1)^3/(24n^2) \le 10^{-5} \quad \Leftrightarrow \quad n^2 \ge 2/\left[24\left(10^{-5}\right)\right] \quad \Leftrightarrow \quad n \ge 91.3$ $\Rightarrow \quad n \ge 92$. Take $n = 92$ for M_n.

23. **(a)** Using the CAS, we differentiate $f(x) = e^{\cos x}$ twice, and find that $f''(x) = e^{\cos x}(\sin^2 x - \cos x)$.

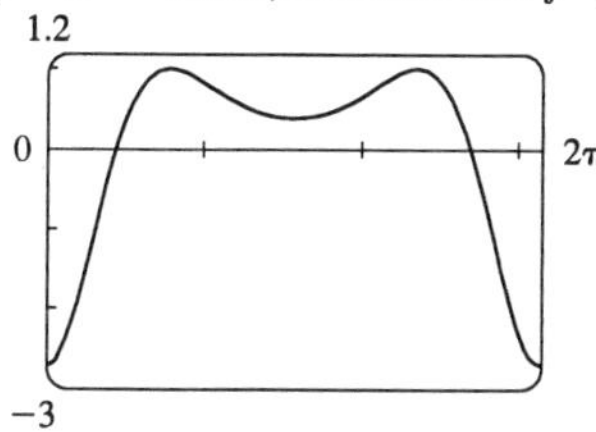

From the graph, we see that $|f''(x)| < 2.8$ on $[0, 2\pi]$. Other possible upper bounds for $|f''(x)|$ are $K = 3$ or $K = e$ (the actual maximum value.)

(b) A CAS gives $M_{10} \approx 7.954926518$. (In Maple, use `student[middlesum]`.)

(c) Using Theorem 5 for the Midpoint Rule, with $K = 2.8$, we get $|E_M| \le \dfrac{2.8(2\pi - 0)^3}{24 \cdot 10^2} \approx 0.287$.

(d) A CAS gives $I \approx 7.954926521$.

(e) The actual error is only about 3×10^{-9}, much less than the estimate in part (c).

(f) We use the CAS to differentiate twice more: $f^{(4)}(x) = e^{\cos x}(\sin^4 x - 6\sin^2 x \cos x + 3 - 7\sin^2 x + \cos x)$

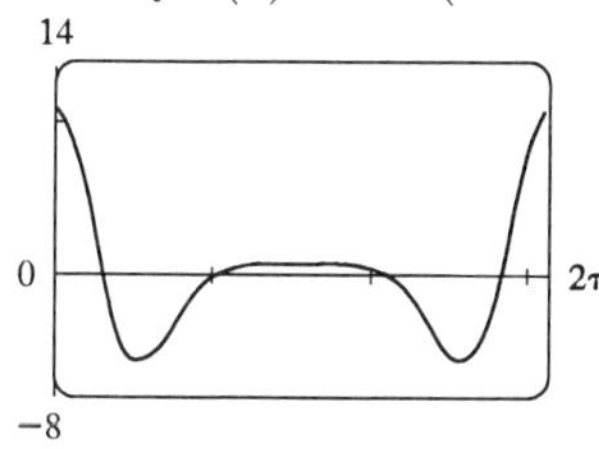

From the graph, it appears that $\left|f^{(4)}(x)\right| < 10.9$ on $[0, 2\pi]$. Another possible upper bound for $\left|f^{(4)}(x)\right|$ is $4e$ (the actual maximum value.)

(g) A CAS gives $S_{10} \approx 7.953789422$. (In Maple, use `student[simpson]`.)

(h) Using Theorem 7 with $K = 10.9$, we get $|E_S| \le \dfrac{10.9(2\pi - 0)^5}{180 \cdot 10^4} \approx 0.0593$.

(i) The actual error is about $7.954926521 - 7.953789427 \approx 0.00114$. This is quite a bit smaller than the estimate in part (h), though the difference is not nearly as great as it was in the case of the Midpoint Rule.

(j) To ensure that $|E_S| \le 0.0001$, we use Theorem 7: $|E_S| \le \dfrac{10.9(2\pi)^5}{180 \cdot n^4} \le 0.0001 \quad \Leftrightarrow \quad \dfrac{10.9(2\pi)^5}{180 \cdot 0.0001} \le n^4$ $\Leftrightarrow \quad n^4 \ge 5{,}929{,}981 \quad \Leftrightarrow \quad n \ge 49.4$. So we must take $n \ge 50$ to ensure that $|I - S_n| \le 0.0001$.

25. $\int_0^1 x^3\,dx = \left[\frac{1}{4}x^4\right]_0^1 = 0.25.\ f(x) = x^3.$

$n = 4$: $L_4 = \frac{1}{4}\left[0^3 + \left(\frac{1}{4}\right)^3 + \left(\frac{1}{2}\right)^3 + \left(\frac{3}{4}\right)^3\right] = 0.140625$

$R_4 = \frac{1}{4}\left[\left(\frac{1}{4}\right)^3 + \left(\frac{1}{2}\right)^3 + \left(\frac{3}{4}\right)^3 + 1^3\right] = 0.390625,$

$T_4 = \frac{1}{4\cdot 2}\left[0^3 + 2\left(\frac{1}{4}\right)^3 + 2\left(\frac{1}{2}\right)^3 + 2\left(\frac{3}{4}\right)^3 + 1^3\right] = \frac{17}{64} = 0.265625,$

$M_4 = \frac{1}{4}\left[\left(\frac{1}{8}\right)^3 + \left(\frac{3}{8}\right)^3 + \left(\frac{5}{8}\right)^3 + \left(\frac{7}{8}\right)^3\right] = 0.2421875$

$E_L = \int_0^1 x^3\,dx - L_4 = \frac{1}{4} - 0.140625 = 0.109375,\ E_R = \frac{1}{4} - 0.390625 = -0.140625,$

$E_T = \frac{1}{4} - 0.265625 = -0.015625,\ E_M = \frac{1}{4} - 0.2421875 = 0.0078125$

$n = 8$: $L_8 = \frac{1}{8}\left[f(0) + f\left(\frac{1}{8}\right) + f\left(\frac{2}{8}\right) + \cdots + f\left(\frac{7}{8}\right)\right] \approx 0.191406$

$R_8 = \frac{1}{8}\left[f\left(\frac{1}{8}\right) + f\left(\frac{2}{8}\right) + \cdots + f\left(\frac{7}{8}\right) + f(1)\right] \approx 0.316406$

$T_8 = \frac{1}{8\cdot 2}\left[f(0) + 2\left(f\left(\frac{1}{8}\right) + f\left(\frac{2}{8}\right) + \cdots + f\left(\frac{7}{8}\right)\right) + f(1)\right] \approx 0.253906$

$M_8 = \frac{1}{8}\left[f\left(\frac{1}{16}\right) + f\left(\frac{3}{16}\right) + \cdots + f\left(\frac{13}{16}\right) + f\left(\frac{15}{16}\right)\right] = 0.248047$

$E_L \approx \frac{1}{4} - 0.191406 \approx 0.058594,\ E_R \approx \frac{1}{4} - 0.316406 \approx -0.066406,$

$E_T \approx \frac{1}{4} - 0.253906 \approx -0.003906,\ E_M \approx \frac{1}{4} - 0.248047 \approx 0.001953.$

$n = 16$: $L_{16} = \frac{1}{16}\left[f(0) + f\left(\frac{1}{16}\right) + f\left(\frac{2}{16}\right) + \cdots + f\left(\frac{15}{16}\right)\right] \approx 0.219727$

$R_{16} = \frac{1}{16}\left[f\left(\frac{1}{16}\right) + f\left(\frac{2}{16}\right) + \cdots + f\left(\frac{15}{16}\right) + f(1)\right] \approx 0.282227$

$T_{16} = \frac{1}{16\cdot 2}\left(f(0) + 2\left[f\left(\frac{1}{16}\right) + f\left(\frac{2}{16}\right) + \cdots + f\left(\frac{15}{16}\right)\right] + f(1)\right) \approx 0.250977$

$M_{16} = \frac{1}{16}\left[f\left(\frac{1}{32}\right) + f\left(\frac{3}{32}\right) + \cdots + f\left(\frac{31}{32}\right)\right] \approx 0.249512$

$E_L \approx \frac{1}{4} - 0.219727 \approx 0.030273,\ E_R \approx \frac{1}{4} - 0.282227 \approx -0.032227,$

$E_T \approx \frac{1}{4} - 0.250977 \approx -0.000977,\ E_M \approx \frac{1}{4} - 0.249512 \approx 0.000488.$

n	L_n	R_n	T_n	M_n
4	0.140625	0.390625	0.265625	0.242188
8	0.191406	0.316406	0.253906	0.248047
16	0.219727	0.282227	0.250977	0.249512

n	E_L	E_R	E_T	E_M
4	0.109375	−0.140625	−0.015625	0.007813
8	0.058594	−0.066406	−0.003906	0.001953
16	0.030273	−0.032227	−0.000977	0.000488

Observations:

1. E_L and E_R are always opposite in sign, as are E_T and E_M.
2. As n is doubled, E_L and E_R are decreased by about a factor of 2, and E_T and E_M are decreased by a factor of about 4.
3. The Midpoint approximation is about twice as accurate as the Trapezoidal approximation.
4. All the approximations become more accurate as the value of n increases.
5. The Midpoint and Trapezoidal approximations are much more accurate than the endpoint approximations.

27. $\int_1^4 \sqrt{x}\,dx = \left[\frac{2}{3}x^{3/2}\right]_1^4 = \frac{2}{3}(8-1) = \frac{14}{3} \approx 4.666667$

$n = 6$: $\Delta x = (4-1)/6 = \frac{1}{2}$

$$T_6 = \frac{1}{2 \cdot 2}\left[\sqrt{1} + 2\sqrt{1.5} + 2\sqrt{2} + 2\sqrt{2.5} + 2\sqrt{3} + 2\sqrt{3.5} + \sqrt{4}\right] \approx 4.661488$$

$$M_6 = \frac{1}{2}\left[\sqrt{1.25} + \sqrt{1.75} + \sqrt{2.25} + \sqrt{2.75} + \sqrt{3.25} + \sqrt{3.75}\right] \approx 4.669245$$

$$S_6 = \frac{1}{2 \cdot 3}\left[\sqrt{1} + 4\sqrt{1.5} + 2\sqrt{2} + 4\sqrt{2.5} + 2\sqrt{3} + 4\sqrt{3.5} + \sqrt{4}\right] \approx 4.666563$$

$E_T \approx \frac{14}{3} - 4.661488 \approx 0.005178$, $E_M \approx \frac{14}{3} - 4.669245 \approx -0.002578$,
$E_S \approx \frac{14}{3} - 4.666563 \approx 0.000104$.

$n = 12$: $\Delta x = (4-1)/12 = \frac{1}{4}$

$$T_{12} = \frac{1}{4 \cdot 2}(f(1) + 2[f(1.25) + f(1.5) + \cdots + f(3.5) + f(3.75)] + f(4)) \approx 4.665367$$

$$M_{12} = \frac{1}{4}[f(1.125) + f(1.375) + f(1.625) + \cdots + f(3.875)] \approx 4.667316$$

$$S_{12} = \frac{1}{4 \cdot 3}[f(1) + 4\,f(1.25) + 2\,f(1.5) + 4\,f(1.75) + \cdots + 4\,f(3.75) + f(4)] \approx 4.666659$$

$E_T \approx \frac{14}{3} - 4.665367 \approx 0.001300$, $E_M \approx \frac{14}{3} - 4.667316 \approx -0.000649$,
$E_S \approx \frac{14}{3} - 4.666659 \approx 0.000007$.

Note: These errors were computed more precisely and then rounded to six places. That is, they were not computed by comparing the rounded values of T_n, M_n, and S_n with the rounded value of the actual definite integral.

n	T_n	M_n	S_n
6	4.661488	4.669245	4.666563
12	4.665367	4.667316	4.666659

n	E_T	E_M	E_S
6	0.005178	−0.002578	0.000104
12	0.001300	−0.000649	0.000007

Observations:

1. E_T and E_M are opposite in sign and decrease by a factor of about 4 as n is doubled.
2. The Simpson's approximation is much more accurate than the Midpoint and Trapezoidal approximations, and seems to decrease by a factor of about 16 as n is doubled.

29. $\int_1^{3.2} y\,dx \approx \frac{0.2}{2}[4.9 + 2(5.4) + 2(5.8) + 2(6.2) + 2(6.7) + 2(7.0)$
$+ 2(7.3) + 2(7.5) + 2(8.0) + 2(8.2) + 2(8.3) + 8.3] = 15.4$

31. $\Delta t = 1\text{ min} = \frac{1}{60}\text{ h}$, so distance $= \int_0^{1/6} v(t)dt \approx \frac{1/60}{3}[40 + 4(42) + 2(45) + 4(49) + 2(52)$
$+ 4(54) + 2(56) + 4(57) + 2(57) + 4(55) + 56] \approx 8.6\text{ mi}.$

33. $\Delta x = (4-0)/4 = 1$

(a) $T_4 = \frac{1}{2}[f(0) + 2f(1) + 2f(2) + 2f(3) + f(4)] \approx \frac{1}{2}[0 + 2(3) + 2(5) + 2(3) + 1] = 11.5$

(b) $M_4 = 1 \cdot [f(0.5) + f(1.5) + f(2.5) + f(3.5)] \approx 1 + 4.5 + 4.5 + 2 = 12$

(c) $S_4 = \frac{1}{3}[f(0) + 4f(1) + 2f(2) + 4f(3) + f(4)] \approx \frac{1}{3}[0 + 4(3) + 2(5) + 4(3) + 1] = 11.\overline{6}$

35. Volume $= \pi \int_0^2 \left(\sqrt[3]{1+x^3}\right)^2 dx = \pi \int_0^2 \left(1+x^3\right)^{2/3} dx$. $V \approx \pi \cdot S_{10}$ where $f(x) = \left(1+x^3\right)^{2/3}$ and $\Delta x = (2-0)/10 = \frac{1}{5}$. Therefore

$$V \approx \pi \cdot S_{10} = \pi \tfrac{1}{5 \cdot 3}\left[f(0) + 4f(0.2) + 2f(0.4) + 4f(0.6) + 2f(0.8) + 4f(1) + 2f(1.2) + 4f(1.4) + 2f(1.6) + 4f(1.8) + f(2)\right] \approx 12.325078.$$

37. Since the Trapezoidal and Midpoint approximations on the interval $[a,b]$ are the sums of the Trapezoidal and Midpoint approximations on the subintervals $[x_{i-1}, x_i]$, $i = 1, 2, \ldots, n$, we can focus our attention on one such interval. The condition $f''(x) < 0$ for $a \le x \le b$ means that the graph of f is concave down as in Figure 5. In that figure, T_n is the area of the trapezoid $AQRD$, $\int_a^b f(x)dx$ is the area of the region $AQPRD$, and M_n is the area of the trapezoid $ABCD$, so $T_n < \int_a^b f(x)dx < M_n$. In general, the condition $f'' < 0$ implies that the graph of f on $[a,b]$ lies above the chord joining the points $(a, f(a))$ and $(b, f(b))$. Thus $\int_a^b f(x)dx > T_n$. Since M_n is the area under a tangent to the graph, and since $f'' < 0$ implies that the tangent lies above the graph, we also have $M_n > \int_a^b f(x)dx$. Thus $T_n < \int_a^b f(x)dx < M_n$.

39. $T_n = \frac{1}{2}\Delta x[f(x_0) + 2f(x_1) + \cdots + 2f(x_{n-1}) + f(x_n)]$ and $M_n = \Delta x[f(\overline{x}_1) + f(\overline{x}_2) + \cdots + f(\overline{x}_{n-1}) + f(\overline{x}_n)]$, where $\overline{x}_i = \frac{1}{2}(x_{i-1} + x_i)$. Now

$$T_{2n} = \tfrac{1}{2}\left(\tfrac{1}{2}\Delta x\right)[f(x_0) + 2f(\overline{x}_1) + 2f(x_1) + 2f(\overline{x}_2) + 2f(x_2) + \cdots + 2f(\overline{x}_{n-1}) + 2f(x_{n-1}) + 2f(\overline{x}_n) + f(x_n)],$$ so

$$\tfrac{1}{2}(T_n + M_n) = \tfrac{1}{4}\Delta x[f(x_0) + 2f(x_1) + \cdots + 2f(x_{n-1}) + f(x_n)] + \tfrac{1}{4}\Delta x[2f(\overline{x}_1) + 2f(\overline{x}_2) + \cdots + 2f(\overline{x}_{n-1}) + 2f(\overline{x}_n)] = T_{2n}.$$

EXERCISES 8.9

1. The area under the graph of $y = 1/x^3 = x^{-3}$ between $x = 1$ and $x = t$ is $A(t) = \int_1^t x^{-3}\,dx = \left[-\frac{1}{2}x^{-2}\right]_1^t = \frac{1}{2} - 1/(2t^2)$. So the area for $1 \le x \le 10$ is $A(10) = 0.5 - 0.005 = 0.495$, the area for $1 \le x \le 100$ is $A(100) = 0.5 - 0.00005 = 0.49995$, and the area for $1 \le x \le 1000$ is $A(1000) = 0.5 - 0.0000005 = 0.4999995$. The total area under the curve for $x \ge 1$ is $\int_1^\infty x^{-3}\,dx = \lim_{t\to\infty}\left[-\frac{1}{2}t^{-2} - \left(-\frac{1}{2}\right)\right] = 0 - \left(-\frac{1}{2}\right) = \frac{1}{2}$.

3. $\displaystyle\int_2^\infty \frac{dx}{\sqrt{x+3}} = \lim_{t\to\infty}\int_2^t \frac{dx}{\sqrt{x+3}} = \lim_{t\to\infty}\left[2\sqrt{x+3}\right]_2^t = \lim_{t\to\infty}\left(2\sqrt{t+3} - 2\sqrt{5}\right) = \infty$. Divergent

5. $\displaystyle\int_{-\infty}^1 \frac{dx}{(2x-3)^2} = \lim_{t\to-\infty}\frac{1}{2}\int_t^1 \frac{2\,dx}{(2x-3)^2} = \lim_{t\to-\infty}\frac{1}{2}\left[-\frac{1}{2x-3}\right]_t^1 = \lim_{t\to-\infty}\left[\frac{1}{2} + \frac{1}{2(2t-3)}\right] = \frac{1}{2}$

7. $\int_{-\infty}^\infty x\,dx = \int_{-\infty}^0 x\,dx + \int_0^\infty x\,dx$. $\int_{-\infty}^0 x\,dx = \lim_{t\to-\infty}\left[\frac{1}{2}x^2\right]_t^0 = \lim_{t\to-\infty}\left(-\frac{1}{2}t^2\right) = -\infty$. Divergent

9. $\int_0^\infty e^{-x}\,dx = \lim_{t\to\infty}\int_0^t e^{-x}\,dx = \lim_{t\to\infty}\left[-e^{-x}\right]_0^t = \lim_{t\to\infty}\left(-e^{-t} + 1\right) = 1$

11. $\int_{-\infty}^{\infty} xe^{-x^2}\,dx = \int_{-\infty}^{0} xe^{-x^2}\,dx + \int_{0}^{\infty} xe^{-x^2}\,dx$, $\int_{-\infty}^{0} xe^{-x^2}\,dx = \lim\limits_{t\to-\infty} -\frac{1}{2}\left[e^{-x^2}\right]_t^0 = \lim\limits_{t\to-\infty} -\frac{1}{2}\left(1 - e^{-t^2}\right) = -\frac{1}{2}$,

and $\int_0^{\infty} xe^{-x^2}\,dx = \lim\limits_{t\to\infty} -\frac{1}{2}\left[e^{-x^2}\right]_0^t = \lim\limits_{t\to\infty} -\frac{1}{2}\left(e^{-t^2} - 1\right) = \frac{1}{2}$. Therefore $\int_{-\infty}^{\infty} xe^{-x^2}\,dx = -\frac{1}{2} + \frac{1}{2} = 0$.

13. $$\int_0^{\infty} \frac{dx}{(x+2)(x+3)} = \lim_{t\to\infty} \int_0^t \left[\frac{1}{x+2} - \frac{1}{x+3}\right] dx = \lim_{t\to\infty}\left[\ln\left(\frac{x+2}{x+3}\right)\right]_0^t = \lim_{t\to\infty}\left[\ln\left(\frac{t+2}{t+3}\right) - \ln\tfrac{2}{3}\right]$$
$$= \ln 1 - \ln\tfrac{2}{3} = -\ln\tfrac{2}{3}$$

15. $\int_0^{\infty} \cos x\,dx = \lim\limits_{t\to\infty} [\sin x]_0^t = \lim\limits_{t\to\infty} \sin t$, which does not exist. Divergent

17. $$\int_0^{\infty} \frac{5\,dx}{2x+3} = \frac{5}{2}\lim_{t\to\infty}\int_0^t \frac{2\,dx}{2x+3} = \tfrac{5}{2}\lim_{t\to\infty}[\ln(2x+3)]_0^t = \tfrac{5}{2}\lim_{t\to\infty}[\ln(2t+3) - \ln 3] = \infty.$$ Divergent

19. $$\int_{-\infty}^{1} xe^{2x}\,dx = \lim_{t\to-\infty}\int_t^1 xe^{2x}\,dx = \lim_{t\to-\infty}\left[\tfrac{1}{2}xe^{2x} - \tfrac{1}{4}e^{2x}\right]_t^1 \quad \text{(by parts)}$$
$$= \lim_{t\to-\infty}\left[\tfrac{1}{2}e^2 - \tfrac{1}{4}e^2 - \tfrac{1}{2}te^{2t} + \tfrac{1}{4}e^{2t}\right] = \tfrac{1}{4}e^2 - 0 + 0 = \tfrac{1}{4}e^2,$$

since $\lim\limits_{t\to-\infty} te^{2t} = \lim\limits_{t\to-\infty}\dfrac{t}{e^{-2t}} \overset{\text{H}}{=} \lim\limits_{t\to-\infty}\dfrac{1}{-2e^{-2t}} = \lim\limits_{t\to-\infty} -\frac{1}{2}e^{2t} = 0$.

21. $$\int_1^{\infty}\frac{\ln x}{x}\,dx = \lim_{t\to\infty}\left[\frac{(\ln x)^2}{2}\right]_1^t = \lim_{t\to\infty}\frac{(\ln t)^2}{2} = \infty.$$ Divergent

23. $$\int_{-\infty}^{\infty}\frac{x\,dx}{1+x^2} = \int_{-\infty}^{0}\frac{x\,dx}{1+x^2} + \int_0^{\infty}\frac{x\,dx}{1+x^2}$$ and
$$\int_{-\infty}^{0}\frac{x\,dx}{1+x^2} = \lim_{t\to-\infty}\left[\tfrac{1}{2}\ln(1+x^2)\right]_t^0 = \lim_{t\to-\infty}\left[0 - \tfrac{1}{2}\ln(1+t^2)\right] = -\infty.$$ Divergent

25. Integrate by parts with $u = \ln x$, $dv = dx/x^2 \quad\Rightarrow\quad du = dx/x$, $v = -1/x$.
$$\int_1^{\infty}\frac{\ln x}{x^2}\,dx = \lim_{t\to\infty}\int_1^t \frac{\ln x}{x^2}\,dx = \lim_{t\to\infty}\left[-\frac{\ln x}{x} - \frac{1}{x}\right]_1^t = \lim_{t\to\infty}\left[-\frac{\ln t}{t} - \frac{1}{t} + 0 + 1\right]$$
$$= -0 - 0 + 0 + 1 = 1, \text{ since } \lim_{t\to\infty}\frac{\ln t}{t} \overset{\text{H}}{=} \lim_{t\to\infty}\frac{1/t}{1} = 0.$$

27. $$\int_0^3 \frac{dx}{\sqrt{x}} = \lim_{t\to0^+}\int_t^3\frac{dx}{\sqrt{x}} = \lim_{t\to0^+}\left[2\sqrt{x}\right]_t^3 = \lim_{t\to0^+}\left(2\sqrt{3} - 2\sqrt{t}\right) = 2\sqrt{3}$$

29. $$\int_{-1}^{0}\frac{dx}{x^2} = \lim_{t\to0^-}\int_{-1}^{t}\frac{dx}{x^2} = \lim_{t\to0^-}\left[\frac{-1}{x}\right]_{-1}^t = \lim_{t\to0^-}\left[-\frac{1}{t} + \frac{1}{-1}\right] = \infty.$$ Divergent

31. $$\int_{-2}^{3}\frac{dx}{x^4} = \int_{-2}^{0}\frac{dx}{x^4} + \int_0^3\frac{dx}{x^4} \text{ and } \int_{-2}^{0}\frac{dx}{x^4} = \lim_{t\to0^-}\left[-\tfrac{1}{3}x^{-3}\right]_{-2}^t = \lim_{t\to0^-}\left[-\frac{1}{3t^3} - \frac{1}{24}\right] = \infty.$$ Divergent

33. $$\int_4^5 \frac{dx}{(5-x)^{2/5}} = \lim_{t\to5^-}\left[-\tfrac{5}{3}(5-x)^{3/5}\right]_4^t = \lim_{t\to5^-}\left[-\tfrac{5}{3}(5-t)^{3/5} + \tfrac{5}{3}\right] = 0 + \tfrac{5}{3} = \tfrac{5}{3}$$

35. $\int_{\pi/4}^{\pi/2}\tan^2 x\,dx = \lim\limits_{t\to\pi/2^-}\int_{\pi/4}^{t}(\sec^2 x - 1)dx = \lim\limits_{t\to\pi/2^-}[\tan x - x]_{\pi/4}^t$

$= \frac{\pi}{4} - 1 + \lim\limits_{t\to\pi/2^-}(\tan t - t) = \infty$. Divergent

37. $\int_0^{\pi} \sec x\,dx = \int_0^{\pi/2} \sec x\,dx + \int_{\pi/2}^{\pi} \sec x\,dx$. $\int_0^{\pi/2} \sec x\,dx = \lim_{t\to\pi/2^-} \int_0^t \sec x\,dx$

$$= \lim_{t\to\pi/2^-} [\ln|\sec x + \tan x|]_0^t = \lim_{t\to\pi/2^-} \ln|\sec t + \tan t| = \infty. \text{ Divergent}$$

39. $\displaystyle\int_{-2}^{2} \frac{dx}{x^2-1} = \int_{-2}^{-1} \frac{dx}{x^2-1} + \int_{-1}^{0} \frac{dx}{x^2-1} + \int_{0}^{1} \frac{dx}{x^2-1} + \int_{1}^{2} \frac{dx}{x^2-1}$, and

$$\int \frac{dx}{x^2-1} = \int \frac{dx}{(x-1)(x+1)} = \frac{1}{2}\ln\left|\frac{x-1}{x+1}\right| + C, \text{ so}$$

$$\int_0^1 \frac{dx}{x^2-1} = \lim_{t\to1^-}\left[\frac{1}{2}\ln\left|\frac{x-1}{x+1}\right|\right]_0^t = \lim_{t\to1^-}\frac{1}{2}\ln\left|\frac{t-1}{t+1}\right| = -\infty. \text{ Divergent}$$

41. Integrate by parts with $u = \ln x$, $dv = x\,dx$:

$$\int_0^1 x\ln x\,dx = \lim_{t\to0^+}\int_t^1 x\ln x\,dx = \lim_{t\to0^+}\left[\tfrac{1}{2}x^2\ln x - \tfrac{1}{4}x^2\right]_t^1 = -\tfrac{1}{4} - \lim_{t\to0^+}\tfrac{1}{2}t^2\ln t$$

$$= -\frac{1}{4} - \frac{1}{2}\lim_{t\to0^+}\frac{\ln t}{1/t^2} \overset{\text{H}}{=} -\frac{1}{4} - \frac{1}{2}\lim_{t\to0^+}\frac{1/t}{-2/t^3} = -\tfrac{1}{4} + \tfrac{1}{4}\lim_{t\to0^+} t^2 = -\tfrac{1}{4}$$

43.

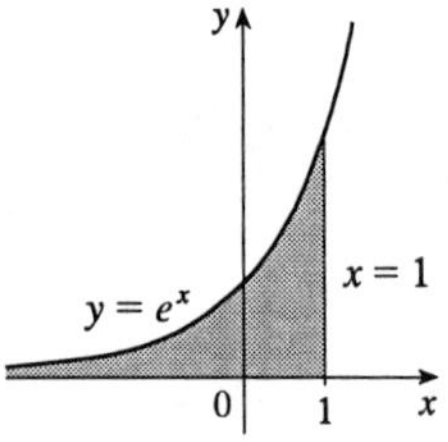

y
$y = e^x$
$x = 1$
0 1 x

Area $= \int_{-\infty}^1 e^x\,dx = \lim_{t\to-\infty}[e^x]_t^1 = e - \lim_{t\to-\infty} e^t = e$

45.

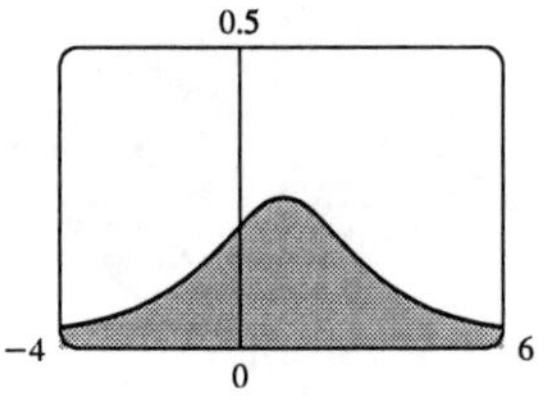

0.5
−4 0 6

$$\text{Area} = \int_{-\infty}^{\infty}\frac{dx}{x^2-2x+5} = \int_{-\infty}^{0}\frac{dx}{(x-1)^2+4} + \int_{0}^{\infty}\frac{dx}{(x-1)^2+4}$$

$$= \lim_{t\to-\infty}\left[\frac{1}{2}\tan^{-1}\left(\frac{x-1}{2}\right)\right]_t^0 + \lim_{t\to\infty}\left[\frac{1}{2}\tan^{-1}\left(\frac{x-1}{2}\right)\right]_0^t$$

$$= \tfrac{1}{2}\tan^{-1}\left(-\tfrac{1}{2}\right) - \tfrac{1}{2}\left(-\tfrac{\pi}{2}\right) + \tfrac{1}{2}\left(\tfrac{\pi}{2}\right) - \tfrac{1}{2}\tan^{-1}\left(-\tfrac{1}{2}\right) = \tfrac{\pi}{2}.$$

47.

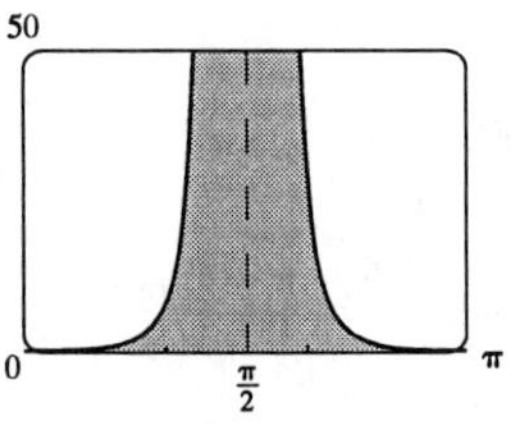

50
0 $\frac{\pi}{2}$ π

Area $= \int_0^{\pi} \tan^2 x \sec^2 x\,dx$

$= \int_0^{\pi/2} \tan^2 x\sec^2 x\,dx + \int_{\pi/2}^{\pi} \tan^2 x \sec^2 x\,dx.$

But $\int_0^{\pi/2} \tan^2 x\sec^2 x\,dx = \lim_{t\to\pi/2^-}\left[\tfrac{1}{3}\tan^3 x\right]_0^t = \infty$,

so the area is infinite.

49. $\dfrac{\sin^2 x}{x^2} \le \dfrac{1}{x^2}$ on $[1,\infty)$. $\displaystyle\int_1^{\infty}\frac{dx}{x^2}$ is convergent by Example 4, so $\displaystyle\int_1^{\infty}\frac{\sin^2 x}{x^2}\,dx$ is convergent by the Comparison Theorem.

51. For $x \geq 1$, $x + e^{2x} > e^{2x} > 0 \quad \Rightarrow \quad \dfrac{1}{x + e^{2x}} \leq \dfrac{1}{e^{2x}} = e^{-2x}$ on $[1, \infty)$.

$\int_1^\infty e^{-2x}\,dx = \lim\limits_{t\to\infty}\left[-\frac{1}{2}e^{-2x}\right]_1^t = \lim\limits_{t\to\infty}\left[-\frac{1}{2}e^{-2t} + \frac{1}{2}e^{-2}\right] = \frac{1}{2}e^{-2}$. Therefore $\int_1^\infty e^{-2x}\,dx$ is convergent, and by the Comparison Theorem, $\displaystyle\int_1^\infty \frac{dx}{x + e^{2x}}$ is also convergent.

53. $\dfrac{1}{x \sin x} \geq \dfrac{1}{x}$ on $\left(0, \frac{\pi}{2}\right]$ since $0 \leq \sin x \leq 1$. $\displaystyle\int_0^{\pi/2} \frac{dx}{x} = \lim_{t\to 0^+}\int_t^{\pi/2}\frac{dx}{x} = \lim_{t\to 0^+}[\ln x]_t^{\pi/2}$. But $\ln t \to -\infty$ as $t \to 0^+$, so $\displaystyle\int_0^{\pi/2}\frac{dx}{x}$ is divergent, and by the Comparison Theorem, $\displaystyle\int_0^{\pi/2}\frac{dx}{x\sin x}$ is also divergent.

55.
$$\int_0^\infty \frac{dx}{\sqrt{x}(1+x)} = \int_0^1 \frac{dx}{\sqrt{x}(1+x)} + \int_1^\infty \frac{dx}{\sqrt{x}(1+x)} = \lim_{t\to 0^+}\int_t^1 \frac{dx}{\sqrt{x}(1+x)} + \lim_{t\to\infty}\int_1^t \frac{dx}{\sqrt{x}(1+x)}$$
$$= \lim_{t\to 0^+}\int_{\sqrt{t}}^1 \frac{2\,du}{1+u^2} + \lim_{t\to\infty}\int_1^{\sqrt{t}} \frac{2\,du}{1+u^2}\ \left[\text{put } u = \sqrt{x},\ x = u^2\right] = \lim_{t\to 0^+}\left[2\tan^{-1}u\right]_{\sqrt{t}}^1 + \lim_{t\to\infty}\left[2\tan^{-1}u\right]_1^{\sqrt{t}}$$
$$= \lim_{t\to 0^+}\left[2\left(\tfrac{\pi}{4}\right) - 2\tan^{-1}\sqrt{t}\right] + \lim_{t\to\infty}\left[2\tan^{-1}\sqrt{t} - 2\left(\tfrac{\pi}{4}\right)\right] = \tfrac{\pi}{2} - 0 + 2\left(\tfrac{\pi}{2}\right) - \tfrac{\pi}{2} = \pi$$

57. If $p = 1$, then $\displaystyle\int_0^1 \frac{dx}{x^p} = \lim_{t\to 0^+}[\ln x]_t^1 = \infty$. Divergent. If $p \neq 1$, then $\displaystyle\int_0^1 \frac{dx}{x^p} = \lim_{t\to 0^+}\int_t^1 \frac{dx}{x^p}$ (Note that the integral is not improper if $p < 0$) $\displaystyle = \lim_{t\to 0^+}\left[\frac{x^{-p+1}}{-p+1}\right]_t^1 = \lim_{t\to 0^+}\frac{1}{1-p}\left[1 - \frac{1}{t^{p-1}}\right]$. If $p > 1$, then $p - 1 > 0$, so $\dfrac{1}{t^{p-1}} \to \infty$ as $t \to 0^+$, and the integral diverges.

Finally, if $p < 1$, then $\displaystyle\int_0^1 \frac{dx}{x^p} = \frac{1}{1-p}\left[\lim_{t\to 0^+}\left(1 - t^{1-p}\right)\right] = \frac{1}{1-p}$.

Thus the integral converges if and only if $p < 1$, and in that case its value is $\dfrac{1}{1-p}$.

59. First suppose $p = -1$. Then

$\displaystyle\int_0^1 x^p \ln x\,dx = \int_0^1 \frac{\ln x}{x}\,dx = \lim_{t\to 0^+}\int_t^1 \frac{\ln x}{x}\,dx = \lim_{t\to 0^+}\left[\tfrac{1}{2}(\ln x)^2\right]_t^1 = -\tfrac{1}{2}\lim_{t\to 0^+}(\ln t)^2 = -\infty$, so the integral diverges. Now suppose $p \neq -1$. Then integration by parts gives

$\displaystyle\int x^p \ln x\,dx = \frac{x^{p+1}}{p+1}\ln x - \int \frac{x^p}{p+1}\,dx = \frac{x^{p+1}}{p+1}\ln x - \frac{x^{p+1}}{(p+1)^2} + C$. If $p < -1$, then $p + 1 < 0$, so

$\displaystyle\int_0^1 x^p \ln x\,dx = \lim_{t\to 0^+}\left[\frac{x^{p+1}}{p+1}\ln x - \frac{x^{p+1}}{(p+1)^2}\right]_t^1 = \frac{-1}{(p+1)^2} - \left(\frac{1}{p+1}\right)\lim_{t\to 0^+}\left[t^{p+1}\left(\ln t - \frac{1}{p+1}\right)\right] = \infty$. If $p > -1$, then $p + 1 > 0$ and

$$\int_0^1 x^p \ln x\,dx = \frac{-1}{(p+1)^2} - \left(\frac{1}{p+1}\right)\lim_{t\to 0^+}\frac{\ln t - 1/(p+1)}{t^{-(p+1)}}$$
$$\overset{\text{H}}{=} \frac{-1}{(p+1)^2} - \left(\frac{1}{p+1}\right)\lim_{t\to 0^+}\frac{1/t}{-(p+1)t^{-(p+2)}} = \frac{-1}{(p+1)^2} + \frac{1}{(p+1)^2}\lim_{t\to 0^+}t^{p+1} = \frac{-1}{(p+1)^2}.$$

Thus the integral converges to $-\dfrac{1}{(p+1)^2}$ if $p > -1$ and diverges otherwise.

61. **(a)** $\int_{-\infty}^{\infty} x\,dx = \int_{-\infty}^{0} x\,dx + \int_0^{\infty} x\,dx$, and $\int_0^{\infty} x\,dx = \lim_{t\to\infty}\int_0^t x\,dx = \lim_{t\to\infty}\left[\frac{1}{2}t^2 - \frac{1}{2}(0^2)\right] = \infty$, so the integral is divergent.

(b) $\int_{-t}^{t} x\,dx = \left[\frac{1}{2}x^2\right]_{-t}^{t} = \frac{1}{2}t^2 - \frac{1}{2}t^2 = 0$, so $\lim_{t\to\infty}\int_{-t}^{t} x\,dx = 0$. Therefore $\int_{-\infty}^{\infty} x\,dx \neq \lim_{t\to\infty}\int_{-t}^{t} x\,dx$.

63. Volume $= \displaystyle\int_1^{\infty} \pi\left(\frac{1}{x}\right)^2 dx = \pi \lim_{t\to\infty}\int_1^t \frac{dx}{x^2} = \pi\lim_{t\to\infty}\left[-\frac{1}{x}\right]_1^t = \pi\lim_{t\to\infty}\left(1-\frac{1}{t}\right) = \pi < \infty.$

65. Work $= \displaystyle\int_R^{\infty} F\,dr = \lim_{t\to\infty}\int_R^t \frac{GmM}{r^2}\,dr = \lim_{t\to\infty} GmM\left(\frac{1}{R}-\frac{1}{t}\right) = \frac{GmM}{R}$. The initial kinetic energy provides the work, so $\frac{1}{2}mv_0^2 = \dfrac{GmM}{R} \quad\Rightarrow\quad v_0 = \sqrt{\dfrac{2GM}{R}}$.

67. **(a)** $F(s) = \displaystyle\int_0^{\infty} f(t)e^{-st}\,dt = \int_0^{\infty} e^{-st}\,dt = \lim_{n\to\infty}\left[-\frac{e^{-st}}{s}\right]_0^n = \lim_{n\to\infty}\left(\frac{e^{-sn}}{-s}+\frac{1}{s}\right)$. This converges to $\dfrac{1}{s}$ only if $s>0$. Therefore $F(s) = \dfrac{1}{s}$ with domain $\{s \mid s>0\}$.

(b)
$$F(s) = \int_0^{\infty} f(t)e^{-st}\,dt = \int_0^{\infty} e^t e^{-st}\,dt = \lim_{n\to\infty}\int_0^n e^{t(1-s)}\,dt$$
$$= \lim_{n\to\infty}\left[\frac{1}{1-s}e^{t(1-s)}\right]_0^n = \lim_{n\to\infty}\left(\frac{e^{(1-s)n}}{1-s} - \frac{1}{1-s}\right).$$
This converges only if $1-s<0 \quad\Rightarrow\quad s>1$, in which case $F(s) = \dfrac{1}{s-1}$ with domain $\{s \mid s>1\}$.

(c) $F(s) = \int_0^{\infty} f(t)e^{-st}\,dt = \lim_{n\to\infty}\int_0^n te^{-st}\,dt$. Use integration by parts: let $u=t$, $dv = e^{-st}\,dt \quad\Rightarrow$ $du = dt$, $v = -\dfrac{e^{-st}}{s}$. Then $F(s) = \displaystyle\lim_{n\to\infty}\left[-\frac{t}{s}e^{-st} - \frac{1}{s^2}e^{-st}\right]_0^n = \lim_{n\to\infty}\left(\frac{-n}{se^{sn}} - \frac{1}{s^2e^{sn}} + 0 + \frac{1}{s^2}\right) = \frac{1}{s^2}$ only if $s>0$. Therefore $F(s) = \dfrac{1}{s^2}$ and the domain of F is $\{s \mid s>0\}$.

69. $G(s) = \int_0^{\infty} f'(t)e^{-st}\,dt$. Integrate by parts with $u = e^{-st}$, $dv = f'(t)dt \quad\Rightarrow\quad du = -se^{-st}$, $v = f(t)$:
$G(s) = \lim_{n\to\infty}\left[f(t)e^{-st}\right]_0^n + s\int_0^{\infty} f(t)e^{-st}\,dt = \lim_{n\to\infty} f(n)e^{-sn} - f(0) + sF(s)$.
But $0 \le f(t) \le Me^{at} \quad\Rightarrow\quad 0 \le f(t)e^{-st} \le Me^{at}e^{-st}$ and $\lim_{t\to\infty} Me^{t(a-s)} = 0$ for $s>a$. So by the Squeeze Theorem, $\lim_{t\to\infty} f(t)e^{-st} = 0$ for $s>a \quad\Rightarrow\quad G(s) = 0 - f(0) + sF(s) = sF(s) - f(0)$ for $s>a$.

71. Use integration by parts: let $u=x$, $dv = xe^{-x^2}\,dx \quad\Rightarrow\quad du = dx$, $v = -\frac{1}{2}e^{-x^2}$. So
$\displaystyle\int_0^{\infty} x^2e^{-x^2}\,dx = \lim_{t\to\infty}\left[-\frac{x}{2}e^{-x^2}\right]_0^t + \frac{1}{2}\int_0^{\infty} e^{-x^2}\,dx = \lim_{t\to\infty} -\frac{t}{2e^{t^2}} + \frac{1}{2}\int_0^{\infty} e^{-x^2}\,dx = \frac{1}{2}\int_0^{\infty} e^{-x^2}\,dx.$
(The limit is 0 by l'Hospital's Rule.)

73. For the first part of the integral, let $x = 2\tan\theta \quad \Rightarrow \quad dx = 2\sec^2\theta\, d\theta$.

$\displaystyle\int \frac{1}{\sqrt{x^2+4}}\,dx = \int \sec\theta = \ln|\sec\theta + \tan\theta|$. But $\tan\theta = \frac{1}{2}x$, and

$\sec\theta = \sqrt{1+\tan^2\theta} = \sqrt{1+\frac{1}{4}x^2} = \frac{1}{2}\sqrt{x^2+4}$. So

$$\int_0^\infty \left(\frac{1}{\sqrt{x^2+4}} - \frac{C}{x+2}\right)dx = \lim_{t\to\infty}\left[\ln\left|\frac{\sqrt{x^2+4}}{2} + \frac{x}{2}\right| - C\ln|x+2|\right]_0^t$$

$$= \lim_{t\to\infty}\ln\left(\frac{\sqrt{t^2+4}+t}{2(t+2)^C}\right) - (\ln 1 - C\ln 2) = \ln\left(\lim_{t\to\infty}\frac{t+\sqrt{t^2+4}}{(t+2)^C}\right) + \ln 2^{C-1}.$$

By l'Hospital's Rule, $\displaystyle\lim_{t\to\infty}\frac{t+\sqrt{t^2+4}}{(t+2)^C} = \lim_{t\to\infty}\frac{1+t/\sqrt{t^2+4}}{C(t+2)^{C-1}} = \frac{2}{C\lim\limits_{t\to\infty}(t+2)^{C-1}}$.

If $C < 1$, we get ∞ and the interval diverges. If $C = 1$, we get 2, so the original integral converges to $\ln 2 + \ln 2^0 = \ln 2$. If $C > 1$, we get 0, so the original integral diverges to $-\infty$.

75. We integrate by parts with $u = \dfrac{1}{\ln(1+x+t)}$, $dv = \sin t\,dt$, so $du = \dfrac{-1}{(1+x+t)[\ln(1+x+t)]^2}$ and $v = -\cos t$. The integral becomes

$$I = \int_0^\infty \frac{\sin t\,dt}{\ln(1+x+t)} = \lim_{b\to\infty}\left(\frac{-\cos t}{\ln(1+x+t)}\bigg|_0^b - \int_0^b \frac{\cos t\,dt}{(1+x+t)[\ln(1+x+t)]^2}\right)$$

$$= \lim_{b\to\infty}\frac{-\cos b}{\ln(1+x+b)} + \frac{1}{\ln(1+x)} + \int_0^\infty \frac{-\cos t\,dt}{(1+x+t)[\ln(1+x+t)]^2}$$

$$= \frac{1}{\ln(1+x)} + J, \text{ where } J = \int_0^\infty \frac{-\cos t\,dt}{(1+x+t)[\ln(1+x+t)]^2}.$$

Now $-1 \le -\cos t \le 1$ for all t; in fact, the inequality is strict except at isolated points. So

$$-\int_0^\infty \frac{dt}{(1+x+t)[\ln(1+x+t)]^2} < J < \int_0^\infty \frac{dt}{(1+x+t)[\ln(1+x+t)]^2} \quad \Leftrightarrow$$

$$-\frac{1}{\ln(1+x)} < J < \frac{1}{\ln(1+x)} \quad \Leftrightarrow \quad 0 < I < \frac{2}{\ln(1+x)}.$$

REVIEW EXERCISES FOR CHAPTER 8

1. False. Since the numerator has a higher degree than the denominator,

$$\frac{x(x^2+4)}{x^2-4} = x + \frac{8x}{x^2-4} = x + \frac{A}{x+2} + \frac{B}{x-2}.$$

3. False. It can be put in the form $\dfrac{A}{x} + \dfrac{B}{x^2} + \dfrac{C}{x-4}$.

5. False. This is an improper integral, since the denominator vanishes at $x = 1$.

$$\int_0^4 \frac{x}{x^2-1}\,dx = \int_0^1 \frac{x}{x^2-1}\,dx + \int_1^4 \frac{x}{x^2-1}\,dx \text{ and } \int_0^1 \frac{x}{x^2-1}\,dx = \lim_{t\to 1^-} \int_0^t \frac{x}{x^2-1}\,dx$$

$$= \lim_{t\to 1^-} \left[\tfrac{1}{2}\ln|x^2-1|\right]_0^t = \lim_{t\to 1^-} \tfrac{1}{2}\ln|t^2-1| = \infty. \text{ So the integral diverges.}$$

7. False. See Exercise 61 in Section 8.9.

9. $\displaystyle\int \frac{x-1}{x+1}\,dx = \int \left[1 - \frac{2}{x+1}\right] dx = x - 2\ln|x+1| + C$

11. Let $u = \arctan x$. Then $du = dx/(1+x^2)$, so $\displaystyle\int \frac{(\arctan x)^5}{1+x^2}\,dx = \int u^5\,du = \tfrac{1}{6}u^6 + C = \tfrac{1}{6}(\arctan x)^6 + C$.

13. Let $u = \sin x$. Then $\displaystyle\int \frac{\cos x\,dx}{e^{\sin x}} = \int e^{-u}\,du = -e^{-u} + C = -\frac{1}{e^{\sin x}} + C$.

15. Use integration by parts with $u = \ln x$, $dv = x^4\,dx \quad\Rightarrow\quad du = dx/x$, $v = x^5/5$:

$\int x^4 \ln x\,dx = \frac{1}{5}x^5 \ln x - \frac{1}{5}\int x^4\,dx = \frac{1}{5}x^5 \ln x - \frac{1}{25}x^5 + C = \frac{1}{25}x^5(5\ln x - 1) + C$.

17. Let $u = x^2$. Then $du = 2x\,dx$, so $\int x\sin(x^2)dx = \frac{1}{2}\int \sin u\,du = -\frac{1}{2}\cos u + C = -\frac{1}{2}\cos(x^2) + C$.

19. $\displaystyle\int \frac{dx}{2x^2-5x+2} = \int \left[\frac{-2/3}{2x-1} + \frac{1/3}{x-2}\right] dx = -\tfrac{1}{3}\ln|2x-1| + \tfrac{1}{3}\ln|x-2| + C = \tfrac{1}{3}\ln\left|\frac{x-2}{2x-1}\right| + C$.

21. Let $u = \sec x$. Then $du = \sec x \tan x\,dx$, so $\int \tan^7 x \sec^3 x\,dx = \int \tan^6 x \sec^2 x \sec x \tan x\,dx$

$= \int (u^2-1)^3 u^2\,du = \int (u^8 - 3u^6 + 3u^4 - u^2)du = \frac{1}{9}u^9 - \frac{3}{7}u^7 + \frac{3}{5}u^5 - \frac{1}{3}u^3 + C$

$= \frac{1}{9}\sec^9 x - \frac{3}{7}\sec^7 x + \frac{3}{5}\sec^5 x - \frac{1}{3}\sec^3 x + C$.

23. Let $u = \sqrt{1+2x}$. Then $x = \frac{1}{2}(u^2-1)$, $dx = u\,du$, so $\displaystyle\int \frac{dx}{\sqrt{1+2x}+3} = \int \frac{u\,du}{u+3}$

$$= \int \left[1 - \frac{3}{u+3}\right] du = u - 3\ln|u+3| + C = \sqrt{1+2x} - 3\ln\left(\sqrt{1+2x}+3\right) + C.$$

25. $u = \sqrt{x} \quad\Rightarrow\quad du = \dfrac{dx}{2\sqrt{x}} \quad\Rightarrow\quad \displaystyle\int \frac{e^{\sqrt{x}}\,dx}{\sqrt{x}} = 2\int e^u\,du = 2e^u + C = 2e^{\sqrt{x}} + C$

27. Let $x = \sec\theta$. Then

$$\int \frac{dx}{(x^2-1)^{3/2}} = \int \frac{\sec\theta\tan\theta}{\tan^3\theta}\,d\theta = \int \frac{\sec\theta}{\tan^2\theta}\,d\theta = \int \frac{\cos\theta\,d\theta}{\sin^2\theta} = -\frac{1}{\sin\theta} + C = -\frac{x}{\sqrt{x^2-1}} + C.$$

29. $\displaystyle\int \frac{dx}{x^3+x} = \int \left(\frac{1}{x} - \frac{x}{x^2+1}\right) dx = \ln|x| - \tfrac{1}{2}\ln(x^2+1) + C$

31. $\int \cot^2 x\,dx = \int (\csc^2 x - 1)dx = -\cot x - x + C$

33. $\displaystyle\int \frac{2x^2+3x+11}{x^3+x^2+3x-5}\,dx = \int\left(\frac{2}{x-1} - \frac{1}{x^2+2x+5}\right)dx = 2\ln|x-1| - \int \frac{dx}{(x+1)^2+4}$

$$= 2\ln|x-1| - \frac{1}{2}\tan^{-1}\left(\frac{x+1}{2}\right) + C$$

35. Let $u = \cot 4x$. Then $du = -4\csc^2 4x\,dx \quad\Rightarrow\quad \int \csc^4 4x\,dx = \int(\cot^2 4x + 1)\csc^2 4x\,dx$

$= \int(u^2+1)\left(-\frac{1}{4}\,du\right) = -\frac{1}{4}\left(\frac{1}{3}u^3 + u\right) + C = -\frac{1}{12}(\cot^3 4x + 3\cot 4x) + C.$

37. Let $u = \ln x$. Then $\displaystyle\int \frac{\ln(\ln x)}{x}\,dx = \int \ln u\,du$. Now use parts with $w = \ln u$, $dv = du \quad\Rightarrow\quad dw = du/u$,

$v = u \quad\Rightarrow\quad \int \ln u\,du = u\ln u - u + C = (\ln x)[\ln(\ln x) - 1] + C.$

39. Let $u = 2x+1$. Then $du = 2\,dx \quad\Rightarrow$

$$\int \frac{dx}{\sqrt{4x^2+4x+5}} = \int \frac{(1/2)du}{\sqrt{u^2+4}} = \frac{1}{2}\int \frac{2\sec^2\theta\,d\theta}{2\sec\theta} \quad \text{(put } u = 2\tan\theta,\ du = 2\sec^2\theta\,d\theta\text{)}$$

$$= \tfrac{1}{2}\int \sec\theta\,d\theta = \tfrac{1}{2}\ln|\sec\theta + \tan\theta| + C_1 = \frac{1}{2}\ln\left|\frac{\sqrt{u^2+4}}{2} + \frac{u}{2}\right| + C_1$$

$$= \tfrac{1}{2}\ln\left(u + \sqrt{u^2+4}\right) + C = \tfrac{1}{2}\ln\left(2x+1+\sqrt{4x^2+4x+5}\right) + C.$$

41. $\int(\cos x + \sin x)^2\cos 2x\,dx = \int(\cos^2 x + 2\sin x\cos x + \sin^2 x)\cos 2x\,dx$

$= \int(1+\sin 2x)\cos 2x\,dx = \int \cos 2x\,dx + \frac{1}{2}\int \sin 4x\,dx = \frac{1}{2}\sin 2x - \frac{1}{8}\cos 4x + C.$

Or: $\int(\cos x + \sin x)^2\cos 2x\,dx = \int(\cos x + \sin x)^2(\cos^2 x - \sin^2 x)dx$

$= \int(\cos x + \sin x)^3(\cos x - \sin x)dx = \frac{1}{4}(\cos x + \sin x)^4 + C_2.$

43. $\int_0^{\pi/2}\cos^3 x\sin 2x\,dx = \int_0^{\pi/2} 2\cos^4 x\sin x\,dx = \left[-\frac{2}{5}\cos^5 x\right]_0^{\pi/2} = \frac{2}{5}$

45. $\displaystyle\int_0^3 \frac{dx}{x^2-x-2} = \int_0^3 \frac{dx}{(x+1)(x-2)} = \int_0^2 \frac{dx}{(x+1)(x-2)} + \int_2^3 \frac{dx}{(x+1)(x-2)}$, and

$$\int_2^3 \frac{dx}{x^2-x-2} = \lim_{t\to 2^+}\int_t^3\left[\frac{-1/3}{x+1} + \frac{1/3}{x-2}\right]dx = \lim_{t\to 2^+}\left[\frac{1}{3}\ln\left|\frac{x-2}{x+1}\right|\right]_t^3 = \lim_{t\to 2^+}\left[\tfrac{1}{3}\ln\tfrac{1}{4} - \frac{1}{3}\ln\left|\frac{t-2}{t+1}\right|\right] = \infty.$$

Divergent

47. $\displaystyle\int_0^1 \frac{t^2-1}{t^2+1}\,dt = \int_0^1\left[1 - \frac{2}{t^2+1}\right]dt = \left[t - 2\tan^{-1}t\right]_0^1 = \left(1 - 2\cdot\tfrac{\pi}{4}\right) - 0 = 1 - \tfrac{\pi}{2}$

49. $\displaystyle\int_0^\infty \frac{dx}{(x+2)^4} = \lim_{t\to\infty}\left[\frac{-1}{3(x+2)^3}\right]_0^t = \lim_{t\to\infty}\left[\frac{1}{3\cdot 2^3} - \frac{1}{3(t+2)^3}\right] = \frac{1}{24}$

51. Let $u = \ln x$. Then

$$\int_1^e \frac{dx}{x\sqrt{\ln x}} = \lim_{t\to 1^+}\int_t^e \frac{dx}{x\sqrt{\ln x}} = \lim_{t\to 1^+}\int_{\ln t}^1 \frac{du}{\sqrt{u}} = \lim_{t\to 1^+}\left[2\sqrt{u}\right]_{\ln t}^1 = \lim_{t\to 1^+}\left(2 - 2\sqrt{\ln t}\right) = 2.$$

53. Let $u = \sqrt{x} + 2$. Then $x = (u-2)^2$, $dx = 2(u-2)du$, so

$$\int_1^4 \frac{\sqrt{x}\,dx}{\sqrt{x}+2} = \int_3^4 \frac{2(u-2)^2\,du}{u} = \int_3^4 \left[2u - 8 + \frac{8}{u}\right]du = \left[u^2 - 8u + 8\ln u\right]_3^4$$

$$= (16 - 32 + 8\ln 4) - (9 - 24 + 8\ln 3) = -1 + 8\ln 4 - 8\ln 3 = 8\ln\tfrac{4}{3} - 1.$$

55. Let $u = 2x + 1$. Then $\displaystyle\int_{-\infty}^{\infty} \frac{dx}{4x^2 + 4x + 5} = \int_{-\infty}^{\infty} \frac{\frac{1}{2}\,du}{u^2 + 4} = \frac{1}{2}\int_{-\infty}^{0} \frac{du}{u^2+4} + \frac{1}{2}\int_0^{\infty} \frac{du}{u^2+4}$

$= \frac{1}{2}\lim\limits_{t\to-\infty}\left[\frac{1}{2}\tan^{-1}\left(\frac{1}{2}u\right)\right]_t^0 + \frac{1}{2}\lim\limits_{t\to\infty}\left[\frac{1}{2}\tan^{-1}\left(\frac{1}{2}u\right)\right]_0^t = \frac{1}{4}\left[0 - \left(-\frac{\pi}{2}\right)\right] + \frac{1}{4}\left[\frac{\pi}{2} - 0\right] = \frac{\pi}{4}.$

57. Let $x = \sec\theta$. Then $\displaystyle\int_1^2 \frac{\sqrt{x^2-1}}{x}\,dx = \int_0^{\pi/3} \frac{\tan\theta}{\sec\theta}\sec\theta\tan\theta\,d\theta = \int_0^{\pi/3}\tan^2\theta\,d\theta$

$= \int_0^{\pi/3}(\sec^2\theta - 1)d\theta = [\tan\theta - \theta]_0^{\pi/3} = \sqrt{3} - \frac{\pi}{3}.$

59. $\int_0^\infty e^{ax}\cos bx\,dx = \lim\limits_{t\to\infty}\int_0^t e^{ax}\cos bx\,dx$. Integrate by parts twice:

$$\int e^{ax}\cos bx\,dx = \frac{1}{b}e^{ax}\sin bx - \frac{a}{b}\int e^{ax}\sin bx\,dx = \frac{1}{b}e^{ax}\sin bx + \frac{a}{b^2}e^{ax}\cos bx - \frac{a^2}{b^2}\int e^{ax}\cos bx\,dx, \text{ so}$$

$$\left(1 + \frac{a^2}{b^2}\right)\int e^{ax}\cos bx\,dx = \frac{1}{b}e^{ax}\sin bx + \frac{a}{b^2}e^{ax}\cos bx + C_1. \text{ Thus}$$

$$\int e^{ax}\cos bx\,dx = \frac{e^{ax}}{a^2+b^2}(b\sin bx + a\cos bx) + C. \text{ Now}$$

$$\int_0^\infty e^{ax}\cos bx\,dx = \lim_{t\to\infty}\left[\frac{e^{ax}}{a^2+b^2}(a\cos bx + b\sin bx)\right]_0^t = \lim_{t\to\infty}\frac{e^{at}}{a^2+b^2}(a\cos bt + b\sin bt) - \frac{a}{a^2+b^2}.$$

If $a \geq 0$, the limit does not exist and the integral is divergent. If $a < 0$, the limit is 0 (since $|e^{at}\cos bt| \leq e^{at}$ and $|e^{at}\sin bt| \leq e^{at}$), so the integral converges to $-a/(a^2+b^2)$.

61. We first make the substitution $t = x + 1$, so $\ln(x^2 + 2x + 2) = \ln\left[(x+1)^2 + 1\right] = \ln(t^2+1)$. Then we use parts with $u = \ln(t^2+1)$, $dv = dt$:

$$\int \ln(t^2+1)dt = t\ln(t^2+1) - \int\frac{t(2t)dt}{t^2+1} = t\ln(t^2+1) - 2\int\frac{t^2\,dt}{t^2+1}$$

$$= t\ln(t^2+1) - 2\int\left(1 - \frac{1}{t^2+1}\right)dt = t\ln(t^2+1) - 2t + 2\arctan t + C$$

$$= (x+1)\ln(x^2+2x+2) - 2x + 2\arctan(x+1) + K.$$

[Alternately, we could have integrated by parts immediately with $u = \ln(x^2 + 2x + 2)$]
Notice from the graph that $f = 0$ where F has a horizontal tangent. Also, F is always increasing, and $f \geq 0$.

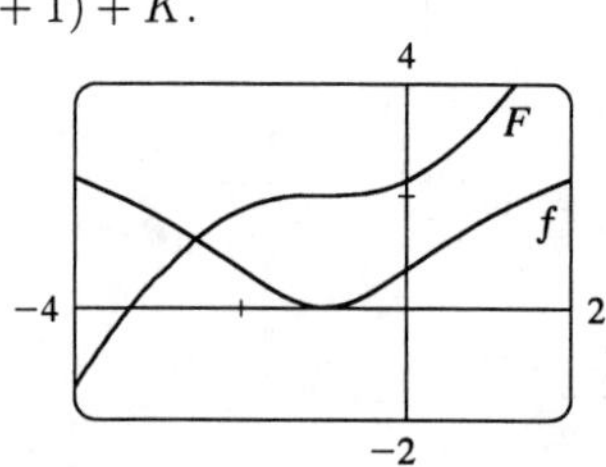

63. Let $u = e^x$. Then $du = e^x\,dx$, so $\int e^x\sqrt{1-e^{2x}}\,dx = \int\sqrt{1-u^2}\,du = \frac{1}{2}u\sqrt{1-u^2} + \frac{1}{2}\sin^{-1}u + C$

(Formula 30) $= \frac{1}{2}\left[e^x\sqrt{1-e^{2x}} + \sin^{-1}(e^x)\right] + C.$

65. Let $u = x + \frac{1}{2}$. Then $du = dx$, so $\int \sqrt{x^2 + x + 1}\,dx = \int \sqrt{\left(x + \frac{1}{2}\right)^2 + \frac{3}{4}}\,dx$

$= \int \sqrt{u^2 + \left(\frac{\sqrt{3}}{2}\right)^2}\,du = \frac{1}{2}u\sqrt{u^2 + \frac{3}{4}} + \frac{3}{8}\ln\left|u + \sqrt{u^2 + \frac{3}{4}}\right| + C$ (Formula 21)

$= \dfrac{2x+1}{4}\sqrt{x^2 + x + 1} + \frac{3}{8}\ln\left|x + \frac{1}{2} + \sqrt{x^2 + x + 1}\right| + C.$

67. **(a)** $\dfrac{d}{du}\left[-\dfrac{1}{u}\sqrt{a^2 - u^2} - \sin^{-1}\left(\dfrac{u}{a}\right) + C\right] = \dfrac{1}{u^2}\sqrt{a^2 - u^2} + \dfrac{1}{\sqrt{a^2 - u^2}} - \dfrac{1}{\sqrt{1 - u^2/a^2}} \cdot \dfrac{1}{a}$

$= \left(a^2 - u^2\right)^{-1/2}\left[\dfrac{1}{u^2}\left(a^2 - u^2\right) + 1 - 1\right] = \dfrac{\sqrt{a^2 - u^2}}{u^2}.$

(b) Let $u = a\sin\theta \quad\Rightarrow\quad du = a\cos\theta\,d\theta$, $a^2 - u^2 = a^2(1 - \sin^2\theta) = a^2\cos^2\theta$.

$\displaystyle\int \frac{\sqrt{a^2 - u^2}}{u^2}\,du = \int \frac{a^2\cos^2\theta}{a^2\sin^2\theta}\,d\theta = \int \frac{1 - \sin^2\theta}{\sin^2\theta}\,d\theta$

$\displaystyle = \int (\csc^2\theta - 1)\,d\theta = -\cot\theta - \theta + C = -\frac{\sqrt{a^2 - u^2}}{u} - \sin^{-1}\left(\frac{u}{a}\right) + C.$

69. $f(x) = \sqrt{1 + x^4}$, $\Delta x = \dfrac{b - a}{n} = \dfrac{1 - 0}{10} = \dfrac{1}{10}$

(a) $T_{10} = \frac{0.1}{2}[f(0) + 2\,f(0.1) + 2\,f(0.2) + \cdots + 2\,f(0.8) + 2\,f(0.9) + f(1)] \approx 1.090608$

(b) $M_{10} = 0.1\left[f\left(\frac{1}{20}\right) + f\left(\frac{3}{20}\right) + f\left(\frac{5}{20}\right) + \cdots + f\left(\frac{19}{20}\right)\right] \approx 1.088840$

(c) $S_{10} = \frac{0.1}{3}\left[f(0) + 4\,f(0.1) + 2\,f(0.2) + 4\,f(0.3) + 2\,f(0.4) + 4\,f(0.5)\right.$
$\left. + 2\,f(0.6) + 4\,f(0.7) + 2\,f(0.8) + 4\,f(0.9) + f(1)\right] \approx 1.089429$

71. $f(x) = \left(1 + x^4\right)^{1/2}$, $f'(x) = \frac{1}{2}\left(1 + x^4\right)^{-1/2}(4x^3) = 2x^3\left(1 + x^4\right)^{-1/2}$, $f''(x) = (2x^6 + 6x^2)\left(1 + x^4\right)^{-3/2}$. Thus $|f''(x)| \le 8 \cdot 1^{-3/2} = 8$ on $[0, 1]$. By taking $K = 8$, we find that the error in Exercise 69(a) is bounded by $K\dfrac{(b-a)^3}{12n^2} = \dfrac{8}{1200} = \dfrac{1}{150} < 0.0067$, and in (b) by $K\dfrac{(b-a)^3}{24n^2} = \dfrac{1}{300} = 0.00\overline{3}$.

73. **(a)** $f(x) = \sin(\sin x)$. A CAS gives

$f^{(4)}(x) = \sin(\sin x)\cos^4 x + 6\cos(\sin x)\cos^2 x\sin x$
$+ 3\sin(\sin x) + \sin(\sin x)\cos^2 x + \cos(\sin x)\sin x.$

From the graph, we see that $f^{(4)}(x) < 3.8$ for $x \in [0, \pi]$.

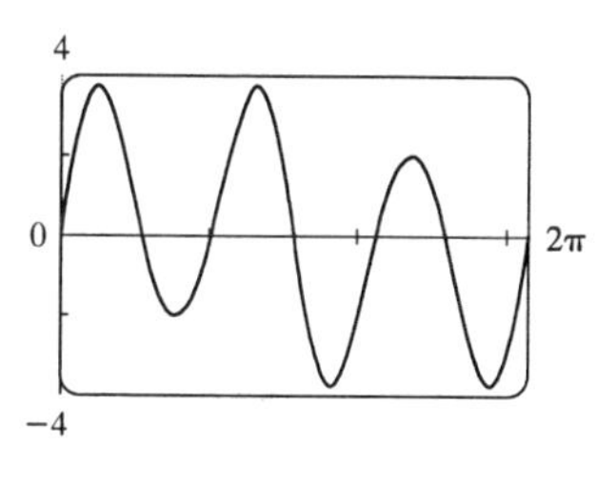

(b) We use Simpson's Rule with $f(x) = \sin(\sin x)$ and $\Delta x = \frac{\pi}{10}$:

$\int_0^\pi f(x)dx \approx \frac{\pi}{10 \cdot 3}\left[f(0) + 4f\left(\frac{\pi}{10}\right) + 2f\left(\frac{2\pi}{10}\right) + \cdots\right.$
$\left. + 2f\left(\frac{8\pi}{10}\right) + 4f\left(\frac{9\pi}{10}\right) + f(\pi)\right] \approx 1.7867$

From part (a), we know that $f^{(4)}(x) < 3.8$ on $[0, \pi]$, so we use Theorem 8.8.7 with $K = 3.8$, and estimate the error as $|E_S| \le \dfrac{3.8(\pi - 0)^5}{180(10)^4} \approx 0.000646$.

(c) If we want the error to be less than 0.00001, we must have $|E_S| \le \dfrac{3.8\pi^5}{180n^4} \le 0.00001$, so

$n^4 \ge \dfrac{3.8\pi^5}{180(0.00001)} \approx 646{,}041.5 \quad\Rightarrow\quad n \ge 28.35$. Since n must be even for Simpson's Rule, we must have $n \ge 30$ to ensure the desired accuracy.

75. $\dfrac{x^3}{x^5+2} \le \dfrac{x^3}{x^5} = \dfrac{1}{x^2}$ for x in $[1,\infty)$. $\displaystyle\int_1^\infty \frac{1}{x^2}\,dx$ is convergent by (8.9.2) with $p = 2 > 1$. Therefore $\displaystyle\int_1^\infty \frac{x^3}{x^5+2}\,dx$ is convergent by the Comparison Theorem.

77. For x in $\left[0, \frac{\pi}{2}\right]$, $0 \le \cos^2 x \le \cos x$. For x in $\left[\frac{\pi}{2}, \pi\right]$, $\cos x \le 0 \le \cos^2 x$. Thus

$$\text{area} = \int_0^{\pi/2}(\cos x - \cos^2 x)dx + \int_{\pi/2}^{\pi}(\cos^2 x - \cos x)dx$$
$$= \left[\sin x - \tfrac{1}{2}x - \tfrac{1}{4}\sin 2x\right]_0^{\pi/2} + \left[\tfrac{1}{2}x + \tfrac{1}{4}\sin 2x - \sin x\right]_{\pi/2}^{\pi} = \left[\left(1 - \tfrac{\pi}{4}\right) - 0\right] + \left[\tfrac{\pi}{2} - \left(\tfrac{\pi}{4} - 1\right)\right] = 2.$$

79. Using the formula for disks, the volume is

$$V = \int_0^{\pi/2} \pi[f(x)]^2\,dx = \pi\int_0^{\pi/2}(\cos^2 x)^2\,dx = \pi\int_0^{\pi/2}\left[\tfrac{1}{2}(1+\cos 2x)\right]^2 dx = \tfrac{\pi}{4}\int_0^{\pi/2}(1 + \cos^2 2x + 2\cos 2x)dx$$
$$= \tfrac{\pi}{4}\int_0^{\pi/2}\left[1 + \tfrac{1}{2}(1+\cos 4x) + 2\cos 2x\right]dx = \tfrac{\pi}{4}\left[\tfrac{3}{2}x + \tfrac{1}{2}\left(\tfrac{1}{4}\sin 4x\right) + 2\left(\tfrac{1}{2}\sin 2x\right)\right]_0^{\pi/2}$$
$$= \tfrac{\pi}{4}\left[\left(\tfrac{3\pi}{4} + \tfrac{1}{8}\cdot 0 + 0\right) - 0\right] = \tfrac{3\pi^2}{16}.$$

81. For $n \ge 0$, $\int_0^\infty x^n\,dx = \lim\limits_{t\to\infty}\left[x^{n+1}/(n+1)\right]_0^t = \infty$. For $n < 0$, $\int_0^\infty x^n\,dx = \int_0^1 x^n\,dx + \int_1^\infty x^n\,dx$. Both integrals are improper. By (8.9.2), the second integral diverges if $-1 \le n < 0$. By Exercise 8.9.57, the first integral diverges if $n \le -1$. Thus $\int_0^\infty x^n\,dx$ is divergent for all values of n.

83. By the Fundamental Theorem of Calculus,

$$\int_0^\infty f'(x)dx = \lim_{t\to\infty}\int_0^t f'(x)dx = \lim_{t\to\infty}\left[f(t) - f(0)\right] = \lim_{t\to\infty} f(t) - f(0) = 0 - f(0) = -f(0).$$

85. Let $u = 1/x \;\Rightarrow\; x = 1/u \;\Rightarrow\; dx = -(1/u^2)du$.

$$\int_0^\infty \frac{\ln x}{1+x^2}\,dx = \int_\infty^0 \frac{\ln(1/u)}{1+1/u^2}\left(-\frac{du}{u^2}\right) = \int_\infty^0 \frac{-\ln u}{u^2+1}(-du) = \int_\infty^0 \frac{\ln u}{1+u^2}\,du$$
$$= -\int_0^\infty \frac{\ln u}{1+u^2}\,du. \text{ Therefore } \int_0^\infty \frac{\ln x}{1+x^2}\,dx = -\int_0^\infty \frac{\ln x}{1+x^2}\,dx = 0.$$

APPLICATIONS PLUS (after Chapter 8)

1. **(a)** Coefficient of inequality $= \dfrac{\text{area between Lorenz curve and straight line}}{\text{area under straight line}}$

$$= \frac{\int_0^1 [x - L(x)]dx}{\int_0^1 x\,dx} = \frac{\int_0^1 [x - L(x)]dx}{\left[x^2/2\right]_0^1} = \frac{\int_0^1 [x - L(x)]dx}{1/2} = 2\int_0^1 [x - L(x)]dx$$

(b) $L(x) = \frac{5}{12}x^2 + \frac{7}{12}x \quad \Rightarrow \quad L\left(\frac{1}{2}\right) = \frac{5}{48} + \frac{7}{24} = \frac{19}{48} = 0.3958\overline{3}$, so the bottom 50% of the households receive about 40% of the income.

Coefficient of inequality $= 2\int_0^1 \left[x - \frac{5}{12}x^2 - \frac{7}{12}x\right]dx = 2\int_0^1 \frac{5}{12}(x - x^2)dx = \frac{5}{6}\left(\frac{1}{2}x^2 - \frac{1}{3}x^3\right)\Big|_0^1 = \frac{5}{36}$

(c) Coefficient of inequality $= 2\int_0^1 [x - L(x)]dx = 2\int_0^1 \left(x - \dfrac{5x^3}{4 + x^2}\right)dx$

$$= 2\int_0^1 \left[x - \left(5x - \frac{20x}{x^2 + 4}\right)\right]dx = 2\int_0^1 \left(-4x + \frac{20x}{x^2 + 4}\right)dx$$

$$= 2\left[-2x^2 + 10\ln(x^2 + 4)\right]_0^1 = 2(-2 + 10\ln 5 - 10\ln 4)$$

$$= -4 + 20\ln\tfrac{5}{4} \approx 0.46$$

3. **(a)** The tangent to the curve $y = f(x)$ at $x = x_0$ has the equation $y - f(x_0) = f'(x_0)(x - x_0)$. The y-intercept of this tangent line is $f(x_0) - f'(x_0)x_0$. Thus L is the distance from the point $(0, f(x_0) - f'(x_0)x_0)$ to the point $(x_0, f'(x_0))$. That is, $L^2 = x_0^2 + [f'(x_0)]^2 x_0^2$, so $[f'(x_0)]^2 = \dfrac{L^2 - x_0^2}{x_0^2}$ and $f'(x_0) = -\dfrac{\sqrt{L^2 - x_0^2}}{x_0}$ for each $0 < x_0 < L$.

(b) $\dfrac{dy}{dx} = -\dfrac{\sqrt{L^2 - x^2}}{x} \quad \Rightarrow$

$$y = \int -\frac{\sqrt{L^2 - x^2}}{x}\,dx = \int \frac{-L\cos\theta\, L\cos\theta\, d\theta}{L\sin\theta} \quad (\text{where } x = L\sin\theta)$$

$$= L\int \frac{\sin^2\theta - 1}{\sin\theta}\,d\theta = L\int(\sin\theta - \csc\theta)d\theta = -L\cos\theta + L\ln|\csc\theta + \cot\theta| + C$$

$$= -\sqrt{L^2 - x^2} + L\ln\left(\frac{L}{x} + \frac{\sqrt{L^2 - x^2}}{x}\right) + C$$

When $x = L$, $0 = y = -0 + L\ln(1 + 0) + C$, so $C = 0$. Therefore

$$y = -\sqrt{L^2 - x^2} + L\ln\left(\frac{L + \sqrt{L^2 - x^2}}{x}\right).$$

5. **(a)** $F = ma = m\dfrac{dv}{dt}$, so by the Substitution Rule we have

$$\int_{t_0}^{t_1} F(t)dt = \int_{t_0}^{t_1} m\left(\frac{dv}{dt}\right)dt = m\int_{v_0}^{v_1} dv = mv\big|_{v_0}^{v_1} = mv_1 - mv_0 = p(t_1) - p(t_0).$$

(b) **(i)** We have $v_1 = 110\,\text{mi/h} = \frac{110(5280)}{3600}\,\text{ft/s} = 161.\overline{3}\,\text{ft/s}$, $v_0 = -90\,\text{mi/h} = -132\,\text{ft/s}$, and $m = \frac{5/16}{32} = \frac{5}{512}$. So the change in momentum is

$p(t_1) - p(t_0) = mv_1 - mv_0 = \frac{5}{512}\left[161.\overline{3} - (-132)\right] = \frac{1466.\overline{6}}{512} \approx 2.86\,\text{slug-ft/s}.$

(ii) From part (a) and part (b)(i) we have $\int_0^{0.01} F(t)dt = p(0.01) - p(0) \approx 2.86$, so the average force over the interval $[0, 0.01] = \frac{1}{0.01}\int_0^{0.01} F(t)dt \approx \frac{1}{0.01}(2.86) = 286\,\text{lb}$.

7. **(a)** Here we have a differential equation of the form $\dfrac{dv}{dt} = kv$, so by Theorem 3.5.2, the solution is $v(t) = v(0)e^{kt}$. In this case $k = -\frac{1}{10}$ and $v(0) = 100\,\text{ft/s}$, so $v(t) = 100e^{-t/10}$. We are interested in the time that the ball takes to travel 280 ft, so we find the distance function

$s(t) = \int_0^t v(x)dx = \int_0^t 100e^{-x/10}\,dx = 100\left[-10e^{-x/10}\right]_0^t = -1000\left(e^{-t/10} - 1\right) = 1000\left(1 - e^{-t/10}\right).$

Now we set $s(t) = 280$ and solve for t: $280 = 1000\left(1 - e^{-t/10}\right) \;\Rightarrow\; 1 - e^{-t/10} = \frac{7}{25} \;\Rightarrow$ $-\frac{1}{10}t = \ln\left(1 - \frac{7}{25}\right) \approx -0.3285$, so $t \approx 3.285$ seconds.

(b) Let x be the distance of the shortstop from home plate. We calculate the time for the ball to reach home plate as a function of x, then differentiate with respect to x to find the value of x which corresponds to the minimum time.

The total time that it takes the ball to reach home is the sum of the times of the two throws, plus the relay time $\left(\frac{1}{2}\,\text{s}\right)$. The distance from the fielder to the shortstop is $280 - x$, so to find the time t_1 taken by the first throw, we solve the equation $s_1(t_1) = 280 - x \;\Leftrightarrow\; 1 - e^{-t_1/10} = \dfrac{280 - x}{1000} \;\Leftrightarrow\; t_1 = -10\ln\dfrac{720 + x}{1000}$.

We find the time t_2 taken by the second throw if the shortstop throws with velocity w, since we see that this velocity varies in the rest of the problem. We use $v = we^{-t/10}$ and isolate t_2 in the equation

$s(t_2) = 10w\left(1 - e^{-t_2/10}\right) = x \;\Leftrightarrow\; e^{-t_2/10} = 1 - \dfrac{x}{10w} \;\Leftrightarrow\; t_2 = -10\ln\dfrac{10w - x}{10w}$, so the total time is $t_w(x) = \dfrac{1}{2} - 10\left[\ln\dfrac{720 + x}{1000} + \ln\dfrac{10w - x}{10w}\right]$. To find the minimum, we differentiate:

$\dfrac{dt_w}{dx} = -10\left[\dfrac{1}{720 + x} - \dfrac{1}{10w - x}\right]$, which changes from negative to positive when $720 + x = 10w - x$ $\Leftrightarrow\; x = 5w - 360$. So by the First Derivative Test, t_w has a minimum at this distance from the shortstop to home plate. So if the shortstop throws at $w = 105$ ft/s, the minimum time is

$t_{105}(165) = -10\ln\frac{720 + 165}{1000} + \frac{1}{2} - 10\ln\frac{1050 - 165}{1050} \approx 3.431$ seconds. This is longer than the time taken in part (a), so in this case the manager should encourage a direct throw.

If $w = 115$ ft/s, the minimum time is $t_{115}(215) = -10\ln\frac{720 + 215}{1000} + \frac{1}{2} - 10\ln\frac{1150 - 215}{1150} \approx 3.242$ seconds. This is less than the time taken in part (a), so in this case, the manager should encourage a relayed throw.

(c) In general, the minimum time is

$$t_w(5w-360)=\frac{1}{2}-10\left[\ln\frac{360+5w}{1000}+\ln\frac{360+5w}{10w}\right]=\frac{1}{2}-10\ln\frac{(w+72)^2}{400w}.$$

We want to find out when this is about 3.285 seconds, the same time as the direct throw.

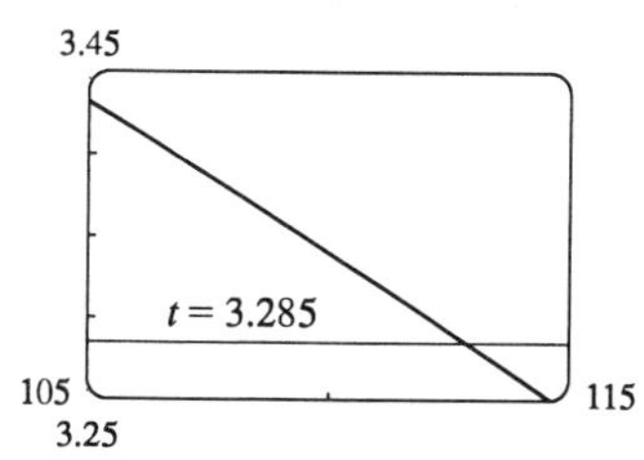

From the graph, we estimate that this is the case for $w\approx 112.8$ ft/s. So if the shortstop can throw the ball with this velocity, then a relayed throw takes the same time as a direct throw.

9. **(a)** $|VP|=9+x\cos\alpha$, $|PT|=35-(4+x\sin\alpha)=31-x\sin\alpha$, and $|PB|=(4+x\sin\alpha)-10=x\sin\alpha-6$.

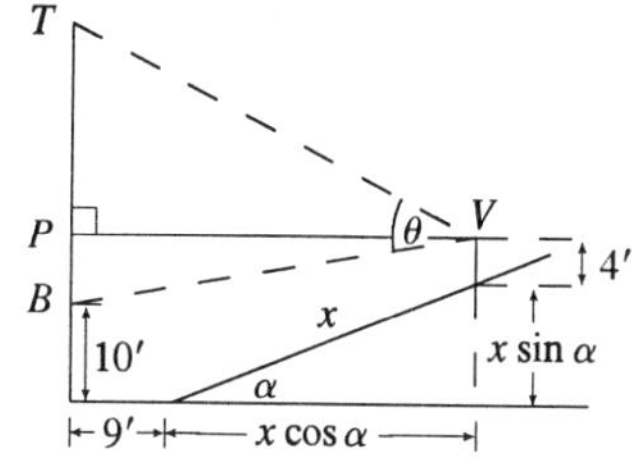

So using the Pythagorean Theorem, we have

$$|VT|=\sqrt{|VP|^2+|PT|^2}$$
$$=\sqrt{(9+x\cos\alpha)^2+(31-x\sin\alpha)^2}=a, \text{ and}$$

$|VB|=\sqrt{|VP|^2+|PB|^2}=\sqrt{(9+x\cos\alpha)^2+(x\sin\alpha-6)^2}=b$. Using the Law of Cosines on $\triangle VBT$, we get $25^2=a^2+b^2-2ab\cos\theta \quad\Leftrightarrow$

$$\cos\theta=\frac{a^2+b^2-625}{2ab}\quad\Leftrightarrow\quad\theta=\arccos\left(\frac{a^2+b^2-625}{2ab}\right), \text{ as required.}$$

(b) From the graph, it appears that the value of x which maximizes θ is $x\approx 8.25$ ft. The row closest to this value of x is the fourth row, at $x=9$ ft, and from the graph, the viewing angle in this row seems to be about 0.85 radians, or about 49°.

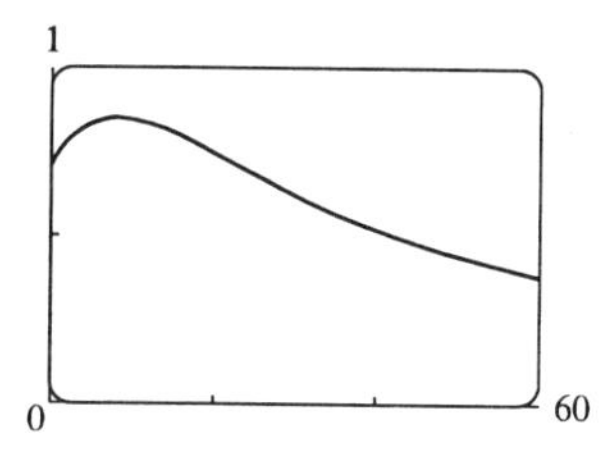

(c) With a CAS, we type in the definition of θ (calling it T), substitute in the proper values of a and b in terms of x and $\alpha=20°=\frac{\pi}{9}$ radians, and then use the differentiation command (`diff` in Maple) to find the derivative. We use a numerical root finder (`fsolve` in Maple) and find that the root of the equation $d\theta/dx=0$ is $x\approx 8.25306209$, as approximated above.

(d) From the graph in part (b), it seems that the average value of the function on the interval $[0,60]$ is about 0.6. We can use a CAS to approximate $\frac{1}{60}\int_0^{60}\theta(x)dx\approx 0.625\approx 36°$. (The calculation is faster if we reduce the number of digits of accuracy required.) The minimum value is $\theta(60)\approx 0.38$ and, from part (b), the maximum value is about 0.85.

CHAPTER NINE

EXERCISES 9.1

1. $\dfrac{dy}{dx} = y^2 \;\Rightarrow\; \dfrac{dy}{y^2} = dx\ (y \neq 0) \;\Rightarrow\; \displaystyle\int \frac{dy}{y^2} = \int dx \;\Rightarrow\; -\frac{1}{y} = x + C \;\Rightarrow\; -y = \frac{1}{x+C} \;\Rightarrow\; y = \frac{-1}{x+C}$, and $y = 0$ is also a solution.

3. $yy' = x \;\Rightarrow\; \int y\,dy = \int x\,dx \;\Rightarrow\; \dfrac{y^2}{2} = \dfrac{x^2}{2} + C_1 \;\Rightarrow\; y^2 = x^2 + 2C_1 \;\Rightarrow\; x^2 - y^2 = C$ (where $C = -2C_1$). This represents a family of hyperbolas.

5. $x^2y' + y = 0 \;\Rightarrow\; \dfrac{dy}{dx} = -\dfrac{y}{x^2} \;\Rightarrow\; \displaystyle\int \frac{dy}{y} = \int \frac{-dx}{x^2}\ (y \neq 0) \;\Rightarrow\; \ln|y| = \frac{1}{x} + K \;\Rightarrow$ $|y| = e^K e^{1/x} \;\Rightarrow\; y = Ce^{1/x}$, where now we allow C to be any constant.

7. $\dfrac{du}{dt} = e^{u+2t} = e^u e^{2t} \;\Rightarrow\; \int e^{-u}\,du = \int e^{2t}\,dt \;\Rightarrow\; -e^{-u} = \frac{1}{2}e^{2t} + C_1 \;\Rightarrow\; e^{-u} = -\frac{1}{2}e^{2t} + C$ (where $C = -C_1$ and the right-hand side is positive, since $e^{-u} > 0$) $\Rightarrow\; -u = \ln\left(C - \frac{1}{2}e^{2t}\right) \;\Rightarrow$ $u = -\ln\left(C - \frac{1}{2}e^{2t}\right)$

9. $e^y y' = \dfrac{3x^2}{1+y}$, $y(2) = 0$. $\displaystyle\int e^y(1+y)dy = \int 3x^2\,dx \;\Rightarrow\; ye^y = x^3 + C$. $y(2) = 0$, so $0 = 2^3 + C$ and $C = -8$. Thus $ye^y = x^3 - 8$.

11. $xe^{-t}\dfrac{dx}{dt} = t$, $x(0) = 1$. $\int x\,dx = \int te^t dt \;\Rightarrow\; \frac{1}{2}x^2 = (t-1)e^t + C$. $x(0) = 1$, so $\frac{1}{2} = (0-1)e^0 + C$ and $C = \frac{3}{2}$. Thus $x^2 = 2(t-1)e^t + 3 \;\Rightarrow\; x = \sqrt{2(t-1)e^t + 3}$.

13. $\dfrac{du}{dt} = \dfrac{2t+1}{2(u-1)}$, $u(0) = -1$. $\int 2(u-1)du = \int (2t+1)dt \;\Rightarrow\; u^2 - 2u = t^2 + t + C$. $u(0) = -1$ so $(-1)^2 - 2(-1) = 0^2 + 0 + C$ and $C = 3$. Thus $u^2 - 2u = t^2 + t + 3$; the quadratic formula gives $u = 1 - \sqrt{t^2 + t + 4}$.

15. Let $y = f(x)$. Then $\dfrac{dy}{dx} = x^3 y$ and $y(0) = 1$. $\dfrac{dy}{y} = x^3\,dx$ (if $y \neq 0$), so $\displaystyle\int \frac{dy}{y} = \int x^3\,dx$ and $\ln|y| = \frac{1}{4}x^4 + C$; $y(0) = 1 \;\Rightarrow\; C = 0$, so $\ln|y| = \frac{1}{4}x^4$, $|y| = e^{x^4/4}$ and $y = f(x) = e^{x^4/4}$ [since $y(0) = 1$].

17. $\dfrac{dy}{dx} = 4x^3 y$, $y(0) = 7$. $\dfrac{dy}{y} = 4x^3\,dx$ (if $y \neq 0$) $\Rightarrow\; \displaystyle\int \frac{dy}{y} = \int 4x^3\,dx \;\Rightarrow\; \ln|y| = x^4 + C \;\Rightarrow$ $y = Ae^{x^4}$; $y(0) = 7 \;\Rightarrow\; A = 7 \;\Rightarrow\; y = 7e^{x^4}$.

19. $y' = e^{x-y}$, $y(0) = 1$.

So $\dfrac{dy}{dx} = e^x e^{-y} \;\Leftrightarrow\; \displaystyle\int e^y dy = \int e^x dx \;\Leftrightarrow$ $e^y = e^x + C$. From the initial condition, we must have $e^1 = e^0 + C \;\Rightarrow C = e - 1$. So the solution is $e^y = e^x + e - 1 \Rightarrow\; y = \ln(e^x + e - 1)$.

3
-3
0
3

21. $\dfrac{dy}{dx} = \dfrac{\sin x}{\sin y}$, $y(0) = \dfrac{\pi}{2}$. So $\int \sin y\,dy = \int \sin x\,dx \quad \Leftrightarrow \quad -\cos y = -\cos x + C \quad \Leftrightarrow \quad \cos y = \cos x - C$.

From the initial condition, we need $\cos\frac{\pi}{2} = \cos 0 - C \quad \Rightarrow \quad 0 = 1 - C \quad \Rightarrow \quad C = 1$, so the solution is $\cos y = \cos x - 1$. Note that we cannot take $\cos^{-1}$ of both sides, since that would unnecessarily restrict the solution to the case where $-1 \le \cos x - 1 \quad \Leftrightarrow \quad 0 \le \cos x$, since $\cos^{-1}$ is defined only on $[-1, 1]$. Instead we plot the graph using Maple's `plots[implicitplot]` or Mathematica's `Plot[Evaluate[⋯]]`.

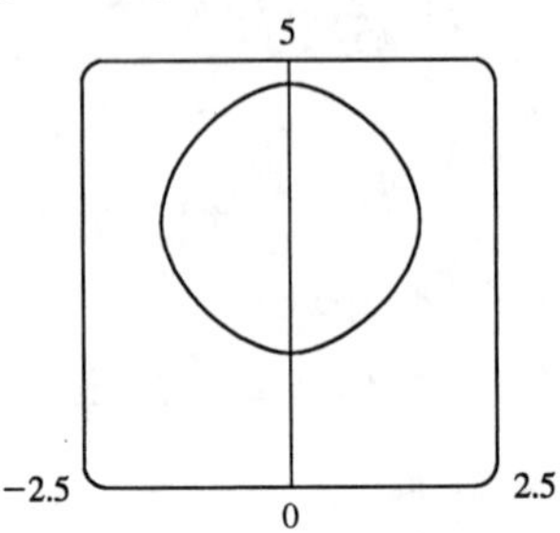

23. (a)

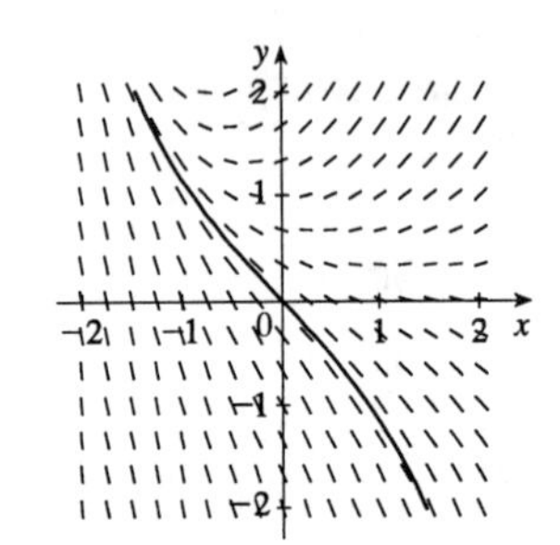

(b)

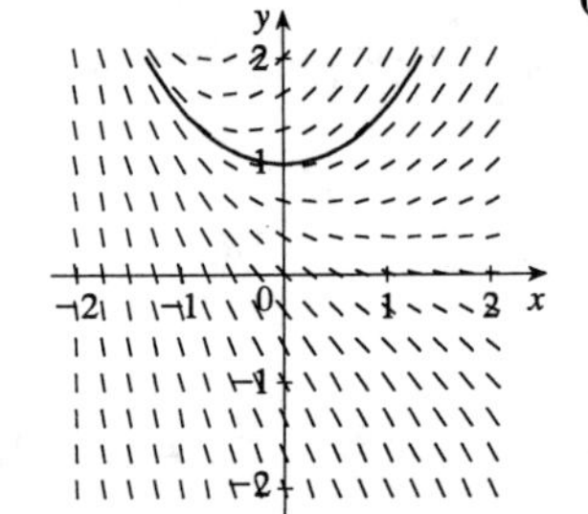

(c)

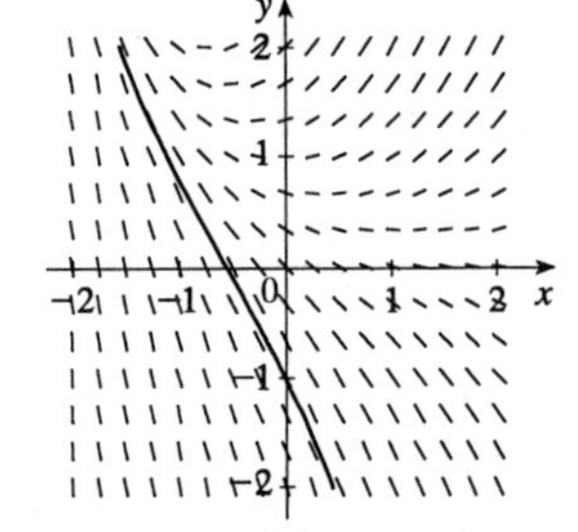

25. $y' = x - y$

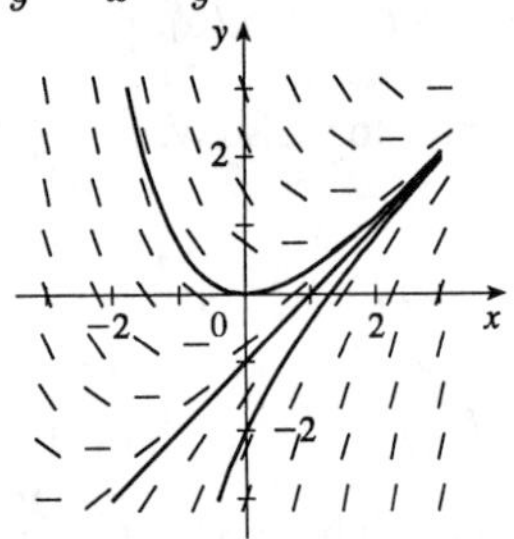

27. In Maple, we can use either `directionfield` (in Maple's share library) or `plots[fieldplot]` to plot the direction field. To plot the solution, we can either use the initial-value option in `directionfield`, or actually solve the equation. In *Mathematica*, we use `PlotVectorField` for the direction field, and the `Plot[Evaluate[⋯]]` construction to plot the solution, which is $y = e^{(1-\cos 2x)/2}$.

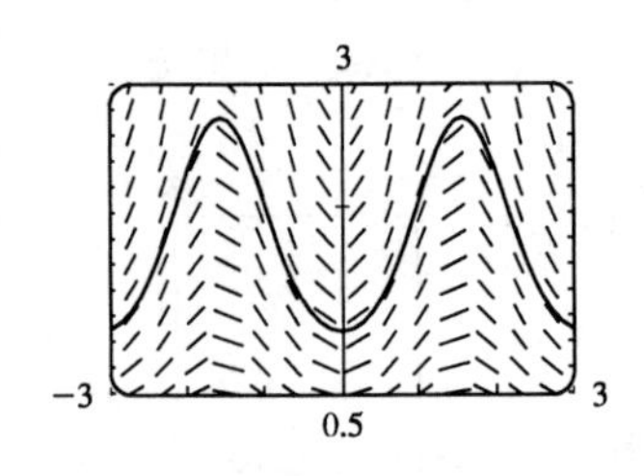

29. **(a)** Let $y(t)$ be the amount of salt (in kg) after t minutes. Then $y(0) = 15$. The amount of liquid in the tank is 1000 L at all times, so the concentration at time t (in minutes) is $y(t)/1000$ kg/L and $\frac{dy}{dt} = -\left[\frac{y(t)}{1000}\,\frac{\text{kg}}{\text{L}}\right]\left(10\,\frac{\text{L}}{\text{min}}\right) = -\frac{y(t)}{100}\,\frac{\text{kg}}{\text{min}}$. $\int \frac{dy}{y} = -\frac{1}{100}\int dt \Rightarrow \ln y = -\frac{t}{100} + C$, and $y(0) = 15 \Rightarrow \ln 15 = C$, so $\ln y = \ln 15 - \frac{t}{100}$. It follows that $\ln\left(\frac{y}{15}\right) = -\frac{t}{100}$ and $\frac{y}{15} = e^{-t/100}$, so $y = 15e^{-t/100}$ kg.

(b) After 20 minutes, $y = 15e^{-20/100} = 15e^{-0.2} \approx 12.3$ kg.

31. $\frac{dx}{dt} = k(a-x)(b-x)$, $a \neq b$. $\int \frac{dx}{(a-x)(b-x)} = \int k\,dt \Rightarrow \frac{1}{b-a}\int\left(\frac{1}{a-x} - \frac{1}{b-x}\right)dx = \int k\,dt$ $\Rightarrow \frac{1}{b-a}(-\ln|a-x| + \ln|b-x|) = kt + C \Rightarrow \ln\left|\frac{b-x}{a-x}\right| = (b-a)(kt+C)$. Here the concentrations $[A] = a - x$ and $[B] = b - x$ cannot be negative, so $\frac{b-x}{a-x} \geq 0$ and $\left|\frac{b-x}{a-x}\right| = \frac{b-x}{a-x}$. We now have $\ln\left(\frac{b-x}{a-x}\right) = (b-a)(kt+C)$. Since $x(0) = 0$, $\ln\left(\frac{b}{a}\right) = (b-a)C$. Hence $\ln\left(\frac{b-x}{a-x}\right) = (b-a)kt + \ln\left(\frac{b}{a}\right)$, $\frac{b-x}{a-x} = \frac{b}{a}e^{(b-a)kt}$, and $x = \frac{b\left[e^{(b-a)kt} - 1\right]}{be^{(b-a)kt}/a - 1} = \frac{ab\left[e^{(b-a)kt} - 1\right]}{be^{(b-a)kt} - a}$ moles/L.

33. **(a)** Let $P(t)$ be the world population in the year t. Then $dP/dt = 0.02P$, so $\int (1/P)dP = \int 0.02\,dt$ and $\ln P = 0.02t + C \Rightarrow P(t) = Ae^{0.02t}$. $P(1986) = 5 \times 10^9 \Rightarrow P(t) = 5 \times 10^9 e^{0.02(t-1986)}$.

(b) **(i)** The predicted population in 2000 is $P(2000) = 5e^{0.28} \times 10^9 \approx 6.6$ billion.

(ii) The predicted population in 2100 is $P(2100) = 5e^{2.28} \times 10^9 \approx 49$ billion.

(iii) The predicted population in 2500 is $P(2500) = 5e^{10.28} \times 10^9 \approx 146$ trillion.

(c) According to this model, in 2000 the area per person will be $\frac{1.8 \times 10^{15}}{6.6 \times 10^9} \approx 270{,}000\ \text{ft}^2$. In 2100 it will be $\frac{1.8 \times 10^{15}}{49 \times 10^9} \approx 37{,}000\ \text{ft}^2$, and in 2500 it will be $\frac{1.8 \times 10^{15}}{146 \times 10^{12}} \approx 12\ \text{ft}^2$. (!)

35. **(a)** Our assumption is that $\frac{dy}{dt} = ky(1-y)$, where y is the fraction of the population that has heard the rumor.

(b) Take $M = 1$ in (11) to get $y = \frac{y_0}{y_0 + (1-y_0)e^{-kt}}$.

(c) Let t be the number of hours since 8 A.M. Then $y_0 = y(0) = \frac{80}{1000} = 0.08$ and $y(4) = \frac{1}{2}$, so $\frac{1}{2} = y(4) = \frac{0.08}{0.08 + 0.92e^{-4k}}$. Thus $0.08 + 0.92e^{-4k} = 0.16$, $e^{-4k} = \frac{0.08}{0.92} = \frac{2}{23}$, and $e^{-k} = \left(\frac{2}{23}\right)^{1/4}$, so $y = \frac{0.08}{0.08 + 0.92(2/23)^{t/4}} = \frac{2}{2 + 23(2/23)^{t/4}}$ and $\left(\frac{2}{23}\right)^{t/4} = \frac{2}{23}\cdot\frac{1-y}{y}$ or $\left(\frac{2}{23}\right)^{t/4-1} = \frac{1-y}{y}$. It follows that $\frac{t}{4} - 1 = \frac{\ln[(1-y)/y]}{\ln(2/23)}$, so $t = 4\left[1 + \frac{\ln[(1-y)/y]}{\ln(2/23)}\right]$. When $y = 0.9$, $\frac{1-y}{y} = \frac{1}{9}$, so $t = 4\left(1 - \frac{\ln 9}{\ln 23}\right) \approx 7.6$ h or 7 h 36 min. Thus 90% of the population will have heard the rumor by 3:36 P.M..

37. y increases most rapidly when y' is maximal, that is, when $y'' = 0$. But $y' = ky(M - y) \quad \Rightarrow$

$y'' = ky'(M - y) + ky(-y') = ky'(M - 2y) = k^2 y(M - y)(M - 2y)$. Since $0 < y < M$, we see that $y'' = 0$

$\Leftrightarrow \quad y = M/2$.

39. At $t = 0$, the exponential model $y = e^{0.1t}$ has derivative $y' = 0.1e^0 = 0.1$. From the original differential equation, the logistic model has derivative $y' = ky(M - y)$. At $t = 0$, this is equal to $ky_0(M - y_0) = 9k$. So the two derivatives are equal at $t = 0$ if $9k = 0.1 \quad \Leftrightarrow \quad k = 0.1/9 = \frac{1}{90}$.

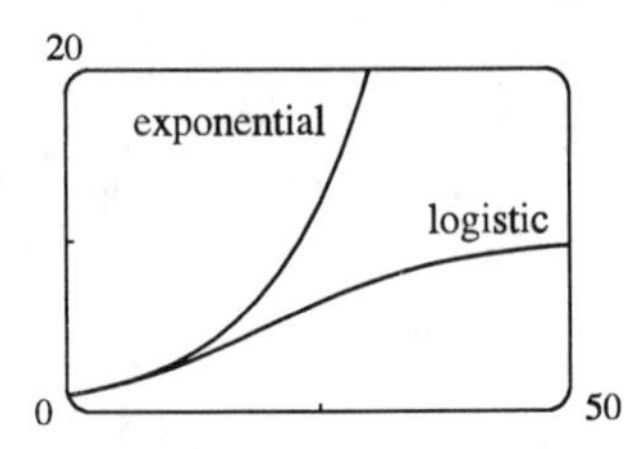

We graph both models, and see that for small values of t they agree closely, but for large values of t, the exponential model increases rapidly, while the logistic model levels off and approaches the line $y = 10$.

41. **(a)** The rate of growth of the area is jointly proportional to $\sqrt{A(t)}$ and $M - A(t)$; that is, the rate is proportional to the product of those two quantities. So for some constant k, $dA/dt = k\sqrt{A}(M - A)$. We are interested in the maximum of the function dA/dt, so we differentiate, using the Chain Rule and then substituting for dA/dt from the differential equation:

$$\frac{d}{dt}\left(\frac{dA}{dt}\right) = k\left[\tfrac{1}{2}A^{-1/2}(M - A)\frac{dA}{dt} + \sqrt{A}(-1)\frac{dA}{dt}\right] = \tfrac{1}{2}kA^{-1/2}\frac{dA}{dt}[(M - A) - 2A]$$

$= \frac{1}{2}k^2(M - A)(M - 3A)$. This is 0 when $M - A = 0$ [this situation never actually occurs, since the graph of $A(t)$ is asymptotic to the line $y = M$, as in the logistic model] and when $M - 3A = 0$ $\Leftrightarrow \quad A(t) = M/3$. This represents a maximum by the First Derivative Test, since $\frac{d}{dt}\left(\frac{dA}{dt}\right)$ goes from positive to negative when $A(t) = M/3$.

(b) To solve the differential equation, we separate variables in our expression for dA/dt and integrate:

$\int \frac{dA}{\sqrt{A}(M - A)} = \int k\,dt = kt + C_1$. To evaluate the LHS, we make the substitution $x = \sqrt{A}$, so $A = x^2$ and $dA = 2x\,dx$. The LHS becomes

$$2\int \frac{dx}{M - x^2} = \int \left(\frac{1/\sqrt{M}}{\sqrt{M} + x} + \frac{1/\sqrt{M}}{\sqrt{M} - x}\right)dx \quad \text{(difference of squares; partial fractions)}$$

$$= \frac{1}{\sqrt{M}}\left(\ln\left|\sqrt{M} + x\right| - \ln\left|\sqrt{M} - x\right|\right) = \frac{1}{\sqrt{M}}\ln\frac{\sqrt{M} + \sqrt{A}}{\sqrt{M} - \sqrt{A}}.$$

So, multiplying by $\sqrt{M}$ and exponentiating both sides of the equation, we get $\dfrac{\sqrt{M}+\sqrt{A}}{\sqrt{M}-\sqrt{A}} = Ce^{\sqrt{M}kt}$, where $C = e^{\sqrt{M}C_1}$. Solving for A: $\sqrt{M}+\sqrt{A} = Ce^{\sqrt{M}kt}\left(\sqrt{M}-\sqrt{A}\right) \quad \Leftrightarrow$

$\sqrt{A}\left(Ce^{\sqrt{M}kt}+1\right) = \sqrt{M}\left(Ce^{\sqrt{M}kt}-1\right) \quad \Leftrightarrow \quad A = M\left(\dfrac{Ce^{\sqrt{M}kt}-1}{Ce^{\sqrt{M}kt}+1}\right)^2.$

To get C in terms of the initial area A_0 and the maximum area M, we substitute $t = 0$ and $A = A_0$:

$A_0 = M\left(\dfrac{C-1}{C+1}\right)^2 \quad \Leftrightarrow \quad (C+1)\sqrt{A_0} = (C-1)\sqrt{M} \quad \Leftrightarrow \quad C = \dfrac{\sqrt{M}+\sqrt{A_0}}{\sqrt{M}-\sqrt{A_0}}$. (Notice that if $A_0 = 0$, then $C = 1$.)

43. $RI + LI'(t) = V, I(0) = 0$. $LI' = V - RI \;\Rightarrow\; L\dfrac{dI}{dt} = V - RI \;\Rightarrow\; \displaystyle\int \frac{L\,dI}{V-RI} = \int dt \;\Rightarrow$

$-\dfrac{L}{R}\ln|V-RI| = t + C \quad\Rightarrow\quad V - RI = Ae^{-Rt/L} \quad\Rightarrow\quad I = \dfrac{V}{R} - \dfrac{A}{R}e^{-Rt/L}.$

$I(0) = 0 \quad\Rightarrow\quad 0 = \dfrac{V}{R} - \dfrac{A}{R}\cdot e^0 \quad\Rightarrow\quad A = V$. So $I = \dfrac{V}{R}\left(1 - e^{-Rt/L}\right)$.

EXERCISES 9.2

1. $y' + x^2y = y^2$ is not linear since it cannot be put into the standard linear form (1).

3. $xy' = x - y \quad\Rightarrow\quad xy' + y = x \quad\Rightarrow\quad y' + \dfrac{1}{x}y = 1$, which is in the standard linear form (1), and thus this differential equation is linear.

5. $I(x) = e^{\int -3\,dx} = e^{-3x}$. Multiplying the differential equation by $I(x)$ gives $e^{-3x}y' - 3e^{-3x}y = e^{-2x} \quad\Rightarrow$ $(e^{-3x}y)' = e^{-2x} \quad\Rightarrow\quad y = e^{3x}\left[\int(e^{-2x})dx + C\right] = Ce^{3x} - \frac{1}{2}e^{x}$.

7. $I(x) = e^{\int -2x\,dx} = e^{-x^2}$. Multiplying the differential equation by $I(x)$ gives $e^{-x^2}(y' - 2xy) = xe^{-x^2} \quad\Rightarrow$ $\left(e^{-x^2}y\right)' = xe^{-x^2} \quad\Rightarrow\quad y = e^{x^2}\left(\int xe^{-x^2}\,dx + C\right) = Ce^{x^2} - \frac{1}{2}$.

9. $y' - y\tan x = \dfrac{\sin 2x}{\cos x} = 2\sin x$, so $I(x) = e^{\int -\tan x\,dx} = e^{\ln|\cos x|} = \cos x$ (since $-\frac{\pi}{2} < x < \frac{\pi}{2}$.) Multiplying the differential equation by $I(x)$ gives $(y' - y\tan x)\cos x = \dfrac{\sin 2x}{\cos x}\cos x \quad\Rightarrow\quad [y\cos x]' = \sin 2x \quad\Rightarrow$ $y\cos x = \int \sin 2x\,dx + C = -\frac{1}{2}\cos 2x + C = \frac{1}{2} - \cos^2 x + C \quad\Rightarrow\quad y = \dfrac{\sec x}{2} - \cos x + C\sec x$.

11. $I(x) = e^{\int 2x\,dx} = e^{x^2}$. Multiplying the differential equation by $I(x)$ gives $e^{x^2}y' + 2xe^{x^2}y = x^2e^{x^2} \quad\Rightarrow$ $\left(e^{x^2}y\right)' = x^2e^{x^2}$. Thus

$y = e^{-x^2}\left[\int x^2e^{x^2}\,dx + C\right] = e^{-x^2}\left[\frac{1}{2}xe^{x^2} - \int \frac{1}{2}e^{x^2}\,dx + C\right] = \frac{1}{2}x + Ce^{-x^2} - e^{-x^2}\int \frac{1}{2}e^{x^2}\,dx.$

13. $I(\theta) = e^{\int -\tan\theta\, d\theta} = e^{-\ln(\sec\theta)} = \cos\theta$. Multiplying the differential equation by $I(\theta)$ gives $\cos\theta(dy/d\theta) - y\sin\theta = \cos\theta \Rightarrow (y\cos\theta)' = \cos\theta \Rightarrow y\cos\theta = \int \cos\theta\, d\theta \Rightarrow$ $y\cos\theta = \sin\theta + C \Rightarrow y = \tan\theta + C\sec\theta$.

15. $I(x) = e^{\int dx} = e^x$. Multiplying the differential equation by $I(x)$ gives $e^x y' + e^x y = e^x(x + e^x) \Rightarrow$ $(e^x y)' = e^x(x + e^x)$. Thus $y = e^{-x}\left[\int e^x(x + e^x)dx + C\right] = e^{-x}\left[xe^x - e^x + \frac{e^{2x}}{2} + C\right] = x - 1 + \frac{e^x}{2} + \frac{C}{e^x}$.
But $0 = y(0) = -1 + \frac{1}{2} + C$, so $C = \frac{1}{2}$, and the solution to the initial-value problem is $y = x - 1 + \frac{1}{2}e^x + \frac{1}{2}e^{-x} = x - 1 + \cosh x$.

17. $I(x) = e^{\int -2x\, dx} = e^{-x^2}$. Multiplying the differential equation by $I(x)$ gives $e^{-x^2}y' - 2xe^{-x^2}y = 2x \Rightarrow$ $\left(e^{-x^2}y\right)' = 2x \Rightarrow y = e^{x^2}\left[\int 2x\, dx + C\right] = x^2e^{x^2} + Ce^{x^2}$. But $3 = y(0) = C$, so the solution to the initial-value problem is $y = (x^2 + 3)e^{x^2}$.

19. $y' + 2\frac{y}{x} = \frac{\cos x}{x^2}$ $(x \neq 0)$, so $I(x) = e^{\int (2/x)dx} = x^2$. Multiplying the differential equation by $I(x)$ gives $x^2y' + 2xy = \cos x \Rightarrow (x^2y)' = \cos x \Rightarrow y = x^{-2}\left[\int \cos x\, dx + C\right] = x^{-2}(\sin x + C)$ $(x \neq 0)$. But $0 = y(\pi) = C$, so the solution to the initial-value problem is $y = (\sin x)/x^2$.

21. $y' + \frac{1}{x}y = \cos x$ $(x \neq 0)$, so $I(x) = e^{\int (1/x)dx} = e^{\ln|x|} = x$ (for $x > 0$). Multiplying the differential equation by $I(x)$ gives $xy' + y = x\cos x \Rightarrow (xy)' = x\cos x$. Thus,

$$y = \frac{1}{x}\left[\int x\cos x\, dx + C\right] = \frac{1}{x}[x\sin x + \cos x + C]$$

$= \sin x + \frac{\cos x}{x} + \frac{C}{x}$. The solutions are asymptotic to the y-axis (except for $C = -1$). In fact, for $C > -1$, $y \to \infty$ as $x \to 0^+$, whereas for $C < -1$, $y \to -\infty$ as $x \to 0^+$. As x gets larger, the solutions approximate $y = \sin x$ more closely. The graphs for larger C lie above those for smaller C. The distance between the graphs lessens as x increases.

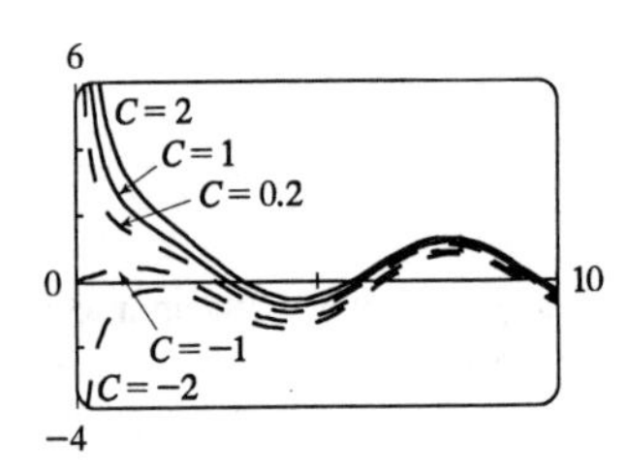

23. Setting $u = y^{1-n}$, $\frac{du}{dx} = (1-n)y^{-n}\frac{dy}{dx}$ or $\frac{dy}{dx} = \frac{y^n}{1-n}\frac{du}{dx} = \frac{u^{n/(1-n)}}{1-n}\frac{du}{dx}$. Then the Bernoulli differential equation becomes $\frac{u^{n/(1-n)}}{1-n}\frac{du}{dx} + P(x)u^{1/(1-n)} = Q(x)u^{n/(1-n)}$ or $\frac{du}{dx} + (1-n)P(x)u = Q(x)(1-n)$.

25. Here $n = 3$, $P(x) = \frac{2}{x}$, $Q(x) = \frac{1}{x^2}$ and setting $u = y^{-2}$, u satisfies $u' - \frac{4u}{x} = -\frac{2}{x^2}$. Then $I(x) = e^{\int -4/x\, dx} = x^{-4}$ and $u = x^4\left(\int -\frac{2}{x^6}\, dx + C\right) = x^4\left(\frac{2}{5x^5} + C\right) = Cx^4 + \frac{2}{5x}$. Thus $y = \pm\left(Cx^4 + \frac{2}{5x}\right)^{-1/2}$.

27. **(a)** $2\frac{dI}{dt}+10I=40$ or $\frac{dI}{dt}+5I=20$. Then the integrating factor is $e^{\int 5\,dt}=e^{5t}$. Multiplying the differential equation by the integrating factor gives $e^{5t}\frac{dI}{dt}+5Ie^{5t}=20e^{5t}$ $\Rightarrow$ $\left(e^{5t}I\right)'=20e^{5t}$ $\Rightarrow$ $I(t)=e^{-5t}\left[\int 20e^{5t}\,dt+C\right]=4+Ce^{-5t}$. But $0=I(0)=4+C$ so $I(t)=4-4e^{-5t}$.

(b) $I(0.1)=4-4e^{-0.5}\approx 1.57$ A.

29. $5\frac{dQ}{dt}+20Q=60$ with $Q(0)=0$ C. Then the integrating factor is $e^{\int 4\,dt}=e^{4t}$ and multiplying the differential equation by the integrating factor gives $e^{4t}\frac{dQ}{dt}+4e^{4t}Q=12e^{4t}$ $\Rightarrow$ $\left(e^{4t}Q\right)'=12e^{4t}$ $\Rightarrow$ $Q(t)=e^{-4t}\left[\int 12e^{4t}\,dt+C\right]=3+Ce^{-4t}$. But $0=Q(0)=3+C$ so $Q(t)=3\left(1-e^{-4t}\right)$ is the charge at time t and $I=dQ/dt=12e^{-4t}$ is the current at time t.

31. $\frac{dP}{dt}+kP=kM$ so $I(t)=e^{\int k\,dt}=e^{kt}$. Multiplying the differential equation by $I(t)$ gives $e^{kt}\frac{dP}{dt}+kPe^{kt}=kMe^{kt}$ $\Rightarrow$ $\left(e^{kt}P\right)'=kMe^{kt}$ $\Rightarrow$ $P(t)=e^{-kt}\left(\int kMe^{kt}dt+C\right)=M+Ce^{-kt}$, $k>0$. Furthermore it is reasonable to assume $0\le P(0)\le M$, so $-M\le C\le 0$.

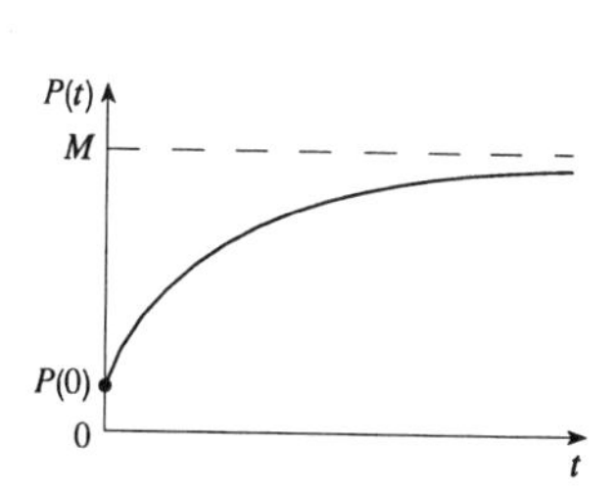

33. **(a)** $\frac{dv}{dt}+\frac{c}{m}v=g$ and $I(t)=e^{\int (c/m)dt}=e^{(c/m)t}$ and multiplying the differential equation by $I(t)$ gives $e^{(c/m)t}\frac{dv}{dt}+\frac{vce^{(c/m)t}}{m}=ge^{(c/m)t}$ $\Rightarrow$ $\left[e^{(c/m)t}v\right]'=ge^{(c/m)t}$. Hence $v(t)=e^{-(c/m)t}\left[\int ge^{(c/m)t}\,dt+K\right]=\frac{mg}{c}+Ke^{-(c/m)t}$. But the object is dropped from rest so $v(0)=0$ and $K=-\frac{mg}{c}$. Thus the velocity at time t is $v(t)=\frac{mg}{c}\left[1-e^{-(c/m)t}\right]$.

(b) $\lim_{t\to\infty} v(t)=mg/c$.

(c) $s(t)=\int v(t)\,dt=\frac{mg}{c}\left[t+\frac{m}{c}e^{-(c/m)t}\right]+c_1$ where $c_1=s(0)-\frac{m^2g}{c^2}$, $s(0)$ is the initial position so $s(0)=0$ and $s(t)=\frac{mg}{c}\left[t+\frac{m}{c}e^{-(c/m)t}\right]-\frac{m^2g}{c^2}$.

EXERCISES 9.3

1. $y = 2x+1 \Longrightarrow L = \displaystyle\int_{-1}^{3}\sqrt{1+\left(\frac{dy}{dx}\right)^2}\,dx = \int_{-1}^{3}\sqrt{1+2^2}\,dx = \sqrt{5}\,[3-(-1)] = 4\sqrt{5}.$

The arc length can be calculated using the distance formula, since the curve is a line segment, so

$L = [\text{distance from } (-1,-1) \text{ to } (3,7)] = \sqrt{[3-(-1)]^2+[7-(-1)]^2} = \sqrt{80} = 4\sqrt{5}.$

3. $x = e^t\cos t,\ y = e^t\sin t,\ 0 \le t \le \pi.$

$\left(\frac{dx}{dt}\right)^2 + \left(\frac{dy}{dt}\right)^2 = \left[e^t(\cos t - \sin t)\right]^2 + \left[e^t(\sin t + \cos t)\right]^2 = e^{2t}\left(2\cos^2 t + 2\sin^2 t\right) = 2e^{2t}.$

$L = \int_0^\pi \sqrt{2}e^t\,dt = \sqrt{2}\left(e^\pi - 1\right).$

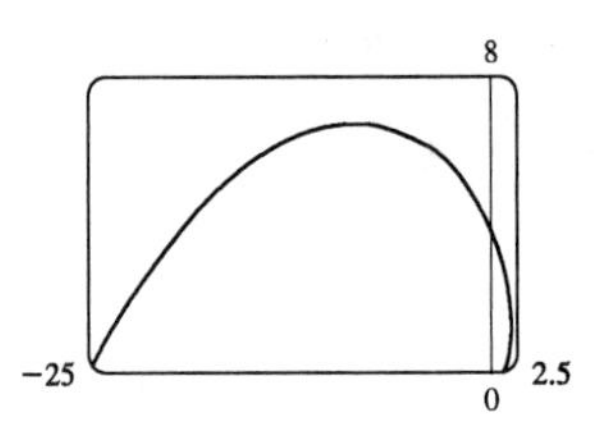

5.

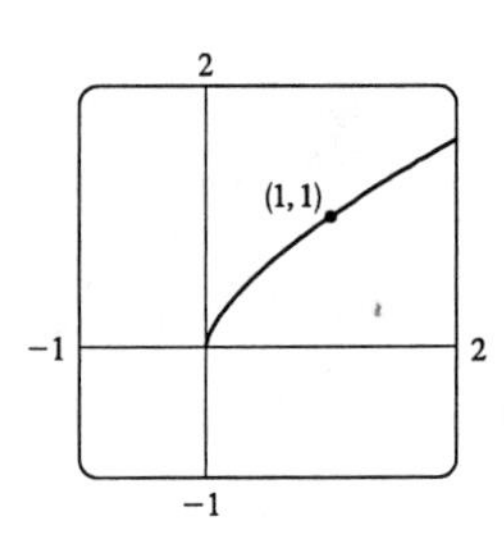

$x = y^{3/2} \Longrightarrow 1 + (dx/dy)^2 = 1 + \left(\frac{3}{2}y^{1/2}\right)^2 = 1 + \frac{9}{4}y.$

$L = \int_0^1 \sqrt{1+\frac{9}{4}y}\,dy = \frac{4}{9}\cdot\frac{2}{3}\left[\left(1+\frac{9}{4}y\right)^{3/2}\right]_0^1 = \frac{8}{27}\left(\frac{13\sqrt{13}}{8}-1\right) = \frac{13\sqrt{13}-8}{27}.$

7. $y = \frac{1}{3}\left(x^2+2\right)^{3/2} \Rightarrow dy/dx = \frac{1}{2}\left(x^2+2\right)^{1/2}(2x) = x\sqrt{x^2+2} \Rightarrow$
$1 + (dy/dx)^2 = 1 + x^2\left(x^2+2\right) = \left(x^2+1\right)^2.$ So $L = \int_0^1 \left(x^2+1\right)dx = \left[\frac{1}{3}x^3+x\right]_0^1 = \frac{4}{3}.$

9. $y = \dfrac{x^4}{4} + \dfrac{1}{8x^2} \Rightarrow \dfrac{dy}{dx} = x^3 - \dfrac{1}{4x^3} \Rightarrow 1 + \left(\dfrac{dy}{dx}\right)^2 = 1 + x^6 - \dfrac{1}{2} + \dfrac{1}{16x^6} = x^6 + \dfrac{1}{2} + \dfrac{1}{16x^6}.$
So $L = \int_1^3 \left(x^3 + \frac{1}{4}x^{-3}\right)dx = \left[\frac{1}{4}x^4 - \frac{1}{8}x^{-2}\right]_1^3 = \left(\frac{81}{4} - \frac{1}{72}\right) - \left(\frac{1}{4} - \frac{1}{8}\right) = \frac{181}{9}.$

11. $y = x^3 \Longrightarrow 1 + (y')^2 = 1 + \left(3x^2\right)^2 = 1 + 9x^4 \Longrightarrow L = \int_0^1 \sqrt{1+9x^4}\,dx.$ Let
$f(x) = \sqrt{1+9x^4}.$ Then by Simpson's Rule with $n = 10$,
$L \approx \frac{1/10}{3}[f(0) + 4f(0.1) + 2f(0.2) + 4f(0.3) + \cdots + 2f(0.8) + 4f(0.9) + f(1)] \approx 1.548.$

13. $y = \sin x \Longrightarrow 1 + (y')^2 = 1 + \cos^2 x \Longrightarrow L = \int_0^\pi \sqrt{1 + \cos^2 x}\, dx$. Let $f(x) = \sqrt{1 + \cos^2 x}$. Then

$L \approx \frac{\pi/10}{3}\left[f(0) + 4f\left(\frac{\pi}{10}\right) + 2f\left(\frac{2\pi}{10}\right) + 4f\left(\frac{3\pi}{10}\right) + \cdots + 2f\left(\frac{8\pi}{10}\right) + 4f\left(\frac{9\pi}{10}\right) + f(\pi)\right] \approx 3.820.$

15. **(a)**

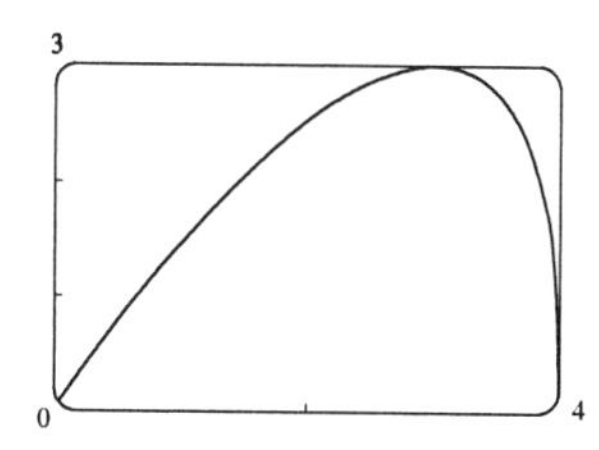

(b)

Let $f(x) = y = x\sqrt[3]{4 - x}$. The polygon with one side is just the line segment joining the points $(0, f(0)) = (0, 0)$ and $(4, f(4)) = (4, 0)$, and its length is 4. The polygon with two sides joins the points $(0, 0)$, $(2, f(2)) = \left(2, 2\sqrt[3]{2}\right)$ and $(4, 0)$. Its length is

$\sqrt{(2-0)^2 + \left(2\sqrt[3]{2} - 0\right)^2} + \sqrt{(4-2)^2 + \left(0 - 2\sqrt[3]{2}\right)^2} = 2\sqrt{4 + 2^{8/3}} \approx 6.43$. Similarly, the inscribed polygon with four sides joins the points $(0, 0)$, $\left(1, \sqrt[3]{3}\right)$, $\left(2, 2\sqrt[3]{2}\right)$, $(3, 3)$, and $(4, 0)$, so its length is

$\sqrt{1 + \left(\sqrt[3]{3}\right)^2} + \sqrt{1 + \left(2\sqrt[3]{2} - \sqrt[3]{3}\right)^2} + \sqrt{1 + \left(3 - 2\sqrt[3]{2}\right)^2} + \sqrt{1 + 9} \approx 7.50.$

(c) Using the arc length formula with $\dfrac{dy}{dx} = x\left[\frac{1}{3}(4 - x)^{-2/3}(-1)\right] + \sqrt[3]{4 - x} = \dfrac{12 - 4x}{3(4 - x)^{2/3}}$, the length of the curve is $L = \displaystyle\int_0^4 \sqrt{1 + \left(\frac{dy}{dx}\right)^2}\, dx = \int_0^4 \sqrt{1 + \left[\frac{12 - 4x}{3(4 - x)^{2/3}}\right]^2}\, dx.$

(d) According to a CAS, the length of the curve is $L \approx 7.7988$. The actual value is larger than any of the approximations in part (b). This is always true, since any approximating straight line between two points on the curve is shorter than the length of the curve between the two points.

17. $x = t^3 \Longrightarrow dx/dt = 3t^2$ and $y = t^4 \Longrightarrow dy/dt = 4t^3 \Longrightarrow$

$L = \int_0^1 \sqrt{9t^4 + 16t^6}\, dt = \int_0^1 t^2\sqrt{9 + 16t^2}\, dt = \frac{205}{128} - \frac{81 \ln 3}{512} \approx 1.428.$

19. $y = \ln(\cos x) \Longrightarrow y' = \dfrac{1}{\cos x}(-\sin x) = -\tan x \Longrightarrow 1 + (y')^2 = 1 + \tan^2 x = \sec^2 x.$

So $L = \int_0^{\pi/4} \sec x\, dx = [\ln(\sec x + \tan x)]_0^{\pi/4} = \ln\left(\sqrt{2} + 1\right) \approx 0.881.$

21. **(a)**

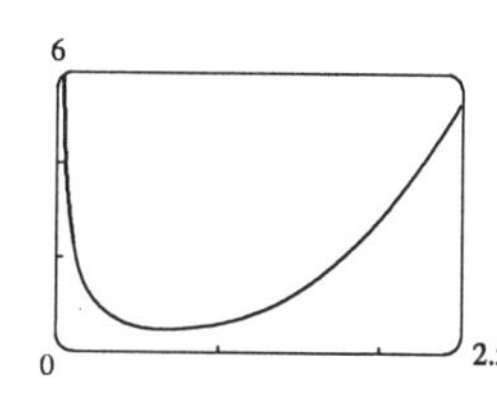

(b) $1 + \left(\dfrac{dy}{dx}\right)^2 = x^4 + \dfrac{1}{2} + \dfrac{1}{16x^4}$,

$$\begin{aligned} s(x) &= \int_1^x \left[t^2 + 1/\left(4t^2\right)\right] dt \\ &= \left[\tfrac{1}{3}t^3 - 1/(4t)\right]_1^x \\ &= \tfrac{1}{3}x^3 - 1/(4x) - \left(\tfrac{1}{3} - \tfrac{1}{4}\right) \\ &= \tfrac{1}{3}x^3 - 1/(4x) - \tfrac{1}{12} \text{ for } x > 0. \end{aligned}$$

(c)

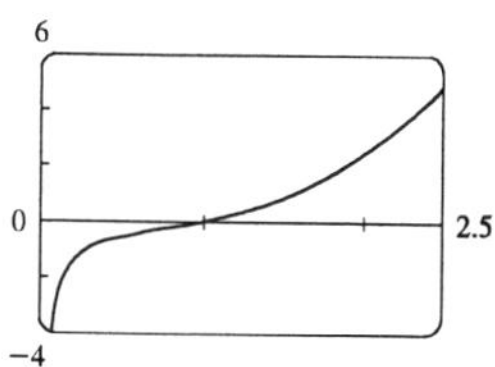

23. **(a)** $y = a\cosh\left(\frac{x}{a}\right) \Rightarrow y' = \sinh\left(\frac{x}{a}\right) \Rightarrow 1 + (y')^2 = 1 + \sinh^2\left(\frac{x}{a}\right) = \cosh^2\left(\frac{x}{a}\right)$.

So $L = \int_{-b}^{b}\cosh\left(\frac{x}{a}\right)dx = \left[a\sinh\left(\frac{x}{a}\right)\right]_{-b}^{b} = 2a\sinh\left(\frac{b}{a}\right)$.

(b) From part (a), we have an expression involving the distance between the poles ($2b$), the height of the lowest point of the wire (a), and the length of the wire (L).

For this problem, we substitute $L = 56$ and $b = \frac{1}{2}(50) = 25$ into that expression: $56 = 2a\sinh(25/a)$. It would be difficult to solve this equation exactly, so we use a machine to graph the line $y = 56$ and the curve $y = 2x\sinh(25/x)$, and estimate the x-coordinate of the point of intersection. From the graph, it appears that the root of the equation is $x \approx 30$ ft. So the lowest point of the wire is about 30 ft above the ground.

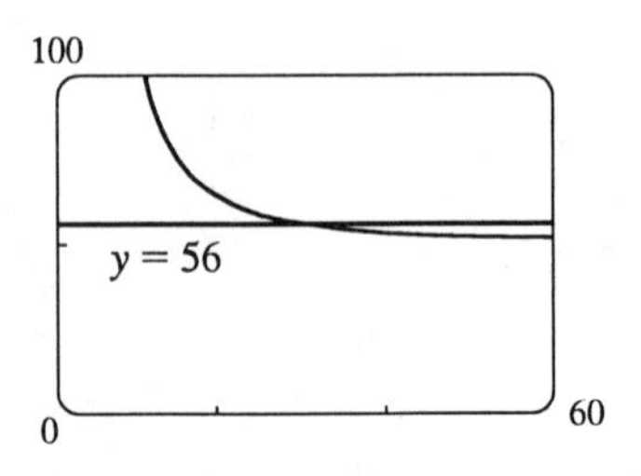

25. The sine wave has amplitude 1 and period 14, since it goes through two periods in a distance of 28 in., so its equation is $y = 1\sin\left(\frac{2\pi}{14}x\right) = \sin\left(\frac{\pi}{7}x\right)$. The width w of the flat metal sheet needed to make the panel is the arc length of the sine curve from $x = 0$ to $x = 28$. We set up the integral to evaluate w using the arc length formula with $\frac{dy}{dx} = \frac{\pi}{7}\cos\left(\frac{\pi}{7}x\right)$: $L = \int_0^{28}\sqrt{1 + \left[\frac{\pi}{7}\cos\left(\frac{\pi}{7}x\right)\right]^2}\,dx = 2\int_0^{14}\sqrt{1 + \left[\frac{\pi}{7}\cos\left(\frac{\pi}{7}x\right)\right]^2}\,dx$. This integral would be very difficult to evaluate exactly, so we use a CAS, and find that $L \approx 29.36$ inches.

27. $x = a\sin\theta,\ y = b\cos\theta,\ 0 \le \theta \le 2\pi$.

$$\left(\frac{dx}{d\theta}\right)^2 + \left(\frac{dy}{d\theta}\right)^2 = (a\cos\theta)^2 + (-b\sin\theta)^2 = a^2\cos^2\theta + b^2\sin^2\theta = a^2\left(1 - \sin^2\theta\right) + b^2\sin^2\theta$$

$$= a^2 - \left(a^2 - b^2\right)\sin^2\theta = a^2 - c^2\sin^2\theta = a^2\left(1 - \frac{c^2}{a^2}\sin^2\theta\right)$$

$$= a^2\left(1 - e^2\sin^2\theta\right)$$

So $L = 4\int_0^{\pi/2}\sqrt{a^2\left(1 - e^2\sin^2\theta\right)}\,d\theta$ (by symmetry) $= 4a\int_0^{\pi/2}\sqrt{1 - e^2\sin^2\theta}\,d\theta$.

29. **(a)** Notice that $0 \le t \le 2\pi$ does not give the complete curve because $x(0) \ne x(2\pi)$. In fact, we must take $t \in [0, 4\pi]$ in order to obtain the complete curve, since the first term in each of the parametric equations has period 2π and the second has period $\frac{4\pi}{11}$, and the least common integer multiple of these two numbers is 4π.

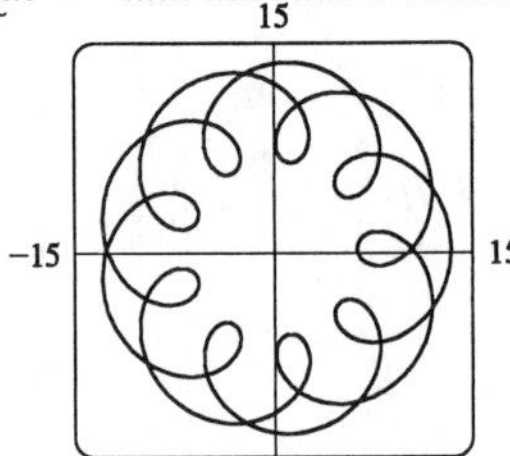

(b) For $x = 11\cos t - 4\cos(11t/2)$ and $y = 11\sin t - 4\sin(11t/2)$, we can show that $L = 11\int_0^{4\pi}\sqrt{5 - 4\cos(9t/2)}\,dt \approx 294$.

EXERCISES 9.4

1. $y = \sqrt{x} \quad \Rightarrow \quad 1 + \left(\dfrac{dy}{dx}\right)^2 = 1 + \left(\dfrac{1}{2\sqrt{x}}\right)^2 = 1 + \dfrac{1}{4x}$. So

$$S = \int_4^9 2\pi y \sqrt{1 + \left(\frac{dy}{dx}\right)^2}\, dx = \int_4^9 2\pi\sqrt{x}\sqrt{1 + \frac{1}{4x}}\, dx = 2\pi \int_4^9 \sqrt{x + \tfrac{1}{4}}\, dx$$

$$= 2\pi\left[\tfrac{2}{3}\left(x + \tfrac{1}{4}\right)^{3/2}\right]_4^9 = \tfrac{4\pi}{3}\left[\tfrac{1}{8}(4x+1)^{3/2}\right]_4^9 = \tfrac{\pi}{6}\left(37\sqrt{37} - 17\sqrt{17}\right).$$

3. $y = x^3 \quad \Rightarrow \quad y' = 3x^2$. So $S = \displaystyle\int_0^2 2\pi y\sqrt{1 + (y')^2}\, dx = 2\pi \int_0^2 x^3\sqrt{1 + 9x^4}\, dx$ (Let $u = 1 + 9x^4$, so $du = 36x^3\, dx$) $= \frac{2\pi}{36}\int_1^{145}\sqrt{u}\, du = \frac{\pi}{18}\left[\frac{2}{3}u^{3/2}\right]_1^{145} = \frac{\pi}{27}\left(145\sqrt{145} - 1\right)$.

5. $y = \sin x \quad \Rightarrow \quad 1 + \left(\dfrac{dy}{dx}\right)^2 = 1 + \cos^2 x$. So

$$S = 2\pi\int_0^\pi \sin x\,\sqrt{1 + \cos^2 x}\, dx = 2\pi\int_{-1}^1 \sqrt{1 + u^2}\, du \quad (\text{put } u = -\cos x \Rightarrow du = \sin x\, dx)$$

$$= 4\pi\int_0^1 \sqrt{1 + u^2}\, du = 4\pi\int_0^{\pi/4} \sec^3\theta\, d\theta \quad (\text{put } u = \tan\theta \quad \Rightarrow \quad du = \sec^2\theta\, d\theta)$$

$$= 2\pi[\sec\theta\tan\theta + \ln|\sec\theta + \tan\theta|]_0^{\pi/4} = 2\pi\left[\sqrt{2} + \ln\left(\sqrt{2} + 1\right)\right].$$

7. $y = \cosh x \quad \Rightarrow \quad 1 + \left(\dfrac{dy}{dx}\right)^2 = 1 + \sinh^2 x = \cosh^2 x$. So $S = 2\pi\int_0^1 \cosh x \cosh x\, dx$

$$= 2\pi\int_0^1 \tfrac{1}{2}(1 + \cosh 2x)dx = \pi\left[x + \tfrac{1}{2}\sinh 2x\right]_0^1 = \pi\left(1 + \tfrac{1}{2}\sinh 2\right) \text{ or } \pi\left[1 + \tfrac{1}{4}(e^2 - e^{-2})\right].$$

9. $x = \frac{1}{3}(y^2 + 2)^{3/2} \quad \Rightarrow \quad dx/dy = \frac{1}{2}(y^2 + 2)^{1/2}(2y) = y\sqrt{y^2 + 2} \quad \Rightarrow$

$1 + (dx/dy)^2 = 1 + y^2(y^2 + 2) = (y^2 + 1)^2$. So

$$S = 2\pi\int_1^2 y(y^2 + 1)dy = 2\pi\left[\tfrac{1}{4}y^4 + \tfrac{1}{2}y^2\right]_1^2 = 2\pi\left(4 + 2 - \tfrac{1}{4} - \tfrac{1}{2}\right) = \tfrac{21\pi}{2}.$$

11. $y = \sqrt[3]{x} \quad \Rightarrow \quad x = y^3 \quad \Rightarrow \quad 1 + (dx/dy)^2 = 1 + 9y^4$. So

$$S = 2\pi\int_1^2 x\sqrt{1 + (dx/dy)^2}\, dy = 2\pi\int_1^2 y^3\sqrt{1 + 9y^4}\, dy = \frac{2\pi}{36}\int_1^2 \sqrt{1 + 9y^4}\, 36y^3\, dy$$

$$= \tfrac{\pi}{18}\left[\tfrac{2}{3}(1 + 9y^4)^{3/2}\right]_1^2 = \tfrac{\pi}{27}\left(145\sqrt{145} - 10\sqrt{10}\right).$$

13. $x = e^{2y} \quad \Rightarrow \quad 1 + (dx/dy)^2 = 1 + 4e^{4y}$. So

$$S = 2\pi\int_0^{1/2} e^{2y}\sqrt{1 + (2e^{2y})^2}\, dy = 2\pi\int_2^{2e}\sqrt{1 + u^2}\,\tfrac{1}{4}\, du \quad [\text{put } u = 2e^{2y},\ du = 4e^{2y}\, dy]$$

$$= \tfrac{\pi}{2}\int_2^{2e}\sqrt{1 + u^2}\, du = \tfrac{\pi}{2}\left[\tfrac{1}{2}u\sqrt{1 + u^2} + \tfrac{1}{2}\ln\left|u + \sqrt{1 + u^2}\right|\right]_2^{2e} \quad [\text{put } u = \tan\theta \text{ or use Formula 21}]$$

$$= \tfrac{\pi}{2}\left[e\sqrt{1 + 4e^2} + \tfrac{1}{2}\ln\left(2e + \sqrt{1 + 4e^2}\right) - \sqrt{5} - \tfrac{1}{2}\ln\left(2 + \sqrt{5}\right)\right]$$

$$= \frac{\pi}{4}\left[2e\sqrt{1 + 4e^2} - 2\sqrt{5} + \ln\left(\frac{2e + \sqrt{1 + 4e^2}}{2 + \sqrt{5}}\right)\right].$$

15. $x = \frac{1}{2\sqrt{2}}(y^2 - \ln y) \Rightarrow \dfrac{dx}{dy} = \dfrac{1}{2\sqrt{2}}\left(2y - \dfrac{1}{y}\right) \Rightarrow 1 + \left(\dfrac{dx}{dy}\right)^2 = 1 + \dfrac{1}{8}\left(2y - \dfrac{1}{y}\right)^2$

$= 1 + \dfrac{1}{8}\left(4y^2 - 4 + \dfrac{1}{y^2}\right) = \dfrac{1}{8}\left(4y^2 + 4 + \dfrac{1}{y^2}\right) = \left[\dfrac{1}{2\sqrt{2}}\left(2y + \dfrac{1}{y}\right)\right]^2$. So

$S = 2\pi \displaystyle\int_1^2 \frac{1}{2\sqrt{2}}(y^2 - \ln y)\frac{1}{2\sqrt{2}}\left(2y + \frac{1}{y}\right)dy = \frac{\pi}{4}\int_1^2 \left(2y^3 + y - 2y\ln y - \frac{\ln y}{y}\right)dy$

$= \frac{\pi}{4}\left[\frac{1}{2}y^4 + \frac{1}{2}y^2 - y^2\ln y + \frac{1}{2}y^2 - \frac{1}{2}(\ln y)^2\right]_1^2 = \frac{\pi}{8}\left[y^4 + 2y^2 - 2y^2\ln y - (\ln y)^2\right]_1^2$

$= \frac{\pi}{8}\left[16 + 8 - 8\ln 2 - (\ln 2)^2 - 1 - 2\right] = \frac{\pi}{8}\left[21 - 8\ln 2 - (\ln 2)^2\right]$.

17. $S = 2\pi\int_0^1 x^4\sqrt{1 + (4x^3)^2}\,dx = 2\pi\int_0^1 x^4\sqrt{16x^6 + 1}\,dx$

$\approx 2\pi\frac{1/10}{3}[f(0) + 4f(0.1) + 2f(0.2) + 4f(0.3) + 2f(0.4) + 4f(0.5) + 2f(0.6)$

$+ 4f(0.7) + 2f(0.8) + 4f(0.9) + f(1)] \approx 3.44$. Here $f(x) = x^4\sqrt{16x^6 + 1}$.

19. The curve $8y^2 = x^2(1 - x^2)$ actually consists of two loops in the region described by the inequalities $|x| \le 1$, $|y| \le \frac{\sqrt{2}}{8}$. (The maximum value of $|y|$ is attained when $|x| = \frac{1}{\sqrt{2}}$.) If we consider the loop in the region $x \ge 0$, the surface area S it generates when rotated about the x-axis is calculated as follows: $16y\dfrac{dy}{dx} = 2x - 4x^3$, so

$\left(\dfrac{dy}{dx}\right)^2 = \left(\dfrac{x - 2x^3}{8y}\right)^2 = \dfrac{x^2(1 - 2x^2)^2}{64y^2} = \dfrac{x^2(1 - 2x^2)^2}{8x^2(1 - x^2)} = \dfrac{(1 - 2x^2)^2}{8(1 - x^2)}$ for $x \ne 0, \pm 1$. The formula also holds for $x = 0$ by continuity. $1 + \left(\dfrac{dy}{dx}\right)^2 = 1 + \dfrac{(1 - 2x^2)^2}{8(1 - x^2)} = \dfrac{9 - 12x^2 + 4x^4}{8(1 - x^2)} = \dfrac{(3 - 2x^2)^2}{8(1 - x^2)}$. So

$S = 2\pi\displaystyle\int_0^1 \frac{\sqrt{x^2(1 - x^2)}}{2\sqrt{2}} \cdot \frac{3 - 2x^2}{2\sqrt{2}\sqrt{1 - x^2}}\,dx = \frac{\pi}{4}\int_0^1 x(3 - 2x^2)dx = \frac{\pi}{4}\left[\frac{3}{2}x^2 - \frac{1}{2}x^4\right]_0^1 = \frac{\pi}{4}\left(\frac{3}{2} - \frac{1}{2}\right) = \frac{\pi}{4}$.

21. $S = 2\pi\displaystyle\int_1^\infty y\sqrt{1 + \left(\frac{dy}{dx}\right)^2}\,dx = 2\pi\int_1^\infty \frac{1}{x}\sqrt{1 + \frac{1}{x^4}}\,dx = 2\pi\int_1^\infty \frac{\sqrt{x^4 + 1}}{x^3}\,dx$

$> 2\pi\displaystyle\int_1^\infty \frac{x^2}{x^3}\,dx = 2\pi\int_1^\infty \frac{dx}{x} = 2\pi\lim_{t\to\infty}[\ln x]_1^t = 2\pi\lim_{t\to\infty}\ln t = \infty$.

23. $x = t^3, y = t^2, 0 \le t \le 1$. $\left(\dfrac{dx}{dt}\right)^2 + \left(\dfrac{dy}{dt}\right)^2 = (3t^2)^2 + (2t)^2 = 9t^4 + 4t^2$.

$S = \displaystyle\int_0^1 2\pi y\sqrt{\left(\frac{dx}{dt}\right)^2 + \left(\frac{dy}{dt}\right)^2}\,dt = \int_0^1 2\pi t^2\sqrt{9t^4 + 4t^2}\,dt$

$= 2\pi\displaystyle\int_4^{13} \frac{u - 4}{9}\sqrt{u}\left(\tfrac{1}{18}du\right)$ (where $u = 9t^2 + 4$) $= \frac{\pi}{81}\left[\frac{2}{5}u^{5/2} - \frac{8}{3}u^{3/2}\right]_4^{13} = \frac{2\pi}{1215}\left(247\sqrt{13} + 64\right)$

25. $x = a\cos^3\theta, y = a\sin^3\theta, 0 \le \theta \le \frac{\pi}{2}$

$\left(\dfrac{dx}{d\theta}\right)^2 + \left(\dfrac{dy}{d\theta}\right)^2 = (-3a\cos^2\theta\sin\theta)^2 + (3a\sin^2\theta\cos\theta)^2 = 9a^2\sin^2\theta\cos^2\theta$

$S = \int_0^{\pi/2} 2\pi a\sin^3\theta\, 3a\sin\theta\cos\theta\,d\theta = 6\pi a^2\int_0^{\pi/2}\sin^4\theta\cos\theta\,d\theta = \frac{6}{5}\pi a^2[\sin^5\theta]_0^{\pi/2} = \frac{6}{5}\pi a^2$

27. $\left(\dfrac{dx}{dt}\right)^2 + \left(\dfrac{dy}{dt}\right)^2 = (6t)^2 + (6t^2)^2 = 36t^2(1+t^2) \quad \Rightarrow$

$$S = \int_0^5 2\pi x\sqrt{(dx/dt)^2 + (dy/dt)^2}\,dt = \int_0^5 2\pi(3t^2)6t\sqrt{1+t^2}\,dt = 18\pi\int_0^5 t^2\sqrt{1+t^2}\,2t\,dt$$

$$= 18\pi\int_1^{26}(u-1)\sqrt{u}\,du \quad (\text{where } u = 1+t^2) \quad = 18\pi\int_1^{26}\left(u^{3/2} - u^{1/2}\right)du = 18\pi\left[\tfrac{2}{5}u^{5/2} - \tfrac{2}{3}u^{3/2}\right]_1^{26}$$

$$= 18\pi\left[\left(\tfrac{2}{5}\cdot 676\sqrt{26} - \tfrac{2}{3}\cdot 26\sqrt{26}\right) - \left(\tfrac{2}{5} - \tfrac{2}{3}\right)\right] = \tfrac{24}{5}\pi\left(949\sqrt{26}+1\right)$$

29. $x = a\cos\theta,\ y = b\sin\theta,\ 0 \le \theta \le 2\pi.$

$$(dx/d\theta)^2 + (dy/d\theta)^2 = (-a\sin\theta)^2 + (b\cos\theta)^2 = a^2\sin^2\theta + b^2\cos^2\theta = a^2(1-\cos^2\theta) + b^2\cos^2\theta$$

$$= a^2 - (a^2-b^2)\cos^2\theta = a^2 - c^2\cos^2\theta = a^2\left(1 - \frac{c^2}{a^2}\cos^2\theta\right) = a^2(1 - e^2\cos^2\theta)$$

(a) $S = \displaystyle\int_0^\pi 2\pi b\sin\theta\, a\sqrt{1-e^2\cos^2\theta}\,d\theta = 2\pi ab\int_{-e}^{e}\sqrt{1-u^2}\left(\frac{1}{e}\right)du$ [where $u = -e\cos\theta$, $du = e\sin\theta\,d\theta$]

$$= \frac{4\pi ab}{e}\int_0^e (1-u^2)^{1/2}\,du = \frac{4\pi ab}{e}\int_0^{\sin^{-1}e}\cos^2 v\,dv \ (\text{where } u = \sin v) \ = \frac{2\pi ab}{e}\int_0^{\sin^{-1}e}(1+\cos 2v)dv$$

$$= \frac{2\pi ab}{e}\left[v + \tfrac{1}{2}\sin 2v\right]_0^{\sin^{-1}e} = \frac{2\pi ab}{e}\left[v + \sin v\cos v\right]_0^{\sin^{-1}e} = \frac{2\pi ab}{e}\left(\sin^{-1}e + e\sqrt{1-e^2}\right).$$

But $\sqrt{1-e^2} = \sqrt{1 - \dfrac{c^2}{a^2}} = \sqrt{\dfrac{a^2-c^2}{a^2}} = \sqrt{\dfrac{b^2}{a^2}} = \dfrac{b}{a}$, so $S = \dfrac{2\pi ab}{e}\sin^{-1}e + 2\pi b^2$.

(b) $S = \int_{-\pi/2}^{\pi/2} 2\pi a\cos\theta\, a\sqrt{1-e^2\cos^2\theta}\,d\theta$

$$= 4\pi a^2\int_0^{\pi/2}\cos\theta\,\sqrt{(1-e^2) + e^2\sin^2\theta}\,d\theta$$

$$= \frac{4\pi a^2(1-e^2)}{e}\int_0^{\pi/2}\frac{e}{\sqrt{1-e^2}}\cos\theta\sqrt{1 + \left(\frac{e\sin\theta}{\sqrt{1-e^2}}\right)^2}\,d\theta$$

$$= \frac{4\pi a^2(1-e^2)}{e}\int_0^{e/\sqrt{1-e^2}}\sqrt{1+u^2}\,du \quad \left(\text{where } u = \frac{e\sin\theta}{\sqrt{1-e^2}}\right)$$

$$= \frac{4\pi a^2(1-e^2)}{e}\int_0^{\sin^{-1}e}\sec^3 v\,dv \quad (\text{where } u = \tan v,\ du = \sec^2 v\,dv)$$

$$= \frac{2\pi a^2(1-e^2)}{e}\left[\sec v\tan v + \ln|\sec v + \tan v|\right]_0^{\sin^{-1}e}$$

$$= \frac{2\pi a^2(1-e^2)}{e}\left[\frac{1}{\sqrt{1-e^2}}\frac{e}{\sqrt{1-e^2}} + \ln\left|\frac{1}{\sqrt{1-e^2}} + \frac{e}{\sqrt{1-e^2}}\right|\right]$$

$$= 2\pi a^2 + \frac{2\pi a^2(1-e^2)}{e}\ln\sqrt{\frac{1+e}{1-e}} = 2\pi a^2 + \frac{2\pi b^2}{e}\frac{1}{2}\ln\left(\frac{1+e}{1-e}\right) \quad \left(\text{since } 1-e^2 = \frac{b^2}{a^2}\right)$$

$$= 2\pi\left[a^2 + \frac{b^2}{2e}\ln\frac{1+e}{1-e}\right]$$

31. In the derivation of (4), we computed a typical contribution to the surface area to be $2\pi \frac{y_{i-1} + y_i}{2} |P_{i-1}P_i|$, the area of a frustum of a cone. When $f(x)$ is not necessarily positive, the approximations $y_i = f(x_i) \approx f(x_i^*)$ and $y_{i-1} = f(x_{i-1}) \approx f(x_i^*)$ must be replaced by $y_i = |f(x_i)| \approx |f(x_i^*)|$ and $y_{i-1} = |f(x_{i-1})| \approx |f(x_i^*)|$. Thus $2\pi \frac{y_{i-1} + y_i}{2} |P_{i-1}P_i| \approx 2\pi |f(x_i^*)| \sqrt{1 + [f'(x_i^*)]^2}\, \Delta x_i$. Continuing with the rest of the derivation as before, we obtain $S = \int_a^b 2\pi |f(x)| \sqrt{1 + [f'(x)]^2}\, dx$.

33. For the upper semicircle, $f(x) = \sqrt{r^2 - x^2}$, $f'(x) = -x/\sqrt{r^2 - x^2}$. The surface area generated is

$$S_1 = \int_{-r}^{r} 2\pi \left(r - \sqrt{r^2 - x^2}\right) \sqrt{1 + \frac{x^2}{r^2 - x^2}}\, dx = 4\pi \int_0^r \left(r - \sqrt{r^2 - x^2}\right) \frac{r}{\sqrt{r^2 - x^2}} dx$$

$$= 4\pi \int_0^r \left(\frac{r^2}{\sqrt{r^2 - x^2}} - r\right) dx.$$

For the lower semicircle, $f(x) = -\sqrt{r^2 - x^2}$ and $f'(x) = \frac{x}{\sqrt{r^2 - x^2}}$, so $S_2 = 4\pi \int_0^r \left(\frac{r^2}{\sqrt{r^2 - x^2}} + r\right) dx$.

Thus the total area is $S = S_1 + S_2 = 8\pi \int_0^r \left(\frac{r^2}{\sqrt{r^2 - x^2}}\right) dx = 8\pi \left[r^2 \sin^{-1}\left(\frac{x}{r}\right)\right]_0^r = 8\pi r^2 \left(\frac{\pi}{2}\right) = 4\pi^2 r^2$.

EXERCISES 9.5

1. $m_1 = 4$, $m_2 = 8$; $P_1(-1, 2)$, $P_2(2, 4)$. $m = m_1 + m_2 = 12$. $M_x = 4 \cdot 2 + 8 \cdot 4 = 40$;
$M_y = 4 \cdot (-1) + 8 \cdot 2 = 12$; $\overline{x} = M_y/m = 1$ and $\overline{y} = M_x/m = \frac{10}{3}$, so the center of mass is $(\overline{x}, \overline{y}) = \left(1, \frac{10}{3}\right)$.

3. $m = m_1 + m_2 + m_3 = 4 + 2 + 5 = 11$. $M_x = 4 \cdot (-2) + 2 \cdot 4 + 5 \cdot (-3) = -15$;
$M_y = 4 \cdot (-1) + 2 \cdot (-2) + 5 \cdot 5 = 17$, $(\overline{x}, \overline{y}) = \left(\frac{17}{11}, -\frac{15}{11}\right)$.

5. $A = \int_0^2 x^2\, dx = \left[\frac{1}{3}x^3\right]_0^2 = \frac{8}{3}$, $\overline{x} = A^{-1} \int_0^2 x \cdot x^2\, dx = \frac{3}{8}\left[\frac{1}{4}x^4\right]_0^2 = \frac{3}{8} \cdot 4 = \frac{3}{2}$,
$\overline{y} = A^{-1} \int_0^2 \frac{1}{2}(x^2)^2\, dx = \frac{3}{8} \cdot \frac{1}{2}\left[\frac{1}{5}x^5\right]_0^2 = \frac{3}{16} \cdot \frac{32}{5} = \frac{6}{5}$. Centroid $(\overline{x}, \overline{y}) = \left(\frac{3}{2}, \frac{6}{5}\right) = (1.5, 1.2)$.

7. $A = \int_{-1}^2 (3x + 5) dx = \left[\frac{3}{2}x^2 + 5x\right]_{-1}^2 = (6 + 10) - \left(\frac{3}{2} - 5\right) = 16 + \frac{7}{2} = \frac{39}{2}$,
$\overline{x} = A^{-1} \int_{-1}^2 x(3x + 5) dx = \frac{2}{39} \int_{-1}^2 (3x^2 + 5x) dx = \frac{2}{39}\left[x^3 + \frac{5}{2}x^2\right]_{-1}^2$
$= \frac{2}{39}\left[(8 + 10) - \left(-1 + \frac{5}{2}\right)\right] = \frac{2}{39}\left(\frac{36 - 3}{2}\right) = \frac{11}{13}$,
$\overline{y} = A^{-1} \int_{-1}^2 \frac{1}{2}(3x + 5)^2\, dx = \frac{1}{39} \int_{-1}^2 (9x^2 + 30x + 25) dx = \frac{1}{39}[3x^3 + 15x^2 + 25x]_{-1}^2$
$= \frac{1}{39}[(24 + 60 + 50) - (-3 + 15 - 25)] = \frac{147}{39} = \frac{49}{13}$. $(\overline{x}, \overline{y}) = \left(\frac{11}{13}, \frac{49}{13}\right)$.

9. By symmetry, $\overline{x} = 0$ and $A = 2\int_0^{\pi/4} \cos 2x\, dx = \sin 2x\big|_0^{\pi/4} = 1$,

$\overline{y} = A^{-1}\int_{-\pi/4}^{\pi/4} \frac{1}{2}\cos^2 2x\, dx = \int_0^{\pi/4}\cos^2 2x\, dx = \frac{1}{2}\int_0^{\pi/4}(1+\cos 4x)dx = \frac{1}{2}\left[x + \frac{1}{4}\sin 4x\right]_0^{\pi/4}$

$= \frac{1}{2}\left(\frac{\pi}{4} + \frac{1}{4}\cdot 0\right) = \frac{\pi}{8}$. $(\overline{x},\overline{y}) = \left(0, \frac{\pi}{8}\right)$.

11. $A = \int_0^1 e^x\, dx = [e^x]_0^1 = e - 1$,

$\overline{x} = \dfrac{1}{A}\displaystyle\int_0^1 xe^x\, dx = \dfrac{1}{e-1}[xe^x - e^x]_0^1$ (integration by parts) $= \dfrac{1}{e-1}[0-(-1)] = \dfrac{1}{e-1}$,

$\overline{y} = \dfrac{1}{A}\displaystyle\int_0^1 \dfrac{(e^x)^2}{2}\, dx = \dfrac{1}{e-1}\cdot \frac{1}{4}[e^{2x}]_0^1 = \dfrac{1}{4(e-1)}(e^2-1) = \dfrac{e+1}{4}$. $(\overline{x},\overline{y}) = \left(\dfrac{1}{e-1}, \dfrac{e+1}{4}\right)$.

13. $A = \int_0^1(\sqrt{x} - x)dx = \left[\frac{2}{3}x^{3/2} - \frac{1}{2}x^2\right]_0^1 = \frac{2}{3} - \frac{1}{2} = \frac{1}{6}$,

$\overline{x} = A^{-1}\int_0^1 x(\sqrt{x} - x)dx = 6\int_0^1(x^{3/2} - x^2)dx = 6\left[\frac{2}{5}x^{5/2} - \frac{1}{3}x^3\right]_0^1 = 6\left(\frac{2}{5} - \frac{1}{3}\right) = \frac{2}{5}$,

$\overline{y} = A^{-1}\int_0^1 \frac{1}{2}\left[(\sqrt{x})^2 - x^2\right]dx = 3\int_0^1(x - x^2)dx = 3\left[\frac{1}{2}x^2 - \frac{1}{3}x^3\right]_0^1 = 3\left(\frac{1}{2} - \frac{1}{3}\right) = \frac{1}{2}$.

$(\overline{x},\overline{y}) = \left(\frac{2}{5}, \frac{1}{2}\right) = (0.4, 0.5)$.

15. $A = \int_0^{\pi/4}(\cos x - \sin x)dx = [\sin x + \cos x]_0^{\pi/4} = \sqrt{2} - 1$,

$\overline{x} = A^{-1}\displaystyle\int_0^{\pi/4} x(\cos x - \sin x)dx = A^{-1}[x(\sin x + \cos x) + \cos x - \sin x]_0^{\pi/4}$ [integration by parts]

$= A^{-1}\left(\frac{\pi}{4}\sqrt{2} - 1\right) = \dfrac{\frac{1}{4}\pi\sqrt{2} - 1}{\sqrt{2} - 1}$,

$\overline{y} = A^{-1}\displaystyle\int_0^{\pi/4}\frac{1}{2}\left(\cos^2 x - \sin^2 x\right)dx = \dfrac{1}{2A}\int_0^{\pi/4}\cos 2x\, dx = \dfrac{1}{4A}[\sin 2x]_0^{\pi/4} = \dfrac{1}{4A} = \dfrac{1}{4(\sqrt{2}-1)}$.

$(\overline{x},\overline{y}) = \left(\dfrac{\pi\sqrt{2} - 4}{4(\sqrt{2}-1)}, \dfrac{1}{4(\sqrt{2}-1)}\right)$.

17. By symmetry, $M_y = 0$ and $\overline{x} = 0$. $A = \frac{1}{2}bh = \frac{1}{2}\cdot 2\cdot 2 = 2$.

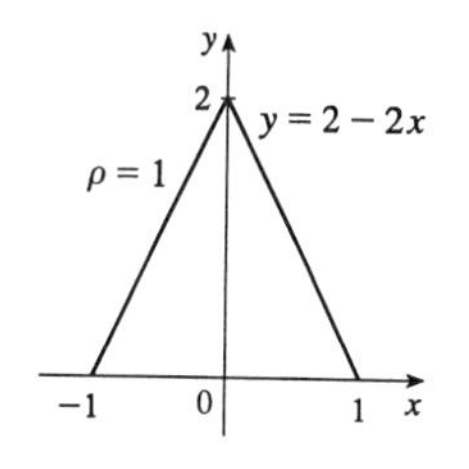

$M_x = 2\rho\int_0^1 \frac{1}{2}(2-2x)^2\, dx = 4\int_0^1(1-x)^2\, dx$

$= 4\left[-\frac{1}{3}(1-x)^3\right]_0^1 = 4\cdot\frac{1}{3} = \frac{4}{3}$.

$\overline{y} = \dfrac{1}{\rho A}M_x = \dfrac{2}{3}$. $(\overline{x},\overline{y}) = \left(0, \frac{2}{3}\right)$.

19. By symmetry, $M_y = 0$ and $\overline{x} = 0$. $A =$ area of triangle $+$ area of square $= 1 + 4 = 5$, so $m = \rho A = 4\cdot 5 = 20$.

$M_x = \rho\cdot 2\int_0^1 \frac{1}{2}\left[(1-x)^2 - (-2)^2\right]dx = 4\int_0^1(x^2 - 2x - 3)dx$

$= 4\left[\frac{1}{3}x^3 - x^2 - 3x\right]_0^1 = 4\left(\frac{1}{3} - 1 - 3\right) = 4\left(-\frac{11}{3}\right) = -\frac{44}{3}$. $\overline{y} = M_x/m = \frac{1}{20}\left(-\frac{44}{3}\right) = -\frac{11}{15}$. $(\overline{x},\overline{y}) = \left(0, -\frac{11}{15}\right)$.

21. Choose x- and y-axes so that the base (one side of the triangle) lies along the x-axis with the other vertex along the positive y-axis as shown. From geometry, we know the medians intersect at a point $\frac{2}{3}$ of the way from each vertex (along the median) to the opposite side.

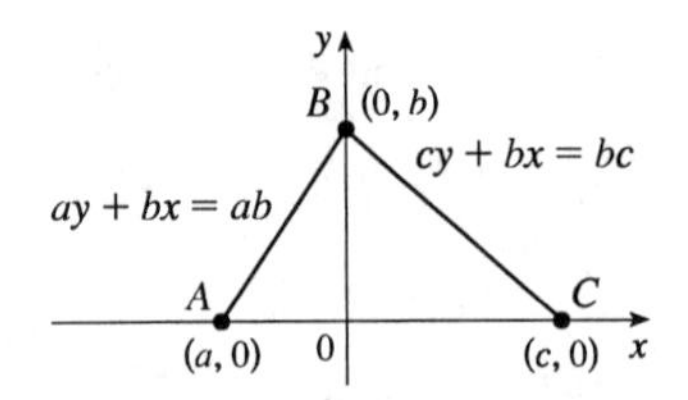

The median from B goes to the midpoint $\left(\frac{1}{2}(a+c), 0\right)$ of side AC, so the point of intersection of the medians is $\left(\frac{2}{3}\cdot\frac{1}{2}(a+c), \frac{1}{3}b\right) = \left(\frac{1}{3}(a+c), \frac{1}{3}b\right)$. This can also be verified by finding the equations of two medians, and solving them simultaneously to find their point of intersection. Now let us compute the location of the centroid of the triangle. The area is $A = \frac{1}{2}(c-a)b$.

$$\overline{x} = \frac{1}{A}\left[\int_a^0 x\cdot\frac{b}{a}(a-x)dx + \int_0^c x\cdot\frac{b}{c}(c-x)dx\right] = \frac{1}{A}\left[\frac{b}{a}\int_a^0 (ax-x^2)dx + \frac{b}{c}\int_0^c (cx-x^2)dx\right]$$
$$= \frac{b}{Aa}\left[\tfrac{1}{2}ax^2 - \tfrac{1}{3}x^3\right]_a^0 + \frac{b}{Ac}\left[\tfrac{1}{2}cx^2 - \tfrac{1}{3}x^3\right]_0^c = \frac{b}{Aa}\left[-\tfrac{1}{2}a^3 + \tfrac{1}{3}a^3\right] + \frac{b}{Ac}\left[\tfrac{1}{2}c^3 - \tfrac{1}{3}c^3\right]$$
$$= \frac{2}{a(c-a)}\cdot\frac{-a^3}{6} + \frac{2}{c(c-a)}\cdot\frac{c^3}{6} = \frac{1}{3(c-a)}(c^2-a^2) = \frac{a+c}{3}, \text{ and}$$
$$\overline{y} = \frac{1}{A}\left[\int_a^0 \tfrac{1}{2}\left(\frac{b}{a}(a-x)\right)^2 dx + \int_0^c \tfrac{1}{2}\left(\frac{b}{c}(c-x)\right)^2 dx\right]$$
$$= \frac{1}{A}\left[\frac{b^2}{2a^2}\int_a^0 (a^2-2ax+x^2)dx + \frac{b^2}{2c^2}\int_0^c (c^2-2cx+x^2)dx\right]$$
$$= \frac{1}{A}\left[\frac{b^2}{2a^2}\left[a^2x - ax^2 + \tfrac{1}{3}x^3\right]_a^0 + \frac{b^2}{2c^2}\left[c^2x - cx^2 + \tfrac{1}{3}x^3\right]_0^c\right]$$
$$= \frac{1}{A}\left[\frac{b^2}{2a^2}\left(-a^3+a^3-\tfrac{1}{3}a^3\right) + \frac{b^2}{2c^2}\left(c^3-c^3+\tfrac{1}{3}c^3\right)\right] = \frac{1}{A}\left[\frac{b^2}{6}(-a+c)\right] = \frac{2}{(c-a)b}\cdot\frac{(c-a)b^2}{6} = \frac{b}{3}.$$

Thus $(\overline{x}, \overline{y}) = \left(\frac{a+c}{3}, \frac{b}{3}\right)$ as claimed.

Remarks: Actually the computation of $\overline{y}$ is all that is needed. By considering each side of the triangle in turn to be the base, we see that the centroid is $\frac{1}{3}$ of the way from each side to the opposite vertex and must therefore be the intersection of the medians.

The computation of $\overline{y}$ in this problem (and many others) can be simplified by using horizontal rather than vertical approximating rectangles. If the length of a thin rectangle at coordinate y is $\ell(y)$, then its area is $\ell(y)\Delta y$, its mass is $\rho\ell(y)\Delta y$, and its moment about the x-axis is $\Delta M_x = \rho y\ell(y)\Delta y$. Thus $M_x = \int \rho y\ell(y)dy$ and $\overline{y} = \dfrac{\int \rho y\ell(y)dy}{\rho A} = \dfrac{1}{A}\int y\ell(y)dy$. In this problem, $\ell(y) = \dfrac{c-a}{b}(b-y)$ by similar triangles, so

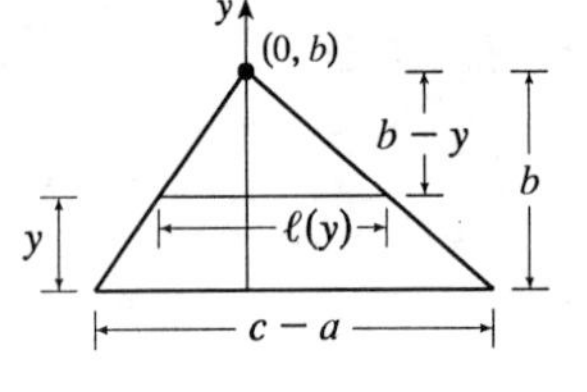

$$\overline{y} = \frac{1}{A}\int_0^b \frac{c-a}{b}y(b-y)dy = \frac{2}{b^2}\int_0^b (by-y^2)dy = \frac{2}{b^2}\left[\tfrac{1}{2}by^2 - \tfrac{1}{3}y^3\right]_0^b = \frac{2}{b^2}\cdot\frac{b^3}{6} = \frac{b}{3}.$$

Notice that only one integral is needed when this method is used.

Since the position of a centroid is independent of density when the density is constant, we will assume for convenience that $\rho = 1$ in Exercises 23 and 25.

23. Divide the lamina into two triangles and one rectangle with respective masses of 2, 2 and 4, so that the total mass is 8. Using the result of Exercise 21, the triangles have centroids $\left(-1, \frac{2}{3}\right)$ and $\left(1, \frac{2}{3}\right)$. The centroid of the rectangle (its center) is $\left(0, -\frac{1}{2}\right)$. So, using Formulas 5 and 7, we have

$\overline{y} = \dfrac{\sum m_i y_i}{m} = \frac{2}{8}\left(\frac{2}{3}\right) + \frac{2}{8}\left(\frac{2}{3}\right) + \frac{4}{8}\left(-\frac{1}{2}\right) = \frac{1}{12}$, and $\overline{x} = 0$, since the lamina is symmetric about the line $x = 0$.

Therefore $(\overline{x}, \overline{y}) = \left(0, \frac{1}{12}\right)$.

25. Suppose first that the large rectangle were complete, so that its mass would be $6 \cdot 3 = 18$. Its centroid would be $\left(1, \frac{3}{2}\right)$. The mass removed from this object to create the one being studied is 3. The centroid of the cut-out piece is $\left(\frac{3}{2}, \frac{3}{2}\right)$. Therefore, for the actual lamina, whose mass is 15, $\overline{x} = \frac{18}{15}(1) - \frac{3}{15}\left(\frac{3}{2}\right) = \frac{9}{10}$, and $\overline{y} = \frac{3}{2}$, since the lamina is symmetric about the line $y = \frac{3}{2}$. Therefore $(\overline{x}, \overline{y}) = \left(\frac{9}{10}, \frac{3}{2}\right)$.

27. A cone of height h and radius r can be generated by rotating a right triangle about one of its legs as shown. By Exercise 21, $\overline{x} = \frac{1}{3}r$, so by the Theorem of Pappus, the volume of the cone is

$V = Ad = \frac{1}{2}rh \cdot 2\pi\left(\frac{1}{3}r\right) = \frac{1}{3}\pi r^2 h.$

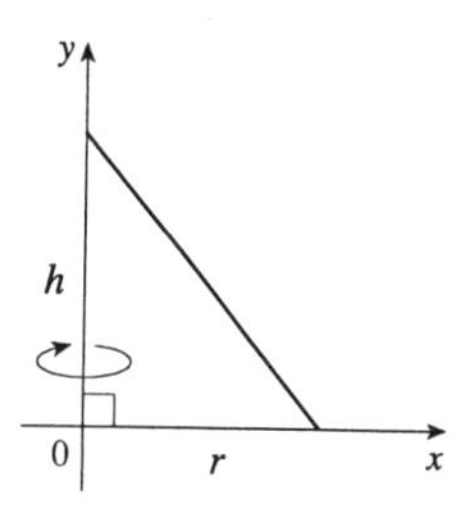

29. Suppose the region lies between two curves $y = f(x)$ and $y = g(x)$ where $f(x) \geq g(x)$, as illustrated in Figure 9. Take a partition P by points x_i with $a = x_0 < x_1 < \cdots < x_n = b$ and choose x_i^* to be the midpoint of the ith subinterval; that is, $x_i^* = \frac{1}{2}(x_{i-1} + x_i)$. Then the centroid of the ith approximating rectangle R_i is its center $C_i = \left(x_i^*, \frac{1}{2}[f(x_i^*) + g(x_i^*)]\right)$. Its area is $[f(x_i^*) - g(x_i^*)]\Delta x_i$, so its mass is $\rho[f(x_i^*) - g(x_i^*)]\Delta x_i$. Thus

$M_y(R_i) = \rho[f(x_i^*) - g(x_i^*)]\Delta x_i \cdot x_i^* = \rho x_i^*[f(x_i^*) - g(x_i^*)]\Delta x_i$ and

$M_x(R_i) = \rho[f(x_i^*) - g(x_i^*)]\Delta x_i \cdot \frac{1}{2}[f(x_i^*) + g(x_i^*)] = \rho \cdot \frac{1}{2}\left[f(x_i^*)^2 - g(x_i^*)^2\right]\Delta x_i$.

Summing over i and taking the limit as $\|P\| \to 0$, we get

$M_y = \lim\limits_{\|P\| \to 0} \sum\limits_i \rho x_i^*[f(x_i^*) - g(x_i^*)]\Delta x_i = \rho \int_a^b x[f(x) - g(x)]\,dx$ and

$M_x = \lim\limits_{\|P\| \to 0} \sum\limits_i \rho \cdot \frac{1}{2}\left[f(x_i^*)^2 - g(x_i^*)^2\right]\Delta x_i = \rho \int_a^b \frac{1}{2}\left[f(x)^2 - g(x)^2\right]dx$. Thus

$$\overline{x} = \frac{M_y}{m} = \frac{M_y}{\rho A} = \frac{1}{A}\int_a^b x[f(x) - g(x)]\,dx \text{ and } \overline{y} = \frac{M_x}{m} = \frac{M_x}{\rho A} = \frac{1}{A}\int_a^b \tfrac{1}{2}\left[f(x)^2 - g(x)^2\right]dx.$$

EXERCISES 9.6

1. **(a)** $P = \rho g d = (1000\,\text{kg/m}^3)(9.8\,\text{m/s}^2)(1\,\text{m}) = 9800\,\text{Pa} = 9.8\,\text{kPa}$

(b) $F = \rho g d A = PA = (9800\,\text{N/m}^2)(2\,\text{m}^2) = 1.96 \times 10^4\,\text{N}$

(c) $F = \int_0^1 \rho g x \cdot 1\,dx = 9800 \int_0^1 x\,dx = 4900x^2\big|_0^1 = 4.90 \times 10^3\,\text{N}$

3. $F = \int_0^{10} \rho g x \cdot 2\sqrt{100 - x^2}\,dx = 9.8 \times 10^3 \int_0^{10} \sqrt{100 - x^2}\,2x\,dx$
$= 9.8 \times 10^3 \int_{100}^0 u^{1/2}(-du)$ (put $u = 100 - x^2$)
$= 9.8 \times 10^3 \int_0^{100} u^{1/2}\,du = 9.8 \times 10^3 \left[\frac{2}{3}u^{3/2}\right]_0^{100}$
$= \frac{2}{3} \cdot 9.8 \times 10^6 \approx 6.5 \times 10^6\,\text{N}$

5. $F = \displaystyle\int_{-r}^{r} \rho g(x + r) \cdot 2\sqrt{r^2 - x^2}\,dx = \rho g \int_{-r}^{r} \sqrt{r^2 - x^2}\,2x\,dx + 2\rho g r \int_{-r}^{r} \sqrt{r^2 - x^2}\,dx$. The first integral is 0 because the integrand is an odd function. The second integral can be interpreted as the area of a semicircular disk with radius r, or we could make the trigonometric substitution $x = r\sin\theta$. Continuing:
$F = \rho g \cdot 0 + 2\rho g r \cdot \frac{1}{2}\pi r^2 = \rho g \pi r^3 = 1000 g \pi r^3\,\text{N}$ (SI units assumed).

7. $F = \displaystyle\int_0^6 \delta x \cdot \frac{2x}{3}\,dx = \left[\frac{2}{9}\delta x^3\right]_0^6$
$= 48\delta \approx 48 \times 62.5 = 3000\,\text{lb}$

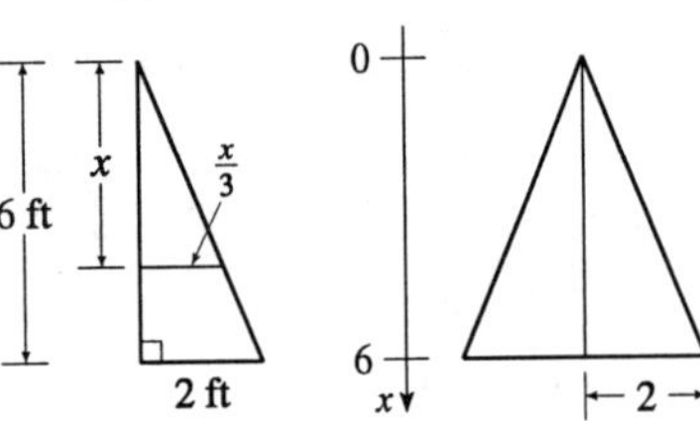

9. $F = \int_2^6 \delta(x - 2)\frac{2}{3}x\,dx = \frac{2}{3}\delta \int_2^6 (x^2 - 2x)dx$
$= \frac{2}{3}\delta\left[\frac{1}{3}x^3 - x^2\right]_2^6 = \frac{2}{3}\delta\left[36 - \left(-\frac{4}{3}\right)\right] = \frac{224}{9}\delta$
$\approx 1.56 \times 10^3\,\text{lb}$

11. $F = \int_0^8 \delta x \cdot (12 + x)dx = \delta \int_0^8 (12x + x^2)dx$
$= \delta\left[6x^2 + \dfrac{x^3}{3}\right]_0^8 = \delta\left(384 + \frac{512}{3}\right)$
$= (62.5)\frac{1664}{3} \approx 3.47 \times 10^4\,\text{lb}$

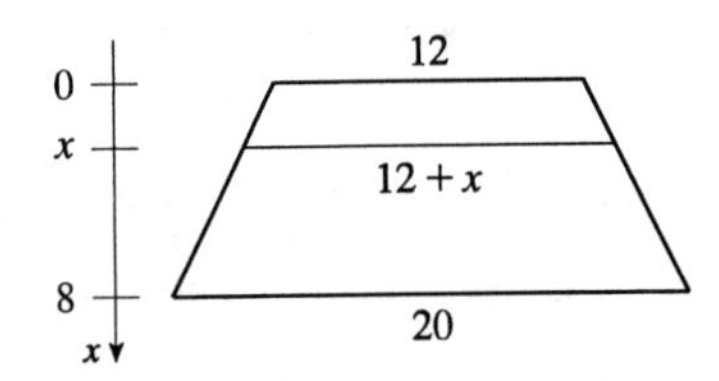

13. $F = \displaystyle\int_0^{4\sqrt{3}} \rho g\left(4\sqrt{3} - x\right)\frac{2x}{\sqrt{3}}\,dx = 8\rho g \int_0^{4\sqrt{3}} x\,dx - \frac{2\rho g}{\sqrt{3}} \int_0^{4\sqrt{3}} x^2\,dx$
$= 4\rho g\left[x^2\right]_0^{4\sqrt{3}} - \dfrac{2\rho g}{3\sqrt{3}}\left[x^3\right]_0^{4\sqrt{3}} = 192\rho g - \dfrac{2\rho g}{3\sqrt{3}} 64 \cdot 3\sqrt{3}$
$= 192\rho g - 128\rho g = 64\rho g \approx 64(840)(9.8) \approx 5.27 \times 10^5\,\text{N}$

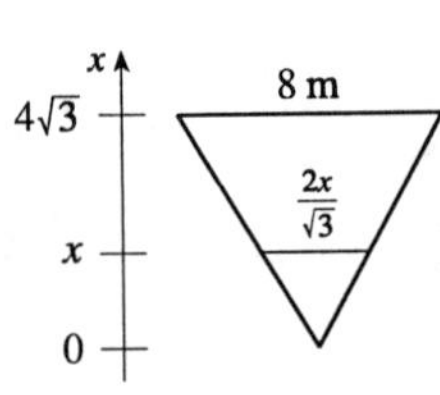

15. **(a)** $F = \rho g d A \approx (1000)(9.8)(0.8)(0.2)^2 \approx 314\,\text{N}$

(b) $F = \int_{0.8}^1 \rho g x(0.2)dx = 0.2\rho g\left[\frac{1}{2}x^2\right]_{0.8}^1 = (0.2\rho g)(0.18) = 0.036\rho g \approx 353\,\text{N}$

17. $F = \int_0^2 \rho g x \cdot 3 \cdot \sqrt{2}\,dx = 3\sqrt{2}\rho g \int_0^2 x\,dx$
$= 3\sqrt{2}\rho g\left[\frac{1}{2}x^2\right]_0^2 = 6\sqrt{2}\rho g$
$\approx 8.32 \times 10^4\text{ N}$

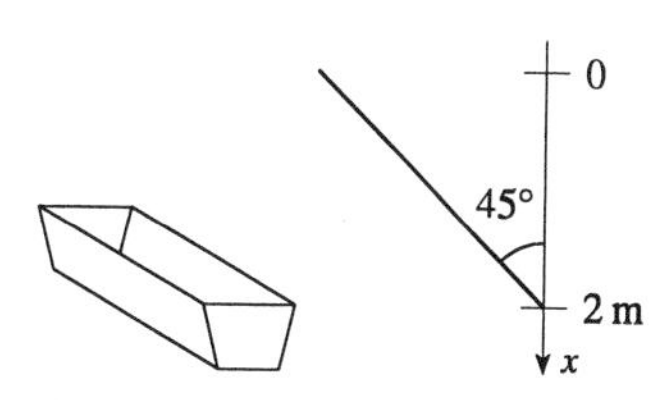

19. Assume that the pool is filled with water.

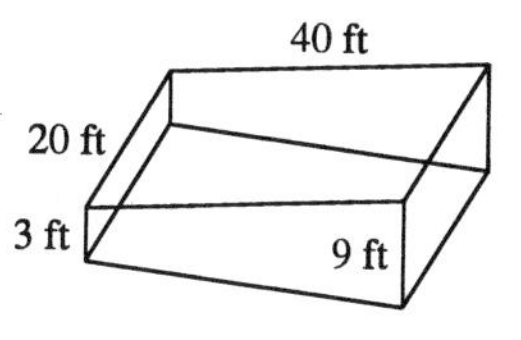

(a) $F = \int_0^3 \delta x\, 20\,dx = 20\delta\left[\frac{1}{2}x^2\right]_0^3 = 20\delta \cdot \frac{9}{2} = 90\delta \approx 5625\text{ lb} \approx 5.63 \times 10^3\text{ lb}$

(b) $F = \int_0^9 \delta x 20\,dx = 20\delta\left[\frac{1}{2}x^2\right]_0^9 = 810\delta \approx 50625\text{ lb} \approx 5.06 \times 10^4\text{ lb.}$

(c) $F = \displaystyle\int_0^3 \delta x\, 40\,dx + \int_3^9 \delta x(40)\frac{9-x}{6}\,dx = 40\delta\left[\tfrac{1}{2}x^2\right]_0^3 + \tfrac{20}{3}\delta\int_3^9 (9x - x^2)dx$
$= 180\delta + \frac{20}{3}\delta\left[\frac{9}{2}x^2 - \frac{1}{3}x^3\right]_3^9 = 180\delta + \frac{20}{3}\delta\left[\left(\frac{729}{2} - 243\right) - \left(\frac{81}{2} - 9\right)\right]$
$= 780\delta \approx 4.88 \times 10^4\text{ lb}$

(d) $F = \displaystyle\int_3^9 \delta x\, 20\frac{\sqrt{409}}{3}\,dx$
$= \frac{1}{3}\left(20\sqrt{409}\right)\delta\left[\frac{1}{2}x^2\right]_3^9$
$= \frac{1}{3} \cdot 10\sqrt{409}\delta(81 - 9)$
$\approx 3.03 \times 10^5\text{ lb}$

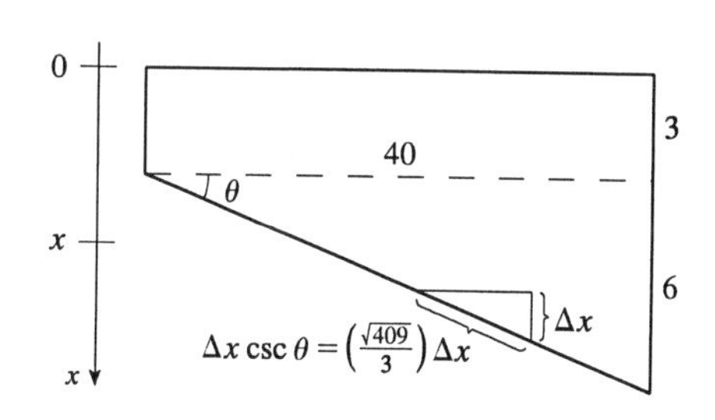

21. $\overline{x} = A^{-1}\int_a^b xw(x)dx$ (Equation 1) $\Rightarrow$ $A\overline{x} = \int_a^b xw(x)dx$ $\Rightarrow$ $(\rho g\overline{x})A = \int_a^b \rho g x w(x)dx = F$ by Exercise 20.

EXERCISES 9.7

1. $C(2000) = C(0) + \int_0^{2000} C'(x)dx = 1{,}500{,}000 + \int_0^{2000}(0.006x^2 - 1.5x + 8)dx$
$= 1{,}500{,}000 + \left[0.002x^3 - 0.75x^2 + 8x\right]_0^{2000} = \$14{,}516{,}000$

3. $C(5000) - C(3000) = \int_{3000}^{5000}(140 - 0.5x + 0.012x^2)dx = [140x - 0.25x^2 + 0.004x^3]_{3000}^{5000}$
$= 494{,}450{,}000 - 106{,}170{,}000 = \$388{,}280{,}000$

5. $p(x) = 20 = \dfrac{1000}{x+20} \quad \Rightarrow \quad x + 20 = 50 \quad \Rightarrow \quad x = 30.$

Consumer surplus $= \displaystyle\int_0^{30} [p(x) - 20]dx = \int_0^{30}\left(\frac{1000}{x+20} - 20\right)dx = [1000\ln(x+20) - 20x]_0^{30}$
$= 1000\ln\left(\frac{50}{20}\right) - 600 = 1000\ln\left(\frac{5}{2}\right) - 600 \approx \$316.29.$

7. $P = p(x) = 10 = 5 + \frac{1}{10}\sqrt{x} \quad \Rightarrow \quad 50 = \sqrt{x} \quad \Rightarrow \quad x = 2500.$

Producer surplus $= \int_0^{2500}[P - p(x)]dx = \int_0^{2500}\left(10 - 5 - \frac{1}{10}\sqrt{x}\right)dx = \left[5x - \frac{1}{15}x^{3/2}\right]_0^{2500} \approx \$4166.67$

9. The demand function is linear, with slope $\frac{-10}{100}$ and $p(1000) = 450$. So its equation is $p - 450 = -\frac{1}{10}(x - 1000)$ or $p = -\frac{1}{10}x + 550$. A selling price of \$400 $\Rightarrow$ $400 = -\frac{1}{10}x + 550$ $\Rightarrow$ $x = 1500.$

Consumer surplus $= \int_0^{1500}\left(550 - \frac{1}{10}x - 400\right)dx = \left[150x - \frac{1}{20}x^2\right]_0^{1500} = \$112{,}500$

11. Pretend that it is five years later. Then the fund will start in five years and continue for 15 years, so the present value (five years from now) is $\int_5^{20} 12{,}000e^{-0.06t}\,dt = -\frac{12{,}000}{0.06}\left[e^{-0.06t}\right]_5^{20} = \frac{12{,}000}{0.06}\left(e^{-0.3} - e^{-1.2}\right) \approx \$87{,}924.80.$

13. **(a)** $f(t) = A$, so present value $= \int_0^\infty Ae^{-rt}\,dt = \lim_{x\to\infty}\int_0^x Ae^{-rt}\,dt = \lim_{x\to\infty} -(A/r)\left[e^{-rt}\right]_0^x$
$= \lim_{x\to\infty} -(A/r)[e^{-rx} - 1] = A/r$ (since $r > 0$, $e^{-rx} \to 0$ as $x \to \infty$.)

(b) $r = 0.08$, $A = 5000$, so present value $= \frac{5000}{0.08} = \$62{,}500$ [by part (a)].

15. $f(8) - f(4) = \int_4^8 f'(t)dt = \int_4^8 \sqrt{t}\,dt = \frac{2}{3}t^{3/2}\Big|_4^8 = \frac{2}{3}\left(16\sqrt{2} - 8\right) = \frac{16(2\sqrt{2}-1)}{3} \approx \$9.75$ million

17. $F = \dfrac{\pi PR^4}{8\eta\ell} = \dfrac{\pi(4000)(0.008)^4}{8(0.027)(2)} \approx 1.19 \times 10^{-4}\ \text{cm}^3/\text{s}$

19. $\int_0^{12} c(t)dt = \int_0^{12}\frac{1}{4}t(12 - t)dt = \left[\frac{3}{2}t^2 - \frac{1}{12}t^3\right]_0^{12} = \frac{144}{2} = 72\ \text{mg}\cdot\text{s/L}$. Therefore,
$F = A/72 = \frac{8}{72} = \frac{1}{9}\ \text{L/s} = \frac{60}{9}\ \text{L/min}.$

REVIEW EXERCISES FOR CHAPTER 9

1. $y^2\dfrac{dy}{dx} = x + \sin x \quad\Rightarrow\quad \displaystyle\int y^2\,dy = \int (x + \sin x)dx \quad\Rightarrow\quad \frac{y^3}{3} = \frac{x^2}{2} - \cos x + C \quad\Rightarrow$

$y^3 = \frac{3}{2}x^2 - 3\cos x + K$ (where $K = 3C$) $\quad\Rightarrow\quad y = \sqrt[3]{\frac{3}{2}x^2 - 3\cos x + K}$

3. $y' = \dfrac{1}{x^2y - 2x^2 + y - 2} \quad\Rightarrow\quad \dfrac{dy}{dx} = \dfrac{1}{(x^2+1)(y-2)} \quad\Rightarrow\quad \displaystyle\int (y-2)dy = \int \frac{dx}{x^2+1} \quad\Rightarrow$

$\frac{1}{2}y^2 - 2y = \tan^{-1}x + K \quad\Rightarrow\quad y = 2 \pm \sqrt{2\tan^{-1}x + C}$, where $C = 4 + 2K$.

5. $xyy' = \ln x$, $y(1) = 2$. $y\,dy = \dfrac{\ln x}{x}\,dx \quad\Rightarrow\quad \displaystyle\int y\,dy = \int \frac{\ln x}{x}\,dx$ (Make the substitution $u = \ln x$; then $du = dx/x$.) So $\int y\,dy = \int u\,du \;\Rightarrow\; \frac{1}{2}y^2 = \frac{1}{2}u^2 + C \;\Rightarrow\; \frac{1}{2}y^2 = \frac{1}{2}(\ln x)^2 + C$. $y(1) = 2 \;\Rightarrow$ $\frac{1}{2}2^2 = \frac{1}{2}(\ln 1)^2 + C = C \quad\Leftrightarrow\quad C = 2$. Therefore, $\frac{1}{2}y^2 = \frac{1}{2}(\ln x)^2 + 2$, or $y = \sqrt{(\ln x)^2 + 4}$. The negative square root is inadmissible, since $y(1) > 0$.

7. Since it's linear, $I(x) = e^{\int 2x\,dx} = e^{x^2}$ and multiplying the differential equation by $I(x)$ gives

$e^{x^2}y' + 2xe^{x^2}y = 2x^3e^{x^2} \quad\Rightarrow\quad \left(e^{x^2}y\right)' = 2x^3e^{x^2} \quad\Rightarrow$

$y(x) = e^{-x^2}\left(\int 2x^3e^{x^2}\,dx + C\right) = e^{-x^2}\left(x^2e^{x^2} - e^{x^2} + C\right) = x^2 - 1 + Ce^{-x^2}$.

9. $y' = (2+y)(1+x^2) \quad\Rightarrow\quad \dfrac{dy}{2+y} = (1+x^2)dx \quad\Rightarrow\quad \ln|2+y| = x + \dfrac{x^3}{3} + c_1 \quad\Rightarrow\quad 2 + y = ke^{x+x^3/3}$

and the solution is $y(x) = ke^{x+x^3/3} - 2$.

11. Since the equation is linear, let $I(x) = e^{\int dx} = e^x$. Then multiplying by $I(x)$ gives $e^xy' + e^xy = \sqrt{x} \;\Rightarrow$ $(e^xy)' = \sqrt{x} \quad\Rightarrow\quad y(x) = e^{-x}\left(\int \sqrt{x}\,dx + c\right) = e^{-x}\left(\frac{2}{3}x^{3/2} + c\right)$. But $3 = y(0) = c$, so the solution to the initial-value problem is $y(x) = e^{-x}\left(\frac{2}{3}x^{3/2} + 3\right)$.

13. $3x = 2(y-1)^{3/2}$, $2 \le y \le 5$. $x = \frac{2}{3}(y-1)^{3/2}$, so $dx/dy = (y-1)^{1/2}$ and the arc length formula gives

$$L = \int_2^5 \sqrt{1 + (dx/dy)^2}\,dy = \int_2^5 \sqrt{1 + (y-1)}\,dy = \int_2^5 \sqrt{y}\,dy = \left[\tfrac{2}{3}y^{3/2}\right]_2^5 = \tfrac{2}{3}\left(5\sqrt{5} - 2\sqrt{2}\right).$$

15. **(a)** $y = \frac{1}{6}x^3 + \dfrac{1}{2x}$, $1 \le x \le 2 \quad\Rightarrow\quad y' = \dfrac{1}{2}\left(x^2 - \dfrac{1}{x^2}\right) \quad\Rightarrow\quad (y')^2 = \dfrac{1}{4}\left(x^4 - 2 + \dfrac{1}{x^4}\right) \quad\Rightarrow$

$$1 + (y')^2 = \frac{1}{4}\left(x^4 + 2 + \frac{1}{x^4}\right) = \frac{1}{4}\left(x^2 + \frac{1}{x^2}\right)^2 \quad\Rightarrow$$

$$L = \int_1^2 \sqrt{1 + (y')^2}\,dy = \frac{1}{2}\int_1^2 \left(x^2 + \frac{1}{x^2}\right)dx = \frac{1}{2}\left[\frac{x^3}{3} - \frac{1}{x}\right]_1^2 = \frac{1}{2}\left(\frac{17}{6}\right) = \frac{17}{12}.$$

(b) $$S = \int_1^2 2\pi y\sqrt{1 + \left(\frac{dy}{dx}\right)^2}\,dx = 2\pi\int_1^2 \left(\frac{x^3}{6} + \frac{1}{2x}\right)\frac{1}{2}\left(x^2 + \frac{1}{x^2}\right)dx$$

$$= \pi\int_1^2 \left(\tfrac{1}{6}x^5 + \tfrac{2}{3}x + \tfrac{1}{2}x^{-3}\right)dx = \pi\left[\tfrac{1}{36}x^6 + \tfrac{1}{3}x^2 - \tfrac{1}{4}x^{-2}\right]_1^2$$

$$= \pi\left[\left(\tfrac{64}{36} + \tfrac{4}{3} - \tfrac{1}{16}\right) - \left(\tfrac{1}{36} + \tfrac{1}{3} - \tfrac{1}{4}\right)\right] = \tfrac{47\pi}{16}$$

17. $y = \frac{1}{x^2}$, $1 \le x \le 2$. $\frac{dy}{dx} = -\frac{2}{x^3}$, so $1 + \left(\frac{dy}{dx}\right)^2 = 1 + \frac{4}{x^6}$. $L = \int_1^2 \sqrt{1 + \frac{4}{x^6}}\,dx$. By Simpson's Rule with $n = 10$, $L \approx \frac{1/10}{3}[f(1) + 4f(1.1) + 2f(1.2) + 4f(1.3) + 2f(1.4) + 4f(1.5)$
$+ 2f(1.6) + 4f(1.7) + 2f(1.8) + 4f(1.9) + f(2)] \approx 1.297$. Here $f(x) = \sqrt{1 + 4/x^6}$.

19. The loop lies between $x = 0$ and $x = 3a$ and is symmetric about the x-axis. We can assume without loss of generality that $a > 0$, since if $a = 0$, the graph is the parallel lines $x = 0$ and $x = 3a$, so there is no loop. The upper half of the loop is given by $y = \frac{1}{3\sqrt{a}}\sqrt{x}(3a - x) = \sqrt{a}x^{1/2} - \frac{x^{3/2}}{3\sqrt{a}}$, $0 \le x \le 3a$. The desired surface area is twice the area generated by the upper half of the loop, that is, $S = 2(2\pi)\int_0^{3a} x\sqrt{1 + \left(\frac{dy}{dx}\right)^2}\,dx$.

$$\frac{dy}{dx} = \frac{\sqrt{a}}{2}x^{-1/2} - \frac{x^{1/2}}{2\sqrt{a}} \quad\Rightarrow\quad 1 + \left(\frac{dy}{dx}\right)^2 = \frac{a}{4x} + \frac{1}{2} + \frac{x}{4a}. \text{ Therefore}$$

$$S = 2(2\pi)\int_0^{3a} x\left(\frac{\sqrt{a}}{2}x^{-1/2} + \frac{x^{1/2}}{2\sqrt{a}}\right)dx = 2\pi\int_0^{3a}\left(\sqrt{a}x^{1/2} + \frac{x^{3/2}}{\sqrt{a}}\right)dx$$

$$= 2\pi\left[\frac{2\sqrt{a}}{3}x^{3/2} + \frac{2}{5\sqrt{a}}x^{5/2}\right]_0^{3a} = 2\pi\left[\frac{2\sqrt{a}}{3}3a\sqrt{3a} + \frac{2}{5\sqrt{a}}9a^2\sqrt{3a}\right] = \frac{56\sqrt{3}\pi a^2}{5}.$$

21. $A = \int_{-2}^{1}[(4 - x^2) - (x + 2)]dx = \int_{-2}^{1}(2 - x - x^2)dx = \left[2x - \frac{1}{2}x^2 - \frac{1}{3}x^3\right]_{-2}^{1}$
$= \left(2 - \frac{1}{2} - \frac{1}{3}\right) - \left(-4 - 2 + \frac{8}{3}\right) = \frac{9}{2} \quad\Rightarrow$
$\overline{x} = A^{-1}\int_{-2}^{1} x(2 - x - x^2)dx = \frac{2}{9}\int_{-2}^{1}(2x - x^2 - x^3)dx = \frac{2}{9}\left[x^2 - \frac{1}{3}x^3 - \frac{1}{4}x^4\right]_{-2}^{1}$
$= \frac{2}{9}\left[\left(1 - \frac{1}{3} - \frac{1}{4}\right) - \left(4 + \frac{8}{3} - 4\right)\right] = -\frac{1}{2}$ and
$\overline{y} = A^{-1}\int_{-2}^{1}\frac{1}{2}\left[(4 - x^2)^2 - (x + 2)^2\right]dx = \frac{1}{9}\int_{-2}^{1}(x^4 - 9x^2 - 4x + 12)dx$
$= \frac{1}{9}\left[\frac{1}{5}x^5 - 3x^3 - 2x^2 + 12x\right]_{-2}^{1} = \frac{1}{9}\left[\left(\frac{1}{5} - 3 - 2 + 12\right) - \left(-\frac{32}{5} + 24 - 8 - 24\right)\right] = \frac{12}{5}$.
So $(\overline{x}, \overline{y}) = \left(-\frac{1}{2}, \frac{12}{5}\right)$.

23. The equation of the line passing through $(0, 0)$ and $(3, 2)$ is $y = \frac{2}{3}x$. $A = \frac{1}{2}\cdot 3\cdot 2 = 3$. Therefore,
$\overline{x} = \frac{1}{3}\int_0^3 x\left(\frac{2}{3}x\right)dx = \frac{2}{27}\left[x^3\right]_0^3 = 2$, and $\overline{y} = \frac{1}{3}\int_0^3 \frac{1}{2}\left(\frac{2}{3}x\right)^2 dx = \frac{2}{81}\left[x^3\right]_0^3 = \frac{2}{3}$. $(\overline{x}, \overline{y}) = \left(2, \frac{2}{3}\right)$.
Or: Use Exercise 9.4.21.

25. The centroid of this circle, $(1, 0)$, travels a distance $2\pi(1)$ when the lamina is rotated about the y-axis. The area of the circle is $\pi(1)^2$. So by the Theorem of Pappus, $V = A2\pi\overline{x} = \pi(1)^2 2\pi(1) = 2\pi^2$.

27. As in Example 1 of Section 9.5, $F = \int_0^2 \rho g x(5 - x)dx = \rho g\left[\frac{5}{2}x^2 - \frac{1}{3}x^3\right]_0^2 = \rho g\frac{22}{3} = \frac{22}{3}\delta \approx \frac{22}{3}\cdot 62.5 \approx 458$ lb.

29. $x = 100 \quad\Rightarrow\quad P = 2000 - 0.1(100) - 0.01(100)^2 = 1890$

Consumer surplus $= \int_0^{100}[p(x) - P]dx = \int_0^{100}(2000 - 0.1x - 0.01x^2 - 1890)dx$
$= \left[110x - 0.05x^2 - \frac{0.01}{3}x^3\right]_0^{100} = 11{,}000 - 500 - \frac{10{,}000}{3} \approx \$7166.67$

31. **(a)** $\dfrac{dL}{dt} \propto L_\infty - L \quad\Rightarrow\quad \dfrac{dL}{dt} = k(L_\infty - L) \quad\Rightarrow\quad \displaystyle\int \frac{dL}{L_\infty - L} = \int k\,dt \quad\Rightarrow\quad -\ln|L_\infty - L| = kt + C$

$\Rightarrow \quad L_\infty - L = Ae^{-kt} \quad\Rightarrow\quad L = L_\infty - Ae^{-kt}$. At $t = 0$, $L = L(0) = L_\infty - A \quad\Rightarrow$

$A = L_\infty - L(0) \quad\Rightarrow\quad L(t) = L_\infty - [L_\infty - L(0)]e^{-kt}$

(b) $L_\infty = 53$ cm, $L(0) = 10$ cm and $k = 0.2$. So $L(t) = 53 - (53 - 10)e^{-0.2t} = 53 - 43e^{-0.2t}$.

33. Let P be the population and I be the number of infected people. The rate of spread dI/dt is jointly proportional to I and to $P - I$, so for some constant k, $\dfrac{dI}{dt} = kI(P - I) \quad\Rightarrow\quad I = \dfrac{I_0 P}{I_0 + (P - I_0)e^{-kPt}}$ (from the discussion of logistic growth in Section 9.1).

Now, measuring t in days, we substitute $t = 7$, $P = 5000$, $I_0 = 160$ and $I(7) = 1200$ to find k:

$1200 = \dfrac{160 \cdot 5000}{160 + (5000 - 160)e^{-5000 \cdot 7 \cdot k}} \quad\Leftrightarrow\quad k \approx 0.00006448$. So, putting $I = 5000 \times 80\% = 4000$, we solve for t: $4000 = \dfrac{160 \cdot 5000}{160 + (5000 - 160)e^{-0.00006448 \cdot 5000 \cdot t}} \quad\Leftrightarrow\quad 160 + 4840e^{-0.3224t} = 200 \quad\Leftrightarrow$

$-0.3224t = \ln \dfrac{40}{4840} \quad\Leftrightarrow\quad t \approx 14.9$. So it takes about 15 days for 80% of the population to be infected.

35. **(a)** We are given that $V = \frac{1}{3}\pi r^2 h$, $dV/dt = 60{,}000\pi$ ft³/h, and $r = 1.5h = \frac{3}{2}h$. So $V = \frac{1}{3}\pi\left(\frac{3}{2}h\right)^2 \quad\Rightarrow$

$\dfrac{dV}{dt} = \frac{3}{4}\pi \cdot 3h^2 \dfrac{dh}{dt} = \frac{9}{4}\pi h^2 \dfrac{dh}{dt}$. Therefore, $\dfrac{dh}{dt} = \dfrac{4(dV/dt)}{9\pi h^2} = \dfrac{240{,}000\pi}{9\pi h^2} = \dfrac{80{,}000}{3h^2}$ (★) $\quad\Rightarrow$

$\int 3h^2\,dh = \int 80{,}000\,dt \quad\Rightarrow\quad h^3 = 80{,}000t + C$. When $t = 0$, $h = 60$. Therefore, $C = 60^3 = 216{,}000$, so $h^3 = 80{,}000t + 216{,}000$. Let $h = 100$. Then $100^3 = 1{,}000{,}000 = 80{,}000t + 216{,}000 \quad\Rightarrow$

$80{,}000t = 784{,}000 \quad\Rightarrow\quad t = 9.8$, so the time required is 9.8 hours.

(b) The floor area of the silo is $F = \pi \cdot 200^2 = 40{,}000\pi$ ft², and the area of the base of the pile is $A = \pi r^2 = \pi\left(\frac{3}{2}h\right)^2 = \frac{9\pi}{4}h^2 = 8100\pi$ ft². So the area of the floor which is not covered is $F - A = 31{,}900\pi \approx 100{,}000$ ft².

Now $A = \frac{9\pi}{4}h^2 \quad\Rightarrow\quad \dfrac{dA}{dt} = \dfrac{9\pi}{4} \cdot 2h\dfrac{dh}{dt}$, and from (★) in part (a) we know that when $h = 60$,

$\dfrac{dh}{dt} = \dfrac{80{,}000}{3(60)^2} = \dfrac{200}{27}\dfrac{\text{ft}}{\text{h}}$. Therefore $\dfrac{dA}{dt} = \dfrac{9\pi}{4}(2)(60)\left(\dfrac{200}{27}\right) \approx 6283\,\dfrac{\text{ft}^2}{\text{h}}$.

(c) At $h = 90$ ft, $\dfrac{dV}{dt} = 60{,}000\pi - 20{,}000\pi = 40{,}000\pi$ ft³/h. From (★) in (a),

$\dfrac{dh}{dt} = \dfrac{4(dV/dt)}{9\pi h^2} = \dfrac{4(40{,}000\pi)}{9\pi h^2} = \dfrac{160{,}000}{9h^2} \quad\Rightarrow\quad \int 9h^2\,dh = \int 160{,}000\,dt \quad\Rightarrow$

$3h^3 = 160{,}000t + C$. When $t = 0$, $h = 90$; therefore, $C = 3 \cdot 729{,}000 = 2{,}187{,}000$. So

$3h^3 = 160{,}000t + 2{,}187{,}000$. At the top, $h = 100$, so $3(100)^3 = 160{,}000t + 2{,}187{,}000 \quad\Rightarrow$

$t = \dfrac{813{,}000}{160{,}000} \approx 5.1$. The pile reaches the top after about 5.1 h.

PROBLEMS PLUS (after Chapter 9)

1. By symmetry, the problem can be reduced to finding the line $x = c$ such that the shaded area is one-third of the area of the quarter-circle. The equation of the circle is $y = \sqrt{49 - x^2}$, so we require that $\int_0^c \sqrt{49 - x^2}\, dx = \frac{1}{3} \cdot \frac{1}{4}\pi(7)^2 \quad \Leftrightarrow$ $\left[\frac{1}{2}x\sqrt{49 - x^2} + \frac{49}{2}\sin^{-1}(x/7)\right]_0^c = \frac{49}{12}\pi$ (Formula 30) $\Leftrightarrow$ $\frac{1}{2}c\sqrt{49 - c^2} + \frac{49}{2}\sin^{-1}(c/7) = \frac{49}{12}\pi$.

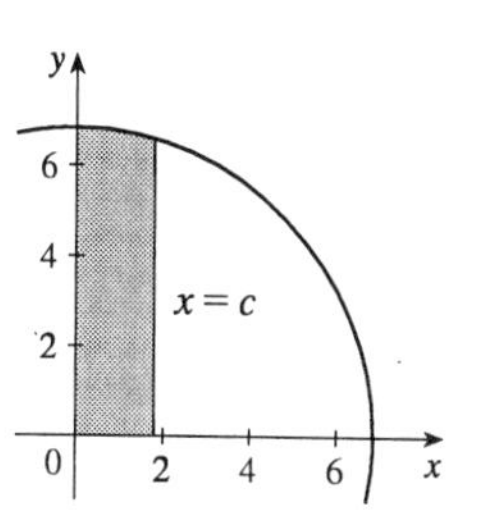

This equation would be difficult to solve exactly, so we plot the left-hand side as a function of c, and find that the equation holds for $c \approx 1.85$. So the cuts should be made at distances of about 1.85 inches from the center of the pizza.

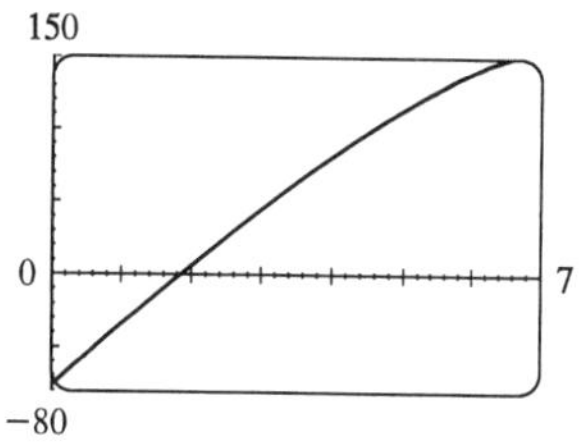

3.

$$\begin{aligned} f'(x) &= \lim_{h\to 0} \frac{f(x+h) - f(x)}{h} \\ &= \lim_{h\to 0} \frac{f(x)[f(h) - 1]}{h} \quad [\text{since } f(x+h) = f(x)f(h)] \\ &= f(x) \lim_{h\to 0} \frac{f(h) - 1}{h} \\ &= f(x) \lim_{h\to 0} \frac{f(h) - f(0)}{h - 0} \\ &= f(x)f'(0) \\ &= f(x) \end{aligned}$$

Therefore, $f'(x) = f(x)$ for all x. In Leibniz notation, we can either solve the differential equation $\frac{dy}{dx} = y$ $\Rightarrow \int \frac{dy}{y} = \int dx \Rightarrow \ln|y| = x + C \Rightarrow y = Ae^x$, or we can use Theorem 4.5.2 to get the same result. Now $f(0) = 1 \Rightarrow A = 1 \Rightarrow f(x) = e^x$.

5. First we show that $x(1 - x) \le \frac{1}{4}$ for all x. Let $f(x) = x(1 - x) = x - x^2$. Then $f'(x) = 1 - 2x$. This is 0 when $x = \frac{1}{2}$ and $f'(x) > 0$ for $x < \frac{1}{2}$, $f'(x) < 0$ for $x > \frac{1}{2}$, so the absolute maximum of f is $f\left(\frac{1}{2}\right) = \frac{1}{4}$. Thus $x(1 - x) \le \frac{1}{4}$ for all x.

Now suppose that the given assertion is false, that is, $a(1 - b) > \frac{1}{4}$ and $b(1 - a) > \frac{1}{4}$. Multiply these inequalities: $a(1 - b)b(1 - a) > \frac{1}{16} \Rightarrow [a(1 - a)][b(1 - b)] > \frac{1}{16}$. But we know that $a(1 - a) \le \frac{1}{4}$ and $b(1 - b) \le \frac{1}{4} \Rightarrow [a(1 - a)][b(1 - b)] \le \frac{1}{16}$. Thus we have a contradiction, so the given assertion is proved.

7. Let $F(x) = \int_a^x f(t)dt - \int_x^b f(t)dt$. Then $F(a) = -\int_a^b f(t)dt$ and $F(b) = \int_a^b f(t)dt$. Also F is continuous by the Fundamental Theorem. So by the Intermediate Value Theorem, there is a number c in $[a, b]$ such that $F(c) = 0$. For that number c, we have $\int_a^c f(t)dt = \int_c^b f(t)dt$.

9. The given integral represents the difference of the shaded areas, which appears to be 0. It can be calculated by integrating with respect to either x or y, so we find x in terms of y for each curve:

$y = \sqrt[3]{1-x^7} \Rightarrow x = \sqrt[7]{1-y^3}$ and $y = \sqrt[7]{1-x^3} \Rightarrow x = \sqrt[3]{1-y^7}$, so

$$\int_0^1 \left(\sqrt[3]{1-y^7} - \sqrt[7]{1-y^3}\right) dy = \int_0^1 \left(\sqrt[7]{1-x^3} - \sqrt[3]{1-x^7}\right) dx.$$

But this equation is of the form $z = -z$. So $\int_0^1 \left(\sqrt[3]{1-x^7} - \sqrt[7]{1-x^3}\right) dx = 0$.

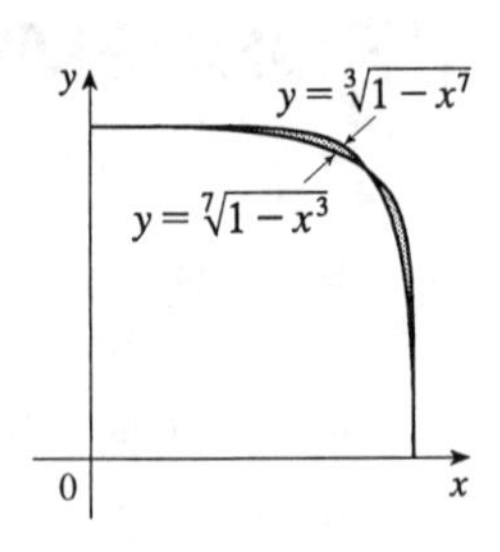

11. First we find the domain explicitly. $5x^2 \geq x^4 + 4 \quad \Leftrightarrow \quad x^4 - 5x^2 + 4 \leq 0 \quad \Leftrightarrow \quad (x^2-4)(x^2-1) \leq 0$ $\Leftrightarrow \quad (x-2)(x-1)(x+1)(x+2) \leq 0 \quad \Leftrightarrow \quad x \in [-2,-1]$ or $[1,2]$. Therefore, the domain is $\{x \mid 5x^2 \geq x^4 + 4\} = [-2,-1] \cup [1,2]$. Now f is decreasing on its domain, since $f(x) = 3x - 2x^3 \quad \Rightarrow$ $f'(x) = 3 - 6x^2 \leq -3$. So f's maximum value is either $f(-2)$ or $f(1)$. $f(-2) = 10$ and $f(1) = 1$, so the maximum value of f is 10.

13. Recall that $\cos A \cos B = \frac{1}{2}[\cos(A+B) + \cos(A-B)]$. So

$$\begin{aligned} f(x) &= \textstyle\int_0^\pi \cos t \cos(x-t)\,dt = \frac{1}{2}\int_0^\pi [\cos(t+x-t) + \cos(t-x+t)]\,dt \\ &= \textstyle\frac{1}{2}\int_0^\pi [\cos x + \cos(2t-x)]\,dt = \frac{1}{2}\left[t\cos x + \frac{1}{2}\sin(2t-x)\right]_0^\pi \\ &= \textstyle\frac{\pi}{2}\cos x + \frac{1}{4}\sin(2\pi - x) - \frac{1}{4}\sin(-x) = \frac{\pi}{2}\cos x + \frac{1}{4}\sin(-x) - \frac{1}{4}\sin(-x) \\ &= \textstyle\frac{\pi}{2}\cos x. \end{aligned}$$

The minimum of $\cos x$ on this domain is -1, so the minimum value of $f(x)$ is $f(\pi) = -\frac{\pi}{2}$.

15. $0 < a < b$. $\displaystyle\int_0^1 [bx + a(1-x)]^t\,dx = \int_a^b \frac{u^t}{(b-a)}\,du$ [put $u = bx + a(1-x)$]

$$= \left[\frac{u^{t+1}}{(t+1)(b-a)}\right]_a^b = \frac{b^{t+1} - a^{t+1}}{(t+1)(b-a)}.\ \text{Now let } y = \lim_{t\to 0}\left[\frac{b^{t+1} - a^{t+1}}{(t+1)(b-a)}\right]^{1/t}.$$

Then $\ln y = \displaystyle\lim_{t\to 0}\left[\frac{1}{t}\ln\frac{b^{t+1} - a^{t+1}}{(t+1)(b-a)}\right]$. This limit is of the form $\frac{0}{0}$, so we can apply l'Hospital's Rule to get

$$\ln y = \lim_{t\to 0}\left[\frac{b^{t+1}\ln b - a^{t+1}\ln a}{b^{t+1} - a^{t+1}} - \frac{1}{t+1}\right] = \frac{b\ln b - a\ln a}{b-a} - 1 = \frac{b\ln b}{b-a} - \frac{a\ln a}{b-a} - \ln e = \ln\frac{b^{b/(b-a)}}{ea^{a/(b-a)}}.$$

Therefore, $y = e^{-1}\left(\dfrac{b^b}{a^a}\right)^{1/(b-a)}$.

17. In accordance with the hint, we let $I_k = \int_0^1 (1-x^2)^k\,dx$, and we find an expression for I_{k+1} in terms of I_k. We integrate I_{k+1} by parts with $u = (1-x^2)^{k+1} \quad \Rightarrow \quad du = (k+1)(1-x^2)^k(-2x)$, $dv = dx \quad \Rightarrow \quad v = x$, and then split the remaining integral into identifiable quantities:

$$\begin{aligned} I_{k+1} &= x(1-x^2)^{k+1}\Big|_0^1 + 2(k+1)\textstyle\int_0^1 x^2(1-x^2)^k\,dx = (2k+2)\int_0^1 (1-x^2)^k[1-(1-x^2)]\,dx \\ &= (2k+2)(I_k - I_{k+1}).\ \text{So } I_{k+1}[1 + (2k+2)] = (2k+2)I_k \end{aligned}$$

$\Rightarrow \quad I_{k+1} = \dfrac{2k+2}{2k+3}I_k$. Now to complete the proof, we use induction: $I_0 = 1 = \dfrac{2^0(0!)^2}{1!}$, so the formula holds for $n = 0$. Now suppose it holds for $n = k$. Then

$$I_{k+1} = \frac{2k+2}{2k+3} I_k = \frac{2k+2}{2k+3}\left[\frac{2^{2k}(k!)^2}{(2k+1)!}\right] = \frac{2(k+1)2^{2k}(k!)^2}{(2k+3)(2k+1)!} = \frac{2(k+1)}{2k+2} \cdot \frac{2(k+1)2^{2k}(k!)^2}{(2k+3)(2k+1)!}$$

$$= \frac{[2(k+1)]^2 2^{2k}(k!)^2}{(2k+3)(2k+2)(2k+1)!} = \frac{2^{2(k+1)}[(k+1)!]^2}{[2(k+1)+1]!}.$$

So by induction, the formula holds for all integers $n \geq 0$.

19. We use the Fundamental Theorem of Calculus to differentiate the given equation:

$[f(x)]^2 = 100 + \int_0^x \left([f(t)]^2 + [f'(t)]^2\right) dt \quad \Rightarrow \quad 2f(x)f'(x) = [f(x)]^2 + [f'(x)]^2 \quad \Rightarrow$

$[f(x)]^2 + [f'(x)]^2 - 2f(x)f'(x) = [f(x) - f'(x)]^2 = 0 \quad \Leftrightarrow \quad f(x) = f'(x)$. We can solve this as a separable equation, or else use Theorem 4.5.2, which says that the only solutions are $f(x) = Ce^x$. Now $[f(0)]^2 = 100$, so $f(0) = C = \pm 10$, and hence $f(x) = \pm 10e^x$ are the only functions satisfying the given equation.

21. To find the height of the pyramid, we use similar triangles. The first figure shows a cross-section of the pyramid passing through the top and through two opposite corners of the square base. Now $|BD| = b$, since it is a radius of the sphere, which has diameter $2b$ since it is tangent to the opposite sides of the square base. Also, $|AD| = b$ since $\triangle ADB$ is isosceles. So the height is $|AB| = \sqrt{b^2 + b^2} = \sqrt{2}b$.

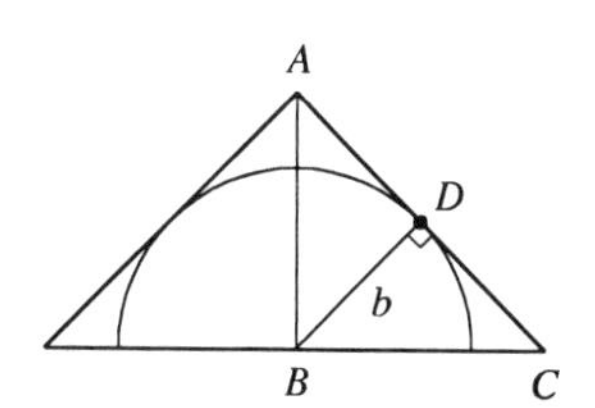

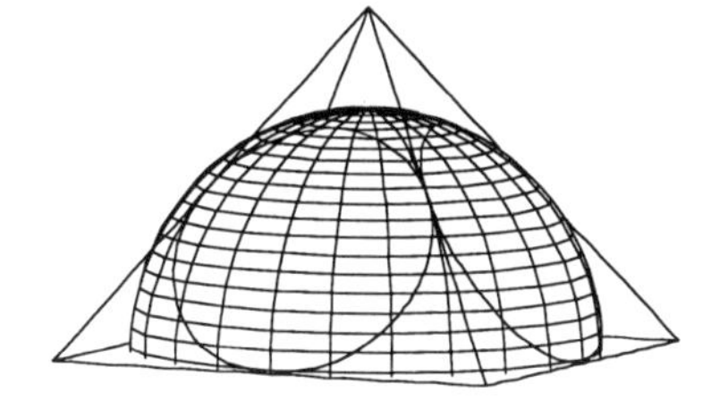

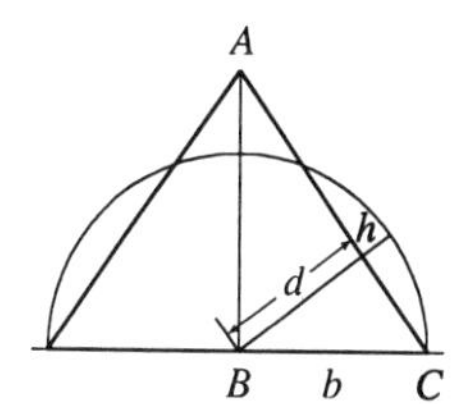

We observe that the shared volume is equal to half the volume of the sphere, minus the sum of the four equal volumes (caps of the sphere) cut off by the triangular faces of the pyramid. See Exercise 7.2.51 for a derivation of the formula for the volume of a cap of a sphere. To use the formula, we need to find the perpendicular distance h of each triangular face from the surface of the sphere. We first find the distance d from the center of the sphere to one of the triangular faces. The third figure shows a cross-section of the pyramid through the top and through the midpoints of opposite sides of the square base. From similar triangles we find that

$\frac{d}{b} = \frac{|AB|}{|AC|} = \frac{\sqrt{2}b}{\sqrt{b^2 + \left(\sqrt{2}b\right)^2}} \quad \Rightarrow \quad d = \frac{\sqrt{2}b^2}{\sqrt{3b^2}} = \frac{\sqrt{6}}{3}b$. So $h = b - \frac{\sqrt{6}}{3}b = \frac{3-\sqrt{6}}{3}b$. So, using the formula from Exercise 5.2.49 with $r = b$, we find that the volume of each of the caps is

$\pi\left(\frac{3-\sqrt{6}}{3}b\right)^2\left(b - \frac{3-\sqrt{6}}{3\cdot 3}b\right) = \frac{15-6\sqrt{6}}{9} \cdot \frac{6+\sqrt{6}}{9}\pi b^3 = \left(\frac{2}{3} - \frac{7}{27}\sqrt{6}\right)\pi b^3$. So, using our first observation, the shared volume is $V = \frac{1}{2}\left(\frac{4}{3}\pi b^3\right) - 4\left(\frac{2}{3} - \frac{7}{27}\sqrt{6}\right)\pi b^3 = \left(\frac{28}{27}\sqrt{6} - 2\right)\pi b^3$.

CHAPTER TEN

EXERCISES 10.1

1. **(a)** A sequence is an ordered list of numbers. It can also be defined as a function whose domain is the set of positive integers.

(b) The terms a_n approach 8 as n becomes large. In fact, we can make a_n as close to 8 as we like by taking n sufficiently large.

(c) The terms a_n become large as n becomes large.

3. The first six terms of $a_n = \dfrac{n}{2n+1}$ are: $\dfrac{1}{3}, \dfrac{2}{5}, \dfrac{3}{7}, \dfrac{4}{9}, \dfrac{5}{11}, \dfrac{6}{13}$. It appears that the sequence is approaching $\dfrac{1}{2}$.

$$\lim_{n\to\infty} \frac{n}{2n+1} = \lim_{n\to\infty} \frac{1}{2+1/n} = \frac{1}{2}$$

5. The numerators are all 1 and the denominators are powers of 2, so $a_n = \dfrac{1}{2^n}$.

7. In each term, the numerator is 2 greater than the number of the term, and the denominator is the square of the number that is 3 greater than the term number. So $a_n = \dfrac{n+2}{(n+3)^2}$.

9. $\displaystyle\lim_{n\to\infty} \frac{1}{5^n} = \lim_{n\to\infty} \left(\frac{1}{5}\right)^n = 0$ by Equation 6 with $r = \dfrac{1}{5}$. Convergent

11. $\displaystyle\lim_{n\to\infty} \frac{n^2-1}{n^2+1} = \lim_{n\to\infty} \frac{1-1/n^2}{1+1/n^2} = 1$. Convergent

13. $\{a_n\}$ diverges since $\dfrac{n^2}{n+1} = \dfrac{n}{1+1/n} \to \infty$ as $n \to \infty$.

15. $\{a_n\} = \left\{\cos\frac{\pi}{2}, \cos\pi, \cos\frac{3\pi}{2}, \cos 2\pi, \cos\frac{5\pi}{2}, \cos 3\pi, \cos\frac{7\pi}{2}, \cos 4\pi, \ldots\right\}$

$= \{0, -1, 0, 1, 0, -1, 0, 1, \ldots\}$

This sequence oscillates among 0, -1, and 1 and so diverges.

17. $a_n = \dfrac{\pi^n}{3^n} = \left(\dfrac{\pi}{3}\right)^n$, so $\{a_n\}$ diverges by Equation 6 with $r = \frac{\pi}{3} > 1$.

19. $\displaystyle\lim_{x\to\infty} \frac{\ln(x^2)}{x} = \lim_{x\to\infty} \frac{2\ln x}{x} \overset{\text{H}}{=} \lim_{x\to\infty} \frac{2/x}{1} = 0$, so by Theorem 2, $\left\{\dfrac{\ln(n^2)}{n}\right\}$ converges to 0.

21. $b_n = \sqrt{n+2} - \sqrt{n} = \left(\sqrt{n+2} - \sqrt{n}\right)\dfrac{\sqrt{n+2}+\sqrt{n}}{\sqrt{n+2}+\sqrt{n}} = \dfrac{2}{\sqrt{n+2}+\sqrt{n}} < \dfrac{2}{2\sqrt{n}} = \dfrac{1}{\sqrt{n}} \to 0$ as $n \to \infty$. So by the Squeeze Theorem with $a_n = 0$ and $c_n = 1/\sqrt{n}$, $\left\{\sqrt{n+2} - \sqrt{n}\right\}$ converges to 0.

23. $\displaystyle\lim_{x\to\infty} \frac{x}{2^x} \overset{\text{H}}{=} \lim_{x\to\infty} \frac{1}{(\ln 2)\, 2^x} = 0$, so by Theorem 2, $\{n2^{-n}\}$ converges to 0.

25. $0 \leq \dfrac{\cos^2 n}{2^n} \leq \dfrac{1}{2^n}$ [since $0 \leq \cos^2 n \leq 1$], so since $\lim\limits_{n\to\infty} \dfrac{1}{2^n} = 0$, $\left\{\dfrac{\cos^2 n}{2^n}\right\}$ converges to 0 by the Squeeze Theorem.

27.

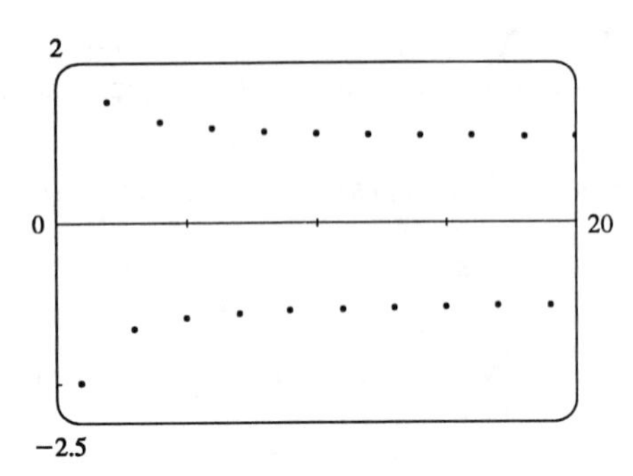

From the graph, we see that the sequence $\left\{(-1)^n \dfrac{n+1}{n}\right\}$ is divergent, since it oscillates between 1 and -1 (approximately).

29.

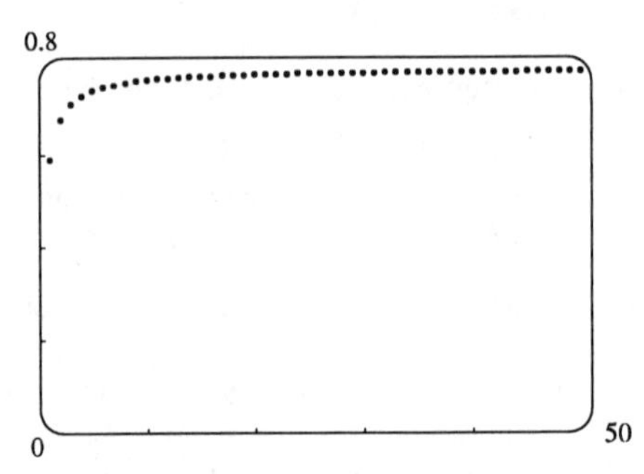

From the graph, it appears that the sequence converges to about 0.78.

$\lim\limits_{n\to\infty} \dfrac{2n}{2n+1} = \lim\limits_{n\to\infty} \dfrac{2}{2+1/n} = 1$, so

$\lim\limits_{n\to\infty} \arctan\left(\dfrac{2n}{2n+1}\right) = \arctan 1 = \dfrac{\pi}{4}$.

31.

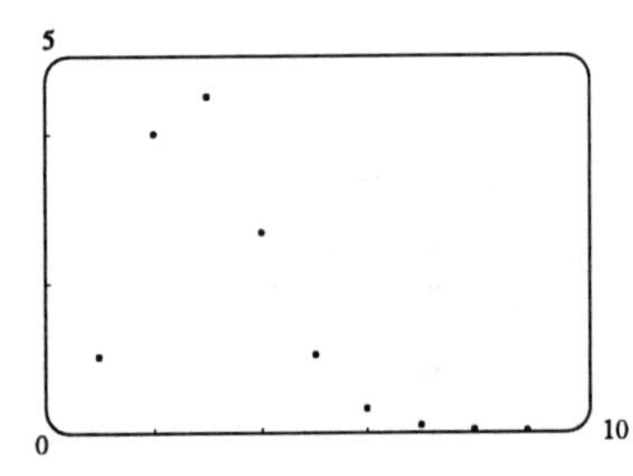

From the graph, it appears that the sequence converges to 0.

$$\begin{aligned} 0 < a_n &= \frac{n^3}{n!} = \frac{n}{n} \cdot \frac{n}{(n-1)} \cdot \frac{n}{(n-2)} \cdot \frac{1}{(n-3)} \cdot \cdots \cdot \frac{1}{3} \cdot \frac{1}{2} \cdot \frac{1}{1} \\ &\leq \frac{n^2}{(n-1)(n-2)(n-3)} \quad (\text{for } n \geq 4) \\ &= \frac{1/n}{(1-1/n)(1-2/n)(1-3/n)} \to 0 \text{ as } n \to \infty \end{aligned}$$

So by the Squeeze Theorem, $\{n^3/n!\}$ converges to 0.

33. **(a)** $a_1 = 1$, $a_2 = 4 - a_1 = 4 - 1 = 3$, $a_3 = 4 - a_2 = 4 - 3 = 1$, $a_4 = 4 - a_3 = 4 - 1 = 3$, $a_5 = 4 - a_4 = 4 - 3 = 1$. Since the terms of the sequence alternate between 1 and 3, the sequence is divergent.

(b) $a_1 = 2$, $a_2 = 4 - a_1 = 4 - 2 = 2$, $a_3 = 4 - a_2 = 4 - 2 = 2$. Since all of the terms are 2, $\lim\limits_{n\to\infty} a_n = 2$ and hence, the sequence is convergent.

35. **(a)** Let a_n be the number of rabbit pairs in the nth month. Clearly $a_1 = 1 = a_2$. In the nth month, each pair that is 2 or more months old (that is, a_{n-2} pairs) will have a pair of children to add to the a_{n-1} pairs already present. Thus, $a_n = a_{n-1} + a_{n-2}$, so that $\{a_n\} = \{f_n\}$, the Fibonacci sequence.

(b) $a_n = \dfrac{f_{n+1}}{f_n} \Rightarrow a_{n-1} = \dfrac{f_n}{f_{n-1}} = \dfrac{f_{n-1} + f_{n-2}}{f_{n-1}} = 1 + \dfrac{f_{n-2}}{f_{n-1}} = 1 + \dfrac{1}{f_{n-1}/f_{n-2}} = 1 + \dfrac{1}{a_{n-2}}$. If $L = \lim\limits_{n\to\infty} a_n$, then $L = \lim\limits_{n\to\infty} a_{n-1}$ and $L = \lim\limits_{n\to\infty} a_{n-2}$, so L must satisfy $L = 1 + \dfrac{1}{L} \Rightarrow$ $L^2 - L - 1 = 0 \Rightarrow L = \frac{1+\sqrt{5}}{2}$ (since L must be positive).

37. $3(n+1) + 5 > 3n + 5$ so $\dfrac{1}{3(n+1)+5} < \dfrac{1}{3n+5} \Leftrightarrow a_{n+1} < a_n$, so $\{a_n\}$ is decreasing.

39. $\left\{\dfrac{n-2}{n+2}\right\}$ is increasing since $a_n < a_{n+1} \Leftrightarrow \dfrac{n-2}{n+2} < \dfrac{(n+1)-2}{(n+1)+2} \Leftrightarrow$
$(n-2)(n+3) < (n+2)(n-1) \Leftrightarrow n^2+n-6 < n^2+n-2 \Leftrightarrow -6 < -2$, which is true.

41. Since $\{a_n\}$ is a decreasing sequence, $a_n > a_{n+1}$ for all $n \geq 1$. Because all of its terms lie between 5 and 8, $\{a_n\}$ is a bounded sequence. By the Monotonic Sequence Theorem, $\{a_n\}$ is convergent, that is, $\{a_n\}$ has a limit L. L must be less than 8 since $\{a_n\}$ is decreasing, so $5 \leq L < 8$.

43. We show by induction that $\{a_n\}$ is increasing and bounded above by 3.
Let P_n be the proposition that $a_{n+1} > a_n$ and $0 < a_n < 3$. Clearly P_1 is true. Assume that P_n is true. Then
$a_{n+1} > a_n \Rightarrow \dfrac{1}{a_{n+1}} < \dfrac{1}{a_n} \Rightarrow -\dfrac{1}{a_{n+1}} > -\dfrac{1}{a_n}$.
Now $a_{n+2} = 3 - \dfrac{1}{a_{n+1}} > 3 - \dfrac{1}{a_n} = a_{n+1} \Leftrightarrow P_{n+1}$. This proves that $\{a_n\}$ is increasing and bounded above by 3, so $1 = a_1 < a_n < 3$, that is, $\{a_n\}$ is bounded, and hence convergent by the Monotonic Sequence Theorem.
If $L = \lim\limits_{n\to\infty} a_n$, then $\lim\limits_{n\to\infty} a_{n+1} = L$ also, so L must satisfy $L = 3 - 1/L \Rightarrow L^2 - 3L + 1 = 0 \Rightarrow$
$L = \frac{3\pm\sqrt{5}}{2}$. But $L > 1$, so $L = \frac{3+\sqrt{5}}{2}$.

45. $(0.8)^n < 0.000001 \Rightarrow \ln(0.8)^n < \ln(0.000001) \Rightarrow n\ln(0.8) < \ln(0.000001) \Rightarrow$
$n > \dfrac{\ln(0.000001)}{\ln(0.8)} \Rightarrow n > 61.9$, so n must be at least 62 to satisfy the given inequality.

47. **(a)** First we show that $a > a_1 > b_1 > b$.
$a_1 - b_1 = \dfrac{a+b}{2} - \sqrt{ab} = \frac{1}{2}\left(a - 2\sqrt{ab} + b\right) = \frac{1}{2}\left(\sqrt{a} - \sqrt{b}\right)^2 > 0$ (since $a > b$) $\Rightarrow a_1 > b_1$.
Also $a - a_1 = a - \frac{1}{2}(a+b) = \frac{1}{2}(a-b) > 0$ and $b - b_1 = b - \sqrt{ab} = \sqrt{b}\left(\sqrt{b} - \sqrt{a}\right) < 0$, so $a > a_1 > b_1 > b$. In the same way we can show that $a_1 > a_2 > b_2 > b_1$ and so the given assertion is true for $n = 1$. Suppose it is true for $n = k$, that is, $a_k > a_{k+1} > b_{k+1} > b_k$. Then

$$\begin{aligned} a_{k+2} - b_{k+2} = \tfrac{1}{2}(a_{k+1} + b_{k+1}) - \sqrt{a_{k+1}b_{k+1}} &= \tfrac{1}{2}\left(a_{k+1} - 2\sqrt{a_{k+1}b_{k+1}} + b_{k+1}\right) \\ &= \tfrac{1}{2}\left(\sqrt{a_{k+1}} - \sqrt{b_{k+1}}\right)^2 > 0, \end{aligned}$$

$$a_{k+1} - a_{k+2} = a_{k+1} - \tfrac{1}{2}(a_{k+1} + b_{k+1}) = \tfrac{1}{2}(a_{k+1} - b_{k+1}) > 0, \text{ and}$$

$b_{k+1} - b_{k+2} = b_{k+1} - \sqrt{a_{k+1}b_{k+1}} = \sqrt{b_{k+1}}\left(\sqrt{b_{k+1}} - \sqrt{a_{k+1}}\right) < 0 \Rightarrow$
$a_{k+1} > a_{k+2} > b_{k+2} > b_{k+1}$, so the assertion is true for $n = k+1$. Thus, it is true for all n by mathematical induction.

(b) From part (a) we have $a > a_n > a_{n+1} > b_{n+1} > b_n > b$, which shows that both sequences, $\{a_n\}$ and $\{b_n\}$, are monotonic and bounded. So they are both convergent by the Monotonic Sequence Theorem.

(c) Let $\lim\limits_{n\to\infty} a_n = \alpha$ and $\lim\limits_{n\to\infty} b_n = \beta$. Then $\lim\limits_{n\to\infty} a_{n+1} = \lim\limits_{n\to\infty} \dfrac{a_n + b_n}{2} \Rightarrow \alpha = \dfrac{\alpha+\beta}{2} \Rightarrow$
$2\alpha = \alpha + \beta \Rightarrow \alpha = \beta$.

EXERCISES 10.2

1. **(a)** A sequence is an ordered list of numbers whereas a series is the *sum* of a list of numbers.

(b) A series is convergent if the sequence of partial sums is a convergent sequence. A series is divergent if it is not convergent.

3.

n	s_n
1	3.33333
2	4.44444
3	4.81481
4	4.93827
5	4.97942
6	4.99314
7	4.99771
8	4.99924
9	4.99975
10	4.99992
11	4.99997
12	4.99999

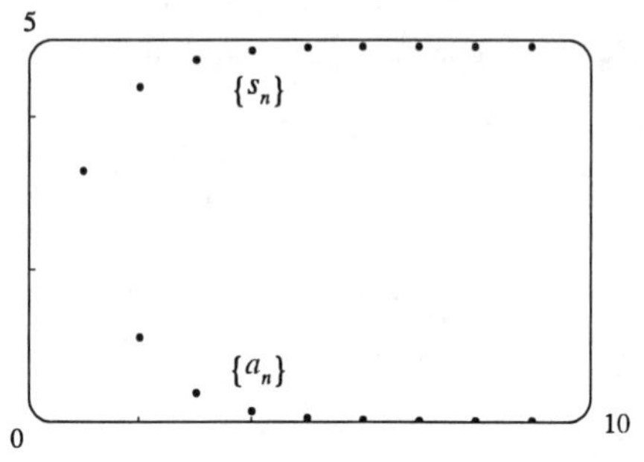

From the graph, it seems that the series converges. In fact, it is a geometric series with $a = \frac{10}{3}$ and $r = \frac{1}{3}$, so its sum is $\sum_{n=1}^{\infty} \frac{10}{3^n} = \frac{10/3}{1 - 1/3} = 5$. Note that the dot corresponding to $n = 1$ is part of both $\{a_n\}$ and $\{s_n\}$.

5.

n	s_n
1	0.50000
2	1.16667
3	1.91667
4	2.71667
5	3.55000
6	4.40714
7	5.28214
8	6.17103
9	7.07103
10	7.98012

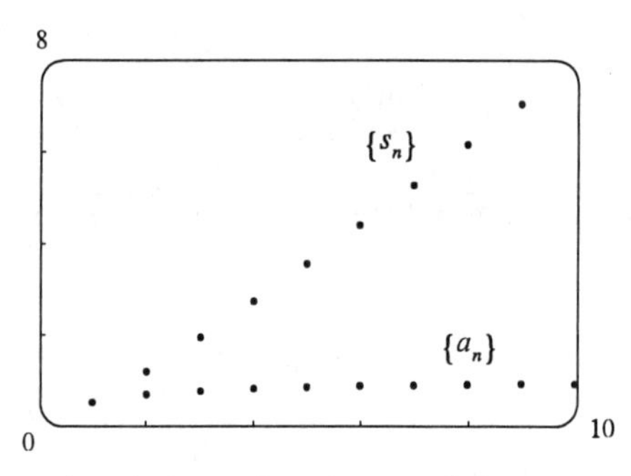

The series diverges, since its terms do not approach 0.

7.

n	s_n
1	0.64645
2	0.80755
3	0.87500
4	0.91056
5	0.93196
6	0.94601
7	0.95581
8	0.96296
9	0.96838
10	0.97259

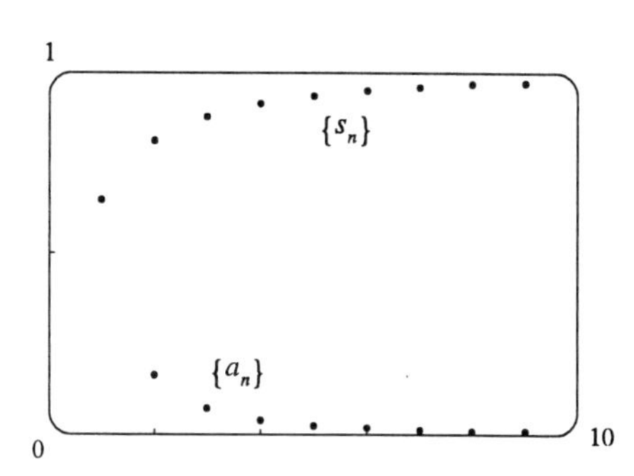

From the graph, it seems that the series converges to 1. To find the sum, we write

$$s_n = \sum_{i=1}^{n}\left(\frac{1}{i^{1.5}} - \frac{1}{(i+1)^{1.5}}\right) = \left(1 - \frac{1}{2^{1.5}}\right) + \left(\frac{1}{2^{1.5}} - \frac{1}{3^{1.5}}\right)$$
$$+ \left(\frac{1}{3^{1.5}} - \frac{1}{4^{1.5}}\right) + \cdots + \left(\frac{1}{n^{1.5}} - \frac{1}{(n+1)^{1.5}}\right) = 1 - \frac{1}{(n+1)^{1.5}}.$$

So the sum is $\lim_{n\to\infty} s_n = 1$.

9. **(a)** $\lim_{n\to\infty} a_n = \lim_{n\to\infty} \dfrac{2n}{3n+1} = \dfrac{2}{3}$, so the *sequence* $\{a_n\}$ is convergent by (10.1.1).

(b) Since $\lim_{n\to\infty} a_n = \frac{2}{3} \neq 0$, the *series* $\sum_{n=1}^{\infty} a_n$ is divergent by the Test for Divergence (7).

11. $4 + \frac{8}{5} + \frac{16}{25} + \frac{32}{125} + \cdots$ is a geometric series with $a = 4$ and $r = \frac{2}{5}$. Since $|r| = \frac{2}{5} < 1$, the series converges to $\frac{4}{1-2/5} = \frac{4}{3/5} = \frac{20}{3}$.

13. $\displaystyle\sum_{n=1}^{\infty} \frac{(-3)^{n-1}}{4^n} = \frac{1}{4}\sum_{n=1}^{\infty}\left(-\frac{3}{4}\right)^{n-1}$. The latter series is geometric with $a = 1$ and $r = -\frac{3}{4}$. Since $|r| = \frac{3}{4} < 1$, it converges to $\frac{1}{1-(-3/4)} = \frac{4}{7}$. Thus, the given series converges to $\left(\frac{1}{4}\right)\left(\frac{4}{7}\right) = \frac{1}{7}$.

15. For $\sum_{n=1}^{\infty} 3^{-n}8^{n+1} = \sum_{n=1}^{\infty} 8\left(\frac{8}{3}\right)^n$, $a = \frac{64}{3}$ and $r = \frac{8}{3} > 1$, so the series diverges.

17. $\sum_{n=1}^{\infty}\left[2\,(0.1)^n + (0.2)^n\right] = 2\sum_{n=1}^{\infty}(0.1)^n + \sum_{n=1}^{\infty}(0.2)^n$. These are convergent geometric series and so by Theorem 8, the sum is also convergent. $2\left(\dfrac{0.1}{1-0.1}\right) + \dfrac{0.2}{1-0.2} = \dfrac{2}{9} + \dfrac{1}{4} = \dfrac{17}{36}$

19. $\lim_{n\to\infty} a_n = \lim_{n\to\infty} \dfrac{n}{\sqrt{1+n^2}} = \lim_{n\to\infty} \dfrac{1}{\sqrt{1+1/n^2}} = 1 \neq 0$, so the series diverges by the Test for Divergence.

21. Converges. $s_n = \sum_{i=1}^{n} \frac{1}{i(i+2)} = \sum_{i=1}^{n} \left(\frac{1/2}{i} - \frac{1/2}{i+2}\right)$ (using partial fractions) $= \frac{1}{2}\sum_{i=1}^{n}\left(\frac{1}{i} - \frac{1}{i+2}\right)$. The latter sum is

$$\left(1-\tfrac{1}{3}\right)+\left(\tfrac{1}{2}-\tfrac{1}{4}\right)+\left(\tfrac{1}{3}-\tfrac{1}{5}\right)+\cdots+\left(\frac{1}{n-1}-\frac{1}{n+1}\right)+\left(\frac{1}{n}-\frac{1}{n+2}\right)$$
$$= 1+\frac{1}{2}-\frac{1}{n+1}-\frac{1}{n+2} \quad \text{(telescoping series).}$$

Thus, $\sum_{n=1}^{\infty} \frac{1}{n(n+2)} = \frac{1}{2}\lim_{n\to\infty}\left(1+\frac{1}{2}-\frac{1}{n+1}-\frac{1}{n+2}\right) = \frac{1}{2}\left(1+\frac{1}{2}\right) = \frac{3}{4}$.

23. Converges. $\sum_{n=1}^{\infty} \frac{3^n+2^n}{6^n} = \sum_{n=1}^{\infty}\left(\frac{3^n}{6^n}+\frac{2^n}{6^n}\right) = \sum_{n=1}^{\infty}\left[\left(\tfrac{1}{2}\right)^n+\left(\tfrac{1}{3}\right)^n\right] = \frac{1/2}{1-1/2}+\frac{1/3}{1-1/3} = 1+\frac{1}{2} = \frac{3}{2}$

25. Converges. $s_n = \left(\sin 1 - \sin \frac{1}{2}\right) + \left(\sin\frac{1}{2} - \sin\frac{1}{3}\right) + \cdots + \left(\sin\frac{1}{n} - \sin\frac{1}{n+1}\right) = \sin 1 - \sin\frac{1}{n+1}$, so $\sum_{n=1}^{\infty}\left(\sin\frac{1}{n} - \sin\frac{1}{n+1}\right) = \lim_{n\to\infty} s_n = \sin 1 - \sin 0 = \sin 1$.

27. $\lim_{n\to\infty} a_n = \lim_{n\to\infty} \arctan n = \frac{\pi}{2} \neq 0$, so the series diverges by the Test for Divergence.

29. $0.\overline{5} = 0.5 + 0.05 + 0.005 + \cdots = \frac{0.5}{1-0.1} = \frac{5}{9}$

31. $0.\overline{307} = 0.307 + 0.000307 + 0.000000307 + \cdots = \frac{0.307}{1-0.001} = \frac{307}{999}$

33. $\sum_{n=0}^{\infty} (x-3)^n$ is a geometric series with $r = x-3$, so it converges whenever $|x-3| < 1 \Leftrightarrow -1 < x-3 < 1 \Leftrightarrow 2 < x < 4$, and the sum is $\frac{1}{1-(x-3)} = \frac{1}{4-x}$.

35. $\sum_{n=0}^{\infty}\left(\frac{1}{x}\right)^n$ is geometric with $r = \frac{1}{x}$, so it converges whenever $\left|\frac{1}{x}\right| < 1 \Leftrightarrow |x| > 1 \Leftrightarrow x > 1$ or $x < -1$, and the sum is $\frac{1}{1-1/x} = \frac{x}{x-1}$.

37. After defining f, We use `convert(f,parfrac);` in Maple, `Apart` in Mathematica, or `Expand Rational` and `Simplify` in Derive to find that the general term is $\frac{1}{(4n+1)(4n-3)} = -\frac{1/4}{4n+1} + \frac{1/4}{4n-3}$. So the nth partial sum is

$$s_n = \sum_{k=1}^{n}\left(-\frac{1/4}{4k+1}+\frac{1/4}{4k-3}\right) = \frac{1}{4}\left(\frac{1}{4k-3}-\frac{1}{4k+1}\right)$$
$$= \frac{1}{4}\left[\left(1-\frac{1}{5}\right)+\left(\frac{1}{5}-\frac{1}{9}\right)+\left(\frac{1}{9}-\frac{1}{13}\right)+\cdots+\left(\frac{1}{4n-3}-\frac{1}{4n+1}\right)\right] = \frac{1}{4}\left(1-\frac{1}{4n+1}\right)$$

The series converges to $\lim_{n\to\infty} s_n = \frac{1}{4}$. This can be confirmed by directly computing the sum using `sum(f,1..infinity);` (in Maple), `Sum[f,{n,1,Infinity}]` (in Mathematica), or `Calculus Sum` (from 1 to ∞) and `Simplify` (in Derive).

39. For $n=1$, $a_1=0$ since $s_1=0$. For $n>1$,

$$a_n = s_n - s_{n-1} = \frac{n-1}{n+1} - \frac{(n-1)-1}{(n-1)+1} = \frac{(n-1)\,n-(n+1)\,(n-2)}{(n+1)\,n} = \frac{2}{n\,(n+1)}.$$

Also, $\sum\limits_{n=1}^{\infty} a_n = \lim\limits_{n\to\infty} s_n = \lim\limits_{n\to\infty} \dfrac{1-1/n}{1+1/n} = 1.$

41. **(a)** The first step in the chain occurs when the local government spends D dollars. The people who receive it spend a fraction c of those D dollars, that is, Dc dollars. Those who receive the Dc dollars spend a fraction c of it, that is, Dc^2 dollars. Continuing in this way, we see that the total spending after n transactions is

$S_n = D + Dc + Dc^2 + \cdots + Dc^{n\text{-}1} = \dfrac{D\,(1-c^n)}{1-c}$ by (3).

(b)
$$\lim_{n\to\infty} S_n = \lim_{n\to\infty} \frac{D\,(1-c^n)}{1-c} = \frac{D}{1-c} \lim_{n\to\infty} (1-c^n)$$
$$= \frac{D}{1-c} \quad (\text{since } 0<c<1 \;\Rightarrow\; \lim_{n\to\infty} c^n = 0)$$
$$= \frac{D}{s} \quad (\text{since } c+s=1) \quad = kD \quad (\text{since } k = 1/s).$$

If $c = 0.8$, then $s = 1-c = 0.2$ and the multiplier is $k = 1/s = 5$.

43. $\sum\limits_{n=2}^{\infty} (1+c)^{-n}$ is a geometric series with $a=(1+c)^{-2}$ and $r=(1+c)^{-1}$, so the series converges when $\left|(1+c)^{-1}\right|<1 \;\Rightarrow\; |1+c|>1 \;\Rightarrow\; 1+c>1$ or $1+c<-1 \;\Rightarrow\; c>0$ or $c<-2$. We calculate the sum of the series and set it equal to 2: $\dfrac{(1+c)^{-2}}{1-(1+c)^{-1}} = 2 \;\Leftrightarrow\; \left(\dfrac{1}{1+c}\right)^2 = 2-2\left(\dfrac{1}{1+c}\right) \;\Leftrightarrow$ $1 = 2\,(1+c)^2 - 2\,(1+c) = 0 \;\Leftrightarrow\; 2c^2+2c-1=0 \;\Leftrightarrow\; c = \frac{-2\pm\sqrt{12}}{4} = \frac{\pm\sqrt{3}-1}{2}$. However, the negative root is inadmissible because $-2 < \frac{-\sqrt{3}-1}{2} < 0$. So $c = \frac{\sqrt{3}-1}{2}$.

45.

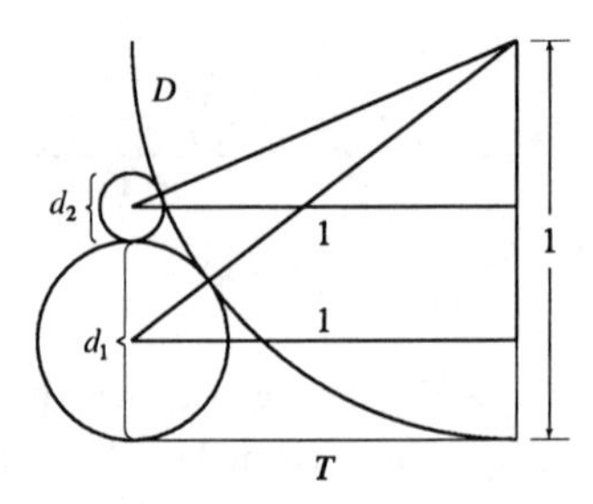

Let d_n be the diameter of C_n. We draw lines from the centers of the C_i to the center of D (or C), and using the Pythagorean Theorem, we can write $1^2 + \left(1 - \frac{1}{2}d_1\right)^2 = \left(1 + \frac{1}{2}d_1\right)^2 \Leftrightarrow$
$1 = \left(1 + \frac{1}{2}d_1\right)^2 - \left(1 - \frac{1}{2}d_1\right)^2 = 2d_1(\text{difference of squares}) \Rightarrow d_1 = \frac{1}{2}$. Similarly,
$1 = \left(1 + \frac{1}{2}d_2\right)^2 - \left(1 - d_1 - \frac{1}{2}d_2\right)^2 = 2d_2 + 2d_1 - d_1^2 - d_1 d_2 = (2 - d_1)(d_1 + d_2) \Leftrightarrow$
$d_2 = \dfrac{1}{2 - d_1} - d_1 = \dfrac{(1 - d_1)^2}{2 - d_1}$, $1 = \left(1 + \frac{1}{2}d_3\right)^2 - \left(1 - d_1 - d_2 - \frac{1}{2}d_3\right)^2 \Leftrightarrow d_3 = \dfrac{[1 - (d_1 + d_2)]^2}{2 - (d_1 + d_2)}$, and

in general, $d_{n+1} = \dfrac{\left(1 - \sum\limits_{i=1}^{n} d_i\right)^2}{2 - \sum\limits_{i=1}^{n} d_i}$. If we actually calculate d_2 and d_3 from the formulas above, we find that

they are $\frac{1}{6} = \frac{1}{2 \cdot 3}$ and $\frac{1}{12} = \frac{1}{3 \cdot 4}$ respectively, so we suspect that in general, $d_n = \dfrac{1}{n(n+1)}$. To prove this, we use induction: assume that for all $k \le n$, $d_k = \dfrac{1}{k(k+1)} = \dfrac{1}{k} - \dfrac{1}{k+1}$. Then

$\sum\limits_{i=1}^{n} d_i = 1 - \dfrac{1}{n+1} = \dfrac{n}{n+1}$ (telescoping sum). Substituting this into our formula for d_{n+1}, we get

$$d_{n+1} = \frac{\left[1 - \dfrac{n}{n+1}\right]^2}{2 - \left(\dfrac{n}{n+1}\right)} = \frac{\dfrac{1}{(n+1)^2}}{\dfrac{n+2}{n+1}} = \frac{1}{(n+1)(n+2)},$$ and the induction is complete.

Now, we observe that the partial sums $\sum\limits_{i=1}^{n} d_i$ of the diameters of the circles approach 1 as $n \to \infty$; that is,

$\sum\limits_{n=1}^{\infty} a_n = \sum\limits_{n=1}^{\infty} \dfrac{1}{n(n+1)} = 1$, which is what we wanted to prove.

47. The series $1 - 1 + 1 - 1 + 1 - 1 + \cdots$ diverges (geometric series with $r = -1$) so we cannot say $0 = 1 - 1 + 1 - 1 + 1 - 1 + \cdots$.

49. Suppose on the contrary that $\sum (a_n + b_n)$ converges. Then by Theorem 8(iii), so would $\sum [(a_n + b_n) - a_n] = \sum b_n$, a contradiction.

51. The partial sums $\{s_n\}$ form an increasing sequence, since $s_n - s_{n-1} = a_n > 0$ for all n. Also, the sequence $\{s_n\}$ is bounded since $s_n \le 1000$ for all n. So by Theorem 10.1.7 , the sequence of partial sums converges, that is, the series $\sum a_n$ is convergent.

53. **(a)** At the first step, only the interval $\left(\frac{1}{3},\frac{2}{3}\right)$ (length $\frac{1}{3}$) is removed. At the second step, we remove the intervals $\left(\frac{1}{9},\frac{2}{9}\right)$ and $\left(\frac{7}{9},\frac{8}{9}\right)$, which have a total length of $2\cdot\left(\frac{1}{3}\right)^2$. At the third step, we remove 2^2 intervals, each of length $\left(\frac{1}{3}\right)^3$. In general, at the nth step we remove 2^{n-1} intervals, each of length $\left(\frac{1}{3}\right)^n$, for a length of $2^{n-1}\cdot\left(\frac{1}{3}\right)^n=\frac{1}{3}\left(\frac{2}{3}\right)^{n-1}$. Thus, the total length of all removed intervals is $\sum\limits_{n=1}^{\infty}\frac{1}{3}\left(\frac{2}{3}\right)^{n-1}=\frac{1/3}{1-2/3}=1$(geometric series with $a=\frac{1}{3}$ and $r=\frac{2}{3}$). Notice that at the nth step, the leftmost interval that is removed is $\left(\left(\frac{1}{3}\right)^n,\left(\frac{2}{3}\right)^n\right)$, so we never remove 0, and 0 is in the Cantor set. Also, the rightmost interval removed is $\left(1-\left(\frac{2}{3}\right)^n,1-\left(\frac{1}{3}\right)^n\right)$, so 1 is never removed. Some other numbers in the Cantor set are $\frac{1}{3}$, $\frac{2}{3}$, $\frac{1}{9}$, $\frac{2}{9}$, $\frac{7}{9}$, and $\frac{8}{9}$.

(b) The area removed at the first step is $\frac{1}{9}$; at the second step, $8\cdot\left(\frac{1}{9}\right)^2$; at the third step, $(8)^2\cdot\left(\frac{1}{9}\right)^3$. In general, the area removed at the nth step is $(8)^{n-1}\left(\frac{1}{9}\right)^n=\frac{1}{9}\left(\frac{8}{9}\right)^{n-1}$, so the total area of all removed squares is $\sum\limits_{n=1}^{\infty}\frac{1}{9}\left(\frac{8}{9}\right)^{n-1}=\frac{1/9}{1-8/9}=1$.

55. **(a)** $\displaystyle\sum_{n=1}^{\infty}\frac{n}{(n+1)!}\quad\Rightarrow\quad s_1=\frac{1}{1\cdot 2}=\frac{1}{2},\ s_2=\frac{1}{2}+\frac{2}{1\cdot 2\cdot 3}=\frac{5}{6},\ s_3=\frac{5}{6}+\frac{3}{1\cdot 2\cdot 3\cdot 4}=\frac{23}{24},$

$\displaystyle s_4=\frac{23}{24}+\frac{4}{1\cdot 2\cdot 3\cdot 4\cdot 5}=\frac{119}{120}$. The denominators are $(n+1)!$, so a guess would be $\displaystyle s_n=\frac{(n+1)!-1}{(n+1)!}$.

(b) For $n=1$, $s_1=\dfrac{1}{2}=\dfrac{2!-1}{2!}$, so the formula holds for $n=1$. Assume $s_k=\dfrac{(k+1)!-1}{(k+1)!}$. Then

$$\begin{aligned} s_{k+1} &= \frac{(k+1)!-1}{(k+1)!}+\frac{k+1}{(k+2)!}=\frac{(k+1)!-1}{(k+1)!}+\frac{k+1}{(k+1)!\,(k+2)} \\ &= \frac{(k+2)!-(k+2)+k+1}{(k+2)!}=\frac{(k+2)!-1}{(k+2)!} \end{aligned}$$

Thus, the formula is true for $n=k+1$. So by induction, the guess is correct.

(c) $\displaystyle\lim_{n\to\infty}s_n=\lim_{n\to\infty}\frac{(n+1)!-1}{(n+1)!}=\lim_{n\to\infty}\left[1-\frac{1}{(n+1)!}\right]=1$ and so $\displaystyle\sum_{n=0}^{\infty}\frac{n}{(n+1)!}=1$.

EXERCISES 10.3

1.

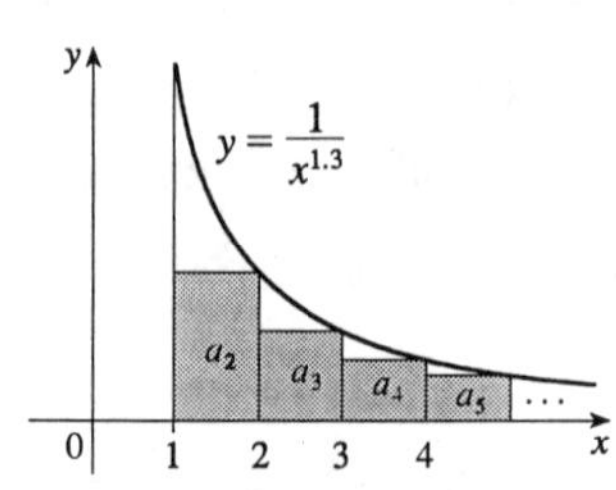

The picture shows that $a_2 = \frac{1}{2^{1.3}} < \int_1^2 \frac{1}{x^{1.3}}\,dx$, $a_3 = \frac{1}{3^{1.3}} < \int_2^3 \frac{1}{x^{1.3}}\,dx$, and so on, so

$\sum_{n=2}^{\infty} \frac{1}{n^{1.3}} < \int_1^{\infty} \frac{1}{x^{1.3}}\,dx$. The integral converges by (8.9.2) with $p = 1.3 > 1$, so the series converges.

3. **(a)** We cannot say anything about $\sum a_n$. If $a_n > b_n$ for all n and $\sum b_n$ is convergent, then $\sum a_n$ could be convergent or divergent.

(b) If $a_n < b_n$ for all n, then $\sum a_n$ is convergent. [This is part (a) of the Comparison Test.]

5. $\sum_{n=1}^{\infty} n^b$ is a p-series with $p = -b$. $\sum_{n=1}^{\infty} b^n$ is a geometric series. By (1), the p-series is convergent if $p > 1$. In this case, $\sum_{n=1}^{\infty} n^b = \sum_{n=1}^{\infty} \left(1/n^{-b}\right)$, so $-b > 1 \iff b < -1$ are the values for which the series converge. A geometric series $\sum_{n=1}^{\infty} ar^{n-1}$ converges if $|r| < 1$, so $\sum_{n=1}^{\infty} b^n$ converges if $|b| < 1 \iff -1 < b < 1$.

7. $f(x) = xe^{-x^2}$ is continuous and positive on $[1, \infty)$, and since $f'(x) = e^{-x^2}\left(1 - 2x^2\right) < 0$ for $x > 1$, f is decreasing as well. Thus, we can use the Integral Test.

$\int_1^{\infty} xe^{-x^2}\,dx = \lim_{t\to\infty} \left[-\frac{1}{2}e^{-x^2}\right]_1^t = 0 - \left(-\frac{e^{-1}}{2}\right) = \frac{1}{2e}$. Since the integral converges, the series converges.

9. $f(x) = \frac{x}{x^2+1}$ is continuous and positive on $[1, \infty)$, and since $f'(x) = \frac{1-x^2}{\left(x^2+1\right)^2} < 0$ for $x > 1$, f is also decreasing. Using the Integral Test, $\int_1^{\infty} \frac{x}{x^2+1}\,dx = \lim_{t\to\infty} \left[\frac{\ln\left(x^2+1\right)}{2}\right]_1^t = \infty$, so the series diverges.

11. $f(x) = \frac{1}{x \ln x}$ is continuous and positive on $[2, \infty)$, and also decreasing since $f'(x) = -\frac{1 + \ln x}{x^2 (\ln x)^2} < 0$ for $x > 2$, so we can use the Integral Test. $\int_2^{\infty} \frac{1}{x \ln x}\,dx = \lim_{t\to\infty} \left[\ln(\ln x)\right]_2^t = \lim_{t\to\infty} \left[\ln(\ln t) - \ln(\ln 2)\right] = \infty$, so the series diverges.

13. $\frac{1}{n^3+n^2} < \frac{1}{n^3}$ since $n^3 + n^2 > n^3$ for all n, and since $\sum_{n=1}^{\infty} \frac{1}{n^3}$ is a convergent p-series ($p = 3 > 1$),

$\sum_{n=1}^{\infty} \frac{1}{n^3+n^2}$ also converges by the Comparison Test [part (a)].

15. $\dfrac{1+5^n}{4^n} > \dfrac{5^n}{4^n} = \left(\dfrac{5}{4}\right)^n$. $\displaystyle\sum_{n=0}^{\infty}\left(\frac{5}{4}\right)^n$ is a divergent geometric series ($|r| = \frac{5}{4} > 1$), so $\displaystyle\sum_{n=0}^{\infty}\frac{1+5^n}{4^n}$ diverges by the Comparison Test.

17. $\dfrac{3}{n(n+3)} < \dfrac{3}{n^2}$. $\displaystyle\sum_{n=1}^{\infty}\frac{3}{n^2} = 3\sum_{n=1}^{\infty}\frac{1}{n^2}$ is a convergent p- series ($p = 2 > 1$), so $\displaystyle\sum_{n=1}^{\infty}\frac{3}{n(n+3)}$ converges by the Comparison Test.

19. Use the Limit Comparison Test with $a_n = \dfrac{1}{1+\sqrt{n}}$ and $b_n = \dfrac{1}{\sqrt{n}}$: $\displaystyle\lim_{n\to\infty}\frac{a_n}{b_n} = \lim_{n\to\infty}\frac{\sqrt{n}}{1+\sqrt{n}} = 1 > 0$. Since $\displaystyle\sum_{n=1}^{\infty}\frac{1}{\sqrt{n}}$ is a divergent p- series ($p = \frac{1}{2} \le 1$), $\displaystyle\sum_{n=1}^{\infty}\frac{1}{1+\sqrt{n}}$ also diverges.

21. Use the Limit Comparison Test with $a_n = \sin\left(\dfrac{1}{n}\right)$ and $b_n = \dfrac{1}{n}$:

$\displaystyle\lim_{n\to\infty}\frac{a_n}{b_n} = \lim_{n\to\infty}\frac{\sin(1/n)}{1/n} = \lim_{\theta\to 0}\frac{\sin\theta}{\theta} = 1 > 0$. Since $\displaystyle\sum_{n=1}^{\infty} b_n$ is the divergent harmonic series,

$\displaystyle\sum_{n=1}^{\infty}\sin\left(\frac{1}{n}\right)$ also diverges.

23. We have already shown (in Exercise 11) that when $p = 1$ the series $\displaystyle\sum_{n=2}^{\infty}\frac{1}{n(\ln n)^p}$ diverges, so assume $p \ne 1$.

$f(x) = \dfrac{1}{x(\ln x)^p}$ is continuous and positive on $[2,\infty)$, and $f'(x) = -\dfrac{p+\ln x}{x^2(\ln x)^{p+1}} < 0$ if $x > e^{-p}$, so that f

is eventually decreasing and we can use the Integral Test. $\displaystyle\int_2^{\infty}\frac{1}{x(\ln x)^p}\,dx = \lim_{t\to\infty}\left[\frac{(\ln x)^{1-p}}{1-p}\right]_2^t$ (for $p \ne 1$)

$= \displaystyle\lim_{t\to\infty}\left[\frac{(\ln t)^{1-p}}{1-p}\right] - \frac{(\ln 2)^{1-p}}{1-p}$.

This limit exists whenever $1 - p < 0 \quad\Leftrightarrow\quad p > 1$, so the series converges for $p > 1$.

25. (a) $f(x) = \dfrac{1}{x^2}$ is positive and continuous and $f'(x) = \dfrac{-2}{x^3}$ is negative for $x > 1$, and so the Integral Test applies. $\displaystyle\sum_{n=1}^{\infty}\frac{1}{n^2} \approx s_{10} = \frac{1}{1^2} + \frac{1}{2^2} + \frac{1}{3^2} + \cdots + \frac{1}{10^2} \approx 1.549768$.

$R_{10} \le \displaystyle\int_{10}^{\infty}\frac{1}{x^2}\,dx = \lim_{t\to\infty}\left[\frac{-1}{x}\right]_{10}^t = \lim_{t\to\infty}\left(-\frac{1}{t} + \frac{1}{10}\right) = \frac{1}{10}$, so the error is at most 0.1.

(b) $s_{10} + \displaystyle\int_{11}^{\infty}\frac{1}{x^2}\,dx \le s \le s_{10} + \int_{10}^{\infty}\frac{1}{x^2}\,dx \quad\Rightarrow\quad s_{10} + \frac{1}{11} \le s \le s_{10} + \frac{1}{10} \quad\Rightarrow$
$1.549768 + 0.090909 = 1.640677 \le s \le 1.549768 + 0.1 = 1.649768$, so we get $s \approx 1.64522$ (the average of 1.640677 and 1.649768) with error ≤ 0.005 (the maximum of $1.649768 - 1.64522$ and $1.64522 - 1.640677$, rounded up).

(c) $R_n \le \displaystyle\int_n^{\infty}\frac{1}{x^2}\,dx = \frac{1}{n}$. So $R_n < 0.001$ if $\dfrac{1}{n} < \dfrac{1}{1000} \quad\Leftrightarrow\quad n > 1000$.

27. $f(x) = x^{-3/2}$ is positive and continuous and $f'(x) = -\frac{3}{2}x^{-5/2}$ is negative for $x > 1$, so the Integral Test applies. From the end of Example 7, we see that the error is at most half the length of the interval. From (4), the interval is $\left(s_n + \int_{n+1}^{\infty} f(x)\,dx, s_n + \int_n^{\infty} f(x)\,dx\right)$, so its length is $\int_n^{\infty} f(x)\,dx - \int_{n+1}^{\infty} f(x)\,dx$. Thus, we need n such that

$$\begin{aligned} 0.01 &> \tfrac{1}{2}\left(\int_n^{\infty} x^{-3/2}\,dx - \int_{n+1}^{\infty} x^{-3/2}\,dx\right) \\ &= \frac{1}{2}\left(\lim_{t\to\infty}\left[\frac{-2}{\sqrt{x}}\right]_n^t - \lim_{t\to\infty}\left[\frac{-2}{\sqrt{x}}\right]_{n+1}^t\right) \\ &= \frac{1}{\sqrt{n}} - \frac{1}{\sqrt{n+1}} \quad \Leftrightarrow \quad n > 13.08 \end{aligned}$$

Again from the end of Example 7, we approximate s by the midpoint of this interval. In general, the midpoint is $\frac{1}{2}\left[\left(s_n + \int_{n+1}^{\infty} f(x)\,dx\right) + \left(s_n + \int_n^{\infty} f(x)\,dx\right)\right] = s_n + \frac{1}{2}\left(\int_{n+1}^{\infty} f(x)\,dx + \int_n^{\infty} f(x)\,dx\right)$. So using $n = 14$, we have $s \approx s_{14} + \frac{1}{2}\left(\int_{14}^{\infty} x^{-3/2}\,dx + \int_{15}^{\infty} x^{-3/2}\,dx\right) = 2.0872 + \frac{1}{\sqrt{14}} + \frac{1}{\sqrt{15}} \approx 2.6127$. Any larger value of n will also work. For instance, $s \approx s_{30} + \frac{1}{\sqrt{30}} + \frac{1}{\sqrt{31}} \approx 2.6124$.

29. $\displaystyle\sum_{n=1}^{10} \frac{1}{n^4+n^2} = \frac{1}{2} + \frac{1}{20} + \frac{1}{90} + \cdots + \frac{1}{10{,}100} \approx 0.567975$. Now $\dfrac{1}{n^4+n^2} < \dfrac{1}{n^4}$, so using the reasoning and notation of Example 7, the error is $R_{10} \le T_{10} = \displaystyle\sum_{n=11}^{\infty} \frac{1}{n^4} \le \int_{10}^{\infty} \frac{dx}{x^4} = \lim_{t\to\infty}\left[-\frac{x^{-3}}{3}\right]_{10}^t = \frac{1}{3000} = 0.000\overline{3}$.

31. **(a)** From the figure, $a_2 + a_3 + \cdots + a_n \le \int_1^n f(x)\,dx$, so with $f(x) = \dfrac{1}{x}$,

$$\frac{1}{2} + \frac{1}{3} + \frac{1}{4} + \cdots + \frac{1}{n} \le \int_1^n \frac{1}{x}\,dx = \ln n.$$

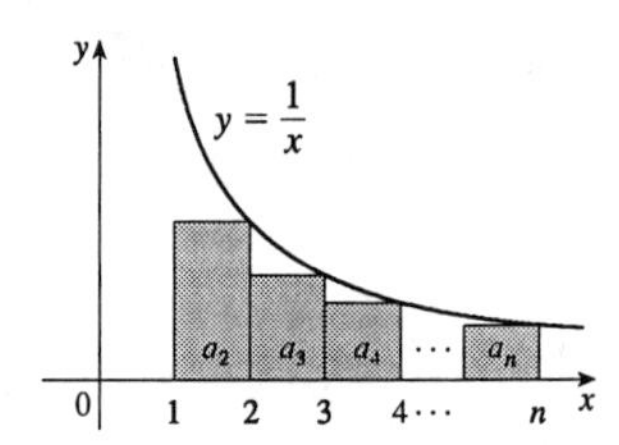

Thus, $s_n = 1 + \dfrac{1}{2} + \dfrac{1}{3} + \dfrac{1}{4} + \cdots + \dfrac{1}{n} \le 1 + \ln n$.

(b) By part (a), $s_{10^6} \le 1 + \ln 10^6 \approx 14.82 < 15$ and $s_{10^9} \le 1 + \ln 10^9 \approx 21.72 < 22$.

33. Since $\dfrac{d_n}{10^n} \le \dfrac{9}{10^n}$ for each n, and since $\displaystyle\sum_{n=1}^{\infty} \frac{9}{10^n}$ is a convergent geometric series ($|r| = \frac{1}{10} < 1$), $0.d_1d_2d_3\ldots = \displaystyle\sum_{n=1}^{\infty} \frac{d_n}{10^n}$ will always converge by the Comparison Test.

35. Yes. Since $\sum a_n$ converges, its terms approach 0 as $n \to \infty$, so $\displaystyle\lim_{n\to\infty} \frac{\sin a_n}{a_n} = 1$ by Theorem 3.4.4. Thus, $\sum \sin a_n$ converges by the Limit Comparison Test.

EXERCISES 10.4

1. **(a)** An alternating series is a series whose terms are alternately positive and negative.

(b) An alternating series $\sum_{n=1}^{\infty} (-1)^{n-1} b_n$ converges if $0 < b_{n+1} \le b_n$ for all n and $\lim_{n \to \infty} b_n = 0$. (This is the Alternating Series Test.)

(c) The error involved in using the partial sum s_n as an approximation to the total sum s is the remainder $R_n = s - s_n$ and the size of the error is smaller than b_{n+1}, that is, $|R_n| \le b_{n+1}$. (This is the Alternating Series Estimation Theorem.)

3. $\sum_{n=1}^{\infty} (-1)^{n-1} \frac{3}{n+4}$. $b_n = \frac{3}{n+4} > 0$ and $b_{n+1} \le b_n$ for all n; $\lim_{n\to\infty} b_n = 0$, so the series converges by the Alternating Series Test.

5. $\sum_{n=1}^{\infty} (-1)^{n+1} \frac{n}{5n+1}$. $\lim_{n\to\infty} \frac{n}{5n+1} = \frac{1}{5}$, so $\lim_{n\to\infty} (-1)^{n+1} \frac{n}{5n+1}$ does not exist and the series diverges by the Test for Divergence.

7. $\sum_{n=1}^{\infty} (-1)^n \frac{n}{n^2+1}$. $b_n = \frac{n}{n^2+1} > 0$ for all n. $b_{n+1} \le b_n \Leftrightarrow \frac{n+1}{(n+1)^2+1} \le \frac{n}{n^2+1} \Leftrightarrow$ $(n+1)(n^2+1) \le \left[(n+1)^2+1\right] n \Leftrightarrow n^3+n^2+n+1 \le n^3+2n^2+2n \Leftrightarrow 0 \le n^2+n-1$, which is true for all $n \ge 1$. Also, $\lim_{n\to\infty} b_n = \lim_{n\to\infty} \frac{n}{n^2+1} = \lim_{n\to\infty} \frac{1/n}{1+1/n^2} = 0$. Therefore, the series converges by the Alternating Series Test.

9. $\sum_{n=1}^{\infty} \frac{(-1)^{n-1}}{n} = 1 - \frac{1}{2} + \frac{1}{3} - \frac{1}{4} + \cdots + \frac{1}{49} - \frac{1}{50} + \frac{1}{51} - \frac{1}{52} + \cdots$. The 50th partial sum of this series is an underestimate, since $\sum_{n=1}^{\infty} \frac{(-1)^{n-1}}{n} = s_{50} + \left(\frac{1}{51} - \frac{1}{52}\right) + \left(\frac{1}{53} - \frac{1}{54}\right) + \cdots$, and the terms in parentheses are all positive. The result can be seen geometrically in Figure 1.

11. If $p > 0$, $\frac{1}{(n+1)^p} \le \frac{1}{n^p}$ and $\lim_{n\to\infty} \frac{1}{n^p} = 0$, so the series converges by the Alternating Series Test. If $p \le 0$, $\lim_{n\to\infty} \frac{(-1)^{n-1}}{n^p}$ does not exist, so the series diverges by the Test for Divergence.

Thus, $\sum_{n=1}^{\infty} \frac{(-1)^{n-1}}{n^p}$ converges $\Leftrightarrow$ $p > 0$.

13. $b_7 = 2^7/7! \approx 0.025 > 0.01$ and $b_8 = 2^8/8! \approx 0.006 < 0.01$, so by the Alternating Series Estimation Theorem, $n = 7$. (That is, since the 8th term is less than the desired error, we need to add the first 7 terms to get the sum to the desired accuracy.)

15.

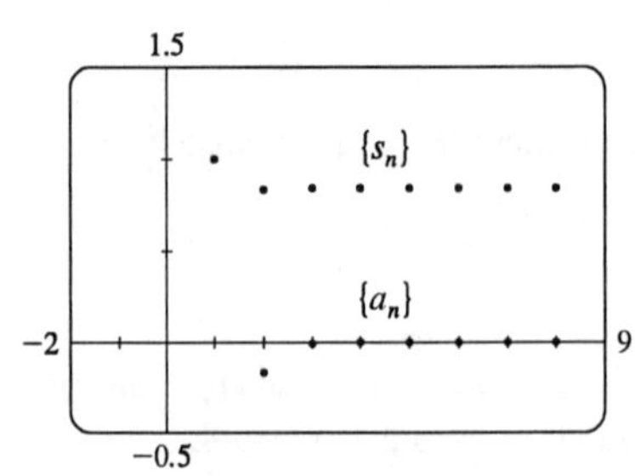

The graph gives us an estimate for the sum of the series $\sum_{n=1}^{\infty} \frac{(-1)^{n-1}}{(2n-1)!}$ of 0.84.

$$b_5 = \frac{1}{(2 \cdot 5 - 1)!} = \frac{1}{362{,}880} \approx 0.000003 < 0.00001, \text{ so } \sum_{n=1}^{\infty} \frac{(-1)^{n-1}}{(2n-1)!} \approx \sum_{n=1}^{4} \frac{(-1)^{n-1}}{(2n-1)!} \approx 0.8415.$$

17. $b_6 = \frac{1}{2^6 6!} = \frac{1}{46{,}080} \approx 0.000022 < 0.00001$, so $\sum_{n=0}^{\infty} \frac{(-1)^n}{2^n n!} \approx \sum_{n=0}^{5} \frac{(-1)^n}{2^n n!} \approx 0.6065.$

19. Using the Ratio Test, $\lim_{n\to\infty} \left| \frac{a_{n+1}}{a_n} \right| = \lim_{n\to\infty} \left| \frac{(-3)^{n+1}/(n+1)^3}{(-3)^n/n^3} \right| = 3 \lim_{n\to\infty} \left(\frac{n}{n+1} \right)^3 = 3 > 1$, so the series diverges.

21. $\sum_{n=1}^{\infty} \frac{1}{2n+1}$ diverges (use the Integral Test or the Limit Comparison Test with $b_n = 1/n$).

23. $\left| \frac{\sin 2n}{n^2} \right| \leq \frac{1}{n^2}$ and $\sum_{n=1}^{\infty} \frac{1}{n^2}$ converges (p-series, $p = 2 > 1$), so $\sum_{n=1}^{\infty} \frac{\sin 2n}{n^2}$ converges absolutely by the Comparison Test.

25. $\lim_{n\to\infty} \left| \frac{a_{n+1}}{a_n} \right| = \lim_{n\to\infty} \frac{(n+2)\,5^{n+1} / \left[(n+1)\,3^{2(n+1)}\right]}{(n+1)\,5^n / (n 3^{2n})} = \lim_{n\to\infty} \frac{5n\,(n+2)}{9\,(n+1)^2} = \frac{5}{9} < 1$, so the series converges absolutely by the Ratio Test.

27. $\lim_{n\to\infty} \left| \frac{a_{n+1}}{a_n} \right| = \lim_{n\to\infty} \frac{(n+3)!/\left[(n+1)!10^{n+1}\right]}{(n+2)!/(n!10^n)} = \frac{1}{10} \lim_{n\to\infty} \frac{n+3}{n+1} = \frac{1}{10} < 1$, so the series converges absolutely by the Ratio Test.

29. By the recursive definition, $\lim_{n\to\infty} \left| \frac{a_{n+1}}{a_n} \right| = \lim_{n\to\infty} \left| \frac{5n+1}{4n+3} \right| = \frac{5}{4} > 1$, so the series diverges by the Ratio Test.

31. **(a)** $\lim_{n\to\infty}\left|\dfrac{1/(n+1)^3}{1/n^3}\right| = \lim_{n\to\infty}\dfrac{n^3}{(n+1)^3} = \lim_{n\to\infty}\dfrac{1}{(1+1/n)^3} = 1$. Inconclusive.

(b) $\lim_{n\to\infty}\left|\dfrac{(n+1)}{2^{n+1}}\cdot\dfrac{2^n}{n}\right| = \lim_{n\to\infty}\dfrac{n+1}{2n} = \lim_{n\to\infty}\left(\dfrac{1}{2}+\dfrac{1}{2n}\right) = \dfrac{1}{2}$. Conclusive (convergent).

(c) $\lim_{n\to\infty}\left|\dfrac{(-3)^n}{\sqrt{n+1}}\cdot\dfrac{\sqrt{n}}{(-3)^{n-1}}\right| = 3\lim_{n\to\infty}\sqrt{\dfrac{n}{n+1}} = 3\lim_{n\to\infty}\sqrt{\dfrac{1}{1+1/n}} = 3$. Conclusive (divergent).

(d) $\lim_{n\to\infty}\left|\dfrac{\sqrt{n+1}}{1+(n+1)^2}\cdot\dfrac{1+n^2}{\sqrt{n}}\right| = \lim_{n\to\infty}\left[\sqrt{1+\dfrac{1}{n}}\cdot\dfrac{1/n^2+1}{1/n^2+(1+1/n)^2}\right] = 1$. Inconclusive.

33. **(a)** $\lim_{n\to\infty}\left|\dfrac{a_{n+1}}{a_n}\right| = \lim_{n\to\infty}\dfrac{|x|^{n+1}/(n+1)!}{|x|^n/n!} = |x|\lim_{n\to\infty}\dfrac{1}{n+1} = 0$, so by the Ratio Test the series converges for all x.

(b) Since the series of part (a) always converges, we must have $\lim_{n\to\infty}\dfrac{x^n}{n!} = 0$ by Theorem 10.2.6.

35. **(a)** $s_5 = \sum_{n=1}^{5}\dfrac{1}{n2^n} = \dfrac{1}{2}+\dfrac{1}{8}+\dfrac{1}{24}+\dfrac{1}{64}+\dfrac{1}{160} = \dfrac{661}{960} \approx 0.68854$. Now the ratios

$r_n = \dfrac{a_{n+1}}{a_n} = \dfrac{n2^n}{(n+1)\,2^{n+1}} = \dfrac{n}{2\,(n+1)}$ form an increasing sequence, since

$r_{n+1}-r_n = \dfrac{n+1}{2\,(n+2)} - \dfrac{n}{2\,(n+1)} = \dfrac{(n+1)^2 - n\,(n+2)}{2\,(n+1)\,(n+2)} = \dfrac{1}{2\,(n+1)\,(n+2)} > 0$. So by Exercise 34(b), the error is less than $\dfrac{a_6}{1-\lim_{n\to\infty} r_n} = \dfrac{1/(6\cdot 2^6)}{1-1/2} = \dfrac{1}{192} \approx 0.00521$.

(b) The error in using s_n as an approximation to the sum is $R_n = \dfrac{a_{n+1}}{1-\frac{1}{2}} = \dfrac{2}{(n+1)\,2^{n+1}}$. We want

$R_n < 0.00005 \quad\Leftrightarrow\quad \dfrac{1}{(n+1)\,2^n} < 0.00005 \quad\Leftrightarrow\quad (n+1)\,2^n > 20{,}000$. To find such an n we can use

trial and error or a graph. We calculate $(11+1)\,2^{11} = 24{,}576$, so $s_{11} = \sum_{n=1}^{11}\dfrac{1}{n2^n} \approx 0.693109$ is within

0.00005 of the actual sum.

EXERCISES 10.5

Note: "R" stands for "radius of convergence" and "I" stands for "interval of convergence" in this section.

1. A power series is a series of the form $\sum_{n=0}^{\infty} c_n x^n = c_0 + c_1 x + c_2 x^2 + c_3 x^3 + \cdots$, where x is a variable and the c_n's are constants called the coefficients of the series.
More generally, a series of the form $\sum_{n=0}^{\infty} c_n (x-a)^n = c_0 + c_1 (x-a) + c_2 (x-a)^2 + \cdots$ is called a power series in $(x-a)$ or a power series centered at a or a power series about a.

3. **(a)** We are given that the power series $\sum_{n=0}^{\infty} c_n x^n$ is convergent for $x = 4$. So by Theorem 3 it must converge for at least $-4 < x \le 4$. In particular, it converges when $x = -2$, that is, $\sum_{n=0}^{\infty} c_n (-2)^n$ is convergent.

(b) It does not follow that $\sum_{n=0}^{\infty} c_n (-4)^n$ is necessarily convergent. [See the comments after Theorem 3 about convergence at the endpoint of an interval. An example is $c_n = (-1)^n / (n4^n)$.]

5. If $a_n = \dfrac{x^n}{n+2}$, then $\lim\limits_{n\to\infty} \left|\dfrac{a_{n+1}}{a_n}\right| = \lim\limits_{n\to\infty} \left|\dfrac{x^{n+1}}{n+3} \cdot \dfrac{n+2}{x^n}\right| = |x| \lim\limits_{n\to\infty} \dfrac{n+2}{n+3} = |x| < 1$ for convergence (by the Ratio Test), and $R = 1$. When $x = 1$, the series is $\sum\limits_{n=0}^{\infty} \dfrac{1}{n+2}$ which diverges (Integral Test or Comparison Test), and when $x = -1$, it is $\sum\limits_{n=0}^{\infty} \dfrac{(-1)^n}{n+2}$ which converges (Alternating Series Test), so $I = [-1, 1)$.

7. If $a_n = \dfrac{x^n}{n!}$, then $\lim\limits_{n\to\infty} \left|\dfrac{a_{n+1}}{a_n}\right| = \lim\limits_{n\to\infty} \left|\dfrac{x^{n+1}/(n+1)!}{x^n/n!}\right| = |x| \lim\limits_{n\to\infty} \dfrac{1}{n+1} = 0 < 1$ for all x. So, by the Ratio Test, $R = \infty$, and $I = (-\infty, \infty)$.

9. If $a_n = \dfrac{(-1)^n x^n}{n2^n}$, then $\lim\limits_{n\to\infty} \left|\dfrac{a_{n+1}}{a_n}\right| = \lim\limits_{n\to\infty} \left|\dfrac{x^{n+1}/\left[(n+1)\,2^{n+1}\right]}{x^n/(n2^n)}\right| = \left|\dfrac{x}{2}\right| \lim\limits_{n\to\infty} \dfrac{n}{n+1} = \left|\dfrac{x}{2}\right| < 1$ for convergence, so $|x| < 2$ and $R = 2$. When $x = 2$, $\sum\limits_{n=1}^{\infty} \dfrac{(-1)^n x^n}{n2^n} = \sum\limits_{n=1}^{\infty} \dfrac{(-1)^n}{n}$, which converges by the Alternating Series Test. When $x = -2$, $\sum\limits_{n=1}^{\infty} \dfrac{(-1)^n x^n}{n2^n} = \sum\limits_{n=1}^{\infty} \dfrac{1}{n}$, which diverges (harmonic series), so $I = (-2, 2]$.

11. If $a_n = \dfrac{n}{4^n}(2x-1)^n$, then

$$\left|\frac{a_{n+1}}{a_n}\right| = \left|\frac{(n+1)(2x-1)^{n+1}}{4^{n+1}} \cdot \frac{4^n}{n(2x-1)^n}\right| = \left|\frac{2x-1}{4}\left(1+\frac{1}{n}\right)\right| \to \tfrac{1}{2}\left|x-\tfrac{1}{2}\right| \text{ as } n\to\infty.$$

For convergence, $\frac{1}{2}\left|x - \frac{1}{2}\right| < 1 \;\Rightarrow\; \left|x-\frac{1}{2}\right| < 2 \;\Rightarrow\; R = 2$ and $-2 < x - \frac{1}{2} < 2 \;\Rightarrow\; -\frac{3}{2} < x < \frac{5}{2}$. If $x = -\frac{3}{2}$, the series becomes $\sum\limits_{n=0}^{\infty} \dfrac{n}{4^n}(-4)^n = \sum\limits_{n=0}^{\infty} (-1)^n n$, which is divergent by the Test for Divergence. If $x = \frac{5}{2}$, the series is $\sum\limits_{n=0}^{\infty} \dfrac{n}{4^n} 4^n = \sum\limits_{n=0}^{\infty} n$, also divergent by the Test for Divergence. So $I = \left(-\frac{3}{2}, \frac{5}{2}\right)$.

13. If $a_n = \dfrac{(-1)^n (x-1)^n}{\sqrt{n}}$, then

$$\lim_{n\to\infty}\left|\frac{a_{n+1}}{a_n}\right| = \lim_{n\to\infty}\left|\frac{(x-1)^{n+1}}{\sqrt{n+1}}\cdot\frac{\sqrt{n}}{(x-1)^n}\right| = |x-1|\lim_{n\to\infty}\sqrt{\frac{n}{n+1}}$$
$$= |x-1| < 1 \text{ for convergence, or } 0 < x < 2, \text{ and } R = 1.$$

When $x = 0$, $\displaystyle\sum_{n=1}^{\infty}\frac{(-1)^n (x-1)^n}{\sqrt{n}} = \sum_{n=1}^{\infty}\frac{1}{\sqrt{n}}$ which is a divergent p-series ($p = \frac{1}{2} \leq 1$). When $x = 2$, the series is $\displaystyle\sum_{n=1}^{\infty}\frac{(-1)^n}{\sqrt{n}}$, which converges by the Alternating Series Test. So $I = (0, 2]$.

15. If $a_n = \dfrac{2^n (x-3)^n}{n+3}$, then

$$\lim_{n\to\infty}\left|\frac{a_{n+1}}{a_n}\right| = \lim_{n\to\infty}\left|\frac{2^{n+1}(x-3)^{n+1}}{n+4}\cdot\frac{n+3}{2^n (x-3)^n}\right| = 2|x-3|\lim_{n\to\infty}\frac{n+3}{n+4}$$
$$= 2|x-3| < 1 \text{ for convergence, or}$$

$|x-3| < \frac{1}{2} \;\Leftrightarrow\; \frac{5}{2} < x < \frac{7}{2}$, and $R = \frac{1}{2}$. When $x = \frac{5}{2}$, $\displaystyle\sum_{n=0}^{\infty}\frac{2^n (x-3)^n}{n+3} = \sum_{n=0}^{\infty}\frac{(-1)^n}{n+3}$ which converges by the Alternating Series Test. When $x = \frac{7}{2}$, $\displaystyle\sum_{n=0}^{\infty}\frac{2^n (x-3)^n}{n+3} = \sum_{n=0}^{\infty}\frac{1}{n+3}$, similar to the harmonic series, which diverges. So $I = \left[\frac{5}{2}, \frac{7}{2}\right)$.

17. If $a_n = n!\,(2x-1)^n$, then
$\displaystyle\lim_{n\to\infty}\left|\frac{a_{n+1}}{a_n}\right| = \lim_{n\to\infty}\left|\frac{(n+1)!\,(2x-1)^{n+1}}{n!\,(2x-1)^n}\right| = \lim_{n\to\infty}(n+1)|2x-1| \to \infty$ as $n\to\infty$ for all $x \neq \frac{1}{2}$. Since the series diverges for all $x \neq \frac{1}{2}$, $R = 0$ and $I = \left\{\frac{1}{2}\right\}$.

19. If $a_n = \dfrac{(n!)^k}{(kn)!}x^n$, then

$$\lim_{n\to\infty}\left|\frac{a_{n+1}}{a_n}\right| = \lim_{n\to\infty}\frac{[(n+1)!]^k\,(kn)!}{(n!)^k\,[k(n+1)]!}|x|$$
$$= \lim_{n\to\infty}\frac{(n+1)^k}{(kn+k)(kn+k-1)\cdots(kn+2)(kn+1)}|x|$$
$$= \lim_{n\to\infty}\left[\frac{(n+1)}{(kn+1)}\frac{(n+1)}{(kn+2)}\cdots\frac{(n+1)}{(kn+k)}\right]|x|$$
$$= \lim_{n\to\infty}\left[\frac{n+1}{kn+1}\right]\lim_{n\to\infty}\left[\frac{n+1}{kn+2}\right]\cdots\lim_{n\to\infty}\left[\frac{n+1}{kn+k}\right]|x|$$
$$= \left(\frac{1}{k}\right)^k|x| < 1$$

$\Leftrightarrow\;|x| < k^k$ for convergence, and the radius of convergence is $R = k^k$.

21. **(a)** If $a_n = \dfrac{(-1)^n x^{2n+1}}{n!\,(n+1)!2^{2n+1}}$, then $\lim\limits_{n\to\infty}\left|\dfrac{a_{n+1}}{a_n}\right| = \left(\dfrac{x}{2}\right)^2 \lim\limits_{n\to\infty}\dfrac{1}{(n+1)(n+2)} = 0$ for all x. So $J_1(x)$ converges for all x; the domain is $(-\infty,\infty)$.

(b),(c) The initial terms of $J_1(x)$ up to $n=5$ are $a_0 = \dfrac{x}{2}$, $a_1 = -\dfrac{x^3}{16}$, $a_2 = \dfrac{x^5}{384}$, $a_3 = -\dfrac{x^7}{18{,}432}$,

$a_4 = \dfrac{x^9}{1{,}474{,}560}$, and $a_5 = -\dfrac{x^{11}}{176{,}947{,}200}$.

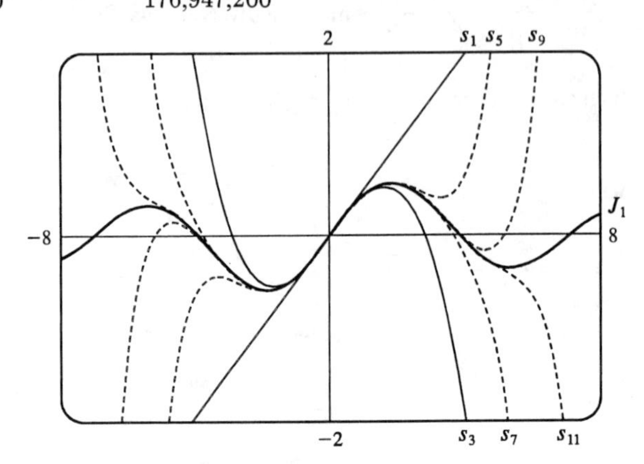

The partial sums seem to approximate $J_1(x)$ well near the origin, but as $|x|$ increases, we need to take a large number of terms to get a good approximation.

23. $s_{2n-1} = 1 + 2x + x^2 + 2x^3 + \cdots + x^{2n-2} + 2x^{2n-1} = (1+2x)\left(1 + x^2 + x^4 + \cdots + x^{2n-2}\right)$

$$= (1+2x)\frac{1-x^{2n}}{1-x^2} \quad \text{[by (10.2.3) with } r = x^2\text{]} \to \frac{1+2x}{1-x^2} \text{ as } n\to\infty \text{ [by (10.2.4)],}$$

when $|x|<1$. Also $s_{2n} = s_{2n-1} + x^{2n} \to \dfrac{1+2x}{1-x^2}$ since $x^{2n}\to 0$ for $|x|<1$. Therefore, $s_n \to \dfrac{1+2x}{1-x^2}$ since s_{2n} and s_{2n-1} both approach $\dfrac{1+2x}{1-x^2}$ as $n\to\infty$. Thus, the interval of convergence is $(-1,1)$ and $f(x) = \dfrac{1+2x}{1-x^2}$.

25. $\sum (c_n + d_n)\,x^n = \sum c_n x^n + \sum d_n x^n$ on the interval $(-2,2)$, since both series converge there. So the radius of convergence must be at least 2. Now since $\sum c_n x^n$ has $R=2$, it must diverge either at $x=-2$ or at $x=2$. So by Exercise 10.2.49, $\sum (c_n + d_n)\,x^n$ diverges either at $x=-2$ or at $x=2$, and so its radius of convergence is 2.

EXERCISES 10.6

Note: "R" stands for "radius of convergence" and "I" stands for "interval of convergence" in this section.

1. If $f(x) = \sum_{n=0}^{\infty} c_n x^n$ has radius of convergence 10, then $f'(x) = \sum_{n=1}^{\infty} n c_n x^{n-1}$ also has radius of convergence 10 by Theorem 2.

3. $f(x) = \dfrac{1}{1+x} = \dfrac{1}{1-(-x)} = \sum_{n=0}^{\infty} (-x)^n = \sum_{n=0}^{\infty} (-1)^n x^n$ with $|-x| < 1 \Leftrightarrow |x| < 1$, so $R = 1$ and $I = (-1, 1)$.

5. $f(x) = \dfrac{1}{1+4x^2} = \sum_{n=0}^{\infty} (-1)^n (4x^2)^n$ (substituting $4x^2$ for x in the series from Exercise 3)

$= \sum_{n=0}^{\infty} (-1)^n 4^n x^{2n}$, with $|4x^2| < 1 \Leftrightarrow x^2 < \frac{1}{4} \Leftrightarrow |x| < \frac{1}{2}$, so $R = \frac{1}{2}$ and $I = \left(-\frac{1}{2}, \frac{1}{2}\right)$.

7. $f(x) = \dfrac{x}{x-3} = 1 + \dfrac{3}{x-3} = 1 - \dfrac{1}{1-x/3} = 1 - \sum_{n=0}^{\infty} \left(\dfrac{x}{3}\right)^n = -\sum_{n=1}^{\infty} \left(\dfrac{x}{3}\right)^n$ (since the 0th term of the series is 1). For convergence, $\dfrac{|x|}{3} < 1 \Leftrightarrow |x| < 3$, so $R = 3$ and $I = (-3, 3)$.

Another Method: $\dfrac{x}{x-3} = -\dfrac{x}{3(1-x/3)} = -\dfrac{x}{3}\sum_{n=0}^{\infty} \left(\dfrac{x}{3}\right)^n = -\sum_{n=0}^{\infty} \dfrac{x^{n+1}}{3^{n+1}} = -\sum_{n=1}^{\infty} \dfrac{x^n}{3^n}$

9. $f(x) = \dfrac{1}{(1+x)^2} = -\dfrac{d}{dx}\left(\dfrac{1}{1+x}\right) = -\dfrac{d}{dx}\left(\sum_{n=0}^{\infty} (-1)^n x^n\right)$ (from Exercise 3)

$= \sum_{n=1}^{\infty} (-1)^{n+1} n x^{n-1}$ [from Theorem 2(a)] $= \sum_{n=0}^{\infty} (-1)^n (n+1) x^n$ with $R = 1$.

11. $f(x) = \dfrac{1}{(1+x)^3} = -\dfrac{1}{2}\dfrac{d}{dx}\left[\dfrac{1}{(1+x)^2}\right] = -\dfrac{1}{2}\dfrac{d}{dx}\left[\sum_{n=0}^{\infty} (-1)^n (n+1) x^n\right]$ (from Exercise 9)

$= -\frac{1}{2}\sum_{n=1}^{\infty} (-1)^n (n+1) n x^{n-1} = \frac{1}{2}\sum_{n=0}^{\infty} (-1)^n (n+2)(n+1) x^n$ with $R = 1$.

13. $f(x) = \ln(5-x) = -\displaystyle\int \dfrac{dx}{5-x} = -\dfrac{1}{5}\int \dfrac{dx}{1-x/5}$

$= -\dfrac{1}{5}\displaystyle\int \left[\sum_{n=0}^{\infty} \left(\dfrac{x}{5}\right)^n\right] dx = C - \dfrac{1}{5}\sum_{n=0}^{\infty} \dfrac{x^{n+1}}{5^n (n+1)} = C - \sum_{n=1}^{\infty} \dfrac{x^n}{n5^n}$

Putting $x = 0$, we get $C = \ln 5$. The series converges for $|x/5| < 1 \Leftrightarrow |x| < 5$, so $R = 5$.

15. $f(x) = \ln(3+x) = \int \frac{dx}{3+x} = \frac{1}{3}\int \frac{dx}{1+x/3} = \frac{1}{3}\int \sum_{n=0}^{\infty} (-1)^n \left(\frac{x}{3}\right)^n dx$ (from Exercise 3)

$$= C + \frac{1}{3}\sum_{n=0}^{\infty} \frac{(-1/3)^n}{n+1} x^{n+1} = \ln 3 + \frac{1}{3}\sum_{n=1}^{\infty} \frac{(-1/3)^{n-1}}{n} x^n \quad [C = f(0) = \ln 3]$$

$$= \ln 3 + \sum_{n=1}^{\infty} \frac{(-1)^{n-1}}{n3^n} x^n \text{ with } R = 3.$$

The terms of the series are $a_0 = \ln 3, a_1 = \frac{x}{3}, a_2 = -\frac{x^2}{18}, a_3 = \frac{x^3}{81}, a_4 = -\frac{x^4}{324}, a_5 = \frac{x^5}{1215}, \ldots$.

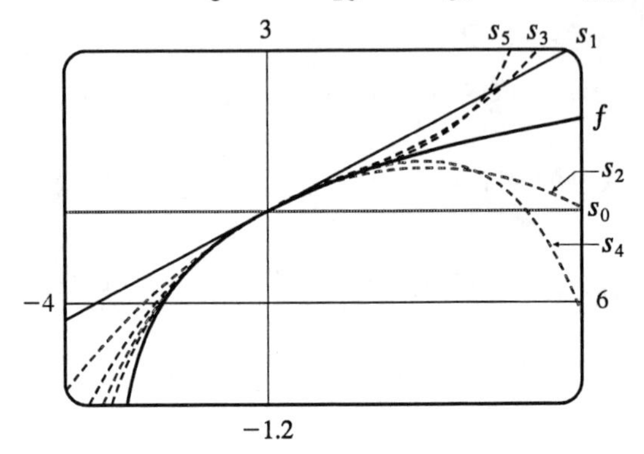

As n increases, $s_n(x)$ approximates f better on the interval of convergence, which is $(-3, 3)$.

17. $f(x) = \ln\left(\frac{1+x}{1-x}\right) = \ln(1+x) - \ln(1-x) = \int \frac{dx}{1+x} + \int \frac{dx}{1-x}$

$$= \int \left[\sum_{n=0}^{\infty} (-1)^n x^n + \sum_{n=0}^{\infty} x_n\right] dx = \int \sum_{n=0}^{\infty} 2x^{2n}\, dx = \sum_{n=0}^{\infty} \frac{2x^{2n+1}}{2n+1} + C$$

But $f(0) = \ln\frac{1}{1} = 0$, so $C = 0$ and we have $f(x) = \sum_{n=0}^{\infty} \frac{2x^{2n+1}}{2n+1}$ with $R = 1$. If $x = \pm 1$, then $f(x) = \pm 2\sum_{n=0}^{\infty} \frac{1}{2n+1}$, which both diverge by the Limit Comparison Test with $b_n = \frac{1}{n}$.

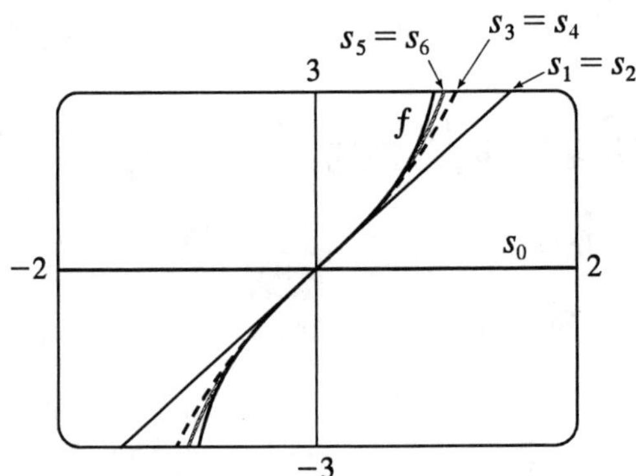

As n increases, $s_n(x)$ approximates f better on the interval of convergence, which is $(-1, 1)$.

19. $\displaystyle\int \frac{dx}{1+x^4} = \int \sum_{n=0}^{\infty} (-1)^n x^{4n}\, dx = C + \sum_{n=0}^{\infty} \frac{(-1)^n x^{4n+1}}{4n+1}$ with $R = 1$.

21. By Example 7, $\arctan x = \displaystyle\sum_{n=0}^{\infty} (-1)^n \frac{x^{2n+1}}{2n+1}$, so

$$\int \frac{\arctan x}{x}\, dx = \int \sum_{n=0}^{\infty} (-1)^n \frac{x^{2n}}{2n+1}\, dx = C + \sum_{n=0}^{\infty} (-1)^n \frac{x^{2n+1}}{(2n+1)^2} \text{ with } R = 1.$$

23. We use the representation $\displaystyle\int \frac{dx}{1+x^4} = C + \sum_{n=0}^{\infty} \frac{(-1)^n x^{4n+1}}{4n+1}$ from Exercise 19 with $C = 0$.

So $\displaystyle\int_0^{0.2} \frac{dx}{1+x^4} = \left[x - \frac{x^5}{5} + \frac{x^9}{9} - \frac{x^{13}}{13} + \cdots\right]_0^{0.2} = 0.2 - \frac{0.2^5}{5} + \frac{0.2^9}{9} - \frac{0.2^{13}}{13} + \cdots$. Since the series is alternating, the error in the n th-order approximation is less than the first neglected term, by The Alternating Series Estimation Theorem. If we use only the first two terms of the series, then the error is at most $0.2^9/9 \approx 5.7 \times 10^{-8}$. So, to six decimal places, $\displaystyle\int_0^{0.2} \frac{dx}{1+x^4} \approx 0.2 - \frac{0.2^5}{5} = 0.199936$.

25. We substitute x^4 for x in Example 7, and find that

$$\begin{aligned}\int x^2 \tan^{-1}\left(x^4\right) dx &= \int x^2 \sum_{n=0}^{\infty} (-1)^n \frac{\left(x^4\right)^{2n+1}}{2n+1}\, dx \\ &= \int \sum_{n=0}^{\infty} (-1)^n \frac{x^{8n+6}}{2n+1}\, dx = C + \sum_{n=0}^{\infty} (-1)^n \frac{x^{8n+7}}{(2n+1)(8n+7)}\end{aligned}$$

So $\int_0^{1/3} x^2 \tan^{-1}\left(x^4\right) dx = \left[\dfrac{x^7}{7} - \dfrac{x^{15}}{45} + \cdots\right]_0^{1/3} = \dfrac{1}{7 \cdot 3^7} - \dfrac{1}{45 \cdot 3^{15}} + \cdots$. The series is alternating, so if we use only one term, the error is at most $1/\left(45 \cdot 3^{15}\right) \approx 1.5 \times 10^{-9}$. So $\int_0^{1/3} x^2 \tan^{-1}\left(x^4\right) dx \approx 1/\left(7 \cdot 3^7\right) \approx 0.000065$ to six decimal places.

27. Using the result of Example 6, $\ln(1-x) = -\displaystyle\sum_{n=1}^{\infty} \frac{x^n}{n}$, with $x = -0.1$, we have

$\ln 1.1 = \ln[1 - (-0.1)] = 0.1 - \dfrac{0.01}{2} + \dfrac{0.001}{3} - \dfrac{0.0001}{4} + \dfrac{0.00001}{5} - \cdots$. The series is alternating, so if we use only the first four terms, the error is at most $\dfrac{0.00001}{5} = 0.000002$. So

$\ln 1.1 \approx 0.1 - \dfrac{0.01}{2} + \dfrac{0.001}{3} - \dfrac{0.0001}{4} \approx 0.09531$.

29. **(a)** $J_0(x) = \sum_{n=0}^{\infty} \frac{(-1)^n x^{2n}}{2^{2n}(n!)^2}$, $J_0'(x) = \sum_{n=1}^{\infty} \frac{(-1)^n 2n x^{2n-1}}{2^{2n}(n!)^2}$, and

$J_0''(x) = \sum_{n=1}^{\infty} \frac{(-1)^n 2n(2n-1) x^{2n-2}}{2^{2n}(n!)^2}$, so

$$x^2 J_0''(x) + x J_0'(x) + x^2 J_0(x) = \sum_{n=1}^{\infty} \frac{(-1)^n 2n(2n-1) x^{2n}}{2^{2n}(n!)^2} + \sum_{n=1}^{\infty} \frac{(-1)^n 2n x^{2n}}{2^{2n}(n!)^2} + \sum_{n=0}^{\infty} \frac{(-1)^n x^{2n+2}}{2^{2n}(n!)^2}$$

$$= \sum_{n=1}^{\infty} \frac{(-1)^n 2n(2n-1) x^{2n}}{2^{2n}(n!)^2} + \sum_{n=1}^{\infty} \frac{(-1)^n 2n x^{2n}}{2^{2n}(n!)^2} + \sum_{n=1}^{\infty} \frac{(-1)^{n-1} x^{2n}}{2^{2n-2}[(n-1)!]^2}$$

$$= \sum_{n=1}^{\infty} (-1)^n \left[\frac{2n(2n-1) + 2n - 2^2 n^2}{2^{2n}(n!)^2}\right] x^{2n} = \sum_{n=1}^{\infty} (-1)^n \left[\frac{4n^2 - 2n + 2n - 4n^2}{2^{2n}(n!)^2}\right] x^{2n} = 0$$

(b) $$\int_0^1 J_0(x)\,dx = \int_0^1 \left[\sum_{n=0}^{\infty} \frac{(-1)^n x^{2n}}{2^{2n}(n!)^2}\right] dx = \int_0^1 \left(1 - \frac{x^2}{4} + \frac{x^4}{64} - \frac{x^6}{2304} + \cdots\right) dx$$

$$= \left[x - \frac{x^3}{3 \cdot 4} + \frac{x^5}{5 \cdot 64} - \frac{x^7}{7 \cdot 2304} + \cdots\right]_0^1 = 1 - \frac{1}{12} + \frac{1}{320} - \frac{1}{16{,}128} + \cdots$$

Since $\frac{1}{16{,}128} \approx 0.000062$, it follows from The Alternating Series Estimation Theorem that, correct to three decimal places, $\int_0^1 J_0(x)\,dx \approx 1 - \frac{1}{12} + \frac{1}{320} \approx 0.920$.

31. **(a)** $f(x) = \sum_{n=0}^{\infty} \frac{x^n}{n!} \Rightarrow f'(x) = \sum_{n=1}^{\infty} \frac{n x^{n-1}}{n!} = \sum_{n=1}^{\infty} \frac{x^{n-1}}{(n-1)!} = \sum_{n=0}^{\infty} \frac{x^n}{n!} = f(x)$.

(b) By Theorem 4.5.2, the only solution to the differential equation $df(x)/dx = f(x)$ is $f(x) = Ke^x$, but $f(0) = 1$, so $K = 1$ and $f(x) = e^x$.
Or: We could solve the equation $df(x)/dx = f(x)$ as a separable differential equation.

33. If $a_n = \frac{x^n}{n^2}$, then by the Ratio Test, $\lim_{n\to\infty} \left|\frac{a_{n+1}}{a_n}\right| = |x| \lim_{n\to\infty} \left(\frac{n}{n+1}\right)^2 = |x| < 1$ for convergence, so $R = 1$.
When $x = \pm 1$, $\sum_{n=1}^{\infty} \left|\frac{x^n}{n^2}\right| = \sum_{n=1}^{\infty} \frac{1}{n^2}$ which is a convergent p-series ($p = 2 > 1$), so the interval of convergence for f is $[-1, 1]$. By Theorem 2, the radii of convergence of f' and f'' are both 1, so we need only check the endpoints. $f(x) = \sum_{n=1}^{\infty} \frac{x^n}{n^2} \Rightarrow f'(x) = \sum_{n=1}^{\infty} \frac{n x^{n-1}}{n^2} = \sum_{n=0}^{\infty} \frac{x^n}{n+1}$, and this series diverges for $x = 1$ (harmonic series) and converges for $x = -1$ (Alternating Series Test), so the interval of convergence is $[-1, 1)$.
$f''(x) = \sum_{n=1}^{\infty} \frac{n x^{n-1}}{n+1}$ diverges at both 1 and -1 (Test for Divergence) since $\lim_{n\to\infty} \frac{n}{n+1} = 1 \neq 0$, so its interval of convergence is $(-1, 1)$.

EXERCISES 10.7

1. Using Theorem 5 with $\sum_{n=0}^{\infty} b_n (x-5)^n$, $b_n = \dfrac{f^{(n)}(a)}{n!}$, so $b_8 = \dfrac{f^{(8)}(5)}{8!}$.

3.

n	$f^{(n)}(x)$	$f^{(n)}(0)$
0	$\cos x$	1
1	$-\sin x$	0
2	$-\cos x$	-1
3	$\sin x$	0
4	$\cos x$	1
$\cdots$	$\cdots$	$\cdots$

$$\begin{aligned}\cos x &= f(0) + f'(0)x + \frac{f''(0)}{2!}x^2 \\ &\quad + \frac{f^{(3)}(0)}{3!}x^3 + \frac{f^{(4)}(0)}{4!}x^4 + \cdots \\ &= 1 - \frac{x^2}{2!} + \frac{x^4}{4!} - \cdots = \sum_{n=0}^{\infty} \frac{(-1)^n x^{2n}}{(2n)!}\end{aligned}$$

If $a_n = \dfrac{(-1)^n x^{2n}}{(2n)!}$, then $\lim\limits_{n\to\infty}\left|\dfrac{a_{n+1}}{a_n}\right| = x^2 \lim\limits_{n\to\infty} \dfrac{1}{(2n+2)(2n+1)} = 0 < 1$ for all x. So $R = \infty$ (Ratio Test).

5.

n	$f^{(n)}(x)$	$f^{(n)}(0)$
0	$(1+x)^{-2}$	1
1	$-2(1+x)^{-3}$	-2
2	$2\cdot 3(1+x)^{-4}$	$2\cdot 3$
3	$-2\cdot 3\cdot 4(1+x)^{-5}$	$-2\cdot 3\cdot 4$
4	$2\cdot 3\cdot 4\cdot 5(1+x)^{-6}$	$2\cdot 3\cdot 4\cdot 5$
$\cdots$	$\cdots$	$\cdots$

So $f^{(n)}(0) = (-1)^n (n+1)!$ and

$$\frac{1}{(1+x)^2} = \sum_{n=0}^{\infty} \frac{(-1)^n (n+1)!}{n!} x^n = \sum_{n=0}^{\infty} (-1)^n (n+1) x^n$$

If $a_n = (-1)^n (n+1) x^n$, then $\lim\limits_{n\to\infty}\left|\dfrac{a_{n+1}}{a_n}\right| = |x| < 1$ for convergence, so $R = 1$.

7. Clearly, $f^{(n)}(x) = e^x$, so $f^{(n)}(3) = e^3$ and $e^x = \sum_{n=0}^{\infty} \dfrac{e^3}{n!}(x-3)^n$. If $a_n = \dfrac{e^3}{n!}(x-3)^n$, then $\lim\limits_{n\to\infty}\left|\dfrac{a_{n+1}}{a_n}\right| = \lim\limits_{n\to\infty} \dfrac{|x-3|}{n+1} = 0$ for all x, so $R = \infty$.

9.

n	$f^{(n)}(x)$	$f^{(n)}(1)$
0	x^{-1}	1
1	$-x^{-2}$	-1
2	$2x^{-3}$	2
3	$-3\cdot 2x^{-4}$	$-3\cdot 2$
4	$4\cdot 3\cdot 2x^{-5}$	$4\cdot 3\cdot 2$
$\cdots$	$\cdots$	$\cdots$

So $f^{(n)}(1) = (-1)^n n!$, and $\dfrac{1}{x} = \sum_{n=0}^{\infty} \dfrac{(-1)^n n!}{n!}(x-1)^n = \sum_{n=0}^{\infty} (-1)^n (x-1)^n$. If $a_n = (-1)^n (x-1)^n$ then $\lim\limits_{n\to\infty}\left|\dfrac{a_{n+1}}{a_n}\right| = |x-1| < 1$ for convergence, so $0 < x < 2$ and $R = 1$.

11.

n	$f^{(n)}(x)$	$f^{(n)}\left(\frac{\pi}{4}\right)$
0	$\sin x$	$\sqrt{2}/2$
1	$\cos x$	$\sqrt{2}/2$
2	$-\sin x$	$-\sqrt{2}/2$
3	$-\cos x$	$-\sqrt{2}/2$
4	$\sin x$	$\sqrt{2}/2$
$\cdots$	$\cdots$	$\cdots$

$$\begin{aligned}\sin x &= f\left(\tfrac{\pi}{4}\right) + f'\left(\tfrac{\pi}{4}\right)\left(x-\tfrac{\pi}{4}\right) + \frac{f''\left(\frac{\pi}{4}\right)}{2!}\left(x-\tfrac{\pi}{4}\right)^2 \\ &\quad + \frac{f^{(3)}\left(\frac{\pi}{4}\right)}{3!}\left(x-\tfrac{\pi}{4}\right)^3 + \frac{f^{(4)}\left(\frac{\pi}{4}\right)}{4!}\left(x-\tfrac{\pi}{4}\right)^4 + \cdots \\ &= \tfrac{\sqrt{2}}{2}\left[1 + \left(x-\tfrac{\pi}{4}\right) - \tfrac{1}{2!}\left(x-\tfrac{\pi}{4}\right)^2 \right. \\ &\quad \left. - \tfrac{1}{3!}\left(x-\tfrac{\pi}{4}\right)^3 + \tfrac{1}{4!}\left(x-\tfrac{\pi}{4}\right)^4 + \cdots\right] \\ &= \frac{\sqrt{2}}{2}\sum_{n=0}^{\infty}\frac{(-1)^{n(n-1)/2}\left(x-\frac{\pi}{4}\right)^n}{n!}\end{aligned}$$

If $a_n = \dfrac{(-1)^{n(n-1)/2}\left(x-\frac{\pi}{4}\right)^n}{n!}$, then $\lim\limits_{n\to\infty}\left|\dfrac{a_{n+1}}{a_n}\right| = \lim\limits_{n\to\infty}\dfrac{\left|x-\frac{\pi}{4}\right|}{n+1} = 0 < 1$ for all x, so $R=\infty$.

13. If $f(x) = \cos x$, then by Formula 9 with $a=0$, $|R_n(x)| \leq \dfrac{\left|f^{(n+1)}(x)\right|}{(n+1)!}|x|^{n+1}$. But $f^{(n+1)}(x) = \pm\sin x$ or $\pm\cos x$. In each case, $\left|f^{(n+1)}(x)\right| \leq 1$, so $|R_n(x)| \leq \dfrac{1}{(n+1)!}|x|^{n+1} \to 0$ as $n\to\infty$ by Equation 10. So $\lim\limits_{n\to\infty} R_n(x) = 0$ and, by Theorem 8, the series in Exercise 3 represents $\cos x$ for all x.

15. $e^x = \displaystyle\sum_{n=0}^{\infty}\frac{x^n}{n!} \Rightarrow e^{3x} = \sum_{n=0}^{\infty}\frac{(3x)^n}{n!} = \sum_{n=0}^{\infty}\frac{3^n x^n}{n!}$, with $R=\infty$.

17. $\cos x = \displaystyle\sum_{n=0}^{\infty}\frac{(-1)^n x^{2n}}{(2n)!} \Rightarrow x^2\cos x = x^2\sum_{n=0}^{\infty}\frac{(-1)^n x^{2n}}{(2n)!} = \sum_{n=0}^{\infty}\frac{(-1)^n x^{2n+2}}{(2n)!}$, $R=\infty$

19. $\sin x = \displaystyle\sum_{n=0}^{\infty}\frac{(-1)^n x^{2n+1}}{(2n+1)!} \Rightarrow x\sin\left(\frac{x}{2}\right) = x\sum_{n=0}^{\infty}\frac{(-1)^n (x/2)^{2n+1}}{(2n+1)!} = \sum_{n=0}^{\infty}\frac{(-1)^n x^{2n+2}}{(2n+1)!2^{2n+1}}$, with $R=\infty$.

21. $\sin^2 x = \frac{1}{2}[1-\cos 2x] = \dfrac{1}{2}\left[1 - \displaystyle\sum_{n=0}^{\infty}\frac{(-1)^n(2x)^{2n}}{(2n)!}\right] = 2^{-1}\left[1 - 1 - \displaystyle\sum_{n=1}^{\infty}\frac{(-1)^n(2x)^{2n}}{(2n)!}\right]$

$$= \sum_{n=1}^{\infty}\frac{(-1)^{n+1}2^{2n-1}x^{2n}}{(2n)!}, \text{ with } R=\infty.$$

23.

n	$f^{(n)}(x)$	$f^{(n)}(0)$
0	$(1+x)^{1/2}$	1
1	$\frac{1}{2}(1+x)^{-1/2}$	$\frac{1}{2}$
2	$-\frac{1}{4}(1+x)^{-3/2}$	$-\frac{1}{4}$
3	$\frac{3}{8}(1+x)^{-5/2}$	$\frac{3}{8}$
4	$-\frac{15}{16}(1+x)^{-7/2}$	$-\frac{15}{16}$
...	...	...

So $f^{(n)}(0) = \dfrac{(-1)^{n-1}\, 1 \cdot 3 \cdot 5 \cdot \cdots \cdot (2n-3)}{2^n}$ for $n \geq 2$, and $\sqrt{1+x} = 1 + \dfrac{x}{2} + \displaystyle\sum_{n=2}^{\infty} \dfrac{(-1)^{n-1}\, 1 \cdot 3 \cdot 5 \cdot \cdots \cdot (2n-3)}{2^n n!} x^n$. If $a_n = \dfrac{(-1)^{n-1}\, 1 \cdot 3 \cdot 5 \cdot \cdots \cdot (2n-3)}{2^n n!} x^n$, then $\displaystyle\lim_{n\to\infty} \left| \frac{a_{n+1}}{a_n} \right| = \frac{|x|}{2} \lim_{n\to\infty} \frac{2n-1}{n+1} = |x| < 1$ for convergence, so $R = 1$.

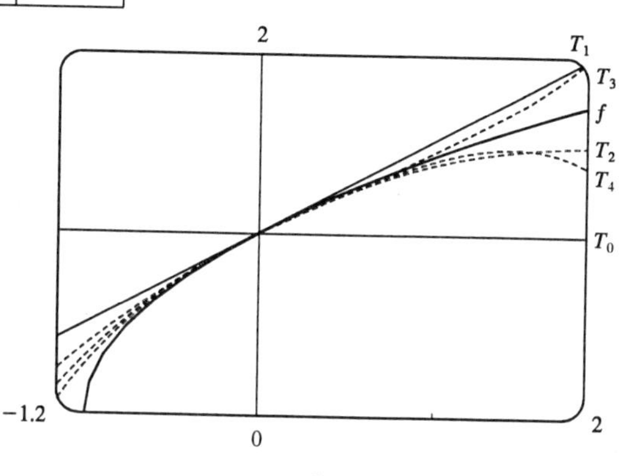

25. $f(x) = (1+x)^{-3} = -\dfrac{1}{2}\dfrac{d}{dx}\left[\dfrac{1}{(1+x)^2}\right] = -\dfrac{1}{2}\dfrac{d}{dx}\left[\displaystyle\sum_{n=0}^{\infty} (-1)^n (n+1) x^n\right]$ (from Exercise 5)

$$= -\frac{1}{2} \sum_{n=1}^{\infty} (-1)^n n(n+1) x^{n-1} = \sum_{n=1}^{\infty} \frac{(-1)^{n+1} n(n+1) x^{n-1}}{2}$$

$$= \sum_{n=0}^{\infty} \frac{(-1)^n (n+1)(n+2) x^n}{2}$$ with $R = 1$ as in Exercise 5.

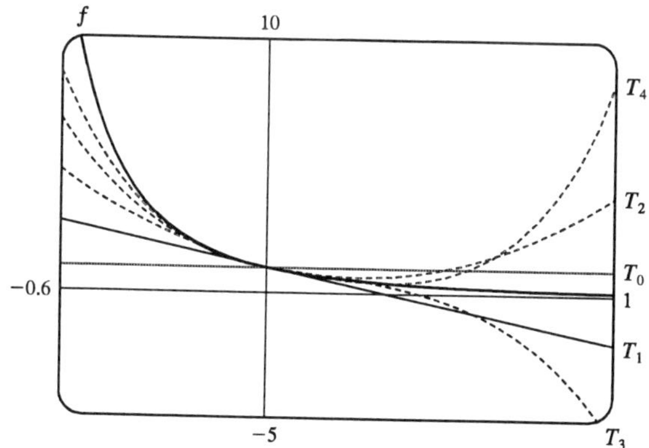

27. $\ln(1+x) = \int \frac{dx}{1+x} = \int \sum_{n=0}^{\infty} (-1)^n x^n \, dx = C + \sum_{n=0}^{\infty} (-1)^n \frac{x^{n+1}}{n+1} = \sum_{n=1}^{\infty} \frac{(-1)^{n-1} x^n}{n}$ with $C = 0$ and $R = 1$, so $\ln(1.1) = \sum_{n=1}^{\infty} \frac{(-1)^{n-1}(0.1)^n}{n}$. This is an alternating series with $b_5 = \frac{(0.1)^5}{5} = 0.000002$, so to five decimal places, $\ln(1.1) \approx \sum_{n=1}^{4} \frac{(-1)^{n-1}(0.1)^n}{n} \approx 0.09531$.

29. $\int \sin(x^2)\, dx = \int \sum_{n=0}^{\infty} (-1)^n \frac{(x^2)^{2n+1}}{(2n+1)!} dx = \int \sum_{n=0}^{\infty} \frac{(-1)^n x^{4n+2}}{(2n+1)!} dx = C + \sum_{n=0}^{\infty} \frac{(-1)^n x^{4n+3}}{(4n+3)(2n+1)!}$

31. Using the series from Exercise 23 and substituting x^3 for x, we get

$$\begin{aligned}\int \sqrt{x^3+1}\, dx &= \int \left[1 + \frac{x^3}{2} + \sum_{n=2}^{\infty} \frac{(-1)^{n-1} 1 \cdot 3 \cdot 5 \cdot \cdots \cdot (2n-3)}{2^n n!} x^{3n}\right] dx \\ &= C + x + \frac{x^4}{8} + \sum_{n=2}^{\infty} \frac{(-1)^{n-1} 1 \cdot 3 \cdot 5 \cdot \cdots \cdot (2n-3)}{2^n n! (3n+1)} x^{3n+1}\end{aligned}$$

33. Using our series from Exercise 29, we get

$\int_0^1 \sin(x^2)\, dx = \sum_{n=0}^{\infty} \left[\frac{(-1)^n x^{4n+3}}{(4n+3)(2n+1)!}\right]_0^1 = \sum_{n=0}^{\infty} \frac{(-1)^n}{(4n+3)(2n+1)!}$ and $|c_3| = \frac{1}{75{,}600} < 0.000014$, so by the Alternating Series Estimation Theorem, we have $\sum_{n=0}^{2} \frac{(-1)^n}{(4n+3)(2n+1)!} \approx \frac{1}{3} - \frac{1}{42} + \frac{1}{1320} \approx 0.310$ (correct to three decimal places).

35. We first find a series representation for $f(x) = (1+x)^{-1/2}$, and then substitute.

n	$f^{(n)}(x)$	$f^{(n)}(0)$
0	$(1+x)^{-1/2}$	1
1	$-\frac{1}{2}(1+x)^{-3/2}$	$-\frac{1}{2}$
2	$\frac{3}{4}(1+x)^{-5/2}$	$\frac{3}{4}$
3	$-\frac{15}{8}(1+x)^{-7/2}$	$-\frac{15}{8}$
$\cdots$	$\cdots$	$\cdots$

$\frac{1}{\sqrt{1+x}} = 1 - \frac{x}{2} + \frac{3}{4}\left(\frac{x^2}{2!}\right) - \frac{15}{8}\left(\frac{x^3}{3!}\right) + \cdots \Rightarrow \frac{1}{\sqrt{1+x^3}} = 1 - \frac{1}{2}x^3 + \frac{3}{8}x^6 - \frac{5}{16}x^9 + \cdots \Rightarrow$

$\int_0^{0.1} \frac{dx}{\sqrt{1+x^3}} = \left[x - \frac{1}{8}x^4 + \frac{3}{56}x^7 - \frac{1}{32}x^{10} + \cdots\right]_0^{0.1} \approx (0.1) - \frac{1}{8}(0.1)^4$, by the Alternating Series Estimation Theorem, since $\frac{3}{56}(0.1)^7 \approx 0.0000000054 < 10^{-8}$, which is the maximum desired error. Therefore, $\int_0^{0.1} \frac{dx}{\sqrt{1+x^3}} \approx 0.09998750$.

37.
$$\begin{aligned}\lim_{x\to 0} \frac{x - \tan^{-1} x}{x^3} &= \lim_{x\to 0} \frac{x - \left(x - \frac{1}{3}x^3 + \frac{1}{5}x^5 - \frac{1}{7}x^7 + \cdots\right)}{x^3} = \lim_{x\to 0} \frac{\frac{1}{3}x^3 - \frac{1}{5}x^5 + \frac{1}{7}x^7 - \cdots}{x^3} \\ &= \lim_{x\to 0} \left(\frac{1}{3} - \frac{1}{5}x^2 + \frac{1}{7}x^4 - \cdots\right) = \frac{1}{3}\end{aligned}$$

since power series are continuous functions.

39. $\displaystyle \lim_{x\to 0}\frac{\sin x - x + \frac{1}{6}x^3}{x^5} = \lim_{x\to 0}\frac{\left(x - \frac{1}{3!}x^3 + \frac{1}{5!}x^5 - \frac{1}{7!}x^7 + \cdots\right) - x + \frac{1}{6}x^3}{x^5}$

$$= \lim_{x\to 0}\frac{\frac{1}{5!}x^5 - \frac{1}{7!}x^7 + \cdots}{x^5} = \lim_{x\to 0}\left(\frac{1}{5!} - \frac{x^2}{7!} + \frac{x^4}{9!} - \cdots\right) = \frac{1}{5!} = \frac{1}{120}$$

since power series are continuous functions.

41. As in Example 8(a), we have $e^{-x^2} = 1 - \dfrac{x^2}{1!} + \dfrac{x^4}{2!} - \dfrac{x^6}{3!} + \cdots$ and we know that

$\cos x = 1 - \dfrac{x^2}{2!} + \dfrac{x^4}{4!} - \cdots$ from Equation 16. Therefore,

$e^{-x^2}\cos x = \left(1 - x^2 + \frac{1}{2}x^4 - \cdots\right)\left(1 - \frac{1}{2}x^2 + \frac{1}{24}x^4 - \cdots\right)$. Writing only the terms with degree ≤ 4, we get

$e^{-x^2}\cos x = 1 - \frac{1}{2}x^2 + \frac{1}{24}x^4 - x^2 + \frac{1}{2}x^4 + \frac{1}{2}x^4 + \cdots = 1 - \frac{3}{2}x^2 + \frac{25}{24}x^4 + \cdots$.

43.

$$\begin{array}{r|l} & -x + \frac{1}{2}x^2 - \frac{1}{3}x^3 + \cdots \\ \hline 1 + x + \frac{1}{2}x^2 + \frac{1}{6}x^3 + \cdots & -x - \frac{1}{2}x^2 - \frac{1}{3}x^3 - \cdots \\ & -x - x^2 - \frac{1}{2}x^3 - \cdots \\ \hline & \frac{1}{2}x^2 + \frac{1}{6}x^3 - \cdots \\ & \frac{1}{2}x^2 + \frac{1}{2}x^3 + \cdots \\ \hline & -\frac{1}{3}x^3 + \cdots \\ & -\frac{1}{3}x^3 + \cdots \\ \hline & \cdots \end{array}$$

From Example 6 in Section 10.6, we have $\ln(1 - x) = -x - \frac{1}{2}x^2 - \frac{1}{3}x^3 - \cdots$, $|x| < 1$. Therefore,

$y = \dfrac{\ln(1 - x)}{e^x} = \dfrac{-x - \frac{1}{2}x^2 - \frac{1}{3}x^3 - \cdots}{1 + x + \frac{1}{2}x^2 + \frac{1}{6}x^3 + \cdots}$. So by the long division above,

$\dfrac{\ln(1 - x)}{e^x} = -x + \dfrac{x^2}{2} - \dfrac{x^3}{3} + \cdots$, $|x| < 1$.

45. $\displaystyle \sum_{n=0}^{\infty}(-1)^n\frac{x^{4n}}{n!} = \sum_{n=0}^{\infty}\frac{\left(-x^4\right)^n}{n!} = e^{-x^4}$, by (11).

47. $\displaystyle \sum_{n=0}^{\infty}\frac{(-1)^n\pi^{2n+1}}{4^{2n+1}(2n+1)!} = \sum_{n=0}^{\infty}\frac{(-1)^n\left(\frac{\pi}{4}\right)^{2n+1}}{(2n+1)!} = \sin\frac{\pi}{4} = \frac{1}{\sqrt{2}}$, by (15).

49. $\displaystyle \sum_{n=0}^{\infty}\frac{x^{n+1}}{(n+1)!} = \frac{x}{1!} + \frac{x^2}{2!} + \frac{x^3}{3!} + \cdots = \left(1 + \frac{x}{1!} + \frac{x^2}{2!} + \frac{x^3}{3!} + \cdots\right) - 1 = e^x - 1$, by (11).

EXERCISES 10.8

1. The general binomial series in (2) is

$$(1+x)^k = \sum_{n=0}^{\infty} \binom{k}{n} x^n = 1 + kx + \frac{k(k-1)}{2!}x^2 + \frac{k(k-1)(k-2)}{3!}x^3 + \cdots.$$

$$\begin{aligned}(1+x)^{1/2} &= \sum_{n=0}^{\infty} \binom{\frac{1}{2}}{n} x^n = 1 + \left(\frac{1}{2}\right)x + \frac{\left(\frac{1}{2}\right)\left(-\frac{1}{2}\right)}{2!}x^2 + \frac{\left(\frac{1}{2}\right)\left(-\frac{1}{2}\right)\left(-\frac{3}{2}\right)}{3!}x^3 + \cdots \\ &= 1 + \frac{x}{2} - \frac{x^2}{2^2 \cdot 2!} + \frac{1 \cdot 3 \cdot x^3}{2^3 \cdot 3!} - \frac{1 \cdot 3 \cdot 5 \cdot x^4}{2^4 \cdot 4!} + \cdots \\ &= 1 + \frac{x}{2} + \sum_{n=2}^{\infty} \frac{(-1)^{n-1}\, 1 \cdot 3 \cdot 5 \cdot \cdots \cdot (2n-3)\, x^n}{2^n \cdot n!},\ R = 1\end{aligned}$$

3.

$$\begin{aligned}[1+(2x)]^{-4} &= 1 + (-4)(2x) + \frac{(-4)(-5)}{2!}(2x)^2 + \frac{(-4)(-5)(-6)}{3!}(2x)^3 + \cdots \\ &= 1 + \sum_{n=1}^{\infty} \frac{(-1)^n\, 2^n \cdot 4 \cdot 5 \cdot 6 \cdot \cdots \cdot (n+3)}{n!} x^n \\ &= 1 + \sum_{n=1}^{\infty} \frac{(-1)^n\, 2^n \cdot 2 \cdot 3 \cdot 4 \cdot 5 \cdot 6 \cdot \cdots \cdot (n+1)(n+2)(n+3)}{2 \cdot 3 \cdot n!} x^n \\ &= \sum_{n=0}^{\infty} (-1)^n \frac{2^n (n+1)(n+2)(n+3)}{6} x^n,\ |2x| < 1 \quad \Leftrightarrow \quad |x| < \tfrac{1}{2}, \text{ so } R = \tfrac{1}{2}.\end{aligned}$$

5.

$$\begin{aligned}\left[1+\left(-x^4\right)\right]^{1/4} &= 1 + \left(\frac{1}{4}\right)\left(-x^4\right) + \frac{\left(\frac{1}{4}\right)\left(-\frac{3}{4}\right)}{2!}\left(-x^4\right)^2 + \frac{\left(\frac{1}{4}\right)\left(-\frac{3}{4}\right)\left(-\frac{7}{4}\right)}{3!}\left(-x^4\right)^3 + \cdots \\ &= 1 - \frac{x^4}{4} - \sum_{n=2}^{\infty} \frac{3 \cdot 7 \cdot 11 \cdot \cdots \cdot (4n-5)}{4^n \cdot n!} x^{4n}, \text{ with } R = 1.\end{aligned}$$

7.

$$\begin{aligned}\frac{1}{\sqrt[3]{8+x}} &= (8+x)^{-1/3} = 8^{-1/3}\left(1 + \frac{x}{8}\right)^{-1/3} = \frac{1}{2}\left(1 + \frac{x}{8}\right)^{-1/3} \\ &= \frac{1}{2}\left[1 + \left(-\frac{1}{3}\right)\left(\frac{x}{8}\right) + \frac{\left(-\frac{1}{3}\right)\left(-\frac{4}{3}\right)}{2!}\left(\frac{x}{8}\right)^2 + \cdots\right] \\ &= \frac{1}{2}\left[1 + \sum_{n=1}^{\infty} \frac{(-1)^n\, 1 \cdot 4 \cdot 7 \cdot \cdots \cdot (3n-2)}{3^n \cdot n!\, 8^n} x^n\right] \text{ and } \left|\frac{x}{8}\right| < 1 \quad \Leftrightarrow \quad |x| < 8, \text{ so } R = 8.\end{aligned}$$

The three Taylor polynomials are $T_1(x) = \frac{1}{2} - \frac{1}{48}x$, $T_2(x) = \frac{1}{2} - \frac{1}{48}x + \frac{1}{576}x^2$, and $T_3(x) = \frac{1}{2} - \frac{1}{48}x + \frac{1}{576}x^2 - \frac{4 \cdot 7}{2 \cdot 27 \cdot 6 \cdot 512}x^3 = \frac{1}{2} - \frac{1}{48}x + \frac{1}{576}x^2 - \frac{7}{41{,}472}x^3$.

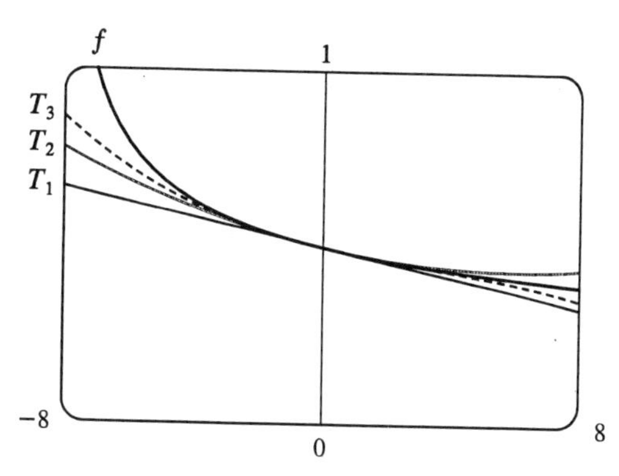

9. **(a)** $\left[1+\left(-x^2\right)\right]^{-1/2} = 1 + \left(-\frac{1}{2}\right)\left(-x^2\right) + \frac{\left(-\frac{1}{2}\right)\left(-\frac{3}{2}\right)}{2!}\left(-x^2\right)^2 + \frac{\left(-\frac{1}{2}\right)\left(-\frac{3}{2}\right)\left(-\frac{5}{2}\right)}{3!}\left(-x^2\right)^3 + \cdots$

$$= 1 + \sum_{n=1}^{\infty} \frac{1\cdot 3\cdot 5\cdot\cdots\cdot(2n-1)}{2^n\cdot n!}x^{2n}$$

(b) $\sin^{-1}x = \int \frac{1}{\sqrt{1-x^2}}\,dx = C + x + \sum_{n=1}^{\infty} \frac{1\cdot 3\cdot 5\cdot\cdots\cdot(2n-1)}{(2n+1)\,2^n\cdot n!}x^{2n+1}$

$$= x + \sum_{n=1}^{\infty} \frac{1\cdot 3\cdot 5\cdot\cdots\cdot(2n-1)}{(2n+1)\,2^n\cdot n!}x^{2n+1} \text{ since } 0 = \sin^{-1}0 = C.$$

11. **(a)** $[1+(-x)]^{-2} = 1 + (-2)(-x) + \frac{(-2)(-3)}{2!}(-x)^2 + \frac{(-2)(-3)(-4)}{3!}(-x)^3 + \cdots$

$$= 1 + 2x + 3x^2 + 4x^3 + \cdots = \sum_{n=0}^{\infty}(n+1)\,x^n,$$

so $\frac{x}{(1-x)^2} = \sum_{n=0}^{\infty}(n+1)\,x^{n+1} = \sum_{n=1}^{\infty} nx^n.$

(b) With $x = \frac{1}{2}$ in part (a), we have $\sum_{n=1}^{\infty}\frac{n}{2^n} = \frac{\frac{1}{2}}{\left(1-\frac{1}{2}\right)^2} = 2.$

13. **(a)** $\left(1+x^2\right)^{1/2} = 1 + \left(\frac{1}{2}\right)x^2 + \frac{\left(\frac{1}{2}\right)\left(-\frac{1}{2}\right)}{2!}\left(x^2\right)^2 + \frac{\left(\frac{1}{2}\right)\left(-\frac{1}{2}\right)\left(-\frac{3}{2}\right)}{3!}\left(x^2\right)^3 + \cdots$

$$= 1 + \frac{x^2}{2} + \sum_{n=2}^{\infty}\frac{(-1)^{n-1}\,1\cdot 3\cdot 5\cdot\cdots\cdot(2n-3)}{2^n\cdot n!}x^{2n}$$

(b) The coefficient of x^{10} (corresponding to $n=5$) in the above Maclaurin series is $\frac{f^{(10)}(0)}{10!}$, so

$$\frac{f^{(10)}(0)}{10!} = \frac{(-1)^4\cdot 1\cdot 3\cdot 5\cdot 7}{2^5\cdot 5!} \quad\Rightarrow\quad f^{(10)}(0) = 10!\left(\frac{1\cdot 3\cdot 5\cdot 7}{2^5\cdot 5!}\right) = 99{,}225.$$

15. **(a)** $g(x) = \sum_{n=0}^{\infty} \binom{k}{n} x^n \Rightarrow g'(x) = \sum_{n=1}^{\infty} \binom{k}{n} n x^{n-1}$, so

$$\begin{aligned}
(1+x)\,g'(x) &= (1+x) \sum_{n=1}^{\infty} \binom{k}{n} n x^{n-1} \\
&= \sum_{n=1}^{\infty} \binom{k}{n} n x^{n-1} + \sum_{n=1}^{\infty} \binom{k}{n} n x^{n} \\
&= \sum_{n=0}^{\infty} \binom{k}{n+1} (n+1)\, x^{n} + \sum_{n=0}^{\infty} \binom{k}{n} n x^{n} \qquad \left[\begin{array}{c}\text{Replace } n \text{ with } n+1 \\ \text{in the first series}\end{array}\right] \\
&= \sum_{n=0}^{\infty} (n+1) \frac{k(k-1)(k-2)\cdots(k-n+1)(k-n)}{(n+1)!} x^n \\
&\qquad + \sum_{n=0}^{\infty} \left[(n) \frac{k(k-1)(k-2)\cdots(k-n+1)}{n!} \right] x^n \\
&= \sum_{n=0}^{\infty} \frac{(n+1)\,k(k-1)(k-2)\cdots(k-n+1)}{(n+1)!} \left[(k-n)+n\right] x^n \\
&= k \sum_{n=0}^{\infty} \frac{k(k-1)(k-2)\cdots(k-n+1)}{n!} x^n \\
&= k \sum_{n=0}^{\infty} \binom{k}{n} x^n \\
&= k g(x)
\end{aligned}$$

Thus, $g'(x) = \dfrac{kg(x)}{1+x}$.

(b) $h(x) = (1+x)^{-k} g(x) \Rightarrow$

$$\begin{aligned}
h'(x) &= -k(1+x)^{-k-1} g(x) + (1+x)^{-k} g'(x) \qquad \text{[Product Rule]} \\
&= -k(1+x)^{-k-1} g(x) + (1+x)^{-k} \frac{kg(x)}{1+x} \qquad \text{[from part (a)]} \\
&= -k(1+x)^{-k-1} g(x) + k(1+x)^{-k-1} g(x) \\
&= 0
\end{aligned}$$

(c) From part (b) we see that $h(x)$ must be constant for $x \in (-1,1)$, so $h(x) = h(0) = 1$ for $x \in (-1,1)$. Thus, $h(x) = 1 = (1+x)^{-k} g(x) \Leftrightarrow g(x) = (1+x)^k$ for $x \in (-1,1)$.

EXERCISES 10.9

1. **(a)**

n	$f^{(n)}(x)$	$f^{(n)}(0)$	$T_n(x)$
0	$\cos x$	1	1
1	$-\sin x$	0	1
2	$-\cos x$	-1	$1 - \frac{1}{2}x^2$
3	$\sin x$	0	$1 - \frac{1}{2}x^2$
4	$\cos x$	1	$1 - \frac{1}{2}x^2 + \frac{1}{24}x^4$
5	$-\sin x$	0	$1 - \frac{1}{2}x^2 + \frac{1}{24}x^4$
6	$-\cos x$	-1	$1 - \frac{1}{2}x^2 + \frac{1}{24}x^4 - \frac{1}{720}x^6$

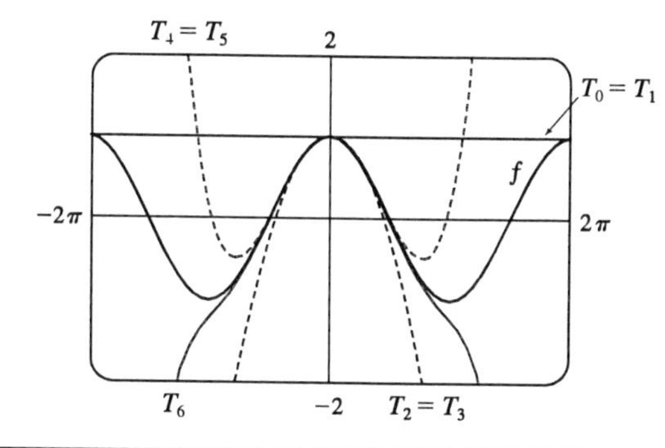

(b)

x	f	$T_0 = T_1$	$T_2 = T_3$	$T_4 = T_5$	T_6
$\frac{\pi}{4}$	0.7071	1	0.6916	0.7074	0.7071
$\frac{\pi}{2}$	0	1	-0.2337	0.0200	-0.0009
π	-1	1	-3.9348	0.1239	-1.2114

(c) As n increases, $T_n(x)$ is a good approximation to $f(x)$ on a larger and larger interval.

3.

n	$f^{(n)}(x)$	$f^{(n)}\left(\frac{\pi}{6}\right)$
0	$\sin x$	$\frac{1}{2}$
1	$\cos x$	$\frac{\sqrt{3}}{2}$
2	$-\sin x$	$-\frac{1}{2}$
3	$-\cos x$	$-\frac{\sqrt{3}}{2}$

$$T_3(x) = \sum_{n=0}^{3} \frac{f^{(n)}\left(\frac{\pi}{6}\right)}{n!}\left(x - \frac{\pi}{6}\right)^n$$

$$= \frac{1}{2} + \frac{\sqrt{3}}{2}\left(x - \frac{\pi}{6}\right) - \frac{1}{4}\left(x - \frac{\pi}{6}\right)^2 - \frac{\sqrt{3}}{12}\left(x - \frac{\pi}{6}\right)^3$$

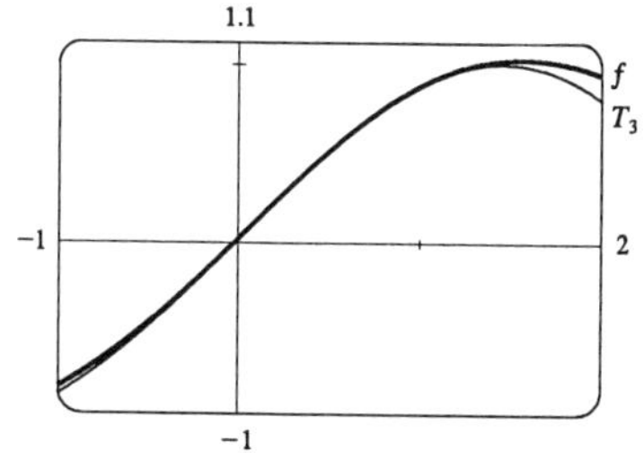

5.

n	$f^{(n)}(x)$	$f^{(n)}(0)$
0	$\tan x$	0
1	$\sec^2 x$	1
2	$2\sec^2 x \tan x$	0
3	$4\sec^2 x \tan^2 x + 2\sec^4 x$	2
4	$8\sec^2 x \tan^3 x + 16\sec^4 x \tan x$	0

$$T_4(x) = \sum_{n=0}^{4} \frac{f^{(n)}(0)}{n!} x^n = x + \frac{2x^3}{3!}$$
$$= x + \frac{x^3}{3}$$

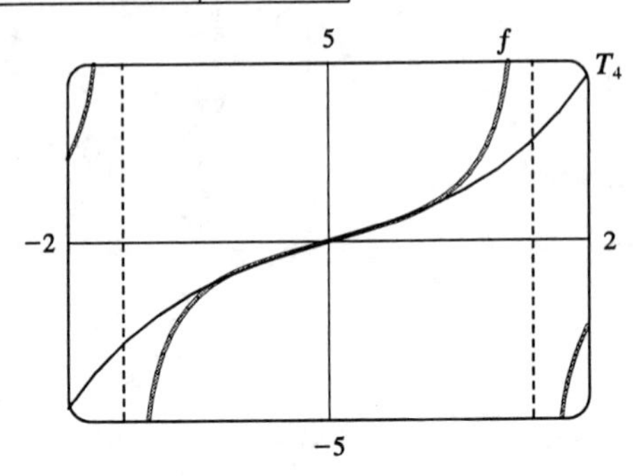

7.

n	$f^{(n)}(x)$	$f^{(n)}\left(\frac{\pi}{3}\right)$
0	$\sec x$	2
1	$\sec x \tan x$	$2\sqrt{3}$
2	$\sec x \tan^2 x + \sec^3 x$	14
3	$\sec x \tan^3 x$ $+5\sec^3 x \tan x$	$46\sqrt{3}$

$$T_3(x) = \sum_{n=0}^{3} \frac{f^{(n)}\left(\frac{\pi}{3}\right)}{n!} \left(x - \tfrac{\pi}{3}\right)^n$$
$$= 2 + 2\sqrt{3}\left(x - \tfrac{\pi}{3}\right) + 7\left(x - \tfrac{\pi}{3}\right)^2 + \tfrac{23\sqrt{3}}{3}\left(x - \tfrac{\pi}{3}\right)^3$$

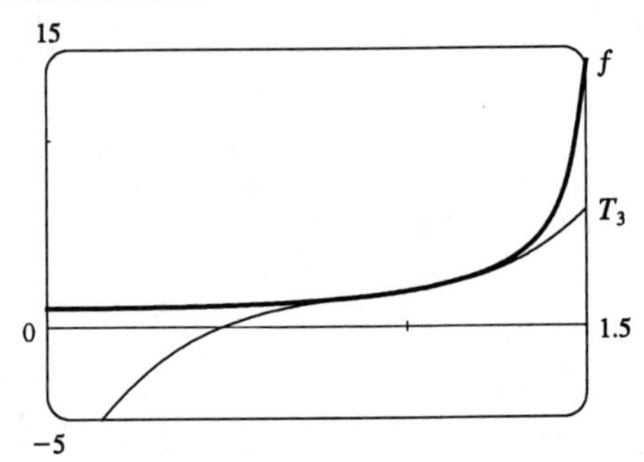

9. In Maple, we can find the Taylor polynomials by the following method: first define `f:=sec(x);` and then set `T2:=convert(taylor(f,x=0,3),polynom);`, `T4:=convert(taylor(f,x=0,5),polynom);`, etc. (The third argument in the `taylor` function is one more than the degree of the desired polynomial). We must `convert` to the type `polynom` because the output of the `taylor` function contains an error term which we do not want. In Mathematica, we use `Tn:=Normal[Series[f,{x,0,n}]]`, with `n=2, 4`, etc. Note that in Mathematica, the "degree" argument is the same as the degree of the desired polynomial. In Derive, author $\sec x$, then enter `Calculus,Taylor,8,0;` and then simplify the expression. The eighth Taylor polynomial is $T_8(x) = 1 + \frac{1}{2}x^2 + \frac{5}{24}x^4 + \frac{61}{720}x^6 + \frac{277}{8064}x^8$.

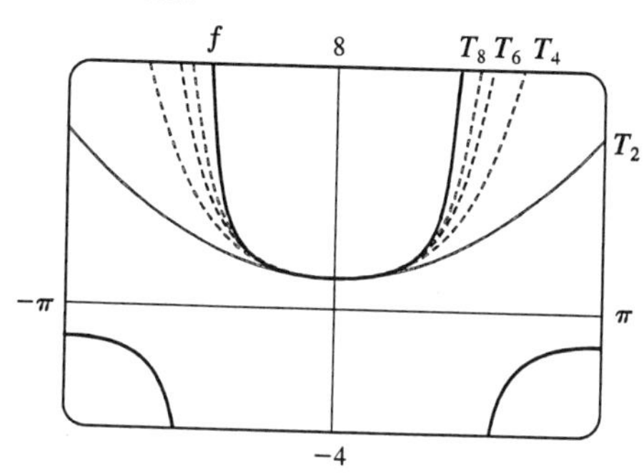

11.

$$f(x) = (1+x)^{1/2} \qquad f(0) = 1$$

$$f'(x) = \tfrac{1}{2}(1+x)^{-1/2} \qquad f'(0) = \tfrac{1}{2}$$

$$f''(x) = -\tfrac{1}{4}(1+x)^{-3/2} = -\frac{1}{4(1+x)^{3/2}}$$

(a) $(1+x)^{1/2} \approx T_1(x) = 1 + \frac{1}{2}x$

(b) By Taylor's Inequality, the remainder is $|R_1(x)| \le \frac{M}{2!}|x|^2$, where $|f''(x)| \le M$. Now $0 \le x \le 0.1 \;\Rightarrow\; 0 \le x^2 \le 0.01$, and letting $x = 0$ gives $M = 0.25$, so $|R_1(x)| \le \frac{0.25}{2}(0.01) = 0.00125$.

(c)

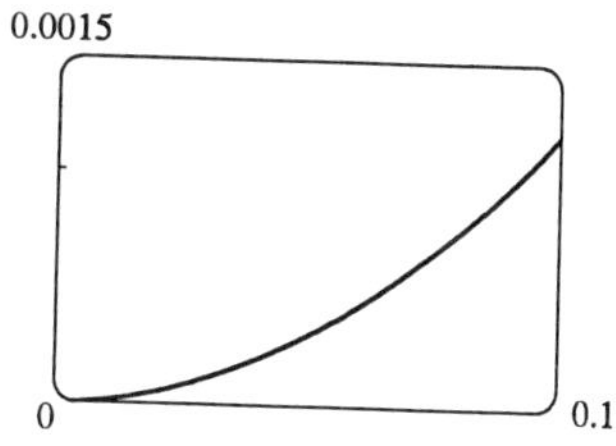

From the graph of $|R_1(x)| = \left|\sqrt{1+x} - \left(1 + \frac{1}{2}x\right)\right|$, it seems that the error is at most 0.0012 on $[0, 0.1]$.

13.

$$f(x)=\sin x \qquad f\left(\tfrac{\pi}{4}\right)=\tfrac{\sqrt{2}}{2} \qquad f^{(4)}(x)=\sin x \qquad f^{(4)}\left(\tfrac{\pi}{4}\right)=\tfrac{\sqrt{2}}{2}$$
$$f'(x)=\cos x \qquad f'\left(\tfrac{\pi}{4}\right)=\tfrac{\sqrt{2}}{2} \qquad f^{(5)}(x)=\cos x \qquad f^{(5)}\left(\tfrac{\pi}{4}\right)=\tfrac{\sqrt{2}}{2}$$
$$f''(x)=-\sin x \qquad f''\left(\tfrac{\pi}{4}\right)=-\tfrac{\sqrt{2}}{2} \qquad f^{(6)}(x)=-\sin x$$
$$f'''(x)=-\cos x \qquad f'''\left(\tfrac{\pi}{4}\right)=-\tfrac{\sqrt{2}}{2}$$

(a) $\sin x \approx T_5(x)$
$$=\tfrac{\sqrt{2}}{2}+\tfrac{\sqrt{2}}{2}\left(x-\tfrac{\pi}{4}\right)-\tfrac{\sqrt{2}}{4}\left(x-\tfrac{\pi}{4}\right)^2-\tfrac{\sqrt{2}}{12}\left(x-\tfrac{\pi}{4}\right)^3+\tfrac{\sqrt{2}}{48}\left(x-\tfrac{\pi}{4}\right)^4+\tfrac{\sqrt{2}}{240}\left(x-\tfrac{\pi}{4}\right)^5$$

(b) $|R_5(x)| \le \dfrac{M}{6!}\left|x-\tfrac{\pi}{4}\right|^6$, where $\left|f^{(6)}(x)\right| \le M$. Now $0 \le x \le \frac{\pi}{2} \Rightarrow \left(x-\frac{\pi}{4}\right)^6 \le \left(\frac{\pi}{4}\right)^6$, and letting $x=\frac{\pi}{2}$ gives $M=1$, so $|R_5(x)| \le \frac{1}{6!}\left(\frac{\pi}{4}\right)^6=\frac{1}{720}\left(\frac{\pi}{4}\right)^6 \approx 0.00033$.

(c)

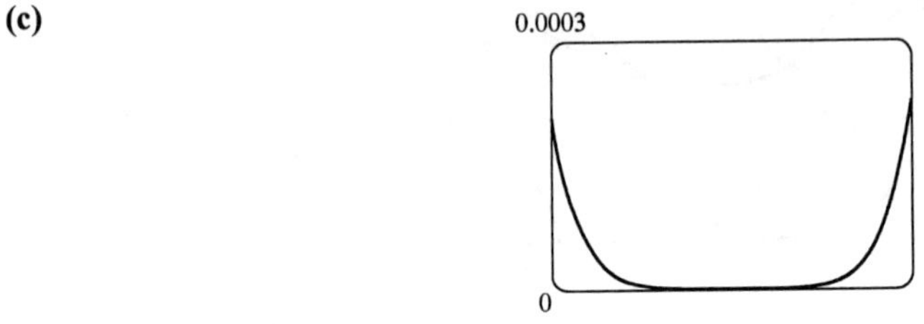

From the graph of $|R_5(x)|=|\sin x - T_5(x)|$, it seems that the error is less than 0.00026 on $\left[0,\frac{\pi}{2}\right]$.

15.

$$f(x)=e^{x^2} \qquad f(0)=1$$
$$f'(x)=e^{x^2}(2x) \qquad f'(0)=0$$
$$f''(x)=e^{x^2}\left(2+4x^2\right) \qquad f''(0)=2$$
$$f'''(x)=e^{x^2}\left(12x+8x^3\right) \qquad f'''(0)=0$$
$$f^{(4)}(x)=e^{x^2}\left(12+48x^2+16x^4\right)$$

(a) $e^{x^2} \approx T_3(x)=1+\frac{2}{2!}x^2=1+x^2$

(b) $|R_3(x)| \le \dfrac{M}{4!}|x|^4$, where $\left|f^{(4)}(x)\right| \le M$. Now $0 \le x \le 0.1 \Rightarrow x^4 \le (0.1)^4$, and letting $x=0.1$ gives $|R_3(x)| \le \dfrac{e^{0.01}(12+0.48+0.0016)}{24}(0.1)^4 < 0.00006$.

(c)

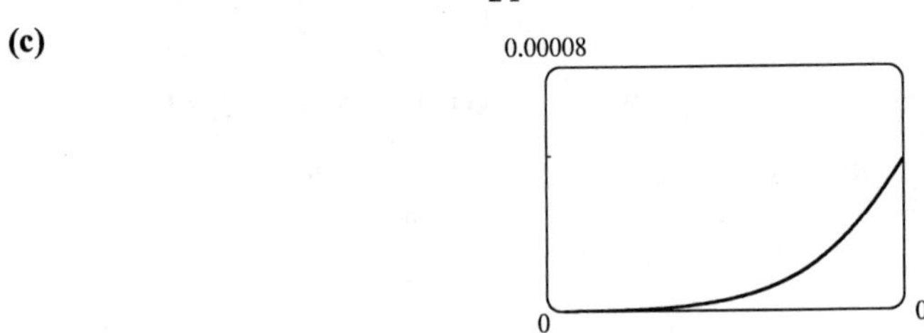

From the graph of $|R_3(x)|=\left|e^{x^2}-\left(1+x^2\right)\right|$, it appears that the error is less than 0.000051 on $[0,0.1]$.

17. From Exercise 3, $\sin x = \frac{1}{2} + \frac{\sqrt{3}}{2}\left(x - \frac{\pi}{6}\right) - \frac{1}{4}\left(x - \frac{\pi}{6}\right)^2 - \frac{\sqrt{3}}{12}\left(x - \frac{\pi}{6}\right)^3 + R_3(x)$, where $R_3(x) \leq \frac{M}{4!}\left|x - \frac{\pi}{6}\right|^4$ with $\left|f^{(4)}(x)\right| = |\sin x| \leq M = 1$. Now $35^\circ = \left(\frac{\pi}{6} + \frac{\pi}{36}\right)$ radians, so the error is $\left|R_3\left(\frac{\pi}{36}\right)\right| \leq \frac{\left(\frac{\pi}{36}\right)^4}{4!} < 0.000003$. Therefore, to five decimal places, $\sin 35^\circ \approx \frac{1}{2} + \frac{\sqrt{3}}{2}\left(\frac{\pi}{36}\right) - \frac{1}{4}\left(\frac{\pi}{36}\right)^2 - \frac{\sqrt{3}}{12}\left(\frac{\pi}{36}\right)^3 \approx 0.57358$.

19. All derivatives of e^x are e^x, so $|R_n(x)| \leq \frac{e^x}{(n+1)!}|x|^{n+1}$, where $0 < x < 0.1$. Letting $x = 0.1$, $R_n(0.1) \leq \frac{e^{0.1}}{(n+1)!}(0.1)^{n+1} < 0.00001$, and by trial and error we find that $n = 3$ satisfies this inequality since $R_3(0.1) < 0.0000046$. Thus, by adding the three terms of the Maclaurin series for e^x corresponding to $n = 0$, 1, and 2, we can estimate $e^{0.1}$ to within 0.00001.

21. $\sin x = x - \frac{1}{3!}x^3 + \frac{1}{5!}x^5 - \cdots$. By the Alternating Series Estimation Theorem, the error in the approximation $\sin x = x - \frac{1}{3!}x^3$ is less than $\left|\frac{1}{5!}x^5\right| < 0.01 \Leftrightarrow \left|x^5\right| < 120(0.01) \Leftrightarrow |x| < (1.2)^{1/5} \approx 1.037$.

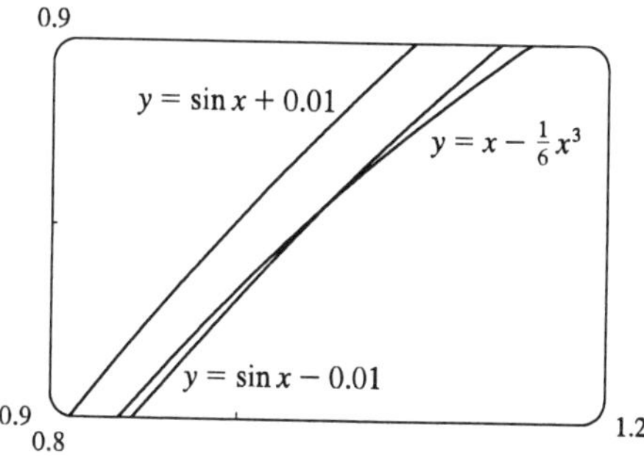

The curves intersect at $x \approx 1.043$, so the graph confirms our estimate. Since both the sine function and the given approximation are odd functions, we only need to check the estimate for $x > 0$.

23. Let $s(t)$ be the position function of the car, and for convenience set $s(0) = 0$. The velocity of the car is $v(t) = s'(t)$ and the acceleration is $a(t) = s''(t)$, so the second degree Taylor polynomial is $T_2(t) = s(0) + v(0)t + \frac{a(0)}{2}t^2 = 20t + t^2$. We estimate the distance travelled during the next second to be $s(1) \approx T_2(1) = 20 + 1 = 21$ m. The function $T_2(t)$ would not be accurate over a full minute, since the car could not possibly maintain an acceleration of 2 m/s^2 for that long (if it did, its final speed would be $140\text{ m/s} \approx 315$ mi/h!)

25. $E = \dfrac{q}{D^2} - \dfrac{q}{(D+d)^2} = \dfrac{q}{D^2} - \dfrac{q}{D^2(1+d/D)^2} = \dfrac{q}{D^2}\left[1 - \left(1 + \dfrac{d}{D}\right)^{-2}\right]$.

We use the Binomial Series to expand $(1 + d/D)^{-2}$:

$$\begin{aligned} E &= \frac{q}{D^2}\left[1 - \left(1 - 2\left(\frac{d}{D}\right) + \frac{2 \cdot 3}{2!}\left(\frac{d}{D}\right)^2 - \frac{2 \cdot 3 \cdot 4}{3!}\left(\frac{d}{D}\right)^3 + \cdots\right)\right] \\ &= \frac{q}{D^2}\left[2\left(\frac{d}{D}\right) - 3\left(\frac{d}{D}\right)^2 + 4\left(\frac{d}{D}\right)^3 - \cdots\right] \approx 2qd \cdot \frac{1}{D^3} \end{aligned}$$

when D is much larger than d, that is, when P is far away from the dipole.

27. **(a)** If the water is deep, then $2\pi d/L$ is large, and we know that $\tanh x \to 1$ as $x \to \infty$. So we can approximate $\tanh(2\pi d/L) \approx 1$, and so $v^2 \approx gL/(2\pi) \Leftrightarrow v \approx \sqrt{gL/(2\pi)}$.

(b) From the calculations at right, the first term in the Maclaurin series of $\tanh x$ is x, so if the water is shallow, we can approximate

$$\begin{aligned} f(x) &= \tanh x & f(0) &= 0 \\ f'(x) &= \operatorname{sech}^2 x & f'(0) &= 1 \\ f''(x) &= -2\operatorname{sech}^2 x \tanh x & f''(0) &= 0 \\ f'''(x) &= 2\operatorname{sech}^2 x\,(3\tanh^2 x - 1) & f'''(0) &= -2 \end{aligned}$$

$\tanh \dfrac{2\pi d}{L} \approx \dfrac{2\pi d}{L}$, and so $v^2 \approx \dfrac{gL}{2\pi} \cdot \dfrac{2\pi d}{L} \Leftrightarrow v \approx \sqrt{gd}$.

(c) Since $\tanh x$ is an odd function, its Maclaurin series is alternating, so the error in the approximation $\tanh \dfrac{2\pi d}{L} \approx \dfrac{2\pi d}{L}$ is less than the first neglected term, which is $\dfrac{|f'''(0)|}{3!}\left(\dfrac{2\pi d}{L}\right)^3 = \dfrac{1}{3}\left(\dfrac{2\pi d}{L}\right)^3$. If $L > 10d$, then $\dfrac{1}{3}\left(\dfrac{2\pi d}{L}\right)^3 < \dfrac{1}{3}\left(2\pi \cdot \dfrac{1}{10}\right)^3 = \dfrac{\pi^3}{375}$, so the error in the approximation $v^2 = gd$ is less than $\dfrac{gL}{2\pi} \cdot \dfrac{\pi^3}{375} \approx 0.0132gL$.

29. Using Taylor's Formula with $n = 1$, $a = x_n$, $x = r$, we get $f(r) = f(x_n) + f'(x_n)(r - x_n) + R_1(x)$, where $R_1(x) = \frac{1}{2}f''(z)(r - x_n)^2$ and z lies between x_n and r. But r is a root, so $f(r) = 0$ and Taylor's Formula becomes $0 = f(x_n) + f'(x_n)(r - x_n) + \frac{1}{2}f''(z)(r - x_n)^2$. Taking the first two terms to the left side and dividing by $f'(x_n)$, we have $x_n - r - \dfrac{f(x_n)}{f'(x_n)} = \dfrac{1}{2}\dfrac{f''(z)}{f'(x_n)}|x_n - r|^2$. By the formula for Newton's Method, we have $|x_{n+1} - r| = \left|x_n - \dfrac{f(x_n)}{f'(x_n)} - r\right| = \dfrac{1}{2}\dfrac{|f''(z)|}{|f'(x_n)|}|x_n - r|^2 \le \dfrac{M}{2K}|x_n - r|^2$ since $|f''(z)| \le M$ and $|f'(x_n)| \ge K$.

REVIEW EXERCISES FOR CHAPTER 10

1. False. See the warning in Note 2 after Theorem 10.2.6.

3. False. For example, take $a_n = (-1)^n/(n6^n)$.

5. False, since $\lim\limits_{n\to\infty}\left|\dfrac{a_{n+1}}{a_n}\right| = \lim\limits_{n\to\infty}\left|\dfrac{n^3}{(n+1)^3}\right| = \lim\limits_{n\to\infty}\dfrac{1}{(1+1/n)^3} = 1$.

7. False. See the remarks after Example 4 in Section 10.3.

9. True. See Example 8 in Section 10.1.

11. True. By Theorem 10.7.5 the coefficient of x^3 is $\dfrac{f'''(0)}{3!} = \dfrac{1}{3} \quad\Rightarrow\quad f'''(0) = 2$.

Or: Use Theorem 10.6.2 to differentiate f three times.

13. False. For example, let $a_n = b_n = (-1)^n$. Then $\{a_n\}$ and $\{b_n\}$ are divergent, but $a_nb_n = 1$, so $\{a_nb_n\}$ is convergent.

15. True by Theorem 10.4.1. $\left[\sum(-1)^n a_n \text{ is absolutely convergent and hence convergent.}\right]$

17. $\lim\limits_{n\to\infty}\dfrac{n}{2n+5} = \lim\limits_{n\to\infty}\dfrac{1}{2+5/n} = \dfrac{1}{2}$ and the sequence is convergent.

19. $\{2n+5\}$ is divergent since $2n+5\to\infty$ as $n\to\infty$.

21. $\{\sin n\}$ is divergent since $\lim\limits_{n\to\infty}\sin n$ does not exist.

23. $\left\{\left(1+\dfrac{3}{n}\right)^{4n}\right\}$ is convergent. Let $y = \left(1+\dfrac{3}{x}\right)^{4x}$. Then

$$\lim_{x\to\infty}\ln y = \lim_{x\to\infty}4x\ln(1+3/x) = \lim_{x\to\infty}\frac{\ln(1+3/x)}{1/(4x)} \overset{\text{H}}{=} \lim_{x\to\infty}\frac{\dfrac{1}{1+3/x}\left(-\dfrac{3}{x^2}\right)}{-1/(4x^2)}$$

$$= \lim_{x\to\infty}\frac{12}{1+3/x} = 12, \text{ so } \lim_{x\to\infty}y = \lim_{n\to\infty}\left(1+\frac{3}{n}\right)^{4n} = e^{12}.$$

25. Use the Limit Comparison Test with $a_n = \dfrac{n^2}{n^3+1}$ and $b_n = \dfrac{1}{n}$. $\lim\limits_{n\to\infty}\dfrac{a_n}{b_n} = \lim\limits_{n\to\infty}\dfrac{n^2/(n^3+1)}{1/n}$ $= \lim\limits_{n\to\infty}\dfrac{1}{1+1/n^3} = 1$. Since $\sum\limits_{n=1}^{\infty}\dfrac{1}{n}$ (the harmonic series) diverges, $\sum\limits_{n=1}^{\infty}\dfrac{n^2}{n^3+1}$ diverges also.

27. An alternating series with $a_n = 1/(n^{1/4})$, $a_n > 0$ for all n, and $a_n > a_{n+1}$. $\lim\limits_{n\to\infty}a_n = \lim\limits_{n\to\infty}\left[1/(n^{1/4})\right] = 0$, so the series converges by the Alternating Series Test.

29. $\lim\limits_{n\to\infty}\left|\dfrac{a_{n+1}}{a_n}\right| = \lim\limits_{n\to\infty}\dfrac{4^{n+1}}{(n+1)3^{n+1}}\cdot\dfrac{n3^n}{4^n} = \frac{4}{3}\lim\limits_{n\to\infty}\dfrac{n}{n+1} = \frac{4}{3} > 1$ so the series diverges by the Ratio Test.

31. $\dfrac{|\sin n|}{1+n^2} \le \dfrac{1}{1+n^2} < \dfrac{1}{n^2}$ and since $\sum\limits_{n=1}^{\infty}\dfrac{1}{n^2}$ converges (p-series with $p = 2 > 1$), so does $\sum\limits_{n=1}^{\infty}\dfrac{|\sin n|}{1+n^2}$ by the Comparison Test.

33. $\lim\limits_{n\to\infty}\left|\dfrac{a_{n+1}}{a_n}\right| = \lim\limits_{n\to\infty}\dfrac{1\cdot3\cdot5\cdot\cdots\cdot(2n-1)(2n+1)}{5^{n+1}(n+1)!}\cdot\dfrac{5^n\,n!}{1\cdot3\cdot5\cdot\cdots\cdot(2n-1)} = \lim\limits_{n\to\infty}\dfrac{2n+1}{5(n+1)}$ $= \frac{2}{5} < 1$, so the series converges by the Ratio Test.

35. Convergent geometric series. $\sum\limits_{n=1}^{\infty} \frac{2^{2n+1}}{5^n} = 2\sum\limits_{n=1}^{\infty} \frac{4^n}{5^n} = 2\left(\frac{4/5}{1-4/5}\right) = 8.$

37. $\sum\limits_{n=1}^{\infty}\left[\tan^{-1}(n+1) - \tan^{-1}n\right] = \lim\limits_{n\to\infty}\left[\left(\tan^{-1}2 - \tan^{-1}1\right) + \left(\tan^{-1}3 - \tan^{-1}2\right) + \cdots + \left(\tan^{-1}(n+1) - \tan^{-1}n\right)\right]$

$= \lim\limits_{n\to\infty}\left[\tan^{-1}(n+1) - \tan^{-1}1\right] = \frac{\pi}{2} - \frac{\pi}{4} = \frac{\pi}{4}$

39. $1.2 + 0.0\overline{345} = \frac{12}{10} + \frac{345/10{,}000}{1 - 1/1000} = \frac{12}{10} + \frac{345}{9990} = \frac{4111}{3330}$

41. $\sum\limits_{n=1}^{\infty} \frac{(-1)^{n+1}}{n^5} = 1 - \frac{1}{32} + \frac{1}{243} - \frac{1}{1024} + \frac{1}{3125} - \frac{1}{7776} + \frac{1}{16{,}807} - \frac{1}{32{,}768} + \cdots.$

4ince $\frac{1}{32{,}768} < 0.000031$, $\sum\limits_{n=1}^{\infty} \frac{(-1)^{n+1}}{n^5} \approx \sum\limits_{n=1}^{7} \frac{(-1)^{n+1}}{n^5} \approx 0.9721.$

43. $\sum\limits_{n=1}^{\infty} \frac{1}{2+5^n} \approx \sum\limits_{n=1}^{8} \frac{1}{2+5^n} \approx 0.18976224$. To estimate the error, note that $\frac{1}{2+5^n} < \frac{1}{5^n}$, so the remainder term is $R_8 = \sum\limits_{n=9}^{\infty} \frac{1}{2+5^n} < \sum\limits_{n=9}^{\infty} \frac{1}{5^n} = \frac{1/5^9}{1-1/5} \approx 6.4 \times 10^{-7}$ (geometric series with $a = 1/5^9$ and $r = \frac{1}{5}$).

45. Use the Limit Comparison Test. $\lim\limits_{n\to\infty}\left|\frac{\left(\frac{n+1}{n}\right)a_n}{a_n}\right| = \lim\limits_{n\to\infty} \frac{n+1}{n} = \lim\limits_{n\to\infty}\left(1 + \frac{1}{n}\right) = 1 > 0$. Since $\sum |a_n|$ is convergent, so is $\sum\left|\left(\frac{n+1}{n}\right)a_n\right|$ by the Limit Comparison Test.

47. $\lim\limits_{n\to\infty}\left|\frac{a_{n+1}}{a_n}\right| = \lim\limits_{n\to\infty}\left|\frac{x^{n+1}}{3^{n+1}(n+1)^3} \cdot \frac{3^n n^3}{x^n}\right| = \frac{|x|}{3} \lim\limits_{n\to\infty}\left(\frac{n}{n+1}\right)^3 = \frac{|x|}{3} < 1$ for convergence (Ratio Test) $\Rightarrow$ $|x| < 3$ and the radius of convergence is 3. When $x = \pm 3$, $\sum\limits_{n=1}^{\infty} |a_n| = \sum\limits_{n=1}^{\infty} \frac{1}{n^3}$ which is a convergent p-series ($p = 3 > 1$), so the interval of convergence is $[-3, 3]$.

49. $\lim\limits_{n\to\infty}\left|\frac{a_{n+1}}{a_n}\right| = \lim\limits_{n\to\infty}\left|\frac{2^{n+1}(x-3)^{n+1}}{\sqrt{n+4}} \cdot \frac{\sqrt{n+3}}{2^n(x-3)^n}\right| = 2|x-3| \lim\limits_{n\to\infty}\sqrt{\frac{n+3}{n+4}} = 2|x-3| < 1 \quad\Leftrightarrow$ $|x-3| < \frac{1}{2}$ so the radius of convergence is $\frac{1}{2}$. For $x = \frac{7}{2}$ the series becomes $\sum\limits_{n=0}^{\infty} \frac{1}{\sqrt{n+3}} = \sum\limits_{n=3}^{\infty} \frac{1}{n^{1/2}}$ which diverges ($p = \frac{1}{2} < 1$), but for $x = \frac{5}{2}$ we get $\sum\limits_{n=0}^{\infty} \frac{(-1)^n}{\sqrt{n+3}}$ which is a convergent alternating series, so the interval of convergence is $\left[\frac{5}{2}, \frac{7}{2}\right)$.

51.

$f(x) = \sin x$	$f\left(\frac{\pi}{6}\right) = \frac{1}{2}$
$f'(x) = \cos x$	$f'\left(\frac{\pi}{6}\right) = \frac{\sqrt{3}}{2}$
$f''(x) = -\sin x$	$f''\left(\frac{\pi}{6}\right) = -\frac{1}{2}$
$f'''(x) = -\cos x$	$f'''\left(\frac{\pi}{6}\right) = -\frac{\sqrt{3}}{2}$
$f^{(4)}(x) = \sin x$	$f^{(4)}\left(\frac{\pi}{6}\right) = \frac{1}{2}$
$\cdots$	$\cdots$

$f^{(2n)}\left(\frac{\pi}{6}\right) = (-1)^n \cdot \frac{1}{2}$ and

$f^{(2n+1)}\left(\frac{\pi}{6}\right) = (-1)^n \cdot \frac{\sqrt{3}}{2}$.

$$\sin x = \sum_{n=0}^{\infty} \frac{f^{(n)}\left(\frac{\pi}{6}\right)}{n!}\left(x - \frac{\pi}{6}\right)^n$$

$$= \sum_{n=0}^{\infty} \frac{(-1)^n}{2(2n)!}\left(x - \frac{\pi}{6}\right)^{2n} + \sum_{n=0}^{\infty} \frac{(-1)^n\sqrt{3}}{2(2n+1)!}\left(x - \frac{\pi}{6}\right)^{2n+1}$$

53. $\frac{1}{1+x} = \frac{1}{1-(-x)} = \sum\limits_{n=0}^{\infty}(-1)^n x^n$ for $|x| < 1 \quad\Rightarrow\quad \frac{x^2}{1+x} = \sum\limits_{n=0}^{\infty}(-1)^n x^{n+2}$ with $R = 1$.

55. $\dfrac{1}{1-x} = \displaystyle\sum_{n=0}^{\infty} x^n$ for $|x| < 1 \quad \Rightarrow \quad \ln(1-x) = -\displaystyle\int \frac{dx}{1-x} = -\int \sum_{n=0}^{\infty} x^n \, dx = C - \sum_{n=0}^{\infty} \frac{x^{n+1}}{n+1}$.

$\ln(1-0) = C - 0 \quad \Rightarrow \quad C = 0 \quad \Rightarrow \quad \ln(1-x) = -\displaystyle\sum_{n=0}^{\infty} \frac{x^{n+1}}{n+1} = \sum_{n=1}^{\infty} \frac{-x^n}{n}$ with $R = 1$.

57. $\sin x = \displaystyle\sum_{n=0}^{\infty} \frac{(-1)^n x^{2n+1}}{(2n+1)!} \quad \Rightarrow \quad \sin(x^4) = \sum_{n=0}^{\infty} \frac{(-1)^n (x^4)^{2n+1}}{(2n+1)!} = \sum_{n=0}^{\infty} \frac{(-1)^n x^{8n+4}}{(2n+1)!}$ for all x, so the radius of convergence is ∞.

59. $(16-x)^{-1/4} = \frac{1}{2}\left(1 - \frac{1}{16}x\right)^{-1/4} = \dfrac{1}{2}\left[1 + \left(-\dfrac{1}{4}\right)\left(-\dfrac{x}{16}\right) + \dfrac{\left(-\frac{1}{4}\right)\left(-\frac{5}{4}\right)}{2!}\left(-\dfrac{x}{16}\right)^2 + \cdots\right]$

$= \dfrac{1}{2} + \displaystyle\sum_{n=1}^{\infty} \frac{1 \cdot 5 \cdot 9 \cdot \cdots \cdot (4n-3)}{2 \cdot 4^n \cdot n! \cdot 16^n} x^n = \frac{1}{2} + \sum_{n=1}^{\infty} \frac{1 \cdot 5 \cdot 9 \cdot \cdots \cdot (4n-3)}{2^{6n+1}\, n!} x^n$ for $\left|-\dfrac{x}{16}\right| < 1 \quad \Rightarrow \quad R = 16$.

61. $e^x = \displaystyle\sum_{n=0}^{\infty} \frac{x^n}{n!}$ so $\dfrac{e^x}{x} = \dfrac{1}{x} + \displaystyle\sum_{n=1}^{\infty} \frac{x^{n-1}}{n!}$ and $\displaystyle\int \frac{e^x}{x}\, dx = C + \ln|x| + \sum_{n=1}^{\infty} \frac{x^n}{n \cdot n!}$

63. **(a)**

$f(x) = x^{1/2}$ $\qquad f(1) = 1$

$f'(x) = \frac{1}{2}x^{-1/2}$ $\qquad f'(1) = \frac{1}{2}$

$f''(x) = -\frac{1}{4}x^{-3/2}$ $\qquad f''(1) = -\frac{1}{4}$

$f'''(x) = \frac{3}{8}x^{-5/2}$ $\qquad f'''(1) = \frac{3}{8}$

$f^{(4)}(x) = -\frac{15}{16}x^{-7/2}$

$\sqrt{x} \approx T_3(x) = 1 + \frac{1}{2}(x-1) - \frac{1}{8}(x-1)^2 + \frac{1}{16}(x-1)^3$

(b)

1.5

T_3

f

0 $\qquad$ 2

(c) By Taylor's Formula,

$R_3(x) = \dfrac{f^{(4)}(z)}{4!}(x-1)^4 = -\dfrac{5(x-1)^4}{128 z^{7/2}}$,

with z between x and 1. If $0.9 \le x \le 1.1$ then $0 \le |x-1| \le 0.1$ and $z^{7/2} > (0.9)^{7/2}$ so

$|R_3(x)| < \dfrac{5(0.1)^4}{128(0.9)^{7/2}} < 0.000006$.

(d)

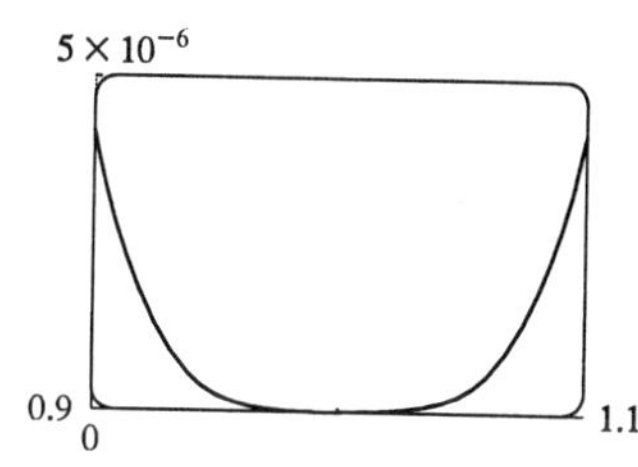

It appears that the error is less than 5×10^{-6} on $(0.9, 1.1)$.

65. $e^x = \displaystyle\sum_{n=0}^{\infty} \frac{x^n}{n!} \quad \Rightarrow \quad e^{-1/x^2} = \sum_{n=0}^{\infty} \frac{\left(-1/x^2\right)^n}{n!} = 1 - \frac{1}{x^2} + \frac{1}{2x^4} - \cdots \quad \Rightarrow$

$x^2\left(1 - e^{-1/x^2}\right) = x^2\left(\dfrac{1}{x^2} - \dfrac{1}{2x^4} + \cdots\right) = 1 - \dfrac{1}{2x^2} + \cdots \to 1$ as $x \to \infty$.

PROBLEMS PLUS (after Chapter 10)

1. It would be far too much work to compute 15 derivatives of f. The key idea is to remember that $f^{(n)}(0)$ occurs in the coefficient of x^n in the Maclaurin series of f. We start with the Maclaurin series for sin: $\sin x = x - \dfrac{x^3}{3!} + \dfrac{x^5}{5!} - \cdots$. Then $\sin(x^3) = x^3 - \dfrac{x^9}{3!} + \dfrac{x^{15}}{5!} - \cdots$ and so the coefficient of x^{15} is $\dfrac{f^{(15)}(0)}{15!} = \dfrac{1}{5!}$. Therefore, $f^{(15)}(0) = \dfrac{15!}{5!} = 6 \cdot 7 \cdot 8 \cdot 9 \cdot 10 \cdot 11 \cdot 12 \cdot 13 \cdot 14 \cdot 15 = 10{,}897{,}286{,}400$.

3. **(a)** From Formula 14a in Appendix D, with $x = y = \theta$, we get $\tan 2\theta = \dfrac{2\tan\theta}{1 - \tan^2\theta}$, so $\cot 2\theta = \dfrac{1 - \tan^2\theta}{2\tan\theta}$ $\Rightarrow$ $2\cot 2\theta = \dfrac{1 - \tan^2\theta}{\tan\theta} = \cot\theta - \tan\theta$. Replacing θ by $\frac{1}{2}x$, we get $2\cot x = \cot\frac{1}{2}x - \tan\frac{1}{2}x$, or $\tan\frac{1}{2}x = \cot\frac{1}{2}x - 2\cot x$.

(b) From part (a) we have $\tan\dfrac{x}{2^n} = \cot\dfrac{x}{2^n} - 2\cot\dfrac{x}{2^{n-1}}$, so the nth partial sum of the given series is

$$\begin{aligned}
s_n &= \frac{\tan(x/2)}{2} + \frac{\tan(x/4)}{4} + \frac{\tan(x/8)}{8} + \cdots + \frac{\tan(x/2^n)}{2^n} \\
&= \left[\frac{\cot(x/2)}{2} - \cot x\right] + \left[\frac{\cot(x/4)}{4} - \frac{\cot(x/2)}{2}\right] + \left[\frac{\cot(x/8)}{8} - \frac{\cot(x/4)}{4}\right] \\
&\qquad + \cdots + \left[\frac{\cot(x/2^n)}{2^n} - \frac{\cot(x/2^{n-1})}{2^{n-1}}\right] \\
&= -\cot x + \frac{\cot(x/2^n)}{2^n} \quad \text{(telescoping sum).}
\end{aligned}$$

Now $\dfrac{\cot(x/2^n)}{2^n} = \dfrac{\cos(x/2^n)}{2^n \sin(x/2^n)} = \dfrac{\cos(x/2^n)}{x} \cdot \dfrac{x/2^n}{\sin(x/2^n)} \to \dfrac{1}{x} \cdot 1 = \dfrac{1}{x}$ as $n \to \infty$ since $\dfrac{x}{2^n} \to 0$ for $x \neq 0$. Therefore, if $x \neq 0$ and $x \neq n\pi$, then $\displaystyle\sum_{n=1}^{\infty} \frac{1}{2^n}\tan\frac{x}{2^n} = \lim_{n\to\infty}\left(-\cot x + \frac{1}{2^n}\cot\frac{x}{2^n}\right) = -\cot x + \frac{1}{x}$.

If $x = 0$, then all terms in the series are 0, so the sum is 0.

5. **(a)** At each stage, each side is replaced by four shorter sides, each of length $\frac{1}{3}$ of the side length at the preceding stage. Writing s_0 and ℓ_0 for the number of sides and the length of the side of the initial triangle, we generate the following table.

$s_0 = 3$	$\ell_0 = 1$
$s_1 = 3 \cdot 4$	$\ell_1 = \frac{1}{3}$
$s_2 = 3 \cdot 4^2$	$\ell_2 = \frac{1}{3^2}$
$s_3 = 3 \cdot 4^3$	$\ell_3 = \frac{1}{3^3}$
$\cdots$	$\cdots$

In general, we have $s_n = 3 \cdot 4^n$ and $\ell_n = \left(\frac{1}{3}\right)^n$, so the length of the perimeter at the nth stage of construction is $p_n = s_n \ell_n = 3 \cdot 4^n \cdot \left(\frac{1}{3}\right)^n = 3 \cdot \left(\frac{4}{3}\right)^n$.

(b) $p_n = \dfrac{4^n}{3^{n-1}} = 4\left(\frac{4}{3}\right)^{n-1}$. Since $\frac{4}{3} > 1$, $p_n \to \infty$ as $n \to \infty$.

(c) The area of each of the small triangles added at a given stage is one-ninth of the area of the triangle added at the preceding stage. Let a be the area of the original triangle. Then the area a_n of each of the small triangles added at stage n is $a_n = a \cdot \dfrac{1}{9^n} = \dfrac{a}{9^n}$. Since a small triangle is added to each side at every stage, it follows that the total area A_n added to the figure at the nth stage is

$A_n = s_{n-1} \cdot a_n = 3 \cdot 4^{n-1} \cdot \dfrac{a}{9^n} = a \cdot \dfrac{4^{n-1}}{3^{2n-1}}$. Then the total area enclosed by the snowflake curve is

$A = a + A_1 + A_2 + A_3 + \cdots = a + a \cdot \dfrac{1}{3} + a \cdot \dfrac{4}{3^3} + a \cdot \dfrac{4^2}{3^5} + a \cdot \dfrac{4^3}{3^7} + \cdots$.

After the first term, this is a geometric series with common ratio $\frac{4}{9}$, so $A = a + \dfrac{a/3}{1 - \frac{4}{9}} = a + \dfrac{a}{3} \cdot \dfrac{9}{5} = \dfrac{8a}{5}$.

But the area of the original equilateral triangle with side 1 is $a = \frac{1}{2} \cdot 1 \cdot \sin\frac{\pi}{3} = \frac{\sqrt{3}}{4}$.

So the area enclosed by the snowflake curve is $\frac{8}{5} \cdot \frac{\sqrt{3}}{4} = \frac{2\sqrt{3}}{5}$.

7. (a) Let $a = \arctan x$ and $b = \arctan y$. Then, from the endpapers,

$\tan(a-b) = \dfrac{\tan a - \tan b}{1 + \tan a \tan b} = \dfrac{\tan(\arctan x) - \tan(\arctan y)}{1 + \tan(\arctan x)\tan(\arctan y)} \quad \Rightarrow \quad \tan(a-b) = \dfrac{x-y}{1+xy} \quad \Rightarrow$

$\arctan x - \arctan y = a - b = \arctan \dfrac{x-y}{1+xy}$ since $-\dfrac{\pi}{2} < \arctan x - \arctan y < \dfrac{\pi}{2}$.

(b) From part (a) we have $\arctan\frac{120}{119} - \arctan\frac{1}{239} = \arctan \dfrac{\frac{120}{119} - \frac{1}{239}}{1 + \frac{120}{119} \cdot \frac{1}{239}} = \arctan \dfrac{\frac{28{,}561}{28{,}441}}{\frac{28{,}561}{28{,}441}} = \arctan 1 = \frac{\pi}{4}$.

(c) Replacing y by $-y$ in the formula of part (a), we get $\arctan x + \arctan y = \arctan \dfrac{x+y}{1-xy}$. So

$$4 \arctan \tfrac{1}{5} = 2\left(\arctan\tfrac{1}{5} + \arctan\tfrac{1}{5}\right) = 2 \arctan \frac{\frac{1}{5} + \frac{1}{5}}{1 - \frac{1}{5} \cdot \frac{1}{5}} = 2 \arctan \tfrac{5}{12}$$

$$= \arctan\tfrac{5}{12} + \arctan\tfrac{5}{12} = \arctan \frac{\frac{5}{12} + \frac{5}{12}}{1 - \frac{5}{12} \cdot \frac{5}{12}} = \arctan\tfrac{120}{119}.$$

Thus, from part (b), we have $4 \arctan\frac{1}{5} - \arctan\frac{1}{239} = \arctan\frac{120}{119} - \arctan\frac{1}{239} = \frac{\pi}{4}$.

(d) From Example 7 in Section 10.9 we have $\arctan x = x - \dfrac{x^3}{3} + \dfrac{x^5}{5} - \dfrac{x^7}{7} + \dfrac{x^9}{9} - \dfrac{x^{11}}{11} + \cdots$, so

$\arctan\dfrac{1}{5} = \dfrac{1}{5} - \dfrac{1}{3 \cdot 5^3} + \dfrac{1}{5 \cdot 5^5} - \dfrac{1}{7 \cdot 5^7} + \dfrac{1}{9 \cdot 5^9} - \dfrac{1}{11 \cdot 5^{11}} + \cdots$. This is an alternating series and the size of the terms decreases to 0, so by Theorem 10.5.1, the sum lies between s_5 and s_6, that is, $0.197395560 < \arctan\frac{1}{5} < 0.197395562$.

(e) From the series in part (d) we get $\arctan\dfrac{1}{239} = \dfrac{1}{239} - \dfrac{1}{3 \cdot 239^3} + \dfrac{1}{5 \cdot 239^5} - \cdots$. The third term is less than 2.6×10^{-13}, so by Theorem 10.5.1 we have, to nine decimal places, $\arctan\frac{1}{239} \approx s_2 \approx 0.004184076$. Thus $0.004184075 < \arctan\frac{1}{239} < 0.004184077$.

(f) From part (c) we have $\pi = 16 \arctan\frac{1}{5} - 4 \arctan\frac{1}{239}$, so from parts (d) and (e) we have

$16(0.197395560) - 4(0.004184077) < \pi < 16(0.197395562) - 4(0.004184075) \quad \Rightarrow$

$3.141592652 < \pi < 3.141592692$. So, to 7 decimal places, $\pi \approx 3.1415927$.

9. We start with the geometric series $\sum_{n=0}^{\infty} x^n = \dfrac{1}{1-x}$, $|x| < 1$, and differentiate:

$$\sum_{n=1}^{\infty} nx^{n-1} = \frac{d}{dx}\left(\sum_{n=0}^{\infty} x^n\right) = \frac{d}{dx}\left(\frac{1}{1-x}\right) = \frac{1}{(1-x)^2} \text{ for } |x| < 1 \quad \Rightarrow \quad \sum_{n=1}^{\infty} nx^n = x\sum_{n=1}^{\infty} nx^{n-1} = \frac{x}{(1-x)^2}$$

for $|x| < 1$. Differentiate again: $\displaystyle\sum_{n=1}^{\infty} n^2x^{n-1} = \frac{d}{dx}\frac{x}{(1-x)^2} = \frac{(1-x)^2 - x \cdot 2(1-x)(-1)}{(1-x)^4} = \frac{x+1}{(1-x)^3} \quad \Rightarrow$

$$\sum_{n=1}^{\infty} n^2x^n = \frac{x^2+x}{(1-x)^3} \quad \Rightarrow \quad \sum_{n=1}^{\infty} n^3x^{n-1} = \frac{d}{dx}\frac{x^2+x}{(1-x)^3}$$

$$= \frac{(1-x)^3(2x+1) - (x^2+x)3(1-x)^2(-1)}{(1-x)^6} = \frac{x^2+4x+1}{(1-x)^4} \quad \Rightarrow \quad \sum_{n=1}^{\infty} n^3x^n = \frac{x^3+4x^2+x}{(1-x)^4}, \ |x| < 1.$$

The radius of convergence is 1 because that is the radius of convergence for the geometric series we started with.

If $x = \pm 1$, the series is $\sum n^3(\pm 1)^n$, which diverges by the Test For Divergence, so the interval of convergence is

$(-1, 1)$.

11. $u = 1 + \dfrac{x^3}{3!} + \dfrac{x^6}{6!} + \dfrac{x^9}{9!} + \cdots$, $v = x + \dfrac{x^4}{4!} + \dfrac{x^7}{7!} + \dfrac{x^{10}}{10!} + \cdots$, $w = \dfrac{x^2}{2!} + \dfrac{x^5}{5!} + \dfrac{x^8}{8!} + \cdots$.

The key idea is to differentiate: $\dfrac{du}{dx} = \dfrac{3x^2}{3!} + \dfrac{6x^5}{6!} + \dfrac{9x^8}{9!} + \cdots = \dfrac{x^2}{2!} + \dfrac{x^5}{5!} + \dfrac{x^8}{8!} + \cdots = w$.

Similarly, $\dfrac{dv}{dx} = 1 + \dfrac{x^3}{3!} + \dfrac{x^6}{6!} + \dfrac{x^9}{9!} + \cdots = u$, and $\dfrac{dw}{dx} = x + \dfrac{x^4}{4!} + \dfrac{x^7}{7!} + \dfrac{x^{10}}{10!} + \cdots = v$.

So $u' = w$, $v' = u$, and $w' = v$. Now differentiate the left hand side of the desired equation:

$$\frac{d}{dx}\left(u^3 + v^3 + w^3 - 3uvw\right) = 3u^2u' + 3v^3v' + 3w^3w' - 3(u'vw + uv'w + uvw')$$

$$= 3u^2w + 3v^2u + 3w^2v - 3(vw^2 + u^2w + uv^2) = 0 \quad \Rightarrow \quad u^3 + v^3 + w^3 - 3uvw = C.$$

To find the value of the constant C, we put $x = 1$ in the equation and get $1^3 + 0 + 0 - 3(1 \cdot 0 \cdot 0) = C \quad \Rightarrow$

$C = 1$, so $u^3 + v^3 + w^3 - 3uvw = 1$.

13. If L is the length of a side of the equilateral triangle, then the area is $A = \frac{1}{2}L \cdot \frac{\sqrt{3}}{2}L = \frac{\sqrt{3}}{4}L^2$ and so $L^2 = \frac{4}{\sqrt{3}}A$. Let r be the radius of one of the circles when there are n rows of circles. The figure shows that

$$L = \sqrt{3}r + r + (n-2)(2r) + r + \sqrt{3}r$$
$$= r\left(2n - 2 + 2\sqrt{3}\right), \text{ so } r = \frac{L}{2\left(n + \sqrt{3} - 1\right)}.$$

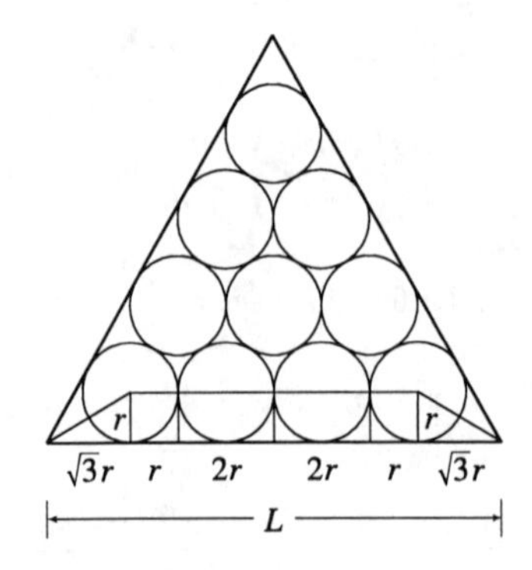

The number of circles is $1 + 2 + \cdots + n = \dfrac{n(n+1)}{2}$ and so the total area of the circles is

$$A_n = \frac{n(n+1)}{2}\pi r^2 = \frac{n(n+1)}{2}\pi\frac{L^2}{4\left(n+\sqrt{3}-1\right)^2} = \frac{n(n+1)}{2}\pi\frac{4A/\sqrt{3}}{4\left(n+\sqrt{3}-1\right)^2} = \frac{n(n+1)}{\left(n+\sqrt{3}-1\right)^2}\frac{\pi A}{2\sqrt{3}}$$

$$\Rightarrow \quad \frac{A_n}{A} = \frac{n(n+1)}{\left(n+\sqrt{3}-1\right)^2}\frac{\pi}{2\sqrt{3}} = \frac{1+1/n}{\left[1+\left(\sqrt{3}-1\right)/n\right]^2}\frac{\pi}{2\sqrt{3}} \to \frac{\pi}{2\sqrt{3}} \text{ as } n \to \infty.$$

15. Call the series S. We group the terms according to the number of digits in their denominators:

$$S = \underbrace{\left(1 + \frac{1}{2} + \cdots + \frac{1}{8} + \frac{1}{9}\right)}_{g_1} + \underbrace{\left(\frac{1}{11} + \cdots + \frac{1}{99}\right)}_{g_2} + \underbrace{\left(\frac{1}{111} + \cdots + \frac{1}{999}\right)}_{g_3} + \cdots$$

Now in the group g_n, there are 9^n terms, since we have 9 choices for each of the n digits in the denominator.

Furthermore, each term in g_n is less than $\dfrac{1}{10^{n-1}}$. So $g_n < 9^n \cdot \dfrac{1}{10^{n-1}} = 9\left(\dfrac{9}{10}\right)^{n-1}$.

Now $\displaystyle\sum_{n=1}^{\infty} 9\left(\frac{9}{10}\right)^{n-1}$ is a geometric series with $a = 9$ and $r = \frac{9}{10} < 1$. Therefore, by the Comparison Test,

$$S = \sum_{n=1}^{\infty} g_n < \sum_{n=1}^{\infty} 9\left(\frac{9}{10}\right)^{n-1} = \frac{9}{1 - \frac{9}{10}} = 90.$$

CHAPTER ELEVEN

EXERCISES 11.1

1. **(a)**

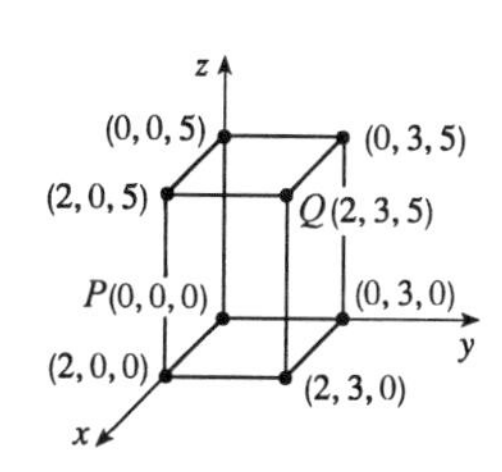

(b) $|PQ| = \sqrt{(2-0)^2 + (3-0)^2 + (5-0)^2}$
$= \sqrt{38}$

3. **(a)**

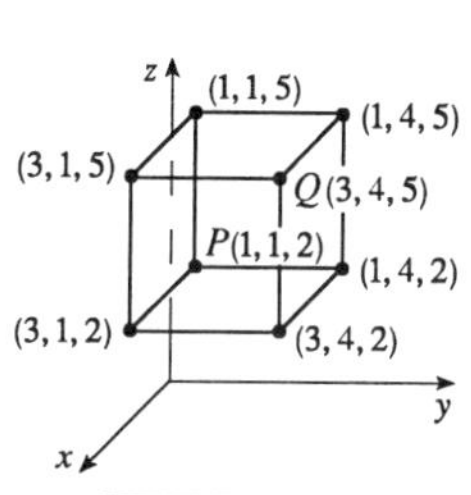

(b) $|PQ| = \sqrt{(3-1)^2 + (4-1)^2 + (5-2)^2}$
$= \sqrt{22}$

5. $|AB| = \sqrt{(3-2)^2 + (3-1)^2 + (4-0)^2} = \sqrt{21}$
$|BC| = \sqrt{(5-3)^2 + (4-3)^2 + (3-4)^2} = \sqrt{6}$
$|CA| = \sqrt{(5-2)^2 + (4-1)^2 + (3-0)^2} = \sqrt{27} = 3\sqrt{3}$
Since no two of the sides are equal in length the triangle isn't isosceles. But $|AB|^2 + |BC|^2 = 27 = |CA|^2$, so the triangle is a right triangle.

7. $|AB| = \sqrt{[5-(-2)]^2 + (4-6)^2 + (-3-1)^2} = \sqrt{49+4+16} = \sqrt{69}$
$|BC| = \sqrt{(2-5)^2 + (-6+4)^2 + [4-(-3)]^2} = \sqrt{9+100+49} = \sqrt{158}$
$|CA| = \sqrt{(-2-2)^2 + [6-(-6)]^2 + (1-4)^2} = \sqrt{16+144+9} = \sqrt{169} = 13$
Since no two sides are of equal length and since $|AB|^2 + |BC|^2 \neq |CA|^2$ the triangle is neither isosceles nor a right triangle.

9. $|PQ| = \sqrt{(0-1)^2 + (3-2)^2 + (7-3)^2} = \sqrt{18} = 3\sqrt{2}$
$|PR| = \sqrt{(3-1)^2 + (5-2)^2 + (11-3)^2} = \sqrt{4+9+64} = \sqrt{77}$
$|QR| = \sqrt{(3-0)^2 + (5-3)^2 + (11-7)^2} = \sqrt{9+4+16} = \sqrt{29}$
Since the sum of the two shortest distances isn't equal to the longest distance, the points aren't collinear. To show that $\sqrt{18} + \sqrt{29} \neq \sqrt{77}$, assume they are equal. Then squaring both sides gives $18 + 29 + 2\sqrt{(18)(29)} = 77$; solving for the radical and squaring again gives $(18)(29) = (15)^2$ or $522 = 225$ which of course is not true, so the values can't be equal.

11. $(x-0)^2+(y-1)^2+[z-(-1)]^2=4^2$ or $x^2+(y-1)^2+(z+1)^2=16$

13. $(x+6)^2+(y+1)^2+(z-2)^2=12$

15. Completing the squares in the equation gives

$(x^2+2x+1)+(y^2+8y+16)+(z^2-4z+4)=28+1+16+4 \quad\Rightarrow$

$(x+1)^2+(y+4)^2+(z-2)^2=49 \quad\Rightarrow\quad C(-1,-4,2)$, and $r=7$.

17. $(x^2+x+\frac{1}{4})+(y^2-2y+1)+(z^2+6z+9)=2+\frac{1}{4}+1+9 \quad\Rightarrow\quad (x+\frac{1}{2})^2+(y-1)^2+(z+3)^2=\frac{49}{4}$

$\Rightarrow\quad C(-\frac{1}{2},1,-3)$, and $r=\frac{7}{2}$.

19. $(x^2-x+\frac{1}{4})+y^2+z^2=0+\frac{1}{4}\Rightarrow(x-\frac{1}{2})^2+(y-0)^2+(z-0)^2=\frac{1}{4}\quad\Rightarrow\quad C(\frac{1}{2},0,0)$, $r=\frac{1}{2}$.

21. Call the given point Q and show that $|P_1Q|=|QP_2|=\frac{1}{2}|P_1P_2|$.

$|P_1P_2|=\sqrt{(x_2-x_1)^2+(y_2-y_1)^2+(z_2-z_1)^2}$

$|P_1Q|=\sqrt{\left[\frac{1}{2}(x_1+x_2)-x_1\right]^2+\left[\frac{1}{2}(y_1+y_2)-y_1\right]^2+\left[\frac{1}{2}(z_1+z_2)-z_1\right]^2}$

$=\frac{1}{2}\sqrt{(x_2-x_1)^2+(y_2-y_1)^2+(z_2-z_1)^2}=\frac{1}{2}|P_1P_2|$

Similarly $|QP_2|=\frac{1}{2}|P_1P_2|$.

23. From Exercise 21, the midpoints of sides AB, BC and CA are respectively $P_1(-\frac{1}{2},1,4)$, $P_2(1,\frac{1}{2},5)$ and $P_3(\frac{5}{2},\frac{3}{2},4)$. Then the lengths of the medians are:

$|AP_2|=\sqrt{0^2+(\frac{1}{2}-2)^2+(5-3)^2}=\sqrt{\frac{9}{4}+4}=\sqrt{\frac{25}{4}}=\frac{5}{2}$,

$|BP_3|=\sqrt{(\frac{5}{2}+2)^2+(\frac{3}{2})^2+(4-5)^2}=\sqrt{\frac{81}{4}+\frac{9}{4}+1}=\sqrt{\frac{94}{4}}=\frac{1}{2}\sqrt{94}$, and

$|CP_1|=\sqrt{(-\frac{1}{2}-4)^2+(1-1)^2+(4-5)^2}=\sqrt{\frac{81}{4}+1}=\frac{1}{2}\sqrt{85}$.

25. **(a)** Since the sphere touches the xy-plane, its radius is the distance from its center, $(2,-3,6)$, to the xy-plane, namely 6. Therefore $r=6$ and the equation of the sphere is $(x-2)^2+(y+3)^2+(z-6)^2=36$.

(b) Here $r=$ distance from center to yz-plane $=2$. Therefore, the equation is

$(x-2)^2+(y+3)^2+(z-6)^2=4$.

(c) Here $r=$ distance from center to xz-plane $=3$ Therefore, the equation is

$(x-2)^2+(y+3)^2+(z-6)^2=9$.

27. We need to find a set of points $\{P(x,y,z)\mid |AP|=|BP|\}$.

$\sqrt{(x+1)^2+(y-5)^2+(z-3)^2}=\sqrt{(x-6)^2+(y-2)^2+(z+2)^2}\quad\Rightarrow$

$(x+1)^2+(y-5)^2+(z-3)^2=(x-6)^2+(y-2)^2+(z+2)^2\quad\Rightarrow$

$x^2+2x+1+y^2-10y+25+z^2-6z+9=x^2-12x+36+y^2-4y+4+z^2+4z+4\quad\Rightarrow$

$14x-6y-10z=9$. Thus the set of points is a plane perpendicular to the line segment joining A and B (since this plane must contain the perpendicular bisector of the line segment AB).

29. A plane parallel to the yz-plane and 9 units in front of it.

31. A half-space containing all points to the right of the plane $y = 2$.

33. A plane perpendicular to the xz-plane and intersecting the xz-plane in the line $x = z$, $y = 0$.

35. A right circular cylinder with radius 1 and axis the z-axis.

37. All points outside the sphere with radius 1 and center $(0, 0, 0)$.

39. Completing the square in z gives $x^2 + y^2 + (z^2 - 2z + 1) < 3 + 1$ or

$x^2 + y^2 + (z-1)^2 < 4$, all points inside the sphere with radius two and center $(0, 0, 1)$.

41. In the xy-plane the equation $xy = 1$ represents a hyperbola with center at the origin. Since z can assume any value, the region in $\mathbb{R}^3$ is a hyperbolic cylinder.

43. All points on and between the two horizontal planes $z = 2$ and $z = -2$.

45. $y < 0$

47. $r < \sqrt{x^2 + y^2 + z^2} < R$, or $r^2 < x^2 + y^2 + z^2 < R^2$

49. **(a)** To find the x- and y-coordinates of the point P, we project it onto L_2 and project the resulting point A onto the x- and y-axes.
To find the z-coordinate, we project P onto either the xz-plane or the yz-plane (using our knowledge of its x- or y-coordinate) and then project the resulting point onto the z-axis. (Or, we could draw a line parallel to AO from P to the z-axis.)
The coordinates of P are $(2, 1, 4)$.

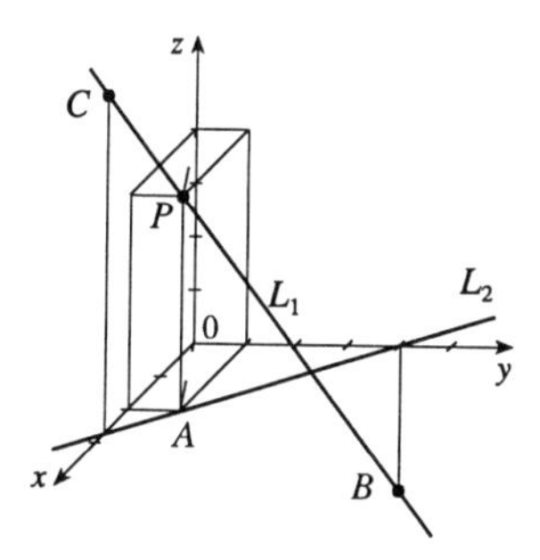

(b) A is the intersection of L_1 and L_2, B is directly below the y-intercept of L_2, and C is directly above the x-intercept of L_2.

EXERCISES 11.2

1. $\overrightarrow{PQ} = \overrightarrow{OQ} - \overrightarrow{OP} = \langle 6, 5, 0 \rangle - \langle 1, 3, 2 \rangle = \langle 5, 2, -2 \rangle$

3. $\mathbf{a} = \langle 2, 0, -2 \rangle$

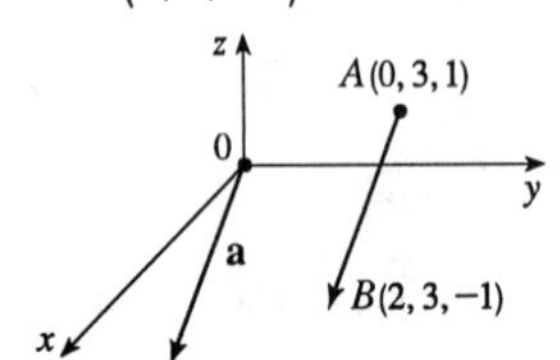

5. $\langle 1, 0, 1 \rangle + \langle 0, 0, 1 \rangle = \langle 1, 0, 2 \rangle$

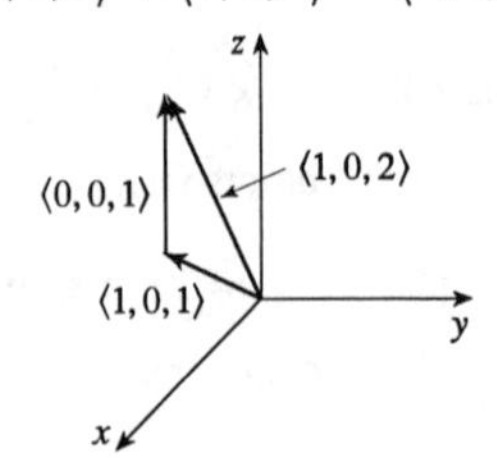

7. $|\mathbf{a}| = \sqrt{2^2 + (-3)^2 + 6^2} = \sqrt{49} = 7$ $\qquad \mathbf{a} + \mathbf{b} = \langle 3, -2, 10 \rangle$

$\mathbf{a} - \mathbf{b} = \langle 1, -4, 2 \rangle$ $\qquad 2\mathbf{a} = \langle 4, -6, 12 \rangle$

$3\mathbf{a} + 4\mathbf{b} = \langle 6, -9, 18 \rangle + \langle 4, 4, 16 \rangle = \langle 10, -5, 34 \rangle$

9. $|\mathbf{a}| = \sqrt{1^2 + 1^2 + 1^2} = \sqrt{3}$ $\qquad \mathbf{a} + \mathbf{b} = 3\mathbf{i} + 4\mathbf{k}$

$\mathbf{a} - \mathbf{b} = \mathbf{i} + \mathbf{j} + \mathbf{k} - 2\mathbf{i} + \mathbf{j} - 3\mathbf{k} = -\mathbf{i} + 2\mathbf{j} - 2\mathbf{k}$ $\qquad 2\mathbf{a} = 2\mathbf{i} + 2\mathbf{j} + 2\mathbf{k}$

$3\mathbf{a} + 4\mathbf{b} = (3\mathbf{i} + 3\mathbf{j} + 3\mathbf{k}) + (8\mathbf{i} - 4\mathbf{j} + 12\mathbf{k}) = 11\mathbf{i} - \mathbf{j} + 15\mathbf{k}$

11. $|\langle -2, 4, 3 \rangle| = \sqrt{(-2)^2 + 4^2 + 3^2} = \sqrt{29}$. Thus $\mathbf{u} = \frac{1}{\sqrt{29}}\langle -2, 4, 3 \rangle = \left\langle -\frac{2}{\sqrt{29}}, \frac{4}{\sqrt{29}}, \frac{3}{\sqrt{29}} \right\rangle$.

13. $|\mathbf{i} + \mathbf{j}| = \sqrt{1^2 + 1^2} = \sqrt{2}$. Thus $\mathbf{u} = \frac{1}{\sqrt{2}}(\mathbf{i} + \mathbf{j}) = \frac{1}{\sqrt{2}}\mathbf{i} + \frac{1}{\sqrt{2}}\mathbf{j}$.

15. $\mathbf{a} \cdot \mathbf{b} = (2)(3)\cos\frac{\pi}{3} = 6 \cdot \frac{1}{2} = 3$

17. $\mathbf{a} \cdot \mathbf{b} = (4)(-2) + (7)(1) + (-1)(4) = -5$

19. $\mathbf{a} \cdot \mathbf{b} = (2)(1) + (3)(-3) + (-4)(1) = -11$

21. $\mathbf{a} \cdot \mathbf{i} = \langle a_1, a_2, a_3 \rangle \cdot \langle 1, 0, 0 \rangle = (a_1)(1) + (a_2)(0) + (a_3)(0) = a_1$. Similarly $\mathbf{a} \cdot \mathbf{j} = (a_1)(0) + (a_2)(1) + (a_3)(0) = a_2$ and $\mathbf{a} \cdot \mathbf{k} = (a_1)(0) + (a_2)(0) + (a_3)(1) = a_3$.

23. $|\mathbf{a}| = \sqrt{1^2 + 2^2 + 2^2} = 3$, $|\mathbf{b}| = \sqrt{3^2 + 4^2 + 0^2} = 5$, $\mathbf{a} \cdot \mathbf{b} = 3 + 8 + 0 = 11$, $\cos\theta = \frac{11}{3 \cdot 5}$, so $\theta = \cos^{-1}\frac{11}{15} \approx 43°$.

25. $|\mathbf{a}| = \sqrt{36 + 4 + 9} = 7$, $|\mathbf{b}| = \sqrt{3}$, $\mathbf{a} \cdot \mathbf{b} = 6 - 2 - 3 = 1$, $\cos\theta = \dfrac{1}{7\sqrt{3}}$, so $\theta = \cos^{-1}\dfrac{1}{7\sqrt{3}} \approx 85°$.

27. Let a, b and c be the angles at vertices A, B and C respectively. Then a is the angle between vectors $\overrightarrow{AB}$ and $\overrightarrow{AC}$, b is the angle between vectors $\overrightarrow{BA}$ and $\overrightarrow{BC}$, and c is the angle between vectors $\overrightarrow{CA}$ and $\overrightarrow{CB}$. Thus

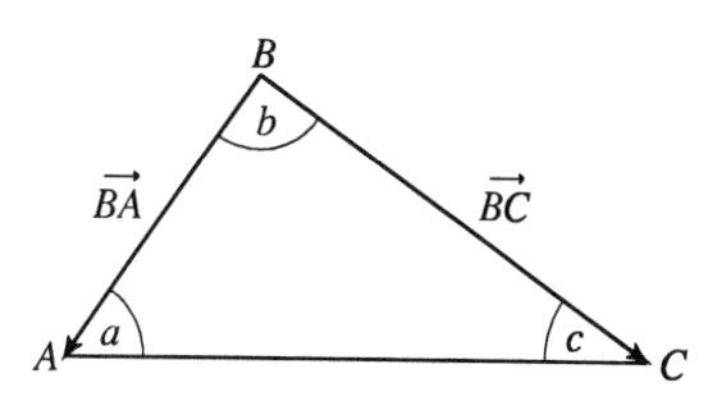

$$\cos a = \frac{\overrightarrow{AB}\cdot\overrightarrow{AC}}{\left|\overrightarrow{AB}\right|\left|\overrightarrow{AC}\right|} = \langle 5,-1,2\rangle\cdot\langle -2,-4,-3\rangle\frac{1}{\sqrt{30\cdot 29}} = \frac{1}{\sqrt{870}}(-10+4-6) = -\frac{12}{\sqrt{870}} \text{ and}$$

$a = \cos^{-1}\left(-\frac{12}{\sqrt{870}}\right) \approx 114°$. Similarly

$$\cos b = \frac{\overrightarrow{BA}\cdot\overrightarrow{BC}}{\left|\overrightarrow{BA}\right|\left|\overrightarrow{BC}\right|} = \frac{1}{\sqrt{30\cdot 83}}\langle -5,1,-2\rangle\cdot\langle -7,-3,-5\rangle = \frac{1}{\sqrt{2490}}(35-3+10) = \frac{42}{\sqrt{2490}}, \text{ so}$$

$b = \cos^{-1}\frac{42}{\sqrt{2490}} \approx 33°$, and

$$\cos c = \frac{\overrightarrow{CA}\cdot\overrightarrow{CB}}{\left|\overrightarrow{CA}\right|\left|\overrightarrow{CB}\right|} = \frac{1}{\sqrt{29\cdot 83}}\langle 2,4,3\rangle\cdot\langle 7,3,5\rangle = \frac{1}{\sqrt{2407}}(14+12+15) = \frac{41}{\sqrt{2407}}, \text{ so}$$

$c = \cos^{-1}\frac{41}{\sqrt{2407}} \approx 33°$.

Alternate Solution: Apply the Law of Cosines three times as follows: $\cos a = \dfrac{\left|\overrightarrow{BC}\right|^2 - \left|\overrightarrow{AB}\right|^2 - \left|\overrightarrow{AC}\right|^2}{2\left|\overrightarrow{AB}\right|\left|\overrightarrow{AC}\right|}$,

$$\cos b = \frac{\left|\overrightarrow{AC}\right|^2 - \left|\overrightarrow{AB}\right|^2 - \left|\overrightarrow{BC}\right|^2}{2\left|\overrightarrow{AB}\right|\left|\overrightarrow{BC}\right|}, \text{ and } \cos c = \frac{\left|\overrightarrow{AB}\right|^2 - \left|\overrightarrow{AC}\right|^2 - \left|\overrightarrow{BC}\right|^2}{2\left|\overrightarrow{AC}\right|\left|\overrightarrow{BC}\right|}.$$

29. $\mathbf{a}\cdot\mathbf{b} = -2+16+(-15) \neq 0$, so **a** and **b** are not orthogonal. Also since **a** isn't a scalar multiple of **b**, **a** and **b** aren't parallel.

31. $\mathbf{a}\cdot\mathbf{b} = 3+(-1)+(-2) = 0$, so **a** and **b** are orthogonal.

33. For the two vectors to be orthogonal, we need $0 = \langle x,1,2\rangle\cdot\langle 3,4,x\rangle = 3x+4+2x = 5x+4 \Leftrightarrow x = -\frac{4}{5}$.

35. Let $\mathbf{a} = a_1\mathbf{i} + a_2\mathbf{j} + a_3\mathbf{k}$ be a vector orthogonal to both $\mathbf{i}+\mathbf{j}$ and $\mathbf{i}+\mathbf{k}$. Then, by definition, $a_1 + a_2 = 0$ and $a_1 + a_3 = 0$, so $a_1 = -a_2 = -a_3$. Furthermore **a** is to be a unit vector, so $1 = a_1^2 + a_2^2 + a_3^2 = 3a_1^2$ implies $a_1 = \pm\frac{1}{\sqrt{3}}$. Thus $\mathbf{a} = \frac{1}{\sqrt{3}}\mathbf{i} - \frac{1}{\sqrt{3}}\mathbf{j} - \frac{1}{\sqrt{3}}\mathbf{k}$ and $\mathbf{a} = -\frac{1}{\sqrt{3}}\mathbf{i} + \frac{1}{\sqrt{3}}\mathbf{j} + \frac{1}{\sqrt{3}}\mathbf{k}$ are the two such unit vectors.

37. $|\langle 1,2,2\rangle| = \sqrt{1+4+4} = 3$, so $\cos\alpha = \frac{1}{3}$, $\cos\beta = \frac{2}{3}$ and $\cos\gamma = \frac{2}{3}$, while $\alpha = \cos^{-1}\frac{1}{3} \approx 71°$ and $\beta = \gamma = \cos^{-1}\frac{2}{3} \approx 48°$.

39. $|-8\mathbf{i}+3\mathbf{j}+2\mathbf{k}| = \sqrt{64+9+4} = \sqrt{77}$, so $\cos\alpha = -\frac{8}{\sqrt{77}}$, $\cos\beta = \frac{3}{\sqrt{77}}$ and $\cos\gamma = \frac{2}{\sqrt{77}}$, while $\alpha = \cos^{-1}\frac{-8}{\sqrt{77}} \approx 156°$, $\beta = \cos^{-1}\frac{3}{\sqrt{77}} \approx 70°$ and $\gamma = \cos^{-1}\frac{2}{\sqrt{77}} \approx 77°$.

41. $|\langle 2, 1.2, 0.8\rangle| = \sqrt{4 + 1.44 + 0.64} = \frac{5}{5}\sqrt{6.08} = \frac{\sqrt{152}}{5}$, so $\cos\alpha = \frac{10}{\sqrt{152}} = \frac{5}{\sqrt{38}}$, $\cos\beta = \frac{6}{\sqrt{152}} = \frac{3}{\sqrt{38}}$ and $\cos\gamma = \frac{4}{\sqrt{152}} = \frac{2}{\sqrt{38}}$, while $\alpha = \cos^{-1}\frac{5}{\sqrt{38}} \approx 36°$, $\beta = \cos^{-1}\frac{3}{\sqrt{38}} \approx 61°$ and $\gamma = \cos^{-1}\frac{2}{\sqrt{38}} \approx 71°$.

43. $|\mathbf{a}| = \sqrt{4+9} = \sqrt{13}$. The scalar projection of **b** onto **a** is $\text{comp}_\mathbf{a}\,\mathbf{b} = \dfrac{\mathbf{a}\cdot\mathbf{b}}{|\mathbf{a}|} = \dfrac{2\cdot 4 + 3\cdot 1}{\sqrt{13}} = \dfrac{11}{\sqrt{13}}$.

The vector projection of **b** onto **a** is $\text{proj}_\mathbf{a}\,\mathbf{b} = \dfrac{\mathbf{a}\cdot\mathbf{b}}{|\mathbf{a}|^2}\mathbf{a} = \frac{11}{\sqrt{13}} \cdot \frac{1}{\sqrt{13}}\langle 2, 3\rangle = \frac{11}{13}\langle 2, 3\rangle = \langle \frac{22}{13}, \frac{33}{13}\rangle$.

45. $|\mathbf{a}| = \sqrt{16+4+0} = 2\sqrt{5}$ so the scalar projection of **b** onto **a** is $\text{comp}_\mathbf{a}\,\mathbf{b} = \dfrac{\mathbf{a}\cdot\mathbf{b}}{|\mathbf{a}|} = \dfrac{1}{2\sqrt{5}}(4+2+0) = \dfrac{3}{\sqrt{5}}$.

The vector projection of **b** onto **a** is $\text{proj}_\mathbf{a}\,\mathbf{b} = \dfrac{\mathbf{a}\cdot\mathbf{b}}{|\mathbf{a}|^2}\mathbf{a} = \frac{3}{\sqrt{5}} \cdot \frac{1}{2\sqrt{5}}\langle 4, 2, 0\rangle = \frac{1}{5}\langle 6, 3, 0\rangle = \langle \frac{6}{5}, \frac{3}{5}, 0\rangle$.

47. $|\mathbf{a}| = \sqrt{1+0+1} = \sqrt{2}$ so the scalar projection of **b** onto **a** is $\text{comp}_\mathbf{a}\,\mathbf{b} = \dfrac{\mathbf{a}\cdot\mathbf{b}}{|\mathbf{a}|} = \frac{1}{\sqrt{2}}(1+0+0) = \frac{1}{\sqrt{2}}$ while the vector projection is $\text{proj}_\mathbf{a}\,\mathbf{b} = \dfrac{\mathbf{a}\cdot\mathbf{b}}{|\mathbf{a}|^2}\mathbf{a} = \frac{1}{\sqrt{2}} \cdot \frac{1}{\sqrt{2}}(\mathbf{i}+\mathbf{k}) = \frac{1}{2}(\mathbf{i}+\mathbf{k})$.

49. $(\text{orth}_\mathbf{a}\,\mathbf{b})\cdot\mathbf{a} = (\mathbf{b} - \text{proj}_\mathbf{a}\mathbf{b})\cdot\mathbf{a} = \mathbf{b}\cdot\mathbf{a} - (\text{proj}_\mathbf{a}\,\mathbf{b})\cdot\mathbf{a} = \mathbf{b}\cdot\mathbf{a} - \dfrac{\mathbf{a}\cdot\mathbf{b}}{|\mathbf{a}|^2}\mathbf{a}\cdot\mathbf{a} = \mathbf{b}\cdot\mathbf{a} - \dfrac{\mathbf{a}\cdot\mathbf{b}}{|\mathbf{a}|^2}|\mathbf{a}|^2$

$= \mathbf{b}\cdot\mathbf{a} - \mathbf{a}\cdot\mathbf{b} = 0$. So they are orthogonal by (7).

51. $\text{comp}_\mathbf{a}\,\mathbf{b} = \dfrac{\mathbf{a}\cdot\mathbf{b}}{|\mathbf{a}|} = 2 \quad\Leftrightarrow\quad \mathbf{a}\cdot\mathbf{b} = 2|\mathbf{a}| = 2\sqrt{10}$. If $\mathbf{b} = \langle b_1, b_2, b_3\rangle$ then we need $3b_1 + 0b_2 - 1b_3 = 2\sqrt{10}$.

One possible solution is obtained by taking $b_1 = 0$, $b_2 = 0$, $b_3 = -2\sqrt{10}$.

In general, $\mathbf{b} = \langle s, t, 3s - 2\sqrt{10}\rangle$, $s, t \in \mathbb{R}$.

53. Here $\mathbf{D} = (4-2)\mathbf{i} + (9-3)\mathbf{j} + (15-0)\mathbf{k} = 2\mathbf{i} + 6\mathbf{j} + 15\mathbf{k}$ so $W = \mathbf{F}\cdot\mathbf{D} = 20 + 108 - 90 = 38$ joules.

55. $W = |\mathbf{F}||\mathbf{D}|\cos\theta = (25)(10)\cos 20° \approx 235$ ft-lb

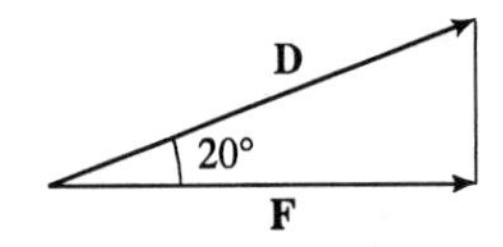

57. $|\mathbf{r} - \mathbf{r}_0|$ is the distance between the points (x, y, z) and (x_0, y_0, z_0), so the set of points is a sphere with radius 1 and center (x_0, y_0, z_0).

Alternate Method: $|\mathbf{r} - \mathbf{r}_0| = 1 \quad\Leftrightarrow\quad \sqrt{(x-x_0)^2 + (y-y_0)^2 + (z-z_0)^2} = 1 \quad\Leftrightarrow$

$(x-x_0)^2 + (y-y_0)^2 + (z-z_0)^2 = 1$, which is the equation of a sphere with radius 1 and center (x_0, y_0, z_0).

59. For convenience, consider the unit cube positioned so that its back left corner is at the origin, and its edges lie along the coordinate axes. The diagonal of the cube that begins at the origin and ends at $(1,1,1)$ has vector representation $\langle 1,1,1\rangle$. The angle θ between this vector and the vector of the edge which also begins at the origin and runs along the x-axis $\left[\text{that is, } \langle 1,0,0\rangle\right]$ is given by $\cos\theta = \dfrac{\langle 1,1,1\rangle \cdot \langle 1,0,0\rangle}{|\langle 1,1,1\rangle||\langle 1,0,0\rangle|} = \dfrac{1}{\sqrt{3}} \quad\Rightarrow$ $\theta = \cos^{-1}\frac{1}{\sqrt{3}} \approx 55°$.

61. Consider the H-C-H combination consisting of the sole carbon atom and the two hydrogen atoms that are at $(1,0,0)$ and $(0,1,0)$ (or any H-C-H combination, for that matter). Vector representations of the line segments emanating from the carbon atom and extending to these two hydrogen atoms are $\langle 1-\frac{1}{2}, 0-\frac{1}{2}, 0-\frac{1}{2}\rangle = \langle \frac{1}{2}, -\frac{1}{2}, -\frac{1}{2}\rangle$ and $\langle 0-\frac{1}{2}, 1-\frac{1}{2}, 0-\frac{1}{2}\rangle = \langle -\frac{1}{2}, \frac{1}{2}, -\frac{1}{2}\rangle$. The bond angle, θ, is therefore given by $\cos\theta = \dfrac{\langle \frac{1}{2}, -\frac{1}{2}, -\frac{1}{2}\rangle \cdot \langle -\frac{1}{2}, \frac{1}{2}, -\frac{1}{2}\rangle}{|\langle \frac{1}{2}, -\frac{1}{2}, -\frac{1}{2}\rangle||\langle -\frac{1}{2}, \frac{1}{2}, -\frac{1}{2}\rangle|} = \dfrac{-\frac{1}{4}-\frac{1}{4}+\frac{1}{4}}{\sqrt{\frac{3}{4}}\sqrt{\frac{3}{4}}} = -\dfrac{1}{3} \quad\Rightarrow\quad \theta = \cos^{-1}\left(-\frac{1}{3}\right) \approx 109.5°$.

EXERCISES 11.3

1. **(a)** Since $\mathbf{b}\times\mathbf{c}$ is a vector, the dot product $\mathbf{a}\cdot(\mathbf{b}\times\mathbf{c})$ is meaningful and is a scalar.

(b) $\mathbf{b}\cdot\mathbf{c}$ is a scalar, so $\mathbf{a}\times(\mathbf{b}\cdot\mathbf{c})$ is meaningless, as the cross product is defined only for two *vectors*.

(c) Since $\mathbf{b}\times\mathbf{c}$ is a vector, the cross product $\mathbf{a}\times(\mathbf{b}\times\mathbf{c})$ is meaningful and results in another vector.

(d) $\mathbf{a}\cdot\mathbf{b}$ is a scalar, so the cross product $(\mathbf{a}\cdot\mathbf{b})$ is meaningless.

(e) Since $(\mathbf{a}\cdot\mathbf{b})$ and $(\mathbf{c}\cdot\mathbf{d})$ are both scalars, the cross product $(\mathbf{a}\cdot\mathbf{b})\times(\mathbf{c}\cdot\mathbf{d})$ is meaningless.

(f) $\mathbf{a}\times\mathbf{b}$ and $\mathbf{c}\times\mathbf{d}$ are both vectors, so the dot product $(\mathbf{a}\times\mathbf{b})\cdot(\mathbf{c}\times\mathbf{d})$ is meaningful and is a scalar.

3. If we sketch $\mathbf{u}$ and $\mathbf{v}$ starting from the same initial point, we see that the angle between them is $30°$, so $|\mathbf{u}\times\mathbf{v}| = |\mathbf{u}||\mathbf{v}|\sin 30° = (6)(8)\left(\frac{1}{2}\right) = 24$. By the right-hand rule, $\mathbf{u}\times\mathbf{v}$ is directed into the page.

5. The magnitude of the torque is

$|\tau| = |\mathbf{r}\times\mathbf{F}| = |\mathbf{r}||\mathbf{F}|\sin\theta = (0.18\,\text{m})(60\,\text{N})\sin(180 - (70+10))° = 10.8\sin 100° \approx 10.6\,\text{J}$.

7. $\mathbf{a}\times\mathbf{b}=\begin{vmatrix}\mathbf{i}&\mathbf{j}&\mathbf{k}\\-2&3&4\\3&0&1\end{vmatrix}=\begin{vmatrix}3&4\\0&1\end{vmatrix}\mathbf{i}-\begin{vmatrix}-2&4\\3&1\end{vmatrix}\mathbf{j}+\begin{vmatrix}-2&3\\3&0\end{vmatrix}\mathbf{k}=3\mathbf{i}+14\mathbf{j}-9\mathbf{k}$

9. $\mathbf{a}\times\mathbf{b}=\begin{vmatrix}\mathbf{i}&\mathbf{j}&\mathbf{k}\\1&1&1\\1&1&-1\end{vmatrix}=\begin{vmatrix}1&1\\1&-1\end{vmatrix}\mathbf{i}-\begin{vmatrix}1&1\\1&-1\end{vmatrix}\mathbf{j}+\begin{vmatrix}1&1\\1&1\end{vmatrix}\mathbf{k}=-2\mathbf{i}+2\mathbf{j}$

11. $\begin{vmatrix}\mathbf{i}&\mathbf{j}&\mathbf{k}\\1&-1&1\\0&4&4\end{vmatrix}=\begin{vmatrix}-1&1\\4&4\end{vmatrix}\mathbf{i}-\begin{vmatrix}1&1\\0&4\end{vmatrix}\mathbf{j}+\begin{vmatrix}1&-1\\0&4\end{vmatrix}\mathbf{k}=-8\mathbf{i}-4\mathbf{j}+4\mathbf{k}$ and this cross product is orthogonal to both of the original vectors. So two unit vectors orthogonal to both are $\pm\dfrac{\langle -8,-4,4\rangle}{\sqrt{64+16+16}}=\pm\dfrac{\langle -8,-4,4\rangle}{4\sqrt{6}}$ or $\left\langle -\frac{2}{\sqrt{6}},-\frac{1}{\sqrt{6}},\frac{1}{\sqrt{6}}\right\rangle$ and $\left\langle \frac{2}{\sqrt{6}},\frac{1}{\sqrt{6}},-\frac{1}{\sqrt{6}}\right\rangle$.

13. The vectors corresponding to $\overrightarrow{AB}$ and $\overrightarrow{AD}$ are $\mathbf{a}=\langle 3,-1,0\rangle$ and $\mathbf{b}=\langle 2,-2,0\rangle$. The area of the parallelogram with the given vertices is $|\mathbf{a}\times\mathbf{b}|=\left\|\begin{matrix}\mathbf{i}&\mathbf{j}&\mathbf{k}\\3&-1&0\\2&-2&0\end{matrix}\right\|=|(0)\mathbf{i}-(0)\mathbf{j}+(-6+2)\mathbf{k}|=|-4\mathbf{k}|=4.$

15. **(a)** A vector orthogonal to the plane through the points P, Q and R is a vector orthogonal to both $\overrightarrow{PQ}$ and $\overrightarrow{PR}$ and thus is $\overrightarrow{PQ}\times\overrightarrow{PR}$. Here $\overrightarrow{PQ}=\langle -1,2,0\rangle$ and $\overrightarrow{PR}=\langle -1,0,3\rangle$, so
$\overrightarrow{PQ}\times\overrightarrow{PR}=\langle (2)(3)-(0)(0),(0)(-1)-(-1)(3),(-1)(0)-(2)(-1)\rangle=\langle 6,3,2\rangle$
(or any nonzero scalar multiple of $\langle 6,3,2\rangle$) is the desired vector.

(b) From (a), $\left|\overrightarrow{PQ}\times\overrightarrow{PR}\right|=|\langle 6,3,2\rangle|=\sqrt{36+9+4}=7$, so the area of the triangle is $\frac{7}{2}$.

17. Using the notation of (1), $\mathbf{r}=\langle 0,0.3,0\rangle$ and $\mathbf{F}$ has direction $\langle 0,3,-4\rangle$. The angle θ between them can be determined by $\langle 0,0.3,0\rangle\cdot\langle 0,3,-4\rangle=|\langle 0,0.3,0\rangle||\langle 0,3,-4\rangle|\cos\theta\ \Rightarrow\ 0.9=(0.3)(5)\cos\theta\ \Rightarrow$ $\cos\theta=0.6\ \Rightarrow\ \theta\approx 53.1^\circ$ and $\sin\theta=0.8$. Then $|\tau|=|\mathbf{r}||\mathbf{F}|\sin\theta\ \Rightarrow\ 100=(0.3)|\mathbf{F}|(0.8)\ \Rightarrow$ $|\mathbf{F}|\approx 417\text{ N}$.

19. $\mathbf{a}\cdot(\mathbf{b}\times\mathbf{c})=\begin{vmatrix}1&0&6\\2&3&-8\\8&-5&6\end{vmatrix}=1\begin{vmatrix}3&-8\\-5&6\end{vmatrix}-0+6\begin{vmatrix}2&3\\8&-5\end{vmatrix}=(18-40)+6(-10-24)=-226.$ Thus the volume is $|-226|=226$ cubic units.

21. $\mathbf{a}=\overrightarrow{PQ}=\langle 1,-1,2\rangle$, $\mathbf{b}=\overrightarrow{PR}=\langle 3,0,6\rangle$ and $\mathbf{c}=\overrightarrow{PS}=\langle 2,-2,-3\rangle$.
$\mathbf{a}\cdot(\mathbf{b}\times\mathbf{c})=\begin{vmatrix}1&-1&2\\3&0&6\\2&-2&-3\end{vmatrix}=1\begin{vmatrix}0&6\\-2&-3\end{vmatrix}-(-1)\begin{vmatrix}3&6\\2&-3\end{vmatrix}+2\begin{vmatrix}3&0\\2&-2\end{vmatrix}=12-21-12=-21$, and the volume is 21 cubic units.

23. $\mathbf{a}\cdot(\mathbf{b}\times\mathbf{c}) = \begin{vmatrix} 2 & 3 & 1 \\ 1 & -1 & 0 \\ 7 & 3 & 2 \end{vmatrix} = 2\begin{vmatrix} -1 & 0 \\ 3 & 2 \end{vmatrix} - 3\begin{vmatrix} 1 & 0 \\ 7 & 2 \end{vmatrix} + 1\begin{vmatrix} 1 & -1 \\ 7 & 3 \end{vmatrix} = -4 - 6 + 10 = 0$, which says that the volume of the parallelepiped determined by **a**, **b** and **c** is 0, and thus these three vectors are coplanar.

25. **(a)**

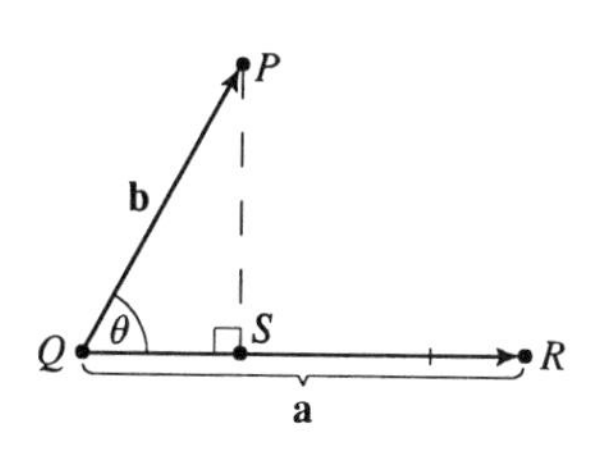

The distance between a point and a line is the length of the perpendicular from the point to the line, here $\left|\overrightarrow{PS}\right| = d$. But referring to triangle PQS,
$d = \left|\overrightarrow{PS}\right| = \left|\overrightarrow{QP}\right| \sin\theta = |\mathbf{b}|\sin\theta$. But θ is the angle between $\overrightarrow{QP} = \mathbf{b}$ and $\overrightarrow{QR} = \mathbf{a}$. Thus by the definition,
$\sin\theta = \dfrac{|\mathbf{a}\times\mathbf{b}|}{|\mathbf{a}||\mathbf{b}|}$ and so $d = |\mathbf{b}|\sin\theta = \dfrac{|\mathbf{b}||\mathbf{a}\times\mathbf{b}|}{|\mathbf{a}||\mathbf{b}|} = \dfrac{|\mathbf{a}\times\mathbf{b}|}{|\mathbf{a}|}$.

(b) $\mathbf{a} = \overrightarrow{QR} = \langle -1, -2, -1\rangle$ and $\mathbf{b} = \overrightarrow{QP} = \langle 1, -5, -7\rangle$. Then
$\mathbf{a}\times\mathbf{b} = \langle (-2)(-7) - (-1)(-5), (-1)(1) - (-1)(-7), (-1)(-5) - (-2)(1)\rangle = \langle 9, -8, 7\rangle$. Thus the distance is $d = \dfrac{|\mathbf{a}\times\mathbf{b}|}{|\mathbf{a}|} = \frac{1}{\sqrt{6}}\sqrt{81 + 64 + 49} = \sqrt{\frac{194}{6}} = \sqrt{\frac{97}{3}}$.

27.
$$\begin{aligned}
(\mathbf{a}-\mathbf{b})\times(\mathbf{a}+\mathbf{b}) &= (\mathbf{a}-\mathbf{b})\times\mathbf{a} + (\mathbf{a}-\mathbf{b})\times\mathbf{b} && \text{by Property 3}\\
&= \mathbf{a}\times\mathbf{a} + (-\mathbf{b})\times\mathbf{a} + \mathbf{a}\times\mathbf{b} + (-\mathbf{b})\times\mathbf{b} && \text{by Property 4}\\
&= (\mathbf{a}\times\mathbf{a}) - (\mathbf{b}\times\mathbf{a}) + (\mathbf{a}\times\mathbf{b}) - (\mathbf{b}\times\mathbf{b}) && \text{by Property 2}\\
&= \mathbf{0} - (\mathbf{b}\times\mathbf{a}) + (\mathbf{a}\times\mathbf{b}) - \mathbf{0}\\
&= (\mathbf{a}\times\mathbf{b}) + (\mathbf{a}\times\mathbf{b}) && \text{by Property 1}\\
&= 2(\mathbf{a}\times\mathbf{b})
\end{aligned}$$

29. $\mathbf{a}\times(\mathbf{b}\times\mathbf{c}) + \mathbf{b}\times(\mathbf{c}\times\mathbf{a}) + \mathbf{c}\times(\mathbf{a}\times\mathbf{b})$
$= [(\mathbf{a}\cdot\mathbf{c})\mathbf{b} - (\mathbf{a}\cdot\mathbf{b})\mathbf{c}] + [(\mathbf{b}\cdot\mathbf{a})\mathbf{c} - (\mathbf{b}\cdot\mathbf{c})\mathbf{a}] + [(\mathbf{c}\cdot\mathbf{b})\mathbf{a} - (\mathbf{c}\cdot\mathbf{a})\mathbf{b}]$ (by Exercise 28)
$= (\mathbf{a}\cdot\mathbf{c})\mathbf{b} - (\mathbf{a}\cdot\mathbf{b})\mathbf{c} + (\mathbf{a}\cdot\mathbf{b})\mathbf{c} - (\mathbf{b}\cdot\mathbf{c})\mathbf{a} + (\mathbf{b}\cdot\mathbf{c})\mathbf{a} - (\mathbf{a}\cdot\mathbf{c})\mathbf{b} = \mathbf{0}$

31. **(a)** No. If $\mathbf{a}\cdot\mathbf{b} = \mathbf{a}\cdot\mathbf{c}$, then $\mathbf{a}\cdot(\mathbf{b}-\mathbf{c}) = 0$, so **a** is perpendicular to $\mathbf{b}-\mathbf{c}$, which can happen if $\mathbf{b}\neq\mathbf{c}$. For example, let $\mathbf{a} = \langle 1, 1, 1\rangle$, $\mathbf{b} = \langle 1, 0, 0\rangle$ and $\mathbf{c} = \langle 0, 1, 0\rangle$.

(b) No. If $\mathbf{a}\times\mathbf{b} = \mathbf{a}\times\mathbf{c}$ then $\mathbf{a}\times(\mathbf{b}-\mathbf{c}) = \mathbf{0}$, which implies that **a** is parallel to $\mathbf{b}-\mathbf{c}$ which of course can happen if $\mathbf{b}\neq\mathbf{c}$.

(c) Yes. Since $\mathbf{a}\cdot\mathbf{c} = \mathbf{a}\cdot\mathbf{b}$, **a** is perpendicular to $\mathbf{b}-\mathbf{c}$, by part (a). From part (b), **a** is also parallel to $\mathbf{b}-\mathbf{c}$. Thus since $\mathbf{a}\neq\mathbf{0}$ but is both parallel and perpendicular to $\mathbf{b}-\mathbf{c}$, we have $\mathbf{b}-\mathbf{c} = \mathbf{0}$ or $\mathbf{b} = \mathbf{c}$.

EXERCISES 11.4

1. $\mathbf{r}_0 = 3\mathbf{i} - \mathbf{j} + 8\mathbf{k}$ and $\mathbf{v} = \mathbf{a}$ so the vector equation is
$\mathbf{r} = (3\mathbf{i} - \mathbf{j} + 8\mathbf{k}) + t(2\mathbf{i} + 3\mathbf{j} + 5\mathbf{k}) = (3 + 2t)\mathbf{i} + (-1 + 3t)\mathbf{j} + (8 + 5t)\mathbf{k}$, and the parametric equations are
$x = 3 + 2t, y = -1 + 3t, z = 8 + 5t$.

3. $\mathbf{r} = (\mathbf{j} + 2\mathbf{k}) + t(6\mathbf{i} + 3\mathbf{j} + 2\mathbf{k}) = (6t)\mathbf{i} + (1 + 3t)\mathbf{j} + (2 + 2t)\mathbf{k}$ is the vector equation, while
$x = 6t$, $y = 1 + 3t$, $z = 2 + 2t$ are the parametric equations.

5. The parallel vector is $\mathbf{v} = \langle 6 - 2, 0 - 1, 3 - 8 \rangle = \langle 4, -1, -5 \rangle$ so the direction numbers are $a = 4$, $b = -1$, $c = -5$. Letting $P_0 = (2, 1, 8)$, the parametric equations are $x = 2 + 4t$, $y = 1 - t$, $z = 8 - 5t$ and symmetric equations are $\dfrac{x - 2}{4} = \dfrac{y - 1}{-1} = \dfrac{z - 8}{-5}$.

7. $\mathbf{v} = \langle 0, 1, -5 \rangle$ and letting $P_0 = (3, 1, -1)$, the parametric equations are $x = 3$, $y = 1 + t$, $z = -1 - 5t$ while the symmetric equations are $x = 3, y - 1 = \dfrac{z + 1}{-5}$. Notice here that the direction number $a = 0$, so rather than writing $\dfrac{x - 3}{0}$ in the symmetric equation we write the equation $x = 3$ separately.

9. $\mathbf{v} = \left\langle \frac{1}{3}, 4, -9 \right\rangle$ and letting $P_0 = \left(-\frac{1}{3}, 1, 1\right)$, the parametric equations are $x = -\frac{1}{3} + \frac{1}{3}t$, $y = 1 + 4t$, $z = 1 - 9t$ while the symmetric equations are $\dfrac{x + 1/3}{1/3} = \dfrac{y - 1}{4} = \dfrac{z - 1}{-9}$.

11. Direction vectors of the lines are respectively $\mathbf{v}_1 = \langle 6, 9, 12 \rangle$ and $\mathbf{v}_2 = \langle 4, 6, 8 \rangle$ and since $\mathbf{v}_1 = \frac{3}{2}\mathbf{v}_2$ the direction vectors and thus the lines are parallel.

13. **(a)** A direction vector of the line with given parametric equations is $\mathbf{v} = \langle 2, 3, -7 \rangle$ and the desired parallel line must also have $\mathbf{v}$ as a direction vector. Here $P_0 = (0, 2, -1)$ so the symmetric equations for the line are $\dfrac{x}{2} = \dfrac{y - 2}{3} = \dfrac{z + 1}{-7}$.

(b) The line intersects the xy-plane when $z = 0$, so we need $\dfrac{x}{2} = \dfrac{y - 2}{3} = \dfrac{1}{-7}$ or
$x = -\frac{2}{7}$, $y = \frac{11}{7}$. Thus the point of intersection with the xy-plane is $\left(-\frac{2}{7}, \frac{11}{7}, 0\right)$. Similarly for the yz- and xz-planes, we need respectively $x = 0$ and $y = 0$ or $0 = \dfrac{y - 2}{3} = \dfrac{z + 1}{-7}$ and $\dfrac{x}{2} = -\dfrac{2}{3} = \dfrac{z + 1}{-7}$ or $y = 2$, $z = -1$ and $x = -\frac{4}{3}$, $z = \frac{11}{3}$. Thus the line intersects the yz-plane at $(0, 2, -1)$ and the xz-plane at $\left(-\frac{4}{3}, 0, \frac{11}{3}\right)$.

15. The lines aren't parallel since the direction vectors $\langle 2, 4, -3 \rangle$ and $\langle 1, 3, 2 \rangle$ aren't parallel so we check to see if the lines intersect. The parametric equations of the lines are L_1: $x = 4 + 2t$, $y = -5 + 4t$, $z = 1 - 3t$ and L_2: $x = 2 + s$, $y = -1 + 3s$, $z = 2s$. For the lines to intersect we must be able to find one value of t and one value of s satisfying the following three equations: $4 + 2t = 2 + s$, $-5 + 4t = -1 + 3s$, $1 - 3t = 2s$. Solving the first two equations we get $t = -5$, $s = -8$ and checking we see that these values don't satisfy the third equation. Thus L_1 and L_2 aren't parallel and don't intersect, so they must be skew lines.

17. Since the direction vectors are $\mathbf{v}_1 = \langle -6, 9, -3 \rangle$ and $\mathbf{v}_2 = \langle 2, -3, 1 \rangle$, we have $\mathbf{v}_1 = -3\mathbf{v}_2$ so the lines are parallel.

19. Setting $a = 7$, $b = 1$, $c = 4$, $x_0 = 1$, $y_0 = 4$, $z_0 = 5$ in Equation 6 gives $7(x-1) + 1(y-4) + 4(z-5) = 0$ or $7x + y + 4z = 31$ to be the equation of the plane.

21. Setting $a = 15$, $b = 9$, $c = -12$, $x_0 = 1$, $y_0 = 2$, $z_0 = 3$ in Equation 6 gives $15(x-1) + 9(y-2) - 12(z-3) = 0$ or $5x + 3y - 4z = -1$ to be the equation of the plane.

23. Since the two planes are parallel, they will have the same normal vectors. Thus $\mathbf{n} = \langle 1, 1, -1 \rangle$ and the equation of the plane is $1(x-6) + 1(y-5) + 1(z-2) = 0$ or $x + y - z = 13$.

25. The equation is $3(x+1) - 4(y-3) - 6(z+8) = 0$ or $3x - 4y - 6z = 33$.

27. Here the vectors $\mathbf{a} = \langle 1, 1, 1 \rangle$ and $\mathbf{b} = \langle 1, 2, 3 \rangle$ lie in the plane, so $\mathbf{a} \times \mathbf{b}$ is a normal vector to the plane. Thus $\mathbf{n} = \mathbf{a} \times \mathbf{b} = \langle 3-2, 1-3, 2-1 \rangle = \langle 1, -2, 1 \rangle$ and the equation of the plane is $x - 2y + z = 0$.

29. $\mathbf{a} = \langle -1, -2, -1 \rangle$ and $\mathbf{b} = \langle 3, 1, 9 \rangle$ so a normal vector to the plane is $\mathbf{n} = \mathbf{a} \times \mathbf{b} = \langle -18+1, -3+9, -1+6 \rangle = \langle -17, 6, 5 \rangle$ and the equation of the plane is $-17(x-1) + 6(y-0) + 5(z+3) = 0$ or $-17x + 6y + 5z = -32$.

31. To find the equation of the plane we must first find two nonparallel vectors in the plane, then their cross product will be a normal vector to the plane. But since the given line lies in the plane, its direction vector $\mathbf{a} = \langle 2, -3, -1 \rangle$ is one vector in the plane. To find another nonparallel vector $\mathbf{b}$ which lies in the plane pick any point on the line [say $(1, 2, 3)$, found by setting $t = 0$] and let $\mathbf{b}$ be the vector connecting this point to the given point in the plane. (But beware; we should first check that the given point is not on the given line. If it were on the line, the plane wouldn't be uniquely determined. What would $\mathbf{n}$ then be when we set $\mathbf{n} = \mathbf{a} \times \mathbf{b}$?) Here $\mathbf{b} = \langle 0, 4, -7 \rangle$ so $\mathbf{n} = \mathbf{a} \times \mathbf{b} = \langle 21+4, 0+14, 8-0 \rangle = \langle 25, 14, 8 \rangle$ and the equation of the plane is $25(x-1) + 14(y-7) + 8(z+4) = 0$ or $25x + 14y + 8z = 77$.

33. $(0, 0, 0)$ is a point on $x = y = z$. $\langle 1, 1, 1 \rangle$ is the direction of this line, and thus also of the plane. $\langle 0-0, 1-0, 2-0 \rangle = \langle 0, 1, 2 \rangle$ is also a vector in the plane. Therefore, $\mathbf{n} = \langle 1, 1, 1 \rangle \times \langle 0, 1, 2 \rangle = \langle 2-1, -2+0, 1-0 \rangle = \langle 1, -2, 1 \rangle$. Choosing $(x_0, y_0, z_0) = (0, 0, 0)$, the equation of the plane is, by Equation 7, $x - 2y + z = 1 \cdot 0 - 2 \cdot 0 + 1 \cdot 0 \quad \Leftrightarrow \quad x - 2y + z = 0$.

35. Substituting the parametric equations of the line into the equation of the plane gives $x + y + z = 1 + t + 2t + 3t = 1 \quad \Rightarrow \quad t = 0$. This value of t corresponds to the point of intersection $(1, 0, 0)$, obtained by substitution of $t = 0$ into the equations of the line.

37. Substituting the parametric equations of the line into the equation of the plane gives $2x + y - z + 5 = 2(1+2t) + (-1) - t + 5 = 0 \quad \Rightarrow \quad 3t + 6 = 0 \quad \Rightarrow \quad t = -2$. Therefore, the point of intersection is $x = 1 + 2(-2) = -3$, $y = -1$ and $z = -2$ and the point of intersection is $(-3, -1, -2)$.

39. Setting $x = 0$, we see that $(0, 1, 0)$ satisfies the equations of both planes, so that they do in fact have a line of intersection. $\mathbf{v} = \mathbf{n}_1 \times \mathbf{n}_2 = \langle 1, 1, 1 \rangle \times \langle 1, 0, 1 \rangle = \langle 1, 0, -1 \rangle$ is the direction of this line. Therefore, direction numbers of the intersecting line are 1, 0, -1.

41. The normal vectors to the planes are respectively $\mathbf{n}_1 = \langle 1, 0, 1 \rangle$ and $\mathbf{n}_2 = \langle 0, 1, 1 \rangle$. Thus the normal vectors and consequently the planes aren't parallel. Furthermore $\mathbf{n}_1 \cdot \mathbf{n}_2 = 1 \neq 0$ so the planes aren't perpendicular. Letting θ be the angle between the two planes, we have $\cos\theta = \dfrac{\mathbf{n}_1 \cdot \mathbf{n}_2}{|\mathbf{n}_1||\mathbf{n}_2|} = \dfrac{1}{\sqrt{2}\sqrt{2}} = \dfrac{1}{2}$ and $\theta = \cos^{-1}\frac{1}{2} = 60°$.

43. The respective normals are $\mathbf{n}_1 = \langle 1, 4, -3 \rangle$ and $\mathbf{n}_2 = \langle -3, 6, 7 \rangle$ so the normals (and thus the planes) fail to be parallel. But $\mathbf{n}_1 \cdot \mathbf{n}_2 = -3 + 24 - 21 = 0$ so the normals and thus the planes are perpendicular.

45. The normals are $\mathbf{n}_1 = \langle 2, 4, -2 \rangle$ and $\mathbf{n}_2 = \langle -3, -6, 3 \rangle$ respectively. So $\mathbf{n}_1 = -\frac{2}{3}\mathbf{n}_2$. The normals, and hence the planes, are parallel. The planes are not the same because $-\frac{3}{2}(1) \neq 10$.

47. **(a)** To find a point on the line of intersection, set one of the variables equal to a constant, say $z = 0$. (This will only work if the line of intersection crosses the xy-plane, otherwise try setting x or y equal to 0.) Then the equations of the planes reduce to $x + y = 2$ and $3x - 4y = 6$. Solving these two equations gives $x = 2$, $y = 0$. So a point on the line of intersection is $(2, 0, 0)$. The direction of the line is $\mathbf{v} = \mathbf{n}_1 \times \mathbf{n}_2 = \langle 5 - 4, -3 - 5, -4 - 3 \rangle = \langle 1, -8, -7 \rangle$, and symmetric equations for the line are $x - 2 = \dfrac{y}{-8} = \dfrac{z}{-7}$.

(b) The angle between the planes satisfies $\cos\theta = \dfrac{\mathbf{n}_1 \cdot \mathbf{n}_2}{|\mathbf{n}_1||\mathbf{n}_2|} = \dfrac{3 - 4 - 5}{\sqrt{3}\sqrt{50}} = -\dfrac{\sqrt{6}}{5}$. Therefore $\theta = \cos^{-1}\left(-\frac{\sqrt{6}}{5}\right) \approx 119°$ (or 61°).

49. Setting $x = 0$, the equations of the two planes become $z = y$ and $5y + z = -1$, which intersect at $y = -\frac{1}{6}$ and $z = -\frac{1}{6}$. Thus we can choose $(x_0, y_0, z_0) = \left(0, -\frac{1}{6}, -\frac{1}{6}\right)$. The direction of the line of intersection is $\mathbf{v} = \mathbf{n}_1 \times \mathbf{n}_2 = \langle 2, -5, -1 \rangle \times \langle 1, 1, -1 \rangle = \langle 6, 1, 7 \rangle$. Parametric equations for this line are, by Equation 2, $x = 6t$, $y = -\frac{1}{6} + t$, $z = -\frac{1}{6} + 7t$.

51. The plane contains all perpendicular bisectors of the line segment joining $(1, 1, 0)$ and $(0, 1, 1)$. All of these bisectors pass through the midpoint of this segment $\left(\dfrac{1}{2}, \dfrac{1+1}{2}, \dfrac{1}{2}\right) = \left(\dfrac{1}{2}, 1, \dfrac{1}{2}\right)$. The direction of this line segment $\langle 1 - 0, 1 - 1, 0 - 1 \rangle = \langle 1, 0, -1 \rangle$ is perpendicular to the plane so that we can choose this to be $\mathbf{n}$. Therefore the equation of the plane is $1\left(x - \frac{1}{2}\right) + 0(y - 1) - 1\left(z - \frac{1}{2}\right) = 0 \quad \Leftrightarrow \quad x = z$.

53. A direction vector for the line of intersection is $\mathbf{a} = \mathbf{n}_1 \times \mathbf{n}_2 = \langle 1, 1, -1 \rangle \times \langle 2, -1, 3 \rangle = \langle 2, -5, -3 \rangle$ and $\mathbf{a}$ is parallel to the desired plane. Another vector parallel to the plane is the vector connecting any point on the line of intersection to the given point $(-1, 2, 1)$ in the plane. Setting $x = 0$, the equation of the planes reduce to $y - z = 2$ and $-y + 3z = 1$ with simultaneous solution $y = \frac{7}{2}$ and $z = \frac{3}{2}$. So a point on the line is $\left(0, \frac{7}{2}, \frac{3}{2}\right)$ and another vector parallel to the plane is $\left\langle -1, -\frac{3}{2}, -\frac{1}{2} \right\rangle$. Then a normal vector to the plane is $\mathbf{n} = \langle 2, -5, -3 \rangle \times \left\langle -1, -\frac{3}{2}, -\frac{1}{2} \right\rangle = \langle -2, 4, -8 \rangle$ and an equation of the plane is $-2(x+1) + 4(y-2) - 8(z-1) = 0$ or $x - 2y + 4z = -1$.

55. The plane contains the points $(a, 0, 0)$, $(0, b, 0)$ and $(0, 0, c)$. Thus the vectors $\mathbf{a} = \langle -a, b, 0 \rangle$ and $\mathbf{b} = \langle -a, 0, c \rangle$ lie in the plane and $\mathbf{n} = \mathbf{a} \times \mathbf{b} = \langle bc - 0, 0 + ac, 0 + ab \rangle = \langle bc, ac, ab \rangle$ is a normal vector to the plane. The equation of the plane is therefore $bcx + acy + abz = abc + 0 + 0$ or $bcx + acy + abz = abc$. Notice that if $a \neq 0$, $b \neq 0$ and $c \neq 0$ then we can rewrite the equation as $\frac{x}{a} + \frac{y}{b} + \frac{z}{c} = 1$. This is a good equation to remember!

57. Two vectors which are perpendicular to the required line are the normal of the given plane, $\langle 1, 1, 1 \rangle$, and a direction vector for the given line, $\langle 1, -1, 2 \rangle$. So a direction vector for the required line is $\langle 1, 1, 1 \rangle \times \langle 1, -1, 2 \rangle = \langle 3, -1, -2 \rangle$. So L is given by $\langle x, y, z \rangle = \langle 0, 1, 2 \rangle + t\langle 3, -1, -2 \rangle$ or parametric equations $x = 3t$, $y = 1 - t$, $z = 2 - 2t$.

59. Let P_i have normal vector $\mathbf{n}_i$, then $\mathbf{n}_1 = \langle 4, -2, 6 \rangle$, $\mathbf{n}_2 = \langle 4, -2, -2 \rangle$, $\mathbf{n}_3 = \langle -6, 3, -9 \rangle$, $\mathbf{n}_4 = \langle 2, -1, -1 \rangle$. Then $\mathbf{n}_1 = \frac{-2}{3}\mathbf{n}_3$, so $\mathbf{n}_1$ and $\mathbf{n}_3$ are parallel, and hence P_1 and P_3 are parallel, similarly P_2 and P_4 are parallel because $\mathbf{n}_2 = 2\mathbf{n}_4$. However $\mathbf{n}_1$ and $\mathbf{n}_2$ are not parallel. $\left(0, 0, \frac{1}{2}\right)$ lies on P_1, but not on P_3, so they are not the same plane, but both P_2 and P_4 contain the point $(0, 0, -3)$, so they are identical.

61. Let $Q = (2, 2, 0)$ and $R = (3, -1, 5)$, points on the line corresponding to $t = 0$ and $t = 1$. Let $P = (1, 2, 3)$, then $\mathbf{a} = \overrightarrow{QR} = \langle 1, -3, 5 \rangle$, $\mathbf{b} = \overrightarrow{QP} = \langle -1, 0, 3 \rangle$. The distance is

$$d = \frac{|\mathbf{a} \times \mathbf{b}|}{|\mathbf{a}|} = \frac{|\langle 1, -3, 5 \rangle \times \langle -1, 0, 3 \rangle|}{|\langle 1, -3, 5 \rangle|} = \frac{|\langle -9, -8, -3 \rangle|}{|\langle 1, -3, 5 \rangle|} = \frac{\sqrt{9^2 + 8^2 + 3^2}}{\sqrt{1^2 + 3^2 + 5^2}} = \frac{\sqrt{154}}{\sqrt{35}} = \sqrt{\frac{22}{5}}$$

63. $D = \dfrac{1}{\sqrt{1 + 4 + 4}}[(1)(2) + (-2)(8) + (-2)(5) - 1] = \dfrac{25}{3}$

65. Put $y = z = 0$ in the equation of the first plane, to get the point $(-1, 0, 0)$ on the plane. Because the planes are parallel, the distance, D, between them is the distance from $(-1, 0, 0)$ to the second plane. Using the formula from Example 8, $D = \dfrac{|3(-1) + 6(0) - 3(0) - 4|}{\sqrt{3^2 + 6^2 + (-3)^2}} = \dfrac{7}{3\sqrt{6}}$.

67. The distance between two parallel planes is the same as the distance between a point on one of the planes and the other plane. Let $P_0 = (x_0, y_0, z_0)$ be a point on the plane given by $ax + by + cz = d_1$. Then $ax_0 + by_0 + cz_0 = d_1$ and the distance between P_0 and the plane given by $ax + by + cz = d_2$ is $D = \dfrac{1}{\sqrt{a^2 + b^2 + c^2}} |ax_0 + by_0 + cz_0 - d_2| = \dfrac{|d_1 - d_2|}{\sqrt{a^2 + b^2 + c^2}}$.

69. L_1: $x = y = z \quad \Rightarrow \quad x = y$ (1). L_2: $x + 1 = y/2 = z/3 \quad \Rightarrow \quad x + 1 = y/2$ (2). The solution of (1) and (2) is $x = y = -2$. However, when $x = -2$, $x = z \quad \Rightarrow \quad z = -2$, but $x + 1 = z/3 \quad \Rightarrow \quad z = -3$, a contradiction. Hence the lines do not intersect. For L_1, $\mathbf{v}_1 = \langle 1, 1, 1 \rangle$, and for L_2, $\mathbf{v}_2 = \langle 1, 2, 3 \rangle$, so the lines are not parallel. Thus the lines are skew lines. If two lines are skew, they can be viewed as lying in two parallel planes and so the distance between the skew lines would be the same as the distance between these parallel planes. The common normal vector to the planes must be perpendicular to both $\langle 1, 1, 1 \rangle$ and $\langle 1, 2, 3 \rangle$, the direction vectors of the two lines. So set $\mathbf{n} = \langle 1, 1, 1 \rangle \times \langle 1, 2, 3 \rangle = \langle 3 - 2, -3 + 1, 2 - 1 \rangle = \langle 1, -2, 1 \rangle$. From above, we know that $(-2, -2, -2)$ and $(-2, -2, -3)$ are points of L_1 and L_2 respectively. So in the notation of Equation 7, $d_1 = 1(-2) - 2(-2) + 1(-2) = 0$ and $d_2 = 1(-2) - 2(-2) + 1(-3) = -1$. By Exercise 67, the distance between these two skew lines is $D = \dfrac{|0 - (-1)|}{\sqrt{1 + 4 + 1}} = \dfrac{1}{\sqrt{6}}$.

Alternate solution (without reference to planes): A vector which is perpendicular to both of the lines is $\mathbf{n} = \langle 1, 1, 1 \rangle \times \langle 1, 2, 3 \rangle = \langle 1, -2, 1 \rangle$. Pick any point on each of the lines, say $(-2, -2, -2)$ and $(-2, -2, -3)$, and form the vector $\mathbf{b} = \langle 0, 0, 1 \rangle$ connecting the two points. The distance between the two skew lines is the absolute value of the scalar projection of $\mathbf{b}$ along $\mathbf{n}$, that is, $D = \dfrac{|\mathbf{n} \cdot \mathbf{b}|}{|\mathbf{n}|} = \dfrac{|1 \cdot 0 - 2 \cdot 0 + 1 \cdot 1|}{\sqrt{1 + 4 + 1}} = \dfrac{1}{\sqrt{6}}$.

71. If $a \neq 0$ then $ax + by + cz = d \quad \Rightarrow \quad ax - d + by + cz = 0 \quad \Rightarrow \quad a(x - d/a) + b(y - 0) + c(z - 0) = 0$ which by (6) is the scalar equation of the plane through the point $(d/a, 0, 0)$ with normal vector $\langle a, b, c \rangle$. Similarly if $b \neq 0$ (or if $c \neq 0$) the equation of the plane can be rewritten as $a(x - 0) + b(y - d/b) + c(z - 0) = 0$ [or as $a(x - 0) + b(y - 0) + c(z - d/c) = 0$] which by (6) is the scalar equation of a plane through the point $(0, d/b, 0)$ [or the point $(0, 0, d/c)$] with normal vector $\langle a, b, c \rangle$.

EXERCISES 11.5

1. The trace in any plane $x = k$ is given by $z^2 - y^2 = 1 - k^2$, $x = k$ whose graph is a hyperbola. The trace in any plane $y = k$ is the circle given by $x^2 + z^2 = 1 + k^2$, $y = k$, and the trace in any plane $z = k$ is the hyperbola given by $x^2 - y^2 = 1 - k^2$, $z = k$. Thus the surface is a hyperboloid of one sheet with axis the y-axis.

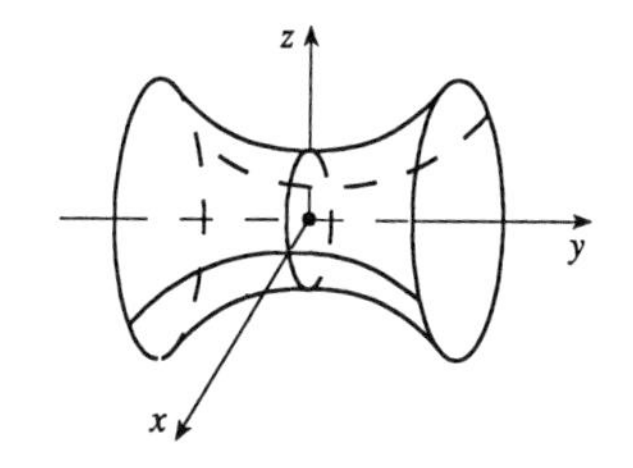

3. Traces: $x = k$, $9y^2 + 36z^2 = 36 - 4k^2$, an ellipse for $|k| < 3$; $y = k$, $4x^2 + 36z^2 = 36 - 9k^2$, an ellipse for $|k| < 2$; $z = k$, $4x^2 + 9y^2 = 36(1 - k^2)$, an ellipse for $|k| < 1$. Thus the surface is an ellipsoid with center at the origin and axes along the x-, y- and z-axes.

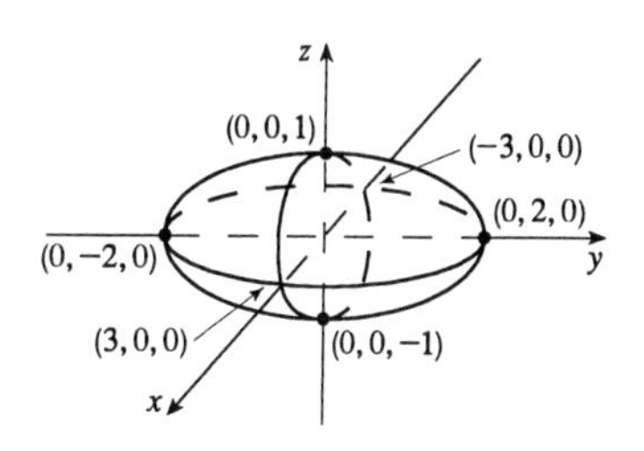

5. Traces: $x = k$, $4z^2 - y^2 = 1 + k^2$, a hyperbola; $y = k$, $4z^2 - x^2 = 1 + k^2$, a hyperbola; $z = k$, $-x^2 - y^2 = 1 - 4k^2$ or $x^2 + y^2 = 4k^2 - 1$, a circle for $k > \frac{1}{2}$ or $k < -\frac{1}{2}$. Thus the surface is a hyperboloid of two sheets with axis the z-axis.

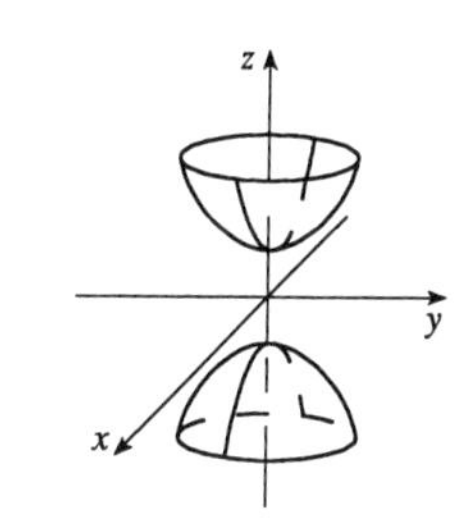

7. Traces: $x = k$, $z = y^2$, a parabola; $y = k$, $z = k^2$, a line; $z = k$, $y^2 = k$ or $y = \pm\sqrt{k}$, two parallel lines for $k > 0$. Thus the surface is a parabolic cylinder opening upward.

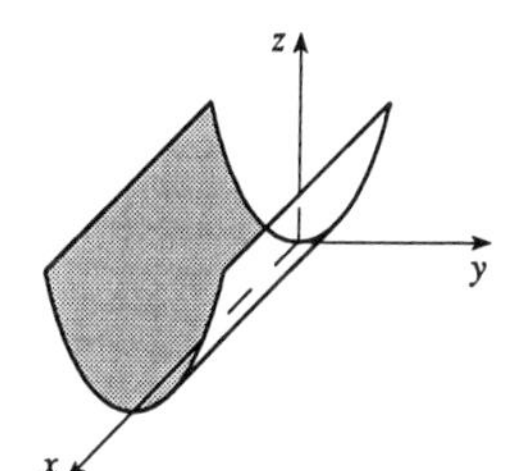

9. Traces: $x = k$, $y^2 = k^2 + z^2$ or $y^2 - z^2 = k^2$, a hyperbola for $k \neq 0$ and two intersecting lines for $k = 0$; $y = k$, $x^2 + z^2 = k^2$, a circle for $k \neq 0$; $z = k$, $y^2 = x^2 + k^2$ or $y^2 - x^2 = k^2$, a hyperbola for $k \neq 0$ and two intersecting lines for $k = 0$. Thus the surface is a cone (right circular) with axis the y-axis and vertex the origin.

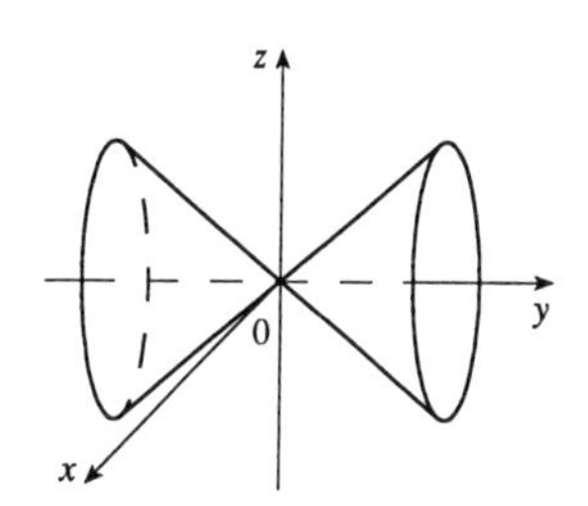

11. Traces: $x = k$, $k^2 + 4z^2 - y = 0$ or $y - k^2 = 4z^2$, a parabola;
$y = k$, $x^2 + 4z^2 = k$, an ellipse for $k > 0$;
$z = k$, $x^2 + 4k^2 - y = 0$ or $y - 4k^2 = x^2$, a parabola.
Thus the surface is an elliptic paraboloid with axis the y-axis and vertex the origin.

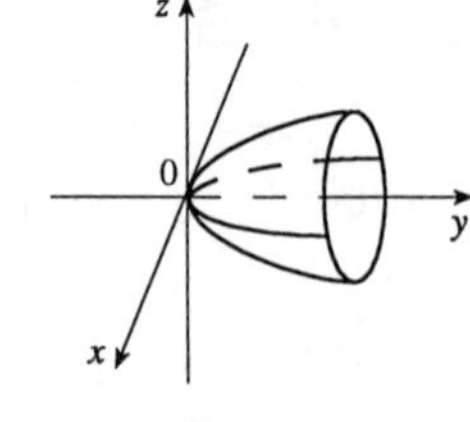

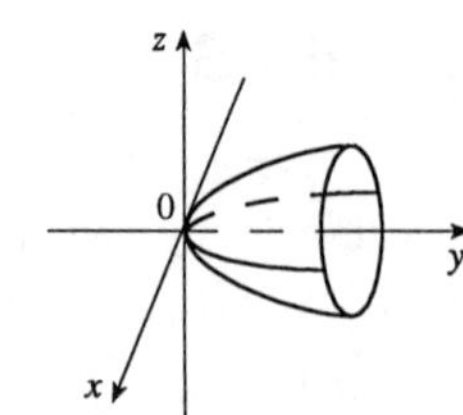

13. Traces: $x = k \quad \Rightarrow \quad \dfrac{y^2}{9} + \dfrac{z^2}{1} = 1$, ellipses;
$y = k$, $|k| \le 3 \quad \Rightarrow \quad 9z^2 = 9 - k^2 \quad \Rightarrow \quad z = \pm\sqrt{1 - (k^2/9)}$,
pairs of lines; $z = k$, $|k| \le 1 \quad \Rightarrow \quad y^2 = 9(1 - k^2) \quad \Rightarrow$
$y = \pm 3\sqrt{1 - k^2}$, pairs of lines. This is the equation of an elliptic cylinder, centered at the origin, whose axis is the x-axis.

15. Traces: $x = k \quad \Rightarrow \quad y = z^2 - k^2$, parabolas;
$y = k \quad \Rightarrow \quad k = z^2 - x^2$, hyperbolas on the z-axis for $k > 0$, and hyperbolas on the x-axis for $k < 0$; $z = k \quad \Rightarrow \quad y = k^2 - x^2$, parabolas.
Thus, $\dfrac{y}{1} = \dfrac{z^2}{1^2} - \dfrac{x^2}{1^2}$ is a hyperbolic paraboloid.

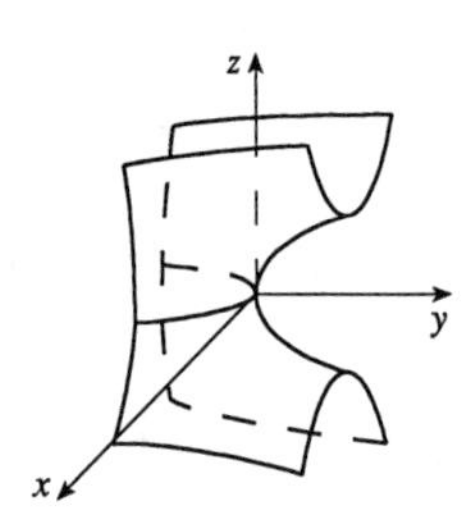

17. This is the equation of an ellipsoid: $x^2 + 4y^2 + 9z^2 = x^2 + \dfrac{y^2}{(1/2)^2} + \dfrac{z^2}{(1/3)^2} = 1$, with x-intercepts ± 1, y-intercepts $\pm\frac{1}{2}$ and z-intercepts $\pm\frac{1}{3}$. So the major axis is the x-axis and the only possible graph is VII.

19. This is the equation of a hyperboloid of one sheet, with $a = b = c = 1$. Since the minus sign is in front of the y term, the axis of the hyperboloid is the y-axis, hence the correct graph is II.

21. There are no real values of x and z that satisfy this equation for $y < 0$, so this surface does not extend to the left of the xz-plane. The surface intersects the plane $y = k > 0$ in an ellipse. Notice that y occurs to the first power whereas x and z occur to the second power. So the surface is an elliptic paraboloid with axis the y-axis. Its graph is VI.

23. This surface is a cylinder because the variable y is missing from the equation. The intersection of the surface and the xz-plane is an ellipse. So the graph is VIII.

25. $z^2 = 3x^2 + 4y^2 - 12$ or $3x^2 + 4y^2 - z^2 = 12$

or $\dfrac{x^2}{4} + \dfrac{y^2}{3} - \dfrac{z^2}{12} = 1$

or $\dfrac{x^2}{2^2} + \dfrac{y^2}{\left(\sqrt{3}\right)^2} - \dfrac{z^2}{\left(\sqrt{12}\right)^2} = 1$

represents a hyperboloid of one sheet with axis the z-axis.

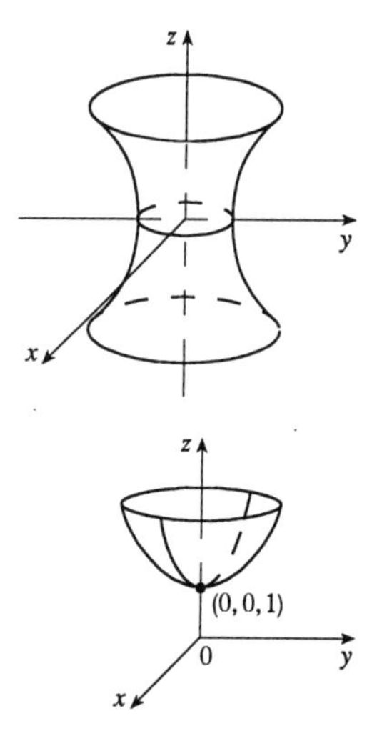

27. $z = x^2 + y^2 + 1$ or $z - 1 = x^2 + y^2$, a circular paraboloid with axis the z-axis and vertex $(0, 0, 1)$.

29. Completing the square in all three variables gives

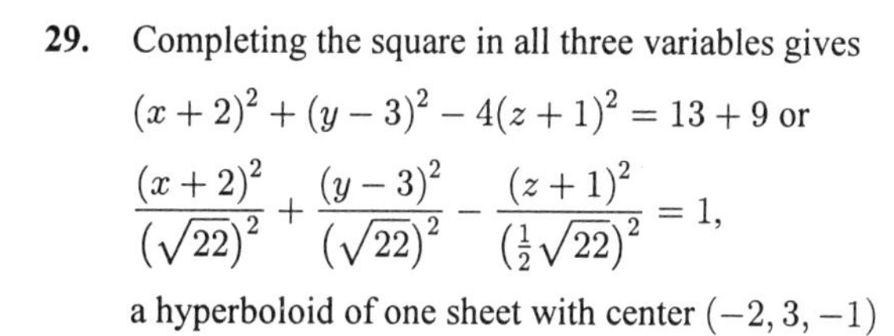

$(x+2)^2 + (y-3)^2 - 4(z+1)^2 = 13 + 9$ or

$\dfrac{(x+2)^2}{\left(\sqrt{22}\right)^2} + \dfrac{(y-3)^2}{\left(\sqrt{22}\right)^2} - \dfrac{(z+1)^2}{\left(\frac{1}{2}\sqrt{22}\right)^2} = 1$,

a hyperboloid of one sheet with center $(-2, 3, -1)$ and axis the vertical line $y = 3$, $x = -2$.

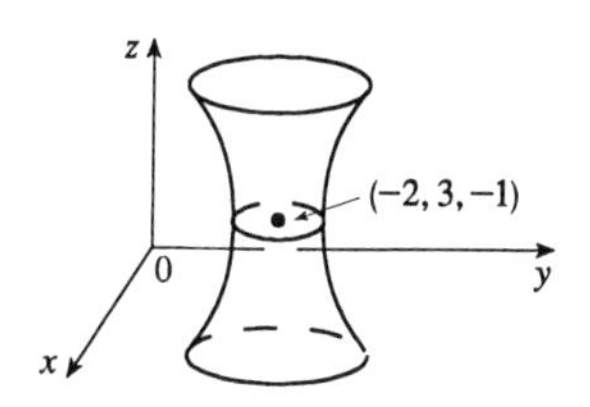

31. $x^2 + 4y^2 = 100$ or $\dfrac{x^2}{10^2} + \dfrac{y^2}{5^2} = 1$,

an elliptic cylinder with axis the z-axis.

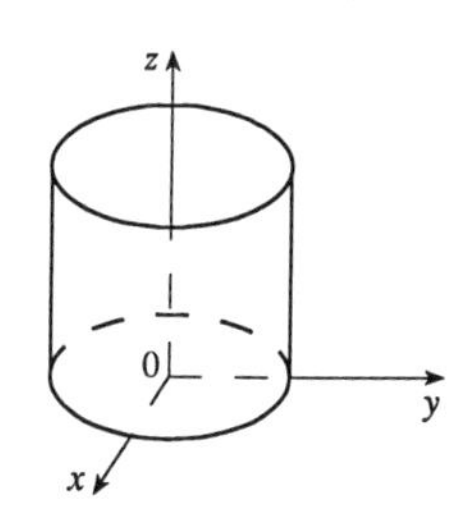

33. Completing the square in y gives

$x^2 - (y-2)^2 + z = 4 - 4 = 0$

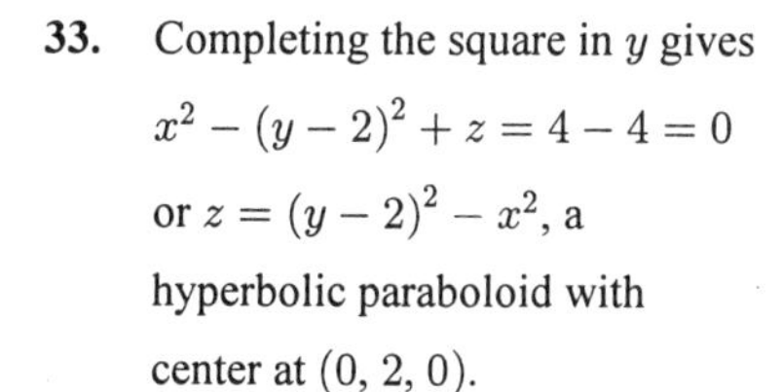

or $z = (y-2)^2 - x^2$, a hyperbolic paraboloid with center at $(0, 2, 0)$.

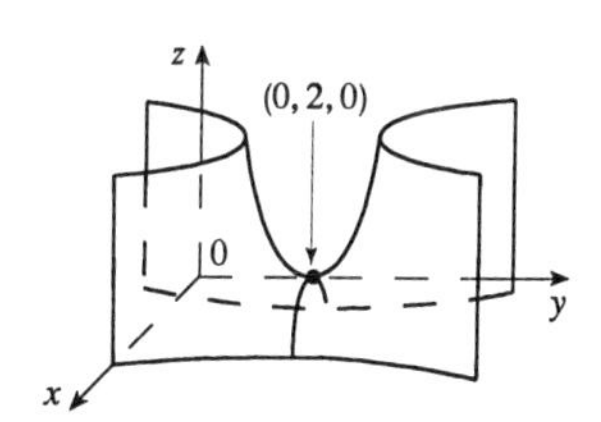

35.

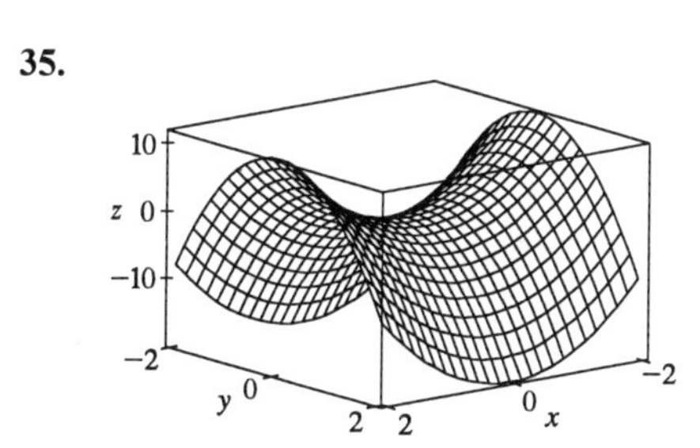

37.

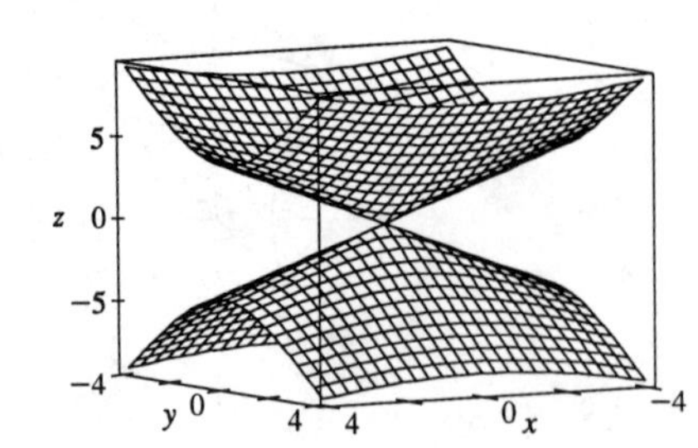

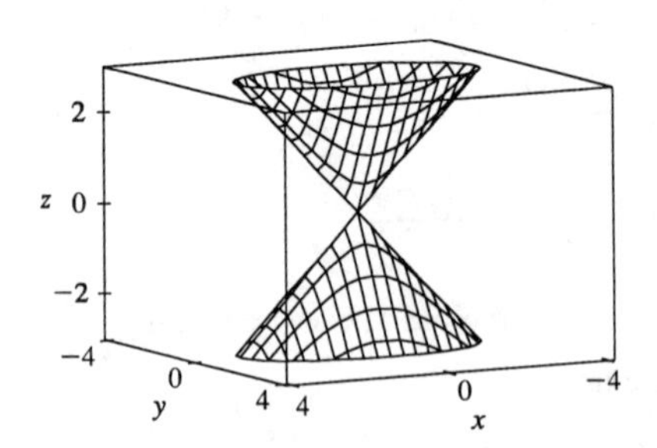

To restrict the z-range as in the second graph, we can use the option `view=-2..2` in Maple's `plot3d` command, or `PlotRange -> {-2,2}` in Mathematica's `Plot3D` command.

39.

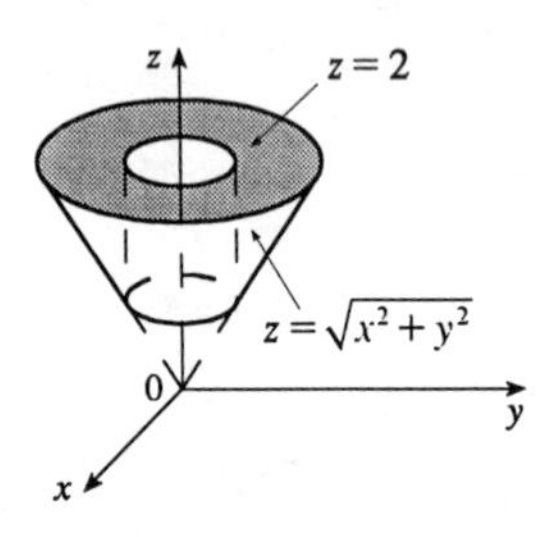

41. The surface is a paraboloid of revolution (circular paraboloid) with vertex at the origin, axis the y-axis and opens to the right. Thus the trace in the yz-plane is also a parabola: $y = z^2$, $x = 0$. The equation is $y = x^2 + z^2$.

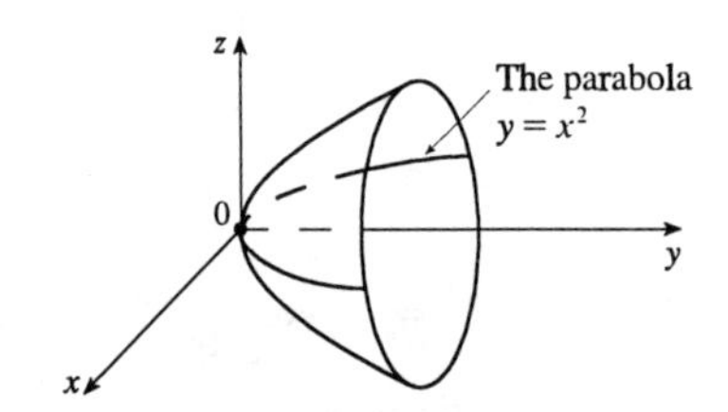

43. Let $P = (x, y, z)$ be an arbitrary point equidistant from $(-1, 0, 0)$ and the plane $x = 1$. Then the distance from P to $(-1, 0, 0)$ is $\sqrt{(x+1)^2 + y^2 + z^2}$ and the distance from P to the plane $x = 1$ is $|x - 1|/\sqrt{1^2} = |x - 1|$ (by the formula in Example 11.4.8). So $|x - 1| = \sqrt{(x+1)^2 + y^2 + z^2} \quad \Leftrightarrow \quad (x-1)^2 = (x+1)^2 + y^2 + z^2$ $\Leftrightarrow \quad x^2 - 2x + 1 = x^2 + 2x + 1 + y^2 + z^2 \quad \Leftrightarrow \quad -4x = y^2 + z^2$. Thus the collection of all such points P is a circular paraboloid with vertex at the origin, axis the x-axis, which opens in the negative direction.

45. If (a, b, c) satisfies $z = y^2 - x^2$, then $c = b^2 - a^2$. L_1: $x = a + t$, $y = b + t$, $z = c + 2(b - a)t$, L_2: $x = a + t$, $y = b - t$, $z = c - 2(b + a)t$. Substitute the parametric equations of L_1 into the equation of the hyperbolic paraboloid in order to find the points of intersection: $z = y^2 - x^2 \quad \Rightarrow$
$c + 2(b - a)t = (b + t)^2 - (a + t)^2 = b^2 - a^2 + 2(b - a)t \quad \Rightarrow \quad c = b^2 - a^2$. As this is true for all values of t, L_1 lies on $z = y^2 - x^2$. Performing similar operations with L_2 gives: $z = y^2 - x^2 \quad \Rightarrow$
$c - 2(b + a)t = (b - t)^2 - (a + t)^2 = b^2 - a^2 - 2(b + a)t \quad \Rightarrow \quad c = b^2 - a^2$. This tells us that all of L_2 also lies on $z = y^2 - x^2$.

47. The curve of intersection looks like a bent ellipse. The projection of this onto the xy-plane is the set of x- and y-coordinates of points that satisfy $x^2 + y^2 = z = 1 - y^2$. That is, points in the xy-plane that satisfy $x^2 + y^2 = 1 - y^2 \quad \Leftrightarrow \quad x^2 + 2y^2 = 1$

$\Leftrightarrow \quad x^2 + \dfrac{y^2}{(1/\sqrt{2})^2} = 1$, which is the equation of an ellipse.

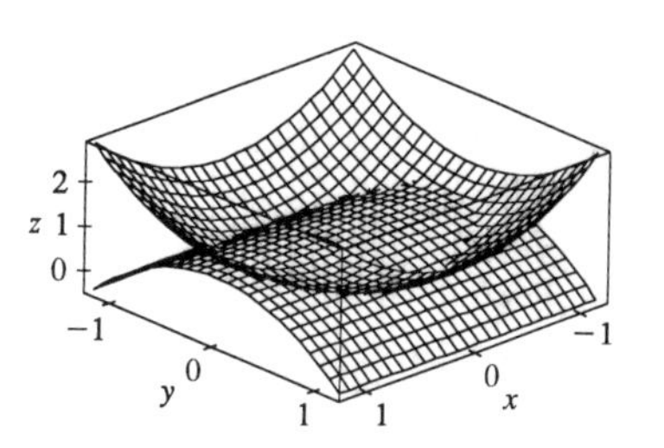

EXERCISES 11.6

1. $x = \cos 4t$, $y = t$, $z = \sin 4t$. At any point (x, y, z) on the curve, $x^2 + z^2 = \cos^2 4t + \sin^2 4t = 1$. So the curve lies on a circular cylinder with axis the y-axis. Since $y = t$, this is a helix. So the graph is V.

3. $x = t$, $y = 1/(1 + t^2)$, $z = t^2$. Note that y and z are positive for all t. The curve passes through $(0, 1, 0)$ when $t = 0$. As $t \to \infty$, $(x, y, z) \to (\infty, 0, \infty)$, and as $t \to -\infty$, $(x, y, z) \to (-\infty, 0, \infty)$. So the graph is I.

5. $x = \cos t$, $y = \sin t$, $z = \sin 5t$. $\quad x^2 + y^2 = \cos^2 t + \sin^2 t = 1$, so the curve lies on a circular cylinder, axis the z-axis. Each of x, y and z is periodic, and at $t = 0$ and $t = 2\pi$ the curve passes through the same point, so the curve repeats itself and the graph is IV.

7. The corresponding parametric equations are $x = t$, $y = -t$, $z = 2t$, which are the parametric equations of a line through the origin and with direction vector $\langle 1, -1, 2 \rangle$.

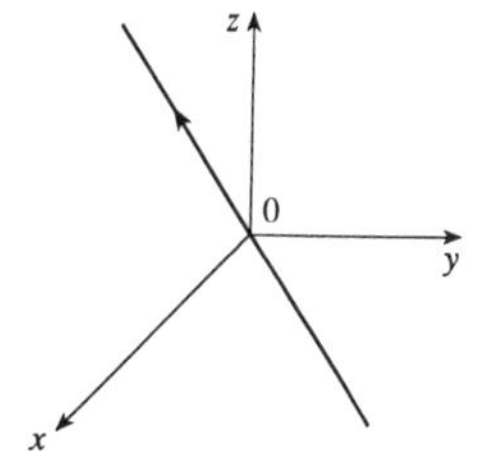

9. The parametric equations give $x^2 + z^2 = \sin^2 t + \cos^2 t = 1$, $y = 3$, which is a circle of radius 1, center $(0, 3, 0)$ in the plane $y = 3$.

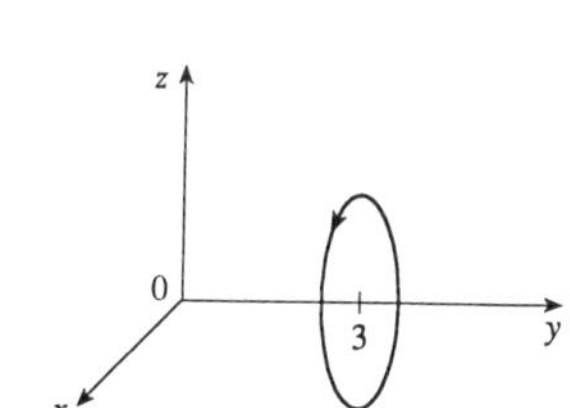

11. Eliminating the parameter t by substituting $z = t$ into $x = t^4 + 1$ gives $x = z^4 + 1$, which is a fourth-degree curve in the xz-plane that opens along the positive x-axis with vertex $(1, 0, 0)$.

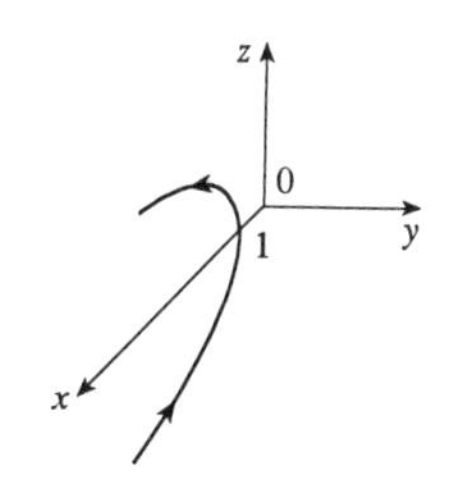

13. The parametric equations are $x = t^2$, $y = t^4$, $z = t^6$. These are positive for $t \neq 0$ and 0 when $t = 0$. So the curve lies entirely in the first quadrant. The projection of the graph onto the xy-plane is $y = x^2$, $y > 0$, a half parabola. On the xz-plane $z = x^3$, $z > 0$, a half cubic, and the yz-plane, $y^3 = z^2$.

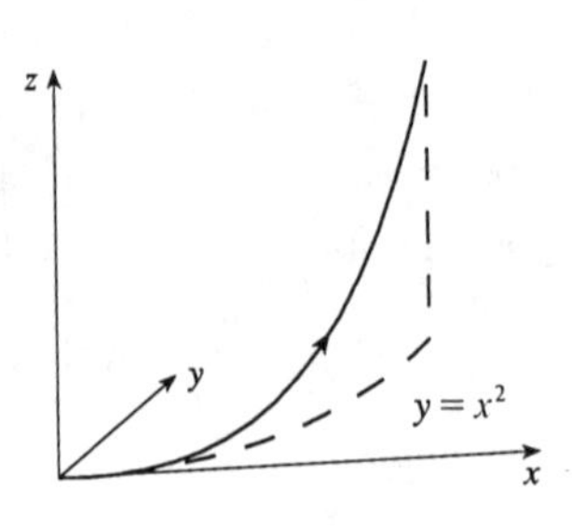

15. If $x = t\cos t$, $y = t\sin t$, and $z = t$, then
$x^2 + y^2 = t^2\cos^2 t + t^2\sin^2 t = t^2 = z^2$,
so the curve lies on the cone $z^2 = x^2 + y^2$.
Thus the curve is a spiral on this cone.

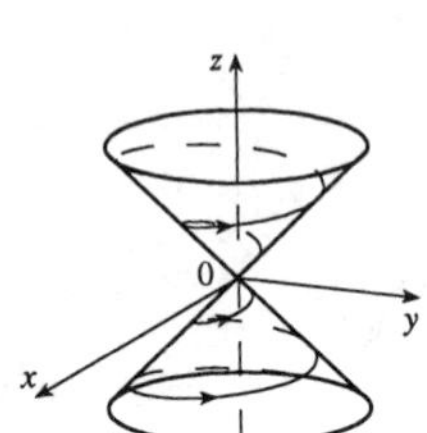

17. $\mathbf{r}(t) = \langle \sin t, \cos t, t^2 \rangle$

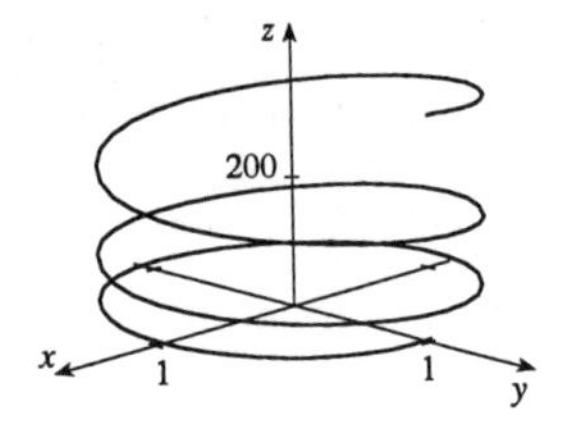

19. $\mathbf{r}(t) = \langle \sqrt{t}, t, t^2 - 2 \rangle$

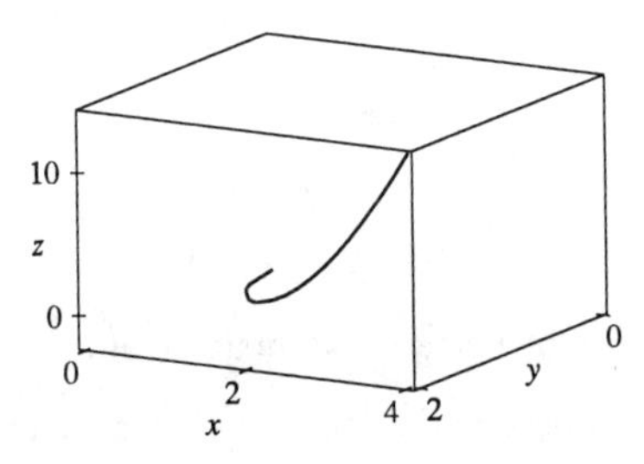

21.

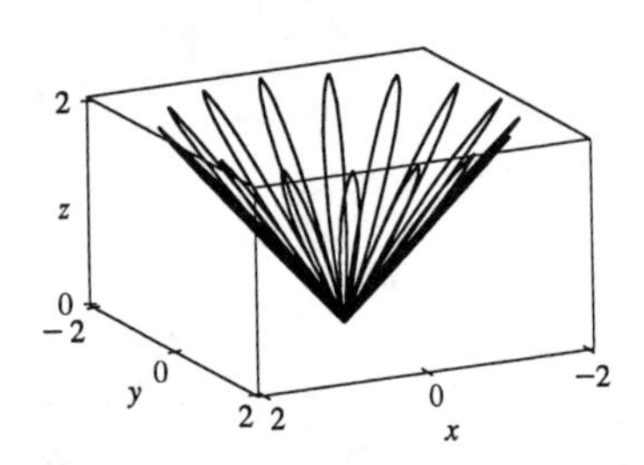

$x = (1 + \cos 16t)\cos t$, $y = (1 + \cos 16t)\sin t$, $z = 1 + \cos 16t$.
At any point on the graph,

$$x^2 + y^2 = (1 + \cos 16t)^2 \cos^2 t + (1 + \cos 16t)^2 \sin^2 t$$
$$= (1 + \cos 16t)^2 = z^2,$$ so the graph lies on the cone $x^2 + y^2 = z^2$. From the graph at left, we see that this curve looks like the projection of a leaved two-dimensional curve onto a cone.

23. $\lim_{t\to 0} \langle t, \cos t, 2 \rangle = \left\langle \lim_{t\to 0} t, \lim_{t\to 0} \cos t, \lim_{t\to 0} 2 \right\rangle = \langle 0, 1, 2 \rangle$

25. $\lim_{t\to 1} \sqrt{t+3} = 2$, $\lim_{t\to 1} \dfrac{t-1}{t^2-1} = \lim_{t\to 1} \dfrac{1}{t+1} = \dfrac{1}{2}$, $\lim_{t\to 1}\left(\dfrac{\tan t}{t}\right) = \tan 1$

Thus the given limit equals $\langle 2, \frac{1}{2}, \tan 1 \rangle$.

27. The domain of $\mathbf{r}$ is $\mathbb{R}$ and $\mathbf{r}'(t) = \langle 1, 2t, 3t^2 \rangle$.

29. Since $\tan t$ and $\sec t$ aren't defined for odd multiples of $\frac{\pi}{2}$, the domain of $\mathbf{r}$ is $\left\{ t \mid t \neq (2n+1)\frac{\pi}{2}, n \text{ an integer} \right\}$.
$\mathbf{r}'(t) = (\sec^2 t)\mathbf{j} + (\sec t \tan t)\mathbf{k}$.

31. We need $4 - t^2 > 0$ and $1 + t \geq 0$, so the domain of $\mathbf{r}$ is $\{t \mid -1 \leq t < 2\}$.

$$\mathbf{r}'(t) = -\frac{2t}{4-t^2}\mathbf{i} + \frac{1}{2\sqrt{1+t}}\mathbf{j} - 12e^{3t}\mathbf{k}.$$

33. The domain of $\mathbf{r}$ is $\mathbb{R}$ and $\mathbf{r}'(t) = 0 + \mathbf{b} + 2\mathbf{c}t = \mathbf{b} + 2t\mathbf{c}$ by Theorem 5 #1.

35. **(a), (c)**

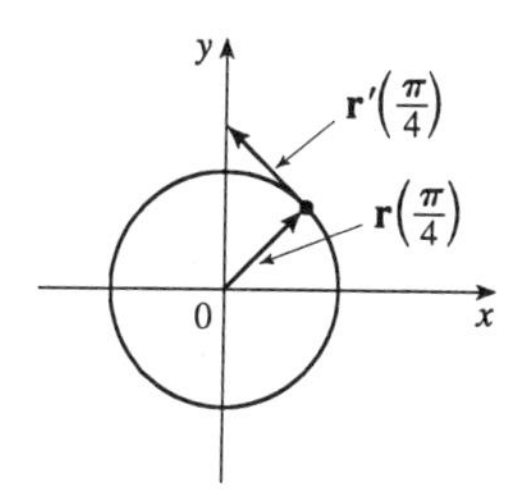

(b) $\mathbf{r}'(t) = \langle -\sin t, \cos t \rangle$

37. Since $(x-1)^2 = t^2 = y$, the curve is a parabola.

(a), (c)

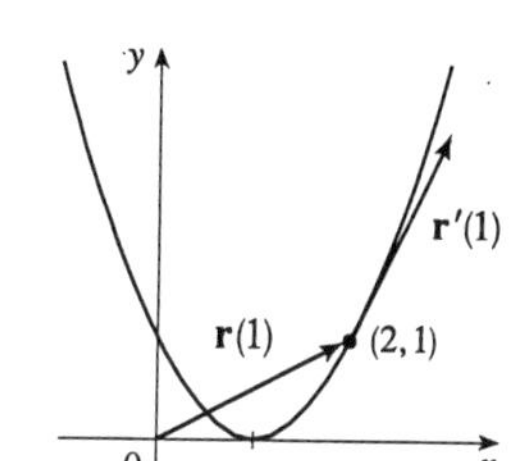

(b) $\mathbf{r}'(t) = \mathbf{i} + 2t\mathbf{j}$

39. $x^{-2} = e^{-2t} = y$, so $y = 1/x^2 > 0$.

(a), (c)

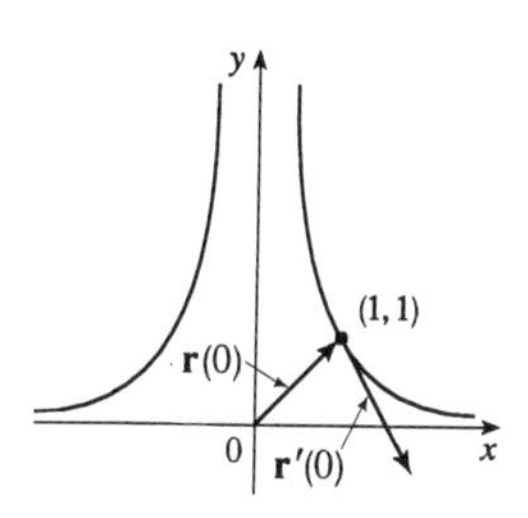

(b) $\mathbf{r}'(t) = e^t\mathbf{i} - 2e^{-2t}\mathbf{j}$

41. $\mathbf{r}'(t) = \langle 2, 6t, 12t^2 \rangle$, $\mathbf{r}(1) = \langle 2, 3, 4 \rangle$,

$\mathbf{r}'(1) = \langle 2, 6, 12 \rangle$. Thus

$$\mathbf{T}(1) = \frac{\mathbf{r}'(1)}{|\mathbf{r}'(1)|} = \frac{1}{\sqrt{188}}\langle 2, 6, 12 \rangle = \left\langle \frac{1}{\sqrt{46}}, \frac{3}{\sqrt{46}}, \frac{6}{\sqrt{46}} \right\rangle$$

43. $\mathbf{r}'(t) = \mathbf{i} + 2\cos t\,\mathbf{j} - 3\sin t\,\mathbf{k}$, $\mathbf{r}'(\frac{\pi}{6}) = \mathbf{i} + \sqrt{3}\mathbf{j} - \frac{3}{2}\mathbf{k}$. Thus

$$\mathbf{T}(\tfrac{\pi}{6}) = \frac{1}{\sqrt{25/4}}\left(\mathbf{i} + \sqrt{3}\mathbf{j} - \tfrac{3}{2}\mathbf{k}\right) = \tfrac{2}{5}\mathbf{i} + \tfrac{2\sqrt{3}}{5}\mathbf{j} - \tfrac{3}{5}\mathbf{k}.$$

45. The vector equation of the curve is $\mathbf{r}(t) = t\mathbf{i} + t^2\mathbf{j} + t^3\mathbf{k}$, so $\mathbf{r}'(t) = \mathbf{i} + 2t\mathbf{j} + 3t^2\mathbf{k}$. At the point $(1, 1, 1)$, $t = 1$, so the tangent vector here is $\mathbf{i} + 2\mathbf{j} + 3\mathbf{k}$. The tangent line goes through the point $(1, 1, 1)$ and has direction vector $\mathbf{i} + 2\mathbf{j} + 3\mathbf{k}$. Thus parametric equations are $x = 1 + t$, $y = 1 + 2t$, $z = 1 + 3t$.

47. $\mathbf{r}(t) = \langle t\cos 2\pi t, t\sin 2\pi t, 4t \rangle$, $\mathbf{r}'(t) = \langle \cos 2\pi t - 2\pi t\sin 2\pi t, \sin 2\pi t + 2\pi t\cos 2\pi t, 4 \rangle$. At $\left(0, \frac{1}{4}, 1\right)$, $t = \frac{1}{4}$ and $\mathbf{r}'(\frac{1}{4}) = \langle 0 - \frac{\pi}{2}, 1 + 0, 4 \rangle = \langle -\frac{\pi}{2}, 1, 4 \rangle$. Thus the parametric equations of the tangent line are $x = -\frac{\pi}{2}t$, $y = \frac{1}{4} + t$, $z = 1 + 4t$.

49. $\mathbf{r}(t) = \langle t, \sqrt{2}\cos t, \sqrt{2}\sin t\rangle \quad \Rightarrow$

$\mathbf{r}'(t) = \langle 1, -\sqrt{2}\sin t, \sqrt{2}\cos t\rangle$.

At $\left(\frac{\pi}{4}, 1, 1\right)$, $t = \frac{\pi}{4}$ and $\mathbf{r}'\left(\frac{\pi}{4}\right) = \langle 1, -1, 1\rangle$.

Thus the parametric equations of the tangent line are

$x = \frac{\pi}{4} + t, y = 1 - t, z = 1 + t$.

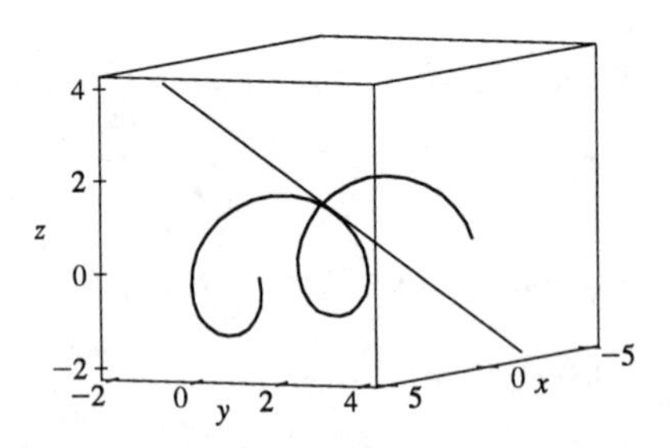

51. The angle of intersection of the two curves is the angle between the two tangent vectors to the curves at the point of intersection. Since $\mathbf{r}_1'(t) = \langle 1, 2t, 3t^2\rangle$ and $t = 0$ at $(0,0,0)$, $\mathbf{r}_1'(0) = \langle 1, 0, 0\rangle$ is a tangent vector to $\mathbf{r}_1$ at $(0,0,0)$. Also $\mathbf{r}_2' = \langle \cos t, 2\cos t, 1\rangle$ so $\mathbf{r}_2'(0) = \langle 1, 2, 1\rangle$ is a tangent vector to $\mathbf{r}_2$ at $(0,0,0)$. If θ is the angle between these two tangent vectors, then $\cos\theta = \frac{1}{\sqrt{1}\sqrt{6}}\langle 1,0,0\rangle \cdot \langle 1,2,1\rangle = \frac{1}{\sqrt{6}}$ and $\theta = \cos^{-1}\frac{1}{\sqrt{6}} \approx 66°$.

53. If the point $(1, 4, 0)$ lies on the curve, then $1 - 3t = y = 4 \quad \Rightarrow \quad t = -1$. The point which corresponds to $t = -1$ is $(t^2, 1 - 3t, 1 + t^3) = \left((-1)^2, 1 - 3(-1), 1 + (-1)^3\right) = (1, 4, 0)$. For the next point, $-8 = y = 1 - 3t \quad \Rightarrow \quad t = 3$. The point which corresponds to $t = 3$ is $(3^2, 1 - 3(3), 1 + 3^3) = (9, -8, 28)$. If the last point is on the curve, then $7 = y = 1 - 3t \quad \Rightarrow \quad t = -2$. But the point for $t = -2$ is $\left((-2)^2, 1 - 3(-2), 1 + (-2)^3\right) = (4, 7, -7) \neq (4, 7, -6)$. So this point is not on the curve.

55.

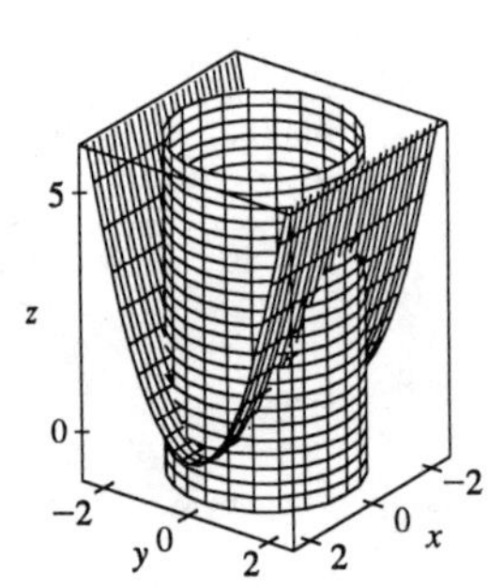

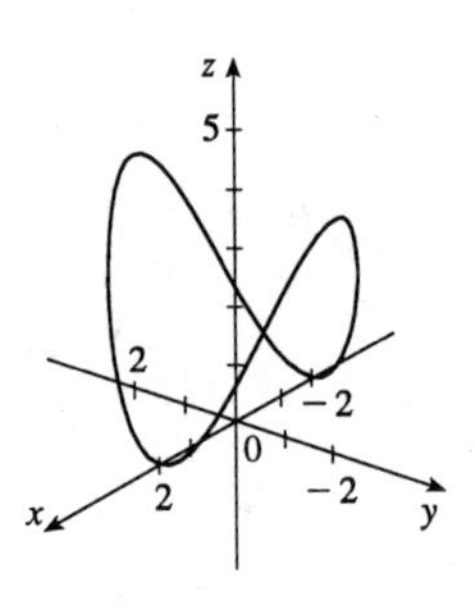

$x = 2\cos t \quad \Rightarrow \quad y^2 = 4 - 4\cos^2 t = 4\sin^2 t \quad \Rightarrow \quad y = \pm 2\sin t$ and $z = x^2 = 4\cos^2 t$. We choose the + sign, so parametric equations for the curve of intersection are $x = 2\cos t$, $y = 2\sin t$, $z = 4\cos^2 t$.

57. $\int_0^1 (t\mathbf{i} + t^2\mathbf{j} + t^3\mathbf{k})dt = \left(\int_0^1 t\,dt\right)\mathbf{i} + \left(\int_0^1 t^2\,dt\right)\mathbf{j} + \left(\int_0^1 t^3\,dt\right)\mathbf{k} = \left[\frac{t^2}{2}\right]_0^1\mathbf{i} + \left[\frac{t^3}{3}\right]_0^1\mathbf{j} + \left[\frac{t^4}{4}\right]_0^1\mathbf{k} = \frac{1}{2}\mathbf{i} + \frac{1}{3}\mathbf{j} + \frac{1}{4}\mathbf{k}$

59. $\int_0^{\pi/4}(\cos 2t\,\mathbf{i} + \sin 2t\,\mathbf{j} + t\sin t\,\mathbf{k})dt = \left[\frac{1}{2}\sin 2t\,\mathbf{i} - \frac{1}{2}\cos 2t\,\mathbf{j}\right]_0^{\pi/4} + \left[\left[-t\cos t\right]_0^{\pi/4} + \int_0^{\pi/4}\cos t\,dt\right]\mathbf{k}$

$$= \tfrac{1}{2}\mathbf{i} + \tfrac{1}{2}\mathbf{j} + \left[-\tfrac{\pi}{4}\cos\tfrac{\pi}{4} + \sin\tfrac{\pi}{4}\right]\mathbf{k} = \tfrac{1}{2}\mathbf{i} + \tfrac{1}{2}\mathbf{j} + \tfrac{1}{\sqrt{2}}\left(1 - \tfrac{\pi}{4}\right)\mathbf{k} = \tfrac{1}{2}\mathbf{i} + \tfrac{1}{2}\mathbf{j} + \tfrac{4-\pi}{4\sqrt{2}}\mathbf{k}$$

61. $\mathbf{r}(t) = \dfrac{t^3}{3}\mathbf{i} + t^4\mathbf{j} - \dfrac{t^3}{3}\mathbf{k} + \mathbf{c}$ where $\mathbf{c}$ is a constant vector. But $\mathbf{j} = \mathbf{r}(0) = (0)\mathbf{i} + (0)\mathbf{j} - (0)\mathbf{k} + \mathbf{c}$. Thus $\mathbf{c} = \mathbf{j}$ and $\mathbf{r}(t) = \dfrac{t^3}{3}\mathbf{i} + (t^4 + 1)\mathbf{j} - \dfrac{t^3}{3}\mathbf{k}$.

63. Let $\mathbf{u}(t) = \langle u_1(t), u_2(t), u_3(t) \rangle$ and $\mathbf{v}(t) = \langle v_1(t), v_2(t), v_3(t) \rangle$. In each part of this problem the basic procedure is to use Equation 1 and then analyze the individual component functions using the limit properties we have already developed for real-valued functions.

(a) $\lim_{t\to a} \mathbf{u}(t) + \lim_{t\to a} \mathbf{v}(t) = \left\langle \lim_{t\to a} u_1(t), \lim_{t\to a} u_2(t), \lim_{t\to a} u_3(t) \right\rangle + \left\langle \lim_{t\to a} v_1(t), \lim_{t\to a} v_2(t), \lim_{t\to a} v_3(t) \right\rangle$ and the limits of these component functions must each exist since the vector functions both possess limits as $t \to a$. Then adding the two vectors and using the addition property of limits for real-valued functions we have that

$$\lim_{t\to a} \mathbf{u}(t) + \lim_{t\to a} \mathbf{v}(t) = \left\langle \lim_{t\to a} u_1(t) + \lim_{t\to a} v_1(t), \lim_{t\to a} u_2(t) + \lim_{t\to a} v_2(t), \lim_{t\to a} u_3(t) + \lim_{t\to a} v_3(t) \right\rangle$$

$$= \left\langle \lim_{t\to a}[u_1(t) + v_1(t)], \lim_{t\to a}[u_2(t) + v_2(t)], \lim_{t\to a}[u_3(t) + v_3(t)] \right\rangle$$

$$= \lim_{t\to a} \langle u_1(t) + v_1(t), u_2(t) + v_2(t), u_3(t) + v_3(t) \rangle \quad \text{[using (1) backward]} \quad = \lim_{t\to a}[\mathbf{u}(t) + \mathbf{v}(t)].$$

(b) $\lim_{t\to a} c\mathbf{u}(t) = \lim_{t\to a} \langle cu_1(t), cu_2(t), cu_3(t) \rangle = \left\langle \lim_{t\to a} cu_1(t), \lim_{t\to a} cu_2(t), \lim_{t\to a} cu_3(t) \right\rangle$

$$= \left\langle c\lim_{t\to a} u_1(t), c\lim_{t\to a} u_2(t), c\lim_{t\to a} u_3(t) \right\rangle = c\left\langle \lim_{t\to a} u_1(t), \lim_{t\to a} u_2(t), \lim_{t\to a} u_3(t) \right\rangle$$

$$= c\lim_{t\to a} \langle u_1(t), u_2(t), u_3(t) \rangle = c\lim_{t\to a} \mathbf{u}(t)$$

(c) $\lim_{t\to a} \mathbf{u}(t) \cdot \lim_{t\to a} \mathbf{v}(t) = \left\langle \lim_{t\to a} u_1(t), \lim_{t\to a} u_2(t), \lim_{t\to a} u_3(t) \right\rangle \cdot \left\langle \lim_{t\to a} v_1(t), \lim_{t\to a} v_2(t), \lim_{t\to a} v_3(t) \right\rangle$

$$= \left[\lim_{t\to a} u_1(t)\right]\left[\lim_{t\to a} v_1(t)\right] + \left[\lim_{t\to a} u_2(t)\right]\left[\lim_{t\to a} v_2(t)\right] + \left[\lim_{t\to a} u_3(t)\right]\left[\lim_{t\to a} v_3(t)\right]$$

$$= \lim_{t\to a} u_1(t)v_1(t) + \lim_{t\to a} u_2(t)v_2(t) + \lim_{t\to a} u_3(t)v_3(t)$$

$$= \lim_{t\to a} [u_1(t)v_1(t) + u_2(t)v_2(t) + u_3(t)v_3(t)] = \lim_{t\to a} [\mathbf{u}(t) \cdot \mathbf{v}(t)]$$

(d) $\lim_{t\to a} \mathbf{u}(t) \times \lim_{t\to a} \mathbf{v}(t) = \left\langle \lim_{t\to a} u_1(t), \lim_{t\to a} u_2(t), \lim_{t\to a} u_3(t) \right\rangle \times \left\langle \lim_{t\to a} v_1(t), \lim_{t\to a} v_2(t), \lim_{t\to a} v_3(t) \right\rangle$

$$= \left\langle \left[\lim_{t\to a} u_2(t)\right]\left[\lim_{t\to a} v_3(t)\right] - \left[\lim_{t\to a} u_3(t)\right]\left[\lim_{t\to a} v_2(t)\right], \left[\lim_{t\to a} u_3(t)\right]\left[\lim_{t\to a} v_1(t)\right] - \left[\lim_{t\to a} u_1(t)\right]\left[\lim_{t\to a} v_3(t)\right], \right.$$

$$\left. \left[\lim_{t\to a} u_1(t)\right]\left[\lim_{t\to a} v_2(t)\right] - \left[\lim_{t\to a} u_2(t)\right]\left[\lim_{t\to a} v_1(t)\right] \right\rangle$$

$$= \left\langle \lim_{t\to a}[u_2(t)v_3(t) - u_3(t)v_2(t)], \lim_{t\to a}[u_3(t)v_1(t) - u_1(t)v_3(t)], \lim_{t\to a} [u_1(t)v_2(t) - u_2(t)v_1(t)] \right\rangle$$

$$= \lim_{t\to a} \langle u_2(t)v_3(t) - u_3(t)v_2(t), u_3(t)v_1(t) - u_1(t)v_3(t), u_1(t)v_2(t) - u_2(t)v_1(t) \rangle = \lim_{t\to a} [\mathbf{u}(t) \times \mathbf{v}(t)]$$

For Exercises 65-68, let $\mathbf{u}(t) = \langle u_1(t), u_2(t), u_3(t) \rangle$ ***and*** $\mathbf{v}(t) = \langle v_1(t), v_2(t), v_3(t) \rangle$. ***In each of these exercises, the procedure is to apply Theorem 4 so the corresponding properties of derivatives of real-valued functions can be used.***

65. $$\frac{d}{dt}[\mathbf{u}(t) + \mathbf{v}(t)] = \frac{d}{dt}\langle u_1(t) + v_1(t), u_2(t) + v_2(t), u_3(t) + v_3(t) \rangle$$
$$= \left\langle \frac{d}{dt}[u_1(t) + v_1(t)], \frac{d}{dt}[u_2(t) + v_2(t)], \frac{d}{dt}[u_3(t) + v_3(t)] \right\rangle$$
$$= \langle u_1'(t) + v_1'(t), u_2'(t) + v_2'(t), u_3'(t) + v_3'(t) \rangle$$
$$= \langle u_1'(t), u_2'(t), u_3'(t) \rangle + \langle v_1'(t), v_2'(t), v_3'(t) \rangle = \mathbf{u}'(t) + \mathbf{v}'(t).$$

67. $$\frac{d}{dt}[\mathbf{u}(t) \times \mathbf{v}(t)] = \frac{d}{dt}\langle u_2(t)v_3(t) - u_3(t)v_2(t), u_3(t)v_1(t) - u_1(t)v_3(t), u_1(t)v_2(t) - u_2(t)v_1(t) \rangle$$
$$= \langle u_2'v_3(t) + u_2(t)v_3'(t) - u_3'(t)v_2(t) - u_3(t)v_2'(t),$$
$$u_3'(t)v_1(t) + u_3(t)v_1'(t) - u_1'(t)v_3(t) - u_1(t)v_3'(t),$$
$$u_1'(t)v_2(t) + u_1(t)v_2'(t) - u_2'(t)v_1(t) - u_2(t)v_1'(t) \rangle$$
$$= \langle u_2'(t)v_3(t) - u_3'(t)v_2(t), u_3'(t)v_1(t) - u_1'(t)v_3(t), u_1'(t)v_2(t) - u_2'(t)v_1(t) \rangle$$
$$+ \langle u_2(t)v_3'(t) - u_3(t)v_2'(t), u_3(t)v_1'(t) - u_1(t)v_3'(t), u_1(t)v_2'(t) - u_2(t)v_1'(t) \rangle$$
$$= \mathbf{u}'(t) \times \mathbf{v}(t) + \mathbf{u}(t) \times \mathbf{v}'(t)$$

Alternate Solution: Let $\mathbf{r}(t) = \mathbf{u}(t) \times \mathbf{v}(t)$, then

$$\mathbf{r}(t+h) - \mathbf{r}(t) = [\mathbf{u}(t+h) \times \mathbf{v}(t+h)] - [\mathbf{u}(t) \times \mathbf{v}(t)]$$
$$= [\mathbf{u}(t+h) \times \mathbf{v}(t+h)] - [\mathbf{u}(t) \times \mathbf{v}(t)] + [\mathbf{u}(t+h) \times \mathbf{v}(t)] - [\mathbf{u}(t+h) \times \mathbf{v}(t)]$$
$$= \mathbf{u}(t+h) \times [\mathbf{v}(t+h) - \mathbf{v}(t)] + [\mathbf{u}(t+h) - \mathbf{u}(t)] \times \mathbf{v}(t).$$

(Be careful of the order of the cross product.) Dividing through by h and taking the limit as $h \to 0$ we have

$$\mathbf{r}'(t) = \lim_{h \to 0} \frac{\mathbf{u}(t+h) \times [\mathbf{v}(t+h) - \mathbf{v}(t)]}{h} + \lim_{h \to 0} \frac{[\mathbf{u}(t+h) - \mathbf{u}(t)] \times \mathbf{v}(t)}{h}$$
$$= \mathbf{u}(t) \times \mathbf{v}'(t) + \mathbf{u}'(t) \times \mathbf{v}(t)$$ by Exercise 63(a) and Definition 3.

69. $D_t[\mathbf{u}(t) \cdot \mathbf{v}(t)] = \mathbf{u}'(t) \cdot \mathbf{v}(t) + \mathbf{u}(t) \cdot \mathbf{v}'(t)$ by Theorem 5 #4
$$= (-4t\mathbf{j} + 9t^2\mathbf{k}) \cdot (t\mathbf{i} + \cos t\,\mathbf{j} + \sin t\,\mathbf{k}) + (\mathbf{i} - 2t^2\mathbf{j} + 3t^3\mathbf{k}) \cdot (\mathbf{i} - \sin t\,\mathbf{j} + \cos t\,\mathbf{k})$$
$$= -4t\cos t + 9t^2 \sin t + 1 + 2t^2 \sin t + 3t^3 \cos t$$
$$= 1 - 4t\cos t + 11t^2 \sin t + 3t^3 \cos t$$

71. $\frac{d}{dt}[\mathbf{r}(t) \times \mathbf{r}'(t)] = \mathbf{r}'(t) \times \mathbf{r}'(t) + \mathbf{r}(t) \times \mathbf{r}''(t)$ by Equation 5 #5. But $\mathbf{r}'(t) \times \mathbf{r}'(t) = \mathbf{0}$. Thus
$$\frac{d}{dt}[\mathbf{r}(t) \times \mathbf{r}'(t)] = \mathbf{r}(t) \times \mathbf{r}''(t).$$

73. $$\frac{d}{dt}|\mathbf{r}(t)| = \frac{d}{dt}[\mathbf{r}(t) \cdot \mathbf{r}(t)]^{1/2} = \tfrac{1}{2}[\mathbf{r}(t) \cdot \mathbf{r}(t)]^{-1/2}[2\mathbf{r}(t) \cdot \mathbf{r}'(t)] = \frac{\mathbf{r}(t) \cdot \mathbf{r}'(t)}{|\mathbf{r}(t)|}$$

75. Since $\mathbf{u}(t) = \mathbf{r}(t) \cdot [\mathbf{r}'(t) \times \mathbf{r}''(t)]$,
$$\mathbf{u}'(t) = \mathbf{r}'(t) \cdot [\mathbf{r}'(t) \times \mathbf{r}''(t)] + \mathbf{r}(t) \cdot \frac{d}{dt}[\mathbf{r}'(t) \times \mathbf{r}''(t)]$$
$$= 0 + \mathbf{r}(t) \cdot [\mathbf{r}''(t) \times \mathbf{r}''(t) + \mathbf{r}'(t) \times \mathbf{r}'''(t)] \qquad [\text{since } \mathbf{r}'(t) \perp \mathbf{r}'(t) \times \mathbf{r}''(t)]$$
$$= \mathbf{r}(t) \cdot [\mathbf{r}'(t) \times \mathbf{r}'''(t)] \qquad [\text{since } \mathbf{r}''(t) \times \mathbf{r}''(t) = \mathbf{0}]$$

EXERCISES 11.7

1. $\mathbf{r}'(t) = \langle 2, 3\cos t, -3\sin t\rangle$, $|\mathbf{r}'(t)| = \sqrt{4 + 9\cos^2 t + 9\sin^2 t} = \sqrt{13}$

$$L = \int_a^b \sqrt{13}\, dt = \sqrt{13}(b - a)$$

3. $\mathbf{r}'(t) = \langle 6, 6\sqrt{2}t, 6t^2\rangle$, $|\mathbf{r}'(t)| = 6\sqrt{1 + 2t^2 + t^4} = 6(1 + t^2)$

$$L = \int_0^1 6(1 + t^2)dt = \left[\frac{6(t + t^3)}{3}\right]_0^1 = \frac{24}{3} = 8$$

5. The point $(2, 4, 8)$ corresponds to $t = 2$, so by Equation 2, $L = \int_0^2 \sqrt{(1)^2 + (2t)^2 + (3t^2)^2}\, dt$.

If $f(t) = \sqrt{1 + 4t^2 + 9t^4}$, then Simpson's Rule gives

$$L \approx \frac{2 - 0}{10 \cdot 3}[f(0) + 4f(0.2) + 2f(0.4) + \cdots + 4f(1.8) + f(2)] \approx 9.5706.$$

7. $\mathbf{r}'(t) = e^t(\cos t + \sin t)\mathbf{i} + e^t(\cos t - \sin t)\mathbf{j}$,

$ds/dt = |\mathbf{r}'(t)| = e^t\sqrt{(\cos t + \sin t)^2 + (\cos t - \sin t)^2} = e^t\sqrt{2\cos^2 t + 2\sin^2 t} = \sqrt{2}e^t$

$s(t) = \int_0^t |\mathbf{r}'(u)|du = \int_0^t \sqrt{2}e^u\, du = \sqrt{2}(e^t - 1) \quad\Rightarrow\quad \frac{1}{\sqrt{2}}s + 1 = e^t \quad\Rightarrow\quad t(s) = \ln\left(\frac{1}{\sqrt{2}}s + 1\right)$.

Therefore, $\mathbf{r}(t(s)) = \left(\frac{1}{\sqrt{2}}s + 1\right)\left[\sin\left(\ln\left(\frac{1}{\sqrt{2}}s + 1\right)\right)\mathbf{i} + \cos\left(\ln\left(\frac{1}{\sqrt{2}}s + 1\right)\right)\mathbf{j}\right]$.

9. $|\mathbf{r}'(t)| = \sqrt{(3\cos t)^2 + 16 + (-3\sin t)^2} = \sqrt{9 + 16} = 5$ and $s(t) = \int_0^t |\mathbf{r}'(u)|du = \int_0^t 5\, du = 5t \quad\Rightarrow$

$t(s) = \frac{1}{5}s$. Therefore, $\mathbf{r}(t(s)) = 3\sin\left(\frac{1}{5}s\right)\mathbf{i} + \frac{4}{5}s\mathbf{j} + 3\cos\left(\frac{1}{5}s\right)\mathbf{k}$.

11. **(a)** $\mathbf{T}(t) = \dfrac{\mathbf{r}'(t)}{|\mathbf{r}'(t)|} = \dfrac{1}{\sqrt{16 + 9}}\langle 4\cos 4t, 3, -4\sin 4t\rangle = \dfrac{1}{5}\langle 4\cos 4t, 3, -4\sin 4t\rangle$

$\mathbf{N}(t) = \dfrac{\mathbf{T}'(t)}{|\mathbf{T}'(t)|} = \dfrac{5}{16 \cdot 5}\langle -16\sin 4t, 0, -16\cos 4t\rangle = \langle -\sin 4t, 0, -\cos 4t\rangle$

(b) $\kappa(t) = \dfrac{|\mathbf{T}'(t)|}{|\mathbf{r}'(t)|} = \dfrac{16}{5 \cdot 5} = \dfrac{16}{25}$

13. **(a)** $\mathbf{T}(t) = \dfrac{\mathbf{r}'(t)}{|\mathbf{r}'(t)|} = \dfrac{1}{\sqrt{2\sin^2 t + 2\cos^2 t}}\langle -\sqrt{2}\sin t, \cos t, \cos t\rangle = \frac{1}{\sqrt{2}}\langle -\sqrt{2}\sin t, \cos t, \cos t\rangle$

$\mathbf{N}(t) = \dfrac{\mathbf{T}'(t)}{|\mathbf{T}'(t)|} = \dfrac{1}{\sqrt{2\cos^2 t + 2\sin^2 t}}\langle -\sqrt{2}\cos t, -\sin t, -\sin t\rangle = \frac{1}{\sqrt{2}}\langle -\sqrt{2}\cos t, -\sin t, -\sin t\rangle$

(b) $\kappa(t) = \dfrac{|\mathbf{T}'(t)|}{|\mathbf{r}'(t)|} = \dfrac{1}{\sqrt{2}}$

15. $\mathbf{r}'(t) = \mathbf{j} - 2t\mathbf{k}$, $\mathbf{r}''(t) = -2\mathbf{k}$, $|\mathbf{r}'(t)|^3 = (4t^2 + 1)^{3/2}$, $|\mathbf{r}'(t) \times \mathbf{r}''(t)| = |-2\mathbf{i}| = 2$,

$$\kappa(t) = \frac{|\mathbf{r}'(t) \times \mathbf{r}''(t)|}{|\mathbf{r}'(t)|^3} = \frac{2}{(4t^2 + 1)^{3/2}}$$

17. $\mathbf{r}'(t) = \langle 6t^2, -6t, 6\rangle$, $\mathbf{r}''(t) = \langle 12t, -6, 0\rangle$, $|\mathbf{r}'(t)|^3 = 6^3(t^4+t^2+1)^{3/2}$,

$|\mathbf{r}'(t)\times\mathbf{r}''(t)| = |36\langle 1, 2t, t^2\rangle| = 36\sqrt{1+4t^2+t^4}$,

$$\kappa(t) = \frac{|\mathbf{r}'(t)\times\mathbf{r}''(t)|}{|\mathbf{r}'(t)|^3} = \frac{36\sqrt{1+4t^2+t^4}}{6^3(t^4+t^2+1)^{3/2}} = \frac{\sqrt{1+4t^2+t^4}}{6(t^4+t^2+1)^{3/2}}$$

19. $\mathbf{r}'(t) = \langle \cos t, -\sin t, \cos t\rangle$, $\mathbf{r}''(t) = \langle -\sin t, -\cos t, -\sin t\rangle$, $|\mathbf{r}'(t)|^3 = \left(\sqrt{\cos^2 t + 1}\right)^3$,

$|\mathbf{r}'(t)\times\mathbf{r}''(t)| = |\langle 1, 0, -1\rangle| = \sqrt{2}$, $\kappa(t) = \dfrac{|\mathbf{r}'(t)\times\mathbf{r}''(t)|}{|\mathbf{r}'(t)|^3} = \dfrac{\sqrt{2}}{(1+\cos^2 t)^{3/2}}$

21. $f'(x) = 3x^2$, $f''(x) = 6x$, $\kappa(x) = \dfrac{|f''(x)|}{\left[1+(f'(x))^2\right]^{3/2}} = \dfrac{6|x|}{[1+9x^4]^{3/2}}$

23. $y' = \cos x$, $y'' = -\sin x$, $\kappa(x) = \dfrac{|y''(x)|}{\left[1+(y'(x))^2\right]^{3/2}} = \dfrac{|\sin x|}{[1+\cos^2 x]^{3/2}}$

25. Since $y' = y'' = e^x$, the curvature is $\kappa(x) = \dfrac{|y''(x)|}{\left[1+(y'(x))^2\right]^{3/2}} = \dfrac{e^x}{(1+e^{2x})^{3/2}} = e^x\left[1+e^{2x}\right]^{-3/2} \Rightarrow$

$\kappa'(x) = e^x[1+e^{2x}]^{-3/2} + e^x\left(-\frac{3}{2}\right)[1+e^{2x}]^{-5/2}(2e^{2x}) = e^x\dfrac{1+e^{2x}-3e^{2x}}{(1+e^{2x})^{5/2}} = e^x\dfrac{1-2e^{2x}}{(1+e^{2x})^{5/2}}$. Then when $\kappa'(x) = 0$, we must have $1-2e^{2x} = 0$ or $e^{2x} = \frac{1}{2}$ or $x = -\frac{1}{2}\ln 2$. And since $1-2e^{2x} > 0$ for $x < -\frac{1}{2}\ln 2$ and $1-2e^{2x} < 0$ for $x > -\frac{1}{2}\ln 2$, the maximum curvature is attained at the point $\left(-\frac{1}{2}\ln 2, e^{(-\ln 2)/2}\right) = \left(-\frac{1}{2}\ln 2, \frac{1}{\sqrt{2}}\right)$.

27. $y = x^4 \quad\Rightarrow\quad y' = 4x^3$, $y'' = 12x^2$, and

$$\kappa(x) = \frac{|y''|}{\left[1+(y')^2\right]^{3/2}} = \frac{12x^2}{(1+16x^6)^{3/2}}.$$

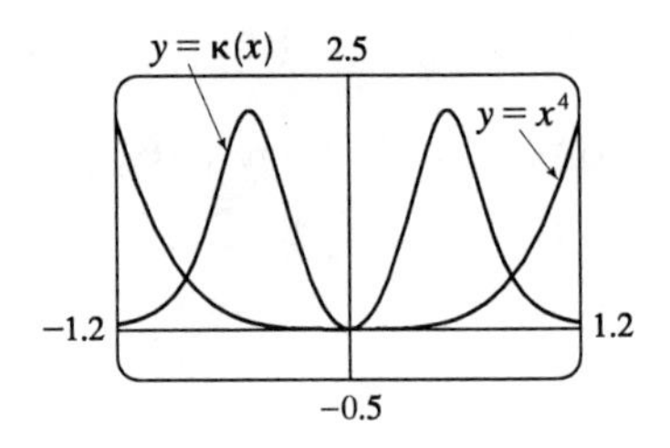

The appearance of the two humps in this graph is perhaps a little surprising, but it is explained by the fact that $y = x^4$ is very flat around the origin, and so here the curvature is zero.

29. $\kappa(t) = \dfrac{|\dot{x}\ddot{y} - \ddot{x}\dot{y}|}{(\dot{x}^2+\dot{y}^2)^{3/2}} = \dfrac{|(3t^2)(2)-(6t)(2t)|}{(9t^4+4t^2)^{3/2}} = \dfrac{6t^2}{(t^2)^{3/2}(9t^2+4)^{3/2}} = \dfrac{6t^2}{|t|^3(9t^2+4)^{3/2}} = \dfrac{6}{|t|(9t^2+4)^{3/2}}$

31. $\left(1, \frac{2}{3}, 1\right)$ corresponds to $t = 1$. $\mathbf{T}(t) = \dfrac{\mathbf{r}'(t)}{|\mathbf{r}'(t)|} = \dfrac{\langle 2t, 2t^2, 1\rangle}{\sqrt{4t^2+4t^4+1}} = \dfrac{\langle 2t, 2t^2, 1\rangle}{2t^2+1}$, so $\mathbf{T}(1) = \left\langle \frac{2}{3}, \frac{2}{3}, \frac{1}{3}\right\rangle$.

$\mathbf{T}'(t) = -4t(2t^2+1)^{-2}\langle 2t, 2t^2, 1\rangle + (2t^2+1)^{-1}\langle 2, 4t, 0\rangle$ (By Theorem 11.6.5)

$= (2t^2+1)^{-2}\langle -8t^2+4t^2+2, -8t^3+8t^3+4t, -4t\rangle = 2(2t^2+1)^{-2}\langle 1-2t^2, 2t, -2t\rangle$

$$\mathbf{N}(t) = \frac{\mathbf{T}'(t)}{|\mathbf{T}'(t)|} = \frac{2(2t^2+1)^{-2}\langle 1-2t^2, 2t, -2t\rangle}{2(2t^2+1)^{-2}\sqrt{(1-2t^2)^2+(2t)^2+(-2t)^2}} = \frac{\langle 1-2t^2, 2t, -2t\rangle}{\sqrt{1-4t^2+4t^4+8t^2}} = \frac{\langle 1-2t^2, 2t, -2t\rangle}{1+2t^2}$$

$\mathbf{N}(1) = \left\langle -\frac{1}{3}, \frac{2}{3}, -\frac{2}{3}\right\rangle$ and $\mathbf{B}(1) = \mathbf{T}(1)\times\mathbf{N}(1) = \left\langle -\frac{4}{9}-\frac{2}{9}, -\left(-\frac{4}{9}+\frac{1}{9}\right), \frac{4}{9}+\frac{2}{9}\right\rangle = \left\langle -\frac{2}{3}, \frac{1}{3}, \frac{2}{3}\right\rangle$.

33. $t = \pi$ corresponds to $(0, \pi, -2)$. $\mathbf{T}(t) = \dfrac{\mathbf{r}'(t)}{|\mathbf{r}'(t)|} = \dfrac{\langle 6\cos 3t, 1, -6\sin 3t\rangle}{\sqrt{36\cos^2 3t + 1 + 36\sin^2 3t}} = \frac{1}{\sqrt{37}}\langle 6\cos 3t, 1, -6\sin 3t\rangle$.

$\mathbf{T}(\pi) = \frac{1}{\sqrt{37}}\langle -6, 1, 0\rangle$ is a normal vector for the normal plane, and so $\langle -6, 1, 0\rangle$ is also normal. Thus an equation for the plane is $-6(x-0) + 1(y-\pi) + 0(z+2) = 0$ or $y - 6x = \pi$.

$\mathbf{T}'(t) = \frac{1}{\sqrt{37}}\langle -18\sin 3t, 0, -18\cos 3t\rangle \quad\Rightarrow\quad |\mathbf{T}'(t)| = \dfrac{\sqrt{18^2\sin^2 3t + 18^2\cos^2 3t}}{\sqrt{37}} = \dfrac{18}{\sqrt{37}} \quad\Rightarrow$

$\mathbf{N}(t) = \langle -\sin 3t, 0, -\cos 3t\rangle$. So $\mathbf{B}(\pi) = \frac{1}{\sqrt{37}}\langle -6, 1, 0\rangle \times \langle 0, 0, 1\rangle = \frac{1}{\sqrt{37}}\langle 1, 6, 0\rangle$. Since $\mathbf{B}(\pi)$ is a normal to the osculating plane, so is $\langle 1, 6, 0\rangle$ and an equation for the plane is $1(x-0) + 6(y-\pi) + 0(z+2) = 0$ or $x + 6y = 6\pi$.

35. The ellipse is given by the parametric equations $x = 2\cos t$, $y = 3\sin t$, so using the result from Exercise 28,

$$\kappa(t) = \frac{|\dot{x}\ddot{y} - \ddot{x}\dot{y}|}{\left[\dot{x}^2 + \dot{y}^2\right]^{3/2}} = \frac{|(-2\sin t)(-3\sin t) - (3\cos t)(-2\cos t)|}{(4\sin^2 t + 9\cos^2 t)^{3/2}} = \frac{6}{(4\sin^2 t + 9\cos^2 t)^{3/2}}.$$

At $(2, 0)$, $t = 0$. Now $\kappa(0) = \frac{6}{27} = \frac{2}{9}$, so the radius of the osculating circle is $1/\kappa(0) = \frac{9}{2}$ and its center is $\left(-\frac{5}{2}, 0\right)$. Its equation is therefore $\left(x + \frac{5}{2}\right)^2 + y^2 = \frac{81}{4}$.

At $(0, 3)$, $t = \frac{\pi}{2}$, and $\kappa\left(\frac{\pi}{2}\right) = \frac{6}{8} = \frac{3}{4}$. So the radius of the osculating circle is $\frac{4}{3}$ and its center is $\left(0, \frac{5}{3}\right)$. Hence its equation is $x^2 + \left(y - \frac{5}{3}\right)^2 = \frac{16}{9}$.

5

−7.5

2.5

−5

37. The tangent vector is normal to the normal plane, and the vector $\langle 6, 6, -8\rangle$ is normal to the given plane. But $\mathbf{T}(t) \parallel \mathbf{r}'(t)$ and $\langle 6, 6, -8\rangle \parallel \langle 3, 3, -4\rangle$, so we need to find t such that $\mathbf{r}'(t) \parallel \langle 3, 3, -4\rangle$. $\mathbf{r}(t) = \langle t^3, 3t, t^4\rangle$ $\Rightarrow$ $\mathbf{r}'(t) = \langle 3t^2, 3, 4t^3\rangle \parallel \langle 3, 3, -4\rangle$ when $t = -1$. So the planes are parallel at the point $\mathbf{r}(-1) = (-1, -3, 1)$.

39. $\kappa = \left|\dfrac{d\mathbf{T}}{ds}\right| = \left|\dfrac{d\mathbf{T}/dt}{ds/dt}\right| = \dfrac{|d\mathbf{T}/dt|}{ds/dt}$ and $\mathbf{N} = \dfrac{d\mathbf{T}/dt}{|d\mathbf{T}/dt|}$, so $\kappa\mathbf{N} = \dfrac{\left|\dfrac{d\mathbf{T}}{dt}\right|\dfrac{d\mathbf{T}}{dt}}{\left|\dfrac{d\mathbf{T}}{dt}\right|\dfrac{ds}{dt}} = \dfrac{d\mathbf{T}/dt}{ds/dt} = \dfrac{d\mathbf{T}}{ds}$ by the Chain Rule.

41. **(a)** $|\mathbf{B}| = 1 \quad\Rightarrow\quad \mathbf{B}\cdot\mathbf{B} = 1 \quad\Rightarrow\quad \dfrac{d}{ds}(\mathbf{B}\cdot\mathbf{B}) = 0 \quad\Rightarrow\quad 2\dfrac{d\mathbf{B}}{ds}\cdot\mathbf{B} = 0 \quad\Rightarrow\quad \dfrac{d\mathbf{B}}{ds} \perp \mathbf{B}$

(b) $\mathbf{B} = \mathbf{T}\times\mathbf{N} \quad\Rightarrow\quad \dfrac{d\mathbf{B}}{ds} = \dfrac{d}{ds}(\mathbf{T}\times\mathbf{N}) = \dfrac{d}{dt}(\mathbf{T}\times\mathbf{N})\dfrac{1}{ds/dt} = \dfrac{d}{dt}(\mathbf{T}\times\mathbf{N})\dfrac{1}{|\mathbf{r}'(t)|}$

$$= [(\mathbf{T}'\times\mathbf{N}) + (\mathbf{T}\times\mathbf{N}')]\frac{1}{|\mathbf{r}'(t)|} = \left[\left(\mathbf{T}'\times\frac{\mathbf{T}'}{|\mathbf{T}'|}\right) + (\mathbf{T}\times\mathbf{N}')\right]\frac{1}{|\mathbf{r}'(t)|} = \frac{\mathbf{T}\times\mathbf{N}'}{|\mathbf{r}'(t)|} \quad\Rightarrow\quad \frac{d\mathbf{B}}{ds} \perp \mathbf{T}.$$

(c) $\mathbf{B} = \mathbf{T}\times\mathbf{N} \quad\Rightarrow\quad \mathbf{T}\perp\mathbf{N}$, $\mathbf{B}\perp\mathbf{T}$ and $\mathbf{B}\perp\mathbf{N}$. So $\mathbf{B}$, $\mathbf{T}$ and $\mathbf{N}$ form an orthogonal set of vectors in the three-dimensional space $\mathbb{R}^3$, which makes them a basis for this space. From parts (a) and (b), $d\mathbf{B}/ds$ is perpendicular to both $\mathbf{B}$ and $\mathbf{T}$, so $d\mathbf{B}/ds$ is parallel to $\mathbf{N}$. Therefore, $d\mathbf{B}/ds = -\tau(s)\mathbf{N}$, where $\tau(s)$ is a scalar.

43. Let $\mathbf{r}_1(t)$, $a \le t \le b$ and $\mathbf{r}_2(u)$, $\alpha \le u \le \beta$ be two parametrizations of a piecewise smooth curve C where $t = g(u)$, $g'(u) > 0$ (so that as u increases, so does t, and conversely,) $g(\alpha) = a$, and $g(\beta) = b$. Then by (3), $L_1 = \int_a^b |\mathbf{r}_1'(t)|dt = \int_\alpha^\beta |\mathbf{r}_1'(g(u))|g'(u)du = \int_\alpha^\beta |\mathbf{r}_1'(g(u))g'(u)|du$ using the substitution $t = g(u)$ and noting $g'(u) > 0$. But since $\mathbf{r}_1(t)$ and $\mathbf{r}_2(u)$ are parametrizations of the same curve C and $t = g(u)$ where g is a strictly increasing function, $\mathbf{r}_1(g(u)) = \mathbf{r}_2(u)$ for each $u \in [\alpha, \beta]$. Thus $L_1 = \int_\alpha^\beta |\mathbf{r}_1'(g(u))g'(u)|du$

$= \int_\alpha^\beta \left|\frac{d}{du}\mathbf{r}_1(g(u))\right|du = \int_\alpha^\beta |\mathbf{r}_2'(u)|du = L_2$. So the arc length is independent of parametrization.

45. **(a)** $\mathbf{r}' = s'\mathbf{T} \;\Rightarrow\; \mathbf{r}'' = s''\mathbf{T} + s'\mathbf{T}' = s''\mathbf{T} + s'\dfrac{d\mathbf{T}}{ds}s' = s''\mathbf{T} + \kappa(s')^2\mathbf{N}$ by the first Serret-Frenet formula.

(b) Using part (a), we have

$$\mathbf{r}' \times \mathbf{r}'' = (s'\mathbf{T}) \times \left[s''\mathbf{T} + \kappa(s')^2\mathbf{N}\right] = [(s'\mathbf{T}) \times (s''\mathbf{T})] + \left[(s'\mathbf{T}) \times \left(\kappa(s')^2\mathbf{N}\right)\right]$$
$$= (s's'')(\mathbf{T} \times \mathbf{T}) + \kappa(s')^3(\mathbf{T} \times \mathbf{N}) = \mathbf{0} + \kappa(s')^3\mathbf{B} = \kappa(s')^3\mathbf{B}.$$

(c) Using part (a), we have

$$\mathbf{r}''' = \left[s''\mathbf{T} + \kappa(s')^2\mathbf{N}\right]' = s'''\mathbf{T} + s''\mathbf{T}' + \kappa'(s')^2\mathbf{N} + 2\kappa s's''\mathbf{N} + \kappa(s')^2\mathbf{N}'$$
$$= s'''\mathbf{T} + s''\frac{d\mathbf{T}}{ds}s' + \kappa'(s')^2\mathbf{N} + 2\kappa s's''\mathbf{N} + \kappa(s')^2\frac{d\mathbf{N}}{ds}s'$$
$$= s'''\mathbf{T} + s''s'\kappa\mathbf{N} + \kappa'(s')^2\mathbf{N} + 2\kappa s's''\mathbf{N} + \kappa(s')^3(-\kappa\mathbf{T} + \tau\mathbf{B}) \quad \text{(by the second formula)}$$
$$= \left[s''' - \kappa^2(s')^3\right]\mathbf{T} + \left[3\kappa s's'' + \kappa'(s')^2\right]\mathbf{N} + \kappa\tau(s')^3\mathbf{B}.$$

(d) Using parts (b) and (c) and the facts that $\mathbf{B} \cdot \mathbf{T} = 0$, $\mathbf{B} \cdot \mathbf{N} = 0$, and $\mathbf{B} \cdot \mathbf{B} = 1$, we get

$$\frac{(\mathbf{r}' \times \mathbf{r}'') \cdot \mathbf{r}'''}{|\mathbf{r}' \times \mathbf{r}''|^2} = \frac{\kappa(s')^3\mathbf{B} \cdot \left\{\left[s''' - \kappa^2(s')^3\right]\mathbf{T} + \left[3\kappa s's'' + \kappa'(s')^2\right]\mathbf{N} + \kappa\tau(s')^3\mathbf{B}\right\}}{\left|\kappa(s')^3\mathbf{B}\right|^2} = \frac{\kappa(s')^3\kappa\tau(s')^3}{\left[\kappa(s')^3\right]^2} = \tau.$$

47. $\mathbf{r} = \langle t, \frac{1}{2}t^2, \frac{1}{3}t^3\rangle \;\Rightarrow\; \mathbf{r}' = \langle 1, t, t^2\rangle$, $\mathbf{r}'' = \langle 0, 1, 2t\rangle$, $\mathbf{r}''' = \langle 0, 0, 2\rangle \;\Rightarrow\; \mathbf{r}' \times \mathbf{r}'' = \langle t^2, -2t, 1\rangle \;\Rightarrow$

$$\tau = \frac{(\mathbf{r}' \times \mathbf{r}'') \cdot \mathbf{r}'''}{|\mathbf{r}' \times \mathbf{r}''|^2} = \frac{\langle t^2, -2t, 1\rangle \cdot \langle 0, 0, 2\rangle}{t^4 + 4t^2 + 1} = \frac{2}{t^4 + 4t^2 + 1}$$

49. For one helix the vector equation is $\mathbf{r}(t) = \langle 10\cos t, 10\sin t, 34t/(2\pi)\rangle$ because the radius of each helix is 10 angstroms and z increases by 34 angstroms for each increase of 2π in t. Therefore

$$L = \int_0^{2\pi \cdot 2.9 \times 10^8} |\mathbf{r}'(t)|dt = \int_0^{2\pi \cdot 2.9 \times 10^8} \sqrt{(-10\sin t)^2 + (10\cos t)^2 + \left(\frac{34}{2\pi}\right)^2}\,dt$$
$$= \int_0^{2\pi \cdot 2.9 \times 10^8} \sqrt{100 + \left(\frac{34}{2\pi}\right)^2}\,dt = \sqrt{100 + \left(\frac{34}{2\pi}\right)^2}(2\pi)2.9 \times 10^8 \approx 2.07 \times 10^{10} \text{ angstroms} \approx 2 \text{ m}.$$

This is the approximate length of each helix in a DNA molecule.

EXERCISES 11.8

1. $\mathbf{r}(t) = \langle t^2 - 1, t \rangle \Rightarrow$

$\mathbf{v}(t) = \mathbf{r}'(t) = \langle 2t, 1 \rangle$,

$\mathbf{a}(t) = \mathbf{r}''(t) = \langle 2, 0 \rangle$,

$|\mathbf{v}(t)| = \sqrt{4t^2 + 1}$

At $t = 1$:

$\mathbf{v}(1) = \langle 2, 1 \rangle$

$\mathbf{a}(1) = \langle 2, 0 \rangle$

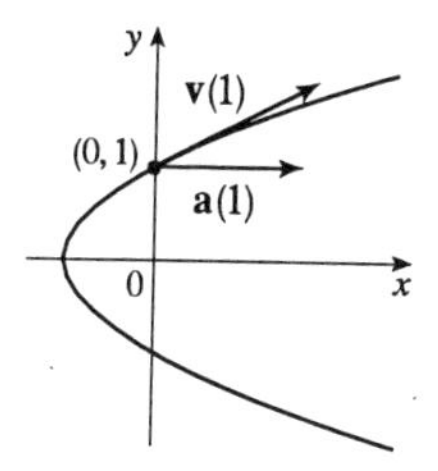

3. $\mathbf{r}(t) = e^t\mathbf{i} + e^{-t}\mathbf{j} \Rightarrow$

$\mathbf{v}(t) = e^t\mathbf{i} - e^{-t}\mathbf{j}$,

$\mathbf{a}(t) = e^t\mathbf{i} + e^{-t}\mathbf{j}$,

$|\mathbf{v}(t)| = \sqrt{e^{2t} + e^{-2t}} = e^{-t}\sqrt{e^{4t} + 1}$

At $t = 0$:

$\mathbf{v}(0) = \mathbf{i} - \mathbf{j}$,

$\mathbf{a}(0) = \mathbf{i} + \mathbf{j}$

Since $x = e^t$, $t = \ln x$ and $y = e^{-t} = e^{-\ln x} = 1/x$, and $x > 0$, $y > 0$.

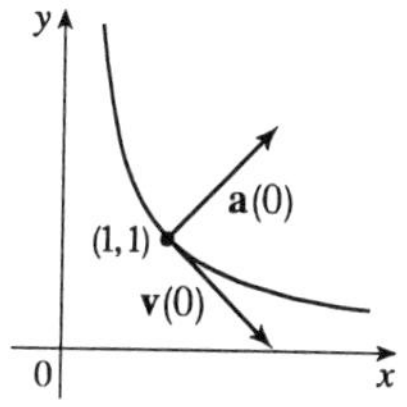

5. $\mathbf{r}(t) = \langle \sin t, t, \cos t \rangle \Rightarrow$

$\mathbf{v}(t) = \langle \cos t, 1, -\sin t \rangle$, $\mathbf{v}(0) = \langle 1, 1, 0 \rangle$

$\mathbf{a}(t) = \langle -\sin t, 0, -\cos t \rangle$, $\mathbf{a}(0) = \langle 0, 0, -1 \rangle$

$|\mathbf{v}(t)| = \sqrt{\cos^2 t + 1 + \sin^2 t} = \sqrt{2}$

Since $x^2 + z^2 = 1$, $y = t$, the path of the particle is a helix about the y-axis.

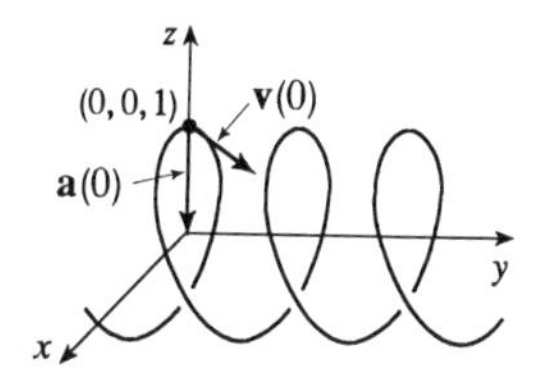

7. $\mathbf{r}(t) = \langle t^3, t^2 + 1, t^3 - 1 \rangle \Rightarrow \mathbf{v}(t) = \langle 3t^2, 2t, 3t^2 \rangle$, $\mathbf{a}(t) = \langle 6t, 2, 6t \rangle$,

$|\mathbf{v}(t)| = \sqrt{9t^4 + 4t^2 + 9t^4} = \sqrt{18t^4 + 4t^2} = |t|\sqrt{18t^2 + 4}$

9. $\mathbf{r}(t) = \langle 1/t, 1, t^2 \rangle \Rightarrow \mathbf{v}(t) = \langle -t^{-2}, 0, 2t \rangle$, $\mathbf{a}(t) = \langle 2t^{-3}, 0, 2 \rangle$, $|\mathbf{v}(t)| = \sqrt{t^{-4} + 4t^2} = \dfrac{1}{t^2}\sqrt{4t^6 + 1}$

11. $\mathbf{r}(t) = e^t\langle \cos t, \sin t, t \rangle \Rightarrow$

$\mathbf{v}(t) = e^t\langle \cos t, \sin t, t \rangle + e^t\langle -\sin t, \cos t, 1 \rangle = e^t\langle \cos t - \sin t, \sin t + \cos t, t + 1 \rangle$

$\mathbf{a}(t) = e^t\langle \cos t - \sin t - \sin t - \cos t, \sin t + \cos t + \cos t - \sin t, t + 1 + 1 \rangle = e^t\langle -2\sin t, 2\cos t, t + 2 \rangle$

$|\mathbf{v}(t)| = e^t\sqrt{\cos^2 t + \sin^2 t - 2\cos t\sin t + \sin^2 t + \cos^2 t + 2\sin t\cos t + t^2 + 2t + 1} = e^t\sqrt{t^2 + 2t + 3}$

13. $\mathbf{a}(t) = \mathbf{k} \Rightarrow \mathbf{v}(t) = \int \mathbf{k}\, dt = t\mathbf{k} + \mathbf{c}_1$ and $\mathbf{i} - \mathbf{j} = \mathbf{v}(0) = 0\mathbf{k} + \mathbf{c}_1$, so $\mathbf{c}_1 = \mathbf{i} - \mathbf{j}$ and $\mathbf{v}(t) = \mathbf{i} - \mathbf{j} + t\mathbf{k}$.

$\mathbf{r}(t) = \int(\mathbf{i} - \mathbf{j} + t\mathbf{k})dt = t\mathbf{i} - t\mathbf{j} + \frac{1}{2}t^2\mathbf{k} + \mathbf{c}_2$. But $\mathbf{0} = \mathbf{r}(0) = \mathbf{0} + \mathbf{c}_2$, so $\mathbf{c}_2 = \mathbf{0}$ and $\mathbf{r}(t) = t\mathbf{i} - t\mathbf{j} + \frac{1}{2}t^2\mathbf{k}$.

15. **(a)** $\mathbf{a}(t) = \mathbf{i} + 2\mathbf{j} + 2t\mathbf{k} \quad \Rightarrow$

$\mathbf{v}(t) = \int (\mathbf{i} + 2\mathbf{j} + 2t\mathbf{k})dt = t\mathbf{i} + 2t\mathbf{j} + t^2\mathbf{k} + \mathbf{c}_1$,

and $\mathbf{0} = \mathbf{v}(0) = \mathbf{0} + \mathbf{c}_1$, so $\mathbf{c}_1 = \mathbf{0}$ and $\mathbf{v}(t) = t\mathbf{i} + 2t\mathbf{j} + t^2\mathbf{k}$.

$\mathbf{r}(t) = \int (t\mathbf{i} + 2t\mathbf{j} + t^2\mathbf{k})dt = \frac{1}{2}t^2\mathbf{i} + t^2\mathbf{j} + \frac{1}{3}t^3\mathbf{k} + \mathbf{c}_2$.

But $\mathbf{i} + \mathbf{k} = \mathbf{r}(0) = \mathbf{0} + \mathbf{c}_2$, so $\mathbf{c}_2 = \mathbf{i} + \mathbf{k}$ and

$\mathbf{r}(t) = \left(1 + \frac{1}{2}t^2\right)\mathbf{i} + t^2\mathbf{j} + \left(1 + \frac{1}{3}t^3\right)\mathbf{k}$.

(b)

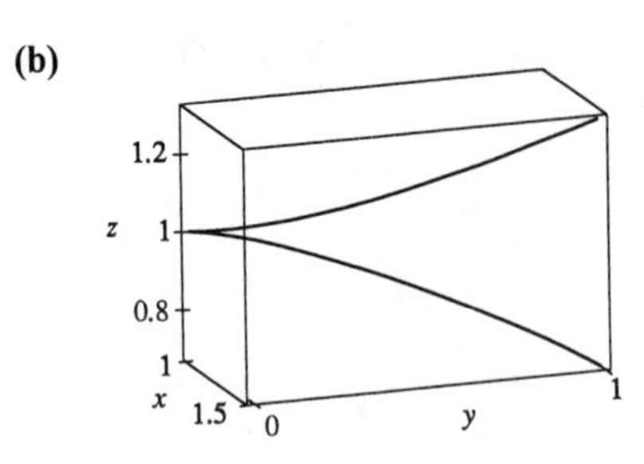

17. $\mathbf{r}(t) = \langle t^2, 5t, t^2 - 16t\rangle \quad \Rightarrow \quad \mathbf{v}(t) = \langle 2t, 5, 2t - 16\rangle$,

$|\mathbf{v}(t)| = \sqrt{4t^2 + 25 + 4t^2 - 64t + 256} = \sqrt{8t^2 - 64t + 281}$ and

$\frac{d}{dt}|\mathbf{v}(t)| = \frac{1}{2}(8t^2 - 64t + 281)^{-1/2}(16t - 64)$. This is zero if and only if the numerator is zero, that is, $16t - 64 = 0$ or $t = 4$. Since $\frac{d}{dt}|\mathbf{v}(t)| < 0$ for $t < 4$ and $\frac{d}{dt}|\mathbf{v}(t)| > 0$ for $t > 4$, the minimum speed of $\sqrt{153}$ is attained at $t = 4$ units of time.

19. $|\mathbf{F}(t)| = 20\,\mathrm{N}$ in the direction of the positive z-axis, so $\mathbf{F}(t) = 20\mathbf{k}$. Also $m = 4\,\mathrm{kg}$, $\mathbf{r}(0) = \mathbf{0}$ and $\mathbf{v}(0) = \mathbf{i} - \mathbf{j}$. Since $20\mathbf{k} = \mathbf{F}(t) = 4\mathbf{a}(t)$, $\mathbf{a}(t) = 5\mathbf{k}$. Then $\mathbf{v}(t) = 5t\mathbf{k} + \mathbf{c}_1$ where $\mathbf{c}_1 = \mathbf{i} - \mathbf{j}$ so $\mathbf{v}(t) = \mathbf{i} - \mathbf{j} + 5t\mathbf{k}$ and the speed is $|\mathbf{v}(t)| = \sqrt{1 + 1 + 25t^2} = \sqrt{25t^2 + 2}$. Also $\mathbf{r}(t) = t\mathbf{i} - t\mathbf{j} + \frac{5}{2}t^2\mathbf{k} + \mathbf{c}_2$ and $\mathbf{0} = \mathbf{r}(0)$, so $\mathbf{c}_2 = \mathbf{0}$ and $\mathbf{r}(t) = t\mathbf{i} - t\mathbf{j} + \frac{5}{2}t^2\mathbf{k}$.

21. $\mathbf{r}'(t) = (1 - \cos t)\mathbf{i} + (\sin t)\mathbf{j}$, $|\mathbf{r}'(t)| = \sqrt{1 - 2\cos t + 1} = \sqrt{2(1 - \cos t)}$, $\mathbf{r}''(t) = (\sin t)\mathbf{i} + (\cos t)\mathbf{j}$. Thus

$a_T = \dfrac{\sin t}{\sqrt{2(1 - \cos t)}}$ and

$$a_N = \frac{|(\cos t - \cos^2 t - \sin^2 t)\mathbf{k}|}{\sqrt{2(1 - \cos t)}} = \frac{\sqrt{[(\cos t) - 1]^2}}{\sqrt{2}\sqrt{1 - \cos t}} = \frac{1}{\sqrt{2}}\sqrt{\frac{(1 - \cos t)^2}{1 - \cos t}} = \frac{\sqrt{1 - \cos t}}{\sqrt{2}}.$$

23. $\mathbf{r}'(t) = 3t^2\mathbf{i} + 2t\mathbf{j} + \mathbf{k}$, $|\mathbf{r}'(t)| = \sqrt{9t^4 + 4t^2 + 1}$, $\mathbf{r}''(t) = 6t\mathbf{i} + 2\mathbf{j}$. Thus $a_T = \dfrac{18t^3 + 4t}{\sqrt{9t^4 + 4t^2 + 1}}$ and

$$a_N = \frac{|-2\mathbf{i} + 6t\mathbf{j} + (6t^2 - 12t^2)\mathbf{k}|}{\sqrt{9t^4 + 4t^2 + 1}} = \frac{\sqrt{4 + 36t^2 + 36t^4}}{\sqrt{9t^4 + 4t^2 + 1}} = \frac{2\sqrt{9t^4 + 9t^2 + 1}}{\sqrt{9t^4 + 4t^2 + 1}}.$$

25. $\mathbf{r}'(t) = e^t\mathbf{i} + \sqrt{2}\mathbf{j} - e^{-t}\mathbf{k}$, $|\mathbf{r}(t)| = \sqrt{e^{2t} + 2 + e^{-2t}} = e^t + e^{-t}$, $\mathbf{r}''(t) = e^t\mathbf{i} + e^{-t}\mathbf{k}$. Then

$a_T = \dfrac{e^{2t} - e^{-2t}}{e^t + e^{-t}} = e^t - e^{-t} = 2\sinh t$ and

$$a_N = \frac{\left|\sqrt{2}e^{-t}\mathbf{i} - 2\mathbf{j} - \sqrt{2}e^t\mathbf{k}\right|}{e^t + e^{-t}} = \frac{\sqrt{2(e^{-2t} + 2 + e^{2t})}}{e^t + e^{-t}} = \sqrt{2}\frac{e^t + e^{-t}}{e^t + e^{-t}} = \sqrt{2}.$$

27. If the engines are turned off at time t, then the spacecraft will continue to travel in the direction of $\mathbf{v}(t)$, so we need a t such that for some scalar $s > 0$, $\mathbf{r}(t) + s\mathbf{v}(t) = \langle 6, 4, 9\rangle$. $\mathbf{v}(t) = \mathbf{r}'(t) = \mathbf{i} + \frac{1}{t}\mathbf{j} + \frac{8t}{(t^2 + 1)^2}\mathbf{k} \quad \Rightarrow$

$\mathbf{r}(t) + s\mathbf{v}(t) = \left\langle 3 + t + s, 2 + \ln t + \frac{s}{t}, 7 - \frac{4}{t^2 + 1} + \frac{8st}{(t^2 + 1)^2}\right\rangle \quad \Rightarrow \quad 3 + t + s = 6 \quad \Rightarrow \quad s = 3 - t$, so

$7 - \dfrac{4}{t^2 + 1} + \dfrac{8(3 - t)t}{(t^2 + 1)^2} = 9 \quad \Leftrightarrow \quad \dfrac{24t - 12t^2 - 4}{(t^2 + 1)^2} = 2 \quad \Leftrightarrow \quad t^4 + 8t^2 - 12t + 3 = 0$. It is easily seen that $t = 1$ is a root of this polynomial. Also $2 + \ln 1 + \frac{3 - 1}{1} = 4$, so $t = 1$ is the desired solution.

29. With $r = (r\cos\theta)\mathbf{i} + (r\sin\theta)\mathbf{j}$ and $\mathbf{h} = \alpha\mathbf{k}$ where $\alpha > 0$,

(a) $\mathbf{h} = \mathbf{r}\times\mathbf{r}' = [(r\cos\theta)\mathbf{i} + (r\sin\theta)\mathbf{j}] \times \left[\left(r'\cos\theta - r\sin\theta\,\frac{d\theta}{dt}\right)\mathbf{i} + \left(r'\sin\theta + r\cos\theta\,\frac{d\theta}{dt}\right)\mathbf{j}\right]$

$$= \left[rr'\cos\theta\sin\theta + r^2\cos^2\theta\,\frac{d\theta}{dt} - rr'\cos\theta\sin\theta + r^2\sin^2\theta\,\frac{d\theta}{dt}\right]\mathbf{k} = r^2\frac{d\theta}{dt}\mathbf{k}$$

(b) Since $\mathbf{h} = \alpha\mathbf{k}$, $\alpha > 0$, $\alpha = |\mathbf{h}|$. But by (a), $\alpha = |\mathbf{h}| = r^2\dfrac{d\theta}{dt}$.

(c) $A(t) = \dfrac{1}{2}\displaystyle\int_{\theta_0}^{\theta} |\mathbf{r}|^2\,d\theta = \dfrac{1}{2}\int_{t_0}^{t} r^2\frac{d\theta}{dt}\,dt$ in polar coordinates. Thus, by the Fundamental Theorem of Calculus, $\dfrac{dA}{dt} = \dfrac{r^2}{2}\dfrac{d\theta}{dt}$.

(d) $\dfrac{dA}{dt} = \dfrac{r^2}{2}\dfrac{d\theta}{dt} = \dfrac{h}{2}$ = constant since $\mathbf{h}$ is a constant vector and $h = |\mathbf{h}|$.

31. From Exercise 30, $T^2 = \dfrac{4\pi^2}{GM}a^3$. $T \approx 365.25 \text{ days} \times 24\cdot 60^2\,\dfrac{\text{seconds}}{\text{day}} \approx 3.1558\times 10^7$ seconds. Therefore

$$a^3 = \frac{GMT^2}{4\pi^2} \approx \frac{(6.67\times 10^{-11})(1.99\times 10^{30})(3.1558\times 10^7)^2}{4\pi^2} \approx 3.348\times 10^{33}\text{ m}^3 \quad\Rightarrow\quad a \approx 1.496\times 10^{11}\text{ m}.$$

Thus, the length of the major axis of the earth's orbit (that is, $2a$) is approximately $2.99\times 10^{11}\text{ m} = 2.99\times 10^8$ km.

REVIEW EXERCISES FOR CHAPTER 11

1. By Properties of the Dot Product, this is true.

3. True. If θ is the angle between $\mathbf{u}$ and $\mathbf{v}$, then by the definition of cross product, $|\mathbf{u}\times\mathbf{v}| = |\mathbf{u}||\mathbf{v}|\sin\theta = |\mathbf{v}||\mathbf{u}|\sin\theta = |\mathbf{v}\times\mathbf{u}|$.
(Or, by Properties of the Cross Product, $|\mathbf{u}\times\mathbf{v}| = |-\mathbf{v}\times\mathbf{u}| = |-1||\mathbf{v}\times\mathbf{u}| = |\mathbf{v}\times\mathbf{u}|$.)

5. Property 2 of the Cross Product tells us that this is true.

7. This is true by Theorem 11.3.6.

9. This is true; the angle between $\mathbf{u}\times\mathbf{v}$ and $\mathbf{u}$ is $90°$.

11. If $|\mathbf{u}| = 1$, $|\mathbf{v}| = 1$ and θ is the angle between these two vectors (so $0 \le \theta \le \pi$), then by the definition of cross product, $|\mathbf{u}\times\mathbf{v}| = |\mathbf{u}||\mathbf{v}|\sin\theta = \sin\theta$, which is equal to 1 if and only if $\theta = \frac{\pi}{2}$ (that is, the two vectors are orthogonal). Therefore, the assertion that the cross product of two unit vectors is a unit vector is false.

13. This is false because by 11.5.7, $\dfrac{x^2}{1} + \dfrac{y^2}{1} = 1$ is the equation of a circular cylinder.

15. $|AB| = \sqrt{9+16+144} = \sqrt{169} = 13$, $|BC| = \sqrt{1+1+36} = \sqrt{38}$, $|CA| = \sqrt{4+25+36} = \sqrt{65}$

17. Completing the squares gives $(x+2)^2+(y+3)^2+(z-5)^2=-2+4+9+25=36$. Thus the circle is centered at $(-2,-3,5)$ and has radius 6.

19. $6\mathbf{a}-5\mathbf{c}=(6-0)\mathbf{i}+(6-5)\mathbf{j}+(-12+25)\mathbf{k}=6\mathbf{i}+\mathbf{j}+13\mathbf{k}$

21. $\mathbf{a}\cdot\mathbf{b}=(1)(3)+(1)(-2)+(-2)(1)=-1$

23. $\mathbf{b}\times\mathbf{c}=\begin{vmatrix}\mathbf{i} & \mathbf{j} & \mathbf{k}\\ 3 & -2 & 1\\ 0 & 1 & -5\end{vmatrix}=9\mathbf{i}+15\mathbf{j}+3\mathbf{k}$, $|\mathbf{b}\times\mathbf{c}|=3\sqrt{9+25+1}=3\sqrt{35}$

25. $\mathbf{c}\times\mathbf{c}=\mathbf{0}$ for any $\mathbf{c}$.

27. $\cos\theta=\dfrac{\mathbf{a}\cdot\mathbf{b}}{|\mathbf{a}||\mathbf{b}|}=\dfrac{-1}{\sqrt{6}\sqrt{14}}=\dfrac{-1}{2\sqrt{21}}$ and $\theta=\cos^{-1}\frac{-1}{2\sqrt{21}}\approx 96°$.

29. The scalar projection is $\text{comp}_{\mathbf{a}}\,\mathbf{b}=|\mathbf{b}|\cos\theta=\mathbf{a}\cdot\mathbf{b}/|\mathbf{a}|=-\frac{1}{\sqrt{6}}$.

31. We need $4x+3x-28=0$ or $x=4$.

33. **(a)** $(\mathbf{u}\times\mathbf{v})\cdot\mathbf{w}=\mathbf{u}\cdot(\mathbf{v}\times\mathbf{w})=2$

(b) $\mathbf{u}\cdot(\mathbf{w}\times\mathbf{v})=\mathbf{u}\cdot[-(\mathbf{v}\times\mathbf{w})]=-\mathbf{u}\cdot(\mathbf{v}\times\mathbf{w})=-2$

(c) $\mathbf{v}\cdot(\mathbf{u}\times\mathbf{w})=(\mathbf{v}\times\mathbf{u})\cdot\mathbf{w}=-(\mathbf{u}\times\mathbf{v})\cdot\mathbf{w}=-2$

(d) $(\mathbf{u}\times\mathbf{v})\cdot\mathbf{v}=\mathbf{u}\cdot(\mathbf{v}\times\mathbf{v})=\mathbf{u}\cdot\mathbf{0}=0$

35. Determine the vectors $\overrightarrow{PQ}=\langle a_1,a_2,a_3\rangle$ and $\overrightarrow{PR}=\langle b_1,b_2,b_3\rangle$. If there is a scalar t such that $\langle a_1,a_2,a_3\rangle=t\langle b_1,b_2,b_3\rangle$, then the vectors are parallel and the points must all lie on the same line. Alternatively, if $\overrightarrow{PQ}\times\overrightarrow{PR}=\mathbf{0}$, then $\overrightarrow{PQ}$ and $\overrightarrow{PR}$ are parallel, so P, Q, and R are collinear. Thirdly, an algebraic method is to determine the equation of the line joining two of the points, and then check whether or not the third point is on that line.

37. For simplicity, consider a unit cube positioned with its back left corner at the origin. Vector representations of the diagonals joining the points $(0,0,0)$ to $(1,1,1)$ and $(1,0,0)$ to $(0,1,1)$ are $\langle 1,1,1\rangle$ and $\langle -1,1,1\rangle$ respectively. Let θ be the angle between these two vectors.
$\langle 1,1,1\rangle\cdot\langle -1,1,1\rangle=-1+1+1=1=|\langle 1,1,1\rangle||\langle -1,1,1\rangle|\cos\theta=3\cos\theta \quad\Rightarrow\quad \cos\theta=\frac{1}{3} \quad\Rightarrow$
$\theta=\cos^{-1}\frac{1}{3}\approx 71°$ (or 109°).

39. $\overrightarrow{AB}=\langle 1,0,-1\rangle$, $\overrightarrow{AC}=\langle 0,4,3\rangle$, so

(a) a vector perpendicular to the plane is $\overrightarrow{AB}\times\overrightarrow{AC}=\langle 0+4,-(3+0),4-0\rangle=\langle 4,-3,4\rangle$.

(b) $\frac{1}{2}\left|\overrightarrow{AB}\times\overrightarrow{AC}\right|=\frac{1}{2}\sqrt{16+9+16}=\frac{\sqrt{41}}{2}$

41. Let F_1 be the magnitude of the force directed 20° away from the direction of shore, and let F_2 be the magnitude of the other force. Separating these forces into components parallel to the direction of the resultant force and perpendicular to it gives $F_1 \cos 20° + F_2 \cos 30° = 255$ (1), and $F_1 \sin 20° - F_2 \sin 30° = 0 \Rightarrow$ $F_1 = F_2 \dfrac{\sin 30°}{\sin 20°}$ (2). Substituting (2) into (1) gives $F_2(\sin 30° \cot 20° + \cos 30°) = 255 \Rightarrow F_2 \approx 114\,\text{N}$. Substituting this into (2) gives $F_1 \approx 166\,\text{N}$.

43. $x = 1 + 2t$, $y = 2 - t$, $z = 4 + 3t$

45. $\mathbf{v} = \langle 4, -3, 5 \rangle$ so $x = 1 + 4t$, $y = -3t$, $z = 1 + 5t$.

47. $(x + 4) + 2(y - 1) + 5(z - 2) = 0$ or $x + 2y + 5z = 8$.

49. Substitution of the parametric equations into the equation of the plane gives $2x - y + z = 2(2 - t) - (1 + 3t) + 4t = 2 \Rightarrow -t + 3 = 2 \Rightarrow t = 1$. When $t = 1$, the parametric equations give $x = 2 - 1 = 1$, $y = 1 + 3 = 4$ and $z = 4$. Therefore, the point of intersection is $(1, 4, 4)$.

51. Since the direction vectors $\langle 2, 3, 4 \rangle$ and $\langle 6, -1, 2 \rangle$ aren't parallel, neither are the lines. For the lines to intersect, the three equations $1 + 2t = -1 + 6s$, $2 + 3t = 3 - s$, $3 + 4t = -5 + 2s$ must be satisfied simultaneously. Solving the first two equations gives $t = \frac{1}{5}$, $s = \frac{2}{5}$ and checking we see these values don't satisfy the third equation. Thus the lines aren't parallel and they don't intersect, so they must be skew.

53. By Exercise 11.4.67, $D = \dfrac{|2 - 24|}{\sqrt{26}} = \dfrac{22}{\sqrt{26}}$.

55. A plane through the x-axis intersecting the yz-plane in the line $y = z$, $x = 0$.

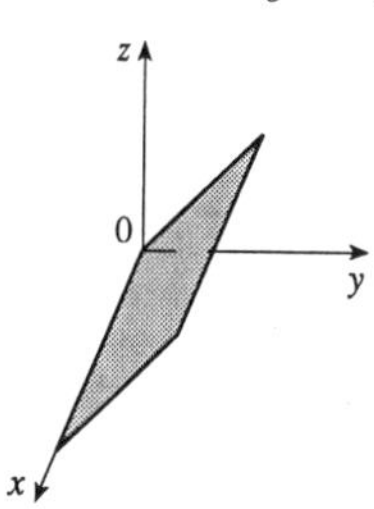

57. A circular paraboloid with vertex the origin and axis the y-axis.

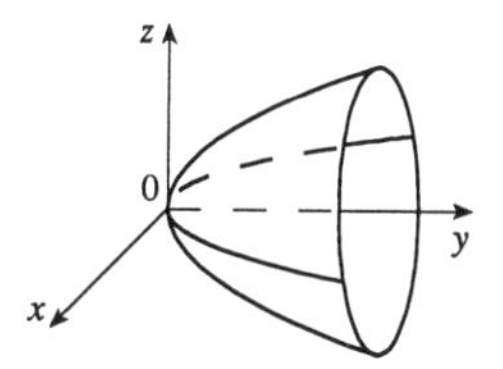

59. A (right elliptical) cone with vertex at the origin and axis the x-axis.

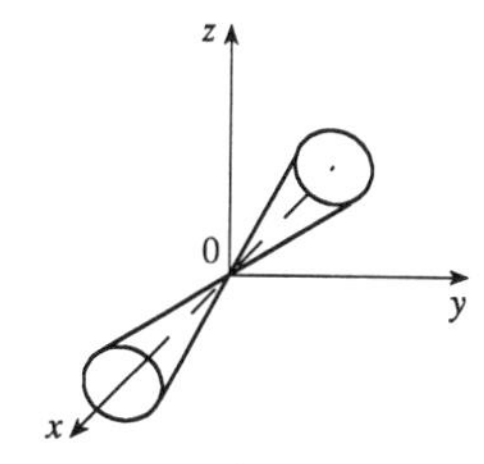

61. A hyperboloid of two sheets with axis the y-axis. For $|y| > 2$, traces parallel to the xz-plane are circles.

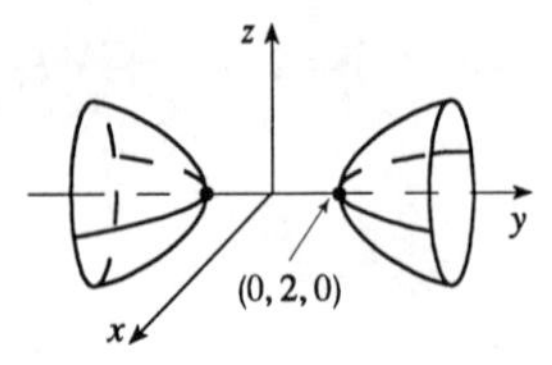

63. $4x^2 + y^2 = 16 \quad \Leftrightarrow \quad \dfrac{x^2}{4} + \dfrac{y^2}{16} = 1.$

The equation of the ellipsoid is

$\dfrac{x^2}{4} + \dfrac{y^2}{16} + \dfrac{z^2}{c^2} = 1$, since the horizontal trace in the plane $z = 0$ must be the original ellipse. The traces of the ellipsoid in the yz-plane must be circles since the surface is obtained by rotation about the x-axis. Therefore, $c^2 = 16$ and an equation of the ellipsoid is

$$\frac{x^2}{4} + \frac{y^2}{16} + \frac{z^2}{16} = 1 \quad \Leftrightarrow \quad 4x^2 + y^2 + z^2 = 16.$$

65. **(a)** Since $x = 2$ and $y^2 + z^2 = 1$, the curve is a circle in the plane $x = 2$ with center $(2, 0, 0)$ and radius 1.

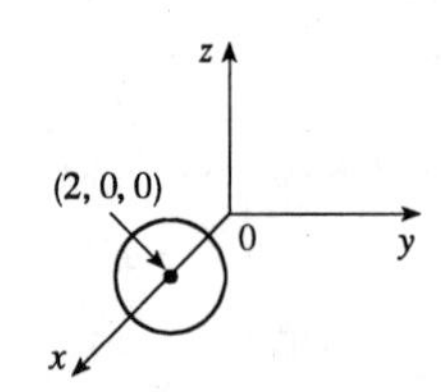

(b) $\mathbf{r}'(t) = \cos t\,\mathbf{j} - \sin t\,\mathbf{k} \quad \Rightarrow$

$\mathbf{r}''(t) = -\sin t\,\mathbf{j} - \cos t\,\mathbf{k}$

67. $\int_0^1 [(t + t^2)\mathbf{i} + (2 + t^3)\mathbf{j} + t^4\mathbf{k}]dt = \left[\left(\frac{1}{2}t^2 + \frac{1}{3}t^3\right)\mathbf{i} + \left(2t + \frac{1}{4}t^4\right)\mathbf{j} + \left(\frac{1}{5}t^5\right)\mathbf{k}\right]_0^1 = \frac{5}{6}\mathbf{i} + \frac{9}{4}\mathbf{j} + \frac{1}{5}\mathbf{k}$

69. $t = 1$ at $(1, 4, 2)$ and $t = 4$ at $(2, 1, 17)$, so

$$L = \int_1^4 \sqrt{\frac{1}{4t} + \frac{16}{t^4} + 4t^2}\,dt$$

$$\approx \frac{4-1}{3 \cdot 4}\left[\sqrt{\frac{1}{4} + 16 + 4} + 4 \cdot \sqrt{\frac{1}{4 \cdot 7/4} + \frac{16}{(7/4)^4} + 4\left(\frac{7}{4}\right)^2} + 2 \cdot \sqrt{\frac{1}{4 \cdot 10/4} + \frac{16}{(10/4)^4} + 4\left(\frac{10}{4}\right)^2}\right.$$

$$\left. + 4 \cdot \sqrt{\frac{1}{4 \cdot 13/4} + \frac{16}{(13/4)^4} + 4\left(\frac{13}{4}\right)^2} + \sqrt{\frac{1}{4 \cdot 4} + \frac{16}{4^4} + 4 \cdot 4^2}\right]$$

$\approx 15.9241.$

71. The angle of intersection of the two curves, θ, is the angle between their respective tangents at the point of intersection. For both curves the point $(1, 0, 0)$ occurs when $t = 0$. $\mathbf{r}_1'(t) = -\sin t\,\mathbf{i} + \cos t\,\mathbf{j} + \mathbf{k} \Rightarrow$ $\mathbf{r}_1'(0) = \mathbf{j} + \mathbf{k}$ and $\mathbf{r}_2'(t) = \mathbf{i} + 2t\mathbf{j} + 3t^2\mathbf{k} \quad \Rightarrow \quad \mathbf{r}_2'(0) = \mathbf{i}$. $\mathbf{r}_1'(0) \cdot \mathbf{r}_2'(0) = (\mathbf{j} + \mathbf{k}) \cdot \mathbf{i} = 0$. Therefore, the curves intersect at right angles to each other, that is, $\theta = \frac{\pi}{2}$.

73. **(a)** $\mathbf{T}(t) = \dfrac{\langle t^2, t, 1\rangle}{\sqrt{t^4 + t^2 + 1}}$

(b) $\mathbf{T}'(t) = -\frac{1}{2}(t^4 + t^2 + 1)^{-3/2}(4t^3 + 2t)\langle t^2, t, 1\rangle + (t^4 + t^2 + 1)^{-1/2}\langle 2t, 1, 0\rangle$

$$= \frac{-2t^3 - t}{(t^4 + t^2 + 1)^{3/2}}\langle t^2, t, 1\rangle + \frac{1}{(t^4 + t^2 + 1)^{1/2}}\langle 2t, 1, 0\rangle$$

$$= \frac{\langle -2t^5 - t^3, -2t^4 - t^2, -2t^3 - t\rangle + \langle 2t^5 + 2t^3 + 2t, t^4 + t^2 + 1, 0\rangle}{(t^4 + t^2 + 1)^{3/2}} = \frac{\langle 2t, -t^4 + 1, -2t^3 - t\rangle}{(t^4 + t^2 + 1)^{3/2}}$$

$|\mathbf{T}'(t)| = \dfrac{\sqrt{4t^2 + t^8 - 2t^4 + 1 + 4t^6 + 4t^4 + t^2}}{(t^4 + t^2 + 1)^{3/2}} = \dfrac{\sqrt{t^8 + 4t^6 + 2t^4 + 5t^2}}{(t^4 + t^2 + 1)^{3/2}}$, and

$\mathbf{N}(t) = \dfrac{\langle 2t, 1 - t^4, -2t^3 - t\rangle}{\sqrt{t^8 + 4t^6 + 2t^4 + 5t^2}}$.

(c) $\kappa(t) = \dfrac{|\mathbf{T}'(t)|}{|\mathbf{r}'(t)|} = \dfrac{\sqrt{t^8 + 4t^6 + 2t^4 + 5t^2}}{(t^4 + t^2 + 1)^2}$

75. $y' = 4x^3$, $y'' = 12x^2$ and $\kappa(x) = \dfrac{|12x^2|}{(1 + 16x^6)^{3/2}}$ so $\kappa(1) = \dfrac{12}{(17)^{3/2}}$.

77. $\mathbf{r}(t) = \langle \sin 2t, t, \cos 2t\rangle \Rightarrow \mathbf{r}'(t) = \langle 2\cos 2t, 1, -2\sin 2t\rangle \Rightarrow \mathbf{T}(t) = \frac{1}{\sqrt{5}}\langle 2\cos 2t, 1, -2\sin 2t\rangle \Rightarrow$ $\mathbf{T}'(t) = \frac{1}{\sqrt{5}}\langle -4\sin 2t, 0, -4\cos 2t\rangle \Rightarrow \mathbf{N}(t) = \langle -\sin 2t, 0, -\cos 2t\rangle$. So $\mathbf{N} = \mathbf{N}(\pi) = \langle 0, 0, -1\rangle$ and $\mathbf{B} = \mathbf{T} \times \mathbf{N} = \frac{1}{\sqrt{5}}\langle -1, 2, 0\rangle$. So a normal to the osculating plane is $\langle -1, 2, 0\rangle$ and the equation is $-1(x - 0) + 2(y - \pi) + 0(z - 1) = 0$ or $x - 2y + 2\pi = 0$.

79. $\mathbf{v}(t) = \int(t\mathbf{i} + \mathbf{j} + t^2\mathbf{k})dt = \frac{1}{2}t^2\mathbf{i} + t\mathbf{j} + \frac{1}{3}t^3\mathbf{k} + \mathbf{c}_1$, but $\mathbf{i} + 2\mathbf{j} + \mathbf{k} = \mathbf{v}(0) = \mathbf{0} + \mathbf{c}_1$ so $\mathbf{c}_1 = \mathbf{i} + 2\mathbf{j} + \mathbf{k}$ and $\mathbf{v}(t) = \left(1 + \frac{1}{2}t^2\right)\mathbf{i} + (2 + \mathrm{t})\mathbf{j} + \left(1 + \frac{1}{3}t^3\right)\mathbf{k}$. $\mathbf{r}(t) = \int \mathbf{v}(t)dt = \left(t + \frac{1}{6}t^3\right)\mathbf{i} + \left(2t + \frac{1}{2}t^2\right)\mathbf{j} + \left(t + \frac{1}{12}t^4\right)\mathbf{k} + \mathbf{c}_2$. But $\mathbf{r}(0) = \mathbf{0}$ so $\mathbf{c}_2 = \mathbf{0}$ and $\mathbf{r}(t) = \left(t + \frac{1}{6}t^3\right)\mathbf{i} + \left(2t + \frac{1}{2}t^2\right)\mathbf{j} + \left(t + \frac{1}{12}t^4\right)\mathbf{k}$.

81. **(a)** Instead of proceeding directly, we use Theorem 11.6.5 #3:

$\mathbf{r}(t) = t\mathbf{R}(t) \Rightarrow \mathbf{v} = \mathbf{r}'(t) = \mathbf{R}(t) + t\mathbf{R}'(t) = \cos\omega t\,\mathbf{i} + \sin\omega t\,\mathbf{j} + t\mathbf{v}_d$

(b) Using the same method as in part (a) and starting with $\mathbf{v} = \mathbf{R}(t) + t\mathbf{R}'(t)$, we have

$\mathbf{a} = \mathbf{v}' = \mathbf{R}'(t) + \mathbf{R}'(t) + t\mathbf{R}''(t) = 2\mathbf{R}'(t) + t\mathbf{R}''(t) = 2\mathbf{v}_d + t\mathbf{a}_d$.

(c) Here we have $\mathbf{r}(t) = e^{-t}\cos\omega t\,\mathbf{i} + e^{-t}\sin\omega t\,\mathbf{j} = e^{-t}\mathbf{R}(t)$. So, as in parts (a) and (b),

$\mathbf{v} = \mathbf{r}'(t) = e^{-t}\mathbf{R}'(t) - e^{-t}\mathbf{R}(t) = e^{-t}[\mathbf{R}'(t) - \mathbf{R}(t)] \Rightarrow$

$\mathbf{a} = \mathbf{v}' = e^{-t}[\mathbf{R}''(t) - \mathbf{R}'(t)] - e^{-t}[\mathbf{R}'(t) - \mathbf{R}(t)] = e^{-t}[\mathbf{R}''(t) - 2\mathbf{R}'(t) + \mathbf{R}(t)] = e^{-t}\mathbf{a}_d - 2e^{-t}\mathbf{v}_d + e^{-t}\mathbf{R}$.

Thus the Coriolis acceleration (the "extra" terms not involving $\mathbf{a}_d$) is $-2e^{-t}\mathbf{v}_d + e^{-t}\mathbf{R}$.

APPLICATIONS PLUS (after Chapter 11)

1. **(a)** $\mathbf{r}(t) = R\cos\omega t\,\mathbf{i} + R\sin\omega t\,\mathbf{j} \quad\Rightarrow\quad \mathbf{v} = \mathbf{r}'(t) = -\omega R\sin\omega t\,\mathbf{i} + \omega R\cos\omega t\,\mathbf{j}$, so

$\mathbf{r} = R\,(\cos\omega t\,\mathbf{i} + \sin\omega t\,\mathbf{j})$ and $\mathbf{v} = \omega R(-\sin\omega t\,\mathbf{i} + \cos\omega t\,\mathbf{j})$.

$\mathbf{v}\cdot\mathbf{r} = \omega R^2(-\cos\omega t\sin\omega t + \sin\omega t\cos\omega t) = 0$, so $\mathbf{v}\perp\mathbf{r}$. Since $\mathbf{r}$ points along a radius of the circle, and $\mathbf{v}\perp\mathbf{r}$, $\mathbf{v}$ is tangent to the circle. Because it is a velocity vector, $\mathbf{v}$ points in the direction of motion.

(b) In (a), we wrote $\mathbf{v}$ in the form $\omega R\mathbf{u}$, where $\mathbf{u}$ is the unit vector $-\sin\omega t\,\mathbf{i} + \cos\omega t\,\mathbf{j}$. Clearly $|\mathbf{v}| = \omega R|\mathbf{u}| = \omega R$. At speed ωR, the particle completes one revolution, a distance $2\pi R$, in time $T = \dfrac{2\pi R}{\omega R} = \dfrac{2\pi}{\omega}$.

(c) $\mathbf{a} = \dfrac{d\mathbf{v}}{dt} = -\omega^2 R\cos\omega t\,\mathbf{i} - \omega^2 R\sin\omega t\,\mathbf{j} = -\omega^2 R\,(\cos\omega t\,\mathbf{i} + \sin\omega t\,\mathbf{j})$, so $\mathbf{a} = -\omega^2\mathbf{r}$. This shows that $\mathbf{a}$ is proportional to $\mathbf{r}$ and points in the opposite direction (toward the origin). Also, $|\mathbf{a}| = \omega^2|\mathbf{r}| = \omega^2 R$.

(d) By Newton's Second Law (see Section 11.8 of the text), $\mathbf{F} = m\mathbf{a}$, so

$$|\mathbf{F}| = m|\mathbf{a}| = mR\omega^2 = \frac{m(\omega R)^2}{R} = \frac{m|\mathbf{v}|^2}{R}.$$

3. **(a)** The projectile reaches maximum height when $0 = \dfrac{dy}{dt} = \dfrac{d}{dt}\left[(v_0\sin\alpha)t - \frac{1}{2}gt^2\right] = v_0\sin\alpha - gt$; that is, when $t = (v_0\sin\alpha)/g$ and $y = (v_0\sin\alpha)\left(\dfrac{v_0\sin\alpha}{g}\right) - \frac{1}{2}g\left(\dfrac{v_0\sin\alpha}{g}\right)^2 = \dfrac{v_0^2\sin^2\alpha}{2g}$. This is the maximum height attained when the projectile is fired with an angle of elevation α. This maximum height is largest when $\alpha = \frac{\pi}{2}$. In that case, $\sin\alpha = 1$ and the maximum height is $v_0^2/(2g)$.

(b) Let $R = v_0^2/g$. We are asked to consider the parabola $x^2 + 2Ry - R^2 = 0$ which can be rewritten as $y = -\dfrac{1}{2R}x^2 + \dfrac{R}{2}$. The points on or inside this parabola are those for which $-R \le x \le R$ and $0 \le y \le \dfrac{-1}{2R}x^2 + \dfrac{R}{2}$. When the projectile is fired at angle of elevation α, the points (x, y) along its path satisfy the relations $x = (v_0\cos\alpha)t$ and $y = (v_0\sin\alpha)t - \frac{1}{2}gt^2$, where $0 \le t \le (2v_0\sin\alpha)/g$ (as in Example 5 in Section 11.8). Thus $|x| \le |(v_0\cos\alpha)(2v_0\sin\alpha)/g| = |(v_0^2/g)\sin 2\alpha| \le |v_0^2/g| = |R|$. This shows that $-R \le x \le R$.

For t in the specified range, we also have $y = t\left(v_0\sin\alpha - \frac{1}{2}gt\right) = \frac{1}{2}gt\left(\dfrac{2v_0\sin\alpha}{g} - t\right) \ge 0$ and

$$y = (v_0\sin\alpha)\frac{x}{v_0\cos\alpha} - \frac{g}{2}\left(\frac{x}{v_0\cos\alpha}\right)^2 = (\tan\alpha)x - \frac{g}{2v_0^2\cos^2\alpha}x^2 = \frac{-1}{2R\cos^2\alpha}x^2 + (\tan\alpha)x.$$ Thus

$$y - \left(\frac{-1}{2R}x^2 + \frac{R}{2}\right) = \frac{-1}{2R\cos^2\alpha}x^2 + \frac{1}{2R}x^2 + (\tan\alpha)x - \frac{R}{2} = \frac{x^2}{2R}\left(1 - \frac{1}{\cos^2\alpha}\right) + (\tan\alpha)x - \frac{R}{2}$$

$$= \frac{x^2(1 - \sec^2\alpha) + 2R(\tan\alpha)x - R^2}{2R} = \frac{-(\tan^2\alpha)x^2 + 2R(\tan\alpha)x - R^2}{2R} = \frac{-[(\tan\alpha)x - R]^2}{2R} \le 0.$$

We have shown that every target that can be hit by the projectile lies on or inside the parabola $y = -\frac{1}{2R}x^2 + \frac{R}{2}$. Now let (a, b) be any point on or inside the parabola $y = -\frac{1}{2R}x^2 + \frac{R}{2}$. Then $-R \le a \le R$ and $0 \le b \le -\frac{1}{2R}a^2 + \frac{R}{2}$. We seek an angle α such that (a, b) lies in the path of the projectile; that is, we wish to find an angle α such that $b = \frac{-1}{2R\cos^2\alpha}a^2 + (\tan\alpha)a$ or equivalently $b = \frac{-1}{2R}(\tan^2\alpha + 1)a^2 + (\tan\alpha)a$. Rearranging this equation we get $\frac{a^2}{2R}\tan^2\alpha - a\tan\alpha + \left(\frac{a^2}{2R} + b\right) = 0$ or $a^2(\tan\alpha)^2 - 2aR(\tan\alpha) + (a^2 + 2bR) = 0$ $(\bigstar)$. This quadratic equation for $\tan\alpha$ has real solutions exactly when the discriminant is nonnegative. Now $B^2 - 4AC \ge 0 \quad \Leftrightarrow \quad (-2aR)^2 - 4a^2(a^2 + 2bR) \ge 0 \quad \Leftrightarrow \quad 4a^2(R^2 - a^2 - 2bR) \ge 0 \quad \Leftrightarrow$ $-a^2 - 2bR + R^2 \ge 0 \quad \Leftrightarrow \quad b \le \frac{1}{2R}(R^2 - a^2) \quad \Leftrightarrow \quad b \le \frac{-1}{2R}a^2 + \frac{R}{2}$. This condition is satisfied since (a, b) is on or inside the parabola $y = \frac{-1}{2R}x^2 + \frac{R}{2}$. It follows that (a, b) lies in the path of the projectile when $\tan\alpha$ satisfies $(\bigstar)$, that is, when

$$\tan\alpha = \frac{2aR \pm \sqrt{4a^2(R^2 - a^2 - 2bR)}}{2a^2} = \frac{R \pm \sqrt{R^2 - 2bR - a^2}}{a}.$$

(c)

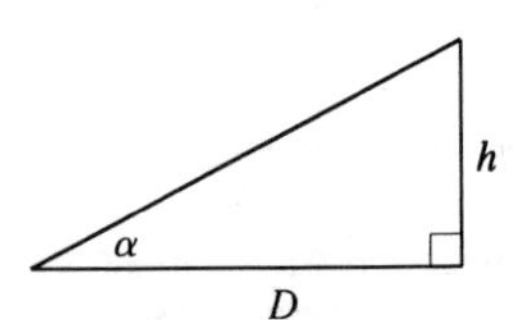

If the gun is pointed at a target with height h at a distance D downrange, then $\tan\alpha = h/D$. When the projectile reaches a distance D downrange (remember we are assuming that it doesn't hit the ground first,) we have $D = x = (v_0\cos\alpha)t$, so $t = \frac{D}{v_0\cos\alpha}$ and $y = (v_0\sin\alpha)t - \frac{1}{2}gt^2 = D\tan\alpha - \frac{gD^2}{2v_0^2\cos^2\alpha}$.

Meanwhile, the target, whose x-coordinate is also D, has fallen from height h to height $h - \frac{1}{2}gt^2 = D\tan\alpha - \frac{gD^2}{2v_0^2\cos^2\alpha}$. Thus the projectile hits the target.

5. **(a)** $m\frac{d^2\mathbf{R}}{dt^2} = -mg\mathbf{j} - k\frac{d\mathbf{R}}{dt} \quad \Rightarrow \quad \frac{d}{dt}\left(m\frac{d\mathbf{R}}{dt} + k\mathbf{R} + mgt\mathbf{j}\right) = \mathbf{0} \quad \Rightarrow \quad m\frac{d\mathbf{R}}{dt} + k\mathbf{R} + mgt\mathbf{j} = \mathbf{c}$ ($\mathbf{c}$ is a constant vector in the xy-plane). At $t = 0$, this says $m\mathbf{v}(0) + k\mathbf{R}(0) = \mathbf{c}$. Since $\mathbf{v}(0) = \mathbf{v}_0$ and $\mathbf{R}(0) = \mathbf{0}$, we have $\mathbf{c} = m\mathbf{v}_0$. Therefore $\frac{d\mathbf{R}}{dt} + \frac{k}{m}\mathbf{R} + gt\mathbf{j} = \mathbf{v}_0$, or $\frac{d\mathbf{R}}{dt} + \frac{k}{m}\mathbf{R} = \mathbf{v}_0 - gt\mathbf{j}$.

(b) Multiplying by $e^{(k/m)t}$ gives $e^{(k/m)t}\dfrac{d\mathbf{R}}{dt} + \dfrac{k}{m}e^{(k/m)t}\mathbf{R} = e^{(k/m)t}\mathbf{v}_0 - gte^{(k/m)t}\mathbf{j}$ or $\dfrac{d}{dt}\left(e^{(k/m)t}\mathbf{R}\right) = e^{(k/m)t}\mathbf{v}_0 - gte^{(k/m)t}\mathbf{j}$. Integrating gives

$e^{(k/m)t}\mathbf{R} = \dfrac{m}{k}e^{(k/m)t}\mathbf{v}_0 - \left(\dfrac{mg}{k}te^{(k/m)t} - \dfrac{m^2g}{k^2}e^{(k/m)t}\right)\mathbf{j} + \mathbf{b}$ for some constant vector $\mathbf{b}$. Setting $t = 0$ yields the relation $\mathbf{R}(0) = \dfrac{m}{k}\mathbf{v}_0 + \dfrac{m^2g}{k^2}\mathbf{j} + \mathbf{b}$, so $\mathbf{b} = -\dfrac{m}{k}\mathbf{v}_0 - \dfrac{m^2g}{k^2}\mathbf{j}$. Thus

$e^{(k/m)t}\mathbf{R} = \dfrac{m}{k}\left[e^{(k/m)t} - 1\right]\mathbf{v}_0 - \left[\dfrac{mg}{k}te^{(k/m)t} - \dfrac{m^2g}{k^2}\left(e^{(k/m)t} - 1\right)\right]\mathbf{j}$ and

$\mathbf{R}(t) = \dfrac{m}{k}\left[1 - e^{-kt/m}\right]\mathbf{v}_0 + \dfrac{mg}{k}\left[\dfrac{m}{k}\left(1 - e^{-kt/m}\right) - t\right]\mathbf{j}$.

7. (a) $F(x) = \begin{cases} 1 & \text{if } x \le 0 \\ \sqrt{1-x^2} & \text{if } 0 < x < \frac{1}{\sqrt{2}} \\ -x + \sqrt{2} & \text{if } x \ge \frac{1}{\sqrt{2}} \end{cases} \Rightarrow F'(x) = \begin{cases} 0 & \text{if } x < 0 \\ -x/\sqrt{1-x^2} & \text{if } 0 < x < \frac{1}{\sqrt{2}} \\ -1 & \text{if } x > \frac{1}{\sqrt{2}} \end{cases} \Rightarrow$

$F''(x) = \begin{cases} 0 & \text{if } x < 0 \\ -1/(1-x^2)^{3/2} & \text{if } 0 < x < \frac{1}{\sqrt{2}} \\ 0 & \text{if } x > \frac{1}{\sqrt{2}} \end{cases}$ since

$\dfrac{d}{dx}\left[-x\left(1-x^2\right)^{-1/2}\right] = -\left(1-x^2\right)^{-1/2} - x^2\left(1-x^2\right)^{-3/2} = -\left(1-x^2\right)^{-3/2}$.

Now $\lim\limits_{x\to 0^+}\sqrt{1-x^2} = 1 = F(0)$ and $\lim\limits_{x\to(1/\sqrt{2})^-}\sqrt{1-x^2} = \frac{1}{\sqrt{2}} = F\left(\frac{1}{\sqrt{2}}\right)$, so F is continuous. Also, since $\lim\limits_{x\to 0^+}F'(x) = 0 = \lim\limits_{x\to 0^-}F'(x)$ and $\lim\limits_{x\to(1/\sqrt{2})^-}F'(x) = -1 = \lim\limits_{x\to(1/\sqrt{2})^+}F'(x)$, F' is continuous. But $\lim\limits_{x\to 0^+}F''(x) = -1 \ne 0 = \lim\limits_{x\to 0^-}F''(x)$, so F'' is not continuous at $x = 0$. $\left(\text{The same is true at } x = \frac{1}{\sqrt{2}}.\right)$ So F does not have continuous curvature.

(b) Set $P(x) = ax^5 + bx^4 + cx^3 + dx^2 + ex + f$. The continuity conditions on P are $P(0) = 0$, $P(1) = 1$, $P'(0) = 0$ and $P'(1) = 1$. Also the curvature must be continuous. For $x \le 0$ and $x \ge 1$ $\kappa(x) = 0$, elsewhere $\kappa(x) = \dfrac{|P''(x)|}{\left(1 + [P'(x)]^2\right)^{3/2}}$, so we need $P''(0) = 0$ and $P''(1) = 0$. The conditions $P(0) = P'(0) = P''(0) = 0$ imply that $d = e = f = 0$. The other conditions imply that $a + b + c = 1$, $5a + 4b + 3c = 1$, and $10a + 6b + 3c = 0$. From these, we find that $a = 3$, $b = -8$, and $c = 6$. Therefore $P(x) = 3x^5 - 8x^4 + 6x^3$. Since there was no solution with $a = 0$, this could not have been done with a polynomial of degree 4.

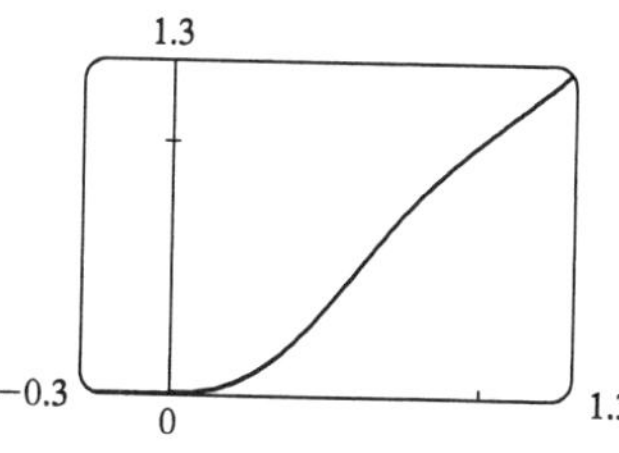

CHAPTER TWELVE

EXERCISES 12.1

1. **(a)** $f(2,1) = 4 - 1 + 4(2)(1) - 7(2) + 10 = 7$

(b) $f(-3,5) = 9 - 25 + 4(-3)(5) + 21 + 10 = -45$

(c) $f(x+h,y) = (x+h)^2 - y^2 + 4(x+h)y - 7(x+h) + 10$
$= x^2 + 2xh + h^2 - y^2 + 4xy + 4hy - 7x - 7h + 10$

(d) $f(x,y+k) = x^2 - (y+k)^2 + 4x(y+k) - 7x + 10 = x^2 - y^2 - 2ky - k^2 + 4xy + 4xk - 7x + 10$

(e) $f(x,x) = x^2 - x^2 + 4x^2 - 7x + 10 = 4x^2 - 7x + 10$

3. **(a)** $F(1,1) = 3(1)(1)/(1+2) = 1$

(b) $F(-1,2) = 3(-1)(2)/(1+8) = -\frac{2}{3}$

(c) $F(t,1) = 3t/(t^2+2)$

(d) $F(-1,y) = 3(-1)y/(1+2y^2) = -3y/(1+2y^2)$

(e) $F(x,x^2) = 3xx^2\Big/\left[x^2 + 2(x^2)^2\right] = 3x^3/(x^2+2x^4) = 3x/(1+2x^2)$

5. $D = \mathbb{R}^2$ and the range is $\mathbb{R}$.

7. $x + y \neq 0$ so $D = \{(x,y) \mid x + y \neq 0\}$. Since $2/(x+y)$ can't be zero, the range is $\{z \mid z \neq 0\}$.

9. $D = \mathbb{R}^2$ since the exponential function is defined everywhere and the range is $\{z \mid z > 0\}$.

11. For the logarithmic function to be defined, we need $x - y + z > 0$. Thus $D = \{(x,y,z) \mid x + z > y\}$ and the range is $\mathbb{R}$.

13. $D = \mathbb{R}^3$ and the range is $\mathbb{R}$.

15. $y - 2x \geq 0$ so $D = \{(x,y) \mid y \geq 2x\}$.

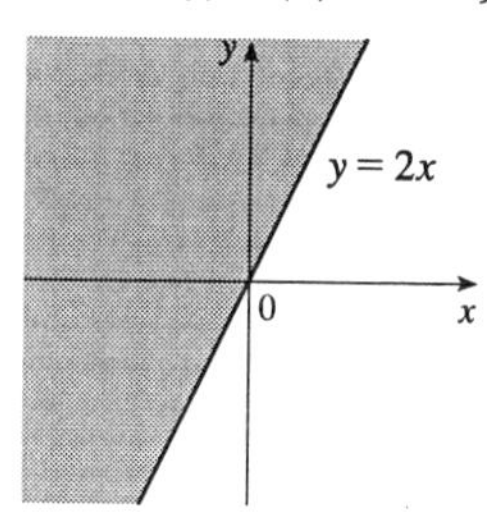

17. $x + 2y \neq 0$ and $9 - x^2 - y^2 \geq 0$, so
$D = \left\{(x,y) \mid y \neq -\frac{1}{2}x \text{ and } x^2 + y^2 \leq 9\right\}$.

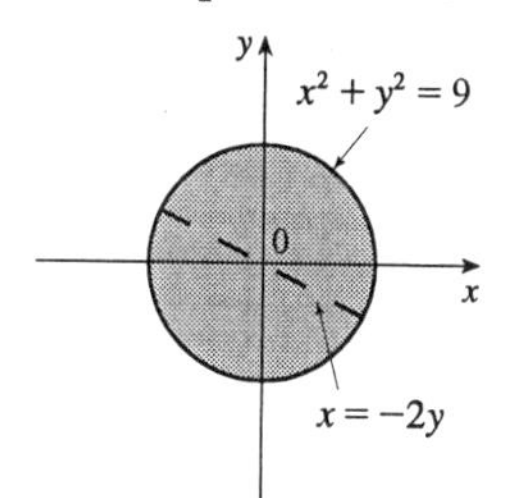

19. $D = \{(x,y) \mid x^2 + y \geq 0\} = \{(x,y) \mid y \geq -x^2\}$

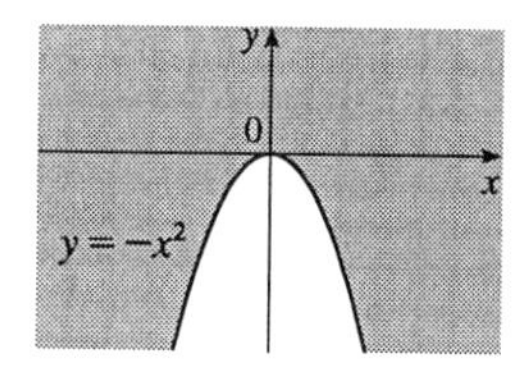

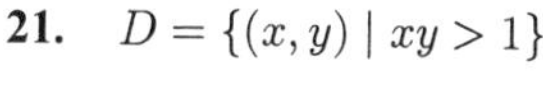

21. $D = \{(x,y) \mid xy > 1\}$

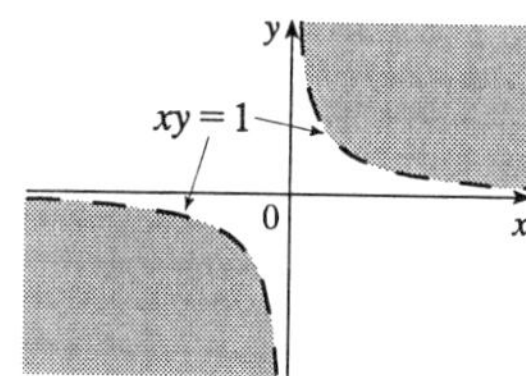

23. $D = \{(x, y) \mid y \neq \frac{\pi}{2} + n\pi, n \text{ an integer}\}$

25. $D = \{(x, y) \mid -1 \leq x + y \leq 1\}$
$= \{(x, y) \mid -1 - x \leq y \leq 1 - x\}$

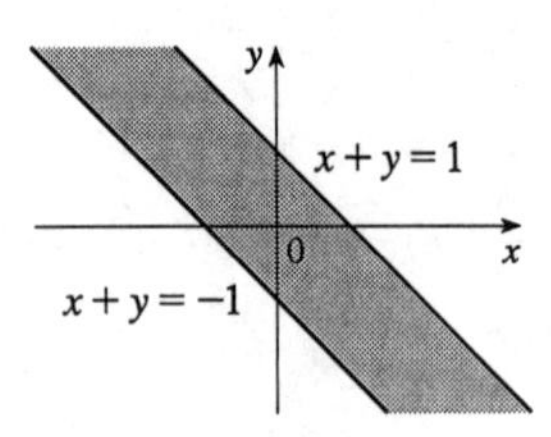

27. Since $\sin y > 0$ implies $2n\pi < y < (2n + 1)\pi$, n an integer,
$D = \{(x, y) \mid x > 0 \text{ and } 2n\pi < y < (2n + 1)\pi, n \text{ an integer}\}$.

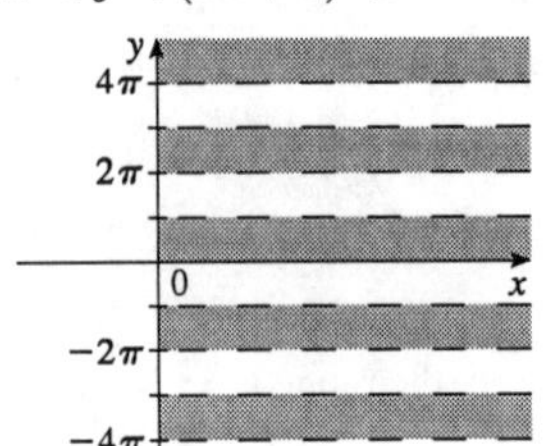

29. $D = \{(x, y, z) \mid x^2 + y^2 + z^2 \leq 1\}$ (the points inside or on the sphere of radius 1, center the origin).

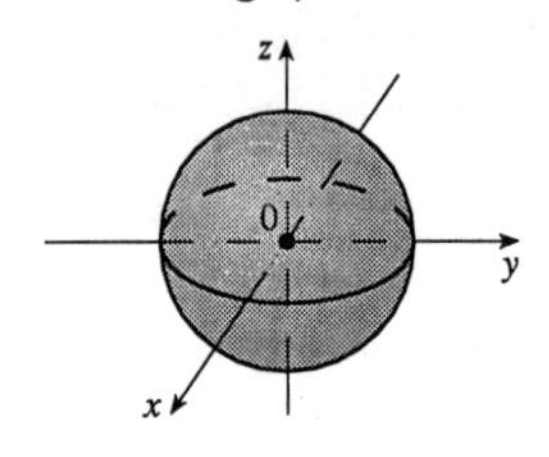

31. $z = 3$, a horizontal plane through the point $(0, 0, 3)$.

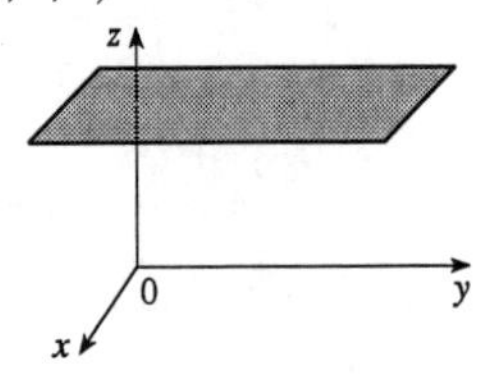

33. $z = 1 - x - y$ or $x + y + z = 1$, a plane with intercepts 1, 1, and 1.

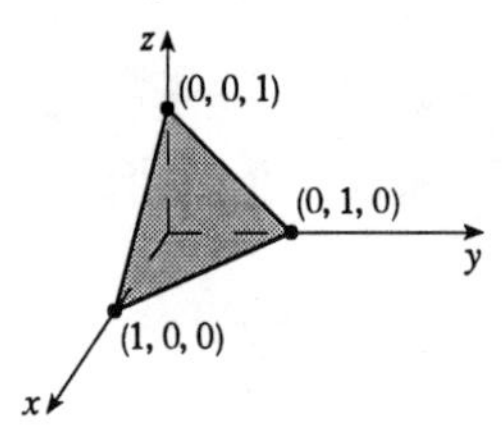

35. $z = x^2 + 9y^2$, an elliptic paraboloid with vertex the origin.

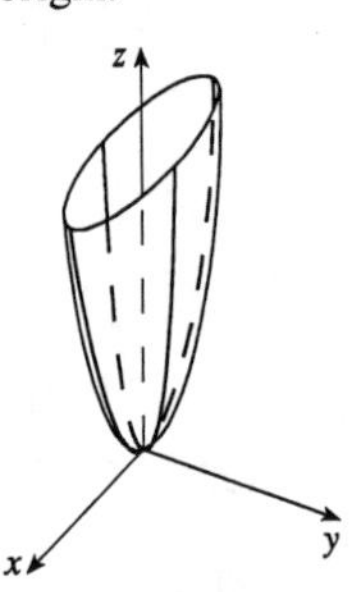

37. $z = \sqrt{x^2 + y^2}$, the top half of a right circular cone.

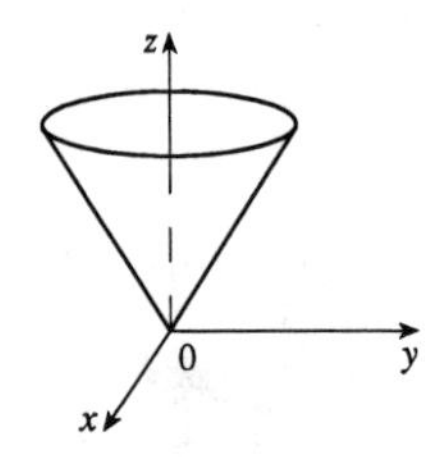

39. $z = y^2 - x^2$, a hyperbolic paraboloid.

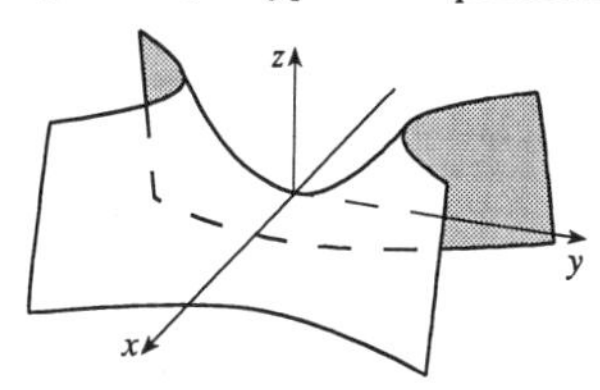

41. $z = 1 - x^2$, a parabolic cylinder.

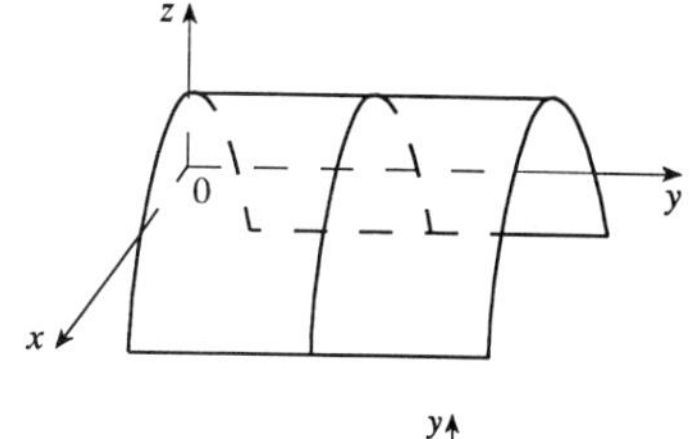

43. The level curves are $xy = k$. For $k = 0$ the curves are the coordinate axis; if $k > 0$, they are hyperbolas in the first and third quadrants; if $k < 0$, they are hyperbolas in the second and fourth quadrants.

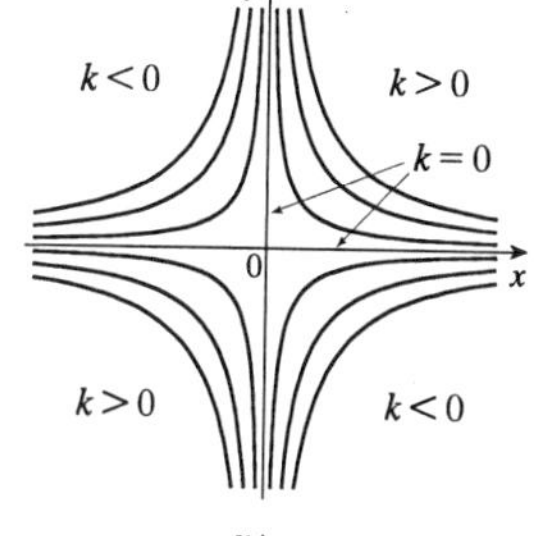

45. $k = x^2 + 9y^2$, a family of ellipses with major axis the x-axis. (Or, if $k = 0$, the origin.)

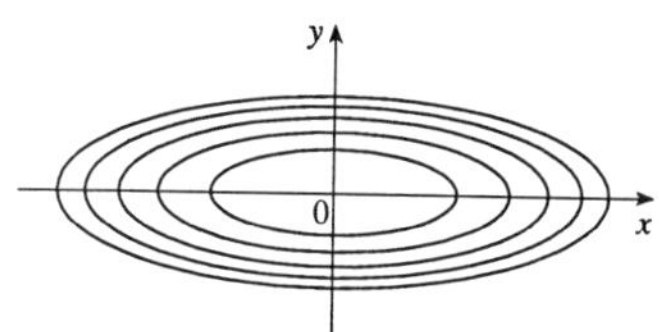

47. $k = x/y$ is a family of lines without the point $(0, 0)$.

49. 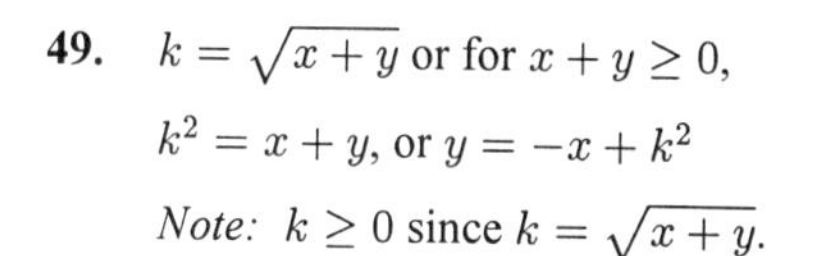
$k = \sqrt{x + y}$ or for $x + y \geq 0$,
$k^2 = x + y$, or $y = -x + k^2$
Note: $k \geq 0$ since $k = \sqrt{x + y}$.

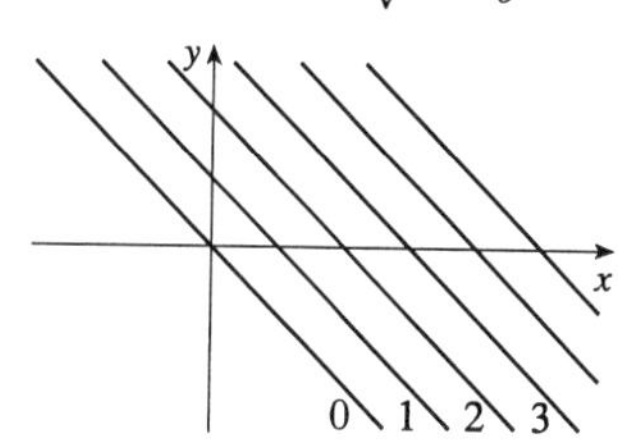

51. $k = x - y^2$, or $x - k = y^2$, a family of parabolas with vertex $(k, 0)$.

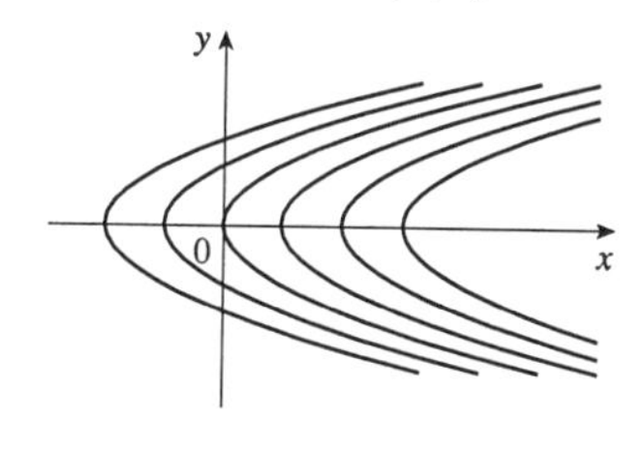

53. $k = x + 3y + 5z$ is a family of parallel planes with normal vector $\langle 1, 3, 5 \rangle$.

55. $k = x^2 - y^2 + z^2$ are the equations of the level surfaces. For $k = 0$, the surface is a right circular cone with vertex the origin and axis the y-axis. For $k > 0$, we have a family of hyperboloids of one sheet with axis the y-axis. For $k < 0$, we have a family of hyperboloids of two sheets with axis the y-axis.

57. The isothermals are given by $k = 100/(1 + x^2 + 2y^2)$ or $x^2 + 2y^2 = (100 - k)/k$ $(0 < k \leq 100)$, a family of ellipses.

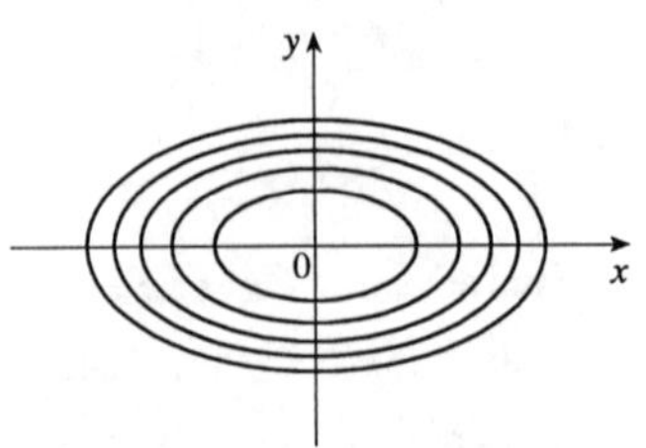

59. **(a)** B **(b)** III

Reasons: This function is constant on any circle centered at the origin, a description which matches only B and III.

61. **(a)** F **(b)** V

Reasons: $f(x, y) \to \infty$ as $(x, y) \to (0, 0)$, a condition satisfied only by F and V.

63. **(a)** D **(b)** IV

Reasons: This function is periodic in both x and y, with period 2π in each variable.

65. $f(x, y) = x^3 + y^3$

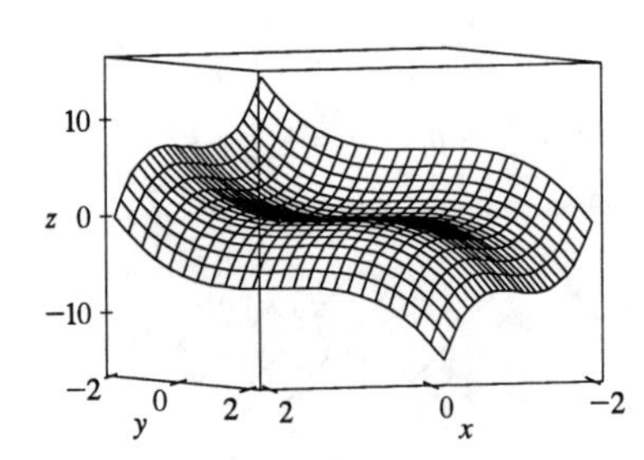

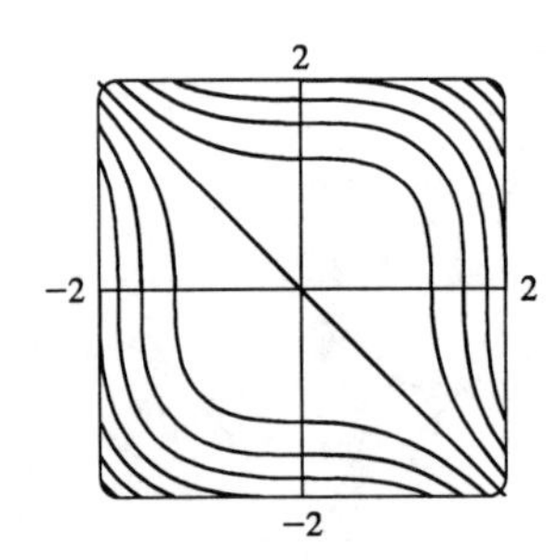

Note that the function is 0 along the line $y = -x$.

67. $f(x, y) = xy^2 - x^3$

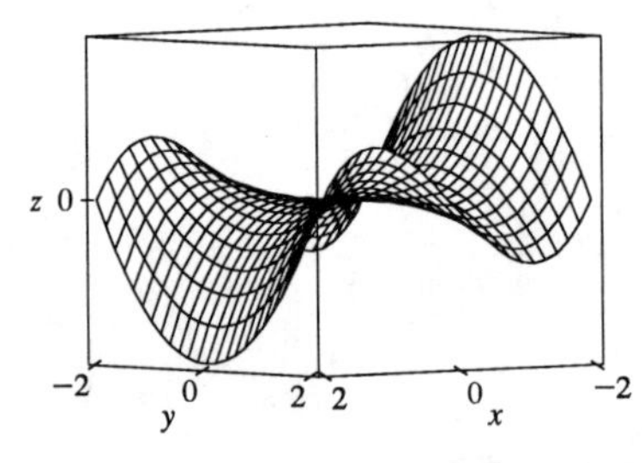

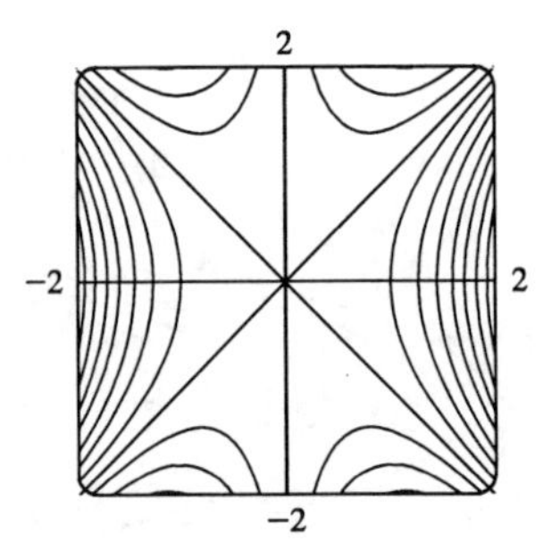

The cross-sections parallel to the yz-plane (such as the left-front trace in the graph above) are parabolas; those parallel to the xz-plane (such as the right-front trace) are cubic curves. The surface is called a monkey saddle because a monkey sitting on the surface near the origin has places for both legs and tail to rest.

69. $f(x,y) = e^{ax^2+by^2}$. We start with the case $a = b = 1$. This gives a graph whose level curves are circles centered at the origin.

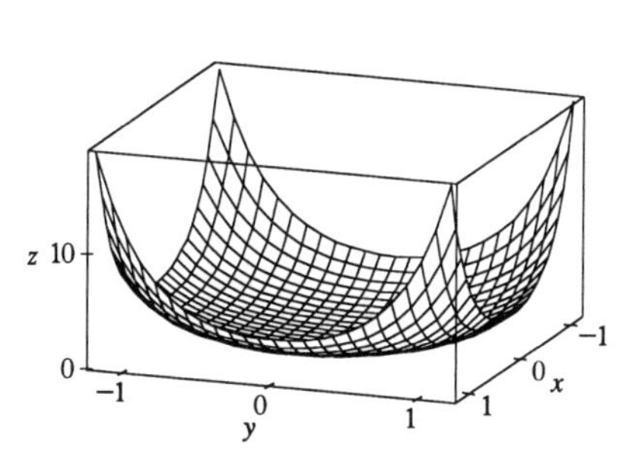

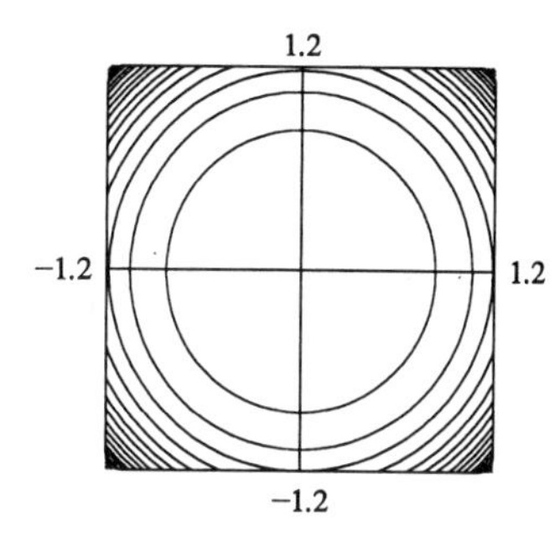

$a = 1, b = 1$

If we increase the ratio a/b, the level curves become ellipses whose eccentricity increases as a/b increases.

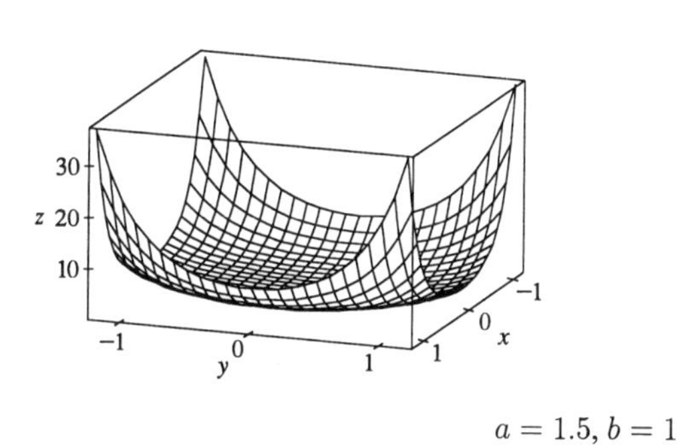

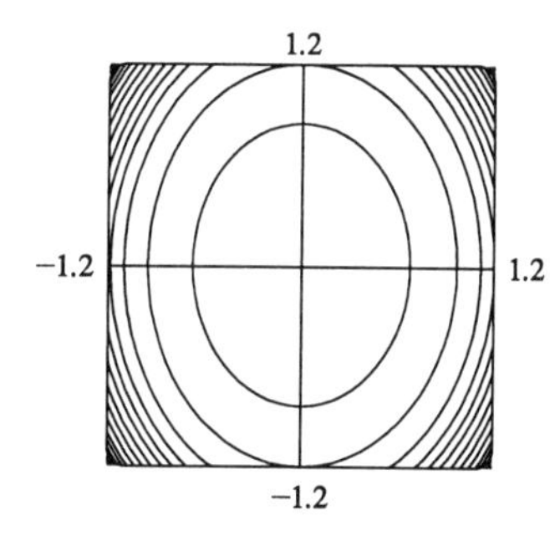

$a = 1.5, b = 1$

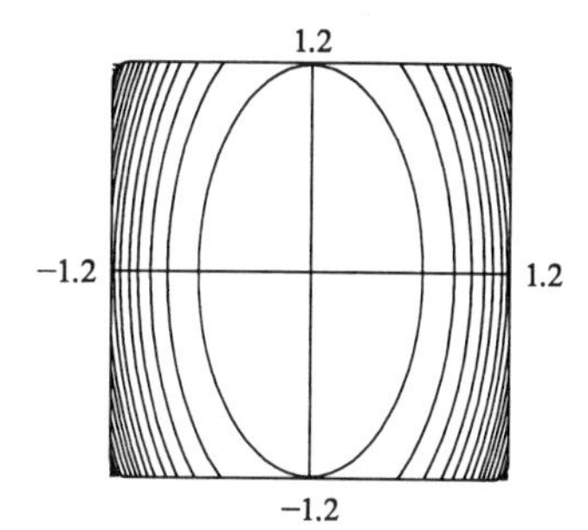

$a = 1.5, b = 0.5$

If one of a and b is 0, the graph is a cylinder. Note that in general, if we interchange a and b, the graph is rotated by $90°$ about the z-axis.

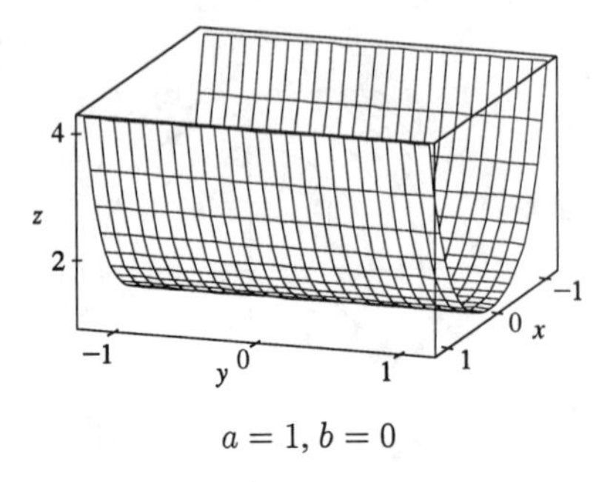

$a = 1, b = 0$

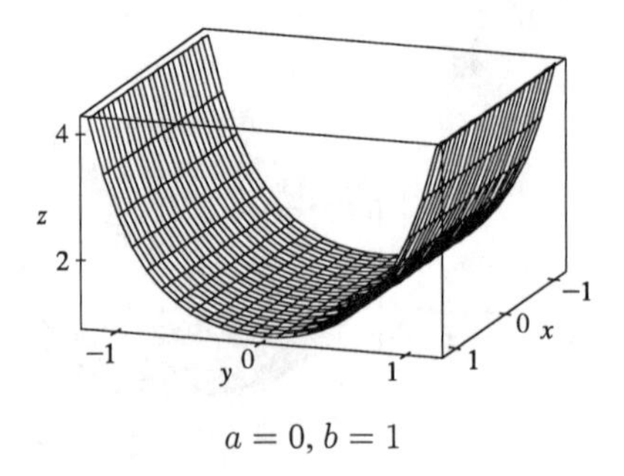

$a = 0, b = 1$

If a is positive and b is negative, the graph is saddle-shaped near the point $(0, 0, 1)$.

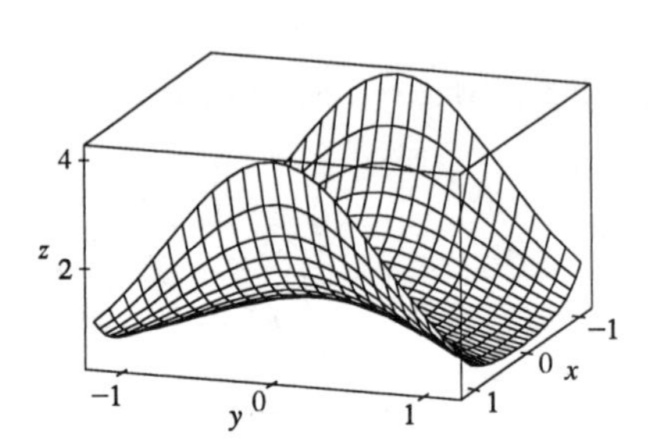

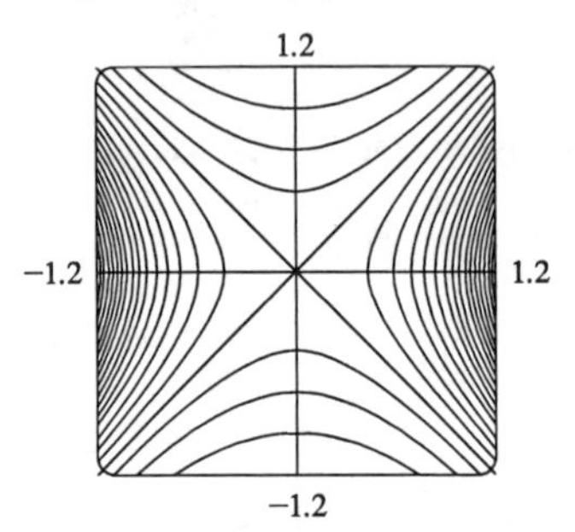

$a = 1, b = -1$

If a and b are both negative, the graph has a bump at the origin, which gets narrower in the x-direction as a decreases, and narrower in the y-direction as b decreases.

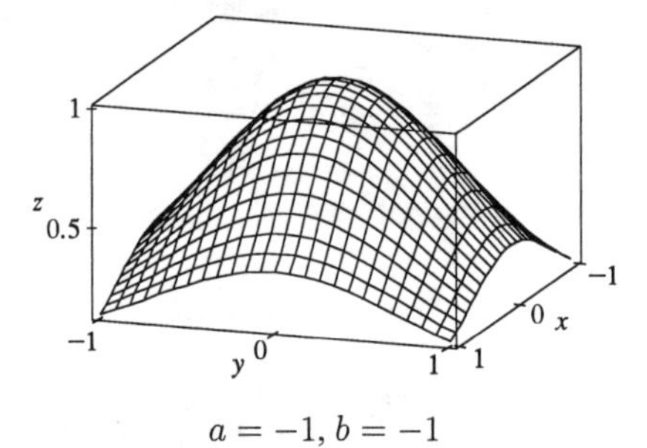

$a = -1, b = -1$

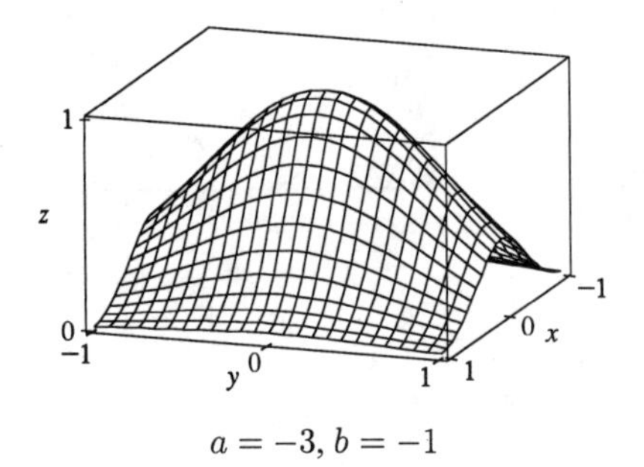

$a = -3, b = -1$

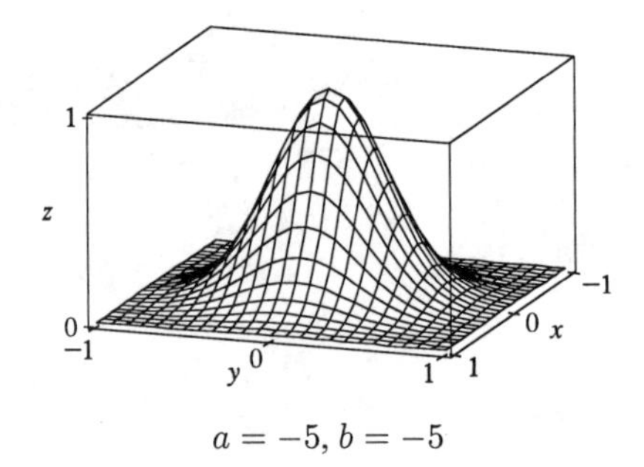

$a = -5, b = -5$

EXERCISES 12.2

1. The function is a polynomial, so the limit equals $(2^2)(3^2) - 2(2)(3^5) + 3(3) = -927$.

3. Since this is a rational function defined at $(0,0)$, the limit equals $(0+0-5)/(2-0) = -\frac{5}{2}$.

5. The product of two functions continuous at (π,π), so the limit equals $\pi \sin[(\pi+\pi)/4] = \pi$.

7. Let $f(x,y) = (x-y)/(x^2+y^2)$. First approach $(0,0)$ along the x-axis. Then $f(x,0) = x/x^2 = 1/x$ and $\lim\limits_{x\to 0} f(x,0)$ doesn't exist. Thus $\lim\limits_{(x,y)\to(0,0)} f(x,y)$ doesn't exist.

9. Let $f(x,y) = 8x^2y^2/(x^4+y^4)$. Approaching $(0,0)$ along the x-axis gives $f(x,y) \to 0$ as $(x,y) \to (0,0)$ along the x-axis. Approaching $(0,0)$ along the line $y = x$, $f(x,x) = 8x^4/2x^4 = 4$ for $x \neq 0$, so along this line $f(x,y) \to 4$ as $(x,y) \to (0,0)$. Thus the limit doesn't exist.

11. Let $f(x,y) = 2xy/(x^2+2y^2)$. As $(x,y) \to (0,0)$ along the x-axis, $f(x,y) \to 0$. But as $(x,y) \to (0,0)$ along the line $y = x$, $f(x,x) = 2x^2/3x^2$, so $f(x,y) \to \frac{2}{3}$ as $(x,y) \to (0,0)$ along this line. So the limit doesn't exist.

13. We can show that the limit along any line through $(0,0)$ is 0 and that the limits along the paths $x = y^2$ and $y = x^2$ are also 0. So we suspect that the limit exists and equals 0. Let $\epsilon > 0$ be given. We need to find $\delta > 0$ such that $\left|xy/\sqrt{x^2+y^2} - 0\right| < \epsilon$ whenever $0 < \sqrt{x^2+y^2} < \delta$ or $|xy|/\sqrt{x^2+y^2} < \epsilon$ whenever $0 < \sqrt{x^2+y^2} < \delta$. But $|x| = \sqrt{x^2} \leq \sqrt{x^2+y^2}$ so $|xy|/\sqrt{x^2+y^2} \leq |y| = \sqrt{y^2} \leq \sqrt{x^2+y^2}$. Thus choose $\delta = \epsilon$ and let $0 < \sqrt{x^2+y^2} < \delta = \epsilon$, then $\left|xy/\sqrt{x^2+y^2} - 0\right| \leq \sqrt{x^2+y^2} < \delta = \epsilon$. Hence by definition, $\lim\limits_{(x,y)\to(0,0)} xy/\sqrt{x^2+y^2} = 0$.

Or: Use the Squeeze Theorem. $0 \leq \left|\dfrac{xy}{\sqrt{x^2+y^2}}\right| \leq |x|$ since $|y| \leq \sqrt{x^2+y^2}$, and $|x| \to 0$ as $(x,y) \to (0,0)$.

15. Let $f(x,y) = \dfrac{2x^2y}{x^4+y^2}$. Then $f(x,0) = 0$ for $x \neq 0$, so $f(x,y) \to 0$ as $(x,y) \to (0,0)$ along the x-axis. But $f(x,x^2) = \dfrac{2x^4}{2x^4} = 1$ for $x \neq 0$, so $f(x,y) \to 1$ as $(x,y) \to (0,0)$ along the parabola $y = x^2$. Thus the limit doesn't exist.

17. $$\lim_{(x,y)\to(0,0)} \frac{x^2+y^2}{\sqrt{x^2+y^2+1}-1} = \lim_{(x,y)\to(0,0)} \frac{(x^2+y^2)\left(\sqrt{x^2+y^2+1}+1\right)}{x^2+y^2} = \lim_{(x,y)\to(0,0)} \left[\sqrt{x^2+y^2+1}+1\right] = 2$$

19. Let $f(x,y) = (xy-x)/(x^2+y^2-2y+1)$. Then $f(0,y) = 0$ for $y \neq 1$, so $f(x,y) \to 0$ as $(x,y) \to (0,1)$ along the y-axis. But $f(x,x+1) = x(x+1-1)/\left(x^2+(x+1-1)^2\right) = \frac{1}{2}$ for $x \neq 0$ so $f(x,y) \to 1/2$ as $(x,y) \to (0,1)$ along the line $y = x+1$. Thus the limit doesn't exist.

21. $\displaystyle\lim_{(x,y,z)\to(1,2,3)} \frac{xz^2-y^2z}{xyz-1} = \frac{1\cdot 3^2 - 2^2\cdot 3}{1\cdot 2\cdot 3-1} = -\frac{3}{5}$ since the function is continuous at $(1,2,3)$.

23. Let $f(x,y,z) = (x^2 - y^2 - z^2)/(x^2 + y^2 + z^2)$. Then $f(x,0,0) = 1$ for $x \neq 0$ and $f(0,y,0) = -1$ for $y \neq 0$, so as $(x,y,z) \to (0,0,0)$ along the x-axis, $f(x,y,z) \to 1$ but as $(x,y,z) \to (0,0,0)$ along the y-axis, $f(x,y,z) \to -1$. Thus the limit doesn't exist.

25. Let $f(x,y,z) = \dfrac{xy + yz^2 + xz^2}{x^2 + y^2 + z^4}$. Then $f(x,0,0) = 0/x^2 = 0$ for $x \neq 0$, so as $(x,y,z) \to (0,0,0)$ along the x-axis, $f(x,y,z) \to 0$. But $f(x,x,0) = x^2/(2x^2) = \frac{1}{2}$ for $x \neq 0$, so as $(x,y,z) \to (0,0,0)$ along the line $y = x$, $z = 0$, $f(x,y,z) \to \frac{1}{2}$. Thus the limit does not exist.

27.

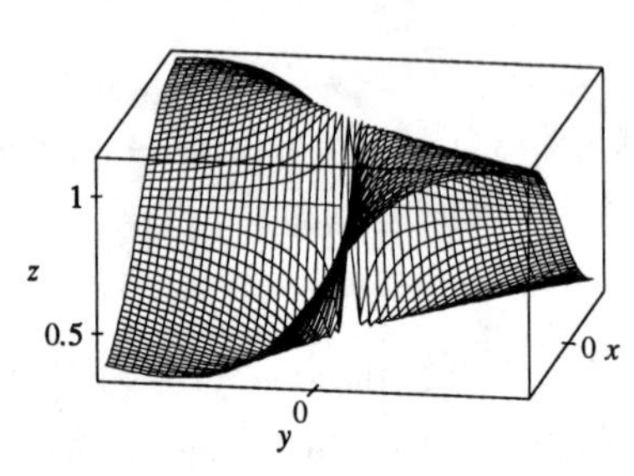

From the ridges on the graph, we see that as $(x,y) \to (0,0)$ along the lines under the two ridges, $f(x,y)$ approaches different values. So the limit does not exist.

29. $h(x,y) = g(f(x,y)) = g(x^4 + x^2y^2 + y^4) = e^{-(x^4+x^2y^2+y^4)}\cos(x^4 + x^2y^2 + y^4)$. Since f is a polynomial it is continuous throughout $\mathbb{R}^2$ and g is the product of two functions, both of which are continuous on $\mathbb{R}$, h is continuous on $\mathbb{R}^2$ by Theorem 5.

31. $h(x,y) = g(f(x,y)) = (2x + 3y - 6)^2 + \sqrt{2x + 3y - 6}$. Since f is a polynomial, it is continuous on $\mathbb{R}^2$ and g is continuous on its domain $\{t \mid t \geq 0\}$. Thus h is continuous on its domain $D = \{(x,y) \mid 2x + 3y - 6 \geq 0\}$ $= \left\{(x,y) \mid y \geq -\frac{2}{3}x + 2\right\}$ which consists of all points on and above right of the line $y = -\frac{2}{3}x + 2$.

33. $F(x,y)$ is a rational function and thus is continuous on its domain $D = \{(x,y) \mid x^2 + y^2 - 1 \neq 0\}$, that is, F is continuous except on the circle $x^2 + y^2 = 1$.

35. $F(x,y) = g(f(x,y))$ where $f(x,y) = x^4 - y^4$, a polynomial so continuous on $\mathbb{R}^2$ and $g(t) = \tan t$, continuous on its domain $\left\{t \mid t \neq (2n+1)\frac{\pi}{2}, n \text{ an integer}\right\}$. Thus F is continuous on its domain $D = \left\{(x,y) \mid x^4 - y^4 \neq (2n+1)\frac{\pi}{2}, n \text{ an integer}\right\}$.

37. $G(x,y) = g(x,y)f(x,y)$ where $g(x,y) = e^{xy}$ and $f(x,y) = \sin(x+y)$ both of which are continuous on $\mathbb{R}^2$. Thus G is continuous on $\mathbb{R}^2$.

39. $G(x,y) = g_1(f_1(x,y)) - g_2(f_2(x,y))$ where $f_1(x,y) = x + y$ and $f_2(x,y) = x - y$ both of which are polynomials so continuous on $\mathbb{R}^2$ and $g_1(t) = \sqrt{t}$, $g_2(s) = \sqrt{s}$ both of which are continuous on their respective domains $\{t \mid t \geq 0\}$ and $\{s \mid s \geq 0\}$. Thus $g_1 \circ f_1$ is continuous on its domain $D_1 = \{(x,y) \mid x + y \geq 0\} = \{(x,y) \mid y \geq -x\}$ and $g_2 \circ f_2$ is continuous on its domain $D_2 = \{(x,y) \mid x - y \geq 0\} = \{(x,y) \mid y \leq x\}$. Then G, being the difference of these two composite functions, is continuous on its domain $D = D_1 \cap D_2 = \{(x,y) \mid -x \leq y \leq x\} = \{(x,y) \mid |y| \leq x\}$.

41. $f(x,y,z) = xg(f(y,z))$ where $f(y,z) = yz$, continuous on $\mathbb{R}^2$ and $g(t) = \ln t$, continuous on its domain $\{t \mid t > 0\}$. Since $h(x) = x$ is continuous on $\mathbb{R}$, $f(x,y,z)$ is continuous on its domain $D = \{(x,y,z) \mid yz > 0\}$.

In Exercises 43 and 45 each f is a piecewise defined function where each first piece is a rational function defined everywhere except at the origin. Thus each f is continuous on $\mathbb{R}^2$ except possibly at the origin. So for each we need only check $\lim\limits_{(x,y)\to(0,0)} f(x,y)$.

43. Letting $z = \sqrt{2}x$, $\lim\limits_{(x,y)\to(0,0)} \dfrac{2x^2 - y^2}{2x^2 + y^2} = \lim\limits_{(z,y)\to(0,0)} \dfrac{z^2 - y^2}{z^2 + y^2}$ which doesn't exist by Example 1. Thus f is not continuous at $(0,0)$ and the largest set on which f is continuous is $\{(x,y) \mid (x,y) \neq (0,0)\}$.

45. Since $x^2 \leq 2x^2 + y^2$, we have $\left|x^2y^3/(2x^2+y^2)\right| \leq |y^3|$. We know that $|y^3| \to 0$ as $(x,y) \to (0,0)$. So, by the Squeeze Theorem, $\lim\limits_{(x,y)\to(0,0)} f(x,y) = \lim\limits_{(x,y)\to(0,0)} \dfrac{x^2y^3}{2x^2+y^2} = 0$. But $f(0,0) = 1$, so f is discontinuous at $(0,0)$. For $(x,y) \neq (0,0)$, $f(x,y)$ is equal to a rational function and is therefore continuous. Therefore f is continuous on the set $\{(x,y) \mid (x,y) \neq (0,0)\}$.

47. **(a)** Let $\epsilon > 0$ be given. We need to find $\delta > 0$ such that $|x - a| < \epsilon$ whenever $0 < \sqrt{(x-a)^2 + (y-b)^2} < \delta$. But $|x - a| = \sqrt{(x-a)^2} \leq \sqrt{(x-a)^2 + (y-b)^2}$. Thus setting $\delta = \epsilon$ and letting $0 < \sqrt{(x-a)^2 + (y-b)^2} < \delta$, we have $|x - a| \leq \sqrt{(x-a)^2 + (y-b)^2} < \delta = \epsilon$. Hence, by Definition 1, $\lim\limits_{(x,y)\to(a,b)} x = a$.

(b) The argument is the same as in (a) with the roles of x and y interchanged.

(c) Let $\epsilon > 0$ be given and set $\delta = \epsilon$. Then $|f(x,y) - L| = |c - c| = 0 \leq \sqrt{(x-a)^2 + (y-b)^2} < \delta = \epsilon$ whenever $0 < \sqrt{(x-a)^2 + (y-b)^2} < \delta$. Thus by Definition 1, $\lim\limits_{(x,y)\to(a,b)} c = c$.

49. Since $|\mathbf{x} - \mathbf{a}|^2 = |\mathbf{x}|^2 + |\mathbf{a}|^2 - 2|\mathbf{x}||\mathbf{a}|\cos\theta \geq |\mathbf{x}|^2 + |\mathbf{a}|^2 - 2|\mathbf{x}||\mathbf{a}| = (|\mathbf{x}| - |\mathbf{a}|)^2$, we have $||\mathbf{x}| - |\mathbf{a}|| \leq |\mathbf{x} - \mathbf{a}|$. Let $\epsilon > 0$ be given and set $\delta = \epsilon$. Then whenever $0 < |\mathbf{x} - \mathbf{a}| < \delta$, $||\mathbf{x}| - |\mathbf{a}|| \leq |\mathbf{x} - \mathbf{a}| < \delta = \epsilon$. Hence $\lim\limits_{\mathbf{x}\to\mathbf{a}} |\mathbf{x}| = |\mathbf{a}|$ and $f(\mathbf{x}) = |\mathbf{x}|$ is continuous on $\mathbb{R}^n$.

EXERCISES 12.3

1. $f(x,y) = 16 - 4x^2 - y^2 \quad\Rightarrow\quad f_x(x,y) = -8x$ and $f_y(x,y) = -2y \quad\Rightarrow\quad f_x(1,2) = -8$ and $f_y(1,2) = -4$. The graph of f is the paraboloid $z = 16 - 4x^2 - y^2$ and the vertical plane $y = 2$ intersects it in the parabola $z = 12 - 4x^2$, $y = 2$, (the curve C_1 in the first figure). The slope of the tangent line to this parabola at $(1,2,8)$ is $f_x(1,2) = -8$. Similarly the plane $x = 1$ intersects the paraboloid in the parabola $z = 12 - y^2$, $x = 1$, (the curve C_2 in the second figure) and the slope of the tangent line at $(1,2,8)$ is $f_y(1,2) = -4$.

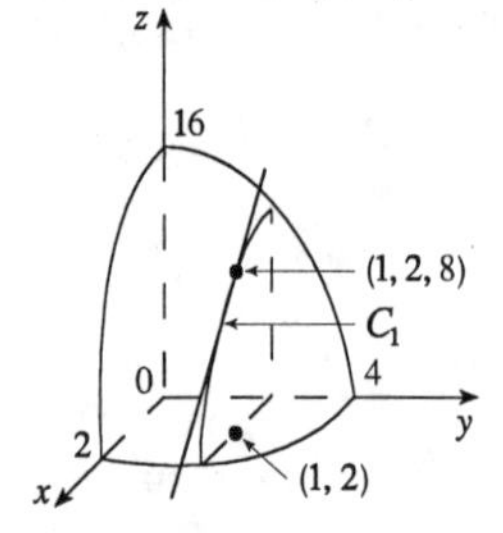

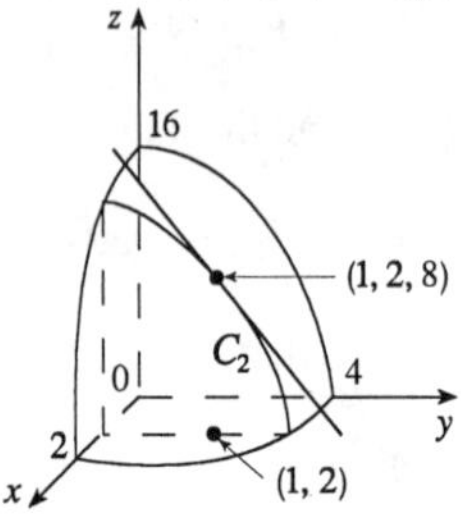

3. $f(x,y) = x^3y^5 \quad\Rightarrow\quad f_x(x,y) = 3x^2y^5,\ f_x(3,-1) = -27$

5. $f(x,y) = xe^{-y} + 3y \quad\Rightarrow\quad \dfrac{\partial f}{\partial y} = x(-1)e^{-y} + 3,\ \dfrac{\partial f}{\partial y}(1,0) = -1 + 3 = 2$

7. $z = \dfrac{x^3+y^3}{x^2+y^2} \quad\Rightarrow\quad \dfrac{\partial z}{\partial x} = \dfrac{3x^2(x^2+y^2) - (x^3+y^3)(2x)}{(x^2+y^2)^2} = \dfrac{x^4 + 3x^2y^2 - 2xy^3}{(x^2+y^2)^2}$,

$\dfrac{\partial z}{\partial y} = \dfrac{3y^2(x^2+y^2) - (x^3+y^3)(2y)}{(x^2+y^2)^2} = \dfrac{3x^2y^2 + y^4 - 2yx^3}{(x^2+y^2)^2}$.

9. $xy + yz = xz \quad\Rightarrow\quad \dfrac{\partial}{\partial x}(xy+yz) = \dfrac{\partial}{\partial x}(xz) \quad\Leftrightarrow\quad y + y\dfrac{\partial z}{\partial x} = z + x\dfrac{\partial z}{\partial x} \quad\Leftrightarrow\quad (y-x)\dfrac{\partial z}{\partial x} = z - y$, so $\dfrac{\partial z}{\partial x} = \dfrac{z-y}{y-x}$. $\dfrac{\partial}{\partial y}(xy+yz) = \dfrac{\partial}{\partial y}(xz) \quad\Leftrightarrow\quad x + z + y\dfrac{\partial z}{\partial y} = x\dfrac{\partial z}{\partial y} \quad\Leftrightarrow\quad (y-x)\dfrac{\partial z}{\partial y} = -(x+z)$, so $\dfrac{\partial z}{\partial y} = \dfrac{x+z}{x-y}$.

11. $x^2 + y^2 - z^2 = 2x(y+z) \quad\Leftrightarrow\quad \dfrac{\partial}{\partial x}\left(x^2+y^2-z^2\right) = \dfrac{\partial}{\partial x}[2x(y+z)] \quad\Leftrightarrow$

$2x - 2z\dfrac{\partial z}{\partial x} = 2(y+z) + 2x\dfrac{\partial z}{\partial x} \quad\Leftrightarrow\quad 2(x+z)\dfrac{\partial z}{\partial x} = 2(x-y-z)$, so $\dfrac{\partial z}{\partial x} = \dfrac{x-y-z}{x+z}$.

$\dfrac{\partial}{\partial y}\left(x^2+y^2-z^2\right) = \dfrac{\partial}{\partial y}[2x(y+z)] \quad\Leftrightarrow\quad 2y - 2z\dfrac{\partial z}{\partial y} = 2x\left(1 + \dfrac{\partial z}{\partial y}\right) \quad\Leftrightarrow\quad 2(x+z)\dfrac{\partial z}{\partial y} = 2(y-x)$, so $\dfrac{\partial z}{\partial y} = \dfrac{y-x}{x+z}$.

13. $f(x,y,z) = xyz \quad\Rightarrow\quad f_y(x,y,z) = xz$, so $f_y(0,1,2) = 0$.

15. $u = xy + yz + zx \quad\Rightarrow\quad u_x = y + z,\ u_y = x + z,\ u_z = y + x$

17. $f(x,y) = x^3y^5 - 2x^2y + x \quad\Rightarrow\quad f_x(x,y) = 3x^2y^5 - 4xy + 1,\ f_y(x,y) = 5x^3y^4 - 2x^2$

19. $f(x,y) = x^4 + x^2y^2 + y^4 \quad\Rightarrow\quad f_x(x,y) = 4x^3 + 2xy^2,\ f_y(x,y) = 2x^2y + 4y^3$

21. $f(x,y) = \dfrac{x-y}{x+y} \quad\Rightarrow\quad f_x(x,y) = \dfrac{(1)(x+y)-(x-y)(1)}{(x+y)^2} = \dfrac{2y}{(x+y)^2}$,

$f_y(x,y) = \dfrac{(-1)(x+y)-(x-y)(1)}{(x+y)^2} = -\dfrac{2x}{(x+y)^2}$

23. $f(x,y) = e^x \tan(x-y) \quad\Rightarrow\quad f_x(x,y) = e^x \tan(x-y) + e^x \sec^2(x-y) = e^x[\tan(x-y) + \sec^2(x-y)]$,

$f_y(x,y) = e^x[\sec^2(x-y)](-1) = -e^x \sec^2(x-y)$

25. $f(u,v) = \tan^{-1}\left(\dfrac{u}{v}\right) \quad\Rightarrow\quad f_u(u,v) = \dfrac{1}{1+(u/v)^2}\left(\dfrac{1}{v}\right) = \dfrac{1}{v}\left(\dfrac{v^2}{u^2+v^2}\right) = \dfrac{v}{u^2+v^2}$,

$f_v(u,v) = \dfrac{1}{1+(u/v)^2}\left(-\dfrac{u}{v^2}\right) = -\dfrac{u}{v^2}\left(\dfrac{v^2}{u^2+v^2}\right) = -\dfrac{u}{u^2+v^2}$

27. $g(x,y) = y\tan(x^2y^3) \quad\Rightarrow\quad g_x(x,y) = [y\sec^2(x^2y^3)](2xy^3) = 2xy^4\sec^2(x^2y^3)$,

$g_y(x,y) = \tan(x^2y^3) + [y\sec^2(x^2y^3)](3x^2y^2) = \tan(x^2y^3) + 3x^2y^3\sec^2(x^2y^3)$

29. $z = \ln\left(x + \sqrt{x^2+y^2}\right) \quad\Rightarrow$

$$\frac{\partial z}{\partial x} = \frac{1}{x+\sqrt{x^2+y^2}}\left[1 + \tfrac{1}{2}(x^2+y^2)^{-1/2}(2x)\right] = \frac{\left(\sqrt{x^2+y^2}+x\right)/\sqrt{x^2+y^2}}{\left(x+\sqrt{x^2+y^2}\right)} = \frac{1}{\sqrt{x^2+y^2}},$$

$$\frac{\partial z}{\partial y} = \frac{1}{x+\sqrt{x^2+y^2}}\left(\frac{1}{2}\right)\left(x^2+y^2\right)^{-1/2}(2y) = \frac{y}{x\sqrt{x^2+y^2}+x^2+y^2}$$

31. $f(x,y) = \displaystyle\int_x^y e^{t^2}\,dt$. By the Fundamental Theorem of Calculus, Part I, $\dfrac{d}{dx}\displaystyle\int_a^x f(t)dt = f(x)$ for f continuous.

Thus $f_x(x,y) = \dfrac{\partial}{\partial x}\displaystyle\int_x^y e^{t^2}\,dt = \dfrac{\partial}{\partial x}\left(-\int_y^x e^{t^2}\,dt\right) = -e^{x^2}$ and $f_y(x,y) = \dfrac{\partial}{\partial y}\displaystyle\int_x^y e^{t^2}\,dt = e^{y^2}$.

33. $f(x,y,z) = x^2yz^3 + xy - z \quad\Rightarrow\quad f_x(x,y,z) = 2xyz^3 + y$, $f_y(x,y,z) = x^2z^3 + x$, $f_z(x,y,z) = 3x^2yz^2 - 1$

35. $f(x,y,z) = x^{yz} \quad\Rightarrow\quad f_x(x,y,z) = yzx^{yz-1}$. By Theorem 4.4.5, $f_y(x,y,z) = x^{yz}\ln(x^z) = zx^{yz}\ln x$ and by symmetry $f_z(x,y,z) = yx^{yz}\ln x$.

37. $u = z\sin\left(\dfrac{y}{x+z}\right) \quad\Rightarrow\quad u_x = z\cos\left(\dfrac{y}{x+z}\right)\left[-y(x+z)^{-2}\right] = \dfrac{-yz}{(x+z)^2}\cos\left(\dfrac{y}{x+z}\right)$,

$u_y = z\cos\left(\dfrac{y}{x+z}\right)\left(\dfrac{1}{x+z}\right) = \dfrac{z}{x+z}\cos\left(\dfrac{y}{x+z}\right)$,

$u_z = \sin\left(\dfrac{y}{x+z}\right) + z\cos\left(\dfrac{y}{x+z}\right)\left[-y(x+z)^{-2}\right] = \sin\left(\dfrac{y}{x+z}\right) - \dfrac{yz}{(x+z)^2}\cos\left(\dfrac{y}{x+z}\right)$

39. $u = xy^2z^3\ln(x+2y+3z) \quad\Rightarrow$

$u_x = y^2z^3\ln(x+2y+3z) = xy^2z^3\left(\dfrac{1}{x+2y+3z}\right) = y^2z^3\left[\ln(x+2y+3z) + \dfrac{x}{x+2y+3z}\right]$,

$u_y = 2xyz^3\ln(x+2y+3z) + xy^2z^3\left(\dfrac{1}{x+2y+3z}\right)(2) = 2xyz^3\left[\ln(x+2y+3z) + \dfrac{y}{x+2y+3z}\right]$, and by symmetry, $u_z = 3xy^2z^2\left[\ln(x+2y+3z) + \dfrac{z}{x+2y+3z}\right]$.

41. $f(x,y,z,t) = \dfrac{x-y}{z-t} \quad \Rightarrow \quad f_x(x,y,z,t) = \dfrac{1}{z-t}$, $f_y(x,y,z,t) = -\dfrac{1}{z-t}$,
$f_z(x,y,z,t) = (x-y)(-1)(z-t)^{-2} = \dfrac{y-x}{(z-t)^2}$, and $f_t(x,y,z,t) = (x-y)(-1)(z-t)^{-2}(-1) = \dfrac{x-y}{(z-t)^2}$.

43. $u = \sqrt{x_1^2 + x_2^2 + \cdots + x_n^2}$. For each $i = 1, \ldots, n$,
$u_{x_i} = \frac{1}{2}(x_1^2 + x_2^2 + \cdots + x_n^2)^{-1/2}(2x_i) = \dfrac{x_i}{\sqrt{x_1^2 + x_2^2 + \cdots + x_n^2}}$.

45. $f(x,y) = x^2 - xy + 2y^2 \quad \Rightarrow$

$$f_x(x,y) = \lim_{h\to 0} \frac{f(x+h,y) - f(x,y)}{h} = \lim_{h\to 0} \frac{(x+h)^2 - (x+h)y + 2y^2 - (x^2 - xy + 2y^2)}{h}$$
$$= \lim_{h\to 0} \frac{h(2x - y + h)}{h} = \lim_{h\to 0}(2x - y + h) = 2x - y,$$

$$f_y(x,y) = \lim_{h\to 0} \frac{f(x,y+h) - f(x,y)}{h} = \lim_{h\to 0} \frac{x^2 - x(y+h) + 2(y+h)^2 - (x^2 - xy + 2y^2)}{h}$$
$$= \lim_{h\to 0} \frac{h(4y - x + 2h)}{h} = \lim_{h\to 0}(4y - x + 2h) = 4y - x$$

47. $f(x,y) = x^2 + y^2 + x^2y \quad \Rightarrow \quad f_x = 2x + 2xy$, $f_y = 2y + x^2$

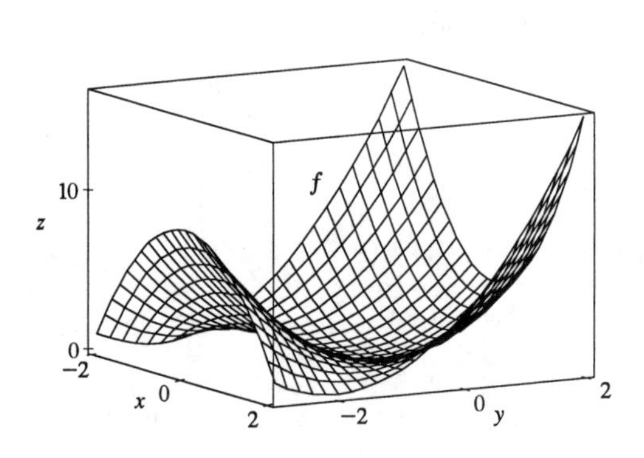

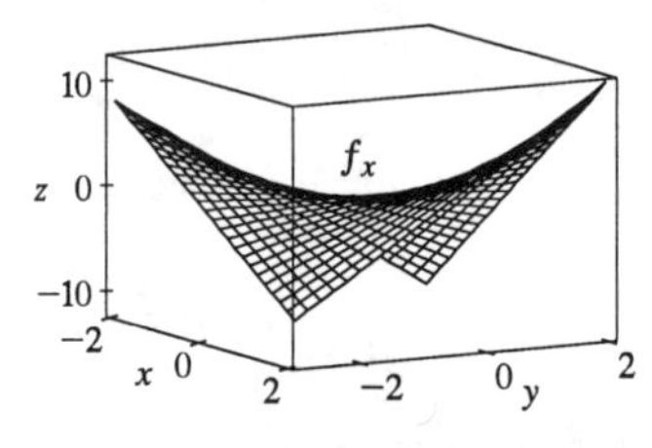

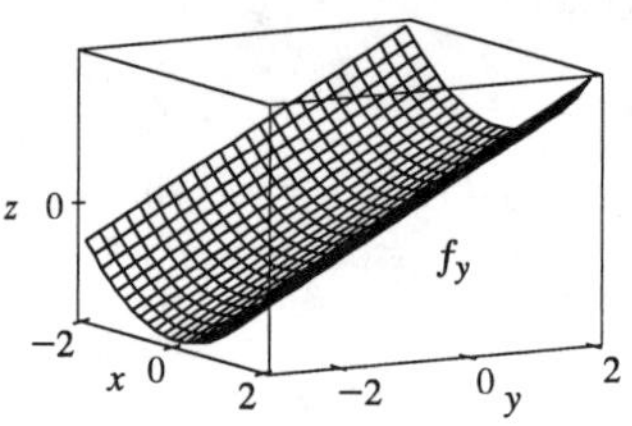

Note that the traces of f in planes parallel to the xz-plane are parabolas which open downward for $y < -1$ and upward for $y > -1$, and the traces of f_x in these planes are straight lines, which have negative slopes for $y < -1$ and positive slopes for $y > -1$. The traces of f in planes parallel to the yz-plane are parabolas which always open upward, and the traces of f_y in these planes are straight lines with positive slopes.

49. First of all, if we start at the point $(3, -3)$ and move in the positive y-direction, we see that both b and c decrease, while a increases. Both b and c have a low point at about $(3, -1.5)$, while a is 0 at this point. So a is definitely the graph of f_y, and one of b and c is the graph of f. To see which is which, we start at the point $(-3, -1.5)$ and move in the positive x-direction. b traces out a line with negative slope, while c traces out a parabola opening downward. This tells us that b is the x-derivative of c. So c is the graph of f, b is the graph of f_x, and a is the graph of f_y.

51. $z = f(x) + g(y) \quad \Rightarrow \quad \dfrac{\partial z}{\partial x} = f'(x),\ \dfrac{\partial z}{\partial y} = g'(y)$

53. $z = f(x+y)$. Let $u = x + y$. Then $\dfrac{\partial z}{\partial x} = \dfrac{df}{du}\dfrac{\partial u}{\partial x} = \dfrac{df}{d(x+y)} = f'(x+y)$,

$$\frac{\partial z}{\partial y} = \frac{df}{du}\frac{\partial u}{\partial y} = \frac{df}{d(x+y)} = f'(x+y).$$

55. $z = f\left(\dfrac{x}{y}\right)$. Let $u = \dfrac{x}{y}$. Then $\dfrac{\partial u}{\partial x} = \dfrac{1}{y}$ and $\dfrac{\partial u}{\partial y} = -\dfrac{x}{y^2}$. Hence $\dfrac{\partial z}{\partial x} = \dfrac{df}{du}\dfrac{\partial u}{\partial x} = \dfrac{df/d(x/y)}{y} = \dfrac{f'(x/y)}{y}$ and

$$\frac{\partial z}{\partial y} = \frac{df}{d(x/y)}\left(-\frac{x}{y^2}\right) = -x\left[\frac{df/d(x/y)}{y^2}\right] = -\frac{xf'(x/y)}{y^2}.$$

57. $f(x,y) = x^2y + x\sqrt{y} \quad \Rightarrow \quad f_x = 2xy + \sqrt{y},\ f_y = x^2 + \dfrac{x}{2\sqrt{y}}$. Thus $f_{xx} = 2y,\ f_{xy} = 2x + \dfrac{1}{2\sqrt{y}}$,

$f_{yx} = 2x + \dfrac{1}{2\sqrt{y}}$ and $f_{yy} = -\dfrac{x}{4y^{3/2}}$.

59. $z = (x^2+y^2)^{3/2} \quad \Rightarrow \quad z_x = \frac{3}{2}(x^2+y^2)^{1/2}(2x) = 3x(x^2+y^2)^{1/2}$ and $z_y = 3y(x^2+y^2)^{1/2}$. Thus

$$z_{xx} = 3\left(x^2+y^2\right)^{1/2} + 3x\left(x^2+y^2\right)^{-1/2}\left(\tfrac{1}{2}\right)(2x) = \frac{3(x^2+y^2)+3x^2}{\sqrt{x^2+y^2}} = \frac{3(2x^2+y^2)}{\sqrt{x^2+y^2}}$$ and

$z_{xy} = 3x\left(\frac{1}{2}\right)(x^2+y^2)^{-1/2}(2y) = \dfrac{3xy}{\sqrt{x^2+y^2}}$. By symmetry $z_{yx} = \dfrac{3xy}{\sqrt{x^2+y^2}}$ and $z_{yy} = \dfrac{3(x^2+2y^2)}{\sqrt{x^2+y^2}}$.

61. $z = t\sin^{-1}x \quad \Rightarrow \quad z_x = t\dfrac{1}{\sqrt{1-\left(\sqrt{x}\right)^2}}\left(\tfrac{1}{2}\right)x^{-1/2} = \dfrac{t}{2\sqrt{x-x^2}},\ z_t = \sin^{-1}\sqrt{x}$. Thus

$z_{xx} = \frac{1}{2}t\left(-\frac{1}{2}\right)(x-x^2)^{-3/2}(1-2x) = \dfrac{t(2x-1)}{4(x-x^2)^{3/2}},\ z_{xt} = \dfrac{1}{2\sqrt{x-x^2}}$,

$z_{tx} = \dfrac{1}{\sqrt{1-\left(\sqrt{x}\right)^2}}\left(\tfrac{1}{2}x^{-1/2}\right) = \dfrac{1}{2\sqrt{x-x^2}}$, and $z_{tt} = 0$.

63. $u = x^5y^4 - 3x^2y^3 + 2x^2 \quad \Rightarrow \quad u_x = 5x^4y^4 - 6xy^3 + 4x,\ u_{xy} = 20x^4y^3 - 18xy^2$ and

$u_y = 4x^5y^3 - 9x^2y^2,\ u_{yx} = 20x^4y^3 - 18xy^2$. Thus $u_{xy} = u_{yx}$.

65. $u = \sin^{-1}\left(xy^2\right) \quad \Rightarrow \quad u_x = \dfrac{1}{\sqrt{1-\left(xy^2\right)^2}}\left(y^2\right) = y^2\sqrt{1-x^2y^4}$,

$u_{xy} = 2y(1-x^2y^4)^{-1/2} + y^2\left(-\frac{1}{2}\right)(1-x^2y^4)^{-3/2}(-4x^2y^3) = \dfrac{2y(1-x^2y^4)+2x^2y^5}{(1-x^2y^4)^{3/2}} = \dfrac{2y}{(1-x^2y^4)^{3/2}}$, and

$u_y = \dfrac{1}{\sqrt{1-\left(xy^2\right)^2}}(2xy) = \dfrac{2xy}{\sqrt{1-x^2y^4}},\ u_{yx} = \dfrac{2y\sqrt{1-x^2y^4} - 2xy\left(\frac{1}{2}\right)(1-x^2y^4)^{-1/2}(-2xy^4)}{1-x^2y^4}$

$$= \frac{2y - 2x^2y^5 + 2x^2y^5}{(1-x^2y^4)^{3/2}} = \frac{2y}{(1-x^2y^4)^{3/2}}.$$

Thus $u_{xy} = u_{yx}$.

67. $f(x,y) = x^2y^3 - 2x^4y \quad \Rightarrow \quad f_x = 2xy^3 - 8x^3y,\ f_{xx} = 2y^3 - 24x^2y,\ f_{xxx} = -48xy$

69. $f(x,y,z) = x^5 + x^4y^4z^3 + yz^2 \quad \Rightarrow \quad f_x = 5x^4 + 4x^3y^4z^3,\ f_{xy} = 16x^3y^3z^3$, and $f_{xyz} = 48x^3y^3z^2$

71. $z = x\sin y \quad\Rightarrow\quad \dfrac{\partial z}{\partial x} = \sin y,\ \dfrac{\partial^2 z}{\partial y\,\partial x} = \cos y$, and $\dfrac{\partial^3 z}{\partial y^2\,\partial x} = -\sin y$.

73. $u = \ln(x + 2y^2 + 3z^3) \quad\Rightarrow\quad \dfrac{\partial u}{\partial z} = \dfrac{1}{x + 2y^2 + 3z^3}(9z^2) = \dfrac{9z^2}{x + 2y^2 + 3z^3}$,

$\dfrac{\partial^2 u}{\partial y\,\partial z} = -9z^2\left(x + 2y^2 + 3z^3\right)^{-2}(4y) = -\dfrac{36yz^2}{(x + 2y^2 + 3z^3)^2}$, and $\dfrac{\partial^3 u}{\partial x\,\partial y\,\partial z} = \dfrac{72yz^2}{(x + 2y^2 + 3z^3)^3}$.

75. $u = e^{-\alpha^2 k^2 t}\sin kx \quad\Rightarrow\quad u_x = ke^{-\alpha^2 k^2 t}\cos kx,\ u_{xx} = -k^2 e^{-\alpha^2 k^2 t}\sin kx$, and $u_t = -\alpha^2 k^2 e^{-\alpha^2 k^2 t}\sin kx$. Thus $\alpha^2 u_{xx} = u_t$.

77. $u = \dfrac{1}{\sqrt{x^2 + y^2 + z^2}} \quad\Rightarrow\quad u_x = \left(-\tfrac{1}{2}\right)(x^2 + y^2 + z^2)^{-3/2}(2x) = -x(x^2 + y^2 + z^2)^{-3/2}$ and

$u_{xx} = -(x^2 + y^2 + z^2)^{-3/2} - x\left(-\tfrac{3}{2}\right)(x^2 + y^2 + z^2)^{-5/2}(2x) = \dfrac{2x^2 - y^2 - z^2}{(x^2 + y^2 + z^2)^{5/2}}$. By symmetry,

$u_{yy} = \dfrac{2y^2 - x^2 - z^2}{(x^2 + y^2 + z^2)^{5/2}}$ and $u_{zz} = \dfrac{2z^2 - x^2 - y^2}{(x^2 + y^2 + z^2)^{5/2}}$. Thus

$$u_{xx} + u_{yy} + u_{zz} = \frac{2x^2 - y^2 - z^2 + 2y^2 - x^2 - z^2 + 2z^2 - x^2 - y^2}{(x^2 + y^2 + z^2)^{5/2}} = 0.$$

79. Let $v = x + at,\ w = x - at$. Then $u_t = \dfrac{\partial[f(v) + g(w)]}{\partial t} = \dfrac{df(v)}{dv}\dfrac{\partial v}{\partial t} + \dfrac{dg(w)}{dw}\dfrac{\partial w}{\partial t} = af'(v) - ag'(w)$ and

$u_{tt} = \dfrac{\partial[af'(v) - ag'(w)]}{\partial t} = a[af''(v) + ag''(w)] = a^2(f''(v) + g''(w))$. Similarly by using the Chain Rule we have $u_x = f'(v) + g'(w)$ and $u_{xx} = f''(v) + g''(w)$. Thus $u_{tt} = a^2 u_{xx}$.

81. $z_x = e^y + ye^x,\ z_{xx} = ye^x,\ \partial^3 z/\partial x^3 = ye^x$. By symmetry $z_y = xe^y + e^x,\ z_{yy} = xe^y$ and $\partial^3 z/\partial y^3 = xe^y$. Then $\partial^3 z/\partial x\partial y^2 = e^y$ and $\partial^3 z/\partial x^2\partial y = e^x$. Thus $z = xe^y + ye^x$ satisfies the given partial differential equation.

83. $f(x_1, \ldots, x_n) = \left(x_1^2 + \cdots + x_n^2\right)^{(2-n)/2} \quad\Rightarrow\quad \dfrac{\partial f}{\partial x_i} = \left(1 - \dfrac{n}{2}\right)2x_i\left(x_1^2 + \cdots + x_n^2\right)^{-n/2},\ 1 \le i \le n \quad\Rightarrow$

$\dfrac{\partial^2 f}{\partial x_i^2} = 2\left(1 - \dfrac{n}{2}\right)\left(x_1^2 + \cdots + x_n^2\right)^{-n/2} - (2n)\left(1 - \dfrac{n}{2}\right)\left(x_i^2\right)\left(x_1^2 + \cdots + x_n^2\right)^{-(2+n)/2},\ 1 \le i \le n$. Therefore

$$\begin{aligned}\frac{\partial^2 f}{\partial x_1^2} + \cdots + \frac{\partial^2 f}{\partial x_n^2} &= \sum_{i=1}^{n}\left[(2 - n)\left(x_1^2 + \cdots + x_n^2\right)^{-n/2} - n(2 - n)\left(x_i^2\right)\left(x_1^2 + \cdots + x_n^2\right)^{-(2+n)/2}\right] \\ &= n(2 - n)\left(x_1^2 + \cdots + x_n^2\right)^{-n/2} - n(2 - n)\left(x_1^2 + \cdots + x_n^2\right)\left(x_1^2 + \cdots + x_n^2\right)^{-(2+n)/2} \\ &= n(2 - n)\left(x_1^2 + \cdots + x_n^2\right)^{-n/2} - n(2 - n)\left(x_1^2 + \cdots + x_n^2\right)^{-n/2} = 0.\end{aligned}$$

85. By the Chain Rule, taking the partial derivative of both sides with respect to R_1 gives

$\dfrac{\partial R^{-1}}{\partial R}\dfrac{\partial R}{\partial R_1} = \dfrac{\partial[(1/R_1) + (1/R_2) + (1/R_3)]}{\partial R_1}$ or $-R^{-2}\dfrac{\partial R}{\partial R_1} = -R_1^{-2}$. Thus $\dfrac{\partial R}{\partial R_1} = \dfrac{R^2}{R_1^2}$.

87. $\dfrac{\partial K}{\partial m} = \tfrac{1}{2}V^2,\ \dfrac{\partial K}{\partial V} = mV,\ \dfrac{\partial^2 K}{\partial V^2} = m$. Thus $\dfrac{\partial K}{\partial m}\cdot\dfrac{\partial^2 K}{\partial V^2} = \tfrac{1}{2}V^2 m = K$.

89. $f_x(x, y) = x + 4y \quad\Rightarrow\quad f_{xy}(x, y) = 4$ and $f_y(x, y) = 3x - y \quad\Rightarrow\quad f_{yx}(x, y) = 3$. Since f_{xy} and f_{yx} are continuous everywhere but $f_{xy}(x, y) \ne f_{yx}(x, y)$, Clairaut's Theorem implies that such a function $f(x, y)$ does not exist.

91. By the geometry of partial derivatives, the slope of the tangent line is $f_x(1,2)$. By implicit differentiation of $4x^2 + 2y^2 + z^2 = 16$, we get $8x + 2z(\partial z/\partial x) = 0 \quad \Rightarrow \quad \partial z/\partial x = -4x/z$, so when $x = 1$ and $z = 2$ we have $\partial z/\partial x = -2$. So the slope is $f_x(1,2) = -2$. Thus the tangent line is given by $z - 2 = -2(x-1)$, $y = 2$. Taking the parameter to be $t = x - 1$, we can write parametric equations for this line: $x = 1 + t$, $y = 2$, $z = 2 - 2t$.

93. By Clairaut's Theorem, $f_{xyy} = (f_{xy})_y = (f_{yx})_y = f_{yxy} = (f_y)_{xy} = (f_y)_{yx} = f_{yyx}$.

95. Let $g(x) = f(x,0) = x(x^2)^{-3/2}e^0 = x|x|^{-3}$. But we are using the point $(1,0)$, so near $(1,0)$, $g(x) = x^{-2}$. Then $g'(x) = -2x^{-3}$ and $g'(1) = -2$, so using (1) we have $f_x(1,0) = g'(1) = -2$.

97. **(a)**

(b) For $(x,y) \neq (0,0)$,

$$f_x(x,y) = \frac{(3x^2y - y^3)(x^2+y^2) - (x^3y - xy^3)(2x)}{(x^2+y^2)^2} = \frac{x^4y + 4x^2y^3 - y^5}{(x^2+y^2)^2}$$

and by symmetry

$$f_y(x,y) = \frac{x^5 - 4x^3y^2 - xy^4}{(x^2+y^2)^2}.$$

(c) $f_x(0,0) = \lim\limits_{h\to 0} \dfrac{f(h,0) - f(0,0)}{h} = \lim\limits_{h\to 0} \dfrac{(0/h^2) - 0}{h} = 0$ and $f_y(0,0) = \lim\limits_{h\to 0} \dfrac{f(0,h) - f(0,0)}{h} = 0$.

(d) Using (3), $f_{xy}(0,0) = \dfrac{\partial f_x}{\partial y} = \lim\limits_{h\to 0} \dfrac{f_x(0,h) - f_x(0,0)}{h} = \lim\limits_{h\to 0} \dfrac{(-h^5 - 0)/h^4}{h} = -1$ while by (2),

$$f_{yx}(0,0) = \frac{\partial f_y}{\partial x} = \lim_{h\to 0} \frac{f_y(h,0) - f_y(0,0)}{h} = \lim_{h\to 0} \frac{h^5/h^4}{h} = 1.$$

(e) For $(x,y) \neq (0,0)$, we use a CAS to compute that $f_{xy}(x,y) = \dfrac{x^6 + 9x^4y^2 - 4x^2y^4 + 4y^6}{(x^2+y^2)^3}$. Now as $(x,y) \to (0,0)$ along the x-axis, $f_{xy}(x,y) \to 1$ while as $(x,y) \to (0,0)$ along the y-axis, $f_{xy}(x,y) \to 4$. Thus f_{xy} isn't continuous at $(0,0)$ and Clairaut's Theorem doesn't apply, so there is no contradiction. The graphs of f_{xy} and f_{yx} are identical except at the origin, where we observe the discontinuity.

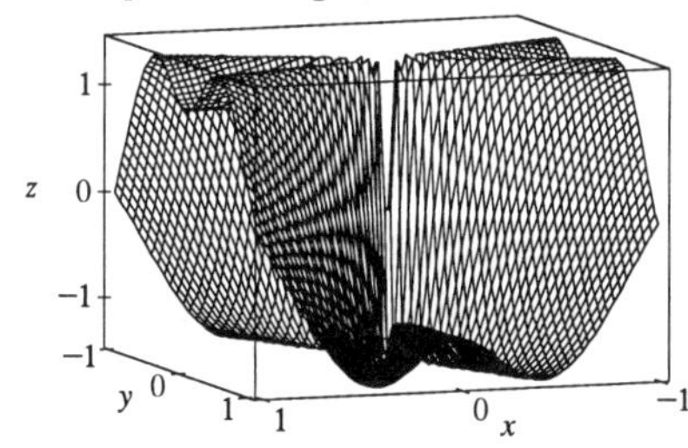

EXERCISES 12.4

1. $z = f(x,y) = x^2 + 4y^2 \quad\Rightarrow\quad f_x(x,y) = 2x$, $f_y(x,y) = 8y$, $f_x(2,1) = 4$, $f_y(2,1) = 8$. Thus the equation of the tangent plane is $z - 8 = 4(x-2) + 8(y-1)$ or $4x + 8y - z = 8$.

3. $z = f(x,y) = 5 + (x-1)^2 + (y+2)^2 \quad\Rightarrow\quad f_x(x,y) = 2(x-1)$, $f_y(x,y) = 2(y+2)$, $f_x(2,0) = 2$, $f_y(2,0) = 4$ and the equation is $z - 10 = 2(x-2) + 4y$ or $2x + 4y - z = -6$.

5. $z = f(x,y) = \ln(2x+y) \quad\Rightarrow\quad f_x(x,y) = \dfrac{2}{2x+y}$, $f_y(x,y) = \dfrac{1}{2x+y}$, $f_x(-1,3) = 2$, $f_y(-1,3) = 1$. Thus the equation of the tangent plane is $z = 2(x+1) + (y-3)$ or $2x + y - z = 1$.

7. $z = f(x,y) = xy$, so $f_x(-1,2) = 2$, $f_y(-1,2) = -1$ and the equation of the tangent plane is $z + 2 = 2(x+1) + (-1)(y-2)$ or $2x - y - z = -2$.

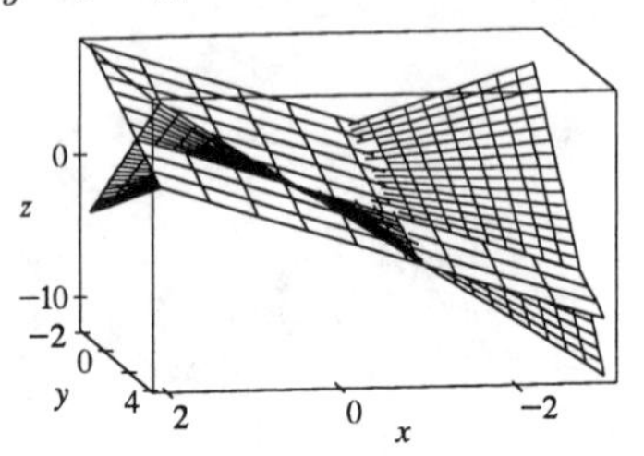

9. $f(x,y) = e^{-(x^2+y^2)/15}(\sin^2 x + \cos^2 y)$. A CAS gives $f_x = -\frac{2}{15}e^{-(x^2+y^2)/15}(x\sin^2 x + x\cos^2 y - 15\sin x\cos x)$ and $f_y = -\frac{2}{15}e^{-(x^2+y^2)/15}(y\sin^2 x + y\cos^2 y + 15\sin y\cos y)$. We use the CAS to evaluate these at $(2,3)$, and then substitute the results into Equation 2 in order to plot the tangent plane.

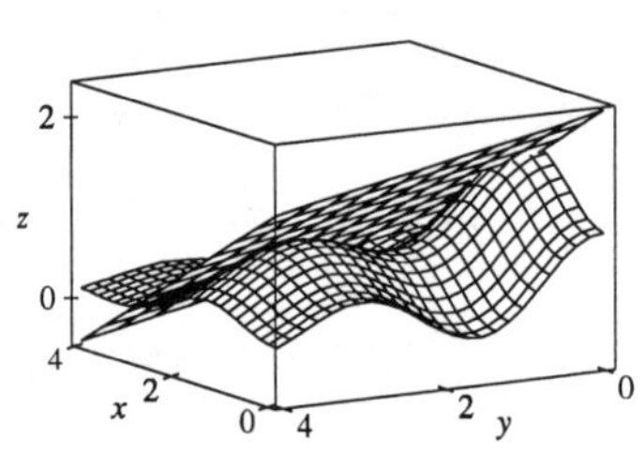

11. $z = x^2y^3 \quad\Rightarrow\quad dz = \dfrac{\partial z}{\partial x}dx + \dfrac{\partial z}{\partial y}dy = 2xy^3dx + 3x^2y^2dy$

13. $u = e^x\cos xy \quad\Rightarrow\quad du = e^x(\cos xy - y\sin xy)dx - (xe^x\sin xy)dy$

15. $w = x^2y + y^2z \quad\Rightarrow\quad dw = \dfrac{\partial w}{\partial x}dx + \dfrac{\partial w}{\partial y}dy + \dfrac{\partial w}{\partial z}dz = 2xy\,dx + \left(x^2 + 2yz\right)dy + y^2dz$

17. $w = \ln\sqrt{x^2+y^2+z^2} \quad\Rightarrow$

$$dw = \left(\frac{1}{2}\right)\frac{2x(x^2+y^2+z^2)^{-1/2}dx + 2y(x^2+y^2+z^2)^{-1/2}dy + 2z(x^2+y^2+z^2)^{-1/2}dz}{(x^2+y^2+z^2)^{1/2}} = \frac{x\,dx + y\,dy + z\,dz}{x^2+y^2+z^2}$$

19. $\Delta x = 0.05$, $\Delta y = 0.1$, $z = 5x^2 + y^2$, $z_x = 10x$, $z_y = 2y$. Thus when $x = 1$, $y = 2$, $dz = (10)(0.05) + (4)(0.1) = 0.9$, while $\Delta z = f(1.05, 2.1) - f(1,2) = 5(1.05)^2 + (2.1)^2 - 5 - 4 = 0.9225$.

21. $f(x,y) = \sqrt{20 - x^2 - 7y^2} \quad\Rightarrow\quad f_x = -\dfrac{x}{\sqrt{20-x^2-7y^2}}$ and $f_y = -\dfrac{7y}{\sqrt{20-x^2-7y^2}}$. Since $f(2,1) = \sqrt{20-4-7} = 3$, we set $(a,b) = (2,1)$. Then $\Delta x = -0.05$, $\Delta y = 0.08$. Thus $f(1.95, 1.08) \approx f(2,1) + dz = 3 + \left(-\frac{2}{3}\right)(-0.05) + \left(-\frac{7}{3}\right)(0.08) = 2.84\overline{6}$.

23. $f(x,y,z) = x^2y^3z^4 \quad\Rightarrow\quad f_x = 2xy^3z^4$, $f_y = 3x^2y^2z^4$ and $f_z = 4x^2y^3z^3$. Since $f(1,1,3) = 81$, we set $(a,b,c) = (1,1,3)$. Then $\Delta x = 0.05$, $\Delta y = -0.1$, $\Delta z = 0.01$, and so $f(1.05, 0.9, 3.01) \approx f(1,1,3) + dw = 81 + (162)(0.05) + (243)(-0.1) + (108)(0.01) = 65.88$.

25. Let $w = f(x,y,z) = x\sqrt{y-z^3} \quad\Rightarrow\quad f_x = \sqrt{y-z^3}$, $f_y = \dfrac{x}{2\sqrt{y-z^3}}$, and $f_z = -\dfrac{3xz^2}{2\sqrt{y-z^3}}$. Then $f(9,10,1) = 27$, so we set $(a,b,c) = (9,10,1)$. Then $\Delta x = -0.06$, $\Delta y = -0.01$, and $\Delta z = 0.01$. Thus $8.94\sqrt{9.99-(1.01)^3} \approx 27 + (3)(-0.06) + \frac{9}{6}(-0.01) + \left(-\frac{27}{6}\right)(0.01) = 26.76$.

27. Let $z = f(x,y) = \sqrt{x}e^y \quad\Rightarrow\quad f_x = \dfrac{e^y}{2\sqrt{x}}$, $f_y = \sqrt{x}e^y$. Now $f(1,0) = 1$, so we set $(a,b) = (1,0)$, $\Delta x = -0.01$, $\Delta y = 0.02$. Thus $\sqrt{0.99}e^{0.02} \approx 1 + \frac{1}{2}(-0.01) + 1(0.02) = 1.015$.

29. $dA = \dfrac{\partial A}{\partial x}\,dx + \dfrac{\partial A}{\partial y}\,dy = y\,dx + x\,dy$ and $|\Delta x| \le 0.1$, $|\Delta y| \le 0.1$. Thus the maximum error in the area is about $dA = 24(0.1) + 30(0.1) = 5.4\text{ cm}^2$.

31. The volume of a can is $V = \pi r^2 h$ and $\Delta V \approx dV$ is an estimate of the amount of tin. Here $dV = 2\pi rh\,dr + \pi r^2\,dh$, so $\Delta V \approx dV = 2\pi(48)(-0.04) + \pi(16)(-0.08) \approx -16.08\text{ cm}^3$. Thus the amount of tin is about 16 cm^3.

33. The area of the rectangle is $A = xy$, and $\Delta A \approx dA$ is an estimate of the area of paint in the stripe. Here $dA = y\,dx + x\,dy$, so $\Delta A \approx dA = (100)\left(\frac{1}{2}\right) + (200)\left(\frac{1}{2}\right) = 150\text{ ft}^2$. Thus there are approximately 150 ft^2 of paint in the stripe.

35. By the Chain Rule, taking partial derivatives of both sides with respect to R_1 gives

$$\frac{\partial R^{-1}}{\partial R}\frac{\partial R}{\partial R_1} = \frac{\partial[(1/R_1)+(1/R_2)+(1/R_3)]}{\partial R_1} \quad\Rightarrow\quad -R^{-2}\frac{\partial R}{\partial R_1} = -R_1^{-2} \quad\Rightarrow\quad \frac{\partial R}{\partial R_1} = \frac{R^2}{R_1^2},$$

and by symmetry, $\dfrac{\partial R}{\partial R_2} = \dfrac{R^2}{R_2^2}$, $\dfrac{\partial R}{\partial R_3} = \dfrac{R^2}{R_3^2}$. When $R_1 = 25$, $R_2 = 40$ and $R_3 = 50$, $\dfrac{1}{R} = \dfrac{17}{200} \quad\Leftrightarrow\quad R = \dfrac{200}{17}$ ohms. Since the possible error for each R_i is 0.5%, the maximum error of R is attained by setting $\Delta R_i = 0.005R_i$. So

$$\begin{aligned}\Delta R \approx dR &= \frac{\partial R}{\partial R_1}\Delta R_1 + \frac{\partial R}{\partial R_2}\Delta R_2 + \frac{\partial R}{\partial R_3}\Delta R_3 = (0.005)R^2\left[\frac{1}{R_1} + \frac{1}{R_2} + \frac{1}{R_3}\right] = (0.005)R \\ &= \tfrac{1}{17} \approx 0.059\text{ ohms.}\end{aligned}$$

37.

$$\begin{aligned}\Delta z &= f(a+\Delta x, b+\Delta y) - f(a,b) = (a+\Delta x)^2 + (b+\Delta y)^2 - (a^2+b^2) \\ &= a^2 + 2a\Delta x + (\Delta x)^2 + b^2 + 2b\Delta y + (\Delta y)^2 - a^2 - b^2 = 2a\Delta x + (\Delta x)^2 + 2b\Delta y + (\Delta y)^2.\end{aligned}$$

But $f_x(a,b) = 2a$ and $f_y(a,b) = 2b$ and so $\Delta z = f_x(a,b)\Delta x + f_y(a,b)\Delta y + \Delta x\Delta x + \Delta y\Delta y$, which is Definition 12 with $\epsilon_1 = \Delta x$ and $\epsilon_2 = \Delta y$. Hence f is differentiable.

39. To show that f is continuous at (a,b) we need to show that $\displaystyle\lim_{(x,y)\to(a,b)} f(x,y) = f(a,b)$ or equivalently $\displaystyle\lim_{(\Delta x,\Delta y)\to(0,0)} f(a+\Delta x, b+\Delta y) = f(a,b)$. Since f is differentiable at (a,b), $f(a+\Delta x, b+\Delta y) - f(a,b) = \Delta z = f_x(a,b)\Delta x + f_y(a,b)\Delta y + \epsilon_1\Delta x + \epsilon_2\Delta y$, where ϵ_1 and $\epsilon_2 \to 0$ as $(\Delta x,\Delta y) \to (0,0)$. Thus $f(a+\Delta x, b+\Delta y) = f(a,b) + f_x(a,b)\Delta x + f_y(a,b)\Delta y + \epsilon_1\Delta x + \epsilon_2\Delta y$. Taking the limit of both sides as $(\Delta x,\Delta y) \to (0,0)$ gives $\displaystyle\lim_{(\Delta x,\Delta y)\to(0,0)} f(a+\Delta x, b+\Delta y) = f(a,b)$. Thus f is continuous at (a,b).

EXERCISES 12.5

1. $z = x^2 + y^2,\ x = t^3,\ y = 1 + t^2 \quad \Rightarrow \quad \dfrac{dz}{dt} = 2x\dfrac{dx}{dt} + 2y\dfrac{dy}{dt} = (2t^3)(3t^2) + 2(1+t^2)(2t) = 6t^5 + 4t^3 + 4t$

3. $z = \ln(x + y^2),\ x = \sqrt{1+t},\ y = 1 + \sqrt{t} \quad \Rightarrow$

$$\frac{dz}{dt} = \frac{1}{(x+y^2)}\frac{1}{2\sqrt{1+t}} + \frac{1}{(x+y^2)}2y\frac{1}{2\sqrt{t}} = \frac{1}{\sqrt{1+t}+1+\sqrt{t}}\left(\frac{1}{2\sqrt{1+t}} + \frac{1+\sqrt{t}}{\sqrt{t}}\right)$$

5. $w = xy^2z^3,\ x = \sin t,\ y = \cos t,\ z = 1 + e^{2t} \quad \Rightarrow \quad \dfrac{dw}{dt} = y^2z^3(\cos t) + 2xyz^3(-\sin t) + 3xy^2z^2(2e^{2t})$

7. $z = x^2 \sin y,\ x = s^2 + t^2,\ y = 2st \quad \Rightarrow \quad \dfrac{\partial z}{\partial s} = (2x\sin y)(2s) + (x^2\cos y)(2t) = 4sx\sin y + 2tx^2\cos y,$

$\dfrac{\partial z}{\partial t} = (2x\sin y)(2t) + (x^2\cos y)(2s) = 4xt\sin y + 2sx^2\cos y$

9. $z = x^2 - 3x^2y^3,\ x = se^t,\ y = se^{-t} \quad \Rightarrow$

$\partial z/\partial s = (2x - 6xy^3)(e^t) + (-9x^2y^2)(e^{-t}) = (2x - 6xy^3)e^t - 9x^2y^2e^{-t},$

$\partial z/\partial t = (2x - 6xy^3)(se^t) + (-9x^2y^2)(-se^{-t}) = (2x - 6xy^3)se^t + 9x^2y^2se^{-t}$

11. $z = 2^{x-3y},\ x = s^2t,\ y = st^2 \quad \Rightarrow \quad \partial z/\partial s = (z\ln 2)(2st) + z(-3\ln 2)(t^2) = (2^{x-3y}\ln 2)(2st - 3t^2),$

$\partial z/\partial t = (z\ln 2)(s^2) + z(-3\ln 2)(2st) = (2^{x-3y}\ln 2)(s^2 - 6st)$

13. $u = f(x,y),\ x = x(r,s,t),\ y = y(r,s,t) \quad \Rightarrow$

$$\frac{\partial u}{\partial r} = \frac{\partial u}{\partial x}\frac{\partial x}{\partial r} + \frac{\partial u}{\partial y}\frac{\partial y}{\partial r},\ \frac{\partial u}{\partial s} = \frac{\partial u}{\partial x}\frac{\partial x}{\partial s} + \frac{\partial u}{\partial y}\frac{\partial y}{\partial s},\ \frac{\partial u}{\partial t} = \frac{\partial u}{\partial x}\frac{\partial x}{\partial t} + \frac{\partial u}{\partial y}\frac{\partial y}{\partial t}$$

15. $v = f(p,q,r),\ p = p(x,y,z),\ q = q(x,y,z),\ r = r(x,y,z) \quad \Rightarrow$

$$\frac{\partial v}{\partial x} = \frac{\partial v}{\partial p}\frac{\partial p}{\partial x} + \frac{\partial v}{\partial q}\frac{\partial q}{\partial x} + \frac{\partial v}{\partial r}\frac{\partial r}{\partial x},\ \frac{\partial v}{\partial y} = \frac{\partial v}{\partial p}\frac{\partial p}{\partial y} + \frac{\partial v}{\partial q}\frac{\partial q}{\partial y} + \frac{\partial v}{\partial r}\frac{\partial r}{\partial y},\ \frac{\partial v}{\partial z} = \frac{\partial v}{\partial p}\frac{\partial p}{\partial z} + \frac{\partial v}{\partial q}\frac{\partial q}{\partial z} + \frac{\partial v}{\partial r}\frac{\partial r}{\partial z}$$

17. $w = x^2 + y^2 + z^2,\ x = st,\ y = s\cos t,\ z = s\sin t \quad \Rightarrow$

$\dfrac{\partial w}{\partial s} = \dfrac{\partial w}{\partial x}\dfrac{\partial x}{\partial s} + \dfrac{\partial w}{\partial y}\dfrac{\partial y}{\partial s} + \dfrac{\partial w}{\partial z}\dfrac{\partial z}{\partial s} = 2xt + 2y\cos t + 2z\sin t$. When $s = 1,\ t = 0$, we have $x = 0,\ y = 1$ and $z = 0$, so $\partial w/\partial s = 2\cos 0 = 2$. Similarly $\partial w/\partial t = 2xs + 2y(-s\sin t) + 2z(s\cos t) = 0 + (-2)\sin 0 + 0 = 0$, when $s = 1,\ t = 0$.

19. $z = y^2\tan x,\ x = t^2uv,\ y = u + tv^2 \quad \Rightarrow$

$\partial z/\partial t = (y^2\sec^2 x)2tuv + (2y\tan x)v^2,\ \partial z/\partial u = (y^2\sec^2 x)t^2v + 2y\tan x,$

$\partial z/\partial v = (y^2\sec^2 x)t^2u + (2y\tan x)2tv$. When $t = 2,\ u = 1$ and $v = 0$, we have $x = 0,\ y = 1$, so $\partial z/\partial t = 0$, $\partial z/\partial u = 0,\ \partial z/\partial v = 4$.

21. $u = \dfrac{x+y}{y+z},\ x = p + r + t,\ y = p - r + t,\ z = p + r - t \quad \Rightarrow$

$$\frac{\partial u}{\partial p} = \frac{1}{y+z} + \frac{(y+z)-(x+y)}{(y+z)^2} - \frac{x+y}{(y+z)^2} = \frac{(y+z)+(z-x)-(x+y)}{(y+z)^2} = 2\frac{z-x}{(y+z)^2} = 2\frac{-2t}{4p^2} = -\frac{t}{p^2},$$

$$\frac{\partial u}{\partial r} = \frac{1}{y+z} + \frac{z-x}{(y+z)^2}(-1) - \frac{x+y}{(y+z)^2} = 0, \text{ and}$$
$$\frac{\partial u}{\partial t} = \frac{1}{y+z} + \frac{z-x}{(y+z)^2} + \frac{x+y}{(y+z)^2} = 2\frac{y+z}{(y+z)^2} = \frac{2}{2p} = \frac{1}{p}.$$

23. $x^2 - xy + y^3 = 8$, so let $F(x,y) = x^2 - xy + y^3 - 8 = 0$. Then $\dfrac{dy}{dx} = -\dfrac{F_x}{F_y} = \dfrac{-(2x-y)}{-x+3y^2} = \dfrac{y-2x}{3y^2-x}$.

25. $2y^2 + \sqrt[3]{xy} = 3x^2 + 17$, so let $F(x,y) = 2y^2 + \sqrt[3]{xy} - 3x^2 - 17 = 0$. Then
$$\frac{dy}{dx} = -\frac{y\big/\left[3(xy)^{2/3}\right] - 6x}{4y + x\big/\left[3(xy)^{2/3}\right]} = \frac{18x - x^{-2/3}y^{1/3}}{12y + x^{1/3}y^{-2/3}}.$$

27. Let $F(x,y,z) = xy + yz - xz = 0$. Then $\dfrac{\partial z}{\partial x} = -\dfrac{F_x}{F_z} = -\dfrac{y-z}{y-x} = \dfrac{z-y}{y-x}$, $\dfrac{\partial z}{\partial y} = -\dfrac{F_y}{F_z} = -\dfrac{x+z}{y-x} = \dfrac{x+z}{x-y}$.

29. $x^2 + y^2 - z^2 = 2x(y+z)$. Let $F(x,y,z) = x^2 + y^2 - z^2 - 2x(y+z) = 0$. Then
$$\frac{\partial z}{\partial x} = -\frac{F_x}{F_z} = -\frac{2x-2y-2z}{-2z-2x} = \frac{x-y-z}{z+x},\quad \frac{\partial z}{\partial y} = -\frac{F_y}{F_z} = -\frac{2y-2x}{-2z-2x} = \frac{y-x}{z+x}$$

31. Let $F(x,y,z) = xe^y + yz + ze^x = 0$. Then $\dfrac{\partial z}{\partial x} = -\dfrac{F_x}{F_z} = -\dfrac{e^y + ze^x}{y+e^x}$, $\dfrac{\partial z}{\partial y} = -\dfrac{F_y}{F_z} = -\dfrac{xe^y + z}{y+e^x}$.

33. $dr/dt = -1.2$, $dh/dt = 3$, $V = \pi r^2 h$ and $dV/dt = 2\pi rh(dr/dt) + \pi r^2(dh/dt)$. Thus when $r = 80$ and $h = 150$, $dV/dt = (-28{,}800)\pi + (19{,}200)\pi = -9600\pi \text{ cm}^3/\text{s}$.

35. **(a)** $V = \ell wh$, so by the Chain Rule, $\dfrac{dV}{dt} = \dfrac{\partial V}{\partial \ell}\dfrac{d\ell}{dt} + \dfrac{\partial V}{\partial w}\dfrac{dw}{dt} + \dfrac{\partial V}{\partial h}\dfrac{dh}{dt} = wh\dfrac{d\ell}{dt} + \ell h\dfrac{dw}{dt} + \ell w\dfrac{dh}{dt}$
$= 2\cdot 2\cdot 2 + 1\cdot 2\cdot 2 + 1\cdot 2\cdot(-3) = 6 \text{ m}^3/\text{s}$.

(b) $S = 2(\ell w + \ell h + wh)$, so by the Chain Rule,
$$\frac{dS}{dt} = \frac{\partial S}{\partial \ell}\frac{d\ell}{dt} + \frac{\partial S}{\partial w}\frac{dw}{dt} + \frac{\partial S}{\partial h}\frac{dh}{dt} = 2(w+h)\frac{d\ell}{dt} + 2(\ell+h)\frac{dw}{dt} + 2(\ell+w)\frac{dh}{dt}$$
$$= 2(2+2)2 + 2(1+2)2 + 2(1+2)(-3) = 10 \text{ m}^2/\text{s}.$$

(c) $L^2 = \ell^2 + w^2 + h^2 \quad\Rightarrow\quad 2L\dfrac{dL}{dt} = 2\ell\dfrac{d\ell}{dt} + 2w\dfrac{dw}{dt} + 2h\dfrac{dh}{dt} = 2(1)(2) + 2(2)(2) + 2(2)(-3) = 0$
$\Rightarrow\quad dL/dt = 0 \text{ m/s}$.

37. $\dfrac{dP}{dt} = 0.05$, $\dfrac{dT}{dt} = 0.15$, $V = 8.31\dfrac{T}{P}$ and $\dfrac{dV}{dt} = \dfrac{8.31}{P}\dfrac{dT}{dt} - 8.31\dfrac{T}{P^2}\dfrac{dP}{dt}$. Thus when $P = 20$ and $T = 320°$,
$$\frac{dV}{dt} = 8.31\left[\frac{0.15}{20} - \frac{(0.05)(320)}{400}\right] \approx -0.27 \text{ L/s}.$$

39. **(a)** Using the Chain Rule, $\dfrac{\partial z}{\partial r} = \dfrac{\partial z}{\partial x}\cos\theta + \dfrac{\partial z}{\partial y}\sin\theta$, $\dfrac{\partial z}{\partial \theta} = \dfrac{\partial z}{\partial x}(-r\sin\theta) + \dfrac{\partial z}{\partial y}r\cos\theta$.

(b) $\left(\dfrac{\partial z}{\partial r}\right)^2 = \left(\dfrac{\partial z}{\partial x}\right)^2\cos^2\theta + 2\dfrac{\partial z}{\partial x}\dfrac{\partial z}{\partial y}\cos\theta\sin\theta + \left(\dfrac{\partial z}{\partial y}\right)^2\sin^2\theta$,

$\left(\dfrac{\partial z}{\partial \theta}\right)^2 = \left(\dfrac{\partial z}{\partial x}\right)^2 r^2\sin^2\theta - 2\dfrac{\partial z}{\partial x}\dfrac{\partial z}{\partial y}r^2\cos\theta\sin\theta + \left(\dfrac{\partial z}{\partial y}\right)^2 r^2\cos^2\theta$. Thus

$$\left(\frac{\partial z}{\partial r}\right)^2 + \frac{1}{r^2}\left(\frac{\partial z}{\partial \theta}\right)^2 = \left[\left(\frac{\partial z}{\partial x}\right)^2 + \left(\frac{\partial z}{\partial y}\right)^2\right](\cos^2\theta + \sin^2\theta) = \left(\frac{\partial z}{\partial x}\right)^2 + \left(\frac{\partial z}{\partial y}\right)^2.$$

41. Let $u = x - y$. Then $\dfrac{\partial z}{\partial x} = \dfrac{dz}{du}\dfrac{\partial u}{\partial x} = \dfrac{dz}{du}$ and $\dfrac{\partial z}{\partial y} = \dfrac{dz}{du}(-1)$. Thus $\dfrac{\partial z}{\partial x} + \dfrac{\partial z}{\partial y} = 0$.

43. Let $u = x + at$, $v = x - at$. Then $z = f(u) + g(v)$, so $\partial z/\partial u = f'(u)$ and $\partial z/\partial v = g'(v)$. Thus

$\dfrac{\partial z}{\partial t} = \dfrac{\partial z}{\partial u}\dfrac{\partial u}{\partial t} + \dfrac{\partial z}{\partial v}\dfrac{\partial v}{\partial t} = af'(u) - ag'(v)$ and $\dfrac{\partial^2 z}{\partial t^2} = a\dfrac{\partial}{\partial t}[f'(u) - g'(v)] = a\left(\dfrac{df'(u)}{du}\dfrac{\partial u}{\partial t} - \dfrac{dg'(v)}{dv}\dfrac{\partial v}{\partial t}\right)$

$= a^2 f''(u) + a^2 g''(v)$. Similarly $\dfrac{\partial z}{\partial x} = f'(u) + g'(v)$ and $\dfrac{\partial^2 z}{\partial x^2} = f''(u) + g''(v)$. Thus $\dfrac{\partial^2 z}{\partial t^2} = a^2\dfrac{\partial^2 z}{\partial x^2}$.

45. $\dfrac{\partial z}{\partial s} = \dfrac{\partial z}{\partial x}2s + \dfrac{\partial z}{\partial y}2r$. Then

$$\frac{\partial^2 z}{\partial r\partial s} = \frac{\partial}{\partial r}\left(\frac{\partial z}{\partial x}2s\right) + \frac{\partial}{\partial r}\left(\frac{\partial z}{\partial y}2r\right)$$

$$= \frac{\partial^2 z}{\partial x^2}\frac{\partial x}{\partial r}2s + \frac{\partial}{\partial y}\left(\frac{\partial z}{\partial x}\right)\frac{\partial y}{\partial r}2s + \frac{\partial z}{\partial x}\frac{\partial}{\partial r}(2s) + \frac{\partial^2 z}{\partial y^2}\frac{\partial y}{\partial r}2r + \frac{\partial}{\partial x}\left(\frac{\partial z}{\partial y}\right)\frac{\partial x}{\partial r}2r + \frac{\partial z}{\partial y}2$$

$$= 4rs\frac{\partial^2 z}{\partial x^2} + \frac{\partial^2 z}{\partial y\partial x}4s^2 + 0 + 4rs\frac{\partial^2 z}{\partial y^2} + \frac{\partial^2 z}{\partial x\partial y}4r^2 + 2\frac{\partial z}{\partial y}.$$

And by the continuity of the partials, $\dfrac{\partial^2 z}{\partial r\partial s} = 4rs\dfrac{\partial^2 z}{\partial x^2} + 4rs\dfrac{\partial^2 z}{\partial y^2} + (4r^2 + 4s^2)\dfrac{\partial^2 z}{\partial x\partial y} + 2\dfrac{\partial z}{\partial y}$.

47. $\dfrac{\partial z}{\partial r} = \dfrac{\partial z}{\partial x}\cos\theta + \dfrac{\partial z}{\partial y}\sin\theta$ and $\dfrac{\partial z}{\partial \theta} = -\dfrac{\partial z}{\partial x}r\sin\theta + \dfrac{\partial z}{\partial y}r\cos\theta$. Then

$$\frac{\partial^2 z}{\partial r^2} = \cos\theta\left(\frac{\partial^2 z}{\partial x^2}\cos\theta + \frac{\partial^2 z}{\partial y\partial x}\sin\theta\right) + \sin\theta\left(\frac{\partial^2 z}{\partial y^2}\sin\theta + \frac{\partial^2 z}{\partial x\partial y}\cos\theta\right)$$

$$= \cos^2\theta\,\frac{\partial^2 z}{\partial x^2} + 2\cos\theta\sin\theta\,\frac{\partial^2 z}{\partial x\partial y} + \sin^2\theta\,\frac{\partial^2 z}{\partial y^2} \text{ and}$$

$$\frac{\partial^2 z}{\partial \theta^2} = -r\cos\theta\,\frac{\partial z}{\partial x} + (-r\sin\theta)\left(\frac{\partial^2 z}{\partial x^2}(-r\sin\theta) + \frac{\partial^2 z}{\partial y\partial x}r\cos\theta\right)$$

$$-r\sin\theta\,\frac{\partial z}{\partial y} + r\cos\theta\left(\frac{\partial^2 z}{\partial y^2}r\cos\theta + \frac{\partial^2 z}{\partial x\partial y}(-r\sin\theta)\right)$$

$$= -r\cos\theta\,\frac{\partial z}{\partial x} - r\sin\theta\,\frac{\partial z}{\partial y} + r^2\sin^2\theta\,\frac{\partial^2 z}{\partial x^2} - 2r^2\cos\theta\sin\theta\,\frac{\partial^2 z}{\partial x\partial y} + r^2\cos^2\theta\,\frac{\partial^2 z}{\partial y^2}.\text{ Thus}$$

$$\frac{\partial^2 z}{\partial r^2} + \frac{1}{r^2}\frac{\partial^2 z}{\partial \theta^2} + \frac{1}{r}\frac{\partial z}{\partial r} = (\cos^2\theta + \sin^2\theta)\frac{\partial^2 z}{\partial x^2} + (\sin^2\theta + \cos^2\theta)\frac{\partial^2 z}{\partial y^2} - \frac{1}{r}\cos\theta\,\frac{\partial z}{\partial x}$$

$$-\frac{1}{r}\sin\theta\,\frac{\partial z}{\partial y} + \frac{1}{r}\left(\cos\theta\,\frac{\partial z}{\partial x} + \sin\theta\,\frac{\partial z}{\partial y}\right) = \frac{\partial^2 z}{\partial x^2} + \frac{\partial^2 z}{\partial y^2} \text{ as desired.}$$

49. Differentiating both sides of $f(tx, ty) = t^n f(x, y)$ with respect to t using the Chain Rule, we get

$$\frac{\partial}{\partial t}f(tx, ty) = \frac{\partial}{\partial t}[t^n f(x, y)] \quad\Leftrightarrow\quad \frac{\partial}{\partial(tx)}f(tx, ty)\cdot\frac{\partial(tx)}{\partial t} + \frac{\partial}{\partial(ty)}f(tx, ty)\cdot\frac{\partial(ty)}{\partial t}$$

$= x\dfrac{\partial}{\partial(tx)}f(tx, ty) + y\dfrac{\partial}{\partial(ty)}f(tx, ty) = nt^{n-1}f(x, y)$. Setting $t = 1$: $x\dfrac{\partial}{\partial x}f(x, y) + y\dfrac{\partial}{\partial y}f(x, y) = nf(x, y)$.

51. Differentiating both sides of $f(tx, ty) = t^n f(x, y)$ with respect to x using the Chain Rule, we get

$$\frac{\partial}{\partial x}f(tx, ty) = \frac{\partial}{\partial x}[t^n f(x, y)] \quad\Leftrightarrow\quad \frac{\partial}{\partial(tx)}f(tx, ty)\cdot\frac{\partial(tx)}{\partial x} + \frac{\partial}{\partial(ty)}f(tx, ty)\cdot\frac{\partial(ty)}{\partial x} = t^n\frac{\partial}{\partial x}f(x, y) \quad\Leftrightarrow$$

$tf_x(tx, ty) = t^n f_x(x, y)$. Thus $f_x(tx, ty) = t^{n-1}f_x(x, y)$.

EXERCISES 12.6

1. $f(x,y) = x^2y^3 + 2x^4y \quad \Rightarrow \quad D_{\mathbf{u}}f(x,y) = (2xy^3 + 8x^3y)\cos\frac{\pi}{3} + (3x^2y^2 + 2x^4)\sin\frac{\pi}{3}$. Thus $D_{\mathbf{u}}f(1,-2) = (-16-16)\left(\frac{1}{2}\right) + (12+2)\left(\frac{\sqrt{3}}{2}\right) = 7\sqrt{3} - 16$.

3. $f(x,y) = y^x \quad \Rightarrow \quad D_{\mathbf{u}}f(x,y) = (y^x \ln y)\cos\frac{\pi}{2} + (xy^{x-1})\sin\frac{\pi}{2} = xy^{x-1}$. Thus $D_{\mathbf{u}}f(1,2) = (1)(2)^{1-1} = 1$.

5. $f(x,y) = x^3 - 4x^2y + y^2$

(a) $\nabla f(x,y) = f_x\mathbf{i} + f_y\mathbf{j} = (3x^2 - 8xy)\mathbf{i} + (2y - 4x^2)\mathbf{j}$

(b) $\nabla f(0,-1) = -2\mathbf{j}$

(c) $\nabla f(0,-1)\cdot\mathbf{u} = -\frac{8}{5}$

7. $f(x,y,z) = xy^2z^3$

(a) $\nabla f(x,y,z) = f_x\mathbf{i} + f_y\mathbf{j} + f_z\mathbf{k} = y^2z^3\mathbf{i} + 2xyz^3\mathbf{j} + 3xy^2z^2\mathbf{k}$

(b) $\nabla f(1,-2,1) = 4\mathbf{i} - 4\mathbf{j} + 12\mathbf{k}$

(c) $\nabla f(1,-2,1)\cdot\mathbf{u} = \frac{1}{\sqrt{3}}(20) = \frac{20}{\sqrt{3}}$

9. $f(x,y) = \sqrt{x-y} \quad \Rightarrow \quad \nabla f(x,y) = \left\langle \frac{1}{2}(x-y)^{-1/2}, -\frac{1}{2}(x-y)^{-1/2}\right\rangle$, $\nabla f(5,1) = \left\langle \frac{1}{4}, -\frac{1}{4}\right\rangle$, and a unit vector in the direction of $\mathbf{v}$ is $\mathbf{u} = \left\langle \frac{12}{13}, \frac{5}{13}\right\rangle$, so $D_{\mathbf{u}}f(5,1) = \nabla f(5,1)\cdot\mathbf{u} = \frac{12}{52} - \frac{5}{52} = \frac{7}{52}$.

11. $g(x,y) = xe^{xy} \quad \Rightarrow \quad \nabla g(x,y) = \langle e^{xy}(1+xy), x^2e^{xy}\rangle$, $\nabla g(-3,0) = \langle 1,9\rangle$, $\mathbf{u} = \left\langle \frac{2}{\sqrt{13}}, \frac{3}{\sqrt{13}}\right\rangle$ and $D_{\mathbf{u}}g(-3,0) = \frac{2}{\sqrt{13}} + \frac{27}{\sqrt{13}} = \frac{29}{\sqrt{13}}$.

13. $f(x,y,z) = \sqrt{xyz} \quad \Rightarrow \quad \nabla f(x,y,z) = \frac{1}{2}(xyz)^{-1/2}\langle yz, xz, xy\rangle$, $\nabla f(2,4,2) = \left\langle 1, \frac{1}{2}, 1\right\rangle$, $\mathbf{u} = \left\langle \frac{2}{3}, \frac{1}{3}, -\frac{2}{3}\right\rangle$ and $D_{\mathbf{u}}f(2,4,2) = 1\cdot\frac{2}{3} + \frac{1}{2}\cdot\frac{1}{3} + 1\left(-\frac{2}{3}\right) = \frac{1}{6}$.

15. $g(x,y,z) = x\tan^{-1}(y/z) \quad \Rightarrow \quad \nabla g(x,y,z) = \left\langle \tan^{-1}(y/z), xz/(y^2+z^2), -xy/(y^2+z^2)\right\rangle$, $\nabla g(1,2,-2) = \left\langle -\frac{\pi}{4}, -\frac{1}{4}, -\frac{1}{4}\right\rangle$, $\mathbf{u} = \frac{1}{\sqrt{3}}\langle 1,1,-1\rangle$ and $D_{\mathbf{u}}g(1,2,-2) = \frac{(-\pi)(1)}{4\sqrt{3}} + \frac{(-1)(1)}{4\sqrt{3}} + \frac{(-1)(-1)}{4\sqrt{3}} = -\frac{\pi}{4\sqrt{3}}$.

17. $f(x,y) = xe^{-y} + 3y \quad \Rightarrow \quad \nabla f(x,y) = \langle e^{-y}, 3 - xe^{-y}\rangle$, $\nabla f(1,0) = \langle 1,2\rangle$ is the direction and the maximum rate is $|\nabla f(1,0)| = \sqrt{5}$.

19. $f(x,y) = \sqrt{x^2+2y} \quad \Rightarrow \quad \nabla f(x,y) = \left\langle \dfrac{x}{\sqrt{x^2+2y}}, \dfrac{1}{\sqrt{x^2+2y}}\right\rangle$. Thus the maximum rate of change is $|\nabla f(4,10)| = \frac{\sqrt{17}}{6}$ in the direction $\left\langle \frac{2}{3}, \frac{1}{6}\right\rangle$ or $\langle 4,1\rangle$.

21. $f(x,y) = \cos(3x+2y) \quad \Rightarrow \quad \nabla f(x,y) = \langle -3\sin(3x+2y), -2\sin(3x+2y)\rangle$, so the maximum rate of change is $\left|\nabla f\left(\frac{\pi}{6}, -\frac{\pi}{8}\right)\right| = \sqrt{\frac{13}{2}}$ in the direction $\left\langle -\frac{3\sqrt{2}}{2}, -\sqrt{2}\right\rangle$ or $\langle -3,-2\rangle$.

23. As in the proof of Theorem 15, $D_{\mathbf{u}}f = |\nabla f|\cos\theta$. Since the minimum value of $\cos\theta$ is -1 occurring when $\theta = \pi$, the minimum value of $D_{\mathbf{u}}f$ is $-|\nabla f|$ occurring when $\theta = \pi$, that is when **u** is in the opposite direction of ∇f (assuming $\nabla f \neq \mathbf{0}$).

25. $T = \dfrac{k}{\sqrt{x^2+y^2+z^2}}$ and $120 = T(1,2,2) = \dfrac{k}{3}$ so $k = 360$.

(a) $\mathbf{u} = \dfrac{\langle 1,-1,1\rangle}{\sqrt{3}}$,

$$D_{\mathbf{u}}T(1,2,2) = \nabla T(1,2,2)\cdot\mathbf{u} = \left[-360\left(x^2+y^2+z^2\right)^{-3/2}\langle x,y,z\rangle\right]_{(1,2,2)}\cdot\mathbf{u}$$
$$= -\tfrac{40}{3}\langle 1,2,2\rangle\cdot\tfrac{1}{\sqrt{3}}\langle 1,-1,1\rangle = -\tfrac{40}{3\sqrt{3}}$$

(b) From (a), $\nabla T = -360(x^2+y^2+z^2)^{-3/2}\langle x,y,z\rangle$, and since $\langle x,y,z\rangle$ is the position vector of the point (x,y,z), the vector $-\langle x,y,z\rangle$, and thus ∇T, always points toward the origin.

27. $\nabla V(x,y,z) = \langle 10x-3y+yz, xz-3x, xy\rangle$, $\nabla V(3,4,5) = \langle 38,6,12\rangle$

(a) $D_{\mathbf{u}}V(3,4,5) = \langle 38,6,12\rangle\cdot\frac{1}{\sqrt{3}}\langle 1,1,-1\rangle = \frac{32}{\sqrt{3}}$

(b) $\langle 38,6,12\rangle$ or $\langle 19,3,6\rangle$.

(c) $|\nabla V(x,y,z)| = \sqrt{(10x-3y+yz)^2+(xz-3x)^2+x^2y^2}$ in general or at $(3,4,5)$ is $|\nabla V(3,4,5)| = \sqrt{1624} = 2\sqrt{406}$.

29. A unit vector in the direction of $\overrightarrow{AB}$ is **i** and a unit vector in the direction of $\overrightarrow{AC}$ is **j**. Thus $D_{\overrightarrow{AB}}f(1,3) = f_x(1,3) = 3$ and $D_{\overrightarrow{AC}}f(1,3) = f_y(1,3) = 26$. Therefore $\nabla f(1,3) = \langle f_x(1,3), f_y(1,3)\rangle = \langle 3,26\rangle$ and by definition $D_{\overrightarrow{AD}}f(1,3) = \nabla f\cdot\mathbf{u}$ where **u** is a unit vector in the direction of $\overrightarrow{AD}$, which is $\left\langle\frac{5}{13},\frac{12}{13}\right\rangle$. Therefore $D_{\overrightarrow{AD}}f(1,3) = \langle 3,26\rangle\cdot\left\langle\frac{5}{13},\frac{12}{13}\right\rangle = 3\cdot\frac{5}{13}+26\cdot\frac{12}{13} = \frac{327}{13}$.

31. $$\nabla(au+bv) = \left\langle\frac{\partial(au+bv)}{\partial x},\frac{\partial(au+bv)}{\partial y}\right\rangle = \left\langle a\frac{\partial u}{\partial x}+b\frac{\partial v}{\partial x}, a\frac{\partial u}{\partial y}+b\frac{\partial v}{\partial y}\right\rangle = a\left\langle\frac{\partial u}{\partial x},\frac{\partial u}{\partial y}\right\rangle + b\left\langle\frac{\partial v}{\partial x},\frac{\partial v}{\partial y}\right\rangle$$
$$= a\nabla u + b\nabla v$$

33. $$\nabla\left(\frac{u}{v}\right) = \left\langle\frac{v\dfrac{\partial u}{\partial x}-u\dfrac{\partial v}{\partial x}}{v^2},\frac{v\dfrac{\partial u}{\partial y}-u\dfrac{\partial v}{\partial y}}{v^2}\right\rangle = \frac{v\left\langle\dfrac{\partial u}{\partial x},\dfrac{\partial u}{\partial y}\right\rangle - u\left\langle\dfrac{\partial v}{\partial x},\dfrac{\partial v}{\partial y}\right\rangle}{v^2} = \frac{v\nabla u - u\nabla v}{v^2}$$

35. $F(x,y,z) = 4x^2+y^2+z^2$, $\nabla F(2,2,2) = \langle 16,4,4\rangle$, so

(a) the equation of the tangent plane is $16x+4y+4z = 48$ or $4x+y+z = 12$, and

(b) the normal line is given by $\dfrac{x-2}{16} = \dfrac{y-2}{4} = \dfrac{z-2}{4}$ or $\dfrac{x-2}{4} = y-2 = z-2$.

37. $\nabla F(x,y,z) = \langle 2x-2y+4z, 2y-2x, -2z+4x\rangle$, $\nabla F(1,0,1) = \langle 6,-2,2\rangle$

(a) $6x-2y+2z = 8$ or $3x-y+z = 4$ **(b)** $\dfrac{x-1}{3} = -y = z-1$

39. $F(x, y, z) = -z + xe^y \cos z$, $\nabla F(x, y, z) = \langle e^y \cos z, xe^y \cos z, -1 - xe^y \sin z \rangle$, $\nabla F(1, 0, 0) = \langle 1, 1, -1 \rangle$

(a) $x + y - z = 1$

(b) $x - 1 = y = -z$

41. $\nabla F(x, y, z) = \langle y + z,\ x + z,\ y + x \rangle$, $\nabla F(1, 1, 1) = \langle 2, 2, 2 \rangle$, so the equation of the tangent plane is $2x + 2y + 2z = 6$ or $x + y + z = 3$, and the normal line is given by $x - 1 = y - 1 = z - 1$ or $x = y = z$.

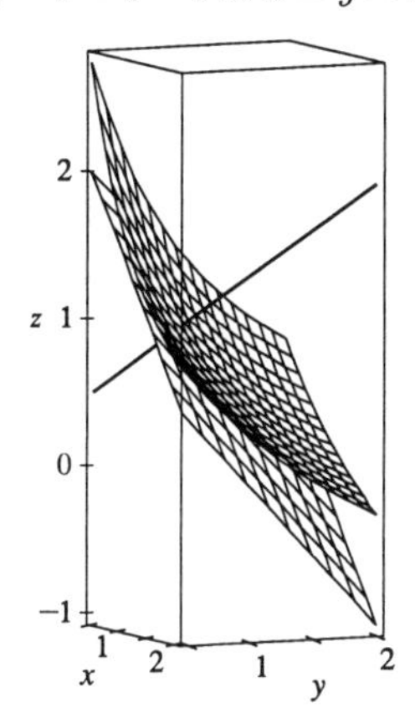

43. $\nabla f(x, y) = \langle 2x, 8y \rangle$, $\nabla f(2, 1) = \langle 4, 8 \rangle$. The tangent line has equation $\nabla f(2, 1) \cdot \langle x - 2, y - 1 \rangle = 0 \quad \Rightarrow$ $4(x - 2) + 8(y - 1) = 0$, which simplifies to $x + 2y = 4$.

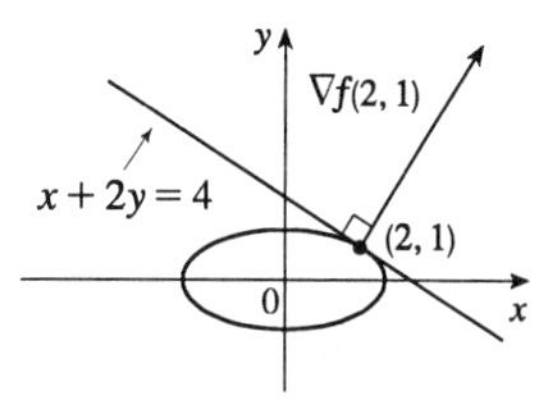

45. $\nabla F(x_0, y_0, z_0) = \left\langle \dfrac{2x_0}{a^2}, \dfrac{2y_0}{b^2}, \dfrac{2z_0}{c^2} \right\rangle$. Thus an equation of the tangent plane at (x_0, y_0, z_0) is

$\dfrac{2x_0}{a^2}x + \dfrac{2y_0}{b^2}y + \dfrac{2z_0}{c^2}z = 2\left(\dfrac{x_0^2}{a^2} + \dfrac{y_0^2}{b^2} + \dfrac{z_0^2}{c^2}\right) = 2(1) = 2$ since (x_0, y_0, z_0) is a point on the ellipsoid. Hence $\dfrac{x_0}{a^2}x + \dfrac{y_0}{b^2}y + \dfrac{z_0}{c^2}z = 1$ is an equation of the tangent plane.

47. $\nabla F(x_0, y_0, z_0) = \left\langle \dfrac{2x_0}{a^2}, \dfrac{2y_0}{b^2}, \dfrac{-1}{c} \right\rangle$, so an equation of the tangent plane is

$\dfrac{2x_0}{a^2}x + \dfrac{2y_0}{b^2}y - \dfrac{1}{c}z = \dfrac{2x_0^2}{a^2} + \dfrac{2y_0^2}{b^2} - \dfrac{z_0}{c}$ or $\dfrac{2x_0}{a^2}x + \dfrac{2y_0}{b^2}y = \dfrac{z}{c} + 2\left(\dfrac{x_0^2}{a^2} + \dfrac{y_0^2}{b^2}\right) - \dfrac{z_0}{c}$. But $\dfrac{z_0}{c} = \dfrac{x_0^2}{a^2} + \dfrac{y_0^2}{b^2}$, so the equation can be written as $\dfrac{2x_0}{a^2}x + \dfrac{2y_0}{b^2}y = \dfrac{z + z_0}{c}$.

49. $\nabla f(x_0, y_0, z_0) = \langle 2x_0, -2y_0, 4z_0 \rangle$ and the given line has direction numbers 2, 4, 6, so

$\langle 2x_0, -2y_0, 4z_0 \rangle = k\langle 2, 4, 6 \rangle$ or $x_0 = k$, $y_0 = -2k$ and $z_0 = \frac{3}{2}k$. But $x_0^2 - y_0^2 + 2z_0^2 = 1$ or $\left(1 - 4 + \frac{9}{2}\right)k^2 = 1$, so $k = \pm\sqrt{\frac{2}{3}} = \pm\frac{\sqrt{6}}{3}$ and there are two such points: $\left(\pm\frac{\sqrt{6}}{3}, \mp\frac{2\sqrt{6}}{3}, \pm\frac{\sqrt{6}}{2}\right)$.

51. Let (x_0, y_0, z_0) be a point on the cone [other than $(0, 0, 0)$]. Then the equation of the tangent plane to the cone at this point is $2x_0x + 2y_0y - 2z_0z = 2(x_0^2 + y_0^2 - z_0^2)$. But $x_0^2 + y_0^2 = z_0^2$ so the tangent plane is given by $x_0x + y_0y - z_0z = 0$, a plane which always contains the origin.

53. Let (x_0, y_0, z_0) be a point on the surface. Then the equation of the tangent plane at the point is $\dfrac{x}{2\sqrt{x_0}} + \dfrac{y}{2\sqrt{y_0}} + \dfrac{z}{2\sqrt{z_0}} = \dfrac{\sqrt{x_0} + \sqrt{y_0} + \sqrt{z_0}}{2}$. But $\sqrt{x_0} + \sqrt{y_0} + \sqrt{z_0} = \sqrt{c}$, so the equation is $\dfrac{x}{\sqrt{x_0}} + \dfrac{y}{\sqrt{y_0}} + \dfrac{z}{\sqrt{z_0}} = \sqrt{c}$. The x-, y-, and z-intercepts are $\sqrt{cx_0}$, $\sqrt{cy_0}$ and $\sqrt{cz_0}$ respectively. (The x-intercept is found by setting $y = z = 0$ and solving the resulting equation for x, and the y- and z-intercepts are found similarly.) So the sum of the intercepts is $\sqrt{c}\left(\sqrt{x_0} + \sqrt{y_0} + \sqrt{z_0}\right) = c$, a constant.

55. If $f(x, y, z) = z - x^2 - y^2$ and $g(x, y, z) = 4x^2 + y^2 + z^2$, then the tangent line is perpendicular to both ∇f and ∇g at $(-1, 1, 2)$. The vector $\mathbf{v} = \nabla f \times \nabla g$ will therefore be parallel to the tangent line. We have: $\nabla f(x, y, z) = \langle -2x, -2y, 1 \rangle \quad \Rightarrow \quad \nabla f(-1, 1, 2) = \langle 2, -2, 1 \rangle$, and $\nabla g(x, y, z) = \langle 8x, 2y, 2z \rangle \quad \Rightarrow$ $\nabla g(-1, 1, 2) = \langle -8, 2, 4 \rangle$. Hence $\mathbf{v} = \nabla f \times \nabla g = \begin{vmatrix} \mathbf{i} & \mathbf{j} & \mathbf{k} \\ 2 & -2 & 1 \\ -8 & 2 & 4 \end{vmatrix} = -10\mathbf{i} - 16\mathbf{j} - 12\mathbf{k}$. Parametric equations are: $x = -1 - 10t$, $y = 1 - 16t$, $z = 2 - 12t$.

57. **(a)** The direction of the normal line of F is given by ∇F, and that of G by ∇G. Assuming that $\nabla F \neq 0 \neq \nabla G$, the two normal lines are be perpendicular at P if $\nabla F \cdot \nabla G = 0$ at $P \quad \Leftrightarrow$ $\langle \partial F/\partial x, \partial F/\partial y, \partial F/\partial z \rangle \cdot \langle \partial G/\partial x, \partial G/\partial y, \partial G/\partial z \rangle = 0$ at $P \quad \Leftrightarrow \quad F_x G_x + F_y G_y + F_z G_z = 0$ at P.

(b) Here $F = x^2 + y^2 - z^2$ and $G = x^2 + y^2 + z^2 - r^2$, so $\nabla F \cdot \nabla G = \langle 2x, 2y, -2z \rangle \cdot \langle 2x, 2y, 2z \rangle = 4x^2 + 4y^2 - 4z^2 = 4F = 0$, since the point $\langle x, y, z \rangle$ lies on the graph of $F = 0$. To see that this is true without using calculus, note that $G = 0$ is the equation of a sphere centered at the origin and $F = 0$ is the equation of a right circular cone with vertex at the origin (which is generated by lines through the origin). At any point of intersection, the sphere's normal line (which passes through the origin) lies on the cone, and thus is perpendicular to the cone's normal line. So the surfaces with equations $F = 0$ and $G = 0$ are everywhere orthogonal.

59. Let $\mathbf{u} = \langle a, b \rangle$ and $\mathbf{v} = \langle c, d \rangle$. Then we know that at the given point, $D_\mathbf{u} f = \nabla f \cdot \mathbf{u} = af_x + bf_y$ and $D_\mathbf{v} f = \nabla f \cdot \mathbf{v} = cf_x + df_y$. But these are just two linear equations in the two unknowns f_x and f_y, and since $\mathbf{u}$ and $\mathbf{v}$ are not parallel, we can solve the equations to find $\nabla f = \langle f_x, f_y \rangle$ at the given point. In fact, $\nabla f = \left\langle \dfrac{dD_\mathbf{u} f - bD_\mathbf{v} f}{ad - bc}, \dfrac{aD_\mathbf{v} f - cD_\mathbf{u} f}{ad - bc} \right\rangle$.

EXERCISES 12.7

1. $f(x,y) = x^2 + y^2 + 4x - 6y \quad \Rightarrow \quad f_x = 2x + 4$, $f_y = 2y - 6$, $f_{xx} = f_{yy} = 2$, $f_{xy} = 0$. Then $f_x = 0$ and $f_y = 0$ implies $(x,y) = (-2,3)$ and $D(-2,3) = 4 > 0$, so $f(-2,3) = -13$ is a local minimum.

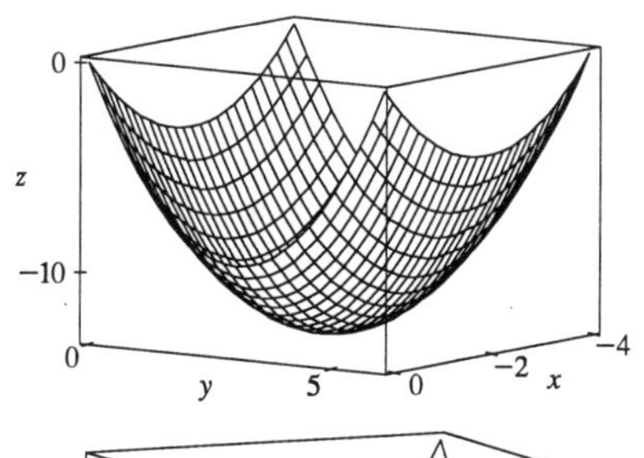

3. $f(x,y) = 2x^2 + y^2 + 2xy + 2x + 2y \quad \Rightarrow$
$f_x = 4x + 2y + 2$, $f_y = 2y + 2x + 2$,
$f_{xx} = 4$, $f_{yy} = 2$, $f_{xy} = 2$.
Then $f_x = 0$ and $f_y = 0$ implies $2x = 0$, so the critical point is $(0,-1)$. $D(0,-1) = 8 - 4 > 0$, so $f(0,-1) = -1$ is a local minimum.

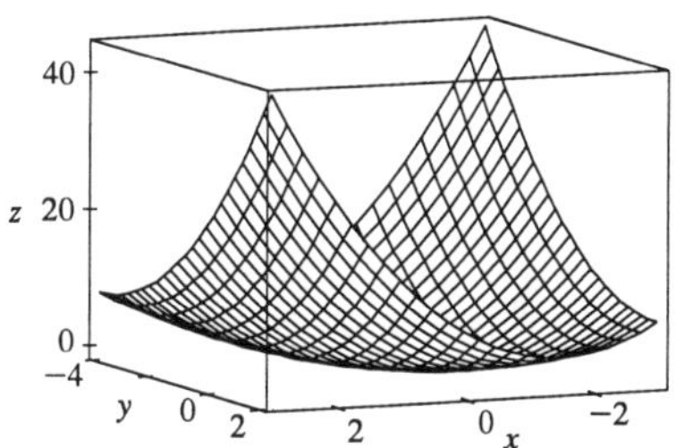

5. $f(x,y) = x^2 + y^2 + x^2y + 4 \quad \Rightarrow \quad f_x = 2x + 2xy$, $f_y = 2y + x^2$, $f_{xx} = 2 + 2y$, $f_{yy} = 2$, $f_{xy} = 2x$. Then $f_y = 0$ implies $y = -\frac{1}{2}x^2$, substituting into $f_x = 0$ gives $2x - x^3 = 0$ so $x = 0$ or $x = \pm\sqrt{2}$. Thus the critical points are $(0,0)$, $\left(\sqrt{2},-1\right)$ and $\left(-\sqrt{2},-1\right)$. Now $D(0,0) = 4$, $D\left(\sqrt{2},-1\right) = -8 = D\left(-\sqrt{2},-1\right)$, $f_{xx}(0,0) = 2$, $f_{xx}\left(\pm\sqrt{2},-1\right) = 0$. Thus $f(0,0) = 4$ is a local minimum and $\left(\pm\sqrt{2},-1\right)$ are saddle points.

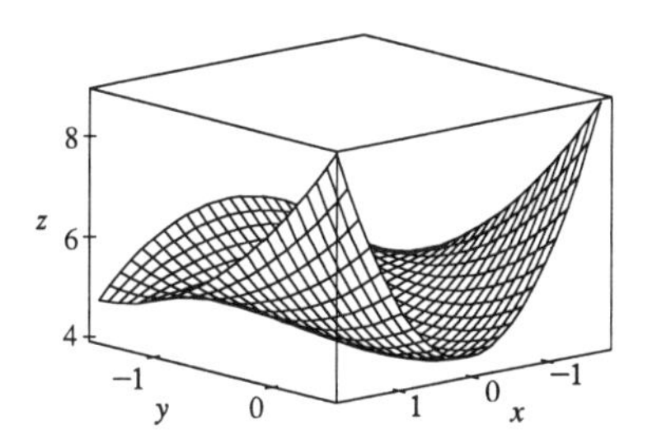

7. $f(x,y) = x^3 - 3xy + y^3 \quad \Rightarrow \quad f_x = 3x^2 - 3y$, $f_y = 3y^2 - 3x$, $f_{xx} = 6x$, $f_{yy} = 6y$, $f_{xy} = -3$. Then $f_x = 0$ implies $x^2 = y$ and substituting into $f_y = 0$ gives $x = 0$ or $x = 1$. Thus the critical points are $(0,0)$ and $(1,1)$. Now $D(0,0) = -9 < 0$ so $(0,0)$ is a saddle point and $D(1,1) = 36 - 9 > 0$ while $f_{xx}(1,1) = 6$ so $f(1,1) = -1$ is a local minimum.

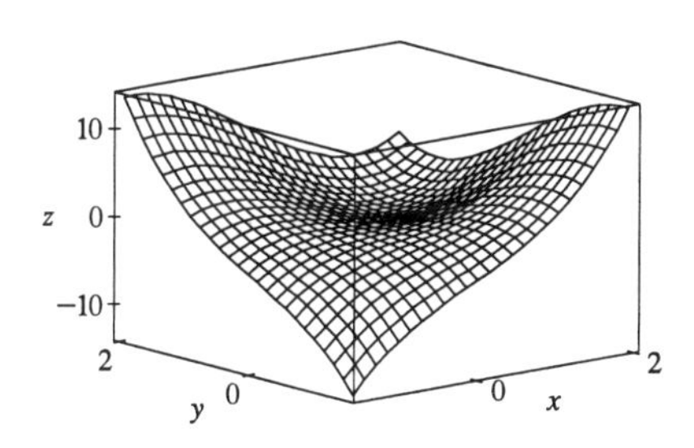

9. $f(x,y) = xy - 2x - y \quad \Rightarrow \quad f_x = y - 2$, $f_y = x - 1$, $f_{xx} = f_{yy} = 0$, $f_{xy} = 1$ and the only critical point is $(1,2)$. Now $D(1,2) = -1$ so $(1,2)$ is a saddle point and f has no local maximum or minimum.

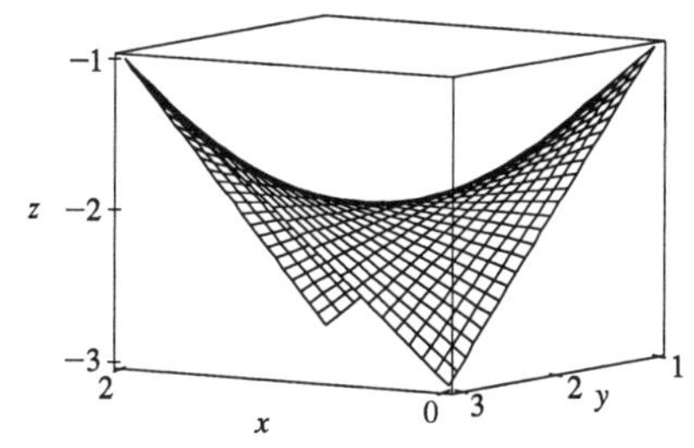

11. $f(x,y) = \dfrac{x^2y^2 - 8x + y}{xy} \quad \Rightarrow \quad f_x = y - x^{-2}$, $f_y = x + 8y^{-2}$, $f_{xx} = 2x^{-3}$, $f_{yy} = -16y^{-3}$ and $f_{xy} = 1$. Then $f_x = 0$ implies $y = x^{-2}$, substituting into $f_y = 0$ gives $x + 8x^4 = 0$ so $x = 0$ or $x = -\frac{1}{2}$ but $(0,y)$ is not in the domain of f. Thus the only critical point is $\left(-\frac{1}{2}, 4\right)$. Then $f_{xx}\left(-\frac{1}{2},4\right) = -16$ and $D\left(-\frac{1}{2},4\right) = 4 - 1 > 0$ so $f\left(-\frac{1}{2},4\right) = -6$ is a local maximum.

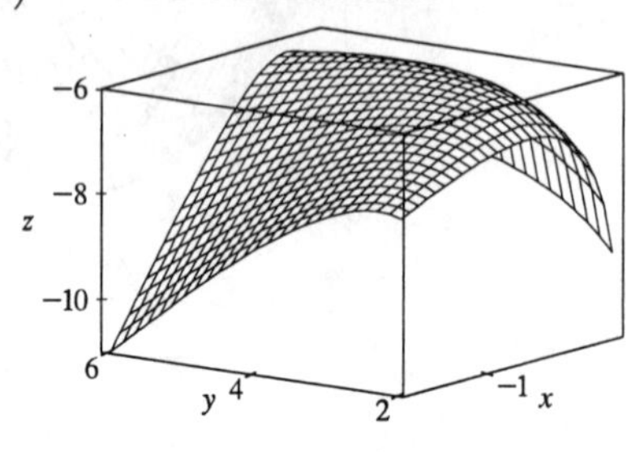

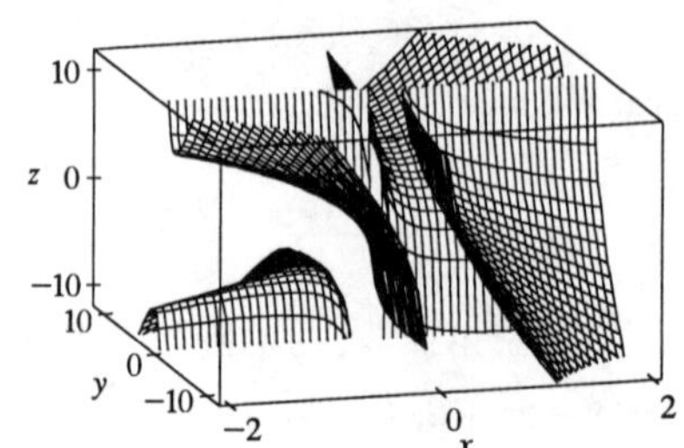

13. $f(x,y) = e^x \cos y \quad \Rightarrow \quad f_x = e^x \cos y$, $f_y = -e^x \sin y$.
Now $f_x = 0$ implies $\cos y = 0$ or $y = \frac{\pi}{2} + n\pi$ for n an integer.
But $\sin\left(\frac{\pi}{2} + n\pi\right) \neq 0$, so there are no critical points.

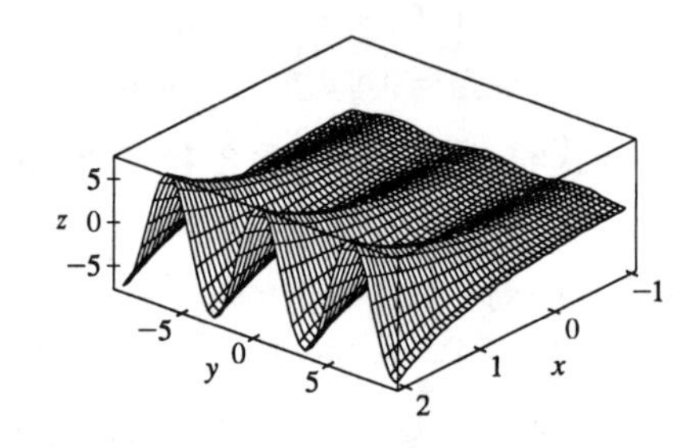

15. $f(x,y) = x \sin y \quad \Rightarrow \quad f_x = \sin y$, $f_y = x \cos y$, $f_{xx} = 0$, $f_{yy} = -x \sin y$ and $f_{xy} = \cos y$. Then $f_x = 0$ if and only if $y = n\pi$, n an integer, and substituting into $f_y = 0$ requires $x = 0$ for each of these y-values. Thus the critical points are $(0, n\pi)$, n an integer. But $D(0,n\pi) = -\cos^2(n\pi) < 0$ so each critical point is a saddle point.

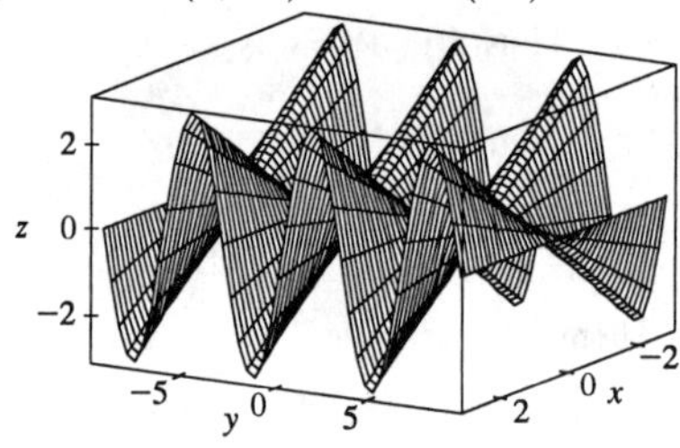

17. $f(x, y) = 3x^2y + y^3 - 3x^2 - 3y^2 + 2$

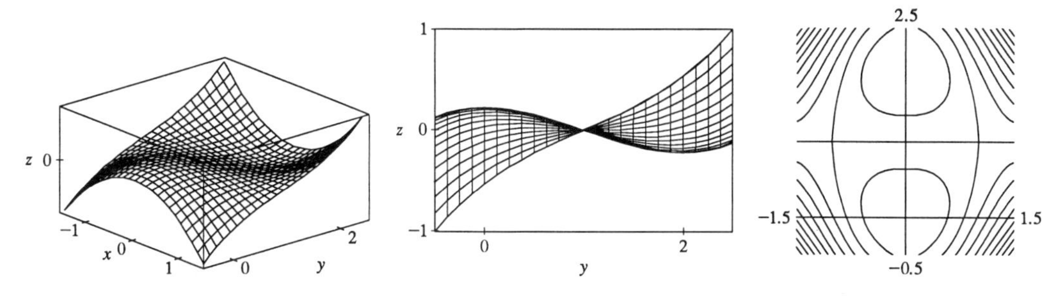

From the graphs, it appears that f has a local maximum $f(0, 0) \approx 2$ and a local minimum $f(0, 2) \approx -2$. There appear to be saddle points near $(\pm 1, 1)$.

$f_x = 6xy - 6x$, $f_y = 3x^2 + 3y^2 - 6y$. Then $f_x = 0$ implies $x = 0$ or $y = 1$ and when $x = 0$, $f_y = 0$ implies $y = 0$ or $y = 2$; when $y = 1$, $f_y = 0$ implies $x^2 = 1$ or $x = \pm 1$. Thus the critical points are $(0, 0)$, $(0, 2)$, $(\pm 1, 1)$. Now $f_{xx} = 6y - 6$, $f_{yy} = 6y - 6$ and $f_{xy} = 6x$, so $D(0, 0) = D(0, 2) = 36 > 0$ while $D(\pm 1, 1) = -36 < 0$ and $f_{xx}(0, 0) = -6$, $f_{xx}(0, 2) = 6$. Hence $(\pm 1, 1)$ are saddle points while $f(0, 0) = 2$ is a local maximum and $f(0, 2) = -2$ is a local minimum.

19. $f(x, y) = \sin x + \sin y + \sin(x + y)$, $0 \le x \le 2\pi$, $0 \le y \le 2\pi$

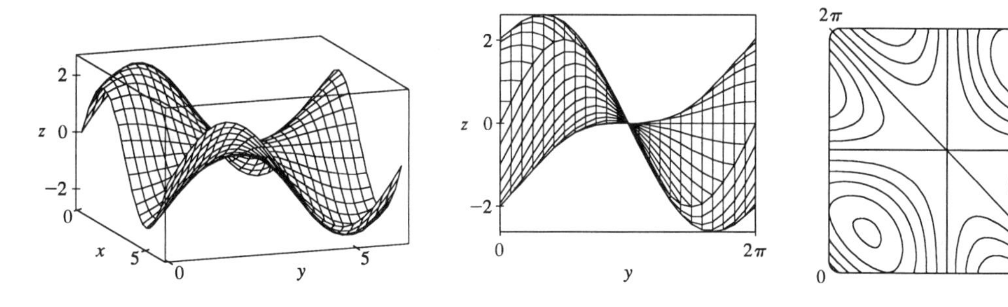

From the graphs it appears that f has a local maximum at about $(1, 1)$ with value ≈ 2.6, a local minimum at about $(5, 5)$ with value ≈ -2.6, and a saddle point at about $(3, 3)$.

$f_x = \cos x + \cos(x + y)$, $f_y = \cos y + \cos(x + y)$, $f_{xx} = -\sin x - \sin(x + y)$, $f_{yy} = -\sin y - \sin(x + y)$, $f_{xy} = -\sin(x + y)$. Setting $f_x = 0$ and $f_y = 0$ and subtracting gives $\cos x - \cos y = 0$ or $\cos x = \cos y$. Thus $x = y$ or $x = 2\pi - y$. If $x = y$, $f_x = 0$ becomes $\cos x + \cos 2x = 0$ or $2\cos^2 x + \cos x - 1 = 0$, a quadratic in $\cos x$. Thus $\cos x = -1$ or $\frac{1}{2}$ and $x = \pi$, $\frac{\pi}{3}$, or $\frac{5\pi}{3}$, yielding the critical points (π, π), $\left(\frac{\pi}{3}, \frac{\pi}{3}\right)$ and $\left(\frac{5\pi}{3}, \frac{5\pi}{3}\right)$. Similarly if $x = 2\pi - y$, $f_x = 0$ becomes $(\cos x) + 1 = 0$ and the resulting critical point is (π, π). Now $D(x, y) = \sin x \sin y + \sin x \sin(x + y) + \sin y \sin(x + y)$. So $D(\pi, \pi) = 0$ and (3) doesn't apply. $D\left(\frac{\pi}{3}, \frac{\pi}{3}\right) = \frac{9}{4} > 0$ and $f_{xx}\left(\frac{\pi}{3}, \frac{\pi}{3}\right) < 0$ so $f\left(\frac{\pi}{3}, \frac{\pi}{3}\right) = \frac{3\sqrt{3}}{2}$ is a local maximum while $D\left(\frac{5\pi}{3}, \frac{5\pi}{3}\right) = \frac{9}{4} > 0$ and $f_{xx}\left(\frac{5\pi}{3}, \frac{5\pi}{3}\right) > 0$, so $f\left(\frac{5\pi}{3}, \frac{5\pi}{3}\right) = -\frac{3\sqrt{3}}{2}$ is a local minimum.

21. $f(x,y) = x^4 - 5x^2 + y^2 + 3x + 2 \quad\Rightarrow\quad f_x(x,y) = 4x^3 - 10x + 3$ and $f_y(x,y) = 2y$. $f_y = 0 \quad\Rightarrow\quad y = 0$, and the graph of f_x shows that the roots of $f_x = 0$ are approximately $x = -1.714$, 0.312 and 1.402. (Alternatively, we could have used a calculator or a CAS to find these roots.) So to three decimal places, the critical points are $(-1.714, 0)$, $(1.402, 0)$, and $(0.312, 0)$. Now since $f_{xx} = 12x^2 - 10$, $f_{xy} = 0$, $f_{yy} = 2$, and $D = 24x^2 - 20$, we have $D(-1.714, 0) > 0$, $f_{xx}(-1.714, 0) > 0$, $D(1.402, 0) > 0$, $f_{xx}(1.402, 0) > 0$, and $D(0.312, 0) < 0$. Therefore $f(-1.714, 0) \approx -9.200$ and $f(1.402, 0) \approx 0.242$ are local minima, and $(0.312, 0)$ is a saddle point. The lowest point on the graph is approximately $(-1.714, 0, -9.200)$.

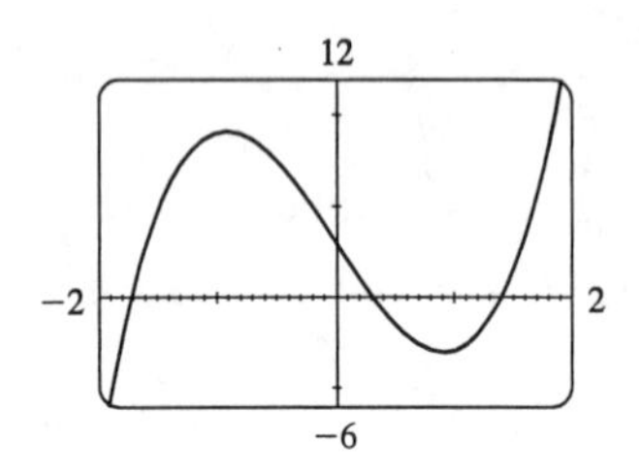

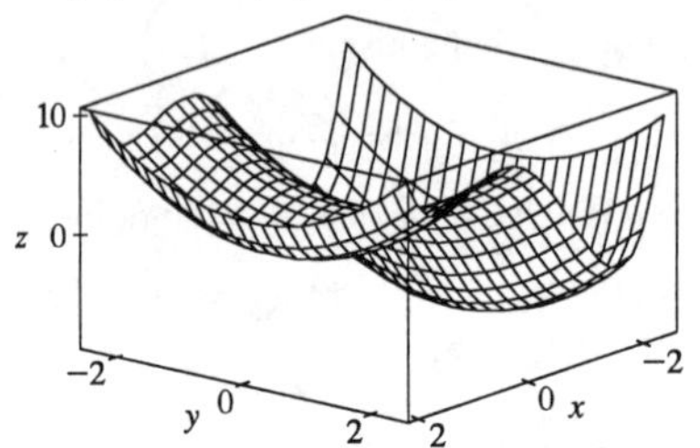

23. $f(x,y) = 2x + 4x^2 - y^2 + 2xy^2 - x^4 - y^4 \quad\Rightarrow\quad f_x(x,y) = 2 + 8x + 2y^2 - 4x^3$, $f_y(x,y) = -2y + 4xy - 4y^3$. Now $f_y = 0 \quad\Leftrightarrow\quad 2y(2y^2 - 2x + 1) = 0 \quad\Leftrightarrow\quad y = 0$ or $y^2 = x - \frac{1}{2}$. The first of these implies that $f_x = -4x^3 + 8x + 2$, and the second implies that $f_x = 2 + 8x + 2\left(x - \frac{1}{2}\right) - 4x^3 = -4x^3 + 10x + 1$. From the graphs, we see that the first possibility for f_x has roots at approximately -1.267, -0.259, and 1.526, and the second has a root at approximately 1.629 (the negative roots do not give critical points, since $y^2 = x - \frac{1}{2}$ must be positive). So to three decimal places, f has critical points at $(-1.267, 0)$, $(-0.259, 0)$, $(1.526, 0)$, and $(1.629, \pm 1.063)$. Now since $f_{xx} = 8 - 12x^2$, $f_{xy} = 4y$, $f_{yy} = 4x - 12y^2$, and $D = (8 - 12x^2)(4x - 12y^2) - 16y^2$, we have $D(-1.267, 0) > 0$, $f_{xx}(-1.267, 0) > 0$, $D(-0.259, 0) < 0$, $D(1.526, 0) < 0$, $D(1.629, \pm 1.063) > 0$, and $f_{xx}(1.629, \pm 1.063) < 0$. Therefore, to three decimal places, $f(-1.267, 0) \approx 1.310$ and $f(1.629, \pm 1.063) \approx 8.105$ are local maxima, and $(-0.259, 0)$ and $(1.526, 0)$ are saddle points. The highest points on the graph are approximately $(1.629, \pm 1.063, 8.105)$.

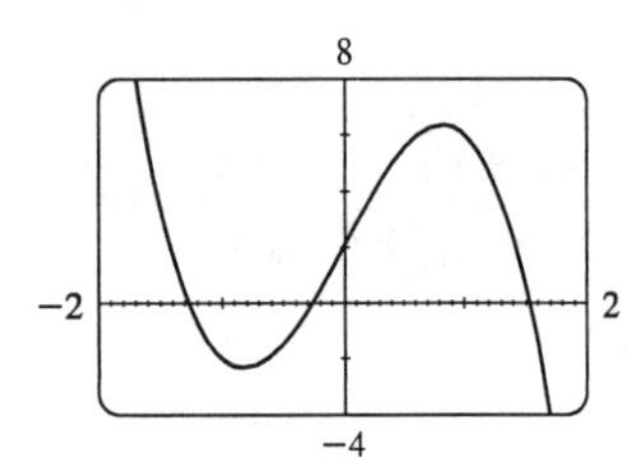

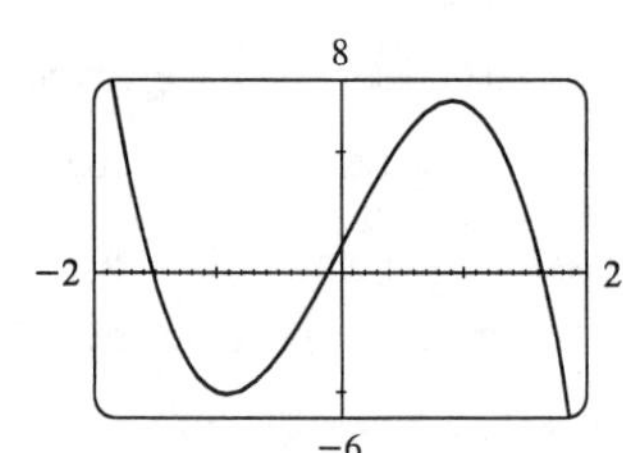

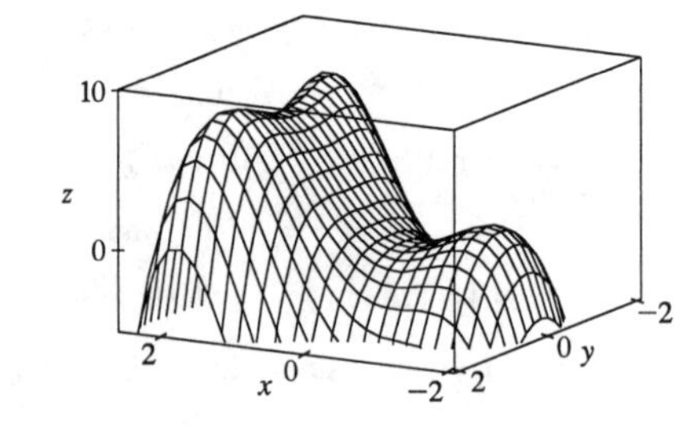

25. Since f is a polynomial it is continuous on D, so an absolute maximum and minimum exist. Here $f_x = -3$, $f_y = 4$ so there are no critical points inside D. Thus the absolute extrema must both occur on the boundary. Along L_1, $y = 0$ and $f(x,0) = 5 - 3x$, a decreasing function in x, so the maximum value is $f(0,0) = 5$ and the minimum value is $f(4,0) = -7$. Along L_2, $x = 4$ and $f(4,y) = -7 + 4y$, an increasing function in y, so the minimum value is $f(4,0) = -7$ and the maximum value is $f(4,5) = 13$. Along L_3, $y = \frac{5}{4}x$ and $f\left(x, \frac{5}{4}x\right) = 5 + 2x$, an increasing function in x, so the minimum value is $f(0,0) = 5$ and the maximum value is $f(4,5) = 13$. Thus the absolute minimum of f on D is $f(4,0) = -7$ and the absolute maximum is $f(4,5) = 13$.

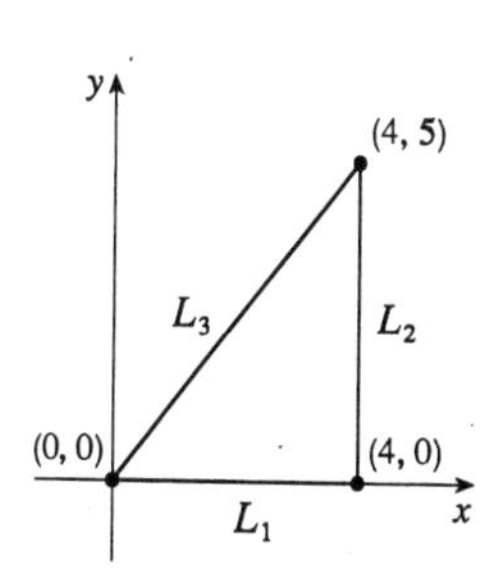

27. In Exercise 5, we found the critical points of f; only $(0,0)$ with $f(0,0) = 4$ is in D. On L_1: $y = -1$, $f(x,-1) = 5$, a constant. On L_2: $x = 1$, $f(1,y) = y^2 + y + 5$, a quadratic in y which attains its maximum at $(1,1)$, $f(1,1) = 7$ and its minimum at $\left(1, -\frac{1}{2}\right)$, $f\left(1, -\frac{1}{2}\right) = \frac{17}{4}$. On L_3: $f(x,1) = 2x^2 + 5$ which attains its maximum at $(-1,1)$ and $(1,1)$ with $f(\pm 1, 1) = 7$ and its minimum at $(0,1)$, $f(0,1) = 5$. On L_4: $f(-1,y) = y^2 + y + 5$ with maximum at $(-1,1)$, $f(-1,1) = 7$ and minimum at $\left(-1, -\frac{1}{2}\right)$, $f\left(-1, -\frac{1}{2}\right) = \frac{17}{4}$. Thus the absolute maximum is attained at both $(\pm 1, 1)$ with $f(\pm 1, 1) = 7$ and the absolute minimum on D is attained at $(0,0)$ with $f(0,0) = 4$.

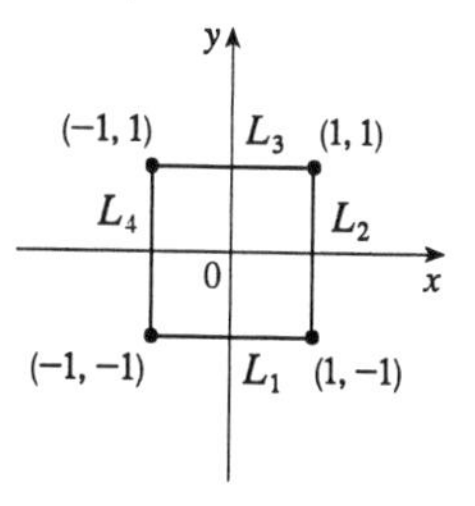

29. $f_x(x,y) = y - 1$ and $f_y(x,y) = x - 1$ and so the critical point is $(1,1)$ (in D), where $f(1,1) = 0$. Along L_1: $y = 4$, so $f(x,4) = 1 + 4x - x - 4 = 3x - 3$, $-2 \le x \le 2$, which is an increasing function and has a maximum value when $x = 2$ where $f(2,4) = 3$ and a minimum of $f(-2,4) = -9$. Along L_2: $y = x^2$, so let $g(x) = f(x, x^2) = x^3 - x^2 - x + 1$.Then $g'(x) = 3x^2 - 2x - 1 = 0 \Leftrightarrow x = -\frac{1}{3}$ or $x = 1$. $f\left(-\frac{1}{3}, \frac{1}{9}\right) = \frac{32}{27}$ and $f(1,1) = 0$. As a result, the absolute maximum and minimum values of f on D are $f(2,4) = 3$ and $f(-2,4) = -9$.

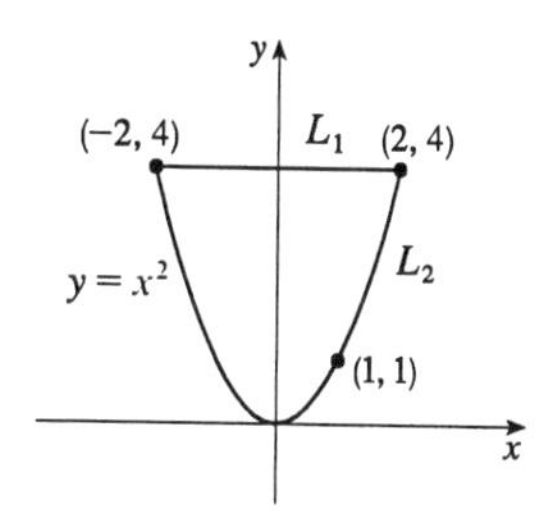

31. $f_x(x,y) = 6x^2$ and $f_y(x,y) = 4y^3$. And so $f_x = 0$ and $f_y = 0$ only occur when $x = y = 0$. Hence, the only critical point inside the disk is at $x = y = 0$ where $f(0,0) = 0$. Now on the circle $x^2 + y^2 = 1$, $y^2 = 1 - x^2$ so let $g(x) = f(x,y) = 2x^3 + (1 - x^2)^2 = x^4 + 2x^3 - 2x^2 + 1$, $-1 \le x \le 1$. Then $g'(x) = 4x^3 + 6x^2 - 4x = 0$ $\Rightarrow$ $x = 0, -2$, or $\frac{1}{2}$. $f(0, \pm 1) = g(0) = 1$, $f\left(\frac{1}{2}, \pm\frac{\sqrt{3}}{2}\right) = g\left(\frac{1}{2}\right) = \frac{13}{16}$, and $(-2, -3)$ is not in D. Checking the endpoints, we get $f(-1,0) = g(-1) = -2$ and $f(1,0) = g(1) = 2$. Thus the absolute maximum and minimum of f on D are $f(1,0) = 2$ and $f(-1,0) = -2$.

Another method: On the boundary $x^2 + y^2 = 1$ we can write $x = \cos\theta$, $y = \sin\theta$, so $f(\cos\theta, \sin\theta) = 2\cos^3\theta + \sin^4\theta$, $0 \le \theta \le 2\pi$.

33. $f(x,y) = -(x^2 - 1)^2 - (x^2 y - x - 1)^2$ $\Rightarrow$ $f_x(x,y) = -2(x^2 - 1)(2x) - 2(x^2 y - x - 1)(2xy - 1)$ and $f_y(x,y) = -2(x^2 y - x - 1)x^2$. Setting $f_y(x,y) = 0$ gives either $x = 0$ or $x^2 y - x - 1 = 0$. There are no critical points for $x = 0$, since $f_x(0,y) = -2$, so we set $x^2 y - x - 1 = 0$ $\Leftrightarrow$ $y = \dfrac{x+1}{x^2}$ $(x \ne 0)$, so

$$f_x\left(x, \frac{x+1}{x^2}\right) = -2(x^2 - 1)(2x) - 2\left(x^2 \frac{x+1}{x^2} - x - 1\right)\left(2x\frac{x+1}{x^2} - 1\right) = -4x(x^2 - 1).$$ Therefore $f_x(x,y) = f_y(x,y) = 0$ at the points $(1,2)$ and $(-1,0)$. To classify these critical points, we calculate $f_{xx}(x,y) = -12x^2 - 12x^2y^2 + 12xy + 4y + 2$, $f_{yy}(x,y) = -2x^4$, and $f_{xy}(x,y) = -8x^3 y + 6x^2 + 4x$. In order to use the Second Derivatives Test we calculate $D(-1,0) = f_{xx}(-1,0) f_{yy}(-1,0) - [f_{xy}(-1,0)]^2 = 16 > 0$, $f_{xx}(-1,0) = -10 < 0$, $D(1,2) = 16 > 0$, and $f_{xx}(1,2) = -26 < 0$, so both $(-1,0)$ and $(1,2)$ give local maxima.

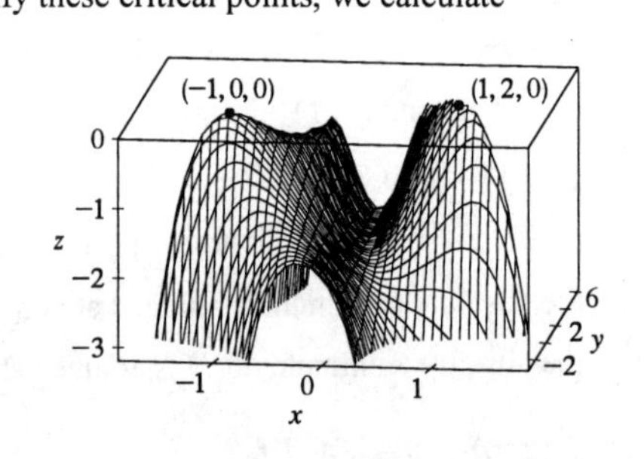

35. Here $d = \sqrt{x^2 + y^2 + z^2}$ with $z = \frac{1}{3}(4 - x - 2y)$. So minimize $d^2 = x^2 + y^2 + \frac{1}{9}(4 - x - 2y)^2 = f(x,y)$. Then $f_x = 2x - \frac{2}{9}(4 - x - 2y) = \frac{1}{9}(20x + 4y - 8)$, $f_y = \frac{1}{9}(26y + 4x - 16)$. Solving the equations $20x + 4y - 8 = 0$ and $26y + 4x - 16 = 0$, we get $x = \frac{2}{7}$, $y = \frac{4}{7}$, so the only critical point is $\left(\frac{2}{7}, \frac{4}{7}\right)$. Since the absolute minimum has to occur at a critical point, the point $\left(\frac{2}{7}, \frac{4}{7}, \frac{6}{7}\right)$ is is the closest point to the origin $\left[\text{or check } D\left(\frac{2}{7}, \frac{4}{7}\right)\right]$.

37. $d = \sqrt{(x-2)^2 + (y+2)^2 + (z-3)^2}$, where $z = \frac{1}{3}(6x + 4y - 2)$, so we minimize $d^2 = f(x,y) = (x-2)^2 + (y+2)^2 + \left(2x + \frac{4}{3}y - \frac{11}{3}\right)^2$. Then $f_x = 10x + \frac{16}{3}y - \frac{56}{3}$ and $f_y = \frac{50}{9}y + \frac{16}{3}x - \frac{52}{9}$. Solving $50y + 48x = 52$ and $16y + 30x = 56$ simultaneously gives $x = \frac{164}{61}$, $y = -\frac{94}{61}$. The absolute minimum must occur at a critical point. Thus $d^2 = \left(\frac{42}{61}\right)^2 + \left(\frac{28}{61}\right)^2 + \left(-\frac{21}{61}\right)^2$ or $d = \frac{7}{\sqrt{61}}$.

39. Minimize $d^2 = x^2 + y^2 + z^2 = x^2 + y^2 + xy + 1$. Then $f_x = 2x + y$, $f_y = 2y + x$ so the critical point is $(0, 0)$ and $D(0, 0) = 4 - 1 > 0$ with $f_{xx}(0, 0) = 2$ so this is a minimum. Thus $z^2 = 1$ or $z = \pm 1$ and the points on the surface are $(0, 0, \pm 1)$.

41. $x + y + z = 100$, so maximize $f(x, y) = xy(100 - x - y)$. $f_x = 100y - 2xy - y^2$, $f_y = 100x - x^2 - 2xy$, $f_{xx} = -2y$, $f_{yy} = -2x$, $f_{xy} = 100 - 2x - 2y$. Then $f_x = 0$ implies $y = 0$ or $y = 100 - 2x$. Substituting $y = 0$ into $f_y = 0$ gives $x = 0$ or $x = 100$ and substituting $y = 100 - 2x$ into $f_y = 0$ gives $3x^2 - 100x = 0$ so $x = 0$ or $\frac{100}{3}$. Thus the critical points are $(0, 0)$, $(100, 0)$, $(0, 100)$ and $\left(\frac{100}{3}, \frac{100}{3}\right)$.
$D(0, 0) = D(100, 0) = D(0, 100) = -10{,}000$ while $D\left(\frac{100}{3}, \frac{100}{3}\right) = \frac{10{,}000}{3}$ and $f_{xx}\left(\frac{100}{3}, \frac{100}{3}\right) = -\frac{200}{3} < 0$. Thus $(0, 0)$, $(100, 0)$ and $(0, 100)$ are saddle points whereas $f\left(\frac{100}{3}, \frac{100}{3}\right)$ is a local maximum. Thus the numbers are $x = y = z = \frac{100}{3}$.

43. Maximize $f(x, y) = xy(36 - 9x^2 - 36y^2)^{1/2}/2$ with (x, y, z) in first octant. Then

$$f_x = \frac{y(36 - 9x^2 - 36y^2)^{1/2}}{2} + \frac{-9x^2y(36 - 9x^2 - 36y^2)^{-1/2}}{2} = \frac{(36y - 18x^2y - 36y^3)}{2(36 - 9x^2 - 36y^2)^{1/2}} \text{ and}$$

$$f_y = \frac{36x - 9x^3 - 72xy^2}{2(36 - 9x^2 - 36y^2)^{1/2}}.$$ Setting $f_x = 0$ gives $y = 0$ or $y^2 = \dfrac{2 - x^2}{2}$ but $y > 0$, so only the latter solution applies. Substituting this y into $f_y = 0$ gives $x^2 = \frac{4}{3}$ or $x = \frac{2}{\sqrt{3}}$, $y = \frac{1}{\sqrt{3}}$ and then $z^2 = (36 - 12 - 12)/4 = 3$. That this gives a maximum volume follows from the geometry. This maximum volume is
$V = (2x)(2y)(2z) = 8\left(\frac{2}{\sqrt{3}}\right)\left(\frac{1}{\sqrt{3}}\right)\left(\sqrt{3}\right) = \frac{16}{\sqrt{3}}$.

45. Maximize $f(x, y) = \dfrac{xy}{3}(6 - x - 2y)$, then the maximum volume is $V = xyz$.
$f_x = \frac{1}{3}(6y - 2xy - y^2) = \dfrac{y}{3}(6 - 2x - 2y)$ and $f_y = \dfrac{x}{3}(6 - x - 4y)$. Setting $f_x = 0$ and $f_y = 0$ gives the critical point $(2, 1)$ which geometrically must yield a maximum. Thus the volume of the largest such box is $V = (2)(1)\left(\frac{2}{3}\right) = \frac{4}{3}$.

47. Let the dimensions be x, y, and z; then $4x + 4y + 4z = c$ and the volume is
$V = xyz = xy\left[\frac{1}{4}c - x - y\right] = \frac{1}{4}cxy - x^2y - xy^2$, $x > 0$, $y > 0$. Then $V_x = \frac{1}{4}cy - 2xy - y^2$ and $V_y = \frac{1}{4}cx - x^2 - 2xy$, so $V_x = 0 = V_y$ when $2x + y = \frac{1}{4}c$ and $x + 2y = \frac{1}{4}c$. Solving, we get $x = \frac{1}{12}c$, $y = \frac{1}{12}c$ and $z = \frac{1}{4}c - x - y = \frac{1}{12}c$. From the geometrical nature of the problem, this critical point must give an absolute maximum. Thus the box is a cube with edge length $\frac{1}{12}c$.

49. Let the dimensions be x, y and z, then minimize $xy + 2(xz + yz)$ if $xyz = 32{,}000 \text{ m}^3$. Then
$f(x, y) = xy + [64{,}000(x + y)/xy] = xy + 64{,}000(x^{-1} + y^{-1})$, $f_x = y - 64{,}000x^{-2}$, $f_y = x - 64{,}000y^{-2}$. And $f_x = 0$ implies $y = 64{,}000/x^2$; substituting into $f_y = 0$ implies $x^3 = 64{,}000$ or $x = 40$ and then $y = 40$. Now $D(x, y) = [(2)(64{,}000)]^2 x^{-3}y^{-3} - 1 > 0$ for $(40, 40)$ and $f_{xx}(40, 40) > 0$ so this is indeed a minimum. Thus the dimensions of the box are $x = y = 40$ cm, $z = 20$ cm.

51. Note that here the variables are m and b, and $f(m,b) = \sum_{i=1}^{n}[y_i - (mx_i + b)]^2$. Then

$f_m = \sum_{i=1}^{n} -2x_i[y_i - (mx_i + b)] = 0$ implies $\sum_{i=1}^{n}(x_iy_i - mx_i^2 - bx_i) = 0$ or $\sum_{i=1}^{n} x_iy_i = m\sum_{i=1}^{n} x_i^2 + b\sum_{i=1}^{n} x_i$ and

$f_b = \sum_{i=1}^{n} -2[y_i - (mx_i + b)] = 0$ implies $\sum_{i=1}^{n} y_i = m\sum_{i=1}^{n} x_i + \sum_{i=1}^{n} b = m\left(\sum_{i=1}^{n} x_i\right) + nb$. Thus we have the

two desired equations. Now $f_{mm} = \sum_{i=1}^{n} 2x_i^2$, $f_{bb} = \sum_{i=1}^{n} 2 = 2n$ and $f_{mb} = \sum_{i=1}^{n} 2x_i$. And $f_{mm}(m,b) > 0$ always

and $D(m,b) = 4n\left(\sum_{i=1}^{n} x_i^2\right) - 4\left(\sum_{i=1}^{n} x_i\right)^2 = 4\left[n\left(\sum_{i=1}^{n} x_i^2\right) - \left(\sum_{i=1}^{n} x_i\right)^2\right] > 0$ always so the solutions of

these two equations do indeed minimize $\sum_{i=1}^{n} d_i^2$.

53.

x_i	y_i	x_i^2	x_iy_i
69	138	4761	9522
65	127	4225	8255
71	178	5041	12,638
73	185	5329	13,505
68	141	4624	9588
63	122	3969	7686
70	158	4900	11,060
67	135	4489	9045
69	145	4761	10,005
70	162	4900	11,340
685	1491	46,999	102,644

From the table, $685m + 10b = 1491$ and
$46{,}999m + 685b = 102{,}644$.
Solving simultaneously gives $m = \frac{1021}{153} \approx 6.67$
and $b = 149.1 - 68.5\left(\frac{1021}{153}\right) \approx -308.01$.
Thus a 6 ft boy is predicted to weigh about
$(6.67)(72) - 308.01$ or 172.2 lbs.

EXERCISES 12.8

1. $f(x,y) = x^2 - y^2$, $g(x,y) = x^2 + y^2 = 1 \Rightarrow \nabla f = \langle 2x, -2y \rangle$, $\lambda \nabla g = \langle 2\lambda x, 2\lambda y \rangle$. Then $2x = 2\lambda x$ implies $x = 0$ or $\lambda = 1$. If $x = 0$, then $x^2 + y^2 = 1$ implies $y = \pm 1$ and if $\lambda = 1$, then $-2y = 2\lambda y$ implies $y = 0$ and thus $x = \pm 1$. Thus the possible points for the extrema of f are $(\pm 1, 0)$, $(0, \pm 1)$. But $f(\pm 1, 0) = 1$ while $f(0, \pm 1) = -1$ so the maximum value of f on $x^2 + y^2 = 1$ is $f(\pm 1, 0) = 1$ and the minimum value is $f(0, \pm 1) = -1$.

3. $f(x,y) = xy$, $g(x,y) = 9x^2 + y^2 = 4 \Rightarrow \nabla f = \langle y, x \rangle$, $\lambda \nabla g = \langle 18\lambda x, 2\lambda y \rangle$. Then $y = 18\lambda x$ implies $(x,y) = (0,0)$ or $\lambda = y/18x$ and $x = 2\lambda y$ implies $(x,y) = (0,0)$ or $\lambda = \dfrac{x}{2y}$. Thus $(x,y) = (0,0)$ or $\dfrac{y}{18x} = \dfrac{x}{2y}$ implies $y^2 = 9x^2$. Now $(x,y) = (0,0)$ doesn't satisfy $g(x,y) = 4$, and when $y^2 = 9x^2$, $g(x,y) = 4$ implies $x^2 = \frac{2}{9}$ or $x = \pm\frac{\sqrt{2}}{3}$. Hence the possible points are $\left(\pm\frac{\sqrt{2}}{3}, \sqrt{2}\right)$, $\left(\pm\frac{\sqrt{2}}{3}, -\sqrt{2}\right)$ and the maximum value of f on the ellipse is $f\left(\frac{\sqrt{2}}{3}, \sqrt{2}\right) = f\left(-\frac{\sqrt{2}}{3}, -\sqrt{2}\right) = \frac{2}{3}$ while the minimum value is $f\left(-\frac{\sqrt{2}}{3}, \sqrt{2}\right) = f\left(\frac{\sqrt{2}}{3}, -\sqrt{2}\right) = -\frac{2}{3}$.

5. $f(x,y,z) = x + 3y + 5z$, $g(x,y,z) = x^2 + y^2 + z^2 = 1 \Rightarrow \nabla f = \langle 1, 3, 5 \rangle$, $\lambda \nabla g = \langle 2\lambda x, 2\lambda y, 2\lambda z \rangle$. Then $\nabla f = \lambda \nabla g$ implies $\lambda = \dfrac{1}{2x} = \dfrac{3}{2y} = \dfrac{5}{2z}$ so $x = \frac{1}{5}z$, $y = \frac{3}{5}z$. Then $x^2 + y^2 + z^2 = 1$ implies $\frac{1}{25}z^2 + \frac{9}{25}z^2 + z^2 = 1$ or $z = \pm\sqrt{\frac{5}{7}}$. Thus the possible points are $\left(\pm\frac{1}{\sqrt{35}}, \pm\frac{3}{\sqrt{35}}, \pm\frac{5}{\sqrt{35}}\right)$ with the maximum being $f\left(\frac{1}{\sqrt{35}}, \frac{3}{\sqrt{35}}, \frac{5}{\sqrt{35}}\right) = \sqrt{35}$ and the minimum being $f\left(-\frac{1}{\sqrt{35}}, -\frac{3}{\sqrt{35}}, -\frac{5}{\sqrt{35}}\right) = -\sqrt{35}$.

7. $f(x,y,z) = xyz$, $g(x,y,z) = x^2 + 2y^2 + 3z^2 = 6 \Rightarrow \nabla f = \langle yz, xz, xy \rangle$, $\lambda \nabla g = \langle 2\lambda x, 4\lambda y, 6\lambda z \rangle$. Then $\nabla f = \lambda \nabla g$ implies $\lambda = (yz)/(2x) = (xz)/(4y) = (xy)/(6z)$ or $x^2 = 2y^2$ and $z^2 = \frac{2}{3}y^2$. Thus $x^2 + 2y^2 + 3z^2 = 6$ implies $6y^2 = 6$ or $y = \pm 1$. Then the possible points are $\left(\sqrt{2}, \pm 1, \sqrt{\frac{2}{3}}\right)$, $\left(\sqrt{2}, \pm 1, -\sqrt{\frac{2}{3}}\right)$, $\left(-\sqrt{2}, \pm 1, \sqrt{\frac{2}{3}}\right)$, $\left(-\sqrt{2}, \pm 1, -\sqrt{\frac{2}{3}}\right)$. And the maximum value of f on the ellipsoid is $\frac{2}{\sqrt{3}}$, occurring when all coordinates are positive or exactly two are negative and the minimum is $-\frac{2}{\sqrt{3}}$ occurring when 1 or 3 of the coordinates are negative.

9. $f(x,y,z) = x^2 + y^2 + z^2$, $g(x,y,z) = x^4 + y^4 + z^4 = 1 \quad\Rightarrow\quad \nabla f = \langle 2x, 2y, 2z \rangle$,
$\lambda \nabla g = \langle 4\lambda x^3, 4\lambda y^3, 4\lambda z^3 \rangle$.
Case 1: If $x \neq 0$, $y \neq 0$ and $z \neq 0$, then $\nabla f = \lambda \nabla g$ implies $\lambda = 1/(2x^2) = 1/(2y^2) = 1/(2z^2)$ or $x^2 = y^2 = z^2$ and $3x^4 = 1$ or $x = \pm\frac{1}{\sqrt[4]{3}}$ giving the points $\left(\pm\frac{1}{\sqrt[4]{3}}, \frac{1}{\sqrt[4]{3}}, \frac{1}{\sqrt[4]{3}}\right)$, $\left(\pm\frac{1}{\sqrt[4]{3}}, -\frac{1}{\sqrt[4]{3}}, \frac{1}{\sqrt[4]{3}}\right)$, $\left(\pm\frac{1}{\sqrt[4]{3}}, \frac{1}{\sqrt[4]{3}}, -\frac{1}{\sqrt[4]{3}}\right)$, $\left(\pm\frac{1}{\sqrt[4]{3}}, -\frac{1}{\sqrt[4]{3}}, -\frac{1}{\sqrt[4]{3}}\right)$ all with an f-value of $\sqrt{3}$.
Case 2: If one of the variables equals zero and the other two are not zero, then the squares of the two nonzero coordinates are equal with common value $\frac{1}{\sqrt{2}}$ and corresponding f value of $\sqrt{2}$.
Case 3: If exactly two of the variables are zero, then the third variable has value ± 1 with the corresponding f value of 1. Thus on $x^4 + y^4 + z^4 = 1$, the maximum value of f is $\sqrt{3}$ and the minimum value is 1.

11. $f(x,y,z,t) = x + y + z + t$, $g(x,y,z,t) = x^2 + y^2 + z^2 + t^2 = 1 \quad\Rightarrow\quad \langle 1,1,1,1 \rangle = \langle 2\lambda x, 2\lambda y, 2\lambda z, 2\lambda t \rangle$ so $\lambda = 1/(2x) = 1/(2y) = 1/(2z) = 1/(2t)$ and $x = y = z = t$. But $x^2 + y^2 + z^2 + t^2 = 1$, so the possible points are $\left(\pm\frac{1}{2}, \pm\frac{1}{2}, \pm\frac{1}{2}, \pm\frac{1}{2}\right)$. Thus the maximum value of f is $f\left(\frac{1}{2}, \frac{1}{2}, \frac{1}{2}, \frac{1}{2}\right) = 2$ and the minimum value is $f\left(-\frac{1}{2}, -\frac{1}{2}, -\frac{1}{2}, -\frac{1}{2}\right) = -2$.

13. $f(x,y,z) = x + 2y$, $g(x,y,z) = x + y + z = 1$, $h(x,y,z) = y^2 + z^2 = 4 \quad\Rightarrow\quad \nabla f = \langle 1, 2, 0 \rangle$, $\lambda \nabla g = \langle \lambda, \lambda, \lambda \rangle$ and $\mu \nabla h = \langle 0, 2\mu y, 2\mu z \rangle$. Then $1 = \lambda$, $2 = \lambda + 2\mu y$ and $0 = \lambda + 2\mu z$ so $\mu y = \frac{1}{2} = -\mu z$ or $y = 1/(2\mu)$, $z = -1/(2\mu)$. Thus $x + y + z = 1$ implies $x = 1$ and $y^2 + z^2 = 4$ implies $\mu = \pm\frac{1}{2\sqrt{2}}$. Then the possible points are $\left(1, \pm\sqrt{2}, \mp\sqrt{2}\right)$ and the maximum value is $f\left(1, \sqrt{2}, -\sqrt{2}\right) = 1 + 2\sqrt{2}$ and the minimum value is $f\left(1, -\sqrt{2}, \sqrt{2}\right) = 1 - 2\sqrt{2}$.

15. $f(x,y,z) = yz + xy$, $g(x,y,z) = xy = 1$, $h(x,y,z) = y^2 + z^2 = 1 \quad\Rightarrow\quad \nabla f = \langle y, x + z, y \rangle$, $\lambda \nabla g = \langle \lambda y, \lambda x, 0 \rangle$, $\mu \nabla h = \langle 0, 2\mu y, 2\mu z \rangle$. Then $y = \lambda y$ implies $\lambda = 1$ [$y \neq 0$ since $g(x,y,z) = 1$], $x + z = \lambda x + 2\mu y$ and $y = 2\mu z$. Thus $\mu = z/(2y) = y/(2z)$ or $y^2 = z^2$, and so $y^2 + z^2 = 1$ implies $y = \pm\frac{1}{\sqrt{2}}$, $z = \pm\frac{1}{\sqrt{2}}$. Then $xy = 1$ implies $x = \pm\sqrt{2}$ and the possible points are $\left(\pm\sqrt{2}, \pm\frac{1}{\sqrt{2}}, \frac{1}{\sqrt{2}}\right)$, $\left(\pm\sqrt{2}, \pm\frac{1}{\sqrt{2}}, -\frac{1}{\sqrt{2}}\right)$. Hence the maximum of f subject to the constraints is $f\left(\pm\sqrt{2}, \pm\frac{1}{\sqrt{2}}, \pm\frac{1}{\sqrt{2}}\right) = \frac{3}{2}$ and the minimum is $f\left(\pm\sqrt{2}, \pm\frac{1}{\sqrt{2}}, \mp\frac{1}{\sqrt{2}}\right) = \frac{1}{2}$.
Note: Since $xy = 1$ is one of the constraints we could have solved the problem by solving $f(y,z) = yz + 1$ subject to $y^2 + z^2 = 1$.

17. $f(x,y) = e^{-xy}$. For the interior of the region, we find the critical points: $f_x = -ye^{-xy}$, $f_y = -xe^{-xy}$, so the only critical point is $(0,0)$, and $f(0,0) = 1$. For the boundary, we use Lagrange multipliers. $g(x,y) = x^2 + 4y^2 = 1 \quad\Rightarrow\quad \lambda \nabla g = \langle 2\lambda x, 8\lambda y \rangle$, so setting $\nabla f = \lambda \nabla g$ we get $-ye^{-xy} = 2\lambda x$ and $-xe^{-xy} = 8\lambda y$. The first of these gives $e^{-xy} = -2\lambda x/y$, and then the second gives $-x(-2\lambda x/y) = 8\lambda y \quad\Rightarrow\quad x^2 = 4y^2$. Solving this last equation with the constraint $x^2 + 4y^2 = 1$ gives $x = \pm\frac{1}{\sqrt{2}}$ and $y = \pm\frac{1}{2\sqrt{2}}$. Now $f\left(\pm\frac{1}{\sqrt{2}}, \mp\frac{1}{2\sqrt{2}}\right) = e^{1/4} \approx 1.284$ and $f\left(\pm\frac{1}{\sqrt{2}}, \pm\frac{1}{2\sqrt{2}}\right) = e^{-1/4} \approx 0.779$. The former are the maxima on the region and the latter are the minima.

19. **(a)** The graphs of $f(x, y) = 3.7$ and $f(x, y) = 350$ seem to be tangent to the circle, and so 3.7 and 350 are the approximate minimum and maximum values of the function $f(x, y)$ subject to the constraint $(x-3)^2 + (y-3)^2 = 9$.

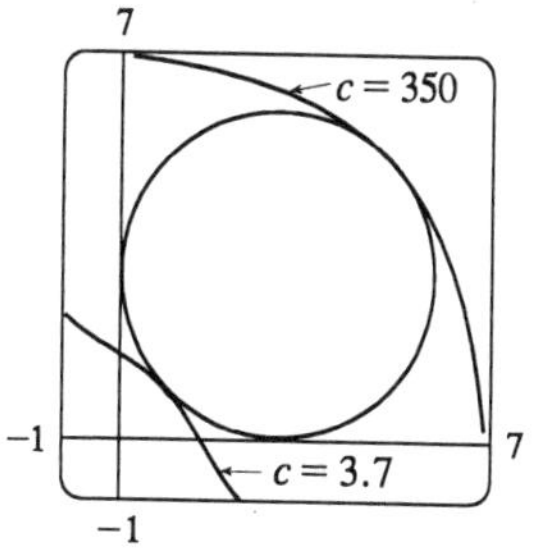

7
$c = 350$
-1
7
$c = 3.7$
-1

(b) Let $g(x, y) = (x-3)^2 + (y-3)^2$. We calculate $f_x(x, y) = 3x^2 + 3y$, $f_y(x, y) = 3y^2 + 3x$, $g_x(x, y) = 2x - 6$, and $g_y(x, y) = 2y - 6$, and use a CAS to search for solutions to the equations $g(x, y) = (x-3)^2 + (y-3)^2 = 9$, $f_x = \lambda g_x$, and $f_y = \lambda g_y$. The solutions are $(x, y) = \left(3 - \frac{3}{2}\sqrt{2}, 3 - \frac{3}{2}\sqrt{2}\right) \approx (0.879, 0.879)$ and $(x, y) = \left(3 + \frac{3}{2}\sqrt{2}, 3 + \frac{3}{2}\sqrt{2}\right) \approx (5.121, 5.121)$. These give $f\left(3 - \frac{3}{2}\sqrt{2}, 3 - \frac{3}{2}\sqrt{2}\right) = \frac{351}{2} - \frac{243}{2}\sqrt{2} \approx 3.673$ and $f\left(3 + \frac{3}{2}\sqrt{2}, 3 + \frac{3}{2}\sqrt{2}\right) = \frac{351}{2} + \frac{243}{2}\sqrt{2} \approx 347.33$, in accordance with part (a).

21. $Q(x, y) = Kx^\alpha y^{1-\alpha}$, $g(x, y) = mx + ny = p \quad \Rightarrow \quad \nabla Q = \langle \alpha K x^{\alpha-1} y^{1-\alpha}, (1-\alpha) K x^\alpha y^{-\alpha} \rangle$, $\lambda \nabla g = \langle \lambda m, \lambda n \rangle$. Then $\alpha K (y/x)^{1-\alpha} = \lambda m$ and $(1-\alpha) K (x/y)^\alpha = \lambda n$ and $mx + ny = p$, so $\alpha K (y/x)^{1-\alpha}/m = (1-\alpha) K (x/y)^\alpha / n$ or $n\alpha/[m(1-\alpha)] = (x/y)^\alpha (x/y)^{1-\alpha}$ or $x = yn\alpha/[m(1-\alpha)]$. Substituting into $mx + ny = p$ gives $y = p(1-\alpha)/n$ and $x = p\alpha/m$ for the maximum production.

23. Let the sides of the rectangle be x and y. Then $f(x, y) = xy$, $g(x, y) = 2x + 2y = p \quad \Rightarrow \quad \nabla f(x, y) = \langle y, x \rangle$, $\lambda \nabla g = \langle 2\lambda, 2\lambda \rangle$. Then $\lambda = \frac{1}{2}y = \frac{1}{2}x$ implies $x = y$ and the rectangle with maximum area is a square with side length $\frac{1}{4}p$.

25. $f(x, y, z) = x^2 + y^2 + z^2$, $g(x, y, z) = x + 2y + 3z = 4$. Then $\nabla f = \langle 2x, 2y, 2z \rangle = \lambda \nabla g = \langle \lambda, 2\lambda, 3\lambda \rangle \quad \Rightarrow$ $x = \frac{1}{2}\lambda$, $y = \lambda$, $z = \frac{3}{2}\lambda$ and $\frac{1}{2}\lambda + 2\lambda + \frac{9}{2}\lambda = 4 \quad \Rightarrow \quad \lambda = \frac{4}{7}$. Hence the point closest to the origin is $\left(\frac{2}{7}, \frac{4}{7}, \frac{6}{7}\right)$.

27. $f(x, y, z) = (x-2)^2 + (y+2)^2 + (z-3)^2$, $g(x, y, z) = 6x + 4y - 3z = 2 \quad \Rightarrow$ $\nabla f = \langle 2(x-2), 2(y+2), 2(z-3) \rangle = \lambda \nabla g = \langle 6\lambda, 4\lambda, -3\lambda \rangle$, so $x = 3\lambda + 2$, $y = 2\lambda - 2$, $z = -\frac{3}{2}\lambda + 3$ and $(18\lambda + 12) + (8\lambda - 8) + \frac{9}{2}\lambda - 9 = 2$ implies $\lambda = \frac{14}{61}$. Thus the shortest distance is $\sqrt{\left(\frac{42}{61}\right)^2 + \left(\frac{28}{61}\right)^2 + \left(-\frac{21}{61}\right)^2} = \frac{7}{\sqrt{61}}$.

29. $f(x, y, z) = x^2 + y^2 + z^2$, $g(x, y, z) = z^2 - xy - 1 = 0 \;\Rightarrow\; \nabla f = \langle 2x, 2y, 2z \rangle = \lambda \nabla g = \langle -\lambda y, -\lambda x, 2\lambda z \rangle$. Then $2z = 2\lambda z$ implies $z = 0$ or $\lambda = 1$. If $z = 0$ then $g(x, y, z) = 1$ implies $xy = -1$ or $x = -1/y$. Thus $2x = -\lambda y$ and $2y = -\lambda x$ imply $\lambda = 2/y^2 = 2y^2$ or $y = \pm 1$, $x = \pm 1$. If $\lambda = 1$, then $2x = -y$ and $2y = -x$ imply $x = y = 0$, so $z = \pm 1$. Hence the possible points are $(\pm 1, \mp 1, 0)$, $(0, 0, \pm 1)$ and the minimum value of f is $f(0, 0, \pm 1) = 1$, so the points closest to the origin are $(0, 0, \pm 1)$.

31. $f(x,y,z) = xyz$, $g(x,y,z) = x+y+z = 100 \quad \Rightarrow \quad \nabla f = \langle yz, xz, xy\rangle = \lambda \nabla g = \langle \lambda, \lambda, \lambda\rangle$. Then $\lambda = yz = xz = xy$ implies $x = y = z = \frac{100}{3}$.

33. If the dimensions are $2x$, $2y$ and $2z$, then $f(x,y,z) = 8xyz$ and $g(x,y,z) = 9x^2 + 36y^2 + 4z^2 = 36 \quad \Rightarrow$ $\nabla f = \langle 8yz, 8xz, 8xy\rangle = \lambda\nabla g = \langle 18\lambda x, 72\lambda y, 8\lambda z\rangle$. Thus $18\lambda x = 8yz$, $72\lambda y = 8xz$, $8\lambda z = 8xy$ so $x^2 = 4y^2$, $z^2 = 9y^2$ and $36y^2 + 36y^2 + 36y^2 = 36$ or $y = \frac{1}{\sqrt{3}}$ $(y > 0)$. Thus the volume of the largest such rectangle is $8\left(\frac{1}{\sqrt{3}}\right)\left(\frac{2}{\sqrt{3}}\right)\left(\frac{3}{\sqrt{3}}\right) = 16\sqrt{3}$.

35. $f(x,y,z) = xyz$, $g(x,y,z) = x + 2y + 3z = 6 \quad \Rightarrow \quad \nabla f = \langle yz, xz, xy\rangle = \lambda\nabla g = \langle \lambda, 2\lambda, 3\lambda\rangle$. Then $\lambda = yz = \frac{1}{2}xz = \frac{1}{3}xy$ implies $x = 2y$, $z = \frac{2}{3}y$. But $2y + 2y + 2y = 6$ so $y = 1$, $x = 2$, $z = \frac{2}{3}$ and the volume is $V = \frac{4}{3}$.

37. $f(x,y,z) = xyz$, $g(x,y,z) = 4(x+y+z) = c \quad \Rightarrow \quad \nabla f = \langle yz, xz, xy\rangle$, $\lambda\nabla g = \langle 4\lambda, 4\lambda, 4\lambda\rangle$. Thus $4\lambda = yz = xz = xy$ or $x = y = z = \frac{1}{12}c$ are the dimensions giving the maximum volume.

39. $f(x,y,z) = xy + 2xz + 2yz$, $g(x,y,z) = xyz = 32{,}000\text{ cm}^3 \quad \Rightarrow$ $\nabla f = \langle 2z + y, 2z + x, 2(x+y)\rangle = \lambda\nabla g = \langle \lambda yz, \lambda xz, \lambda xy\rangle$. Then (1) $\lambda yz = 2z + y$, (2) $\lambda xz = 2z + x$, and (3) $\lambda xy = 2(x+y)$. Now (1) – (2) implies $\lambda z(y - x) = y - x$, so $x = y$ or $\lambda = 1/z$. If $\lambda = 1/z$ then (1) implies $z = 0$ which can't be, so $x = y$. But twice (2) minus (3) together with $x = y$ implies $\lambda y(2x - y) = (4z + 2y) - 4y$ or $\lambda y(2z - y) = 2(2z - y)$ so $z = y/2$ or $\lambda = 2/y$. If $\lambda = 2/y$ then (3) implies $y = 0$ which can't be. Thus $x = y = 2z$ and $\frac{1}{2}y^3 = 32{,}000$ or $y = 40$ and the dimensions which minimize the volume are $x = y = 40$ cm, $z = 20$ cm.

41. We need to find the extrema of $f(x,y,z) = x^2 + y^2 + z^2$ subject to the two constraints $g(x,y,z) = x + y + 2z = 2$ and $h(x,y,z) = x^2 + y^2 - z = 0$. $\nabla f = \langle 2x, 2y, 2z\rangle$, $\lambda\nabla g = \langle \lambda, \lambda, 2\lambda\rangle$ and $\mu\nabla h = \langle 2\mu x, 2\mu y, -\mu\rangle$. Thus we need (1) $2x = \lambda + 2\mu x$, (2) $2y = \lambda + 2\mu y$, (3) $2z = 2\lambda - \mu$, (4) $x + y + 2z = 2$, and (5) $x^2 + y^2 - z = 0$. From (1) and (2), $2(x - y) = 2\mu(x - y)$, so if $x \neq y$, $\mu = 1$. Putting this in (3) gives $2z = 2\lambda - 1$ or $\lambda = z + \frac{1}{2}$, but putting $\mu = 1$ into (1) says $\lambda = 0$. Hence $z + \frac{1}{2} = 0$ or $z = -\frac{1}{2}$. Then (4) and (5) become $x + y - 3 = 0$ and $x^2 + y^2 + \frac{1}{2} = 0$. The last equation cannot be true, so this case gives no solution. So we must have $x = y$. Then (4) and (5) become $2x + 2z = 2$ and $2x^2 - z = 0$ which imply $z = 1 - x$ and $z = 2x^2$. Thus $2x^2 = 1 - x$ or $2x^2 + x - 1 = (2x - 1)(x + 1) = 0$ so $x = \frac{1}{2}$ or $x = -1$. The two points to check are $\left(\frac{1}{2}, \frac{1}{2}, \frac{1}{2}\right)$ and $(-1, -1, 2)$: $f\left(\frac{1}{2}, \frac{1}{2}, \frac{1}{2}\right) = \frac{3}{4}$ and $f(-1,-1,2) = 6$. Thus $\left(\frac{1}{2}, \frac{1}{2}, \frac{1}{2}\right)$ is the point on the ellipse nearest the origin and $(-1,-1,2)$ is the one farthest from the origin.

REVIEW EXERCISES FOR CHAPTER 12

1. True. $f_y(a,b) = \lim\limits_{h\to 0} \dfrac{f(a,b+h) - f(a,b)}{h}$ from (12.3.3). Let $h = y - b$. As $h \to 0$, $y \to b$. Then by substituting, we get $f_y(a,b) = \lim\limits_{y\to b} \dfrac{f(a,y) - f(a,b)}{y - b}$.

3. False. $f_{xy} = \partial^2 f/(\partial y \partial x)$.

5. False. See Example 3 in Section 12.2.

7. True. If f has a local minimum and f is differentiable at (a,b) then by (12.7.2), $f_x(a,b) = 0$ and $f_y(a,b) = 0$ so $\nabla f(a,b) = \langle f_x(a,b), f_y(a,b)\rangle = \langle 0,0\rangle = \mathbf{0}$.

9. False. $\nabla f(x,y) = \langle 0, 1/y\rangle$.

11. True. $\nabla f = \langle \cos x, \cos y\rangle$, so $|\nabla f| = \sqrt{\cos^2 x + \cos^2 y}$. But $|\cos\theta| \le 1$, so $|\nabla f| \le \sqrt{2}$. Now $D_{\mathbf{u}} f(x,y) = \nabla f \cdot \mathbf{u} = |\nabla f||\mathbf{u}|\cos\theta$, but $\mathbf{u}$ is a unit vector, so $|D_{\mathbf{u}} f(x,y)| \le \sqrt{2}\cdot 1\cdot 1 = \sqrt{2}$.

13. $x \neq 1$ and $x + y + 1 > 0$, so $D = \{(x,y) \mid y > -x - 1, x \neq 1\}$.

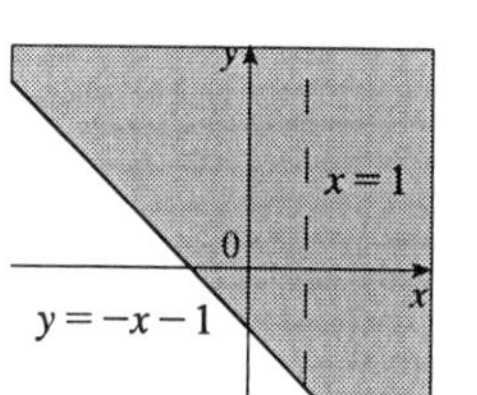

15. $D = \{(x,y) \mid -1 \le x \le 1\}$

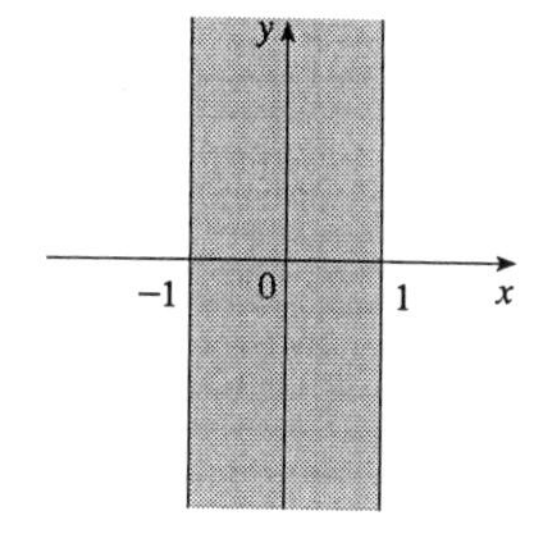

17. $z = f(x,y) = 1 - x^2 - y^2$, a paraboloid with vertex $(0,0,1)$.

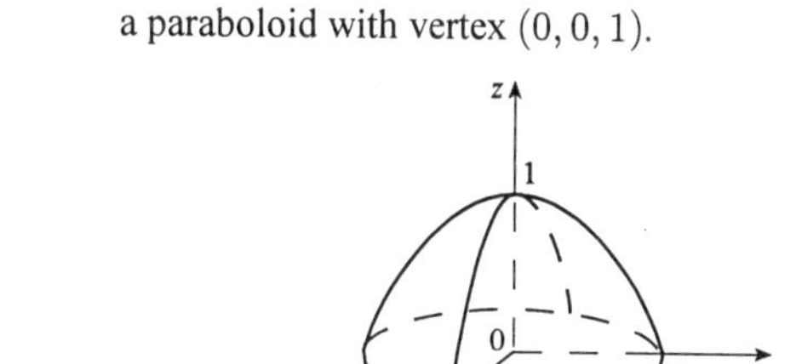

19. Let $k = e^{-c} = e^{-(x^2+y^2)}$ be the level curves, then $-\ln k = c = x^2 + y^2$, so we have a family of concentric circles.

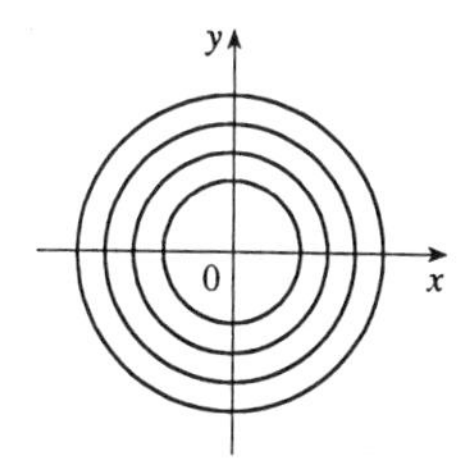

21. Since $0 \le \dfrac{x^2y^2}{x^2+2y^2} \le \dfrac{x^2y^2}{x^2+y^2} \le \dfrac{(x^2+y^2)(x^2+y^2)}{x^2+y^2} = x^2+y^2$, given $\epsilon > 0$, let $\delta = \sqrt{\epsilon}$. Then whenever $0 < \sqrt{x^2+y^2} < \delta$, $\left|\dfrac{x^2y^2}{x^2+2y^2} - 0\right| = \dfrac{x^2y^2}{x^2+2y^2} \le x^2+y^2 < \delta^2 = \epsilon$. Hence $\displaystyle\lim_{(x,y)\to(0,0)} \frac{x^2y^2}{x^2+2y^2} = 0$.

(Or use the Squeeze Theorem.)

23. $f(x,y) = 3x^4 - x\sqrt{y} \quad\Rightarrow\quad f_x = 12x^3 - \sqrt{y}$, $f_y = -\frac{1}{2}xy^{-1/2}$

25. $f(s,t) = e^{2s}\cos\pi t \quad\Rightarrow\quad f_s = 2e^{2s}\cos\pi t$, $f_t = -\pi e^{2s}\sin\pi t$

27. $f(x,y,z) = xy^z \quad\Rightarrow\quad f_x = y^z$, $f_y = xzy^{z-1}$, $f_z = xy^z\ln y$

29. $f(x,y) = x^2y^3 - 2x^4 + y^2 \quad\Rightarrow\quad f_x = 2xy^3 - 8x^3$, $f_y = 3x^2y^2 + 2y$, $f_{xx} = 2y^3 - 24x^2$, $f_{yy} = 6x^2y + 2$, and $f_{xy} = f_{yx} = 6xy^2$.

31. $f(x,y,z) = xy^2z^3 \quad\Rightarrow\quad f_x = y^2z^3$, $f_y = 2xyz^3$, $f_z = 3xy^2z^2$, $f_{xx} = 0$, $f_{yy} = 2xz^3$, $f_{zz} = 6xy^2z$, $f_{xy} = f_{yx} = 2yz^3$, $f_{xz} = f_{zx} = 3y^2z^2$, and $f_{yz} = f_{zy} = 6xyz^2$.

33. $u = x^y \quad\Rightarrow\quad u_x = yx^{y-1}$, $u_y = x^y\ln x$ and $(x/y)u_x + (\ln x)^{-1}u_y = x^y + x^y = 2u$.

35. $z_x(0,1) = 0$, $z_y(0,1) = 6$ and the equation of the tangent plane is $z - 5 = 6(y-1)$ or $z - 6y = -1$.

37. $F(x,y,z) = xy^2z^3$, $\nabla F = \langle y^2z^3, 2xyz^3, 3xy^2z^2\rangle$ and $\nabla F(3,2,1) = \langle 4, 12, 36\rangle$. Thus the equation of the tangent plane is $4x + 12y + 36z = 72$ or $x + 3y + 9z = 18$.

39. $F(x,y,z) = x^2 + 2y^2 - 3z^2$, $F_x = 2x$, $F_y = 4y$, $F_z = -6z$; $F_x(3,2,-1) = 6$, $F_y(3,2,-1) = 8$, $F_z(3,2,-1) = 6$. So the equation of the tangent plane is $6(x-3) + 8(y-2) + 6(z+1) = 0$ or $3x + 4y + 3z = 14$.

41. $F(x,y,z) = x^2 + y^2 + z^2$, $\nabla F(x_0,y_0,z_0) = \langle 2x_0, 2y_0, 2z_0\rangle = k\langle 2,1,-3\rangle$ or $x_0 = k$, $y_0 = \frac{1}{2}k$ and $z_0 = -\frac{3}{2}k$. But $x_0^2 + y_0^2 + z_0^2 = 1$, so $\frac{7}{2}k^2 = 1$ and $k = \pm\sqrt{\frac{2}{7}}$. Hence there are two such points: $\left(\pm\sqrt{\frac{2}{7}}, \pm\frac{1}{\sqrt{14}}, \mp\frac{3}{\sqrt{14}}\right)$.

43. Let $w = f(x,y,z) = x^3\sqrt{y^2+z^2}$, then $f(2,3,4) = 8(5) = 40$ so set $(a,b,c) = (2,3,4)$. Then $\Delta x = -0.02$, $\Delta y = 0.01$, $\Delta z = -0.03$, $f_x = 3x^2\sqrt{y^2+z^2}$, $f_y = yx^3/\sqrt{y^2+z^2}$, $f_z = zx^3/\sqrt{y^2+z^2}$. Thus $(1.98)^3\sqrt{(3.01)^2 + (3.97)^2} \approx 40 + (60)(-0.02) + (24/5)(0.01) + (32/5)(-0.03) = 38.656$.

45. $\dfrac{dw}{dt} = \dfrac{1}{2\sqrt{x}}\left(2e^{2t}\right) + \dfrac{2y}{z}\left(3t^2+4\right) + \dfrac{-y^2}{z^2}(2t) = e^t + \dfrac{2y}{z}\left(3t^2+4\right) - 2t\dfrac{y^2}{z^2}$

47. $\dfrac{\partial z}{\partial x} = 2xf'(x^2-y^2)$, $\dfrac{\partial z}{\partial y} = 1 - 2yf'(x^2-y^2)$ $\left[\text{where } f' = \dfrac{df}{d(x^2-y^2)}\right]$. Then $y\dfrac{\partial z}{\partial x} + x\dfrac{\partial z}{\partial y} = 2xyf'(x^2-y^2) + x - 2xyf'(x^2-y^2) = x$.

49. $\dfrac{\partial z}{\partial x} = \dfrac{\partial z}{\partial u} y + \dfrac{\partial z}{\partial v}\dfrac{-y}{x^2}$ and

$$\frac{\partial^2 z}{\partial x^2} = y\frac{\partial}{\partial x}\left(\frac{\partial z}{\partial u}\right) + \frac{2y}{x^3}\frac{\partial z}{\partial v} + \frac{-y}{x^2}\frac{\partial}{\partial x}\left(\frac{\partial z}{\partial v}\right)$$

$$= \frac{2y}{x^3}\frac{\partial z}{\partial v} + y\left(\frac{\partial^2 z}{\partial u^2}y + \frac{\partial^2 z}{\partial v \partial u}\frac{-y}{x^2}\right) + \frac{-y}{x^2}\left(\frac{\partial^2 z}{\partial v^2}\frac{-y}{x^2} + \frac{\partial^2 z}{\partial u \partial v}y\right)$$

$$= \frac{2y}{x^3}\frac{\partial z}{\partial v} + y^2\frac{\partial^2 z}{\partial u^2} - \frac{2y^2}{x^2}\frac{\partial^2 z}{\partial u \partial v} + \frac{y^2}{x^4}\frac{\partial^2 z}{\partial v^2}.$$ Also $\dfrac{\partial z}{\partial y} = x\dfrac{\partial z}{\partial u} + \dfrac{1}{x}\dfrac{\partial z}{\partial v}$ and

$$\frac{\partial^2 z}{\partial y^2} = x\frac{\partial}{\partial y}\left(\frac{\partial z}{\partial u}\right) + \frac{1}{x}\frac{\partial}{\partial y}\left(\frac{\partial z}{\partial v}\right) = x\left(\frac{\partial^2 z}{\partial u^2}x + \frac{\partial^2 z}{\partial v \partial u}\frac{1}{x}\right) + \frac{1}{x}\left(\frac{\partial^2 z}{\partial v^2}\frac{1}{x} + \frac{\partial^2 z}{\partial u \partial v}x\right)$$

$$= x^2\frac{\partial^2 z}{\partial u^2} + 2\frac{\partial^2 z}{\partial u \partial v} + \frac{1}{x^2}\frac{\partial^2 z}{\partial v^2}.$$ Thus

$$x^2\frac{\partial^2 z}{\partial x^2} - y^2\frac{\partial^2 z}{\partial y^2} = \frac{2y}{x}\frac{\partial z}{\partial v} + x^2y^2\frac{\partial^2 z}{\partial u^2} - 2y^2\frac{\partial^2 z}{\partial u \partial v} + \frac{y^2}{x^2}\frac{\partial^2 z}{\partial v^2} - x^2y^2\frac{\partial^2 z}{\partial u^2} - 2y^2\frac{\partial^2 z}{\partial u \partial v} - \frac{y^2}{x^2}\frac{\partial^2 z}{\partial v^2}$$

$$= \frac{2y}{x}\frac{\partial z}{\partial v} - 4y^2\frac{\partial^2 z}{\partial u \partial v} = 2v\frac{\partial z}{\partial v} - 4uv\frac{\partial^2 z}{\partial u \partial v}$$ since $y = xv = \dfrac{uv}{y}$ or $y^2 = uv$.

51. $\nabla f = \left\langle z^2\sqrt{y}e^{x\sqrt{y}}, \dfrac{xz^2e^{x\sqrt{y}}}{2\sqrt{y}}, 2ze^{x\sqrt{y}}\right\rangle = ze^{x\sqrt{y}}\left\langle z\sqrt{y}, \dfrac{xz}{2\sqrt{y}}, 2\right\rangle$

53. $\nabla f = \langle 1/\sqrt{x}, -2y\rangle$, $\nabla f(1,5) = \langle 1, -10\rangle$, $\mathbf{u} = \frac{1}{5}\langle 3, -4\rangle$. Then $D_{\mathbf{u}} f(1,5) = \frac{43}{5}$.

55. $\nabla f = \langle 2xy, x^2 + 1/\left(2\sqrt{y}\right)\rangle$, $|\nabla f(2,1)| = \left|\langle 4, \frac{9}{2}\rangle\right|$. Thus the maximum rate of change of f at $(2,1)$ is $\frac{\sqrt{145}}{2}$ in the direction $\langle 4, \frac{9}{2}\rangle$.

57. $\dfrac{x^2+y^2}{(x-1)^2+y^2} \to \infty$ as $(x,y) \to (1,0)$ and so $\displaystyle\lim_{(x,y)\to(1,0)} \tan^{-1}\frac{x^2+y^2}{(x-1)^2+y^2} = \frac{\pi}{2}$.

59. $f(x,y) = x^2 - xy + y^2 + 9x - 6y + 10 \quad \Rightarrow$
$f_x = 2x - y + 9$, $f_y = -x + 2y - 6$,
$f_{xx} = 2 = f_{yy}$, $f_{xy} = -1$. Then $f_x = 0$ and $f_y = 0$ imply $y = 1$, $x = -4$. Thus the only critical point is $(-4, 1)$ and $f_{xx}(-4,1) > 0$, $D(-4,1) = 3 > 0$ so $f(-4,1) = -11$ is a local minimum.

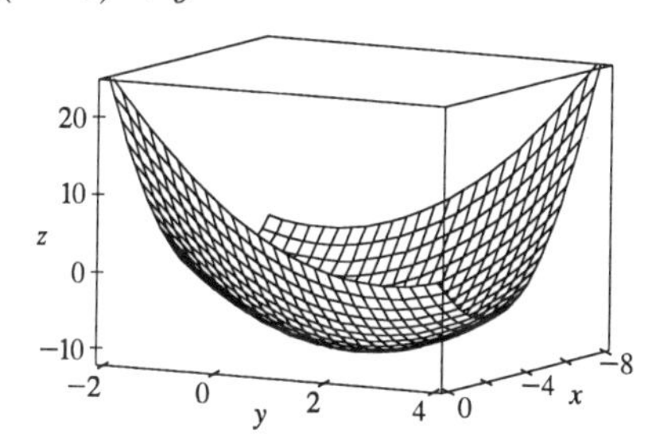

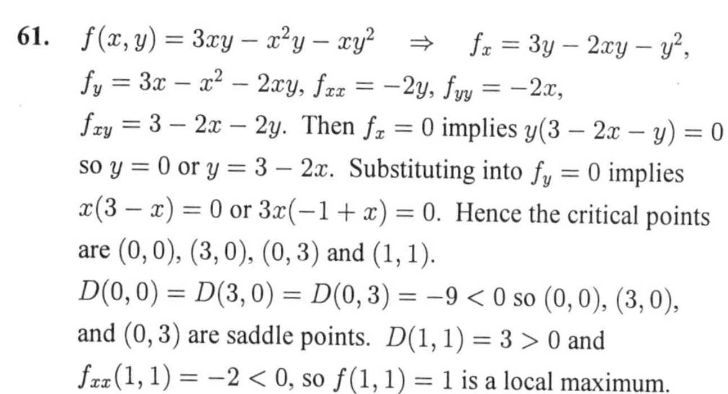

61. $f(x,y) = 3xy - x^2y - xy^2 \quad \Rightarrow \quad f_x = 3y - 2xy - y^2$,
$f_y = 3x - x^2 - 2xy$, $f_{xx} = -2y$, $f_{yy} = -2x$,
$f_{xy} = 3 - 2x - 2y$. Then $f_x = 0$ implies $y(3 - 2x - y) = 0$ so $y = 0$ or $y = 3 - 2x$. Substituting into $f_y = 0$ implies $x(3 - x) = 0$ or $3x(-1 + x) = 0$. Hence the critical points are $(0,0)$, $(3,0)$, $(0,3)$ and $(1,1)$.
$D(0,0) = D(3,0) = D(0,3) = -9 < 0$ so $(0,0)$, $(3,0)$, and $(0,3)$ are saddle points. $D(1,1) = 3 > 0$ and $f_{xx}(1,1) = -2 < 0$, so $f(1,1) = 1$ is a local maximum.

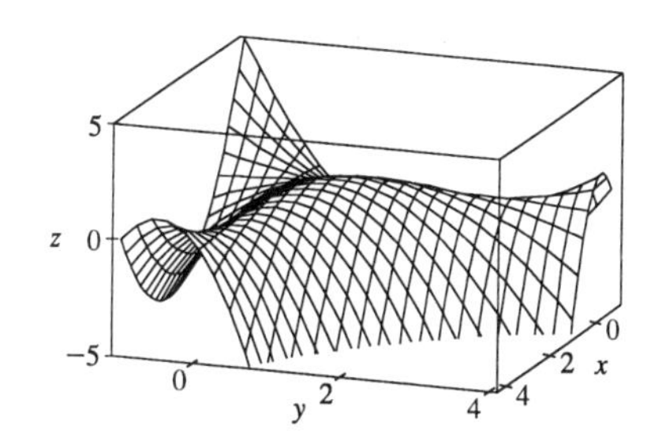

63. First solve inside D. Here $f_x = 4y^2 - 2xy^2 - y^3$, $f_y = 8xy - 2x^2y - 3xy^2$. Then $f_x = 0$ implies $y = 0$ or $y = 4 - 2x$, but $y = 0$ isn't inside D. Substituting $y = 4 - 2x$ into $f_y = 0$ implies $x = 0$, $x = 2$ or $x = 1$, but $x = 0$ isn't inside D, and when $x = 2$, $y = 0$ but $(2, 0)$ isn't inside D. Thus the only critical point inside D is $(1, 2)$ and $f(1, 2) = 4$. Secondly we consider the boundary of D. On L_1, $f(x, 0) = 0$ and so $f = 0$ on L_1. On L_2, $x = -y + 6$ and

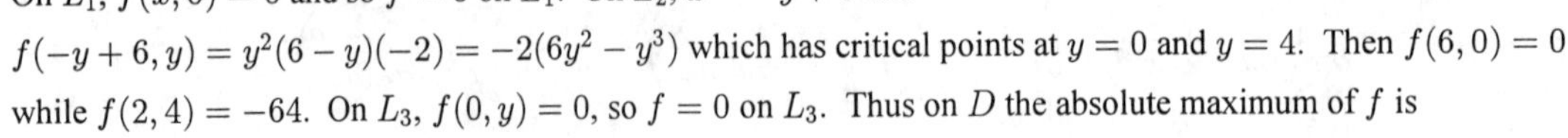

$f(-y + 6, y) = y^2(6 - y)(-2) = -2(6y^2 - y^3)$ which has critical points at $y = 0$ and $y = 4$. Then $f(6, 0) = 0$ while $f(2, 4) = -64$. On L_3, $f(0, y) = 0$, so $f = 0$ on L_3. Thus on D the absolute maximum of f is $f(1, 2) = 4$ while the absolute minimum is $f(2, 4) = -64$.

65. $f(x, y) = x^3 - 3x + y^4 - 2y^2$

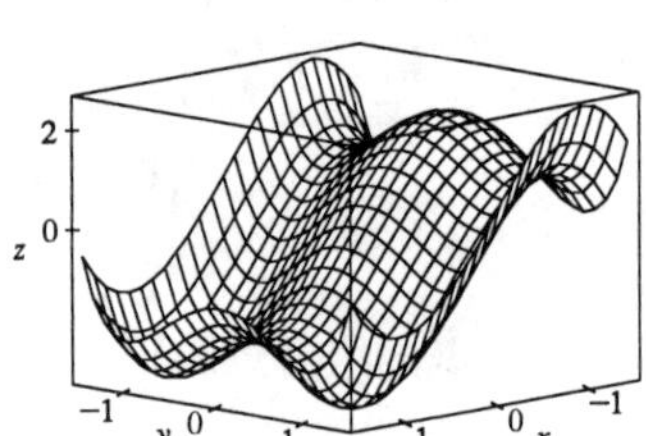

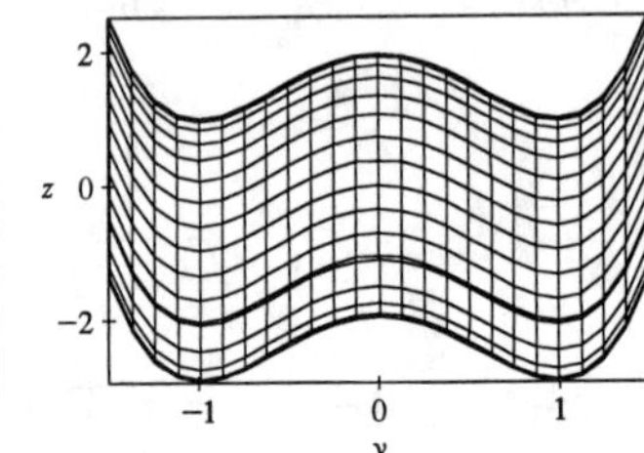

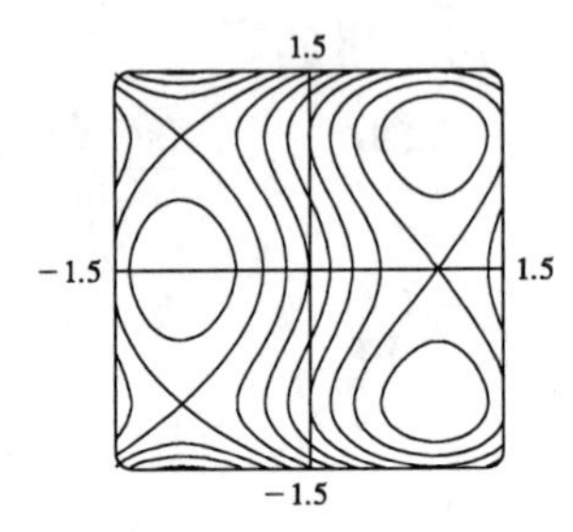

From the graphs, it appears that f has a local maximum $f(-1, 0) \approx 2$, local minima $f(1, \pm 1) \approx -3$, and saddle points at $(-1, \pm 1)$ and $(1, 0)$.

To find the exact quantities, we calculate $f_x = 3x^2 - 3 = 0 \quad \Leftrightarrow \quad x = \pm 1$ and $f_y = 4y^3 - 4y = 0 \quad \Leftrightarrow$ $y = 0, \pm 1$, giving the critical points estimated above. Also $f_{xx} = 6x$, $f_{xy} = 0$, $f_{yy} = 12y^2 - 4$, so using the Second Derivatives Test, $D(-1, 0) = 24 > 0$ and $f_{xx}(-1, 0) = -6 < 0$ indicating a local maximum $f(-1, 0) \approx 2$; $D(1, \pm 1) = 48 > 0$ and $f_{xx}(1, \pm 1) = 6 > 0$ indicating local minima $f(1, \pm 1) = -3$; and $D(-1, \pm 1) = -48$ and $D(1, 0) = -24$, indicating saddle points.

67. $f(x, y) = x^2 y$, $g(x, y) = x^2 + y^2 = 1 \quad \Rightarrow \quad \nabla f = \langle 2xy, x^2 \rangle = \lambda \nabla g = \langle 2\lambda x, 2\lambda y \rangle$. Then $2xy = 2\lambda x$ and $x^2 = 2\lambda y$ imply $\lambda = x^2/(2y)$ and $\lambda = y$ if $x \neq 0$ and $y \neq 0$. Hence $x^2 = 2y^2$. Then $x^2 + y^2 = 1$ implies $3y^2 = 1$ so $y = \pm\frac{1}{\sqrt{3}}$ and $x = \pm\sqrt{\frac{2}{3}}$. [Note if $x = 0$ then $x^2 = 2\lambda y$ implies $y = 0$ and $f(0, 0) = 0$.] Thus the possible points are $\left(\pm\sqrt{\frac{2}{3}}, \pm\frac{1}{\sqrt{3}}\right)$ and the absolute maxima are $f\left(\pm\sqrt{\frac{2}{3}}, \frac{1}{\sqrt{3}}\right) = \frac{2}{3\sqrt{3}}$ while the absolute minima are $f\left(\pm\sqrt{\frac{2}{3}}, -\frac{1}{\sqrt{3}}\right) = -\frac{2}{3\sqrt{3}}$.

69. $f(x, y, z) = x + y + z$, $g(x, y, z) = 1/x + 1/y + 1/z = 1 \quad \Rightarrow$ $\nabla f = \langle 1, 1, 1 \rangle = \lambda \nabla g = \langle -\lambda x^{-2}, -\lambda y^{-2}, -\lambda z^{-2} \rangle$. Thus $\lambda = -x^2 = -y^2 = -z^2$ or $y = \pm x$, $z = \pm x$. Substituting into $1/x + 1/y + 1/z = 1$ gives (1) $3/x = 1$ so $x = 3$, or (2) $1/x = 1$ so $x = 1$, or (3) $-1/x = 1$ so $x = -1$ with the associated points (1) $(3, 3, 3)$, (2) $(1, 1, -1)$ or $(1, -1, 1)$, (3) $(-1, 1, 1)$. Thus the absolute maximum is $f(3, 3, 3) = 9$ and the absolute minimum is $f(1, 1, -1) = f(1, -1, 1) = f(-1, 1, 1) = 1$.

71. $f(x,y,z)=x^2+y^2+z^2$, $g(x,y,z)=xy^2z^3=2 \quad\Rightarrow$
$\nabla f=\langle 2x,2y,2z\rangle=\lambda\nabla g=\langle \lambda y^2z^3, 2\lambda xyz^3, 3\lambda xy^2z^2\rangle$. Since $xy^2z^3=2$, $x\neq 0$, $y\neq 0$ and $z\neq 0$, so (1) $2x=\lambda y^2z^3$, (2) $1=\lambda xz^3$, (3) $2=3\lambda xy^2z$. Then (2) and (3) imply $\dfrac{1}{xz^3}=\dfrac{2}{3xy^2z}$ or $y^2=\frac{2}{3}z^2$ so $y=\pm z\sqrt{\frac{2}{3}}$. Similarly (1) and (3) imply $\dfrac{2x}{y^2z^3}=\dfrac{2}{3xy^2z}$ or $3x^2=z^2$ so $x=\pm\frac{1}{\sqrt{3}}z$. But $xy^2z^3=2$ so x and z must have the same sign, that is, $x=\frac{1}{\sqrt{3}}z$. Thus $g(x,y,z)=2$ implies $\frac{1}{\sqrt{3}}z\left(\frac{2}{3}z^2\right)z^3=2$ or $z=\pm 3^{1/4}$ and the possible points are $\left(\pm 3^{-1/4}, 3^{-1/4}\sqrt{2}, \pm 3^{1/4}\right)$, $\left(\pm 3^{-1/4}, -3^{-1/4}\sqrt{2}, \pm 3^{1/4}\right)$. However at each of these points f takes on the same value, $2\sqrt{3}$. But $(2,1,1)$ also satisfies $g(x,y,z)=2$ and $f(2,1,1)=6>2\sqrt{3}$. Thus f has an absolute minimum value of $2\sqrt{3}$ and no absolute maximum subject to the constraint $xy^2z^3=2$.

Alternate Solution: $g(x,y,z)=xy^2z^3=2$ implies $y^2=\dfrac{2}{xz^3}$, so minimize $f(x,z)=x^2+\dfrac{2}{xz^3}+z^2$. Then $f_x=2x-\dfrac{2}{x^2z^3}$, $f_z=-\dfrac{6}{xz^4}+2z$, $f_{xx}=2+\dfrac{4}{x^3z^3}$, $f_{zz}=\dfrac{24}{xz^5}+2$ and $f_{xz}=\dfrac{6}{x^2z^4}$. Now $f_x=0$ implies $2x^3z^3-2=0$ or $z=1/x$. Substituting into $f_y=0$ implies $-6x^3+2x^{-1}=0$ or $x=\frac{1}{\sqrt[4]{3}}$, so the two critical points are $\left(\pm\frac{1}{\sqrt[4]{3}}, \pm\sqrt[4]{3}\right)$. Then $D\left(\pm\frac{1}{\sqrt[4]{3}}, \pm\sqrt[4]{3}\right)=(2+4)\left(2+\frac{24}{3}\right)-\left(\frac{6}{\sqrt{3}}\right)^2>0$ and $f_{xx}\left(\pm\frac{1}{\sqrt[4]{3}}, \pm\sqrt[4]{3}\right)=6>0$, so each point is a minimum. Finally, $y^2=\dfrac{2}{xz^3}$, so the four points closest to the origin are $\left(\pm\frac{1}{\sqrt[4]{3}}, \frac{\sqrt{2}}{\sqrt[4]{3}}, \pm\sqrt[4]{3}\right)$, $\left(\pm\frac{1}{\sqrt[4]{3}}, -\frac{\sqrt{2}}{\sqrt[4]{3}}, \pm\sqrt[4]{3}\right)$.

73.

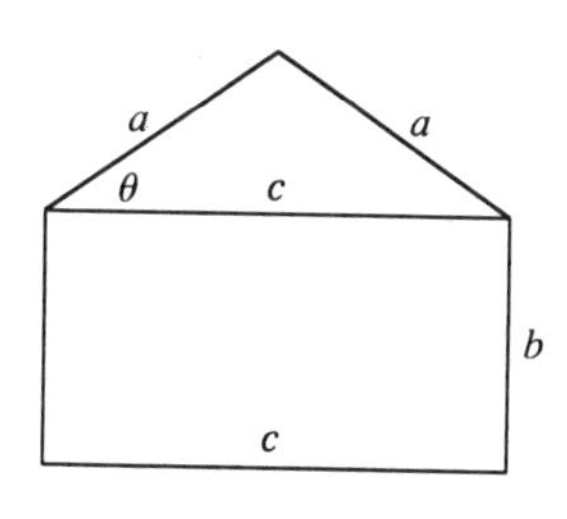

The area of the triangle is $\frac{1}{2}ca\sin\theta$ and the area of the rectangle is bc. Thus, the area of the whole object is $f(a,b,c)=\frac{1}{2}ca\sin\theta+bc$. The perimeter of the object is $g(a,b,c)=2a+2b+c=P$. To simplify $\sin\theta$ in terms of a, b, and c notice that $a^2\sin^2\theta+\left(\frac{1}{2}c\right)^2=a^2$ $\Rightarrow \quad \sin\theta=\dfrac{1}{2a}\sqrt{4a^2-c^2}$. Thus $f(a,b,c)=\dfrac{c}{4}\sqrt{4a^2-c^2}+bc$. (Instead of using θ, we could just have used the Pythagorean Theorem.) As a result, by Lagrange's method, we must find a, b, c, and λ by solving $\nabla f=\lambda\nabla g$ which gives the following equations: (1) $ca(4a^2-c^2)^{-1/2}=2\lambda$, (2) $c=2\lambda$, (3) $\frac{1}{4}(4a^2-c^2)^{1/2}-\frac{1}{4}c^2(4a^2-c^2)^{-1/2}+b=\lambda$, and (4) $2a+2b+c=P$. From (2), $\lambda=\frac{1}{2}c$ and so (1) produces $ca(4a^2-c^2)^{-1/2}=c \quad\Rightarrow\quad (4a^2-c^2)^{1/2}=a$ $\Rightarrow\quad 4a^2-c^2=a^2 \quad\Rightarrow\quad$ (5) $c=\sqrt{3}a$. Similarly, since $(4a^2-c^2)^{1/2}=a$ and $\lambda=\frac{1}{2}c$, (3) gives $\frac{1}{4}a-\dfrac{c^2}{4a}+b=\dfrac{c}{2}$, so from (5), $\dfrac{a}{4}-\dfrac{3a}{4}+b=\dfrac{\sqrt{3}a}{2} \quad\Rightarrow\quad -\dfrac{a}{2}-\dfrac{\sqrt{3}a}{2}=-b \quad\Rightarrow\quad$ (6) $b=\frac{a}{2}\left(1+\sqrt{3}\right)$.
Substituting (5) and (6) into (4) we get: $2a+a\left(1+\sqrt{3}\right)+\sqrt{3}a=P \quad\Rightarrow\quad 3a+2\sqrt{3}a=P \quad\Rightarrow$
$a=\dfrac{P}{3+2\sqrt{3}}=\dfrac{2\sqrt{3}-3}{3}P$ and thus $b=\dfrac{(2\sqrt{3}-3)(1+\sqrt{3})}{6}P=\dfrac{3-\sqrt{3}}{6}P$ and $c=(2-\sqrt{3})P$.

PROBLEMS PLUS (after Chapter 12)

1. Since three-dimensional situations are often difficult to visualize and work with, let us first try to find an analogous problem in two dimensions. The analogue of a cube is a square and the analogue of a sphere is a circle. Thus a similar problem in two dimensions is the following: if five circles with the same radius r are contained in a square of side 1 m so that the circles touch each other and four of the circles touch two sides of the square, find r.

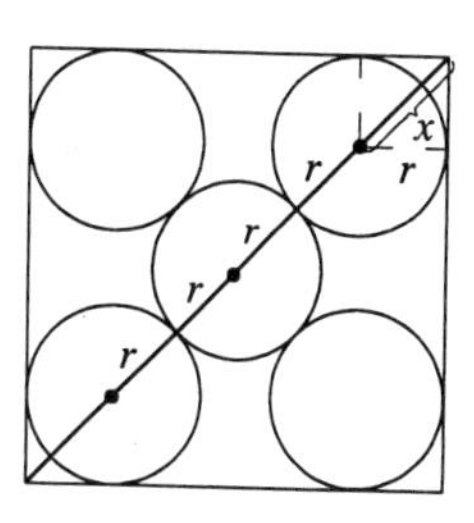

The diagonal of the square is $\sqrt{2}$. The diagonal is also $4r + 2x$. But x is the diagonal of a smaller square of side r. Therefore $x = \sqrt{2}r \quad \Rightarrow \quad \sqrt{2} = 4r + 2x = 4r + 2\sqrt{2}r = \left(4 + 2\sqrt{2}\right)r \quad \Rightarrow \quad r = \dfrac{\sqrt{2}}{4 + 2\sqrt{2}}$.

Let us use these ideas to solve the original three-dimensional problem. The diagonal of the cube is $\sqrt{1^2 + 1^2 + 1^2} = \sqrt{3}$. The diagonal of the cube is also $4r + 2x$ where x is the diagonal of a smaller cube with edge r. Therefore $x = \sqrt{r^2 + r^2 + r^2} = \sqrt{3}r \quad \Rightarrow \quad \sqrt{3} = 4r + 2x = 4r + 2\sqrt{3}r = \left(4 + 2\sqrt{3}\right)r$.

Therefore $r = \dfrac{\sqrt{3}}{4 + 2\sqrt{3}} = \dfrac{2\sqrt{3} - 3}{2}$. The radius of each ball is $\left(\sqrt{3} - \frac{3}{2}\right)$m.

3. We introduce a coordinate system, as shown. Recall that the area of the parallelogram spanned by two vectors is equal to the length of their cross product, so since $\mathbf{u} \times \mathbf{v} = \langle -q, r, 0 \rangle \times \langle -q, 0, p \rangle = \langle pr, pq, qr \rangle$, we have $|\mathbf{u} \times \mathbf{v}| = \sqrt{(pr)^2 + (pq)^2 + (qr)^2}$, and therefore

$$D^2 = \left(\tfrac{1}{2}|\mathbf{u} \times \mathbf{v}|\right)^2 = \tfrac{1}{4}\left[(pr)^2 + (pq)^2 + (qr)^2\right]$$
$$= \left(\tfrac{1}{2}pr\right)^2 + \left(\tfrac{1}{2}pq\right)^2 + \left(\tfrac{1}{2}qr\right)^2 = A^2 + B^2 + C^2.$$

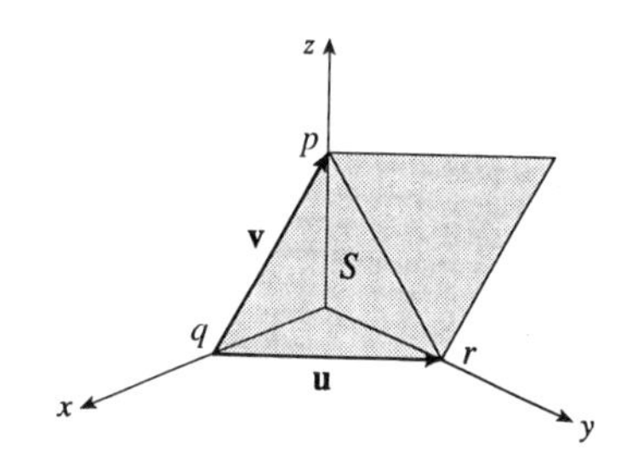

Another Method: We draw a line from S perpendicular to QR, as shown. Now $D = \frac{1}{2}ch$, so $D^2 = \frac{1}{4}c^2h^2$. Substituting $h^2 = p^2 + k^2$, we get $D^2 = \frac{1}{4}c^2(p^2 + k^2) = \frac{1}{4}c^2p^2 + \frac{1}{4}c^2k^2$. But $C = \frac{1}{2}ck$, so $D^2 = \frac{1}{4}c^2p^2 + C^2$. Now substituting $c^2 = q^2 + r^2$ gives $D^2 = \frac{1}{4}p^2q^2 + \frac{1}{4}q^2r^2 + C^2 = A^2 + B^2 + C^2$.

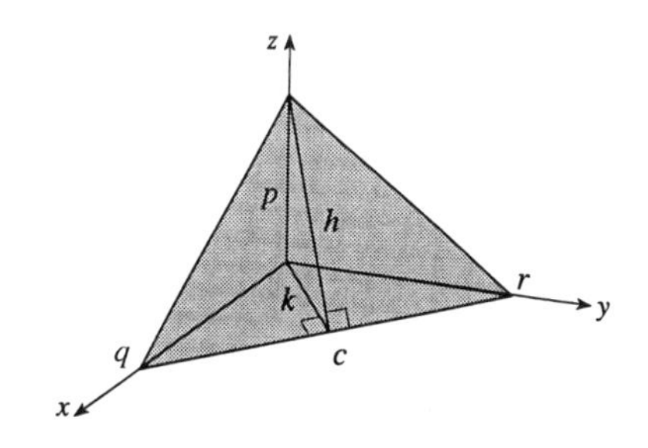

5. The areas of the smaller rectangles are $A_1 = xy$, $A_2 = (L-x)y$, $A_3 = (L-x)(W-y)$, $A_4 = x(W-y)$. For $0 \le x \le L$, $0 \le y \le W$, let

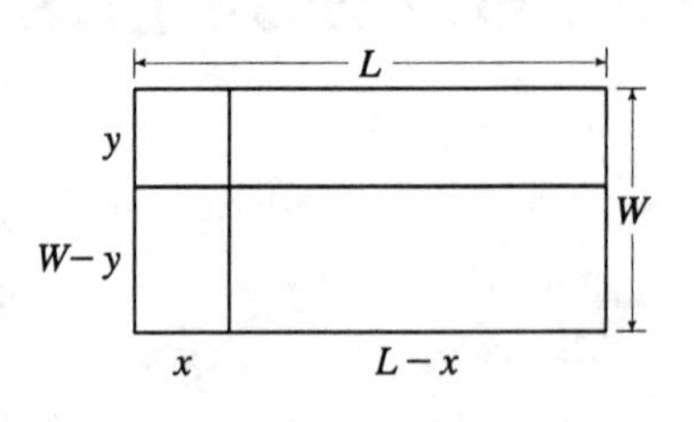

$$\begin{aligned} f(x,y) &= A_1^2 + A_2^2 + A_3^2 + A_4^2 \\ &= x^2y^2 + (L-x)^2y^2 + (L-x)^2(W-y)^2 + x^2(W-y)^2 \\ &= \left[x^2 + (L-x)^2\right]\left[y^2 + (W-y)^2\right]. \end{aligned}$$

Then we need to find the maximum and minimum values of $f(x,y)$. Here

$f_x(x,y) = [2x - 2(L-x)]\left[y^2 + (W-y)^2\right] = 0 \quad \Rightarrow \quad 4x - 2L = 0$ or $x = \frac{1}{2}L$, and

$f_y(x,y) = \left[x^2 + (L-x)^2\right][2y - 2(W-y)] = 0 \quad \Rightarrow \quad 4y - 2W = 0$ or $y = W/2$. Also

$f_{xx} = 4\left[y^2 + (W-y)^2\right]$, $f_{yy} = 4\left[x^2 + (L-x)^2\right]$, and $f_{xy} = (4x-2L)(4y-2W)$. Then

$D = 16\left[y^2 + (W-y)^2\right]\left[x^2 + (L-x)^2\right] - (4x-2L)^2(4y-2W)^2$. Thus when $x = \frac{1}{2}L$ and $y = \frac{1}{2}W$, $D > 0$ and $f_{xx} = 2W^2 > 0$. Thus a minimum of f occurs at $\left(\frac{1}{2}L, \frac{1}{2}W\right)$ and this minimum value is $f\left(\frac{1}{2}L, \frac{1}{2}W\right) = \frac{1}{4}L^2W^2$. There are no other critical points, so the maximum must occur on the boundary. Now along the width of the rectangle let $g(y) = f(0,y) = f(L,y) = L^2\left[y^2 + (W-y)^2\right]$, $0 \le y \le W$. Then $g'(y) = L^2[2y - 2(W-y)] = 0 \quad \Leftrightarrow \quad y = \frac{1}{2}W$. And $g\left(\frac{1}{2}\right) = \frac{1}{2}L^2W^2$. Checking the endpoints we get $g(0) = g(W) = L^2W^2$. Along the length of the rectangle let $h(x) = f(x,0) = f(x,W) = W^2\left[x^2 + (L-x)^2\right]$, $0 \le x \le L$. By symmetry $h'(x) = 0 \quad \Leftrightarrow \quad x = \frac{1}{2}L$ and $h\left(\frac{1}{2}L\right) = \frac{1}{2}L^2W^2$. At the endpoints we have $h(0) = h(L) = L^2W^2$. Therefore L^2W^2 is the maximum value of f. This maximum value of f occurs when the "cutting" lines correspond to sides of the rectangle.

7. Let $g(x,y) = xf\left(\frac{y}{x}\right)$. Then $g_x(x,y) = f\left(\frac{y}{x}\right) + xf'\left(\frac{y}{x}\right)\left(-\frac{y}{x^2}\right) = f\left(\frac{y}{x}\right) - \frac{y}{x}f'\left(\frac{y}{x}\right)$ and $g_y(x,y) = xf'\left(\frac{y}{x}\right)\left(\frac{1}{x}\right) = f'\left(\frac{y}{x}\right)$. Thus the tangent plane at (x_0, y_0, z_0) on the surface has equation

$$z - x_0 f\left(\frac{y_0}{x_0}\right) = \left[f\left(\frac{y_0}{x_0}\right) - y_0x_0^{-1}f'\left(\frac{y_0}{x_0}\right)\right](x - x_0) + f'\left(\frac{y_0}{x_0}\right)(y - y_0) \quad \Rightarrow$$

$\left[f\left(\frac{y_0}{x_0}\right) - y_0x_0^{-1}f'\left(\frac{y_0}{x_0}\right)\right]x + \left[f'\left(\frac{y_0}{x_0}\right)\right]y - z = 0$. But any plane whose equation is of the form $ax + by + cz = 0$ passes through the origin. Thus the origin is the common point of intersection.

9. At $(x_1, y_1, 0)$ the equations of the tangent planes to $z = f(x,y)$ and $z = g(x,y)$ are
P_1: $z - f(x_1, y_1) = f_x(x_1, y_1)(x - x_1) + f_y(x_1, y_1)(y - y_1)$ and
P_2: $z - g(x_1, y_1) = g_x(x_1, y_1)(x - x_1) + g_y(x_1, y_1)(y - y_1)$, respectively. P_1 intersects the xy-plane in the line given by $f_x(x_1, y_1)(x - x_1) + f_y(x_1, y_1)(y - y_1) = -f(x_1, y_1)$, $z = 0$; and P_2 intersects the xy-plane in the line given by $g_x(x_1, y_1)(x - x_1) + g_y(x_1, y_1)(y - y_1) = -g(x_1, y_1)$, $z = 0$. The point $(x_2, y_2, 0)$ is the point of intersection of these two lines, since $(x_2, y_2, 0)$ is the point where the line of intersection of the two tangent planes intersects the xy-plane. Thus (x_2, y_2) is the solution of the simultaneous equations
$f_x(x_1, y_1)(x_2 - x_1) + f_y(x_1, y_1)(y_2 - y_1) = -f(x_1, y_1)$ and
$g_x(x_1, y_1)(x_2 - x_1) + g_y(x_1, y_1)(y_2 - y_1) = -g(x_1, y_1)$. For simplicity, rewrite $f_x(x_1, y_1)$ as f_x and similarly for f_y, g_x, g_y, f and g and solve the equations $(f_x)(x_2 - x_1) + (f_y)(y_2 - y_1) = -f$ and
$(g_x)(x_2 - x_1) + (g_y)(y_2 - y_1) = -g$ simultaneously for $(x_2 - x_1)$ and $(y_2 - y_1)$. Then $y_2 - y_1 = \dfrac{gf_x - fg_x}{g_x f_y - f_x g_y}$
or $y_2 = y_1 - \dfrac{gf_x - fg_x}{f_x g_y - g_x f_y}$ and $(f_x)(x_2 - x_1) + \dfrac{(f_y)(gf_x - fg_x)}{g_x f_y - f_x g_y} = -f$ so

$$x_2 - x_1 = \frac{-f - [(f_y)(gf_x - fg_x)/(g_x f_y - f_x g_y)]}{f_x} = \frac{fg_y - f_y g}{g_x f_y - f_x g_y}.$$ Hence $x_2 = x_1 - \dfrac{fg_y - f_y g}{f_x g_y - g_x f_y}$.

11. **(a)** $\nabla T = \langle -2x, -4y \rangle$, so $\nabla T(4,2) = \langle -8, -8 \rangle$ is the direction of maximum increase at $(4,2)$, or $\left\langle -\frac{1}{\sqrt{2}}, -\frac{1}{\sqrt{2}} \right\rangle$ as a unit vector.

(b) The level curves are ellipses with center the origin. At $(0,0)$, $T = 100$. The path followed by the particle is always perpendicular to these level curves.

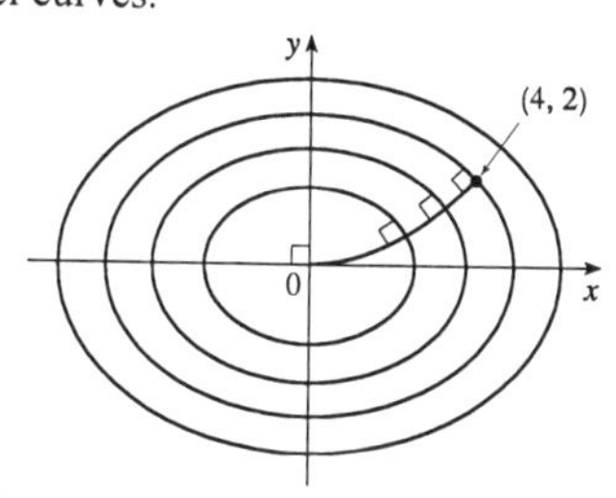

(c) Let the path be given by the function $\mathbf{r}(t) = x(t)\mathbf{i} + y(t)\mathbf{j}$. A tangent vector at each point $(x(t), y(t))$ is given by $\mathbf{r}'(t) = \dfrac{dx}{dt}\mathbf{i} + \dfrac{dy}{dt}\mathbf{j}$. Since the particle seeks maximum temperature increase, the directions of $\mathbf{r}'(t)$ and $\nabla T(x,y) = -2x\mathbf{i} - 4y\mathbf{j}$ are the same at each point of the path. Thus $dx/dt = -2kx$ and $dy/dt = -4ky$, for some constant k. These differential equations represent exponential growth and the solutions are $x(t) = C_1 e^{-2kt}$ and $y(t) = C_2 e^{-4kt}$ [see Section 8.1 or Section 6.5 (or Section 3.5 in the Early Transcendentals version)]. Since the particle starts at $(4,2)$, we have $4 = x(0) = C_1$ and $2 = y(0) = C_2$. Thus, the path is given by $x = 4e^{-2kt}$, $y = 2e^{-4kt}$. Eliminating the parameter t we get $y/2 = e^{-4kt} = (x/4)^2$, so the curve is part of the parabola $y = \frac{1}{8}x^2$.

13. Since we are minimizing the area of the ellipse, and the circle lies above the x-axis, the ellipse will intersect the circle for only one value of y. This y-value must satisfy both the equation of the circle and the equation of the ellipse. Now $\frac{x^2}{a^2} + \frac{y^2}{b^2} = 1 \Rightarrow x^2 = \frac{a^2}{b^2}(b^2 - y^2)$. Substituting into the equation of the circle gives $\frac{a^2}{b^2}(b^2 - y^2) + y^2 - 2y = 0 \Rightarrow$

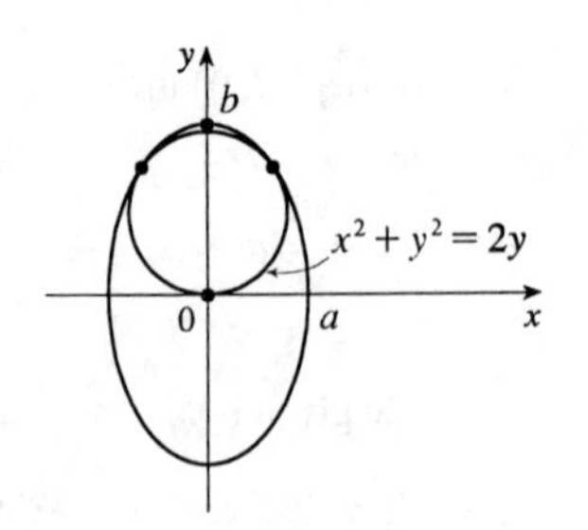

$\left(\frac{b^2 - a^2}{b^2}\right)y^2 - 2y + a^2 = 0$. In order for there to be only one solution to this quadratic equation, the discriminant must be 0, so $4 - 4a^2\frac{b^2 - a^2}{b^2} = 0 \Rightarrow b^2 - a^2b^2 + a^4 = 0$. The area of the ellipse is $A(a, b) = \pi ab$, and we minimize this function subject to the constraint $g(a, b) = b^2 - a^2b^2 + a^4 = 0$. Now $\nabla A = \lambda \nabla g \Leftrightarrow \pi b = \lambda(4a^3 - 2ab^2), \pi a = \lambda(2b - 2ba^2) \Rightarrow$ (1) $\lambda = \frac{\pi b}{2a(2a^2 - b^2)}$, (2) $\lambda = \frac{\pi a}{2b(1 - a^2)}$, (3) $b^2 - a^2b^2 + a^4 = 0$. Comparing (1) and (2) gives $\frac{\pi b}{2a(2a^2 - b^2)} = \frac{\pi a}{2b(1 - a^2)} \Rightarrow$ $2\pi b^2 = 4\pi a^4 \Leftrightarrow a^2 = \frac{1}{\sqrt{2}}b$. Substitute this into (3) to get $b = \frac{3}{\sqrt{2}} \Rightarrow a = \sqrt{\frac{3}{2}}$.

CHAPTER THIRTEEN

EXERCISES 13.1

1. **(a)** $\sum_{i=1}^{2}\sum_{j=1}^{2} f(x_{ij}^*, y_{ij}^*)\Delta A_{ij} = \frac{1}{2}\left[f\left(0, \frac{3}{2}\right) + f(0,2) + f\left(1, \frac{3}{2}\right) + f(1,2)\right]$

$= \frac{1}{2}\left[\left(-\frac{27}{4}\right) + (-12) + \left(1 - \frac{27}{4}\right) + (1-12)\right] = \frac{1}{2}\left(-\frac{71}{2}\right) = -17.75$

(b) $\frac{1}{2}\left[f\left(1, \frac{3}{2}\right) + f(1,2) + f\left(2, \frac{3}{2}\right) + f(2,2)\right] = \frac{1}{2}\left[-\frac{23}{4} + (-11) + \left(-\frac{19}{4}\right) + (-10)\right] = \frac{1}{2}\left(-\frac{63}{2}\right) = -15.75$

(c) $\frac{1}{2}\left[f(0,1) + f\left(0, \frac{3}{2}\right) + f(1,1) + f\left(1, \frac{3}{2}\right)\right] = \frac{1}{2}\left[-3 - \frac{27}{4} - 2 - \frac{23}{4}\right] = -8.75$

(d) $\frac{1}{2}\left[f(1,1) + f\left(1, \frac{3}{2}\right) + f(2,1) + f\left(2, \frac{3}{2}\right)\right] = \frac{1}{2}\left[-2 - \frac{23}{4} - 1 - \frac{19}{4}\right] = -6.75$

3.

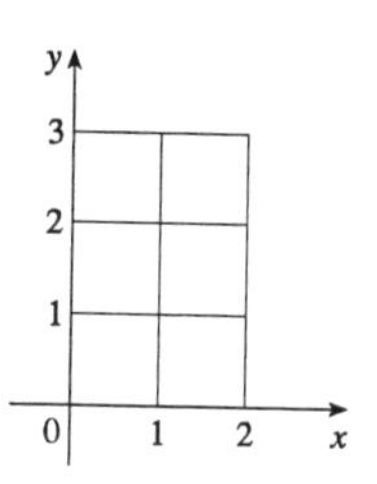

$\Delta A_{ij} = 1$ for $i = 1, 2$, $j = 1, 2, 3$.

$$\iint_R (x^2 + 4y)\,dA \approx (1)\left[f(1,1) + f(1,2) + f(1,3) + f(2,1) + f(2,2) + f(2,3)\right]$$
$$= 5 + 9 + 13 + 8 + 12 + 16 = 63$$

$\|P\| = \sqrt{1+1} = \sqrt{2}$

5.

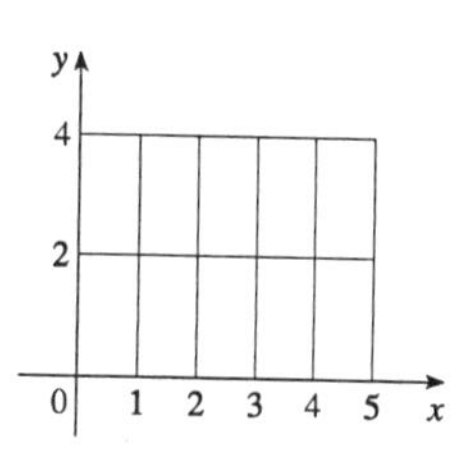

$\Delta A_{ij} = 2$ for $i = 1, 2, 3, 4, 5$, $j = 1, 2$.

$$\iint_R (xy - y^2)\,dA$$
$$\approx (2)\left[f\left(\tfrac{1}{2}, 1\right) + f\left(\tfrac{1}{2}, 3\right) + f\left(\tfrac{3}{2}, 1\right) + f\left(\tfrac{3}{2}, 3\right) + f\left(\tfrac{5}{2}, 1\right) + f\left(\tfrac{5}{2}, 3\right) + f\left(\tfrac{7}{2}, 1\right) + f\left(\tfrac{7}{2}, 3\right) + f\left(\tfrac{9}{2}, 1\right) + f\left(\tfrac{9}{2}, 3\right)\right]$$
$$= (2)\left[-\tfrac{1}{2} + \left(-\tfrac{15}{2}\right) + \tfrac{1}{2} + \left(-\tfrac{9}{2}\right) + \tfrac{3}{2} + \left(-\tfrac{3}{2}\right) + \tfrac{5}{2} + \tfrac{3}{2} + \tfrac{7}{2} + \tfrac{9}{2}\right] = 0$$

and $\|P\| = \sqrt{1+4} = \sqrt{5}$.

7.

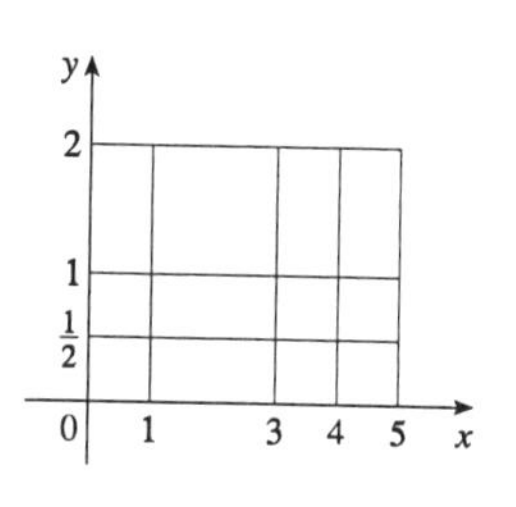

$$\iint_R (x^2 - y^2)\,dA \approx \tfrac{1}{2}f\left(0, \tfrac{1}{2}\right) + \tfrac{1}{2}f(0,1) + 1f(0,2) + 1f\left(1, \tfrac{1}{2}\right) + 1f(1,1) + 2f(1,2) + \tfrac{1}{2}f\left(3, \tfrac{1}{2}\right) + \tfrac{1}{2}f(3,1) + 1f(3,2) + \tfrac{1}{2}f\left(4, \tfrac{1}{2}\right) + \tfrac{1}{2}f(4,1) + 1f(4,2)$$
$$= \tfrac{1}{2}\left(-\tfrac{1}{4}\right) + \tfrac{1}{2}(-1) + 1(-4) + 1\left(\tfrac{3}{4}\right) + 1(0) + 2(-3) + \tfrac{1}{2}\left(\tfrac{35}{4}\right) + \tfrac{1}{2}(8) + (1)(5) + \tfrac{1}{2}\left(\tfrac{63}{4}\right) + \tfrac{1}{2}(15) + 1(12)$$
$$= \tfrac{1}{2}\left(\tfrac{185}{4}\right) + \tfrac{55}{4} - 6 = \tfrac{247}{8},$$

and the length of the longest diagonal is $\|P\| = \sqrt{1+4} = \sqrt{5}$.

9. The values of $f(x,y) = \sqrt{52 - x^2 - y^2}$ get smaller as we move further from the origin, so on any of the subrectangles in the problem, the function will have its largest value at the lower left corner of the subrectangle and its smallest value at the upper right corner, and any other value will lie between these two. So for this partition (and in fact for any partition) $U < V < L$.

11. To calculate the estimates using a programmable calculator, we can use an algorithm similar to that of Exercise 6.2.19. In Maple, we can define the function $f(x, y) = e^{-x^2-y^2}$ (calling it `f`), load the `student` package, and then use the command `middlesum(middlesum(f,x=0..1,m),y=0..1,m);` to get the estimate with $n = m^2$ squares of equal size. Mathematica has no special Riemann sum command, but we can define `f` and then use nested `Sum` commands to calculate the estimates.

n	estimate
1	0.6065
4	0.5694
16	0.5606
64	0.5585
256	0.5579
1024	0.5578

12.

n	estimate
1	0.9922
4	0.9262
16	0.8797
64	0.8660
256	0.8625
1024	0.8616

13. $z = f(x, y) = 4 - 2y \geq 0$ for $0 \leq y \leq 1$.
Thus the integral represents the volume of that part of the rectangular solid $[0, 1] \times [0, 1] \times [0, 4]$ which lies below the plane $z = 4 - 2y$. So
$\iint_R (4 - y)dA = (1)(1)(2) + \frac{1}{2}(1)(1)(2) = 3$.

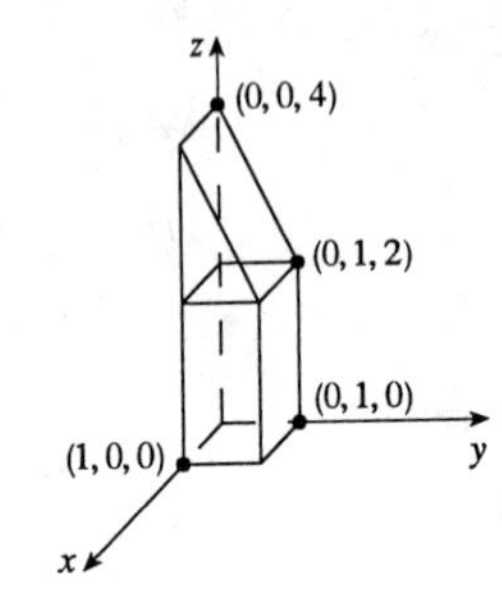

15. For any partition, $\iint_R k\,dA \approx \sum_i \sum_j f(x_{ij}^*, y_{ij}^*)\Delta A_{ij}$ but $f(x_{ij}^*, y_{ij}^*) = k$ always and $\sum_i \sum_j \Delta A_{ij} =$ area of $R = (b-a)(c-d)$. Thus for every partition $\sum_i \sum_j f(x_{ij}^*, y_{ij}^*)\Delta A_{ij} = k \sum_i \sum_j \Delta A_{ij} = k(b-a)(c-d)$ and so as $\|P\| \to 0$ the limit is $k(b-a)(c-d)$.

EXERCISES 13.2

1. $\int_0^2 x^2y^3\,dy = x^2\left[\frac{1}{4}y^4\right]_0^2 = 4x^2$, $\int_0^1 x^2y^3\,dx = y^3\left[\frac{1}{3}x^3\right]_0^1 = \frac{1}{3}y^3$

3. $\int_0^2 xe^{x+y}\,dy = xe^x[e^y]_0^2 = x(e^{x+2} - e^x) = xe^x(e^2 - 1)$, $\int_0^1 xe^{x+y}\,dx = e^y \int_0^1 xe^x\,dx = e^y[xe^x - e^x]_0^1 = e^y$

5. $$\int_0^4 \int_0^2 x\sqrt{y}\,dx\,dy = \int_0^4 \sqrt{y}\left[\tfrac{1}{2}x^2\right]_0^2 dy = \int_0^4 2\sqrt{y}\,dy = \left[\tfrac{4}{3}y^{3/2}\right]_0^4 = \frac{32}{3}$$

7. $\int_{-1}^1 \int_0^1 (x^3y^2 + 3xy^2)dy\,dx = \int_{-1}^1 \left[\frac{1}{4}x^3y^4 + xy^3\right]_{y=0}^{y=1} dx = \int_{-1}^1 \left[\frac{1}{4}x^3 + x\right]dx = \left[\frac{1}{16}x^4 + \frac{1}{2}x^2\right]_{-1}^1 = 0$

Alternate Solution: Applying Fubini's Theorem, the integral equals
$\int_0^1 \int_{-1}^1 (x^3y^2 + 3xy^2)dx\,dy = \int_0^1 \left[\frac{1}{4}y^2x^4 + \frac{3}{2}y^2x^2\right]_{x=-1}^{x=1} dy = \int_0^1 0\,dy = 0$.

9. $$\int_0^3 \int_0^1 \sqrt{x+y}\,dx\,dy = \int_0^3 \left[\tfrac{2}{3}(x+y)^{3/2}\right]_{x=0}^{x=1} dy = \tfrac{2}{3}\int_0^3 \left[(1+y)^{3/2} - y^{3/2}\right]dy$$
$$= \tfrac{2}{3}\left[\tfrac{2}{5}(1+y)^{5/2} - \tfrac{2}{5}y^{5/2}\right]_0^3 = \tfrac{4}{15}\left[32 - 3^{5/2} - 1\right] = \tfrac{4}{15}\left(31 - 9\sqrt{3}\right)$$

11. $\int_0^{\pi/4}\int_0^3 \sin x\,dy\,dx = 3\int_0^{\pi/4}\sin x\,dx = 3[-\cos x]_0^{\pi/4} = 3\left(1-\frac{1}{\sqrt{2}}\right)$

13. $\int_0^{\ln 2}\int_0^{\ln 5} e^{2x-y}\,dx\,dy = \left(\int_0^{\ln 5} e^{2x}\,dx\right)\left(\int_0^{\ln 2} e^{-y}\,dy\right) = \left[\frac{1}{2}e^{2x}\right]_0^{\ln 5}[-e^{-y}]_0^{\ln 2} = \left(\frac{25}{2}-\frac{1}{2}\right)\left(-\frac{1}{2}+1\right) = 6$

15. $\int_1^2\int_0^3 (2y^2-3xy^3)dy\,dx = \int_1^2\left[\frac{2}{3}y^3-\frac{3}{4}xy^4\right]_{y=0}^{y=3}dx = \int_1^2\left(18-\frac{243}{4}x\right)dx = \left[18x-\frac{243}{8}x^2\right]_1^2 = -\frac{585}{8}$

17. $\int_0^{\pi/6}\int_1^4 x\sin y\,dx\,dy = \left(\int_0^{\pi/6}\sin y\,dy\right)\left(\int_1^4 x\,dx\right) = \left(1-\frac{\sqrt{3}}{2}\right)\frac{15}{2} = \frac{15(2-\sqrt{3})}{4}$

19. $\int_0^{\pi/6}\int_0^{\pi/3} x\sin(x+y)dy\,dx = \int_0^{\pi/6}[-x\cos(x+y)]_0^{\pi/3}dx = \int_0^{\pi/6}\left[x\cos x - x\cos\left(x+\frac{\pi}{3}\right)\right]dx$

$= x\left[\sin x - \sin\left(x+\frac{\pi}{3}\right)\right]_0^{\pi/6} - \int_0^{\pi/6}\left[\sin x - \sin\left(x+\frac{\pi}{3}\right)\right]dx$

$= \frac{\pi}{6}\left[\frac{1}{2}-1\right] - \left[-\cos x + \cos\left(x+\frac{\pi}{3}\right)\right]_0^{\pi/6} = -\frac{\pi}{12} - \left[-\frac{\sqrt{3}}{2}+0-\left(-1+\frac{1}{2}\right)\right] = \frac{\sqrt{3}-1}{2} - \frac{\pi}{12}$

21. $\displaystyle\int_0^1\int_1^2 \frac{1}{x+y}\,dx\,dy = \int_0^1 [\ln(x+y)]_1^2\,dy = \int_0^1 [\ln(2+y)-\ln(1+y)]dy$

$= \Big[\left[(2+y)\ln(2+y)-(2+y)\right] - \left[(1+y)\ln(1+y)-(1+y)\right]\Big]_0^1$

$= (3\ln 3) - 3 - (2\ln 2) + 2 - [(2\ln 2 - 2) - (0-1)] = 3\ln 3 - 4\ln 2 = \ln\frac{27}{16}$

23. $z = f(x,y) = 4-x-2y \ge 0$ for $0\le x\le 1$ and $0\le y\le 1$. So the solid is the region in the first octant which lies below the plane $z = 4-x-2y$ and above $[0,1]\times[0,1]$.

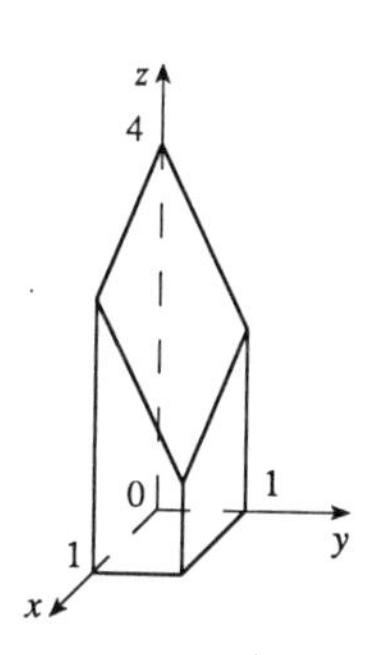

25. $V = \int_1^4\int_{-1}^0 (2x+5y+1)dx\,dy = \int_1^4[x^2+5xy+x]_{-1}^0\,dy = \int_1^4 5y\,dy = \frac{5}{2}y^2\big|_1^4 = \frac{75}{2}$

27. $V = \int_{-2}^2\int_{-1}^1\left(1-\frac{1}{4}x^2-\frac{1}{9}y^2\right)dx\,dy = 4\int_0^2\int_0^1\left(1-\frac{1}{4}x^2-\frac{1}{9}y^2\right)dx\,dy$

$= 4\int_0^2\left[x-\frac{1}{12}x^3-\frac{1}{9}y^2x\right]_0^1dy = 4\int_0^2\left(\frac{11}{12}-\frac{1}{9}y^2\right)dy = 4\left[\frac{11}{12}y-\frac{1}{27}y^3\right]_0^2 = 4\cdot\frac{83}{54} = \frac{166}{27}$

29. Here we need the volume of the solid lying under the surface $z = x\sqrt{x^2+y}$ and above the square $R = [0,1]\times[0,1]$ in the xy-plane.

$V = \int_0^1\int_0^1 x\sqrt{x^2+y}\,dx\,dy = \int_0^1 \frac{1}{3}\left[(x^2+y)^{3/2}\right]_0^1 dy = \frac{1}{3}\int_0^1\left[(1+y)^{3/2}-y^{3/2}\right]dy$

$= \frac{1}{3}\cdot\frac{2}{5}\left[(1+y)^{5/2}-y^{5/2}\right]_0^1 = \frac{4}{15}\left(2\sqrt{2}-1\right)$

31. In the first octant, $z\ge 0 \quad\Rightarrow\quad y\le 3$, so

$V = \int_0^3\int_0^2(9-y^2)dx\,dy = \int_0^3[9x-y^2x]_0^2\,dy = \int_0^3(18-2y^2)dy = \left[18y-\frac{2}{3}y^3\right]_0^3 = 36$

33. In Maple, we can calculate the integral by defining the integrand as `f` and then using the command `int(int(f,x=0..1),y=0..1);`.
In Mathematica, we can use the command `Integrate[Integrate[f,{x,0,1}],{y,0,1}]`.
We find that $\iint_R x^5 y^3 e^{xy}\,dA = 21e - 57 \approx 0.0839$.
We can use `plot3d` (in Maple) or `Plot3d` (in Mathematica) to graph the function.

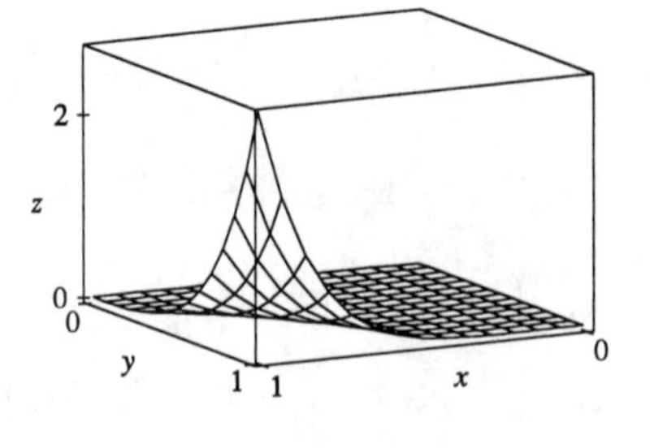

35. $A(R) = 2 \cdot 5 = 10$, so

$$f_{\text{ave}} = \frac{1}{A(R)}\iint\limits_R f(x,y)dA = \frac{1}{10}\int_0^5\int_{-1}^1 x^2 y\,dx\,dy = \frac{1}{10}\int_0^5\left[\frac{x^3}{3}y\right]_{-1}^1 dy = \frac{1}{10}\int_0^5 \frac{2y}{3}\,dy = \frac{1}{10}\left[\frac{y^2}{3}\right]_0^5 = \frac{5}{6}.$$

37. Let $f(x,y) = \dfrac{x-y}{(x+y)^3}$. Then a CAS gives $\int_0^1\int_0^1 f(x,y)dy\,dx = \frac{1}{2}$ and $\int_0^1\int_0^1 f(x,y)\,dx\,dy = -\frac{1}{2}$.

To explain the seeming violation of Fubini's Theorem, note that f has an infinite discontinuity at $(0,0)$ and thus does not satisfy the conditions of Fubini's Theorem. In fact, both iterated integrals involve improper integrals which diverge at their lower limits of integration.

EXERCISES 13.3

1. $\int_0^1\int_0^y x\,dx\,dy = \int_0^1\left[\frac{1}{2}x^2\right]_0^y dy = \int_0^1\left[\frac{1}{2}y^2\right]dy = \frac{1}{6}$

3. $\int_0^2\int_{\sqrt{x}}^3 (x^2+y)dy\,dx = \int_0^2\left[x^2y + \frac{1}{2}y^2\right]_{\sqrt{x}}^3 dx = \int_0^2\left[3x^2 + \frac{9}{2} - x^{5/2} - \frac{1}{2}x\right]dx$

$$= \left[x^3 + \tfrac{9}{2}x - \tfrac{2}{7}x^{7/2} - \tfrac{1}{4}x^2\right]_0^2 = 16\left(1 - \tfrac{\sqrt{2}}{7}\right)$$

5. $\int_0^1\int_0^x \sin(x^2)dy\,dx = \int_0^1 x\sin(x^2)dx = \frac{1}{2}[-\cos(x^2)]_0^1 = \frac{1}{2}(1-\cos 1)$

7. $\int_0^1\int_{x^2}^{\sqrt{x}} xy\,dy\,dx = \int_0^1\left[\frac{1}{2}xy^2\right]_{x^2}^{\sqrt{x}} dx = \frac{1}{2}\int_0^1(x^2 - x^5)dx = \frac{1}{2}\left[\frac{1}{3}x^3 - \frac{1}{6}x^6\right]_0^1 = \frac{1}{12}$

9. $\int_0^1\int_{\sqrt{x}}^{2-x}(x^2 - 2xy)dy\,dx = \int_0^1[x^2y - xy^2]_{\sqrt{x}}^{2-x} dx = \int_0^1\left(-2x^3 + 7x^2 - 4x - x^{5/2}\right)dx$

$$= \left[-\tfrac{1}{2}x^4 + \tfrac{7}{3}x^3 - 2x^2 - \tfrac{2}{7}x^{7/2}\right]_0^1 = -\tfrac{19}{42}$$

11. $\int_1^2\int_y^{y^3} e^{x/y}\,dx\,dy = \int_1^2\left[ye^{x/y}\right]_y^{y^3} dy = \int_1^2\left(ye^{y^2} - ey\right)dy = \left[\frac{1}{2}e^{y^2} - \frac{1}{2}ey^2\right]_1^2 = \frac{1}{2}(e^4 - 4e)$

13. $\int_0^1\int_0^{x^2} x\cos y\,dy\,dx = \int_0^1[x\sin y]_0^{x^2} dx = \int_0^1 x\sin x^2\,dx = -\frac{1}{2}\cos x^2\Big|_0^1 = \frac{1}{2}(1-\cos 1)$

15.

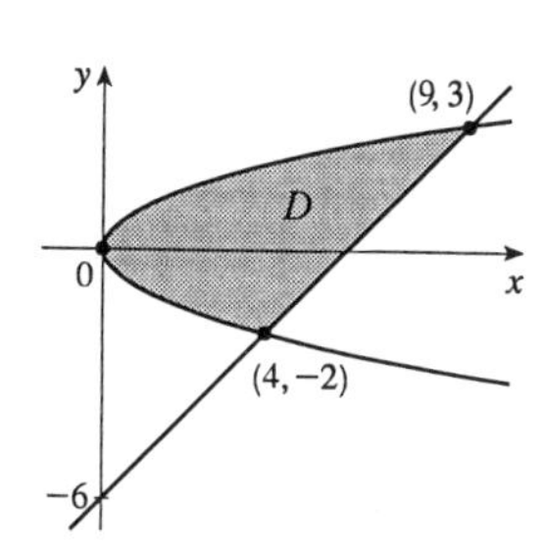

$$\int_{-2}^{3}\int_{y^2}^{y+6} 4y^3\,dx\,dy = \int_{-2}^{3}\left(4y^4 + 24y^3 - 4y^5\right)dy$$
$$= \left[\frac{4y^5}{5} + 6y^4 - \frac{2y^6}{3}\right]_{-2}^{3}$$
$$= 3^4\left(\frac{12}{5}\right) - 16\left(\frac{26}{15}\right)$$
$$= \frac{500}{3}$$

17.

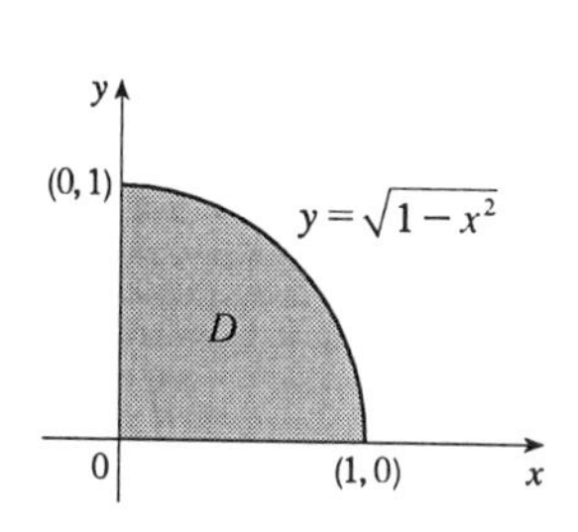

$$\int_0^1\int_0^{\sqrt{1-x^2}} xy\,dy\,dx = \int_0^1 \left[\tfrac{1}{2}xy^2\right]_0^{\sqrt{1-x^2}}dx$$
$$= \int_0^1 \frac{x - x^3}{2}\,dx$$
$$= \frac{1}{2}\left[\frac{x^2}{2} - \frac{x^4}{4}\right]_0^1$$
$$= \frac{1}{8}$$

19.

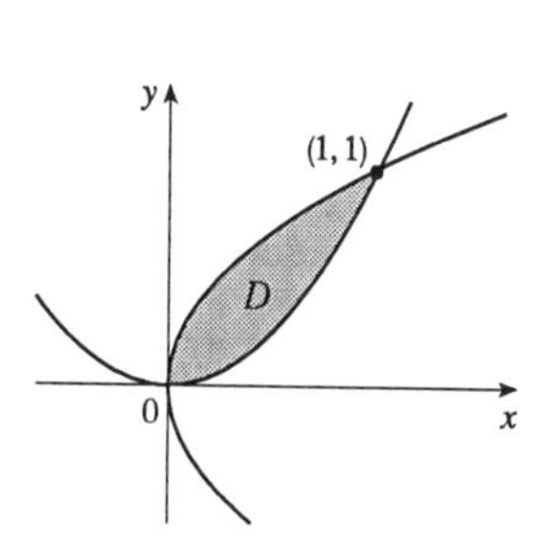

$$V = \int_0^1\int_{x^2}^{\sqrt{x}} (x^2 + y^2)\,dy\,dx$$
$$= \int_0^1 \left(x^{5/2} - x^4 + \tfrac{1}{3}x^{3/2} - \tfrac{1}{3}x^6\right)dx$$
$$= \left[\tfrac{2}{7}x^{7/2} - \tfrac{1}{5}x^5 + \tfrac{2}{15}x^{5/2} - \tfrac{1}{21}x^7\right]_0^1$$
$$= \frac{18}{105} = \frac{6}{35}$$

21.

$$V = \int_1^2\int_1^{7-3y} xy\,dx\,dy = \int_1^2 \left[\frac{yx^2}{2}\right]_1^{7-3y} dy$$
$$= \frac{1}{2}\int_1^2 \left(48y - 42y^2 + 9y^3\right)dy$$
$$= \tfrac{1}{2}\left[24y^2 - 14y^3 + \tfrac{9}{4}y^4\right]_1^2$$
$$= \frac{31}{8}$$

23.

$$V = \textstyle\int_0^2\int_0^{1-x/2} \sqrt{9 - x^2}\,dy\,dx = \int_0^2\left(\sqrt{9 - x^2} - \tfrac{1}{2}x\sqrt{9 - x^2}\right)dx$$
$$= \textstyle\int_0^2 \sqrt{9 - x^2}\,dx + \tfrac{1}{4}\int_0^2\left(-2x\sqrt{9 - x^2}\right)dx$$
$$= \left[\tfrac{1}{2}x\sqrt{9 - x^2} + \tfrac{9}{2}\sin^{-1}(x/3) + \tfrac{1}{6}(9 - x^2)^{3/2}\right]_0^2$$
$$= \sqrt{5} + \tfrac{9}{2}\sin^{-1}\tfrac{2}{3} + \tfrac{5}{6}\sqrt{5} - \tfrac{1}{6}(27)$$
$$= \tfrac{1}{6}\left(11\sqrt{5} - 27\right) + \tfrac{9}{2}\sin^{-1}\tfrac{2}{3}$$

25.

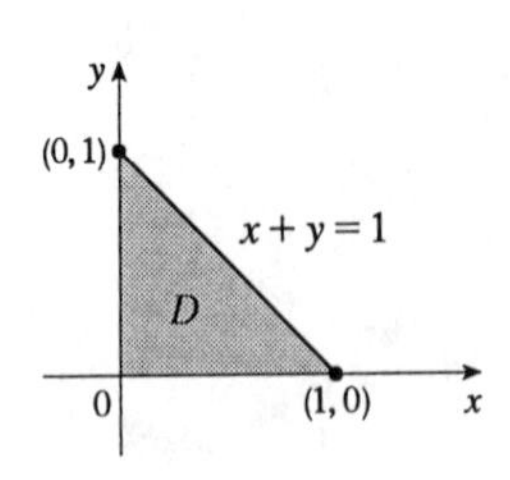

$$V = \int_0^1 \int_0^{1-x} (1 - x - y)\,dy\,dx = \int_0^1 \left[(1-x)^2 - \tfrac{1}{2}(1-x)^2\right] dx = \int_0^1 \tfrac{1}{2}(1-x)^2\,dx = \left[-\tfrac{1}{6}(1-x)^3\right]_0^1 = \frac{1}{6}$$

27.

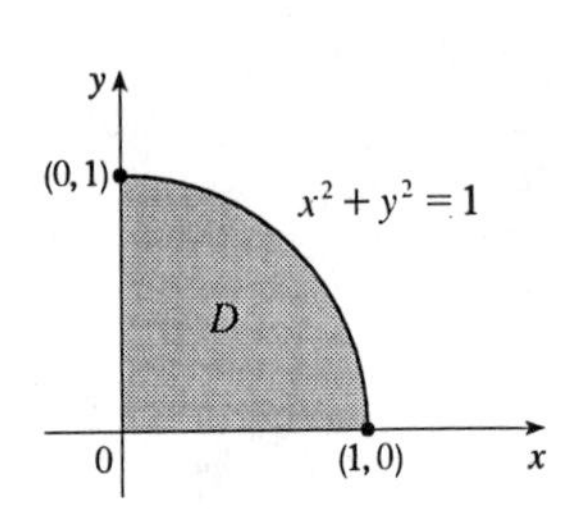

$$V = \int_0^1 \int_0^{\sqrt{1-x^2}} y\,dy\,dx = \int_0^1 \frac{1-x^2}{2}\,dx = \tfrac{1}{2}\left[x - \tfrac{1}{3}x^3\right]_0^1 = \frac{1}{3}$$

29.

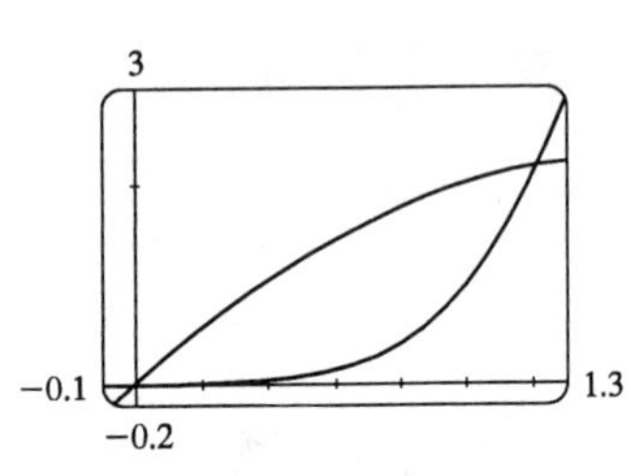

From the graph, it appears that the two curves intersect at $x = 0$ and at $x \approx 1.213$. Thus the desired integral is

$$\iint_D x\,dA \approx \int_0^{1.213} \int_{x^4}^{3x-x^2} x\,dy\,dx = \int_0^{1.213} [xy]_{x^4}^{3x-x^2}\,dx = \left[\left(x^3 - \tfrac{1}{4}x^4\right) - \tfrac{1}{6}x^6\right]_0^{1.213} \approx 0.713$$

31. The two bounding curves $y = x^3 - x$ and $y = x^2 + x$ intersect at the origin and at $x = 2$, with $x^2 + x > x^3 - x$ on $(0, 2)$. Using a CAS, we find that the volume is

$$V = \int_0^2 \int_{x^3-x}^{x^2+x} z\,dy\,dx = \int_0^2 \int_{x^3-x}^{x^2+x} (x^3y^4 + xy^2)\,dy\,dx = \frac{13{,}984{,}735{,}616}{14{,}549{,}535}.$$

33.

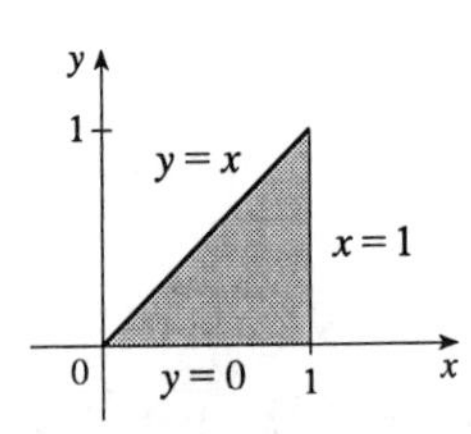

Because the region of integration is

$$D = \{(x, y) \mid 0 \le y \le x, 0 \le x \le 1\} = \{(x, y) \mid y \le x \le 1, 0 \le y \le 1\}$$

we have

$$\int_0^1\int_0^x f(x, y)\,dy\,dx = \iint_D f(x, y)\,dA = \int_0^1\int_y^1 f(x, y)\,dx\,dy$$

35.

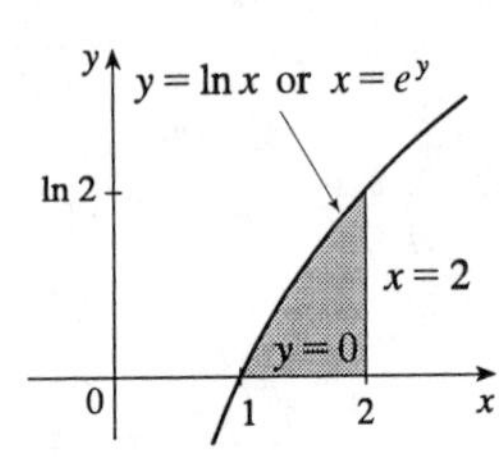

Because the region of integration is

$$D = \{(x, y) \mid 0 \le y \le \ln x, 1 \le x \le 2\} = \{(x, y) \mid e^y \le x \le 2, 0 \le y \le \ln 2\}$$

we have

$$\int_1^2\int_0^{\ln x} f(x, y)\,dy\,dx = \iint_D f(x, y)\,dA = \int_0^{\ln 2}\int_{e^y}^2 f(x, y)\,dx\,dy$$

37.

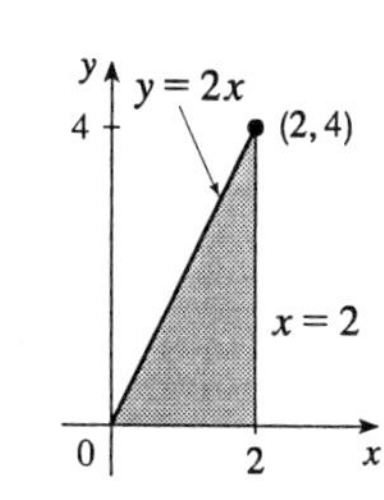

Because the region of integration is

$$D = \{(x,y) \mid y/2 \le x \le 2, 0 \le y \le 4\}$$
$$= \{(x,y) \mid 0 \le y \le 2x, 0 \le x \le 2\}$$

we have

$$\int_0^4 \int_{y/2}^2 f(x,y)\,dx\,dy = \iint_D f(x,y)\,dA = \int_0^2 \int_0^{2x} f(x,y)\,dy\,dx$$

39.

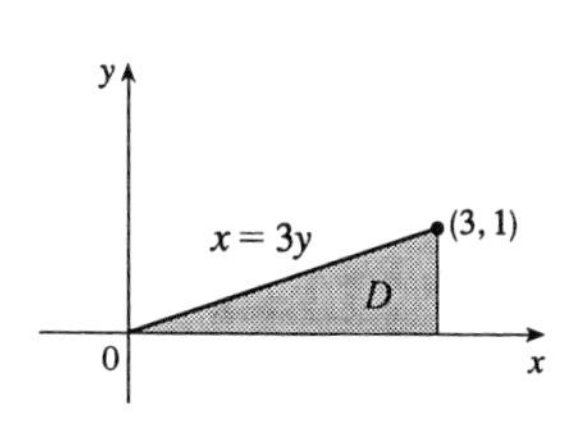

$$\int_0^1 \int_{3y}^3 e^{x^2}\,dx\,dy = \int_0^3 \int_0^{x/3} e^{x^2}\,dy\,dx$$
$$= \int_0^3 \left(\frac{x}{3}\right) e^{x^2}\,dx$$
$$= \tfrac{1}{6} e^{x^2}\Big|_0^3 = \frac{e^9 - 1}{6}$$

41.

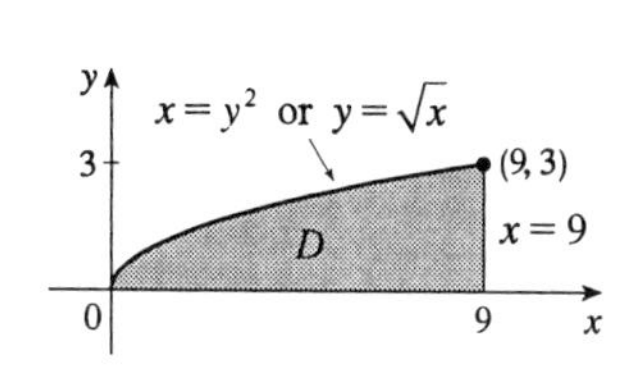

$$\int_0^3 \int_{y^2}^9 y \cos x^2\,dx\,dy = \int_0^9 \int_0^{\sqrt{x}} y \cos x^2\,dy\,dx$$
$$= \int_0^9 \cos x^2 \left[\frac{y^2}{2}\right]_0^{\sqrt{x}} dx$$
$$= \int_0^9 \tfrac{1}{2} x \cos x^2\,dx = \tfrac{1}{4} \sin x^2 \Big|_0^9$$
$$= \tfrac{1}{4} \sin 81$$

43.

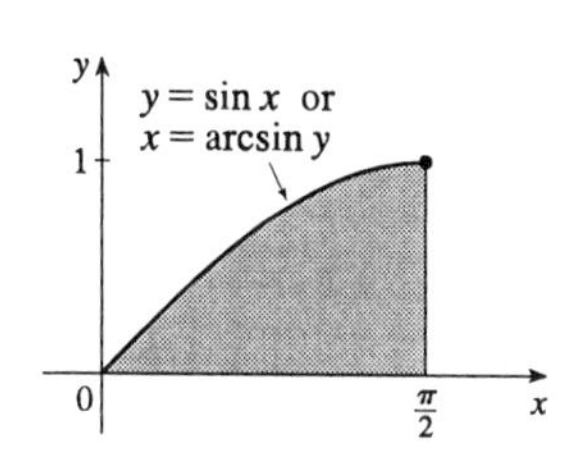

$$\int_0^1 \int_{\arcsin y}^{\pi/2} \cos x\,\sqrt{1 + \cos^2 x}\,dx\,dy = \int_0^{\pi/2} \int_0^{\sin x} \cos x\,\sqrt{1 + \cos^2 x}\,dy\,dx$$
$$= \int_0^{\pi/2} \cos x\,\sqrt{1 + \cos^2 x}\,[y]_0^{\sin x}\,dx$$
$$= \int_0^{\pi/2} \cos x\,\sqrt{1 + \cos^2 x}\,\sin x\,dx \quad \left[\begin{array}{c}\text{Let } u = \cos x,\ du = -\sin x\,dx, \\ dx = du/(-\sin x)\end{array}\right]$$
$$= \int_1^0 -u\sqrt{1 + u^2}\,du = -\tfrac{1}{3}(1 + u^2)^{3/2}\Big|_1^0 = \tfrac{1}{3}\left(\sqrt{8} - 1\right) = \tfrac{1}{3}\left(2\sqrt{2} - 1\right)$$

45. $D = \{(x,y) \mid 0 \le x \le 1, -x + 1 \le y \le 1\} \cup \{(x,y) \mid -1 \le x \le 0, x + 1 \le y \le 1\}$
$\cup \{(x,y) \mid 0 \le x \le 1, -1 \le y \le x - 1\} \cup \{(x,y) \mid -1 \le x \le 0, -1 \le y \le -x - 1\}$, all type I.

$\iint_D x^2\,dA = \int_0^1 \int_{1-x}^1 x^2\,dy\,dx + \int_{-1}^0 \int_{x+1}^1 x^2\,dy\,dx + \int_0^1 \int_{-1}^{x-1} x^2\,dy\,dx + \int_{-1}^0 \int_{-1}^{-x-1} x^2\,dy\,dx$

$= 4\int_0^1 \int_{1-x}^1 x^2\,dy\,dx$ (by symmetry of the regions, and because $f(x,y) = x^2 \ge 0$) $= 4\int_0^1 x^3\,dx = 1$

47. For $D = [0,1] \times [0,1]$, $0 \le \sqrt{x^3 + y^3} \le \sqrt{2}$ and $A(D) = 1$, so $0 \le \iint_D \sqrt{x^3 + y^3}\,dA \le \sqrt{2}$.

49. Since $m \le f(x,y) \le M$, $\iint_D m\,dA \le \iint_D f(x,y)dA \le \iint_D M\,dA$ by (8) $\Rightarrow$
$m\iint_D 1\,dA \le \iint_D f(x,y)\,dA \le M\iint_D 1\,dA$ by (7) $\Rightarrow$ $mA(D) \le \iint_D f(x,y)dA \le MA(D)$ by (10).

51. $\iint_D (x^2 \tan x + y^3 + 4)dA = \iint_D x^2 \tan x\,dA + \iint_D y^3\,dA + \iint_D 4\,dA$. But $x^2 \tan x$ is an odd function and D is symmetric with respect to the y-axis, so $\iint_D x^2 \tan x\,dA = 0$. Similarly, y^3 is an odd function and D is symmetric with respect to the x-axis, so $\iint_D y^3\,dA = 0$. Thus
$\iint_D (x^2 \tan x + y^3 + 4)dA = 4\iint_D dA = 4(\text{area of } D) = 4 \cdot \pi\left(\sqrt{2}\right)^2 = 8\pi$.

EXERCISES 13.4

1. $\left(1, \frac{\pi}{2}\right)$

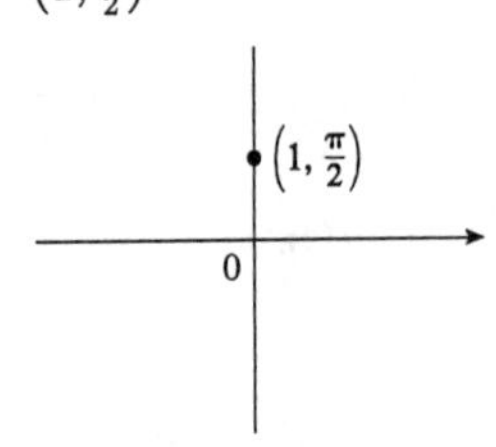

$\left(1, \frac{5\pi}{2}\right), \left(-1, \frac{3\pi}{2}\right)$

3. $\left(-1, \frac{\pi}{5}\right)$

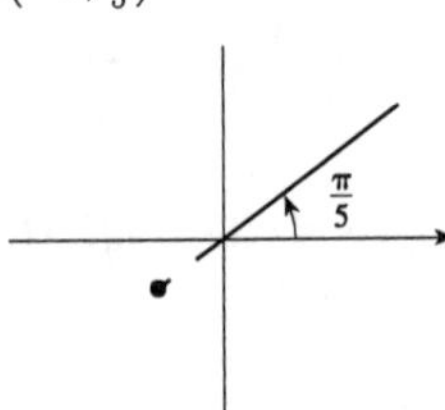

$\left(1, \frac{6\pi}{5}\right), \left(-1, \frac{11\pi}{5}\right)$

5. $(3, 2)$

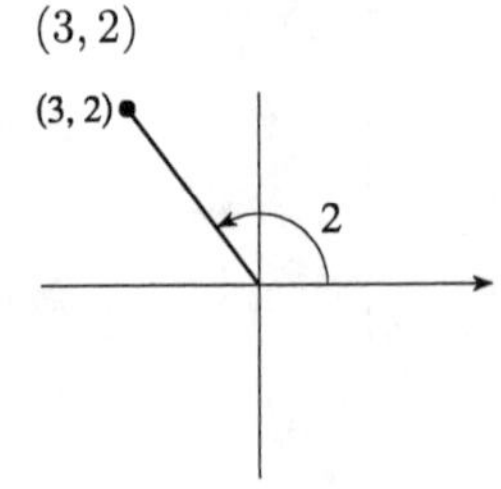

$(3, 2 + 2\pi), (-3, 2 + \pi)$

7.

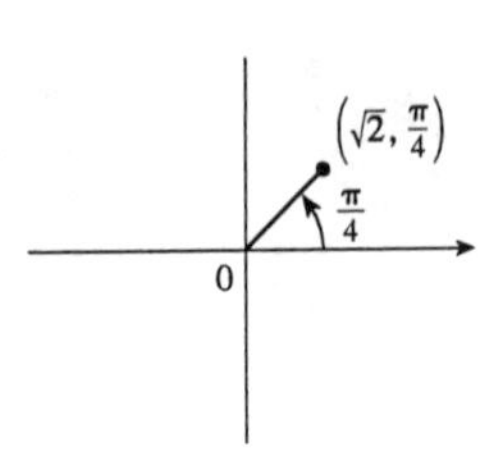

$x = \sqrt{2} \cos \frac{\pi}{4} = 1,$

$y = \sqrt{2} \sin \frac{\pi}{4} = 1$

9.

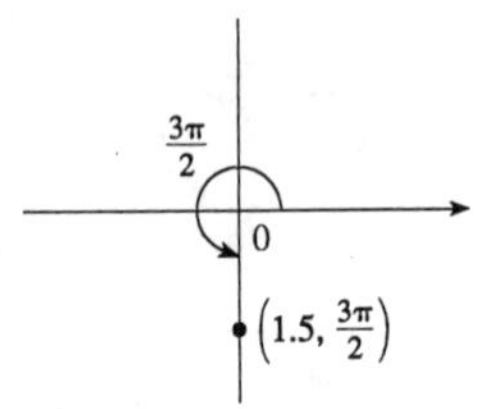

$\left(0, -\frac{3}{2}\right)$

11.

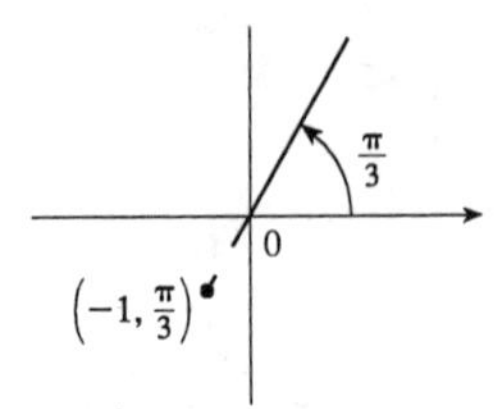

$x = -\cos \frac{\pi}{3} = -\frac{1}{2}$

$y = -\sin \frac{\pi}{3} = -\frac{\sqrt{3}}{2}$

13. $(x, y) = (-1, 1)$, $r = \sqrt{(-1)^2 + 1^2} = \sqrt{2}$, $\tan\theta = y/x = -1$ and (x, y) is in quadrant II, so $\theta = \frac{3\pi}{4}$. Coordinates $\left(\sqrt{2}, \frac{3\pi}{4}\right)$.

15. $(x, y) = \left(2\sqrt{3}, -2\right)$. $r = \sqrt{12 + 4} = 4$, $\tan\theta = y/x = -\frac{1}{\sqrt{3}} \quad \Rightarrow \quad (x, y)$ is in quadrant IV, so $\theta = \frac{11\pi}{6}$. The polar coordinates are $\left(4, \frac{11\pi}{6}\right)$.

17. $r > 1$

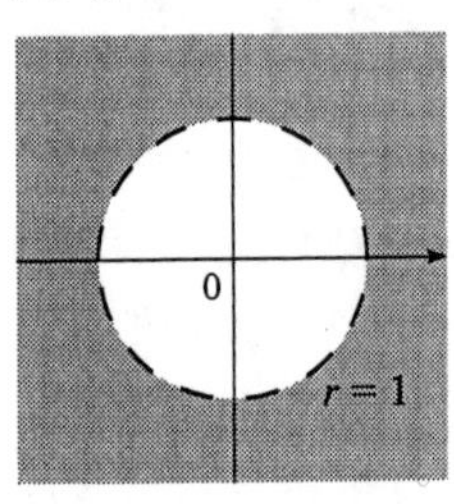

19. $0 \le r \le 2, \frac{\pi}{2} \le \theta \le \pi$

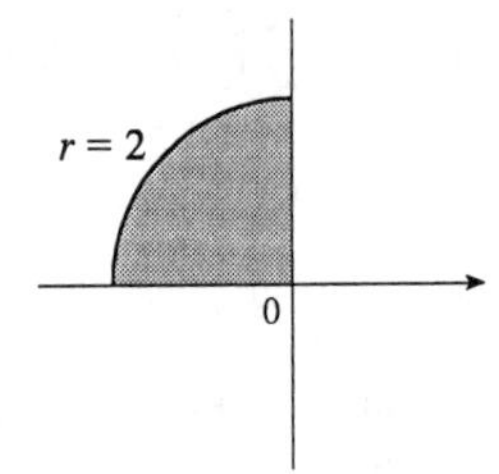

21. $3 < r < 4, -\frac{\pi}{2} \le \theta \le \pi$

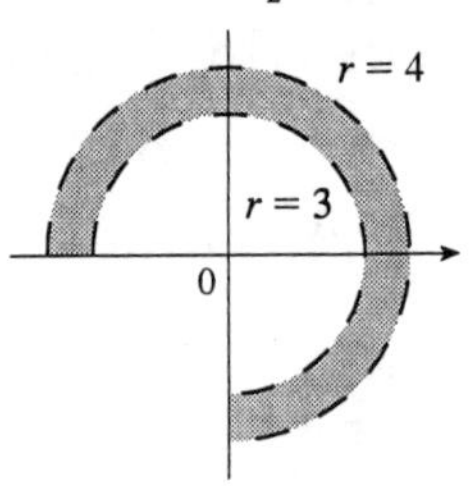

23. $\left(1, \frac{\pi}{6}\right)$ is $\left(\frac{\sqrt{3}}{2}, \frac{1}{2}\right)$ Cartesian and $\left(3, \frac{3\pi}{4}\right)$ is $\left(-\frac{3}{\sqrt{2}}, \frac{3}{\sqrt{2}}\right)$ Cartesian. The square of the distance between them is $\left(\frac{\sqrt{3}}{2} + \frac{\sqrt{3}}{2}\right)^2 + \left(\frac{1}{2} - \frac{3}{\sqrt{2}}\right)^2 = \frac{1}{4}\left(40 + 6\sqrt{6} - 6\sqrt{2}\right)$, so the distance is $\frac{1}{2}\sqrt{40 + 6\sqrt{6} - 6\sqrt{2}}$.

25. Since $y = r\sin\theta$, the equation $r\sin\theta = 2$ becomes $y = 2$.

27. $r = \dfrac{1}{1 - \cos\theta} \quad \Leftrightarrow \quad r - r\cos\theta = 1 \quad \Leftrightarrow \quad r = 1 + r\cos\theta \quad \Leftrightarrow \quad r^2 = (1 + r\cos\theta)^2 \quad \Leftrightarrow$

$x^2 + y^2 = (1 + x)^2 = 1 + 2x + x^2 \quad \Leftrightarrow \quad y^2 = 1 + 2x$

29. $r^2 = \sin 2\theta = 2\sin\theta\cos\theta \quad\Leftrightarrow\quad r^4 = 2r\sin\theta\, r\cos\theta \quad\Leftrightarrow\quad (x^2+y^2)^2 = 2yx$

31. $y = 5 \quad\Leftrightarrow\quad r\sin\theta = 5$

33. $x^2 + y^2 = 25 \quad\Leftrightarrow\quad r^2 = 25 \quad\Leftrightarrow\quad r = 5$

35. $2xy = 1 \quad\Leftrightarrow\quad 2r\cos\theta\, r\sin\theta = 1 \quad\Leftrightarrow\quad r^2\sin 2\theta = 1 \quad\Leftrightarrow\quad r^2 = \csc 2\theta$

37. $r = 5$

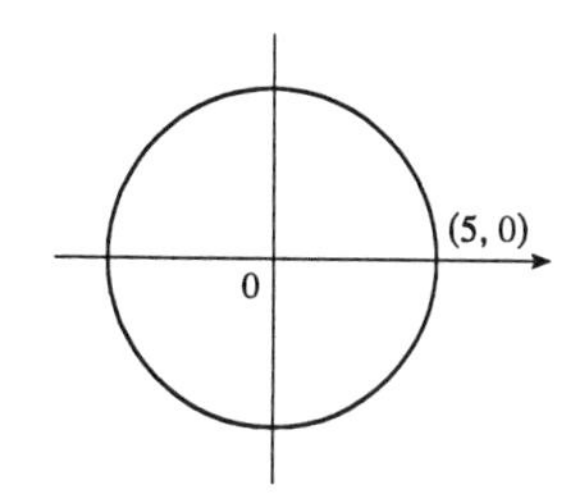

39. $r = 2\sin\theta \quad\Leftrightarrow\quad r^2 = 2r\sin\theta \quad\Leftrightarrow$
$x^2 + y^2 = 2y \quad\Leftrightarrow\quad x^2 + (y-1)^2 = 1$

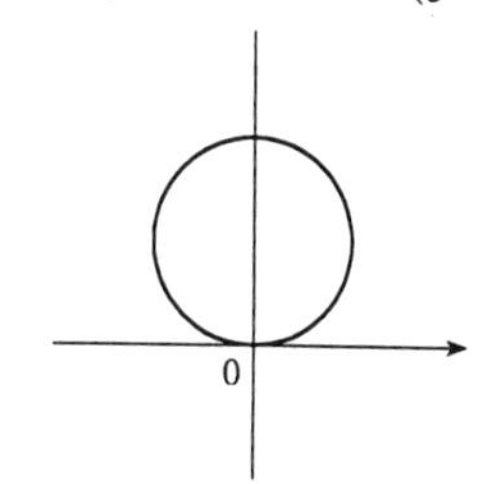

41. $r = -\cos\theta \quad\Leftrightarrow\quad r^2 = -r\cos\theta$
$\Leftrightarrow\ x^2 + y^2 = -x \ \Leftrightarrow\ \left(x + \frac{1}{2}\right)^2 + y^2 = \frac{1}{4}$

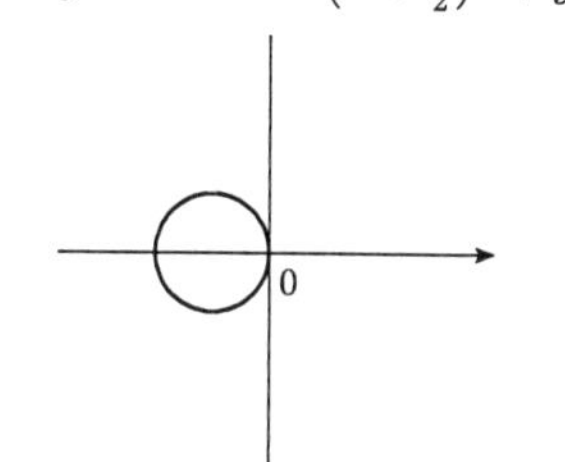

43. $r = 3(1 - \cos\theta)$

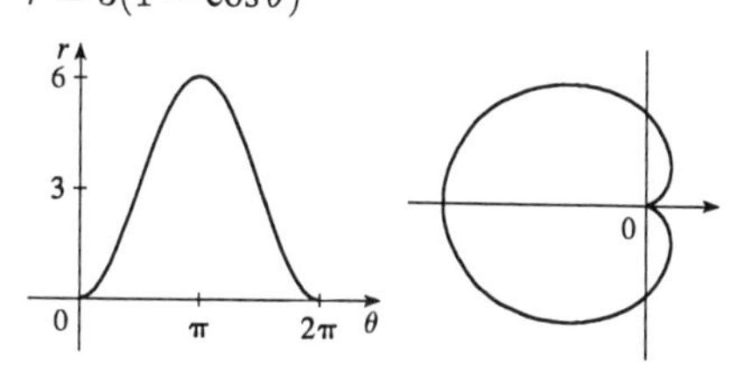

45. $r = \theta, \theta \geq 0$

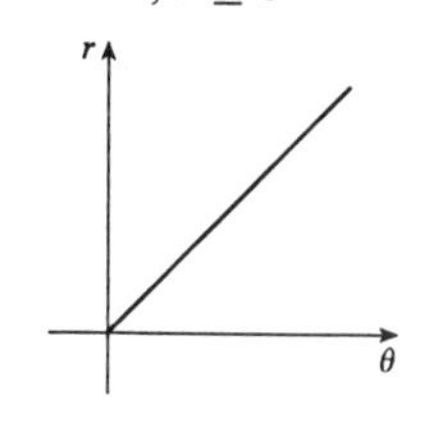

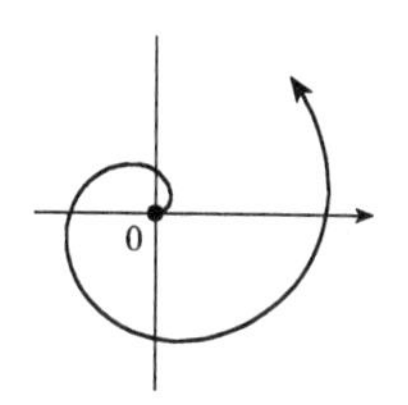

47. $r = 1/\theta$

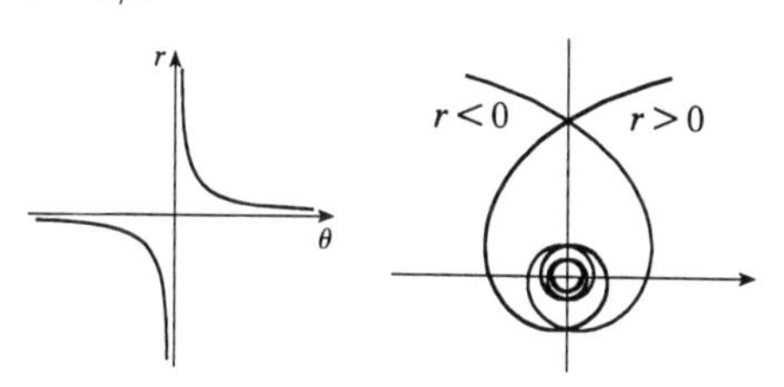

49. $r = 1 - 2\cos\theta$

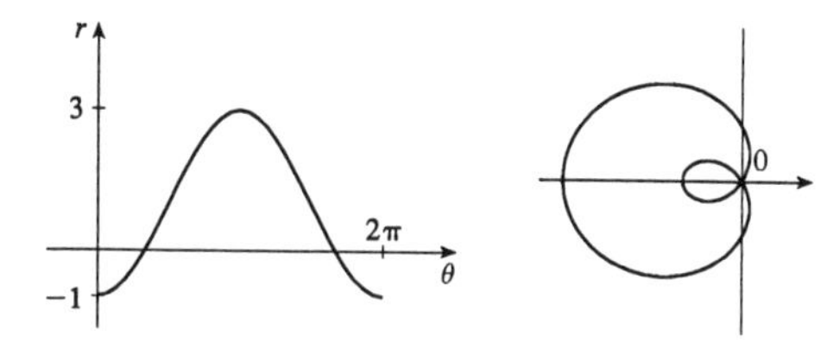

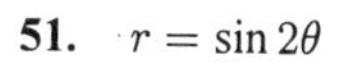

51. $r = \sin 2\theta$

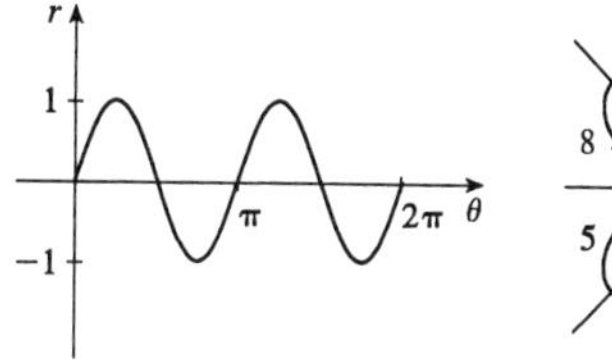

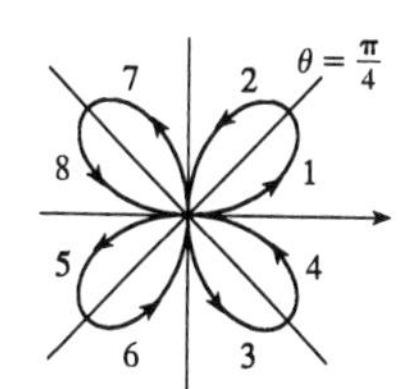

53. $r = 2\cos 4\theta$

55. $r^2 = 4\cos 2\theta$

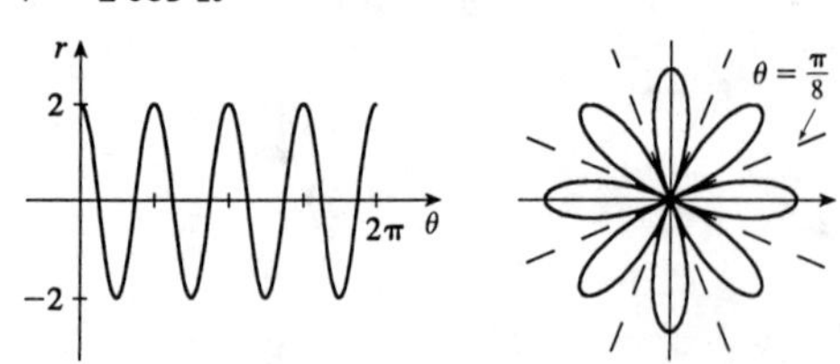

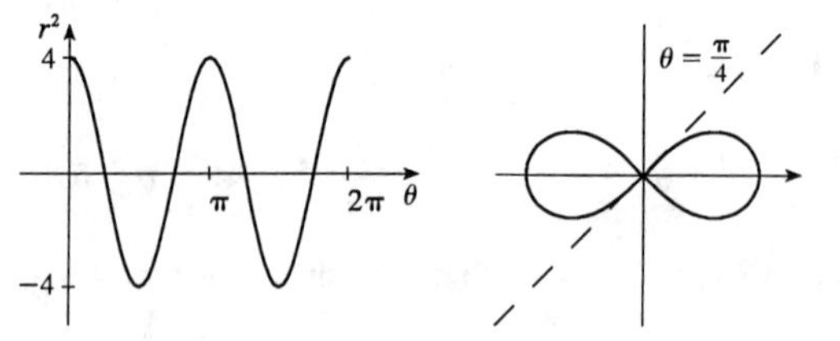

57. $r = 2\cos\left(\frac{3}{2}\theta\right)$

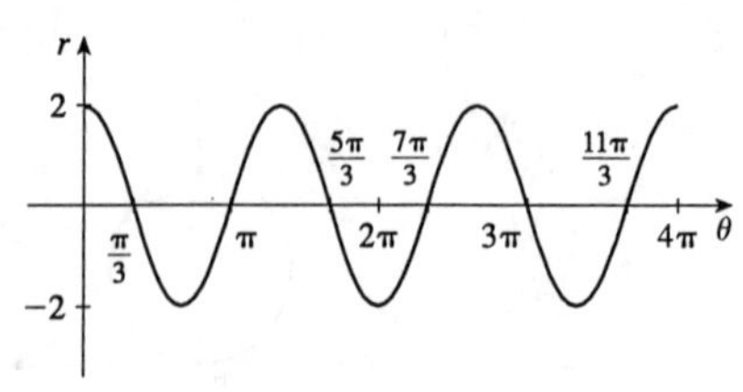

59. $x = r\cos\theta = 4\cos\theta + 2\sec\theta\cos\theta = 4\cos\theta + 2$.
Now, $r \to \infty \;\Rightarrow\; (4 + 2\sec\theta) \to \infty \;\Rightarrow\; \theta \to \frac{\pi}{2}^-$
(since we need only consider $0 \le \theta < 2\pi$), so
$\lim_{r\to\infty} x = \lim_{\theta\to\pi/2^-} (4\cos\theta + 2) = 2$. Also,
$r \to -\infty \;\Rightarrow\; (4 + 2\sec\theta) \to -\infty \;\Rightarrow\; \theta \to \frac{\pi}{2}^+$,
so $\lim_{r\to-\infty} x = \lim_{\theta\to\pi/2^+} (4\cos\theta + 2) = 2$. Therefore
$\lim_{r\to\pm\infty} x = 2 \;\Rightarrow\; x = 2$ is a vertical asymptote.

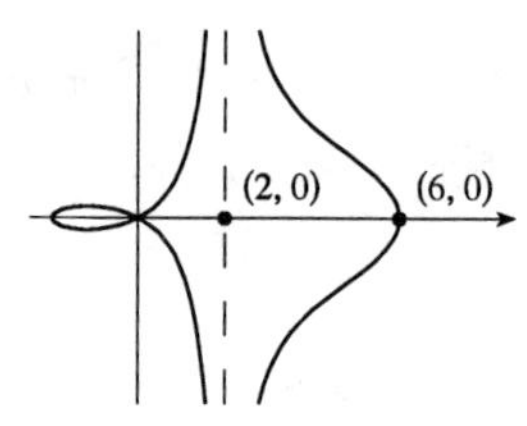

61. To show that $x = 1$ is an asymptote we must prove $\lim_{r\to\pm\infty} x = 1$. $x = r\cos\theta = \sin\theta\tan\theta\cos\theta = \sin^2\theta$. Now,

$r \to \infty \;\Rightarrow\; \sin\theta\tan\theta \to \infty \;\Rightarrow\; \theta \to \frac{\pi}{2}^-$, so $\lim_{r\to\infty} x = \lim_{\theta\to\pi/2^-} \sin^2\theta = 1$. Also, $r \to -\infty \;\Rightarrow$

$\sin\theta\tan\theta \to -\infty \Rightarrow \;\theta \to \pi/2^+$, so $\lim_{r\to-\infty} x = \lim_{\theta\to\pi/2^+} \sin^2\theta = 1$.

Therefore $\lim_{r\to\pm\infty} x = 1 \;\Rightarrow\; x = 1$ is a vertical asymptote.

Also notice that $x = \sin^2\theta \ge 0$ for all θ, and $x = \sin^2\theta \le 1$ for all θ.
And $x \ne 1$, since the curve is not defined at odd multiples of $\frac{\pi}{2}$.
Therefore the curve lies entirely within the vertical strip $0 \le x < 1$.

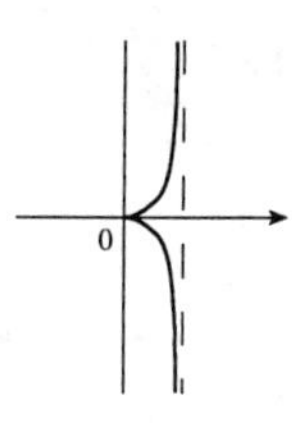

63. **(a)** We see that the curve crosses itself at the origin, where $r = 0$ (in fact the inner loop corresponds to negative r-values,) so we solve the equation of the limaçon for $r = 0 \quad\Leftrightarrow\quad c\sin\theta = -1 \quad\Leftrightarrow$ $\sin\theta = -1/c$. Now if $|c| < 1$, then this equation has no solution and hence there is no inner loop. But if $c < -1$, then on the interval $(0, 2\pi)$ the equation has the two solutions $\theta = \sin^{-1}(-1/c)$ and $\theta = \pi - \sin^{-1}(-1/c)$, and if $c > 1$, the solutions are $\theta = \pi + \sin^{-1}(1/c)$ and $\theta = 2\pi - \sin^{-1}(1/c)$. In each case, $r < 0$ for θ between the two solutions, indicating a loop.

(b) For $0 < c < 1$, the dimple (if it exists) is characterized by the fact that y has a local maximum at $\theta = \frac{3\pi}{2}$. So we determine for what c-values $\dfrac{d^2y}{d\theta^2}$ is negative at $\theta = \frac{3\pi}{2}$, since by the Second Derivative Test this indicates a maximum: $y = r\sin\theta = \sin\theta + c\sin^2\theta \;\Rightarrow\; \dfrac{dy}{d\theta} = \cos\theta + 2c\sin\theta\cos\theta = \cos\theta + c\sin 2\theta$ $\Rightarrow\; \dfrac{d^2y}{d\theta^2} = -\sin\theta + 2c\cos 2\theta$. At $\theta = \frac{3\pi}{2}$, this is equal to $-(-1) + 2c(-1) = 1 - 2c$, which is negative only for $c > \frac{1}{2}$. A similar argument shows that for $-1 < c < 0$, y only has a local *minimum* at $\theta = \frac{\pi}{2}$ (indicating a dimple) for $c < -\frac{1}{2}$.

Note for Exercises 65 and 67: Maple is able to plot polar curves using the `polarplot` command, or using the `coords=polar` option in a regular `plot` command. In Mathematica, use `PolarPlot`. If your graphing device cannot plot polar equations, you must convert to parametric equations. For example, in Exercise 65, $x = r\cos\theta = [1 + 2\sin(\theta/2)]\cos\theta$, $y = r\sin\theta = [1 + 2\sin(\theta/2)]\sin\theta$.

65. $r = 1 + 2\sin(\theta/2)$

The correct parameter interval is $[0, 4\pi]$.

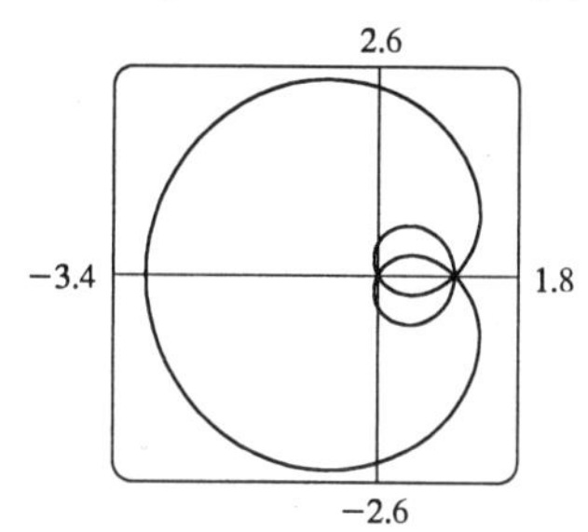

67. $r = \sin(9\theta/4)$

The correct parameter interval is $[0, 8\pi]$.

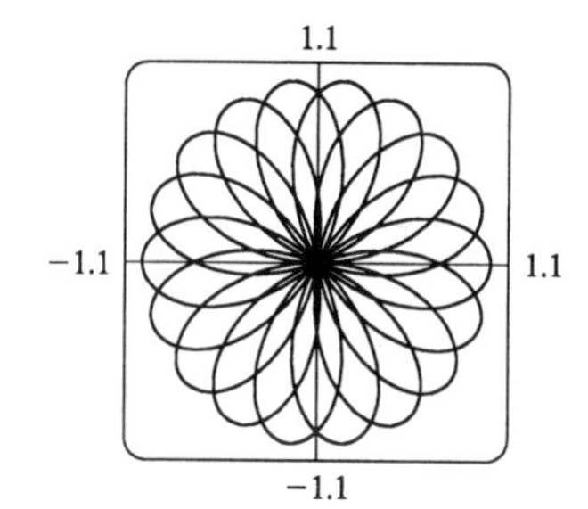

69.

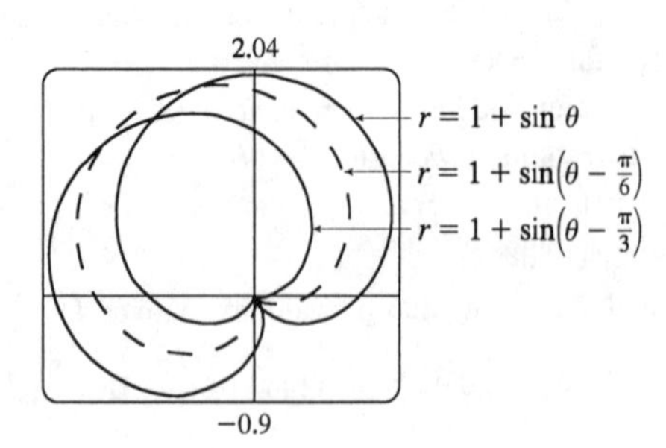

It appears that the graph of $r = 1 + \sin\left(\theta - \frac{\pi}{6}\right)$ is the same shape as the graph of $r = 1 + \sin\theta$, but rotated counterclockwise about the origin by $\frac{\pi}{6}$. Similarly, the graph of $r = 1 + \sin\left(\theta - \frac{\pi}{3}\right)$ is rotated by $\frac{\pi}{3}$. In general, the graph of $r = f(\theta - \alpha)$ is the same shape as that of $r = f(\theta)$, but rotated counterclockwise through α about the origin.

That is, for any point (r_0, θ_0) on the curve $r = f(\theta)$, the point $(r_0, \theta_0 + \alpha)$ is on the curve the curve $r = f(\theta - \alpha)$, since $r_0 = f(\theta_0) = f((\theta_0 + \alpha) - \alpha)$.

71. **(a)** $r = \sin n\theta$. From the graphs, it seems that when n is even, the number of loops in the curve (called a rose) is $2n$, and when n is odd, the number of loops is simply n. This is because in the case of n odd, every point on the graph is traversed twice, due to the fact that

$$r(\theta + \pi) = \sin[n(\theta + \pi)] = \sin n\theta \cos n\pi + \cos n\theta \sin n\pi = \begin{cases} \sin n\theta & n \text{ even} \\ -\sin n\theta & n \text{ odd.} \end{cases}$$

$n = 2$

$n = 3$

$n = 4$

$n = 5$

(b) The graph of $r = |\sin n\theta|$ has $2n$ loops whether n is odd or even, since $r(\theta + \pi) = r(\theta)$.

$n = 2$

$n = 3$

$n = 4$

$n = 5$

73. $r = \dfrac{1 - a\cos\theta}{1 + a\cos\theta}$. We start with $a = 0$, since in this case the curve is simply the circle $r = 1$. As a increases, the graph moves to the left, and its right side becomes flattened. As a increases through about 0.4, the right side seems to grow a dimple, which upon closer investigation (with narrower θ-ranges) seems to appear at $a \approx 0.42$ (the actual value is $\sqrt{2} - 1$.) As $a \to 1$, this dimple becomes more pronounced, and the curve begins to stretch out horizontally, until at $a = 1$ the denominator vanishes at $\theta = \pi$, and the dimple becomes an actual cusp. For $a > 1$ we must choose our parameter interval carefully, since $r \to \infty$ as $1 + a\cos\theta \to 0 \quad \Leftrightarrow$ $\theta \to \pm\cos^{-1}(-1/a)$. As a increases from 1, the curve splits into two parts. The left part has a loop, which grows larger as a increases, and the right part grows broader vertically, and its left tip develops a dimple when $a \approx 2.42$ (actually, $\sqrt{2} + 1$). As a increases, the dimple grows more and more pronounced.

If $a < 0$, we get the same graph as we do for the corresponding positive a-value, but with a rotation through π about the pole, as happened when c was replaced with $-c$ in Exercise 72.

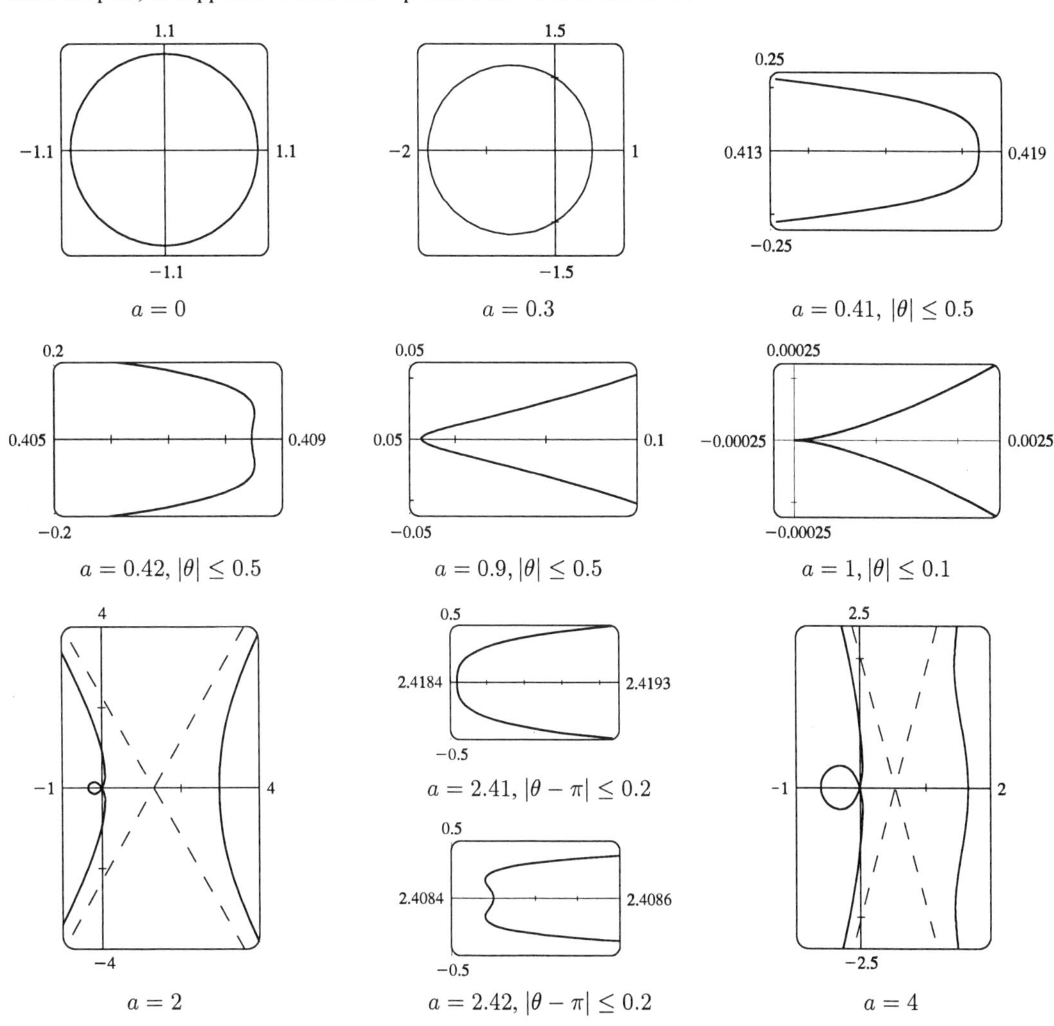

$a = 0$ $\quad$ $a = 0.3$ $\quad$ $a = 0.41,\ |\theta| \le 0.5$

$a = 0.42,\ |\theta| \le 0.5$ $\quad$ $a = 0.9,\ |\theta| \le 0.5$ $\quad$ $a = 1,\ |\theta| \le 0.1$

$a = 2$ $\quad$ $a = 2.41,\ |\theta - \pi| \le 0.2$ $\quad$ $a = 2.42,\ |\theta - \pi| \le 0.2$ $\quad$ $a = 4$

EXERCISES 13.5

1. $\iint_R x\,dA = \int_0^{2\pi}\int_0^5 r^2\cos\theta\,dr\,d\theta = \left(\int_0^{2\pi}\cos\theta\,d\theta\right)\left(\int_0^5 r^2\,dr\right) = 0$

3. $\displaystyle\iint_R xy\,dA = \int_0^{\pi/2}\int_2^5 r^3\cos\theta\sin\theta\,dr\,d\theta = \left(\int_0^{\pi/2}\frac{\sin 2\theta}{2}\,d\theta\right)\left(\int_2^5 r^3\,dr\right) = \frac{1}{2}\cdot\frac{5^4-2^4}{4} = \frac{609}{8}$

5. The circle $r=1$ intersects the cardioid $r = 1+\sin\theta$ when $1 = 1+\sin\theta \quad\Rightarrow\quad \theta = 0$ or $\theta = \pi$, so

$\displaystyle\iint_D \frac{1}{\sqrt{x^2+y^2}}\,dA = \int_0^{\pi}\int_1^{1+\sin\theta}\left(\frac{1}{r}\right)r\,dr\,d\theta = \int_0^{\pi}[r]_1^{1+\sin\theta}\,d\theta = \int_0^{\pi}\sin\theta\,d\theta = [-\cos\theta]_0^{\pi} = 2.$

7. $\int_0^{2\pi}\int_\theta^{2\theta} r^3\,dr\,d\theta = \int_0^{2\pi}\frac{15}{4}\theta^4\,d\theta = \left[\frac{3}{4}\theta^5\right]_0^{2\pi} = 24\pi^5$

9. $A = \int_{-\pi/6}^{\pi/6}\int_0^{\cos 3\theta} r\,dr\,d\theta = \int_{-\pi/6}^{\pi/6}\frac{1}{2}r^2\Big|_0^{\cos 3\theta}\,d\theta = \frac{1}{2}\int_{-\pi/6}^{\pi/6}\cos^2 3\theta\,d\theta$

$= \int_0^{\pi/6}\frac{1}{2}(1+\cos 6\theta)d\theta = \frac{1}{2}\left[\theta + \frac{1}{6}\sin 6\theta\right]_0^{\pi/6} = \frac{\pi}{12}$

11. By symmetry, the two loops of the lemniscate are equal in area, so

$A = 2\int_{-\pi/4}^{\pi/4}\int_0^{2\sqrt{\cos 2\theta}} r\,dr\,d\theta = \int_{-\pi/4}^{\pi/4} 4\cos 2\theta\,d\theta = 8\int_0^{\pi/4}\cos 2\theta\,d\theta = 4\sin 2\theta\big|_0^{\pi/4} = 4.$

13. $3\cos\theta = 1+\cos\theta$ implies $\cos\theta = \frac{1}{2}$, so $\theta = \pm\frac{\pi}{3}$. Then by symmetry

$A = 2\int_0^{\pi/3}\int_{1+\cos\theta}^{3\cos\theta} r\,dr\,d\theta = 2\int_0^{\pi/3}\left[\frac{1}{2}r^2\right]_{1+\cos\theta}^{3\cos\theta}d\theta = \int_0^{\pi/3}(9\cos^2\theta - 1 - 2\cos\theta - \cos^2\theta)d\theta$

$= \int_0^{\pi/3}\left[8\cdot\frac{1}{2}(1+\cos 2\theta) - 2\cos\theta - 1\right]d\theta = [4\theta + 2\sin 2\theta - 2\sin\theta - \theta]_0^{\pi/3} = \pi.$

15. $V = \displaystyle\iint_{x^2+y^2\le 9}(x^2+y^2)dA = \int_0^{2\pi}\int_0^3 r^3\,dr\,d\theta = 2\pi\left(\tfrac{81}{4}\right) = \tfrac{81\pi}{2}$

17.

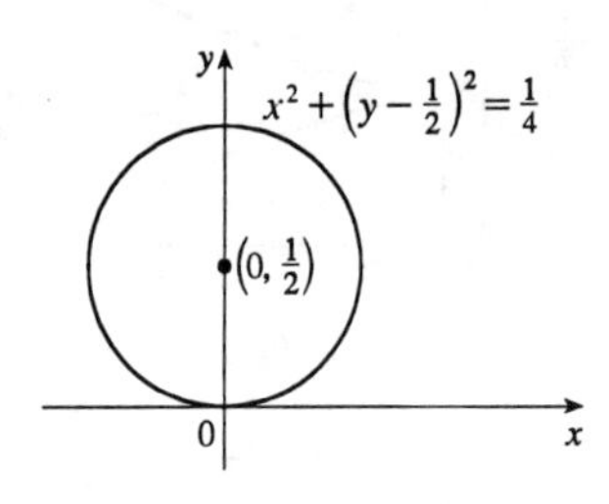

$V = \displaystyle\iint_{x^2+(y-\frac{1}{2})^2\le\frac{1}{4}}(12-6x-4y)dA$

$= \int_0^{\pi}\int_0^{\sin\theta}(12r - 6r^2\cos\theta - 4r^2\sin\theta)dr\,d\theta$

$= \int_0^{\pi}\left(6\sin^2\theta - 2\sin^3\theta\cos\theta - \frac{4}{3}\sin^4\theta\right)d\theta$

$= \left[\frac{5}{2}(\theta - \sin\theta\cos\theta) - \frac{1}{2}\sin^4\theta + \frac{1}{3}\sin^2\theta\cos\theta\right]_0^{\pi} = \frac{5\pi}{2}$

This can also be done without calculus:
the volume of this cylinder is $\pi\left(\frac{1}{4}\right)\left(\frac{12+8}{2}\right) = \frac{5\pi}{2}$.

19. The cone $z = \sqrt{x^2+y^2}$ intersects the sphere $x^2+y^2+z^2 = 1$ when $x^2+y^2+\left(\sqrt{x^2+y^2}\right)^2 = 1$ or $x^2+y^2 = \frac{1}{2}$. So

$V = \displaystyle\iint_{x^2+y^2\le\frac{1}{2}}\left(\sqrt{1-x^2-y^2} - \sqrt{x^2+y^2}\right)dA = \int_0^{2\pi}\int_0^{1/\sqrt{2}}\left(\sqrt{1-r^2} - r\right)r\,dr\,d\theta$

$= \frac{2\pi}{3}\left[-(1-r^2)^{3/2} - r^3\right]_0^{1/\sqrt{2}} = \frac{2\pi}{3}\left(-\frac{1}{\sqrt{2}}+1\right) = \frac{\pi}{3}\left(2-\sqrt{2}\right)$

21. The given solid is the region inside the cylinder $x^2+y^2=4$ between the surfaces $z=\sqrt{64-4x^2-4y^2}$ and $z=-\sqrt{64-4x^2-4y^2}$. So

$$V=\iint_{x^2+y^2\le 4}\left[\sqrt{64-4x^2-4y^2}-\left(-\sqrt{64-4x^2-4y^2}\right)\right]dA=\iint_{x^2+y^2\le 4}2\sqrt{64-4x^2-4y^2}\,dA$$

$$=4\int_0^{2\pi}\int_0^2\sqrt{16-r^2}\,r\,dr\,d\theta=8\pi\left[-\tfrac{1}{3}(16-r^2)^{3/2}\right]_0^2=\tfrac{8\pi}{3}\left(64-12^{3/2}\right)=\tfrac{8\pi}{3}\left(64-24\sqrt{3}\right)$$

23. $V=2\iint_{x^2+y^2\le a^2}\sqrt{a^2-x^2-y^2}\,dA=2\int_0^{2\pi}\int_0^a\sqrt{a^2-r^2}\,r\,dr\,d\theta=\frac{4\pi}{3}\left[-(a^2-r^2)^{3/2}\right]_0^a=\frac{4\pi}{3}a^3.$

25.

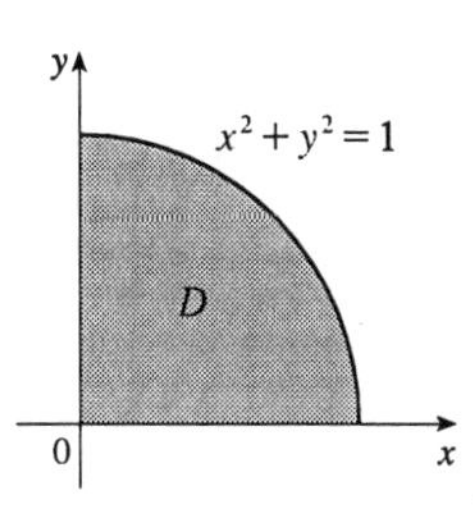

$\int_0^{\pi/2}\int_0^1 re^{r^2}\,dr\,d\theta=\frac{\pi}{2}\left[\frac{1}{2}e^{r^2}\right]_0^1=\frac{1}{4}\pi(e-1)$

27.

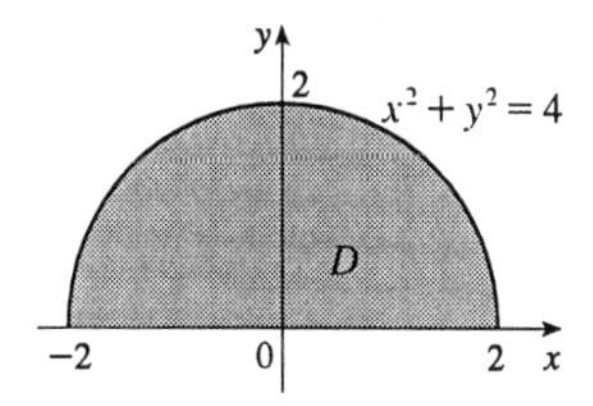

$\int_0^{\pi}\int_0^2(r^4\cos^2\theta\sin^2\theta)r\,dr\,d\theta=\int_0^{\pi}\int_0^2\left(\frac{1}{4}r^5\sin^2 2\theta\right)dr\,d\theta$

$=\frac{8}{3}\int_0^{\pi}\sin^2 2\theta\,d\theta=\frac{8}{12}[2\theta-\sin 2\theta\cos 2\theta]_0^{\pi}=\frac{4\pi}{3}$

29. $$\int_0^1\int_0^{\sqrt{1-x^2}}e^{-(x^2+y^2)^2}\,dy\,dx=\int_0^{\pi/2}\int_0^1 re^{-(r^2)^2}\,dr\,d\theta=\frac{\pi}{2}\int_0^1 re^{-r^4}\,dr\approx 0.587$$

31. $$\int_{1/\sqrt{2}}^1\int_{\sqrt{1-x^2}}^{x}xy\,dy\,dx+\int_1^{\sqrt{2}}\int_0^x xy\,dy\,dx+\int_{\sqrt{2}}^2\int_0^{\sqrt{4-x^2}}xy\,dy\,dx=\int_0^{\pi/4}\int_1^2 r^3\cos\theta\sin\theta\,dr\,d\theta$$

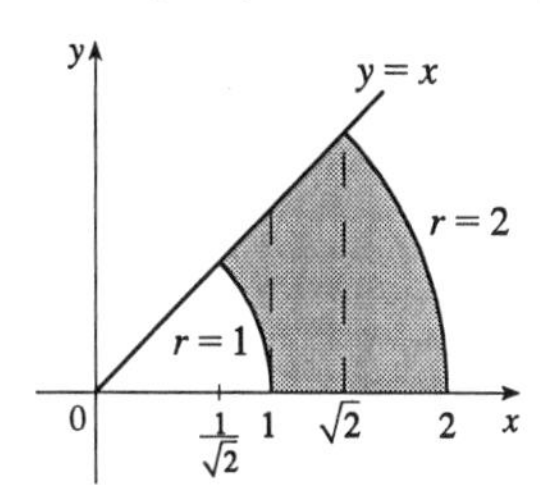

$$=\int_0^{\pi/4}\left[\frac{r^4}{4}\cos\theta\sin\theta\right]_1^2 d\theta$$

$$=\frac{15}{4}\int_0^{\pi/4}\sin\theta\cos\theta\,d\theta$$

$$=\frac{15}{4}\left[\frac{\sin^2\theta}{2}\right]_0^{\pi/4}=\frac{15}{16}$$

33. **(a)** We integrate by parts with $u=x$ and $dv=xe^{-x^2}dx$. Then $du=dx$ and $v=-\frac{1}{2}e^{-x^2}$, so

$\int_0^\infty x^2e^{-x^2}\,dx=\lim_{t\to\infty}\int_0^t x^2e^{-x^2}\,dx=\lim_{t\to\infty}\left(-\frac{1}{2}xe^{-x^2}\Big|_0^t+\int_0^t\frac{1}{2}e^{-x^2}\,dx\right)$

$=\lim_{t\to\infty}\left(-\frac{1}{2}te^{-t^2}\right)+\frac{1}{2}\int_0^\infty e^{-x^2}\,dx=0+\frac{1}{2}\int_0^\infty e^{-x^2}\,dx$ (by l'Hospital's Rule)

$=\frac{1}{4}\int_{-\infty}^\infty e^{-x^2}\,dx$ (since e^{-x^2} is an even function) $=\frac{1}{4}\sqrt{\pi}$ [by Exercise 32(c)].

(b) Let $u=\sqrt{x}$. Then $u^2=x \;\Rightarrow\; dx=2u\,du \;\Rightarrow$

$\int_0^\infty\sqrt{x}e^{-x}\,dx=\lim_{t\to\infty}\int_0^t\sqrt{x}e^{-x}\,dx=\lim_{t\to\infty}\int_0^{\sqrt{t}}ue^{-u^2}2u\,du$

$=2\int_0^\infty u^2e^{-u^2}\,du=2\left(\frac{1}{4}\sqrt{\pi}\right)=\frac{1}{2}\sqrt{\pi}$ [by part (a)]

EXERCISES 13.6

1. $Q = \iint_D (x^2 + 3y^2)dA = \int_0^2\int_1^2 (x^2 + 3y^2)dy\,dx = \int_0^2 (x^2 + 7)dx = 14 + \frac{8}{3} = \frac{50}{3}$ C

3. $m = \int_{-1}^1\int_0^1 x^2\,dy\,dx = \int_{-1}^1 x^2\,dx = \frac{2}{3}$, $\overline{x} = \frac{3}{2}\int_{-1}^1\int_0^1 x^3\,dy\,dx = 0$,
$\overline{y} = \frac{3}{2}\int_{-1}^1\int_0^1 x^2 y\,dy\,dx = \frac{3}{2}\int_{-1}^1 \frac{1}{2}x^2\,dx = \frac{3}{2}\left(\frac{2}{6}\right) = \frac{1}{2}$. Hence $(\overline{x}, \overline{y}) = \left(0, \frac{1}{2}\right)$.

5. $m = \int_0^2\int_{x/2}^{3-x}(x + y)dy\,dx = \int_0^2 \left[x\left(3 - \frac{3}{2}x\right) + \frac{1}{2}(3 - x)^2 - \frac{1}{8}x^2\right]dx = \int_0^2 \left[-\frac{9}{8}x^2 + \frac{9}{2}\right]dx = 6$,
$M_y = \int_0^2\int_{x/2}^{3-x}(x^2 + xy)dy\,dx = \int_0^2 \left[x^2 y + \frac{1}{2}xy^2\right]_{x/2}^{3-x} dx = \int_0^2 \left(\frac{9}{2}x - \frac{9}{8}x^3\right)dx = \frac{9}{2}$, and
$M_x = \int_0^2\int_{x/2}^{3-y}(xy + y^2)dy\,dx = \int_0^2 \left(9 - \frac{9}{2}x\right)dx = 9$. Hence $m = 6$, $(\overline{x}, \overline{y}) = \left(\frac{3}{4}, \frac{3}{2}\right)$.

7. $m = \int_0^1\int_{x^2}^1 xy\,dy\,dx = \int_0^1 \left(\frac{1}{2}x - \frac{1}{2}x^5\right)dx = \frac{1}{4} - \frac{1}{12} = \frac{1}{6}$,
$M_y = \int_0^1\int_{x^2}^1 x^2 y\,dy\,dx = \int_0^1 \left(\frac{1}{2}x^2 - \frac{1}{2}x^6\right)dx = \frac{1}{6} - \frac{1}{14} = \frac{2}{21}$ and
$M_x = \int_0^1\int_{x^2}^1 xy^2\,dy\,dx = \int_0^1 \left(\frac{1}{3}x - \frac{1}{3}x^7\right)dx = \frac{1}{6} - \frac{1}{24} = \frac{1}{8}$. Hence $m = \frac{1}{6}$, $(\overline{x}, \overline{y}) = \left(\frac{4}{7}, \frac{3}{4}\right)$.

9.

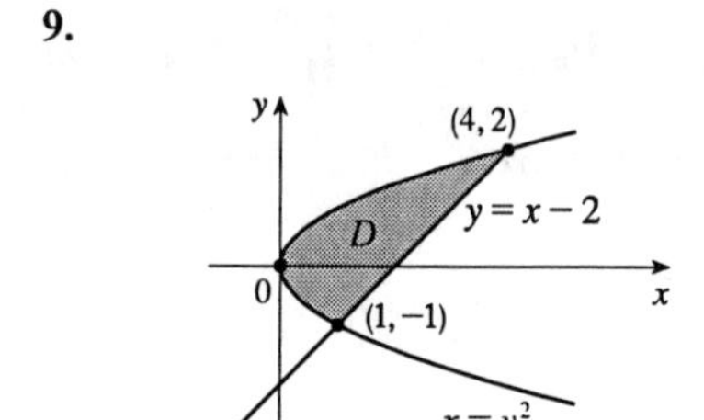

$m = \int_{-1}^2\int_{y^2}^{y+2} 3\,dx\,dy = \int_{-1}^2 (3y + 6 - 3y^2)dy = \frac{27}{2}$,

$M_y = \int_{-1}^2\int_{y^2}^{y+2} 3x\,dx\,dy = \int_{-1}^2 \frac{3}{2}\left[(y + 2)^2 - y^4\right]dy$
$= \left[\frac{1}{2}(y + 2)^3 - \frac{3}{10}y^5\right]_{-1}^2 = \frac{108}{5}$ and

$M_x = \int_{-1}^2\int_{y^2}^{y+2} 3y\,dx\,dy = \int_{-1}^2 (3y^2 + 6y - 3y^3)dy$
$= \left[y^3 + 3y^2 - \frac{3}{4}y^4\right]_{-1}^2 = \frac{27}{4}$.

Hence $m = \frac{27}{2}$, $(\overline{x}, \overline{y}) = \left(\frac{8}{5}, \frac{1}{2}\right)$.

11. $m = \int_0^\pi\int_0^{\sin x} y\,dy\,dx = \int_0^\pi \frac{1}{2}\sin^2 x\,dx = \left[\frac{1}{4}x - \frac{1}{8}\sin 2x\right]_0^\pi = \frac{1}{4}\pi$,
$M_y = \int_0^\pi\int_0^{\sin x} xy\,dy\,dx = \int_0^\pi \frac{1}{2}x\sin^2 x\,dx = \left[\frac{1}{8}x^2 - \frac{1}{8}x\sin 2x - \frac{1}{16}\cos 2x\right]_0^\pi = \frac{1}{8}\pi^2$, and
$M_x = \int_0^\pi\int_0^{\sin x} y^2\,dy\,dx = \int_0^\pi \frac{1}{3}\sin^3 x\,dx = \frac{1}{3}\left[-\cos x + \frac{1}{3}\cos^3 x\right]_0^\pi = \frac{4}{9}$. Hence $m = \frac{\pi}{4}$, $(\overline{x}, \overline{y}) = \left(\frac{\pi}{2}, \frac{16}{9\pi}\right)$.

13. $\rho(x, y) = ky = kr\sin\theta$, $m = \int_0^{\pi/2}\int_0^1 kr^2\sin\theta\,dr\,d\theta = \frac{1}{3}k\int_0^{\pi/2}\sin\theta\,d\theta = \frac{1}{3}k[-\cos\theta]_0^{\pi/2} = \frac{1}{3}k$,
$M_y = \int_0^{\pi/2}\int_0^1 kr^3\sin\theta\cos\theta\,dr\,d\theta = \frac{1}{4}k\int_0^{\pi/2}\sin\theta\cos\theta\,d\theta = \frac{1}{8}k[-\cos 2\theta]_0^{\pi/2} = \frac{1}{8}k$,
$M_x = \int_0^{\pi/2}\int_0^1 kr^3\sin^2\theta\,dr\,d\theta = \frac{1}{4}k\int_0^{\pi/2}\sin^2\theta\,d\theta = \frac{1}{8}k[\theta + \sin 2\theta]_0^{\pi/2} = \frac{\pi}{16}k$. Hence $(\overline{x}, \overline{y}) = \left(\frac{3}{8}, \frac{3\pi}{16}\right)$.

15. Placing the vertex opposite the hypotenuse at $(0,0)$, $\rho(x,y)=k(x^2+y^2)$. Then

$m=\int_0^a\int_0^{a-x}k(x^2+y^2)dy\,dx=k\int_0^a\left[ax^2-x^3+\frac{1}{3}(a-x)^3\right]dx=k\left[\frac{1}{3}ax^3-\frac{1}{4}x^4-\frac{1}{12}(a-x)^4\right]_0^a=\frac{1}{6}ka^4.$

By symmetry,

$M_y=M_x=\int_0^a\int_0^{a-x}ky(x^2+y^2)dy\,dx=k\int_0^a\left[\frac{1}{2}(a-x)^2x^2+\frac{1}{4}(a-x)^4\right]dx$

$=k\left[\frac{1}{6}a^2x^3-\frac{1}{4}ax^4+\frac{1}{10}x^5-\frac{1}{20}(a-x)^5\right]_0^a=\frac{1}{15}ka^5.$

Hence $(\overline{x},\overline{y})=\left(\frac{2}{5}a,\frac{2}{5}a\right)$.

17. $I_x=\int_0^1\int_{x^2}^1 y^2(xy)dy\,dx=\int_0^1\frac{1}{4}(x-x^9)dx=\frac{1}{8}-\frac{1}{40}=\frac{1}{10}$,
$I_y=\int_0^1\int_{x^2}^1 x^3y\,dy\,dx=\int_0^1\frac{1}{2}(x^3-x^7)dx=\frac{1}{8}-\frac{1}{16}=\frac{1}{16}$, $I_0=I_x+I_y=\frac{13}{80}$.

19. $I_x=\int_{-1}^2\int_{y^2}^{y+2}3y^2\,dx\,dy=\int_{-1}^2(3y^3+6y^2-3y^4)dy=\left[\frac{3}{4}y^4+2y^3-\frac{3}{5}y^5\right]_{-1}^2=\frac{189}{20}$,
$I_y=\int_{-1}^2\int_{y^2}^{y+2}3x^2\,dx\,dy=\int_{-1}^2\left[(y+2)^3-y^6\right]dy=\left[\frac{1}{4}(y+2)^4-\frac{1}{7}y^7\right]_{-1}^2=\frac{1269}{28}$, and $I_0=I_x+I_y=\frac{1917}{35}$.

21. $I_x=\int_0^a\int_0^a\rho y^2\,dx\,dy=\rho a\left(\frac{1}{3}a^3\right)=\frac{1}{3}\rho a^4=I_y$ by symmetry, and $m=\rho a^2$ since the lamina is homogeneous.
Hence $\overline{\overline{x}}=\overline{\overline{y}}=\left[\left(\frac{1}{3}\rho a^4\right)/(\rho a^2)\right]^{1/2}=\frac{1}{\sqrt{3}}a.$

23. Since $m=\iint_D\rho(x,y)dA=\rho\iint_D dA=\rho(\text{area of }D)=\rho A(D)$,

$$\overline{x}=\frac{1}{\rho A(D)}\iint_D x\rho\,dA=\frac{1}{A}\int_a^b\int_{g(x)}^{f(x)}x\,dy\,dx=\frac{1}{A}\int_a^b x[f(x)-g(x)]dx \text{ and}$$

$$\overline{y}=\frac{1}{\rho A(D)}\iint_D y\rho\,dA=\frac{1}{A}\int_a^b\int_{g(x)}^{f(x)}y\,dy\,dx=\frac{1}{A}\int_a^b\left[\tfrac{1}{2}y^2\right]_{g(x)}^{f(x)}dx=\frac{1}{A}\int_a^b\tfrac{1}{2}\left[f(x)^2-g(x)^2\right]dx.$$

EXERCISES 13.7

1. Here $z = f(x, y) = 4 - x - 2y$ with $0 \le x^2 + y^2 \le 4$. Thus by (2)

$$A(S) = \iint_D \sqrt{(-1)^2 + (-2)^2 + 1}\, dA = \sqrt{6} \iint_{x^2+y^2 \le 4} dA = \sqrt{6}\pi(2)^2 = 4\sqrt{6}\pi.$$

3. $y^2 + z^2 = 9 \quad \Rightarrow \quad z = \sqrt{9 - y^2}$. $f_x = 0$, $f_y = -y(9 - y^2)^{-1/2} \quad \Rightarrow$

$$A(S) = \int_0^4 \int_0^2 \sqrt{\left[-y(9-y^2)^{-1/2}\right]^2 + 1}\, dy\, dx = \int_0^4 \int_0^2 \sqrt{\frac{y^2}{9 - y^2} + 1}\, dy\, dx$$

$$= \int_0^4 \int_0^2 \frac{3\, dy}{\sqrt{9 - y^2}}\, dx = 3 \int_0^4 \left[\sin^{-1} \frac{y}{3}\right]_0^2 dx = 3\left[\left(\sin^{-1} \tfrac{2}{3}\right) x\right]_0^4 = 12 \sin^{-1} \tfrac{2}{3}$$

5. $z = f(x, y) = y^2 - x^2$ with $1 \le x^2 + y^2 \le 4$. $f_x = -2x$, $f_y = 2y \quad \Rightarrow$

$$A(S) = \iint_D \sqrt{1 + 4x^2 + 4y^2}\, dA = \int_0^{2\pi} \int_1^2 \sqrt{1 + 4r^2}\, r\, dr\, d\theta$$

$$= \tfrac{4\pi}{24} \left[(1 + 4r^2)^{3/2}\right]_1^2 = \tfrac{\pi}{6}\left(17\sqrt{17} - 5\sqrt{5}\right)$$

7. $z = f(x, y) = xy$ with $0 \le x^2 + y^2 \le 1$, so $f_x = y$, $f_y = x \quad \Rightarrow$

$$A(S) = \iint_D \sqrt{y^2 + x^2 + 1}\, dA = \int_0^{2\pi} \int_0^1 \sqrt{r^2 + 1}\, r\, dr\, d\theta = \int_0^{2\pi} \left[\tfrac{1}{3}(r^2 + 1)^{3/2}\right]_0^1 d\theta$$

$$= \int_0^{2\pi} \tfrac{1}{3}\left(2\sqrt{2} - 1\right) d\theta = \tfrac{2\pi}{3}\left(2\sqrt{2} - 1\right).$$

9. $z = \sqrt{a^2 - x^2 - y^2}$, $z_x = -x(a^2 - x^2 - y^2)^{-1/2}$, $z_y = -y(a^2 - x^2 - y^2)^{-1/2}$,

$$A(S) = \iint_D \sqrt{\frac{x^2 + y^2}{a^2 - x^2 - y^2} + 1}\, dA$$

$$= \int_{-\pi/2}^{\pi/2} \int_0^{a\cos\theta} \sqrt{\frac{r^2}{a^2 - r^2} + 1}\, r\, dr\, d\theta$$

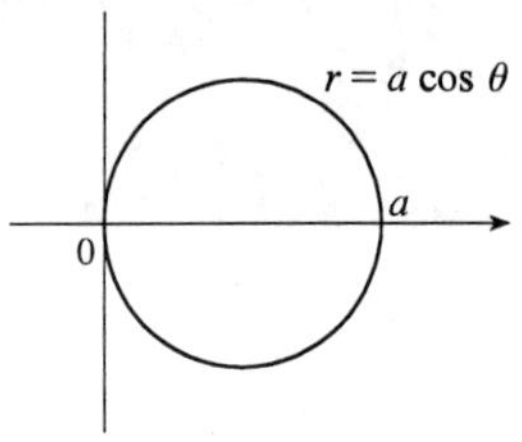

$$= \int_{-\pi/2}^{\pi/2} \int_0^{a\cos\theta} \frac{ar}{\sqrt{a^2 - r^2}}\, dr\, d\theta$$

$$= \int_{-\pi/2}^{\pi/2} \left[-a\sqrt{a^2 - r^2}\right]_0^{a\cos\theta} d\theta$$

$$= \int_{-\pi/2}^{\pi/2} -a\left(\sqrt{a^2 - a^2\cos^2\theta} - a\right) d\theta = 2a^2 \int_0^{\pi/2} \left(1 - \sqrt{1 - \cos^2\theta}\right) d\theta$$

$$= 2a^2 \int_0^{\pi/2} d\theta - 2a^2 \int_0^{\pi/2} \sqrt{\sin^2\theta}\, d\theta = a^2\pi - 2a^2 \int_0^{\pi/2} \sin\theta\, d\theta = a^2(\pi - 2)$$

11. The midpoints of the four squares are $\left(\frac{1}{4},\frac{1}{4}\right)$, $\left(\frac{1}{4},\frac{3}{4}\right)$, $\left(\frac{3}{4},\frac{1}{4}\right)$, and $\left(\frac{3}{4},\frac{3}{4}\right)$, the derivatives of the function $f(x,y)=x^2+y^2$ are $f_x(x,y)=2x$, $f_y(x,y)=2y$, so the Midpoint Rule gives

$$\begin{aligned} A(S) &= \int_0^1\int_0^1 \sqrt{[f_x(x,y)]^2+[f_y(x,y)]^2+1}\,dy\,dx \\ &\approx \frac{1}{4}\Bigg(\sqrt{\left[2\left(\tfrac{1}{4}\right)\right]^2+\left[2\left(\tfrac{1}{4}\right)\right]^2+1}+\sqrt{\left[2\left(\tfrac{1}{4}\right)\right]^2+\left[2\left(\tfrac{3}{4}\right)\right]^2+1} \\ &\qquad +\sqrt{\left[2\left(\tfrac{3}{4}\right)\right]^2+\left[2\left(\tfrac{1}{4}\right)\right]^2+1}+\sqrt{\left[2\left(\tfrac{3}{4}\right)\right]^2+\left[2\left(\tfrac{3}{4}\right)\right]^2+1}\Bigg) \\ &= \frac{1}{4}\left(\sqrt{\frac{3}{2}}+2\sqrt{\frac{7}{2}}+\sqrt{\frac{11}{2}}\right)\approx 1.8279. \end{aligned}$$

13. Since $f(x,y)=x^2+2y$, we have $f_x=2x$, $f_y=2$. We use a CAS to calculate the integral $A(S)=\int_0^1\int_0^1\sqrt{f_x^2+f_y^2+1}\,dy\,dx=\int_0^1\int_0^1\sqrt{4x^2+5}\,dy\,dx=\int_0^1\sqrt{4x^2+5}\,dx$, and find that $A(S)=\frac{3}{2}+\frac{5}{8}\ln 5$.

15. $f(x,y)=1+x^2y^2 \quad\Rightarrow\quad f_x=2xy^2$, $f_y=2x^2y$. We use a CAS (with precision reduced to five significant digits, to speed up the calculation) to estimate the integral $A(S)=\int_{-1}^{1}\int_{-\sqrt{1-x^2}}^{\sqrt{1-x^2}}\sqrt{f_x^2+f_y^2+1}\,dy\,dx=\int_{-1}^{1}\int_{-\sqrt{1-x^2}}^{\sqrt{1-x^2}}\sqrt{4x^2y^4+4x^4y^2+1}\,dy\,dx$, and find that $A(S)\approx 3.3213$.

17. Here $z=f(x,y)=ax+by+c$, $f_x(x,y)=a$, $f_y(x,y)=b$, so by (2),

$$A(S)=\iint_D\sqrt{a^2+b^2+1}\,dA=\sqrt{a^2+b^2+1}\iint_D dA=\sqrt{a^2+b^2+1}\,A(D).$$

EXERCISES 13.8

1. $\int_0^1\int_{-1}^2\int_0^3 xyz^2\,dz\,dx\,dy = \int_0^1\int_{-1}^2 xy(9)dx\,dy = 9\int_0^1 x\left(2-\frac{1}{2}\right)dx = 9\left(\frac{3}{4}\cdot 1^2\right) = \frac{27}{4}$

3. $\int_0^1\int_0^z\int_0^y xyz\,dx\,dy\,dz = \int_0^1\int_0^z\left(\frac{1}{2}y^3z\right)dy\,dz = \int_0^1 \frac{1}{8}z^5\,dz = \frac{1}{48}z^6\Big|_0^1 = \frac{1}{48}$

5. $\int_0^\pi\int_0^2\int_0^{\sqrt{4-z^2}} z\sin y\,dx\,dz\,dy = \int_0^\pi\int_0^2 z\sqrt{4-z^2}\sin y\,dz\,dy$

$$= \int_0^\pi \left[-\tfrac{1}{3}(4-z^2)^{3/2}\right]_0^2 \sin y\,dy = \int_0^\pi \tfrac{8}{3}\sin y\,dy = -\tfrac{8}{3}\cos y\Big|_0^\pi = \tfrac{16}{3}$$

7. $\int_0^1\int_0^{2z}\int_0^{z+2} yz\,dx\,dy\,dz = \int_0^1\int_0^{2z} yz(z+2)dy\,dz = \int_0^1 (2z^4+4z^3)dz = \frac{7}{5}$

9. $\int_0^1\int_0^{x^2}\int_0^{x+2y} y\,dz\,dy\,dx = \int_0^1\int_0^{x^2}(yx+2y^2)dy\,dx = \int_0^1\left[\frac{1}{2}xy^2+\frac{2}{3}y^3\right]_0^{x^2}dx$

$$= \int_0^1\left(\tfrac{1}{2}x^5+\tfrac{2}{3}x^6\right)dx = \left[\tfrac{1}{12}x^6+\tfrac{2}{21}x^7\right]_0^1 = \tfrac{5}{28}$$

11. Here E is the region that lies below the plane with x-, y-, and z-intercepts 1, 2, and 3 respectively, that is, below the plane $2z+6x+3y=6$ and above the region in the xy-plane bounded by the lines $x=0$, $y=0$ and $6x+3y=6$. So

$\iiint_E xy\,dV = \int_0^1\int_0^{2-2x}\int_0^{3-3x-3y/2} xy\,dz\,dy\,dx = \int_0^1\int_0^{2-2x}\left(3xy-3x^2y-\frac{3}{2}xy^2\right)dy\,dx$

$= \int_0^1\left[\frac{3}{2}xy^2-\frac{3}{2}x^2y^2-\frac{1}{2}xy^3\right]_0^{2-2x}dx = \int_0^1(2x-6x^2+6x^3-2x^4)dx = \left[x^2-2x^3+\frac{3}{2}x^4-\frac{2}{5}x^5\right]_0^1 = \frac{1}{10}.$

13.

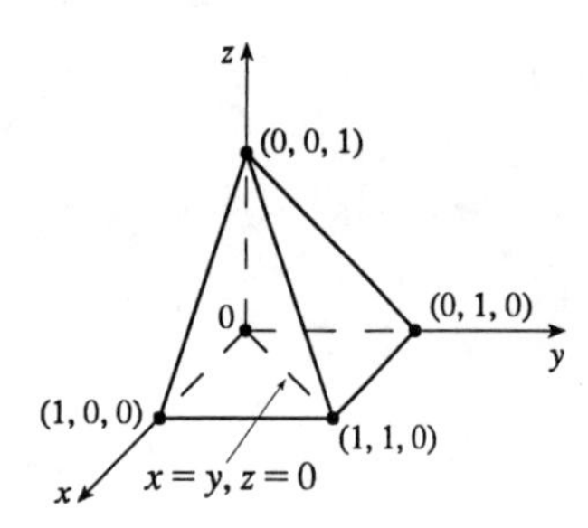

By symmetry $\iiint_E z\,dV = 2\iiint_{E'} z\,dV$ where E' is the part of E to the left [as viewed from $(10,10,0)$] of the plane $x=y$. So

$$\iiint_E z\,dV = \int_0^1\int_y^1\int_0^{1-x} 2z\,dz\,dx\,dy = \int_0^1\int_y^1 (1-x)^2\,dx\,dy$$

$$= \int_0^1\left[-\tfrac{1}{3}(1-x)^3\right]_y^1 dy = \int_0^1 \tfrac{1}{3}(1-y)^3\,dy$$

$$= \tfrac{1}{12}(1-y)^4\Big|_0^1 = \tfrac{1}{12}.$$

15.

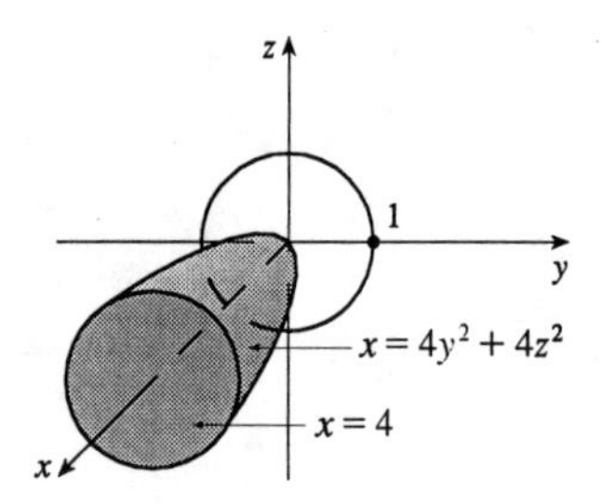

The projection E on the yz-plane is the disk $y^2+z^2\le 1$. Using polar coordinates $y=r\cos\theta$ and $z=r\sin\theta$, we get

$$\iiint_E x\,dV = \iint_D\left[\int_{4y^2+4z^2}^4 x\,dx\right]dA$$

$$= \frac{1}{2}\iint_D\left[4^2-(4y^2+4z^2)^2\right]dA = 8\int_0^{2\pi}\int_0^1(1-r^4)r\,dr\,d\theta$$

$$= 8\int_0^{2\pi}d\theta\int_0^1(r-r^5)d\theta = 8(2\pi)\left[\tfrac{1}{2}r^2-\tfrac{1}{6}r^6\right]_0^1 = \tfrac{16\pi}{3}.$$

17. The plane $2x + 3y + 6z = 12$ intersects the xy-plane when $2x + 3y + 6(0) = 12 \quad \Rightarrow \quad y = 4 - \frac{2}{3}x$. So $E = \left\{(x, y, z) \mid 0 \le x \le 6, 0 \le y \le 4 - \frac{2}{3}x, 0 \le z \le \frac{1}{6}(12 - 2x - 3y)\right\}$ and

$$V = \int_0^6 \int_0^{4-\frac{2}{3}x} \int_0^{(12-2x-3y)/6} dz\,dy\,dx = \tfrac{1}{6}\int_0^6 \int_0^{4-\frac{2}{3}x} (12 - 2x - 3y)dy\,dx$$

$$= \frac{1}{6}\int_0^6 \left[\frac{(12-2x)^2}{3} - \frac{3}{2}\frac{12-2x}{9}\right] dx = \tfrac{1}{36}\int_0^6 (12-2x)^2\,dx = \left[\tfrac{1}{36}\left(-\tfrac{1}{6}\right)(12-2x)^3\right]_0^6 = 8$$

19.

$$V = \int_0^1 \int_{-\sqrt{x}}^{\sqrt{x}} \int_0^{1-x} dz\,dy\,dx = \int_0^1 \int_{-\sqrt{x}}^{\sqrt{x}} (1-x)dy\,dx$$

$$= \int_0^1 2\sqrt{x}(1-x)dx = \int_0^1 2\left(\sqrt{x} - x^{3/2}\right)dx$$

$$= 2\left(\tfrac{2}{3} - \tfrac{2}{5}\right) = \tfrac{8}{15}$$

21. **(a)** The wedge can be described as the region

$$D = \left\{(x, y, z) \mid y^2 + z^2 \le 1, 0 \le x \le 1, 0 \le y \le x\right\}$$

$$= \left\{(x, y, z) \mid 0 \le x \le 1, 0 \le y \le x, 0 \le z \le \sqrt{1-y^2}\right\}.$$

So the integral expressing the volume of the wedge is

$\iiint_D dV = \int_0^1 \int_0^x \int_0^{\sqrt{1-y^2}} dz\,dy\,dx.$

(b) A CAS gives $\int_0^1 \int_0^x \int_0^{\sqrt{1-y^2}} dz\,dy\,dx = \frac{\pi}{4} - \frac{1}{3}$.

(Or use Formulas 30 and 87.)

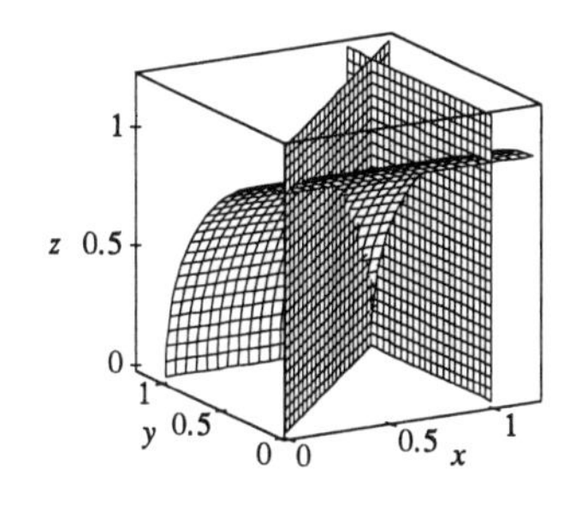

23. Note that $\Delta V_{ijk} = \left(\frac{1}{2}\right)^3 = \frac{1}{8}$, so the Midpoint Rule gives

$$\iiint_B f(x,y,z)dV \approx \tfrac{1}{8}\left[f\left(\tfrac14,\tfrac14,\tfrac14\right) + f\left(\tfrac14,\tfrac14,\tfrac34\right) + f\left(\tfrac14,\tfrac34,\tfrac14\right) + f\left(\tfrac34,\tfrac14,\tfrac14\right)\right.$$
$$\left. + f\left(\tfrac14,\tfrac34,\tfrac34\right) + f\left(\tfrac34,\tfrac14,\tfrac34\right) + f\left(\tfrac34,\tfrac34,\tfrac14\right) + f\left(\tfrac34,\tfrac34,\tfrac34\right)\right]$$
$$= \tfrac{1}{8}\left[e^{-3\left(\frac14\right)^2} + 3e^{-2\left(\frac14\right)^2-\left(\frac34\right)^2} + 3e^{-\left(\frac14\right)^2-2\left(\frac34\right)^2} + e^{-3\left(\frac34\right)^2}\right] \approx 0.42968.$$

The norm of this partition is the longest diagonal of the eight sub-boxes. Of course, all diagonals have the same length: $\|P\| = \sqrt{\left(\frac12\right)^2 + \left(\frac12\right)^2 + \left(\frac12\right)^2} = \frac{\sqrt{3}}{2}$.

25. $E = \left\{(x, y, z) \mid 0 \le x \le 1,\right.$
$\left. 0 \le z \le 1 - x, 0 \le y \le 2 - 2z\right\}$

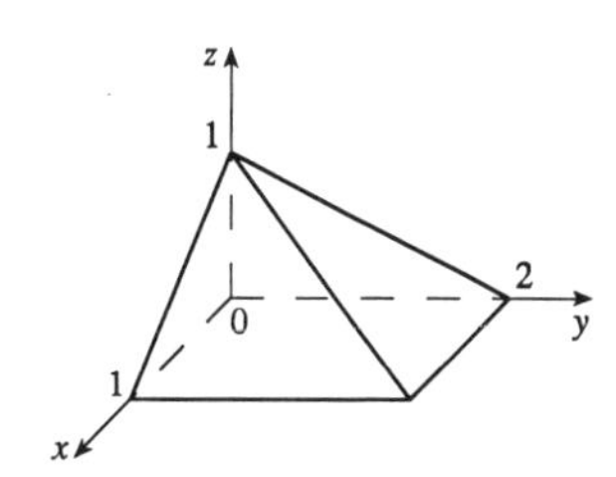

27.

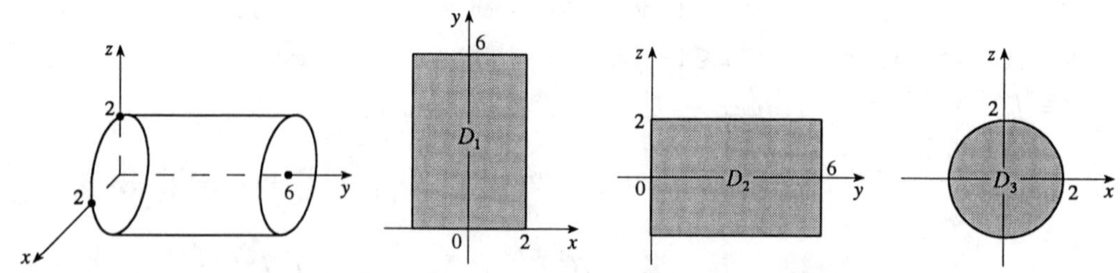

If D_1, D_2, D_3 are the projections of E on the xy-, yz-, and xz-planes, then
$D_1 = \{(x, y) \mid -2 \le x \le 2, 0 \le y \le 6\}$, $D_2 = \{(y, z) \mid -2 \le z \le 2, 0 \le y \le 6\}$,
$D_3 = \{(x, z) \mid x^2 + z^2 \le 4\}$. Therefore

$$E = \left\{(x, y, z) \mid -\sqrt{4 - x^2} \le z \le \sqrt{4 - x^2}, -2 \le x \le 2, 0 \le y \le 6\right\}$$
$$= \left\{(x, y, z) \mid -\sqrt{4 - z^2} \le x \le \sqrt{4 - z^2}, -2 \le z \le 2, 0 \le y \le 6\right\}.$$

$$\iiint_E f(x, y, z)dV = \int_{-2}^{2}\int_0^6\int_{-\sqrt{4-x^2}}^{\sqrt{4-x^2}} f(x, y, z)dz\,dy\,dx = \int_0^6\int_{-2}^{2}\int_{-\sqrt{4-x^2}}^{\sqrt{4-x^2}} f(x, y, z)dz\,dx\,dy$$
$$= \int_0^6\int_{-2}^{2}\int_{-\sqrt{4-z^2}}^{\sqrt{4-z^2}} f(x, y, z)dx\,dz\,dy = \int_{-2}^{2}\int_0^6\int_{-\sqrt{4-z^2}}^{\sqrt{4-z^2}} f(x, y, z)dx\,dy\,dz$$
$$= \int_{-2}^{2}\int_{-\sqrt{4-x^2}}^{\sqrt{4-x^2}}\int_0^6 f(x, y, z)dy\,dz\,dx = \int_{-2}^{2}\int_{-\sqrt{4-z^2}}^{\sqrt{4-z^2}}\int_0^6 f(x, y, z)dy\,dx\,dz.$$

29.

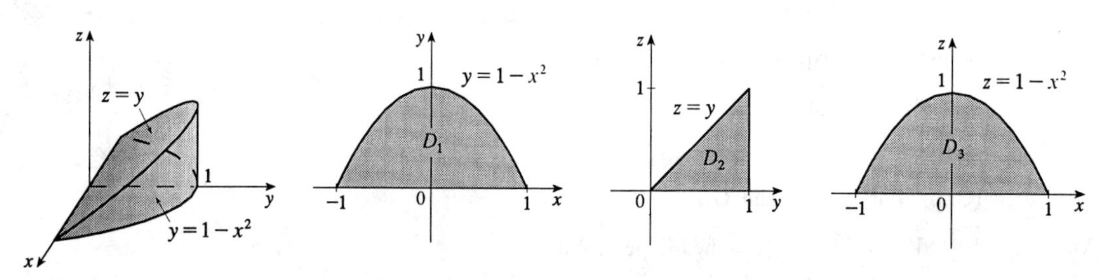

If D_1, D_2, and D_3 are the projections of E on the xy-, yz-, and xz-planes, then
$D_1 = \{(x, y) \mid -1 \le x \le 1, 0 \le y \le 1 - x^2\} = \left\{(x, y) \mid 0 \le y \le 1, -\sqrt{1-y} \le x \le \sqrt{1-y}\right\}$,
$D_2 = \{(y, z) \mid 0 \le y \le 1, 0 \le z \le y\} = \{(y, z) \mid 0 \le z \le 1, z \le y \le 1\}$,
$D_3 = \{(x, z) \mid -1 \le x \le 1, 0 \le z \le 1 - x^2\} = \left\{(x, z) \mid 0 \le z \le 1, -\sqrt{1-z} \le x \le \sqrt{1-z}\right\}$. Therefore

$$E = \left\{(x, y, z) \mid -1 \le x \le 1, 0 \le y \le 1 - x^2, 0 \le z \le y\right\}$$
$$= \left\{(x, y, z) \mid 0 \le y \le 1, -\sqrt{1-y} \le x \le \sqrt{1-y}, 0 \le z \le y\right\}$$
$$= \left\{(x, y, z) \mid 0 \le y \le 1, 0 \le z \le y, -\sqrt{1-y} \le x \le \sqrt{1-y}\right\}$$
$$= \left\{(x, y, z) \mid 0 \le z \le 1, z \le y \le 1, -\sqrt{1-y} \le x \le \sqrt{1-y}\right\}$$
$$= \left\{(x, y, z) \mid -1 \le x \le 1, 0 \le z \le 1 - x^2, z \le y \le 1 - x^2\right\}$$
$$= \left\{(x, y, z) \mid 0 \le z \le 1, -\sqrt{1-z} \le x \le \sqrt{1-z}, z \le y \le 1 - x^2\right\}.$$

Then

$$\iiint_E f(x,y,z)\,dV = \int_{-1}^{1}\int_0^{1-x^2}\int_0^{y} f(x,y,z)\,dz\,dy\,dx = \int_0^1\int_{-\sqrt{1-y}}^{\sqrt{1-y}}\int_0^{y} f(x,y,z)\,dz\,dx\,dy$$
$$= \int_0^1\int_0^{y}\int_{-\sqrt{1-y}}^{\sqrt{1-y}} f(x,y,z)\,dx\,dz\,dy = \int_0^1\int_z^1\int_{-\sqrt{1-y}}^{\sqrt{1-y}} f(x,y,z)\,dx\,dy\,dz$$
$$= \int_{-1}^{1}\int_0^{1-x^2}\int_z^{1-x^2} f(x,y,z)\,dy\,dz\,dx = \int_0^1\int_{-\sqrt{1-z}}^{\sqrt{1-z}}\int_z^{1-x^2} f(x,y,z)\,dy\,dx\,dz.$$

31.

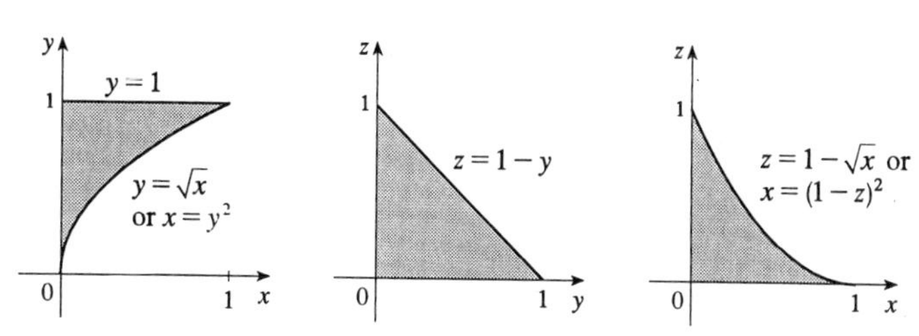

The diagrams show the projections of E on the xy-, yz-, and xz-planes. Therefore

$$\int_0^1\int_{\sqrt{x}}^1\int_0^{1-y} f(x,y,z)\,dz\,dy\,dx = \int_0^1\int_0^{y^2}\int_0^{1-y} f(x,y,z)\,dz\,dx\,dy$$
$$= \int_0^1\int_0^{1-z}\int_0^{y^2} f(x,y,z)\,dx\,dy\,dz = \int_0^1\int_0^{1-y}\int_0^{y^2} f(x,y,z)\,dx\,dz\,dy$$
$$= \int_0^1\int_0^{1-\sqrt{x}}\int_{\sqrt{x}}^{1-z} f(x,y,z)\,dy\,dz\,dx = \int_0^1\int_0^{(1-z)^2}\int_{\sqrt{x}}^{1-z} f(x,y,z)\,dy\,dx\,dz.$$

33.

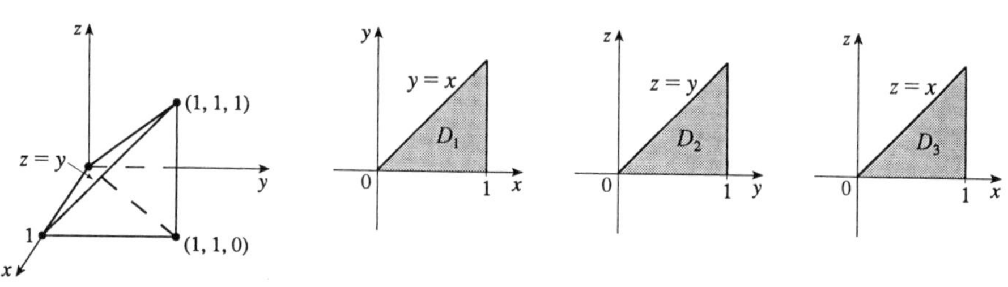

$\int_0^1\int_y^1\int_0^y f(x,y,z)\,dz\,dx\,dy = \iiint_E f(x,y,z)\,dV$ where $E = \{(x,y,z) \mid 0 \le z \le y,\ y \le x \le 1,\ 0 \le y \le 1\}$. If D_1, D_2, and D_3 are the projections of E on the xy-, yz- and xz-planes then

$D_1 = \{(x,y) \mid 0 \le y \le 1,\ y \le x \le 1\} = \{(x,y) \mid 0 \le x \le 1,\ 0 \le y \le x\}$,

$D_2 = \{(y,z) \mid 0 \le y \le 1,\ 0 \le z \le y\} = \{(y,z) \mid 0 \le z \le 1,\ z \le y \le 1\}$, and

$D_3 = \{(x,z) \mid 0 \le x \le 1,\ 0 \le z \le x\} = \{(x,z) \mid 0 \le z \le 1,\ z \le x \le 1\}$. Thus we also have

$$E = \{(x,y,z) \mid 0 \le x \le 1,\ 0 \le y \le x,\ 0 \le z \le y\} = \{(x,y,z) \mid 0 \le y \le 1,\ 0 \le z \le y,\ y \le x \le 1\}$$
$$= \{(x,y,z) \mid 0 \le z \le 1,\ z \le y \le 1,\ y \le x \le 1\} = \{(x,y,z) \mid 0 \le x \le 1,\ 0 \le z \le x,\ z \le y \le x\}$$
$$= \{(x,y,z) \mid 0 \le z \le 1,\ z \le x \le 1,\ z \le y \le x\}.$$ Then

$$\int_0^1\int_y^1\int_0^y f(x,y,z)\,dz\,dx\,dy = \int_0^1\int_0^x\int_0^y f(x,y,z)\,dz\,dy\,dx = \int_0^1\int_0^y\int_y^1 f(x,y,z)\,dx\,dz\,dy$$
$$= \int_0^1\int_z^1\int_y^1 f(x,y,z)\,dx\,dy\,dz = \int_0^1\int_0^x\int_z^x f(x,y,z)\,dy\,dz\,dx = \int_0^1\int_z^1\int_z^x f(x,y,z)\,dy\,dx\,dz.$$

35. $m = \int_0^1 \int_0^{x^2} \int_0^{x+2y} 2\,dz\,dy\,dx = 2\int_0^1 \int_0^{x^2} (x+2y)dy\,dx = 2\int_0^1 (x^3 + x^4)dx = \frac{9}{10}$,
$M_{yz} = \int_0^1 \int_0^{x^2} \int_0^{x+2y} 2x\,dz\,dy\,dx = \int_0^1 \int_0^{x^2} 2(x^2 + 2xy)dy\,dx = \int_0^1 2(x^4 + x^5)dx = \frac{11}{15}$, $M_{xz} = \frac{5}{14}$, and
$M_{xy} = \int_0^1 \int_0^{x^2} \int_0^{x+2y} 2z\,dz\,dy\,dx = \int_0^1 \int_0^{x^2} 2(x+2y)^2/2\,dy\,dx = \int_0^1 2\left[\frac{1}{2}x^2 y + xy^2 + \frac{2}{3}y^3\right]_0^{x^2} dx$
$= \int_0^1 2\left(\frac{1}{2}x^4 + x^5 + \frac{2}{3}x^6\right)dx = \frac{76}{105}$. Hence $(\overline{x}, \overline{y}, \overline{z}) = \left(\frac{22}{27}, \frac{25}{63}, \frac{152}{189}\right)$.

37. $m = \int_0^a \int_0^a \int_0^a (x^2 + y^2 + z^2)dx\,dy\,dz = \int_0^a \int_0^a \left[\frac{1}{3}a^3 + a(y^2 + z^2)\right]dy\,dz = \int_0^a \left[\frac{2}{3}a^4 + a^2 z^2\right]dz = \frac{2}{3}a^5 + \frac{1}{3}a^5 = a^5$,
$M_{yz} = \int_0^a \int_0^a \int_0^a [x^3 + x(y^2 + z^2)]dx\,dy\,dz = \int_0^a \int_0^a \left[\frac{1}{4}a^4 + \frac{1}{2}a^2(y^2 + z^2)\right]dy\,dz$
$= \int_0^a \left[\frac{1}{4}a^5 + \frac{1}{6}a^5 + \frac{1}{2}a^3 z^2\right]dz = \frac{1}{4}a^6 + \frac{1}{3}a^6 = \frac{7}{12}a^6 = M_{xz} = M_{xy}$ by symmetry of E and $\rho(x, y, z)$.
Hence $(\overline{x}, \overline{y}, \overline{z}) = \left(\frac{7}{12}a, \frac{7}{12}a, \frac{7}{12}a\right)$.

39. **(a)** $m = \int_0^1 \int_0^{\sqrt{1-x^2}} \int_0^y (1 + x + y + z)dz\,dy\,dx$

(b) $(\overline{x}, \overline{y}, \overline{z}) = \Big(m^{-1}\int_0^1 \int_0^{\sqrt{1-x^2}} \int_0^y x(1 + x + y + z)dz\,dy\,dx, m^{-1}\int_0^1 \int_0^{\sqrt{1-x^2}} \int_0^y y(1 + x + y + z)dz\,dy\,dx,$
$m^{-1}\int_0^1 \int_0^{\sqrt{1-x^2}} \int_0^y z(1 + x + y + z)dz\,dy\,dx\Big)$

(c) $I_z = \int_0^1 \int_0^{\sqrt{1-x^2}} \int_0^y (x^2 + y^2)(1 + x + y + z)dz\,dy\,dx$

41. **(a)** $m = \int_{-1}^1 \int_{-\sqrt{1-y^2}}^{\sqrt{1-y^2}} \int_{4y^2+4z^2}^4 (x^2 + y^2 + z^2)dx\,dz\,dy$

(b) $(\overline{x}, \overline{y}, \overline{z})$ where $\overline{x} = m^{-1}\int_{-1}^1 \int_{-\sqrt{1-y^2}}^{\sqrt{1-y^2}} \int_{4y^2+4z^2}^4 x(x^2 + y^2 + z^2)dx\,dz\,dy$,
$\overline{y} = m^{-1}\int_{-1}^1 \int_{-\sqrt{1-y^2}}^{\sqrt{1-y^2}} \int_{4y^2+4z^2}^4 y(x^2 + y^2 + z^2)dx\,dz\,dy$, and
$\overline{z} = m^{-1}\int_{-1}^1 \int_{-\sqrt{1-y^2}}^{\sqrt{1-y^2}} \int_{4y^2+4z^2}^4 z(x^2 + y^2 + z^2)dx\,dz\,dy$

(c) $I_z = \int_{-1}^1 \int_{-\sqrt{1-y^2}}^{\sqrt{1-y^2}} \int_{4y^2+4z^2}^4 (x^2 + y^2)(x^2 + y^2 + z^2)dx\,dz\,dy$

43. Using the formulas in the solution to Exercise 39, a CAS gives

(a) $m = \dfrac{3\pi}{32} + \dfrac{11}{24}$ **(b)** $(\overline{x}, \overline{y}, \overline{z}) = \left(\dfrac{28}{9\pi + 44}, \dfrac{30\pi + 128}{45\pi + 220}, \dfrac{45\pi + 208}{135\pi + 660}\right)$ **(c)** $I_z = \dfrac{68 + 15\pi}{240}$

45. $I_x = \int_0^L \int_0^L \int_0^L k(y^2 + z^2)dz\,dy\,dx = kL\int_0^L \left[y^2 L + \frac{1}{3}L^3\right]dy = kL\left[\frac{1}{3}L^4 + \frac{1}{3}L^4\right] = \frac{2}{3}kL^5$.
By symmetry, $I_x = I_y = I_z = \frac{2}{3}kL^5$.

47. $V(E) = L^3$, $f_{\text{ave}} = \dfrac{1}{L^3}\displaystyle\int_0^L \int_0^L \int_0^L xyz\,dx\,dy\,dz = \dfrac{1}{L^3}\int_0^L x^2\,dx \int_0^L y^2\,dy \int_0^L z^2\,dz = \dfrac{1}{L^3}\dfrac{L^2}{2}\dfrac{L^2}{2}\dfrac{L^2}{2} = \dfrac{L^3}{8}$

49. The triple integral will attain its maximum when the integrand $1 - x^2 - 2y^2 - 3z^2$ is positive in the region E and negative everywhere else. For if E contains some region F where the integrand is negative, the integral could be increased by excluding F from E, and if E fails to contain some part G of the region where the integrand is positive, the integral could be increased by including G in E. So we require that $x^2 + 2y^2 + 3z^2 \leq 1$. This describes the region bounded by the ellipsoid $x^2 + 2y^2 + 3z^2 = 1$.

EXERCISES 13.9

1. $r = 3$, $\theta = \frac{\pi}{2}$, $z = 1$ so $x = 0$, $y = 3$ and the point is $(0, 3, 1)$.

3. $x = 2\cos\frac{4\pi}{3} = -1$, $z = 8$, $y = 2\sin\frac{4\pi}{3} = -\sqrt{3}$ so the point is $\left(-1, -\sqrt{3}, 8\right)$.

5. $x = 3\cos 0 = 3$, $y = 3\sin 0 = 0$ and $z = -6$ so the point is $(3, 0, -6)$.

7. $r^2 = (-1)^2 + (0)^2 = 1$ so $r = 1$; $z = 0$; $\tan\theta = 0$ so $\theta = 0$ or π. But $x = -1$ so $\theta = \pi$ and the point is $(1, \pi, 0)$.

9. $r^2 = 4$ so $r = 2$, $\tan\theta = \frac{1}{\sqrt{3}}$ so $\theta = \frac{\pi}{6}$ and $z = 4$. Thus the point in cylindrical coordinates is $\left(2, \frac{\pi}{6}, 4\right)$.

11. $r = \sqrt{4^2 + 4^2} = 4\sqrt{2}$; $z = 4$; $\tan\theta = \frac{4}{4}$, so $\theta = \frac{\pi}{4}$ or $\theta = \frac{5\pi}{4}$, but both x and y are positive, so $\theta = \frac{\pi}{4}$ and the point is $\left(4\sqrt{2}, \frac{\pi}{4}, 4\right)$.

13. $x = (1)\sin 0\cos 0 = 0$, $y = (1)\sin 0\sin 0 = 0$, $z = (1)\cos 0 = 1$ so the point in rectangular coordinates is $(0, 0, 1)$.

15. $x = \sin\frac{\pi}{6}\cos\frac{\pi}{6} = \frac{\sqrt{3}}{4}$, $y = \sin\frac{\pi}{6}\sin\frac{\pi}{6} = \frac{1}{4}$ and $z = \cos\frac{\pi}{6} = \frac{\sqrt{3}}{2}$ so the point is $\left(\frac{\sqrt{3}}{4}, \frac{1}{4}, \frac{\sqrt{3}}{2}\right)$.

17. $x = 4\sin\frac{\pi}{6}\cos\frac{\pi}{4} = 4\left(\frac{1}{2}\right)\frac{1}{\sqrt{2}} = \sqrt{2}$, $y = 4\sin\frac{\pi}{6}\sin\frac{\pi}{4} = \sqrt{2}$ and $z = 4\cos\frac{\pi}{6} = 4\left(\frac{\sqrt{3}}{2}\right) = 2\sqrt{3}$ so the point is $\left(\sqrt{2}, \sqrt{2}, 2\sqrt{3}\right)$.

19. $\rho = \sqrt{9 + 0 + 0} = 3$, $\cos\phi = \frac{0}{3} = 0$ so $\phi = \frac{\pi}{2}$ and $\cos\theta = \dfrac{-3}{3\sin\frac{\pi}{2}} = -1$ so $\theta = \pi$ and the spherical coordinates are $\left(3, \pi, \frac{\pi}{2}\right)$.

21. $\rho = \sqrt{3 + 1} = 2$, $\cos\phi = \frac{1}{2}$ so $\phi = \frac{\pi}{3}$ and $\cos\theta = \dfrac{\sqrt{3}}{2\sin\frac{\pi}{3}} = \dfrac{\sqrt{3}\cdot 2}{2\cdot\sqrt{3}} = 1$ so $\theta = 0$ and the point is $\left(2, 0, \frac{\pi}{3}\right)$.

Note: It is also apparent that $\theta = 0$ since the point is in the xz-plane and $x > 0$.

23. $\rho = \sqrt{1 + 1 + 2} = 2$; $z = -\sqrt{2} = 2\cos\phi$, so $\cos\phi = -\frac{1}{\sqrt{2}}$ which implies that $\phi = \frac{3\pi}{4}$;
$\cos\theta = \dfrac{1}{2\sin\frac{3\pi}{4}} = \dfrac{1}{\sqrt{2}}$, so $\theta = \frac{\pi}{4}$ or $\theta = \frac{7\pi}{4}$, but $x > 0$ and $y < 0$ so $\theta = \frac{7\pi}{4}$. Thus the point is $\left(2, \frac{7\pi}{4}, \frac{3\pi}{4}\right)$.

25. $\rho = \sqrt{r^2 + z^2} = \sqrt{2 + 0} = \sqrt{2}$; $\theta = \frac{\pi}{4}$; $z = 0 = \sqrt{2}\cos\phi$ so $\phi = \frac{\pi}{2}$ and the point is $\left(\sqrt{2}, \frac{\pi}{4}, \frac{\pi}{2}\right)$.

27. $\rho = \sqrt{r^2 + z^2} = \sqrt{4^2 + 4^2} = 4\sqrt{2}$; $\theta = \frac{\pi}{3}$; $z = 4 = 4\sqrt{2}\cos\phi$ so $\cos\phi = \frac{1}{\sqrt{2}} \Rightarrow \phi = \frac{\pi}{4}$ and the point is $\left(4\sqrt{2}, \frac{\pi}{3}, \frac{\pi}{4}\right)$.

29. $z = 2\cos 0 = 2$, $r = \sqrt{\rho^2 - z^2} = \sqrt{2^2 - 2^2} = 0$, (or $r = 2\sin 0 = 0$), $\theta = 0$ and the point is $(0, 0, 2)$.

31. $z = 8\cos\frac{\pi}{2} = 0$, $r = 8\sin\frac{\pi}{2} = 8$, $\theta = \frac{\pi}{6}$ and the point is $\left(8, \frac{\pi}{6}, 0\right)$.

33. Since $r = 3$, $x^2 + y^2 = 9$ and the surface is a cylinder of radius 3 and axis the z-axis.

35. Since $\phi = \frac{\pi}{3}$, the surface is one frustum of the right circular cone with vertex at the origin and axis the positive z-axis.

37. $z = r^2 = x^2 + y^2$ so the surface is a circular paraboloid with vertex at the origin and axis the positive z-axis.

39. $2 = \rho\cos\phi = z$ is a plane through the point $(0, 0, 2)$ and parallel to the xy-plane.

41. Since $\phi = 0$, $x = 0$ and $y = 0$ while $z = \rho \geq 0$. Thus the locus is the positive z-axis including the origin.

43. $r = 2\cos\theta \quad \Rightarrow \quad r^2 = x^2 + y^2 = 2r\cos\theta = 2x \quad \Leftrightarrow \quad (x-1)^2 + y^2 = 1$ which is the equation of a circular cylinder of radius 1, whose axis is the vertical line $x = 1$, $y = 0$, $z = z$.

45. Since $r^2 + z^2 = 25$ and $r^2 = x^2 + y^2$, we have $x^2 + y^2 + z^2 = 25$, a sphere of radius 5 and center at the origin.

47. Since $x^2 = \rho^2 \sin^2\phi \cos^2\theta$ and $z^2 = \rho^2\cos^2\phi$, the equation of the surface in rectangular coordinates is $x^2 + z^2 = 4$. Thus the surface is a right circular cylinder of radius 2 about the y-axis.

49. Since $r^2 - r = 0$, $r = 0$ or $r = 1$. But $x^2 + y^2 = r^2$. Thus the surface consists of the right circular cylinder of radius 1 and axis the z-axis along with the surface given by $x^2 + y^2 = 0$, that is, the z-axis.

51. **(a)** $r^2 = x^2 + y^2$, so $r^2 + z^2 = 16$. **(b)** $\rho^2 = x^2 + y^2 + z^2$, so $\rho^2 = 16$ or $\rho = 4$.

53. **(a)** $r\cos\theta + 2r\sin\theta + 3z = 6$

(b) $\rho\sin\phi\cos\theta + 2\rho\sin\phi\sin\theta + 3\rho\cos\phi = 6$ or $\rho(\sin\phi\cos\theta + 2\sin\phi\sin\theta + 3\cos\phi) = 6$.

55. **(a)** $r^2(\cos^2\theta - \sin^2\theta) - 2z^2 = 4$ or $2z^2 = r^2\cos 2\theta - 4$.

(b) $\rho^2(\sin^2\phi\cos^2\theta - \sin^2\phi\sin^2\theta - 2\cos^2\phi) = 4$ or $\rho^2(\sin^2\phi\cos 2\theta - 2\cos^2\phi) = 4$.

57. **(a)** $r^2 = 2r\sin\theta$ or $r = 2\sin\theta$.

(b) $\rho^2\sin^2\phi(\cos^2\theta + \sin^2\theta) = 2\rho\sin\phi\sin\theta$ or $\rho\sin^2\phi = 2\sin\phi\sin\theta$ or $\rho\sin\phi = 2\sin\theta$.

59. $z = r^2 = x^2 + y^2$ is a circular paraboloid with vertex $(0, 0, 0)$, opening upward. $z = 2 - r^2 \quad \Rightarrow \quad z - 2 = -(x^2 + y^2)$ is a circular paraboloid with vertex $(0, 0, 2)$ opening downward. Thus $r^2 \leq z \leq 2 - r^2$ is the solid region enclosed by these two surfaces.

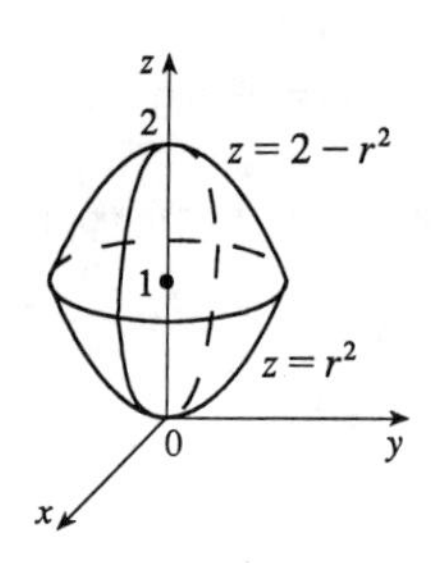

61. $-\frac{\pi}{2} \leq \theta \leq \frac{\pi}{2}$ restricts the solid to the 4 octants in which x is positive. $\rho = \sec\phi \quad \Rightarrow \quad \rho\cos\phi = z = 1$, which is the equation of a horizontal plane. $0 \leq \phi \leq \frac{\pi}{6}$ describes a cone, opening upward. So the solid lies above the cone $\phi = \frac{\pi}{6}$ and below the plane $z = 1$.

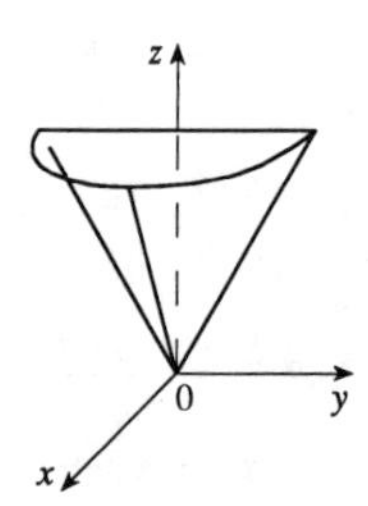

63. $z \ge \sqrt{x^2+y^2}$ because the solid lies above the cone. Squaring both sides of this inequality gives $z^2 \ge x^2+y^2$ $\Rightarrow \quad 2z^2 \ge x^2+y^2+z^2 = \rho^2 \quad \Rightarrow \quad z^2 = \rho^2\cos^2\phi \ge \frac{1}{2}\rho^2 \quad \Rightarrow \quad \cos^2\phi \ge \frac{1}{2}$. The cone opens upward so that the inequality is $\cos\phi \ge \frac{1}{\sqrt{2}}$, or equivalently $0 \le \phi \le \frac{\pi}{4}$. In spherical coordinates the sphere $z = x^2+y^2+z^2$ is $\rho\cos\phi = \rho^2 \quad \Rightarrow \quad \rho = \cos\phi$. $0 \le \rho \le \cos\phi$ because the solid lies below the sphere. The solid can therefore be described as the region in spherical coordinates satisfying $0 \le \rho \le \cos\phi$, $0 \le \phi \le \frac{\pi}{4}$.

65. In cylindrical coordinates, the equation of the cylinder is $r = 3$, $0 \le z \le 10$.
The hemisphere is the upper part of the sphere radius 3, center $(0,0,10)$, equation $r^2 + (z-10)^2 = 3^2$, $z \ge 10$.
In Maple, we can use either the `coords=cylindrical` option in a regular `plot` command, or the `plots[cylinderplot]` command.
In Mathematica, we can use `ParametricPlot3d`.

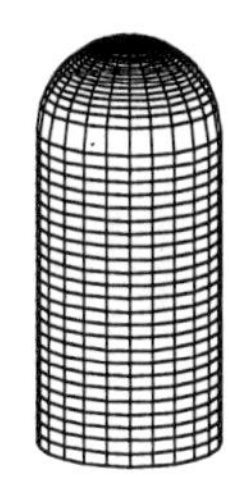

EXERCISES 13.10

1.

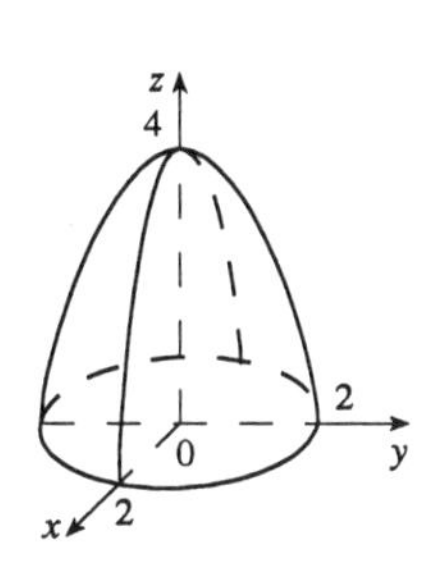

$$\int_0^{2\pi}\int_0^2\int_0^{4-r^2} r\,dz\,dr\,d\theta = \int_0^{2\pi}\int_0^2 \left(4r - r^3\right)dr\,d\theta$$
$$= \int_0^{2\pi} \left[2r^2 - \tfrac{1}{4}r^4\right]_0^2 d\theta$$
$$= \int_0^{2\pi} (8-4)d\theta = 8\pi$$

3.

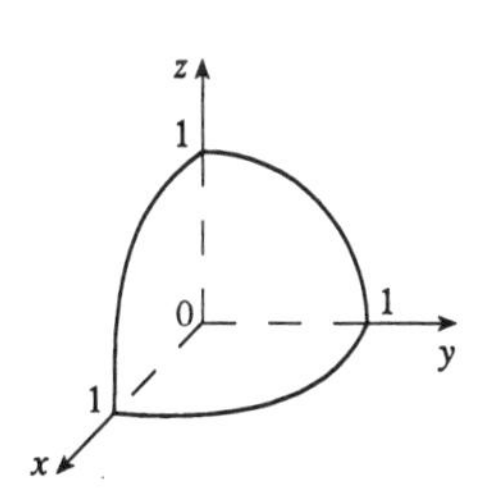

$$\int_0^{\pi/2}\int_0^{\pi/2}\int_0^1 \rho^2\sin\phi\,d\rho\,d\theta\,d\phi = \int_0^{\pi/2}\int_0^{\pi/2} \tfrac{1}{3}\sin\phi\,d\theta\,d\phi$$
$$= \frac{1}{3}\int_0^{\pi/2} \tfrac{\pi}{2}\sin\phi\,d\phi$$
$$= \tfrac{\pi}{6}[-\cos\phi]_0^{\pi/2} = \tfrac{\pi}{6}$$

5. $\iiint_E (x^2+y^2)dV = \int_{-1}^{2}\int_0^{2\pi}\int_0^{2}(r^2)r\,dr\,d\theta\,dz = (3)(2\pi)\left[\frac{1}{4}r^4\right]_0^2 = 24\pi$

7. $\iiint_E y\,dV = \int_0^{2\pi}\int_1^{2}\int_0^{2+r\cos\theta} r^2\sin\theta\,dz\,dr\,d\theta = \int_0^{2\pi}\int_1^2 [2r^2\sin\theta + r^3\cos\theta]dr\,d\theta$

$= \int_0^{2\pi}\left[\frac{1}{3}(2r^3\sin\theta) + \frac{1}{4}(r^4\cos\theta)\right]_1^2 d\theta = \int_0^{2\pi}\left[\frac{14}{3}\sin\theta + \frac{15}{4}\cos\theta\right]d\theta = 0$

9. $\iiint_E x^2\,dV = \int_0^{2\pi}\int_0^1\int_0^{2r} r^2\cos^2\theta\,r\,dz\,dr\,d\theta = \int_0^{2\pi}\int_0^1 \left[r^3\cos^2\theta\,z\right]_0^{2r} dr\,d\theta$

$= \int_0^{2\pi}\int_0^1 2r^4\cos^2\theta\,dr\,d\theta = \int_0^{2\pi}\left[\frac{2}{5}r^5\cos^2\theta\right]_0^1 d\theta = \frac{2}{5}\int_0^{2\pi}\cos^2\theta\,d\theta$

$= \frac{2}{5}\int_0^{2\pi}\frac{1+\cos 2\theta}{2}\,d\theta = \frac{1}{5}\left[\theta + \frac{1}{2}\sin 2\theta\right]_0^{2\pi} = \frac{2\pi}{5}$

11. The paraboloids intersect when $x^2+y^2 = 36 - 3x^2 - 3y^2 \quad\Rightarrow\quad D = \{(x,y)\mid x^2+y^2 \le 9\}$. So, using cylindrical coordinates, $E = \{(r,\theta,z)\mid r^2 \le z \le 36 - r^2, 0 \le r \le 3, 0 \le \theta \le 2\pi\}$ and $V = \int_0^{2\pi}\int_0^3\int_{r^2}^{36-3r^2} r\,dz\,dr\,d\theta = 2\pi\int_0^3 (36r - 4r^3)dr = 2\pi[18r^2 - r^4]_0^3 = 162\pi$.

13. The paraboloid $z = 4x^2 + 4y^2$ intersects the plane $z = a$ when $a = 4x^2+4y^2$ or $x^2+y^2 = \frac{1}{4}a$. So, using cylindrical coordinates, $E = \left\{(r,\theta,z)\mid 0 \le r \le \frac{1}{2}\sqrt{a}, 0 \le \theta \le 2\pi, 4r^2 \le z \le a\right\}$. Thus $m = \int_0^{2\pi}\int_0^{\sqrt{a}/2}\int_{4r^2}^{a} Kr\,dz\,dr\,d\theta = 2\pi K\int_0^{\sqrt{a}/2}(ar - 4r^3)dr = 2\pi K\left[\frac{1}{2}ar^2 - r^4\right]_0^{\sqrt{a}/2} = \frac{1}{8}a^2\pi K$. Since the region is homogeneous and symmetric, $M_{yz} = M_{xz} = 0$ and $M_{xy} = \int_0^{2\pi}\int_0^{\sqrt{a}/2}\int_{4r^2}^{a} Krz\,dz\,dr\,d\theta = 2\pi K\int_0^{\sqrt{a}/2}\left(\frac{1}{2}a^2r - 8r^5\right)dr = 2\pi K\left[\frac{1}{4}a^2r^2 - \frac{4}{3}r^6\right]_0^{\sqrt{a}/2} = \frac{1}{12}a^3\pi K$. Hence $(\overline{x},\overline{y},\overline{z}) = \left(0,0,\frac{2}{3}a\right)$.

15. $\iiint_B (x^2+y^2+z^2)dV = \int_0^{\pi}\int_0^{2\pi}\int_0^1 \rho^4\sin\phi\,d\rho\,d\theta\,d\phi = 2\pi\int_0^{\pi}\frac{1}{5}\sin\phi\,d\phi = \frac{2\pi}{5}[-\cos\phi]_0^{\pi} = \frac{4\pi}{5}$

17. $\iiint_E y^2\,dV = \int_0^{\pi/2}\int_0^{\pi/2}\int_0^1 (\rho^2\sin^2\phi\sin^2\theta)(\rho^2\sin\phi)d\rho\,d\phi\,d\theta$

$= \int_0^{\pi/2}\int_0^{\pi/2}\int_0^1 \rho^4\sin^3\phi\sin^2\theta\,d\rho\,d\phi\,d\theta = \int_0^{\pi/2}\int_0^{\pi/2}\frac{1}{5}\sin^3\phi\sin^2\theta\,d\phi\,d\theta$

$= \int_0^{\pi/2}\frac{2}{15}\sin^2\theta\,d\theta = \frac{2}{15}\left[\frac{1}{2}\theta - \frac{1}{4}\sin 2\theta\right]_0^{\pi/2} = \frac{\pi}{30}$

19. $\iiint_E \sqrt{x^2+y^2+z^2}\,dV = \int_0^{2\pi}\int_0^{\pi/6}\int_0^2 \rho^3\sin\phi\,d\rho\,d\phi\,d\theta$

$= 8\pi\int_0^{\pi/6}\sin\phi\,d\phi = 8\pi[-\cos\phi]_0^{\pi/6} = 8\pi\left(1 - \frac{\sqrt{3}}{2}\right) = 4\pi\left(2 - \sqrt{3}\right)$

21. Since $\rho = 4\cos\phi$ implies $\rho^2 = 4\rho\cos\phi$, the equation is that of a sphere of radius 2 with center at $(0,0,2)$. Thus $V = \int_0^{2\pi}\int_0^{\pi/3}\int_0^{4\cos\phi}\rho^2\sin\phi\,d\rho\,d\phi\,d\theta = 2\pi\int_0^{\pi/3}\sin\phi\left(\frac{64}{3}\cos^3\phi\right)d\phi = \frac{32}{3}\pi[-\cos^4\phi]_0^{\pi/3} = 10\pi$.

23. Placing the center of the base at $(0,0,0)$, $\rho(x,y,z) = K\sqrt{x^2+y^2+z^2}$ is the density function. So

$m = \int_0^{2\pi}\int_0^{\pi/2}\int_0^a K\rho^3 \sin\phi \, d\rho \, d\phi \, d\theta = 2\pi K \int_0^{\pi/2} \frac{1}{4}a^4 \sin\phi \, d\phi = \frac{1}{2}\pi K a^4 [-\cos\phi]_0^{\pi/2} = \frac{1}{2}\pi K a^4.$

25. $I_z = \int_0^{2\pi}\int_0^{\pi/2}\int_0^a (K\rho^3 \sin\phi)(\rho^2 \sin^2\phi) d\rho \, d\phi \, d\theta = 2\pi K \int_0^{\pi/2} \frac{1}{6} a^6 \sin^3\phi \, d\phi$

$= \frac{1}{3}\pi a^6 K \left[-\cos\phi + \frac{1}{3}\cos^3\phi\right]_0^{\pi/2} = \frac{2}{9}\pi K a^6$

27. Place the center of the base at $(0,0,0)$; the density function is $\rho(x,y,z) = K$. By symmetry, the moments of inertia about any two such diameters will be equal, so we just need to find I_x:

$I_x = \int_0^{2\pi}\int_0^{\pi/2}\int_0^a (K\rho^2 \sin\phi)\rho^2(\sin^2\phi \sin^2\theta + \cos^2\phi) d\rho \, d\phi \, d\theta$

$= K\int_0^{2\pi}\int_0^{\pi/2} (\sin^3\phi \sin^2\theta + \sin\phi \cos^2\phi)\left(\frac{1}{5}a^5\right) d\phi \, d\theta$

$= \frac{1}{5}Ka^5 \int_0^{2\pi} \left[\sin^2\theta\left(-\cos\phi + \frac{1}{3}\cos^3\phi\right) + \left(-\frac{1}{3}\cos^3\phi\right)\right]_0^{\pi/2} d\theta = \frac{1}{5}Ka^5 \int_0^{2\pi} \left[\frac{2}{3}\sin^2\theta + \frac{1}{3}\right] d\theta$

$= \frac{1}{5}Ka^5 \left[\frac{2}{3}\left(\frac{1}{2}\theta - \frac{1}{4}\sin 2\theta\right) + \frac{1}{3}\theta\right]_0^{2\pi} = \frac{1}{5}Ka^5 \left[\frac{2}{3}(\pi - 0) + \frac{1}{3}(2\pi - 0)\right] = \frac{4}{15}Ka^5\pi$

29. In spherical coordinates $z = \sqrt{x^2+y^2}$ becomes $\cos\phi = \sin\phi$ or $\phi = \pi/4$. Then

$V = \int_0^{2\pi}\int_0^{\pi/4}\int_0^1 \rho^2 \sin\phi \, d\rho \, d\phi \, d\theta = 2\pi \int_0^{\pi/4} \sin\phi \, d\phi \left(\int_0^1 \rho^2 \, d\rho\right) = \frac{1}{3}\pi\left(2 - \sqrt{2}\right)$,

$M_{xy} = \int_0^{2\pi}\int_0^{\pi/4}\int_0^1 \rho^3 \sin\phi \cos\phi \, d\rho \, d\phi \, d\theta = 2\pi\left[-\frac{1}{4}\cos 2\phi\right]_0^{\pi/4}\left(\frac{1}{4}\right) = \frac{\pi}{8}$ and by symmetry $M_{yz} = M_{xz} = 0$. Hence $(\overline{x}, \overline{y}, \overline{z}) = \left(0, 0, \frac{3}{8(2-\sqrt{2})}\right)$.

31. In cylindrical coordinates the paraboloid is given by $z = r^2$ and the plane by $z = 2r\sin\theta$ and they intersect in the circle $r = 2\sin\theta$. Then $\iiint_E z \, dV = \int_0^{\pi}\int_0^{2\sin\theta}\int_{r^2}^{2r\sin\theta} rz \, dz \, dr \, d\theta = \frac{5\pi}{6}$ (using a CAS).

33. $\int_{-1}^1 \int_{-\sqrt{1-x^2}}^{\sqrt{1-x^2}} \int_{x^2+y^2}^{2-x^2-y^2} (x^2+y^2)^{3/2} \, dz \, dy \, dx = \int_0^{2\pi}\int_0^1\int_{r^2}^{2-r^2} (r^2)^{3/2} \, r \, dz \, dr \, d\theta$

$= 2\pi \int_0^1 \left[r^4 z\right]_{r^2}^{2-r^2} dr = 2\pi \int_0^1 (2r^4 - r^6 - r^6) dr = 4\pi\left[\frac{1}{5}r^5 - \frac{1}{7}r^7\right]_0^1 = \frac{8\pi}{35}$

35. The region of integration E is the top half of the sphere $x^2+y^2+z^2 = 9$. So

$\int_{-3}^3 \int_{-\sqrt{9-x^2}}^{\sqrt{9-x^2}} \int_0^{\sqrt{9-x^2-y^2}} z\sqrt{x^2+y^2+z^2} \, dz \, dy \, dx = \iiint_E z\sqrt{x^2+y^2+z^2} \, dV$

$= \int_0^{2\pi}\int_0^{\pi/2}\int_0^3 (\rho^2 \cos\phi)(\rho^2 \sin\phi) d\rho \, d\phi \, d\theta = 2\pi \int_0^{\pi/2} \frac{243}{5} \cos\phi \sin\phi \, d\phi = \frac{486}{5}\pi\left[-\frac{1}{4}\cos 2\phi\right]_0^{\pi/2} = \frac{243}{5}\pi$

37. (a) From the diagram, $z = r\cot\phi_0$ to $z = \sqrt{a^2 - r^2}$, $r = 0$ to $r = a\sin\phi_0$ (or use $a^2 - r^2 = r^2\cot^2\phi_0$). Thus

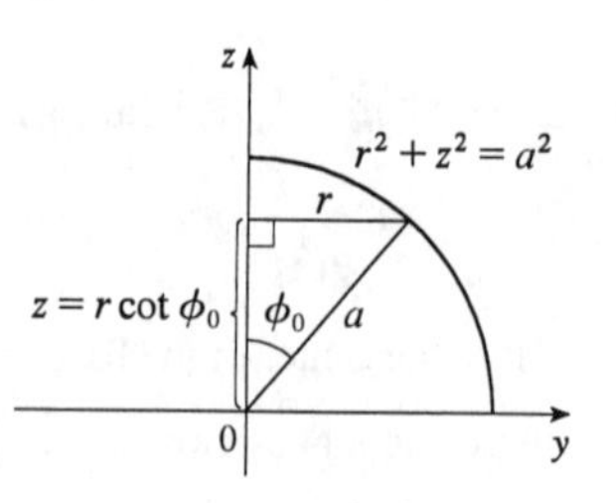

$$V = \int_0^{2\pi}\int_0^{a\sin\phi_0}\int_{r\cot\phi_0}^{\sqrt{a^2-r^2}} r\,dz\,dr\,d\theta$$

$$= 2\pi\int_0^{a\sin\phi_0}\left(r\sqrt{a^2 - r^2} - r^2\cot\phi_0\right)dr$$

$$= \tfrac{2\pi}{3}\left[-(a^2 - r^2)^{3/2} - r^3\cot\phi_0\right]_0^{a\sin\phi_0}$$

$$= \tfrac{2\pi}{3}\left[-(a^2 - a^2\sin^2\phi_0)^{3/2} - a^3\sin^3\phi_0\cot\phi_0 + a^3\right]$$

$$= \tfrac{2}{3}\pi a^3[1 - (\cos^3\phi_0 + \sin^2\phi_0\cos\phi_0)] = \tfrac{2}{3}\pi a^3(1 - \cos\phi_0)$$

(b) The wedge in question is the shaded area rotated from $\theta = \theta_1$ to $\theta = \theta_2$. Letting

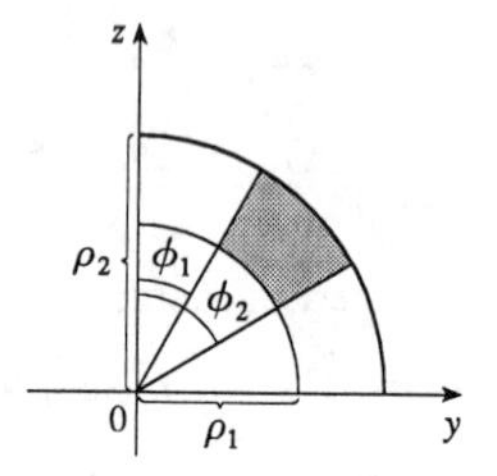

V_{ij} = volume of the region bounded by the sphere of radius ρ_i and the cone with angle ϕ_j ($\theta = \theta_1$ to θ_2)

and letting V be the volume of the wedge, we have

$$V = (V_{22} - V_{21}) - (V_{12} - V_{11})$$

$$= \tfrac{1}{3}(\theta_2 - \theta_1)[\rho_2^3(1 - \cos\phi_2) - \rho_2^3(1 - \cos\phi_1) - \rho_1^3(1 - \cos\phi_2) + \rho_1^3(1 - \cos\phi_1)]$$

$$= \tfrac{1}{3}(\theta_2 - \theta_1)[(\rho_2^3 - \rho_1^3)(1 - \cos\phi_2) - (\rho_2^3 - \rho_1^3)(1 - \cos\phi_1)]$$

$$= \tfrac{1}{3}(\theta_2 - \theta_1)[(\rho_2^3 - \rho_1^3)(\cos\phi_1 - \cos\phi_2)]$$

Or: Show that $V = \int_{\theta_1}^{\theta_2}\int_{\rho_1\sin\phi_1}^{\rho_2\sin\phi_2}\int_{r\cot\phi_2}^{r\cot\phi_1} r\,dz\,dr\,d\theta$.

(c) By the Mean Value Theorem with $f(\rho) = \rho^3$ there exists some $\tilde{\rho}$ with $\rho_1 \le \tilde{\rho} \le \rho_2$ such that $f(\rho_2) - f(\rho_1) = f'(\tilde{\rho})(\rho_2 - \rho_1)$ or $\rho_1^3 - \rho_2^3 = 3\tilde{\rho}^2\Delta\rho$. Similarly there exists ϕ with $\phi_1 \le \tilde{\phi} \le \phi_2$ such that $\cos\phi_2 - \cos\phi_1 = (-\sin\tilde{\phi})\Delta\phi$. Substituting into the result from (b) gives $\Delta V = (\tilde{\rho}^2\Delta\rho)(\theta_2 - \theta_1)(\sin\tilde{\phi})\Delta\phi = \tilde{\rho}^2\sin\tilde{\phi}\,\Delta\rho\,\Delta\phi\,\Delta\theta$.

EXERCISES 13.11

1. $\dfrac{\partial(x,y)}{\partial(u,v)} = \begin{vmatrix} \partial x/\partial u & \partial x/\partial v \\ \partial y/\partial u & \partial y/\partial v \end{vmatrix} = \begin{vmatrix} 1 & -2 \\ 2 & -1 \end{vmatrix} = 1(-1) - 2(-2) = 3$

3. $\dfrac{\partial(x,y)}{\partial(u,v)} = \begin{vmatrix} \partial x/\partial u & \partial x/\partial v \\ \partial y/\partial u & \partial y/\partial v \end{vmatrix} = \begin{vmatrix} 2e^{2u}\cos v & -e^{2u}\sin v \\ 2e^{2u}\sin v & e^{2u}\cos v \end{vmatrix} = 2e^{4u}(\cos^2 v + \sin^2 v) = 2e^{4u}$

5. $\dfrac{\partial(x,y,z)}{\partial(u,v,w)} = \begin{vmatrix} 1 & 1 & 1 \\ 1 & 1 & -1 \\ 1 & -1 & 1 \end{vmatrix} = 1(1-1) - 1(1+1) + 1(-1-1) = -4$

7. S_1: $v = 0$, $0 \le u \le 2$, so $x = u$, $y = 2u$ and $y = 2x$. S_2: $u = 2$, $0 \le v \le 1$, so $x = 2 - 2v$, $y = 4 - v$ and $x = 2y - 6$. S_3: $v = 1$, $0 \le u \le 2$, so $x = u - 2$, $y = 2u - 1$ and $y = 2x + 3$. S_4: $u = 0$, $0 \le v \le 1$, so $x = -2v$, $y = -v$ and $2y = x$.

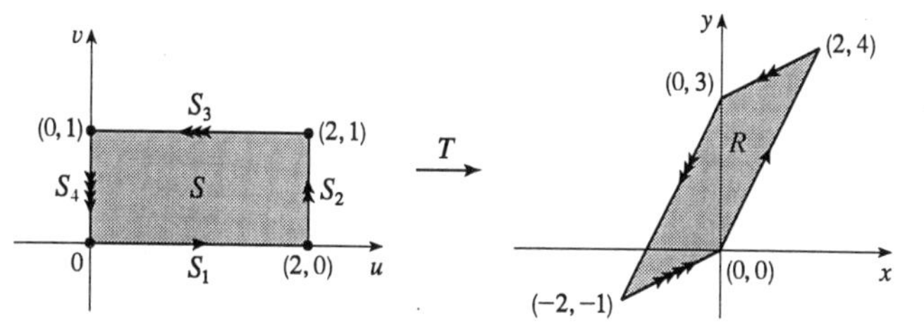

9. S_1: $0 \le u \le 1$, $v = 0$, $x = 4u$, $y = 0$ so $0 \le x \le 4$, $y = 0$ is the image of the first side.

S_2: $u + v = 1$, $x = 4(1 - v) + 3v = 4 - v \Leftrightarrow v = 4 - x$, $y = 4v = 4(4 - x) = 16 - 4x$, $3 \le x \le 4$, so $y = 16 - 4x$, $3 \le x \le 4$ is the image of the second side.

S_3: $u = 0$, $0 \le v \le 1$, $x = 3v$, $y = 4v = \frac{4}{3}x$, $0 \le x \le 3$, so $y = \frac{4}{3}x$, $0 \le x \le 3$ is the image of the third side.

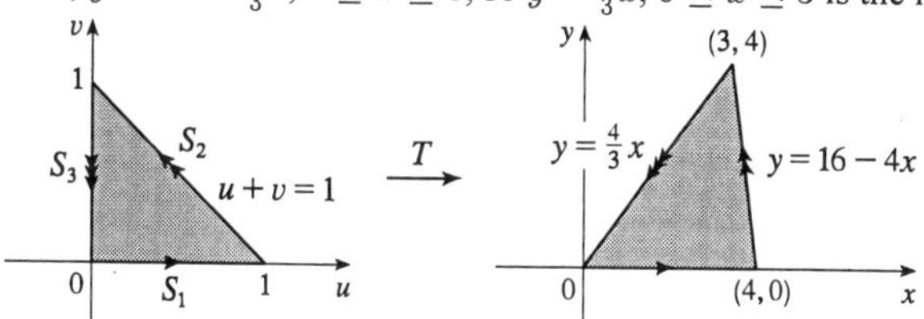

11. $\dfrac{\partial(x,y)}{\partial(u,v)} = \begin{vmatrix} 1/3 & 1/3 \\ -2/3 & 1/3 \end{vmatrix} = \dfrac{1}{3}$ and $3x + 4y = (u + v) + \frac{4}{3}(v - 2u) = \frac{1}{3}(7v - 5u)$. Then S is the region bounded by the lines $u = 0$, $\frac{1}{3}(v - 2) = \frac{1}{3}(u + v) - 2$ or $u = 2$, $\frac{1}{3}(v - 2u) = -\frac{2}{3}(u + v)$ or $v = 0$, and $\frac{1}{3}(v - 2u) = 3 - \frac{2}{3}(u + v)$ or $v = 3$. Thus

$\iint_R (3x + 4y)\,dA = \int_0^3 \int_0^2 \frac{1}{3}(7v - 5u)\left(\frac{1}{3}du\,dv\right) = \frac{1}{9}\int_0^3 (14v - 10)\,dv = \frac{1}{9}(33) = \frac{11}{3}$.

13. $\dfrac{\partial(x,y)}{\partial(u,v)} = \begin{vmatrix} 2 & 0 \\ 0 & 3 \end{vmatrix} = 6$, $x^2 = 4u^2$ and the planar ellipse $9x^2 + 4y^2 \le 36$ is the image of the disk $u^2 + v^2 \le 1$.

Thus $\iint_R x^2\,dA = \iint_{u^2+v^2\le 1} (4u^2)(6)du\,dv = \int_0^{2\pi}\int_0^1 (24r^2\cos^2\theta)r\,dr\,d\theta = 6\pi$.

15. $\dfrac{\partial(x,y)}{\partial(u,v)} = \begin{vmatrix} 1/v & -u/v^2 \\ 0 & 1 \end{vmatrix} = \dfrac{1}{v}$, $xy = u$, $y = x$ is the image of the parabola $v^2 = u$, $y = 3x$ is the image of the parabola $v^2 = 3u$, and the hyperbolas $xy = 1$, $xy = 3$ are the images of the lines $u = 1$ and $u = 3$ respectively.

Thus $\displaystyle\iint_R xy\,dA = \int_1^3\int_{\sqrt{u}}^{\sqrt{3u}} u\left(\frac{1}{v}\right)dv\,du = \int_1^3 u\left(\ln\sqrt{3u} - \ln\sqrt{u}\right)du = \int_1^3 u\ln\sqrt{3}\,du = 4\ln\sqrt{3} = 2\ln 3.$

17. $\dfrac{\partial(x,y,z)}{\partial(u,v,w)} = \begin{vmatrix} a & 0 & 0 \\ 0 & b & 0 \\ 0 & 0 & c \end{vmatrix} = abc$ and the solid enclosed by the ellipsoid is the image of the ball $u^2 + v^2 + w^2 \le 1$. So $\iiint_E dV = \iiint_{u^2+v^2+w^2\le 1} abc\,du\,dv\,dw = (abc)(\text{volume of the ball}) = \frac{4}{3}\pi abc$.

19. Letting $u = 2x - y$ and $v = 3x + y$, we have $x = \frac{1}{5}(u+v)$, $y = \frac{1}{5}(2v - 3u)$. Then $\dfrac{\partial(x,y)}{\partial(u,v)} = \begin{vmatrix} 1/5 & 1/5 \\ -3/5 & 2/5 \end{vmatrix} = \dfrac{1}{5}$ and

$$\iint_R xy\,dA = \int_{-2}^1\int_{-3}^1 \frac{(u+v)(2v-3u)}{25}\left(\frac{1}{5}\right)du\,dv = \frac{1}{125}\int_{-2}^1\int_{-3}^1 (2v^2 - uv - 3u^2)du\,dv$$

$$= \frac{1}{125}\int_{-2}^1 (8v^2 + 4v - 28)dv = -\frac{66}{125}.$$

21. Letting $u = y - x$, $v = y + x$, we have $y = \frac{1}{2}(u+v)$, $x = \frac{1}{2}(v-u)$. Then $\dfrac{\partial(x,y)}{\partial(u,v)} = \begin{vmatrix} -1/2 & 1/2 \\ 1/2 & 1/2 \end{vmatrix} = -\dfrac{1}{2}$ and R is the image of the trapezoidal region with vertices $(-1,1)$, $(-2,2)$, $(2,2)$, and $(1,1)$. Thus

$$\iint_R \cos\frac{y-x}{y+x}\,dA = \int_1^2\int_{-v}^v \left|-\tfrac{1}{2}\right|\cos\frac{u}{v}\,du\,dv = \frac{1}{2}\int_1^2\left[v\sin\frac{u}{v}\right]_{-v}^v dv = \frac{1}{2}\int_1^2 2v\sin(1)dv = \tfrac{3}{2}\sin 1.$$

23.

Let $u = x + y$ and $v = -x + y$. Then $u + v = 2y \Rightarrow y = \frac{1}{2}(u+v)$ and $u - v = 2x \Rightarrow x = \frac{1}{2}(u-v)$.

$\dfrac{\partial(x,y)}{\partial(u,v)} = \begin{vmatrix} 1/2 & -1/2 \\ 1/2 & 1/2 \end{vmatrix} = \dfrac{1}{2}$. Now

$|u| = |x+y| \le |x| + |y| \le 1 \Rightarrow -1 \le u \le 1$, and

$|v| = |-x+y| \le |x| + |y| \le 1 \Rightarrow -1 \le v \le 1$.

R is the image of the square region with vertices $(1,1)$, $(1,-1)$, $(-1,-1)$, and $(-1,1)$. So $\iint_R e^{x+y}\,dA = \frac{1}{2}\int_{-1}^1\int_{-1}^1 e^u\,du\,dv = \frac{1}{2}\cdot 2\cdot e^u\Big|_{-1}^1 = e - e^{-1}$.

REVIEW EXERCISES FOR CHAPTER 13

1. This is true by Fubini's Theorem.

3. $\iint_D \sqrt{4-x^2-y^2}\,dA =$ the volume under the surface $x^2+y^2+z^2=4$ and above the xy-plane $= \frac{1}{2}$(the volume of the sphere $x^2+y^2+z^2=4$) $= \frac{1}{2}\cdot\frac{4}{3}\pi(2)^3 = \frac{16}{3}\pi$.

5. The volume enclosed by the cone $z=\sqrt{x^2+y^2}$ and the plane $z=2$ is, using cylindrical coordinates, $V=\int_0^{2\pi}\int_0^2\int_r^2 r\,dz\,dr\,d\theta \neq \int_0^{2\pi}\int_0^2\int_r^2 dz\,dr\,d\theta$, so the assertion is false.

7. $\int_{-2}^{2}\int_0^4(4x^3+3xy^2)dx\,dy = \int_{-2}^{2}(256+24y^2)dy = 2[256y+8y^3]_0^2 = 1152$

9. $\displaystyle\int_1^2\int_0^{x^2}\frac{1}{x+y}\,dy\,dx = \int_1^2 [\ln|x+y|]_0^{x^2}\,dx = \int_1^2[\ln(1+x)]dx = [(1+x)\ln(1+x)-(1+x)]_1^2 = \ln\left(\tfrac{27}{4}\right)-1$

11. $\int_0^1\int_0^{x^2}\int_0^y y^2z\,dz\,dy\,dx = \int_0^1\int_0^{x^2}\frac{1}{2}y^4\,dy\,dx = \int_0^1\frac{1}{10}x^{10}\,dx = \frac{1}{110}$

13.

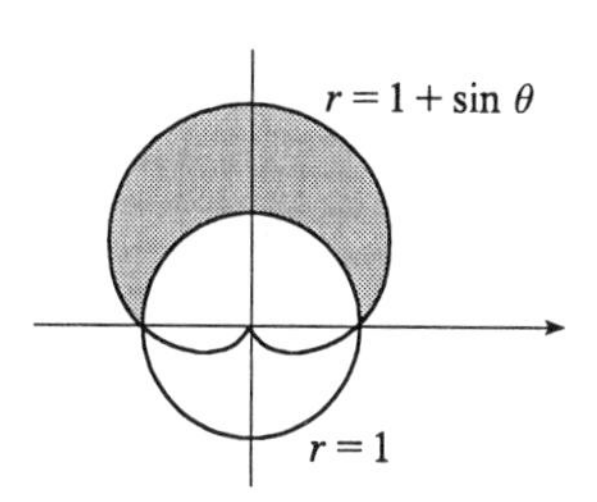

The region whose area is given by $\int_0^{\pi}\int_1^{1+\sin\theta} r\,dr\,d\theta$ is $\{(r,\theta)\mid 0\le\theta\le\pi, 1\le r\le 1+\sin\theta\}$, which is the region outside the circle $r=1$ and inside the cardioid $r=1+\sin\theta$.

15.

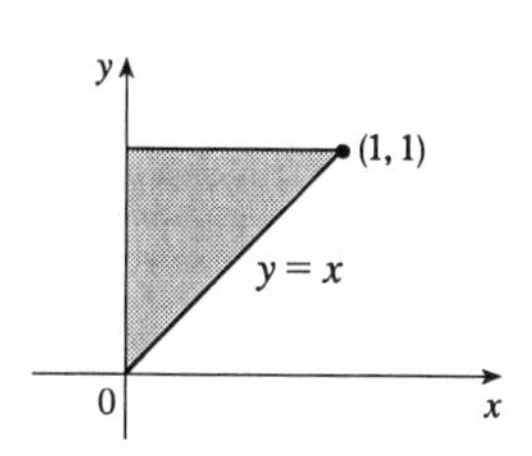

$$\int_0^1\int_x^1 e^{x/y}\,dy\,dx = \int_0^1\int_0^y e^{x/y}\,dx\,dy = \int_0^1(ey-y)dy = \tfrac{1}{2}(e-1)$$

17. $\displaystyle\int_2^4\int_0^1\frac{1}{(x-y)^2}\,dx\,dy = \int_2^4\left(-\frac{1}{y}-\frac{1}{1-y}\right)dy = [-\ln y+\ln|1-y|]_2^4$

$= -\ln 4+\ln 3+\ln 2 = \ln\frac{3}{2}$

19. The curves $y^2=x^3$ and $y=x$ intersect when $x^3=x$, that is when $x=0$ and $x=1$ (note that $x\neq -1$ since $x^3=y^2 \Rightarrow x\ge 0$.) So $\int_0^1\int_{x^{3/2}}^{x} xy\,dy\,dx = \int_0^1\left[\frac{1}{2}x^3-\frac{1}{2}x^4\right]dx = \left[\frac{1}{8}x^4-\frac{1}{10}x^5\right]_0^1 = \frac{1}{40}$.

21. $\int_0^1\int_0^{1-y^2}(xy+2x+3y)dx\,dy = \int_0^1\left[(y+2)\frac{1}{2}(1-y^2)^2+3y(1-y^2)\right]dy$

$= \frac{1}{2}y^5+y^4-4y^3-2y^2+\frac{7}{2}y+1 = \frac{1}{5}+\frac{1}{12}-\frac{2}{3}+\frac{7}{4} = \frac{41}{30}$

23.

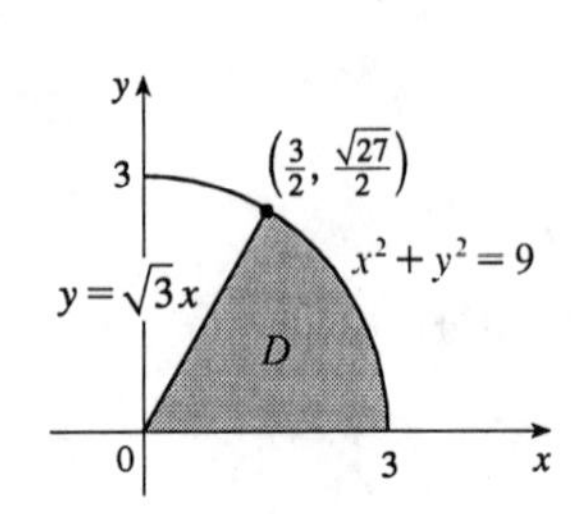

$$\iint_D (x^2+y^2)^{3/2}\,dA = \int_0^{\pi/3}\int_0^3 (r^2)^{3/2}\,r\,dr\,d\theta = \frac{\pi}{3}\frac{3^5}{5} = \frac{81\pi}{5}$$

25. $\iiint_E x^2z\,dV = \int_0^2\int_0^{2x}\int_0^x x^2z\,dz\,dy\,dx = \int_0^2\int_0^{2x} \frac{1}{2}x^4\,dy\,dx = \int_0^2 x^5\,dx = \frac{1}{6}\cdot 2^6 = \frac{32}{3}$

27. $\iiint_E y^2z^2\,dV = \int_{-1}^1\int_{-\sqrt{1-y^2}}^{\sqrt{1-y^2}}\int_0^{1-y^2-z^2} y^2z^2\,dx\,dz\,dy = \int_{-1}^1\int_{-\sqrt{1-y^2}}^{\sqrt{1-y^2}} y^2z^2(1-y^2-z^2)dz\,dy$
$= \int_0^{2\pi}\int_0^1 (r^2\cos^2\theta)(r^2\sin^2\theta)(1-r^2)r\,dr\,d\theta = \int_0^{2\pi}\int_0^1 \frac{1}{4}\sin^2 2\theta(r^5-r^7)dr\,d\theta$
$= \int_0^{2\pi} \frac{1}{8}(1-\cos 4\theta)\left[\frac{1}{6}r^6 - \frac{1}{8}r^8\right]_0^1 d\theta = \frac{1}{192}\left[\theta - \frac{1}{4}\sin 4\theta\right]_0^{2\pi} = \frac{2\pi}{192} = \frac{\pi}{96}$

29. $\iiint_E yz\,dV = \int_{-2}^2\int_0^{\sqrt{4-x^2}}\int_0^y yz\,dz\,dy\,dx = \int_{-2}^2\int_0^{\sqrt{4-x^2}} \frac{1}{2}y^3\,dy\,dx = \int_0^\pi\int_0^2 \frac{1}{2}r^3\sin^3\theta\,r\,dr\,d\theta$
$= \frac{16}{5}\int_0^\pi \sin^3\theta\,d\theta = \frac{16}{5}\left[-\cos\theta + \frac{1}{3}\cos^3\theta\right]_0^\pi = \frac{64}{15}$

31. $V = \int_0^2\int_1^4 (x^2+4y^2)dy\,dx = \int_0^2 (3x^2+84)dx = 176$

33.

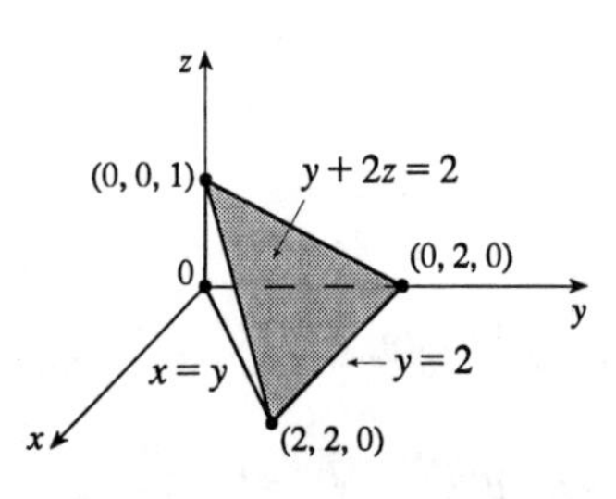

$$V = \int_0^2\int_0^y\int_0^{(2-y)/2} dz\,dx\,dy = \int_0^2\int_0^y \left(1 - \tfrac{1}{2}y\right)dx\,dy = \int_0^2 \left(y - \tfrac{1}{2}y^2\right)dy = \frac{2}{3}$$

35. Using the wedge above the plane $z = 0$ and below the plane $z = mx$ and noting that we have the same volume for $m < 0$ as for $m > 0$ (so use $m > 0$), we have
$V = 2\int_0^{a/3}\int_0^{\sqrt{a^2-9y^2}} mx\,dx\,dy = 2\int_0^{a/3} \frac{1}{2}m(a^2-9y^2)dy = m[a^2y - 3y^3]_0^{a/3} = m\left(\frac{1}{3}a^3 - \frac{1}{9}a^3\right) = \frac{2}{9}ma^3.$

37. $m = \int_0^1\int_0^{1-y^2} y\,dx\,dy = \int_0^1 (y-y^3)dy = \frac{1}{2} - \frac{1}{4} = \frac{1}{4},$
$M_y = \int_0^1\int_0^{1-y^2} xy\,dx\,dy = \int_0^1 \frac{1}{2}y(1-y^2)^2\,dy = -\frac{1}{12}(1-y^2)^3\Big|_0^1 = \frac{1}{12},$
$M_x = \int_0^1\int_0^{1-y^2} y^2\,dx\,dy = \int_0^1 (y^2-y^4)dy = \frac{2}{15}$. Hence $(\overline{x},\overline{y}) = \left(\frac{1}{3}, \frac{8}{15}\right)$.

39. $m = \frac{1}{4}\pi Ka^2$ where K is constant,
$M_y = \iint_{x^2+y^2\le a^2} Kx\,dA = K\int_0^{\pi/2}\int_0^a r^2\cos\theta\,dr\,d\theta = \frac{1}{3}Ka^3\int_0^{\pi/2}\cos\theta\,d\theta = \frac{1}{3}a^3K$, and
$M_x = K\int_0^{\pi/2}\int_0^a r^2\sin\theta\,dr\,d\theta = \frac{1}{3}a^3K$ (by symmetry $M_y = M_x$). Hence the centroid is $(\overline{x},\overline{y}) = \left(\frac{4}{3\pi}a, \frac{4}{3\pi}a\right)$.

41. The equation of the cone with the suggested orientation is $(h-z)=\dfrac{h}{a}\sqrt{x^2+y^2}$, $0\le z\le h$. Then $V=\frac{1}{3}\pi a^2h$ is the volume of one frustum of a cone; by symmetry $M_{yz}=M_{xz}=0$; and

$$M_{xy}=\iint\limits_{x^2+y^2\le a^2}\int_0^{h-(h/a)\sqrt{x^2+y^2}} z\,dz\,dA=\int_0^{2\pi}\int_0^a\int_0^{(h/a)(a-r)} rz\,dz\,dr\,d\theta=\pi\int_0^a r\frac{h^2}{a^2}(a-r)^2\,dr$$

$$=\frac{\pi h^2}{a^2}\int_0^a\left(a^2r-2ar^2+r^3\right)dr=\frac{\pi h^2}{a^2}\left(\frac{a^4}{2}-\frac{2a^4}{3}+\frac{a^4}{4}\right)=\frac{\pi h^2a^2}{12}.$$

Hence the centroid is $(\overline{x},\overline{y},\overline{z})=\left(0,0,\frac{1}{4}h\right)$.

43. $z=1-\frac{1}{2}x-\frac{1}{3}y\quad\Rightarrow\quad\partial z/\partial x=-\frac{1}{2}$ and $\partial z/\partial y=-\frac{1}{3}$. The projection onto the xy-plane of the part of the given plane in the first octant is the triangular region D bounded by the x- and y-axes and the line $3x+2y=6$, which has intercepts 2 and 3. So $A(D)=\frac{1}{3}\cdot 2\cdot 3=3$. Thus the surface area is

$$A(S)=\iint\limits_D\sqrt{1+\left(-\tfrac{1}{2}\right)^2+\left(-\tfrac{1}{3}\right)^2}\,dA=\tfrac{7}{6}A(D)=\tfrac{7}{6}\cdot 3=\tfrac{7}{2}.$$

45. $r=1+3\cos\theta$

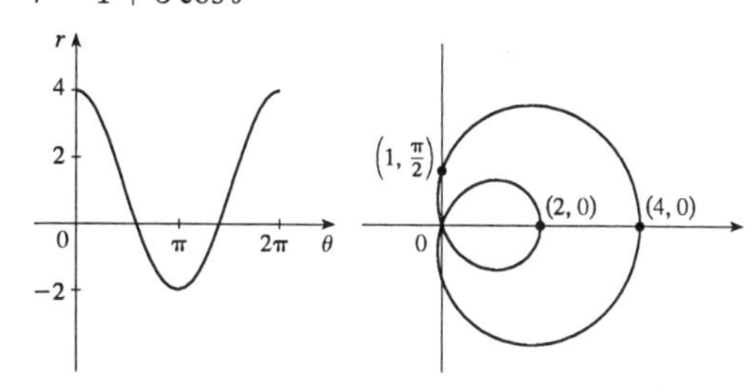

47. $r^2=\sec 2\theta\quad\Rightarrow\quad r^2\cos 2\theta=1\quad\Rightarrow$

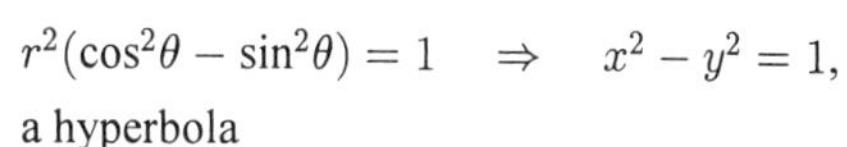

$r^2(\cos^2\theta-\sin^2\theta)=1\quad\Rightarrow\quad x^2-y^2=1$,

a hyperbola

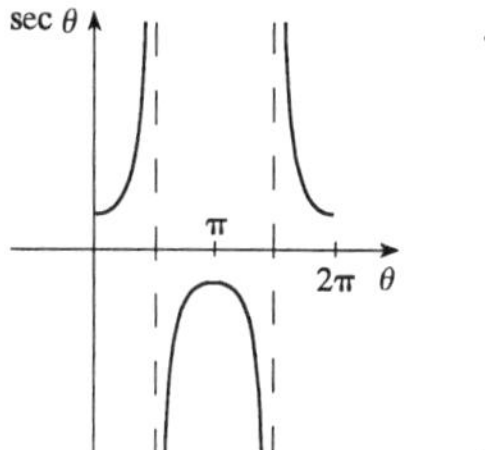

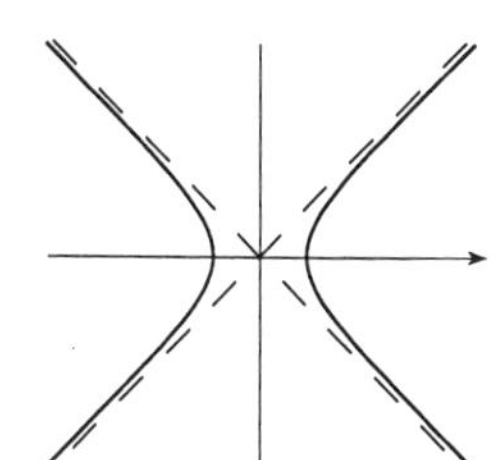

49. $r=2\cos^2(\theta/2)=1+\cos\theta$

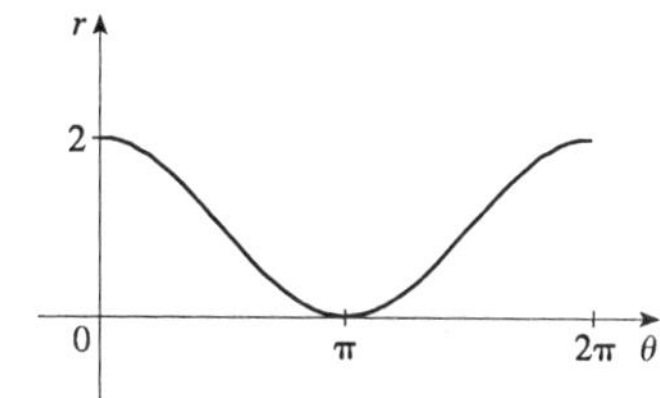

51. $r=\dfrac{1}{1+\cos\theta}\quad\Rightarrow\quad e=1\quad\Rightarrow\quad$ parabola;

$d=1\quad\Rightarrow\quad$ directrix $x=1$ and vertex $\left(\frac{1}{2},0\right)$;

y-intercepts are $\left(1,\frac{\pi}{2}\right)$ and $\left(1,\frac{3\pi}{2}\right)$.

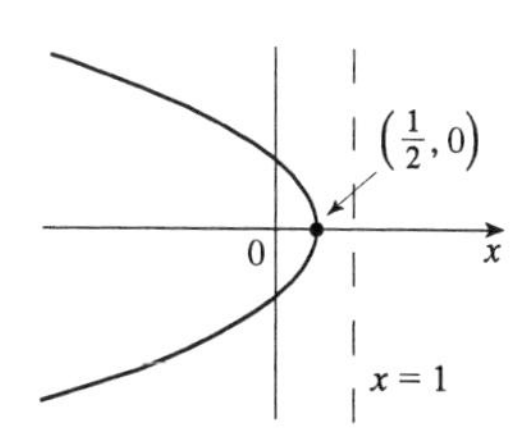

53. $x^2 + y^2 = 4x \quad \Leftrightarrow \quad r^2 = 4r\cos\theta \quad \Leftrightarrow \quad r = 4\cos\theta$

55. $x = 2\cos\frac{\pi}{6} = \sqrt{3}$, $y = 2\sin\frac{\pi}{6} = 1$, $z = 2$, so in rectangular coordinates the point is $\left(\sqrt{3}, 1, 2\right)$. $\rho = \sqrt{3+1+4} = 2\sqrt{2}$, $\theta = \frac{\pi}{6}$, and $\cos\phi = z/\rho = \frac{1}{\sqrt{2}}$, so $\phi = \frac{\pi}{4}$ and the spherical coordinates are $\left(2\sqrt{2}, \frac{\pi}{6}, \frac{\pi}{4}\right)$.

57. $x = 4\sin\frac{\pi}{6}\cos\frac{\pi}{3} = 1$, $y = 4\sin\frac{\pi}{6}\sin\frac{\pi}{3} = \sqrt{3}$, $z = 4\cos\frac{\pi}{6} = 2\sqrt{3}$ so in rectangular coordinates the point is $\left(1, \sqrt{3}, 2\sqrt{3}\right)$. $r^2 = x^2 + y^2 = 4$, $r = 2$, so the cylindrical coordinates are $\left(2, \frac{\pi}{3}, 2\sqrt{3}\right)$.

59. A half-plane including the z-axis and intersecting the xy-plane in the half-line $x = y$, $x > 0$

61. Since $\rho = 3\sec\phi$, $\rho\cos\phi = 3$ or $z = 3$. Thus the surface is a plane parallel to the xy-plane and through the point $(0, 0, 3)$.

63. In cylindrical coordinates: $r^2 + z^2 = 4$. In spherical coordinates: $\rho^2 = 4$ or $\rho = 2$.

65. The resulting surface is a paraboloid of revolution with equation $z = 4x^2 + 4y^2$. Changing to cylindrical coordinates we have $z = 4(x^2 + y^2) = 4r^2$.

67.

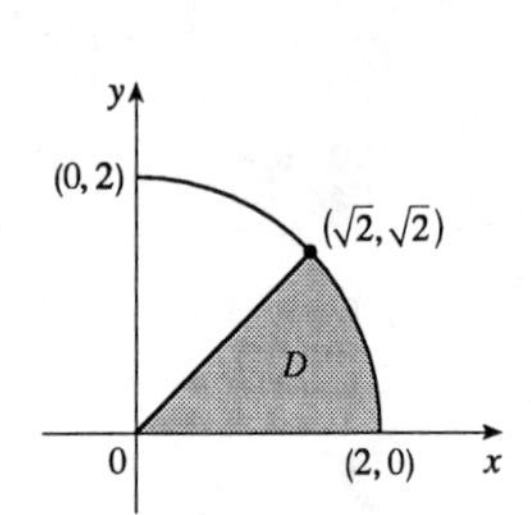

$$\int_0^{\sqrt{2}}\int_y^{\sqrt{4-y^2}} \frac{1}{1+x^2+y^2}\,dx\,dy = \int_0^{\pi/4}\int_0^2 \frac{1}{1+r^2}\,r\,dr\,d\theta$$

$$= \frac{1}{2}\int_0^{\pi/4} \ln\left|1+r^2\right|\Big|_0^2\,d\theta$$

$$= \frac{1}{2}\int_0^{\pi/4} \ln 5\,d\theta = \frac{\pi}{8}\ln 5$$

69. From the graph, it appears that $1 - x^2 = e^x$ at $x \approx -0.71$ and at $x = 0$, with $1 - x^2 > e^x$ on $(-0.71, 0)$. So the desired integral is

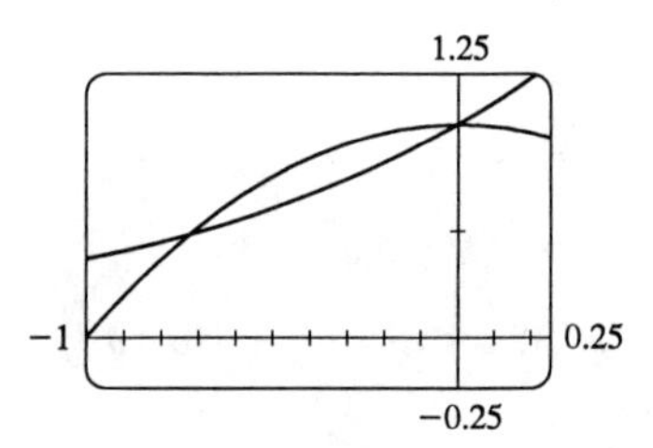

$\iint_D y^2\,dA \approx \int_{-0.71}^0 \int_{e^x}^{1-x^2} y^2\,dy\,dx = \frac{1}{3}\int_{-0.71}^0 \left[(1-x^2)^3 - e^{3x}\right]dx$

$= \frac{1}{3}\left[x - x^3 + \frac{3}{5}x^5 - \frac{1}{7}x^7 - \frac{1}{3}e^{3x}\right]_{-0.71}^0 \approx 0.0512.$

71. Let the tetrahedron be called T. The front face of T is given by the plane $x + \frac{1}{2}y + \frac{1}{3}z = 1$, or $z = 3 - 3x - \frac{3}{2}y$, which intersects the xy-plane in the line $y = 2 - 2x$. So the total mass is

$m = \iiint_T \rho(x,y,z)dV = \int_0^1\int_0^{2-2x}\int_0^{3-3x-\frac{3}{2}y}(x^2+y^2+z^2)dz\,dy\,dx = \frac{7}{5}$. The center of mass is

$(\overline{x}, \overline{y}, \overline{z}) = \left(m^{-1}\iiint_T x\rho(x,y,z)dV, m^{-1}\iiint_T y\rho(x,y,z)dV, m^{-1}\iiint_T z\rho(x,y,z)dV\right) = \left(\frac{4}{21}, \frac{11}{21}, \frac{8}{7}\right).$

73.

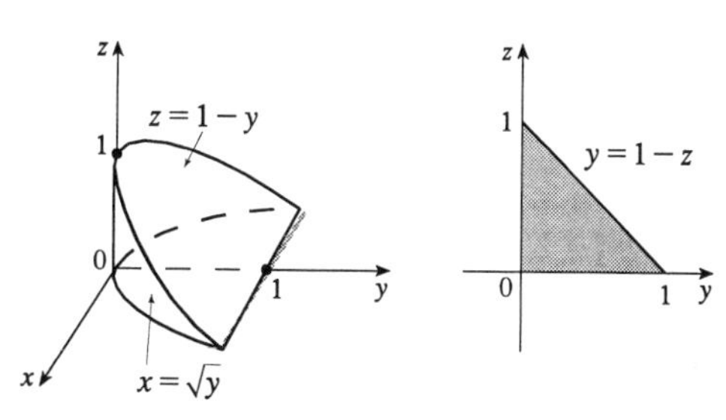

$\int_{-1}^{1}\int_{x^2}^{1}\int_{0}^{1-y} f(x,y,z)dz\,dy\,dx = \int_0^1\int_0^{1-z}\int_{-\sqrt{y}}^{\sqrt{y}} f(x,y,z)dx\,dy\,dz$

75. Since $u = x - y$, $v = x + y$, $x = \frac{1}{2}(u+v)$ and $y = \frac{1}{2}(v-u)$. Thus $\dfrac{\partial(x,y)}{\partial(u,v)} = \begin{vmatrix} 1/2 & 1/2 \\ -1/2 & 1/2 \end{vmatrix} = \dfrac{1}{2}$ and

$$\iint_R \frac{x-y}{x+y}\,dA = \int_2^4\int_{-2}^0 \frac{u}{v}\left(\tfrac{1}{2}\,du\,dv\right) = -\int_2^4 \frac{dv}{v} = -\ln 2.$$

77. Let $u = y - x$ and $v = y + x$ so $x = y - u = (v - x) - u \quad \Rightarrow \quad x = \frac{1}{2}(v-u)$ and

$y = v - \frac{1}{2}(v-u) = \frac{1}{2}(v+u)$. $\left|\dfrac{\partial(x,y)}{\partial(u,v)}\right| = \left|\dfrac{\partial x}{\partial u}\dfrac{\partial y}{\partial v} - \dfrac{\partial x}{\partial v}\dfrac{\partial y}{\partial u}\right| = \left|-\dfrac{1}{2}\left(\dfrac{1}{2}\right) - \dfrac{1}{2}\left(\dfrac{1}{2}\right)\right| = \left|-\dfrac{1}{2}\right| = \dfrac{1}{2}$. R is the image under this transformation of the square with vertices $(u,v) = (0,0)$, $(-2,0)$, $(0,2)$, and $(-2,2)$. So

$$\iint_R xy\,dA = \int_0^2\int_{-2}^0 \frac{v^2-u^2}{4}\left(\frac{1}{2}\right)du\,dv = \frac{1}{8}\int_0^2 \left[v^2u - \tfrac{1}{3}u^3\right]_{-2}^0 dv = \frac{1}{8}\int_0^2 \left(2v^2 - \tfrac{8}{3}\right)dv = \tfrac{1}{8}\left[\tfrac{2}{3}v^3 - \tfrac{8}{3}v\right]_0^2 = 0.$$

This result could have been anticipated by symmetry, since the integrand is an odd function of y and R is symmetric about the x-axis.

79. For each r such that D_r lies within the domain D, $A(D_r) = \pi r^2$, and by the Mean Value Theorem for Double Integrals there exists (x_r, y_r) in D_r such that $f(x_r, y_r) = \dfrac{1}{\pi r^2}\iint_{D_r} f(x,y)dA$. But $\lim\limits_{r\to 0^+}(x_r, y_r) = (a,b)$, so

$\lim\limits_{r\to 0^+} \dfrac{1}{\pi r^2}\iint_{D_r} f(x,y)dA = \lim\limits_{r\to 0^+} f(x_r, y_r) = f(a,b)$ by the continuity of f.

APPLICATIONS PLUS (after Chapter 13)

1. **(a)** The area of a trapezoid is $\frac{1}{2}h(b_1 + b_2)$, where h is the height (the distance between the two parallel sides) and b_1, b_2 are the lengths of the bases (the parallel sides). From the figure in the text, we see that $h = x\sin\theta$, $b_1 = w - 2x$, and $b_2 = w - 2x + 2x\cos\theta$. Therefore the cross-sectional area of the rain gutter is

$$\begin{aligned} A(x,\theta) &= \tfrac{1}{2}x\sin\theta[(w-2x) + (w - 2x + 2x\cos\theta)] = (x\sin\theta)(w - 2x + x\cos\theta) \\ &= wx\sin\theta - 2x^2\sin\theta + x^2\sin\theta\cos\theta,\ \ 0 < x \le \tfrac{1}{2}w,\ 0 < \theta \le \tfrac{\pi}{2} \end{aligned}$$

We look for the critical points of A: $\partial A/\partial x = w\sin\theta - 4x\sin\theta + 2x\sin\theta\cos\theta$ and $\partial A/\partial\theta = wx\cos\theta - 2x^2\cos\theta + x^2(\cos^2\theta - \sin^2\theta)$, so $\partial A/\partial x = 0 \quad\Leftrightarrow$ $\sin\theta(w - 4x + 2x\cos\theta) = 0 \quad\Leftrightarrow\quad \cos\theta = \dfrac{4x - w}{2x} = 2 - \dfrac{w}{2x}$ $\left(0 < \theta \le \frac{\pi}{2} \quad\Rightarrow\quad \sin\theta > 0\right)$. If, in addition, $\partial A/\partial\theta = 0$, then

$$\begin{aligned} 0 &= wx\cos\theta - 2x^2\cos\theta + x^2(2\cos^2\theta - 1) \\ &= wx\left(2 - \frac{w}{2x}\right) - 2x^2\left(2 - \frac{w}{2x}\right) + x^2\left[2\left(2 - \frac{w}{2x}\right)^2 - 1\right] \\ &= 2wx - \tfrac{1}{2}w^2 - 4x^2 + wx + x^2\left[8 - \frac{4w}{x} + \frac{w^2}{2x^2} - 1\right] = -wx + 3x^2 = x(3x - w) \end{aligned}$$

Since $x > 0$, we must have $x = \frac{1}{3}w$, in which case $\cos\theta = \frac{1}{2}$, so $\theta = \frac{\pi}{3}$, $\sin\theta = \frac{\sqrt{3}}{2}$, $k = \frac{\sqrt{3}}{6}w$, $b_1 = \frac{1}{3}w$, $b_2 = \frac{2}{3}w$, and $A = \frac{\sqrt{3}}{12}w^2$. As in Example 12.7.6, we can argue from the physical nature of this problem that we have found a relative maximum of A. Now checking the boundary of A, let $g(\theta) = A(w/2, \theta) = \frac{1}{2}w^2\sin\theta - \frac{1}{2}w^2\sin\theta + \frac{1}{4}w^2\sin\theta\cos\theta = \frac{1}{8}w^2\sin 2\theta$, $0 < \theta \le \frac{\pi}{2}$. Clearly g is maximized when $\sin 2\theta = 1$ in which case $A = \frac{1}{8}w^2$. Also along the line $\theta = \frac{\pi}{2}$, let $h(x) = A\left(x, \frac{\pi}{2}\right) = wx - 2x^2$, $0 < x < \frac{1}{2}w \quad\Rightarrow\quad h'(x) = w - 4x = 0 \quad\Leftrightarrow\quad x = \frac{1}{4}w$, and $h\left(\frac{1}{4}w\right) = w\left(\frac{1}{4}w\right) - 2\left(\frac{1}{4}w\right)^2 = \frac{1}{8}w^2$. Since $\frac{1}{8}w^2 < \frac{\sqrt{3}}{12}w^2$, we conclude that the relative maximum found earlier was an absolute maximum.

(b) If the metal were bent into a semi-circular gutter of radius r, we would have $w = \pi r$ and $A = \frac{1}{2}\pi r^2 = \frac{1}{2}\pi\left(\dfrac{w}{\pi}\right)^2 = \dfrac{w^2}{2\pi}$. Since $\dfrac{w^2}{2\pi} > \dfrac{\sqrt{3}w^2}{12}$, it *would* be better to bend the metal into a gutter with a semi-circular cross-section.

3. **(a)** If $f(P, A)$ is the probability that an individual at A will be infected by an individual at P, and $k\,dA$ is the number of infected individuals in an element of area dA, then $f(P, A)k\,dA$ is the number of infections that should result from exposure of the individual at A to infected people in the element of area dA. Integration over D gives the number of infections of the person at A due to all the infected people in D. In rectangular coordinates (with the origin at the city's center), the exposure of a person at A is

$$E = \iint\limits_D kf(P, A)dA = k\iint\limits_D \frac{20 - d(P, A)}{20}\,dA = k\iint\limits_D \left[1 - \frac{\sqrt{(x - x_0)^2 + (y - y_0)^2}}{20}\right] dx\,dy.$$

(b) If $A = (0, 0)$, then

$$E = k\iint\limits_D \left[1 - \tfrac{1}{20}\sqrt{x^2 + y^2}\right] dx\,dy = k\int_0^{2\pi}\int_0^{10}\left(1 - \frac{r}{20}\right) r\,dr\,d\theta$$

$$= 2\pi k\left[\frac{r^2}{2} - \frac{r^3}{60}\right]_0^{10} = 2\pi k\left(50 - \tfrac{50}{3}\right) = \tfrac{200}{3}\pi k \approx 209k.$$

For A at the edge of the city, it is convenient to use a polar coordinate system centered at A. Then the polar equation for the circular boundary of the city becomes $r = 20\cos\theta$ instead of $r = 10$, and the distance from A to a point P in the city is again r (see the figure.) So

$$E = k\int_{-\pi/2}^{\pi/2}\int_0^{20\cos\theta}\left(1 - \frac{r}{20}\right) r\,dr\,d\theta$$

$$= k\int_{-\pi/2}^{\pi/2}\left[\frac{r^2}{2} - \frac{r^3}{60}\right]_0^{20\cos\theta} d\theta$$

$$= k\int_{-\pi/2}^{\pi/2}\left(200\cos^2\theta - \tfrac{400}{3}\cos^3\theta\right)d\theta$$

$$= 200k\int_{-\pi/2}^{\pi/2}\left[\tfrac{1}{2} + \tfrac{1}{2}\cos 2\theta - \tfrac{2}{3}(1 - \sin^2\theta)\cos\theta\right]d\theta$$

$$= 200k\left[\tfrac{1}{2}\theta + \tfrac{1}{4}\sin 2\theta - \tfrac{2}{3}\sin\theta + \tfrac{2}{3}\cdot\tfrac{1}{3}\sin^3\theta\right]_{-\pi/2}^{\pi/2}$$

$$= 200k\left[\frac{\pi}{4} + 0 - \frac{2}{3} + \frac{2}{9} + \frac{\pi}{4} + 0 - \frac{2}{3} + \frac{2}{9}\right] = 200k\left(\frac{\pi}{2} - \frac{8}{9}\right) \approx 136k.$$

Therefore the risk of infection is much lower at the edge of the city than in the middle, so it is better to live at the edge.

5. **(a)** The mountain comprises a solid conical region C. The work done in lifting a small volume of material ΔV with density $g(P)$ to a height $h(P)$ above sea level is $h(P)g(P)\Delta V$. Summing over the whole mountain we get $W = \iiint_C h(P)g(P)dV$.

(b) Here C is a solid right circular cone with radius $R = 62{,}000$ ft, height $H = 12{,}400$ ft, and density $g(P) = 200\ \text{lb}/\text{ft}^3$ at all points P in C. We use cylindrical coordinates:

$$\begin{aligned}
W &= \int_0^{2\pi}\int_0^H\int_0^{R(1-z/H)} z\cdot 200\, r\,dr\,dz\,d\theta \\
&= 2\pi\int_0^H 200z\left[\tfrac{1}{2}r^2\right]_0^{R(1-z/H)} dz \\
&= 400\pi\int_0^H z\frac{R^2}{2}\left(1-\frac{z}{H}\right)^2 dz \\
&= 200\pi R^2\int_0^H\left(z-\frac{2z^2}{H}+\frac{z^3}{H^2}\right)dz \\
&= 200\pi R^2\left[\frac{z^2}{2}-\frac{2z^3}{3H}+\frac{z^4}{4H^2}\right]_0^H \\
&= 200\pi R^2\left(\frac{H^2}{2}-\frac{2H^2}{3}+\frac{H^2}{4}\right) = \tfrac{50}{3}\pi R^2H^2 = \tfrac{50}{3}\pi(62{,}000)^2(12{,}400)^2 \approx 3.1\times 10^{19}\ \text{ft-lb.}
\end{aligned}$$

$$\frac{r}{R} = \frac{H-z}{H} = 1-\frac{z}{H}$$

7. (a) $mgh = \frac{1}{2}mv^2 + \frac{1}{2}I\omega^2 = \frac{1}{2}(m + I/r^2)v^2$, so

$$v^2 = \frac{2mgh}{m+I/r^2} = \frac{2gh}{1+I^*}.$$

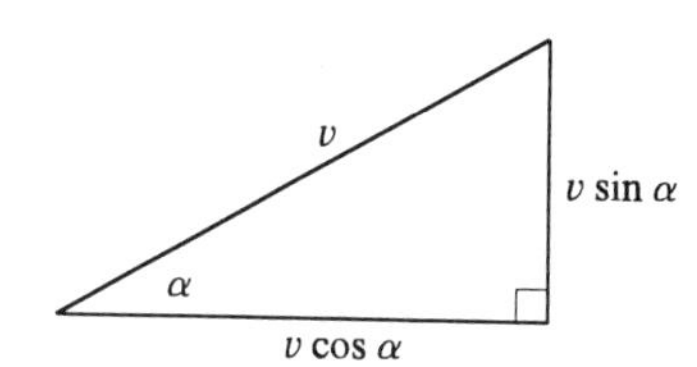

(b) The vertical component of the speed is $v\sin\alpha$, so

$$\frac{dy}{dt} = \sqrt{\frac{2gy}{1+I^*}}\sin\alpha = \sqrt{\frac{2g}{1+I^*}}\sin\alpha\,\sqrt{y}.$$

(c) Solving the separable differential equation, we get $\dfrac{dy}{\sqrt{y}} = \sqrt{\dfrac{2g}{1+I^*}}\sin\alpha\,dt \quad\Rightarrow$

$2\sqrt{y} = \sqrt{\dfrac{2g}{1+I^*}}(\sin\alpha)t + C$. But $y = 0$ when $t = 0$, so $C = 0$ and we have $2\sqrt{y} = \sqrt{\dfrac{2g}{1+I^*}}(\sin\alpha)t$.

Solving for t when $y = h$ gives $T = \dfrac{2\sqrt{h}}{\sin\alpha}\sqrt{\dfrac{1+I^*}{2g}} = \sqrt{\dfrac{2h(1+I^*)}{g\sin^2\alpha}}$.

(d) Assume that the length of each cylinder is ℓ. Then the density of the solid cylinder is $\dfrac{m}{\pi r^2\ell}$, and from (13.8.16), its moment of inertia (using cylindrical coordinates) is

$$I_z = \iiint \frac{m}{\pi r^2\ell}(x^2+y^2)\,dV = \int_0^\ell\int_0^{2\pi}\int_0^r \frac{m}{\pi r^2\ell}R^2R\,dR\,d\theta\,dz = \frac{m}{\pi r^2\ell}2\pi\ell\left[\tfrac{1}{4}R^4\right]_0^r = \frac{mr^2}{2},\ \text{and so}$$

$$I^* = \frac{I_z}{mr^2} = \frac{1}{2}.$$

For the hollow cylinder, we consider its entire mass to lie a distance r from the axis of rotation, so $x^2 + y^2 = r^2$ is a constant. We express the density in terms of mass per unit area as $\rho = \dfrac{m}{2\pi r\ell}$, and then the moment of inertia is calculated as a double integral:

$$I_z = \iint (x^2+y^2)\frac{m}{2\pi r\ell}\,dA = \frac{mr^2}{2\pi r\ell}\iint dA = mr^2,\ \text{so}\ I^* = \frac{I_z}{mr^2} = 1.$$

(e), (f) Before considering the specific cases, we calculate I^* for a partly hollow ball with inner radius a and outer radius r. The volume of such a ball is $\frac{4}{3}\pi(r^3 - a^3) = \frac{4}{3}\pi r^3(1 - b^3)$, and so its density is $\dfrac{m}{\frac{4}{3}\pi r^3(1-b^3)}$.

Using Formula 13.10.4, we get

$$I_z = \iiint (x^2 + y^2)\frac{m}{\frac{4}{3}\pi r^3(1-b^3)}\,dV = \frac{m}{\frac{4}{3}\pi r^3(1-b^3)}\int_a^r \int_0^{2\pi} \int_0^{\pi} (\rho^2 \sin^2\phi)(\rho^2 \sin\phi)\,d\phi\,d\theta\,d\rho$$

$$= \frac{m}{\frac{4}{3}\pi r^3(1-b^3)} \cdot 2\pi\left[-\frac{(2+\sin^2\phi)\cos\phi}{3}\right]_0^{\pi}\left[\frac{\rho^5}{5}\right]_a^r \quad \text{(from the Table of Integrals)}$$

$$= \frac{m}{\frac{4}{3}\pi r^3(1-b^3)} \cdot 2\pi \cdot \frac{4}{3} \cdot \frac{r^5 - a^5}{5} = \frac{2mr^5(1-b^5)}{5r^3(1-b^3)} = \frac{2(1-b^5)mr^2}{5(1-b^3)}$$

Therefore $I^* = \dfrac{2(1-b^5)}{5(1-b^3)}$. Now for the solid ball, we let $a \to 0$, so $b \to 0$ and $I^* \to \dfrac{2}{5}$. For the hollow ball, we let $a \to r$, so $b \to 1$ and we use l'Hospital's Rule: $\lim\limits_{b\to 1} I^* = \dfrac{2}{5}\lim\limits_{b\to 1}\dfrac{-5b^4}{-3b^2} = \dfrac{2}{3}$.

Note: We could instead have calculated $\lim\limits_{b\to 1} I^* = \lim\limits_{b\to 1}\dfrac{2(1-b)(1+b+b^2+b^3+b^4)}{5(1-b)(1+b+b^2)} = \dfrac{2\cdot 5}{5\cdot 3} = \dfrac{2}{3}$.

CHAPTER FOURTEEN

EXERCISES 14.1

1. $\mathbf{F}(x, y) = x\mathbf{i} + y\mathbf{j}$

The length of the vector $x\mathbf{i} + y\mathbf{j}$ is the distance from $(0, 0)$ to (x, y). Flow lines are rays emanating from the origin.

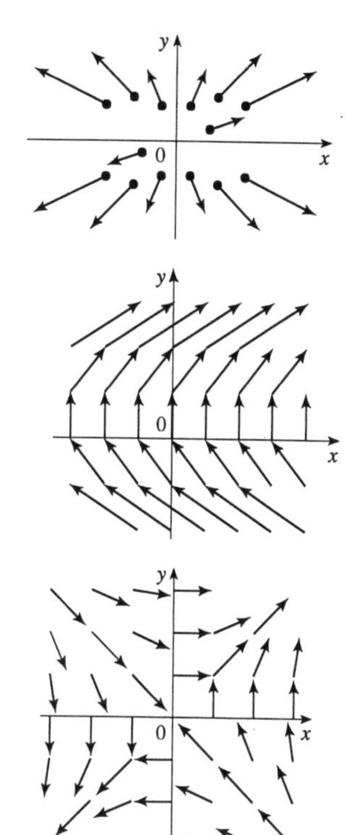

3. $\mathbf{F}(x, y) = y\mathbf{i} + \mathbf{j}$

The length of the vector $y\mathbf{i} + \mathbf{j}$ is $\sqrt{y^2 + 1}$. Flow lines are parabolas opening about the x-axis.

5. $\mathbf{F}(x, y) = \dfrac{y\mathbf{i} + x\mathbf{j}}{\sqrt{x^2 + y^2}}$

The length of the vector $\dfrac{y\mathbf{i} + x\mathbf{j}}{\sqrt{x^2 + y^2}}$ is 1.

7. $\mathbf{F}(x, y, z) = \mathbf{j}$

All vectors in this field are parallel to the y-axis.

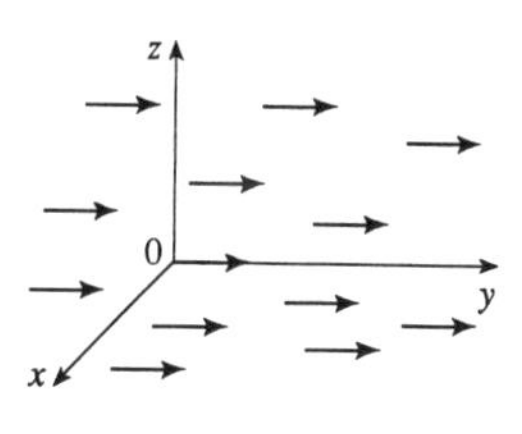

9. $\mathbf{F}(x, y, z) = y\mathbf{j}$

The length of $\mathbf{F}(x, y, z)$ is $|y|$. No vectors emanate from the xz-plane since $y = 0$ there. In each plane $y = b$, all the vectors are identical.

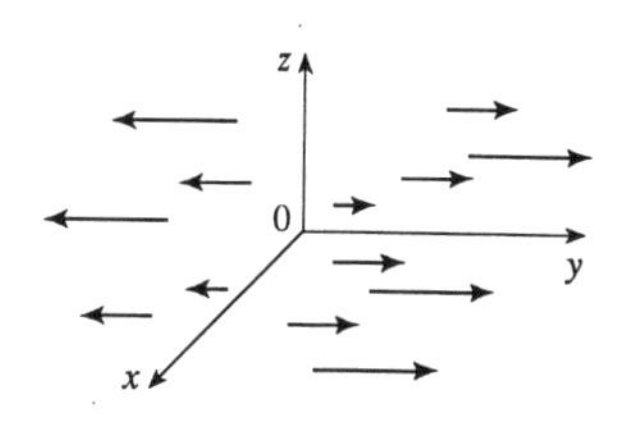

11. $\mathbf{F}(x, y) = \langle y, x \rangle$ corresponds to graph III, since in the first quadrant all the vectors have positive x- and y-components, in the second quadrant all vectors have positive x-components and negative y-components, in the third quadrant all vectors have negative x- and y-components, and in the fourth quadrant all vectors have negative x-components and positive y-components.

13. $\mathbf{F}(x, y) = \langle \sin x, \sin y \rangle$ corresponds to graph II, since the vector field is the same on each square of the form $[2n\pi, 2(n+1)\pi] \times [2m\pi, 2(m+1)\pi]$, m, n any integers.

15. The vector field seems to have very short vectors near the line $y = 2x$. For $\mathbf{F}(x, y) = \langle 0, 0 \rangle$ we must have $y^2 - 2xy = 0$ and $3xy - 6x^2 = 0$. The first equation holds if $y = 0$ or $y = 2x$, and the second holds if $x = 0$ or $y = 2x$. So both equations hold $\left[\text{and thus } \mathbf{F}(x, y) = \mathbf{0}\right]$ along the line $y = 2x$.

17. $\nabla f(x, y) = f_x(x, y)\mathbf{i} + f_y(x, y)\mathbf{j} = (5x^4 - 8xy^3)\mathbf{i} - (12x^2y^2)\mathbf{j}$

19. $\nabla f(x, y) = \langle f_x, f_y \rangle = \langle 3e^{3x} \cos 4y, -4e^{3x} \sin 4y \rangle$

21. $\nabla f(x, y, z) = \langle f_x, f_y, f_z \rangle = \langle y^2, 2xy - z^3, -3yz^2 \rangle$

23. $f(x, y) = x^2 - \frac{1}{2}y^2$, $\nabla f(x, y) = 2x\mathbf{i} - y\mathbf{j}$

The length of $\nabla f(x, y)$ is $\sqrt{4x^2 + y^2}$, and $\nabla f(x, y)$ terminates on the x-axis at the point $(3x, 0)$.

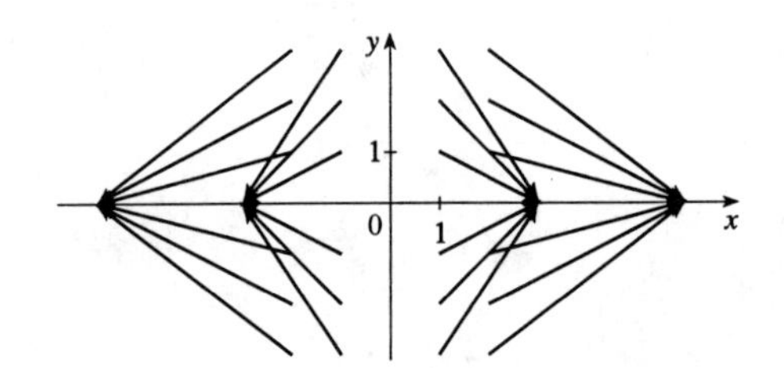

25. **(a)** The flow lines appear to be hyperbolas with equations $y = C/x$.

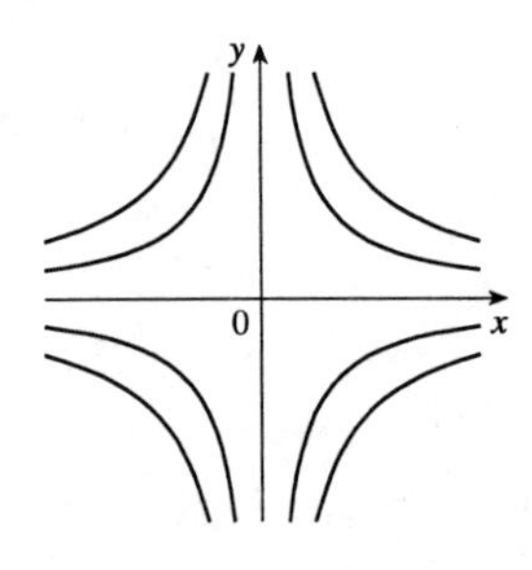

(b) $dx/dt = x \quad \Rightarrow \quad dx/x = dt \quad \Rightarrow$

$\ln|x| = t + C \quad \Rightarrow \quad x = \pm e^{t+C} = Ae^t$

for some constant A.

$dy/dt = -y \quad \Rightarrow \quad dy/y = -dt \quad \Rightarrow$

$\ln|y| = -t + K \quad \Rightarrow \quad y = \pm e^{-t+K} = Be^{-t}$

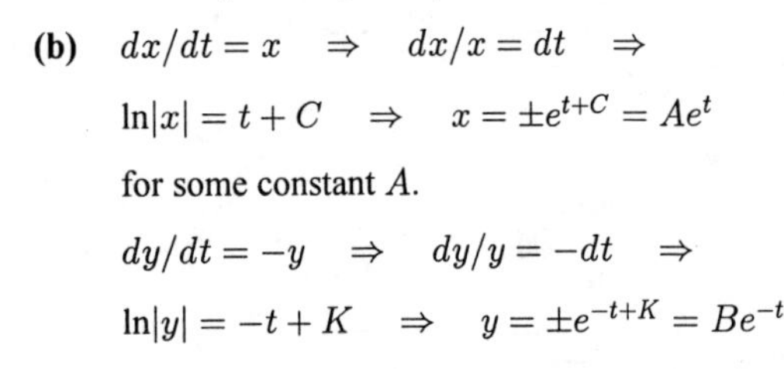

for some constant B. Therefore $xy = Ae^tBe^{-t} = AB = \text{constant}$. If the flow line passes through $(1, 1)$ then $(1)(1) = \text{constant} = 1 \quad \Rightarrow \quad xy = 1 \quad \Rightarrow \quad y = \dfrac{1}{x}$, $x > 0$.

EXERCISES 14.2

1. $\int_C x\,ds = \int_0^1 (t^3)\sqrt{9t^4+1}\,dt = \frac{1}{54}(9t^4+1)^{3/2}\Big|_0^1 = \frac{1}{54}\left(10^{3/2}-1\right)$

3. $x = 4\cos t,\ y = 4\sin t,\ -\frac{\pi}{2} \le t \le \frac{\pi}{2}.$

$$\int_C xy^4\,ds = \int_{-\pi/2}^{\pi/2} \left[(4)^5 \cos t \sin^4 t\right](4)dt = (4)^6\left[\tfrac{1}{5}\sin^5 t\right]_{-\pi/2}^{\pi/2} = \frac{2\cdot 4^6}{5} = 1638.4$$

5. $x = x,\ y = x^2,\ -2 \le x \le 1$. Then

$\int_C (x-2y^2)dy = \int_{-2}^1 (x-2x^4)2x\,dx = \int_{-2}^1 (2x^2-4x^5)dx = \frac{2}{3}[x^3-x^6]_{-2}^1 = 48.$

7. $C = C_1 + C_2$

On C_1: $x = x,\ y = 0,\ 0 \le x \le 2.$

On C_2: $x = x,\ y = 2x-4,\ 2 \le x \le 3$. Then

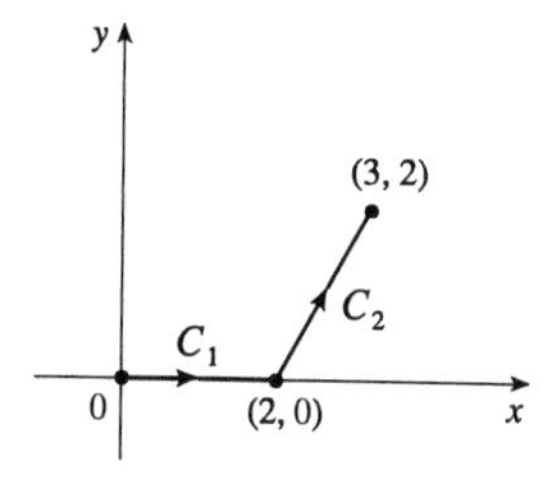

$\int_C xy\,dx + (x-y)dy$

$= \int_{C_1} xy\,dx + (x-y)dy + \int_{C_2} xy\,dx + (x-y)dy$

$= \int_0^2 0\,dx + \int_2^3 [(2x^2-4x) + (-x+4)(2)]dx$

$= \int_2^3 (2x^2-6x+8)dx = \frac{17}{3}.$

9. $\int_C xyz\,ds = \int_0^{\pi/2} (18t\sin t\cos t)\sqrt{4+9}\,dt = 18\sqrt{13}\int_0^{\pi/2} (t\sin t\cos t)dt$

$= 18\sqrt{13}\int_0^{\pi/2} \frac{1}{2}t\sin 2t\,dt = 9\sqrt{13}\left[-\frac{1}{2}t\cos 2t + \frac{1}{4}\sin 2t\right]_0^{\pi/2} = \frac{9\sqrt{13}}{4}\pi$

11. $x = -t+1,\ y = 3t,\ z = 5t+1,\ 0 \le t \le 1.$

$\int_C xy^2z\,ds = \int_0^1 (1-t)(9t^2)(5t+1)\sqrt{1+9+25}\,dt = 9\sqrt{35}\int_0^1 (t^2+4t^3-5t^4)dt = 3\sqrt{35}$

13. $\int_C x^3y^2\,dz = \int_0^1 (8t^3)(t^4)(t^2)(2t)dt = \int_0^1 16t^{10}\,dt = \frac{16}{11}$

15. On C_1: $x = 0,\ y = t,\ z = t,\ 0 \le t \le 1,$

C_2: $x = t,\ y = t+1,\ z = 2t+1,\ 0 \le t \le 1,$

C_3: $x = 1,\ y = 2,\ z = t+3,\ 0 \le t \le 1$. Then

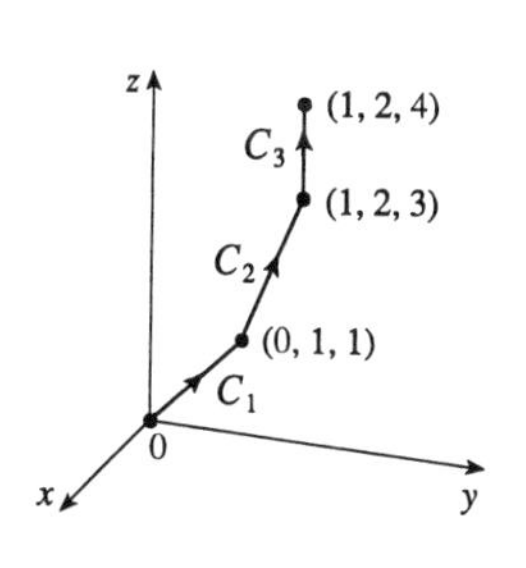

$\int_C z^2\,dx - z\,dy + 2y\,dz$

$= \int_0^1 (0-t+2t)dt + \int_0^1 \left[(2t+1)^2 - (2t+1) + 2(t+1)(2)\right]dt$

$+ \int_0^1 (0+0+4)dt$

$= \frac{1}{2} + \left[\frac{4}{3}t^3 + 3t^2 + 4t\right]_0^1 + 4 = \frac{77}{6}$

17. $\mathbf{F}(\mathbf{r}(t)) = t^{10}\mathbf{i} - t^7\mathbf{j},\ \mathbf{r}'(t) = 3t^2\mathbf{i} + 4t^3\mathbf{j}.\ \int_C \mathbf{F}\cdot d\mathbf{r} = \int_0^1 (3t^{12}-4t^{10})dt = \frac{3}{13} - \frac{4}{11} = -\frac{19}{143}$

19. $\int_C \mathbf{F}\cdot d\mathbf{r} = \int_0^1 \langle \sin t^3, \cos(-t^2), t^4\rangle \cdot \langle 3t^2, -2t, 1\rangle dt$

$= \int_0^1 (3t^2\sin t^3 - 2t\cos t^2 + t^4)dt = \left[-\cos t^3 - \sin t^2 + \frac{1}{5}t^5\right]_0^1 = \frac{6}{5} - \cos 1 - \sin 1$

21. **(a)** $\int_C \mathbf{F} \cdot d\mathbf{r} = \int_0^1 \left\langle e^{t^2-1}, t^5 \right\rangle \cdot \langle 2t, 3t^2 \rangle dt = \int_0^1 \left(2te^{t^2-1} + 3t^7 \right) dt$

$= \left[e^{t^2-1} + \frac{3}{8}t^8 \right]_0^1 = \frac{11}{8} - 1/e$

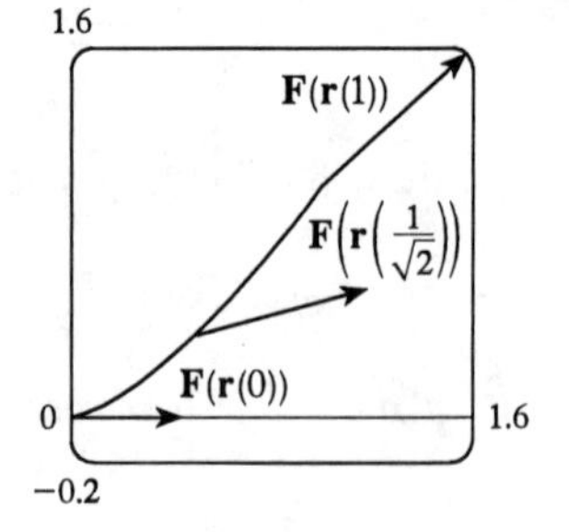

(b) $\mathbf{r}(0) = \mathbf{0}$, $\mathbf{F}(\mathbf{r}(0)) = \langle e^{-1}, 0 \rangle$;

$\mathbf{r}\left(\frac{1}{\sqrt{2}}\right) = \left\langle \frac{1}{2}, \frac{1}{2\sqrt{2}} \right\rangle$, $\mathbf{F}\left(\mathbf{r}\left(\frac{1}{\sqrt{2}}\right)\right) = \left\langle e^{-1/2}, \frac{1}{4\sqrt{2}} \right\rangle$;

$\mathbf{r}(1) = \langle 1, 1 \rangle$, $\mathbf{F}(\mathbf{r}(1)) = \langle 1, 1 \rangle$.

In order to generate the graph with Maple, we use the `PLOT` command (not to be confused with the `plot` command) to define each of the vectors. For example,

`v1:=PLOT(CURVES([[0,0],[evalf(1/E),0]]));` generates the vector from the vector field at the point $(0, 0)$ (but without an arrowhead) and gives it the name `v1`. To show everything on the same screen, we use the `display` command.

In Mathematica, we use `ListPlot` (with the `PlotJoined -> True` option) to generate the vectors, and then `Show` to show everything on the same screen.

23. A calculator or CAS gives $\int_C x \sin y\, ds = \int_1^2 \ln t \sin(e^{-t}) \sqrt{(1/t)^2 + (-e^{-t})^2}\, dt \approx 0.052$.

25. The part of the astroid that lies in the quadrant is parametrized by $x = \cos^3 t$, $y = \sin^3 t$, $0 \le t \le \frac{\pi}{2}$. Now $\dfrac{dx}{dt} = 3\cos^2 t\,(-\sin t)$ and $\dfrac{dy}{dt} = 3\sin^2 t \cos t$, so

$\sqrt{(dx/dt)^2 + (dy/dt)^2} = \sqrt{9\cos^4 t \sin^2 t + 9\sin^4 t \cos^2 t} = 3\cos t \sin t \sqrt{\cos^2 t + \sin^2 t} = 3\cos t \sin t$.

Therefore $\int_C x^3 y^5\, ds = \int_0^{\pi/2} \cos^9 t \sin^{15} t\,(3\cos t \sin t)dt = \frac{945}{16{,}777{,}216}\pi$.

27. **(a)** Along the line $x = -3$, the vectors of $\mathbf{F}$ have positive y-components, so since the path goes upward, the integrand $\mathbf{F} \cdot \mathbf{T}$ is always positive. Therefore $\int_{C_1} \mathbf{F} \cdot d\mathbf{r}$ is positive.

(b) All of the (nonzero) field vectors along the circle with radius 3 are pointed in the clockwise direction, that is, opposite the direction to the path. So $\mathbf{F} \cdot \mathbf{T}$ is negative, and therefore $\int_{C_2} \mathbf{F} \cdot d\mathbf{r}$ is negative.

29. $x = 4\cos t$, $y = 4\sin t$, $-\frac{\pi}{2} \le t \le \frac{\pi}{2}$, $m = \int_C k\, ds = k\int_{-\pi/2}^{\pi/2} \sqrt{4\cos^2 t + 4\sin^2 t}\, dt = 2k(\pi)$,

$\overline{x} = \dfrac{1}{2\pi k} \displaystyle\int_C xk\, dS = \dfrac{1}{2\pi} \int_{-\pi/2}^{\pi/2} 4\cos t\, dt = \dfrac{1}{2\pi}[4\sin t]_{-\pi/2}^{\pi/2} = \dfrac{4}{\pi}$, $\overline{y} = \dfrac{1}{2\pi k} \displaystyle\int_C yk\, dS = \dfrac{1}{2\pi} \int_{-\pi/2}^{\pi/2} 4\sin t\, dt = 0$.

Hence $(\overline{x}, \overline{y}) = \left(\frac{4}{\pi}, 0\right)$.

31. **(a)** $\overline{x} = \dfrac{1}{m}\int_C x\rho(x,y,z)ds$, $\overline{y} = \dfrac{1}{m}\int_C y\rho(x,y,z)ds$, $\overline{z} = \dfrac{1}{m}\int_C z\rho(x,y,z)ds$ where $m = \int_C \rho(x,y,z)ds$.

(b) $m = \int_C k\,ds = k\int_0^{2\pi}\sqrt{4\sin^2 t + 4\cos^2 t + 9}\,dt = k\sqrt{13}\int_0^{2\pi}dt = 2\pi k\sqrt{13}$,

$\overline{x} = \dfrac{1}{2\pi k\sqrt{13}}\displaystyle\int_0^{2\pi} k2\sqrt{13}\sin t\,dt = 0$, $\overline{y} = \dfrac{1}{2\pi k\sqrt{13}}\displaystyle\int_0^{2\pi} k2\sqrt{13}\cos t\,dt = 0$,

$\overline{z} = \dfrac{1}{2\pi k\sqrt{13}}\displaystyle\int_0^{2\pi}\left(k\sqrt{13}\right)(3t)dt = \dfrac{3}{2\pi}\left(2\pi^2\right) = 3\pi$. Hence $(\overline{x},\overline{y},\overline{z}) = (0,0,3\pi)$.

33. From Example 3, $\rho(x,y) = k(1-y)$, $x = \cos t$, $y = \sin t$, and $ds = dt$, $0 \le t \le \pi \quad \Rightarrow$

$$I_x = \int_C y^2\rho(x,y)ds = \int_0^{\pi}\sin^2 t[k(1-\sin t)]dt = k\int_0^{\pi}\left(\sin^2 t - \sin^3 t\right)dt$$

$$= \frac{k}{2}\int_0^{\pi}(1-\cos 2t)dt - k\int_0^{\pi}\left(1-\cos^2 t\right)\sin t\,dt \text{ (let } u = \cos t,\ du = -\sin t \text{ in the second integral)}$$

$$= k\left[\frac{\pi}{2} + \int_1^{-1}\left(1-u^2\right)du\right] = k\left(\frac{\pi}{2} - \frac{4}{3}\right).$$

$I_y = \displaystyle\int_C x^2\rho(x,y)ds = k\int_0^{\pi}\cos^2 t(1-\sin t)dt = \frac{k}{2}\int_0^{\pi}(1+\cos 2t)dt - k\int_0^{\pi}\cos^2 t\sin t\,dt = k\left[\frac{\pi}{2} - \frac{2}{3}\right]$, using the same substitution as above.

35. $W = \int_C \mathbf{F}\cdot d\mathbf{r} = \int_0^{2\pi}\langle t - \sin t, 3 - \cos t\rangle \cdot \langle 1 - \cos t, \sin t\rangle dt$

$= \int_0^{2\pi}(t - t\cos t - \sin t + \sin t\cos t + 3\sin t - \sin t\cos t)dt = 2\pi^2$

37. $W = \int_0^1\langle t^6, -t^5, -t^7\rangle \cdot \langle 2t, -3t^2, 4t^3\rangle dt = \int_0^1(5t^7 - 4t^{10})dt = \frac{5}{8} - \frac{4}{11} = \frac{23}{88}$

39. Let $\mathbf{F} = 185\mathbf{k}$. To parametrize the staircase, let $x = 20\cos t$, $y = 20\sin t$, $z = \frac{90}{6\pi}t = \frac{15}{\pi}t$, $0 \le t \le 6\pi \quad \Rightarrow$

$W = \int_C \mathbf{F}\cdot d\mathbf{r} = \int_0^{6\pi}\langle 0,0,185\rangle \cdot \langle -20\sin t, 20\cos t, \frac{15}{\pi}\rangle dt = (185)\frac{15}{\pi}\int_0^{6\pi}dt = (185)(90) \approx 1.67 \times 10^4$ ft-lb.

41. Use the orientation pictured in the figure. Then since $\mathbf{B}$ is tangent to any circle that lies in the plane perpendicular to the wire, $\mathbf{B} = |\mathbf{B}|\mathbf{T}$ where $\mathbf{T}$ is the unit tangent to the circle C: $x = r\cos\theta$, $y = r\sin\theta$. Thus $\mathbf{B} = |\mathbf{B}|\langle -\sin\theta, \cos\theta\rangle$. Then

$\int_C \mathbf{B}\cdot d\mathbf{r} = \int_0^{2\pi}|\mathbf{B}|\langle -\sin\theta, \cos\theta\rangle \cdot \langle -r\sin\theta, r\cos\theta\rangle d\theta = \int_0^{2\pi}|\mathbf{B}|r\,d\theta = 2\pi r|\mathbf{B}|$. (Note that $|\mathbf{B}|$ here is the magnitude of the field at a distance r from the wire's center.) But by Ampere's Law $\int_C \mathbf{B}\cdot d\mathbf{r} = \mu_0 I$. Hence $|\mathbf{B}| = \mu_0 I/(2\pi r)$.

EXERCISES 14.3

1. $\partial(2x-3y)/\partial y = -3 = \partial(2y-3x)/\partial x$ and the domain of **F** is $\mathbb{R}^2$ which is open and simply-connected, so **F** is conservative. Thus there exists f such that $\nabla f = \mathbf{F}$, that is, $f_x(x,y) = 2x-3y$ and $f_y(x,y) = 2y-3x$. But $f_x(x,y) = 2x-3y$ implies $f(x,y) = x^2 - 3yx + g(y)$ and differentiating both sides of this equation with respect to y gives $f_y(x,y) = -3x + g'(y)$. Thus $2y - 3x = -3x + g'(y)$ so $g'(y) = 2y$ and $g(y) = y^2 + K$ where K is a constant. Hence $f(x,y) = x^2 - 3xy + y^2 + K$ is a potential for **F**.

3. $\partial(x^2+y)/\partial y = 1$, $\partial(x^2)/\partial x = 2x$ and these are not equal, so **F** is not conservative.

5. $\partial(1+4x^3y^3)/\partial y = 12x^3y^2 = \partial(3x^4y^2)/\partial x$ and the domain of **F** is $\mathbb{R}^2$ which is open and simply-connected. Thus **F** is conservative so there exists f such that $\nabla f = \mathbf{F}$. Then $f_x(x,y) = 1 + 4x^3y^3$ implies $f(x,y) = x + x^4y^3 + g(y)$ and $f_y(x,y) = 3x^4y^3 + g'(y)$. But $f_y(x,y) = 3x^4y^2$ implies $g(y) = K$. Hence a potential for **F** is $f(x,y) = x + x^4y^3 + K$.

7. $\partial(e^{2x} + x\sin y)/\partial y = x\cos y$, $\partial(x^2\cos y)/\partial x = 2x\cos y$, so **F** is not conservative.

9. $\partial(ye^x + \sin y)/\partial y = e^x + \cos y = \partial(e^x + x\cos y)/\partial x$ and the domain of **F** is $\mathbb{R}^2$. Hence **F** is conservative so there exists f such that $\nabla f = \mathbf{F}$. Then $f_x(x,y) = ye^x + \sin y$ implies $f(x,y) = ye^x + x\sin y + g(y)$ and $f_y(x,y) = e^x + x\cos y + g'(y)$. But $f_y(x,y) = e^x + x\cos y$ so $g(y) = K$ and $f(x,y) = ye^x + x\sin y + K$ is a potential for **F**.

11. **(a)** $f_x(x,y) = x$ implies $f(x,y) = \frac{1}{2}x^2 + g(y)$ and $f_y(x,y) = g'(y)$. But $f_y(x,y) = y$ so $g(y) = \frac{1}{2}y^2 + K$ and $f(x,y) = \frac{1}{2}x^2 + \frac{1}{2}y^2 + K$ (or set $K = 0$.)

(b) $\int_C \mathbf{F}\cdot d\mathbf{r} = f(3,9) - f(-1,1) = 44$

13. **(a)** $f_x(x,y) = 2xy^3$ implies $f(x,y) = x^2y^3 + g(y)$ and $f_y(x,y) = 3x^2y^2 + g'(y)$. But $f_y(x,y) = 3x^2y^2$ so $f(x,y) = x^2y^3$ (setting $K = 0$).

(b) Since $\mathbf{r}(0) = \langle 0,1\rangle$ and $\mathbf{r}(\frac{\pi}{2}) = \langle 1, \frac{1}{4}(\pi^2+4)\rangle$, $\int_C \mathbf{F}\cdot d\mathbf{r} = f\left(1, \frac{1}{4}(\pi^2+4)\right) - f(0,1) = \frac{1}{64}(\pi^2+4)^3$.

15. **(a)** $f_x(x,y,z) = y$ implies $f(x,y,z) = xy + g(y,z)$ and $f_y(x,y,z) = x + \partial g/\partial y$. But $f_y(x,y,z) = x + z$ so $\partial g/\partial y = z$ and $g(y,z) = yz + h(z)$. Thus $f(x,y,z) = xy + yz + h(z)$ and $f_z(x,y,z) = y + h'(z)$. But $f_z(x,y,z) = y$ so $h'(z) = 0$ or $h(z) = K$. Hence $f(x,y,z) = xy + yz$ (setting $K = 0$).

(b) $\int_C \mathbf{F}\cdot d\mathbf{r} = f(8,3,-1) - f(2,1,4) = 21 - 6 = 15$

17. **(a)** $f_x(x,y,z) = 2xz + \sin y$ implies $f(x,y,z) = x^2z + x\sin y + g(y,z)$ and $f_y(x,y,z) = x\cos y + g_y(y,z)$. But $f_y(x,y,z) = x\cos y$ so $g_y(y,z) = 0$ and $f(x,y,z) = x^2z + x\sin y + h(z)$. Thus $f_z(x,y,z) = x^2 + h'(z)$. But $f_z(x,y,z) = x^2$ so $h'(z) = 0$ and $f(x,y,z) = x^2z + x\sin y$ (setting $K = 0$).

(b) $\mathbf{r}(0) = \langle 1,0,0\rangle$, $\mathbf{r}(2\pi) = \langle 1,0,2\pi\rangle$. Thus $\int_C \mathbf{F}\cdot d\mathbf{r} = f(1,0,2\pi) - f(1,0,0) = 2\pi$.

19. Here $\mathbf{F}(x,y) = (2x\sin y)\mathbf{i} + (x^2\cos y - 3y^2)\mathbf{j}$. Then $f(x,y) = x^2\sin y - y^3$ is a potential for $\mathbf{F}$, that is, $\nabla f = \mathbf{F}$ so $\mathbf{F}$ is conservative and thus its line integral is independent of path. Hence $\int_C 2x\sin y\,dx + (x^2\cos y - 3y^2)dy = \int_C \mathbf{F}\cdot d\mathbf{r} = f(5,1) - f(-1,0) = 25\sin 1 - 1$.

21. Here $\mathbf{F}(x,y) = x^2y^3\mathbf{i} + x^3y^2\mathbf{j}$. $W = \int_C \mathbf{F}\cdot d\mathbf{r}$. Since $\partial(x^2y^3)/\partial y = 3x^2y^2 = \partial(x^3y^2)/\partial x$, there exists f such that $\nabla f = \mathbf{F}$. In fact, $f_x = x^2y^3 \quad\Rightarrow\quad f(x,y) = \frac{1}{3}x^3y^3 + g(y) \quad\Rightarrow\quad f_y = x^3y^2 + g'(y) \quad\Rightarrow\quad g'(y) = 0$, so we can take $f(x,y) = \frac{1}{3}x^3y^3$. Thus $W = \int_C \mathbf{F}\cdot d\mathbf{r} = f(2,1) - f(0,0) = \frac{1}{3}(2^3)(1^3) - 0 = \frac{8}{3}$.

23. We know that if the vector field (call it $\mathbf{F}$) is conservative, then around any closed path C, $\int_C \mathbf{F}\cdot d\mathbf{r} = 0$. But take C to be some circle centered at the origin, oriented counterclockwise. All of the field vectors along C point "against" the direction of C (that is, within 90° of $-C$) so the integral around C will be negative. Therefore the field is not conservative.

25.

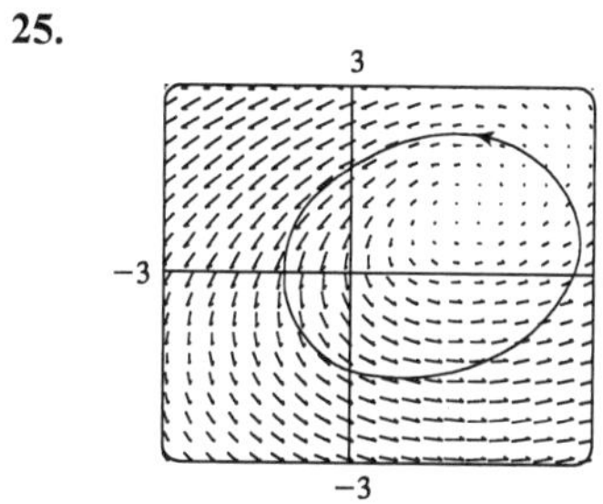

From the graph, it appears that $\mathbf{F}$ is not conservative. For example, any closed curve containing the point $(2,1)$ seems to have many field vectors pointing counterclockwise along it, and none pointing clockwise. So along this path the integral $\int \mathbf{F}\cdot d\mathbf{r} \neq 0$. To confirm our guess, we calculate

$$\frac{\partial}{\partial y}\left(\frac{x-2y}{\sqrt{1+x^2+y^2}}\right) = (x-2y)\left[\frac{-y}{(1+x^2+y^2)^{3/2}}\right] - \frac{2}{\sqrt{1+x^2+y^2}}$$

$$= \frac{-2-2x^2-xy}{(1+x^2+y^2)^{3/2}},\quad \frac{\partial}{\partial x}\left(\frac{x-2}{\sqrt{1+x^2+y^2}}\right) = (x-2)\left[\frac{-x}{(1+x^2+y^2)^{3/2}}\right] + \frac{1}{\sqrt{1+x^2+y^2}}$$

$$= \frac{1+y^2+2x}{(1+x^2+y^2)^{3/2}}.$$ These are not equal, so the field is not conservative, by Theorem 5.

27. Since $\mathbf{F}$ is conservative, there exists a function f such that $\mathbf{F} = \nabla f$, that is, $P = f_x$, $Q = f_y$, and $R = f_z$. Since P, Q and R have continuous first order partial derivatives, Clairaut's Theorem says $\partial P/\partial y = f_{xy} = f_{yx} = \partial Q/\partial x$, $\partial P/\partial z = f_{xz} = f_{zx} = \partial R/\partial x$, and $\partial Q/\partial z = f_{yz} = f_{zy} = \partial R/\partial y$.

29. $D = \{(x,y) \mid x > 0, y > 0\}$ = the first quadrant (excluding the axes).

(a) D is open because around every point in D we can put a disk that lies in D.

(b) D is connected because the straight line segment joining any two points in D lies in D.

(c) D is simply-connected because it's connected and has no holes.

31. $D = \{(x,y) \mid 1 < x^2 + y^2 < 4\}$ = the annular region between the circles with center $(0,0)$ and radii 1 and 2.

(a) D is open.

(b) D is connected.

(c) D is not simply-connected. For example, $x^2 + y^2 = (1.5)^2$ is simple and closed and lies within D but encloses points that are not in D. (Or, D has a hole, so is not simply-connected.)

33. **(a)** $P = -\dfrac{y}{x^2+y^2}$, $\dfrac{\partial P}{\partial y} = \dfrac{y^2-x^2}{(x^2+y^2)^2}$ and $Q = \dfrac{x}{x^2+y^2}$, $\dfrac{\partial Q}{\partial x} = \dfrac{y^2-x^2}{(x^2+y^2)^2}$. Thus $\dfrac{\partial P}{\partial y} = \dfrac{\partial Q}{\partial x}$.

(b) C_1: $x = \cos t$, $y = \sin t$, $0 \le t \le \pi$, C_2: $x = \cos t$, $y = \sin t$, $t = 2\pi$ to $t = \pi$. Then

$$\int_{C_1} \mathbf{F}\cdot d\mathbf{r} = \int_0^{\pi} \frac{(-\sin t)(-\sin t) + (\cos t)(\cos t)}{\cos^2 t + \sin^2 t}\,dt = \int_0^{\pi} dt = \pi \text{ and } \int_{C_2} \mathbf{F}\cdot d\mathbf{r} = \int_{2\pi}^{\pi} dt = -\pi.$$

Since these aren't equal, the line integral of $\mathbf{F}$ isn't independent of path. (Or notice that $\int_{C_3} \mathbf{F}\cdot d\mathbf{r} = \int_0^{2\pi} dt = 2\pi$ where C_3 is the circle $x^2+y^2=1$, and apply the contrapositive of Theorem 3.) This doesn't contradict Theorem 6, since the domain of $\mathbf{F}$, which is $\mathbb{R}^2$ except the origin, isn't simply-connected.

EXERCISES 14.4

1. **(a)**

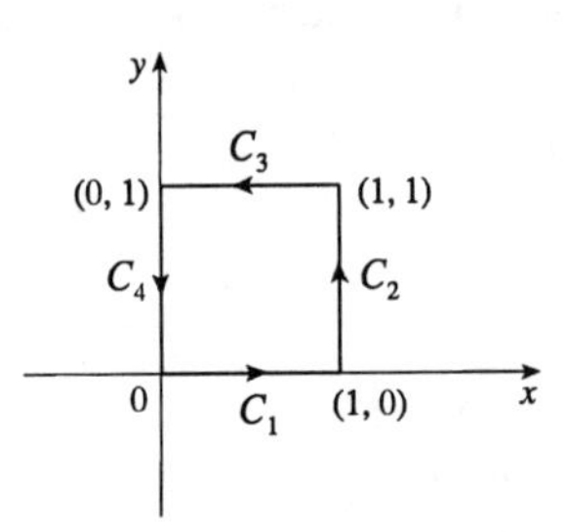

$$\begin{aligned}&\oint_C x^2y\,dx + xy^3\,dy\\ &= \oint_{C_1+C_2+C_3+C_4} x^2y\,dx + xy^3\,dy\\ &= \textstyle\int_0^1 0\,dx + \int_0^1 y^3\,dy + \int_1^0 x^2\,dx + \int_1^0 0\,dy\\ &= \tfrac14 - \tfrac13 = -\tfrac{1}{12}\end{aligned}$$

(b) $\oint_C x^2y\,dx + xy^3\,dy = \int_0^1\int_0^1 (y^3 - x^2)\,dx\,dy = \int_0^1 \left(y^3 - \frac13\right)dy = \frac14 - \frac13 = -\frac{1}{12}$

3. **(a)**

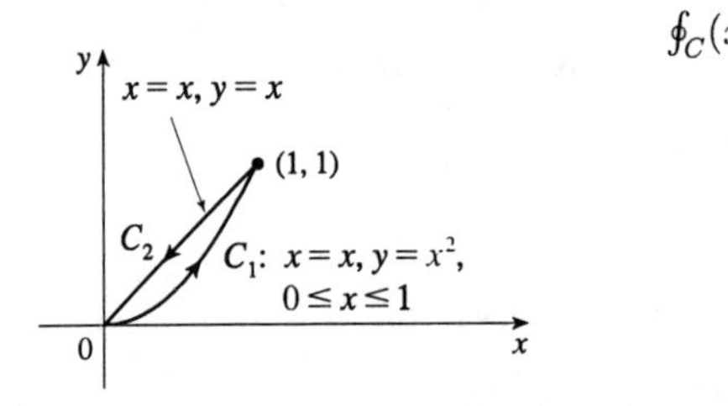

$$\begin{aligned}&\oint_C (x+2y)dx + (x-2y)dy\\ &= \oint_{C_1+C_2} (x+2y)dx + (x-2y)dy\\ &= \textstyle\int_0^1 [x + 2x^2 + (x-2x^2)(2x)]dx\\ &\qquad + \textstyle\int_1^0 [3x + (-x)]dx\\ &= \textstyle\int_0^1 (x + 4x^2 - 4x^3)dx + \int_1^0 2x\,dx\\ &= \left(\tfrac12 + \tfrac43 - 1\right) - 1 = -\tfrac16\end{aligned}$$

(b) $\oint_C (x+2y)dx + (x-2y)dy = \int_0^1\int_{x^2}^{x} (1-2)\,dy\,dx = \int_0^1 (x^2 - x)dx = \frac13 - \frac12 = -\frac16$

5. We can parametrize C as $x = \cos\theta$, $y = \sin\theta$, $0 \le \theta \le 2\pi$. Then the line integral is $\oint_C P\,dx + Q\,dy = \int_0^{2\pi} \cos^4\theta\sin^5\theta(-\sin\theta)d\theta + \int_0^{2\pi}(-\cos^7\theta\sin^6\theta)\cos\theta\,d\theta = -\frac{29\pi}{1024}$, according to a CAS. The double integral is $\iint_D \left(\dfrac{\partial Q}{\partial x} - \dfrac{\partial P}{\partial y}\right) dA = \int_{-1}^{1}\int_{-\sqrt{1-x^2}}^{\sqrt{1-x^2}} (-7x^6y^6 - 5x^4y^4)\,dy\,dx = -\frac{29\pi}{1024}$, verifying Green's Theorem in this case.

7. $\oint_C xy\,dx + y^5\,dy = \int_0^2\int_0^{x/2}(0-x)dy\,dx = \int_0^2\left(-\frac{1}{2}x^2\right)dx = -\frac{4}{3}$

9. $\int_0^1\int_{y^2}^{\sqrt{y}}(2-1)dx\,dy = \int_0^1\left(y^{1/2}-y^2\right)dy = \frac{1}{3}$

11. $\iint_D(0-0)dA = 0$

13. $\displaystyle\iint\limits_{\substack{0\le x^2+y^2\le 4\\ y\ge 0}}(4x-x)dA = 3\int_0^\pi\int_0^2 r^2\cos\theta\,dr\,d\theta = 0$ since $\int_0^\pi \cos\theta\,d\theta = 0$ or $\iint_D 3x\,dA = 3M_y = 0$.

15. $\iint_D(2x-x)dA = \int_0^\pi\int_0^{\sin x} x\,dy\,dx = \int_0^\pi x\sin x\,dx = [-x\cos x + \sin x]_0^\pi = \pi$

17. $\displaystyle\int_C \mathbf{F}\cdot d\mathbf{r} = \int_C(y^2-x^2y)dx + xy^2\,dy = \iint\limits_{\substack{x^2+y^2\le 4\\ 0\le y\le x}}(y^2-2y+x^2)dA = \int_0^{\pi/4}\int_0^2(r^2-2r\sin\theta)r\,dr\,d\theta$

$= \int_0^{\pi/4}\left[4-\frac{16}{3}\sin\theta\right]d\theta = \left[4\theta+\frac{16}{3}\cos\theta\right]_0^{\pi/4} = \pi + \frac{8}{3}\left(\sqrt{2}-2\right)$

19. By Green's Theorem, $W = \int_C \mathbf{F}\cdot d\mathbf{r} = \int_C x(x+y)dx + xy^2\,dy = \iint_D(y^2-x)dy\,dx$ where C is the path described in the question and D is the triangle bounded by C. So

$W = \int_0^1\int_0^{1-x}(y^2-x)dy\,dx = \int_0^1\left[\frac{1}{3}y^3-xy\right]_0^{1-x}dx = \int_0^1\left(\frac{1}{3}(1-x)^3 - x(1-x)\right)dx$

$= \left[-\frac{1}{12}(1-x)^4 - \frac{1}{2}x^2 + \frac{1}{3}x^3\right]_0^1 = \left(-\frac{1}{2}+\frac{1}{3}\right) - \left(-\frac{1}{12}\right) = -\frac{1}{12}.$

21. $A = \oint_C x\,dy = \int_0^{2\pi}(\cos^3 t)(3\sin^2 t\cos t)dt = 3\int_0^{2\pi}(\cos^4 t\sin^2 t)dt$

$= 3\left[-\frac{1}{6}(\sin t\cos^5 t) + \frac{1}{6}\left[\frac{1}{4}(\sin t\cos^3 t) + \frac{3}{8}(\cos t\sin t) + \frac{3}{8}t\right]\right]_0^{2\pi} = 3\left(\frac{1}{6}\right)\left(\frac{6}{8}\pi\right) = \frac{3}{8}\pi$

Or: $\int_0^{2\pi}(\cos^4 t\sin^2 t)dt = \int_0^{2\pi}\frac{1}{8}\left[\frac{1}{2}(1-\cos 4t) + \sin^2 2t\cos 2t\right]dt = \frac{\pi}{8}$

23. **(a)** Using Equation 14.2.8, we write parametric equations of the line segment as $x = (1-t)x_1 + tx_2$, $y = (1-t)y_1 + ty_2$, $0\le t\le 1$. Then $dx = (x_2-x_1)dt$ and $dy = (y_2-y_1)dt$, so

$\int_C x\,dy - y\,dx = \int_0^1 [(1-t)x_1 + tx_2](y_2-y_1)dt + [(1-t)y_1 + ty_2](x_2-x_1)dt$

$= \int_0^1(x_1(y_2-y_1) - y_1(x_2-x_1) + t[(y_2-y_1)(x_2-x_1) - (x_2-x_1)(y_2-y_1)])dt$

$= \int_0^1(x_1y_2 - x_2y_1)dt = x_1y_2 - x_2y_1.$

(b) We apply Green's Theorem to the path $C = C_1\cup C_2\cup\cdots\cup C_n$, where C_i is the line segment that joins (x_i, y_i) to (x_{i+1}, y_{i+1}) for $i = 1, 2, \ldots, n-1$, and C_n is the line segment that joins (x_n, y_n) to (x_1, y_1). From (6), $\frac{1}{2}\int_C x\,dy - y\,dx = \iint_D dA$, where D is the polygon bounded by C. Therefore

area of polygon $= A(D) = \iint_D dA = \frac{1}{2}\int_C x\,dy - y\,dx$

$= \frac{1}{2}\left(\int_{C_1} x\,dy - y\,dx + \int_{C_2} x\,dy - y\,dx + \cdots + \int_{C_{n-1}} x\,dy - y\,dx + \int_{C_n} x\,dy - y\,dx\right).$

To evaluate these integrals we use the formula from (a) to get

$A(D) = \frac{1}{2}[(x_1y_2 - x_2y_1) + (x_2y_3 - x_3y_2) + \cdots + (x_{n-1}y_n - x_ny_{n-1}) + (x_ny_1 - x_1y_n)].$

(c) $A = \frac{1}{2}[(0\cdot 1 - 2\cdot 0) + (2\cdot 3 - 1\cdot 1) + (1\cdot 2 - 0\cdot 3)$

$+ (0\cdot 1 - (-1)\cdot 2) + (-1\cdot 0 - 0\cdot 1)] = \frac{1}{2}(0+5+2+2) = \frac{9}{2}$

25. Here $A = \frac{1}{2}(1)(1) = \frac{1}{2}$ and $C = C_1 + C_2 + C_3$, where C_1: $x = x$, $y = 0$, $0 \le x \le 1$; C_2: $x = x$, $y = 1 - x$, $x = 1$ to $x = 0$; and C_3: $x = 0$, $y = 1$ to $y = 0$. Then

$\overline{x} = \frac{1}{2A}\int_C x^2\,dy = \int_{C_1} x^2\,dy + \int_{C_2} x^2\,dy + \int_{C_3} x^2\,dy = 0 + \int_1^0 (x^2)(-dx) + 0 = \frac{1}{3}$. Similarly,

$\overline{y} = -\frac{1}{2A}\int_C y^2\,dx = \int_{C_1} y^2\,dx + \int_{C_2} y^2\,dx + \int_{C_3} y^2\,dx = 0 + \int_1^0 (1-x)^2(-dx) + 0 = \frac{1}{3}$. Therefore $(\overline{x}, \overline{y}) = \left(\frac{1}{3}, \frac{1}{3}\right)$.

27. By Green's Theorem, $-\frac{1}{3}\rho\oint_C y^3\,dx = -\frac{1}{3}\rho\iint_D(-3y^2)dA = \iint_D y^2\rho\,dA = I_x$ and $\frac{1}{3}\rho\oint_C x^3\,dy = \frac{1}{3}\rho\iint_D(3x^2)dA = \iint_D x^2\rho\,dA = I_y$.

29. Since C is a simple closed path which doesn't pass through or enclose the origin, there exists an open region that doesn't contain the origin but does contain D. Thus $P = -\frac{y}{x^2+y^2}$ and $Q = \frac{x}{x^2+y^2}$ have continuous partials on this open region containing D and we can apply Green's Theorem. But by Exercise 14.3.33(a), $\frac{\partial P}{\partial y} = \frac{\partial Q}{\partial x}$, so $\oint_C \mathbf{F}\cdot d\mathbf{r} = \iint_D 0\,dA = 0$.

31. Using the first part of (6) we have that $\iint_R dx\,dy = A(R) = \int_{\partial R} x\,dy$. But $x = g(u,v)$, and $dy = \frac{\partial h}{\partial u}\,du + \frac{\partial h}{\partial v}\,dv$, and we orient ∂S by taking the positive direction to be that which corresponds, under the mapping, to the positive direction along ∂R, so

$$\int_{\partial R} x\,dy = \int_{\partial S} g(u,v)\left(\frac{\partial h}{\partial u}\,du + \frac{\partial h}{\partial v}\,dv\right) = \int_{\partial S} g(u,v)\frac{\partial h}{\partial u}\,du + g(u,v)\frac{\partial h}{\partial v}\,dv$$

$$= \pm\iint_S\left[\frac{\partial}{\partial u}\left(g(u,v)\frac{\partial h}{\partial v}\right) - \frac{\partial}{\partial v}\left(g(u,v)\frac{\partial h}{\partial u}\right)\right]dA \quad \text{(using Green's Theorem in the } uv\text{-plane)}$$

$$= \pm\iint_S\left(\frac{\partial g}{\partial u}\frac{\partial h}{\partial v} + g(u,v)\frac{\partial^2 h}{\partial u\partial v} - \frac{\partial g}{\partial v}\frac{\partial h}{\partial u} - g(u,v)\frac{\partial^2 h}{\partial v\partial u}\right)dA \quad \text{(using the Chain Rule)}$$

$$= \pm\iint_S\left(\frac{\partial x}{\partial u}\frac{\partial y}{\partial v} - \frac{\partial x}{\partial v}\frac{\partial y}{\partial u}\right)dA \quad \text{(by the equality of mixed partials)} \quad = \pm\iint_S \frac{\partial(x,y)}{\partial(u,v)}\,du\,dv.$$

The sign is chosen to be positive if the orientation that we gave to ∂S corresponds to the usual positive orientation, and it is negative otherwise. In either case, since $A(R)$ is positive, the sign chosen must be the same as the sign of $\frac{\partial(x,y)}{\partial(u,v)}$. Therefore $A(R) = \iint_R dx\,dy = \iint_S \left|\frac{\partial(x,y)}{\partial(u,v)}\right| du\,dv$.

EXERCISES 14.5

1. **(a)** $\operatorname{curl}\mathbf{F} = \nabla\times\mathbf{F} = \begin{vmatrix} \mathbf{i} & \mathbf{j} & \mathbf{k} \\ \partial/\partial x & \partial/\partial y & \partial/\partial z \\ x & y & z \end{vmatrix} = (0-0)\mathbf{i} + (0-0)\mathbf{j} + (0-0)\mathbf{k} = \mathbf{0}$

(b) $\operatorname{div}\mathbf{F} = \nabla\cdot\mathbf{F} = \dfrac{\partial}{\partial x}(x) + \dfrac{\partial}{\partial y}(y) + \dfrac{\partial}{\partial z}(z) = 1+1+1 = 3$

3. **(a)** $\operatorname{curl}\mathbf{F} = \nabla\times\mathbf{F} = \begin{vmatrix} \mathbf{i} & \mathbf{j} & \mathbf{k} \\ \partial/\partial x & \partial/\partial y & \partial/\partial z \\ yz & xz & xy \end{vmatrix} = (x-x)\mathbf{i} + (y-y)\mathbf{j} + (z-z)\mathbf{k} = \mathbf{0}$

(b) $\operatorname{div}\mathbf{F} = \nabla\cdot\mathbf{F} = \dfrac{\partial}{\partial x}(yz) + \dfrac{\partial}{\partial y}(xz) + \dfrac{\partial}{\partial z}(xy) = 0+0+0 = 0$

5. **(a)** $\operatorname{curl}\mathbf{F} = \nabla\times\mathbf{F} = \begin{vmatrix} \mathbf{i} & \mathbf{j} & \mathbf{k} \\ \partial/\partial x & \partial/\partial y & \partial/\partial z \\ 0 & xy & xyz \end{vmatrix} = xz\mathbf{i} - yz\mathbf{j} + y\mathbf{k}$

(b) $\operatorname{div}\mathbf{F} = \nabla\cdot\mathbf{F} = \dfrac{\partial}{\partial x}(0) + \dfrac{\partial}{\partial y}(xy) + \dfrac{\partial}{\partial z}(xyz) = 0 + x + xy = x(1+y)$

7. **(a)** $\nabla\times\mathbf{F} = \begin{vmatrix} \mathbf{i} & \mathbf{j} & \mathbf{k} \\ \partial/\partial x & \partial/\partial y & \partial/\partial z \\ e^{xz} & -2e^{yz} & 3xe^{y} \end{vmatrix} = (3xe^{y} + 2ye^{yz})\mathbf{i} + (xe^{xz} - 3e^{y})\mathbf{j}$

(b) $\nabla\cdot\mathbf{F} = \dfrac{\partial}{\partial x}(e^{xz}) + \dfrac{\partial}{\partial y}(-2e^{yz}) + \dfrac{\partial}{\partial z}(3xe^{y}) = ze^{xz} - 2ze^{yz}$

9. **(a)** $\operatorname{curl}\mathbf{F} = \begin{vmatrix} \mathbf{i} & \mathbf{j} & \mathbf{k} \\ \partial/\partial x & \partial/\partial y & \partial/\partial z \\ xe^{y} & -ze^{-y} & y\ln z \end{vmatrix} = (e^{-y} + \ln z)\mathbf{i} - xe^{y}\mathbf{k}$

(b) $\operatorname{div}\mathbf{F} = \dfrac{\partial}{\partial x}(xe^{y}) + \dfrac{\partial}{\partial y}(-ze^{-y}) + \dfrac{\partial}{\partial z}(y\ln z) = e^{y} + ze^{-y} + \dfrac{y}{z}$

11. $\operatorname{curl}\mathbf{F} = \begin{vmatrix} \mathbf{i} & \mathbf{j} & \mathbf{k} \\ \partial/\partial x & \partial/\partial y & \partial/\partial z \\ y & x & 1 \end{vmatrix} = \mathbf{0}$ and $\mathbf{F}$ is defined on all of $\mathbb{R}^3$ with component functions which have continuous partial derivatives, so by (4), $\mathbf{F}$ is conservative. Thus there exists f such that $\mathbf{F} = \nabla f$. Then $f_x(x,y,z) = y$ implies $f(x,y,z) = xy + g(y,z)$ and $f_y(x,y,z) = x + g_y(y,z)$. But $f_y(x,y,z) = x$, so $g(y,z) = h(z)$ and $f(x,y,z) = xy + h(z)$. Thus $f_z(x,y,z) = h'(z)$ but $f_z(x,y,z) = 1$ so $h(z) = z + k$. Hence a potential for $\mathbf{F}$ is $f(x,y,z) = xy + z + k$.

13. $\operatorname{curl}\mathbf{F} = \begin{vmatrix} \mathbf{i} & \mathbf{j} & \mathbf{k} \\ \partial/\partial x & \partial/\partial y & \partial/\partial z \\ yz & -z^2 & x^2 \end{vmatrix} = 2z\mathbf{i} + (y-2x)\mathbf{j} - z\mathbf{k} \neq \mathbf{0}$. Hence $\mathbf{F}$ isn't conservative.

15. $\operatorname{curl}\mathbf{F} = \begin{vmatrix} \mathbf{i} & \mathbf{j} & \mathbf{k} \\ \partial/\partial x & \partial/\partial y & \partial/\partial z \\ \cos y & \sin x & \tan z \end{vmatrix} = (\cos x - \sin y)\mathbf{k} \neq \mathbf{0}$. Hence $\mathbf{F}$ isn't conservative.

17. Since $\text{curl}\,\mathbf{F} = \begin{vmatrix} \mathbf{i} & \mathbf{j} & \mathbf{k} \\ \partial/\partial x & \partial/\partial y & \partial/\partial z \\ yz & y^2 + xz & xy \end{vmatrix} = (x - x)\mathbf{i} + (y - y)\mathbf{j} + (z - z)\mathbf{k} = \mathbf{0}$, **F** is defined on $\mathbb{R}^3$, and since the partial derivatives of the components of **F** are continuous, **F** is conservative. Thus there exists f such that $\nabla f = \mathbf{F}$. Then $f_x(x, y, z) = yz$ implies $f(x, y, z) = xyz + g(y, z)$ and $f_y(x, y, z) = xz + g_y(y, z)$. But $f_y(x, y, z) = xz + y^2$ so $g(y, z) = \frac{1}{3}y^3 + h(z)$ and $f(x, y, z) = xyz + \frac{1}{3}y^3 + h(z)$. Then $f_z(x, y, z) = xy + h'(z)$. But $f_z(x, y, z) = xy$ so $h(z) = k$. Hence $f(x, y, z) = xyz + \frac{1}{3}y^3 + k$ is a potential for **F**.

19. No. Assume there is such a **G**. Then $\text{div}(\text{curl}\,\mathbf{G}) = y^2 + z^2 + x^2 \neq 0$, which contradicts Theorem 11.

21. $\text{curl}\,\mathbf{F} = \begin{vmatrix} \mathbf{i} & \mathbf{j} & \mathbf{k} \\ \partial/\partial x & \partial/\partial y & \partial/\partial z \\ f(x) & g(y) & h(z) \end{vmatrix} = (0 - 0)\mathbf{i} + (0 - 0)\mathbf{j} + (0 - 0)\mathbf{k} = \mathbf{0}$. Hence $\mathbf{F} = f(x)\mathbf{i} + g(y)\mathbf{j} + h(z)\mathbf{k}$ is irrotational.

Note: **For Exercises 23--29, let $\mathbf{F}(x, y, z) = P_1\mathbf{i} + Q_1\mathbf{j} + R_1\mathbf{k}$ and $\mathbf{G}(x, y, z) = P_2\mathbf{i} + Q_2\mathbf{j} + R_2\mathbf{k}$.**

23.
$$\begin{aligned} \text{div}(\mathbf{F} + \mathbf{G}) &= \frac{\partial(P_1 + P_2)}{\partial x} + \frac{\partial(Q_1 + Q_2)}{\partial y} + \frac{\partial(R_1 + R_2)}{\partial z} \\ &= \left(\frac{\partial P_1}{\partial x} + \frac{\partial Q_1}{\partial y} + \frac{\partial R_1}{\partial z}\right) + \left(\frac{\partial P_2}{\partial x} + \frac{\partial Q_2}{\partial y} + \frac{\partial R_3}{\partial z}\right) = \text{div}\,\mathbf{F} + \text{div}\,\mathbf{G} \end{aligned}$$

25.
$$\begin{aligned} \text{div}(f\mathbf{F}) &= \frac{\partial(fP_1)}{\partial x} + \frac{\partial(fQ_1)}{\partial y} + \frac{\partial(fR_1)}{\partial z} = \left[f\frac{\partial P_1}{\partial x} + P_1\frac{\partial f}{\partial x}\right] + \left[f\frac{\partial Q_1}{\partial y} + Q_1\frac{\partial f}{\partial y}\right] + \left[f\frac{\partial R_1}{\partial z} + R_1\frac{\partial f}{\partial z}\right] \\ &= f\left(\frac{\partial P_1}{\partial x} + \frac{\partial Q_1}{\partial y} + \frac{\partial R_1}{\partial z}\right) + \langle P_1, Q_1, R_1\rangle \cdot \left\langle \frac{\partial f}{\partial x}, \frac{\partial f}{\partial y}, \frac{\partial f}{\partial z} \right\rangle = f\,\text{div}\,\mathbf{F} + \mathbf{F} \cdot \nabla f \end{aligned}$$

27.
$$\begin{aligned} \text{div}(\mathbf{F} \times \mathbf{G}) = \nabla \cdot (\mathbf{F} \times \mathbf{G}) &= \begin{vmatrix} \partial/\partial x & \partial/\partial y & \partial/\partial z \\ P_1 & Q_1 & R_1 \\ P_2 & Q_2 & R_2 \end{vmatrix} = \frac{\partial}{\partial x}\begin{vmatrix} Q_1 & R_1 \\ Q_2 & R_2 \end{vmatrix} - \frac{\partial}{\partial y}\begin{vmatrix} P_1 & R_1 \\ P_2 & R_2 \end{vmatrix} + \frac{\partial}{\partial z}\begin{vmatrix} P_1 & Q_1 \\ P_2 & Q_2 \end{vmatrix} \\ &= \left[Q_1\frac{\partial R_2}{\partial x} + R_2\frac{\partial Q_1}{\partial x} - Q_2\frac{\partial R_1}{\partial x} - R_1\frac{\partial Q_2}{\partial x}\right] - \left[P_1\frac{\partial R_2}{\partial y} + R_2\frac{\partial P_1}{\partial y} - P_2\frac{\partial R_1}{\partial y} - R_1\frac{\partial P_2}{\partial y}\right] \\ &\quad + \left[P_1\frac{\partial Q_2}{\partial z} + Q_2\frac{\partial P_1}{\partial z} - P_2\frac{\partial Q_1}{\partial z} - Q_1\frac{\partial P_2}{\partial z}\right] \\ &= \left[P_2\left(\frac{\partial R_1}{\partial y} - \frac{\partial Q_1}{\partial z}\right) + Q_2\left(\frac{\partial P_1}{\partial z} - \frac{\partial R_1}{\partial x}\right) + R_2\left(\frac{\partial Q_1}{\partial x} - \frac{\partial P_1}{\partial y}\right)\right] \\ &\quad - \left[P_1\left(\frac{\partial R_2}{\partial y} - \frac{\partial Q_2}{\partial z}\right) + Q_1\left(\frac{\partial P_2}{\partial z} - \frac{\partial R_2}{\partial x}\right) + R_1\left(\frac{\partial Q_2}{\partial x} - \frac{\partial P_2}{\partial y}\right)\right] \\ &= \mathbf{G} \cdot \text{curl}\,\mathbf{F} - \mathbf{F} \cdot \text{curl}\,\mathbf{G} \end{aligned}$$

29. $\operatorname{curl}\operatorname{curl}\mathbf{F} = \nabla\times(\nabla\times\mathbf{F}) = \begin{vmatrix} \mathbf{i} & \mathbf{j} & \mathbf{k} \\ \partial/\partial x & \partial/\partial y & \partial/\partial z \\ \partial R_1/\partial y - \partial Q_1/\partial z & \partial P_1/\partial z - \partial R_1/\partial x & \partial Q_1/\partial x - \partial P_1/\partial y \end{vmatrix}$

$$= \left(\frac{\partial^2 Q_1}{\partial y\partial x} - \frac{\partial^2 P_1}{\partial y^2} - \frac{\partial^2 P_1}{\partial z^2} + \frac{\partial^2 R_1}{\partial z\partial x}\right)\mathbf{i} + \left(\frac{\partial^2 R_1}{\partial z\partial y} - \frac{\partial^2 Q_1}{\partial z^2} - \frac{\partial^2 Q_1}{\partial x^2} + \frac{\partial^2 P_1}{\partial x\partial y}\right)\mathbf{j}$$
$$+ \left(\frac{\partial^2 P_1}{\partial x\partial z} - \frac{\partial^2 R_1}{\partial x^2} - \frac{\partial^2 R_1}{\partial y^2} + \frac{\partial^2 Q_1}{\partial y\partial z}\right)\mathbf{k}.$$

Now let's consider $\operatorname{grad}\operatorname{div}\mathbf{F} - \nabla^2\mathbf{F}$ and compare with the above.

$\operatorname{grad}\operatorname{div}\mathbf{F} - \nabla^2\mathbf{F}$

$$= \left[\left(\frac{\partial^2 P_1}{\partial x^2} + \frac{\partial^2 Q_1}{\partial x\partial y} + \frac{\partial^2 R_1}{\partial x\partial z}\right)\mathbf{i} + \left(\frac{\partial^2 P_1}{\partial y\partial x} + \frac{\partial^2 Q_1}{\partial y^2} + \frac{\partial^2 R_1}{\partial y\partial z}\right)\mathbf{j} + \left(\frac{\partial^2 P_1}{\partial z\partial x} + \frac{\partial^2 Q_1}{\partial z\partial y} + \frac{\partial^2 R_1}{\partial z^2}\right)\mathbf{k}\right]$$
$$- \left[\left(\frac{\partial^2 P_1}{\partial x^2} + \frac{\partial^2 P_1}{\partial y^2} + \frac{\partial^2 P_1}{\partial z^2}\right)\mathbf{i} + \left(\frac{\partial^2 Q_1}{\partial x^2} + \frac{\partial^2 Q_1}{\partial y^2} + \frac{\partial^2 Q_1}{\partial z^2}\right)\mathbf{j} + \left(\frac{\partial^2 R_1}{\partial x^2} + \frac{\partial^2 R_1}{\partial y^2} + \frac{\partial^2 R_1}{\partial z^2}\right)\mathbf{k}\right]$$
$$= \left(\frac{\partial^2 Q_1}{\partial x\partial y} + \frac{\partial^2 R_1}{\partial x\partial z} - \frac{\partial^2 P_1}{\partial y^2} - \frac{\partial^2 P_1}{\partial z^2}\right)\mathbf{i} + \left(\frac{\partial^2 P_1}{\partial y\partial x} + \frac{\partial^2 R_1}{\partial y\partial z} - \frac{\partial^2 Q_1}{\partial x^2} - \frac{\partial^2 Q_1}{\partial z^2}\right)\mathbf{j}$$
$$+ \left(\frac{\partial^2 P_1}{\partial z\partial x} + \frac{\partial^2 Q_1}{\partial z\partial y} - \frac{\partial^2 R_1}{\partial x^2} - \frac{\partial^2 R_2}{\partial y^2}\right)\mathbf{k}.$$

Then applying Clairaut's Theorem to reverse the order of differentiation in the second partial derivatives as needed and comparing, we have $\operatorname{curl}\operatorname{curl}\mathbf{F} = \operatorname{grad}\operatorname{div}\mathbf{F} - \nabla^2\mathbf{F}$ as desired.

31. **(a)** $\operatorname{curl} f = \nabla\times f$ is meaningless because f is a scalar field.

(b) $\operatorname{grad} f$ is a vector field.

(c) $\operatorname{div}\mathbf{F}$ is a scalar field.

(d) $\operatorname{curl}(\operatorname{grad} f)$ is a vector field.

(e) $\operatorname{grad}\mathbf{F}$ is meaningless.

(f) $\operatorname{grad}(\operatorname{div}\mathbf{F})$ is a vector field.

(g) $\operatorname{div}(\operatorname{grad} f)$ is a scalar field.

(h) $\operatorname{grad}(\operatorname{div} f)$ is meaningless.

(i) $\operatorname{curl}(\operatorname{curl}\mathbf{F})$ is a vector field.

(j) $\operatorname{div}(\operatorname{div}\mathbf{F})$ is meaningless.

(k) $(\operatorname{grad} f)\times(\operatorname{div}\mathbf{F})$ is meaningless because $\operatorname{div}\mathbf{F}$ is a scalar field.

(l) $\operatorname{div}(\operatorname{curl}(\operatorname{grad} f))$ is a scalar field.

33. $\nabla\cdot\mathbf{r} = \left(\frac{\partial}{\partial x}\mathbf{i} + \frac{\partial}{\partial y}\mathbf{j} + \frac{\partial}{\partial z}\mathbf{k}\right)\cdot(x\mathbf{i} + y\mathbf{j} + z\mathbf{k}) = 1 + 1 + 1 = 3$

35. $\nabla\left(\frac{1}{r}\right) = \nabla\left(\frac{1}{\sqrt{x^2+y^2+z^2}}\right)$

$$= \frac{-\frac{1}{2\sqrt{x^2+y^2+z^2}}(2x)}{x^2+y^2+z^2}\mathbf{i} - \frac{\frac{1}{2\sqrt{x^2+y^2+z^2}}(2y)}{x^2+y^2+z^2}\mathbf{j} - \frac{\frac{1}{2\sqrt{x^2+y^2+z^2}}(2z)}{x^2+y^2+z^2}\mathbf{k}$$
$$= -\frac{x\mathbf{i} + y\mathbf{j} + z\mathbf{k}}{(x^2+y^2+z^2)^{3/2}} = -\frac{\mathbf{r}}{r^3}$$

37. $\nabla \ln r = \nabla \ln\left(x^2+y^2+z^2\right)^{1/2} = \frac{1}{2}\nabla \ln(x^2+y^2+z^2)$

$$= \frac{x}{x^2+y^2+z^2}\mathbf{i} + \frac{y}{x^2+y^2+z^2}\mathbf{j} + \frac{z}{x^2+y^2+z^2}\mathbf{k} = \frac{x\mathbf{i}+y\mathbf{j}+z\mathbf{k}}{x^2+y^2+z^2} = \frac{\mathbf{r}}{r^2}$$

39. If the vector field is $\mathbf{F} = P\mathbf{i} + Q\mathbf{j} + R\mathbf{k}$, then we are assuming here that $R = 0$, that is, the vector field does not vary in the z-direction, so $\dfrac{\partial R}{\partial z} = 0$. Since the x-component of each vector of $\mathbf{F}$ is 0, $P = 0$, so $\dfrac{\partial P}{\partial x} = 0$. But Q is decreasing as y increases, so $\dfrac{\partial Q}{\partial y} < 0$. Hence $\operatorname{div}\mathbf{F} = \dfrac{\partial P}{\partial x} + \dfrac{\partial Q}{\partial y} + \dfrac{\partial R}{\partial z}$ is negative at every point.

41. By (13), $\oint_C f(\nabla g)\cdot\mathbf{n}\,ds = \iint_D \operatorname{div}(f\nabla g)dA = \iint_D [f\operatorname{div}(\nabla g) + \nabla g\cdot\nabla f]dA$ by Exercise 25. But $\operatorname{div}(\nabla g) = \nabla^2 g$. Hence $\iint_D f\nabla^2 g\,dA = \oint_C f(\nabla g)\cdot\mathbf{n}\,ds - \iint_D \nabla g\cdot\nabla f\,dA$.

43. **(a)** We know that $\omega = \dfrac{v}{d}$, and from the diagram $\sin\theta = \dfrac{d}{r} \quad\Rightarrow\quad v = d\omega = (\sin\theta)r\omega = |\boldsymbol{\omega}\times\mathbf{r}|$. But $\mathbf{v}$ is perpendicular to both $\mathbf{w}$ and $\mathbf{r}$, so that $\mathbf{v} = \mathbf{w}\times\mathbf{r}$.

(b) From (a), $\mathbf{v} = \mathbf{w}\times\mathbf{r} = \begin{vmatrix} \mathbf{i} & \mathbf{j} & \mathbf{k} \\ 0 & 0 & \omega \\ x & y & z \end{vmatrix} = (0\cdot z - \omega y)\mathbf{i} + (\omega x - 0\cdot z)\mathbf{j} + (0\cdot y - x\cdot 0)\mathbf{k} = -\omega y\mathbf{i} + \omega x\mathbf{j}$

(c) $\operatorname{curl}\mathbf{v} = \nabla\times\mathbf{v} = \begin{vmatrix} \mathbf{i} & \mathbf{j} & \mathbf{k} \\ \partial/\partial x & \partial/\partial y & \partial/\partial z \\ -\omega y & \omega x & 0 \end{vmatrix}$

$$= \left[\frac{\partial}{\partial y}(0) - \frac{\partial}{\partial z}(\omega x)\right]\mathbf{i} + \left[\frac{\partial}{\partial z}(-\omega y) - \frac{\partial}{\partial x}(0)\right]\mathbf{j} + \left[\frac{\partial}{\partial x}(\omega x) - \frac{\partial}{\partial y}(-\omega y)\right]\mathbf{k}$$
$$= [\omega - (-\omega)]\mathbf{k} = 2\omega\mathbf{k} = 2\mathbf{w}$$

EXERCISES 14.6

1. Letting x and y be the parameters, the parametric equations are $x = x$, $y = y$, $z = \sqrt{1-3x^2-2y^2}$ where $-\frac{1}{\sqrt{3}} \le x \le \frac{1}{\sqrt{3}}$ and $-\frac{1}{\sqrt{2}} \le y \le \frac{1}{\sqrt{2}}$. Then the vector equation of the surface is $\mathbf{r}(x,y) = x\mathbf{i} + y\mathbf{j} + \sqrt{1-3x^2-2y^2}\,\mathbf{k}$.

Alternate Solution: Letting ϕ and θ be the parameters, the parametric equations are $x = \frac{1}{\sqrt{3}}\sin\phi\cos\theta$, $y = \frac{1}{\sqrt{2}}\sin\phi\sin\theta$, $z = \cos\phi$ where $0 \le \phi \le \frac{\pi}{2}$ and $0 \le \theta \le 2\pi$.

Note: There are many parametric representations of a given surface.

3. $x = x$, $y = 6 - 3x^2 - 2z^2$, $z = z$ where $3x^2 + 2z^2 \le 6$ since $y \ge 0$. Then the associated vector equation is $\mathbf{r}(x,y) = x\mathbf{i} + (6 - 3x^2 - 2z^2)\mathbf{j} + z\mathbf{k}$.

5. Since the cone intersects the sphere in the circle $x^2 + y^2 = 2$, $z = 2$ and we want the portion of the sphere above this, we can parametrize the surface as $x = x$, $y = y$, $z = \sqrt{4 - x^2 - y^2}$ where $2 \le x^2 + y^2 \le 4$.

Or: Using spherical coordinates, $x = 2\sin\phi\cos\theta$, $y = 2\sin\phi\,\sin\theta$, $z = 2\cos\phi$ where $0 \le \phi \le \frac{\pi}{4}$ and $0 \le \theta \le 2\pi$.

7. The surface is a disc of radius 4 and center $(0, 0, 5)$. Thus $x = r\cos\theta$, $y = r\sin\theta$, $z = 5$ where $0 \le r \le 4$, $0 \le \theta \le 2\pi$ is a parametric representation of the surface.

Or: In rectangular coordinates we could represent the surface as $x = x$, $y = y$, $z = 5$ where $0 \le x^2 + y^2 \le 16$.

9. $\mathbf{r}(u, v) = u\cos v\,\mathbf{i} + u\sin v\,\mathbf{j} + v\mathbf{k}$. This equation must correspond to graph I, since for fixed v, $\mathbf{r}$ parametrizes a straight line in the plane $z = v$. As v increases, the line rotates and moves upward, generating a spiral ramp.

11. $x = (u - \sin u)\cos v$, $y = (1 - \cos u)\sin v$, $z = u$. This corresponds to graph II: when $u = 0$, $x = y = z = 0$, so $(0, 0, 0)$ is on the surface; and when $u = \pi$, $z = \pi$ and $y = 0$ while x ranges between $-\pi$ and π, giving the upper "seam" on the surface.

13. Using Equations 3, we have the parametrization $x = x$, $y = e^{-x}\cos\theta$, $z = e^{-x}\sin\theta$, $0 \le x \le 3$, $0 \le \theta \le 2\pi$.

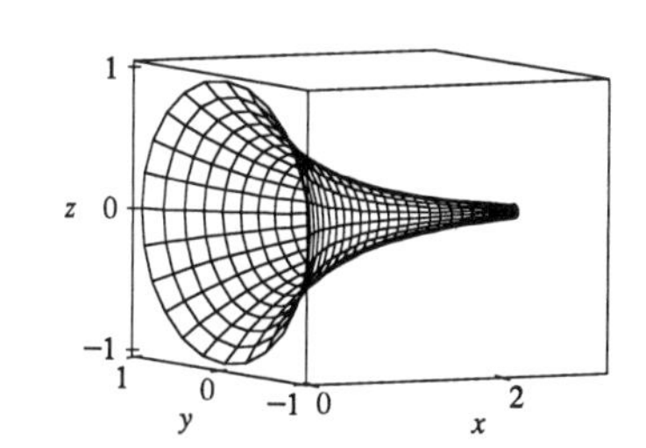

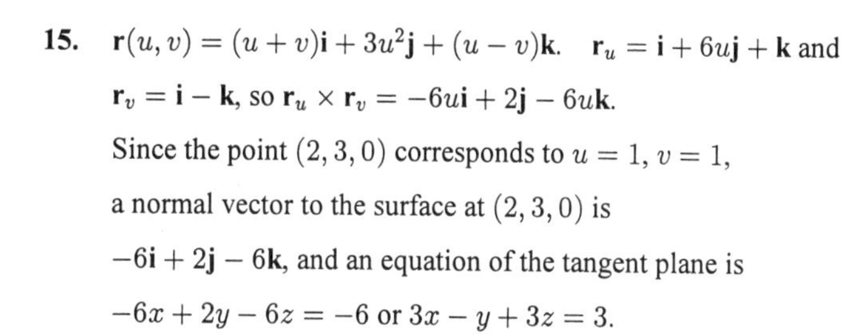

15. $\mathbf{r}(u, v) = (u + v)\mathbf{i} + 3u^2\mathbf{j} + (u - v)\mathbf{k}$. $\mathbf{r}_u = \mathbf{i} + 6u\mathbf{j} + \mathbf{k}$ and $\mathbf{r}_v = \mathbf{i} - \mathbf{k}$, so $\mathbf{r}_u \times \mathbf{r}_v = -6u\mathbf{i} + 2\mathbf{j} - 6u\mathbf{k}$.
Since the point $(2, 3, 0)$ corresponds to $u = 1$, $v = 1$, a normal vector to the surface at $(2, 3, 0)$ is $-6\mathbf{i} + 2\mathbf{j} - 6\mathbf{k}$, and an equation of the tangent plane is $-6x + 2y - 6z = -6$ or $3x - y + 3z = 3$.

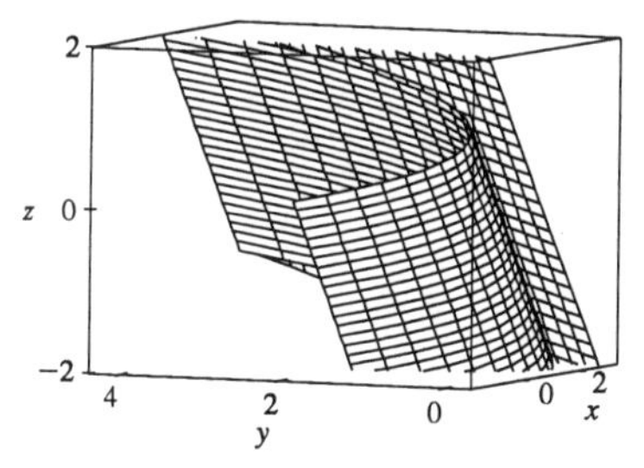

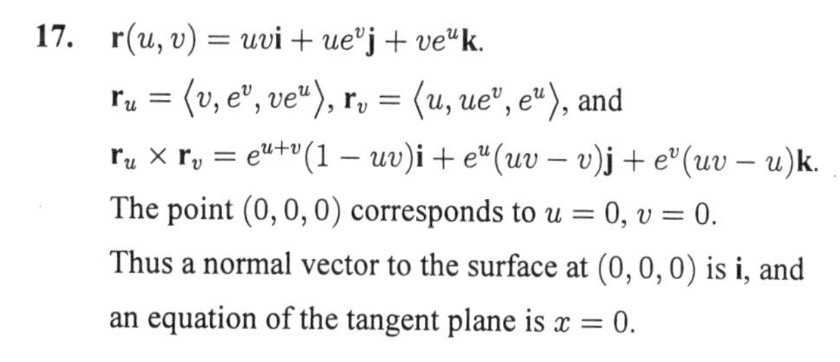

17. $\mathbf{r}(u, v) = uv\mathbf{i} + ue^v\mathbf{j} + ve^u\mathbf{k}$.
$\mathbf{r}_u = \langle v, e^v, ve^u\rangle$, $\mathbf{r}_v = \langle u, ue^v, e^u\rangle$, and
$\mathbf{r}_u \times \mathbf{r}_v = e^{u+v}(1 - uv)\mathbf{i} + e^u(uv - v)\mathbf{j} + e^v(uv - u)\mathbf{k}$.
The point $(0, 0, 0)$ corresponds to $u = 0$, $v = 0$.
Thus a normal vector to the surface at $(0, 0, 0)$ is $\mathbf{i}$, and an equation of the tangent plane is $x = 0$.

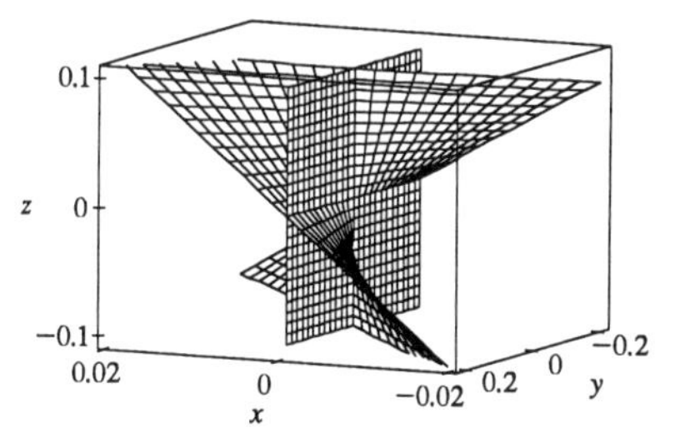

19. Here $z = f(x, y) = 4 - x - 2y$ with $0 \le x^2 + y^2 \le 4$. Thus, by (9),

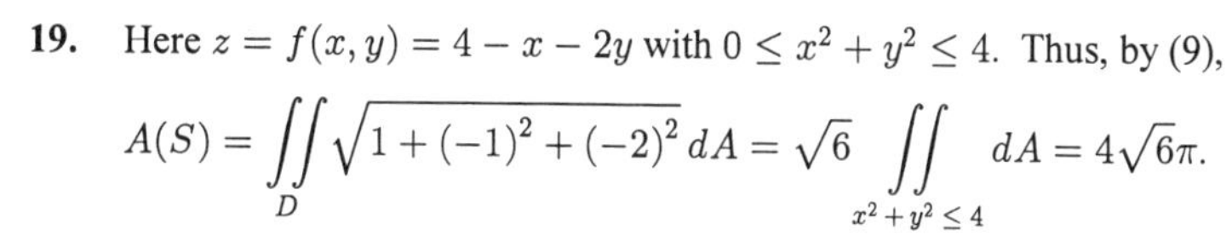

$$A(S) = \iint_D \sqrt{1 + (-1)^2 + (-2)^2}\,dA = \sqrt{6}\iint_{x^2+y^2\le 4} dA = 4\sqrt{6}\pi.$$

21. $z = f(x,y) = y^2 - x^2$ with $1 \le x^2 + y^2 \le 4$. Then

$A(S) = \iint_D \sqrt{1+4x^2+4y^2}\,dA = \int_0^{2\pi}\int_1^2 \sqrt{1+4r^2}\,r\,dr\,d\theta = 4\pi\left(\frac{1}{24}\right)(1+4r^2)^{3/2}\Big|_1^2 = \frac{\pi}{6}\left(17\sqrt{17} - 5\sqrt{5}\right).$

23. A parametric representation of the surface is $x = x$, $y = 4x + z^2$, $z = z$

with $0 \le x \le 1$, $0 \le z \le 1$. Hence $\mathbf{r}_x \times \mathbf{r}_z = 4\mathbf{i} - \mathbf{j} + 2z\mathbf{k}$.

Note: In general, if $y = f(x,z)$ then $\mathbf{r}_z \times \mathbf{r}_x = -\dfrac{\partial f}{\partial x}\mathbf{i} + \mathbf{j} - \dfrac{\partial f}{\partial x}\mathbf{k}$ and

$A(S) = \displaystyle\iint_D \sqrt{1 + \left(\frac{\partial f}{\partial x}\right)^2 + \left(\frac{\partial f}{\partial z}\right)^2}\,dA.$ Then

$A(S) = \int_0^1\int_0^1 \sqrt{17+4z^2}\,dx\,dz = \int_0^1 \sqrt{17+4z^2}dz = \frac{1}{2}\left(z\sqrt{17+4z^2} + \frac{17}{2}\ln\left|2z + \sqrt{4z^2+17}\right|\right)\Big|_0^1$

$= \frac{\sqrt{21}}{2} + \frac{17}{4}\left[\ln\left(2+\sqrt{21}\right) - \ln\sqrt{17}\right].$

25. Let $A(S_1)$ be the surface area of that portion of the surface which lies above the plane $z = 0$, then $A(S) = 2A(S_1)$. Following Example 6, a parametric representation of S_1 is $x = a\sin\phi\cos\theta$, $y = a\sin\phi\sin\theta$, $z = a\cos\phi$ and $|\mathbf{r}_\phi \times \mathbf{r}_\theta| = a^2\sin\phi$. For D, $0 \le \phi \le \frac{\pi}{2}$ and for each fixed ϕ, $\left(x - \frac{1}{2}a\right)^2 + y^2 \le \left(\frac{1}{2}a\right)^2$ or $\left[a\sin\phi\cos\theta - \frac{1}{2}a\right]^2 + a^2\sin^2\phi\sin^2\theta \le (a/2)^2$ implies $a^2\sin^2\phi - a^2\sin\phi\cos\theta \le 0$ or $\sin\phi(\sin\phi - \cos\theta) \le 0$. But $0 \le \phi \le \frac{\pi}{2}$, so $\cos\theta \ge \sin\phi$ or $\sin\left(\frac{\pi}{2} + \theta\right) \ge \sin\phi$ or $\phi - \frac{\pi}{2} \le \theta \le \frac{\pi}{2} - \phi$. Hence $D = \left\{(\phi,\theta) \mid 0 \le \phi \le \frac{\pi}{2},\ \phi - \frac{\pi}{2} \le \theta \le \frac{\pi}{2} - \phi\right\}$. Then

$A(S_1) = \int_0^{\pi/2}\int_{\phi-(\pi/2)}^{(\pi/2)-\phi} a^2\sin\phi\,d\theta\,d\phi = a^2\int_0^{\pi/2}(\pi - 2\phi)\sin\phi\,d\phi = a^2[(-\pi\cos\phi) - 2(-\phi\cos\phi + \sin\phi)]_0^{\pi/2}$

$= a^2(\pi - 2)$. Thus $A(S) = 2a^2(\pi - 2)$.

Alternate Solution: Working on S_1 we could parametrize the portion of the sphere by $x = x$, $y = y$, $z = \sqrt{a^2 - x^2 - y^2}$. Then $|\mathbf{r}_x \times \mathbf{r}_y| = \sqrt{1 + \dfrac{x^2}{a^2 - x^2 - y^2} + \dfrac{y^2}{a^2 - x^2 - y^2}} = \dfrac{a}{\sqrt{a^2 - x^2 - y^2}}$ and

$$A(S_1) = \iint_{0 \le \left(x - \frac{1}{2}a\right)^2 + y^2 \le \left(\frac{1}{2}a\right)^2} \frac{a}{\sqrt{a^2 - x^2 - y^2}}\,dA = \int_{-\pi/2}^{\pi/2}\int_0^{a\cos\theta} \frac{a}{\sqrt{a^2 - r^2}}\,r\,dr\,d\theta$$

$$= \int_{-\pi/2}^{\pi/2} -a\left(a^2 - r^2\right)^{1/2}\Big|_0^{a\cos\theta}\,d\theta = \int_{-\pi/2}^{\pi/2} a^2\left[1 - \left(1 - \cos^2\theta\right)^{1/2}\right]d\theta$$

$$= \int_{-\pi/2}^{\pi/2} a^2(1 - |\sin\theta|)d\theta = 2a^2\int_0^{\pi/2}(1 - \sin\theta)d\theta = 2a^2\left(\tfrac{\pi}{2} - 1\right).$$

Thus $A(S) = 4a^2\left(\frac{\pi}{2} - 1\right) = 2a^2(\pi - 2)$.

Notes: *(1)* Perhaps working in spherical coordinates is the most obvious approach here. However, you must be careful in setting up D.

(2) In the alternate solution, you can avoid having to use $|\sin\theta|$ by working in the first octant and then multiplying by 8. However, if you set up S_1 as above and arrived at $A(S_1) = a^2\pi$, you now see your error.

27. $\mathbf{r}_u = \langle v, 1, 1\rangle$, $\mathbf{r}_v = \langle u, 1, -1\rangle$ and $\mathbf{r}_u \times \mathbf{r}_v = \langle -2, u+v, v-u\rangle$. Then

$$A(S) = \iint_{u^2+v^2\leq 1} \sqrt{4+2u^2+2v^2}\,dA = \int_0^{2\pi}\int_0^1 r\sqrt{4+2r^2}\,dr\,d\theta = 2\pi\left(\tfrac{1}{6}\right)(4+2r^2)^{3/2}\Big|_0^1$$
$$= \tfrac{\pi}{3}\left(6\sqrt{6}-8\right) = \pi\left(2\sqrt{6}-\tfrac{8}{3}\right).$$

29. **(a)** $x = a\sin u\cos v$, $y = b\sin u\sin v$, $z = c\cos u \quad\Rightarrow$

$$\frac{x^2}{a^2}+\frac{y^2}{b^2}+\frac{z^2}{c^2} = (\sin u\cos v)^2 + (\sin u\sin v)^2 + (\cos u)^2$$
$$= \sin^2 u + \cos^2 u = 1$$

and since the ranges of u and v are sufficient to generate the entire graph, the parametric equations represent an ellipsoid.

(b)

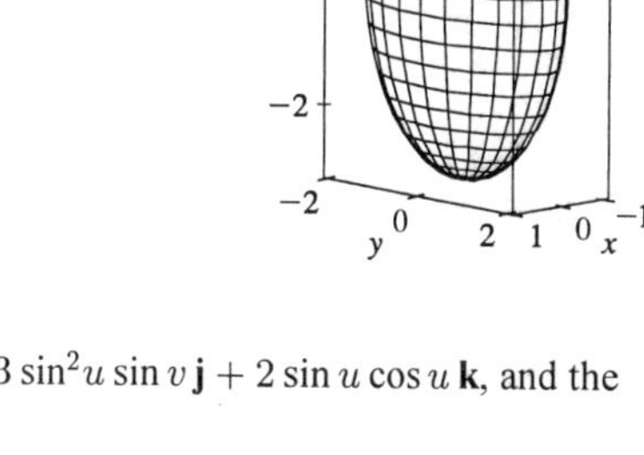

(c) From the parametric equations (with $a = 1$, $b = 2$, and $c = 3$), we calculate $\mathbf{r}_u = \cos u\cos v\,\mathbf{i} + 2\cos u\sin v\,\mathbf{j} - 3\sin u\,\mathbf{k}$ and

$\mathbf{r}_v = -\sin u\sin v\,\mathbf{i} + 2\sin u\cos v\,\mathbf{j}$. So $\mathbf{r}_u \times \mathbf{r}_v = 6\sin^2 u\cos v\,\mathbf{i} + 3\sin^2 u\sin v\,\mathbf{j} + 2\sin u\cos u\,\mathbf{k}$, and the surface area is given by

$$A(S) = \int_0^{2\pi}\int_0^{\pi} |\mathbf{r}_u \times \mathbf{r}_v|\,du\,dv = \int_0^{2\pi}\int_0^{\pi} \sqrt{36\sin^4 u\cos^2 v + 9\sin^4 u\sin^2 v + 4\cos^2 u\sin^2 u}\,du\,dv.$$

31.

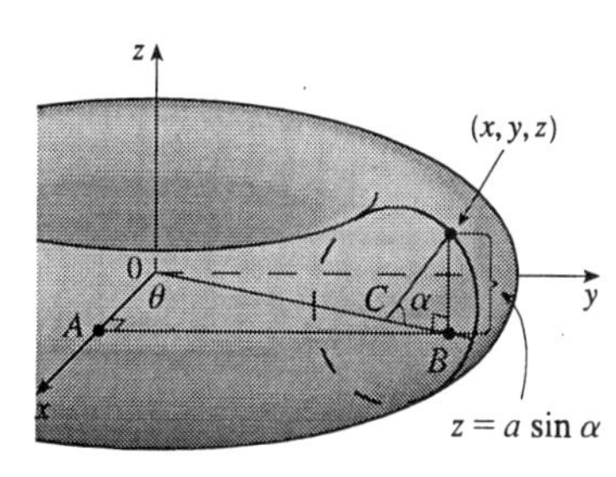

Here $z = a\sin\alpha$, $y = |AB|$, and $x = |OA|$. But $|OB| = |OC| + |CB| = b + a\cos\alpha$ and $\sin\theta = |AB|/|OB|$ so that $y = |OB|\sin\theta = (b + a\cos\alpha)\sin\theta$. Similarly $\cos\theta = |OA|/|OB|$ so $x = (b + a\cos\alpha)\cos\theta$. Hence a parametric representation for the torus is

$x = b\cos\theta + a\cos\alpha\cos\theta$, $y = b\sin\theta + a\cos\alpha\sin\theta$, $z = a\sin\alpha$,

where $0 \leq \alpha \leq 2\pi$, $0 \leq \theta \leq 2\pi$.

EXERCISES 14.7

1. Here $f(x,y,z) = \sqrt{x^2 + 2y^2 + 3z^2}$ and by Definition 1,

$$\iint_S f(x,y,z)dS \approx [f(1,0,0)](4) + [f(-1,0,0)](4) + [f(0,1,0)](4) + [f(0,-1,0)](4) + [f(0,0,1)](4) + [f(0,0,-1)](4)$$
$$= 4\left(1 + 1 + 2\sqrt{2} + 2\sqrt{3}\right) = 8\left(1 + \sqrt{2} + \sqrt{3}\right) \approx 33.170.$$

3. $\mathbf{r}(x,y) = x\mathbf{i} + y\mathbf{j} + (6 - 3x - 2y)\mathbf{k}$, $\mathbf{r}_x \times \mathbf{r}_y = 3\mathbf{i} + 2\mathbf{j} + \mathbf{k}$ (the normal to the plane) and $|\mathbf{r}_x \times \mathbf{r}_y| = \sqrt{14}$. The given plane meets the first octant in the line $3x + 2y = 6$, $z = 0$, $x \geq 0$, $y \geq 0$, so $D = \left\{(x,y) \mid 0 \leq x \leq \frac{1}{3}(6 - 2y), 0 \leq y \leq 3\right\}$. Then $\iint_S y\,dS = \int_0^3\int_0^{(6-2y)/3} y\sqrt{14}\,dx\,dy = \sqrt{14}\int_0^3 \left(2y - \frac{2}{3}y^2\right)dy = 3\sqrt{14}$.

5. $\mathbf{r}(x,z) = x\mathbf{i} + (x^2 + 4z)\mathbf{j} + z\mathbf{k}$, $0 \leq x \leq 2$, $0 \leq z \leq 2$, $|\mathbf{r}_x \times \mathbf{r}_z| = \sqrt{4x^2 + 17}$ (see Exercise 14.6.23.) Then $\iint_S x\,dS = \int_0^2\int_0^2 x\sqrt{4x^2 + 17}\,dx\,dz = 2\left[\frac{1}{12}(4x^2 + 17)^{3/2}\right]_0^2 = \frac{33\sqrt{33} - 17\sqrt{17}}{6}$.

7. Since $z = y + 3$, $|\mathbf{r}_x \times \mathbf{r}_y| = \sqrt{2}$ and $\iint_S yz\,dS = \iint_{x^2+y^2\leq 1} \sqrt{2}y(y+3)dA$
$= \sqrt{2}\int_0^{2\pi}\int_0^1 (r^2\sin^2\theta + 3r\sin\theta)r\,dr\,d\theta = \sqrt{2}\int_0^{2\pi}\left[\frac{1}{4}\sin^2\theta + \sin\theta\right]d\theta = \frac{\pi}{2\sqrt{2}}$.

9. Using spherical coordinates and Example 6 in Section 14.6 we have $\mathbf{r}(\phi,\theta) = 2\sin\phi\cos\theta\,\mathbf{i} + 2\sin\phi\sin\theta\,\mathbf{j} + 2\cos\phi\,\mathbf{k}$ and $|\mathbf{r}_\phi \times \mathbf{r}_\theta| = 4\sin\phi$. Then $\iint_S (x^2z + y^2z)dS = \int_0^{2\pi}\int_0^{\pi/2}(4\sin^2\phi)(2\cos\phi)(4\sin\phi)d\phi\,d\theta = 16\pi\sin^4\phi\Big|_0^{\pi/2} = 16\pi$.

11. Using cylindrical coordinates, $\mathbf{r}(\theta,z) = 3\cos\theta\,\mathbf{i} + 3\sin\theta\,\mathbf{j} + z\mathbf{k}$, $0 \leq \theta \leq 2\pi$, $0 \leq z \leq 2$, and $|\mathbf{r}_\theta \times \mathbf{r}_z| = 3$.
$\iint_S (x^2y + z^2)dS = \int_0^{2\pi}\int_0^2 (27\cos^2\theta\sin\theta + z^2)3\,dz\,d\theta = \int_0^{2\pi}(162\cos^2\theta\sin\theta + 8)d\theta = 16\pi$

13. $\mathbf{r}(u,v) = uv\mathbf{i} + (u+v)\mathbf{j} + (u-v)\mathbf{k}$, $u^2 + v^2 \leq 1$ and $|\mathbf{r}_u \times \mathbf{r}_v| = \sqrt{4 + 2u^2 + 2v^2}$ (see Exercise 14.6.27).
Then $\iint_S yx\,dS = \iint_{u^2+v^2\leq 1}(u^2 - v^2)\sqrt{4 + 2u^2 + 2v^2}\,dA = \int_0^{2\pi}\int_0^1 r^2(\cos^2\theta - \sin^2\theta)\sqrt{4 + 2r^2}\,r\,dr\,d\theta$
$= \left[\int_0^{2\pi}(\cos^2\theta - \sin^2\theta)d\theta\right]\left[\int_0^1 r^3\sqrt{4 + 2r^2}\,dr\right] = 0$ since the first integral is 0.

15. $\mathbf{F}(\mathbf{r}(x,y)) = e^y\mathbf{i} + ye^x\mathbf{j} + x^2y\mathbf{k}$ and $\mathbf{r}_x \times \mathbf{r}_y = -2x\mathbf{i} - 2y\mathbf{j} + \mathbf{k}$. Then $\mathbf{F}(\mathbf{r}(x,y)) \cdot (\mathbf{r}_x \times \mathbf{r}_y) = -2xe^y - 2y^2e^x + x^2y$ and
$\iint_S \mathbf{F} \cdot d\mathbf{S} = \int_0^1\int_0^1(-2xe^y - 2y^2e^x + x^2y)dx\,dy = \int_0^1\left(-e^y - 2ey^2 + \frac{1}{3}y + 2y^2\right)dy = \frac{1}{6}(11 - 10e)$.

17. As in Exercise 3, $D = \left\{(x,y) \mid 0 \leq x \leq 2, 0 \leq y \leq \frac{1}{2}(6 - 3x)\right\}$.
$\iint_S \mathbf{F} \cdot d\mathbf{S} = \int_0^2\int_0^{(6-3x)/2}[x\mathbf{i} + xy\mathbf{j} + x(6 - 3x - 2y)\mathbf{k}] \cdot (3\mathbf{i} + 2\mathbf{j} + \mathbf{k})dy\,dx$
$= \int_0^2\int_0^{(6-3x)/2}(9x - 3x^2)dy\,dx = \int_0^2\left[27x - \frac{45}{2}x^2 + \frac{9}{2}x^3\right]dx = 12.$

19. $\mathbf{F}(\mathbf{r}(\phi,\theta)) = 3\sin\phi\cos\theta\,\mathbf{i} + 3\sin\phi\sin\theta\,\mathbf{j} + 3\cos\phi\,\mathbf{k}$ and
$\mathbf{r}_\phi \times \mathbf{r}_\theta = 9\sin^2\!\phi\cos\theta\,\mathbf{i} + 9\sin^2\!\phi\sin\theta\,\mathbf{j} + 9\sin\phi\cos\phi\,\mathbf{k}$. Then
$\mathbf{F}(\mathbf{r}(\phi,\theta)) \cdot (\mathbf{r}_\phi \times \mathbf{r}_\theta) = 27\sin^3\!\phi\cos^2\!\theta + 27\sin^3\!\phi\sin^2\!\theta + 27\sin\phi\cos^2\!\phi = 27\sin\phi$ and
$\iint_S \mathbf{F}\cdot d\mathbf{S} = \int_0^{2\pi}\!\int_0^{\pi} 27\sin\phi\,d\phi\,d\theta = (2\pi)(54) = 108\pi$.

21. Let S_1 be the paraboloid $y = x^2 + z^2$, $0 \le y \le 1$ and S_2 the disc $x^2 + z^2 \le 1$, $y = 1$. Since S is a closed surface, we use the outward orientation. On S_1: $\mathbf{F}(\mathbf{r}(x,z)) = (x^2+z^2)\mathbf{j} - z\mathbf{k}$ and $\mathbf{r}_x \times \mathbf{r}_z = 2x\mathbf{i} - \mathbf{j} + 2z\mathbf{k}$ (since the $\mathbf{j}$-component must be negative on S_1). Then

$$\iint_{S_1} \mathbf{F}\cdot d\mathbf{S} = \iint_{x^2+z^2\le 1} [-(x^2+z^2) - 2z^2]\,dA = -\int_0^{2\pi}\!\int_0^1 (r^2 + 2r^2\cos^2\theta)r\,dr\,d\theta$$
$$= -\int_0^{2\pi} \tfrac{1}{4}(1 + 2\cos^2\theta)\,d\theta = -\left(\tfrac{\pi}{2} + \tfrac{\pi}{2}\right) = -\pi.$$

On S_2: $\mathbf{F}(\mathbf{r}(x,z)) = \mathbf{j} - z\mathbf{k}$ and $\mathbf{r}_z \times \mathbf{r}_x = \mathbf{j}$. Then $\iint_{S_2} \mathbf{F}\cdot d\mathbf{S} = \iint_{x^2+z^2\le 1} (1)\,dA = \pi$. Hence
$\iint_S \mathbf{F}\cdot d\mathbf{S} = -\pi + \pi = 0$.

23. Here S consists of the six faces of the cube as labeled in the figure.

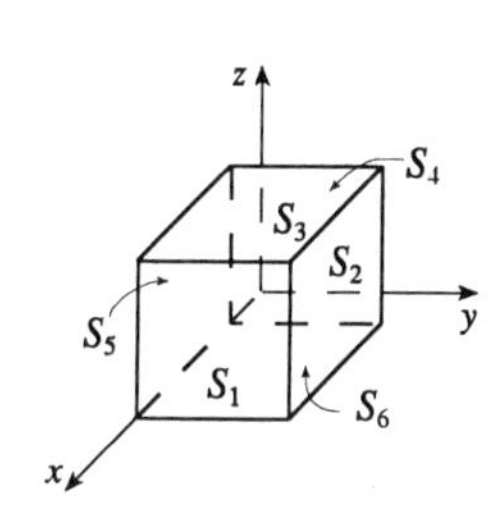

On S_1: $\mathbf{F} = \mathbf{i} + 2y\mathbf{j} + 3z\mathbf{k}$, $\mathbf{r}_y \times \mathbf{r}_z = \mathbf{i}$ and $\iint_{S_1} \mathbf{F}\cdot d\mathbf{S} = \int_{-1}^1\!\int_{-1}^1 dy\,dz = 4$;

S_2: $\mathbf{F} = x\mathbf{i} + 2\mathbf{j} + 3z\mathbf{k}$, $\mathbf{r}_z \times \mathbf{r}_x = \mathbf{j}$ and $\iint_{S_2} \mathbf{F}\cdot d\mathbf{S} = \int_{-1}^1\!\int_{-1}^1 2\,dx\,dz = 8$;

S_3: $\mathbf{F} = x\mathbf{i} + 2y\mathbf{j} + 3\mathbf{k}$, $\mathbf{r}_x \times \mathbf{r}_y = \mathbf{k}$ and $\iint_{S_3} \mathbf{F}\cdot d\mathbf{S} = \int_{-1}^1\!\int_{-1}^1 3\,dx\,dy = 12$;

S_4: $\mathbf{F} = -\mathbf{i} + 2y\mathbf{j} + 3z\mathbf{k}$, $\mathbf{r}_z \times \mathbf{r}_y = -\mathbf{i}$ and $\iint_{S_4} \mathbf{F}\cdot d\mathbf{S} = 4$;

S_5: $\mathbf{F} = x\mathbf{i} - 2\mathbf{j} + 3z\mathbf{k}$, $\mathbf{r}_x \times \mathbf{r}_z = -\mathbf{j}$ and $\iint_{S_5} \mathbf{F}\cdot d\mathbf{S} = 8$;

S_6: $\mathbf{F} = x\mathbf{i} + 2y\mathbf{j} - 3\mathbf{k}$, $\mathbf{r}_y \times \mathbf{r}_x = -\mathbf{k}$ and $\iint_{S_6} \mathbf{F}\cdot d\mathbf{S} = \int_{-1}^1\!\int_{-1}^1 3\,dx\,dy = 12$.

Hence $\iint_S \mathbf{F}\cdot d\mathbf{S} = \sum_{i=1}^6 \iint_{S_i} \mathbf{F}\cdot d\mathbf{S} = 48$.

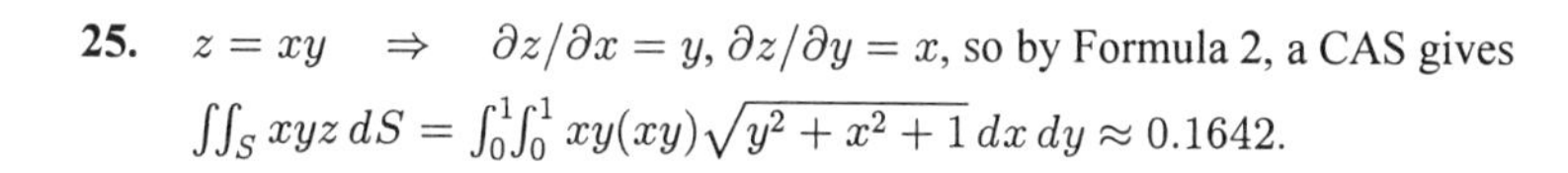

25. $z = xy \quad\Rightarrow\quad \partial z/\partial x = y$, $\partial z/\partial y = x$, so by Formula 2, a CAS gives
$\iint_S xyz\,dS = \int_0^1\!\int_0^1 xy(xy)\sqrt{y^2 + x^2 + 1}\,dx\,dy \approx 0.1642$.

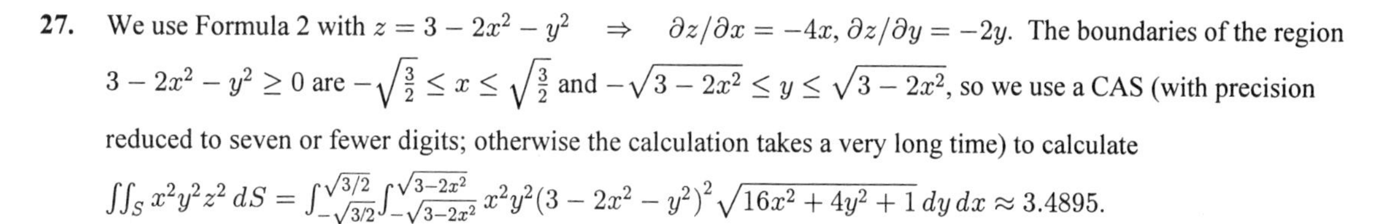

27. We use Formula 2 with $z = 3 - 2x^2 - y^2 \quad\Rightarrow\quad \partial z/\partial x = -4x$, $\partial z/\partial y = -2y$. The boundaries of the region $3 - 2x^2 - y^2 \ge 0$ are $-\sqrt{\frac{3}{2}} \le x \le \sqrt{\frac{3}{2}}$ and $-\sqrt{3 - 2x^2} \le y \le \sqrt{3 - 2x^2}$, so we use a CAS (with precision reduced to seven or fewer digits; otherwise the calculation takes a very long time) to calculate
$\iint_S x^2y^2z^2\,dS = \int_{-\sqrt{3/2}}^{\sqrt{3/2}}\int_{-\sqrt{3-2x^2}}^{\sqrt{3-2x^2}} x^2y^2(3 - 2x^2 - y^2)^2\sqrt{16x^2 + 4y^2 + 1}\,dy\,dx \approx 3.4895$.

29. If S is given by $y = h(x,z)$, then S is also the level surface $f(x,y,z) = y - h(x,z) = 0$.
$\mathbf{n} = \dfrac{\nabla f(x,y,z)}{|\nabla f(x,y,z)|} = \dfrac{-h_x\mathbf{i} + \mathbf{j} - h_z\mathbf{k}}{\sqrt{h_x^2 + 1 + h_z^2}}$, and $-\mathbf{n}$ is the unit normal that points to the left. Now we proceed as in the derivation of (8), using Formula 2 to evaluate

$$\iint_S \mathbf{F}\cdot d\mathbf{S} = \iint_S \mathbf{F}\cdot\mathbf{n}\,dS = \iint_D (P\mathbf{i} + Q\mathbf{j} + R\mathbf{k}) \frac{\dfrac{\partial h}{\partial x}\mathbf{i} - \mathbf{j} + \dfrac{\partial h}{\partial z}\mathbf{k}}{\sqrt{\left(\dfrac{\partial h}{\partial x}\right)^2 + 1 + \left(\dfrac{\partial h}{\partial z}\right)^2}} \sqrt{\left(\frac{\partial h}{\partial x}\right)^2 + 1 + \left(\frac{\partial h}{\partial z}\right)^2}\,dA$$

where D is the projection of $f(x,y,z)$ onto the xz-plane. Therefore

$$\iint_S \mathbf{F}\cdot d\mathbf{S} = \iint_D \left(P\frac{\partial h}{\partial x} - Q + R\frac{\partial h}{\partial z}\right) dA.$$

31. $m = \iint_S K\,dS = K\cdot 4\pi\left(\frac{1}{2}a^2\right) = 2\pi a^2 K$; by symmetry $M_{xz} = M_{yz} = 0$, and
$M_{xy} = \iint_S zK\,dS = K\int_0^{2\pi}\int_0^{\pi/2}(a\cos\phi)(a^2\sin\phi)d\phi\,d\theta = 2\pi K a^3\left[-\frac{1}{4}\cos 2\phi\right]_0^{\pi/2} = \pi K a^3$. Hence
$(\overline{x},\overline{y},\overline{z}) = \left(0,0,\frac{1}{2}a\right)$.

33. **(a)** $I_z = \iint_S (x^2+y^2)\rho(x,y,z)dS$

(b) $I_z = \iint_S (x^2+y^2)\left(10 - \sqrt{x^2+y^2}\right)dS = \iint_{1\le x^2+y^2\le 16} (x^2+y^2)\left(10-\sqrt{x^2+y^2}\right)\sqrt{2}\,dA$

$= \int_0^{2\pi}\int_1^4 \sqrt{2}(10r^3 - r^4)dr\,d\theta = 2\sqrt{2}\pi\left(\frac{4329}{10}\right) = \frac{4329}{5}\sqrt{2}\pi$

35. $\rho(x,y,z) = 1200$, $\mathbf{V} = y\mathbf{i} + \mathbf{j} + z\mathbf{k}$, $\mathbf{F} = \rho\mathbf{V} = (1200)(y\mathbf{i} + \mathbf{j} + z\mathbf{k})$. S is given by
$\mathbf{r}(x,y) = x\mathbf{i} + y\mathbf{j} + \left[9 - \frac{1}{4}(x^2+y^2)\right]\mathbf{k}$, $0 \le x^2+y^2 \le 36$ and $\mathbf{r}_x \times \mathbf{r}_y = \frac{1}{2}x\mathbf{i} + \frac{1}{2}y\mathbf{j} + \mathbf{k}$. Thus the rate of flow is given by

$\iint_S \mathbf{F}\cdot d\mathbf{S} = \iint_{0\le x^2+y^2\le 36} (1200)\left(\frac{1}{2}xy + \frac{1}{2}y + \left[9 - \frac{1}{4}(x^2+y^2)\right]\right)dA$

$= 1200\int_0^6\int_0^{2\pi}\left[\frac{1}{2}r^2\sin\theta\cos\theta + \frac{1}{2}r\sin\theta + 9 - \frac{1}{4}r^2\right]r\,d\theta\,dr = 1200\int_0^6 2\pi\left(9r - \frac{1}{4}r^3\right)dr$

$= (1200)(2\pi)(81) = 194{,}400\pi$.

37. S consists of the hemisphere S_1 given by $z = \sqrt{a^2 - x^2 - y^2}$ and the disk S_2 given by $0 \le x^2+y^2 \le a^2$, $z = 0$.
On S_1: $\mathbf{E} = a\sin\phi\cos\theta\,\mathbf{i} + a\sin\phi\sin\theta\,\mathbf{j} + 2a\cos\phi\,\mathbf{k}$,
$\mathbf{T}_\phi \times \mathbf{T}_\theta = a^2\sin^2\phi\cos\theta\,\mathbf{i} + a^2\sin^2\phi\sin\theta\,\mathbf{j} + a^2\sin\phi\cos\phi\,\mathbf{k}$. Thus
$\iint_{S_1} \mathbf{E}\cdot d\mathbf{S} = \int_0^{2\pi}\int_0^{\pi/2}(a^3\sin^3\phi + 2a^3\sin\phi\cos^2\phi)d\phi\,d\theta$

$= \int_0^{2\pi}\int_0^{\pi/2}(a^3\sin\phi + a^3\sin\phi\cos^2\phi)d\phi\,d\theta = (2\pi)a^3\left(1 + \frac{1}{3}\right) = \frac{8}{3}\pi a^3$.

On S_2: $\mathbf{E} = x\mathbf{i} + y\mathbf{j}$, and $\mathbf{r}_y \times \mathbf{r}_x = -\mathbf{k}$ so $\iint_{S_2} \mathbf{E}\cdot d\mathbf{S} = 0$. Hence the total charge is $q = \epsilon_0\iint_S \mathbf{E}\cdot d\mathbf{S} = \frac{8}{3}\pi a^3\epsilon_0$.

39. $K\nabla u = 6.5(4y\mathbf{j} + 4z\mathbf{k})$. S is given by $\mathbf{r}(x,\theta) = x\mathbf{i} + \sqrt{6}\cos\theta\,\mathbf{j} + \sqrt{6}\sin\theta\,\mathbf{k}$ and since we want the inward heat flow, we use $\mathbf{r}_x \times \mathbf{r}_\theta = -\sqrt{6}\cos\theta\,\mathbf{j} - \sqrt{6}\sin\theta\,\mathbf{k}$. Then the rate of heat flow inward is given by
$\iint_S (-K\nabla u)\cdot d\mathbf{S} = \int_0^{2\pi}\int_0^4 -(6.5)(-24)dx\,d\theta = (2\pi)(156)(4) = 1248\pi$.

EXERCISES 14.8

1. The boundary curve is C: $x^2 + y^2 = 1$, $z = 0$ oriented in the counterclockwise direction. The vector equation of C is $\mathbf{r}(t) = \cos t\,\mathbf{i} + \sin t\,\mathbf{j}$, $0 \le t \le 2\pi$. Then $\mathbf{F}(\mathbf{r}(t)) = \cos t\,\mathbf{j} + e^{\cos t \sin t}\,\mathbf{k}$ and $\mathbf{F}(\mathbf{r}(t)) \cdot \mathbf{r}'(t) = \cos^2 t$. Hence $\iint_S \operatorname{curl} \mathbf{F} \cdot d\mathbf{S} = \oint_C \mathbf{F} \cdot d\mathbf{r} = \int_0^{2\pi} \cos^2 t\, dt = \int_0^{2\pi} \frac{1}{2}(1 + \cos 2t)dt = \pi$.

3. C is the circle $x^2 + z^2 = 1$, $y = 0$ and the vector equation is $\mathbf{r}(t) = \cos t\,\mathbf{i} + \sin t\,\mathbf{k}$, $0 \le t \le 2\pi$ since the surface is oriented toward the xy-plane. Then $\mathbf{F}(\mathbf{r}(t)) = \cos^3 t\,\mathbf{k}$ and $\mathbf{F}(\mathbf{r}(t)) \cdot \mathbf{r}'(t) = \cos^4 t$. Hence
$\iint_S \operatorname{curl} \mathbf{F} \cdot d\mathbf{S} = \oint_C \mathbf{F} \cdot d\mathbf{r} = \int_0^{2\pi} \cos^4 t\, dt = \int_0^{2\pi} \left[\frac{3}{8} + \frac{1}{2}\cos 2t + \frac{1}{8}\cos 4t\right]dt = \frac{3\pi}{4}$.

5. C is the square in the plane $z = -1$. By (4), $\iint_{S_1} \operatorname{curl} \mathbf{F} \cdot d\mathbf{S} = \oint_C \mathbf{F} \cdot d\mathbf{r} = \iint_{S_2} \operatorname{curl} \mathbf{F} \cdot d\mathbf{S}$ where S_1 is the original cube without the bottom and S_2 is the bottom face of the cube.
$\operatorname{curl} \mathbf{F} = x^2 z\,\mathbf{i} + (xy - 2xyz)\mathbf{j} + (y - xz)\mathbf{k}$. For S_2, $\mathbf{n} = -\mathbf{k}$ and $\operatorname{curl} \mathbf{F} \cdot \mathbf{n} = xz - y = -x - y$ on S_2, where $z = -1$. Then $\iint_{S_2} \operatorname{curl} \mathbf{F} \cdot d\mathbf{S} = -\int_{-1}^{1}\int_{-1}^{1}(x + y)dx\,dy = 0$ so $\iint_{S_1} \operatorname{curl} \mathbf{F} \cdot d\mathbf{S} = 0$.

7. $\operatorname{curl} \mathbf{F} = 3x\mathbf{i} + (x - 3y)\mathbf{j} + 2y\mathbf{k}$, $\mathbf{n} = \frac{1}{\sqrt{11}}(3\mathbf{i} + \mathbf{j} + \mathbf{k})$ and

$$\begin{aligned}\oint_C \mathbf{F} \cdot d\mathbf{r} &= \iint_S \operatorname{curl} \mathbf{F} \cdot \mathbf{n}\, dS = \int_0^1\int_0^{3-3x} \frac{1}{\sqrt{11}}[9x + (x - 3y) + 2y]\left(\sqrt{11}\right)dy\,dx \\ &= \int_0^1\int_0^{3-3x}(10x - y)dy\,dx = \int_0^1 \left[10(3x - 3x^2) - \tfrac{1}{2}(3 - 3x)^2\right]dx \\ &= \left[15x^2 - 10x^3 + \tfrac{3}{2}(1 - x^3)\right]_0^1 = \tfrac{7}{2}.\end{aligned}$$

9. The curve of intersection is an ellipse in the plane $z = x + 4$ with unit normal $\mathbf{n} = \frac{1}{\sqrt{2}}(-\mathbf{i} + \mathbf{k})$ and $\operatorname{curl} \mathbf{F} = 5\mathbf{i} + 2\mathbf{j} + 4\mathbf{k}$ so $\operatorname{curl} \mathbf{F} \cdot \mathbf{n} = -\frac{1}{\sqrt{2}}$. Then
$\oint_C \mathbf{F} \cdot d\mathbf{r} = -\iint_S \frac{1}{\sqrt{2}} dS = -\frac{1}{\sqrt{2}}(\text{surface area of planar ellipse}) = -\frac{1}{\sqrt{2}}\pi(2)\left(2\sqrt{2}\right) = -4\pi$.

11. **(a)** The curve of intersection is an ellipse in the plane $x + y + z = 1$ with unit normal $\mathbf{n} = \frac{1}{\sqrt{3}}(\mathbf{i} + \mathbf{j} + \mathbf{k})$, $\operatorname{curl} \mathbf{F} = x^2\mathbf{j} + y^2\mathbf{k}$ and $\operatorname{curl} \mathbf{F} \cdot \mathbf{n} = \frac{1}{\sqrt{3}}(x^2 + y^2)$. Then
$\oint_C \mathbf{F} \cdot d\mathbf{r} = \iint_S \frac{1}{\sqrt{3}}(x^2 + y^2)dS = \iint_{x^2 + y^2 \le 9} (x^2 + y^2)dx\,dy = \int_0^{2\pi}\int_0^3 r^3\, dr\, d\theta = 2\pi\left(\frac{81}{4}\right) = \frac{81\pi}{2}$.

(b)

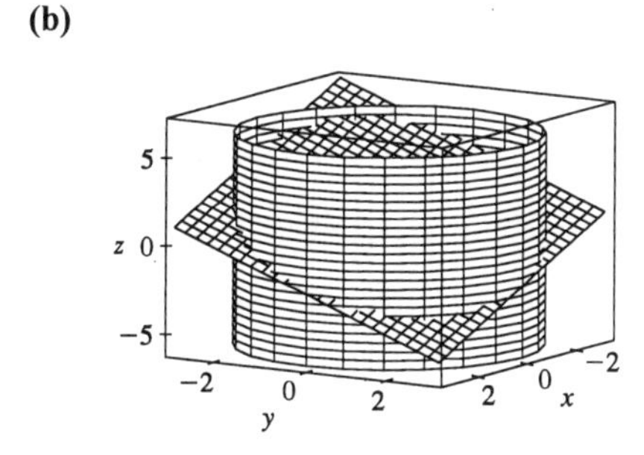

(c) One possible parametrization is $x = 3\cos t$, $y = 3\sin t$, $z = 1 - 3\cos t - 3\sin t$, $0 \le t \le 2\pi$.

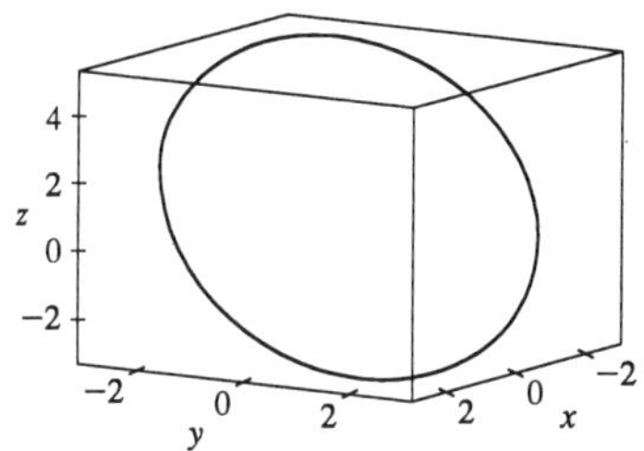

13. The boundary curve C is the circle $x^2+y^2=9$, $z=0$ oriented in the counterclockwise direction as viewed from $(0,0,1)$. Then $\mathbf{r}(t)=3\cos t\,\mathbf{i}+3\sin t\,\mathbf{j}$, $0\le t\le 2\pi$, so $\mathbf{F}(\mathbf{r}(t))=9\sin t\,\mathbf{i}-18\cos t\,\mathbf{k}$ and $\mathbf{F}\cdot\mathbf{r}'(t)=-27\sin^2 t$. Thus $\oint_C \mathbf{F}\cdot d\mathbf{r}=\int_0^{2\pi}(-27\sin^2 t)dt=-27\pi$. Now $\operatorname{curl}\mathbf{F}=-4\mathbf{i}+6\mathbf{j}-3\mathbf{k}$, $\mathbf{r}_x\times\mathbf{r}_y=2x\mathbf{i}+2y\mathbf{j}+\mathbf{k}$, so

$$\iint_S \operatorname{curl}\mathbf{F}\cdot d\mathbf{S}=\iint_{x^2+y^2\le 9}(-8x+12y-3)dA=\int_0^{2\pi}\int_0^3(-8r\cos\theta+12r\sin\theta-3)r\,dr\,d\theta$$
$$=\int_0^3(-3r)(2\pi)dr=-27\pi.$$

15. The x-, y-, and z-intercepts of the plane are all 1, so C consists of the three line segments C_1: $\mathbf{r}_1(t)=(1-t)\mathbf{i}+t\mathbf{j}$, $0\le t\le 1$, C_2: $\mathbf{r}_2(t)=(1-t)\mathbf{j}+t\mathbf{k}$, $0\le t\le 1$, and C_3: $\mathbf{r}_3(t)=t\mathbf{i}+(1-t)\mathbf{k}$, $0\le t\le 1$. Then

$$\oint_C \mathbf{F}\cdot d\mathbf{r}=\int_0^1[t\mathbf{i}+(1-t)\mathbf{k}]\cdot(-\mathbf{i}+\mathbf{j})dt+\int_0^1[(1-t)\mathbf{i}+t\mathbf{j}]\cdot(-\mathbf{j}+\mathbf{k})dt+\int_0^1[(1-t)\mathbf{j}+t\mathbf{k}]\cdot(\mathbf{i}-\mathbf{k})dt$$
$$=\int_0^1(-3t)dt=-\tfrac{3}{2}.$$

Now $\operatorname{curl}\mathbf{F}=-\mathbf{i}-\mathbf{j}-\mathbf{k}$ and $\mathbf{r}_x\times\mathbf{r}_y=\mathbf{i}+\mathbf{j}+\mathbf{k}$. Hence $\iint_S \operatorname{curl}\mathbf{F}\cdot d\mathbf{S}=\int_0^1\int_0^{1-x}(-3)dy\,dx=-\frac{3}{2}$.

17. $\operatorname{curl}\mathbf{F}=\begin{vmatrix}\mathbf{i}&\mathbf{j}&\mathbf{k}\\ \partial/\partial x&\partial/\partial y&\partial/\partial z\\ x^x+z^2&y^y+x^2&z^z+y^2\end{vmatrix}=2y\mathbf{i}+2z\mathbf{j}+2x\mathbf{k}$ and $W=\int_C\mathbf{F}\cdot d\mathbf{r}=\iint_S\operatorname{curl}\mathbf{F}\cdot d\mathbf{S}$.

To parametrize the surface, let $x=2\cos\theta\sin\phi$, $y=2\sin\theta\sin\phi$, $z=2\cos\phi$, so that $\mathbf{r}(\phi,\theta)=2\sin\phi\cos\theta\,\mathbf{i}+2\sin\phi\sin\theta\,\mathbf{j}+2\cos\phi\,\mathbf{k}$, $0\le\phi\le\frac{\pi}{2}$, $0\le\theta\le\frac{\pi}{2}$, and $\mathbf{r}_\phi\times\mathbf{r}_\theta=4\sin^2\phi\cos\theta\,\mathbf{i}+4\sin^2\phi\sin\theta\,\mathbf{j}+4\sin\phi\cos\phi\,\mathbf{k}$. Then $\operatorname{curl}\mathbf{F}(\mathbf{r}(\phi,\theta))=4\sin\phi\sin\theta\,\mathbf{i}+4\cos\phi\,\mathbf{j}+4\sin\phi\cos\theta\,\mathbf{k}$, and $\operatorname{curl}\mathbf{F}\cdot(\mathbf{r}_\phi\times\mathbf{r}_\theta)=16\sin^3\phi\sin\theta\cos\theta+16\cos\phi\sin^2\phi\sin\theta+16\sin^2\phi\cos\phi\cos\theta$. Therefore

$$\iint_S\operatorname{curl}\mathbf{F}\cdot d\mathbf{S}=\iint_D\operatorname{curl}\mathbf{F}\cdot(\mathbf{r}_\phi\times\mathbf{r}_\theta)dA$$
$$=16\left[\int_0^{\pi/2}\sin\theta\cos\theta\,d\theta\right]\left[\int_0^{\pi/2}\sin^3\phi\,d\phi\right]+16\left[\int_0^{\pi/2}\sin\theta\,d\theta\right]\left[\int_0^{\pi/2}\sin^2\phi\cos\phi\,d\phi\right]$$
$$+16\left[\int_0^{\pi/2}\cos\theta\,d\theta\right]\left[\int_0^{\pi/2}\sin^2\phi\cos\phi\,d\phi\right]$$
$$=8\left[-\cos\phi+\tfrac{1}{3}\cos^3\phi\right]_0^{\pi/2}+16(1)\left[\tfrac{1}{3}\sin^3\phi\right]_0^{\pi/2}+16(1)\left[\tfrac{1}{3}\sin^3\phi\right]_0^{\pi/2}$$
$$=8\left[0+1+0-\tfrac{1}{3}\right]+16\left(\tfrac{1}{3}\right)+16\left(\tfrac{1}{3}\right)=\tfrac{16}{3}+\tfrac{16}{3}+\tfrac{16}{3}=16.$$

19. Assume S is centered at the origin with radius a and let H_1 and H_2 be the upper and lower hemispheres, respectively, of S. Then $\iint_S\operatorname{curl}\mathbf{F}\cdot d\mathbf{S}=\iint_{H_1}\operatorname{curl}\mathbf{F}\cdot d\mathbf{S}+\iint_{H_2}\operatorname{curl}\mathbf{F}\cdot d\mathbf{S}=\oint_{C_1}\mathbf{F}\cdot d\mathbf{r}+\oint_{C_2}\mathbf{F}\cdot d\mathbf{r}$ by Stokes' Theorem. But C_1 is the circle $x^2+y^2=a^2$ oriented in the counterclockwise direction while C_2 is the same circle but oriented in the clockwise direction. Hence $\oint_{C_2}\mathbf{F}\cdot d\mathbf{r}=-\oint_{C_1}\mathbf{F}\cdot d\mathbf{r}$ so $\iint_S\operatorname{curl}\mathbf{F}\cdot d\mathbf{S}=0$ as desired.

EXERCISES 14.9

1.

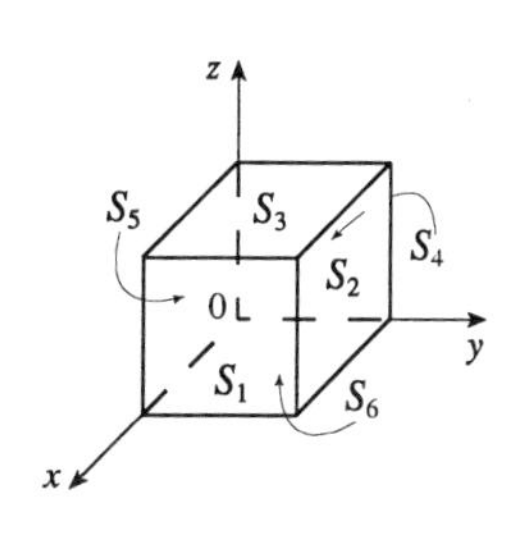

$\text{div}\,\mathbf{F} = 3 + x + 2x = 3 + 3x$, so $\iiint_E \text{div}\,\mathbf{F}\,dV = \int_0^1\int_0^1\int_0^1 (3x+3)dx\,dy\,dz = \frac{9}{2}$ (notice the triple integral is three times the volume of the cube plus three times $\overline{x}$).

To compute $\iint_S \mathbf{F}\cdot d\mathbf{S}$: on

S_1: $\mathbf{n} = \mathbf{i}$, $\mathbf{F} = 3\mathbf{i} + y\mathbf{j} + 2z\mathbf{k}$, and $\iint_{S_1} \mathbf{F}\cdot d\mathbf{S} = \iint_{S_1} 3\,dS = 3$;

S_2: $\mathbf{F} = 3x\mathbf{i} + x\mathbf{j} + 2xz\mathbf{k}$, $\mathbf{n} = \mathbf{j}$ and $\iint_{S_2} \mathbf{F}\cdot d\mathbf{S} = \iint_{S_2} x\,dS = \frac{1}{2}$;

S_3: $\mathbf{F} = 3x\mathbf{i} + xy\mathbf{j} + 2x\mathbf{k}$, $\mathbf{n} = \mathbf{k}$ and $\iint_{S_3} \mathbf{F}\cdot d\mathbf{S} = \iint_{S_3} 2x\,dS = 1$;

S_4: $\mathbf{F} = \mathbf{0}$, $\iint_{S_4} \mathbf{F}\cdot d\mathbf{S} = 0$; S_5: $\mathbf{F} = 3x\mathbf{i} + 2x\mathbf{k}$, $\mathbf{n} = -\mathbf{j}$ and $\iint_{S_5} \mathbf{F}\cdot d\mathbf{S} = \iint_{S_5} 0\,dS = 0$;

S_6: $\mathbf{F} = 3x\mathbf{i} + xy\mathbf{j}$, $\mathbf{n} = -\mathbf{k}$ and $\iint_{S_6} \mathbf{F}\cdot d\mathbf{S} = \iint_{S_6} 0\,dS = 0$. Thus $\iint_S \mathbf{F}\cdot d\mathbf{S} = \frac{9}{2}$.

3. $\text{div}\,\mathbf{F} = \dfrac{\partial}{\partial x}(3y^2z^3) + \dfrac{\partial}{\partial y}(9x^2yz^2) + \dfrac{\partial}{\partial z}(4xy^2) = 9x^2z^2$, so by the Divergence Theorem,

$\iint_S \mathbf{F}\cdot d\mathbf{S} = \iiint_E 9x^2z^2\,dV = \int_{-1}^1\int_{-1}^1\int_{-1}^1 9x^2z^2\,dx\,dy\,dz = 8.$

5. $\iint_S \mathbf{F}\cdot d\mathbf{S} = \iiint_E (-z - z + 2z)dV = 0$

7. $\iint_S \mathbf{F}\cdot d\mathbf{S} = \iiint_E x\,dV = \int_0^1\int_0^{2-2x}\int_0^{2-2x-y} x\,dz\,dy\,dx = \int_0^1\int_0^{2-2x}[x(2-2x) - xy]dy\,dx$

$= \int_0^1 \left[x(2-2x)^2 - \frac{1}{2}x(2-2x)^2\right]dx = \frac{1}{6}$

9. $\iint_S \mathbf{F}\cdot d\mathbf{S} = \iiint_E 3(x^2+y^2+z^2)dV = \int_0^{2\pi}\int_0^{\pi}\int_0^1 3\rho^4 \sin\phi\,d\rho\,d\phi\,d\theta = 2\pi\int_0^{\pi} \frac{3}{5}\sin\phi\,d\phi = \frac{12}{5}\pi$

11. $\iint_S \mathbf{F}\cdot d\mathbf{S} = \iiint_E 2y\,dV = \iint\limits_{x^2+y^2\le 9} \int_{y-3}^0 2y\,dz\,dA = \int_0^{2\pi}\int_0^3\int_{-3+r\sin\theta}^0 (2r^2\sin\theta)dz\,dr\,d\theta$

$= \int_0^{2\pi}\int_0^3 (6r^2\sin\theta - 2r^3\sin^2\theta)dr\,d\theta = \int_0^{2\pi}\left[54\sin\theta - \frac{81}{2}\sin^2\theta\right]d\theta = -\frac{81}{2}\pi$

13. $\iint_S \mathbf{F}\cdot d\mathbf{S} = \iiint_E (x^2+y^2+z)dV = \int_0^{2\pi}\int_1^2\int_1^3 (r^2+z)r\,dz\,dr\,d\theta = 2\pi\int_1^2 (2r^3+4r)dr = 27\pi$

15. $\iint_S \mathbf{F}\cdot d\mathbf{S} = \iiint_E \sqrt{3-x^2}\,dV = \int_{-1}^1\int_{-1}^1\int_0^{2-x^4-y^4} \sqrt{3-x^2}\,dz\,dy\,dx = \frac{341}{60}\sqrt{2} + \frac{81}{20}\sin^{-1}\left(\frac{\sqrt{3}}{3}\right)$

17. For S_1 we have $\mathbf{n} = -\mathbf{k}$, so $\mathbf{F}\cdot\mathbf{n} = \mathbf{F}\cdot(-\mathbf{k}) = -x^2z - y^2 = -y^2$ (since $z = 0$ on S_1). So if D is the unit disk, we get $\iint_{S_1} \mathbf{F}\cdot d\mathbf{S} = \iint_{S_1} \mathbf{F}\cdot\mathbf{n}\,dS = \iint_D (-y^2)dA = -\int_0^{2\pi}\int_0^1 r^2\sin^2\theta\,r\,dr\,d\theta = -\frac{1}{4}\pi$. Now since S_2 is closed, we can use the Divergence Theorem. Since

$\text{div}\,\mathbf{F} = \dfrac{\partial}{\partial x}(z^2x) + \dfrac{\partial}{\partial y}\left(\frac{1}{3}y^3 + \tan z\right) + \dfrac{\partial}{\partial z}(x^2z + y^2) = z^2 + y^2 + x^2$, we use spherical coordinates to get

$\iint_{S_2} \mathbf{F}\cdot d\mathbf{S} = \iiint_E \text{div}\,\mathbf{F}\,dV = \int_0^{2\pi}\int_0^{\pi/2}\int_0^1 \rho^2\cdot\rho^2\sin\phi\,d\rho\,d\phi\,d\theta = \frac{2}{5}\pi$. Finally

$\iint_S \mathbf{F}\cdot d\mathbf{S} = \iint_{S_2} \mathbf{F}\cdot d\mathbf{S} - \iint_{S_1} \mathbf{F}\cdot d\mathbf{S} = \frac{2}{5}\pi - \left(-\frac{1}{4}\pi\right) = \frac{13}{20}\pi.$

19. Since $\dfrac{\mathbf{x}}{|\mathbf{x}|^3} = \dfrac{x\mathbf{i} + y\mathbf{j} + z\mathbf{k}}{(x^2+y^2+z^2)^{3/2}}$ and $\dfrac{\partial}{\partial x}\left[\dfrac{x}{(x^2+y^2+z^2)^{3/2}}\right] = \dfrac{(x^2+y^2+z^2)-3x^2}{(x^2+y^2+z^2)^{5/2}}$ with similar expressions for $\dfrac{\partial}{\partial y}\left[\dfrac{y}{(x^2+y^2+z^2)^{3/2}}\right]$ and $\dfrac{\partial}{\partial z}\left[\dfrac{z}{(x^2+y^2+z^2)^{3/2}}\right]$, we have

$\operatorname{div}\left(\dfrac{\mathbf{x}}{|\mathbf{x}|^3}\right) = \dfrac{3(x^2+y^2+z^2)-3(x^2+y^2+z^2)}{(x^2+y^2+z^2)^{5/2}} = 0$, except at $(0,0,0)$ where it is undefined.

21. $\iint_S \mathbf{a}\cdot\mathbf{n}\,dS = \iiint_E \operatorname{div}\mathbf{a}\,dV = 0$ since $\operatorname{div}\mathbf{a} = 0$.

23. $\iint_S \operatorname{curl}\mathbf{F}\cdot d\mathbf{S} = \iiint_E \operatorname{div}(\operatorname{curl}\mathbf{F})dV = 0$ by Theorem 14.5.11.

25. $\iint_S (f\nabla g)\cdot\mathbf{n}\,dS = \iiint_E \operatorname{div}(f\nabla g)dV = \iiint_E (f\nabla^2 g + \nabla g\cdot\nabla f)dV$ by Exercise 14.5.25.

REVIEW EXERCISES FOR CHAPTER 14

1. False; $\operatorname{div}\mathbf{F}$ is a scalar field.

3. True, by (14.5.3) and the fact that $\operatorname{div}\mathbf{0} = 0$.

5. False. See Exercise 14.3.33. [But the assertion is true if D is simply-connected; see (14.3.6).]

7. True. Apply the Divergence Theorem and use the fact that $\operatorname{div}\mathbf{F} = 0$.

9. $x = \frac{1}{2}y^2$, $y = y$, $\int_C y\,dS = \int_0^2 y\sqrt{y^2+1}\,dy = \frac{1}{3}\left(5\sqrt{5}-1\right)$

11. $\int_C x^3 z\,ds = \int_0^{\pi/2}(16\sin^3 t\cos t)\sqrt{5}\,dt = 4\sqrt{5}\sin^4 t\Big|_0^{\pi/2} = 4\sqrt{5}$

13. $x = \cos t$, $y = \sin t$, $0 \le t \le 2\pi$ and $\int_C x^3 y\,dx - x\,dy = \int_0^{2\pi}(-\cos^3 t\sin^2 t - \cos^2 t)dt = -\pi$

Or: Since C is a simple closed curve, apply Green's Theorem giving

$\displaystyle\iint_{x^2+y^2\le 1}(-1-x^3)dA = \int_0^1\int_0^{2\pi}(-r - r^4\cos^3\theta)d\theta = -\pi.$

15. C_1: $x = t$, $y = t$, $z = 2t$, $0 \le t \le 1$;

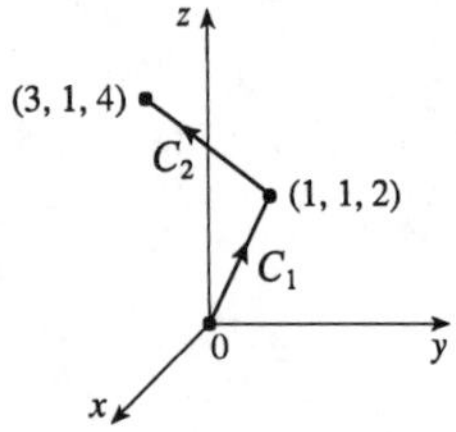

C_2: $x = 1 + 2t$, $y = 1$, $z = 2 + 2t$, $0 \le t \le 1$. Then

$\int_C y\,dx + z\,dy + x\,dz = \int_0^1 5t\,dt + \int_0^1 (4+4t)dt = \frac{17}{2}$.

17. $\mathbf{F}(\mathbf{r}(t)) = (2t+t^2)\mathbf{i} + t^4\mathbf{j} + 4t^4\mathbf{k}$, $\mathbf{F}\cdot\mathbf{r}'(t) = 4t + 2t^2 + 2t^5 + 16t^7$ and

$\int_C \mathbf{F}\cdot d\mathbf{r} = \int_0^1 (4t + 2t^2 + 2t^5 + 16t^7)dt = 5.$

19. $\dfrac{\partial(\sin y)}{\partial y} = \cos y$ and $\dfrac{\partial(x\cos y + \sin y)}{\partial x} = \cos y$ and the domain of $\mathbf{F}$ is $\mathbb{R}^2$ so $\mathbf{F}$ is conservative. Hence there exists f such that $\nabla f = \mathbf{F}$. Then $f_x(x,y) = \sin y$ implies $f(x,y) = x\sin y + g(y)$ and $f_y(x,y) = x\cos y + g'(y)$. But $f_y(x,y) = x\cos y + \sin y$ so $g'(y) = \sin y$ and $f(x,y) = x\sin y - \cos y + K$ is a potential for $\mathbf{F}$.

21. Since $\dfrac{\partial(2x + y^2 + 3x^2 y)}{\partial y} = 2y + 3x^2 = \dfrac{\partial(2xy + x^3 + 3y^2)}{\partial x}$ and the domain of $\mathbf{F}$ is $\mathbb{R}^2$, $\mathbf{F}$ is conservative. Furthermore $f(x,y) = x^2 + xy^2 + x^3 y + y^3 + K$ is a potential for $\mathbf{F}$. Then $\int_C \mathbf{F}\cdot d\mathbf{r} = f(\pi,0) - f(0,0) = \pi^2$.

23.

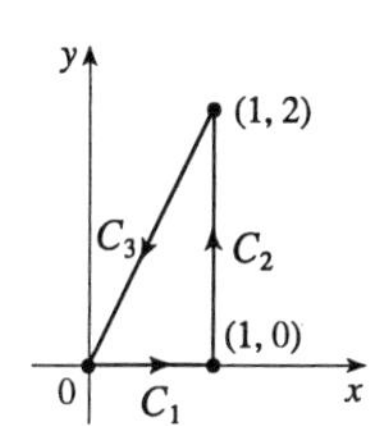

C_1: $0 \le x \le 1$, $y = 0$; C_2: $x = 1$, $0 \le y \le 2$; C_3: $x = x$, $y = 2x$, $x = 1$ to $x = 0$. Then $\oint_C xy\,dx + x^2\,dy = \int_0^1 0\,dx + \int_0^2 (0+1)dy + \int_1^0 (2x^2 + 2x^2)dx = \frac{2}{3}$. And $\iint_D (2x - x)dA = \int_0^1\int_0^{2x} x\,dy\,dx = \frac{2}{3}$.

25. $\displaystyle\int_C x^2 y\,dx - xy^2\,dy = \iint_{x^2+y^2\le 4} (-y^2 - x^2)dA = -\int_0^{2\pi}\int_0^2 r^3\,dr\,d\theta = -8\pi$

27. If we assume there is such a vector field $\mathbf{G}$, then $\operatorname{div}(\operatorname{curl}\mathbf{G}) = 2 + 3z - 2xz$. But $\operatorname{div}(\operatorname{curl}\mathbf{F}) = 0$ for all vector fields $\mathbf{F}$. Thus such a $\mathbf{G}$ cannot exist.

29. For any piecewise-smooth simple closed plane curve C, bounding a region D, we can apply Green's Theorem to $\mathbf{F}(x,y) = f(x)\mathbf{i} + g(y)\mathbf{j}$ to get $\displaystyle\int_C f(x)dx + g(y)dy = \iint_D \left[\frac{\partial}{\partial x}g(y) - \frac{\partial}{\partial y}f(x)\right]dA = \iint_D 0\,dA = 0$.

31. $\nabla^2 f = 0$ means that $\dfrac{\partial^2 f}{\partial x^2} + \dfrac{\partial^2 f}{\partial y^2} = 0$. Now if $\mathbf{F} = f_y\mathbf{i} - f_x\mathbf{j}$ and C is any closed path in D, then applying Green's Theorem, we get

$$\int_C \mathbf{F}\cdot d\mathbf{r} = \int_C f_y\,dx - f_x\,dy = \iint_D \left[\frac{\partial}{\partial x}(-f_x) - \frac{\partial}{\partial y}(f_y)\right]dA = -\iint_D (f_{xx} + f_{yy})dA = -\iint_D 0\,dA = 0$$

Therefore the line integral is independent of path by Theorem 14.3.3.

33. $z = f(x,y) = x^2 + 2y$ with $0 \le x \le 1$, $0 \le y \le 2x$. Thus $A(S) = \iint_D \sqrt{1 + 4x^2 + 4}\,dA = \int_0^1\int_0^{2x} \sqrt{5 + x^2}\,dy\,dx = \int_0^1 2x\sqrt{5 + x^2}\,dx = \frac{2}{3}\left(6\sqrt{6} - 5\sqrt{5}\right)$.

35. $z = f(x, y) = x^2 + y^2$ with $0 \le x^2 + y^2 \le 4$ so $\mathbf{r}_x \times \mathbf{r}_y = -2x\mathbf{i} - 2y\mathbf{j} + \mathbf{k}$ (using upward orientation). Then

$$\iint_S z\,dS = \iint_{x^2+y^2\le 4} (x^2 + y^2)\sqrt{4x^2 + 4y^2 + 1}\,dA = \int_0^{2\pi}\int_0^2 r^3\sqrt{1 + 4r^2}\,dr\,d\theta = \tfrac{1}{60}\pi\left(391\sqrt{17} + 1\right).$$

(Substitute $u = 1 + 4r^2$ and use tables.)

37. Since the sphere bounds a simple solid region, the Divergence Theorem applies and

$\iint_S \mathbf{F}\cdot d\mathbf{S} = \iiint_E (z - 2)dV = \iiint_E z\,dV - 2\iiint_E dV = m\overline{z} - 2\left(\frac{4}{3}\pi 2^3\right) = -\frac{64}{3}\pi.$

Alternate Solution: $\mathbf{F}(\mathbf{r}(\phi,\theta)) = 4\sin\phi\cos\theta\cos\phi\,\mathbf{i} - 4\sin\phi\sin\theta\,\mathbf{j} + 6\sin\phi\cos\theta\,\mathbf{k}$,

$\mathbf{r}_\phi \times \mathbf{r}_\theta = 4\sin^2\!\phi\cos\theta\,\mathbf{i} + 4\sin^2\!\phi\sin\theta\,\mathbf{j} + 4\sin\phi\cos\phi\,\mathbf{k}$, and

$\mathbf{F}\cdot(\mathbf{r}_\phi \times \mathbf{r}_\theta) = 16\sin^3\!\phi\cos^2\!\theta\cos\phi - 16\sin^3\!\phi\sin^2\!\theta + 24\sin^2\!\phi\cos\phi\cos\theta$. Then

$$\iint_S \mathbf{F}\cdot d\mathbf{S} = \int_0^{2\pi}\int_0^{\pi}(16\sin^3\!\phi\cos\phi\cos^2\!\theta - 16\sin^3\!\phi\sin^2\!\theta + 24\sin^2\!\phi\cos\phi\cos\theta)d\phi\,d\theta$$
$$= \int_0^{2\pi}\tfrac{4}{3}(-16\sin^2\!\theta)d\theta = -\tfrac{64}{3}\pi.$$

39. Since $\operatorname{curl}\mathbf{F} = \mathbf{0}$, $\iint_S(\operatorname{curl}\mathbf{F})\cdot d\mathbf{S} = 0$. And C: $\mathbf{r}(t) = \cos t\,\mathbf{i} + \sin t\,\mathbf{j}, 0 \le t \le 2\pi$ and

$\oint_C \mathbf{F}\cdot d\mathbf{r} = \int_0^{2\pi}(-\cos^2\!t\sin t + \sin^2\!t\cos t)dt = \frac{1}{3}\cos^3\!t + \frac{1}{3}\sin^3\!t\Big|_0^{2\pi} = 0.$

41. The surface is given by $x + y + z = 1$ or $z = 1 - x - y$, $0 \le x \le 1$, $0 \le y \le 1 - x$ and $\mathbf{r}_x \times \mathbf{r}_y = \mathbf{i} + \mathbf{j} + \mathbf{k}$.
Then $\oint_C \mathbf{F}\cdot d\mathbf{r} = \iint_S \operatorname{curl}\mathbf{F}\cdot d\mathbf{S} = \iint_D(-y\mathbf{i} - z\mathbf{j} - x\mathbf{k})\cdot(\mathbf{i} + \mathbf{j} + \mathbf{k})dA = \iint_D(-1)dA = -(\text{area of D}) = -\frac{1}{2}$.

43. $\iiint_E \operatorname{div}\mathbf{F}\,dV = \iiint_{x^2+y^2+z^2\le 1} 3\,dV = 3(\text{volume of sphere}) = 4\pi$. Then

$\mathbf{F}(\mathbf{r}(\phi,\theta))\cdot(\mathbf{r}_\phi \times \mathbf{r}_\theta) = \sin^3\!\phi\cos^2\!\theta + \sin^3\!\phi\sin^2\!\theta + \sin\phi\cos^2\!\phi = \sin\phi$ and

$\iint_S \mathbf{F}\cdot d\mathbf{S} = \int_0^{2\pi}\int_0^{\pi}\sin\phi\,d\phi\,d\theta = (2\pi)(2) = 4\pi.$

45. Because $\operatorname{curl}\mathbf{F} = \mathbf{0}$, $\mathbf{F}$ is conservative, and if $f(x,y,z) = x^3yz - 3xy + z^2$, then $\nabla f = \mathbf{F}$. Hence
$\int_C \mathbf{F}\cdot d\mathbf{r} = \int_C \nabla f\cdot d\mathbf{r} = f(0,3,0) - f(0,0,2) = 0 - 4 = -4.$

47. By the Divergence Theorem, $\iint_S \mathbf{F}\cdot\mathbf{n}\,dS = \iiint_E \operatorname{div}\mathbf{F}\,dV = 3(\text{volume of } E) = 3(8 - 1) = 21.$

PROBLEMS PLUS (after Chapter 14)

1. Since $|xy| < 1$, except at $(1,1)$, the formula for the sum of a geometric series gives $\dfrac{1}{1-xy} = \sum_{n=0}^{\infty} (xy)^n$, so

$$\int_0^1 \int_0^1 \frac{1}{1-xy}\,dx\,dy = \int_0^1 \int_0^1 \sum_{n=0}^{\infty} (xy)^n\,dx\,dy = \sum_{n=0}^{\infty} \int_0^1 \int_0^1 (xy)^n\,dx\,dy = \sum_{n=0}^{\infty} \left[\int_0^1 x^n\,dx\right]\left[\int_0^1 y^n\,dy\right]$$

$$= \sum_{n=0}^{\infty} \frac{1}{n+1} \cdot \frac{1}{n+1} = \sum_{n=0}^{\infty} \frac{1}{(n+1)^2} = \frac{1}{1^2} + \frac{1}{2^2} + \frac{1}{3^2} + \cdots = \sum_{n=1}^{\infty} \frac{1}{n^2}.$$

3. **(a)** Since $|xyz| < 1$ except at $(1,1,1)$, the formula for the sum of a geometric series gives

$\dfrac{1}{1-xyz} = \sum_{n=0}^{\infty} (xyz)^n$, so

$$\int_0^1 \int_0^1 \int_0^1 \frac{1}{1-xyz}\,dx\,dy\,dz = \int_0^1 \int_0^1 \int_0^1 \sum_{n=0}^{\infty} (xyz)^n\,dx\,dy\,dz = \sum_{n=0}^{\infty} \int_0^1 \int_0^1 \int_0^1 (xyz)^n\,dx\,dy\,dz$$

$$= \sum_{n=0}^{\infty} \left[\int_0^1 x^n\,dx\right]\left[\int_0^1 y^n\,dy\right]\left[\int_0^1 z^n\,dz\right] = \sum_{n=0}^{\infty} \frac{1}{n+1} \cdot \frac{1}{n+1} \cdot \frac{1}{n+1}$$

$$= \sum_{n=0}^{\infty} \frac{1}{(n+1)^3} = \frac{1}{1^3} + \frac{1}{2^3} + \frac{1}{3^3} + \cdots = \sum_{n=1}^{\infty} \frac{1}{n^3}.$$

(b) Since $|-xyz| < 1$, except at $(1,1,1)$, the formula for the sum of a geometric series gives

$\dfrac{1}{1+xyz} = \sum_{n=0}^{\infty} (-xyz)^n$, so

$$\int_0^1 \int_0^1 \int_0^1 \frac{1}{1+xyz}\,dx\,dy\,dz = \int_0^1 \int_0^1 \int_0^1 \sum_{n=0}^{\infty} (-xyz)^n\,dx\,dy\,dz = \sum_{n=0}^{\infty} \int_0^1 \int_0^1 \int_0^1 (-xyz)^n\,dx\,dy\,dz$$

$$= \sum_{n=0}^{\infty} (-1)^n \left[\int_0^1 x^n\,dx\right]\left[\int_0^1 y^n\,dy\right]\left[\int_0^1 z^n\,dz\right] = \sum_{n=0}^{\infty} (-1)^n \frac{1}{n+1} \cdot \frac{1}{n+1} \cdot \frac{1}{n+1}$$

$$= \sum_{n=0}^{\infty} \frac{(-1)^n}{(n+1)^3} = \frac{1}{1^3} - \frac{1}{2^3} + \frac{1}{3^3} - \cdots = \sum_{n=1}^{\infty} \frac{(-1)^{n-1}}{n^3}.$$

To evaluate this sum, we first write out a few terms: $s = 1 - \frac{1}{2^3} + \frac{1}{3^3} - \frac{1}{4^3} + \frac{1}{5^3} - \frac{1}{6^3} \approx 0.8998$.

Notice that $a_7 = \frac{1}{7^3} < 0.003$. By the Alternating Series Estimation Theorem (in Section 10.4), we have $|s - s_6| \le a_7 < 0.003$. This error of 0.003 will not affect the second decimal place, so we have $s \approx 0.90$.

5.

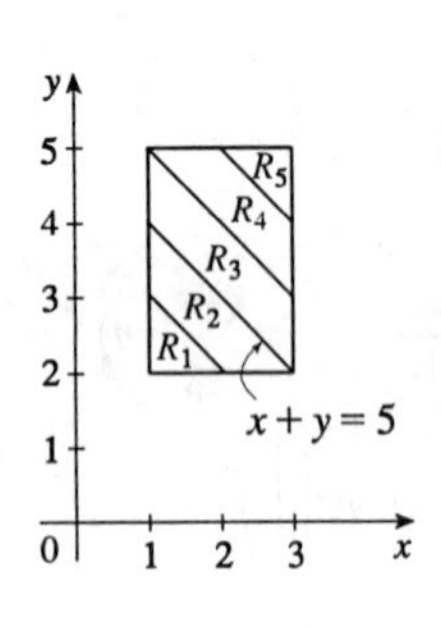

Let $R = \bigcup_{i=1}^{5} R_i$, where

$R_i = \{(x,y) \mid x + y \geq i + 2, x + y < i + 3, 1 \leq x \leq 3, 2 \leq y \leq 5\}$.

$\iint_R [\![x+y]\!]\,dA = \sum_{i=1}^{5} \iint_{R_i} [\![x+y]\!]\,dA = \sum_{i=1}^{5} [\![x+y]\!] \iint_{R_i} dA$,

since $[\![x+y]\!] = \text{constant} = i + 2$ for $(x,y) \in R_i$. Therefore

$$\begin{aligned}\iint_R [\![x+y]\!]\,dA &= \textstyle\sum_{i=1}^{5} (i+2)[A(R_i)] \\ &= 3A(R_1) + 4A(R_2) + 5A(R_3) + 6A(R_4) + 7A(R_5) \\ &= 3\left(\tfrac{1}{2}\right) + 4\left(\tfrac{3}{2}\right) + 5(2) + 6\left(\tfrac{3}{2}\right) + 7\left(\tfrac{1}{2}\right) = 30.\end{aligned}$$

7. $$f_{\text{ave}} = \frac{1}{b-a}\int_a^b f(x)dx = \frac{1}{1-0}\int_0^1\left[\int_x^1 \cos(t^2)dt\right]dx = \int_0^1\int_x^1 \cos(t^2)dt\,dx$$

$$= \int_0^1\int_0^t \cos(t^2)dx\,dt \quad \text{(changing the order of integration)}$$

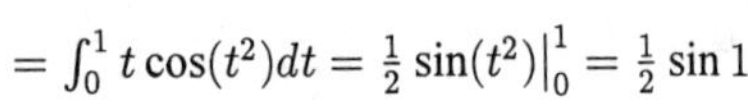

$$= \int_0^1 t\cos(t^2)dt = \tfrac{1}{2}\sin(t^2)\Big|_0^1 = \tfrac{1}{2}\sin 1$$

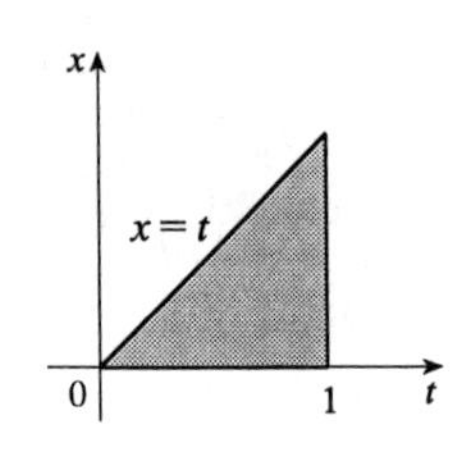

9. $\int_0^x\int_0^y\int_0^z f(t)dt\,dz\,dy = \iiint_E f(t)dV$, where

$E = \{(t,z,y) \mid 0 \leq t \leq z, 0 \leq z \leq y, 0 \leq y \leq x\}$.

If we let D be the projection of E on the yt-plane

then $D = \{(y,t) \mid 0 \leq t \leq x, t \leq y \leq x\}$. And

we see from the diagram that

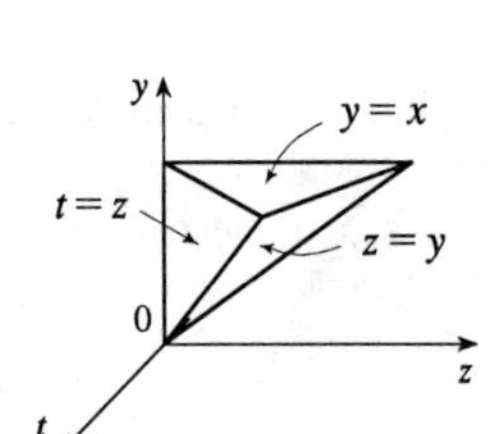

$E = \{(t,z,y) \mid t \leq z \leq y, t \leq y \leq x, 0 \leq t \leq x\}$. So

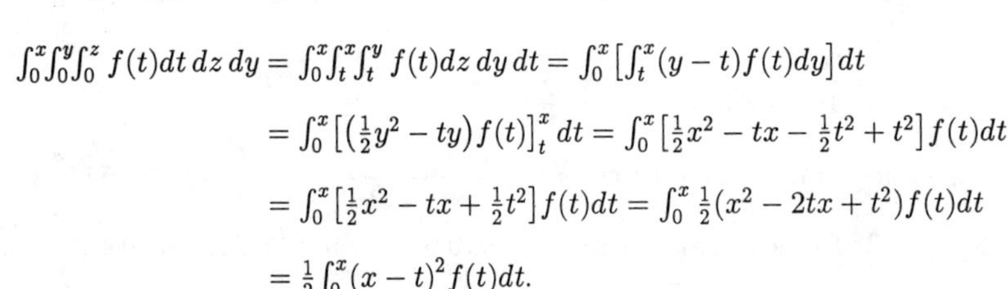

$$\begin{aligned}\int_0^x\int_0^y\int_0^z f(t)dt\,dz\,dy &= \int_0^x\int_t^x\int_t^y f(t)dz\,dy\,dt = \int_0^x\left[\int_t^x (y-t)f(t)dy\right]dt \\ &= \int_0^x\left[\left(\tfrac{1}{2}y^2 - ty\right)f(t)\right]_t^x dt = \int_0^x\left[\tfrac{1}{2}x^2 - tx - \tfrac{1}{2}t^2 + t^2\right]f(t)dt \\ &= \int_0^x\left[\tfrac{1}{2}x^2 - tx + \tfrac{1}{2}t^2\right]f(t)dt = \int_0^x \tfrac{1}{2}(x^2 - 2tx + t^2)f(t)dt \\ &= \tfrac{1}{2}\int_0^x (x-t)^2 f(t)dt.\end{aligned}$$

11. Let $u = \mathbf{a} \cdot \mathbf{r}$, $v = \mathbf{b} \cdot \mathbf{r}$, $w = \mathbf{c} \cdot \mathbf{r}$, where $\mathbf{a} = \langle a_1, a_2, a_3 \rangle$, $\mathbf{b} = \langle b_1, b_2, b_3 \rangle$, $\mathbf{c} = \langle c_1, c_2, c_3 \rangle$. Under this change of variables, E corresponds to the rectangular box $0 \le u \le \alpha$, $0 \le v \le \beta$, $0 \le w \le \gamma$. So, by the Change of Variables Theorem, $\int_0^\gamma \int_0^\beta \int_0^\alpha uvw\, du\, dv\, dw = \iiint_E (\mathbf{a} \cdot \mathbf{r})(\mathbf{b} \cdot \mathbf{r})(\mathbf{c} \cdot \mathbf{r}) \left| \dfrac{\partial(u,v,w)}{\partial(x,y,z)} \right| dV$. But

$$\left| \frac{\partial(u,v,w)}{\partial(x,y,z)} \right| = \left\| \begin{matrix} a_1 & a_2 & a_3 \\ b_1 & b_2 & b_3 \\ c_1 & c_2 & c_3 \end{matrix} \right\| = |\mathbf{a} \cdot \mathbf{b} \times \mathbf{c}| \quad \Rightarrow$$

$$\begin{aligned} \iiint_E (\mathbf{a} \cdot \mathbf{r})(\mathbf{b} \cdot \mathbf{r})(\mathbf{c} \cdot \mathbf{r}) dV &= \frac{1}{|\mathbf{a} \cdot \mathbf{b} \times \mathbf{c}|} \int_0^\gamma \int_0^\beta \int_0^\alpha uvw\, du\, dv\, dw \\ &= \frac{1}{|\mathbf{a} \cdot \mathbf{b} \times \mathbf{c}|} \left(\frac{\alpha^2}{2} \right) \left(\frac{\beta^2}{2} \right) \left(\frac{\gamma^2}{2} \right) \\ &= \frac{(\alpha\beta\gamma)^2}{8|\mathbf{a} \cdot \mathbf{b} \times \mathbf{c}|} \end{aligned}$$

13. Let S_1 be the portion of $\Omega(S)$ between $S(a)$ and S, and let ∂S_1 be its boundary. Also let S_L be the lateral surface of S_1 [that is, the surface of S_1 except S and $S(a)$]. Applying the Divergence Theorem we have $\iint_{\partial S_1} \dfrac{\mathbf{r} \cdot \mathbf{n}}{r^3}\, dS = \iiint_{S_1} \nabla \cdot \dfrac{\mathbf{r}}{r^3}\, dV$. But

$$\begin{aligned} \nabla \cdot \frac{\mathbf{r}}{r^3} &= \left\langle \frac{\partial}{\partial x}, \frac{\partial}{\partial y}, \frac{\partial}{\partial z} \right\rangle \cdot \left\langle \frac{x}{(x^2+y^2+z^2)^{3/2}}, \frac{y}{(x^2+y^2+z^2)^{3/2}}, \frac{z}{(x^2+y^2+z^2)^{3/2}} \right\rangle \\ &= \frac{(x^2+y^2+z^2-3x^2) + (x^2+y^2+z^2-3y^2) + (x^2+y^2+z^2-3z^2)}{(x^2+y^2+z^2)^{5/2}} = 0 \quad \Rightarrow \end{aligned}$$

$\iint_{\partial S_1} \dfrac{\mathbf{r} \cdot \mathbf{n}}{r^3}\, dS = \iiint_{S_1} 0\, dV = 0$. On the other hand, notice that for the surfaces of ∂S_1 other than $S(a)$ and S, $\mathbf{r} \cdot \mathbf{n} = 0 \quad \Rightarrow$

$$0 = \iint_{\partial S_1} \frac{\mathbf{r} \cdot \mathbf{n}}{r^3}\, dS = \iint_S \frac{\mathbf{r} \cdot \mathbf{n}}{r^3}\, dS + \iint_{S(a)} \frac{\mathbf{r} \cdot \mathbf{n}}{r^3}\, dS + \iint_{S_L} \frac{\mathbf{r} \cdot \mathbf{n}}{r^3}\, dS = \iint_S \frac{\mathbf{r} \cdot \mathbf{n}}{r^3}\, dS + \iint_{S(a)} \frac{\mathbf{r} \cdot \mathbf{n}}{r^3}\, dS \quad \Rightarrow$$

$\iint_S \dfrac{\mathbf{r} \cdot \mathbf{n}}{r^3}\, dS = -\iint_{S(a)} \dfrac{\mathbf{r} \cdot \mathbf{n}}{r^3}\, dS$. Notice that on $S(a)$, $r = a \quad \Rightarrow \quad \mathbf{n} = -\dfrac{\mathbf{r}}{r} = -\dfrac{\mathbf{r}}{a}$ and $\mathbf{r} \cdot \mathbf{r} = r^2 = a^2$, so that $-\iint_{S(a)} \dfrac{\mathbf{r} \cdot \mathbf{n}}{r^3}\, dS = \iint_{S(a)} \dfrac{\mathbf{r} \cdot \mathbf{r}}{a^4}\, dS = \iint_{S(a)} \dfrac{a^2}{a^4}\, dS = \dfrac{1}{a^2} \iint_{S(a)} dS = \dfrac{\text{area of } S(a)}{a^2} = |\Omega(S)|$. Therefore $|\Omega(S)| = \iint_S \dfrac{\mathbf{r} \cdot \mathbf{n}}{r^3}\, dS$.

CHAPTER FIFTEEN

EXERCISES 15.1

1. The auxiliary equation is $r^2 - 3r + 2 = (r-2)(r-1) = 0$, so $y = c_1e^x + c_2e^{2x}$.

3. The auxiliary equation is $3r^2 - 8r - 3 = (3r+1)(r-3) = 0$, so $y = c_1e^{-x/3} + c_2e^{3x}$.

5. The auxiliary equation is $r^2 + 2r + 10 = 0 \quad\Rightarrow\quad r = -1 \pm 3i$, so $y = e^{-x}(c_1\cos 3x + c_2\sin 3x)$.

7. The auxiliary equation is $r^2 - 1 = (r-1)(r+1) = 0$, so $y = c_1e^x + c_2e^{-x}$.

9. The auxiliary equation is $r^2 + 25 = 0 \quad\Rightarrow\quad r = \pm 5i$, so $y = c_1\cos 5x + c_2\sin 5x$.

11. The auxiliary equation is $2r^2 + r = r(2r+1) = 0$, so $y = c_1 + c_2e^{-x/2}$.

13. The auxiliary equation is $r^2 - r + 2 = 0 \quad\Rightarrow\quad r = \frac{1}{2}\left(1 \pm \sqrt{7}i\right)$, so $y = e^{x/2}\left[c_1\cos\left(\frac{\sqrt{7}}{2}x\right) + c_2\sin\left(\frac{\sqrt{7}}{2}x\right)\right]$.

15. The auxiliary equation is $r^2 + 2r - 1 = 0 \quad\Rightarrow\quad r = -1 \pm \sqrt{2}$, so $y = c_1e^{(-1+\sqrt{2})x} + c_2e^{(-1-\sqrt{2})x}$.

17. The auxiliary equation is $2r^2 + r + 3 = 0 \quad\Rightarrow\quad r = \frac{1}{4}\left(-1 \pm \sqrt{23}i\right)$, so

$y = e^{-x/4}\left[c_1\cos\left(\frac{\sqrt{23}}{4}x\right) + c_2\sin\left(\frac{\sqrt{23}}{4}x\right)\right]$.

19. $r^2 - 8r + 16 = (r-4)^2 = 0$ so $y = c_1e^{4x} + c_2xe^{4x}$.

The graphs are all asymptotic to the x-axis as $x \to -\infty$, and as $x \to \infty$ the solutions tend to $\pm\infty$.

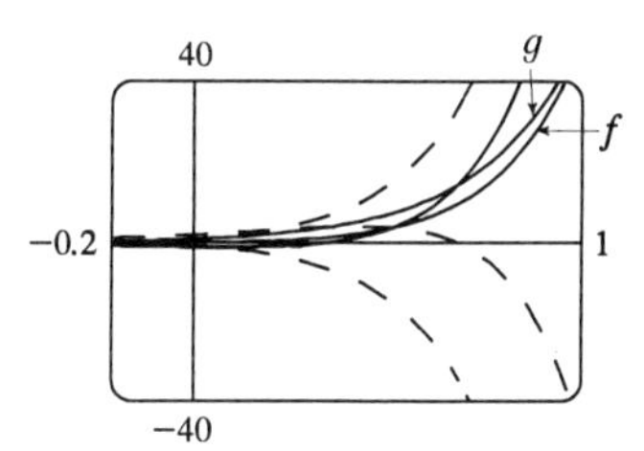

21. $r^2 + 3r - 4 = (r+4)(r-1) = 0$ so the general solution is $y = c_1e^x + c_2e^{-4x}$. Then $2 = y(0) = c_1 + c_2$ and $-3 = y'(0) = c_1 - 4c_2$ so $c_1 = 1$, $c_2 = 1$ and the solution to the initial-value problem is $y = e^x + e^{-4x}$.

23. $r^2 - 2r + 2 = 0 \quad\Rightarrow\quad r = 1 \pm i$ and the general solution is $y = e^x(c_1\cos x + c_2\sin x)$. But $1 = y(0) = c_1$ and $2 = y'(0) = c_1 + c_2$ so the solution to the initial-value problem is $y = e^x(\cos x + \sin x)$.

25. $r^2 - 2r - 3 = (r-3)(r+1) = 0$ so the general solution is $y = c_1e^{-x} + c_2e^{3x}$. However the conditions are given at $x = 1$ so rewrite the general solution as $y = k_1e^{-(x-1)} + k_2e^{3(x-1)}$. Then $3 = y(1) = k_1 + k_2$ and $1 = y'(1) = -k_1 + 3k_2$ so $k_1 = 2$, $k_2 = 1$ and the solution to the initial-value problem is $y = 2e^{-(x-1)} + e^{3(x-1)}$.

27. $r^2 + 9 = 0 \quad\Rightarrow\quad r = \pm 3i$ and the general solution is $y = c_1\cos 3x + c_2\sin 3x$. But $0 = y\left(\frac{\pi}{3}\right) = -c_1$ and $1 = y'\left(\frac{\pi}{3}\right) = -3c_2$, so the solution to the initial-value problem is $y = -\frac{1}{3}\sin 3x$.

29. $r^2 + 4r + 4 = (r+2)^2 = 0$ so the general solution is $y = c_1 e^{-2x} + c_2 x e^{-2x}$. Then $0 = y(0) = c_1$, $3 = y(1) = c_2 e^{-2}$ so $c_2 = 3e^2$ and the solution of the boundary-value problem is $y = 3xe^{-2x+2}$.

31. $r^2 + 1 = 0 \Rightarrow r = \pm i$ and the general solution is $y = c_1 \cos x + c_2 \sin x$. But $1 = y(0) = c_1$ and $0 = y(\pi) = -c_1$ so there is no solution.

33. $r^2 - r - 2 = (r-2)(r+1) = 0$ so the general solution is $y = c_1 e^{-x} + c_2 e^{2x}$. Then $1 = y(-1) = c_1 e + c_2 e^{-2}$ and $0 = y(1) = c_1 e^{-1} + c_2 e^2$ so $c_1 = \dfrac{e^5}{e^6 - 1}$ and $c_2 = \dfrac{e^2}{1 - e^6}$ so the solution to the boundary-value problem is

$$y = \frac{e^5}{e^6 - 1} e^{-x} + \frac{e^2}{1 - e^6} e^{2x} = \frac{1}{e^6 - 1}\left[e^{5-x} - e^{2(1+x)}\right].$$

35. $r^2 + 4r + 13 = 0 \Rightarrow r = -2 \pm 3i$ and the general solution is $y = e^{-2x}(c_1 \cos 3x + c_2 \sin 3x)$. But $2 = y(0) = c_1$ and $1 = y\left(\frac{\pi}{2}\right) = e^{-\pi}(-c_2)$, so the solution to the boundary-value problem is $y = e^{-2x}(2\cos 3x - e^{\pi} \sin 3x)$.

37. **(a)** *Case 1* ($\lambda = 0$): $y'' + \lambda y = 0 \Rightarrow y'' = 0$ which has an auxiliary equation $r^2 = 0 \Rightarrow r = 0 \Rightarrow y = c_1 + c_2 x$ where $y(0) = 0$ and $y(L) = 0$. Thus, $0 = y(0) = c_1$ and $0 = y(L) = c_2 L \Rightarrow c_1 = c_2 = 0$. Thus, $y = 0$.

Case 2 ($\lambda < 0$): $y'' + \lambda y = 0$ has auxiliary equation $r^2 = -\lambda \Rightarrow r = \pm\sqrt{-\lambda}$ (distinct and real since $\lambda < 0$) $\Rightarrow y = c_1 e^{\sqrt{-\lambda}x} + c_2 e^{-\sqrt{-\lambda}x}$ where $y(0) = 0$ and $y(L) = 0$. Thus, $0 = y(0) = c_1 + c_2$ ($\bigstar$) and $0 = y(L) = c_1 e^{\sqrt{-\lambda}L} + c_2 e^{-\sqrt{-\lambda}L}$ ($\dagger$).

Multiplying ($\bigstar$) by $e^{\sqrt{-\lambda}L}$ and subtracting ($\dagger$) gives $c_2\left(e^{\sqrt{-\lambda}L} - e^{-\sqrt{-\lambda}L}\right) = 0 \Rightarrow c_2 = 0$ and thus $c_1 = 0$ from ($\bigstar$). Thus, $y = 0$ for the cases $\lambda = 0$ and $\lambda < 0$.

(b) $y'' + \lambda y = 0$ has an auxiliary equation $r^2 + \lambda = 0 \Rightarrow r = \pm i\sqrt{\lambda} \Rightarrow$ $y = c_1 \cos\sqrt{\lambda}x + c_2 \sin\sqrt{\lambda}x$ where $y(0) = 0$ and $y(L) = 0$. Thus, $0 = y(0) = c_1$ and $0 = y(L) = c_2 \sin\sqrt{\lambda}L$ since $c_1 = 0$. Since we cannot have a trivial solution, $c_2 \neq 0$ and thus $\sin\sqrt{\lambda}L = 0 \Rightarrow \sqrt{\lambda}L = n\pi$ where n is an integer $\Rightarrow \lambda = n^2\pi^2/L^2$ and $y = c_2 \sin(n\pi x/L)$ where n is an integer.

EXERCISES 15.2

1. The auxiliary equation is $r^2 - r - 6 = (r-3)(r+2) = 0$, so the complementary solution is $y_c(x) = c_1e^{3x} + c_2e^{-2x}$. Try the particular solution $y_p(x) = A\cos 3x + B\sin 3x$, so $y_p' = 3B\cos 3x - 3A\sin 3x$ and $y_p'' = -9A\cos 3x - 9B\sin 3x$. Substitution gives $(-9A\cos 3x - 9B\sin 3x) - (3B\cos 3x - 3A\sin 3x) - 6(A\cos 3x + B\sin 3x) = \cos 3x \quad \Rightarrow$ $\cos 3x = (-15A - 3B)\cos 3x + (3A - 15B)\sin 3x$. Hence $-15A - 3B = 1$ and $3A - 15B = 0 \quad \Rightarrow$ $A = -\frac{5}{78}$, $B = -\frac{1}{78}$ and the general solution is $y(x) = y_c + y_p = c_1e^{3x} + c_2e^{-2x} - \frac{5}{78}\cos 3x - \frac{1}{78}\sin 3x$.

3. The complementary solution is $y_c(x) = e^{2x}(c_1x + c_2)$, so try a particular solution $y_p(x) = Ae^{-x}$. Then $y_p'' = y_p' = y_p = Ae^{-x}$and substitution into the differential equation gives $Ae^{-x} + 4Ae^{-x} + 4Ae^{-x} = e^{-x} \quad \Rightarrow$ $A = \frac{1}{9}$. Hence the general solution is $y(x) = e^{2x}(c_1x + c_2) + \frac{1}{9}e^{-x}$.

5. The complementary solution is $y_c(x) = c_1\cos 6x + c_2\sin 6x$. Try $y_p(x) = Ax^2 + Bx + C$. Then $y_p' = 2Ax + B$ and $y_p'' = 2A$; substitution gives $2A + 36(Ax^2 + Bx + C) = 2x^2 - x$. $A = \frac{1}{18}$, $B = -\frac{1}{36}$, $C = -\frac{1}{324}$. The general solution is $y(x) = c_1\cos 6x + c_2\sin 6x + \frac{1}{18}x^2 - \frac{1}{36}x - \frac{1}{324}$.

7. Since the roots of $r^2 - 2r + 5 = 0$ are $1 \pm 2i$, $y_c(x) = e^x(c_1\cos 2x + c_2\sin 2x)$. For $y'' - 2y' + 5y = x$ try $y_{p_1}(x) = Ax + B$. Then $0 - 2A + 5(Ax + B) = x$, so $y_{p_1}(x) = \frac{1}{5}x + \frac{2}{25}$. For $y'' - 2y' + 5y = \sin 3x$ try $y_{p_2}(x) = A\cos 3x + B\sin 3x$. Then $y_{p_2}' = -3A\sin 3x + 3B\cos 3x$ and $y_{p_2}'' = -9A\cos 3x - 9B\sin 3x$. Substituting into the differential equation gives

$-9A\cos 3x - 9B\sin 3x + 6A\sin 3x - 6B\cos 3x + 5A\cos 3x + 5B\sin 3x = \sin 3x$. Thus $(-9A - 6B + 5A) = 0$ and $(-9B + 6A + 5B) = 1$, so $A = \frac{3}{26}$ and $B = -\frac{1}{13}$. Hence the general solution is $y(x) = e^x(c_1\cos 2x + c_2\sin 2x) + \frac{1}{5}x + \frac{2}{25} + \frac{3}{26}\cos 3x - \frac{1}{13}\sin 3x$. But $1 = y(0) = c_1 + \frac{2}{25} + \frac{3}{26} \quad \Rightarrow$ $c_1 = \frac{523}{650}$, $2 = y'(0) = c_1 + 2c_2 + \frac{1}{5} - \frac{3}{13} \quad \Rightarrow \quad c_2 = \frac{797}{1300}$. Thus the solution to the initial-value problem is $y(x) = e^x\left[\frac{523}{650}\cos 2x + \frac{797}{1300}\sin 2x\right] + \frac{1}{5}x + \frac{2}{25} + \frac{3}{26}\cos 3x - \frac{1}{13}\sin 3x$.

9. $y_c(x) = c_1e^x + c_2e^{-x}$. Try $y_p(x) = (Ax + B)e^{3x}$. Then $y_p' = e^{3x}(A + 3Ax + 3B)$ and $y_p'' = e^{3x}(3A + 3A + 9Ax + 9B)$. Substitution into the differential equation gives $e^{3x}[9Ax + 9B + 6A - (Ax + B)] = xe^{3x}$, so $A = \frac{1}{8}$, $B = -\frac{3}{32}$ and the general solution is $y(x) = c_1e^x + c_2e^{-x} + \left(\frac{1}{8}x - \frac{3}{32}\right)e^{3x}$. But $0 = y(0) = c_1 + c_2 - \frac{3}{32}$, $1 = y'(0) = c_1 - c_2 - \frac{9}{32} + \frac{1}{8}$ so the solution to the initial-value problem is $y(x) = \frac{5}{8}e^x - \frac{17}{32}e^{-x} + e^{3x}\left(\frac{1}{8}x - \frac{3}{32}\right)$.

11. $y_c(x) = c_1 e^{-x/4} + c_2 e^{-x}$. Try $y_p(x) = Ae^x$. Then $10Ae^x = e^x$, so $A = \frac{1}{10}$ and the general solution is $y(x) = c_1 e^{-x/4} + c_2 e^{-x} + \frac{1}{10}e^x$.

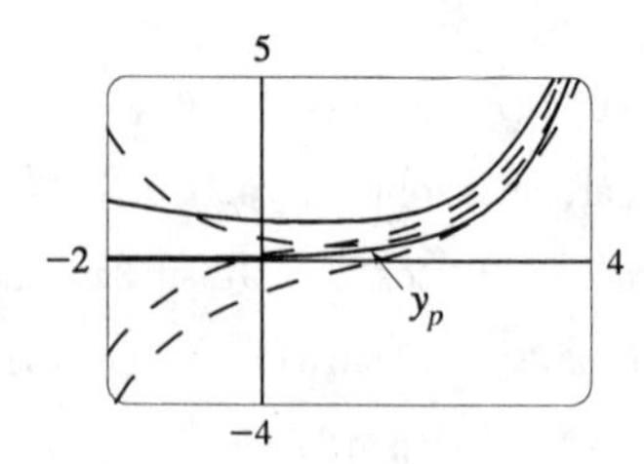

The solutions are all composed of exponential curves and with the exception of the particular solution (which approaches 0 as $x \to -\infty$), they all approach either ∞ or $-\infty$ as $x \to -\infty$. As $x \to \infty$, all solutions are asymptotic to $y_p = \frac{1}{10}e^x$.

13. Since the roots of the auxiliary equation are complex, we need just try $y_p(x) = (Ax^4 + Bx^3 + Cx^2 + Dx + E)e^{2x}$.

15. Since $y_c(x) = e^x(c_1 \cos x + c_2 \sin x)$ we try $y_p(x) = xe^x(A \cos x + B \sin x)$.

Note: **Solving Equations (7) and (9) in The Method of Variation of Parameters gives**

$$u_1' = -\frac{Gy_2}{a(y_1y_2' - y_2y_1')} \quad \textbf{and } u_2' = \frac{Gy_1}{a(y_1y_2' - y_2y_1')}$$

We will use these equations rather than resolving the system in each of the remaining exercises in this section.

17. **(a)** The complementary solution is $y_c(x) = c_1 \cos 2x + c_2 \sin 2x$. A particular solution is of the form $y_p(x) = Ax + B$. Thus, $4Ax + 4B = x \quad \Rightarrow \quad A = \frac{1}{4}$ and $B = 0 \quad \Rightarrow \quad y_p(x) = \frac{1}{4}x$. Thus, the general solution is $y = y_c + y_p = c_1 \cos 2x + c_2 \sin 2x + \frac{1}{4}x$.

(b) In (a), $y_c(x) = c_1 \cos 2x + c_2 \sin 2x$, so set $y_1 = \cos 2x$, $y_2 = \sin 2x$. Then $y_1y_2' - y_2y_1' = 2\cos^2 2x + 2\sin^2 2x = 2$ so $u_1' = -\frac{1}{2}x \sin 2x \quad \Rightarrow$ $u_1(x) = -\frac{1}{2}\int x \sin 2x\, dx = -\frac{1}{4}\left(-x \cos 2x + \frac{1}{2}\sin 2x\right)$ (by parts) and $u_2' = \frac{1}{2}x \cos 2x \quad \Rightarrow$ $u_2(x) = \frac{1}{2}\int x \cos 2x\, dx = \frac{1}{4}\left(x \sin 2x + \frac{1}{2}\cos 2x\right)$ (by parts). Hence $y_p(x) = -\frac{1}{4}\left(-x \cos 2x + \frac{1}{2}\sin 2x\right)\cos 2x + \frac{1}{4}\left(x \sin 2x + \frac{1}{2}\cos 2x\right)\sin 2x = \frac{1}{4}x$. Thus $y(x) = y_c(x) + y_p(x) = c_1 \cos 2x + c_2 \sin 2x + \frac{1}{4}x$.

19. **(a)** $r^2 - r = r(r-1) = 0 \quad \Rightarrow \quad r = 0, 1$, so the complementary solution is $y_c(x) = c_1 e^x + c_2 x e^x$.
A particular solution is of the form $y_p(x) = Ae^{2x}$. Thus $4Ae^{2x} - 4Ae^{2x} + Ae^{2x} = e^{2x} \quad \Rightarrow \quad Ae^{2x} = e^{2x}$
$\Rightarrow \quad A = 1 \quad \Rightarrow \quad y_p(x) = e^{2x}$. So a general solution is $y(x) = y_c(x) + y_p(x) = c_1 e^x + c_2 x e^x + e^{2x}$.

(b) From (a), $y_c(x) = c_1 e^x + c_2 x e^x$, so set $y_1 = e^x$, $y_2 = xe^x$. Then, $y_1 y_2' - y_2 y_1' = e^{2x}(1+x) - xe^{2x} = e^{2x}$
and so $u_1' = -xe^x \quad \Rightarrow \quad u_1(x) = -\int xe^x\,dx = -(x-1)e^x$ (by parts) and $u_2' = e^x \quad \Rightarrow$
$u_2(x) = \int e^x\,dx = e^x$. Hence $y_p(x) = (1-x)e^{2x} + xe^{2x} = e^{2x}$ and the general solution is
$y(x) = y_c(x) + y_p(x) = c_1 e^x + c_2 x e^x + e^{2x}$.

21. As in Example 6, $y_c(x) = c_1 \sin x + c_2 \cos x$, so set $y_1 = \sin x$, $y_2 = \cos x$. Then
$y_1 y_2' - y_2 y_1' = -\sin^2 x - \cos^2 x = -1$, so $u_1' = -\dfrac{\sec x \cos x}{-1} = 1 \Rightarrow u_1(x) = x$ and $u_2' = \dfrac{\sec x \sin x}{-1} = -\tan x$
$\Rightarrow \quad u_2(x) = -\int \tan x\,dx = \ln|\cos x| = \ln(\cos x)$ on $0 < x < \frac{\pi}{2}$. Hence $y_p(x) = x\sin x + \cos x \ln(\cos x)$ and
the general solution is $y(x) = (c_1 + x)\sin x + [c_2 + \ln(\cos x)]\cos x$.

23. $y_1 = e^x$, $y_2 = e^{2x}$ and $y_1 y_2' - y_2 y_1' = e^{3x}$. So $u_1' = \dfrac{-e^{2x}}{(1+e^{-x})e^{3x}} = -\dfrac{e^{-x}}{1+e^{-x}}$ and
$u_1(x) = \displaystyle\int -\frac{e^{-x}}{1+e^{-x}}\,dx = \ln(1+e^{-x})$. $u_2' = \dfrac{e^x}{(1+e^{-x})e^{3x}} = \dfrac{e^x}{e^{3x}+e^{2x}}$ so
$u_2(x) = \displaystyle\int \frac{e^x}{e^{3x}+e^{2x}}\,dx = \ln\left(\frac{e^x+1}{e^x}\right) - e^{-x} = \ln(1+e^{-x}) - e^{-x}$. Hence
$y_p(x) = e^x \ln(1+e^{-x}) + e^{2x}[\ln(1+e^{-x}) - e^{-x}]$ and the general solution is
$y(x) = [c_1 + \ln(1+e^{-x})]e^x + [c_2 - e^{-x} + \ln(1+e^{-x})]e^{2x}$.

25. $y_1 = e^{-x}$, $y_2 = e^x$ and $y_1 y_2' - y_2 y_1' = 2$. So $u_1' = -\dfrac{e^x}{2x}$, $u_2' = \dfrac{e^{-x}}{2x}$ and $y_p(x) = -e^{-x}\displaystyle\int \frac{e^x}{2x}\,dx + e^x \int \frac{e^{-x}}{2x}\,dx$.
Hence the general solution is $y(x) = \left(c_1 - \displaystyle\int \frac{e^x}{2x}\,dx\right)e^{-x} + \left(c_2 + \displaystyle\int \frac{e^{-x}}{2x}\,dx\right)e^x$.

EXERCISES 15.3

1. By Hooke's Law $k(0.6) = 20$ so $k = \frac{100}{3}$ is the spring constant and the differential equation is $3x'' + \frac{100}{3}x = 0$. The general solution is $x(t) = c_1 \cos\left(\frac{10}{3}t\right) + c_2 \sin\left(\frac{10}{3}t\right)$. But $0 = x(0) = c_1$ and $1.2 = x'(0) = \frac{10}{3}c_2$, so the position of the mass after t seconds is $x(t) = 0.36 \sin\left(\frac{10}{3}t\right)$.

3. $k(0.5) = 6$ or $k = 12$ is the spring constant, so the initial-value problem is $2x'' + 14x' + 12x = 0$, $x(0) = 1$, $x'(0) = 0$. The general solution is $x(t) = c_1 e^{-6t} + c_2 e^{-t}$. But $1 = x(0) = c_1 + c_2$ and $0 = x'(0) = -6c_1 - c_2$. Thus the position is given by $x(t) = -\frac{1}{5}e^{-6t} + \frac{6}{5}e^{-t}$.

5. For critical damping we need $c^2 - 4mk = 0$ or $m = c^2/(4k) = 14^2/(4 \cdot 12) = \frac{49}{12}$ kg.

7. The differential equation is $mx'' + kx = F_0 \cos \omega_0 t$ and $\omega_0 \neq \omega = \sqrt{k/m}$. Here the auxiliary equation is $mr^2 + k = 0$ with roots $\pm\sqrt{k/m}i = \pm\omega i$ so $x_c(t) = c_1 \cos \omega t + c_2 \sin \omega t$. Since $\omega_0 \neq \omega$, try $x_p(t) = A \cos \omega_0 t + B \sin \omega_0 t$. Then we need $(m)(-\omega_0^2)(A \cos \omega_0 t + B \sin \omega_0 t) + k(A \cos \omega_0 t + B \sin \omega_0 t) = F_0 \cos \omega_0 t$ or $A(k - m\omega_0^2) = F_0$ and $B(k - m\omega_0^2) = 0$. Hence $B = 0$ and $A = \dfrac{F_0}{k - m\omega_0^2} = \dfrac{F_0}{m(\omega^2 - \omega_0^2)}$ since $\omega^2 = \dfrac{k}{m}$. Thus the motion of the mass is given by $x(t) = c_1 \cos \omega t + c_2 \sin \omega t + \dfrac{F_0}{m(\omega^2 - \omega_0^2)} \cos \omega_0 t$.

9. Here the initial-value problem for the charge is $Q'' + 20Q' + 500Q = 12$, $Q(0) = Q'(0) = 0$. Then $Q_c(t) = e^{-10t}(c_1 \cos 20t + c_2 \sin 20t)$ and try $Q_p(t) = A \Rightarrow 500A = 12$ or $A = \frac{3}{125}$. The general solution is $Q(t) = e^{-10t}(c_1 \cos 20t + c_2 \sin 20t) + \frac{3}{125}$. But $0 = Q(0) = c_1 + \frac{3}{125}$ and $Q'(t) = I(t) = e^{-10t}[(-10c_1 + 20c_2)\cos 20t + (-10c_2 - 20c_1)\sin 20t]$ but $0 = Q'(0) = -10c_1 + 20c_2$. Thus the charge is $Q(t) = -\frac{1}{250}e^{-10t}(6 \cos 20t + 3 \sin 20t) + \frac{3}{125}$ and the current is $I(t) = e^{-10t}\left(\frac{3}{5}\right)\sin 20t$.

11. As in Exercise 9, $Q_c(t) = e^{-10t}(c_1 \cos 20t + c_2 \sin 20t)$ but $E(t) = 12 \sin 10t$ so try $Q_p(t) = A \cos 10t + B \sin 10t$. Substituting into the differential equation gives $(-100A + 200B + 500A)\cos 10t + (-100B - 200A + 500B)\sin 10t = 12 \sin 10t$, $\Rightarrow$ $400A + 200B = 0$ and $400B - 200A = 12$. Thus $A = -\frac{3}{250}$, $B = \frac{3}{125}$ and the general solution is $Q(t) = e^{-10t}(c_1 \cos 20t + c_2 \sin 20t) - \frac{3}{250} \cos 10t + \frac{3}{125} \sin 10t$. But $0 = Q(0) = c_1 - \frac{3}{250}$ so $c_1 = \frac{3}{250}$. Also $Q'(t) = \frac{3}{25}\sin 10t + \frac{6}{25}\cos 10t + e^{-10t}[(-10c_1 + 20c_2)\cos 20t + (-10c_2 - 20c_1)\sin 20t]$ and $0 = Q'(0) = \frac{6}{25} - 10c_1 + 20c_2$ so $c_2 = -\frac{3}{500}$. Hence the charge is given by $Q(t) = e^{-10t}\left[\frac{3}{250} \cos 20t - \frac{3}{500} \sin 20t\right] - \frac{3}{250} \cos 10t + \frac{3}{125} \sin 10t$.

13. $x(t) = A \cos(\omega t + \delta) \Leftrightarrow x(t) = A[\cos \omega t \cos \delta - \sin \omega t \sin \delta] \Leftrightarrow x(t) = A\left(\dfrac{c_1}{A} \cos \omega t + \dfrac{c_2}{A} \sin \omega t\right)$ where $\cos \delta = c_1/A$ and $\sin \delta = -c_2/A \Leftrightarrow x(t) = c_1 \cos \omega t + c_2 \sin \omega t$. (Note that $\cos^2\delta + \sin^2\delta = 1 \Rightarrow c_1^2 + c_2^2 = A^2$.)

EXERCISES 15.4

1. Let $y(x)=\sum_{n=0}^{\infty}a_nx^n$. Then $\sum_{n=0}^{\infty}na_nx^{n-1}-6\sum_{n=0}^{\infty}a_nx^n=0 \quad\Rightarrow\quad \sum_{n=1}^{\infty}na_nx^{n-1}-6\sum_{n=0}^{\infty}a_nx^n=0$.

Replacing n by $n+1$ in the first sum gives $\sum_{n=0}^{\infty}[(n+1)a_{n+1}-6a_n]x^n=0$. Thus the recurrence relation is $a_{n+1}=\dfrac{6a_n}{n+1}$, $n=0,1,2,\ldots$. Then $a_1=6a_0$, $a_2=\dfrac{6a_1}{2}=\dfrac{6^2a_0}{2}$, $a_3=\dfrac{6a_2}{3}=\dfrac{6^3a_0}{2\cdot3},\ldots,$ $a_n=\dfrac{6a_{n-1}}{n}=\dfrac{6^na_0}{n!}$. Thus the solution is $y(x)=\displaystyle\sum_{n=0}^{\infty}a_0\frac{6^n}{n!}x^n=\sum_{n=0}^{\infty}\left[a_0\frac{(6x)^n}{n!}\right]=a_0e^{6x}$.

3. Assuming $y(x)=\sum_{n=0}^{\infty}a_nx^n$, we have $y'(x)=\sum_{n=1}^{\infty}na_nx^{n-1}=\sum_{n=0}^{\infty}(n+1)a_{n+1}x^n$ and $-x^2y=-\sum_{n=0}^{\infty}a_nx^{n+2}=-\sum_{n=2}^{\infty}a_{n-2}x^n$. Hence the differential equation becomes $\sum_{n=0}^{\infty}(n+1)a_{n+1}x^n-\sum_{n=2}^{\infty}a_{n-2}x^n=0$ or $a_1=2a_2x+\sum_{n=2}^{\infty}[(n+1)a_{n+1}-a_{n-2}]x^n=0$. Equating coefficients gives $a_1=a_2=0$ and $a_{n+1}=\dfrac{a_{n-2}}{n+1}$ for $n=2,3,\ldots$. But $a_1=0$, so $a_4=0$ and $a_7=0$ and in general $a_{3n+1}=0$. Similarly $a_2=0$ so $a_{3n+2}=0$. Finally $a_3=\dfrac{a_0}{3}$, $a_6=\dfrac{a_3}{6}=\dfrac{a_0}{6\cdot3}=\dfrac{a_0}{3^2\cdot2!}$, $a_9=\dfrac{a_6}{9}=\dfrac{a_0}{9\cdot6\cdot3}=\dfrac{a_0}{3^3\cdot3!},\ldots,$ and $a_{3n}=\dfrac{a_0}{3^n\cdot n!}$. Thus the solution is

$$y(x)=\sum_{n=0}^{\infty}a_nx^n=\sum_{n=0}^{\infty}a_{3n}x^{3n}=a_0\sum_{n=0}^{\infty}\frac{(x^3/3)^n}{n!}=a_0e^{x^3/3}.$$

5. Assuming $y(x)=\sum_{n=0}^{\infty}a_nx^n$, $y''(x)=\sum_{n=2}^{\infty}n(n-1)a_nx^{n-2}=\sum_{n=0}^{\infty}(n+2)(n+1)a_{n+2}x^n$, $3xy'(x)=3x\sum_{n=0}^{\infty}na_nx^{n-1}=\sum_{n=0}^{\infty}3na_nx^n$ and the differential equation becomes $\sum_{n=0}^{\infty}(n+2)(n+1)a_{n+2}+\sum_{n=0}^{\infty}3na_nx^n+\sum_{n=0}^{\infty}3a_nx^n=0 \quad\Rightarrow$ $\sum_{n=0}^{\infty}[(n+2)(n+1)a_{n+2}+3(n+1)a_n]x^n=0$. Thus the recurrence relation is $a_{n+2}=-\dfrac{3a_n}{n+2}$ for $n=0,1,2,\ldots$. Given a_0 and a_1, $a_2=-\dfrac{3a_0}{2}$, $a_4=-\dfrac{3a_2}{4}=(-1)^2\dfrac{3^2a_0}{2^2\cdot2!}$, $a_6=-\dfrac{3a_4}{6}=(-1)^3\dfrac{3^3a_0}{2^3\cdot3!},\ldots,$ $a_{2n}=(-1)^n\dfrac{3^na_0}{2^nn!}=(-1)^n\left(\dfrac{3}{2}\right)^n\dfrac{a_0}{n!}$ and $a_3=-\dfrac{3a_1}{3}$, $a_5=-\dfrac{3a_3}{5}=(-1)^2\dfrac{3^2a_1}{5\cdot3}$, $a_7=-\dfrac{3a_5}{7}=(-1)^3\dfrac{3^3a_1}{7\cdot5\cdot3}$, $\ldots,$ $a_{2n+1}=(-1)^n\dfrac{3^na_1}{(2n+1)(2n-1)\cdot\cdots\cdot5\cdot3}=(-1)^n\dfrac{3^na_12^nn!}{(2n+1)!}=(-1)^n\dfrac{6^nn!a_1}{(2n+1)!}$ because $(2n+1)(2n-1)\cdot\cdots\cdot5\cdot3=\dfrac{(2n+1)!}{2^n\cdot n!}$. Thus the solution is

$$y(x)=\sum_{n=0}^{\infty}a_{2n}x^{2n}+\sum_{n=0}^{\infty}a_{2n+1}x^{2n+1}=a_0\sum_{n=0}^{\infty}(-1)^n\frac{(3x^2/2)^n}{n!}+a_1\sum_{n=0}^{\infty}\left[(-1)^n\frac{n!6^nx^{2n+1}}{(2n+1)!}\right]$$

$$=a_0e^{-3x^2/2}+a_1\sum_{n=0}^{\infty}\left[(-1)^n\frac{6^nn!x^{2n+1}}{(2n+1)!}\right].$$

7. Let $y(x) = \sum_{n=0}^{\infty} a_n x^n$. Then $y''(x) = \sum_{n=0}^{\infty}(n+2)(n+1)a_{n+2}x^n$, $-xy'(x) = -\sum_{n=0}^{\infty} na_n x^n$ and the differential equation becomes $\sum_{n=0}^{\infty}[(n+2)(n+1)a_{n+2} - (n+1)a_n]x^n = 0$. Thus the recurrence relation is $a_{n+2} = \dfrac{a_n}{n+2}$ for $n = 0, 1, 2, \ldots$. But $a_0 = y(0) = 1$ so $a_2 = \dfrac{1}{2}$, $a_4 = \dfrac{a_2}{4} = \dfrac{1}{2\cdot 4}$, $a_6 = \dfrac{a_4}{6} = \dfrac{1}{2\cdot 4\cdot 6}, \ldots,$ $a_{2n} = \dfrac{1}{2^n\, n!}$. Also $a_1 = y'(0) = 0$ and by the recurrence relation $a_{2n+1} = 0$ for $n = 0, 1, 2, \ldots$. Thus the solution to the initial-value problem is $y(x) = \displaystyle\sum_{n=0}^{\infty} a_n x^n = \sum_{n=0}^{\infty} \frac{x^{2n}}{2^n n!} = \sum_{n=0}^{\infty} \frac{(x^2/2)^n}{n!} = e^{x^2/2}$.

9. Let $y(x) = \sum_{n=0}^{\infty} a_n x^n$. Then $y''(x) = \sum_{n=0}^{\infty} n(n-1)a_n x^{n-2} = \sum_{n=-1}^{\infty}(n+3)(n+2)a_{n+3}x^{n+1}$ $= 2a_2 + \sum_{n=0}^{\infty}(n+3)(n+2)a_{n+3}x^{n+1}$ and the differential equation becomes $2a_2 + \sum_{n=0}^{\infty}[(n+3)(n+2)a_{n+3} + (n+1)a_n]x^{n+1} = 0$. Then $a_2 = 0$ and the recurrence relation is $a_{n+3} = -\dfrac{(n+1)a_n}{(n+3)(n+2)}$, $n = 0, 1, 2, \ldots$. But $a_0 = y(0) = 0 = a_2$ and by the recurrence relation $a_{3n} = a_{3n+2} = 0$ for $n = 0, 1, 2, \ldots$. Also $a_1 = y'(0) = 1$ so

$a_4 = -\dfrac{2}{4\cdot 3}$, $a_7 = -\dfrac{5a_4}{7\cdot 6} = (-1)^2 \dfrac{2\cdot 5}{7\cdot 6\cdot 4\cdot 3} = (-1)^2 \dfrac{2^2 5^2}{7!}, \ldots, a_{3n+1} = (-1)^n \dfrac{2^2 5^2 \cdots (3n-1)^2}{(3n+1)!}$. Thus the solution is $y(x) = \displaystyle\sum_{n=0}^{\infty} a_n x^n = x + \sum_{n=0}^{\infty}\left[(-1)^n \frac{2^2 5^2 \cdots (3n-1)^2 x^{3n+1}}{(3n+1)!}\right]$.

REVIEW EXERCISES FOR CHAPTER 15

1. True. See Theorem 15.1.3.

3. True. $\cosh x$ and $\sinh x$ are linearly independent solutions of this linear homogeneous equation.

5. The auxiliary equation is $r^2 - 6r + 34 = 0$ with roots $r = 3 \pm 5i$. Thus the solution is $y(x) = e^{3x}(c_1 \cos 5x + c_2 \sin 5x)$.

7. The auxiliary equation is $2r^2 + r - 1 = 0$ with roots $r_1 = -1$, $r_2 = \frac{1}{2}$. Thus the solution is $y(x) = c_1 e^{-x} + c_2 e^{x/2}$.

9. $y_c(x) = e^{-x}(c_1 + c_2 x)$ and try $y_p(x) = A\cos 3x + B\sin 3x$. Then $-9A\cos 3x - 9B\sin 3x - 6A\sin 3x + 6B\cos 3x + A\cos 3x + B\sin 3x = \sin 3x \quad\Rightarrow\quad 6B - 8A = 0$ and $-6A - 8B = 1 \quad\Rightarrow\quad A = -\frac{3}{50}$, $B = -\frac{2}{25}$ and the general solution is $y(x) = e^{-x}(c_1 + c_2 x) - \frac{3}{50}\cos 3x - \frac{2}{25}\sin 3x$.

11. $y_c(x) = c_1 \cos\left(\frac{3}{2}x\right) + c_2 \sin\left(\frac{3}{2}x\right)$ and try $y_p(x) = Ax^2 + Bx + C$. Then $8A + 9Ax^2 + 9Bx + 9C = 2x^2 - 3$ $\Rightarrow\quad A = \frac{2}{9}$, $B = 0$, $C = -\frac{43}{81}$, and the general solution is $y(x) = c_1 \cos\left(\frac{3}{2}x\right) + c_2 \sin\left(\frac{3}{2}x\right) + \frac{2}{9}x^2 - \frac{43}{81}$.

13. $y_c(x) = c_1 e^x + c_2 e^{2x}$ so try $y_p(x) = Axe^{2x}$. Then $(4Ax + 4A - 6Ax - 3A)e^{2x} + 2Axe^{2x} = e^{2x} \quad\Rightarrow$ $A = 1$. Thus the general solution is $y(x) = c_1 e^x + c_2 e^{2x} + xe^{2x}$.

15. The auxiliary equation is $r^2 + 6r = 0$ and the general solution is $y(x) = c_1 + c_2 e^{-6x} = k_1 + k_2 e^{-6(x-1)}$. But $3 = y(1) = k_1 + k_2$ and $12 = y'(1) = -6k^2$. Thus $k_2 = -2$, $k_1 = 5$ and the solution is $y(x) = 5 - 2e^{-6(x-1)}$.

17. The auxiliary equation is $r^2 - 5r + 4 = 0$ and the general solution is $y(x) = c_1 e^x + c_2 e^{4x}$. But $0 = y(0) = c_1 + c_2$ and $1 = y'(0) = c_1 + 4c_2$, so the solution is $y(x) = \frac{1}{3}(e^{4x} - e^x)$.

19. Let $y(x) = \sum_{n=0}^{\infty} a_n x^n$. Then $y''(x) = \sum_{n=0}^{\infty} n(n-1)a_n x^{n-2} = \sum_{n=0}^{\infty}(n+2)(n+1)a_{n+2}x^n$ and the differential equation becomes $\sum_{n=0}^{\infty}[(n+2)(n+1)a_{n+2} + (n+1)a_n]x^n = 0$. Thus the recurrence relation is $a_{n+2} = -a_n/(n+2)$ for $n = 0, 1, 2, \ldots$. But $a_0 = y(0) = 0$, so $a_{2n} = 0$ for $n = 0, 1, 2, \ldots$. Also $a_1 = y'(0) = 1$, so $a_3 = -\dfrac{1}{3}$, $a_5 = \dfrac{(-1)^2}{3 \cdot 5}$, $a_7 = \dfrac{(-1)^3}{3 \cdot 5 \cdot 7} = \dfrac{(-1)^3 2^3 3!}{7!}$, $\ldots$, $a_{2n+1} = \dfrac{(-1)^n 2^n\, n!}{(2n+1)!}$ for $n = 0, 1, 2, \ldots$. Thus the solution to the initial-value problem is $y(x) = \displaystyle\sum_{n=0}^{\infty} a_n x^n = \sum_{n=0}^{\infty} \frac{(-1)^n 2^n\, n!\, x^{2n+1}}{(2n+1)!}$.

21. Here the initial-value problem is $2Q'' + 40Q' + 400Q = 12$, $Q(0) = 0.01$, $Q'(0) = 0$. Then $Q_c(t) = e^{-10t}(c_1 \cos 10t + c_2 \sin 10t)$ and we try $Q_p(t) = A$. Thus the general solution is $Q(t) = e^{-10t}(c_1 \cos 10t + c_2 \sin 10t) + \frac{3}{100}$. But $0.01 = Q(0) = c_1 + 0.03$ and $0 = Q'(0) = -10c_1 + 10c_2$, so $c_1 = -0.02 = c_2$. Hence the charge is given by $Q(t) = -0.02e^{-10t}(\cos 10t + \sin 10t) + 0.03$.

23. **(a)** Since we are assuming that the earth is a solid sphere of uniform density, we can calculate the density ρ as follows: $\rho = \dfrac{\text{mass of earth}}{\text{volume of earth}} = \dfrac{M}{\frac{4}{3}\pi R^3}$. If V_r is the volume of the portion of the earth which lies within a distance r of the center, then $V_r = \frac{4}{3}\pi r^3$ and $M_r = \rho V_r = \dfrac{Mr^3}{R^3}$. Thus $F_r = -\dfrac{GM_r m}{r^2} = -\dfrac{GMm}{R^3} r$.

(b) The particle is acted upon by a varying gravitational force during its motion. By Newton's Second Law of Motion, $m\dfrac{d^2y}{dt^2} = F_y = -\dfrac{GMm}{R^3}y$, so $y''(t) = -k^2 y(t)$ where $k^2 = \dfrac{GM}{R^3}$. At the surface, $-mg = F_R = -\dfrac{GMm}{R^2}$, so $g = \dfrac{GM}{R^2}$. Therefore $k^2 = \dfrac{g}{R}$.

(c) The differential equation $y'' + k^2 y = 0$ has auxiliary equation $r^2 + k^2 = 0$. (This is the r of Section 15.1, not the r measuring distance from the earth's center.) The roots of the auxiliary equation are $\pm ik$, so by (15.1.11), the general solution of our differential equation for t is $y(t) = c_1 \cos kt + c_2 \sin kt$. It follows that $y'(t) = -c_1 k \sin kt + c_2 k \cos kt$. Now $y(0) = R$ and $y'(0) = 0$, so $c_1 = R$ and $c_2 k = 0$. Thus $y(t) = R\cos kt$ and $y'(t) = -kR\sin kt$. This is simple harmonic motion (see Section 15.3) with amplitude R, frequency k, and phase angle 0. The period is $T = 2\pi/k$. $R \approx 3960$ mi $= 3960 \cdot 5280$ ft and $g = 32\text{ ft/s}^2$, so $k = \sqrt{g/R} \approx 1.24 \times 10^{-3}\,s^{-1}$ and $T = 2\pi/k \approx 5079\text{ s} \approx 85$ min.

(d) $y(t) = 0 \quad \Leftrightarrow \quad \cos kt = 0 \quad \Leftrightarrow \quad kt = \frac{\pi}{2} + \pi n$ for some integer $n \quad \Rightarrow$
$y'(t) = -kR\sin\left(\frac{\pi}{2} + \pi n\right) = \pm kR$. Thus the particle passes through the center of the earth with speed $kR \approx 4.899\text{ mi/s} \approx 17{,}600\text{ mi/h}$.

EXERCISES A

1. $|5 - 23| = |-18| = 18$

3. $|-\pi| = \pi$ because $\pi > 0$.

5. $\left|\sqrt{5} - 5\right| = -\left(\sqrt{5} - 5\right) = 5 - \sqrt{5}$ because $\sqrt{5} - 5 < 0$.

7. For $x < 2$, $x - 2 < 0$, so $|x - 2| = -(x - 2) = 2 - x$.

9. $|x + 1| = \begin{cases} x + 1 & \text{for } x + 1 \geq 0 \quad \Leftrightarrow \quad x \geq -1 \\ -(x + 1) & \text{for } x + 1 < 0 \quad \Leftrightarrow \quad x < -1 \end{cases}$

11. $|x^2 + 1| = x^2 + 1$ (since $x^2 + 1 \geq 0$ for all x).

13. $2x + 7 > 3 \quad \Leftrightarrow \quad 2x > -4$
$\Leftrightarrow \quad x > -2$, so $x \in (-2, \infty)$.

15. $1 - x \leq 2 \quad \Leftrightarrow \quad -x \leq 1$
$\Leftrightarrow \quad x \geq -1$, so $x \in [-1, \infty)$.

17. $2x + 1 < 5x - 8 \quad \Leftrightarrow \quad 9 < 3x$
$\Leftrightarrow \quad 3 < x$, so $x \in (3, \infty)$.

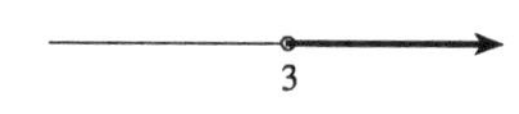

19. $-1 < 2x - 5 < 7 \quad \Leftrightarrow \quad 4 < 2x < 12$
$\Leftrightarrow \quad 2 < x < 6$, so $x \in (2, 6)$.

21. $0 \leq 1 - x < 1 \quad \Leftrightarrow \quad -1 \leq -x < 0$
$\Leftrightarrow \quad 1 \geq x > 0$, so $x \in (0, 1]$.

23. $4x < 2x + 1 \leq 3x + 2$. So $4x < 2x + 1 \quad \Leftrightarrow$
$2x < 1 \quad \Leftrightarrow \quad x < \frac{1}{2}$, and $2x + 1 \leq 3x + 2$
$\Leftrightarrow \quad -1 \leq x$. Thus $x \in \left[-1, \frac{1}{2}\right)$.

25. $1 - x \geq 3 - 2x \geq x - 6$. So $1 - x \geq 3 - 2x \quad \Leftrightarrow \quad x \geq 2$, and $3 - 2x \geq x - 6 \quad \Leftrightarrow \quad 9 \geq 3x \quad \Leftrightarrow$
$3 \geq x$. Thus $x \in [2, 3]$.

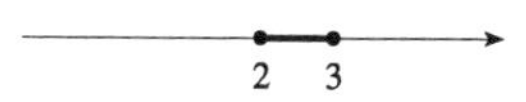

27. $(x - 1)(x - 2) > 0$. *Case 1:* $x - 1 > 0 \quad \Leftrightarrow \quad x > 1$, and $x - 2 > 0 \quad \Leftrightarrow \quad x > 2$, so $x \in [1, \infty)$.
Case 2: $x - 1 < 0 \quad \Leftrightarrow \quad x < 1$, and $x - 2 < 0 \quad \Leftrightarrow \quad x < 2$, so $x \in (-\infty, 1)$. Thus the solution set is
$(-\infty, 1) \cup (2, \infty)$.

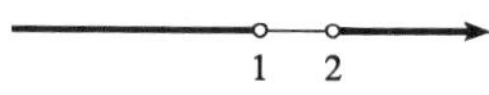

29. $2x^2 + x \leq 1 \quad \Leftrightarrow \quad 2x^2 + x - 1 \leq 0 \quad \Leftrightarrow \quad (2x - 1)(x + 1) \leq 0$. *Case 1:* $2x - 1 \geq 0 \quad \Leftrightarrow \quad x \geq \frac{1}{2}$, and
$x + 1 \leq 0 \quad \Leftrightarrow \quad x \leq -1$, which is impossible. *Case 2:* $2x - 1 \leq 0 \quad \Leftrightarrow \quad x \leq \frac{1}{2}$, and $x + 1 \geq 0 \quad \Leftrightarrow$
$x \geq -1$, so $x \in \left[-1, \frac{1}{2}\right]$. Thus the solution set is $\left[-1, \frac{1}{2}\right]$.

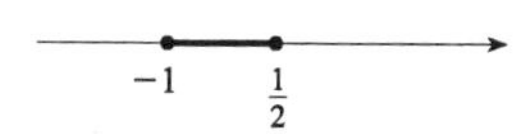

31. $x^2 + x + 1 > 0 \quad \Leftrightarrow \quad x^2 + x + \frac{1}{4} + \frac{3}{4} > 0 \quad \Leftrightarrow \quad \left(x + \frac{1}{2}\right)^2 + \frac{3}{4} > 0$. But since $\left(x + \frac{1}{2}\right)^2 \geq 0$ for every real x, the original inequality will be true for all real x as well. Thus, the solution set is $(-\infty, \infty)$.

33. $x^2 < 3 \quad \Leftrightarrow \quad x^2 - 3 < 0 \quad \Leftrightarrow \quad \left(x - \sqrt{3}\right)\left(x + \sqrt{3}\right) < 0$. *Case 1:* $x > \sqrt{3}$ and $x < -\sqrt{3}$, which is impossible. *Case 2:* $x < \sqrt{3}$ and $x > -\sqrt{3}$. Thus the solution set is $\left(-\sqrt{3}, \sqrt{3}\right)$.

Another Method: $x^2 < 3 \quad \Leftrightarrow \quad |x| < \sqrt{3} \quad \Leftrightarrow \quad -\sqrt{3} < x < \sqrt{3}$.

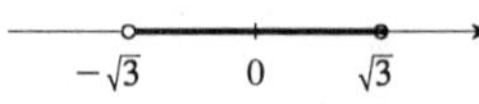

35. $x^3 - x^2 \leq 0 \quad \Leftrightarrow \quad x^2(x - 1) \leq 0$. Since $x^2 \geq 0$ for all x, the inequality is satisfied when $x - 1 \leq 0 \quad \Leftrightarrow \quad x \leq 1$. Thus the solution set is $(-\infty, 1]$.

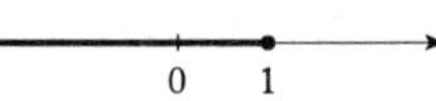

37. $x^3 > x \quad \Leftrightarrow \quad x^3 - x > 0 \quad \Leftrightarrow \quad x(x^2 - 1) > 0 \quad \Leftrightarrow \quad x(x - 1)(x + 1) > 0$. Constructing a table:

Interval	x	$x - 1$	$x + 1$	$x(x-1)(x+1)$
$x < -1$	−	−	−	−
$-1 < x < 0$	−	−	+	+
$0 < x < 1$	+	−	+	−
$x > 1$	+	+	+	+

Since $x^3 > x$ when the last column is positive, the solution set is $(-1, 0) \cup (1, \infty)$.

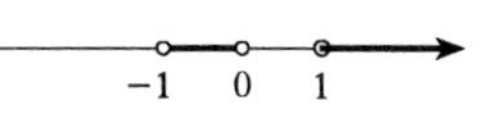

39. $1/x < 4$. This is clearly true for $x < 0$. So suppose $x > 0$. then $1/x < 4 \quad \Leftrightarrow \quad 1 < 4x \quad \Leftrightarrow \quad \frac{1}{4} < x$. Thus the solution set is $(-\infty, 0) \cup \left(\frac{1}{4}, \infty\right)$.

41. Multiply both sides by x. *Case 1:* If $x > 0$, then $4/x < x \quad \Leftrightarrow \quad 4 < x^2 \quad \Leftrightarrow \quad 2 < x$. *Case 2:* If $x < 0$, then $4/x < x \quad \Leftrightarrow \quad 4 > x^2 \quad \Leftrightarrow \quad -2 < x < 0$. Thus the solution set is $(-2, 0) \cup (2, \infty)$.

43. $\dfrac{2x + 1}{x - 5} < 3$. *Case 1:* If $x - 5 > 0$ (that is, $x > 5$), then $2x + 1 < 3(x - 5) \quad \Leftrightarrow \quad 16 < x$, so $x \in (16, \infty)$.

Case 2: If $x - 5 < 0$ (that is, $x < 5$), then $2x + 1 > 3(x - 5) \quad \Leftrightarrow \quad 16 > x$, so in this case $x \in (-\infty, 5)$.

Combining the two cases, the solution set is $(-\infty, 5) \cup (16, \infty)$.

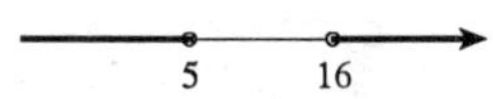

45. $\dfrac{x^2 - 1}{x^2 + 1} \geq 0$. Since $x^2 + 1 \geq 0$ for all real x, this inequality will hold whenever $x^2 - 1 \geq 0 \quad \Leftrightarrow$ $(x - 1)(x + 1) \geq 0$. *Case 1:* $x \geq 1$ and $x \geq -1$, so $x \in [1, \infty)$. *Case 2:* $x \leq 1$ and $x \leq -1$, so $x \in (-\infty, -1]$. Thus the solution set is $(-\infty, -1] \cup [1, \infty)$.

Another Method: $x^2 \geq 1 \quad \Leftrightarrow \quad |x| \geq 1 \quad \Leftrightarrow \quad x \geq 1$ or $x \leq -1$.

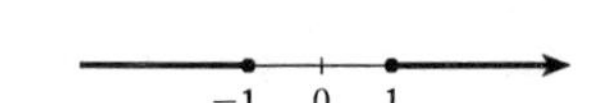

47. $C = \frac{5}{9}(F - 32) \quad \Rightarrow \quad F = \frac{9}{5}C + 32$. So $50 \le F \le 95 \quad \Rightarrow \quad 50 \ge \frac{9}{5}C + 32 \le 95 \quad \Rightarrow \quad 18 \le \frac{9}{5}C \le 63$ $\quad \Rightarrow \quad 10 \le C \le 35$. So the interval is $[10, 35]$.

49. **(a)** Let T represent the temperature in degrees Celsius and h the height in km. $T = 20$ when $h = 0$ and T decreases by $10°$ C for every km. Thus $T = 20 - 10h$ when $0 \le h \le 12$.

(b) From (a), $T = 20 - 10h \quad \Rightarrow \quad h = 2 - T/10$. So $0 \le h \le 5 \quad \Rightarrow \quad 0 \le 2 - T/10 \le 5 \quad \Rightarrow$ $-2 \le -T/10 \le 3 \quad \Rightarrow \quad -30 \le T \le 20$. Thus the range of temperatures to be expected is $[-30, 20]$.

51. $|2x| = 3 \quad \Leftrightarrow \quad$ either $2x = 3$ or $2x = -3 \quad \Leftrightarrow \quad x = \frac{3}{2}$ or $x = -\frac{3}{2}$.

53. $|x + 3| = |2x + 1| \quad \Leftrightarrow \quad$ either $x + 3 = 2x + 1$ or $x + 3 = -(2x + 1)$. In the first case, $x = 2$, and in the second case, $3x = -4 \quad \Leftrightarrow \quad x = -\frac{4}{3}$.

55. By (6), Property 5, $|x| < 3 \quad \Leftrightarrow \quad -3 < x < 3$, so $x \in (-3, 3)$.

57. $|x - 4| < 1 \quad \Leftrightarrow \quad -1 < x - 4 < 1 \quad \Leftrightarrow \quad 3 < x < 5$, so $x \in (3, 5)$.

59. $|x + 5| \ge 2 \quad \Leftrightarrow \quad x + 5 \ge 2$ or $x + 5 \le -2 \quad \Leftrightarrow \quad x \ge -3$ or $x \le -7$, so $x \in (-\infty, -7] \cup [-3, \infty)$.

61. $|2x - 3| \le 0.4 \quad \Leftrightarrow \quad -0.4 \le 2x - 3 \le 0.4 \quad \Leftrightarrow \quad 2.6 \le 2x \le 3.4 \quad \Leftrightarrow \quad 1.3 \le x \le 1.7$, so $x \in [1.3, 1.7]$.

63. $1 \le |x| \le 4$. So either $1 \le x \le 4$ or $1 \le -x \le 4 \quad \Leftrightarrow \quad -1 \ge x \ge -4$. Thus $x \in [-4, -1] \cup [1, 4]$.

65. $|x| > |x - 1|$. Since $|x|, |x - 1| \ge 0$, $|x| > |x - 1| \quad \Leftrightarrow \quad |x|^2 > |x - 1|^2 \quad \Leftrightarrow$ $x^2 > (x - 1)^2 = x^2 - 2x + 1 \quad \Leftrightarrow \quad 0 > -2x + 1 \quad \Leftrightarrow \quad x > \frac{1}{2}$, so $x \in \left(\frac{1}{2}, \infty\right)$.

67. $\left|\dfrac{x}{2 + x}\right| < 1 \quad \Leftrightarrow \quad \left(\dfrac{x}{2 + x}\right)^2 < 1 \quad \Leftrightarrow \quad x^2 < (2 + x)^2 \quad \Leftrightarrow \quad x^2 < 4 + 4x + x^2 \quad \Leftrightarrow \quad 0 < 4 + 4x$ $\quad \Leftrightarrow \quad -1 < x$, so $x \in (-1, \infty)$.

69. $a(bx - c) \ge bc \quad \Leftrightarrow \quad bx - c \ge \dfrac{bc}{a} \quad \Leftrightarrow \quad bx \ge \dfrac{bc}{a} + c = \dfrac{bc + ac}{a} \quad \Leftrightarrow \quad x \ge \dfrac{bc + ac}{ab}$

71. $ax + b < c \quad \Leftrightarrow \quad ax < c - b \quad \Leftrightarrow \quad x > \dfrac{c - b}{a}$ (since $a < 0$)

73. $|(x + y) - 5| = |(x - 2) + (y - 3)| \le |x - 2| + |y - 3| < 0.01 + 0.04 = 0.05$

75. If $a < b$ then $a + a < a + b$ and $a + b < b + b$. So $2a < a + b < 2b$. Dividing by 2, $a < \frac{1}{2}(a + b) < b$.

77. $|ab| = \sqrt{(ab)^2} = \sqrt{a^2 b^2} = \sqrt{a^2}\sqrt{b^2} = |a||b|$

79. If $0 < a < b$, then $a \cdot a < a \cdot b$ and $a \cdot b < b \cdot b$ [using (2), Rule 3]. So $a^2 < ab < b^2$ and hence $a^2 < b^2$.

81. Observe that the sum, difference and product of two integers is always an integer. Let the rational numbers be represented by $r = m/n$ and $s = p/q$ (where m, n, p and q are integers with $n \ne 0$, $q \ne 0$). Now $r + s = \dfrac{m}{n} + \dfrac{p}{q} = \dfrac{mq + pn}{nq}$, but $mq + pn$ and nq are both integers, so $\dfrac{mq + pn}{nq} = r + s$ is a rational number by definition. Similarly, $r - s = \dfrac{m}{n} - \dfrac{p}{q} = \dfrac{mq - pn}{nq}$ is a rational number. Finally, $r \cdot s = \dfrac{m}{n} \cdot \dfrac{p}{q} = \dfrac{mp}{nq}$ but mp and nq are both integers, so $\dfrac{mp}{nq} = r \cdot s$ is a rational number by definition.

EXERCISES B

1. From the Distance Formula (1) with $x_1 = 1$, $x_2 = 4$, $y_1 = 1$, $y_2 = 5$, we find the distance to be $\sqrt{(4-1)^2 + (5-1)^2} = \sqrt{3^2 + 4^2} = \sqrt{25} = 5$.

3. $\sqrt{(-1-6)^2 + [3-(-2)]^2} = \sqrt{(-7)^2 + 5^2} = \sqrt{74}$

5. $\sqrt{(4-2)^2 + (-7-5)^2} = \sqrt{2^2 + (-12)^2} = \sqrt{148} = 2\sqrt{37}$

7. From (2), the slope is $\dfrac{11-5}{4-1} = \dfrac{6}{3} = 2$.

9. $m = \dfrac{-6-3}{-1-(-3)} = -\dfrac{9}{2}$

11. Since $|AC| = \sqrt{(-4-0)^2 + (3-2)^2} = \sqrt{(-4)^2 + 1^2} = \sqrt{17}$ and $|BC| = \sqrt{[-4-(-3)]^2 + [3-(-1)]^2} = \sqrt{(-1)^2 + 4^2} = \sqrt{17}$, the triangle has two sides of equal length, and so is isosceles.

13. Label the points A, B, C, and D respectively. Then

$|AB| = \sqrt{[4-(-2)]^2 + (6-9)^2} = \sqrt{6^2 + (-3)^2} = 3\sqrt{5}$,

$|BC| = \sqrt{(1-4)^2 + (0-6)^2} = \sqrt{(-3)^2 + (-6)^2} = 3\sqrt{5}$,

$|CD| = \sqrt{(-5-1)^2 + (3-0)^2} = \sqrt{(-6)^2 + 3^2} = 3\sqrt{5}$, and

$|DA| = \sqrt{[-2-(-5)]^2 + (9-3)^2} = \sqrt{3^2 + 6^2} = 3\sqrt{5}$. So all sides are of equal length. Moreover, $m_{AB} = \dfrac{6-9}{4-(-2)} = -\dfrac{1}{2}$, $m_{BC} = \dfrac{0-6}{1-4} = 2$, $m_{CD} = \dfrac{3-0}{-5-1} = -\dfrac{1}{2}$, and $m_{DA} = \dfrac{9-3}{-2-(-5)} = 2$, so the sides are perpendicular. Thus, it is a square.

15. The slope of the line segment AB is $\dfrac{4-1}{7-1} = \dfrac{1}{2}$, the slope of CD is $\dfrac{7-10}{-1-5} = \dfrac{1}{2}$, the slope of BC is $\dfrac{10-4}{5-7} = -3$, and the slope of DA is $\dfrac{1-7}{1-(-1)} = -3$. So AB is parallel to CD and BC is parallel to DA. Hence $ABCD$ is a parallelogram.

17. $x = 3$

19. $xy = 0 \quad \Leftrightarrow \quad x = 0$ or $y = 0$

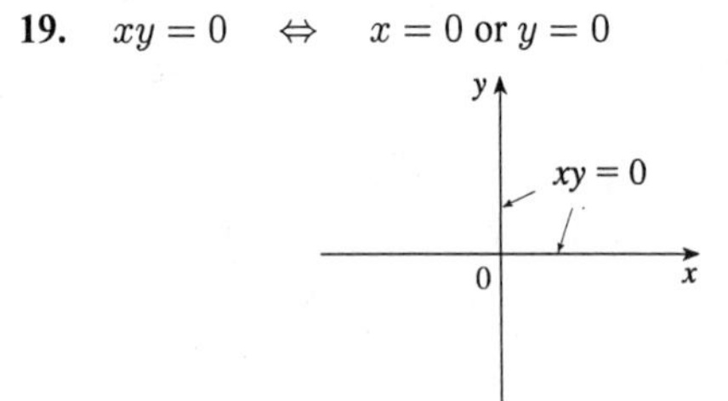

21. From (3), the equation of the line is $y - (-3) = 6(x - 2)$ or $y = 6x - 15$.

23. $y - 7 = \frac{2}{3}(x - 1)$ or $2x - 3y + 19 = 0$

25. The slope is $m = \dfrac{6 - 1}{1 - 2} = -5$, so the equation of the line is $y - 1 = -5(x - 2)$ or $5x + y = 11$.

27. From (4), the equation is $y = 3x - 2$.

29. Since the line passes through $(1, 0)$ and $(0, -3)$, its slope is $m = \dfrac{-3 - 0}{0 - 1} = 3$, so its equation is $y = 3x - 3$.

31. Since $m = 0$, $y - 5 = 0(x - 4)$ or $y = 5$.

33. Putting the line $x + 2y = 6$ into its slope-intercept form $y = -\frac{1}{2}x + 3$, we see that this line has slope $-\frac{1}{2}$. So we want the line of slope $-\frac{1}{2}$ that passes through the point $(1, -6)$: $y - (-6) = -\frac{1}{2}(x - 1) \quad \Leftrightarrow \quad y = -\frac{1}{2}x - \frac{11}{2}$ or $x + 2y + 11 = 0$.

35. $2x + 5y + 8 = 0 \quad \Leftrightarrow \quad y = -\frac{2}{5}x - \frac{8}{5}$. Since this line has slope $-\frac{2}{5}$, a line perpendicular to it would have slope $\frac{5}{2}$, so the required line is $y - (-2) = \frac{5}{2}[x - (-1)] \quad \Leftrightarrow \quad y = \frac{5}{2}x + \frac{1}{2}$ or $5x - 2y + 1 = 0$.

37. $x + 3y = 0 \quad \Leftrightarrow \quad y = -\frac{1}{3}x$, so the slope is $-\frac{1}{3}$ and the y-intercept is 0.

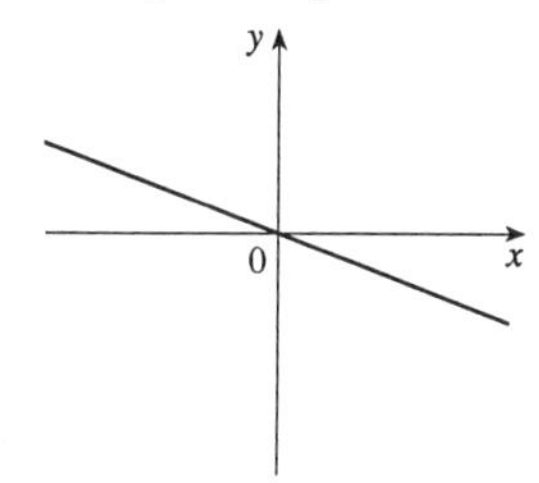

39. $y = -2$ is a horizontal line with slope 0 and y-intercept -2.

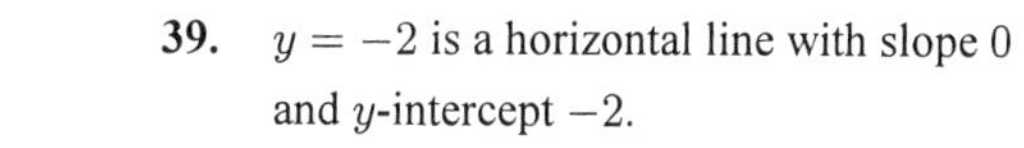

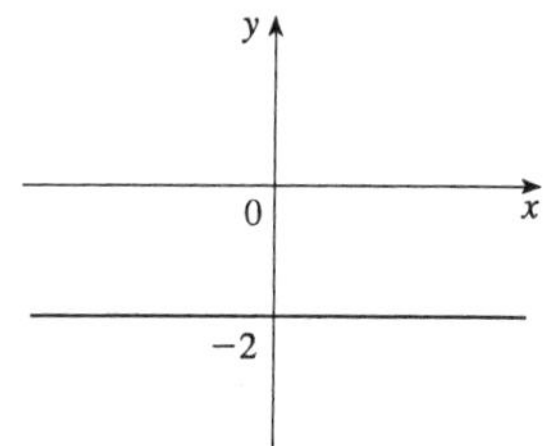

41. $3x - 4y = 12 \quad \Leftrightarrow \quad y = \frac{3}{4}x - 3$, so the slope is $\frac{3}{4}$ and the y-intercept is -3.

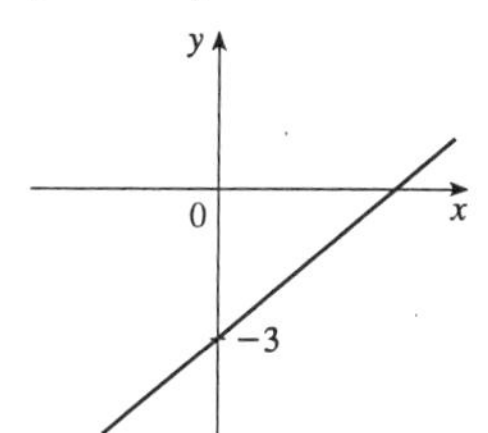

43. $\{(x, y) \mid x < 0\}$

45. $\{(x,y) \mid xy < 0\} = \{(x,y) \mid x < 0 \text{ and } y > 0\} \cup \{(x,y) \mid x > 0 \text{ and } y < 0\}$

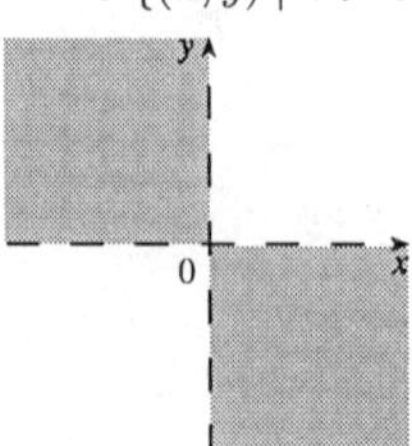

47. $\{(x,y) \mid |x| \le 2\} = \{(x,y) \mid -2 \le x \le 2\}$

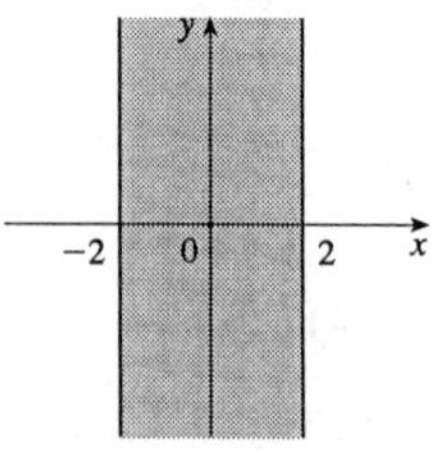

49. $\{(x,y) \mid 0 \le y \le 4, x \le 2\}$

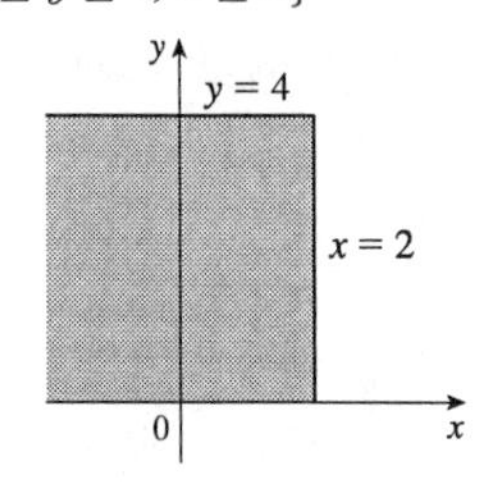

51. $\{(x,y) \mid 1 + x \le y \le 1 - 2x\}$

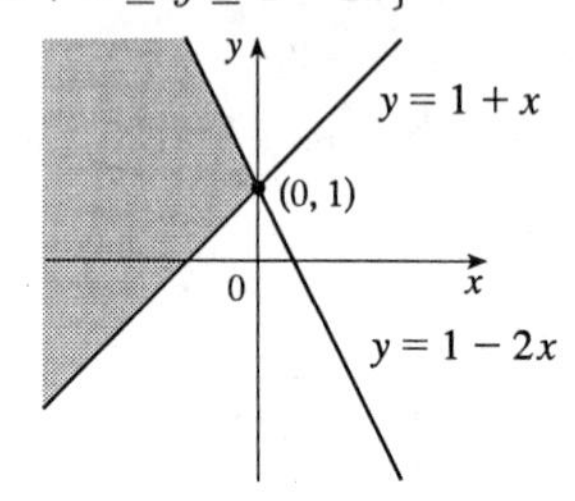

53. Let $P(0,y)$ be a point on the y-axis. The distance from P to $(5,-5)$ is $\sqrt{(5-0)^2 + (-5-y)^2} = \sqrt{5^2 + (y+5)^2}$. The distance from P to $(1,1)$ is $\sqrt{(1-0)^2 + (1-y)^2} = \sqrt{1^2 + (y-1)^2}$. We want these distances to be equal:
$\sqrt{5^2 + (y+5)^2} = \sqrt{1^2 + (y-1)^2} \quad \Leftrightarrow \quad 5^2 + (y+5)^2 = 1^2 + (y-1)^2 \quad \Leftrightarrow$
$25 + (y^2 + 10y + 25) = 1 + (y^2 - 2y + 1) \;\Leftrightarrow\; 12y = -48 \;\Leftrightarrow\; y = -4$. So the desired point is $(0,-4)$.

55. Using the midpoint formula of Exercise 54, we get

(a) $\left(\frac{1+7}{2}, \frac{3+15}{2}\right) = (4,9)$ **(b)** $\left(\frac{-1+8}{2}, \frac{6-12}{2}\right) = \left(\frac{7}{2}, -3\right)$

57. $2x - y = 4 \quad \Leftrightarrow \quad y = 2x - 4 \quad \Rightarrow \quad m_1 = 2$ and $6x - 2y = 10 \quad \Leftrightarrow \quad 2y = 6x - 10 \quad \Leftrightarrow \quad y = 3x - 5$ $\Rightarrow \quad m_2 = 3$. Since $m_1 \ne m_2$, the two lines are not parallel [by 6(a)]. To find the point of intersection:
$2x - 4 = 3x - 5 \quad \Leftrightarrow \quad x = 1 \quad \Rightarrow \quad y = -2$. Thus, the point of intersection is $(1,-2)$.

59. The slope of the segment AB is $\frac{-2-4}{7-1} = -1$, so its perpendicular bisector has slope 1. The midpoint of AB is $\left(\frac{1+7}{2}, \frac{4-2}{2}\right) = (4,1)$, so the equation of the perpendicular bisector is $y - 1 = 1(x-4)$ or $y = x - 3$.

61. **(a)** Since the x-intercept is a, the point $(a,0)$ is on the line, and similarly since the y-intercept is b, $(0,b)$ is on the line. Hence the slope of the line is $m = \frac{b-0}{0-a} = -\frac{b}{a}$. Substituting into $y = mx + b$ gives
$y = -\frac{b}{a}x + b \quad \Leftrightarrow \quad y + \frac{b}{a}x = b \quad \Leftrightarrow \quad \frac{y}{b} + \frac{x}{a} = 1$.

(b) Letting $a = 6$ and $b = -8$ gives $\frac{y}{-8} + \frac{x}{6} = 1 \quad \Leftrightarrow \quad 6y - 8x = -48 \quad \Leftrightarrow \quad 8x - 6y - 48 = 0 \quad \Leftrightarrow$
$4x - 3y - 24 = 0$.

EXERCISES C

1. From (1), the equation is $(x-3)^2+(y+1)^2=25$.

3. The equation has the form $x^2+y^2=r^2$. Since $(4,7)$ lies on the circle, we have $4^2+7^2=r^2 \quad \Rightarrow \quad r^2=65$. So the required equation is $x^2+y^2=65$.

5. $x^2+y^2-4x+10y+13=0 \quad \Leftrightarrow \quad x^2-4x+y^2+10y=-13 \quad \Leftrightarrow$
$(x^2-4x+4)+(y^2+10y+25)=-13+4+25=16 \quad \Leftrightarrow \quad (x-2)^2+(y+5)^2=4^2$. Thus, we have a circle with center $(2,-5)$ and radius 4.

7. $x^2+y^2+x=0 \quad \Leftrightarrow \quad \left(x^2+x+\frac{1}{4}\right)+y^2=\frac{1}{4} \quad \Leftrightarrow \quad \left(x+\frac{1}{2}\right)^2+y^2=\left(\frac{1}{2}\right)^2$. Thus, we have a circle with center $\left(-\frac{1}{2},0\right)$ and radius $\frac{1}{2}$.

9. $2x^2+2y^2-x+y=1 \quad \Leftrightarrow \quad 2\left(x^2-\frac{1}{2}x+\frac{1}{16}\right)+2\left(y^2+\frac{1}{2}y+\frac{1}{16}\right)=1+\frac{1}{8}+\frac{1}{8} \quad \Leftrightarrow$
$2\left(x-\frac{1}{4}\right)^2+2\left(y+\frac{1}{4}\right)^2=\frac{5}{4} \quad \Leftrightarrow \quad \left(x-\frac{1}{4}\right)^2+\left(y+\frac{1}{4}\right)^2=\frac{5}{8}$. Thus, we have a circle with center $\left(\frac{1}{4},-\frac{1}{4}\right)$ and radius $\frac{\sqrt{5}}{2\sqrt{2}}=\frac{\sqrt{10}}{4}$.

11. $y=-x^2$. Parabola

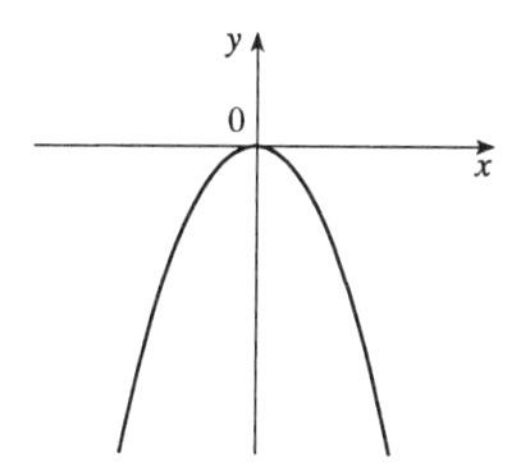

13. $x^2+4y^2=16 \quad \Leftrightarrow \quad \dfrac{x^2}{16}+\dfrac{y^2}{4}=1$. Ellipse

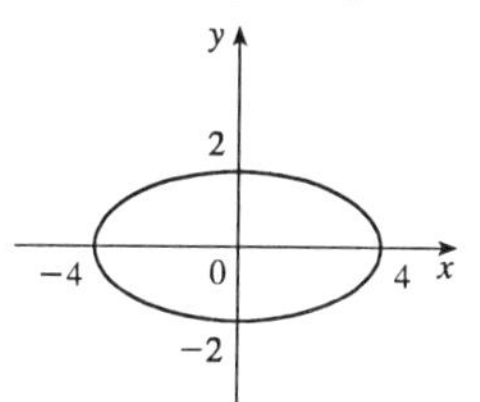

15. $16x^2-25y^2=400 \quad \Leftrightarrow \quad \dfrac{x^2}{25}-\dfrac{y^2}{16}=1$.
Hyperbola

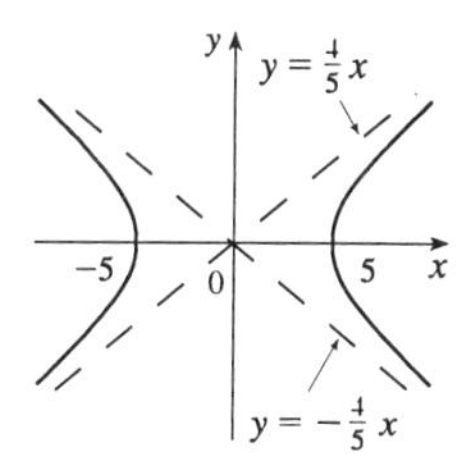

17. $4x^2+y^2=1 \quad \Leftrightarrow \quad \dfrac{x^2}{1/4}+y^2=1$. Ellipse

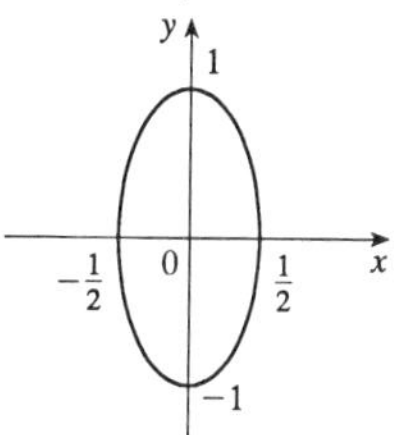

19. $x=y^2-1$. Parabola with vertex at $(-1,0)$

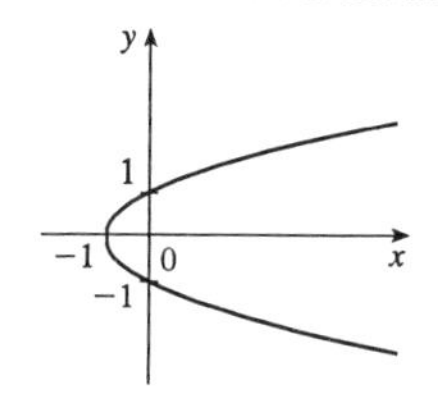

21. $9y^2-x^2=9 \quad \Leftrightarrow \quad y^2-\dfrac{x^2}{9}=1$. Hyperbola

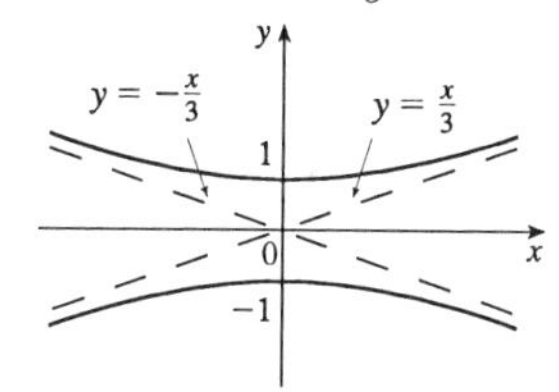

23. $xy = 4$. Hyperbola

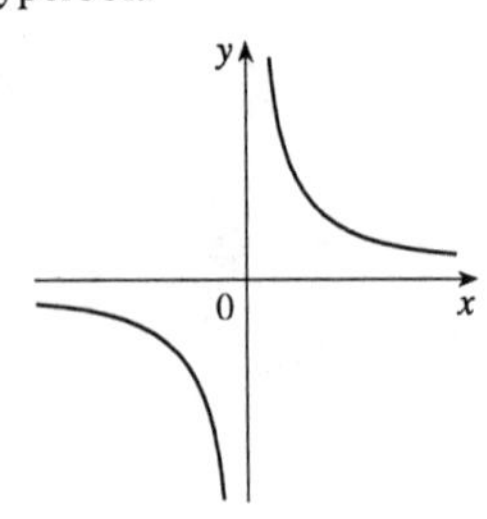

25. $9(x-1)^2 + 4(y-2)^2 = 36 \quad \Leftrightarrow$
$\dfrac{(x-1)^2}{4} + \dfrac{(y-2)^2}{9} = 1.$
Ellipse centered at $(1, 2)$

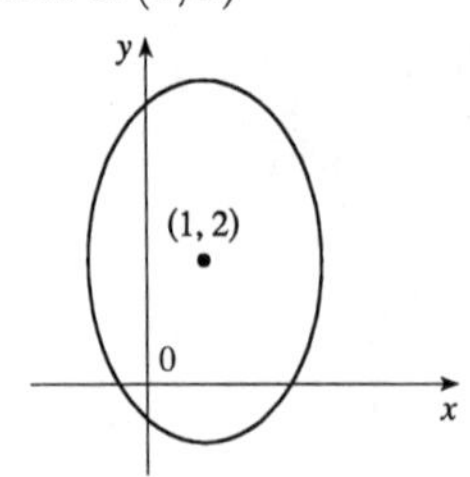

27. $y = x^2 - 6x + 13 = \left(x^2 - 6x + 9\right) + 4$
$= (x-3)^2 + 4$. Parabola with vertex at $(3, 4)$

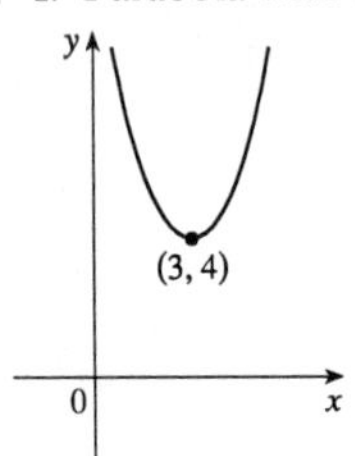

29. $x = -y^2 + 4$. Parabola with vertex at $(4, 0)$

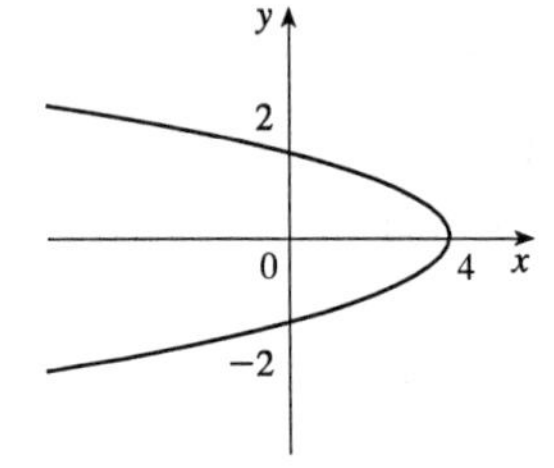

31. $x^2 + 4y^2 - 6x + 5 = 0 \quad \Leftrightarrow$
$(x^2 - 6x + 9) + 4y^2 = -5 + 9 = 4 \quad \Leftrightarrow$
$\dfrac{(x-3)^2}{4} + y^2 = 1$. Ellipse centered at $(3, 0)$

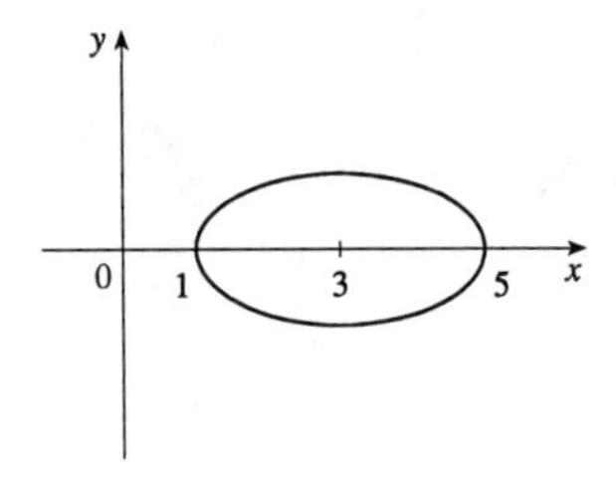

33. $y = 3x$ and $y = x^2$ intersect where $3x = x^2$
$\Leftrightarrow \quad 0 = x^2 - 3x = x(x-3)$,
that is, at $(0, 0)$ and $(3, 9)$.

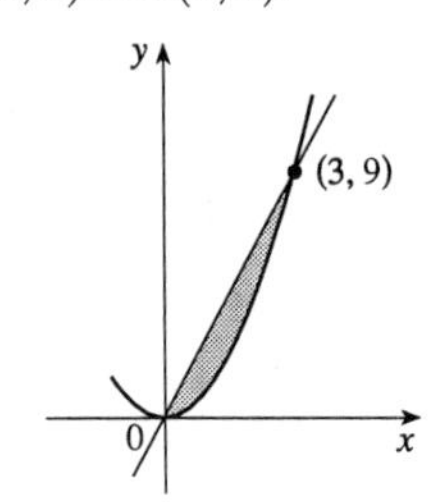

35. The parabola must have an equation of the form $y = a(x-1)^2 - 1$. Substituting $x = 3$ and $y = 3$ into the equation gives $3 = a(3-1)^2 - 1$, so $a = 1$, and the equation is $y = (x-1)^2 - 1 = x^2 - 2x$.
Note that using the other point $(-1, 3)$ would have given the same value for a, and hence the same equation.

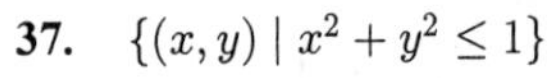

37. $\{(x, y) \mid x^2 + y^2 \leq 1\}$

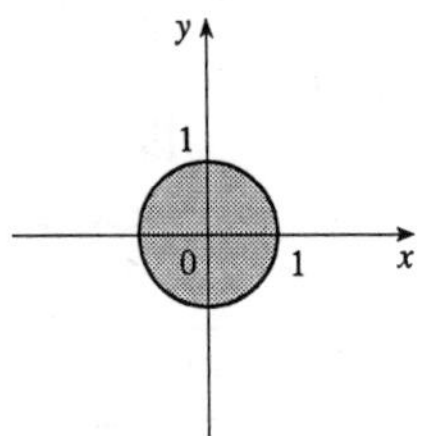

39. $\{(x, y) \mid y \geq x^2 - 1\}$

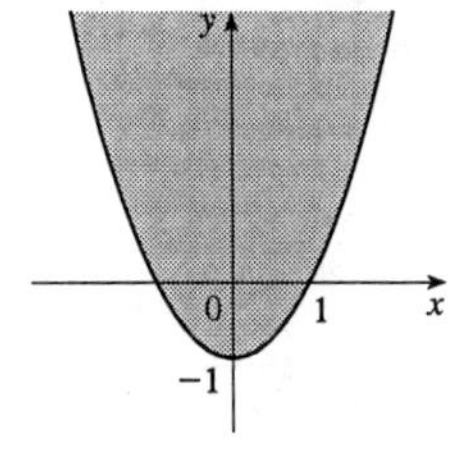

EXERCISES D

1. $210° = 210\left(\frac{\pi}{180}\right) = \frac{7\pi}{6}$ rad

3. $9° = 9\left(\frac{\pi}{180}\right) = \frac{\pi}{20}$ rad

5. $900° = 900\left(\frac{\pi}{180}\right) = 5\pi$ rad

7. 4π rad $= 4\pi\left(\frac{180}{\pi}\right) = 720°$

9. $\frac{5\pi}{12}$ rad $= \frac{5\pi}{12}\left(\frac{180}{\pi}\right) = 75°$

11. $-\frac{3\pi}{8}$ rad $= -\frac{3\pi}{8}\left(\frac{180}{\pi}\right) = -67.5°$

13. Using Formula 3, $a = r\theta = \frac{36\pi}{12} = 3\pi$ cm.

15. Using Formula 3, $\theta = \frac{1}{1.5} = \frac{2}{3}$ rad $= \frac{2}{3}\left(\frac{180}{\pi}\right) = \left(\frac{120}{\pi}\right)^{\circ}$.

17.

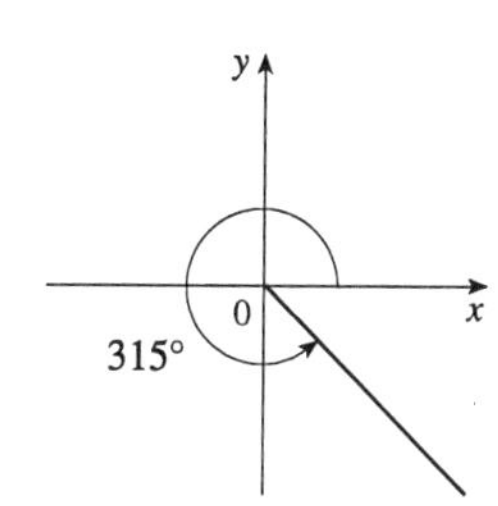

19.

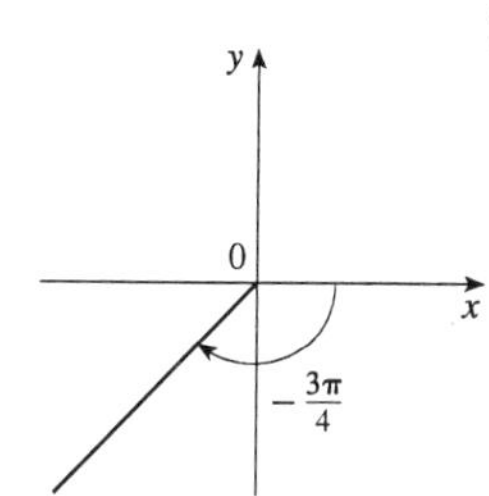

21.

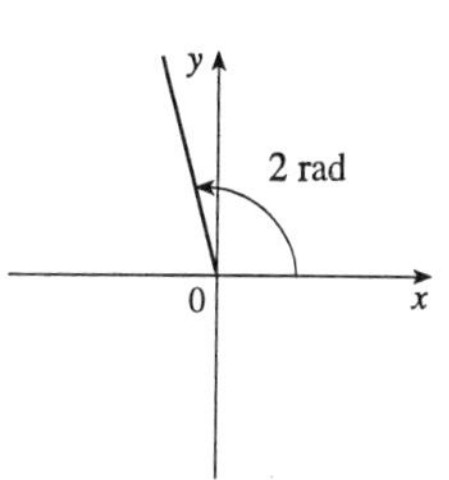

23.

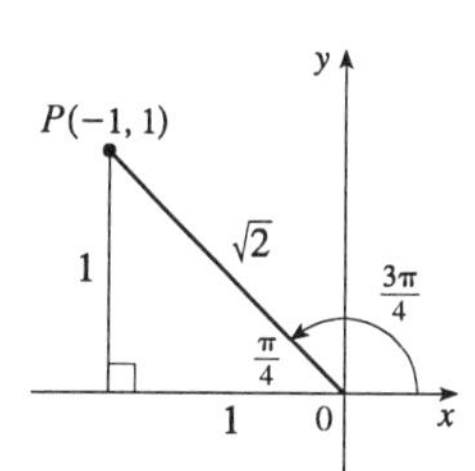

From the diagram we see that a point on the terminal line is $P(-1,1)$. Therefore taking $x = -1$, $y = 1$, $r = \sqrt{2}$ in the definitions of the trigonometric ratios, we have

$\sin\frac{3\pi}{4} = \frac{1}{\sqrt{2}}$, $\cos\frac{3\pi}{4} = -\frac{1}{\sqrt{2}}$, $\tan\frac{3\pi}{4} = -1$,

$\csc\frac{3\pi}{4} = \sqrt{2}$, $\sec\frac{3\pi}{4} = -\sqrt{2}$, and $\cot\frac{3\pi}{4} = -1$.

25.

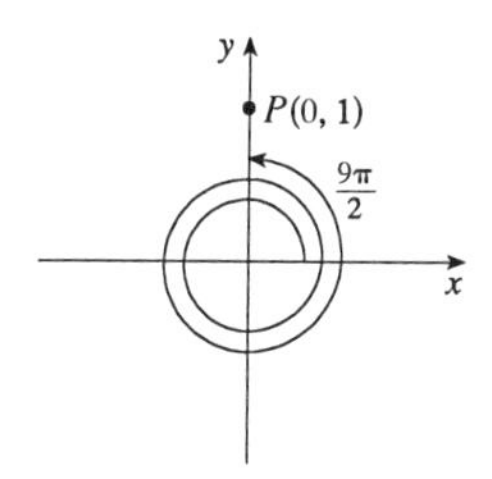

From the diagram we see that a point on the terminal line is $P(0,1)$. Therefore taking $x = 0$, $y = 1$, $r = 1$ in the definitions of the trigonometric ratios, we have $\sin\frac{9\pi}{2} = 1$, $\cos\frac{9\pi}{2} = 0$, $\tan\frac{9\pi}{2} = y/x$ is undefined since $x = 0$, $\csc\frac{9\pi}{2} = 1$, $\sec\frac{9\pi}{2} = r/x$ is undefined since $x = 0$, and $\cot\frac{9\pi}{2} = 0$.

27.

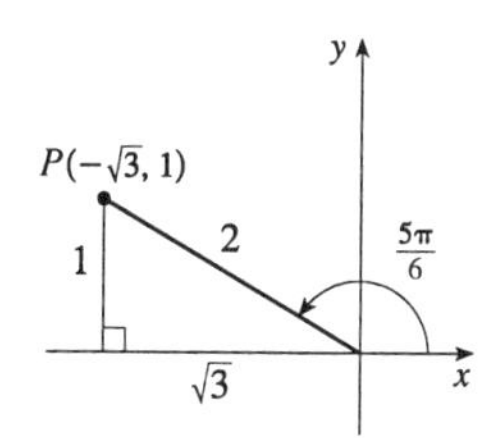

Using Figure 8 we see that a point on the terminal line is $P\left(-\sqrt{3},1\right)$. Therefore taking $x = -\sqrt{3}$, $y = 1$, $r = 2$ in the definitions of the trigonometric ratios, we have

$\sin\frac{5\pi}{6} = \frac{1}{2}$, $\cos\frac{5\pi}{6} = -\frac{\sqrt{3}}{2}$, $\tan\frac{5\pi}{6} = -\frac{1}{\sqrt{3}}$,

$\csc\frac{5\pi}{6} = 2$, $\sec\frac{5\pi}{6} = -\frac{2}{\sqrt{3}}$, and $\cot\frac{5\pi}{6} = -\sqrt{3}$.

29. $\sin\theta = y/r = \frac{3}{5} \quad\Rightarrow\quad y = 3$, $r = 5$, and $x = \sqrt{r^2 - y^2} = 4$ (since $0 < \theta < \frac{\pi}{2}$). Therefore taking $x = 4$, $y = 3$, $r = 5$ in the definitions of the trigonometric ratios, we have $\cos\theta = \frac{4}{5}$, $\tan\theta = \frac{3}{4}$, $\csc\theta = \frac{5}{3}$, $\sec\theta = \frac{5}{4}$, and $\cot\theta = \frac{4}{3}$.

31. $\frac{\pi}{2} < \phi < \pi \quad \Rightarrow \quad \phi$ is in the second quadrant, where x is negative and y is positive.
Therefore $\sec\phi = r/x = -1.5 = -\frac{3}{2} \quad \Rightarrow \quad r = 3$, $x = -2$, and $y = \sqrt{r^2 - x^2} = \sqrt{5}$. Taking $x = -2$, $y = \sqrt{5}$, and $r = 3$ in the definitions of the trigonometric ratios, we have $\sin\phi = \frac{\sqrt{5}}{3}$, $\cos\phi = -\frac{2}{3}$, $\tan\phi = -\frac{\sqrt{5}}{2}$, $\csc\phi = \frac{3}{\sqrt{5}}$, and $\cot\theta = -\frac{2}{\sqrt{5}}$.

33. $\pi < \beta < 2\pi$ means that β is in the third or fourth quadrant where y is negative. Also since $\cot\beta = x/y = 3$ which is positive, x must also be negative. Therefore $\cot\beta = x/y = \frac{3}{1} \quad \Rightarrow \quad x = -3$, $y = -1$, and $r = \sqrt{x^2 + y^2} = \sqrt{10}$. Taking $x = -3$, $y = -1$ and $r = \sqrt{10}$ in the definitions of the trigonometric ratios, we have $\sin\beta = -\frac{1}{\sqrt{10}}$, $\cos\beta = -\frac{3}{\sqrt{10}}$, $\tan\beta = \frac{1}{3}$, $\csc\beta = -\sqrt{10}$, and $\sec\beta = -\frac{\sqrt{10}}{3}$.

35. $\sin 35° = \dfrac{x}{10} \quad \Rightarrow \quad x = 10\sin 35° \approx 5.73576$ cm

37. $\tan\frac{2\pi}{5} = \dfrac{x}{8} \quad \Rightarrow \quad x = 8\tan\frac{2\pi}{5} \approx 24.62147$ cm

39. **(a)** From the diagram we see that

$\sin\theta = \dfrac{y}{r} = \dfrac{a}{c}$, and $\sin(-\theta) = -\dfrac{a}{c} = -\sin\theta$.

(b) Again from the diagram we see that

that $\cos\theta = \dfrac{x}{r} = \dfrac{b}{c} = \cos(-\theta)$.

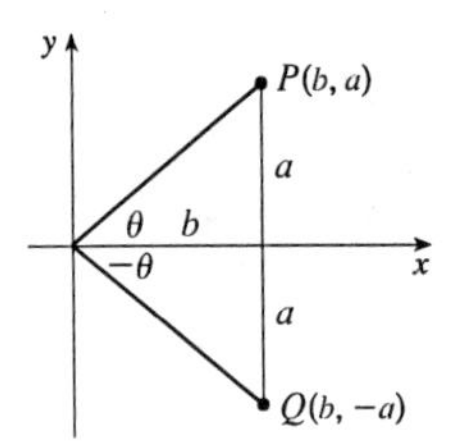

41. **(a)** Using (12a) and (13a), we have $\frac{1}{2}[\sin(x+y) + \sin(x-y)]$
$= \frac{1}{2}[\sin x\cos y + \cos x\sin y + \sin x\cos y - \cos x\sin y] = \frac{1}{2}(2\sin x\cos y) = \sin x\cos y$.

(b) This time, using (12b) and (13b), we have $\frac{1}{2}[\cos(x+y) + \cos(x-y)]$
$= \frac{1}{2}[\cos x\cos y - \sin x\sin y + \cos x\cos y + \sin x\sin y] = \frac{1}{2}(2\cos x\cos y) = \cos x\cos y$.

(c) Again using (12b) and (13b), we have $\frac{1}{2}[\cos(x-y) - \cos(x+y)]$
$= \frac{1}{2}[\cos x\cos y + \sin x\sin y - \cos x\cos y + \sin x\sin y] = \frac{1}{2}(2\sin x\sin y) = \sin x\sin y$.

43. Using (12a), $\sin\left(\frac{\pi}{2} + x\right) = \sin\frac{\pi}{2}\cos x + \cos\frac{\pi}{2}\sin x = 1\cdot\cos x + 0\cdot\sin x = \cos x$.

45. Using (6), $\sin\theta\cot\theta = \sin\theta\cdot\dfrac{\cos\theta}{\sin\theta} = \cos\theta$.

47. $\sec y - \cos y = \dfrac{1}{\cos y} - \cos y$ [by (6)] $= \dfrac{1-\cos^2 y}{\cos y} = \dfrac{\sin^2 y}{\cos y}$ [by (7)] $= \dfrac{\sin y}{\cos y}\sin y = \tan y\sin y$ [by (6)]

49. $\cot^2\theta + \sec^2\theta = \dfrac{\cos^2\theta}{\sin^2\theta} + \dfrac{1}{\cos^2\theta}$ [by (6)] $= \dfrac{\cos^2\theta\cos^2\theta + \sin^2\theta}{\sin^2\theta\cos^2\theta} = \dfrac{(1-\sin^2\theta)(1-\sin^2\theta) + \sin^2\theta}{\sin^2\theta\cos^2\theta}$ [by (7)]
$= \dfrac{1 - \sin^2\theta + \sin^4\theta}{\sin^2\theta\cos^2\theta} = \dfrac{\cos^2\theta + \sin^4\theta}{\sin^2\theta\cos^2\theta}$ [by (7)] $= \dfrac{1}{\sin^2\theta} + \dfrac{\sin^2\theta}{\cos^2\theta} = \csc^2\theta + \tan^2\theta$ [by (6)]

51. Using (14a), we have $\tan 2\theta = \tan(\theta + \theta) = \dfrac{\tan\theta + \tan\theta}{1 - \tan\theta\tan\theta} = \dfrac{2\tan\theta}{1 - \tan^2\theta}$.

53. Using (15a) and (16a), $\sin x\sin 2x + \cos x\cos 2x = \sin x(2\sin x\cos x) + \cos x(2\cos^2 x - 1)$
$= 2\sin^2 x\cos x + 2\cos^3 x - \cos x = 2(1 - \cos^2 x)\cos x + 2\cos^3 x - \cos x$ [by (7)]
$= 2\cos x - 2\cos^3 x + 2\cos^3 x - \cos x = \cos x$.

55. $\dfrac{\sin\phi}{1-\cos\phi} = \dfrac{\sin\phi}{1-\cos\phi}\cdot\dfrac{1+\cos\phi}{1+\cos\phi} = \dfrac{\sin\phi(1+\cos\phi)}{1-\cos^2\phi} = \dfrac{\sin\phi(1+\cos\phi)}{\sin^2\phi}$ [by (7)] $= \dfrac{1+\cos\phi}{\sin\phi}$

$= \dfrac{1}{\sin\phi} + \dfrac{\cos\phi}{\sin\phi} = \csc\phi + \cot\phi$ [by (6)]

57. Using (12a), $\sin 3\theta + \sin\theta = \sin(2\theta+\theta) + \sin\theta = \sin 2\theta\cos\theta + \cos 2\theta\sin\theta + \sin\theta$

$= \sin 2\theta\cos\theta + (2\cos^2\theta - 1)\sin\theta + \sin\theta$ [by (16a)] $= \sin 2\theta\cos\theta + 2\cos^2\theta\sin\theta - \sin\theta + \sin\theta$

$= \sin 2\theta\cos\theta + \sin 2\theta\cos\theta$ [by (15a)] $= 2\sin 2\theta\cos\theta$.

59. Since $\sin x = \frac{1}{3}$ we can label the opposite side as having length 1, the hypotenuse as having length 3, and use the Pythagorean Theorem to get that the adjacent side has length $\sqrt{8}$. Then, from the diagram, $\cos x = \frac{\sqrt{8}}{3}$. Similarly we have that $\sin y = \frac{3}{5}$.

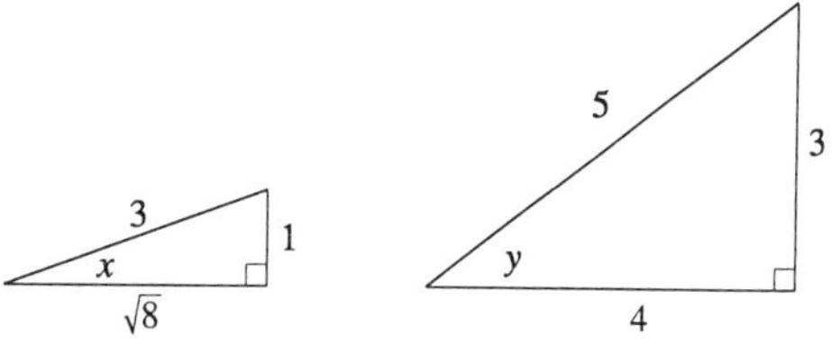

Now use (12a): $\sin(x+y) = \sin x\cos y + \cos x\sin y = \frac{1}{3}\cdot\frac{4}{5} + \frac{\sqrt{8}}{3}\cdot\frac{3}{5} = \frac{4}{15} + \frac{3\sqrt{8}}{15} = \frac{4+6\sqrt{2}}{15}$.

61. Using (13b) and the values for $\cos x$ and $\sin y$ obtained in Exercise 59, we have

$\cos(x-y) = \cos x\cos y + \sin x\sin y = \frac{\sqrt{8}}{3}\cdot\frac{4}{5} + \frac{1}{3}\cdot\frac{3}{5} = \frac{8\sqrt{2}+3}{15}$.

63. Using (15a) and the value for $\sin y$ obtained in Exercise 59, we have

$\sin 2y = 2\sin y\cos y = \dfrac{2\sin y}{\sec y} = 2\left(\frac{3}{5}\right)\left(\frac{4}{5}\right) = \frac{24}{25}$.

65. $2\cos x - 1 = 0 \quad\Leftrightarrow\quad \cos x = \frac{1}{2} \quad\Rightarrow\quad x = \frac{\pi}{3}, \frac{5\pi}{3}$

67. $2\sin^2 x = 1 \quad\Leftrightarrow\quad \sin^2 x = \frac{1}{2} \quad\Leftrightarrow\quad \sin x = \pm\frac{1}{\sqrt{2}} \quad\Rightarrow\quad x = \frac{\pi}{4}, \frac{3\pi}{4}, \frac{5\pi}{4}, \frac{7\pi}{4}$.

69. Using (15a), $\sin 2x = \cos x \;\Rightarrow\; 2\sin x\cos x - \cos x = 0 \;\Leftrightarrow\; \cos x(2\sin x - 1) = 0 \;\Leftrightarrow\; \cos x = 0$ or $2\sin x - 1 = 0 \;\Rightarrow\; x = \frac{\pi}{2}, \frac{3\pi}{2}$ or $\sin x = \frac{1}{2} \;\Rightarrow\; x = \frac{\pi}{6}$ or $\frac{5\pi}{6}$. Therefore the solutions are $x = \frac{\pi}{6}, \frac{\pi}{2}, \frac{5\pi}{6}, \frac{3\pi}{2}$.

71. $\sin x = \tan x \quad\Leftrightarrow\quad \sin x - \tan x = 0 \quad\Leftrightarrow\quad \sin x - \dfrac{\sin x}{\cos x} = 0 \quad\Leftrightarrow\quad \sin x\left(1 - \dfrac{1}{\cos x}\right) = 0 \quad\Leftrightarrow$

$\sin x = 0$ or $1 - \dfrac{1}{\cos x} = 0 \quad\Rightarrow\quad x = 0, \pi, 2\pi$ or $1 = \dfrac{1}{\cos x} \quad\Rightarrow\quad \cos x = 1 \quad\Rightarrow\quad x = 0, 2\pi$. Therefore the solutions are $x = 0, \pi, 2\pi$.

73. We know that $\sin x = \frac{1}{2}$ when $x = \frac{\pi}{6}$ or $\frac{5\pi}{6}$, and from Figure 13(a), we see that $\sin x \le \frac{1}{2} \quad\Rightarrow\quad 0 \le x \le \frac{\pi}{6}$ or $\frac{5\pi}{6} \le x \le 2\pi$.

75. $\tan x = -1$ when $x = \frac{3\pi}{4}, \frac{7\pi}{4}$, and $\tan x = 1$ when $x = \frac{\pi}{4}$ or $\frac{5\pi}{4}$. From Figure 14 we see that $-1 < \tan x < 1$ $\Rightarrow\quad 0 \le x < \frac{\pi}{4}, \frac{3\pi}{4} < x < \frac{5\pi}{4}$, and $\frac{7\pi}{4} < x \le 2\pi$.

77. $y = \cos\left(x - \frac{\pi}{3}\right)$. We start with the graph of $y = \cos x$ and shift it $\frac{\pi}{3}$ units to the right.

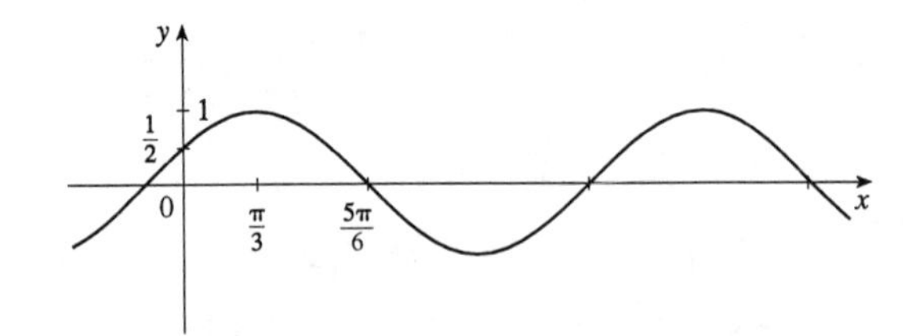

79. $y = \frac{1}{3}\tan\left(x - \frac{\pi}{2}\right)$. We start with the graph of $y = \tan x$, shift it $\frac{\pi}{2}$ units to the right and compress it to $\frac{1}{3}$ of its original vertical size.

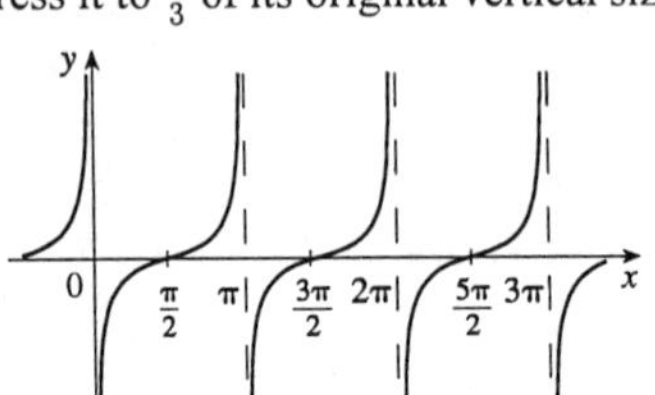

81. $y = |\sin x|$. We start with the graph of $y = \sin x$ and reflect the parts below the x-axis about the x-axis.

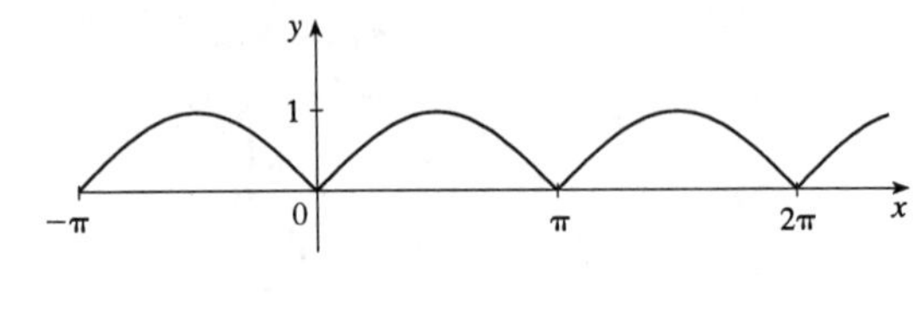

83. From the figure we see that $x = b\cos\theta$, $y = b\sin\theta$, and from the distance formula we have that the distance c from (x, y) to $(a, 0)$ is $c = \sqrt{(x-a)^2 + (y-0)^2} \quad\Rightarrow\quad c^2 = (b\cos\theta - a)^2 + (b\sin\theta)^2$
$= b^2\cos^2\theta - 2ab\cos\theta + a^2 + b^2\sin^2\theta = a^2 + b^2(\cos^2\theta + \sin^2\theta) - 2ab\cos\theta = a^2 + b^2 - 2ab\cos\theta$ [by (7)].

85. Using the Law of Cosines, we have $c^2 = 1^2 + 1^2 - 2(1)(1)\cos(\alpha - \beta) = 2[1 - \cos(\alpha - \beta)]$. Now, using the distance formula, $c^2 = |AB|^2 = (\cos\alpha - \cos\beta)^2 + (\sin\alpha - \sin\beta)^2$. Equating these two expressions for c^2, we get $2[1 - \cos(\alpha - \beta)] = \cos^2\alpha + \sin^2\alpha + \cos^2\beta + \sin^2\beta - 2\cos\alpha\cos\beta - 2\sin\alpha\sin\beta \quad\Rightarrow$
$1 - \cos(\alpha - \beta) = 1 - \cos\alpha\cos\beta - \sin\alpha\sin\beta \quad\Rightarrow\quad \cos(\alpha - \beta) = \cos\alpha\cos\beta + \sin\alpha\sin\beta$.

87. In Exercise 86 we used the subtraction formula for cosine to prove the addition formula for cosine. Using that formula with $x = \frac{\pi}{2} - \alpha$, $y = \beta$, we get $\cos\left[\left(\frac{\pi}{2} - \alpha\right) + \beta\right] = \cos\left(\frac{\pi}{2} - \alpha\right)\cos\beta - \sin\left(\frac{\pi}{2} - \alpha\right)\sin\beta \quad\Rightarrow$
$\cos\left[\frac{\pi}{2} - (\alpha - \beta)\right] = \cos\left(\frac{\pi}{2} - \alpha\right)\cos\beta - \sin\left(\frac{\pi}{2} - \alpha\right)\sin\beta$. Now we use the identities given in the problem to get $\sin(\alpha - \beta) = \sin\alpha\cos\beta - \cos\alpha\sin\beta$.

89. Using the formula derived in Exercise 88, the area of the triangle is $\frac{1}{2}(10)(3)\sin 107° \approx 14.34457$.

EXERCISES E

1. Let S_n be the statement that $2^n > n$.

1. S_1 is true because $2^1 = 2 > 1$.
2. Assume S_k is true, that is, $2^k > k$. Then $2^{k+1} = 2^k \cdot 2 > k \cdot 2 = 2k$ (since $2^k > k$). But $k > 1 \quad \Rightarrow$ $k + k > k + 1 \quad \Rightarrow \quad 2k > k + 1$, so that $2^{k+1} > 2k > k + 1$, which shows that S_{k+1} is true.
3. Therefore, by mathematical induction, $2^n > n$ for every positive integer n.

3. Let S_n be the statement that $(1 + x)^n \geq 1 + nx$.

1. S_1 is true because $(1 + x)^1 = 1 + (1)x$.
2. Assume S_k is true, that is, $(1 + x)^k \geq 1 + kx$. Then $(1 + x)^{k+1} = (1 + x)^k(1 + x) \geq (1 + kx)(1 + x)$ [since $(1 + x)^k \geq 1 + kx$] $= 1 + x + kx + kx^2 \geq 1 + x + kx = 1 + (k + 1)x$, which shows that S_{k+1} is true.
3. Therefore, by mathematical induction, $(1 + x)^n \geq 1 + nx$ for every positive integer n.

5. Let S_n be the statement that $7^n - 1$ is divisible by 6.

1. S_1 is true because $7^1 - 1 = 6$ is divisible by 6.
2. Assume S_k is true, that is, $7^k - 1$ is divisible by 6; in other words $7^k - 1 = 6m$ for some positive integer m. Then $7^{k+1} - 1 = 7^k \cdot 7 - 1 = (6m + 1) \cdot 7 - 1 = 6(7m + 1)$, which is divisible by 6, so S_{k+1} is true.
3. Therefore, by mathematical induction, $7^n - 1$ is divisible by 6 for every positive integer n.

7. Let S_n be the statement that $1 + 3 + 5 + \cdots + (2n - 1) = n^2$.

1. S_1 is true because $[2(1) - 1] = 1 = 1^2$.
2. Assume S_k is true, that is $1 + 3 + 5 + \cdots + (2k - 1) = k^2$. Then $1 + 3 + 5 + \cdots + [2(k + 1) - 1] = 1 + 3 + 5 + \cdots + (2k - 1) + (2k + 1) = k^2 + (2k + 1) = (k + 1)^2$, which shows that S_{k+1} is true.
3. Therefore, by mathematical induction, $1 + 3 + 5 + \cdots + (2n - 1) = n^2$ for every positive integer n.

9. Let S_n be the statement that $\dfrac{1}{2} + \dfrac{1}{6} + \dfrac{1}{12} + \cdots + \dfrac{1}{n(n+1)} = \dfrac{n}{n+1}$.

1. S_1 is true because $\dfrac{1}{1(1+1)} = \dfrac{1}{1+1}$.
2. Assume S_k is true, that is, $\dfrac{1}{2} + \dfrac{1}{6} + \dfrac{1}{12} + \cdots + \dfrac{1}{k(k+1)} = \dfrac{k}{k+1}$. Then
$$\frac{1}{2} + \frac{1}{6} + \frac{1}{12} + \cdots + \frac{1}{(k+1)[(k+1)+1]} = \frac{1}{2} + \frac{1}{6} + \frac{1}{12} + \cdots + \frac{1}{(k+1)(k+2)}$$
$$= \frac{1}{2} + \frac{1}{6} + \frac{1}{12} + \cdots + \frac{1}{k(k+1)} + \frac{1}{(k+1)(k+2)} = \frac{k}{k+1} + \frac{1}{(k+1)(k+2)} = \frac{k(k+2)+1}{(k+1)(k+2)}$$
$$= \frac{k^2+2k+1}{(k+1)(k+2)} = \frac{(k+1)(k+1)}{(k+1)(k+2)} = \frac{(k+1)}{(k+1)+1}$$
which shows that S_{k+1} is true.
3. Therefore, by mathematical induction, $\dfrac{1}{2} + \dfrac{1}{6} + \dfrac{1}{12} + \cdots + \dfrac{1}{n(n+1)} = \dfrac{n}{n+1}$ for every positive integer n.

EXERCISES G

1. The computer results are from Maple, with `Digits:=16`. The last column shows the values of the sixth-degree Taylor polynomial for $f(x) = \csc^2 x - x^{-2}$ near $x = 0$. Note that the second arrangement of Taylor's polynomial is easier to use with a calculator:

$$T_6(x) = \tfrac{1}{3} + \tfrac{1}{15}x^2 + \tfrac{2}{189}x^4 + \tfrac{1}{675}x^6 = \left[\left(\tfrac{1}{675}x^2 + \tfrac{2}{189}\right)x^2 + \tfrac{1}{15}\right]x^2 + \tfrac{1}{3}$$

x	$f(x)_{\text{calculator}}$	$f(x)_{\text{computer}}$	$f(x)_{\text{Taylor}}$
0.1	0.33400107	0.3340010596845	0.33400106
0.01	0.333341	0.33334000010	0.33334000
0.001	0.3334	0.333333400	0.33333340
0.0001	0.34	0.3333334	0.33333333
0.00001	2.0	0.33334	0.33333333
0.000001	100 or 200	0.334	0.33333333
0.0000001	10,000 or 20,000	0.4	0.33333333
0.00000001	1,000,000	0	0.33333333

We see that the calculator results start to deteriorate seriously at $x = 0.0001$, and for smaller x, they are entirely meaningless. The different results "100 or 200" etc. depended on whether we calculated $\left[(\sin x)^2\right]^{-1}$ or $\left[(\sin x)^{-1}\right]^2$. With Maple, the result is off by more than 10% when $x = 0.0000001$ (compare with the calculator result!) A detailed analysis reveals that the values of the function are always greater than $\frac{1}{3}$, but the computer eventually gives results less than $\frac{1}{3}$.

The polynomial $T_6(x)$ was obtained by patient simplification of the expression for $f(x)$, starting with $\sin^2(x) = \frac{1}{2}(1 - \cos 2x)$, where $\cos 2x = 1 - \dfrac{(2x)^2}{2!} + \dfrac{(2x)^4}{4!} - \cdots - \dfrac{(2x)^{10}}{10!} + R_{12}(x)$.
Consequently, the exact value of the limit is $T_6(0) = \frac{1}{3}$. It can also be obtained by several applications of l'Hospital's Rule to the expression $f(x) = \dfrac{x^2 - \sin^2 x}{x^2 \sin^2 x}$ with intermediate simplifications.

3. From $f(x) = \dfrac{x^{25}}{(1.0001)^x}$ (we may assume $x > 0$; why?), we have $\ln f(x) = 25 \ln x - x \ln(1.0001)$ and $\dfrac{f'(x)}{f(x)} = \dfrac{25}{x} - \ln(1.0001)$. This derivative, as well as the derivative $f'(x)$ itself, is positive for

$0 < x < x_0 = \dfrac{25}{\ln(1.0001)} \approx 249{,}971.015$, and negative for $x > x_0$. Hence the maximum value of $f(x)$ is

$f(x_0) = \dfrac{x_0^{25}}{(1.0001)^{x_0}}$, a number too large to be calculated directly. Using decimal logarithms,

$\log_{10} f(x_0) \approx 124.08987757$, so that $f(x_0) \approx 1.229922 \times 10^{124}$. The actual value of the limit is $\lim_{x\to\infty} f(x) = 0$; it would be wasteful and inelegant to use l'Hospital's Rule twenty-five times since we can transform $f(x)$ into

$f(x) = \left[\dfrac{x}{(1.0001)^{x/25}}\right]^{25}$, and the inside expression needs just one application of l'Hospital's Rule to give 0.

5. For $f(x) = \ln\ln x$ with $x \in [a, b]$, $a = 10^9$, and $b = 10^9 + 1$, we need $f'(x) = \dfrac{1}{x \ln x}$, $f''(x) = -\dfrac{\ln x + 1}{x^2 (\ln x)^2}$.

(a) $f'(b) < D < f'(a)$, where $f'(a) \approx 4.8254942434 \times 10^{-11}$, $f'(b) \approx 4.8254942383 \times 10^{-11}$.

(b) Let us estimate $f'(b) - f'(a) = (b-a)f''(c_1) = f''(c_1)$. Since f'' increases (its absolute value decreases), we have $|f'(b) - f'(a)| < |f''(a)| \approx 5.0583 \times 10^{-20}$.

7. **(a)** The 11-digit calculator value of $192 \sin \frac{\pi}{96}$ is 6.2820639018, while the value (on the same device) of p before rationalization is 6.282063885, which is 1.68×10^{-8} less than the trigonometric result.

(b) $p = \dfrac{96}{\sqrt{2+\sqrt{3}} \cdot \sqrt{2+\sqrt{2+\sqrt{3}}} \cdot \sqrt{2+\sqrt{2+\sqrt{2+\sqrt{3}}}} \cdot \sqrt{2+\sqrt{2+\sqrt{2+\sqrt{2+\sqrt{3}}}}}}$, but of course we can avoid repetitious calculations by storing intermediate results in a memory: $p_1 = \sqrt{2+\sqrt{3}}$, $p_2 = \sqrt{2+p_1}$, $p_3 = \sqrt{2+p_2}$, $p_4 = \sqrt{2+p_3}$, and so $p = \dfrac{96}{p_1 p_2 p_3 p_4}$. According to this formula, a calculator gives $p \approx 6.2820639016$, which is within 2×10^{-10} of the trigonometric result. With `Digits:=16;` Maple gives $p \approx 6.282063901781030$ before rationalization (off the trig result by about 1.1×10^{-14}) and $p \approx 6.282063901781018$ after rationalization (error of about 1.7×10^{-15}), a gain of about 1 digit of accuracy for rationalizing. If we set `Digits:=100;`, the difference between Maple's calculation of $192 \sin \frac{\pi}{96}$ and the radical is only about 4×10^{-99}.

9. **(a)** Let $A = \left[\frac{1}{2}\left(27q + \sqrt{729q^2 + 108p^3}\right)\right]^{1/3}$ and $B = \left[\frac{1}{2}\left(27q - \sqrt{729q^2 + 108p^3}\right)\right]^{1/3}$. Then $A^3 + B^3 = 27q$ and $AB = \frac{1}{4}\left[729q^2 - (729q^2 + 108p^3)\right]^{1/3} = -3p$. Substitute into the formula $A + B = \dfrac{A^3 + B^3}{A^2 - AB + B^2}$ where we replace B by $-\dfrac{3p}{A}$:

$x = \frac{1}{3}(A+B) = \dfrac{27q/3}{\left[\frac{1}{2}\left(27q + \sqrt{729q^2 + 108p^3}\right)\right]^{2/3} + 3p + 9p^2\left[\frac{1}{2}\left(27q + \sqrt{729q^2 + 108p^3}\right)\right]^{-2/3}}$ which almost yields the given formula; since replacing q by $-q$ results in replacing x by $-x$, a simple discussion of the cases $q > 0$ and $q < 0$ allows us to replace q by $|q|$ in the denominator, so that it involves only positive numbers. The problems mentioned in the introduction to this exercise have disappeared.

(b) A direct attack works best here. To save space, let $\alpha = 2 + \sqrt{5}$, so we can rationalize, using $\alpha^{-1} = -2 + \sqrt{5}$ and $\alpha - \alpha^{-1} = 4$ (check it!):

$u = \dfrac{4}{\alpha^{2/3} + 1 + \alpha^{-2/3}} \cdot \dfrac{\alpha^{1/3} - \alpha^{-1/3}}{\alpha^{1/3} - \alpha^{-1/3}} = \dfrac{4\left(\alpha^{1/3} - \alpha^{-1/3}\right)}{\alpha - \alpha^{-1}} = \alpha^{1/3} - \alpha^{-1/3}$ and we cube the expression for u: $u^3 = \alpha - 3\alpha^{1/3} + 3\alpha^{-1/3} - \alpha^{-1} = 4 - 3u$, $u^3 + 3u - 4 = (u-1)(u^2 + u + 4) = 0$, so that the only real root is $u = 1$. A check using the formula from part (a): $p = 3$, $q = -4$, so $729q^2 + 108p^3 = 14{,}580 = 54^2 \times 5$, and $x = \dfrac{36}{\left(54 + 27\sqrt{5}\right)^{2/3} + 9 + 81\left(54 + 27\sqrt{5}\right)^{-2/3}}$, which simplifies to the given form after reduction by 9.

11. Proof that $\lim_{n\to\infty} a_n = 0$: From $1 \le e^{1-x} \le e$ it follows that $x^n \le e^{1-x}x^n \le x^n e$, and integration gives $\frac{1}{n+1} = \int_0^1 x^n\,dx \le \int_0^1 e^{1-x}x^n\,dx \le \int_0^1 x^n e\,dx = \frac{e}{n+1}$, that is, $\frac{1}{n+1} \le a_n \le \frac{e}{n+1}$, and since $\lim_{n\to\infty}\frac{1}{n+1} = \lim_{n\to\infty}\frac{e}{n+1} = 0$, it follows from the Squeeze Theorem that $\lim_{n\to\infty} a_n = 0$. Of course, the expression $1/(n+1)$ on the left side could have been replaced by 0 and the proof would still be correct.

Calculations: Using the formula $a_n = \left[e - 1 - \left(\frac{1}{1!} + \frac{1}{2!} + \cdots + \frac{1}{n!}\right)\right]n!$ with an 11-digit pocket calculator:

n	a_n	n	a_n	n	a_n
0	1.7182818284	7	0.1404151360	14	−5.07636992
1	0.7182818284	8	0.1233210880	15	−77.1455488
2	0.4365636568	9	0.1098897920	16	−1235.3287808
3	0.3096909704	10	0.0988979200	17	−21001.589274
4	0.2387638816	11	0.0878771200	18	−378029.60693
5	0.1938194080	12	0.0545254400	19	−7182563.5317
6	0.1629164480	13	−0.2911692800	20	−143651271.63

It is clear that the values calculated from the direct reduction formula will diverge to $-\infty$. If we instead calculate a_n using the reduction formula in Maple (with `Digits:=16`), we get some odd results : $a_{20} = -1000$, $a_{28} = 10^{14}$, $a_{29} = 0$, and $a_{30} = 10^{17}$, for example. But for larger n, the results are at least small and positive (for example, $a_{1000} \approx 0.001$.) For $n > 32{,}175$, we get the delightful `object too large` error message. If, instead of using the reduction formula, we integrate directly with Maple, the results are much better.

13. We can start by expressing e^x and e^{-x} in terms of $E(x) = (e^x - 1)/x$ $(x \ne 0)$, where $E(0) = 1$ to make E continuous at 0 (by L'Hospital's Rule). Namely, $e^x = 1 + xE(x)$, $e^{-x} = 1 - xE(-x)$ and $\sinh x = \frac{1 + xE(x) - [1 - xE(-x)]}{2} = \frac{1}{2}x[E(x) + E(-x)]$, where the addition involves only positive numbers $E(x)$ and $E(-x)$, thus presenting no loss of accuracy due to subtraction.

Another form, which calls the function E only once: we write

$\sinh x = \frac{(e^x)^2 - 1}{2e^x} = \frac{[1 + xE(x)]^2 - 1}{2[1 + xE(x)]} = \frac{x\left[1 + \frac{1}{2}|x|E(|x|)\right]E(|x|)}{1 + |x|E(|x|)}$, taking advantage of the fact that $\frac{\sinh x}{x}$ is an even function, so replacing x by $|x|$ does not change its value.

EXERCISES H

1. $(3+2i)+(7-3i)=(3+7)+(2-3)i=10-i$

3. $(3-i)(4+i)=12+3i-4i-(-1)=13-i$

5. $\overline{12+7i}=12-7i$

7. $\dfrac{2+3i}{1-5i}=\dfrac{2+3i}{1-5i}\cdot\dfrac{1+5i}{1+5i}=\dfrac{2+10i+3i+15(-1)}{1-25(-1)}=\dfrac{-13+13i}{26}=-\frac{1}{2}+\frac{1}{2}i$

9. $\dfrac{1}{1+i}=\dfrac{1}{1+i}\cdot\dfrac{1-i}{1-i}=\dfrac{1-i}{1-(-1)}=\dfrac{1-i}{2}=\frac{1}{2}-\frac{1}{2}i$

11. $i^3=i^2\cdot i=(-1)\,i=-i$

13. $\sqrt{-25}=\sqrt{25}\,i=5i$

15. $\overline{3+4i}=3-4i$, $|3+4i|=\sqrt{3^2+4^2}=\sqrt{25}=5$

17. $\overline{-4i}=\overline{0-4i}=0+4i=4i$, $|-4i|=\sqrt{0^2+(-4)^2}=4$

19. $4x^2+9=0 \quad\Leftrightarrow\quad 4x^2=-9 \quad\Leftrightarrow\quad x^2=-\frac{9}{4} \quad\Leftrightarrow\quad x=\pm\sqrt{-\frac{9}{4}}=\pm\sqrt{\frac{9}{4}}i=\pm\frac{3}{2}i.$

21. By the quadratic formula, $x^2-8x+17=0 \quad\Leftrightarrow\quad x=\dfrac{8\pm\sqrt{8^2-4(1)(17)}}{2(1)}=\dfrac{8\pm\sqrt{-4}}{2}=\dfrac{8\pm 2i}{2}=4\pm i.$

23. By the quadratic formula, $z^2+z+2=0 \quad\Leftrightarrow\quad z=\dfrac{-1\pm\sqrt{1-4(1)(2)}}{2(1)}=\dfrac{-1\pm\sqrt{-7}}{2}=-\frac{1}{2}\pm\frac{\sqrt{7}}{2}i.$

25. $r=\sqrt{(-3)^2+3^2}=3\sqrt{2}$, $\tan\theta=\frac{3}{-3}=-1 \quad\Rightarrow\quad \theta=\frac{3}{4}\pi$ (since the given number is in the second quadrant). Therefore $-3+3i=3\sqrt{2}\left(\cos\frac{3\pi}{4}+i\sin\frac{3\pi}{4}\right)$.

27. $r=\sqrt{3^2+4^2}=5$, $\tan\theta=\frac{4}{3} \quad\Rightarrow\quad \theta=\tan^{-1}\frac{4}{3}$ (since the given number is in the second quadrant). Therefore $3+4i=5\left[\cos\left(\tan^{-1}\frac{4}{3}\right)+i\sin\left(\tan^{-1}\frac{4}{3}\right)\right]$.

29. For $z=\sqrt{3}+i$, $r=\sqrt{\left(\sqrt{3}\right)^2+1^2}=2$, and $\tan\theta=\frac{1}{\sqrt{3}} \quad\Rightarrow\quad \theta=\frac{\pi}{6}$ so that $z=2\left(\cos\frac{\pi}{6}+i\sin\frac{\pi}{6}\right)$. For $w=1+\sqrt{3}i$, $r=2$, and $\tan\theta=\sqrt{3} \quad\Rightarrow\quad \theta=\frac{\pi}{3}$ so that $w=2\left(\cos\frac{\pi}{3}+i\sin\frac{\pi}{3}\right)$. Therefore
$zw=2\cdot 2\left[\cos\left(\frac{\pi}{6}+\frac{\pi}{3}\right)+i\sin\left(\frac{\pi}{6}+\frac{\pi}{3}\right)\right]=4\left(\cos\frac{\pi}{2}+i\sin\frac{\pi}{2}\right)$,
$z/w=\frac{2}{2}\left[\cos\left(\frac{\pi}{6}-\frac{\pi}{3}\right)+i\sin\left(\frac{\pi}{6}-\frac{\pi}{3}\right)\right]=\cos\left(-\frac{\pi}{6}\right)+i\sin\left(-\frac{\pi}{6}\right)$, and $1=1+0i=\cos 0+i\sin 0 \quad\Rightarrow$
$1/z=\frac{1}{2}\left[\cos\left(0-\frac{\pi}{6}\right)+i\sin\left(0-\frac{\pi}{6}\right)\right]=\frac{1}{2}\left[\cos\left(-\frac{\pi}{6}\right)+i\sin\left(-\frac{\pi}{6}\right)\right]$.

31. For $z=2\sqrt{3}-2i$, $r=4$, $\tan\theta=\frac{-2}{2\sqrt{3}}=-\frac{1}{\sqrt{3}} \quad\Rightarrow\quad \theta=-\frac{\pi}{6} \quad\Rightarrow\quad z=4\left[\cos\left(-\frac{\pi}{6}\right)+i\sin\left(-\frac{\pi}{6}\right)\right]$. For $w=-1+i$, $r=\sqrt{2}$, $\tan\theta=\frac{1}{-1}=-1 \quad\Rightarrow\quad \theta=\frac{3\pi}{4} \quad\Rightarrow\quad z=\sqrt{2}\left(\cos\frac{3\pi}{4}+i\sin\frac{3\pi}{4}\right)$. Therefore
$zw=4\sqrt{2}\left[\cos\left(-\frac{\pi}{6}+\frac{3\pi}{4}\right)+i\sin\left(-\frac{\pi}{6}+\frac{3\pi}{4}\right)\right]=4\sqrt{2}\left(\cos\frac{7\pi}{12}+i\sin\frac{7\pi}{12}\right)$,
$z/w=\frac{4}{\sqrt{2}}\left[\cos\left(-\frac{\pi}{6}-\frac{3\pi}{4}\right)+i\sin\left(-\frac{\pi}{6}-\frac{3\pi}{4}\right)\right]=\frac{4}{\sqrt{2}}\left[\cos\left(-\frac{11\pi}{12}\right)+i\sin\left(-\frac{11\pi}{12}\right)\right]=2\sqrt{2}\left(\cos\frac{13\pi}{12}+i\sin\frac{13\pi}{12}\right)$,
and $1=1+0i=\cos 0+i\sin 0 \quad\Rightarrow\quad 1/z=\frac{1}{4}\left[\cos\left(0-\left(-\frac{\pi}{6}\right)\right)+i\sin\left(0-\left(-\frac{\pi}{6}\right)\right)\right]=\frac{1}{4}\left(\cos\frac{\pi}{6}+i\sin\frac{\pi}{6}\right)$.

33. For $z = 1 + i$, $r = \sqrt{2}$, $\tan\theta = \frac{1}{1} = 1 \quad\Rightarrow\quad \theta = \frac{\pi}{4} \quad\Rightarrow\quad 1 + i = \sqrt{2}\left(\cos\frac{\pi}{4} + i\sin\frac{\pi}{4}\right)$. So by De Moivre's Theorem,
$$(1+i)^{20} = \left[\sqrt{2}\left(\cos\tfrac{\pi}{4} + i\sin\tfrac{\pi}{4}\right)\right]^{20} = \left(2^{1/2}\right)^{20}\left(\cos\tfrac{20\pi}{4} + i\sin\tfrac{20\pi}{4}\right) = 2^{10}(\cos 5\pi + i\sin 5\pi)$$
$$= 2^{10}[-1 + i(0)] = -2^{10} = -1024.$$

35. For $z = 2\sqrt{3} + 2i$, $r = 4$, $\tan\theta = \frac{2}{2\sqrt{3}} = \frac{1}{\sqrt{3}} \quad\Rightarrow\quad \theta = \frac{\pi}{6} \quad\Rightarrow\quad 2\sqrt{3} + 2i = 4\left(\cos\frac{\pi}{6} + i\sin\frac{\pi}{6}\right)$. So by De Moivre's Theorem,
$$\left(2\sqrt{3} + 2i\right)^5 = \left[4\left(\cos\tfrac{\pi}{6} + i\sin\tfrac{\pi}{6}\right)\right]^5 = 4^5\left(\cos\tfrac{5\pi}{6} + i\sin\tfrac{5\pi}{6}\right) = 4^5\left[-\tfrac{\sqrt{3}}{2} + i(0.5)\right] = -512\sqrt{3} + 512i.$$

37. $1 = 1 + 0i = \cos 0 + i\sin 0$. Using Equation 3 with $r = 1$, $n = 8$, and $\theta = 0$ we have
$$w_k = 1^{1/8}\left[\cos\left(\frac{0 + 2k\pi}{8}\right) + i\sin\left(\frac{0 + 2k\pi}{8}\right)\right] = \cos\frac{k\pi}{4} + i\sin\frac{k\pi}{4}, \text{ where } k = 0, 1, 2, \ldots, 7.$$

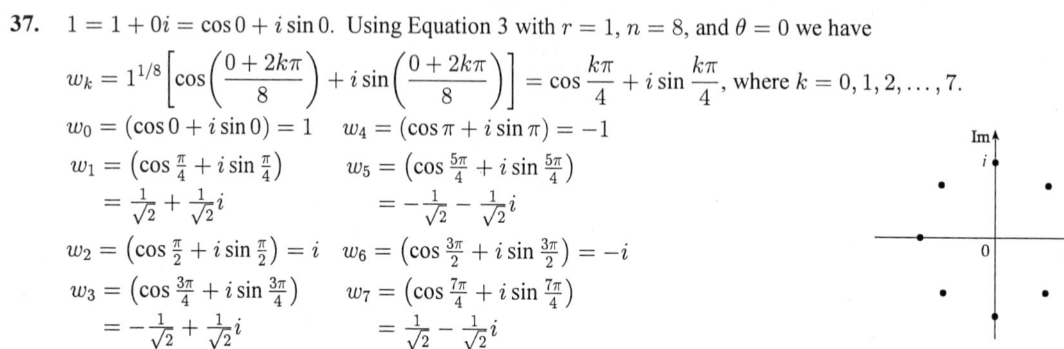

$w_0 = (\cos 0 + i\sin 0) = 1$ $\qquad$ $w_4 = (\cos\pi + i\sin\pi) = -1$

$w_1 = \left(\cos\frac{\pi}{4} + i\sin\frac{\pi}{4}\right) = \frac{1}{\sqrt{2}} + \frac{1}{\sqrt{2}}i$ $\qquad$ $w_5 = \left(\cos\frac{5\pi}{4} + i\sin\frac{5\pi}{4}\right) = -\frac{1}{\sqrt{2}} - \frac{1}{\sqrt{2}}i$

$w_2 = \left(\cos\frac{\pi}{2} + i\sin\frac{\pi}{2}\right) = i$ $\qquad$ $w_6 = \left(\cos\frac{3\pi}{2} + i\sin\frac{3\pi}{2}\right) = -i$

$w_3 = \left(\cos\frac{3\pi}{4} + i\sin\frac{3\pi}{4}\right) = -\frac{1}{\sqrt{2}} + \frac{1}{\sqrt{2}}i$ $\qquad$ $w_7 = \left(\cos\frac{7\pi}{4} + i\sin\frac{7\pi}{4}\right) = \frac{1}{\sqrt{2}} - \frac{1}{\sqrt{2}}i$

39. $0 = 0 + i = \cos\frac{\pi}{2} + i\sin\frac{\pi}{2}$. Using Equation 3 with $r = 1$, $n = 3$, and $\theta = \frac{\pi}{2}$, we have
$$w_k = 1^{1/3}\left[\cos\left(\frac{\pi/2 + 2k\pi}{3}\right) + i\sin\left(\frac{\pi/2 + 2k\pi}{3}\right)\right], \text{ where } k = 0, 1, 2.$$

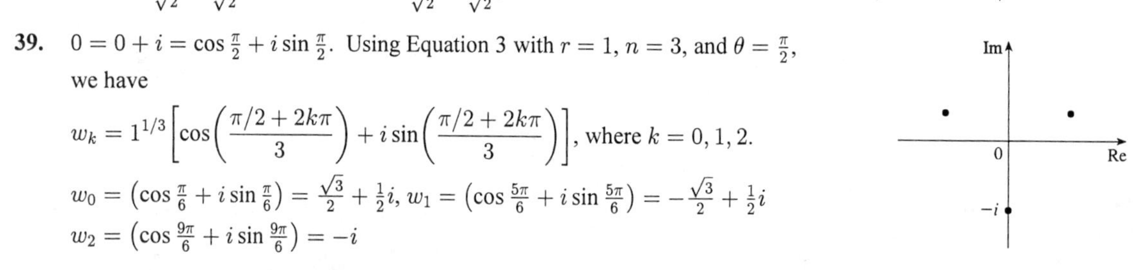

$w_0 = \left(\cos\frac{\pi}{6} + i\sin\frac{\pi}{6}\right) = \frac{\sqrt{3}}{2} + \frac{1}{2}i$, $w_1 = \left(\cos\frac{5\pi}{6} + i\sin\frac{5\pi}{6}\right) = -\frac{\sqrt{3}}{2} + \frac{1}{2}i$

$w_2 = \left(\cos\frac{9\pi}{6} + i\sin\frac{9\pi}{6}\right) = -i$

41. Using Euler's formula (6) with $y = \frac{\pi}{2}$, $e^{i\pi/2} = \cos\frac{\pi}{2} + i\sin\frac{\pi}{2} = i$.

43. Using Euler's formula with $y = \frac{3\pi}{4}$, $e^{i3\pi/4} = \cos\frac{3\pi}{4} + i\sin\frac{3\pi}{4} = -\frac{1}{\sqrt{2}} + \frac{1}{\sqrt{2}}i$.

45. Using Equation 7 with $x = 2$ and $y = \pi$, $e^{2+i\pi} = e^2 e^{i\pi} = e^2(\cos\pi + i\sin\pi) = e^2(-1 + 0) = -e^2$.

47. Take $r = 1$ and $n = 3$ in De Moivre's Theorem to get
$$[1(\cos\theta + i\sin\theta)]^3 = 1^3(\cos 3\theta + i\sin 3\theta) \quad\Rightarrow\quad (\cos\theta + i\sin\theta)^3 = \cos 3\theta + i\sin 3\theta \quad\Rightarrow$$
$$\cos^3\theta + 3(\cos^2\theta)(i\sin\theta) + 3(\cos\theta)(i\sin\theta)^2 + (i\sin\theta)^3 = \cos 3\theta + i\sin 3\theta \quad\Rightarrow$$
$(\cos^3\theta - 3\sin^2\theta\cos\theta) + (3\sin\theta\cos^2\theta - \sin^3\theta)i = \cos 3\theta + i\sin 3\theta$. Equating real and imaginary parts gives $\cos 3\theta = \cos^3\theta - 3\sin^2\theta\cos\theta$ and $\sin 3\theta = 3\sin\theta\cos^2\theta - \sin^3\theta$.

49. $F(x) = e^{rx} = e^{(a+bi)x} = e^{ax+bxi} = e^{ax}(\cos bx + i\sin bx) = e^{ax}\cos bx + i(e^{ax}\sin bx) \quad\Rightarrow$
$$F'(x) = (e^{ax}\cos bx)' + i(e^{ax}\sin bx)' = (ae^{ax}\cos bx - be^{ax}\sin bx) + i(ae^{ax}\sin bx + be^{ax}\cos bx)$$
$$= a[e^{ax}(\cos bx + i\sin bx)] + b[e^{ax}(-\sin bx + i\cos bx)] = ae^{rx} + b\left[e^{ax}\left(i^2\sin bx + i\cos bx\right)\right]$$
$$= ae^{rx} + bi[e^{ax}(\cos bx + i\sin bx)] = ae^{rx} + bie^{rx} = (a + bi)e^{rx} = re^{rx}$$

EXERCISES I

1. $x^2 = -8y$. $4p = -8$, so $p = -2$. The vertex is $(0, 0)$, the focus is $(0, -2)$, and the directrix is $y = 2$.

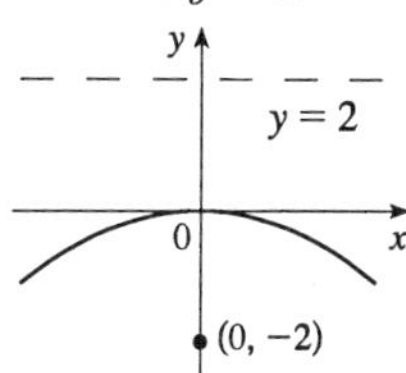

3. $y^2 = x$. $p = \frac{1}{4}$ and the vertex is $(0, 0)$, so the focus is $\left(\frac{1}{4}, 0\right)$, and the directrix is $x = -\frac{1}{4}$.

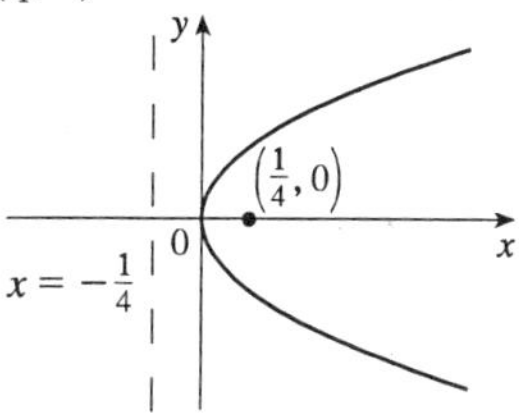

5. $x + 1 = 2(y - 3)^2 \quad \Rightarrow$
$(y - 3)^2 = \frac{1}{2}(x + 1) \quad \Rightarrow \quad p = \frac{1}{8} \quad \Rightarrow$
vertex $(-1, 3)$, focus $\left(-\frac{7}{8}, 3\right)$, directrix $x = -\frac{9}{8}$

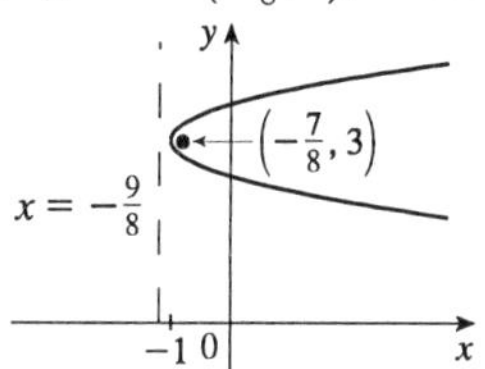

7. $2x + y^2 - 8y + 12 = 0 \quad \Rightarrow$
$(y - 4)^2 = -2(x - 2) \quad \Rightarrow$
$p = -\frac{1}{2} \quad \Rightarrow \quad$ vertex $(2, 4)$,
focus $\left(\frac{3}{2}, 4\right)$, directrix $x = \frac{5}{2}$

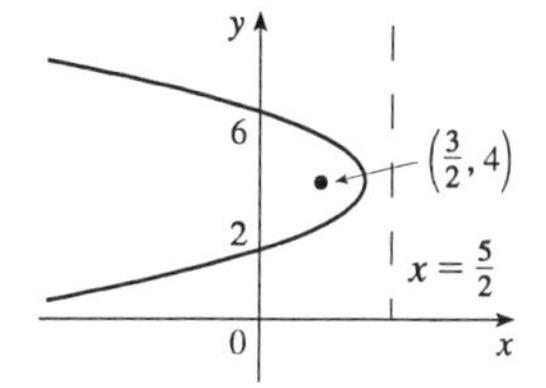

9. $x^2/16 + y^2/4 = 1 \quad \Rightarrow \quad a = 4, b = 2,$
$c = \sqrt{16 - 4} = 2\sqrt{3} \;\Rightarrow\;$ center $(0, 0)$,
vertices $(\pm 4, 0)$, foci $\left(\pm 2\sqrt{3}, 0\right)$

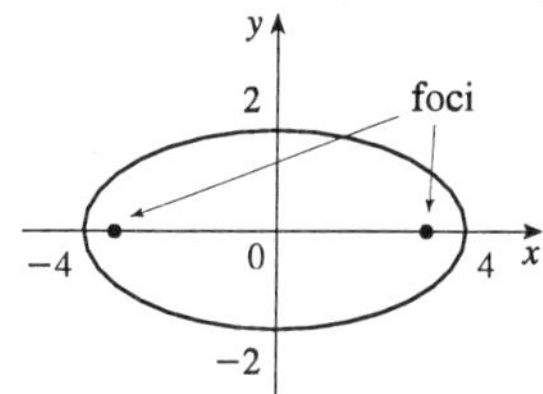

11. $25x^2 + 9y^2 = 225 \quad \Leftrightarrow \quad \frac{1}{9}x^2 + \frac{1}{25}y^2 = 1$
$\Rightarrow \quad a = 5, b = 3, c = 4 \quad \Rightarrow$
center $(0, 0)$, vertices $(0, \pm 5)$, foci $(0, \pm 4)$

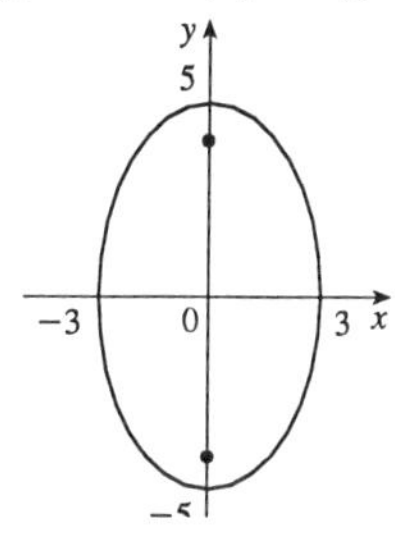

13. $\dfrac{x^2}{144} - \dfrac{y^2}{25} = 1 \quad \Rightarrow \quad a = 12, b = 5,$
$c = \sqrt{144 + 25} = 13 \;\Rightarrow\;$ center $(0, 0)$, vertices $(\pm 12, 0)$, foci $(\pm 13, 0)$, asymptotes $y = \pm \frac{5}{12}x$

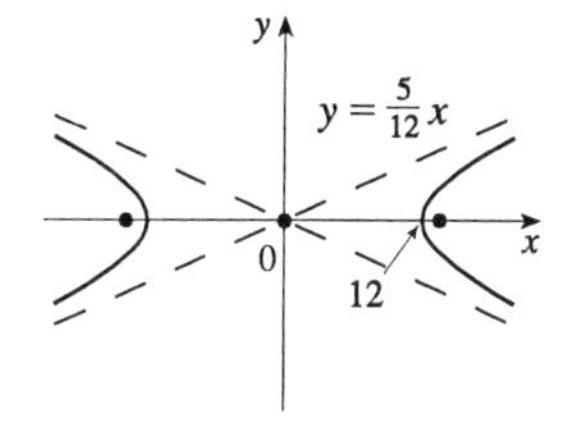

15. $9y^2 - x^2 = 9 \;\Rightarrow\; y^2 - \frac{1}{9}x^2 = 1 \;\Rightarrow\; a = 1,$
$b = 3, c = \sqrt{10} \;\Rightarrow\;$ center $(0, 0)$, vertices $(0, \pm 1)$, foci $\left(0, \pm \sqrt{10}\right)$, asymptotes $y = \pm \frac{1}{3}x$

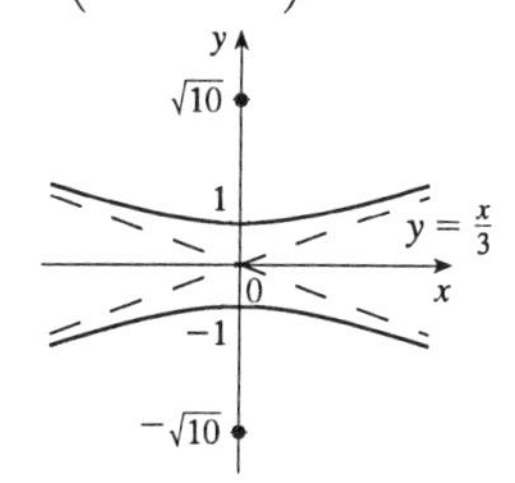

17. $9x^2 - 18x + 4y^2 = 27 \;\Leftrightarrow$
$\dfrac{(x-1)^2}{4} + \dfrac{y^2}{9} = 1 \;\Rightarrow\; a = 3, b = 2,$
$c = \sqrt{5} \;\Rightarrow\;$ center $(1, 0)$,
vertices $(1, \pm 3)$, foci $\left(1, \pm\sqrt{5}\right)$

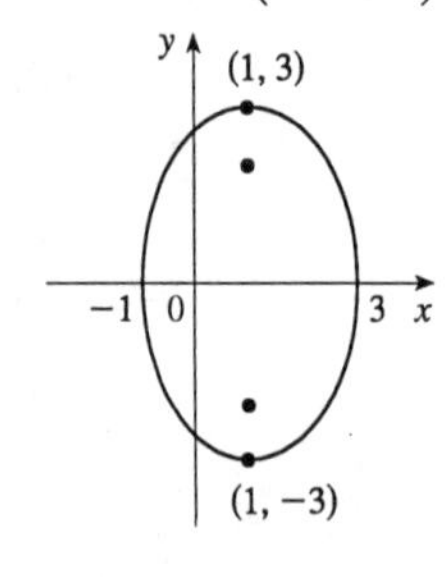

19. $2y^2 - 4y - 3x^2 + 12x = -8 \;\Leftrightarrow$
$\dfrac{(x-2)^2}{6} - \dfrac{(y-1)^2}{9} = 1 \;\Rightarrow\; a = \sqrt{6},$
$b = 3, c = \sqrt{15} \;\Rightarrow\;$ center $(2, 1)$,
vertices $\left(2 \pm \sqrt{6}, 1\right)$, foci $\left(2 \pm \sqrt{15}, 1\right)$,
asymptotes $y - 1 = \pm \frac{3}{\sqrt{6}}(x - 2)$

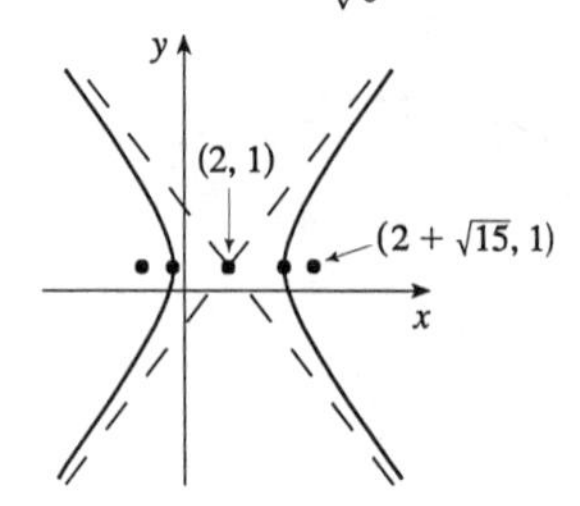

21. Vertex at $(0, 0)$, $p = 3$, opens upward $\;\Rightarrow\; x^2 = 4py = 12y$

23. Vertex at $(2, 0)$, $p = 1$, opens to right $\;\Rightarrow\; y^2 = 4p(x-2) = 4(x-2)$

25. The parabola must have equation $y^2 = 4px$, so $(-4)^2 = 4p(1) \;\Rightarrow\; p = 4 \;\Rightarrow\; y^2 = 16x$.

27. Center $(0, 0)$, $c = 1$, $a = 2 \;\Rightarrow\; b = \sqrt{2^2 - 1^2} = \sqrt{3} \;\Rightarrow\; \frac{1}{4}x^2 + \frac{1}{3}y^2 = 1$

29. Center $(3, 0)$, $c = 1$, $a = 3 \;\Rightarrow\; b = \sqrt{8} = 2\sqrt{2} \;\Rightarrow\; \frac{1}{8}(x-3)^2 + \frac{1}{9}y^2 = 1$

31. Center $(2, 2)$, $c = 2$, $a = 3 \;\Rightarrow\; b = \sqrt{5} \;\Rightarrow\; \frac{1}{9}(x-2)^2 + \frac{1}{5}(y-2)^2 = 1$

33. Center $(0, 0)$, vertical axis, $c = 3$, $a = 1 \;\Rightarrow\; b = \sqrt{8} = 2\sqrt{2} \;\Rightarrow\; y^2 - \frac{1}{8}x^2 = 1$

35. Center $(4, 3)$, horizontal axis, $c = 3$, $a = 2 \;\Rightarrow\; b = \sqrt{5} \;\Rightarrow\; \frac{1}{4}(x-4)^2 - \frac{1}{5}(y-3)^2 = 1$

37. Center $(0, 0)$, horizontal axis, $a = 3$, $\frac{b}{a} = 2 \;\Rightarrow\; b = 6 \;\Rightarrow\; \frac{1}{9}x^2 - \frac{1}{36}y^2 = 1$

39. In Figure 8, we see that the point on the ellipse closest to a focus is the closer vertex (which is a distance $a - c$ from it) while the farthest point is the other vertex (at a distance of $a + c$). So for this lunar orbit, $(a - c) + (a + c) = 2a = (1728 + 110) + (1728 + 314)$, or $a = 1940$; and $(a + c) - (a - c) = 2c = 314 - 110$, or $c = 102$. Thus $b^2 = a^2 - c^2 = 3{,}753{,}196$, and the equation is $\dfrac{x^2}{3{,}763{,}600} + \dfrac{y^2}{3{,}753{,}196} = 1$.

41. **(a)** Set up the coordinate system so that A is $(-200, 0)$ and B is $(200, 0)$.
$|PA| - |PB| = (1200)(980) = 1{,}176{,}000\text{ ft} = \frac{2450}{11}\text{ mi} = 2a \;\Rightarrow\; a = \frac{1225}{11}$, and $c = 200$ so
$b^2 = c^2 - a^2 = \dfrac{3{,}339{,}375}{121} \;\Rightarrow\; \dfrac{121x^2}{1{,}500{,}625} - \dfrac{121y^2}{3{,}339{,}375} = 1.$

(b) Due north of $B \;\Rightarrow\; x = 200 \;\Rightarrow\; \dfrac{(121)(200)^2}{1{,}500{,}625} - \dfrac{121y^2}{3{,}339{,}375} = 1 \;\Rightarrow\; y = \dfrac{133{,}575}{539} \approx 248$ mi

43. The function whose graph is the upper branch of this hyperbola is concave upward. The function is

$y = f(x) = a\sqrt{1+\dfrac{x^2}{b^2}} = \dfrac{a}{b}\sqrt{b^2+x^2}$, so $y' = \dfrac{a}{b}x\left(b^2+x^2\right)^{-1/2}$ and

$y'' = \dfrac{a}{b}\left[\left(b^2+x^2\right)^{-1/2} - x^2\left(b^2+x^2\right)^{-3/2}\right] = ab\left(b^2+x^2\right)^{-3/2} > 0$ for all x, and so f is concave upward.

45. **(a)** ellipse

(b) hyperbola

(c) empty graph (no curve)

(d) In case (a), $a^2 = k$, $b^2 = k - 16$, and $c^2 = a^2 - b^2 = 16$, so the foci are at $(\pm 4, 0)$. In case (b), $k - 16 < 0$, so $a^2 = k$, $b^2 = 16 - k$, and $c^2 = a^2 + b^2 = 16$, and so again the foci are at $(\pm 4, 0)$.

47. Use the parametrization $x = 2\cos t$, $y = \sin t$, $0 \le t \le 2\pi$ to get

$L = 4\int_0^{\pi/2}\sqrt{(dx/dt)^2+(dy/dt)^2}\,dt = 4\int_0^{\pi/2}\sqrt{4\sin^2 t+\cos^2 t}\,dt = 4\int_0^{\pi/2}\sqrt{3\sin^2 t+1}\,dt$. Using Simpson's Rule with $n = 10$, $L \approx \frac{4}{3}\left(\frac{\pi}{20}\right)\left[f(0)+4f\left(\frac{\pi}{20}\right)+2f\left(\frac{\pi}{10}\right)+\cdots+2f\left(\frac{2\pi}{5}\right)+4f\left(\frac{9\pi}{20}\right)+f\left(\frac{\pi}{2}\right)\right]$, with $f(t) = \sqrt{3\sin^2 t+1}$, so $L \approx 9.69$.

49. $\dfrac{x^2}{a^2}+\dfrac{y^2}{b^2} = 1 \;\Rightarrow\; \dfrac{2x}{a^2}+\dfrac{2yy'}{b^2} = 0 \;\Rightarrow\; y' = -\dfrac{b^2x}{a^2y}$ $(y \neq 0)$. Thus the slope of the tangent line at P is $-\dfrac{b^2x_1}{a^2y_1}$. The slope of F_1P is $\dfrac{y_1}{x_1+c}$ and of F_2P is $\dfrac{y_1}{x_1-c}$. By the formula from Problem 21 of Problems Plus after Chapter 3, we have

$$\tan\alpha = \frac{\dfrac{y_1}{x_1+c}+\dfrac{b^2x_1}{a^2y_1}}{1-\dfrac{b^2x_1y_1}{a^2y_1(x_1+c)}} = \frac{a^2y_1^2+b^2x_1(x_1+c)}{a^2y_1(x_1+c)-b^2x_1y_1} = \frac{a^2b^2+b^2cx_1}{c^2x_1y_1+a^2cy_1} \quad \left[\begin{array}{c}\text{using } b^2x_1^2+a^2y_1^2=a^2b^2\\ \text{and } a^2-b^2=c^2\end{array}\right]$$

$$= \frac{b^2(cx_1+a^2)}{cy_1(cx_1+a^2)} = \frac{b^2}{cy_1}, \text{ and}$$

$$\tan\beta = \frac{-\dfrac{y_1}{x_1-c}-\dfrac{b^2x_1}{a^2y_1}}{1-\dfrac{b^2x_1y_1}{a^2y_1(x_1-c)}} = \frac{-a^2y_1^2-b^2x_1(x_1-c)}{a^2y_1(x_1-c)-b^2x_1y_1} = \frac{-a^2b^2+b^2cx_1}{c^2x_1y_1-a^2cy_1} = \frac{b^2(cx_1-a^2)}{cy_1(cx_1-a^2)} = \frac{b^2}{cy_1}.$$

So $\alpha = \beta$.

51. $\dfrac{x^2}{a^2}-\dfrac{y^2}{b^2} = 1 \;\Rightarrow\; y = \pm\dfrac{b}{a}\sqrt{x^2-a^2}$. Now

$$\lim_{x\to\infty}\left[\frac{b}{a}\sqrt{x^2-a^2}-\frac{b}{a}x\right] = \frac{b}{a}\cdot\lim_{x\to\infty}\left(\sqrt{x^2-a^2}-x\right)\frac{\sqrt{x^2-a^2}+x}{\sqrt{x^2-a^2}+x} = \frac{b}{a}\cdot\lim_{x\to\infty}\frac{-a^2}{\sqrt{x^2-a^2}+x} = 0, \text{ which}$$

shows that $y = \dfrac{b}{a}x$ is a slant asymptote.

Similarly $\displaystyle\lim_{x\to\infty}\left[-\frac{b}{a}\sqrt{x^2-a^2}-\left(-\frac{b}{a}x\right)\right] = -\frac{b}{a}\cdot\lim_{x\to\infty}\frac{-a^2}{\sqrt{x^2-a^2}+x} = 0$, so $y = -\dfrac{b}{a}x$ is a slant asymptote.

EXERCISES J

1. $r = \dfrac{ed}{1 + e\cos\theta} = \dfrac{\frac{2}{3} \cdot 3}{1 + \frac{2}{3}\cos\theta} = \dfrac{6}{3 + 2\cos\theta}$

3. $r = \dfrac{ed}{1 + e\sin\theta} = \dfrac{1 \cdot 2}{1 + \sin\theta} = \dfrac{2}{1 + \sin\theta}$

5. $r = 5\sec\theta \quad \Leftrightarrow \quad x = r\cos\theta = 5$, so $r = \dfrac{ed}{1 + e\cos\theta} = \dfrac{4 \cdot 5}{1 + 4\cos\theta} = \dfrac{20}{1 + 4\cos\theta}$

7. Focus $(0, 0)$, vertex $\left(5, \frac{\pi}{2}\right) \quad \Rightarrow \quad$ directrix $y = 10 \quad \Rightarrow \quad r = \dfrac{ed}{1 + e\sin\theta} = \dfrac{10}{1 + \sin\theta}$

9. $e = 3 \quad \Rightarrow \quad$ hyperbola; $ed = 4 \quad \Rightarrow \quad d = \frac{4}{3} \quad \Rightarrow$
directrix $x = \frac{4}{3}$; vertices $(1, 0)$ and $(-2, \pi) = (2, 0)$;
center $\left(\frac{3}{2}, 0\right)$; asymptotes parallel to $\theta = \pm\cos^{-1}\left(-\frac{1}{3}\right)$

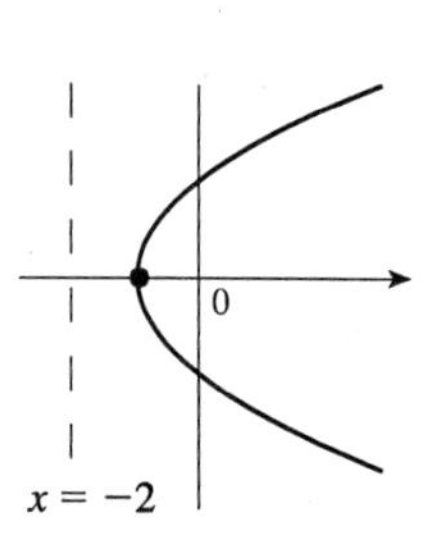

11. $e = 1 \quad \Rightarrow \quad$ parabola; $ed = 2 \quad \Rightarrow \quad d = 2$
$\Rightarrow \quad$ directrix $x = -2$; vertex $(-1, 0) = (1, \pi)$

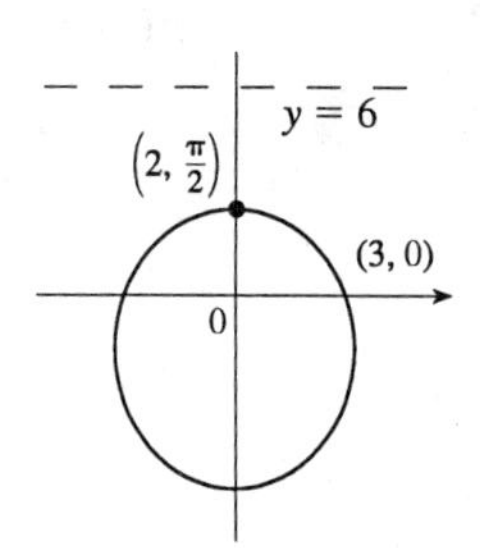

13. $r = \dfrac{3}{1 + \frac{1}{2}\sin\theta} \quad \Rightarrow \quad e = \frac{1}{2} \quad \Rightarrow \quad$ ellipse;
$ed = 3 \quad \Rightarrow \quad d = 6 \quad \Rightarrow \quad$ directrix $y = 6$;
vertices $\left(2, \frac{\pi}{2}\right)$ and $\left(6, \frac{3\pi}{2}\right)$; center $\left(2, \frac{3\pi}{2}\right)$

15. $r = \dfrac{7/2}{1 - \frac{5}{2}\sin\theta} \quad \Rightarrow \quad e = \frac{5}{2} \quad \Rightarrow \quad$ hyperbola;
$ed = \frac{7}{2} \quad \Rightarrow \quad d = \frac{7}{5} \quad \Rightarrow \quad$ directrix $y = -\frac{7}{5}$;
center $\left(\frac{5}{3}, \frac{3\pi}{2}\right)$; vertices $\left(-\frac{7}{3}, \frac{\pi}{2}\right) = \left(\frac{7}{3}, \frac{3\pi}{2}\right)$
and $\left(1, \frac{3\pi}{2}\right)$.

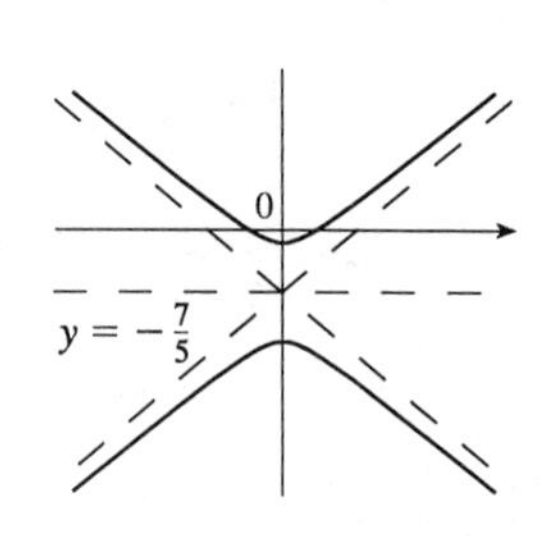

17. **(a)** The equation is $r = \dfrac{1}{4 - 3\cos\theta} = \dfrac{1/4}{1 - \frac{3}{4}\cos\theta}$,

so $e = \frac{3}{4}$ and $ed = \frac{1}{4} \quad \Rightarrow \quad d = \frac{1}{3}$.

The conic is an ellipse, and the equation of its directrix is $x = r\cos\theta = -\frac{1}{3} \quad \Rightarrow \quad r = -\dfrac{1}{3\cos\theta}$.

We must be careful in our choice of parameter values in this equation ($-1 \le \theta \le 1$ works well.)

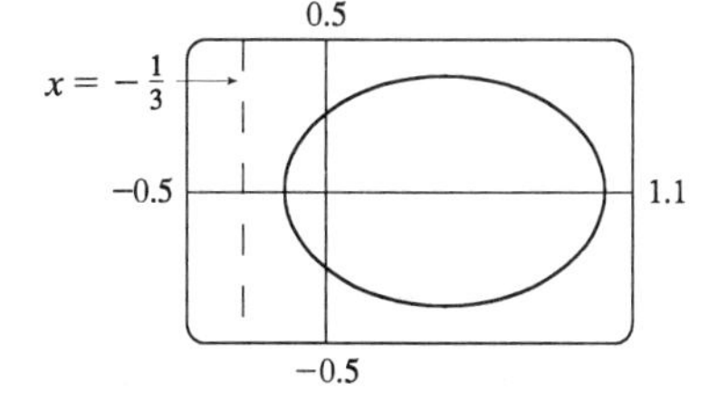

(b) The equation is obtained by replacing θ with $\theta - \frac{\pi}{3}$ in the equation of the original conic (see Example 4), so

$r = \dfrac{1}{4 - 3\cos\left(\theta - \frac{\pi}{3}\right)}$.

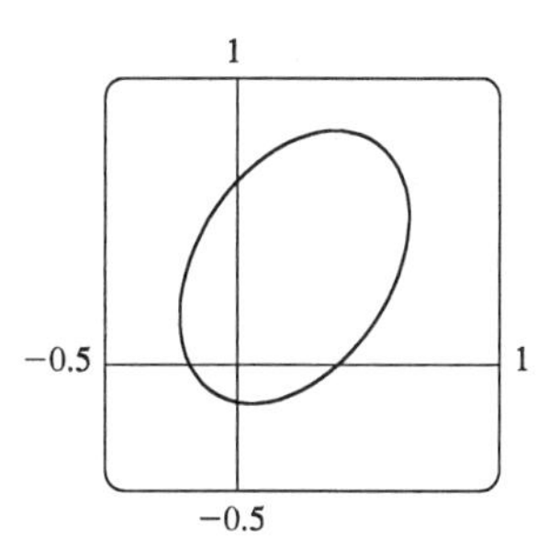

19. For $e < 1$ the curve is an ellipse. It is nearly circular when e is close to 0. As e increases, the graph is stretched out to the right, and grows larger (that is, its right-hand focus moves to the right while its left-hand focus remains at the origin.) At $e = 1$, the curve becomes a parabola with focus at the origin.

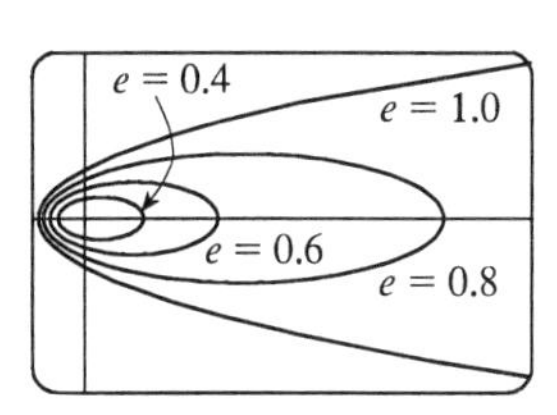

21. $|PF| = e|Pl| \quad \Rightarrow$

$r = e[d - r\cos(\pi - \theta)] = e(d + r\cos\theta) \quad \Rightarrow$

$r(1 - e\cos\theta) = ed \quad \Rightarrow \quad r = \dfrac{ed}{1 - e\cos\theta}$

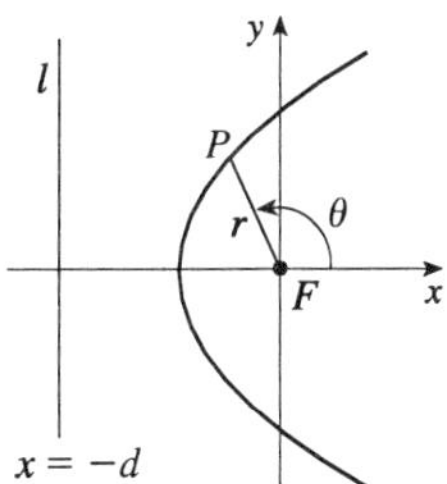

23. $|PF| = e|Pl| \quad \Rightarrow$

$r = e[d - r\sin(\theta - \pi)] = e(d + r\sin\theta)$

$\Rightarrow \quad r(1 - e\sin\theta) = ed \quad \Rightarrow \quad r = \dfrac{ed}{1 - e\sin\theta}$

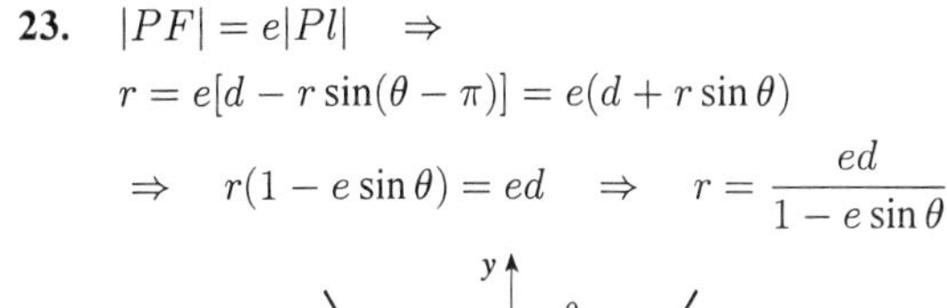

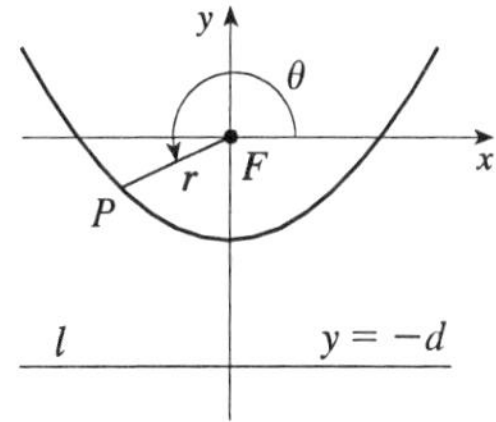

25. **(a)** If the directrix is $x = -d$, then $r = \dfrac{ed}{1 - e\cos\theta}$ [see Figure 8(b)], and, from (4), $a^2 = \dfrac{e^2d^2}{(1-e^2)^2}$ $\Rightarrow$

$ed = a(1-e^2)$. Therefore, $r = \dfrac{a(1-e^2)}{1 - e\cos\theta}$.

(b) $e = 0.017$ and the major axis $= 2a = 2.99 \times 10^8$ $\Rightarrow$ $a = 1.495 \times 10^8$.

Therefore $r = \dfrac{1.495 \times 10^8\left[1 - (0.017)^2\right]}{1 - 0.017\cos\theta} \approx \dfrac{1.49 \times 10^8}{1 - 0.017\cos\theta}$.

27. The minimum distance is at perihelion where $4.6 \times 10^7 = r = a(1-e) = a(1-0.206) = a(0.794)$ $\Rightarrow$ $a = 4.6 \times 10^7/0.794$. So the maximum distance, which is at aphelion, is $r = a(1+e) = \left(4.6 \times 10^7/0.794\right) \times 10^7(1.206) \approx 7.0 \times 10^7$ km.

29. From Exercise 27, we have $e = 0.206$ and $a(1-e) = 4.6 \times 10^7$ km. Thus $a = 4.6 \times 10^7/0.794$. From Exercise 25, we can write the equation of Mercury's orbit as $r = a\dfrac{1-e^2}{1-e\cos\theta}$. So since $\dfrac{dr}{d\theta} = \dfrac{-a(1-e^2)e\sin\theta}{(1-e\cos\theta)^2}$ $\Rightarrow$

$r^2 + \left(\dfrac{dr}{d\theta}\right)^2 = \dfrac{a^2(1-e^2)^2}{(1-e\cos\theta)^2} + \dfrac{a^2(1-e^2)^2e^2\sin^2\theta}{(1-e\cos\theta)^4} = \dfrac{a^2(1-e^2)^2}{(1-e\cos\theta)^4}\left(1 - 2e\cos\theta + e^2\right)$, the length of the orbit is $L = \displaystyle\int_0^{2\pi}\sqrt{r^2 + (dr/d\theta)^2}\,d\theta = a\left(1-e^2\right)\int_0^{2\pi}\frac{\sqrt{1+e^2-2e\cos\theta}}{(1-e\cos\theta)^2}\,d\theta \approx 3.6 \times 10^8$ km.

This seems reasonable, since Mercury's orbit is nearly circular, and the circumference of a circle of radius a is $2\pi a \approx 3.6 \times 10^8$ km.